VIRAL
NANOTECHNOLOGY

VIRAL NANOTECHNOLOGY

Edited by

Yury Khudyakov

Centers for Disease Control and Prevention
Atlanta, Georgia, USA

Paul Pumpens

Latvian Biomedical Research and Study Centre
Riga, Latvia

CRC Press
Taylor & Francis Group
Boca Raton London New York

CRC Press is an imprint of the
Taylor & Francis Group, an **informa** business

CRC Press
Taylor & Francis Group
6000 Broken Sound Parkway NW, Suite 300
Boca Raton, FL 33487-2742

First issued in paperback 2020

© 2016 by Taylor & Francis Group, LLC
CRC Press is an imprint of Taylor & Francis Group, an Informa business

No claim to original U.S. Government works

ISBN-13: 978-1-4665-8352-8 (hbk)
ISBN-13: 978-0-367-65877-9 (pbk)

Visit the Taylor & Francis Web site at
http://www.taylorandfrancis.com

and the CRC Press Web site at
http://www.crcpress.com

Contents

SECTION I Major Concepts

SECTION II Applications

Preface

Nature works as small as it wishes.

Contemplation de la Nature (1764)

Charles Bonnet (1720–1793)

The field of viral nanotechnology is new. As many novel scientific enterprises in their early days, it is frequently confused for something else and under-recognized. To our surprise, when we started developing and discussing the idea of this book, many of our colleagues questioned the sheer existence of such a discipline, interpreting nanotechnology as an elaborate disguise of genetic and protein engineering. The common genealogy in all these disciplines is undeniable. Viral genetic engineering and protein engineering are clear methodological founders and forerunners of viral nanotechnology. However, the focus on application sets viral nanotechnology apart from genetic and protein engineering, both of which focus on constructing.

The transition from molecular technologies of constructing to novel technologies of application is a major event in the field of molecular research over the last two decades. Molecular engineering provides technologies for building molecular modules from genetic elements, while viral nanotechnology explores how these modules can be used. This is similar to the transition from the invention of bricks and other construction materials to advancing the architecture and construction of a variety of buildings and complex structures, including private homes, castles, laboratory facilities, city blocks, and entire cities.

The field of viral nanotechnology is vast and rapidly expanding. The manipulation of atoms and molecules for the fabrication of materials with *novel properties*, which cannot be otherwise obtained, offers opportunities for a multitude of unique applications affecting all human activities. The unique properties of viral proteins, such as their capability to be robustly assembled into well-defined multivalent architectures with a variety of shapes and sizes, which can be easily modified genetically or chemically, serve as a foundation for the major use of viral nanotechnology in the fields of biomedicine, photonics, catalysis, and energy.

The shift from *how to construct* to *how to use* is most crucial, leading to disruptive innovations in all the three areas of medicine—prevention, treatment, and diagnostics—and opens a way to *singularity* in medicine, assuring a total control over health and the fundamental changes in health care. Arnold Schoenberg once wrote in *Theory of Harmony*, "Our noblest impulse, the impulse to know and understand (erkennen), makes it our duty to search." We are confident that this collection of works of eminent researchers will induce such "noblest impulse" to discoveries in the field of viral nanotechnology, promoting the search for unprecedented improvements in human life.

This book, which you hold in your hands, presents a unique opportunity for direct communication with many practitioners of this exciting discipline. We hope that, despite skeptics, you will get this sense of imminence of major breakthroughs, which will convert dreams of controlled health and betterment of humankind into the firm reality of our times.

Yury Khudyakov
Paul Pumpens

Editors

Yury E. Khudyakov, PhD, is chief of the Molecular Epidemiology and Bioinformatics Laboratory, Laboratory Branch (LB), Division of Viral Hepatitis (DVH), Centers for Disease Control and Prevention (CDC), Atlanta, Georgia. He earned his MS in genetics from the Novosibirsk State University, Novosibirsk, Russia, and his PhD in molecular biology from the D.I. Ivanovsky Institute of Virology, Academy of Medical Sciences, Moscow, Russia. He started his research career in the laboratory of gene chemistry at the M.M. Schemyakin Institute of Bioorganic Chemistry, Academy of Sciences, Moscow, Russia. He was a research fellow in the laboratory of viral biochemistry and chief of the Genetic Engineering Section in the laboratory of chemistry of viral nucleic acids and proteins at the D.I. Ivanovsky Institute of Virology, Moscow, Russia. In 1991, he joined the Hepatitis Branch (HB), Division of Viral and Rickettsial Diseases (DVRD)/CDC, as a National Research Council Research Associate, National Academy of Sciences, United States. Since 1996, he has served as chief in the Developmental Diagnostic Unit, Molecular and Immunodiagnostic Section/ HB/DVRD/ NCID/CDC, and later as chief of computational molecular biology activity, and deputy chief of the Developmental Diagnostic Laboratory, LB/DVH/CDC.

Dr. Khudyakov's main research interests are molecular epidemiology of viral diseases, development of new diagnostics and vaccines, molecular biology and evolution of viruses, and bioinformatics.

Dr. Khudyakov has published more than 170 research papers and book chapters. He has edited the books *Artificial DNA*, CRC Press (2002) and *Medicinal Protein Engineering*, CRC Press (2009). He is an author of several issued and pending patents. He is a member of the editorial board for the *Journal of Clinical Virology* and the academic editor for *PLoS ONE*.

Paul Pumpens, PhD, graduated from the Chemical Department of the University of Latvia in 1970 and earned his PhD in molecular biology from the Latvian Academy of Sciences, Riga and DSc from the Institute of Molecular Biology of the USSR Academy of Sciences, Moscow, Russia.

Dr. Pumpens started his research career as a research fellow at the Institute of Organic Synthesis, where he conducted research from 1973 to 1989. He served as head of the Laboratory of Protein Engineering at the Institute of Organic Synthesis (1989–1990), as head of the Department of Protein Engineering at the Institute of Molecular Biology of the Latvian Academy of Sciences (since 1993, the Biomedical Research and Study Centre) in Riga (1990–2002), and as scientific director of the Biomedical Research and Study Centre (2002–2014). He served as a professor of the Biological Department of the University of Latvia from 1999 until 2013.

Dr. Pumpens pioneered genetic engineering research in Latvia. He was one of the first in the world to perform successful cloning of the hepatitis B virus genome and the expression of hepatitis B virus genes in bacterial cells. His major scientific interests are in designing novel recombinant vaccines and diagnostic reagents and development of tools for gene therapy on the basis of virus-like particles.

Dr. Pumpens is an author of more than 300 scientific papers and issues or pending patents.

Contributors

Alaa A.A. Aljabali
Pharmacy Faculty
Yarmouk University
Irbid, Hashemite Kingdom of Jordan

Martin F. Bachmann
Jenner Institute
University of Oxford
Oxford, United Kingdom

and

University Hospital
Bern, Switzerland

Bertrand Bellier
Sorbonne Universités
University Pierre et Marie Curie
Paris, France

Bettina Böttcher
School of Biological Sciences
University of Edinburgh
Edinburgh, United Kingdom

Bryce Chackerian
Department of Molecular Genetics
    and Microbiology
School of Medicine
University of New Mexico
Albuquerque, New Mexico

Charlotte Dalba
Department of Immunology-
    Immunopathology-Immunotherapy
Sorbonne Universités
University Pierre and Marie Curie
Paris, France

Tina Dalianis
Department of Oncology–Pathology
Karolinska Institutet
Cancer Center Karolinska
Stockholm, Sweden

Sandra Diederich
Institute for Novel and Emerging
    Infectious Diseases
Friedrich-Loeffler-Institut
Federal Research Institute for Animal
    Health
Greifswald-Insel Riems, Germany

Trevor Douglas
Department of Chemistry
Indiana University
Bloomington, Indiana

Bogdan Dragnea
Department of Chemistry
Indiana University
Bloomington, Indiana

Diego Espinosa
Department of Molecular Microbiology
    and Immunology
Bloomberg School of Public Health
Johns Hopkins University
Baltimore, Maryland

David J. Evans
Department of Chemistry
University of Hull
Hull, United Kingdom

Jitka Forstová
Faculty of Science
Department of Genetics
    and Microbiology
Charles University in Prague
Prague, Czech Republic

Samson Francis
Department of Molecular and Cellular
    Biochemistry
Indiana University
and
Assembly Biosciences
Bloomington, Indiana

Alma Gedvilaite
Institute of Biotechnology
Vilnius University
Vilnius, Lithuania

Wolfram H. Gerlich
Institute for Medical Virology
Justus Liebig University Giessen
Giessen, Germany

Arin Ghasparian
Virometix AG
Zurich, Switzerland

Xi Jiang
Division of Infectious Diseases
Cincinnati Children's Hospital Medical
    Center
Cincinnati, Ohio

Nicholas Johnson
Animal and Plant Health Agency
Addlestone, United Kingdom

Michael Kann
French National Centre for
    Scientific Research
University of Bordeaux
Bordeaux, France

Andris Kazaks
Department of Structural Biology
Latvian Biomedical Research
    and Study Centre
Riga, Latvia

David Klatzmann
Sorbonne Universités
University Pierre et Marie Curie
Paris, France

Philipp Kolb
Biological Faculty
University of Freiburg
Freiburg, Germany

Tatjana Kozlovska
Latvian Biomedical Research
    and Study Centre
Riga, Latvia

James Lara
Division of Viral Hepatitis
Centers for Disease Control
Atlanta, Georgia

Lye Siang Lee
Department of Molecular and Cellular
    Biochemistry
Indiana University
Bloomington, Indiana

George P. Lomonossoff
Department of Biological Chemistry
John Innes Centre
Norwich, United Kingdom

Kenneth Lundstrom
Pan Therapeutics
Lutry, Switzerland

David R. Milich
Vaccine Research Institute of
    San Diego
and
VLP Biotech, Inc.
San Diego, California

Michael Nassal
Department of Internal Medicine
    2/Molecular Biology
University Hospital Freiburg
Freiburg, Germany

Thi Thai An Nguyen
Department of Internal Medicine
    2/Molecular Biology
University Hospital Freiburg
Freiburg, Germany

and

Laboratory Department
30-4 Hospital
Ho Chi Minh City, Vietnam

Avnish Patel
London School of Hygiene & Tropical
    Medicine
London, United Kingdom

Dustin Patterson
Department of Chemistry
    and Biochemistry
University of Texas at Tyler
Tyler, Texas

David S. Peabody
Department of Molecular Genetics
    and Microbiology
School of Medicine
University of New Mexico
Albuquerque, New Mexico

Darrell Peterson
Department of Biochemistry
Virginia Commonwealth University
Richmond, Virginia

Hadrien Peyret
Department of Biological Chemistry
John Innes Centre
Norwich, United Kingdom

Paul Pumpens
Latvian Biomedical Research
    and Study Centre
Riga, Latvia

Peter Pushko
Medigen, Inc.
Frederick, Maryland

John A. Robinson
Department of Chemistry
University of Zurich
Zurich, Switzerland

David J. Rowlands
Faculty of Biological Sciences
School of Molecular and Cellular
    Biology
and
Astbury Centre for Structural
    Molecular Biology
University of Leeds
Leeds, United Kingdom

Polly Roy
London School of Hygiene & Tropical
    Medicine
London, United Kingdom

Matti Sällberg
Division of Clinical Microbiology
Department of Laboratory Medicine
Karolinska Institutet
Karolinska University Hospital
    Huddinge
Stockholm, Sweden

Kestutis Sasnauskas
Institute of Biotechnology
Vilnius University
Vilnius, Lithuania

Pooja Saxena
Department of Chemistry
Indiana University
Bloomington, Indiana

Benjamin Schwarz
Department of Chemistry
Indiana University
Bloomington, Indiana

Kenza Snoussi
Faculty of Medicine
Department of Infection Biology
University of Tsukuba
Tsukuba, Japan

Hana Španielová
Faculty of Science
Department of Genetics and
    Microbiology
Charles University in Prague
and
Institute of Organic Chemistry and
    Biochemistry
Academy of Sciences of the Czech
    Republic
Prague, Czech Republic

Sam L. Stephen
Faculty of Biological Sciences
School of Molecular and Cellular
    Biology
and
Astbury Centre for Structural
    Molecular Biology
University of Leeds
Leeds, United Kingdom

Nicola J. Stonehouse
Faculty of Biological Sciences
School of Molecular and Cellular
    Biology
and
Astbury Centre for Structural
    Molecular Biology
University of Leeds
Leeds, United Kingdom

Jiřina Suchanová
Faculty of Science
Department of Genetics
    and Microbiology
Charles University in Prague
Prague, Czech Republic

Ming Tan
Division of Infectious Diseases
Cincinnati Children's Hospital Medical
    Center
Cincinnati, Ohio

Kaspars Tars
Latvian Biomedical Research
    and Study Centre
and
Department of Molecular Biology
University of Latvia
Riga, Latvia

Irina Tsvetkova
Department of Chemistry
Indiana University
Bloomington, Indiana

Terrence M. Tumpey
Influenza Division
Centers for Disease Control
    and Prevention
Atlanta, Georgia

Rainer G. Ulrich
Institute for Novel and Emerging
    Infectious Diseases
Friedrich-Loeffler-Institut
Federal Research Institute for Animal
    Health
Greifswald-Insel Riems, Germany

Jelena Vasilevska
Latvian Biomedical Research
    and Study Centre
Riga, Latvia

Andreas Walker
Institute of Virology
Heinrich-Heine-University Düsseldorf
Düsseldorf, Germany

Joseph Che-Yen Wang
Department of Molecular and Cellular
    Biochemistry
Indiana University
Bloomington, Indiana

David Whitacre
Vaccine Research Institute of
    San Diego
and
VLP Biotech, Inc.
San Diego, California

Xiaojun Xu
Division of Viral Hepatitis
Centers for Disease Control
Atlanta, Georgia

Franziska Zabel
University Hospital Zurich
Zurich, Switzerland

and

Saiba GmbH
Rämismühle, Switzerland

Anna Zajakina
Latvian Biomedical Research
    and Study Centre
Riga, Latvia

Fidel Zavala
Department of Molecular Microbiology
    and Immunology
Bloomberg School of Public Health
Johns Hopkins University
Baltimore, Maryland

Andris Zeltins
Laboratory of Plant Virology
Latvian Biomedical Research
    and Study Centre
Riga, Latvia

Adam Zlotnick
Department of Molecular and Cellular
    Biochemistry
Indiana University
Bloomington, Indiana

Aurelija Zvirbliene
Institute of Biotechnology
Vilnius University
Vilnius, Lithuania

Introduction

Yet the spirit of an epoch is reflected not in the arts alone, but in every field of human endeavor, from theology to engineering.

Music in Western Civilization **by Paul Henry Lang**

Nanotechnology is a rapidly progressing field of science that unites a broad range of diverse disciplines from biomedicine, including microbiology, virology, immunology, and vaccinology, to material science, including organic and inorganic chemistry, semiconductor physics, and microfabrication.

Although many definitions are in use, nanotechnology is commonly defined as a knowledge-based manipulation of matter sized from 1 to 100 nm with a special interest in self-assembling and self-regulating systems. Considering that many viral structural proteins match this definition, it is not surprising that nanotechnology embraced and explored various applications for viral capsids and envelopes. At the dawn of genetic engineering in the late 1970s of the twentieth century, macrostructures spontaneously assembled from recombinant viral structural proteins and resembling viral capsids and envelopes were described as *virus-like particles* (*VLPs*). Over recent years, these structures became also known as *virus-like nanoparticles* and *viral nanoparticles* (*VNPs*), reflecting a changing attitude of molecular researchers in the field from observation to technology. This transition resulted in the formulation of *viral nanotechnology* as an independent discipline within *bionanotechnology*, which involves all biological materials.

The subject of viral nanotechnology and its methodological foundation and applications are still being conceptualized. Existing concepts of the field will be certainly challenged and developed further. Nevertheless, the field is already vast and cannot be captured in a single paper or a single book. This volume contains a collection of chapters that just reflect the current state of viral nanotechnology, including basic methodologies (Section I) and selected applications clustered according to VLP structures (Section II). The history of the discipline is still to be written; however, it can be somewhat discerned from the order of applications presented in the book.

Hepatitis B virus (HBV) is one of the very first objects explored by recombinant gene expression technologies (Chapter 10). The initial attempts to express the HBV surface antigen (HBsAg) or envelope protein in *Escherichia coli*, the most popular bacterial host cell for expression of recombinant proteins, resulted, however, in a protein product that only faintly reproduced antigenic properties of the plasma-derived HBsAg. This setback was, however, rapidly overcome by expression of recombinant HBsAg in yeast and mammalian cells. Detection of HBV VLPs in extracts of yeast cells transformed with recombinant plasmids was a stunning demonstration of self-assembly of viral envelope proteins expressed in the heterologous expression system (Chapter 10). Discovery of native immunological properties, including capacity to elicit strong neutralizing immune responses upon immunization, not only prompted vaccine applications of the recombinant HBV VLP (Chapter 10) but re-enforced the notion, common at the time, of importance for the envelope macrostructure to be accurately reproduced in order to model antigenic epitopes. Although experiments with synthetic peptides and recombinant proteins showed later that antigenic epitopes, including conformation-dependent antigenic epitopes, can be efficiently modeled without the actual reproduction of the native envelope macrostructure, self-assembly of viral proteins continues to be an attractive fundamental property guiding a multitude of viral applications (Chapter 2), with VLPs remaining desirable targets for the development of vaccines and diagnostic reagents (Chapter 7).

The early experiences with HBsAg showed that the accuracy of functional reproduction of viral proteins is dependent on the used expression system (Chapter 10). However, selection of the appropriate expression system is not defined by functional reproduction alone but also by technological requirements that make the target nanostructures readily available for applications at low cost (Chapters 16 and 17). Technology of expression of recombinant biomaterials with desirable properties made giant strides since the time of the use of single-cellular bacterial and eukaryotic expression systems that dominated biotechnological applications years ago. Among many, plant expression systems have been recently developed into powerful biological factories for mass-scale production of nanoparticles from a large variety of viruses for vaccine applications (Chapter 16) and for manufacturing novel nanomaterials (Chapter 17).

Applications explored by viral nanotechnology, however, go beyond the selection of expression systems. Transition from functional reproduction of natural structures to rational design of particles with novel properties is a major trend epitomized by viral nanotechnology (Chapter 15). Harnessing of protein folding and assembly of proteins into higher-order structures is essential for application of viral proteins in nanotechnology. Recent advances in resolution of electron cryo-microscopy to the range of 4 Å for icosahedral and helical viral macrostructures and 20 Å for complex polymorphic virions (Chapter 3) and x-ray crystallographic analysis to the atomic resolution of 2 Å (Chapter 4) drastically improved our knowledge of viral architectures and prompted development of highly accurate computational approaches for prediction of protein 3D structures (Chapter 5). Availability of high-quality structural information is key to the rational design of VLP with predetermined properties. The introduction of different functional entities at precisely defined sites of viral structures allows for the development of synthetic VLP (Chapter 29) and extends application of VLP beyond vaccines (Chapter 6).

Nothing in viral nanotechnology has a long history. All is recent, including applications such as nanomedicine (Chapter 8). However, even a short history has its champions. Although nanotechnology has explored structural proteins from almost all major viral families at one time or another (Section II), the contribution of the HBV core protein (HBc) to the field is most long-lasting. Three decades of research made this protein into a legendary model system of viral nanotechnology (Chapters 11 through 14). A robust assembly of HBc into a highly immunogenic structure, which is greatly amenable to introduction of exogenous epitopes, was discovered during the early days of genetic engineering and was immediately explored for the development of a novel generation of vaccines (Chapter 11). Using this model, several unconventional methodologies were introduced for the controlled formation of nanoparticulate structures with predefined properties (Chapter 12) and for boosting immune responses against DNA vaccines (Chapter 14). Unique nanotechnological properties of HBc led researchers to explore core proteins of the other members of Hepadnaviridae, of which HBV is a prototype species. Woodchuck hepatitis virus core (WHc) was used as a carrier for malaria antigenic epitopes. This antigen shares many vaccine-relevant properties of HBc except for HBV-specific immunoreactivity. Thus, the WHc-based malaria vaccine won't immunoreact with an antibody against HBc. Considering that ~15% of human population experienced HBV infection, cross-immunoreactivity with anti-HBc could've compromised efficacy of the HBc-based vaccines, especially in hepatitis B endemic regions of the world (Chapter 13).

To the general scientific community, vaccine development remains the most known application of VLPs (Chapter 9). Indeed, VLPs are natural vaccine targets. Surface geometry of repetitive 3D motives improves avidity binding of VLPs to B-cell receptors and formation of immune complexes, thus promoting strong immune responses that cannot be induced by a single protein (Chapter 7). Viral structural proteins form a variety of particles of different sizes and shapes (Chapter 1). Considering that the VLP macrostructure has a significant effect on immune responses (Chapter 7), VLPs formed from structural proteins of different viruses offer numerous opportunities for novel approaches to vaccines (Section II).

With certain exceptions, viruses are miniature genetic systems capable of encoding only a few different proteins. In order to support the infectious process, viral proteins must be polyfunctional. Structural proteins are responsible for packaging, protection, and delivery of viral genetic material, which is accomplished by binding to cellular receptors, internalization, and disassembly of virions. These proteins also play a key role in evading innate and adaptive immune responses. Polyfunctionality offers a wealth of opportunities and presents a multitude of challenges for nanotechnological applications of viral proteins. Manipulation of the protein structure allows for augmenting or boosting useful functions and suppressing or eliminating the undesirable functions of viral proteins (Section II). Different applications target different viral properties. For example, gene therapy makes use of the nucleic acid–transporting properties of VLPs rather than their capacity to elicit immune responses, while vaccine applications make use of the VLPs' high immunogenicity rather than their binding to cellular receptors and protection of nucleic acids.

Moreover, polyfunctionality of viral proteins presents opportunities for the development of different approaches to the same applications. For example, not all VLP applications to vaccines are alike. VLP can be used directly as a vaccine (Chapters 10 and 19 through 21) or as a carrier of epitopes (Chapters 9, 11 through 13, 20, and 21). In the first case, the immunogenic VLP properties and capacity to elicit the VLP-specific immune response are used to induce protective immunity in vaccinated hosts. In the second case, though, the challenge is to use immunogenic properties of VLP to direct immune response against the inserted epitopes rather than against the carrier itself. Although structural proteins from all viral families can be used as epitope carriers (Section II), phage VLPs are most suitable to carry exogenous epitopes. These VLPs are highly immunogenic and incapable of interacting with antibody against human pathogens (Chapter 15). Antibody binding to the carrier protein reduces efficacy of vaccination against the inserted epitopes, especially in human populations with high prevalence of the carrier-specific antibody (Chapter 13).

VLP-based vaccines are not just dreams of a distant future; they are reality of the present. HBV (Chapter 10) and human papilloma virus (HPV) (Chapters 19 and 20) VLPs are efficient vaccines of today that have already saved many human lives from debilitating and deadly diseases. Both viruses cause malignant transformation of infected cells. HBV vaccine is heralded as the very first anticancer vaccine that prevents the development of liver malignancy. HPV VLPs offer a very potent treatment against HPV-induced tumors. A variety of VLPs from different viral families were generated for cancer therapy (Chapter 27) and evaluated for treatment of various forms of cancer, including melanoma, hepatocellular carcinoma, and malignant tumors of different organs (Chapter 28).

All viruses have a capsid that packages nucleic acids. Many viruses have an additional lipoprotein shell or envelope that covers the capsid. Both capsid and envelope proteins may form VLPs upon production in heterologous expression systems. The envelope VLPs are mainly applied to vaccine development (Chapter 10). However, applications of the capsid VLPs are much broader and not restricted to vaccines against infections with nonenveloped viruses or immunotherapy (Chapters 23, 27, and 28). Their capability to form complexes with exogenous nucleic acids is used for the specific delivery of DNA vaccines and for gene therapy by gene transfer and silencing (Chapter 20).

The range of VLP applications continuously increases, with each passing year bringing new ideas for the biomedical use of protein nanostructures. For example, application of the polyoma- and papilloma-derived VLPs for gene delivery was extended to the use of the same VLPs as nanocontainers for exogenous proteins and peptides with pharmacological properties. Nanocontainers protect the enclosed cargo from proteases and immune responses, preventing its degradation. Such protection helps to sustain a pharmacological concentration of proteins and peptides, which significantly improves their therapeutic effects (Chapter 20). Development of nanoreactors from bacterial microcompartments, or closed capsid-like structures, which encapsulate enzymes and enzymatic pathways, is another extraordinary advancement in nanotechnology that revolutionizes the use of enzymatic reactions, making them considerably more efficient than can be attained by a simple mixing of enzymes (Chapter 22). Although viral nanotechnology is destined to revamp many fields of human endeavor, from electronics to medicine, none of them will be as affected as health care and public health (Chapter 8). In addition to prevention and therapy (Section II), viral nanotechnology will transform diagnostics by devising, for example, VLP-enabled imaging probes for optical and magnetic resonance imaging and for positron-emission tomography (Chapter 23).

Increase in demand for various applications motivates exploration of more complex viral nanostructures. If we look at the developments in the field over the last several years, we will see that viral nanotechnology is progressing from using mainly single-protein VLPs to nanostructures composed of more than one or several proteins (Chapters 18 and 26) and from icosahedral to asymmetrical VLPs (Chapters 24 and 25). These trends do not only reflect significant advancements of molecular technologies, enabling production of such complex protein systems, but point to a rapid conceptual evolution of nanotechnology from needs of yesterday to anticipations of tomorrow.

Nanotechnology is built on actionable ideas. Transition from mere exploring to active improving the world is a major development in scientific thinking of the time, with nanotechnology having been born from this development. The thrust forward experienced by the field of viral nanotechnology in recent years is overwhelming and inspires expectations that people of this generation will see improvement in health and life on a scale unprecedented in the history of humankind.

Yury Khudyakov
Paul Pumpens
Wolfram H. Gerlich

Section I

Major Concepts

1 Introduction to Capsid Architecture

Kaspars Tars

CONTENTS

1.1 MULTIMERIC ORGANIZATION OF VIRAL PARTICLES

As has been recognized by Watson and Crick, regardless of the size, viral genomes cannot code a protein molecule that is big enough to cover the entire genome—for a simple reason that one amino acid is coded by three nucleotides; hence, protein is always smaller than its gene. Therefore, multiple copies of coat protein are needed to enclose the genome. Since in many cases all coat protein monomers are identical, Watson and Crick reasoned that they are presumably engaged in the same protein–protein interactions with each other; that is, each monomer has identical protein environment [1]. Mathematically, there are only two ways to enclose a part of space by identical objects in identical environments. First, objects can be arranged in a helical fashion, thereby enclosing a cylindrical region of space. Second, objects can be arranged in facets of polyhedron, enclosing roughly spherical volume. There are only five regular polyhedrons (also called Platonic solids). Each face of the polyhedron can be divided into three (for tetrahedron, octahedron, and icosahedron), four (for cube), or five (for dodecahedron) identical triangles (Figure 1.1a). Therefore, tetrahedron, having 32 symmetry, can be regarded as built from 12 identical triangular units, cube and octahedron, having 432 symmetry—from 24 units, dodecahedron and icosahedron, both having 532 symmetry (Figure 1.1b)—from 60 units. Therefore, identical subunits enclose the biggest volume in 532 symmetrical arrangement. Viruses utilize both helical and icosahedral arrangement of subunits, although the latter seems to be more common.

1.2 ICOSAHEDRAL CAPSIDS

Icosahedron can be regarded as built from 60 identical units—called icosahedral asymmetric units. Each icosahedral asymmetric unit occupies one-third of an icosahedron face (Figure 1.1c). The word *icosahedral* in the world of viruses case simply implies type of symmetry—since there are no actual lines and planes in virus particles, the arrangement of subunits in principle could be regarded as *icosahedral* or *dodecahedral*, both having 532 symmetry. Morphologically, however, capsids usually look spherical or icosahedral, but rarely (if ever) dodecahedral—this is due to the fact that subunit contacts are usually flat at the threefold symmetry axes and bent at fivefold axes, as explained in Figure 1.2. Also note that pure mathematical icosahedron in addition to 532 symmetry has a number of reflection planes and inversion centers, which are absent in viruses.

Although some viruses indeed utilize strict icosahedral arrangement of capsid, built from chemically and structurally identical monomers, in this case, the obvious limitation is the particle size, which is restricted to 60 subunits. Although capsid size can be increased by simply utilizing larger monomers, a more economical way would be to use more monomers. Indeed, most icosahedral viruses utilize the so-called quasiequivalence principle, first described by Caspar and Klug in 1962 [2], where chemically identical coat protein subunits pack in slightly different environments on icosahedron surface. Each icosahedral asymmetric unit is now divided into a number of smaller, quasiequivalent triangles, a process called triangulation. The consequence is that coat protein monomers are no longer structurally identical, but have slightly different (quasiequivalent) conformations. Due to geometric restrictions, icosahedral particles can be built from T × 60 monomers, where the so-called triangulation number $T = h^2 + hk + k^2$, h and k being any nonnegative integers. To understand the quasiequivalence principle, let us first imagine that coat protein subunits are packed in hexagons, as shown in Figure 1.2a. If we now would attempt to take out one monomer from hexamer (Figure 1.2b) and join the free sides together (Figure 1.2c), we would end up

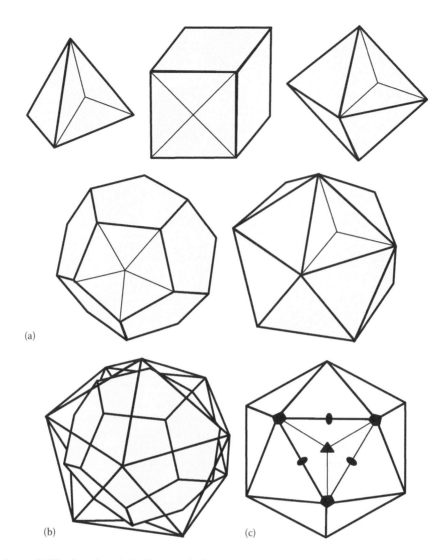

FIGURE 1.1 Polyhedrons. (a) The five platonic bodies: tetrahedron, cube, octahedron, dodecahedron, and icosahedron. The face of each polyhedron can be divided into three, four, or five identical triangles (also called asymmetric units, shown for one face of each polyhedron). (b) Dodecahedron and icosahedron have the same type of symmetry, which can be visualized by placing dodecahedron in icosahedron cage. (c) Icosahedron has a number of symmetry axes, shown for one face. Fivefold symmetry axes pass through vertices, threefold axes through middle of faces, and twofold axes through middle of edges. One icosahedral asymmetric unit is shown in gray.

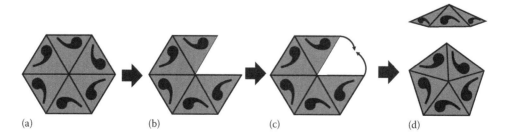

FIGURE 1.2 Hexamers and pentamers. (a) Protein monomers packed in hexameric fashion. (b) If we now take one monomer out and (c) force the two *free* protein sides to interact each with other, the result will be a concave pentamer (d).

with a concave pentamer (Figure 1.2d). Notice that contact surface between monomers in hexamers and pentamers is largely the same, so only minor changes in protein structure would be necessary to accommodate both pentameric and hexameric structures. Next, let us imagine that coat protein monomers are first packed in a planar, hexagonal environment, like honeycomb (Figure 1.3a, note that this is just an imagination and not necessarily the case for real virus assembly). Then, if we replace some coat protein hexamers by regularly interspersed pentamers, this results in curvature of planar surface, eventually leading to closed icosahedral structure (Figure 1.3b and c). h and k indices

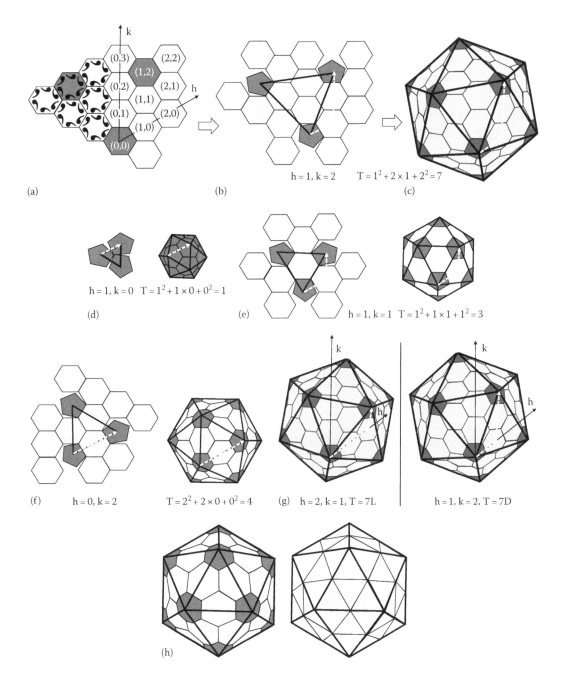

FIGURE 1.3 The quasiequivalence principle. (a) Assume that coat protein subunits (shown as black commas on the left side) are arranged in hexagonal grid. Positions of grid points can be described by two indices, h and k, along two hexagonal axes. If regularly interspersed hexamers (shown in dark gray) are replaced by pentamers (b), the surface begins to curve, until icosahedral object is formed with pentamers at vertices (c). Path from one pentamer to the next one can be described by a vector, represented by two indices, h and k, denoting the number of steps along two axes on hexagonal grid. In the example, illustrated in panels a–c case when h = 1 and k = 2 is shown, resulting in T = 7 symmetry. (d) If h = 1 and h = 0, pentamers are joined directly with no interspersed hexamers, this corresponds to T = 1 lattice, (e) h = 1 and k = 1 corresponds to T = 3 lattice, (f) h = 1 and k = 0 corresponds to T = 4 lattice. (g) if h ≠ k and k ≠ 0, laevo (l) and dextro (d) arrangements are possible, both grids being mirror images of each other, as illustrated for T = 7D and T = 7L cases. (h) There are two alternative ways to show quasiequivalent grid on the surface of icosahedron—by drawing the pentagons and hexagons (pentagon–hexagon system) or by drawing lines, connecting centers of pentagons and hexagons (triangle system). Both systems are shown for T = 4 grid.

then represent the number of steps along two axes in the hexagonal grid to reach the next pentamer from the previous one. In the simplest case (h = 1, k = 0, $T = 1^2 + 1 \times 0 + 0^2 = 1$), the pentamers are directly connected and there are no hexamers at all in the structure—this corresponds to the simplest icosahedron, built from 60 subunits (Figure 1.3d).

The next simplest arrangement corresponds to h = 1 and k = 1, $T = 1^2 + 1 \times 1 + 1^2 = 3$ (Figure 1.3e); hence, icosahedron is now built of 180 chemically identical subunits. By increasing h and k indices, T numbers fall in series 1 (h = 1, k = 0), 3 (h = 1, k = 1), 4 (Figure 1.3f, h = 2, k = 0), 7 (h = 1, k = 2 or h = 2, k = 1), 9 (h = 3, k = 0), and so on.

Consequently, icosahedral lattices, corresponding to some T values like 2, 5, or 6, are geometrically impossible. If h = k or if h = 0 (or k = 0), edges of icosahedron dissect hexagons in two distinct symmetric ways as exemplified in Figure 1.3e and f for cases when h = k = 1 and h = 0, k = 2. In all other cases, that is, when h ≠ k and h ≠ 0 (or k ≠ 0), the icosahedron is called *skewed*, since the edges of icosahedron are not symmetrically disposed with respect to hexagonal lattice lines. Consequently, for all skewed icosahedrons, two different *hands* exist. For example, T = 7 lattice exists in two forms (Figure 1.3g)—T = 7l (h = 1, k = 2) and T = 7d (h = 2, k = 1). Some higher T numbers can be obtained in more than two distinct ways—for example, T = 49 can be

obtained from h = 3, k = 5; h = 5, k = 3 or h = 7, k = 7. The grid points on icosahedron can be shown in two alternative ways (Figure 1.3h)—by showing the actual pentagons and hexagons (pentagon–hexagon system) or by connecting their midpoints (triangular system). In Figure 1.4, T numbers for h and k indices from 0 to 10 are shown. In Figure 1.5 structures of T = 3 bacteriophage MS2 (PDB code 2MS2 [3]) and T = 4 hepatitis B core antigen (PDB code 1QGT [4]) are shown along with their lattices in triangular system.

Another way of describing symmetric and skewed icosahedrons is to define triangulation number as $T = Pf^2$, where $P = h^2 + hk + k^2$, f is the largest common divisor between h and k and h and k—integers without any common factors.

k→ h ↓	1	2	3	4	5	6	7	8	9	10
0	1	4	9	16	25	36	49	64	81	100
1	3	7L	13L	21L	31L	43L	57L	73L	91L	111L
2	7D	12	19L	28L	39L	52L	67L	84L	103L	124L
3	13D	19D	27	37L	49L	63L	79L	97L	117L	139L
4	21D	28D	37D	48	61L	76L	93L	112L	133L	156L
5	31D	39D	49D	61D	75	91L	109L	129L	151L	175L
6	43D	52D	63D	76D	91D	108	127L	148L	171L	196L
7	57D	67D	79D	93D	109D	127D	147	169L	193L	219L
8	73D	84D	97D	112D	129D	148D	169D	192	217L	244L
9	91D	103D	117D	133D	151D	171D	193D	217D	243	271L
10	111D	124D	139D	156D	175D	196D	219D	244D	271D	300

FIGURE 1.4   T numbers of icosahedral viruses for h and k indices from 1 to 10. P = 1 class (h = 0) cases are shown in yellow, P = 3 class (h = k) cases in gray. All other T numbers belong to skew class with D (blue) and L (orange) variants.

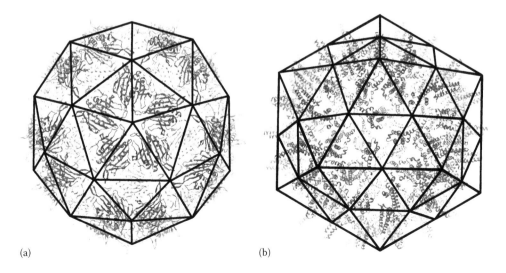

(a)   (b)

FIGURE 1.5   Subunit arrangement in T = 3 and T = 4 capsids. (a) T = 3 capsid of RNA phage MS2. (b) T = 4 capsid of hepatitis B core antigen.

Notice that h and k values using this T number definition are not necessarily the same as in definition $T = h^2 + hk + h^2$. $P = 1$ icosahedrons correspond to symmetric case as described earlier as h = k; $P = 3$ icosahedrons correspond to symmetric case, where h = 0 (or k = 0), whereas icosahedrons with all other P values are skewed.

Viruses utilize a wide variety of T numbers. Some of the simplest viruses have T = 1 particles—for example, plant satellite viruses, like satellite tobacco necrosis virus [5]. On the other extreme, algal dsDNA viruses, like *Paramecium bursaria* chlorella virus 1 (PBCV-1), has T = 169d symmetry [6], *Phaeocystis pouchetii* virus (PpV01) T = 219d symmetry [7]. Mimiviruses are even larger—from cryo-EM reconstruction [8] it has been estimated that the outermost layer of the capsid has h = 19 ± 1 and k = 19 ± 1 (uncertainty comes from difficulty to count capsomeres close to vertices at a relatively low, 65 Å resolution). Hence, the actual T number must be in the range from 972 (if both h and k are 18) to 1200 (if both h and k are 20).

1.3 SPECIAL CASES OF T = 2 AND T = 6

Although, as stated earlier, T = 2 and T = 6 lattices are geometrically impossible, the actual number of coat protein monomers in icosahedral asymmetric unit somewhat unexpectedly can correspond to forbidden numbers like 2 and 6. The *T = 6* phenomenon is observed in papovaviruses (polyoma and papilloma viruses), where both hexavalent and pentavalent positions are occupied by pentamers (Figure 1.6), resulting in 360 subunits in T = 7 lattice instead of expected 420 subunits [9,10]. Monomers of papovaviruses have fairly flexible C-termini, which can exist in a variety of very different conformations, enabling coat protein pentamers to interact with either five or six neighboring pentamers, which is normally not the case for other viruses.

Nucleocapsids of dsRNA viruses, notably rotaviruses and reoviruses, are built from three layers. The outer layers have T = 13 symmetry, while the inner is made from 120 chemically identical monomers, as first demonstrated for bluetongue virus [11]. In this case, two coat protein monomers are occupying one T = 1 lattice position; therefore, the capsid can be regarded as T = 2. Also in this case, the 3D structures of both chemically identical monomers are significantly different.

1.4 PSEUDOSYMMETRY

Many viruses employ more than one coat protein type to build a capsid. For example, capsids of picornaviruses are built from three distinct coat proteins, each encoded by a separate gene. All three proteins (VP1, VP2, and VP3) share little sequence similarity, yet are surprisingly similar in 3D structure [12–15]. The resulting capsid formally has T = 1 symmetry, each equal lattice position being occupied by a single VP1–VP2–VP3 trimer. However, the monomer arrangement is very similar to that of T = 3 capsid, but since the 180 monomers are chemically distinct, it is called P = 3 (**P**seudo-3) symmetry, sometimes also referred to as pT = 3 symmetry.

A more complicated situation is observed in adenoviruses, for which both x-ray and cryo-EM high resolution structures have been determined in recent years [16,17]. Adenoviruses have very icosahedral morphology and even in conventional EM, it is easy to see pentamers and hexamers and even deduce the triangulation number of capsid (Figure 1.7a). However, adenovirus capsid is built from a number of distinct proteins, the major ones being penton base and hexon (Figure 1.7b and c). Penton base forms pentamers and is located at vertices. Hexon is actually a trimer that has hexagonal shape (Figure 1.7d). Each hexon monomer, however, contains two structurally similar domains. Therefore, hexon can be also regarded as built from six similar domains. The resulting symmetry can be described as P = 25. Apart from hexons and pentons, adenoviruses utilize a number of other structural proteins. For example, cement proteins VIII and III hold together pentons and hexons. Similar cement or glue proteins are also observed in other viruses, for example, herpes viruses [18].

1.5 PROLATE ICOSAHEDRONS

Some viruses utilize modified icosahedrons, which can be elongated or twinned. Icosahedral particles can be regarded as built from two caps, connected by a central cylindrical

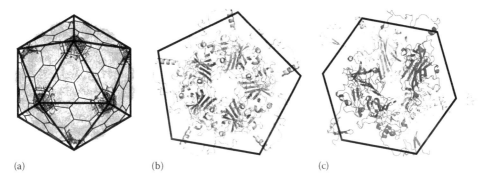

(a) (b) (c)

FIGURE 1.6 Subunit arrangement in papovaviruses. (a) Coat protein monomers are arranged in T = 7l lattice. (b) As expected, pentavalent positions are occupied by five chemically and structurally identical monomers. (c) Hexavalent positions are also occupied by five chemically identical monomers (shown in different colors). Structurally, however, monomers in hexavalent positions differ by their C-terminal parts having quite different conformations.

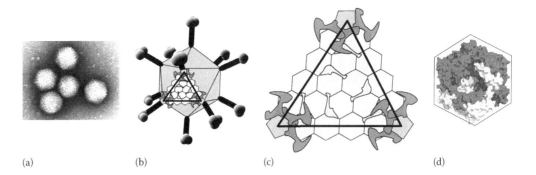

(a)   (b)   (c)   (d)

FIGURE 1.7 Architecture of adenovirus. (a) Adenovirus particle has a very icosahedral morphology, pentameric and hexameric units can be easily seen in conventional EM. (b) Adenoviruses have trimeric spike protein, attached to each penton. (c) Schematic drawing of one icosahedral face of adenovirus. Hexons and pentons are represented as yellow hexagons and orange pentagons, respectively. Glue proteins VIII (blue) and III (green) are shown according to their approximate shape. (d) Although hexon clearly has a very hexagonal shape, it is actually a trimer—three monomers are shown in yellow, red, and blue.

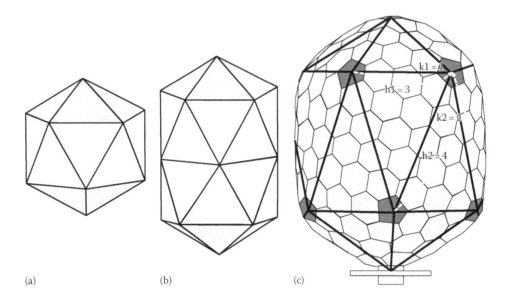

(a)   (b)   (c)

FIGURE 1.8 Prolate icosahedrons. (a) Icosahedron can be regarded as built from two caps (gray) and central body. (b) The central body can be elongated by adding more central sections. (c) Geometry of prolate $T = 4$ phage head can be described by two sets of indices—h1 and k1, describing the vector between two vertices of caps and h2 and k2—describing the vector between two vertices across the central body. The lower pentameric vertex is replaced by a baseplate complex.

body (Figure 1.8a). In case of prolate icosahedrons, the cylindrical section is elongated (Figure 1.8b). Perhaps, the best known prolate icosahedron is that of phage $T = 4$ head, geometry of which is shown in Figure 1.8c. For prolate particles, capsid morphology can be described with two numbers—T number, which describes symmetry of caps much in the same way as for convenient icosahedral viruses and Q number, which describes the length of cylindrical body. The total number of subunits then can be calculated as $30(T + Q)$. Two sets of indices are used—h1 and k1 and h2 and k2. h1 and k1 indices represent a vector between two vertices of icosahedral cap, while h2 and k2—vector between vertices across the elongation. Then, $T = h1^2 + h1k1 + k1^2$ and $Q = h1h2 + h1k2 + k1k2$. If $Q = T$, icosahedron is not prolate but normal. Phage T4 head can be described as $T = 13$, $Q = 20$ [19]. Notice that in contrast to T number, Q number can have any integer value, greater than or equal to T.

1.6 TWINNED ICOSAHEDRONS

Twinned icosahedrons occur in plant viruses belonging to *Geminiviridae* family. In this case, particle is built from two truncated $T = 1$ icosahedrons, each missing a single pentamer (notice that pentamer is a smaller unit compared to cap as discussed earlier). Therefore, the whole particle is built from $2 \times (60 - 5) = 110$ monomers as observed in maize streak virus [20]. In contrast to prolated icosahedrons, twinned icosahedrons have a concavity in the middle (Figure 1.9).

1.7 VIRAL STRUCTURAL PROTEINS THAT DO NOT FOLLOW ICOSAHEDRAL SYMMETRY

Capsids of many icosahedral viruses contain one or several proteins, which do not follow the icosahedral symmetry. For example, $T = 3$ capsid of bacteriophage MS2 is built from

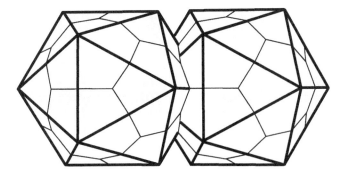

FIGURE 1.9 Geometry of geminiviruses. Two truncated T = 1 icosahedrons, each with a missing pentamer, are joined together. Notice that in contrast to prolate icosahedral particles, geminiviruses have a concavity in the middle.

180 coat protein monomers and a single copy of maturation protein, which is responsible for attachment of phage to bacterial pili [21]. Related phage Qβ in addition to coat and maturation proteins has three to five copies of elongated read-through version of coat protein, which also appears to be necessary for infectivity [22].

Adenoviruses utilize a trimeric fiber (or *spike*) protein, which is bound to each penton base [23]. Spike protein is responsible for attachment to cellular receptors.

In many viruses, notably dsDNA phages, one vertex of icosahedral head is replaced by a portal protein, through which the genome is inserted after assembly. For example, cryo-EM reconstruction of phage P22 has revealed a 12-mer of portal protein sitting at one pentameric vertex [24]. There is strong evidence that besides being the genome packaging motors, portals also participate in correct assembly of capsid by interaction with scaffolding proteins (see the following).

1.8 SCAFFOLDING PROTEINS

Some viruses make use of scaffolds during the assembly. Scaffolding proteins are used to guide correct particle assembly, but are absent in assembled, mature capsids. In most cases, scaffolding proteins are encoded by separate genes, but they can also exist as a part of coat protein, which is cleaved off after capsid assembly, as seen in the case of bacteriophage HK97 [25]. Scaffolding proteins can be internal—that is, structural proteins are assembled around the scaffold or external—in which case structural proteins are assembled inside the scaffold cage. Some viruses use both inner and outer scaffolding proteins—for example, a relatively simple T = 1 bacteriophage phiX174 [26]. In some cases, T numbers can be changed by using different scaffolding proteins. Bacteriophage P4 is a satellite virus, dependent on the presence of host phage P2. P4 satellite lacks a coat protein in its genome, so it utilizes coat protein of P2 host. However, P4 encodes an additional external scaffolding protein sid, which forces P2 coat protein to build T = 4 capsids, while normally P2 builds larger T = 7 capsids from the same coat protein [27]. As a result, there is not enough room for larger P2 genome in T = 4 capsids, so smaller P4 genome is packaged instead.

1.9 HELICAL CAPSIDS

An alternative way to form icosahedral capsids to pack the genome is to enclose a cylindrical volume, packaging coat protein monomers in helical fashion. In contrast to icosahedral capsids, where subunits are related by pure rotation, in helical capsids, both rotation and translation components are combined to produce a screw axis. Those structures are sometimes called *open*, since in principle, any given volume can be enclosed by simply varying the number of monomers. Well-known examples include tobamoviruses (e.g., tobacco mosaic virus), potyviruses (e.g., potato virus X), rhabdoviruses (e.g., rabies virus), and filoviruses (e.g., Ebola virus). Also influenza viruses, which are sometimes regarded as being *irregular*, contain nucleic acid complex with nucleoprotein in double-helical arrangement [28]. Helical symmetry is also present in some complex viruses. For example, dsDNA phages, like bacteriophage T4, have a helical tail.

Morphologically, helical capsids may look rather different—some are very rigid, like tobacco mosaic virus, while others are very flexible, like Ebola virus.

For helical capsids, the most important geometric parameters that determine the shape of virion are axial rise per subunit ρ and the number of monomers per turn μ. The pitch of helix then can be defined as P = ρ × μ. In contrast to icosahedral capsids, helical capsids do not have a strictly determined number of monomers. The length of capsid is largely determined by genome size. Since helical capsids are somewhat heterogenous, they are prone to crystallization, and structural information of intact capsids has to be obtained by other means—x-ray fiber diffraction or cryo-EM. However, coat proteins of some helical capsids also form crystallizable disk-like structures, for which high-resolution structures can be solved, as exemplified by rabies virus nucleoprotein–RNA complex [29]. One of the best studied helical viruses is tobacco mosaic virus, for which high-resolution structure has been obtained by fiber diffraction [30]. The capsid of tobacco mosaic virus forms a rigid rod, which is roughly 300 Å long and 180 Å in diameter (Figure 1.10). Axial rise per subunit is 1.2 Å and there are 16 1/3 monomers per turn of helix.

1.10 ENVELOPED VIRUSES

Nucleocapsids of both helical and icosahedral viruses can have a lipid envelope of cellular origin, in which viral membrane proteins are embedded. In some viruses, like HIV and influenza, membrane forms the outer layer of virion (external membrane), while in some other viruses, for example, flaviviruses and alphaviruses, membrane is surrounded by an outer protein layer (internal membrane). Usually, viral proteins in envelope are not arranged symmetrically; however, there are some notable exceptions.

In alphaviruses, such as Sindbis virus [31], Chickungunya virus [32], and Venezuelan equine encephalitis virus [33], two concentric T = 4 protein layers enclose the membrane in between them. The inner protein layer is represented by

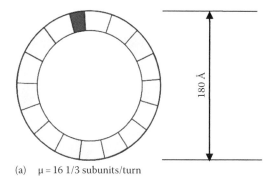

(a)   μ = 16 1/3 subunits/turn

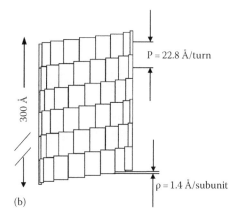

P = 22.8 Å/turn

ρ = 1.4 Å/subunit

(b)

FIGURE 1.10 Geometry of helical nucleocapsid. Schematic representation of helical tobacco mosaic virus capsid from above (a) and from the side (b). There are 16 1/3 coat protein subunits per turn of helix. Axial rise is 1.4 Å/subunit, resulting in helix pitch of 22.8 Å.

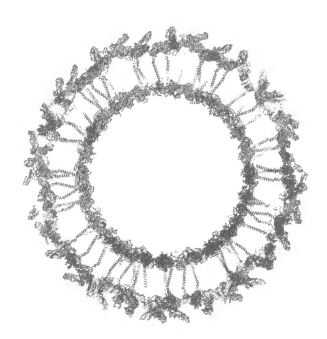

FIGURE 1.11 Organization of Sindbis virus. The T = 4 particle is built of two protein layers—the inner is composed of 240 monomers of coat protein (red), while the outer is composed of 240 copies of heterodimer of two envelope proteins E1 (green) and E2 (blue). Lipid bilayer of cellular origin is located in between the protein layers. E1/E2 proteins are anchored in the membrane with transmembrane helices, which also make contacts with inner capsid.

nucleocapsid protein and the outer by envelope proteins E1 and E2. Both protein layers are interconnected by transmembrane helices (Figure 1.11).

Flaviviruses represent another interesting type of symmetry. Even though the envelope is made of 180 protein monomers, the actual subunit arrangement does not correspond to T = 3 quasisymmetry. As observed in Dengue virus, the particle can be best described as having T = 1 symmetry with three envelope protein copies in each asymmetric unit [34,35].

REFERENCES

1. Crick, F.H. and J.D. Watson, Structure of small viruses. *Nature*, 1956. **177**(4506): 473–475.
2. Caspar, D.L. and A. Klug, Physical principles in the construction of regular viruses. *Cold Spring Harb Symp Quant Biol*, 1962. **27**: 1–24.
3. Valegård, K. et al., The three-dimensional structure of the bacterial virus MS2. *Nature*, 1990. **345**: 36–41.
4. Wynne, S.A., R.A. Crowther, and A.G. Leslie, The crystal structure of the human hepatitis B virus capsid. *Mol Cell*, 1999. **3**(6): 771–780.
5. Jones, T.A. and L. Liljas, Structure of satellite tobacco necrosis virus after crystallographic refinement at 2.5 Å resolution. *J Mol Biol*, 1984. **177**(4): 735–767.
6. Nandhagopal, N. et al., The structure and evolution of the major capsid protein of a large, lipid-containing DNA virus. *Proc Natl Acad Sci USA*, 2002. **99**(23): 14758–14763.
7. Yan, X. et al., The marine algal virus PpV01 has an icosahedral capsid with T = 219 quasisymmetry. *J Virol*, 2005. **79**(14): 9236–9243.
8. Xiao, C. et al., Structural studies of the giant mimivirus. *PLoS Biol*, 2009. **7**(4): e92.
9. Liddington, R.C. et al., Structure of simian virus 40 at 3.8-Å resolution. *Nature*, 1991. **354**(6351): 278–284.
10. Wolf, M. et al., Subunit interactions in bovine papillomavirus. *Proc Natl Acad Sci USA*, 2010. **107**(14): 6298–6303.
11. Grimes, J.M. et al., The atomic structure of the bluetongue virus core. *Nature*, 1998. **395**(6701): 470–478.
12. Hogle, J.M., M. Chow, and D.J. Filman, Three-dimensional structure of poliovirus at 2.9 Å resolution. *Science*, 1985. **229**(4720): 1358–1365.
13. Rossmann, M.G. et al., Structure of a human common cold virus and functional relationship to other picornaviruses. *Nature*, 1985. **317**(6033): 145–153.
14. Luo, M. et al., The atomic structure of Mengo virus at 3.0 Å resolution. *Science*, 1987. **235**(4785): 182–191.
15. Acharya, R. et al., The three-dimensional structure of foot-and-mouth disease virus at 2.9 Å resolution. *Nature*, 1989. **337**(6209): 709–716.
16. Reddy, V.S. et al., Crystal structure of human adenovirus at 3.5 Å resolution. *Science*, 2010. **329**(5995): 1071–1075.
17. Liu, H. et al., Atomic structure of human adenovirus by cryo-EM reveals interactions among protein networks. *Science*, 2010. **329**(5995): 1038–1043.
18. Zhou, Z.H. et al., Seeing the herpesvirus capsid at 8.5 Å. *Science*, 2000. **288**(5467): 877–880.
19. Fokine, A. et al., Molecular architecture of the prolate head of bacteriophage T4. *Proc Natl Acad Sci USA*, 2004. **101**(16): 6003–6008.
20. Zhang, W. et al., Structure of the Maize streak virus geminate particle. *Virology*, 2001. **279**(2): 471–477.

21. Brinton, C.C., P. Gemski, and J. Carnahan, A new type of bacterial pilus genetically controlled by the fertility factor of *E. coli* K12 and its role in chromosome transfer. *Proc Natl Acad Sci USA*, 1964. **52**: 776–783.

22. Rumnieks, J. and K. Tars, Crystal structure of the read-through domain from bacteriophage Qbeta A1 protein. *Protein Sci*, 2011. **20**(10): 1707–1712.

23. van Raaij, M.J. et al., A triple beta-spiral in the adenovirus fibre shaft reveals a new structural motif for a fibrous protein. *Nature*, 1999. **401**(6756): 935–938.

24. Chen, D.H. et al., Structural basis for scaffolding-mediated assembly and maturation of a dsDNA virus. *Proc Natl Acad Sci USA*, 2011. **108**(4): 1355–1360.

25. Huang, R.K. et al., The Prohead-I structure of bacteriophage HK97: Implications for scaffold-mediated control of particle assembly and maturation. *J Mol Biol*, 2011. **408**(3): 541–554.

26. Dokland, T. et al., The role of scaffolding proteins in the assembly of the small, single-stranded DNA virus phiX174. *J Mol Biol*, 1999. **288**(4): 595–608.

27. Dearborn, A.D. et al., Structure and size determination of bacteriophage P2 and P4 procapsids: Function of size responsiveness mutations. *J Struct Biol*, 2012. **178**(3): 215–224.

28. Zheng, W. and Y.J. Tao, Structure and assembly of the influenza A virus ribonucleoprotein complex. *FEBS Lett*, 2013. **587**(8): 1206–1214.

29. Albertini, A.A. et al., Crystal structure of the rabies virus nucleoprotein–RNA complex. *Science*, 2006. **313**(5785): 360–363.

30. Namba, K. and G. Stubbs, Structure of tobacco mosaic virus at 3.6 Å resolution: Implications for assembly. *Science*, 1986. **231**(4744): 1401–1406.

31. Tang, J. et al., Molecular links between the E2 envelope glycoprotein and nucleocapsid core in Sindbis virus. *J Mol Biol*, 2011. **414**(3): 442–459.

32. Sun, S. et al., Structural analyses at pseudo atomic resolution of Chikungunya virus and antibodies show mechanisms of neutralization. *Elife* (Cambridge), 2013. **2**: e00435.

33. Zhang, R. et al., 4.4 Å cryo-EM structure of an enveloped alphavirus Venezuelan equine encephalitis virus. *EMBO J*, 2011. **30**(18): 3854–3863.

34. Li, L. et al., The flavivirus precursor membrane-envelope protein complex: Structure and maturation. *Science*, 2008. **319**(5871): 1830–1834.

35. Zhang, X. et al., Cryo-EM structure of the mature dengue virus at 3.5-Å resolution. *Nat Struct Mol Biol*, 2013. **20**(1): 105–110.

2 Self-Assembling Virus-Like and Virus-Unlike Particles

Adam Zlotnick, Samson Francis, Lye Siang Lee, and Joseph Che-Yen Wang

CONTENTS

2.1 OVERVIEW

Viruses are robust. Their architecture allows them to endure changes in their chemical environments such as temperature, ionic strength, pH, and hydration. They must be stable enough to assemble and protect their genomic cargo yet fragile enough to release that cargo in response to a subtle trigger. In many cases, release of the viral genome is accompanied by the dissociation of the virus.

The evolutionary forces leading to viruses have incorporated nuance not found in designed molecules. The interplay between uniformity, structure, stability, and function has excited the interest of scientists working at the frontier of materials science and biotechnology. Virus-like particles (VLPs) have monodispersity and defined surface chemistry. Their self-assembly can extend from subunits forming nanoparticles to nanoparticles forming high-order hierarchical structures. Both the interior and exterior of viral capsids have been exploited for organization and storage of abiotic cargoes. Viral capsids have been repurposed as nanocontainers for a wide variety of functions including drug encapsidation/delivery, constrained polymer synthesis, and catalysis.

There are three characteristics shared by viruses used in nanotechnology that simplify their use in assembly studies and as materials. They are easily expressed in quantity; the preponderance of bacteriophages on the list is not accidental. They are all simple; capsids are constructed from one or two gene products. The capsid proteins (CPs) of these viruses self-assemble in a controllable manner, though in some cases, a scaffold, nucleic acid (NA), or charged cargo is required. One other virus, bacteriophage M13, has been an important vehicle for virus-based nanotechnology because of its utility in phage display and its ability to form complexes of viruses

[50,146,160]. It is worth considering simplicity, self-assembly, and supramolecular self-assembly in some detail.

In 1956, Crick and Watson observed that "it is a striking fact that almost all small viruses are either rods or spheres" and proposed that "these shells are constructed from a large number of identical protein molecules, of small or moderate size, packed together in a regular manner" [36]. It is altogether striking that these incredibly sophisticated molecular machines actually self-assemble. This was shown with the rodlike (helical) tobacco mosaic virus (TMV) in 1955 and spherical (icosahedral) cowpea chlorotic mottle virus (CCMV) in 1967 [6,52]. A half-century later, TMV and CCMV are still the focus of research on virus assembly and virus modification, and they have been joined by several other viruses including hepatitis B virus (HBV), bacteriophage P22, bacteriophage MS2, and simian virus 40 (SV40).

2.2 SOME VIRUS ASSEMBLY SYSTEMS

The Baltimore system divides viruses by genome and replication mechanism [51]. This generalization contributes to our understanding of virus assembly and structure. dsDNA viruses typically form empty capsids into which dsDNA is pumped. Herpes simplex virus (HSV) is one of the largest demonstrated self-assembly systems. Though HSV's 1000 Å diameter, T = 16 procapsid assembles in the nucleus [12], it has numerous commonalities with the much smaller bacteriophage P22. For the biologically active form, both have portal complexes, associated with the DNA packaging machinery, that are early participants in assembly [27,109]. An in vitro assembled HSV1 capsid comprises 960 copies of VP5 and 320 copies of the heterotrimeric triplex located on quasi-threefold sites [124]. The capsid of P22 is a T = 7 complex of 420 copies of CP. For both P22 and HSV, assembly is induced by addition of scaffold proteins. In P22, scaffold contributes to capsid stability, directs morphology, and accelerates kinetics [17,119,121,122,136]. Without scaffold, P22 CP can form monsters, curlicues of CP that have not quite the right geometry to form a spherical capsid [136,165]; the interactions between CPs in monsters would be slightly stronger than in procapsids if not for the stabilizing and geometrical influence of scaffold protein [120,186]. Remarkably, scaffold incorporation is substoichiometric in vitro as excess scaffold can trap partial capsids [41,119,136,161,165]. In vivo, scaffold expression must also be controlled to allow maximal procapsid assembly [79]. Of particular note, procapsids are labile and cold sensitive [120,135]. These features are also seen in the HSV1. The scaffold protein is necessary for assembly, intermediates can be trapped, procapsids are labile, and they are sensitive to cold [110–112,168]. Both HSV and P22 undergo a maturation step during DNA packaging in which the capsid undergoes conformational change and becomes much more stable [12,134,162]. We note that HSV and P22 both have accessory proteins that may modulate assembly in vivo that are not required in vitro.

Papillomaviruses and polyomaviruses are dsDNA viruses with very different characteristics—they assemble on their genome. Both have T = 7 capsids assembled from 72 pentamers. Human papillomavirus (HPV) particles are the basis for vaccines against cervical cancer [13]. Consider simian virus 40 (SV40). In vivo, SV40 assembles in the host nucleus on a chromatinized form of its genome [69]. Pentamers are recruited to genomic DNA by interaction with the SP1 transcription factor bound to a specific packing signal [56]. Pentamers also interact directly with the NA [140]. SV40 capsid crystal structures show that pentamers interact by exchanging C-terminal domains [98,151], distinctly different from the buried hydrophobic patches that stabilize protein–protein interactions in capsids of most other viruses [4]. Thus, SV40 assembly starts with positively charged subunits and a negatively charged genome, where the genome is roughly preorganized to fit into a spherical compartment. Empty particles are not seen in vivo. In vitro, almost any dsDNA can be packaged into T = 7 capsids by SV40 pentamers [78,140,142]. When dsDNA is titrated with SV40 pentamer, the concentration dependence of the assembly is distinctly sigmoidal [105,106]; the threshold concentration for the assembly indicates a thermodynamic barrier to initiate complex formation [184]. On ssDNA and ssRNA substrates SV40 pentamers assemble with no apparent thermodynamic barrier to yield T = 1 particles and strings of T = 1 particles [82,83], suggesting that the rigidity of the underlying scaffold had a critical role in directing particle morphology.

HBV is another system whose assembly has been studied in vivo and in vitro. In vivo HBV core protein dimers assemble on a complex of viral pregenomic RNA (pgRNA) and viral reverse transcriptase (RT). In the cytoplasm, these RNA-filled particles mature in the cytoplasm to yield HBV cores containing a relaxed circular dsDNA HBV genome [145]. HBV virions from serum predominantly have T = 4 symmetry with a small fraction of smaller T = 3 particles [40,147,150]. A recent study suggests that a large fraction of VLP in the plasma of infected chimpanzees and human patients and the media of cultured cells have no genetic material [115]. Mutants that alter assembly in vivo have similar effects in vitro [22,32,157,175]. Remarkably, mutations that alter the capsid also affect reverse transcription, implying that the capsid plays an active role in reverse transcription [157]. In vitro studies of assembly largely parallel in vivo observations. Empty capsids are readily formed by core protein dimers where the extent of assembly is sensitive to solution conditions [23,171]. In vivo, HBV CP is very specific for viral RNA, though specificity depends on the phosphorylation of CP [55,86,104] and the presence of RT and an RT-binding stem loop on the viral RNA [9,66,85,89]. How HBV CP avoids misassembling on random RNA remains a puzzle, and one hypothesis is that a kinase binds and occludes the RNA-binding domain of HBV to essentially chaperone the assembly in a phosphorylation-dependent manner [26]. When expressed in *Escherichia coli*, capsids pack random RNA at an amount roughly proportional to the charge of the RNA-binding domain [31,99,132,182]; in vitro, RNA binding is nonspecific, independent of phosphorylation [131]. Another puzzle is that though HBV capsids can stably contain a dsDNA genome, unlike SV40, they are unable to assemble on a dsDNA substrate [37,137].

Arguably, the simplest and most common viruses are spherical and carry a plus-sense RNA genome. The bromoviruses brome mosaic virus (BMV) and CCMV provide an excellent example of this class. In vivo, like all plus-sense RNA viruses, bromovirus RNA-dependent RNA polymerases and associated proteins remodel the host's interior to create new cytoplasmic factories for RNA synthesis [2,87,100,144]. While it is known that bromovirus CPs regulate RNA replication [74], the direct connection between RNA factories and capsid assembly remains unclear. Some specificity is based on specific protein–RNA interaction [63,74,114]. In vitro CCMV and BMV both assemble in response to low pH or in the presence of ssRNA [5]. However, CCMV shows no evidence for specificity for viral RNA in vitro [3,71].

Though not covered in this review, it is important to briefly discuss rodlike TMV [155]. TMV was the first virus assembled in vitro [52]. TMV virions can be described as a single start helix with 17.3 copies of CP per turn around an RNA core. In vivo assembly takes place in virus factories that may be involved in intracellular spread. In vitro assembly begins with adsorbing several 20S CP complexes to a specific RNA operator. These complexes are proposed to be helical polymers of about 34 CPs [35]. Subsequent additions are based on a mixture of 20S complexes and smaller aggregates [20]. A critical feature of the subunits, those participating in nucleation and subsequent addition, is that they switch states from an inactive form to an assembly-active form [20,88]. In vitro, TMV CP readily forms disks and larger aggregates as defined by the famous assembly phase diagram [84].

Systems in this preliminary survey were chosen to illustrate specific points. These systems all self-assemble from well-defined subunits. Some require a specific scaffold protein that modulates the stability and geometry of local intersubunit contacts. Others require a template (e.g., RNA) that can also lead to a high local concentration of subunits. Strikingly, where researchers have looked closely enough (e.g., HBV, SV40, TMV), these proteins undergo a conformational transition that is associated with an assembly competent state. Activation helps minimize errors in the assembly, but arguably and more importantly, it prevents assembly of particles that are defective, because they assembled wrongly, in the wrong location, or with the wrong cargo. The common features of virus self-assembly allow

1. Development of general models to describe the biophysics of capsid assembly
2. Diversion of normal assembly to create new intelligently designed structures

2.3 GEOMETRY OF SPHERICAL AND NONSPHERICAL CAPSIDS

Most viruses have a capsid or nucleocapsid component, the focus of this review. Capsids are typically constructed of many copies of a small number of proteins that are arranged with icosahedral or helical symmetry. Many CPs spontaneously self-assemble in vitro, sometimes with the aid of an NA scaffold or a scaffold protein, into a very good facsimile of the in vivo structure.

An icosahedron is a Platonic solid eponymously characterized by its 20 facets. It can also be defined by its 5-3-2 symmetry, a symmetry shared with dodecahedra. As each of the 20 facets of an icosahedron has threefold symmetry, there are $3 \times 20 = 60$ asymmetric units making up an icosahedron; a proteinaceous icosahedron must be built of 60 (or multiples 60) proteins. Caspar and Klug noted that larger icosahedra could be constructed by including more subunits arranged as pentamers and hexamers [18,21], an organization that explains a geometric series of capsid sizes based on a T number indicating the number of quasi-equivalent subunits in the icosahedral asymmetric unit. The structural basis for this geometry is the recognition that each triangular facet can be *cut out* of a hexagonal lattice so that each of the three asymmetric units in the facet comprises one or more quasi-equivalent triangles; thus, $T = h^2 + hk + k^2$ (where h and k are integer indices for a hexagonal lattice). One realization of this lattice is seen in old-style soccer balls. Since a pentagonal dodecahedron has the same symmetry, the same quasi-equivalence applies to viruses with dodecahedral organization such as picornaviruses [95]. A vast library of icosahedral virus structures has been compiled at virus particle explorer database (VIPERDB) [16].

Helical symmetry is ubiquitous not only in biology, in viruses, cytoskeletal proteins, and flagella but also in NAs and polysaccharides (Figure 2.1c). Helical symmetry covers a broad

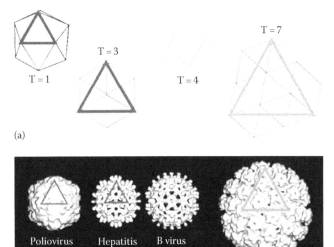

(a)

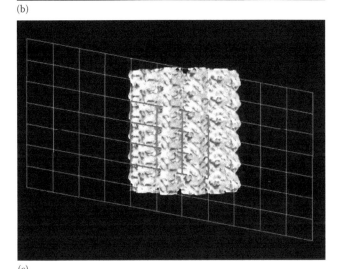

(b)

(c)

FIGURE 2.1 Capsid symmetry. (a) Icosahedral facets cut from a hexagonal lattice. (b) Examples of quasi-equivalence in virus image reconstructions. (c) An example of helical symmetry showing a helix that rises 3 subunits for each 11-subunit turn. (Panels a and b are from Zlotnick, A., *Proc. Natl. Acad. Sci. USA*, 101, 15549, 2004. Panel c is from Tsai, C.J. and Nussinov, R., *Acta Crystallogr. D Biol. Crystallogr.*, 67, 716, 2011.)

array of geometries from rods of stacked rings to a spring-shaped coil with no interaction between adjacent turns. Helical symmetry is a function of the number of intertwined fibers, the number of monomers or bonds per turn, diameter, and the rise per turn [64,92]. These conventions are reviewed and a simplified system has been proposed by Tsai and Nussinov [164].

2.4 SELF-ASSEMBLY OF SPHERICAL AND NONSPHERICAL CAPSIDS

For complex reactions, and any reaction with hundreds of concurrent steps is complex, theory allows the experimentalist to interpret data. Conversely, theory provides predictions that allow the experimentalist to test the accuracy of the theory.

To understand the underlying biophysics of a virus capsid and to manipulate its structure and function require a grasp of the process of capsid assembly. Icosahedral polymers and spherical polymers have a fundamental difference that has substantial effects on the interpretation of their assembly (Figure 2.2). Icosahedral polymers are finite: They have a specific number of subunits in defined environments and, when complete, they have no loose ends [47,75]. Helical polymers are conceptually infinite though (obviously) in practice they have unliganded ends that can freely equilibrate with bulk solvent [117,176]. Infinite polymers come in 1-D (fibers or helices), 2-D (sheets), and 3-D (crystals).

At equilibrium, a solution of finite and infinite polymers looks the same: there is a mixture of free subunits and polymers. The basis of this (pseudo-)critical concentration is very different. For infinite polymers, association of subunits to the polymer is described by the rate equation:

$$-\frac{d[\text{subunit}]}{dt} = f[\text{polymer}_{end}][\text{subunit}] \tag{2.1}$$

where

f is the forward rate constant

polymer_{end} is the concentration of polymer ends to which subunits may associate or dissociate

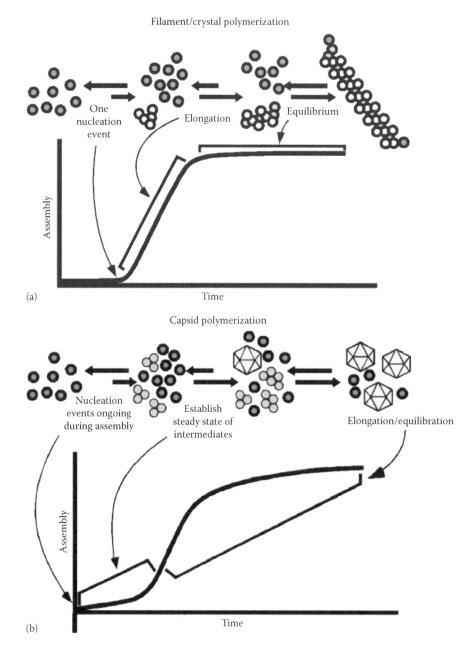

FIGURE 2.2 Assembly of infinite polymers and finite polymers looks similar but has different physics. (a) Assembly of an infinite polymer requires only a single nucleus to initiate a complex with a restricted number of subunits. (b) Assembly of a finite polymer, a capsid, requires one nucleus for each particle of 10s to 100s of subunits. (From Katen, S.P. and Zlotnick, A., *Methods Enzymol.*, 455, 395, 2009.)

Dissociation of subunits is described by:

$$d[subunit] = b[polymer_{end}] \qquad (2.2)$$

where b is the backward rate. Thus, at equilibrium, where the change in subunit concentration, d[subunit]/dt, is zero, the equilibrium constant, the ratio of Equations 2.1 and 2.2, is a critical concentration (K_{crit}) in units of subunit concentration:

$$K_{crit} = \frac{b}{f} = \frac{[polymer_{end}][subunit]}{polymer_{end}} \qquad (2.3)$$

No matter how high the initial concentration of subunit, at equilibrium it cannot exceed K_{crit}.

Assembly of an icosahedron, or any spherical complex (e.g., a micelle), has the appearance of a critical concentration but has a different basis with distinct implications. For a capsid of n subunits, the law of mass action specifies:

$$K_{capsid} = \frac{[subunit]^n}{capsid} \qquad (2.4)$$

For HBV, where n = 120, K_{capsid} is in the unwieldy units of M^{119}. Because of the huge exponent, the concentration of capsid varies nonlinearly with the [subunit]; that is, there is a concentration of subunit above which [capsid] takes off. When n is large, the [subunit] is nearly constant and resembles a critical concentration. This pseudo-critical concentration is approximately ≫ at the point where [capsid] = [subunit], it is directly analogous to a critical micellar concentration [158]. Another important point is that as subunits must be at least trivalent to assemble into a spherical capsid, individual subunit–subunit interactions can be very weak and still lead to a globally stable structure. Indeed there are kinetic advantages to weak association.

Assembly kinetics for infinite and finite polymers also looks the same but has very different physics [47,88,117,131,179]. For both types of reactions, bulk assembly kinetics is sigmoidal. The difference has a simple basis: to grow a typical protein crystal involves about 10^{12} protein molecules and only one nucleus; to use the same amount of protein to grow virus capsids will require about 10^{10} nuclei. The lag phase for infinite polymers (filaments, sheets, and crystals) is the time required for forming that one nucleus. For finite polymers, nucleation occurs continuously throughout the reaction, the lag phase is the time required to build a steady-state population of intermediates, essentially an assembly line [47,75]. Thus, for capsids, the lag phase is directly proportional to the *elongation rate*, the rate of addition of subunits to a growing capsid [61]. Conversely, once the steady state of intermediates is established, the rate of capsid appearance is equal to the rate of nucleation, most easily measured when the reaction is plateauing [47].

Overall, there is solid agreement between approaches to examining assembly, recently elegantly reviewed [125].

Master equation approaches based on differential rate equations [47,76] describe the behavior of large populations, but assume you know all members of the population. Gillespie-type stochastic event simulators provide a detailed view of individuals, but require much the same foreknowledge as master equations [177]. Molecular dynamics (MD) simulations make no such assumptions but are computationally extremely challenging; therefore, most are based on coarse-grain models of the subunits that necessarily include assumptions on geometry, flexibility, and bond strength [46,60,102,113].

Theory for assembly of empty capsids has provided a number of important predictions. Weak association energy will support assembly because of the multivalent nature of subunits. Weak association energy leads to assembly by numerous reversible reactions, allowing mistakes to dissociate, thermodynamic editing [138,139,178]. Completing a capsid may be difficult, in part because the molecular movement of the protein complex partially closes or constricts sites for incoming subunits [113].

2.5 SELF-ASSEMBLY OF HOMOPOLYMERS

CCMV was the first icosahedral virus capsid assembled in vitro and the first infectious nucleocapsid assembled in vitro [6,7]. CP is readily obtained from infected plants. As the virus is not particularly stable at neutral pH, virions are treated with high concentrations calcium at pH 7; the Ca^{2+} precipitates packaged RNA, leaving free dimeric protein in solution [1]. Empty capsids assemble in a pH-dependent manner [5,8]. Neutral pH assembly requires some sort of scaffolding, such as NA, to be discussed in subsequent sections. In vitro assembly of empty particles requires low pH [1,5]. In analytical studies of assembly thermodynamics, the classical pseudo-critical concentration was observed by size exclusion chromatography, indicating that each contact in a tetravalent dimer had an association energy of about −3 kcal/mol [181]. Kinetic studies of assembly showed a complicated series of two phases [70,181]. The first phase correlated with formation of pentamers of dimers and the second phase was due to association of pentamers of dimers and free dimers associating to form T = 3 capsids. One of the peculiarities of CCMV was the observation that the dimer itself appeared to have distinct hinges that adjusted to different quaternary structures [159].

Though HBV CP is all α helix, distantly related to retroviral Gag proteins [185], and CCMV is a classical β barrel, the physics of their assembly is very similar though the biochemistry of the proteins dictates biologically relevant differences. Unlike dimeric CCMV CP, the dimeric assembly domain of HBV CP, Cp149, assembles under near-physiological conditions [23,171]. However, Cp149 assembly is steeply ionic strength dependent. We have proposed that high salt modulates free Cp149 conformation, a hypothesis that is supported by observations that Zn^{2+} and mutations at the intradimer interface can activate capsid assembly [22,93,148,154]. Interestingly, a disulfide at the intradimer interface also strongly modulates assembly and capsid stability while seemingly making no structural change [148,174]. Alternatively, it has been proposed that ionic strength has the effect of suppressing a global electrostatic repulsion [77]. Like CCMV, Cp149 shows the expected pseudo-critical concentration corresponding to about −3 to −4 kcal/mol for each of the four contacts made by the tetravalent dimer. HBV assembly is strongly temperature dependent ($\Delta H > 0$), indicating that it is an entropy-driven reaction [23]. Cp149 assembly kinetics has the expected sigmoidal behavior [131].

The weak association energy seen in CCMV and HBV assembly was anticipated from the theory [179]. When association is strengthened, in both cases we find evidence of defects in normal assembly. For CCMV, at low pH and high ionic strength, virus-unlike particles assembled from 12 pentamers of dimers appear, presumably as a function of insufficient free dimer [159,181]. HBV CP starts accumulating incomplete capsid with some metastable particles lacking one or two icosahedral facets particularly enriched [129]; these incomplete capsids are reminiscent of those predicted by Nguyen et al. [113].

Thus, strong association is not necessary, but is it pathological? In the polyomavirus SV40, strong association energy can lead to kinetic traps. SV40 pentamers are stable under physiological conditions but can assemble and disassemble with the aid of catalytic chaperones [29,30], indicating that the barrier to assembly is kinetic not thermodynamic. However, SV40 pentamers assemble rapidly and quantitatively in the presence of Ca^{2+} or at low pH, resulting in a profusion of polymers with different geometries including aberrant junk, T = 7 capsids, T = 1 capsids, and cylinders [73,141]. These results suggest that SV40 pentamers associated with incorrect geometry are themselves trapped and propagate defective particles. The overall picture with polyomaviruses is that a chaperone, metal, and pH can induce a structural change that allows assembly; abnormal assembly products may arise from the lack of control during assembly or simply the inability to edit out small errors due to high association energy (slow dissociation compared to association) that locks in errors.

2.6 SCAFFOLDED SELF-ASSEMBLY THEORY AND EXPERIMENT

Assembly with scaffold protein or NA adds a layer of complication. The simplest case is an obligate scaffold molecule, a component that is required for each subunit. This example is realized in bacteriophages like P22 and bacteriophage-like viruses such as herpes viruses, which have specific scaffold protein. In vivo, P22 assembly probably is nucleated by the portal complex, which, in the context of the mature particle, is located in place of an icosahedral fivefold; then, scaffold protein and CP thus play an intricate dance, leading to the formation of immature, empty capsids [90,162]. Subsequent to assembly, DNA is pumped in, scaffold exits, and the capsid undergoes a maturation transition. Conceptually, up to 420 scaffold proteins could bind to

a T = 7 particle [27], though in practice the upper limit is about 360 and about 60 is the minimum [121].

The first in vitro observations of assembly kinetics were with P22; these showed that scaffold protein concentration affected the rate of extent of assembly [136]. Correct assembly requires an appropriate amount of scaffold protein. In vivo, too much or too little is toxic [17]. In the absence of scaffold, coat assembly usually leads to irregular particles, indicating the scaffold's structural role [49,118,136,162]. Interestingly, in the absence of scaffold, P22 CP has slightly stronger protein–protein interaction energy, indicating that there is a cost to scaffold-directed geometry and suggesting that scaffold provides a kinetic advantage [186]. Scaffold is flexible protein that binds relatively weakly to capsid via electrostatic interactions [119,123]. In vitro, excess scaffold leads to many incomplete capsids, exactly analogous to the kinetic traps achieved with homopolymer capsid assembly [119,121]. Because scaffold participation is nonstoichiometric, appropriate assembly models must correlate changes in capsid stability with the amount of scaffold [121,186]. A straightforward binding analysis shows the linkage between assembly and scaffold concentration [186].

The assembly may also be scaffolded by polyvalent complexes: NAs, membranes, abiotic polymers (e.g., polysulfostyrene), and nanoparticles. Conceptually, assembly on polyvalent scaffolds may occur by two mechanisms:

1. CPs may bind to the scaffold, forming a disorder micelle, and then rearrange to a capsid.
2. Capsids may assemble concomitant with binding to the scaffold.

Micellar assembly was qualitatively proposed by McPherson [103] and put into a rigorous context by Hagan and coworkers [58,59,80]. MD simulations showed that strong CP–scaffold interactions combined with weak CP–CP interactions led to micellar assembly, whereas strong CP–CP interactions led to stepwise capsid growth, very similar to that seen with empty particles.

2.7 CAPSID ASSEMBLY ON NUCLEIC ACID

NA is a much more complicated scaffold: it is polyvalent and flexible and may have specific sites. CP may affect the structure and organization of packaged NA (Figure 2.3). Conversely, NA can modify the assembly and play a structural role. Indeed, crystal and cryo-electron microsocopy (cryo-EM) structures of the bean pod mottle virus [28], nodaviruses (Flockhouse, Nodamura, and Pariacoto) [50,160,183], and satellite tobacco necrosis virus [91] show highly ordered NA. In the nodaviruses, even heterologous RNA is highly ordered [72,163]. Assembly studies are central to understanding virion formation in a biological setting and critical to packaging nonviral NAs for gene therapy [78,97].

Interactions between CP subunits and NAs may be specific or nonspecific. In general, CPs that simultaneously

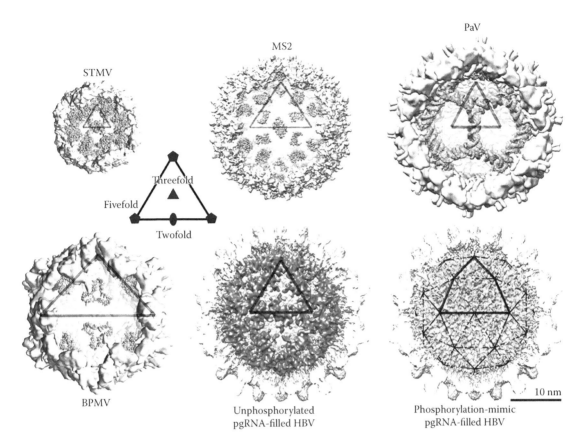

FIGURE 2.3 Ordered RNA in icosahedral viruses. In satellite tobacco mosaic virus (STMV, PDB code 1A34) and Pariacoto virus (PaV, PDB code 1F8V), the coat protein induces double-stranded segments that fit into a groove between segments [91]. In bacteriophage MS2 (PDB code 1ZDH), a specific RNA stem loop binds a specific site at a dimer interface influencing assembly and RNA organization; this image is a structure containing 60 copies of the specific oligonucleotide [166]. In bean pod mottle virus (BPMV, PDB code 1BMV), a trefoil of ssRNA is observed in a nonpolar groove around each icosahedral threefold [28]; sequence analysis suggests it is enriched in purines [101]. In HBV when the arginine-rich C-terminus is unphosphorylated, the nucleic acid (ssRNA to dsDNA) connects from fivefold to fivefold [168]; when three serines are replaced with glutamate to mimic phosphorylation, the RNA is reorganized [168]. The electron microscopy database accession numbers for HBV are EMD-2059 and EMD-2058. Figures were generated using UCSF Chimera [128].

assemble and package NA have a basic segment that enables coassembly. Indeed there are many examples of spontaneous NA packaging concomitant with assembly [6,52,108,142]. Viruses that form shells into which NA is pumped, usually through portals (e.g., bacteriophage P22), generally have an uncharged interior, minimizing CP–NA interaction. Such virions often package polycations to facilitate NA neutralization and condensation. They also will expel their NA in a controlled manner in response to an appropriate shock [48,57]. This generalization raises the question of how picornaviruses, such as poliovirus, which have a relatively neutral interior surface and readily expel their RNA genomes (see [11,65,94,116]), become filled with RNA in the first place.

Assembly on NA is readily observed by electrophoretic mobility shift assay (EMSA), which gives insight into the generalized paths of the assembly [37,71,82,83,131]. It should be noted that electrophoretic mobility is dominated by the external surface of the molecule in question [149]. Consider the EMSA for low-cooperativity assembly where CP binds to NA randomly, irrespective of other binding events. A Gaussian distribution of different numbers of protein bound per NA is expected. When protein concentration increases,

a progressively increasing number of CPs bound per NA in the assembly intermediates will be formed until all NAs are fully encapsidated (Figure 2.4). This leads to a gradual shift of NA migration observed in the EMSAs [131]. Similar results are expected where, in high cooperativity, the binding of one CP to an NA creates adjacent high-affinity sites for the next CP to bind on the same NA. Therefore, when the protein concentration increases, the assembly process is a sequential event completing the intermediates one by one while other NAs remain unbound. As a result, a bimodal distribution of free NA and completely fully encapsidated NA with low concentrations of intermediates is observed in EMSA (Figure 2.4).

In the case of HBV, CPs assemble with high affinity and high cooperativity on ssRNA or ssDNA and produce uniform capsids [37,130]; assembly on dsDNA, however, results in numerous aberrant structures: distorted spiral, string-like capsid fragments or DNA condensates [37]. Although HBV is a dsDNA virus, which suggests that the capsid is capable of accommodating DS NA, the failure of efficiently assembling on dsDNA suggests that the binding energy of the NA appears to be similar to the free energy from the average interaction of an incoming CP dimer for the growing capsid.

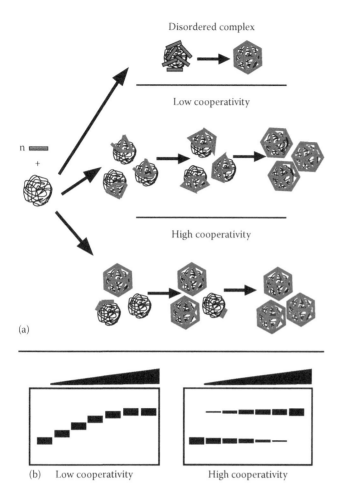

(a)

(b)   Low cooperativity   High cooperativity

FIGURE 2.4 Assembly of a capsid on nucleic acid (NA). (a) In micellar assembly (top path), subunits adsorb to the NA and then rearrange to form a capsid; this path is associated with weak CP–CP interaction [53]. In low-cooperativity assembly, subunits are distributed over many NA polymers but preferentially bind in clusters. In high-cooperativity assembly, an assembly-regulating nucleation step minimizes the number of starts and strong CP–CP association favors addition of CP to NA only where nucleated. (b) Predictions for assembly reactions resolved by native agarose gel electrophoresis. Micellar assembly and low-cooperativity assembly lead to a gradual shift of NA migration through a gel, so long as the CP remains NA associated. High-cooperativity assembly leads to a bimodal distribution of NA into free and encapsidated fractions. (Adapted from Zlotnick, A. et al., *Biophys. J.*, 104, 1595, 2013.)

The heterogeneity of assembly products (capsids, aberrant complexes, or DNA condensates) is likely to represent an energy minimum that is too stable to dissociate in favor of the supramolecular complex(es) at the global minimum [37].

The CP–protein interaction in CCMV, on the other hand, has very weak association energy [53,181]. Interactions between protein and NA are relatively strong [54]. Thus, the assembly reaction must be supported by a scaffold—a viral NA, a heterologous NA, or a synthetic anionic polymer [75]. In vitro assembly of CCMV on ssRNA is a low-cooperativity reaction, which supports formations of spherical particles with various diameters or formation of multiplets [15]. By comparison, assembly on dsDNA, a stiff polymer with a

50 nm persistence length, forms a mixture of rodlike structures with 17 nm in diameter and up to several micrometers in length [14,107].

The polyomavirus SV40, a dsDNA virus, provides a very different example of NA-induced assembly. SV40 assembles from pentameric subunits. As discussed in Section 2.4, SV40 pentamers have high affinity for one another, though with a high kinetic barrier to assembly [30]. SV40 pentamers also have a high affinity for NA [33,96]. In vivo the capsid plays a peculiar role in organizing the highly compacted minichromosome, the circular 5 kB genome with bound nucleosomes [143]. In vitro assembly on dsDNA leads to highly cooperative formation of 50 nm T = 7 particles composed of 72 pentamers, even with very large substrates [78,105]. Surprisingly even short dsDNA substrates, 600-mers, lead to T = 7 particles [106]. However, capsid geometry is strongly affected by the choice of NA substrate. Binding on short ssRNA substrate (≤814 nt), SV40 pentamers form exclusively T = 1 capsid with 22 nm in diameter [82,83]. If RNA is too long to fit in a T = 1 particle, pentamers form multiplets of 22 nm particles and heterogeneous particles of 29–40 nm diameter [83].

The previous examples have been for specific interactions. The effect of specific interaction is probably best dissected for bacteriophages like R17, Qβ, and MS2 [153,172]. In these phages, there are two quasi-equivalent classes of dimer AB and CC, where a stem loop of RNA bound to a dimer induces it to the AB conformer. Assembly can be conceived of as a stepwise process in which AB and CC dimers alternate during assembly. There is substantial evidence that there is a series of stem loops in the genomic RNA to support this mechanism [44,45]. Single-molecule studies with MS2 and also with satellite tobacco necrosis virus indicate that the presence of specific signals dramatically enhances the rate of assembly of formation of a compact structure [10].

Taken together, these data indicate synergism between NA and capsid. In some cases, NA is the scaffold; in other cases, it is the protein. Where information is available, it is clear that NA alters assembly kinetics and path, sometimes with great sensitivity to sequence. NA serves as a scaffold and also a nucleating factor for capsid assembly. The molecular mechanism of how viruses assemble around NAs is based on trade-offs between the stability of protein–protein interaction; a simplified view is that assembly is the sum of the work required to package the NA, the interaction between subunits and the NA, and stability of the capsid [184]. This view suggests that the ability of a virus to assemble on a given NA substrate requires either adapting the substrate conformation to the virus (in the case of most single-stranded NA) or adapting the CP to the substrate [19,43,83]. Specificity and NA secondary structure complicate, but do not fundamentally change this picture [42,45,80,126,127].

2.8 ASSEMBLY ON ABIOTIC SUBSTRATES

The interior and exterior of the viral capsid have been exploited for packaging and storage of nongenomic cargoes. In recent years, viral capsids have been repurposed as nanocontainers for a wide variety of functions including, but not limited to, drug encapsidation/delivery, constrained polymer synthesis, and catalysis.

Several approaches have been pursued to access the interior cavity of these nanocontainers for the deposition and/or synthesis of nongenomic matter. Early examples entailed the use of preassembled capsid structures as starting points to accomplish either incorporation or synthesis of nongenomic entities; some exciting recent work has focused on exploiting the packaging mechanisms of several ssRNA icosahedral viruses. In much the same way they package their genome, they have been demonstrated to engulf pre-prepared cargo such as functionalized metal centers as well as other abiotic substrates with good success and will be the main focus of this discussion. We will not focus on phage display approaches as these have been amply discussed elsewhere [68].

CCMV, first spherical virus assembled in vitro, has also been a leading system for developing abiotic assembly. In the absence of RNA, the positively charged regions within empty CCMV capsids can interact with and accumulate anionic polyoxometalate salts such as tungstates, molybdates, and vanadates; by undergoing template-based mineralization, these complexes give rise to size and shape-constrained inorganic nanoparticles. Mineralization of cationic metal oxide species such as those derived from iron (Fe^{2+}) and cobalt (Co^{2+}) was accomplished by reengineering of the capsid's interior through substitution of the basic N-terminal residues with glutamic acid, confirming the need for complementary electrostatic interactions between capsid and cargo. Similarly, TMV has also been utilized as a template for the fabrication of Ni and Co nanowires, a strategy that takes full advantage of the elongated helical protein cages that form in the absence of viral RNA. It is also noteworthy to report that, in recent work, mineralization of virus particles is now being applied to the exterior of the capsid's protein cage architecture, as demonstrated by Qin et al. who coated the human enterovirus type 71 vaccine strain with calcium phosphate, enhancing the thermostability and immunogenicity of the vaccine [167,170].

BMV and CCMV have also been exploited to accommodate negatively charged metal nanoparticles. However, this has been accomplished through encapsidation of the foreign substrate, assembly, as opposed to the intracapsid synthesis approaches. Encapsidation may require entities of appropriate size. Colloidal gold nanoparticles were first appended (via covalent linkers) with either citrate molecules [39], phosphines, or short DNA strands [25] to confer an overall negative charge, mimicking viral RNA, before assembly with the BMV core protein subunits. These early attempts at functionalizing and encapsidating gold nanoparticles were generally low yielding. Encapsidation efficiencies were greatly improved, however, when the gold core was functionalized with heterobifunctionalized polyethylene glycol (PEG) chains of various lengths, which also allowed for exquisite control over the size of the caged nanoparticle [24]. Further, the overall size of the coated metal nanoparticle was found to influence the triangulation number of the corresponding capsids, yielding T = 1 BMV capsids for 6 nm PEG-coated cores and T = 3 BMV capsids for PEGylated Au cores measuring 12 nm in diameter. Higher-order structures have been observed in the form of large 3-D crystals, grown under the same conditions as native BMV, indicating that there are geometry and stability factors at play regulating BMV assembly [156]. Indeed, crystals of BMV-Au nanoparticles had altered spectroscopic properties, demonstrating an emergent property derived from complex geometry. This strategy was confirmed when applied to HBV, giving rise to T = 3 or T = 4 particles with size-dependent encapsidation efficiencies [62]. Using iron oxide nanotemplates though, Dragnea et al. were able to overcome these limitations, reporting that a viral sphere with a diameter of 41.3 nm could be assembled around a metal oxide core that exceeded the diameter of native BMV's capsid cavity (~20 nm) [67]. The versatility of this approach for the caging of inorganic cores with BMV core protein subunits was further demonstrated by encapsidating cadmium selenide (CdSe) and zinc sulfide (ZnS) quantum dots to generate fluorescent BMV capsids [38].

In a nonionic approach, CCMV was engineered with reactive cysteines on the capsid's exterior, allowing it to be chemoselectively tethered to functionalized gold surface to give ordered 2-D CCMV arrays [81]. Building on this work, a conducting framework of gold-decorated CCMV particles interfaced by conducting organic molecules has also been developed, illustrating its potential use in nanotechnology [133,152,169].

Complex biological supramolecules have been generated using organic substrates. Recent work by the Nolte group revealed that the exterior of CCMV capsids could be decorated with PEG chains albeit at a significant cost to the overall stability of the capsid, which eventually underwent irreversible dissociation [34]. To overcome this instability, a second polymer (polystyrene sulfonate [PSS]) was encapsidated by the PEG-coated protein subunits to give rise to robust 18 nm (T = 1) PSS-CCMV-PEG capsids. While these studies did not employ functionalized polymers, they suggested that the stability of VLPs could be fine-tuned based on their cargoes and exterior decorations. Hammond et al. on the other hand sought to investigate the spontaneous assembly of M13 viruses on polymeric surfaces like linear-polyethylenimine (LPEI) and anionic polyacrylic acid [173]. Their studies concluded that the filamentous M13 viruses could form 2-D monolayer surfaces atop a polyelectrolyte surface that could in turn be used as a scaffold for the nucleation and/or growth of nanoparticles or nanowires.

2.9 FINAL COMMENTS

Virus structures, on their own, are beautifully evolved biological supramolecular entities. Viruses take advantage of host resources and self-assembly to yield huge numbers of uniform complexes on a nanometer scale. The time course of assembly may be as short as milliseconds [82]. Viruses take advantage of their host. Our ability to take advantage of viruses in conjunction with the myriad of nongenomic substrates, chemistries, and nanotechnologies makes viruses invaluable partners for generating new materials with novel properties.

REFERENCES

1. Adolph, K. W. and P. J. G. Butler. 1974. Studies on the assembly of a spherical plant virus. I. States of aggregation of the isolated protein. *J Mol Biol* **88**:327–341.
2. Alvisi, G., V. Madan, and R. Bartenschlager. 2011. Hepatitis C virus and host cell lipids: An intimate connection. *RNA Biol* **8**:258–269.
3. Annamalai, P. and A. L. Rao. 2005. Dispensability of 3′ tRNA-like sequence for packaging cowpea chlorotic mottle virus genomic RNAs. *Virology* **332**:650–658.
4. Bahadur, R. P., F. Rodier, and J. Janin. 2007. A dissection of the protein–protein interfaces in icosahedral virus capsids. *J Mol Biol* **367**:574–590.
5. Bancroft, J. B. 1970. The self-assembly of spherical plant viruses. *Adv Virus Res* **16**:99–134.
6. Bancroft, J. B. and E. Hiebert. 1967. Formation of an infectious nucleoprotein from protein and nucleic acid isolated from a small spherical virus. *Virology* **32**:354–356.
7. Bancroft, J. B., G. J. Hills, and R. Markham. 1967. A study of the self-assembly process in a small spherical virus. Formation of organized structures from protein subunits in vitro. *Virology* **31**:354–379.
8. Bancroft, J. B., G. W. Wagner, and C. E. Bracker. 1968. The self-assembly of a nucleic-acid free pseudo-top component for a small spherical virus. *Virology* **36**:146–149.
9. Bartenschlager, R. and H. Schaller. 1992. Hepadnaviral assembly is initiated by polymerase binding to the encapsidation signal in the viral RNA genome. *EMBO J* **11**:3413–3420.
10. Borodavka, A., R. Tuma, and P. G. Stockley. 2012. Evidence that viral RNAs have evolved for efficient, two-stage packaging. *Proc Natl Acad Sci U S A* **109**:15769–15774.
11. Bostina, M., H. Levy, D. J. Filman, and J. M. Hogle. 2011. Poliovirus RNA is released from the capsid near a twofold symmetry axis. *J Virol* **85**:776–783.
12. Brown, J. C. and W. W. Newcomb. 2011. Herpesvirus capsid assembly: Insights from structural analysis. *Curr Opin Virol* **1**:142–149.
13. Buck, C. B., P. M. Day, and B. L. Trus. 2013. The papillomavirus major capsid protein L1. *Virology* **445**:169–174.
14. Burns, K., S. Mukherjee, T. Keef, J. M. Johnson, and A. Zlotnick. 2010. Altering the energy landscape of virus self-assembly to generate kinetically trapped nanoparticles. *Biomacromolecules* **11**:439–442.
15. Cadena-Nava, R. D., M. Comas-Garcia, R. F. Garmann, A. L. Rao, C. M. Knobler, and W. M. Gelbart. 2012. Self-assembly of viral capsid protein and RNA molecules of different sizes: Requirement for a specific high protein/RNA mass ratio. *J Virol* **86**:3318–3326.
16. Carrillo-Tripp, M., C. M. Shepherd, I. A. Borelli, S. Venkataraman, G. Lander, P. Natarajan, J. E. Johnson, C. L. Brooks 3rd, and V. S. Reddy. 2008. VIPERdb2: An enhanced and web API enabled relational database for structural virology. *Nucleic Acids Res* **37**:D436–D442.
17. Casjens, S. and J. King. 1974. P22 morphogenesis. I: Catalytic scaffolding protein in capsid assembly. *J Supramol Struct* **2**:202–224.
18. Casjens, S. and J. King. 1975. Virus assembly. *Annu Rev Biochem* **44**:555–611.
19. Caspar, D. L. 1980. Movement and self-control in protein assemblies. Quasi-equivalence revisited. *Biophys J* **32**:103–138.
20. Caspar, D. L. and K. Namba. 1990. Switching in the self-assembly of tobacco mosaic virus. *Adv Biophys* **26**:157–185.
21. Caspar, D. L. D. and A. Klug. 1962. Physical principles in the construction of regular viruses. *Cold Spring Harbor Symp Quant Biol* **27**:1–24.
22. Ceres, P., S. J. Stray, and A. Zlotnick. 2004. Hepatitis B virus capsid assembly is enhanced by naturally occurring mutation F97L. *J Virol* **78**:9538–9543.
23. Ceres, P. and A. Zlotnick. 2002. Weak protein–protein interactions are sufficient to drive assembly of hepatitis B virus capsids. *Biochemistry* **41**:11525–11531.
24. Chen, C., M.-C. Daniel, Z. T. Quinkert, M. De, B. Stein, V. D. Bowman, P. R. Chipman, V. M. Rotello, C. C. Kao, and B. Dragnea. 2006. Nanoparticle-templated assembly of viral protein cages. *Nano Lett* **6**:611–615.
25. Chen, C., E.-S. Kwak, B. Stein, C. C. Kao, and B. Dragnea. 2005. Packaging of gold particles in viral capsids. *J Nanosci Nanotechnol* **5**:2029–2033.
26. Chen, C., J. C. Wang, and A. Zlotnick. 2011. A kinase chaperones hepatitis B virus capsid assembly and captures capsid dynamics in vitro. *PLoS Pathog* **7**:e1002388.
27. Chen, D. H., M. L. Baker, C. F. Hryc, F. DiMaio, J. Jakana, W. Wu, M. Dougherty et al. 2011. Structural basis for scaffolding-mediated assembly and maturation of a dsDNA virus. *Proc Natl Acad Sci U S A* **108**:1355–1360.
28. Chen, Z. G., C. Stauffacher, Y. Li, T. Schmidt, W. Bomu, G. Kamer, M. Shanks, G. Lomonossoff, and J. E. Johnson. 1989. Protein–RNA interactions in an icosahedral virus at 3.0 Å resolution. *Science* **245**:154–159.
29. Chromy, L. R., A. Oltman, P. A. Estes, and R. L. Garcea. 2006. Chaperone-mediated in vitro disassembly of polyoma- and papillomaviruses. *J Virol* **80**:5086–5091.
30. Chromy, L. R., J. M. Pipas, and R. L. Garcea. 2003. Chaperone-mediated in vitro assembly of polyomavirus capsids. *Proc Natl Acad Sci U S A* **100**:10477–10482.
31. Chua, P. K., F. M. Tang, J. Y. Huang, C. S. Suen, and C. Shih. 2010. Testing the balanced electrostatic interaction hypothesis of hepatitis B virus DNA synthesis by using an in vivo charge rebalance approach. *J Virol* **84**:2340–2351.
32. Chua, P. K., Y. M. Wen, and C. Shih. 2003. Coexistence of two distinct secretion mutations (P5T and I97L) in hepatitis B virus core produces a wild-type pattern of secretion. *J Virol* **77**:7673–7676.
33. Clever, J., D. A. Dean, and H. Kasamatsu. 1993. Identification of a DNA binding domain in simian virus 40 capsid proteins Vp2 and Vp3. *J Biol Chem* **268**:20877–20883.
34. Comellas-Aragonès, M., A. s. de la Escosura, A. J. Dirks, A. van der Ham, A. Fusté-Cuñé, J. J. L. M. Cornelissen, and R. J. M. Nolte. 2009. Controlled integration of polymers into viral capsids. *Biomacromolecules* **10**:3141–3147.
35. Correia, J. J., S. Shire, D. A. Yphantis, and T. M. Schuster. 1985. Sedimentation equilibrium measurements of the intermediate-size tobacco mosaic virus protein polymers. *Biochemistry* **24**:3292–3297.
36. Crick, F. H. C. and J. D. Watson. 1956. The structure of small viruses. *Nature* **177**:473–475.
37. Dhason, M. S., J. C. Wang, M. F. Hagan, and A. Zlotnick. 2012. Differential assembly of Hepatitis B Virus core protein on single- and double-stranded nucleic acid suggest the dsDNA-filled core is spring-loaded. *Virology* **430**:20–29.
38. Dixit, S. K., N. L. Goicochea, M.-C. Daniel, A. Murali, L. Bronstein, M. De, B. Stein, V. M. Rotello, C. C. Kao, and B. Dragnea. 2006. Quantum dot encapsulation in viral capsids. *Nano Lett* **6**:1993–1999.

39. Dragnea, B., C. Chen, E. S. Kwak, B. Stein, and C. C. Kao. 2003. Gold nanoparticles as spectroscopic enhancers for in vitro studies on single viruses. *J Am Chem Soc* **125**:6374–6375.

40. Dryden, K., S. Wieland, F. Chisari, and M. Yeager. 2002. Structure of native hepatitis B virus particles by electron cryomicroscopy and image reconstruction, pp. 188. Presented at the *American Society for Virology 21st Annual Meeting*. Lexington, KY.

41. Dryden, K. A., S. F. Wieland, C. Whitten-Bauer, J. L. Gerin, F. V. Chisari, and M. Yeager. 2006. Native hepatitis B virions and capsids visualized by electron cryomicroscopy. *Mol Cell* **22**:843–850.

42. Dykeman, E. C., P. G. Stockley, and R. Twarock. 2013. Building a viral capsid in the presence of genomic RNA. *Phys Rev E Stat Nonlin Soft Matter Phys* **87**:022717.

43. Dykeman, E. C., P. G. Stockley, and R. Twarock. 2010. Dynamic allostery controls coat protein conformer switching during MS2 phage assembly. *J Mol Biol* **395**:916–923.

44. Dykeman, E. C., P. G. Stockley, and R. Twarock. 2013. Packaging signals in two single-stranded RNA viruses imply a conserved assembly mechanism and geometry of the packaged genome. *J Mol Biol* **425**:3235–3249.

45. Dykeman, E. C., P. G. Stockley, and R. Twarock. 2014. Solving a Levinthal's paradox for virus assembly identifies a unique antiviral strategy. *Proc Natl Acad Sci U S A* **111**:5361–5366.

46. Elrad, O. M. and M. F. Hagan. 2008. Mechanisms of size control and polymorphism in viral capsid assembly. *Nano Lett* **8**:3850–3857.

47. Endres, D. and A. Zlotnick. 2002. Model-based analysis of assembly kinetics for virus capsids or other spherical polymers. *Biophys J* **83**:1217–1230.

48. Evilevitch, A., L. T. Fang, A. M. Yoffe, M. Castelnovo, D. C. Rau, V. A. Parsegian, W. M. Gelbart, and C. M. Knobler. 2008. Effects of salt concentrations and bending energy on the extent of ejection of phage genomes. *Biophys J* **94**:1110–1120.

49. Fane, B. A. and P. E. Prevelige, Jr. 2003. Mechanism of scaffolding-assisted viral assembly, pp. 259–299. *In* W. Chiu and J. E. Johnson (eds.), *Virus Structure*, vol. 64. Academic Press, San Diego, CA.

50. Fisher, A. J. and J. E. Johnson. 1993. Ordered duplex RNA controls capsid architecture in an icosahedral animal virus. *Nature* **361**:176–179.

51. Flint, S. J., L. W. Enquist, V. R. Racaniello, and A. M. Skalka. 2009. *Principles of Virology*, 3rd edn. ASM Press, Herndon, VA.

52. Fraenkel-Conrat, H. and R. C. Williams. 1955. Reconstitution of active tobacco mosaic virus from its inactive protein and nucleic acid components. *Proc Natl Acad Sci U S A* **41**:690–698.

53. Garmann, R. F., M. Comas-Garcia, A. Gopal, C. M. Knobler, and W. M. Gelbart. 2014. The assembly pathway of an icosahedral single-stranded RNA virus depends on the strength of inter-subunit attractions. *J Mol Biol* **426**:1050–1060.

54. Garmann, R. F., M. Comas-Garcia, M. S. Koay, J. J. Cornelissen, C. M. Knobler, and W. M. Gelbart. 2014. Role of electrostatics in the assembly pathway of a single-stranded RNA virus. *J Virol* **88**:10472–10479.

55. Gazina, E. V., J. E. Fielding, B. Lin, and D. A. Anderson. 2000. Core protein phosphorylation modulates pregenomic RNA encapsidation to different extents in human and duck hepatitis B viruses. *J Virol* **74**:4721–4728.

56. Gordon-Shaag, A., O. Ben-Nun-Shaul, V. Roitman, Y. Yosef, and A. Oppenheim. 2002. Cellular transcription factor Sp1 recruits simian virus 40 capsid proteins to the viral packaging signal, ses. *J Virol* **76**:5915–5924.

57. Grayson, P., A. Evilevitch, M. M. Inamdar, P. K. Purohit, W. M. Gelbart, C. M. Knobler, and R. Phillips. 2006. The effect of genome length on ejection forces in bacteriophage lambda. *Virology* **348**:430–436.

58. Hagan, M. F. 2008. Controlling viral capsid assembly with templating. *Phys Rev E Stat Nonlin Soft Matter Phys* **77**:051904.

59. Hagan, M. F. 2009. A theory for viral capsid assembly around electrostatic cores. *J Chem Phys* **130**:114902.

60. Hagan, M. F. and D. Chandler. 2006. Dynamic pathways for viral capsid assembly. *Biophys J* **91**:42–54.

61. Hagan, M. F. and O. M. Elrad. 2010. Understanding the concentration dependence of viral capsid assembly kinetics—The origin of the lag time and identifying the critical nucleus size. *Biophys J* **98**:1065–1074.

62. He, L., Z. Porterfield, P. van der Schoot, A. Zlotnick, and B. Dragnea. 2013. Hepatitis virus capsid polymorph stability depends on encapsulated cargo size. *ACS Nano* **7**:8447–8454.

63. Hema, M., A. Murali, P. Ni, R. C. Vaughan, K. Fujisaki, I. Tsvetkova, B. Dragnea, and C. C. Kao. 2010. Effects of amino-acid substitutions in the brome mosaic virus capsid protein on RNA encapsidation. *Mol Plant Microbe Interact* **23**:1433–1447.

64. Heymann, J. B., M. Chagoyen, and D. M. Belnap. 2005. Common conventions for interchange and archiving of three-dimensional electron microscopy information in structural biology. *J Struct Biol* **151**:196–207.

65. Hogle, J. M., M. Chow, and D. J. Filman. 1985. Three-dimensional structure of poliovirus at 2.9 Å resolution. *Science* **229**:1358–1365.

66. Hu, J. and M. Boyer. 2006. Hepatitis B virus reverse transcriptase and epsilon RNA sequences required for specific interaction in vitro. *J Virol* **80**:2141–2150.

67. Huang, X., L. M. Bronstein, J. Retrum, C. Dufort, I. Tsvetkova, S. Aniagyei, B. Stein et al. 2007. Self-assembled virus-like particles with magnetic cores. *Nano Lett* **7**:2407–2416.

68. Huang, Y., C. Y. Chiang, S. K. Lee, Y. Gao, E. L. Hu, J. De Yoreo, and A. M. Belcher. 2005. Programmable assembly of nanoarchitectures using genetically engineered viruses. *Nano Lett* **5**:1429–1434.

69. Imperiale, M. J. and E. O. Major. 2007. Polyomaviruses, pp. 2263–2298. *In* D. M. Knipe, D. E. Griffin, R. A. Lamb, M. A. Martin, B. Roizman, and S. E. Straus (eds.), *Fields Virology*, vol. 2. Lippincott Williams & Wilkins, Philadelphia, PA.

70. Johnson, J. M., J. Tang, Y. Nyame, D. Willits, M. J. Young, and A. Zlotnick. 2005. Regulating self-assembly of spherical oligomers. *Nano Lett* **5**:765–770.

71. Johnson, J. M., D. Willits, M. J. Young, and A. Zlotnick. 2004. Interaction with capsid protein alters RNA structure and the pathway for in vitro assembly of cowpea chlorotic mottle virus. *J Mol Biol* **335**:455–464.

72. Johnson, K. N., L. Tang, J. E. Johnson, and L. A. Ball. 2004. Heterologous RNA encapsidated in Pariacoto virus-like particles forms a dodecahedral cage similar to genomic RNA in wild-type virions. *J Virol* **78**:11371–11378.

73. Kanesashi, S. N., K. Ishizu, M. A. Kawano, S. I. Han, S. Tomita, H. Watanabe, K. Kataoka, and H. Handa. 2003. Simian virus 40 VP1 capsid protein forms polymorphic assemblies in vitro. *J Gen Virol* **84**:1899–1905.

74. Kao, C. C., P. Ni, M. Hema, X. Huang, and B. Dragnea. 2011. The coat protein leads the way: An update on basic and applied studies with the brome mosaic virus coat protein. *Mol Plant Pathol* **12**:403–412.

75. Katen, S. P. and A. Zlotnick. 2009. Thermodynamics of virus capsid assembly. *Methods Enzymol* **455**:395–417.

76. Keef, T., C. Micheletti, and R. Twarock. 2006. Master equation approach to the assembly of viral capsids. *J Theor Biol* **242**:713–721.

77. Kegel, W. K. and P. Schoot Pv. 2004. Competing hydrophobic and screened-coulomb interactions in hepatitis B virus capsid assembly. *Biophys J* **86**:3905–3913.

78. Kimchi-Sarfaty, C., O. Ben-Nun-Shaul, D. Rund, A. Oppenheim, and M. M. Gottesman. 2002. In vitro-packaged SV40 pseudovirions as highly efficient vectors for gene transfer. *Hum Gene Ther* **13**:299–310.

79. King, J., C. Hall, and S. Casjens. 1978. Control of the synthesis of phage P22 scaffolding protein is coupled to capsid assembly. *Cell* **15**:551–560.

80. Kivenson, A. and M. F. Hagan. 2010. Mechanisms of capsid assembly around a polymer. *Biophys J* **99**:619–628.

81. Klem, M. T., D. Willits, M. Young, and T. Douglas. 2003. 2-D array formation of genetically engineered viral cages on Au surfaces and imaging by atomic force microscopy. *J Am Chem Soc* **125**:10806–10807.

82. Kler, S., R. Asor, C. Li, A. Ginsburg, D. Harries, A. Oppenheim, A. Zlotnick, and U. Raviv. 2012. RNA encapsidation by SV40-derived nanoparticles follows a rapid two-state mechanism. *J Am Chem Soc* **134**:8823–8830.

83. Kler, S., J. C. Wang, M. Dhason, A. Oppenheim, and A. Zlotnick. 2013. Scaffold properties are a key determinant of the size and shape of self-assembled virus-derived particles. *ACS Chem Biol* **8**:2753–2761.

84. Klug, A. 1999. The tobacco mosaic virus particle: Structure and assembly. *Philos Trans R Soc Lond B Biol Sci* **354**:531–535.

85. Knaus, T. and M. Nassal. 1993. The encapsidation signal on the hepatitis B virus RNA pregenome forms a stem-loop structure that is critical for its function. *Nucleic Acids Res* **21**:3967–3975.

86. Kock, J., M. Nassal, K. Deres, H. E. Blum, and F. von Weizsacker. 2004. Hepatitis B virus nucleocapsids formed by carboxy-terminally mutated core proteins contain spliced viral genomes but lack full-size DNA. *J Virol* **78**:13812–13818.

87. Kopek, B. G., E. W. Settles, P. D. Friesen, and P. Ahlquist. 2010. Nodavirus-induced membrane rearrangement in replication complex assembly requires replicase protein a, RNA templates, and polymerase activity. *J Virol* **84**:12492–12503.

88. Kraft, D. J., W. K. Kegel, and P. van der Schoot. 2012. A kinetic Zipper model and the assembly of tobacco mosaic virus. *Biophys J* **102**:2845–2855.

89. Kramvis, A. and M. C. Kew. 1998. Structure and function of the encapsidation signal of hepadnaviridae. *J Viral Hepat* **5**:357–367.

90. Lander, G. C., R. Khayat, R. Li, P. E. Prevelige, C. S. Potter, B. Carragher, and J. E. Johnson. 2009. The P22 tail machine at subnanometer resolution reveals the architecture of an infection conduit. *Structure* **17**:789–799.

91. Larson, S. B., S. Koszelak, J. Day, A. Greenwood, J. A. Dodds, and A. McPherson. 1993. Double-helical RNA in satellite tobacco mosaic virus. *Nature* **361**:179–182.

92. Lawson, C. L., S. Dutta, J. D. Westbrook, K. Henrick, and H. M. Berman. 2008. Representation of viruses in the remediated PDB archive. *Acta Crystallogr D Biol Crystallogr* **D64**:874–882.

93. Le Pogam, S., T. T. Yuan, G. K. Sahu, S. Chatterjee, and C. Shih. 2000. Low-level secretion of human hepatitis B virus virions caused by two independent, naturally occurring mutations (P5T and L60V) in the capsid protein. *J Virol* **74**:9099–9105.

94. Levy, H. C., M. Bostina, D. J. Filman, and J. M. Hogle. 2010. Catching a virus in the act of RNA release: A novel poliovirus uncoating intermediate characterized by cryo-electron microscopy. *J Virol* **84**:4426–4441.

95. Li, C., J. C. Wang, M. W. Taylor, and A. Zlotnick. 2012. In vitro assembly of an empty picornavirus capsid follows a dodecahedral path. *J Virol* **86**:13062–13069.

96. Li, P. P., A. Nakanishi, D. Shum, P. C. Sun, A. M. Salazar, C. F. Fernandez, S. W. Chan, and H. Kasamatsu. 2001. Simian virus 40 Vp1 DNA-binding domain is functionally separable from the overlapping nuclear localization signal and is required for effective virion formation and full viability. *J Virol* **75**:7321–7329.

97. Li, W., A. Asokan, Z. Wu, T. Van Dyke, N. DiPrimio, J. S. Johnson, L. Govindaswamy et al. 2008. Engineering and selection of shuffled AAV genomes: A new strategy for producing targeted biological nanoparticles. *Mol Ther* **16**:1252–1260.

98. Liddington, R. C., Y. Yan, J. Moulai, R. Sahli, T. L. Benjamin, and S. C. Harrison. 1991. Structure of simian virus 40 at 3.8-Å resolution. *Nature* **354**:278–284.

99. Liu, S., J. He, C. Shih, K. Li, A. Dai, Z. H. Zhou, and J. Zhang. 2010. Structural comparisons of hepatitis B core antigen particles with different C-terminal lengths. *Virus Res* **149**:241–244.

100. Lohmann, V. 2013. Hepatitis C virus RNA replication. *Curr Top Microbiol Immunol* **369**:167–198.

101. MacFarlane, S. A., M. Shanks, J. W. Davies, A. Zlotnick, and G. P. Lomonossoff. 1991. Analysis of the nucleotide sequence of bean pod mottle virus middle component RNA. *Virology* **183**:405–409.

102. Mannige, R. V. and C. L. Brooks, 3rd. 2008. Tilable nature of virus capsids and the role of topological constraints in natural capsid design. *Phys Rev E Stat Nonlin Soft Matter Phys* **77**:051902.

103. McPherson, A. 2005. Micelle formation and crystallization as paradigms for virus assembly. *Bioessays* **27**:447–458.

104. Melegari, M., S. K. Wolf, and R. J. Schneider. 2005. Hepatitis B virus DNA replication is coordinated by core protein serine phosphorylation and HBx expression. *J Virol* **79**:9810–9820.

105. Mukherjee, S., M. Abd-El-Latif, M. Bronstein, O. Ben-nun-Shaul, S. Kler, and A. Oppenheim. 2007. High cooperativity of the SV40 major capsid protein VP1 in virus assembly. *PLoS One* **2**:e765.

106. Mukherjee, S., S. Kler, A. Oppenheim, and A. Zlotnick. 2010. Uncatalyzed assembly of spherical particles from SV40 VP1 pentamers and linear dsDNA incorporates both low and high cooperativity elements. *Virology* **397**:199–204.

107. Mukherjee, S., C. M. Pfeifer, J. M. Johnson, J. Liu, and A. Zlotnick. 2006. Redirecting the coat protein of a spherical virus to assemble into tubular nanostructures. *J Am Chem Soc* **128**:2538–2539.

108. Muriaux, D., J. Mirro, D. Harvin, and A. Rein. 2001. RNA is a structural element in retrovirus particles. *Proc Natl Acad Sci U S A* **98**:5246–5251.

109. Newcomb, W. W., F. L. Homa, and J. C. Brown. 2005. Involvement of the portal at an early step in herpes simplex virus capsid assembly. *J Virol* **79**:10540–10546.

110. Newcomb, W. W., F. L. Homa, D. R. Thomsen, F. P. Booy, B. L. Trus, A. C. Steven, J. V. Spencer, and J. C. Brown. 1996. Assembly of the herpes simplex virus capsid: Characterization of intermediates observed during cell-free capsid formation. *J Mol Biol* **263**:432–446.

111. Newcomb, W. W., F. L. Homa, D. R. Thomsen, B. L. Trus, N. Cheng, A. Steven, F. Booy, and J. C. Brown. 1999. Assembly of the herpes simplex virus procapsid from purified components and identification of small complexes containing the major capsid and scaffolding proteins. *J Virol* **73**:4239–4250.

112. Newcomb, W. W., F. L. Homa, D. R. Thomsen, Z. Ye, and J. C. Brown. 1994. Cell-free assembly of the herpes simple virus capsid. *J Virol* **68**:6059–6063.

113. Nguyen, H. D., V. S. Reddy, and C. L. Brooks, 3rd. 2007. Deciphering the kinetic mechanism of spontaneous self-assembly of icosahedral capsids. *Nano Lett* **7**:338–344.

114. Ni, P., Z. Wang, X. Ma, N. C. Das, P. Sokol, W. Chiu, B. Dragnea, M. Hagan, and C. C. Kao. 2012. An examination of the electrostatic interactions between the N-terminal tail of the brome mosaic virus coat protein and encapsidated RNAs. *J Mol Biol* **419**:284–300.

115. Ning, X., D. Nguyen, L. Mentzer, C. Adams, H. Lee, R. Ashley, S. Hafenstein, and J. Hu. 2011. Secretion of genome-free hepatitis B virus—Single strand blocking model for virion morphogenesis of para-retrovirus. *PLoS Pathog* **7**:e1002255.

116. Nugent, C. I., K. L. Johnson, P. Sarnow, and K. Kirkegaard. 1999. Functional coupling between replication and packaging of poliovirus replicon RNA. *J Virol* **73**:427–435.

117. Oosawa, F. and S. Asakura. 1975. *Thermodynamics of Polymerization of Protein*. Academic Press, London, U.K.

118. Padilla-Meier, G. P. and C. M. Teschke. 2011. Conformational changes in bacteriophage P22 scaffolding protein induced by interaction with coat protein. *J Mol Biol* **410**:226–240.

119. Parent, K. N., S. M. Doyle, E. Anderson, and C. M. Teschke. 2005. Electrostatic interactions govern both nucleation and elongation during phage P22 procapsid assembly. *Virology* **340**:33–45.

120. Parent, K. N., M. M. Suhanovsky, and C. M. Teschke. 2007. Phage P22 procapsids equilibrate with free coat protein subunits. *J Mol Biol* **365**:513–522.

121. Parent, K. N., A. Zlotnick, and C. M. Teschke. 2006. Quantitative analysis of multi-component spherical virus assembly: Scaffolding protein contributes to the global stability of phage P22 procapsids. *J Mol Biol* **359**:1097–1106.

122. Parker, M. H., C. G. Brouillette, and P. E. Prevelige, Jr. 2001. Kinetic and calorimetric evidence for two distinct scaffolding protein binding populations within the bacteriophage P22 procapsid. *Biochemistry* **40**:8962–8970.

123. Parker, M. H. and P. E. Prevelige, Jr. 1998. Electrostatic interactions drive scaffolding/coat protein binding and procapsid maturation in bacteriophage P22. *Virology* **250**:337–349.

124. Pellet, P. E. and B. Roizmann. 2007. Herpesviridae: A brief introduction, pp. 2479–2499. *In* B. N. Fields and D. M. Knipe (eds.), *Fields' Virology*, 2nd edn., vol. 2. Raven Press, New York.

125. Perlmutter, J. D. and M. F. Hagan. 2014. Mechanisms of virus assembly. arXiv **1407**:preprint arXiv:1407.3856.

126. Perlmutter, J. D., M. R. Perkett, and M. F. Hagan. 2014. Pathways for virus assembly around nucleic acids. *J Mol Biol* **426**:3148–3165.

127. Perlmutter, J. D., C. Qiao, and M. F. Hagan. 2013. Viral genome structures are optimal for capsid assembly. *Elife* **2**:e00632.

128. Pettersen, E. F., T. D. Goddard, C. C. Huang, G. S. Couch, D. M. Greenblatt, E. C. Meng, and T. E. Ferrin. 2004. UCSF Chimera—A visualization system for exploratory research and analysis. *J Comput Chem* **25**:1605–1612.

129. Pierson, E. E., D. Z. Keifer, L. Selzer, L. S. Lee, N. C. Contino, J. C. Wang, A. Zlotnick, and M. F. Jarrold. 2014. Detection of late intermediates in virus capsid assembly by charge detection mass spectrometry. *J Am Chem Soc* **136**:3536–3541.

130. Porterfield, J. Z., M. S. Dhason, D. D. Loeb, M. Nassal, S. J. Stray, and A. Zlotnick. 2010. Full-length hepatitis B virus core protein packages viral and heterologous RNA with similarly high levels of cooperativity. *J Virol* **84**:7174–7184.

131. Porterfield, J. Z. and A. Zlotnick. 2010. An overview of capsid assembly kinetics, pp. 131–158. *In* P. G. Stockley and R. Twarock (eds.), *Emerging Topics in Physical Virology*. Imperial College Press, London, U.K.

132. Porterfield, J. Z. and A. Zlotnick. 2010. A simple and general method for determining the protein and nucleic acid content of viruses by UV absorbance. *Virology* **407**:281–288.

133. Portney, N. G., G. Destito, M. Manchester, and M. Ozkan. 2009. Hybrid assembly of CPMV viruses and surface characteristics of different mutants. *Curr Top Microbiol Immunol* **327**:59–69.

134. Prevelige, P. E. and B. A. Fane. 2012. Building the machines: Scaffolding protein functions during bacteriophage morphogenesis. *Adv Exp Med Biol* **726**:325–350.

135. Prevelige, P. E., Jr., J. King, and J. L. Silva. 1994. Pressure denaturation of the bacteriophage P22 coat protein and its entropic stabilization in icosahedral shells. *Biophys J* **66**:1631–1641.

136. Prevelige, P. E., D. Thomas, and J. King. 1993. Nucleation and growth phases in the polymerization of coat and scaffolding subunits into icosahedral procapsid shells. *Biophys J* **64**:824–835.

137. Rabe, B., M. Delaleau, A. Bischof, M. Foss, I. Sominskaya, P. Pumpens, C. Cazenave, M. Castroviejo, and M. Kann. 2009. Nuclear entry of hepatitis B virus capsids involves disintegration to protein dimers followed by nuclear reassociation to capsids. *PLoS Pathog* **5**:e1000563.

138. Rapaport, D. C. 2008. Role of reversibility in viral capsid growth: A paradigm for self-assembly. *Phys Rev Lett* **101**:186101.

139. Rapaport, D. C. 2010. Studies of reversible capsid shell growth. *J Phys Condens Matt* **22**:104115.

140. Roitman-Shemer, V., J. Stokrova, J. Forstova, and A. Oppenheim. 2007. Assemblages of simian virus 40 capsid proteins and viral DNA visualized by electron microscopy. *Biochem Biophys Res Commun* **353**:424–430.

141. Salunke, D., D. L. Caspar, and R. L. Garcea. 1989. Polymorphism in the assembly of polyomavirus capsid protein VP1. *Biophys J* **56**:887–900.

142. Sandalon, Z., N. Dalyot-Herman, A. B. Oppenheim, and A. Oppenheim. 1997. In vitro assembly of SV40 virions and pseudovirions: Vector development for gene therapy. *Hum Gene Ther* **8**:843–849.

143. Saper, G., S. Kler, R. Asor, A. Oppenheim, U. Raviv, and D. Harries. 2013. Effect of capsid confinement on the chromatin organization of the SV40 minichromosome. *Nucleic Acids Res* **41**:1569–1580.

144. Schwartz, M., J. Chen, M. Janda, M. Sullivan, J. den Boon, and P. Ahlquist. 2002. A positive-strand RNA virus replication complex parallels form and function of retrovirus capsids. *Mol Cell* **9**:505–514.

145. Seeger, C., F. Zoulim, and W. S. Mason. 2007. Hepadnaviruses, pp. 2977–3029. *In* D. M. Knipe, D. E. Griffin, R. A. Lamb, M. A. Martin, B. Roizman, and S. E. Straus (eds.), *Fields Virology*, vol. 2. Lippincott Williams & Wilkins, Philadelphia, PA.

146. Seeman, N. C. and A. M. Belcher. 2002. Emulating biology: Building nanostructures from the bottom up. *Proc Natl Acad Sci U S A* **99**:6451–6455.

147. Seitz, S., S. Urban, C. Antoni, and B. Bottcher. 2007. Cryo-electron microscopy of hepatitis B virions reveals variability in envelope capsid interactions. *Embo J* **26**:4160–4167.

148. Selzer, L., S. P. Katen, and A. Zlotnick. 2014. The HBV core protein intra-dimer interface modulates capsid assembly and stability. *Biochemistry* **53**:5496–5504.

149. Serwer, P. and G. A. Griess. 1999. Advances in the separation of bacteriophages and related particles. *J Chromatogr B Biomed Sci Appl* **722**:179–190.

150. Stannard, L. M. and M. Hodgkiss. 1979. Morphological irregularities in Dane particle cores. *J Gen Virol* **45**:509–514.

151. Stehle, T., S. J. Gamblin, Y. Yan, and S. C. Harrison. 1996. The structure of simian virus 40 refined at 3.1 Å resolution. *Structure* **4**:165–182.

152. Steinmetz, N. F., G. Calder, G. P. Lomonossoff, and D. J. Evans. 2006. Plant viral capsids as nanobuilding blocks: Construction of arrays on solid supports. *Langmuir* **22**:10032–10037.

153. Stockley, P. G., R. Twarock, S. E. Bakker, A. M. Barker, A. Borodavka, E. Dykeman, R. J. Ford, A. R. Pearson, S. E. Phillips, N. A. Ranson, and R. Tuma. 2013. Packaging signals in single-stranded RNA viruses: Nature's alternative to a purely electrostatic assembly mechanism. *J Biol Phys* **39**:277–287.

154. Stray, S. J., P. Ceres, and A. Zlotnick. 2004. Zinc ions trigger conformational change and oligomerization of hepatitis B virus capsid protein. *Biochemistry* **43**:9989–9998.

155. Stubbs, G. and A. Kendall. 2012. Helical viruses. *Adv Exp Med Biol* **726**:631–658.

156. Sun, J., C. DuFort, M. C. Daniel, A. Murali, C. Chen, K. Gopinath, B. Stein et al. 2007. Core-controlled polymorphism in virus-like particles. *Proc Natl Acad Sci U S A* **104**:1354–1359.

157. Tan, Z., M. L. Maguire, D. D. Loeb, and A. Zlotnick. 2013. Genetically altering the thermodynamics and kinetics of hepatitis B virus capsid assembly has profound effects on virus replication in cell culture. *J Virol* **87**:3208–3216.

158. Tanford, C. 1980. *The Hydrophobic Effect: Formation of Micelles and Biological Membranes*, 2nd edn. John Wiley & Sons, Inc., New York.

159. Tang, J., J. M. Johnson, K. A. Dryden, M. J. Young, A. Zlotnick, and J. E. Johnson. 2006. The role of subunit hinges and molecular "switches" in the control of viral capsid polymorphism. *J Struct Biol* **154**:59–67.

160. Tang, L., K. N. Johnson, L. A. Ball, T. Lin, M. Yeager, and J. E. Johnson. 2001. The structure of pariacoto virus reveals a dodecahedral cage of duplex RNA. *Nat Struct Biol* **8**:77–83.

161. Teschke, C. M. and D. G. Fong. 1996. Interactions between coat and scaffolding proteins of phage P22 are altered in vitro by amino acid substitutions in coat protein that cause a cold-sensitive phenotype. *Biochemistry* **35**:14831–14840.

162. Teschke, C. M. and K. N. Parent. 2010. 'Let the phage do the work': Using the phage P22 coat protein structures as a framework to understand its folding and assembly mutants. *Virology* **401**:119–130.

163. Tihova, M., K. A. Dryden, T. V. Le, S. C. Harvey, J. E. Johnson, M. Yeager, and A. Schneemann. 2004. Nodavirus coat protein imposes dodecahedral RNA structure independent of nucleotide sequence and length. *J Virol* **78**:2897–2905.

164. Tsai, C. J. and R. Nussinov. 2011. A unified convention for biological assemblies with helical symmetry. *Acta Crystallogr D Biol Crystallogr* **67**:716–728.

165. Tuma, R., H. Tsuruta, K. H. French, and P. E. Prevelige. 2008. Detection of intermediates and kinetic control during assembly of bacteriophage P22 procapsid. *J Mol Biol* **381**:1395–1406.

166. Valegard, K., J. B. Murray, P. G. Stockley, N. J. Stonehouse, and L. Liljas. 1994. Crystal structure of an RNA bacteriophage coat protein-operator complex. *Nature* **371**:623–626.

167. Wang, G., R.-Y. Cao, R. Chen, L. Mo, J.-F. Han, X. Wang, X. Xu et al. 2013. Rational design of thermostable vaccines by engineered peptide-induced virus self-biomineralization under physiological conditions. *Proc Natl Acad Sci U S A* **110**:7619–7624.

168. Wang, J. C., M. S. Dhason, and A. Zlotnick. 2012. Structural organization of pregenomic RNA and the carboxy-terminal domain of the capsid protein of hepatitis B virus. *PLoS Pathog* **8**:e1002919.

169. Wang, Q., T. Lin, J. E. Johnson, and M. G. Finn. 2002. Natural supramolecular building blocks. Cysteine-added mutants of cowpea mosaic virus. *Chem Biol* **9**:813–819.

170. Wang, X., W. Peng, J. Ren, Z. Hu, J. Xu, Z. Lou, X. Li et al. 2012. A sensor-adaptor mechanism for enterovirus uncoating from structures of EV71. *Nat Struct Mol Biol* **19**:424–429.

171. Wingfield, P. T., S. J. Stahl, R. W. Williams, and A. C. Steven. 1995. Hepatitis core antigen produced in *Escherichia coli*: Subunit composition, conformational analysis, and in vitro capsid assembly. *Biochemistry* **34**:4919–4932.

172. Witherell, G. W., J. M. Gott, and O. C. Uhlenbeck. 1991. Specific interaction between RNA phage coat proteins and RNA. *Prog Nucleic Acid Res Mol Biol* **40**:185–220.

173. Yoo, P. J., K. T. Nam, J. Qi, S.-K. Lee, J. Park, A. M. Belcher, and P. T. Hammond. 2006. Spontaneous assembly of viruses on multilayered polymer surfaces. *Nat Mater* **5**:234–240.

174. Yu, X., L. Jin, J. Jih, C. Shih, and Z. H. Zhou. 2013. 3.5A cryoEM structure of hepatitis B virus core assembled from full-length core protein. *PLoS ONE* **8**:e69729.

175. Yuan, T. T., G. K. Sahu, W. E. Whitehead, R. Greenberg, and C. Shih. 1999. The mechanism of an immature secretion phenotype of a highly frequent naturally occurring missense mutation at codon 97 of human hepatitis B virus core antigen. *J Virol* **73**:5731–5740.

176. Zandi, R., P. van der Schoot, D. Reguera, W. Kegel, and H. Reiss. 2006. Classical nucleation theory of virus capsids. *Biophys J* **90**:1939–1948.

177. Zhang, T. and R. Schwartz. 2006. Simulation study of the contribution of oligomer/oligomer binding to capsid assembly kinetics. *Biophys J* **90**:57–64.

178. Zlotnick, A. 2007. Distinguishing reversible from irreversible virus capsid assembly. *J Mol Biol* **366**:14–18.

179. Zlotnick, A. 1994. To build a virus capsid. An equilibrium model of the self assembly of polyhedral protein complexes. *J Mol Biol* **241**:59–67.

180. Zlotnick, A. 2004. Viruses and the physics of soft condensed matter. *Proc Natl Acad Sci U S A* **101**:15549–15550.

181. Zlotnick, A., R. Aldrich, J. M. Johnson, P. Ceres, and M. J. Young. 2000. Mechanism of capsid assembly for an icosahedral plant virus. *Virology* **277**:450–456.

182. Zlotnick, A., N. Cheng, S. J. Stahl, J. F. Conway, A. C. Steven, and P. T. Wingfield. 1997. Localization of the C terminus of the assembly domain of hepatitis B virus capsid protein: Implications for morphogenesis and organization of encapsidated RNA. *Proc Natl Acad Sci U S A* **94**:9556–9561.

183. Zlotnick, A., N. Padmaja, S. Munshi, and J. E. Johnson. 1997. Resolution of space-group ambiguity and structure determination of nodamura virus to 3.3 Å resolution from pseudo-R32 (monoclinic) crystals. *Acta Crystallogr D* **53**:738–746.

184. Zlotnick, A., J. Z. Porterfield, and J. C. Wang. 2013. To build a virus on a nucleic acid substrate. *Biophys J* **104**:1595–1604.

185. Zlotnick, A., S. J. Stahl, P. T. Wingfield, J. F. Conway, N. Cheng, and A. C. Steven. 1998. Shared motifs of the capsid proteins of hepadnaviruses and retroviruses suggest a common evolutionary origin. *FEBS Lett* **431**:301–304.

186. Zlotnick, A., M. M. Suhanovsky, and C. M. Teschke. 2012. The energetic contributions of scaffolding and coat proteins to the assembly of bacteriophage procapsids. *Virology* **428**:64–69.

3 Electron Cryomicroscopy and Image Reconstruction of Viral Nanoparticles

Bettina Böttcher

CONTENTS

3.1 HISTORICAL BACKGROUND

Historically, the developments in electron microscopy and sample preparation have been closely linked to the structure determination of viral nanoparticles. This is partly due to the high availability of some of these viral nanoparticles and to their high symmetry. Tobacco mosaic virus (TMV) was one of the earliest objects that have been imaged with the electron microscope (Kausche et al. 1939). However, sample preparation methods were still crude and relied on drying of the virus from a low-salt buffer. The drying conditions led to bending and distortions of the virus. Furthermore, the lack of contrast made it difficult to recognize features.

Nevertheless, these very early electron micrographs were already sufficient to measure the length and the width of a typical TMV particle. Later, more sophisticated techniques used coating of the surface of TMV with a thin layer of carbon under a shadowing angle (Matthews et al. 1956). This gave much better contrast and also showed some details on the surface of the virus particle. However, the level of detail was still insufficient to measure the correct distance between repeating units or to identify the general building principles of TMV from its subunits.

At the same time, TMV was also popular in x-ray crystallography for advancing x-ray fiber diffraction (Franklin and Holmes 1958), which resulted in the first density functions of TMV. From these density functions, a model of the organization of the subunit arrangement could be derived (Klug and Caspar 1960). Electron microscopy still lacked behind in suitable preparation methods and thus could not resolve these fine details.

Then, in the late 1950s, Brenner and Horne showed the potential of negative staining for the visualization of viruses (Brenner and Horne 1959). Negative staining was a fast and easy preparation method, which preserved the virus structures relatively well and gave a good contrast with lots of image detail. Micrographs of negatively stained TMV showed a high level of detail (Nixon and Woods 1960) that agreed well with the models derived from x-ray diffraction, which gave confidence in the preparation method.

Although a lot could be learned by just looking at these images, much more could be gained from averaging the information and identifying the characteristic, reoccurring structural features. Since computing was still in its infancy, optical methods had been devised, for optimally superposing different images and averaging them photographically (Markham et al. 1963, 1964). While the methods were quite powerful, finding the optimal superposition was still somewhat subjective and relied on recognizing the optimal fit.

To overcome this limitation, Klug et al. decided to make use of the repeating pattern of helical objects such as TMV, which gives rise to regular lines in a diffraction pattern. To observe this diffraction pattern from electron micrographs, they designed an optical device, which allowed generating a diffraction pattern from an electron micrograph in the back focal plane of an optical lens (Klug and Berger 1964). The diffraction pattern of the image could be reconstructed into an image by adding a second imaging lens (Klug and DeRosier 1966). The device did allow not only observing the diffraction pattern but also manipulating it in the back focal plane, for example, by masking, before reconstructing the image with the second lens. Masking was a very powerful tool for reducing the image information to the components, which were regularly packed in a helix and excluding other contributions, which were mainly added by the noise in the background. Thus, the filtered images were much clearer and could be interpreted with much greater confidence than the unfiltered noisy micrographs. Furthermore, differential masking of the diffraction pattern allowed separating the contributions from the front side of a helical arrangement from

the information from the backside. Although this was not yet a 3D image reconstruction, it gave much clearer insights into the 3D building pattern of a helical arrangement. The power of the method was initially demonstrated on TMV and the tail of the bacteriophage T4 (Klug and DeRosier 1966) and then used to elucidate how the structure of bacteriophage T4 changes upon contraction (Krimm and Anderson 1967).

Shortly afterward, DeRosier and Klug (1968) formulated the general principles of image reconstruction in Fourier space and demonstrated its effectiveness on the tail of the bacteriophage T4, for which the helical symmetry of the tail enabled the calculation of a 3D map from a single image. The principle of image reconstruction in Fourier space was later worked out for single particles in general (Crowther et al. 1970b) and developed for the special case of icosahedral symmetry (Crowther 1971). The reconstruction method was applied to various virus structures such as tomato bushy stunt virus, human wart virus (Crowther et al. 1970a), and turnip yellow mosaic virus (Mellema and Amos 1972). The image reconstruction required that the orientation and the origin of a particle relative to the symmetry axes were known. For icosahedral particles, Crowther (1971) came up with a very elegant solution. He showed that in the Fourier transform of a projection, every symmetry axis in the 3D object gives rise to pairs of common lines in the Fourier transform along which the information is the same. The position of the lines changes with the orientation of the object relative to the symmetry axes. Searching for the 37 common lines within a transform of an icosahedral particle was a very elegant and reference-free method for determining the orientation of the particle.

Helical and icosahedral reconstruction methods relied on the fact that the symmetry of the assemblies was well preserved and not distorted by the preparation methods. It soon became apparent that the distortions from perfect symmetry, which were introduced by the staining procedures, severely limited the resolution. This was mainly due to the flattening of the sample during drying and to the uneven staining. So, approaches were devised with which the assemblies could be maintained in an aqueous environment and could be imaged free of stain. First attempts in the early 1970s aimed at lowering the vacuum close to the object and saturating it with water vapor. This allowed imaging wet objects at room temperature. Electron diffraction studies on thin catalase crystals showed that this approach conserved structural details up to 2 Å resolution (Matricardi et al. 1972), which surpassed what could be achieved with stained and dried catalase crystals by almost one order of magnitude. Obviously, retaining the water was essential for obtaining high-resolution structural information.

At about the same time, an alternative route for retaining the water inside the microscope was tested. This approach relied on freezing the sample and imaging it at liquid nitrogen temperature inside the electron microscope (Taylor and Glaeser 1973). At the low temperatures, the vapor pressure of the water dropped below the pressure of the vacuum and thus, evaporation of the sample was efficiently prevented. Freezing proved similarly effective to retain high-resolution features in catalase crystals (Taylor and Glaeser 1974) as using a differential vacuum system in the column to stabilize a wet sample at room temperature. However, from an instrumentational point of view, cold samples were much easier to handle than a differential vacuum system.

The initial preparation method for unstained cold samples simply relied on freezing the sample in liquid nitrogen. Yet, the cooling rates in liquid nitrogen are relatively low and thus, water crystallizes as hexagonal ice. Soon, it became apparent that the growing ice crystals also had a damaging effect onto the biological specimens. So, it was important to keep the frozen water in a state, which was more like a liquid. In the 1980s, it was demonstrated that a vitrified state of water could be created by rapidly cooling water to low temperatures (Bruggeller and Mayer 1980; Dubochet and McDowall 1981). Electron and x-ray diffraction patterns both showed diffuse diffraction rings of this form of water, similarly as expected for a liquid. Dubochet and coworkers were convinced that this amorphous form of water was the best way to preserve the structures of biological specimens for electron microscopy. They developed efficient ways for the preparation (Lepault and Dubochet 1986) of particles in vitrified suspension (Lepault et al. 1983) and for sections of vitrified, thicker objects (McDowall et al. 1983) such as bacteria (Dubochet et al. 1983). Key to the method development was understanding which phase of water forms under which condition and how the different modifications of water change in the electron microscope (Dubochet and McDowall 1981; Dubochet et al. 1988). Many of the investigated specimens were viruses (Adrian et al. 1984; Vogel et al. 1986; Baschong et al. 1988; Dubochet et al. 1994). The level of detail and the structural preservation, which were achieved by vitrification, were stunning and unprecedented.

In the following years, many low-resolution virus structures were determined using the new preparation technique and the image reconstruction methods based on common lines and Fourier reconstruction developed earlier by Crowther. Initially, most of the structures were determined from very few particles (20–100). Nonetheless, making use of the high symmetry of the icosahedral particles gave structures in the range of 20–40 Å resolution. More importantly, the structures were undisturbed by stain or other preparation artifacts. Gradually, the number of particles in the reconstruction increased and reference-based approaches were incorporated into the determination of particle orientations. Later, methods were developed for correcting for the aberrations introduced by the contrast transfer of the electron microscope. By 1999, some 180 different reconstructions of icosahedrally arranged viral nanoparticles ranging from 25 to 140 nm in size were published (reviewed in Baker et al. 1999).

Many of the reconstructions showed capsids in different states and thus gave valuable structural insights into capsid maturation. In some cases, the resemblance of certain features in the reconstructions to crystal structures of isolated subcomponents was intriguing. So finally the crystal structures of these components were fitted into the respective entities in the EM maps, giving the first pseudoatomic models (Wang et al. 1992; Olson et al. 1993; Stewart et al. 1993). Although it was not entirely clear at that time, whether this was scientifically justified, it is now a well-accepted and well-advanced technique to combine EM data with high-resolution structural models to build pseudoatomic models of larger complexes.

Virus particles also proved excellent objects for advancing the methods to obtain higher and higher resolution. The first obstacle was to overcome the limitations which were introduced by the contrast transfer function (CTF) of the electron microscope (Figure 3.1, black). Without correcting for it, the position of the first zero in the CTF, which depends on the defocus, also sets the ultimate limit of the obtainable resolution. Since an underfocus of several micrometers was needed to introduce sufficient phase contrast for determining the orientation of the virus particles, finding a solution to the contrast transfer problem was essential for pushing the resolution limit. A correction algorithm was developed with turnip yellow mosaic virus as a test object (Böttcher and Crowther 1996). Together with a slight increase in the number of particles (180), this already increased the resolution to 15 Å, which was a clear leap compared to what was achieved without correction for the CTF.

Another important technological advance for increasing the resolution was the availability of electron microscopes with a field emission gun. These electron sources were much brighter and had superior spatial and chromatic coherence. As a consequence, high-resolution information was transferred much stronger than in electron microscopes with a conventional thermionic source (Figure 3.1, red and blue).

In addition, faster computers and cheap, large computer storage made it possible to process thousands of particles instead of just a few hundred. So, toward the end of the twentieth century, reconstructions of icosahedral capsids with resolutions below 1 nm were determined. These reconstructions resolved α-helices as rod-shaped features (Böttcher et al. 1997; Conway et al. 1997) and for the first time demonstrated that secondary structural elements could be resolved from the reconstructions of single particles. At that point, the image-processing strategies, which led to the high-resolution structures, were specific to icosahedral symmetry. However, in principle an icosahedral particle is nothing else than a special case of a multicopy object (single particle). So, single-particle image-processing packages were further developed and now include all the necessary symmetry operations to obtain high-resolution structural information from multicopy objects of any symmetry.

In 2008, the 4 Å resolution barrier was surpassed on a rotavirus double-layer particle (Zhang et al. 2008). The 3.8 Å resolution map showed a similar level of detail as the crystal structure, which was published later (McClain et al. 2010). By now, single-particle image-processing methods (Grigorieff 2007) were seriously competing with x-ray crystallography in structure determination. At the end of 2013, 16 structures (Yu et al. 2008, 2011, 2013; Zhang et al. 2008, 2010, 2012, 2013a,b; Liu et al. 2010a; McClain et al. 2010; Wolf et al. 2010;

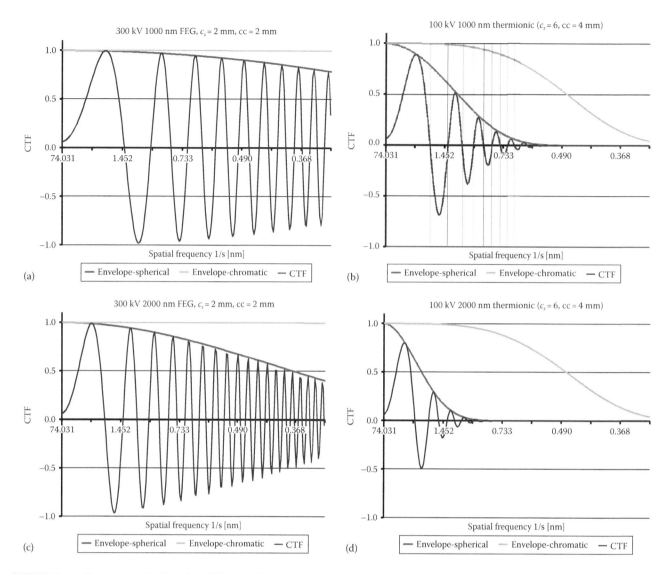

FIGURE 3.1 Contrast transfer function (CTF) for different types of microscopes and for different defocus values. The CTFs (black curves) are modulated by the envelope functions caused by spherical aberration (red curve) and by chromatic aberration (blue curve). (a and c) The CTF for an electron microscope with Schottky field emitter (energy spread 0.7 eV, opening half angle 0.04 mrad) operating at 300 kV and with c_s = 2 mm and cc = 2mm. (b and d) The CTF for a conventional electron microscope with a thermionic electron source (energy spread 1.5 eV, opening half angle 0.25 mrad), and a Biotwin lens (c_s = 6.3 mm and cc = 4 mm) operating at 100 kV. The CTF (black curve) oscillates between +1 and −1 in dependence of the spatial frequency and is dampened by the envelope functions. Different frequency bands are transferred with opposite contrast. In panel (b), the transfer bands with positive contrast transfer are highlighted in yellow and with negative transfer in green. The CTF oscillates faster at larger defocus. The envelope that is imposed by the spherical aberration (red curve) depends on the defocus. In contrast, the envelope that is caused by the chromatic aberration of the electron microscope is independent of the defocus (blue curve). Typically, the main limiting factor is the spherical aberration, while the chromatic aberration is only effective at higher spatial frequencies. Electron microscopes with a thermionic electron source and a large spherical aberration (e.g., Biotwin lens) do not transmit information at 1–2 µm defocus at spatial frequencies >1/7 Å^{-1}, while this range is well transmitted in electron microscopes with a lower spherical aberration (e.g., twin lens or super twin lens) and a Schottky emitter. The diagrams were generated with the Excel sheet created by Stahlberg (2012).

Chen et al. 2011a; Cheng et al. 2011; Settembre et al. 2011) of icosahedral nanoparticles with resolutions better than 4 Å had been deposited in the electron microscopy data bank (EMDB, Henrick et al. 2003; Lawson et al. 2011).

Although computing power was greatly improved and more coherent electron sources were available, high-resolution work still used photographic film as primary detector. Out of the 16 maps of icosahedral nanoparticles that were reconstructed to resolutions better than 4 Å, 15 relied on data from

film as primary detector. Film has a much better transfer of fine details as the conventional CCD cameras (Figure 3.2) that require conversion of the incident electrons into a detectable light signal. In 2009, the first direct detectors for electron microscopy became available (McMullan et al. 2009a,b). It became quickly apparent that they had the potential to outperform film for electrons at 300 kV accelerating voltage. Furthermore, the fast read-out rates of direct detectors allowed recording movies rather than single, high-dose images.

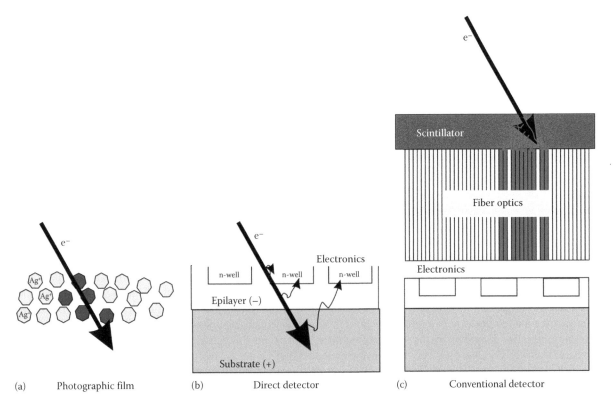

FIGURE 3.2 Detectors in electron microscopes: (a) photographic film contains silver halide grains in a gelatin layer. The size of the grains (gray) is <1 μm and the thickness of the gelatin layer (yellow) is in the order of 25 μm. The incident electrons enter the sensitive layer and change the properties of the grains, which makes them convertible into silver in the subsequent developing process. One electron can change the properties of several grains (amplification of the signal). High-energy electrons are partly back scattered by the film carrier and reenter the sensitive layer at a different position, which leads to an increase in unspecific background. (b) A direct detector consists of several layers. The upper layer contains the electronic read-out circuits and is some 2–4 μm thick. The read-out is located on top of the n-wells in the epilayer (5–25 μm thick). The incident electron generates many hole–electron pairs in the epilayer (signal amplification). The electrons of these electron–hole pairs diffuse toward the n-wells where they change the read-out signal. The substrate layer (up to 500 μm thick) underneath the epilayer acts as a potential barrier, which cannot be entered by the electron–hole pairs. Nevertheless, the substrate layer can be entered by the incident electrons, which have a much higher energy. In the substrate layer, these electrons are partly back scattered and reenter the epilayer at a different position, producing more electron–hole pairs and adding to the background noise. The problem is reduced by back-thinning the substrate layer. The incident electron also passes through the read-out electronics, where it causes damage to the electronics. Therefore, direct detectors have the potential problem of a limited lifetime due to radiation damage. (c) In a conventional electronic detector, the incident electrons pass through a scintillator layer, where they excite emission of many photons across a relatively large area (amplification of the signal). The light signal is passed through fiber optics to the electronic detector, which can be either a CMOS or a CCD detector. Similar as in direct detectors, the photons generate electron–hole pairs that change the read-out signal of the pixels. The photons have much lower energy than the incident electrons in a direct detector. Therefore, a photon typically generates only one single electron–hole pair. Since the electronics of the direct detector is not directly exposed to the high-energy incident electrons, damage to the read-out electronics is not a problem in conventional detectors.

These movies showed beam-induced movement of the particles in the vitrified water (Brilot et al. 2012). The movement degraded the high-resolution image information in a conventional high-dose image. However, in a movie, the independent movement of the particles could be corrected in the individual frames, and thus, the image quality could be greatly improved (Li et al. 2013). Now, it has been demonstrated that with the advent of the new detectors and electron microscopes with automated loading systems for the sample and fully or semi-automated data collection, structures with resolutions better than 4 Å can be acquired within days. This provides a bright future for structure determination of icosahedral nanoparticles and suggests that for these types of assemblies, x-ray crystallography will become largely obsolete.

The analysis of helical particles made a similar rapid progress over the past years. At the beginning of the twenty-first century, researchers made use of iterative helical real-space refinement (Egelman 2000; Sachse et al. 2007). This enabled much more accurate determination of the orientation of small helical patches and thus correcting for small deviations from perfect helical symmetry in the whole helical assembly (e.g., bending of the helical tube). As a result, resolution in helical reconstructions improved fast either by using real-space refinement methods or alternatively by using standard crystallographic procedures. At the end of 2013, there were five helical structures deposited in the electron microscopy database (acetylcholine receptor; Unwin 2005, bacterial L-type flagella;

Maki-Yonekura et al. 2010, and three of TMV; Sachse et al. 2007; Clare and Orlova 2010; Ge and Zhou 2011) with resolutions better than 5 Å.

3.2 SAMPLE PREPARATION

High-resolution structure determination of viral nanoparticles by electron microscopy requires a preparation method, which preserves the whole structure of the nanoparticle in an aqueous surrounding without any staining. The method of choice is vitrification of a particle suspension and has been introduced by Dubochet and coworkers in the 1980s (for review see Dubochet et al. 1988; Dubochet 2012).

Vitrified samples are typically prepared by applying a suspension of nanoparticles onto a sample carrier either directly as small droplets (spraying) or by forming a thin film by removal of most of the sample (e.g., by blotting with filter paper). Afterward, the sample together with its carrier is plunged into a cryogen, which rapidly cools the sample below the vitrification temperature of approximately −137°C. High cooling rates (10^{-4} K/s) support the vitrification of the water, whereas slower cooling rates lead to the formation of hexagonal, crystalline ice. The vitrified state is unstable above temperatures of approximately −120°C (Mayer and Bruggeller 1983) but stable for at least 5 min at temperatures below −137°C (Dubochet et al. 1988). Irradiation with electrons in the electron microscope can also induce devitrification at lower temperatures above −150°C (Heide 1984). To preserve the vitrified state, samples are stored in liquid nitrogen and require permanent cooling during sample handling, sample transfer, and sample imaging.

The success of sample preparation depends on several factors, which include the sample carrier, the method of film formation (spraying vs. blotting), and the choice of cryogen. These factors will be discussed in the following.

3.2.1 SAMPLE CARRIERS

Standard electron microscopy holders accept round grids with a diameter of 3.05 mm, which are available in different materials (e.g., Cu, Cu/Rh, Al, Ti, Au, Mo, Ni, Be). For biological applications, copper or copper/rhodium grids are most commonly used. The grids have a metal surface with either square, round, or hexagonal openings in defined distances. The pitch between openings is given by the mesh size, which is the number of openings per inch. A typical mesh size for the preparation of vitrified samples is 400 mesh, which corresponds to a pitch of 63.5 μm with a typical side length of the opening of 35–40 μm.

Normally, grids are coated with carbon support film (1–40 nm thick), which consists of amorphous carbon that has a high mechanical strength, is inert to most solvents used in biology, and is largely translucent for electrons. These properties make amorphous carbon film an ideal support for electron microscopy. In addition, amorphous carbon has some electrical conductance at liquid nitrogen temperature, which helps to dissipate charges that build up during imaging. Thick carbon

support films (10–40 nm) also have good mechanical stability, which decreases beam-induced movement. However, thick carbon films also scatter electrons much more strongly than thin carbon films and give rise to a grainy background that superimposes with the image of the nanoparticle. The problem is reduced by decreasing the thickness of the carbon support film to 1–2 nm. However, these very thin carbon films are no longer stable on their own over large areas and easily rupture when the sample is applied or when the film is irradiated with electrons. To overcome this problem, a thick carbon support film with holes (typical diameter of the holes 1–5 μm) is used as a carrier, which is coated with an additional, thinner, continuous carbon layer. Now, the thin, continuous carbon film only needs to be self-supporting over the small areas of the holes, where the nanoparticles are imaged.

An alternative to thin carbon film is graphene, which is a different modification of carbon, and has been discovered only a decade ago (Novoselov et al. 2004; Meyer et al. 2007). Graphene is one layer of carbon-atoms thick and forms extended 2D crystals, which are extremely sturdy in respect to their thickness. In addition, the conductivity of graphene per mass is significantly higher than that of amorphous carbon. First experiments have shown that the background added by a graphene support film is indeed minimal and allows the identification of unstained DNA, which is impossible with other support films (Pantelic et al. 2011). Graphene is also sturdy enough to support larger biological complexes (Pantelic et al. 2012). However, the preparation of large self-supporting graphene films is still difficult and thus, it is not yet routinely available as a support film for electron cryomicroscopy. An alternative to grapheme is graphene oxide (Wilson et al. 2009). The additional oxygens at the surface provide a hydrophilic surface that interacts well with the particle suspension. Graphene oxide is easier to produce than graphene and is now commercially available as additional support film on holey carbon-coated grids.

Holey carbon support films are also used without additional continuous carbon support films. In this case, a self-supporting particle suspension with a thickness of approximately 20–100 nm is formed across the holes, where the unsupported nanoparticles can be imaged. This has the advantage that the viral nanoparticles are not deformed by absorption to a support film and that no additional background is added by the support film to the image of the nanoparticles. The drawback is that unsupported holey films are somewhat more labile and give rise to more beam-induced movement during imaging. Despite the increased beam-induced movement, high-resolution (<4 Å) structures have been determined from unsupported viral nanoparticles (e.g., rotavirus [Zhang et al. 2008], cytoplasmic polyhedrosis virus [Yu et al. 2011], or TMV [Ge and Zhou 2011]).

Most applications use holey carbon films either with or without an additional support layer. There are various lab protocols for generating the required holey carbon films, which rely on generating a plastic film with holes as a template. The holes are either generated by condensation of little water droplets on a surface, which exclude the plastic film from these areas (Murray and Ward 1987) or by including bubbles of an immiscible liquid into a solution of plastic, which then give rise to holes in the plastic film after etching (Harris 1962; Bayer and Anderson 1963; Moharir and Prakash 1975; Baumeister and Seredynski 1976). The holey plastic film serves as a template onto which the thick carbon layer is evaporated. Finally, the plastic film is dissolved in organic solvent and a holey carbon film remains. The resulting pattern of holes and the size distributions of holes are irregular. These irregularities are a major obstacle for automated data acquisition schemes, which are more efficient with regular hole patterns and with diameters of the holes that are adapted to the illuminated area or the size of the detector. Regular hole patterns require a controlled pattering algorithm, which is achieved by photolithography either with a consumable soft template that is dissolved (Ermantraut et al. 1998; Chester et al. 2007) or with a hard template and a water-soluble release layer to replicate the pattern onto the carbon film (Quispe et al. 2007). Grids with such patterned holey support films are commercially available from different suppliers with different hole sizes, hole shapes, and hole patterns.

Shortly after evaporation, carbon films are hydrophilic but become more and more hydrophobic and thus water repellent over time. If an aqueous particle suspension is applied onto such hydrophobic carbon film, the suspension interacts very little with the support film and does not wet it properly. Therefore, vitrified samples prepared with hydrophobic grids are often either too dry or the particle suspension is too thick. The hydrophobicity and thus the wettability of the carbon support film can be increased by exposing it to a plasma immediately before use. The plasma is formed by passing an electrical current through a low pressure gas, which leads to a glow discharge. This generates radicals of the gas, which change the surface properties of the exposed surfaces, in this case the properties of the support film. While glow discharge with *air* as gas deposits negative charges onto the amorphous carbon film, an atmosphere of amylamine adds positive charges to the surface (Dubochet et al. 1971). Glow discharge increases the wettability of amorphous carbon and lowers the contact angle between the carbon surface and water droplets (Kutsay et al. 2008). This facilitates easy spreading of a particle suspension and is essential for the formation of a thin, homogeneous particle film. The increase in wettability by glow discharge is reversible and is completely reverted after some 2 h (Kutsay et al. 2008). The choice of positive or negative charges or the use of freshly floated carbon gives a wide range of different surface properties, which have subtle influence on the distribution of the nanoparticles and to some extent, also, on the orientation of the particles on the support film.

Amorphous carbon has a different thermal expansion coefficient to the supporting copper grids. As a consequence, the copper grids shrink much more upon cooling than the carbon support film does, which leads to severe folds in the carbon support film and tensions in the vitrified sample. This phenomenon was described by Booy and coworkers as cryocrinkling of the carbon film (Booy and Pawley 1993). Cryocrinkling is reduced by using a grid material with a lower expansion coefficient such as molybdenum or titanium (Booy and Pawley 1993; Vonck 2000).

3.2.2 FORMATION OF THIN PARTICLE SUSPENSIONS

High-resolution imaging of vitrified viral nanoparticles requires that they are contained in a very thin film of their vitrified buffer on the sample carrier. Ideally, the thickness of the film is approximately the same thickness as the diameter of the particles. Thicker films give reduced contrast and increase the percentage of electrons, which are scattered inelastically or multiple times. Depending on the size of the particle, a useful thickness of a vitrified suspension is approximately 20–200 nm.

3.2.3 BLOTTING AND EVAPORATION

Typically, thin films are formed by applying the sample to a grid with or without support film and then removing excess liquid. The process of film formation is quenched by plunging the grid with the particle suspension into a cryogen (Figure 3.3). In this method, two processes contribute to the removal of the liquid: one is the absorption of the liquid to the filter paper and the other is the evaporation of the water from the sample. The latter gives rise to a decrease in the temperature of the sample and an increase in the concentration of the particles and solutes. Depending on the environment, evaporation is rapid and can be the major driving force of the film formation (Trinick and Cooper 1990). While the increase in the number of particles during evaporation is often advantageous (e.g., for dilute particle suspensions), the increase in the concentration of solutes and especially of salts is unwanted and can interfere with the integrity and/or conformation of the nanoparticles.

The speed of evaporation is influenced by the ambient temperature and the relative humidity in the surroundings (Frederik and Hubert 2005) as well as the concentration of the solutes and the surface area, which is accessible to evaporation during film formation. An experimental setup without control of these parameters in the microenvironment of the sample requires some experience to identify the right moment, when the film has the optimal thickness before quenching the film formation by plunging the grid into the cryogen. To reduce this random factor, many devices control humidity as well as the temperature in the local environment of the sample. By increasing the humidity to almost 100%, evaporation can be virtually stopped. The simplest way to generate an environment with 100% humidity is to bubble air through water before blowing it onto the grid (Dubochet et al. 1988; Cyrklaff et al. 1990). A more sophisticated approach is to surround the sample with a chamber with a fully controlled environment inside (Bellare et al. 1988) (Figure 3.3). The chamber can be humidified with water-soaked sponges, which have a large surface area for exchange of water vapor with the surrounding or by using an ultrasonic nebulizer (Frederik and Hubert 2005). Inside the chamber, it is also possible to control the temperature either by heating or by cooling with a Peltier element. This gives full control of the major environmental parameters that influence the evaporation rate and enables reproducible sample preparation with very little optimization of other factors.

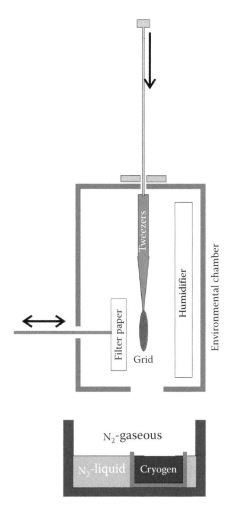

FIGURE 3.3 Schematic drawing of a manual freezing apparatus similar to the Bellare design (Bellare et al. 1988). The grid (red) is held by a pair of tweezers (brown) that is connected to a movable rod. The rod together with the grid and the tweezers can be propelled (gravity, spring, and motor) toward the cryogen. During blotting, the grid is surrounded by an environmental chamber (gray), which is humidified by a humidifier (e.g., wet sponge or ultrasonic nebulizer). The thin film of the sample is formed by coplanar blotting with filter paper (yellow). For blotting, the filter paper can be moved with a rod that is accessible from the outside of the chamber. After film formation, the grid is plunged into the cryogen. The cryogen (ethane, propane, or a mixture of both) is placed in a small vessel that is surrounded by a larger vessel, which contains liquid nitrogen and an atmosphere of cold, dry nitrogen gas.

The other process, which contributes to the formation of the thin film, is the active removal of excess liquid by blotting with filter paper. Again, there are many factors that influence the outcome of the blotting process. They include the type of filter paper, the force with which the filter paper is applied, the blotting time, the contact area and the contact angle between the filter paper and the sample, and the direction of blotting. Blotting can be done either from the same side as the sample was applied to or from the opposite side (Toyoshima 1989) or simultaneously from both sides. Blotting from both sides has the advantage that there is no exposed surface area of the grid,

from which the solvent can evaporate. This makes it unnecessary to control the humidity in the microenvironment of the grid, because the sample is humidified from both sides by the wet filter paper (Cyrklaff et al. 1990; Trinick and Cooper 1990). Devices with double-sided blotting are realized either in self-built plunging devices or in computer-controlled devices such as the FEI-Vitrobot, the Gatan Cryo plunge, or the time-resolved freezing apparatus from White and coworkers (Walker et al. 1995; White et al. 1998). Blotting single sided from the front or back gives similarly good results. Single-sided blotting from the back (opposite to where the sample was applied) minimizes the contact area between nanoparticles and the filter paper. In some cases, this helps to increase the number of particles in the field of view (Toyoshima 1989). Most manual devices use single-sided blotting, but there are also fully computer-controlled, commercial cryoplungers such as the Leica EM GP that support single-sided blotting. In single-sided blotting, the surface-accessible area of the grid is much larger than in double-sided blotting and thus, evaporation is a major factor, which requires a controlled microenvironment for reproducible results as described earlier.

Another important aspect is the contact angle between the sample and the filter paper. In most blotting devices, grid and filter paper are oriented coplanar, which gives the maximal contact across the whole area of the grid. This leads to an even film thickness across the grid. Therefore, coplanar blotting results in the largest possible area with optimal sample thickness, if optimal blotting conditions are found. The disadvantage is that if the blotting conditions are not optimal, the whole grid is likely to be suboptimal. For that reason, a few devices such as the FEI-Vitrobot use a design where the filter paper touches the grid under an angle. This generates a gradient in the strength of the contact between the filter paper and the sample across the whole grid. As a consequence, the thickness of the film is also variable across the grid, leaving some areas with an optimal thickness of the film even if the blotting conditions are not optimized yet. However, after optimization of the conditions, a typical film has still extended areas that are either too thin or too thick for imaging. Furthermore, since mounting of the grid in the device can vary by some ±0.5 mm, the area with the optimal thickness is not always centered, which is problematic for applications such as tomography that requires optimal sample thickness in the center of the grid for recording images at large tilt angles.

In addition to the contact angle between filter paper and grid, blot time and blot force are also important. Blotting too strongly often leads to a completely dry grid. Here, the sample appears to be mechanically ripped off the grid rather than being slowly absorbed into the filter paper. With more moderate blot forces, where blotting is controlled by the absorption of the solution into the filter paper rather than by a mechanical manipulation of the film, useful blotting times are less critical in a controlled microenvironment. Successful blotting times between 1 and 20 s have been reported.

The filter paper itself also influences the result of the blotting process. Filter papers come with different flow rates, which are related to how fast they can blot off the sample from the grid. For forming thin films, filter papers with medium to medium–fast flow rates like Whatman No. 1 (medium flow rate, 11 μm particle retention) or Whatman No. 595 (medium fast, 4–7 μm particle retention) are most commonly used, while filter papers with slow flow rates are uncommon for the formation of thin films in vitrification experiments.

Filter papers also vary in their composition. Standard filter papers have significant amounts of trace elements that can dissolve and enter the sample. In particular, the most commonly used Whatman No. 1 contains 0.13 mg calcium/g filter paper. This calcium is enough to change the conformation of calcium-sensitive assemblies. If sensitivity to trace elements is an issue, ashless filter papers with similar flow rates are a better alternative (e.g., Whatman No. 40, 42, or 43).

A variant of the preparation of thin films by blotting is the bare-grid method. Here, grids are used without further support film (Adrian et al. 1984; Dubochet et al. 1988; Jager 1990). In this method, the excess liquid is blotted off the grids and film formation is followed through binoculars and terminated by plunging the grid into a suitable cryogen (ethane or propane) when the film starts to rupture. Thus, the method uses a mixture of sample removal by blotting and evaporation. The preparation method requires some experience for recognizing the right moment for terminating the process of the film formation. In the bare-grid method, the film thickness varies across the grid, because the thickness partly depends on the gravity-driven flow of the water and thus increases toward the bottom of the grid.

In the bare-grid method, the size of a mesh that is spanned without further support is some 30–50 μm for a 400 mesh grid, which requires a relatively thick film to form a stable, self-supporting particle suspension. The suspension is typically thicker toward the grid bars and thinner in the center of a mesh. A problem of samples prepared by the bare-grid method is that the unsupported suspensions are nonconducting; therefore, they rapidly charge up during irradiation with electrons, which leads to image blurring. Although the bare grid method has been used for structure determination in the past (Adrian et al. 1984; Schatz et al. 1995), it is now no longer considered as suitable for high-resolution studies. Nonetheless, the method might have its merits for very large nanoparticles that get easily distorted by interactions with support films or by the pressure in very thin suspensions.

3.2.4 Spraying

An alternative for removing excess sample by blotting and evaporation is depositing a film of suitable thickness in the form of tiny droplets directly onto the grid. This is achieved by spraying microdroplets onto the support film of a grid while it is plunged into the cryogen (Dubochet et al. 1982). For the formation of a thin film, very small droplets are required, which still cover areas of 1–2 μm². Such droplets are generated by sprayers or *atomizers*. However, on a hydrophobic surface (such as an untreated carbon support film), droplets of this size do not spread and thus, they produce a film, which is far too thick for imaging. To facilitate rapid spreading of the drops after impact, the surface of the grid has to be extremely hydrophilic (glow discharge). But even with all the parameters carefully adjusted, the thickness of the droplets is still variable across each droplet, with a thick center and a thinner outer rim. Furthermore, with the impacting drops spaced some micrometers apart, a significant part of the surface of the grid remains dry, which further reduces the useful area. Therefore, spraying has lost its importance as a standard procedure for preparing thin films for vitrification.

On the other hand, spraying has the potential of starting a reaction by spraying an agent onto the sample just before quenching the reaction by vitrification. Thus, it is an ideal approach for time-resolved electron cryomicroscopy. The concept has been further developed by Berriman and Unwin for imaging the acetylcholine receptor in the open state (Berriman and Unwin 1994). In contrast to the spraying method described previously, Berriman and Unwin first formed the thin film of the acetylcholine receptor by conventional blotting (see previous text) and then sprayed the agent (acetylcholine) onto this preformed, thin film during plunging. This method requires a tight coupling of the action of the sprayer with the plunging of the grid via a photoswitch and enables reactions times of 1–100 ms.

The spreading of droplets on a preformed, thin film is far more efficient than spraying the droplets onto a dry grid. This is likely due to the fact that the already formed film and the impacting droplets have similar surface properties and therefore mix easily. While large particles such as viruses or ferritin remain within the boundaries of the impacting droplet, small solutes, such as ions, diffuse much faster and can be traced outside the impact zone of the drop. Depending on the delay between spraying and freezing, ions can travel several micrometers beyond the boundaries of the drop before the sample is vitrified.

A device that sprays an agent onto the grid allows only the addition of one component and has a very short time interval between the start of a reaction and its quenching, which is determined by the time it takes for the grid after blotting to reach the cryogen. For a more sophisticated control over the time course of a reaction, two mixing chambers are added in front of the spraying device, which allow mixing of different reagents and a free choice of delay times (Walker et al. 1995; White et al. 1998; Lu et al. 2009). Such devices implement a classical quench flow experiment into sample preparation by quenching the reaction by vitrification.

3.2.5 Vitrification

Immediately after forming the thin film by blotting or spraying, the film is stabilized against evaporation by rapid *freezing*. For vitrification, the sample is plunged into a cryogen with a guillotine-like apparatus that is either driven by gravity, accelerated by a spring, moved by pneumatics, or propelled with a motor. The choice of the cryogen is far more critical than the plunging mechanism and is decisive whether hexagonal ice is obtained or a vitrified sample. Vitrification is only achieved when the cooling rates are high and the sample is rapidly (10^4 K/s) cooled below the vitrification temperature. Although liquid nitrogen is suitable for cooling the sample below this temperature, its cooling rate is far too slow. The reason is the Leidenfrost phenomenon, which was discovered in the eighteenth century. It occurs when a *hot* object is emerged into a liquid, which has a boiling point of a least 100 K below the temperature of the hot object. The large difference in temperature causes film boiling at the surface of the hot object and the formation of an insulating gas layer. The heat transfer across this gas layer is slow, and thus, the *hot* object cools down relatively slowly. The boiling temperature of liquid nitrogen is −196°C and that of the *hot* sample is between 0°C and 40°C. This leads to film boiling and thus to a slow heat transfer, which gives the water time to rearrange and to form crystalline ice rather than vitrified water.

To achieve faster cooling rates, it is essential to use a cryogen, which has a higher boiling point but is still liquid below the vitrification temperature. Such cryogens are ethane, propane, or a 60:40 mixture of both (Tivol et al. 2008) (Table 3.1). The cryogen of choice is condensed into a small reservoir that is cooled with a surrounding bath of liquid nitrogen (Figure 3.3). Unfortunately, the melting points of ethane and propane are above the temperature of liquid nitrogen. Therefore, after a while, these cryogens freeze, and have to be melted again to make them ready for a new round of plunge-freezing grids without mechanical damage of the grids by impacting onto the frozen cryogen. To keep propane or ethane stably in the liquid phase, they have to be periodically warmed by condensing more gaseous cryogen into the reservoir, or by thermally insulating the reservoir from the surrounding liquid nitrogen bath or by heating the reservoir with a heater. Alternatively to ethane or propane, a mixture of both can be used, which forms a eutectic system that freezes at around −196°C and hence stays liquid even when it is cooled with liquid nitrogen (Tivol et al. 2008).

After plunging the sample into the cryogen, the sample is immediately vitrified and can be removed from the cryogen for storage or for transfer to the electron microscope. To keep the grids below the devitrification temperature, they are stored in liquid nitrogen and handled either in liquid nitrogen or in an atmosphere of cold, dry nitrogen gas.

TABLE 3.1
Boiling and Melting Points of Some Cryogens

Cryogen	Boiling Point (K)	Melting Point (K)
Nitrogen	77	66
Ethane	185	90
Propane	231	86
60:40 mixture of ethane/propane	185–231	77

3.2.6 STRATEGIES FOR ON-GRID CONCENTRATION OF VIRAL NANOPARTICLES

Often viral nanoparticles cannot be produced with concentrations that give sufficient particles in the field of view for efficient imaging. To obtain 100 particles in a field of view of 1 μm^2 in a film with a thickness of 50 nm, a concentration of approximately 3 μM or 2×10^{15} particles/mL is needed. For many viral nanoparticles, these concentrations are out of reach and require concentrating the nanoparticles in situ on the grid.

Occasionally, more particles than expected from the particle concentration are observed in the field of view. Sometimes, the enrichment can be explained by partial evaporation of the solution in an uncontrolled freezing device. In other cases, this seems to be unlikely, because a humidified environmental chamber has reduced evaporation during sample preparation, suggesting that other processes are involved in concentrating the sample on the grid. One of the likely mechanisms is that the nanoparticles move much slower and therefore enter the filter paper far slower than the buffer (water, salts, and other small molecules). Another effect is that nanoparticles enrich at the air–water interface of drops (Johnson and Gregory 1993). The enrichment is slow (minutes to hours) and depends on the size of the particles and the diameter of the drop. Surface enrichment can be utilized to trap virus particles at the air–water interface and to generate a high local concentration of the nanoparticles, which is transferred by touching the surface of the drop with the grid.

Another process that is utilized for enriching viral nanoparticles is to adsorb them onto a thin support film. For example, adeno-associated virus can is enriched by almost a magnitude, when absorbed onto a carbon film (Kronenberg et al. 2001). Other nanoparticles might be trapped on the grid by using functionalized support films such as *the affinity grid* (Ni-NTA-lipids absorbed to a carbon-coated grid; Kelly et al. 2008, 2010) or streptavidin-functionalized graphene oxide support films (Liu et al. 2010c).

3.3 DATA ACQUISITION

There are two fundamentally different types of objects: one type of object exists in multiple, identical copies, and the other is an object with a unique structure. While for multicopy objects, image information from different particles can be merged into a single representative 3D structure, for unique objects, all structural information has to be obtained from one object. This requires different data acquisition schemes for multicopy objects and for unique objects.

3.3.1 DATA ACQUISITION FOR VIRAL NANOPARTICLES IN MULTIPLE, IDENTICAL COPIES

For multicopy objects, micrographs of many different copies of the object from different areas of the sample are acquired. Each micrograph is taken with the maximal, permissible dose that does not destroy the features at the desired resolution but gives the best possible signal-to-noise ratio. Typically, doses between 5 and 25 electrons/Å^2 are used for vitrified samples. These doses are a compromise between optimal contrast at low spatial frequencies (high dose), which is required for determining the accurate orientations of the particles, and the preservation of high-resolution features (low dose). Systematic dose studies of diffraction patterns of 2D bacteriorhodopsin crystals suggest that the maximal permissible dose that preserves high-resolution features at 3 Å at temperatures of about −175°C is approximately 3 e/Å^2 (Stark et al. 1996) and that low-resolution image information around 40 Å is already significantly reduced at a dose of 30 e/Å^2 (Toyoshima and Unwin 1988; Conway et al. 1993). The optimal signal-to-noise ratio up to 3 Å resolution at an accelerating voltage of 200 kV is obtained at a dose of 10 e/Å^2 (Baker et al. 2010).

3.3.2 Low-Dose Imaging

Low-dose imaging techniques minimize the dose on the object and thus reduce beam damage (Unwin and Henderson 1975). All low-dose acquisition schemes include several steps, which consist of searching, positioning, focusing, and acquisition (Figure 3.4): The first step is to search the area of interest at a low magnification and with a low dose ($\ll 0.5$ e/Å^2). The search step can include information at different magnifications. For example at low magnification (100–400), the whole grid is inspected to identify regions with the most suitable ice thickness and the least surface contamination. Then, at a somewhat higher magnification (1000–5000), meshes are identified with holes that are covered with a homogeneous, thin film of vitrified sample. At this magnification, smaller nanoparticles ($\ll 50$ nm) cannot be recognized as individual particles and the choice of the area is purely based on the properties of the vitrified film.

In the next step, the area of interest is centered as precisely as possible for the later image acquisition. This also happens at low magnification but sometimes at a somewhat higher magnification than the actual search process to allow for more accurate positioning. The third step is focusing and waiting for the drift to settle. This requires a higher dose and a higher magnification than the searching and positioning. Therefore, the image shift capabilities of the microscope are used to deflect the beam away (1–5 μm) from the area of interest. There, the beam is focused to illuminate only a small area, avoiding preexposure of the area of interest. For determining and adjusting the defocus and for measuring the specimen drift, the magnification is also increased to recognize finer details such as the graininess of the supporting carbon film. Typically, a defocus of 1–5 μm underfocus is chosen for unstained vitrified multicopy objects. Drift rates are also measured and should be lower than around 1–2 Å/s depending on the targeted resolution and the exposure time. The final step is recording a micrograph of the area of interest: This requires resetting the image shift to the center of the area of interest and

adjusting the beam diameter and the magnification to the desired values. Finally, the beam is centered on the area of interest that is tightly coordinated with the shutter of the detector in order to irradiate the sample only during image acquisition.

3.3.3 Automated Image Acquisition

Most electron microscopes have assisted low-dose acquisition schemes that control image shifts, beam shifts, changes in magnification, changes in illumination, and the operation of the beam shutter coupled with the detector. Despite these sophisticated schemes, low-dose data acquisition is tiring for the operator and thus limits manual acquisition times to a couple of hours. Fortunately, the different steps of low-dose data acquisition are highly repetitive and structured, and thus are an ideal task for computer-assisted automation. Nowadays, there are several acquisition packages that acquire images of the desired areas without any further user intervention (e.g., leginon, autoem, em-Tools, tom2, SAM, EPU; Zhang et al. 2003; Lei and Frank 2005; Suloway et al. 2005; Shi et al. 2008; Korinek et al. 2011). These acquisition schemes run stably and can acquire thousands of micrographs from a single grid over a couple of days. Furthermore, with electronic detectors, the micrographs are already in an electronic format and thus are immediately ready for further image processing. This allows feeding image information directly into the acquisition process and using it for automatic, accurate alignment and positioning.

3.3.4 Considerations for Acquiring High-Resolution Micrographs

The automated data acquisition on electronic detectors has generated a quantum leap in the availability of image data. The increase in high-quality image data allows researchers to address more demanding projects, which require more images to achieve higher resolution, to identify different conformational states in a population of particles, or to work with less abundant particles.

Many images of a large number of different copies of nanoparticles are a prerequisite for addressing problems that are related to the statistics of the particle population. For example, subpopulations in a noisy dataset can be more reliably identified, in a large dataset ($\gg 10,000$ per subpopulation) than in a small dataset ($\ll 5000$). However, an increase in the amount of data is not necessarily sufficient for obtaining high-resolution (<4 Å) reconstructions. These reconstructions are only obtained if the high-resolution information is present in the micrographs. Various factors reduce the transfer of the high-resolution image information. These factors can be separated into factors that are related to the hardware of the electron microscope (e.g., coherence of the electron source, correction of lens aberrations), the detector (e.g., sensitivity and modulation transfer function of the detector), the stability of the cryoholder, the stability of the environment (e.g., vibrations,

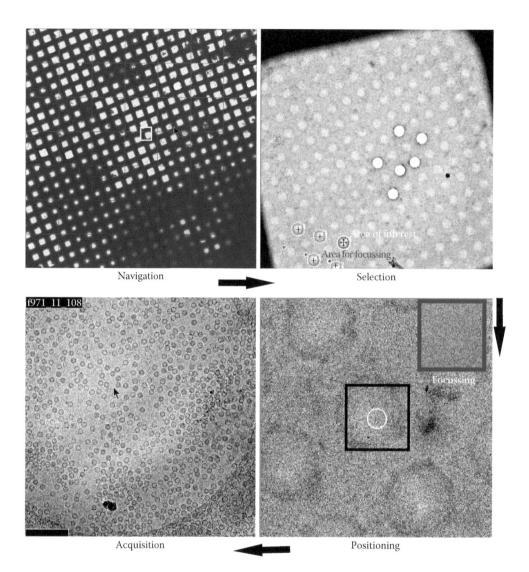

Navigation

Selection

f971 11 108

Acquisition

Positioning

FIGURE 3.4 Semiautomatic low-dose data acquisition. Low-dose data acquisition requires taking images at different magnifications and doses. Upper left panel: A low magnification map of the whole grid allows easy navigation. The grid map also reveals areas with a suitable thickness of the ice. In this example, the ice is too thick for high-resolution data acquisition in the lower right corner of the grid. The grid map is used to identify meshes, which are suitable for further image acquisition. At a magnification of 1000–2000, images of appropriate meshes are acquired (upper right panel), which are used for selecting the areas of interest. The areas of interest are manually selected, avoiding contaminated holes or holes that are not covered with ice. In the upper right panel, some selected areas are outlined with a yellow square and a red circle. The corresponding areas for focusing are highlighted by blue circle. Focusing areas are positioned away from the area of interest. For exact positioning of the area of interest, images at somewhat higher magnification are taken (red outline, lower right panel). These images are cross-correlated with the selected area in the selection image and inform on stage shift and image shift for optimal positioning of the area of interest. This is followed by focusing in the focus area (blue inset, at somewhat higher magnification, typically the same magnification as for the image acquisition). Finally, the image of the area of interest is taken at the desired magnification and the desired dose (lower left panel, black outline). All images were acquired with TVIPS EM-Tools as part of semiautomatic data acquisition.

temperature stability, and absence of electromagnetic fields in the room), the alignment of the electron microscope, and the movement of the particles during image acquisition. While most of these factors cannot be influenced during image acquisition, alignment of the microscope and to some extent detecting the movement of the particles can be addressed during image processing.

It is generally important that the electron microscope is well aligned to recover high-resolution information. Images should be acquired with the sample at the eucentric height where the object is in focus at the optimal lens current of the objective lens. Deviations from the eucentric height lead to changes in the absolute magnification and in image rotation. It is necessary to check and readjust the eucentric height during image acquisition whenever the position has changed by several tens of micrometers (e.g., when moving from one mesh to the next) to keep the eucentric height accurate within 1–2 μm.

Another limiting factor is the electron-beam tilt (Glaeser et al. 2011; Zhang and Hong Zhou 2011), which arises from deviations of the direction of the beam from the optical axis. Electron-beam tilt can be corrected by coma-free alignment (Zemlin et al. 1978) and is minimized by using parallel illumination (Glaeser et al. 2011; Zhang and Hong Zhou 2011). The electron-beam tilt induces a resolution-dependent phase shift that is larger at lower accelerating voltages. The dependence of the phase shift from the spatial frequency is cubic, and thus, it is the main limiting factor of the phase information at large spatial frequencies (high resolution).

Even if the microscope is perfectly aligned and has the appropriate specifications to record high-resolution image information, it is impossible to record high-resolution image information when the object moves during the image acquisition. This movement can either be a directional movement, which is caused by specimen drift, or it is a random movement that is induced by the irradiation with electrons and subsequent charging, beam damage, and heating of the sample. While the directional movement typically ceases after a while, the beam-induced movement occurs during low-dose imaging and is proportional to the applied dose (Chen et al. 2008; Brilot et al. 2012). Its occurrence cannot be avoided by any known imaging or sample preparation technique although it might be somewhat lower at a low-dose rate (1.5 e/$Å^2$ s) (Chen et al. 2008) or on a thicker carbon support film. A way forward for correcting for the beam-induced movement is to record movies rather than single high-dose micrographs. Movies allow correcting for the beam-induced movement of each individual particle in every movie frame (Brilot et al. 2012). The correction requires that frames are recorded fast and that the total imaging time remains short (1–2 s) so that particle movement is predominantly due to beam-induced, random movement rather than to drift-induced directional displacement. With short total recording times, the average over all frames is comparable to a conventional high-dose image that contains the high-contrast information for the accurate determination of the orientation and position of the particles. The dose of the individual frames is typically low and is in the order of 1 e/$Å^2$ or less. Therefore, the signal-to-noise ratio of the individual frames is not sufficient to determine the relative position and orientation of the particles with the accuracy that is required to compensate for the beam-induced movement (Bai et al. 2013). However, the relative alignment of the frames can be improved by computing sliding averages of several frames rather than using individual frames (Brilot et al. 2012) or by using priors about the likely changes in position and orientation in the alignment process (Bai et al. 2013). With the same data, processing of movies gives significantly better resolution than processing the high-dose averages (Bai et al. 2013).

3.3.5 Data Acquisition of Unique Viral Nanoparticles (Tomography)

For a unique object, all the image information for reconstructing the 3D map has to be taken from the same object by recording a tilt series. A tilt range of 180° is required for recovering the complete 3D information. However, it is practically impossible to observe image information over this whole tilt range, because at higher tilt angles, the area of interest becomes occluded by protruding parts of the specimen holder and the thickness of the irradiated sample increases with increasing tilt angle (factor two at 63°). Both factors limit the practical tilt range to approximately 140° (±70°), but often, the tilt range is even lower. This limitation in angular coverage leads to missing information in the z-direction (parallel to the electron beam) of the reconstruction, which is reduced by a double tilt experiment, in which two complete tilt series are acquired with the tilt axes being oriented approximately perpendicular to each other. This requires rotating the specimen in plane by 90° before recording the second tilt series. The in-plane rotation is supported by a special tilt holder and thus can be done inside the electron microscope without remounting the grid.

Similar as for multicopy objects, the total dose with which the object is imaged is closely related to the achievable resolution. Since all the information comes from a single object, the total dose needs to be sufficient to detect features above the noise level at a certain resolution but at the same time needs to be low enough not to destroy these features by beam damage. In the best case, the achievable resolution in a tomogram is around 20 Å (Henderson 2004) but is often considerably worse. This means that in a tomogram, it is not necessary to preserve structural details at spatial frequencies beyond 1/20 $Å^{-1}$ and total doses of 40–100 e/$Å^2$ are typically used for recording the whole tilt series. Therefore, for recording a tilt series in a tilt range of +70° to −70° in 1° steps with a total dose of 100 e/$Å^2$, a dose of only 0.7 e/$Å^2$ is permitted for each tilt image.

Another consideration is that the individual micrographs of a tilt series have to be aligned to a common origin and the orientation of the tilt axis as well as of the actual tilt angle has to be determined for reconstructing the 3D map. Due to the low signal-to-noise ratio in the individual micrographs of the series, the determination of these parameters is often inaccurate when it is solely based on cross-correlation. For improving alignment, electron-dense colloidal gold particles of 5–20 nm diameter are added to the sample as fiducial markers. But even with fiducial markers, accurate alignment of the tilt images still requires 1–2 e/Å per micrograph (Kourkoutis et al. 2012). To further increase the transfer of image information at low spatial frequencies, micrographs are taken with an underfocus of several micrometers (typical values −5 to −10 μm), which is much higher than for high-resolution studies of multicopy objects.

A different aspect of recording the tilt series is how many equally spaced projections n across the whole tilt range of

180° are required to record the complete information for a certain resolution d. The number of tilt images depends on the size of the object D and is given by the Crowther criterion (Crowther et al. 1970b):

$$n = \frac{\pi D}{d}.$$

From this equation, the tilt increment α in degree is derived as

$$\alpha = \frac{180 \cdot d}{\pi D}.$$

Thus, for acquiring the complete information of an object of 100 nm diameter at 20 Å resolution, the required tilt increment is approximately 1°.

To minimize the electron dose on the object, a similar low-dose acquisition scheme is used as for multicopy objects. The area of interest is identified and centered at low magnification. The sample is focused at higher magnification away from the area of interest. For tilted objects, the focus area and the area of interest have to be placed precisely on the tilt axis, because otherwise, the focus in the focus area and in the area of interest would differ. After the micrograph is recorded for a certain tilt angle, the object is tilted to the next tilt position. Often the change in tilt leads to some displacement of the area of interest and therefore, focusing and positioning is repeated. The displacement is minimized, when the sample is exactly at the eucentric height. However, imperfections in the goniometer as well as specimen drift often leave residual displacements. To correct for these displacements, the stage and/or the image is shifted to keep the same focus area centered during the whole tilt series. The procedure of compensating for displacement after tilting is referred to as tracking and is an essential step of tomographic data acquisition.

Nowadays, tomographic image acquisition is fully automated and the procedures take care of tracking, focusing and drift checks as well as of low-dose data acquisition. There are several tomographic acquisition packages, which are either commercially available (e.g., Gatan Tomography package, TVIPS EM-Tools, Xplore 3D of FEI) or have been developed by academics (e.g., TOM, Serial EM, Leginon, UCSF tomo) (Fung et al. 1996; Mastronarde 2005; Suloway et al. 2005; Korinek et al. 2011). The automatic acquisition of a tilt series in small angular intervals, over a large tilt range, can take several hours, especially when low drift rates and accurate tracking and focusing are required.

For recording a high-resolution tilt series of a unique object, there are also stringent requirements for the alignment of the microscope. The most important requirement is that the object is exactly at the eucentric height and that the focus area and the area of interest are exactly on the tilt axis. This is important to minimize displacements and changes in focus during image acquisition. Another important factor is a parallel illumination to minimize changes in magnification with changes in defocus, which can amount to inaccuracies of several angstroms across the micrograph (Fan et al. 1995).

3.4 INSTRUMENTATION

A critical issue for imaging viral nanoparticles at highest possible resolution is the adequate instrumentation, which enables recovering high-resolution image information with a good signal-to-noise ratio. There are several elements in an electron microscope that increase the information transfer at certain spatial frequencies. These elements include emitters, sample holders, energy filters, phase plates, aberration correctors, and detectors. Their importance depends on the targeted resolution and the application and will be discussed in the following.

3.4.1 EMITTERS

The main mechanism of contrast formation in vitrified samples is the phase contrast, which depends on differences in the density in the denser object (typically protein, DNA, RNA) and the surrounding, less dense, vitrified buffer. These differences are small, and thus, the phase contrast is also small. To increase the phase contrast at low spatial frequencies, images are often taken at a considerable defocus (depending on the application between −1 and −10 μm). However, taking highly defocused images also has two major implications: (1) Due to the oscillation of the CTF, certain bands of spatial frequencies are transferred with opposite contrast and others are not transferred at all (Figure 3.1). The frequency-dependent change in contrast leads to a delocalization of the image information and has to be corrected during image processing. (2) The contrast transfer at higher spatial frequencies is dampened by an envelope function (Wade 1992). This envelope function depends on the spatial and chromatic coherence of the electron source, the spherical and chromatic aberration of the electron microscope, the wavelength of the electrons, and the defocus. Conventional thermionic electron sources generate electrons by heating an emitter (e.g., tungsten wire, LaB_6 crystal) to overcome the workfunction. Thus, the emitting surface of the source is relatively large, and therefore, the spatial coherence is low. Also the high temperature that is required to overcome the workfunction leads to a considerable energy spread of the exiting electrons (1.5–4 eV), which contributes to the chromatic aberration. As a consequence of the limited coherence at dose rates of 10–20 e/Å^2 s and a defocus of several micrometers, higher-resolution information (<10 Å) is transferred too weakly to be recovered. Hence, acquisition of data close to atomic resolution requires electron sources with high spatial coherence such as Schottky field emitters or cold field emitters. For both types of emitters, the electron emission is assisted by an external field, which is generated by an additional extraction voltage of a few kilovolts. A field of sufficient strength is only generated at the pointed tip of the electron source, which reduces emission to a very small area of the emitter and thus increases the spatial coherence. For cold field emitters, the emission is achieved by the field alone, which requires an almost atomically sharp tip of the electron source for generating a strong enough field. This is in contrast to Schottky emitters that have a somewhat less pointed tip, and therefore, the extraction voltage on its own is not sufficient to cause emission. In this case, the field only decreases the workfunction of the emitter and the emission is assisted by additional heating of the emitter. In this respect, a Schottky field emitter is not a true field emitter but a field-assisted, thermionic emitter. Similar to cold field emitters, emission only occurs at the tip of the emitter where the field is strong enough.

Although cold field emitters have the best spatial and chromatic coherence, they have been rarely used in transmission electron microscopy of biological specimens. The main reason is that they require a much higher ambient vacuum and contaminate easily, which leads to large fluctuations in their brightness. Cold field emitters require frequent decontamination that is incompatible with high-throughput data acquisition. While the use of field emitters (Schottky or cold) is essential for electron microscopy of vitrified multicopy objects at subnanometer resolution, it is not required for tomographic applications, for which the resolution in a tomogram does not realistically extend beyond 20 Å.

3.4.2 Energy Filters

Often unique objects are considerably larger than multicopy objects, and therefore, the imaged samples can be several hundreds of nanometers thick. In these thick samples, many electrons have undergone inelastic scattering (mean free path of electrons in vitrified buffer at 120 kV is 203 nm; Grimm et al. 1996) and have deposited some of their energy onto the sample. This leads to an energy loss and electrons are no longer properly focused by the lens system of the electron microscope. Image details are blurred and significantly reduced in contrast. A way forward is to remove electrons, which have lost energy from the optical path by an energy filter. This leaves only unscattered and elastically scattered electron in the optical path (zero-loss imaging).

Energy filters are based on the fact that electrons with different energies are differentially deflected in a magnetic field. By using a magnetic field to disperse the electrons according to their energy and a slit aperture, electrons can be selected based on their energy. Energy filters are placed at different positions in the optical path. Some electron microscope manufacturer such as Zeiss and Jeol have developed layouts with the energy filter in-column, whereas Gatan developed a system that is placed post-column and thus can be retrofitted to almost any electron microscope. Both in-column and post-column filers are useful for zero-loss imaging as required for tomographic applications of thicker specimens. By removing electrons from the optical path, energy filters add an additional amplitude contrast (Angert et al. 2000), which also increases the transfer of low spatial frequencies. Therefore, energy filters give an essential improvement of image contrast in tomography, but they are less commonly used for acquiring high-resolution image information of multicopy objects.

3.4.3 Phase Plates

Electron tomography requires strong contrast transfer at low spatial frequencies. Nevertheless, contrast transfer at low spatial frequencies is relatively weak due to the fact that the predominant phase contrast transfer follows a sine function. To increase the transfer at low spatial frequencies, one option is to increase the defocus, which shifts the first maximum of the CTF toward lower spatial frequencies but also introduces contrast reversals at lower spatial frequencies (Figure 3.1). Alternatively, the proportion of amplitude contrast can be increased by using energy filtering (see Section 3.4.2). However, both measures are unsatisfactory for contrast recovery at very low spatial frequencies. To overcome this limitation, phase plates have been proposed, which are similar to Zernike phase plates in light optical microscopy (Zernike 1942). These phase plates shift the phase of the scattered electrons by 90°, which results in optimal contrast transfer at low spatial frequencies (conversion of a sine dependency of contrast transfer into a cosine dependency). Zernike-type phase plates are located in the back-focal plane and are realized by a thin carbon film with a small hole. The hole is aligned in such a way that the unscattered electron beam passes through it and the scattered electrons pass through the carbon, where their phases are shifted by approximately 90° due to the interaction with the carbon film. The smallest spatial frequency that is affected by the phase shift decreases with the size of the hole, and this is in the order of $1/140$–$1/160$ nm^{-1} for retrofitted phase plates (Marko et al. 2011).

Experiments on T4 phage have demonstrated a huge increase in image contrast close to focus and have proven that the phase plate delivers the expected contrast increase for low spatial frequencies (Danev et al. 2010). For image reconstructions of multicopy objects, a significant increase in contrast is observed, and as a consequence, only 30%–50% of particles are required to achieve a certain resolution (Murata et al. 2010). The benefit of the phase plates is demonstrated up to subnanometer resolution.

3.4.4 Spherical Aberration Corrector

Spherical aberration of the objective lens induces an additional frequency-dependent phase shift. For biological low-resolution applications, this is exploited for obtaining more contrast at low spatial frequencies close to focus, by using objective lenses with a very large spherical aberration (e.g., FEI Biotwin lens with a c_s of 6.3 mm). However, such a large spherical aberration also limits the point-to-point resolution of the electron microscope and causes faster defocus-dependent degradation of the contrast transfer at higher spatial frequencies (Figure 3.1). The spherical aberration can be corrected by a c_s-corrector, which allows adjusting c_s values over a wide range. A c_s-value of zero increases the point resolution of the electron microscope and improves contrast transfer at high spatial frequencies. These improvements are essential for material science applications where subangstrom resolution is required. At first glance, the benefits for biological applications, which target much lower resolution (3–4 Å), are less obvious. Here, a c_s of zero also sets the phase shift, which is introduced by electron-beam tilt (e.g., nonoptimal alignment and/or nonparallel illumination) to zero (Zhang and Hong Zhou 2011). Therefore, it is likely that c_s-correctors will improve the resolution of biological applications, for which electron-beam tilt is one of the major limiting factors for obtaining high-resolution image reconstructions.

3.4.5 DETECTORS

For many years, photographic film has been the most efficient detector for recording high-resolution electron micrographs. Film provides an excellent spatial resolution, because it has a large field of view and electrons are detected directly by changing the properties of the fine grain of the film emulsion (Figure 3.2a). Indeed, most of the image reconstructions at resolutions better than 4 Å have been recorded on film (Yu et al. 2008; Zhang et al. 2008, 2010, 2012, 2013a,b; Liu et al. 2010a; Wolf et al. 2010; Chen et al. 2011; Settembre et al. 2011; Yu et al. 2011, 2013). However, the sensitivity and linearity of the response to electrons is poor, the handling is inconvenient, and the data can only be used for further image analysis after time-consuming processing (developing, drying, and scanning of film).

For a long time, an alternative to film has been conventional CCD or CMOS detectors (Figure 3.2c). These detectors measure a light signal, which comes from a scintillator that is excited by the incident electrons. The light signal is transmitted to the sensor either by a fiber-optical coupling or by an optical lens. A major limitation is that the area of the scintillator that is excited by one electron is relatively large. This leads to a delocalization of the information and thus to an unfavorable point spread function, which limits the resolution. The typical quantum efficiencies of a conventional CCD or CMOS detector at 0.5 of the Nyquist frequency are 0.1–0.3, depending on the voltage (better for lower accelerating voltage; quantum efficiency at Nyquist frequency are 0.01–0.07) (Ruskin et al. 2013). To compensate for the delocalization of the signal and the weak contrast transfer close to Nyquist frequency, micrographs are recorded at a higher magnification as would be used for data acquisition on film. Unfortunately, this reduces the field of view to a fraction of what can be imaged with a similar spatial resolution on film.

The advantage of conventional CCD or CMOS detectors is the better linearity in dose response over a wider dynamic range with higher sensitivity. Furthermore, electronic detectors deliver micrographs in a digital format almost instantly. This is a prerequisite for automated, computer-controlled data acquisition such as tomography and automated multicopy particle image acquisition, which relies on feeding back image information into the alignment and positioning process.

More recently, direct CMOS detectors were developed. Other than conventional CCD and CMOS detectors, they do not use a scintillator for converting electrons into a detectable light signal, but directly detect the charge separation caused by the incident electrons in the epilayer of the detector (Figure 3.2b). Thus, the blurring of the signal is less extended and at accelerating voltages above 200 kV, direct detectors outperform conventional CCD or CMOS detectors (Ruskin et al. 2013). A further advantage of the new direct detectors is their

rapid read-out. Rather than recording a single micrograph, it is possible to record the image information as a movie with some 20 frames per second. This allows computational correction of beam-induced movement in the subsequent image processing (Campbell et al. 2012; Bai et al. 2013).

Some of the direct detectors have an even faster read-out that enables counting of individual electrons rather than integrating the signal. Counting increases the linearity of the dose response and thus improves the resolution. Furthermore, by detecting single electrons, the position of the incident electrons can be determined with subpixel resolution, which gives superresolution information beyond the Nyquist frequency (Gatan K2 summit).

3.5 IMAGE PROCESSING

As outlined earlier, there are unique objects and multicopy objects, which have different requirements not only for the data acquisition but also for the subsequent image processing and the expected resolutions. For unique objects, the optimal resolution is determined by the beam damage of the sample and the signal-to-noise ratio at a certain spatial frequency. These considerations suggest that the resolution can probably not exceed 20 Å (Henderson 2004). For multicopy objects, there is no theoretical limitation that prevents achieving atomic resolution. In theory, averaging a few thousand particle images should already be sufficient to reconstruct a structure of a multicopy object at a resolution of approximately 3 Å (Henderson 1995) in the absence of any image degradation. In reality, two to three orders of magnitude more asymmetric units need to be processed and averaged to recover structural information beyond 4 Å resolution. This suggests that electron microscopic images are far from perfect.

Image processing aims at combining the different projections of the object that have been generated by the electron microscope in a consistent way into a 3D map and to correct for the image distortions imposed by the electron microscope and the detector.

3.5.1 IMAGE RECONSTRUCTION OF MULTICOPY OBJECTS

Historically, very different approaches have been used for different types of multicopy objects. For particles with icosahedral and helical symmetry, certain symmetry-related properties of the Fourier transforms have been exploited for determining the particle orientations. These specialized approaches cannot be used for asymmetric multicopy objects, which always require the comparison with other images (e.g., reference images of a noise-free reference or other images of a different view of the object) for determining the relative orientations of the particles. Nowadays, the specialized approaches for helical and icosahedral objects are mainly used for the determination of a first model, which is followed by more general image-processing strategies that are independent of the symmetry.

These general strategies have a common workflow, which includes four steps: (1) Preprocessing: The image data are prepared for further image processing. This includes windowing of the individual particle images from the micrographs and determining the parameters of the CTF of the microscope for every micrograph. (2) Determination of the orientations of the particles: This can be done either ab initio or by comparison of the particle image with a common reference. (3) Image reconstruction: The 3D map of the object is calculated by combining all 2D particle images with the correct orientation into a 3D map. Typically, the determination of the orientations is an iterative process, in which the current best 3D map serves as reference map for the reference-based determination of particle orientations. (4) Postprocessing: The final 3D map is corrected for the attenuation of high spatial frequency information.

3.5.1.1 Preprocessing of the Image Data

Image processing of multicopy objects requires a set of images with equal size, in which each image contains one copy of the multicopy object. To generate this set of particle images, the positions of the particles on the micrographs are determined either by selecting the desired particles manually or automatically, by using certain properties of the local modulation of the gray value distributions or by correlation with reference projections. After identifying the positions of the particles in the micrographs, the particles are extracted into smaller particle images that are somewhat larger than the particle diameter (approximately two times the particle diameter) and contain one particle per image. For further image processing, the particle images are normalized in their gray value distribution.

Automatic approaches to particle selection often incorporate false-positive particle images (e.g., ice contaminations, edge of the illuminated area or edge of the hole), which contain high-contrast features. Therefore, automatically selected datasets require a stringent postsorting process, which uses different parameters for recognizing the particles than the parameters that have been used for the automatic particle selection. One strategy is the reference-free classification in which false-positives typically do not group into well-defined classes and do not align with high accuracy. Particle images from such ill-defined classes are excluded from further processing.

Preprocessing of the particle images often includes further steps such as tight masking of the particles with a smooth mask, band-pass filtration, the inversion of the gray values, and the correction for the CTF. The accurate determination of the CTF (Figure 3.1) and its correction is critical for obtaining high-resolution image reconstructions. Without correction of the CTF, the image information would not average coherently and the gray value distribution of the object would be misrepresented in the final map. A very successful approach to the determination of the CTF is ctffind3 (Mindell and Grigorieff 2003), which divides the whole micrograph into smaller patches, and averages the power spectra of these patches. After a background correction of the averaged power spectra, the cross-correlation to a simulated power spectrum in a given frequency band is maximized. The method works reliably and can determine the parameters of the CTF more accurately than most curve fitting approaches in which the density profile of a background subtracted power spectrum is fitted with the CTF. The determined parameters are incorporated into other programs for further correction of the contrast transfer of the microscope either during preprocessing or later, during image reconstruction.

3.5.1.2 Determination of the Orientations

The extraction of particle images is followed by the determination of the orientations relative to a 3D reference map. This can be done by cross-correlating the particle image to a set of projections of this 3D reference map. The particle image is assigned with the same orientations as the best correlating reference projection. However, for a fine angular sampling, the comparison requires a huge number of reference projections and thus can be computationally ineffective.

Alternatively, a limited number of equally spaced reference projections are computed. In Fourier space, the Fourier transforms of these reference projections define a fixed network of intersecting planes (Figure 3.5). The Fourier transform of the particle image intersects with this network of intersecting planes along certain lines. Along these intersecting lines, the information in the Fourier transforms of the reference projection and in the transform of the particle projection is the same. The position of the intersecting lines depends only on the orientation of the particle and on the orientation of the reference projections. The particle image is assigned with the orientation for which the predicted intersecting lines in the Fourier transforms of the reference projections and in the particle transform have the highest correlation. This method is referred to as cross common line method (Crowther 1971) and has the advantage that finer angular sampling does not require a larger number of reference projections.

The cross common line approach is particularly powerful for icosahedral particles, for which the symmetry gives rise to 60 cross common lines for each transform of a reference projection (Crowther 1971; Crowther et al. 1994). The incorporation of the symmetry in the search also helps to focus the determination of orientations on the icosahedrally related elements. This is, for example, helpful, when the icosahedral organization of a viral nanoparticle is determined, which contains asymmetric components (e.g., packaged genome, attached packaging proteins).

3.5.1.3 Helical Assemblies

A special case of multicopy objects are helical assemblies, for which the asymmetric unit is related by a radial distance to the helical axis, a rotation angle around the helical axis and by an axial rise. Other than for conventional multicopy objects, the particle images are not the whole helical assembly, but overlapping helical segments. For these segments, in projection, a shift along the helical axis has the same effect as a rotation around the helical axis. Therefore, the shift along the helical axis is restricted in search space, whereas full rotation around the helical axis and a limited tilt ($\pm 15°$) of the helical axis out of plane are permitted (Egelman 2000, 2010; Sachse et al. 2007). By identifying the angular orientation and the origin of each helical segment, it is possible to correct for slight deviations from the perfect helical symmetry as would occur

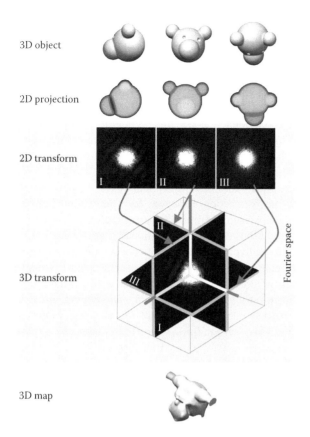

3D object

2D projection

2D transform

3D transform

3D map

Fourier space

FIGURE 3.5 Principle of image reconstruction in Fourier space: The object (upper row) has different orientations in respect to the incident beam. In this example, the object is rotated by 90° around its X- and Y-axes. The microscope generates 2D projections of the object (second row). These 2D projections are transformed into 2D Fourier transforms (third row, labeled I, II, III). The 2D Fourier transforms are section of the 3D transform of the object (fourth row). The sections go through the origin of the transform and have the same relative orientations as the object had relative to the incident beam (in this case perpendicular). The sections intersect along certain lines (blue between I and II, yellow between II and III, and green between I and III). Along these lines, the information of intersecting transforms is the same. These lines are referred to as cross common lines and can be used for determining the orientation of a particle. The image is reconstructed by inverse transforming the 3D Fourier transform that includes all the 2D Fourier transforms into real space (fifth row, surface representation of the reconstruction calculated from the three projections shown in the second row).

by bending of the helical tube. For shorter helical segments, imperfections in the helical symmetry can be more accurately corrected. However, shorter segments also contain less information, and therefore, their orientation is less precisely determined than for longer helical segments. Thus, the resolution of the final reconstruction of a helical assembly depends critically on the chosen length of the helical segments.

3.5.1.4 General Considerations for the Determination of Orientations

The low spatial frequencies ($<1/20$ Å^{-1}) are most important for the determination of the relative orientations of the particle images. Therefore, some search strategies use only

the low-resolution information of the reference map for orientational searches of the full asymmetric unit on a coarse angular grid. After the overall orientation of a particle is established, local searches on a finer angular grid follow, which include the higher spatial frequencies. This strategy helps to find reliable orientations that are not dominated by noise correlation, which often occurs if high-frequency information is included in the iterative refinement.

3.5.1.5 Reconstruction of the 3D Map of the Object

After the orientations of the particle images have been determined, the 3D structure of the particle is reconstructed. Image reconstruction can be done in real space or in Fourier space. Real-space algorithms include weighted back-projection, algebraic reconstruction techniques (ARTs), and statistical image reconstruction techniques (SIRTs).

Alternatively, reconstructions can also be calculated in Fourier space followed by Fourier inversion. Some of the approaches have been specifically adapted for icosahedral nanoparticles or particles with rotational symmetry and use spherical harmonics (Crowther 1971; Navaza 2003; Liu et al. 2008a; Estrozi and Navaza 2010) in Fourier space. In the following, only the two general approaches of Fourier inversion and weighted back-projection will be discussed:

3.5.1.5.1 Fourier Inversion

In electron microscopy, a particle image corresponds to a projection of the 3D particle. The 2D Fourier transform of such a particle projection is a section through the 3D Fourier transform of the particle (Figure 3.5). The orientation of the section in Fourier space depends on the orientation of the particle in real space. All sections go through the origin of the 3D Fourier transform. By adding 2D Fourier transforms of many different views of the particle, the 3D Fourier space is evenly sampled. Since the grid points of the sections and the grid points of the 3D Fourier transform do not necessarily coincide, a proper Fourier interpolation scheme has to be implemented (Grigorieff 2007) to minimize the oversampling of the low spatial frequencies. Finally, the real-space 3D map is calculated by transforming the 3D Fourier transform into real space (Fourier inversion).

The advantage of combining the image data in Fourier space is that image information is represented as a function of spatial frequencies. Therefore, it is simple to manipulate frequency-dependent modulations such as the CTF of the microscope or the frequency-dependent degradation of the image information. In Fourier space, these factors can be conveniently deconvoluted with appropriate frequency-dependent filter functions. Especially the correction of the CTF is effectively done during image reconstruction in Fourier space by multiplying the Fourier transforms of the individual particle images F_i with the respective CTFs c_i and summing the products in the 3D Fourier transform F. The multiplication downweighs weakly transferred data in the averaged transform but misrepresents the amplitudes. In real space, such a map without further correction of the amplitudes would appear somewhat skinny and would overestimate cavities. Therefore, for a quantitative representation of the density of the particle in the 3D map, it is also essential to correct the amplitudes. The amplitudes are restored by dividing the sum of the Fourier components by the summed square of the CTF values. Furthermore, an additional weighting factor w_i can be introduced that weights images based on their properties (e.g., signal-to-noise ratio). Finally, to reduce overrepresentation of globally, weakly transferred data, a Wiener weighting factor f is included (Böttcher and Crowther 1996; Grigorieff 2007):

$$F = \frac{\sum_i c_i w_i F_i}{f \sum_i c_i^2 w_i F_i}.$$

3.5.1.5.2 Back-Projection

Image reconstructions can also be calculated in real space. One of the methods is back-projection, which is the inverse

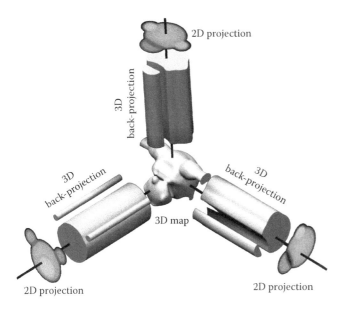

FIGURE 3.6 Image reconstruction in real space by back-projection. The object and its 2D projections are the same as shown in the example in Figure 3.5. The image is back-projected in the opposite direction as it was projected by expanding the gray values of the projection uniformly into the third dimension. The 3D back-projections of the different 2D projections are averaged in a single 3D map. A surface representation of the averaged back-projections is shown in the center of the map (3D map).

of projection (Figure 3.6). While the projection compresses a 3D image into a 2D projection, back-projection smears out the 2D image information into the third dimension. The direction of back-projection is opposite to the direction of projection. To calculate an image reconstruction, the image information from different projections is back-projected into the direction of projection and averaged in the third dimension (Hoppe et al. 1986). Reconstructions by simple back-projection oversample the low-resolution information and thus generate reconstructions that are dominated by the low spatial frequencies. This is counteracted by weighting the image data with a weighting function that depends on the spatial frequency, the angular sampling, and the size of the reconstructed object (for more details, see Penczek 2010). Fourier inversion and back-projection are equivalent methods for reconstructing the 3D volume of the object. Some of the equivalent aspects are summarized in Table 3.2.

3.5.1.5.3 Postprocessing of the Three-Dimensional Map

In the reconstructed map, the signal at higher spatial frequencies is often underestimated due to the modulation transfer function of the detector, the spherical and chromatic aberrations of the electron microscope, uncertainties in the determination of the orientations of the particles, and conformational flexibility of the particles. These factors lead to a fast decay of high-resolution information. Therefore, for a realistic representation of the map, the signal at higher spatial frequencies has to be adjusted. One way forward is to measure small angle x-ray scattering curves of the particle in solution and to use these scattering curves for scaling the amplitudes at different spatial frequencies (Thuman-Commike et al. 1999). Unfortunately, it is not always possible to obtain such experimental scattering curves. Alternatively, Guinier plots (natural logarithm of the average structure factor as a function of spatial frequency^2 $1/d^2$) can be used to estimate the decay of the signal (Rosenthal and Henderson 2003). This follows the assumption that in the absence of any decay, the structure factors at higher spatial frequencies (>1/10 Å^{-1}) depend mainly on the random distribution of atoms and change very little with the spatial frequencies. The decay can be estimated by determining the linear slope of the Guinier plot at these higher spatial frequencies. The slope is comparable to the B-factor in crystallography. Depending on the resolution of the final map, B-factors between 150 Å^2 for high-resolution

TABLE 3.2

Comparison of Image Reconstruction in Fourier Space and in Real Space

Real Space	Fourier Space
Conversion to Fourier space: Fourier transform	Conversion to real space: inverse Fourier transform
2D projection of image	Fourier transform of 2D projection
Back-projected 2D projection	Fourier transform of 2D projection as a section of the 3D Fourier transform
Diameter D of reconstructed object	Thickness of 2D section in the 3D Fourier transform of object: $1/D$
3D image reconstruction of object: back-projections of all image projections averaged in the third dimension	Fourier transform of 3D image reconstruction of object: Fourier transforms of all 2D projection added as sections in the 3D Fourier transform
Weighted back-projection of object (for review: Penczek 2010)	Averaging of Fourier sections with Fourier interpolation scheme (Grigorieff 2007)

EM maps (<4 Å) and 1000 Å^2 for intermediate-resolution EM maps (around 10 Å) have been reported. Subsequently, the signal at higher spatial frequencies in the maps is restored by applying a negative B-factor.

3.5.2 IMAGE PROCESSING OF UNIQUE OBJECTS

For unique objects, all image information is obtained from a tilt series of the same object. Then, this tilt series is used for a tomographic reconstruction of the 3D map of the unique object. Similar to multicopy objects, the image processing of the tomograms consists of several steps that include preparing the images of the tilt series for further processing, aligning the different views of the tilt series to a common origin, determining the tilt angle and the orientation of the tilt axis, and finally reconstructing the 3D map of the object.

The considerations for preparing the raw data for further processing are similar as that for multicopy objects. In the preparation step, the images of the tilt series are cropped to the relevant area and to the same size. The gray value histograms of the individual images of the series are normalized and outliers such as very bright or very dark pixels are replaced by the average density of the surrounding pixels

In the next step, the images of the tilt series are aligned to a common origin and the orientation of the tilt axis and the tilt angle are calculated. The strategies for the alignment of the tilt series are somewhat different from the strategies used for multicopy objects: in principle, the common origin between the images of the tilt series could be determined by the cross-correlation between subsequent images of the tilt series, and the orientation of the tilt axis and the tilt angle are already known from the experimental setup. However, the accuracy with which these parameters are determined is insufficient for calculating high-resolution tomograms. The relative orientation of the images is more precisely determined by using fiducial markers. Such fiducials are colloidal gold particles, which can be easily identified in low-dose images (Fung et al. 1996; Diez et al. 2006) or landmarks in many small high-contrast patches of the image (Brandt et al. 2001; Castano-Diez et al. 2007; Sorzano et al. 2009). The exact position of the gold fiducials is determined by the density profile of the electron-dense marker, whereas the relative positions of the landmarks in patches are determined by cross-correlation. Then the tilt angle, the direction of the tilt axis, and the common origin of the micrographs of the series are determined by solving the transformation matrix and using the relative position of the same fiducials or landmarks in different images as input.

Subsequently, the tilt angle, the tilt axis, and the origin of the images of the tilt series are used to calculate the 3D image reconstruction, which is often done by weighted back-projection (see earlier text, multicopy objects). The image reconstruction is noisy due to the low dose of the individual images of the tilt series and the little overlap in information between subsequent images of the series. Furthermore, the limited tilt leads to missing information in the z-direction and thus an anisotropic representation of details in the reconstruction (Figure 3.7). Features

that are orientated perpendicular to the optical axis of the microscope are more accurately represented than features that have the same direction as the optical axis. For example, surface representations of tomograms of spherical virus particles often show details at the sides of the particle but are lacking the caps at the top and at the bottom (Figure 3.7).

3.5.2.1 Denoising of Tomograms

An important strategy for improving the representation of the image information is the denoising of the tomographic reconstructions. Many different techniques are available. Simple filters such as low-pass filter and block convolution are already quite efficient. In a low-pass filter, the higher spatial frequencies are attenuated, whereas low spatial frequencies are not affected. The filtration is done in Fourier space by multiplying the transform of the reconstruction with a filter function, which contains a spatial frequency-dependent weighting factor. The cutoff frequency for the filter is often chosen within the first transfer band of the CTF of the highest defocused image in the series. This avoids that information with the opposite contrast is included in the image reconstruction and thus makes correcting the image reconstruction for the CTF unnecessary. A filter function with a hard cutoff value in Fourier space introduces ripples at the edges of a high-contrast object in real space. Therefore, the edge of the filter function is typically smooth and gradually changes the weight over a larger frequency band, often with a Gaussian falloff. Generally, the low-pass filter suppresses high spatial frequencies, which are dominated by noise. Block convolution filters have a similar effect as low-pass filters but are computed in real space. In a block convolution, each pixel of the reconstruction is replaced by the average of the pixels in the surrounding subvolume. The size of the subvolume corresponds to the cutoff frequency in the low-pass filter.

Block convolution and low-pass filter only affect the transfer of the information, but they do not change the representation of the transmitted data. This is in contrast to filters that enhance certain features. Examples for such filters are median filters and anisotropic diffusion filters. Median filters work similarly as block convolution filters with the main difference that the pixel in the reconstruction is the median gray value of the surrounding subvolume instead of the average gray value. The median gray value is the value in the middle of the sorted series of gray values of the subvolume. Median filters are computationally more costly than block convolutions, but have the advantage that they enhance edges of features in the tomogram. The effect is further improved by applying the median filter several times iteratively to the tomogram (van der Heide et al. 2007). While edges are enhanced, features that are smaller than half of the box size of the median filter are potentially lost. Another approach for enhancing features in the tomogram is a nonlinear, anisotropic diffusion filter (Frangakis and Hegerl 1999, 2001; Fernandez and Li 2003). This type of filter takes the local gray value gradient into account and thus acts differently on noise and on signal of high- and low-contrast features. Therefore, details in the reconstruction are preserved and noise is efficiently repressed.

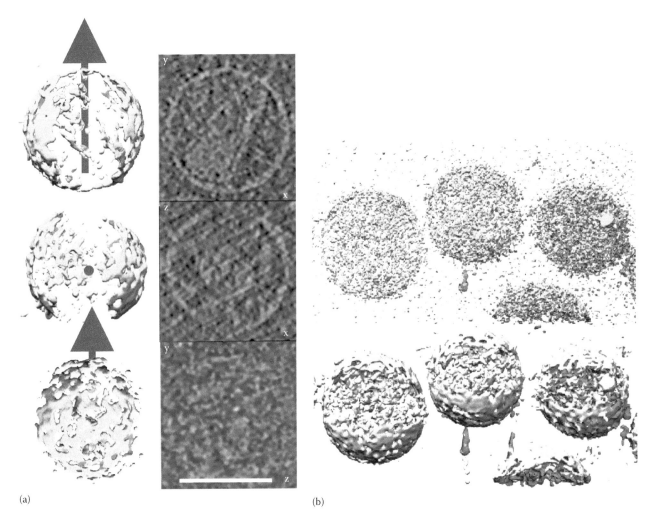

(a)

(b)

FIGURE 3.7 Tomogram of HIV virions (EMDB 1155, Briggs et al. 2006). (a) Left panels show surface representations of median filtered tomogram of a HIV virion; right panels show slices through the center of the tomogram of the virion in the same orientation as in the corresponding left panel. The position of the tilt axis is indicated in red in the surface representations, and the x, y, and z directions are labeled in the corresponding slices in the right panels. The length of the scale bar equals 100 nm. Different levels of detail are resolved in the tomogram, depending on the orientation relative to the tilt axis, demonstrating the effect of anisotropic resolution due to the limited tilt. (b) The top panel shows a surface representation of an unfiltered tomogram of several virions (EMDB 1155). The bottom panel shows a surface representation of the median filtered tomogram. Median filtering clearly reduces noise in the background and enhances the edges of the virions.

While simple filters such as block convolution and low-pass filter do not change the signal-to-noise in the transmitted resolution band, edge-enhancing filters do. Thus, edge-enhancing filters can increase the resolution somewhat. Denoising of tomograms is an essential step for improving the interpretability of the reconstructions and for calculating surface representations of a tomogram (Figure 3.7b).

3.5.2.2 Subtomogram Averaging

The resolution of a unique object in a tomogram is limited by noise. Currently, there are no strategies to overcome this limitation. However, unique objects often contain repeating building blocks, which can be regarded as multicopy objects within the unique object. Examples for such multicopy objects in polymorphic viral nanoparticles are virus surface glycoproteins (Liu et al. 2008b; Huiskonen et al. 2010; Tran et al. 2012; Maurer et al. 2013) and internal structural virus proteins (de Marco et al. 2010; Bharat et al. 2011; Liljeroos et al. 2011; Schur et al. 2013), which both have been studied by subtomogram averaging.

Subtomogram averaging involves similar steps as image processing of isolated multicopy objects with the exception that the data do not consist of individual projections of the multicopy object but of 3D volumes with anisotropic resolution (due to the limited tilt of the tomogram). The workflow for subtomogram averaging consists of the identification of the multicopy object within the tomogram either by identifying the multicopy objects manually or by cross-correlating the tomogram with a template in different orientations. The highest cross-correlation peaks between the tomogram and the template are the likely positions of the multicopy objects within the tomogram. Often additional prior knowledge about the likely position of the object in the tomogram is incorporated for selecting the cross-correlation peaks. For example, if the multicopy object is membrane bound or is located in a certain layer of the capsid, cross-correlation peaks at other places in the tomogram are ignored.

After the positions of the objects in the tomogram have been identified, small subvolumes of equal size are cut out. Ideally, the subvolumes contain one copy of the object each. In the next step, the subvolumes are aligned rotationally and translationally to a common reference volume. For the alignment, the missing wedge of the tomogram is taken into account and cross-correlation is restricted to information that is present in the reference and in the subvolume. Finally, aligned subvolumes are averaged and the whole process of alignment and averaging is repeated with the averaged subvolume as new reference. Often the template-based identification of the multicopy objects in the tomogram is also included in the iterative refinement process, because the accuracy of the localization of the multicopy objects within the tomogram improves with the improving template map.

The average of many aligned copies of a randomly orientated object has isotropic resolution, which is in contrast to the tomogram of the unique object. Furthermore, much higher resolution can be obtained in the subtomogram average, because the information of many objects is averaged. Thus, by averaging, the signal-to-noise ratio can be improved beyond the limit that is dictated by the tolerable dose of the unique object. It has been demonstrated that subnanometer resolution is achievable by subtomogram averaging of some 700,000 copies (Schur et al. 2013) (Figure 3.8).

The processing of subtomograms does not only give structural information of the multicopy object but it also informs where a certain multicopy object is located within a unique object and what its spatial relation to other objects is. This information gives important biological insights into the structural building principles of polymorphic particles.

3.6 EXAMPLES OF ELECTRON CRYOMICROSCOPY OF VIRAL NANOPARTICLES

Over the past 30 years, countless structural studies by electron cryomicroscopy and image processing of diverse viral nanoparticles have been published. These studies covered a wide range of aspects including structure determination of highly symmetrical icosahedral and helical virus particles and determining systematic breaks of their symmetry. These studies gave insights into the structural maturation of viral capsids and the packaging of the genome as well as of its delivery to the host.

3.6.1 HIGH-RESOLUTION STRUCTURAL STUDIES OF ICOSAHEDRAL NANOPARTICLES

For regular icosahedral nanoparticles, electron cryomicroscopy and image processing has emerged as an important alternative to crystallographic studies. The strength of electron cryomicroscopy is that the nanoparticles are studied under almost physiological buffer conditions unhindered by crystal contacts. Until the end of 2013, some 42 structures of viral nanoparticles or viral proteins with resolutions between 3.1 and 5 Å (Table 3.3) have been deposited in the EMDB, which is the unified repository for EM

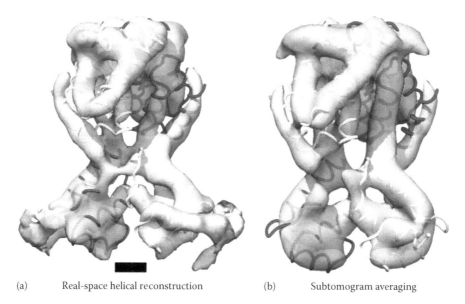

(a) Real-space helical reconstruction (b) Subtomogram averaging

FIGURE 3.8 Surface representations of Mason-Pfizer Monkey virus capsid domain of Gag protein at 7–8 Å resolution with fitted model of the C-terminal (red and green) and N-terminal (yellow and blue) part of the capsid domain (4ARD, Bharat et al. 2012). (a) Real-space reconstruction of the helical assemblies followed by averaging of the independent copies in the asymmetric unit (EMD 2090, Bharat et al. 2012). Averaging of 344,000 units gave a resolution of 6.8 Å. (b) Subtomogram averaging of the capsid domain of the helical tubes. Averaging of 728,000 units gave a resolution of 8.3 Å (EMD 2488, Schur et al. 2013). Both reconstructions show a similar level of detail indicating that subtomogram averaging can deliver subnanometer resolution maps with similar reliability as conventional processing of multicopy objects. The length of the scale bar equals 1 nm.

TABLE 3.3
Maps of Viral Nanoparticles with 3–5 Å Resolution, Which Have Been Deposited in the EMDB until the End of 2013

Virus	EMDB Entry	Publication Year	Resolution	Microscope	Voltage (kV)	Detector	Averaged Number of Particles/Units	Remarks	Citations
Adeno-associated virus	EMD-5681	2013	4.8 Å	FEI Titan Krios	120	Gatan Ultrascan 4k × 4k CCD camera	70,725 particles/4.2 Mio units	Localization of sucrose octa sulfate in difference maps.	Xie et al. (2013)
Adeno-associated virus	EMD-5415	2012	4.5 Å	FEI Titan Krios	120	Gatan Ultrascan 4k × 4k CCD camera	27,312 particles/1.6 Mio units	Localization of an engineered loop. Structural differences to other serotypes might explain the different tropoism of the engineered virus.	Lerch et al. (2012)
Cytoplasmic polyhedrosis virus: Bombyx mori cypovirus 1	EMD-1508	2008	3.9 Å	FEI Polara	300	TVIPS 4k × 4k CCD camera	12,814 particles/768,840 units	The structure of the whole virus was built de novo with O.	Yu et al. (2008)
Cytoplasmic polyhedrosis virus: Bombyx mori cypovirus	EMD-5233	2011	3.9 Å	FEI Titan Krios	300	Gatan Ultrascan 4k × 4k CCD camera	29,000 particles/1.7 Mio units		Cheng et al. (2011)
Cytoplasmic polyhedrosis virus: Bombyx mori cypovirus	EMD-5256	2011	3.1 Å	FEI Titan Krios	300	Film Kodak SO169	28,993 particles/1.7 Mio units	This 3.1 Å resolution map–enabled building of atomic models for all the structural proteins of CPV.	Yu et al. (2011)
Cytoplasmic polyhedrosis virus: Bombyx mori cypovirus	EMD-5376	2012	4.1 Å	FEI Titan Krios	300	Gatan Ultrascan 4k × 4k CCD camera	8,000 particles/480,000 units	The map reveals small-scale structural changes in the capsid that only occur in the transcribing virus.	Yang et al. (2012)
Sulfolobus turreted Icosahedral virus	EMD-5584	2013	4.5 Å	FEI Titan Krios	300	FEI Falcon I Direct detector	8,903 particles/534,180 units	The map reveals the structure of the whole virus and allows model building of the Cα trace.	Veesler et al. (2013)
Penicillium chrysogenum virus	EMD-5600	To be published	4.1 Å	FEI Tecnai F30	300				Luque et al. (2014)
Aquareovirus capsid proteins: Grass carp reovirus	EMD-1653	2010	4.5 Å	FEI Polara	300	TVIPS 4k × 4k CCD camera	15,000 particles/900,000 units	The structural modeling is based on homology modeling and constraints by the EM map.	Cheng et al. (2010)

(Continued)

TABLE 3.3 (Continued)
Maps of Viral Nanoparticles with 3–5 Å Resolution, Which Have Been Deposited in the EMDB until the End of 2013

Virus	EMDB Entry	Publication Year	Resolution	Microscope	Voltage (kV)	Detector	Averaged Number of Particles/Units	Remarks	Citations
Prochlorococcus phage P-SSP7	EMD-1713	2010	4.6 Å	JEOL 3200FSC Energy filter	300	Film Kodak SO163	36,000 particles/2.2 Mio units	The processing of the data without imposed icosahedral symmetry shows the structure of the portal protein at one of the vertices at 9 Å resolution.	Liu et al. (2010b)
Enterobacteria phage P22	EMD-1824 EMD-1826	2011	3.8 and 4 Å	JEOL 3200FSC Energy filter Specimen temperature 4 K	300	Film Kodak SO163	23,400 particles/1.4 Mio units and 18,300 particles/1.1 Mio units	This study compares the structure of the empty procapsid (EMD 1824) with the structure of the infectious virus particle (EMD 1826) at 4 Å resolution. The study reveals the role of the scaffolding protein and shows that the capsid undergoes coordinated conformational changes upon transition from the procapsid into the mature virion.	Chen et al. (2011)
Bordetella phage BPP-1	EMD-5764 EMD-5765 EMD-5766	2013	3.5 Å	FEI Titan Krios	300	Film Kodak SO163	39,549 particles/2.4 Mio units	The map shows a noncircularly permuted Johnson fold topology of MCP.	Zhang et al. (2013)
Dengue virus serotype 1	EMD-2142	2013	4.2 Å for virus and 3.5 Å for averaged heterotetramer	FEI Titan Krios	300	Film Kodak SO163	32,569 particles/2 Mio units and 6 Mio averaged heterotetramers	Three heterotetramers in the asymmetric unit were averaged for improving the resolution of the heterotetramer.	Zhang et al. (2013)
Bovine adenovirus 3	EMD-2272 and EMD-2273	2013	4.5 and 4.5 Å	FEI Titan Krios	300				Cheng et al. (2014)
Human adenovirus type 5	EMD-5172	2010	3.6 Å	FEI Titan Krios	300	Film Kodak SO163	31,815 particles/1.9 Mio units	The map identifies the residues that are important for the interaction between the minor and major proteins. The residues were further validated by mutational studies, which showed a mutation-dependent modulation of the particle stability.	Liu et al. (2010a)
Human adenovirus type 5	EMD-5467	2012	4.2 Å	FEI Titan Krios	300	Gatan Ultrascan 4k × 4k CCD camera	21,000 particles/1.3 Mio units	The map resolves large parts of the fiber at a somewhat lower resolution than the capsid. The Ad5 capsid is pseudotyped with the Ad35 fiber. The map provides insights into the fiber–receptor interaction.	Cao et al. (2012)
Hepatitis B core protein	EMD-2278	2013	3.5 Å	FEI Titan Krios	300	Film Kodak SO163	8,093 particles/485,580 units	The map of the full-length protein Hepatitis B core antigen complements information from the crystal structure of the C-terminally truncated protein (Wynne et al. 1999). The C-terminal linker domain is differently orientated than in the previously published crystal structure.	Yu et al. (2013)
Pseudomonas phage phi6	EMD-2364	2013	4.4 Å	FEI Titan Krios	300	Film Kodak SO163	18,236 particles/1 Mio units	Comparison of the P1 capsid protein in the capsid (EM map) with the crystallized P1 pentamers reveals the likely conformational switch that drives capsid maturation.	Nemecek et al. (2013)

(Continued)

TABLE 3.3 (Continued)
Maps of Viral Nanoparticles with 3–5 Å Resolution, Which Have Been Deposited in the EMDB until the End of 2013

Virus	EMDB Entry	Publication Year	Resolution	Microscope	Voltage (kV)	Detector	Averaged Number of Particles/Units	Remarks	Citations
Dengue virus (mature)	EMD-5499 (partial map) EMD-5520 (full map)	2013	3.6 Å	FEI Titan Krios	300	Film Kodak SO163	9,288 particles/557,280 units	The map shows the mature virus. Comparison to other states gives insights into the pH-dependent maturation process of the capsid.	Zhang et al. (2013)
Dengue virus type 4	EMD-2485	2013	4.1 Å	FEI Titan Krios	300	FEI Falcon 1 direct detector	16,602 particles/1 Mio units	The comparison to structures of other dengue virus serotypes suggests differences in the surface charge distribution, which may explain the differences in the cellular receptor binding.	Kostyuchenko et al. (2014)
Bacteriophage epsilon 15	EMD-5003	2008	4.5 Å	JEOL 3000SFF 4.2 K	300	Film Kodak SO163	36,259 particles/2.1 Mio units	The complete backbone trace of the major capsid protein (gene product 7) was derived from this map.	Jiang et al. (2008)
Bacteriophage epsilon 15	EMD-5678	2013	4.5 Å	JEOL 3000SFF Energy filter	300	Film Kodak SO163	14,000 particles/840,000 units	The resolution and reliability of the maps was validated by comparing multiple independent datasets.	Baker et al. (2013)
Bovine Papilloma virus type 1	EMD-5155 and EMD-5156	2010	4.2 and 3.6 Å after sixfold averaging	FEI Tecnai F30	300	Film Kodak SO163	3,997 particles/239,820 units and 1.4 Mio quasiequivalent units	The structural data suggest a different mode of interaction between the 72 pentameric units in the icosahedral capsid than deduced from an earlier crystal structure (Modis et al. 2002).	Wolf et al. (2010)
Aquareovirus	EMD-5160	2010	3.3 Å	FEI Titan Krios	300	Film Kodak SO163	18,464 particles/1.1 Mio units	The map shows that autocleavage of the membrane protein is important for priming the virus for cell entry. The map shows enough detail to suggest that Lys84 and Glu76 may facilitate the autocleavage in a nucleophilic attack.	Zhang et al. (2010)
Rhesus rotavirus	EMD-5199	2011	3.8 Å	FEI Tecnai F30	300	Film Kodak SO163	4,187 particles/3.2 Mio quasiequivalent units (averaging of the 60 asymmetric units and the 13 quasiequivalent units in the asymmetric unit)	The maps show how the two subfragments of VP4 (VP8* and VP5*) retain their association after proteolytic cleavage in the infectious virion.	Settembre et al. (2011)
Bovine rotavirus	EMD-1461	2008	3.8 Å	FEI Tecnai F30	300	Film Kodak SO163	8,400 particles/6.5 Mio quasiequivalent units	The study was used as proof of concept and for methods development; the comparison of the EM map with the x-ray structure of rotavirus at a similar resolution shows similar clarity and level of detail.	Zhang et al. (2008)
Rotavirus VP6	EMD-1752	2010	3.5 Å	FEI Tecnai F30	300	Film Kodak SO163	7,000 particles (from Zhang et al. 2008)/5.5 Mio quasiequivalent units	Same data were used as in Zhang et al. (2008), but the data were processed with symmetry-adapted functions. The final map shows a clear improvement in resolution.	Estrozi and Navaza (2010)

(Continued)

TABLE 3.3 (Continued)
Maps of Viral Nanoparticles with 3–5 Å Resolution, Which Have Been Deposited in the EMDB until the End of 2013

Virus	EMDB Entry	Publication Year	Resolution	Microscope	Voltage (kV)	Detector	Averaged Number of Particles/Units	Remarks	Citations
Rotavirus VP6	EMD-5487 and EMD-5488	2012	4.9 Å (no movie motion correction) and 4.4 Å (motion corrected)	FEI Tecnai F20	200	Direct Detector DE12 Movies	807 particles/630,000 quasiequivalent units	The study was used as test of performance for direct counting detectors. Furthermore, the advantage of movie processing over processing of single high-dose images was tested; processing of separate movie frames leads to better results.	Campbell et al. (2012)
Rotavirus VP6–VP7 complex	EMD-1609	2009	4.2 Å	FEI Tecnai F30	300	Film Kodak SO163	3,780 particles/2.9 Mio quasiequivalent units	The map was used to build the structure of VP7. The model shows the N-terminal arm, which is missing in the crystal structure (Aoki et al. 2009).	Chen et al. (2009)
Venezuelan equine encephalitis virus	EMD-5275 and EMD-5276	2011	4.8 and 4.4 Å	Jeol 3200FSC Energy filter	300	Gatan Ultrascan 4k × 4k CCD camera	37,000 particles/2.2 Mio units and 8.8 Mio quasiequivalent units	The map resolves features that were missing in the crystal structures of domains of alphavirus subunits. These features are implicated in the fusion, assembly, and budding processes of alphaviruses.	Zhang et al. (2011)
Sputnik virus	EMD-5495 and EMD-5496	2012	3.5 and 3.8 Å	FEI Titan Krios	300	Film Kodak SO163	12,000 particles/720,000 units and 12,000 particles/720,000 units	The structure could be solved, where x-ray crystallography was not feasible. Particles were imaged over thin carbon film.	Zhang et al. (2012)
Tobacco mosaic virus	EMD-1316	2007	4.4 Å	FEI Tecnai 30	200	Film Kodak SO163	135 particles/212,550 units	The map identified better ordered region at lower radii (res 88–109) than previous x-ray fiber diffraction studies (Namba et al. 1989); most likely the reconstruction shows a different stable state of the capsid, which might be relevant for assembly/disassembly.	Sachse et al. (2007)
Tobacco mosaic virus	EMD-1730	2010	4.6 Å	FEI Polara	300	Gatan Ultrascan 4k × 4k CCD camera	260,000 units	The study was done as a proof of principle that conventional CCD cameras are suitable for high-resolution structural studies. The map shows the same solution structure as reported by the other cryo-EM study (Sachse et al. 2007).	Clare and Orlova (2010)
Tobacco mosaic virus	EMD-5185	2011	3.3 Å	FEI Titan Krios	300	Film Kodak SO 163	43,000 helical segments 1.3 Mio units	This work demonstrates that helical reconstruction is possible to atomic resolution. The map shows TMV in its metastable calcium-free assembly state, which is distinctly different from the x-ray fiber diffraction map of TMV (Namba et al. 1989).	Ge and Zhou (2011)

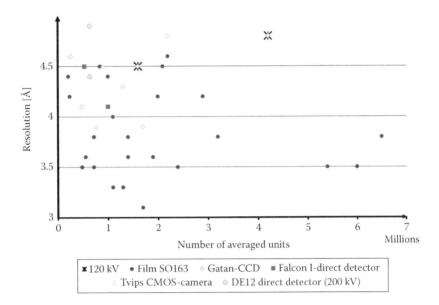

FIGURE 3.9 Plot of the resolution vs. the number of averaged units for maps of viral nanoparticles with 3–5 Å resolution taken with different types of detectors and at different accelerating voltages (see Table 3.3 for details). There is no strong dependency between the resolution and the number of averaged units in this resolution range. Maps with resolution better than 4 Å were predominantly recorded on film.

structures (Lawson et al. 2011). Analysis of the experimental conditions (Figure 3.9) highlights the important factors for achieving high resolution: Most of the structures have been obtained at an accelerating voltage of 300 kV, whereas only four structures have been determined from data recorded at lower voltages (two at 120 kV and two at 200 kV). All apart from one of the highest-resolution structures (<4 Å) were recorded on conventional film. Furthermore, the median number of units, which was averaged in a map with a resolution better than 5 Å, was 1.3 million, which is two to three orders of magnitude larger than what is expected if the only limiting factor would be the scattering efficiency of the object for electrons (Henderson 1995). In general, in the 3–5 Å resolution range, maps do not show a strong dependency of the resolution on the number of averaged units. This implies that other factors such as the structural heterogeneity of the particles, limitations in the imaging system, and beam-induced movement limit the resolution much more than the signal-to-noise ratio of the images. Interestingly, the four structures that have been determined with direct detectors (Falcon I or DE12) do not show any significant reduction in the number of averaged units, which are required for achieving a certain resolution compared to film or conventional CCD or CMOS detectors. This suggests that current image processing either cannot exploit the better signal transfer of the direct detectors at higher spatial frequencies or that the signal transfer of the detector is not the limiting factor. Probably, more studies will be needed to optimize the experimental conditions for the use of direct detectors to fully realize the potential of this new technology.

Maps derived from electron micrographs typically show very clear density with similar if not more detail than x-ray maps of the same object at the same resolution. Since electron microscopy starts to deliver more and more maps with comparable resolution as x-ray crystallography, it becomes important that resolution estimates in crystallography and electron microscopy give the same values for maps with similarly well-resolved features. The determination of the resolution in electron microscopy uses Fourier shell correlation between maps of two independent halves of the data for estimating the resolution (Harauz and Van Heel 1986). The Fourier shell correlation informs how well the halves agree at different spatial frequency and thus also give an estimate of the signal-to-noise ratio at these frequencies. For giving a single resolution value, a cutoff for the correlation (or signal-to-noise ratio) is required to determine the highest spatial frequency that still contains interpretable information. For a long time, there was a controversy which cutoff value is the *correct* one to use. Finally, Rosenthal and Henderson have shown that theoretically a Fourier shell correlation of 0.143 between maps of two independent halves of the noisy data (Rosenthal and Henderson 2003) equals a Fourier shell correlation of 0.5 (and thus a signal-to-noise ratio of 1) between a map of the combined image data and a noise-free reference map. Therefore, they suggested that a map with a resolution determined by a Fourier shell correlation cutoff of 0.143 between the two halves should have a similar quality as an x-ray map with the same resolution. Now that higher-resolution EM maps become available, this theoretical consideration can be tested and indeed the level of detail in EM maps and x-ray maps is most similar at the resolution given by the 0.143 cutoff criterion.

The higher-resolution EM maps contain sufficient information for building atomic models from scratch. They also show sufficient detail of the side chains to understand molecular switching mechanisms. An example for the quality of an EM density map together with the built-in structure

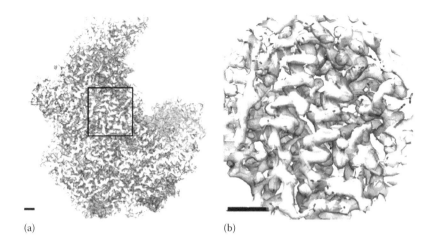

(a)   (b)

FIGURE 3.10 Model of the structural proteins VP1, VP3, and VP5 of Bombyx mori cypovirus capsid (3IXZ) built into the EM map (EMDB 5656) of the virus capsid at 3.1 Å resolution (Yu et al. 2011). (a) Two VP1 (green and cyan), one VP3 (blue) and two VP5 (red and yellow) proteins in the asymmetric unit together with the EM density map (gray). (b) A close-up of VP1 of the area indicated by a square in (a). The Cα backbone of the model is represented as a green ribbon diagram and the side chains are colored by element. The EM map shows distinct densities for the side chains of most residues. In both panels, the length of the scale bar equals 1 nm.

is shown in Figure 3.10 for the cytoplasmic polyhedrosis virus (Yu et al. 2011), which is the map with the highest resolution of a viral nanoparticle reported until the end of 2013. The map shows sufficient detail for ab initio building of the structures of all five proteins in the asymmetric unit. For the refinement of the model, CNS (*C*rystallography and *N*MR *S*ystem; Brunger et al. 1998; Brunger 2007) was used, which is a toolbox designed for model building in crystallography and NMR. To make the EM map accessible to the toolbox, pseudocrystallographic structure factors have been computed by arranging the EM density into a hypothetical crystal. With these pseudocrystallographic structure factors, standard crystallographic refinement procedures could be used for refining the atomic model against the map. Finally, the validity of the derived model was tested by calculating a free *R* factor, which is 27.3 for the model of the asymmetric unit of the cytoplasmic polyhedrosis virus against the EM map.

The somewhat lower-resolution structures (4–5 Å) of viral nanoparticles resolve the secondary structural elements but do not resolve all types of side chains. In many cases, the information of these maps is still sufficient for building Cα traces of the proteins from scratch. However, depending on the resolution of the map, building the Cα trace can be error prone. As an alternative to ab initio model building, homology models can be adapted to the observed density maps (Topf et al. 2008; Cheng et al. 2010; Zhu et al. 2010). The approach uses the cryo-EM map as additional constraint for the homology modeling and thus also allows modeling of conformational changes between the observed structure and the known structure of a homologue. The method is based on molecular dynamics simulations of the model with the density map as a force field that drives the model into the EM map (Topf et al. 2008). For reducing the computational demand of the molecular dynamics simulations, the model refinement can be focused onto areas that do not fit well into

the observed EM map (Zhu et al. 2010). The method is very powerful and works similarly well with lower-resolution maps (resolution ≈ 7 Å) as constraints (Zhu et al. 2010). Therefore, structure-guided homology modeling is an emerging tool for modeling the structure of viral nanoparticles from EM maps at intermediate resolution.

3.6.2 DEALING WITH ASYMMETRIC ATTACHMENTS

Many icosahedral viral nanoparticles have single asymmetric attachments at one of their vertices like, for example, portal proteins that are essential for genome packaging and genome delivery to the host. To understand the function of these portal proteins in greater structural detail, it is necessary to investigate them in the context of the capsid. However, the information of the single portal overlaps with the image information of the icosahedrally ordered capsid that dominates the image information. Icosahedrally averaged capsids only show weak ghosts of the portal proteins at all vertices. Just relaxing the symmetry in the processing of the whole assembly is often not sufficient to determine the unique orientations of the capsids together with the portal at one of the vertices. To overcome this problem, one strategy is to generate an artificial reference, which contains one asymmetric attachment on a single vertex. Such a reference can be generated, by combining the symmetric reconstruction of the capsid with the reconstruction of an isolated attachment added to one of the vertices (Lander et al. 2006). In addition, the ghosts of the attachments at the other vertices are masked off to enhance the asymmetric contribution of the single attachment. This generates an artificial reference model of the whole assembly with the densities of the unique portal and of the capsid being at the same level. Such a reference is suitable for determining the unique orientations of the capsids with attached portal. This approach was quite successful for processing

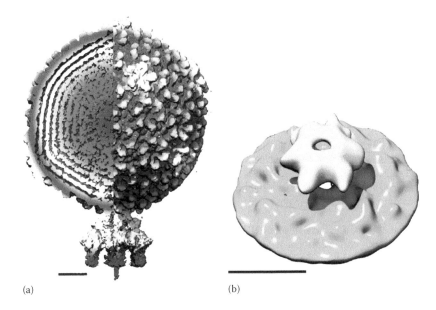

(a)   (b)

FIGURE 3.11 Reconstruction of asymmetric attachments to icosahedral viral nanoparticles: (a) Asymmetric reconstruction of P22 together with its portal protein (EMDB 5348, Tang et al. 2011). The surface is colored according to the radial distance from the center of the capsid. (b) Vertex reconstruction of the hexameric RNA packaging motor in a viral polymerase complex (EMDB 1256, Huiskonen et al. 2007). The hexagonal RNA packaging motor is shown in yellow, and the icosahedral RNA polymerase complex of cystovirus Φ8 is shown in blue. The length of the scale bars equals 10 nm.

bacteriophages such as P22 (Tang et al. 2011) for which the capsid together with its portal protein could be resolved at subnanometer resolution (Figure 3.11a).

For smaller attachments or attachments at several sides of the icosahedral nanoparticle, it is more difficult to identify the position and the structure of the asymmetric contribution. For these assemblies, an alternative approach has been proposed, which aims at only reconstructing the asymmetric attachment at a vertex of the capsid (Briggs et al. 2005). In these vertex reconstructions, the capsids are first processed, assuming full icosahedral symmetry. Then the orientational information of the capsid is used to identify the positions of the vertices. The vertices are extracted and their image information is classified to discern between vertices with and without attachments. For the vertices with attachment, the position and the out-of-plane tilt of the attachment is already known from the capsid orientation. So, the only parameter that is still missing is the relative rotation of the attachment around the vertex axis, which can be determined by angular reconstitution (Schatz et al. 1997). An example is shown for the hexameric packaging motor (about 200 kDa), which is attached to the fivefold vertex of the icosahedral polymerase complex (about 33 MDa) of cystovirus Φ8 (Huiskonen et al. 2007) in Figure 3.11b. Here, a very small attachment can be correctly reconstructed without disturbance by the some 150 times larger icosahedral carrier.

REFERENCES

Adrian, M., J. Dubochet et al. (1984). Cryo-electron microscopy of viruses. *Nature* **308**(5954): 32–36.

Angert, I., E. Majorovits et al. (2000). Zero-loss image formation and modified contrast transfer theory in EFTEM. *Ultramicroscopy* **81**(3–4): 203–222.

Aoki, S. T., E. C. Settembre et al. (2009). Structure of rotavirus outer-layer protein VP7 bound with a neutralizing Fab. *Science* **324**(5933): 1444–1447.

Bai, X. C., I. S. Fernandez et al. (2013). Ribosome structures to near-atomic resolution from thirty thousand cryo-EM particles. *Elife* **2**: e00461.

Baker, L. A., E. A. Smith et al. (2010). The resolution dependence of optimal exposures in liquid nitrogen temperature electron cryomicroscopy of catalase crystals. *J Struct Biol* **169**(3): 431–437.

Baker, M. L., C. F. Hryc et al. (2013). Validated near-atomic resolution structure of bacteriophage epsilon15 derived from cryo-EM and modeling. *Proc Natl Acad Sci USA* **110**(30): 12301–12306.

Baker, T. S., N. H. Olson et al. (1999). Adding the third dimension to virus life cycles: Three-dimensional reconstruction of icosahedral viruses from cryo-electron micrographs. *Microbiol Mol Biol Rev* **63**(4): 862–922, table of contents.

Baschong, W., U. Aebi et al. (1988). Head structure of bacteriophages T2 and T4. *J Ultrastruct Mol Struct Res* **99**(3): 189–202.

Baumeister, W. and J. Seredynski (1976). Preparation of perforated films with pre-determinable hole size distributions. *Micron* **7**(1): 49–54.

Bayer, M. E. and T. F. Anderson (1963). Preparation of holey films for electron microscopy. *Experientia* **19**(8): 433–434.

Bellare, J. R., H. T. Davis et al. (1988). Controlled environment vitrification system: An improved sample preparation technique. *J Electron Microsc Tech* **10**: 87–111.

Berriman, J. and N. Unwin (1994). Analysis of transient structures by cryo-microscopy combined with rapid mixing of spray droplets. *Ultramicroscopy* **56**: 241–252.

Bharat, T. A., N. E. Davey et al. (2012). Structure of the immature retroviral capsid at 8 Å resolution by cryo-electron microscopy. *Nature* **487**(7407): 385–389.

Bharat, T. A., J. D. Riches et al. (2011). Cryo-electron tomography of Marburg virus particles and their morphogenesis within infected cells. *PLoS Biol* **9**(11): e1001196.

Booy, F. P. and J. B. Pawley (1993). Cryo-crinkling: What happens to carbon films on copper grids at low temperature. *Ultramicroscopy* **48**(3): 273–280.

Böttcher, B. and R. A. Crowther (1996). Difference imaging reveals ordered regions of RNA in turnip yellow mosaic virus. *Structure* **4**(4): 387–394.

Böttcher, B., S. A. Wynne et al. (1997). Determination of the fold of the core protein of hepatitis B virus by electron cryomicroscopy. *Nature* **386**(6620): 88–91.

Brandt, S., J. Heikkonen et al. (2001). Automatic alignment of transmission electron microscope tilt series without fiducial markers. *J Struct Biol* **136**(3): 201–213.

Brenner, S. and R. W. Horne (1959). A negative staining method for high resolution electron microscopy of viruses. *Biochim Biophys Acta* **34**: 103–110.

Briggs, J. A., K. Grunewald et al. (2006). The mechanism of HIV-1 core assembly: Insights from three-dimensional reconstructions of authentic virions. *Structure* **14**(1): 15–20.

Briggs, J. A., J. T. Huiskonen et al. (2005). Classification and three-dimensional reconstruction of unevenly distributed or symmetry mismatched features of icosahedral particles. *J Struct Biol* **150**(3): 332–339.

Brilot, A. F., J. Z. Chen et al. (2012). Beam-induced motion of vitrified specimen on holey carbon film. *J Struct Biol* **177**(3): 630–637.

Bruggeller, P. and E. Mayer (1980). Complete vitrification in pure liquid water and dilute aqueous-solutions. *Nature* **288**(5791): 569–571.

Brunger, A. T. (2007). Version 1.2 of the crystallography and NMR system. *Nat Protoc* **2**(11): 2728–2733.

Brunger, A. T., P. D. Adams et al. (1998). Crystallography & NMR system: A new software suite for macromolecular structure determination. *Acta Crystallogr D Biol Crystallogr* **54**(Pt 5): 905–921.

Campbell, M. G., A. Cheng et al. (2012). Movies of ice-embedded particles enhance resolution in electron cryo-microscopy. *Structure* **20**(11): 1823–1828.

Cao, C., X. Dong et al. (2012). Conserved fiber-penton base interaction revealed by nearly atomic resolution cryo-electron microscopy of the structure of adenovirus provides insight into receptor interaction. *J Virol* **86**(22): 12322–12329.

Castano-Diez, D., A. Al-Amoudi et al. (2007). Fiducial-less alignment of cryo-sections. *J Struct Biol* **159**(3): 413–423.

Chen, D. H., M. L. Baker et al. (2011). Structural basis for scaffolding-mediated assembly and maturation of a dsDNA virus. *Proc Natl Acad Sci USA* **108**(4): 1355–1360.

Chen, J. Z., C. Sachse et al. (2008). A dose-rate effect in single-particle electron microscopy. *J Struct Biol* **161**(1): 92–100.

Chen, J. Z., E. C. Settembre et al. (2009). Molecular interactions in rotavirus assembly and uncoating seen by high-resolution cryo-EM. *Proc Natl Acad Sci USA* **106**(26): 10644–10648.

Cheng, L., X. Huang et al. (2014). Cryo-EM structures of two bovine adenovirus type 3 intermediates. *Virology* **450**: 174–181.

Cheng, L., J. Sun et al. (2011). Atomic model of a cypovirus built from cryo-EM structure provides insight into the mechanism of mRNA capping. *Proc Natl Acad Sci USA* **108**(4): 1373–1378.

Cheng, L., J. Zhu et al. (2010). Backbone model of an aquareovirus virion by cryo-electron microscopy and bioinformatics. *J Mol Biol* **397**(3): 852–863.

Chester, D. W., J. F. Klemic et al. (2007). Holey carbon micro-arrays for transmission electron microscopy: A microcontact printing approach. *Ultramicroscopy* **107**(8): 685–691.

Clare, D. K. and E. V. Orlova (2010). 4.6Å Cryo-EM reconstruction of tobacco mosaic virus from images recorded at 300 keV on a 4k × 4k CCD camera. *J Struct Biol* **171**(3): 303–308.

Conway, J. F., N. Chenng et al. (1997). Visualization of a4-helix bundle in the hepatitis B virus capsid by cryo-electron microscopy. *Nature* **368**(6): 91–94.

Conway, J. F., B. L. Trus et al. (1993). The effects of radiation damage on the structure of frozen hydrated HSV-1 capsids. *J Struct Biol* **111**: 222–233.

Crowther, R. A. (1971). Procedures for three-dimensional reconstruction of spherical viruses by Fourier synthesis from electron micrographs. *Phil Trans R Soc* **261**: 221–230.

Crowther, R. A., L. A. Amos et al. (1970a). Three dimensional reconstructions of spherical viruses by Fourier synthesis from electron micrographs. *Nature* **226**(244): 421–425.

Crowther, R. A., D. J. DeRosier et al. (1970b). The reconstruction of a three-dimensional structure from projections and its application to electron microscopy. *Proc R Soc Lond A* **A317**: 319–340.

Crowther, R. A., N. A. Kiselev et al. (1994). Three-dimensional structure of Hepatitis B virus core particles determined by electron cryomicroscopy. *Cell* **77**: 943–950.

Cyrklaff, M., M. Adrian et al. (1990). Evaporation during preparation of unsupported thin vitrified aqueous layers for cryo-electron microscopy. *J Electron Microsc Tech* **16**(4): 351–355.

Danev, R., S. Kanamaru et al. (2010). Zernike phase contrast cryo-electron tomography. *J Struct Biol* **171**(2): 174–181.

de Marco, A., B. Muller et al. (2010). Structural analysis of HIV-1 maturation using cryo-electron tomography. *PLoS Pathog* **6**(11): e1001215.

DeRosier, D. J. and A. Klug (1968). Reconstruction of three dimensional structures from electron micrographs. *Nature* **217**: 130–134.

Diez, D. C., A. Seybert et al. (2006). Tilt-series and electron microscope alignment for the correction of the non-perpendicularity of beam and tilt-axis. *J Struct Biol* **154**(2): 195–205.

Dubochet, J. (2012). Cryo-EM—The first thirty years. *J Microsc* **245**(3): 221–224.

Dubochet, J., M. Adrian et al. (1988). Cryo-electron microscopy of vitrified specimens. *Q Rev Biophys* **21**(2): 129–228.

Dubochet, J., M. Adrian et al. (1994). Structure of intracellular mature vaccinia virus observed by cryoelectron microscopy. *J Virol* **68**(3): 1935–1941.

Dubochet, J., M. Ducommun et al. (1971). A new preparation method for dark-field electron microscopy of biomacromolecules. *J Ultrastruct Res* **35**(1): 147–167.

Dubochet, J., J. Lepault et al. (1982). Electron microscopy of frozen water and aqueous solutions. *J Microsc* **128**: 219–237.

Dubochet, J. and A. McDowall (1981). Vitrification of pure water for electron microscopy. *J Microsc* **124**: 3–4.

Dubochet, J., A. W. McDowall et al. (1983). Electron microscopy of frozen-hydrated bacteria. *J Bacteriol* **155**(1): 381–390.

Egelman, E. H. (2000). A robust algorithm for the reconstruction of helical filaments using single-particle methods. *Ultramicroscopy* **85**(4): 225–234.

Egelman, E. H. (2010). Reconstruction of helical filaments and tubes. *Methods Enzymol* **482**: 167–183.

Ermantraut, E., K. Wohlfahrt et al. (1998). Perforated support foils with pre-defined hole size, shape and arrangement. *Ultramicroscopy* **74**: 75–81.

Estrozi, L. F. and J. Navaza (2010). Ab initio high-resolution single-particle 3D reconstructions: The symmetry adapted functions way. *J Struct Biol* **172**(3): 253–260.

Fan, G. Y., S. J. Young et al. (1995). Conditions for electron tomographic data acquisition. *J Electron Microsc (Tokyo)* **44**(1): 15–21.

Fernandez, J. J. and S. Li (2003). An improved algorithm for anisotropic nonlinear diffusion for denoising cryo-tomograms. *J Struct Biol* **144**(1–2): 152–161.

Frangakis, A. S. and R. Hegerl (1999). Nonlinear anisotropic diffusion in three-dimensional electron microscopy. *Scale-Space Theor Comput Vis* **1682**: 386–397.

Frangakis, A. S. and R. Hegerl (2001). Noise reduction in electron tomographic reconstructions using nonlinear anisotropic diffusion. *J Struct Biol* **135**(3): 239–250.

Franklin, R. E. and K. C. Holmes (1958). Tobacco mosaic virus—Application of the method of isomorphous replacement to the determination of the helical parameters and radial density distribution. *Acta Crystallogr* **11**(3): 213–220.

Frederik, P. M. and D. H. W. Hubert (2005). Cryoelectron microscopy of liposomes. *Methods Enzymol* **391**: 431–448.

Fung, J. C., W. Liu et al. (1996). Toward fully automated high-resolution electron tomography. *J Struct Biol* **116**(1): 181–189.

Ge, P. and Z. H. Zhou (2011). Hydrogen-bonding networks and RNA bases revealed by cryo electron microscopy suggest a triggering mechanism for calcium switches. *Proc Natl Acad Sci USA* **108**(23): 9637–9642.

Glaeser, R. M., D. Typke et al. (2011). Precise beam-tilt alignment and collimation are required to minimize the phase error associated with coma in high-resolution cryo-EM. *J Struct Biol* **174**(1): 1–10.

Grigorieff, N. (2007). FREALIGN: High-resolution refinement of single particle structures. *J Struct Biol* **157**(1): 117–125.

Grimm, R., D. Typke et al. (1996). Determination of the inelastic mean free path in ice by examination of tilted vesicles and automated most probable loss imaging. *Ultramicroscopy* **63**(3–4): 169–179.

Harauz, G. and M. Van Heel (1986). Exact filters for general geometry 3-dimensional reconstruction. *Optik* **73**(4): 146–156.

Harris, W. J. (1962). Holey films for electron microscopy. *Nature* **196**(4853): 499–500.

Heide, H. G. (1984). Observation on ice layers. *Ultramicroscopy* **14**: 271–278.

Henderson, R. (1995). The potential and limitations of neutrons, electrons and X-rays for atomic resolution microscopy of unstained biological molecules. *Q Rev Biophys* **28**(2): 171–193.

Henderson, R. (2004). Realizing the potential of electron cryomicroscopy. *Q Rev Biophys* **37**: 3–13.

Henrick, K., R. Newman et al. (2003). EMDep: A web-based system for the deposition and validation of high-resolution electron microscopy macromolecular structural information. *J Struct Biol* **144**(1–2): 228–237.

Hoppe, W., H. J. Schramm et al. (1986). Three-dimensional electron microscopy of individual biological objects. *Z Naturforsch A* **31**: 645–655.

Huiskonen, J. T., J. Hepojoki et al. (2010). Electron cryotomography of Tula hantavirus suggests a unique assembly paradigm for enveloped viruses. *J Virol* **84**(10): 4889–4897.

Huiskonen, J. T., H. T. Jaalinoja et al. (2007). Structure of a hexameric RNA packaging motor in a viral polymerase complex. *J Struct Biol* **158**(2): 156–164.

Jager, J. (1990). Herstellung von freitragenden Eisfilmen auf Kupfernetzchen zur Einbettung biologischer Makromolekule fur die EM-Beobachtung. *Elektronenmikroskopie* **1**: 24–28.

Jiang, W., M. L. Baker et al. (2008). Backbone structure of the infectious epsilon15 virus capsid revealed by electron cryomicroscopy. *Nature* **451**(7182): 1130–1134.

Johnson, R. P. and D. W. Gregory (1993). Viruses accumulate spontaneously near droplet surfaces: A method to concentrate viruses for electron microscopy. *J Microsc* **171**: 125–136.

Kausche, G. A., E. Pfankuch et al. (1939). The visualisation of herbal viruses in surface microscopes. *Naturwissenschaften* **27**: 292–299.

Kelly, D. F., P. D. Abeyrathne et al. (2008). The affinity grid: A prefabricated EM grid for monolayer purification. *J Mol Biol* **382**(2): 423–433.

Kelly, D. F., D. Dukovski et al. (2010). A practical guide to the use of monolayer purification and affinity grids. *Methods Enzymol* **481**(*Cryo-Em, Part A—Sample Preparation and Data Collection*): 83–107.

Klug, A. and J. E. Berger (1964). An optical method for the analysis of periodicities in electron micrographs, and some observations on the mechanism of negative staining. *J Mol Biol* **10**: 565–569.

Klug, A. and D. L. Caspar (1960). The structure of small viruses. *Adv Virus Res* **7**: 225–325.

Klug, A. and D. J. DeRosier (1966). Optical filtering of electron micrographs: Reconstruction of one-sided images. *Nature* **212**: 29–32.

Korinek, A., F. Beck et al. (2011). Computer controlled cryo-electron microscopy—TOM(2) a software package for high-throughput applications. *J Struct Biol* **175**(3): 394–405.

Kostyuchenko, V. A., P. L. Chew et al. (2014). Near-atomic resolution cryo-electron microscopic structure of dengue serotype 4 virus. *J Virol* **88**(1): 477–482.

Kourkoutis, L. F., J. M. Plitzko et al. (2012). Electron microscopy of biological materials at the nanometer scale. *Annu Rev Mater Res* **42**: 33–58.

Krimm, S. and T. F. Anderson (1967). Structure of normal and contracted tail sheaths of T4 bacteriophage. *J Mol Biol* **27**(2): 197–202.

Kronenberg, S., J. A. Kleinschmidt et al. (2001). Electron cryomicroscopy and image reconstruction of adeno-associated virus type 2 empty capsids. *EMBO Rep* **2**(11): 997–1002.

Kutsay, O., O. Loginova et al. (2008). Surface properties of amorphous carbon films. *Diamond Relat Mater* **17**(7–10): 1689–1691.

Lander, G. C., L. Tang et al. (2006). The structure of an infectious P22 virion shows the signal for headful DNA packaging. *Science* **312**(5781): 1791–1795.

Lawson, C. L., M. L. Baker et al. (2011). EMDataBank.org: Unified data resource for CryoEM. *Nucleic Acids Res* **39**(Database issue): D456–D464.

Lei, J. and J. Frank (2005). Automated acquisition of cryo-electron micrographs for single particle reconstruction on an FEI Tecnai electron microscope. *J Struct Biol* **150**(1): 69–80.

Lepault, J., F. P. Booy et al. (1983). Electron microscopy of frozen biological suspensions. *J Microsc* **129**(Pt 1): 89–102.

Lepault, J. and J. Dubochet (1986). Electron microscopy of frozen hydrated specimens: Preparation and characteristics. *Methods Enzymol* **127**: 719–730.

Lerch, T. F., J. K. O'Donnell et al. (2012). Structure of AAV-DJ, a retargeted gene therapy vector: Cryo-electron microscopy at 4.5 Å resolution. *Structure* **20**(8): 1310–1320.

Li, X., P. Mooney et al. (2013). Electron counting and beam-induced motion correction enable near-atomic-resolution single-particle cryo-EM. *Nat Methods* **10**(6): 584–590.

Liljeroos, L., J. T. Huiskonen et al. (2011). Electron cryotomography of measles virus reveals how matrix protein coats the ribonucleocapsid within intact virions. *Proc Natl Acad Sci USA* **108**(44): 18085–18090.

Liu, H., L. Cheng et al. (2008a). Symmetry-adapted spherical harmonics method for high-resolution 3D single-particle reconstructions. *J Struct Biol* **161**(1): 64–73.

Liu, H., L. Jin et al. (2010a). Atomic structure of human adenovirus by cryo-EM reveals interactions among protein networks. *Science* **329**(5995): 1038–1043.

Liu, J., A. Bartesaghi et al. (2008b). Molecular architecture of native HIV-1 gp120 trimers. *Nature* **455**(7209): 109–113.

Liu, X., Q. Zhang et al. (2010b). Structural changes in a marine podovirus associated with release of its genome into *Prochlorococcus*. *Nat Struct Mol Biol* **17**(7): 830–836.

Liu, Z. F., L. H. Jiang et al. (2010c). A graphene oxide center dot streptavidin complex for biorecognition—Towards affinity purification. *Adv Funct Mater* **20**(17): 2857–2865.

Lu, Z., T. R. Shaikh et al. (2009). Monolithic microfluidic mixing-spraying devices for time-resolved cryo-electron microscopy. *J Struct Biol* **168**(3): 388–395.

Luque, D., J. Gomez-Blanco et al. (2014). Cryo-EM near-atomic structure of a dsRNA fungal virus shows ancient structural motifs preserved in the dsRNA viral lineage. *Proc. Nat. Acad. Sci. USA* **111**: 7641–7646.

Maki-Yonekura, S., K. Yonekura et al. (2010). Conformational change of flagellin for polymorphic supercoiling of the flagellar filament. *Nat Struct Mol Biol* **17**(4): 417–422.

Markham, R., S. Frey et al. (1963). Methods for enhancement of image detail and accentuation of structure in electron microscopy. *Virology* **20**(1): 88–102.

Markham, R., G. J. Hills et al. (1964). Anatomy of tobacco mosaic virus. *Virology* **22**(3): 342–359.

Marko, M., A. Leith et al. (2011). Retrofit implementation of Zernike phase plate imaging for cryo-TEM. *J Struct Biol* **174**(2): 400–412.

Mastronarde, D. N. (2005). Automated electron microscope tomography using robust prediction of specimen movements. *J Struct Biol* **152**(1): 36–51.

Matricardi, V. R., R. C. Moretz et al. (1972). Electron diffraction of wet proteins: Catalase. *Science* **177**(45): 268–270.

Matthews, R. E. F., R. W. Horne et al. (1956). Electron microscope observations of periodicities in the surface structure of tobacco mosaic virus. *Nature* **178**(4534): 635–636.

Maurer, U. E., T. Zeev-Ben-Mordehai et al. (2013). The structure of herpesvirus fusion glycoprotein B-bilayer complex reveals the protein-membrane and lateral protein-protein interaction. *Structure* **21**(8): 1396–1405.

Mayer, E. and P. Bruggeller (1983). Devitrification of glassy water—Evidence for a discontinuity of state. *J Phys Chem* **87**(23): 4744–4749.

McClain, B., E. Settembre et al. (2010). X-ray crystal structure of the rotavirus inner capsid particle at 3.8 Å resolution. *J Mol Biol* **397**(2): 587–599.

McDowall, A. W., J. J. Chang et al. (1983). Electron microscopy of frozen hydrated sections of vitreous ice and vitrified biological samples. *J Microsc* **131**(Pt 1): 1–9.

McMullan, G., S. Chen et al. (2009a). Detective quantum efficiency of electron area detectors in electron microscopy. *Ultramicroscopy* **109**(9): 1126–1143.

McMullan, G., A. R. Faruqi et al. (2009b). Experimental observation of the improvement in MTF from backthinning a CMOS direct electron detector. *Ultramicroscopy* **109**(9): 1144–1147.

Mellema, J. E. and L. A. Amos (1972). Three-dimensional image reconstruction of turnip yellow mosaic virus. *J Mol Biol* **72**: 819–822.

Meyer, J. C., A. K. Geim et al. (2007). The structure of suspended graphene sheets. *Nature* **446**(7131): 60–63.

Mindell, J. A. and N. Grigorieff (2003). Accurate determination of local defocus and specimen tilt in electron microscopy. *J Struct Biol* **142**(3): 334–347.

Modis, Y., B. L. Trus et al. (2002). Atomic model of the papillomavirus capsid. *EMBO J* **21**(18): 4754–4762.

Moharir, A. V. and N. Prakash (1975). Formvar holey films and nets for electron-microscopy. *J Phys E—Sci Instrum* **8**(4): 288–290.

Murata, K., X. Liu et al. (2010). Zernike phase contrast cryo-electron microscopy and tomography for structure determination at nanometer and subnanometer resolutions. *Structure* **18**(8): 903–912.

Murray, J. M. and R. Ward (1987). Preparation of holey carbon films suitable for cryo-electron microscopy. *J Electron Microsc Tech* **5**: 285–290.

Namba, K., R. Pattanayek et al. (1989). Visualization of protein-nucleic acid interaction in a virus: Refined structure of intact tobacco mosaic virus at 2.9Å resolution by X-ray fiber diffraction. *J Mol Biol* **208**: 307–325.

Navaza, J. (2003). On the three-dimensional reconstruction of icosahedral particles. *J Struct Biol* **144**(1–2): 13–23.

Nemecek, D., E. Boura et al. (2013). Subunit folds and maturation pathway of a dsRNA virus capsid. *Structure* **21**(8): 1374–1383.

Nixon, H. L. and R. D. Woods (1960). The structure of tobacco mosaic virus protein. *Virology* **10**(1): 157–159.

Novoselov, K. S., A. K. Geim et al. (2004). Electric field effect in atomically thin carbon films. *Science* **306**(5696): 666–669.

Olson, N. H., P. R. Kolatkar et al. (1993). Structure of human rhinovirus complexed with its receptor molecule. *Proc Natl Acad Sci USA* **90**: 507–511.

Pantelic, R. S., J. C. Meyer et al. (2012). The application of graphene as a sample support in transmission electron microscopy. *Solid State Commun* **152**(15): 1375–1382.

Pantelic, R. S., J. W. Suk et al. (2011). Graphene: Substrate preparation and introduction. *J Struct Biol* **174**(1): 234–238.

Penczek, P. A. (2010). Fundamentals of three-dimensional reconstruction from projections. *Methods Enzymol* **482**: 1–33.

Quispe, J., J. Damiano et al. (2007). An improved holey carbon film for cryo-electron microscopy. *Microsc Microanal* **13**(5): 365–371.

Rosenthal, P. B. and R. Henderson (2003). Optimal determination of particle orientation, absolute hand, and contrast loss in single-particle electron cryomicroscopy. *J Mol Biol* **333**(4): 721–745.

Ruskin, R. S., Z. Yu et al. (2013). Quantitative characterization of electron detectors for transmission electron microscopy. *J Struct Biol* **184**(3): 385–393.

Sachse, C., J. Z. Chen et al. (2007). High-resolution electron microscopy of helical specimens: A fresh look at tobacco mosaic virus. *J Mol Biol* **371**(3): 812–835.

Schatz, M., E. V. Orlova et al. (1995). Structure of *Lumbricus terrestris* hemoglobin at 30 Å resolution determined using angular reconstitution. *J Struct Biol* **114**(1): 28–40.

Schatz, M., E. V. Orlova et al. (1997). Angular reconstitution in three-dimensional electron microscopy: Practical and technical aspects. *Scan Microsc* **11**: 179–193.

Schur, F. K., W. J. Hagen et al. (2013). Determination of protein structure at 8.5Å resolution using cryo-electron tomography and sub-tomogram averaging. *J Struct Biol* **184**(3): 394–400.

Settembre, E. C., J. Z. Chen et al. (2011). Atomic model of an infectious rotavirus particle. *EMBO J* **30**(2): 408–416.

Shi, J., D. R. Williams et al. (2008). A script-assisted microscopy (SAM) package to improve data acquisition rates on FEI Tecnai electron microscopes equipped with Gatan CCD cameras. *J Struct Biol* **164**(1): 166–169.

Sorzano, C. O., C. Messaoudi et al. (2009). Marker-free image registration of electron tomography tilt-series. *BMC Bioinformatics* **10**: 124.

Stahlberg, H. (2012). ctf-simulation, Centre of Cellular Imaging and Nanoanalytics, http://www.c-cina.unibas.ch/download/ctf-simulation, accessed in 2013.

Stark, H., F. Zemlin et al. (1996). Electron radiation damage to protein crystals of bacteriorhodopsin at different temperatures. *Ultramicroscopy* **63**: 75–79.

Stewart, P. L., S. D. Fuller et al. (1993). Difference imaging of adenovirus: Bridging the resolution gap between X-ray crystallography and electron microscopy. *EMBO J* **12**(7): 2589–2599.

Suloway, C., J. Pulokas et al. (2005). Automated molecular microscopy: The new Leginon system. *J Struct Biol* **151**(1): 41–60.

Tang, J., G. C. Lander et al. (2011). Peering down the barrel of a bacteriophage portal: The genome packaging and release valve in p22. *Structure* **19**(4): 496–502.

Taylor, K. A. and R. M. Glaeser (1973). Hydrophilic support films of controlled thickness and composition. *Rev Sci Instrum* **44**(10): 1546–1547.

Taylor, K. A. and R. M. Glaeser (1974). Electron diffraction of frozen, hydrated protein crystals. *Science* **186**(4168): 1036–1037.

Thuman-Commike, P. A., H. Tsuruta et al. (1999). Solution x-ray scattering-based estimation of electron cryomicroscopy imaging parameters for reconstruction of virus particles. *Biophys J* **76**(4): 2249–2261.

Tivol, W. F., A. Briegel et al. (2008). An improved cryogen for plunge freezing. *Microsc Microanal* **14**(5): 375–379.

Topf, M., K. Lasker et al. (2008). Protein structure fitting and refinement guided by cryo-EM density. *Structure* **16**(2): 295–307.

Toyoshima, C. (1989). On the use of holey grids in electron microscopy. *Ultramicroscopy* **30**: 439–444.

Toyoshima, C. and N. Unwin (1988). Contrast transfer for frozen-hydrated specimens—Determination from pairs of defocused images. *Ultramicroscopy* **25**(4): 279–291.

Tran, E. E., M. J. Borgnia et al. (2012). Structural mechanism of trimeric HIV-1 envelope glycoprotein activation. *PLoS Pathog* **8**(7): e1002797.

Trinick, J. and J. Cooper (1990). Concentration of solutes during preparation of aqueous suspensions for cryo-electron microscopy. *J Microsc* **159**: 215–222.

Unwin, N. (2005). Refined structure of the nicotinic acetylcholine receptor at 4Å resolution. *J Mol Biol* **346**(4): 967–989.

Unwin, P. N. and R. Henderson (1975). Molecular structure determination by electron microscopy of unstained crystalline specimens. *J Mol Biol* **94**(3): 425–440.

van der Heide, P., X. P. Xu et al. (2007). Efficient automatic noise reduction of electron tomographic reconstructions based on iterative median filtering. *J Struct Biol* **158**(2): 196–204.

Veesler, D., T. S. Ng et al. (2013). Atomic structure of the 75 MDa extremophile Sulfolobus turreted icosahedral virus determined by CryoEM and X-ray crystallography. *Proc Natl Acad Sci USA* **110**(14): 5504–5509.

Vogel, R. H., S. W. Provencher et al. (1986). Envelope structure of Semliki Forest virus reconstructed from cryo-electron micrographs. *Nature* **320**(6062): 533–535.

Vonck, J. (2000). Parameters affecting specimen flatness of two-dimensional crystals for electron crystallography. *Ultramicroscopy* **85**(3): 123–129.

Wade, R. H. (1992). A brief look at imaging and contrast transfer. *Ultramicroscopy* **46**: 145–156.

Walker, M., J. Trinick et al. (1995). Millisecond time resolution electron cryo-microscopy of the M-ATP transient kinetic state of the acto-myosin ATPase. *Biophys J* **68**(4 Suppl.): 87S–91S.

Wang, G. J., C. Porta et al. (1992). Identification of a Fab interaction footprint site on an icosahedral virus by cryoelectron microscopy and X-ray crystallography. *Nature* **355**(6357): 275–278.

White, H. D., M. L. Walker et al. (1998). A computer-controlled spraying-freezing apparatus for millisecond time-resolution electron cryomicroscopy. *J Struct Biol* **121**(3): 306–313.

Wilson, N. R., P. A. Pandey et al. (2009). Graphene oxide: Structural analysis and application as a highly transparent support for electron microscopy. *ACS Nano* **3**(9): 2547–2556.

Wolf, M., R. L. Garcea et al. (2010). Subunit interactions in bovine papillomavirus. *Proc Natl Acad Sci USA* **107**(14): 6298–6303.

Wynne, S. A., R. A. Crowther et al. (1999). The crystal structure of the human hepatitis B virus capsid. *Mol Cell* **3**(6): 771–780.

Xie, Q., M. Spilman et al. (2013). Electron microscopy analysis of a disaccharide analog complex reveals receptor interactions of adeno-associated virus. *J Struct Biol* **184**(2): 129–135.

Yang, C., G. Ji et al. (2012). Cryo-EM structure of a transcribing cypovirus. *Proc Natl Acad Sci USA* **109**(16): 6118–6123.

Yu, X., P. Ge et al. (2011). Atomic model of CPV reveals the mechanism used by this single-shelled virus to economically carry out functions conserved in multishelled reoviruses. *Structure* **19**(5): 652–661.

Yu, X., L. Jin et al. (2008). 3.88 Å structure of cytoplasmic polyhedrosis virus by cryo-electron microscopy. *Nature* **453**(7193): 415–419.

Yu, X., L. Jin et al. (2013). 3.5Å cryoEM structure of Hepatitis B virus core assembled from full-length core protein. *PLoS One* **8**(9): e69729.

Zemlin, F., K. Weiss et al. (1978). Coma-free alignment of high-resolution electron-microscopes with aid of optical diffractograms. *Ultramicroscopy* **3**(1): 49–60.

Zernike, F. (1942). Phase contrast, a new method for the microscopic observation of transparent objects Part II. *Physica* **9**: 974–986.

Zhang, P., M. J. Borgnia et al. (2003). Automated image acquisition and processing using a new generation of 4K × 4K CCD cameras for cryo electron microscopic studies of macromolecular assemblies. *J Struct Biol* **143**(2): 135–144.

Zhang, R., C. F. Hryc et al. (2011). 4.4 Å cryo-EM structure of an enveloped alphavirus Venezuelan equine encephalitis virus. *EMBO J* **30**(18): 3854–3863.

Zhang, X., P. Ge et al. (2013a). Cryo-EM structure of the mature dengue virus at 3.5-Å resolution. *Nat Struct Mol Biol* **20**(1): 105–110.

Zhang, X., H. Guo et al. (2013b). A new topology of the HK97-like fold revealed in Bordetella bacteriophage by cryoEM at 3.5 Å resolution. *Elife* **2**(0): e01299.

Zhang, X. and Z. Hong Zhou (2011). Limiting factors in atomic resolution cryo electron microscopy: No simple tricks. *J Struct Biol* **175**: 253–263.

Zhang, X., L. Jin et al. (2010). 3.3 Å cryo-EM structure of a nonenveloped virus reveals a priming mechanism for cell entry. *Cell* **141**(3): 472–482.

Zhang, X., E. Settembre et al. (2008). Near-atomic resolution using electron cryomicroscopy and single-particle reconstruction. *Proc Natl Acad Sci USA* **105**(6): 1867–1872.

Zhang, X., S. Sun et al. (2012). Structure of Sputnik, a virophage, at 3.5-Å resolution. *Proc Natl Acad Sci USA* **109**(45): 18431–18436.

Zhu, J., L. Cheng et al. (2010). Building and refining protein models within cryo-electron microscopy density maps based on homology modeling and multiscale structure refinement. *J Mol Biol* **397**(3): 835–851.

4 X-Ray Analysis of Viral Nanoparticles

Kaspars Tars

CONTENTS

4.1 HISTORICAL ASPECTS OF CRYSTALLOGRAPHY

Although the appearance and shape of crystals have caught attention since antiquity, the first scientific description of crystals dates back to 1611 when Johannes Kepler suggested that hexagonal form of snowflakes is due to regular packaging of spherical water particles [1]. Later discoveries of René Just Haüy, in 1784 [2] led to idea that crystals are essentially regular arrays of packed atoms or molecules. During the nineteenth century, principles of crystal composition and symmetry were worked out, but it was not until discovery of x-rays in 1895 by Wilhelm Röntgen [3] before x-ray crystallography became possible. In 1912 in a historical conversation between Max von Laue and Paul Peter Ewald, it was suggested that crystals could be used for diffraction of x-rays, and the first diffraction images using copper sulfate crystals were indeed recorded on photographic film [4]. The first crystal structures of any compound—those of NaCl and KCl—were solved by William Lawrence Bragg already in 1913 [5]. In the following three decades, crystal structures of several organic molecules were solved, notably hexamethylenetetramine [6], long-chain fatty acids [7], phthalocyanine [8], penicillin [9], and vitamin B_{12} [10]. However, due to technical limitations, it was not until 1958 when first high-resolution protein structure of myoglobin was solved [11].

The first viruses for which structural investigations were attempted were of plant origin, since they could be relatively easily obtained from plant biomass in milligram amounts. Large effort was devoted to x-ray fiber diffraction analysis of concentrated gel-like solutions of tobacco mosaic virus (TMV). From obtained diffraction patterns, an overall shape of TMV was suggested already in 1941 [12]. The TMV analysis was continued for several decades, notably by Rosalind Franklin [13,14], and higher-resolution structures were solved by several research groups in 1970s [15–17]. The first true crystals of any virus were those of icosahedral tomato bushy stunt virus (TBSV), obtained by Bawden and Pirie already in 1937 [18], which were further examined by x-ray diffraction [19,20]. However, it was not until several decades later before first high-resolution structure of TBSV emerged, largely due to efforts of Stephen C. Harrison and coworkers. TBSV 30 Å resolution structure was published in 1972 [21], followed by 16 Å structure in 1975 [22], 5.5 Å structure in 1977 [23], and finally, 2.9 Å structure in 1978 [24], which was the first detailed high-resolution structure of any virus. A few years later, structures of southern bean mosaic virus [25] and satellite tobacco necrosis virus [26] were published. As of March 2014, there were 307 virus or virus-like particle entries in PDB, solved by protein x-ray crystallography, representing 37 different virus families. Considering that the total number of protein structures in PDB is almost 100,000, the number of available virus structures remains relatively low, even if 105 available cryo-electron microscopy (cryo-EM)-based models are taken into account. However, 3D structures of many other monomeric or oligomeric (but not icosahedral) viral structural proteins are available in PDB. Additionally, several thousands of structures of numerous nonstructural viral proteins, such as proteinases and polymerases of medically relevant viruses, are available at PDB.

4.2 CRYSTALS

By definition, crystal is a solid material whose constituent atoms, molecules, or ions are arranged in an ordered pattern extending in all three spatial dimensions. Crystals are composed of identical unit cells. Unit cell is the smallest entity of crystal, from which the entire crystal can be made by translation operators only. Unit cell in turn is composed from one or several crystallographic asymmetric units—from which the entire crystal can be constructed using translation and other symmetry operators (rotation, mirror planes, etc.). Asymmetric unit can be composed of one or several molecules that can be related to each other by local symmetry axes, translational operators, or combinations of both. Depending on the symmetry among asymmetric units in the unit cells, crystals are classified in 230 space groups. Natural proteins are chiral molecules made of L-amino acids and can crystallize in only 65 space groups that do not have mirror planes and inversion centers as symmetry elements; otherwise, *mirror* protein molecules made of D-amino acids also had to be present in the crystal. In Figure 4.1, a hypothetical example is shown, in which five molecules form an asymmetric unit and two asymmetric units are related by twofold rotation axis, producing space group P2, which is not common in

reality, but shown here for its simplicity. Local or noncrystallographic symmetry is different from crystallographic symmetry—crystallographic symmetry axes run through the whole crystal and relate asymmetric units, regardless of their distance from the axis. In contrary, local symmetry axes relate molecules only within one asymmetric unit. Also, since molecules within one asymmetric unit have different environments, their structure is also slightly different—so in contrast to crystallographic symmetry, noncrystallographic symmetry is not perfect. In the case of viruses, deviations from icosahedral symmetry due to particle packing in crystal are negligible and can be largely ignored. The concept of crystallographic asymmetric unit is similar to that of icosahedral asymmetric unit, discussed in Chapter 1. Icosahedral asymmetric unit is the smallest entity from which the whole icosahedron can be built and crystallographic asymmetric unit is the smallest entity from which the crystal can be built. In virus crystals, the crystallographic asymmetric unit can contain a single particle or 60 icosahedral asymmetric units, several particles, or, quite frequently, a fraction of a particle. For example, if the virus particle is sitting on a crystallographic twofold symmetry axis in a way that it coincides with the icosahedral twofold symmetry axis, there will be a half of a particle in the asymmetric unit. An icosahedral particle may be sitting on several crystallographic twofold or threefold axes, which coincide with the icosahedral ones—in some cubic space groups, there might be as little as a 1/12th of a particle in the crystallographic asymmetric unit. Since the fivefold symmetry that is present in icosahedral virus particles is not compatible with crystallographic symmetry operators, the fivefold symmetry axis is always noncrystallographic and in virus crystals, there will always be at least five icosahedral asymmetric units related by a local fivefold axis.

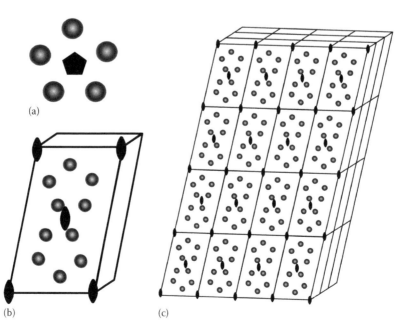

(a)

(b)   (c)

FIGURE 4.1 Example of the composition of crystal. (a) Five molecules (black balls) form asymmetric unit. Five molecules are related by a fivefold noncrystallographic symmetry axis (shown as pentagon). (b) Two asymmetric units form unit cell. Asymmetric units are related by a crystallographic twofold symmetry axis (shown as black ovals). (c) Many unit cells form the crystal.

4.3 CRYSTALLIZATION OF VIRUSES AND VIRUS-LIKE PARTICLES

Assuming that purified virus or virus-like particles are available, the workflow of structure determination (shown schematically in Figure 4.2) begins with crystallization. Crystallization of viruses is in a sense not much different from crystallization of ordinary proteins. The general concept is that the concentration of precipitant—substance that attracts water molecules—is slowly increased. At some point there will be not enough water molecules left to interact with protein, so protein molecules will start interacting with each other. In most cases, precipitate is formed, but sometimes, crystals may form under right conditions. The principle is shown in Figure 4.3 using a simplified phase diagram, where concentrations of protein and precipitant are varied. A good starting concentration for proteins is about 10 mg/mL, but there are reports of using protein concentrations in a range from 2 to 100 mg/mL. Precipitants may be inorganic salts such as ammonium sulfate, polyethylene glycol (PEG) of various molecular weights, and other compounds. In most cases, a buffer solution is added to the crystallization mixture to hold the pH constant. The mixture often also includes additives that help in crystallization by other means than precipitants. Additives may be metal ions,

linkers, or ligands, which help to stabilize protein structure or promote interaction between protein molecules. In general, it is impossible to tell in advance, in which conditions the given protein will crystallize, so frequently hundreds of different conditions should be tried.

Although most crystallization experiments are performed at room temperature (or rather in a crystallization room with a constant +20° temperature), other temperatures, such as +4° or +15°, can be tried. Since many viruses and virus-like particles are fairly stable, it is worth trying temperatures above room temperature, like +30° or +37°. For example, crystals of small RNA phages MS2 and PP7 could be grown only at +37° [27,28].

The two most commonly used crystallization techniques are sitting drop vapor diffusion technique (Figure 4.4) and hanging drop vapor technique. In both cases, protein solution is mixed with precipitant (normally in 1:1 ratio) and placed in sealed compartment above precipitant solution reservoir. Since concentration of precipitant in hanging or sitting drop is lower than in reservoir, the drop will slowly evaporate until the concentration of precipitant in droplet will equal to that of reservoir. Most often, it takes overnight or a few days for the protein crystals to appear, but they can also grow in a matter of hours—or sometimes in a matter of a few months or even a year.

In recent years, use of crystallization robots has simplified protein crystallization significantly. The crystallization

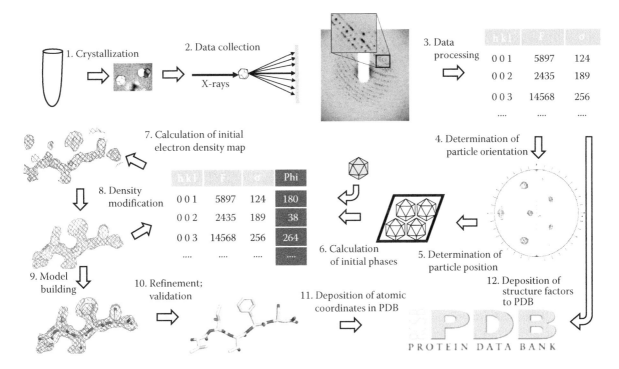

FIGURE 4.2 A general common scheme for virus structure determination. After crystallization (1) diffraction images are collected (2), which are subsequently processed (3)—meaning that each reflection has assigned index h, k, l, amplitude F, and error estimate σ. Determination of particle orientation in the unit cell (4) can be done by self-rotation function, which reveals positions of icosahedral symmetry axes. Determination of particle position in the unit cell (5) usually can be reasoned from the packing considerations. After then, initial phases (red column) can be calculated (6), using similar known structure (represented by orange icosahedron). Initial map (gray mesh) can be calculated (7) by Fourier transform and density modified (8) by cyclic averaging until a map of sufficient quality (blue mesh) is obtained. Atomic model can be built (9) in the map and further refined and validated (10). Finally, the coordinates must be deposited (11) in PDB along with experimental data (12).

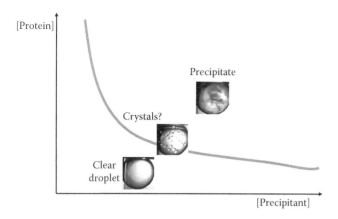

FIGURE 4.3 The phase diagram. In the diagram, two parameters are varied—concentrations of protein and precipitant. If both are low, the droplet stays clear. If both are high, the protein forms precipitate. On the borderline, the protein may form crystals.

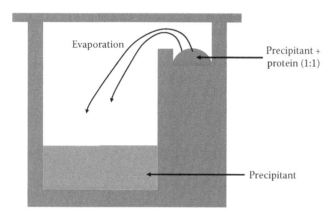

FIGURE 4.4 Protein crystallization by sitting drop vapor technique. In a sealed vial, there are two compartments—one bigger for precipitant and another smaller for precipitant/protein mix (usually in proportions 1:1). Due to the lower concentration of precipitant in the upper compartment, the droplet will start to slowly evaporate until the concentrations of precipitant will be equal. Hanging drop technique is similar, except that the droplet is hanging on the inside of the lid of the vial.

robots not only save time for pipetting but are also able to handle very small volumes down to 0.1 μL. This greatly reduces the amount of protein necessary for crystallization trials, which might be important, if little material is available or if expensive and complicated eukaryotic protein expression systems are used.

While crystallization of viruses is not much different from crystallization of ordinary proteins, there are still some considerations to be kept in mind. First, viruses are most often noncovalent complexes of protein subunits, so under certain conditions, they may disassemble. In particular, this is true for viruses, in which contacts between subunits are enforced by metal ions. Metal-ion-mediated subunit contacts may weaken upon changes in pH, ionic strength, or presence of chelating agents, promoting capsid disassembly. For example, bacteriophage φCb5, which contains stabilizing calcium ions in between subunits, is very salt sensitive, so useful crystals

were obtained in conditions with lower concentrations of salt and buffer than present in standard commercial crystallization screens [29]. Second, those viruses that crystallize in the presence of PEG often (but not always) do so at low PEG concentrations—again lower than present in standard screens. For example, sobemoviruses cocksfoot mottle virus [30] and ryegrass mottle virus [31] were crystallized at PEG 8000 concentrations around 1%, while in most commercially available protein crystallization kits, high MW PEG is found at >10% concentration.

4.4 NEED FOR X-RAYS

In conventional microscopy, small objects are visualized by lenses, which are able to focus the light beam. However, the wavelength of visible light by about three orders of magnitude exceeds the bond lengths between atoms. Since in optical microscopy it is not possible to visualize samples, much smaller than the wavelength of light, for visualization of molecular details, electromagnetic radiation with wavelength comparable to bond lengths between atoms should be used—namely, x-rays with wavelength around 1–2 Å. However, currently, there is no known material that is able to focus hard x-rays; therefore, construction of x-ray microscope, capable to reveal molecular details, is not feasible. That said, x-ray microscopes in fact do exist, but only in soft x-ray range and their spatial resolution currently is limited to about 10 nm.

Since hard x-rays cannot be focused by lenses, the image has to be obtained indirectly by analyzing the diffraction pattern from crystals. In this way, x-ray analysis is fundamentally different from optical and electron microscopy, in which image can be acquired directly.

4.5 INTERACTION OF X-RAYS WITH MATTER AND THE NEED FOR CRYSTALS

When x-ray beam hits a free electron, it starts oscillating, producing spherical secondary x-rays with the same wavelength as primary x-ray beam. The same happens to electrons in atoms and molecules and the secondary x-rays interfere each with other, producing a distinct x-ray scattering pattern, which is dependent on relative positions of all electrons—or in other words, dependent on the structure of molecule. In principle, x-rays interact with any particle—but the intensity of scattered beam is inversely proportional to the square of particles mass. Therefore, atomic nuclei, which are much heavier than electrons, contribute negligibly to scattering. Each scattered x-ray beam has two unique characteristics—amplitude and relative phase (Figure 4.5). If one could measure amplitudes and phases of all scattered x-rays, the calculation of electron density of molecule would be trivial, since electron density and x-ray scattering pattern are Fourier transforms to each other. However, there are two obstacles to direct calculation—first, scattering from a single molecule is far too weak to be observed with the currently available instruments. Second, while values

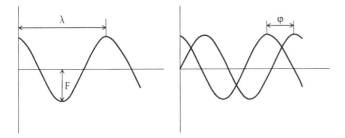

FIGURE 4.5 Electromagnetic wave. Each electromagnetic wave has three essential qualities—wavelength λ, amplitude F, and phase φ, which determines relative shift to another wave.

of amplitudes can be straightforwardly calculated from directly observed intensities, phase of each scattered x-ray beam cannot be measured directly, so information about it has to be acquired indirectly.

The problem of a weak scattering from a single molecule can be overcome by additive scattering from multiple molecules. Scattering from many molecules in random orientations (like in solution) would result in secondary x-rays largely being cancelled out by each other. However, if all molecules could be placed in a certain orientation, scattered x-rays would add up in certain directions, producing measurable spots on the detector. Since in crystals all unit cells (containing one or several molecules of interest) are indeed in the same orientation, crystals are indispensable in determining 3D structures using x-ray technique. The specific directions, in which scattered x-rays are added constructively, are determined by Bragg's law (Figure 4.6). The resulting spots on the detector are usually called *reflections*, because diffracted x-ray beams can be considered as reflected from planes, made by atoms in unit cells. Many different sets of parallel planes exist in crystal, orientations of which are defined by so-called Miller indices (Figure 4.7) and each of them is able to produce a single unique reflection. At a given orientation of crystal

respective to x-ray beam only a few planes produce reflections. Therefore, crystal must be rotated by a small oscillation angle to get the next set of reflections and so on, until all unique reflections at a given resolution are collected. The amount of reflections to be collected depends on the size of unit cell and resolution of the crystal. In the case of viruses, quite often millions of reflections are collected— for example, more than 10 million unique reflections were collected from adenovirus crystals [32].

Resolution of the crystal is defined as the smallest distance between reflecting planes, which produce a measurable reflection. At the same time resolution corresponds to the smallest distance, resolvable in the final electron density maps. High-resolution reflections are located further from the center of the detector; therefore, resolution of the crystal can be seen already after the first images have been acquired. Resolution of the crystal is dependent on how perfectly the unit cells are ordered in the crystal. Resolution is not very dependent on the visual appearance of crystal—although there is some weak tendency that nice-looking crystals produce good diffraction pattern, there are also numerous examples of big crystals with sharp edges displaying very poor resolution and vice versa.

4.6 PHASE PROBLEM

The second problem—lack of direct information about phases (often referred to as *phase problem*)—is of central importance in x-ray crystallography, since if phases are known, the structure of molecule is essentially solved. Information about phases can be obtained in a number of ways. For small molecules (in general, less than 1000 non-hydrogen atoms), so-called direct methods can be applied. In this case, information about phases is extracted directly from observed intensities. Direct methods have been successfully applied for structure determination of a few small proteins [33], but for larger proteins, not speaking about viruses, the method cannot be applied.

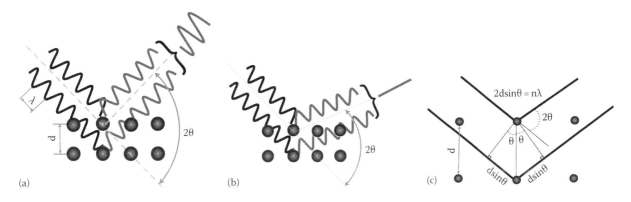

FIGURE 4.6 Reflections and Bragg's law. When primary x-ray beam (black) hits an atom in a crystalline lattice, secondary x-ray beams (red) are scattered in all directions (defined by angle 2θ); however, only in certain directions scattered beams from underlying parallel planes in crystal (separated by distance d) add up by constructive interference (a), producing a measurable signal, while in other directions (b), secondary beams cancel out each other by destructive interference. (c) Directions, in which beams add up, are defined by Bragg's law: $2d\sin\theta = n\lambda$, meaning that the path length difference of each scattered wave should be a multiple of wavelength.

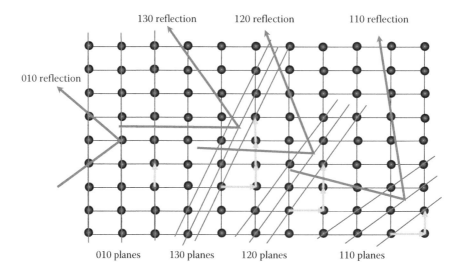

FIGURE 4.7 Planes, reflections, and Miller indices. Since crystal is a regular array of unit cells, many sets of parallel planes can be drawn in all directions. For simplicity, crystal lattice is shown in two dimensions and planes represented by lines. The orientation of each set of planes can be characterized by three Miller indices—for the simplified 2D case, the third index is always zero. According to Bragg's law, each unique set of planes produces one reflection—1,1,0 plane produces 1,1,0 reflection; 1,2,0 plane produces 1,2,0 reflection; and so on.

If a similar protein structure (in general, more than 25% amino acid identity) is already known, it is possible to combine the observed amplitudes with phases of known structure—this is so-called molecular replacement (MR) method, which is becoming increasingly more popular due to high number of structures, already available at Protein Data Bank (PDB). MR is a method of choice for virus structure determination, and as discussed in further sections, in the case of viruses, it can be applied even if similarity of known structure is well below 25% or in some cases indeed without any sequence similarity at all.

Multiple isomorphous replacement (MIR) method was the method of choice for the structure determination of first protein structures. The method depends on introduction of heavy elements in crystals—this can be done by soaking or cocrystallization. In some cases, heavy elements would bind to well-defined places in protein crystal—for example, mercury ion binds to free cysteines and some other metals bind to histidine and acidic residues. Introduction of heavy elements gives rise to small changes in observed amplitudes of reflections. From the observed differences, the positions of heavy elements can be determined by direct methods and used to calculate initial phases. Crystals of at least two different heavy atom derivatives should be prepared to obtain accurate phase information. There are several drawbacks of MIR method. First, crystals frequently are destroyed during treatment with derivatives of heavy elements—and even if the treated crystals look visually undamaged, they often do not diffract well. Second, heavy atoms may not bind to protein. Third, binding of heavy elements may change unit cell dimensions of the crystals or protein structure and the method can no longer be applied. For all those reasons, MIR method is infrequently used in the past decades. However, first struc-

tures of viruses, including TBSV, SBMV, and STNV, were solved using this method [24–26].

Anomalous x-ray scattering (MAD or SAD phasing) currently is a method of choice for structure determination of proteins with little or no similarities in PDB. This method also relies on introduction of heavy atoms in crystals. In this case, amplitudes of reflections with Miller indices h, k, l and -h, -k, -l (so-called Friedel pairs), which are normally identical in the absence of heavy scatterers, become slightly different (phenomenon, known as anomalous signal), and phase information can be extracted from those differences. Anomalous signal is especially strong at a specific wavelength—absorption edge, which is dependent on the particular heavy element and its environment in the crystal. Frequently, three datasets are collected at different wavelengths at and around the absorption edge. A convenient way to introduce heavy scatterer is by exchanging methionine residues in the protein to selenomethionines, which can be done during expression in methionine-auxotroph organism with supplemented selenomethionine. The advantages of anomalous scattering technique using selenomethionine are that (1) there is no doubt about the presence of anomalous scatterer in the crystals, (2) crystals of selenomethionine derivatives usually grow in similar conditions and produce diffraction comparable to native ones, and (3) all the data can be collected from a single crystal. The drawback of the method is that big amount of data has to be collected, preferably from a single crystal, which can be challenging for weak-diffracting and radiation-sensitive crystals. For this reason, anomalous scattering techniques, even if extremely useful in structure determination of proteins, are seldom (if ever) used in structure determination of viruses.

4.7 CRYSTAL TREATMENT, DATA COLLECTION, AND PROCESSING

Data collection from virus crystals brings special challenges, compared to ordinary proteins. This is due to the enormous size of even the simplest viruses. For example, bacteriophage MS2—one of the smallest known viruses—has a molecular weight of 2.5 MDa—roughly 100 times of an average protein of 25 kDa. Molecular weight of adenovirus—the biggest macromolecular structure ever determined at atomic resolution—is 150 MDa. Therefore, unit cell dimensions in the virus crystals are huge, which brings two consequences. First, diffraction from a crystal with big unit cell is weak when compared to diffraction from the crystal of the same size but smaller unit cell. Since the volume of the unit cell of a virus might be by a factor of 100–1000 bigger than that of a typical protein, the x-ray beam intensity, exposure time, and/or crystal size have to be increased proportionally to get a measurable signal. Growing of big crystals can be challenging, while an increase in beam intensity and exposure time often leads to irreversible radiation damage, preventing collection of full dataset of reflections. Second, the distance between recorded reflections on the detector is inversely proportional to the size of unit cells—so diffraction images, collected from virus crystals, are crowded with reflections (see inset in Figure 4.2), sometimes to the extent that they cannot be separated from each other. This problem can be addressed by using small oscillation angles (most frequently 0.1°–0.3° are appropriate for viruses), large detectors, and narrow x-ray beams with low divergence. For all those reasons, it is very difficult to collect useful data from virus crystals at home sources or even weak beamlines at synchrotrons; therefore, modern undulator beamlines, equipped with large detectors, are preferred.

For protein structures, typically 90% or more of theoretically possible reflections at the given resolution have to be collected to get electron density map of good quality. For viruses, however, the figure can be much lower. This is due to the icosahedral symmetry, with prior knowledge that the structure looks the same for all icosahedral asymmetric units; the final electron density map can be improved substantially as discussed in subsequent sections. For example, structure of phage PP7 was solved with only 30% complete data [27].

Due to long exposure times at synchrotron beamlines, radiation damage is a very serious issue for virus crystals. Radiation damage occurs, when incoming x-rays produce free radicals, which react with protein and eventually destroy the crystal. Radiation damage can be greatly reduced by freezing the crystal, since at cryogenic temperatures free radicals cannot travel far. However, to avoid damage during the freezing and to prevent formation of ice crystals, protein crystals first have to be treated by cryoprotectants. Normally, this can be done by soaking protein crystal for a few seconds or minutes in mother liquor, containing a suitable cryoprotectant. The most commonly used cryoprotectant is 25%–35% glycerol, but numerous other cryoprotectants, such as sucrose,

ethylene glycol, or low molecular weight PEG, can be tried. Soaking in cryoprotectant also may damage the crystal by itself; therefore, it is recommended to test various cryoprotectants at different concentrations for new types of crystals. After treatment with cryoprotectant, crystals are picked up with a polymer loop and flash frozen in liquid nitrogen.

Due to long exposure times, radiation damage can be a serious problem even for frozen virus crystals. For regular proteins, normally a full dataset can be collected from a single well-diffracting frozen crystal. In contrast, about 100 frozen adenovirus crystals were used to obtain a 47% complete dataset [32].

Large crystals (typically, larger than 0.3 mm) may be difficult to freeze without affecting the quality of diffraction. Crystals normally are somewhat mosaic—that is, they consist from a number of blocks (not to be confused with unit cells, which are much smaller) in slightly different orientations, which increases the number of diffraction spots in one image. Upon freezing, mosaicity of crystal usually increases, which produces even more spots on the detector—which is already crowded in the case of viruses. Therefore, there is a trade-off among the size of crystal and diffraction quality—although exposure time is shorter for bigger crystals, they tend to be very mosaic when frozen—thereby further increasing the number of spots per image. Also, some virus crystals are extremely fragile and do not tolerate any freezing. Consequently, many virus structures have been solved from data, collected from big unfrozen crystals at a price of quick radiation damage. An extreme example is structure of bluetongue virus—about 1000 crystals were used to collect the diffraction data [34]. In order to prevent rapid drying by evaporation, data from unfrozen crystals are normally collected after the crystal is mounted in a sealed quartz capillary. However, crystals of some viruses are so fragile that they do not tolerate any mechanical handling at all. In this case, crystals can be grown directly in capillary, as successfully demonstrated for phages HK97 [35] and PRD1 [36]. In some cases, it may be beneficial to treat the crystals with dehydrating agents like high molecular weight PEG, thereby lowering the water content and increasing the contacts between virus particles, which may increase the crystal resolution as demonstrated for phage PRD1 [36].

After placement of crystal in the x-ray beam, normally first a couple of images are collected in order to estimate resolution and necessary exposure time. If the quality of diffraction is satisfactory, next the space group and cell dimensions are calculated—this is done automatically by programs (such as MOSFLM [37]) by measuring distances and angles between reflections. Then, the same program suggests optimal data collection strategy after which the actual data collection may be initiated. However, since, as discussed earlier, data collection from virus crystals results in a quick radiation damage, strategy step is often considered as a waste of useful images, which could be obtained from the given crystal before it is dead. Instead *shoot first, think later* strategy (somewhat humorously adopted from Wild West stories) is often used: images are collected starting from a random position of crystal and the orientation parameters figured out later. After the data collection, data have to be indexed and integrated—during which each reflection is assigned to its Miller index, intensity, and estimated error. This is done by programs like MOSFLM, HKL2000 [38], or XDS [39]. After that, scaling and merging is performed in programs like CCP4 suite [40] program SCALA [41] or HKL2000: reflections, belonging to different images, are put on the same scale and reflections, being partially recorded on several images, are merged together. Finally, intensities of reflections are converted (*truncated*) to amplitudes, which can be done, for example, by CCP4 program TRUNCATE [42].

4.8   MOLECULAR REPLACEMENT TECHNIQUE IN DETERMINATION OF VIRUS STRUCTURES

As discussed previously, in MR, experimental amplitudes of unknown structure are combined with phases of related, known structure. However, to be able to perform the combination, both structures have to be in identical unit cells and in identical orientations and positions. Since this is rarely the case in reality, the coordinates of known structure from PDB have to be placed in a virtual crystal unit cell of the same size and symmetry in silico, after which amplitudes and phases can be calculated by inverse Fourier transform. In order to place the known structure in virtual unit cell in the same orientation and position as unknown, special methods have been developed, which use so-called rotation and translation functions. The use of rotation and translation functions depends on Patterson function—essentially an interatomic vector map, which in the case of proteins and viruses is far too noisy to be interpreted by itself, but it can be calculated directly from the observed data by Fourier transform by squaring the structure factor amplitudes and setting all phases to zero. The exact theory behind those techniques is beyond the scope of this book, but roughly—the Patterson functions of known structure in virtual unit cell are calculated in many orientations and positions until the best correlation with Patterson function from observed amplitudes is reached. For determination of particle orientation of viruses, icosahedral symmetry is of great help. Like the structure itself, also the Patterson function does follow the pattern of icosahedral symmetry. By using the so-called self-rotation function, the Patterson map is correlated with itself in various orientations—in such a way it is possible to identify orientations of fivefold, threefold, and twofold symmetry axes of the virus particle in the unit cell. The rotation search is usually done in polar angles by varying phi and psi angles and keeping the third, kappa angle at a constant value, corresponding to the particular symmetry axis—for example, $72°$ for fivefold axis

(since $360°/5 = 72°$). The self-rotation search can be conveniently done in program GLRF [43]. The resulting rotation function map then clearly shows orientations of all symmetry axes of the virion and consequently the orientation of the whole particle (Figure 4.8). After determination of particle orientations, their positions in unit cell have to be determined as well. In the case of viruses, this can frequently be done by simple considerations of particle packaging, taking into account unit cell dimensions and diameter of virion. However, sometimes other techniques, involving translation function, have to be applied, for which programs like Phaser [44] can be used as exemplified by structure determination of phages PRR1 [45] and φCb5 [29]. Sometimes there are two viral particles in crystallographic asymmetric unit in similar orientations. In this case, the translational vector between them can be extracted from Patterson function, calculated at low resolution as done in the case of bacteriophage PP7 [27].

Once the orientation and position of particles has been established, the initial phases can be calculated by combining the phases of the known structure and the observed amplitudes, which in the case of viruses can be conveniently done, for example, in program CNS [46]. In contrast to ordinary proteins, where typically a known structure of at least 25% sequence identity must be used, in the case of viruses, much less similar initial models can be utilized. Initial model can be structure of related virus with low or nondetectable sequence similarity as in the case of bacteriophage φCb5 [29], low-resolution cryo-EM map as in the case of hepatitis E T = 1 virus-like particles [47], or in some cases—even an artificial hollow sphere of uniform density as first demonstrated for canine parvovirus [48] and later also for several other viruses [49].

4.9   DENSITY MODIFICATION

The electron density map, calculated from the initial phases, is usually of very poor quality and uninterpretable before density modification—averaging and solvent flattening.

Since spherical viruses have icosahedral symmetry, the structure is more or less identical in 60 regions in virus particle. Therefore, if there is one particle in crystallographic asymmetric unit, from full dataset, the information about virus structure is essentially present 60 times. The electron density map can be substantially improved by averaging—that is, calculating an average map of all icosahedrally equivalent positions, known to be identical. For viruses, averaged map of good quality can be obtained from initial map, calculated from rather incomplete data. This is quite important, since due to long exposure times and radiation damage, full datasets from virus crystals are difficult to obtain.

Averaging is done in cycles—as shown in Figure 4.9. In case if initial model is only distantly related or has low resolution, phase extension protocol can be applied. In this case, first averaging cycles are performed at low resolution

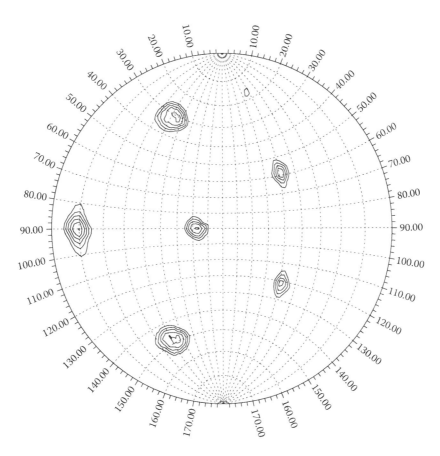

FIGURE 4.8 The self-rotation function. By performing self-rotation function of experimental amplitudes in polar angles phi, psi, and kappa, it is possible to see orientations of icosahedral symmetry axes and hence the particle itself. In the given example, rotation search is performed by varying phi and psi angles from 0° to 180° and keeping the rotation angle kappa constant at 72° (corresponding to fivefold icosahedral axis, since 72° × 5 = 360°). Icosahedron contains six fivefold symmetry axes and all of them show up as clear peaks in the rotation function map. Similar maps can be obtained also for threefold and twofold icosahedral axes by keeping kappa at 120° and 180°, respectively. Data from bacteriophage phiCb5 were used for the calculation of the map.

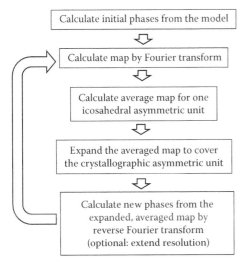

FIGURE 4.9 General scheme for density modification using cyclic averaging.

and slightly higher-resolution data are being added to each next averaging cycle. It is of critical importance that initial phases at low resolution are correct—otherwise, the phase extension protocol fails. If a low-resolution model is used to obtain initial phases, it is critical that low-resolution data are also collected. For structure determination of proteins, low-resolution cutoff for data collection is frequently about 15–20 Å, but this might be not enough for structure determination of viruses. However, collection of low-resolution data below 30–40 Å may be problematic in beamlines with standard setup, since lower-resolution reflections are in the very center of the detector, often behind the beamstop (a small metal piece in between the crystal and the detector, which stops the very strong primary beam, which otherwise may damage the detector). The problem may be solved by simply moving the beamstop closer to the detector, but many automatic goniometers at synchrotrons have nonadjustable beamstops.

Apart from averaging, another useful (but less powerful) tool for electron density modification is solvent flattening—by using this approach, electron density for crystal regions, containing solvent, is set to a low constant value and the same in principle can be done for the interior of viral capsid. The procedure leads to noise reduction in disordered solvent and capsid interior regions, consequently improving the phases.

Density modification can be done by a variety of software, including CCP4 program dm [50] or RAVE [51].

4.10   MODEL BUILDING

Once the obtained averaged map is of sufficiently high quality, atomic model has to be built in the map. If similar structure is available, it may be used as an initial model. If similar model is not available, automatic model building can be tried in programs like BUCCANEER [52]. However, automatic model building is unlikely to work if resolution is lower than about 3.2 Å. If resolution is worse than 3.5 Å, electron densities for many side chains look very similar and unambiguous model building may be difficult, especially for long stretches of small and/or flexible side chains. Tryptophan residues serve as good markers, due to their prominent side chain, which is relatively easily seen even at 4 Å resolution. Other aromatic residues—tyrosine and phenylalanine—also serve as good markers. If this is not enough, data can be collected from crystals with incorporated selenomethionines. When electron density maps from both native and selenomethionine derivative datasets are available, the difference map of both can be calculated. Then, positions of selenium atoms will show up, since selenium contains many more electrons compared to sulfur. This approach was used in model building of phage PRD1 [53].

Viruses with T > 1 have several structurally very similar molecules in the icosahedral asymmetric unit; so after the building of the first molecule, it can be placed in the electron density of other molecules, followed by manual and/or automatic adjustments.

In the past, program O [54] was used to be the gold standard for model building. However, currently COOT [55] is being used more frequently.

4.11   REFINEMENT AND VALIDATION OF VIRUS STRUCTURES

In macromolecular crystallography, refinement is a process of automatic adjustment of atomic coordinates of the model, so it better fits the experimental data. Various restraints are being applied—there are penalties for, for example, unusual bond lengths, bond angles, and close noncovalent contacts. A more precise model then can be used to calculate improved electron density map, which can again be used for improving the model and so on. As a result, both electron density map and the model get improved substantially. In virus crystallography, however, the electron density map is most often as good as possible already after averaging, so the sole purpose of refinement is to improve the model. In most cases, it is assumed that all icosahedral asymmetric units have identical structures, so noncrystallographic symmetry operators, which are subject of refinement for normal proteins, are constrained. However, not all refinement programs support constrained refinement—most frequently, it is done in CNS [46], but now, it is possible also in REFMAC [56]. In some cases, when complete, good-quality, and high-resolution data are available, strict noncrystallographic constraints can be relaxed to take into account deviations from icosahedral symmetry, which do occur due to the crystal contacts [57].

Refinement process is monitored by a number of indicators, the best known are R factor and free R factor (Rfree). R factor reflects fitness of experimentally observed amplitudes of reflections F_{obs} and theoretically calculated amplitudes F_{calc} from the built atomic model ($R = (\Sigma ||F_{obs}| - |F_{calc}||)/\Sigma |F_{obs}|$). R factor of about 0.6 corresponds to a completely random structure, and most x-ray structures in PDB have R < 0.3. Free R factor is calculated for a subset of data (typically 5%), which is not used in refinement. For normal proteins, Rfree is considered a much more reliable indicator when compared to R, since surprisingly low R values can be obtained for partially or even completely wrongly built structures if parameters are overfitted during refinement. As a rule of thumb, for well-refined structures, R factor is typically about 10 times percent points of resolution—so a typical refined protein structure at 2.4 Å resolution will have R factor of 0.24. Rfree is typically by about 2%–5% points higher than R. For virus structures, however, R and Rfree have nearly identical values; therefore, calculation of Rfree can be considered useless. This can be explained by high noncrystallographic symmetry—even if some reflections are omitted and not used in refinement, other reflections at the same resolution still contain the same information.

After the refinement has produced a reasonably good and complete model, it is time for validation. Validation is a process of assessing the model quality, based on various criteria. For example, there should not be more than 5% outliers in Ramachandran plot [58], no clashes between noncovalently bound atoms, no big hydrophobic residues on the surface of protein, etc. After validation, some adjustments to model are normally required, followed by another refinement run, another validation run and so on, until satisfactory model quality is obtained. It is often said that refinement and validation can never be truly completed, since some small errors and issues always remain. After structure deposition in PDB, there are several records in the file header added automatically, identifying potential problems—such as outliers in Ramachandran plot and unusual bond lengths and angles.

4.12   ADVANTAGES AND DISADVANTAGES OF PROTEIN X-RAY CRYSTALLOGRAPHY AND OTHER STRUCTURE DETERMINATION METHODS

Of the nearly 100,000 protein structures, deposited in PDB, about 82,000 are solved by protein x-ray crystallography, 9,000 by nuclear magnetic resonance (NMR), and 500 by cryo-EM. Of more than 400 virus structures available in PDB, about 300 have been solved by crystallography, rest—by cryo-EM.

Solution-state NMR is restricted to analysis of proteins, smaller than 70 kDa, far less than even the smallest viruses. However, certain solid-state NMR techniques have a capability to solve also virus structures in the near future, for example, significant progress has been made toward structure determination of bacteriophage AP205 [59,60].

Crystallography does not have any inherent size restrictions, but the most obvious drawback is that the macromolecule has to be able to form well-diffracting crystals, which is not always the case. Also, crystallography is a pretty static method and does not provide much information about dynamics of protein. Finally, crystals are usually grown in quite nonphysiological conditions, typically using high salt or PEG concentrations, and the conditions cannot be changed easily—for example, to see what structural differences would be observed at different pH values.

Icosahedral symmetry of viruses can bring certain problems in structure determination of virus components by crystallography. In addition to major capsid protein(s), many viruses contain also minor structural proteins, present in one or few copies per virion. For example, capsid of RNA phage MS2 contains 180 copies of coat protein and 1 copy of maturation protein, responsible for the attachment of particles to F pili receptor. Since crystal contacts are maintained only by coat protein, after crystallization, position of A protein in individual phage particles is different. Therefore, A protein does not follow crystal symmetry and no electron density for A protein could be observed [61]. The same is true for any virus component that does not follow icosahedral symmetry. For example, adenovirus contains trimeric spike protein attached to penton bases, located at fivefold symmetry axes. The consequence is that in x-ray structure of adenovirus, the electron density of trimeric spike protein is fivefold averaged, resulting in a largely uninterpretable electron density map [32]. The same in principle applies to viral genome, which is present in one unique copy in virion and therefore cannot follow exact icosahedral symmetry. However, parts of genome can follow approximate icosahedral symmetry and density for stretches of nucleotides can frequently be observed in crystal structures. For example, the recent high-resolution structure of satellite tobacco necrosis virus revealed electron density corresponding to about 57% of viral genome [57]. However, due to random orientation of genome in virus particles in crystal, the exact genome sequence cannot be fitted in the electron density maps, and the density for all bases normally represents the average of all four nucleotides. Despite all mentioned problems, crystallography remains to be the most widely used method to obtain high-resolution structures of proteins, including viruses.

With the advance of free-electron lasers (XFELs), it is now possible to solve 3D structures of proteins from very tiny crystals, barely a few hundred molecules across [62–64]. XFELs produce extremely short and bright x-ray pulses, which destroy any biological sample almost instantly, but the sample still produces useful diffraction pattern just before destruction. In future, using still brighter x-ray sources, it might be possible to obtain useful x-ray scattering images even from single molecules or from single virus particles. The proof-of-principle was demonstrated already in 2011, when 32 nm resolution image was obtained from a single mimivirus particle using XFEL in Linac Coherent Light Source [65].

Cryo-EM has the obvious advantage that no crystals are needed. Also, samples can be studied in physiological or, indeed, almost any arbitrary conditions. Additionally, by cryo-EM, it is in principle possible to study virus components, which do not follow icosahedral symmetry, although this normally reduces the quality of resulting electron density map substantially. For example, position and overall shape of phage MS2 A protein, present in one copy per virion, was determined by cryo-EM at 40 Å resolution [66].

However, the resolution of cryo-EM is rarely comparable to that achieved by crystallography, although lately there are several examples of near-atomic resolution. Also, cryo-EM tends to be more time consuming and less *user friendly* than crystallography.

A powerful tool for structure determination of viruses is combination of cryo-EM and x-ray techniques. Often, whole viruses fail to crystallize, but individual subunits can be crystallized separately and their structure solved. If low-resolution cryo-EM map from the same or related virus is available, structures of subunits can be fitted in the map. In this manner, a pseudoatomic structure can be obtained for the whole particle. For example, recently crystal structure of E1–E2 glycoproteins and neutralizing antibodies of chikungunya virus was fitted in 5.3 Å cryo-EM map [67]. Although the exact interactions between side chains of subunits cannot be precisely identified in this manner, the method can still give a fair idea of structural features of virion and its interactions with receptors or antibodies.

REFERENCES

1. Kepler, J., *Strena seu de Nive Sexangula*. 1611, Frankfurt, Germany: G. Tampach.
2. Haüy, R.J., *Essai d'une théorie sur la structure des cristaux*. 1784, Paris, France: Gogué et Née de La Rochelle.
3. Röntgen, W.C., *Über eine neue Art von Strahlen*. 1896, Würzburg, Germany: Stahel.
4. Friedrich, W., P. Knipping, and M. Laue, Interferenzerscheinungen bei Röntgenstrahlen. *Annalen der Physik*, 1913. 346: 971–988.
5. Bragg, W.L., The structure of some crystals as indicated by their diffraction of X-rays. *Proc. R. Soc. Lond.*, 1913. A89: 248–277.
6. Dickinson, J.G. and A.L. Raymond, The crystal structure of hexamethylene-tetramine. *J. Am. Chem. Soc.*, 1923. 45: 22–29.
7. Müller, A., The X-ray investigation of fatty acids. *J. Chem. Soc., Trans.*, 1923. 123: 2043–2047.
8. Robertson, J.M., An X-ray study of the phthalocyanines. Part II. Quantitative structure determination of the metal-free compound. *J. Chem. Soc.*, 1936. 1936: 1195–1209.

9. Crowfoot, D. et al., X-ray crystallographic investigation of the structure of penicillin, *in* H.T. Clarke, J.R. Johnson, and R. Robinson, eds., *Chemistry of Penicillin*. 1949, Princeton, NJ: Princeton University Press. pp. 310–367.

10. Brink, C. et al., X-ray crystallographic evidence on the structure of vitamin B12. *Nature*, 1954. 174(4443): 1169–1171.

11. Kendrew, J.C. et al., A three-dimensional model of the myoglobin molecule obtained by x-ray analysis. *Nature*, 1958. 181(4610): 662–666.

12. Bernal, J.D. and I. Fankuchen, X-ray and crystallographic studies of plant virus preparations. *J. Gen. Physiol.*, 1941. 25: 111–165.

13. Franklin, R.E. and K.C. Holmes, Tobacco mosaic virus: Application of the method of isomorphous replacement to the determination of the helical parameters and radial density distribution. *Acta Cryst.*, 1958. 11: 213–220.

14. Franklin, R.E., A. Klug, and K.C. Holmes, X-ray diffraction studies of the structure and morphology of tobacco mosaic virus, *in* E.W. Wolstenholme, E.C.P. Millar, and C. Foundation, eds., *Ciba Foundation Symposium on the Nature of Viruses*. 1957, London, U.K.: J. & A. Churchill. pp. 39–55.

15. Holmes, K.C. et al., Structure of tobacco mosaic virus at 6.7Å resolution. *Nature*, 1975. 254(5497): 192–196.

16. Stubbs, G., S. Warren, and K. Holmes, Structure of RNA and RNA binding site in tobacco mosaic virus from 4 map calculated from X-ray fibre diagrams. *Nature*, 1977. 267(5608): 216–221.

17. Bloomer, A.C. et al., Protein disk of tobacco mosaic virus at 2.8 Å resolution showing the interactions within and between subunits. *Nature*, 1978. 276: 362–368.

18. Bawden, F.C. and N.W. Pirie, Crystalline preparations of tomato bushy stunt virus. *Brit. J. Exp. Pathol.*, 1938. 29: 251–263.

19. Bernal, J.D., I. Fankuchen, and D.P. Riley, Structure of the crystals of tomato bushy stunt virus preparations. *Nature*, 1938. 142: 1075.

20. Carlisle, C.H. and K. Dornberger, Some X-ray measurements on single crystals of tomato bushy-stunt virus. *Acta Cryst.*, 1948. 1: 196–200.

21. Harrison, S.C., Structure of tomato bushy stunt virus: Three-dimensional x-ray diffraction analysis at 30 Angstrom resolution. *Cold Spring Harb. Symp. Quant. Biol.*, 1972. 36: 495–501.

22. Harrison, S.C. and A. Jack, Structure of tomato bushy stunt virus. Three-dimensional x-ray diffraction analysis at 16 A resolution. *J. Mol. Biol.*, 1975. 97(2): 173–191.

23. Winkler, F.K. et al., Tomato bushy stunt virus at 5.5-Å resolution. *Nature*, 1977. 265(5594): 509–513.

24. Harrison, S.C. et al., Tomato bushy stunt virus at 2.9 Å resolution. *Nature*, 1978. 276: 368–373.

25. Abad-Zapatero, C. et al., Structure of southern bean mosaic virus at 2.8 Å resolution. *Nature*, 1980. 286: 33–39.

26. Liljas, L. et al., Structure of satellite tobacco necrosis virus at 3.0 Å resolution. *J. Mol. Biol.*, 1982. 159: 93–108.

27. Tars, K. et al., Structure determination of bacteriophage PP7 from *Pseudomonas aeruginosa*: From poor data to a good map. *Acta Crystallogr. D Biol. Crystallogr.*, 2000. 56(Pt 4): 398–405.

28. Valegård, K. et al., Purification, crystallization and preliminary X-ray data of the bacteriophage MS2. *J. Mol. Biol.*, 1986. 190: 587–591.

29. Plevka, P. et al., The structure of bacteriophage phiCb5 reveals a role of the RNA genome and metal ions in particle stability and assembly. *J. Mol. Biol.*, 2009. 391(3): 635–647.

30. Tars, K., A. Zeltins, and L. Liljas, The three-dimensional structure of cocksfoot mottle virus at 2.7 Å resolution. *Virology*, 2003. 310: 287–297.

31. Plevka, P. et al., The three-dimensional structure of ryegrass mottle virus at 2.9 Å resolution. *Virology*, 2007. 369(2): 364–374.

32. Reddy, V.S. et al., Crystal structure of human adenovirus at 3.5 Å resolution. *Science*, 2010. 329(5995): 1071–1075.

33. Uson, I. and G.M. Sheldrick, Advances in direct methods for protein crystallography. *Curr. Opin. Struct. Biol.*, 1999. 9(5): 643–648.

34. Grimes, J.M. et al., The atomic structure of the bluetongue virus core. *Nature*, 1998. 395: 470–478.

35. Wikoff, W.R. et al., Crystallization and preliminary X-ray analysis of the dsDNA bacteriophage HK97 mature empty capsid. *Virology*, 1998. 243: 113–118.

36. Cockburn, J.J. et al., Crystallization of the membrane-containing bacteriophage PRD1 in quartz capillaries by vapour diffusion. *Acta Crystallogr. D Biol. Crystallogr.*, 2003. 59(Pt 3): 538–540.

37. Leslie, A.G.W., Recent changes to the MOSFLM package for processing film and image plate data. *Joint CCP4 + ESF-EAMCB Newsletter on Protein Crystallography*, 1992. 26.

38. Otwinowski, Z. and W. Minor, Processing of X-ray diffraction data collected in oscillation mode, *in* C.W. Carter, Jr. and R.M. Sweet, eds., *Methods in Enzymology*. 1997, New York: Academic Press. pp. 307–326.

39. Kabsch, W., XDS. *Acta Crystallogr. D Biol. Crystallogr.*, 2010. 66(Pt 2): 125–132.

40. Winn, M.D. et al., Overview of the CCP4 suite and current developments. *Acta Crystallogr. D Biol. Crystallogr.*, 2011. 67(Pt 4): 235–242.

41. Evans, P.R., Scala. *Joint CCP4 + ESF-EAMCB Newsletter on Protein Crystallography*, 1997. 33: 22–24.

42. French, S. and K. Wilson, On the treatment of negative intensity observations. *Acta Crystallogr. A*, 1978. 34: 517–525.

43. Tong, L. and M.G. Rossmann, Rotation function calculations with GLRF program. *Methods Enzymol.*, 1997. 276: 594–611.

44. McCoy, A.J. et al., Phaser crystallographic software. *J. Appl. Crystallogr.*, 2007. 40(Pt 4): 658–674.

45. Persson, M., K. Tars, and L. Liljas, The capsid of the small RNA phage PRR1 is stabilized by metal ions. *J. Mol. Biol.*, 2008. 383(4): 914–922.

46. Brünger, A.T. et al., Crystallography and NMR system: A new software suite for macromolecular structure determination. *Acta Cryst. D*, 1998. 54: 905–921.

47. Guu, T.S. et al., Structure of the hepatitis E virus-like particle suggests mechanisms for virus assembly and receptor binding. *Proc. Natl. Acad. Sci. USA*, 2009. 106(31): 12992–12997.

48. Tsao, J. et al., Structure determination of monoclinic canine parvovirus. *Acta Crystallogr. B*, 1992. 48(Pt 1): 75–88.

49. Taka, J. et al., Ab initio crystal structure determination of spherical viruses that exhibit a centrosymmetric location in the unit cell. *Acta Crystallogr. D Biol. Crystallogr.*, 2005. 61(Pt 8): 1099–1106.

50. Cowtan, K.D. and P. Main, Phase combination and cross validation in iterated density-modification calculations. *Acta Crystallogr. D Biol. Crystallogr.*, 1996. 52(Pt 1): 43–48.

51. Kleywegt, G.J. and T.A. Jones, Halloween… Masks and bones, *in* S. Bailey, R. Hubbard, and D. Waller, eds., *From First Map to Final Model. Proceedings of the CCP4 Study Weekend*. 1994, Daresbury, U.K.: SERC Daresbury Laboratory. pp. 59–66.

52. Cowtan, K., The Buccaneer software for automated model building. 1. Tracing protein chains. *Acta Crystallogr. D Biol. Crystallogr.*, 2006. 62(Pt 9): 1002–1011.

53. Abrescia, N.G. et al., Insights into assembly from structural analysis of bacteriophage PRD1. *Nature*, 2004. 432(7013): 68–74.

54. Jones, T.A., M. Bergdoll, and M. Kjeldgaard, O: A macromolecule modeling environment, *in* C. Bugg and S. Ealick, eds., *Crystallographic and Modeling Methods in Molecular Design*. 1990, New York: Springer-Verlag. pp. 189–199.

55. Emsley, P. and K. Cowtan, Coot: Model-building tools for molecular graphics. *Acta Crystallogr. D Biol. Crystallogr.*, 2004. 60(Pt 12 Pt 1): 2126–2132.

56. Murshudov, G.N., A.A. Vagin, and E.J. Dodson, Refinement of macromolecular structures by the maximum-likelihood method. *Acta Crystallogr. D Biol. Crystallogr.*, 1997. 53(Pt 3): 240–255.

57. Larson, S.B., J.S. Day, and A. McPherson, Satellite tobacco mosaic virus refined to 1.4 Å resolution. *Acta Crystallogr. D Biol. Crystallogr.*, 2014. 70(Pt 9): 2316–2330.

58. Kleywegt, G.J. and T.A. Jones, Phi/psi-chology: Ramachandran revisited. *Structure*, 1996. 4: 1395–1400.

59. Barbet-Massin, E. et al., Out-and-back 13C-13C scalar transfers in protein resonance assignment by proton-detected solid-state NMR under ultra-fast MAS. *J. Biomol. NMR*, 2013. 56(4): 379–386.

60. Barbet-Massin, E. et al., Rapid proton-detected NMR assignment for proteins with fast magic angle spinning. *J. Am. Chem. Soc.*, 2014. 136(35): 12489–12497.

61. Valegård, K. et al., The three-dimensional structure of the bacterial virus MS2. *Nature*, 1990. 345: 36–41.

62. Chapman, H.N. et al., Femtosecond X-ray protein nanocrystallography. *Nature*, 2011. 470(7332): 73–77.

63. Barends, T.R. et al., De novo protein crystal structure determination from X-ray free-electron laser data. *Nature*, 2014. 505(7482): 244–247.

64. Redecke, L. et al., Natively inhibited *Trypanosoma brucei* cathepsin B structure determined by using an X-ray laser. *Science*, 2013. 339(6116): 227–230.

65. Seibert, M.M. et al., Single mimivirus particles intercepted and imaged with an X-ray laser. *Nature*, 2011. 470(7332): 78–81.

66. Dent, K.C. et al., The asymmetric structure of an icosahedral virus bound to its receptor suggests a mechanism for genome release. *Structure*, 2013. 21(7): 1225–1234.

67. Sun, S. et al., Structural analyses at pseudo atomic resolution of Chikungunya virus and antibodies show mechanisms of neutralization. *Elife (Cambridge)*, 2013. 2: e00435.

5 Computational Methods for Engineering Protein 3D Nano-Objects

James Lara and Xiaojun Xu

CONTENTS

5.1 INTRODUCTION

Nanotechnology research and development is outlined as a major priority in the National Strategic plans of more than 20 countries around the world [1]. Presently, the United States leads in conducting clinical trials of nanomaterials [2]. Nanostructures are typically between 0.1 and 100 nanometers (nm) in size. This is the scale at which basic functions of biological systems (e.g., protein molecules) operate. However, nanosized materials display unusual physical and chemical properties, that is, quantum properties. Such changes in properties of materials are due to an increase in surface area compared to volume as particles get smaller. Engineering at the nanoscale level is no simple feat, and scientists need to come up with completely different solutions to build from the *bottom-up* rather than using traditional *top-down* manufacturing techniques. This chapter presents a brief overview of

the computational methods used for engineering a group of nanomaterials known as 3D nano-objects,* that is, materials where all three dimensions are at nanometric scale; 3D nano-objects include nanomaterials, such as fullerene, quantum dots, dendrimers, nanoparticles (NPs), and nanopowders, which are among the most important nanomaterials in this group used in the field of medicine, nanomedicine. Concepts and principles of computations as well as methods and problems of computing simulations and modeling 3D nano-objects are discussed in Section 5.2. In Section 5.3, discussion of the computational methods used for simulating such objects is focused on the case problem of protein design.

---

* According to the definition by the International Standard Organization (ISO/TS 80004), a nano-object is an object that at least one of its dimensions is at nanometric scale. Nano-objects are divided into three groups according to the definition (i.e., as 1D, 2D, or 3D nano-object).

In Section 5.4, a brief overview of the role that artificial intelligence (AI) paradigm may have on improving computational simulations and assist in the designing of proteins and materials with improved diagnostic or therapeutic activities is given. Finally, in Section 5.5, conclusions are presented.

5.2 COMPUTATIONAL CHEMISTRY AND NANOTECHNOLOGY

Nanotechnology deals with materials and their application in areas such as material engineering and the manufacture of molecular devices, machines, sensors, etc. Atoms and molecules are considered to be the basic building blocks of fabricating future generations of nanomaterials and molecular devices. For example, molecular strands of DNA* are being proposed as the self-assembling templates for biosensors and detectors [3]. Similarly, synthetic inorganic materials are being proposed as suitable replacement of tissue transplants [4]. Such developments are possible only through the cross correlation and fertilization among many disciplines, among which computational chemistry has been an essential enabling technology for molecular nanotechnology development. Computational chemistry may be defined as the computer simulation of any natural chemistry phenomenon using mathematical and theoretical principles for solving chemical problems. Molecular modeling, a subset of computational chemistry, focuses on predicting the behavior of individual entities (e.g., electrons, atoms, or molecules) within the chemical system. Many types of predictions of the properties of molecules and chemical reactions are possible [5], including

- Molecular energy and structure (thermodynamic stability)
- Vibrational frequencies (infrared and Raman spectra)
- Energies and structure of transition states
- Thermochemical properties
- Magnetic shielding effects (nuclear magnetic resonance [NMR] spectra)
- Pathway reactions, kinetics and mechanisms
- Polarizabilities and hyperpolarizabilities
- Charge distributions and net charge of molecules

The ability to computationally calculate these types of properties for systems in the solid phase, gas phase, or in solution has many applications, including studies focused on the discovery of new chemical objects, mechanisms of chemical reaction exploration, among other endeavors, which are critical for accelerating nanotechnology research and development.

The computational methods used for modeling matter can be divided into four categories, depending on the scale

as well as timescales to which they apply [6]: (1) electronic level, in which matter is regarded as made up of fundamental particles (electrons, protons, etc.) and is described by laws of quantum mechanics (QM); (2) atomistic level, in which matter is made up of atoms and is described by laws of statistical mechanics; (3) mesoscale level, in which matter is regarded as made up of an agglomerate of entities, each comprising a number of atoms, and (4) the continuum level, in which matter is regarded as an object of continuous mass and spatial extent rather than a system of discrete particles and is described by well-known laws of continuum mechanics (e.g., equations of momentum, mass and energy conservation, and constitutive equations, like Ohm's law for electric flow). A wide range of mathematical equations and theoretical principles have been proposed for simulating and modeling vast types of systems, from mineral crystal structures to enzyme structures. Depending on what the focus of a study is (e.g., electronic structure, electron–electron interactions, molecular structure, and atom–atom interactions), different levels of theory handle equation terms or functions in different ways and to different degrees of sophistication. A synopsis of the different approaches and methods is represented in Table 5.1.

Nanotechnology applications for medicine and biotechnology rely heavily on computational chemistry methods to solve problems mainly aimed at finding the most efficient ways for manufacturing nanomaterials, characterizing properties of nanostructures, and harvesting those properties that maximize their effectiveness for the intended application. Some of the computational methods used for modeling the structure of nanomaterials at different levels, specifically, the electronic, atomistic, and mesoscale levels, are briefly overviewed in the next subsections. In-depth discussion of methods used for determining molecular electron structures or other properties of chemical systems can be found in selected Refs. [5–12].

5.2.1 ELECTRONIC SCALE

The *ab initio* or *first principles* methods used for electronic structure calculations are based on QM and offer the most rigorous and accurate molecular modeling methods, because they provide the best mathematical approximation to the actual system. Mathematical calculations implemented in *ab initio* methods are solely based on quantum laws [7], the masses and charges of fundamental particles (e.g., electrons), and the values of fundamental physical constants, such as the speed of light ($c = 2.998 \times 10^8$ m/s) and Planck's constant ($h = 6.626 \times 10^{-34}$ J s). Molecular systems comprising dozens of electrons (approximately 40 electrons) can be very accurately simulated by *ab initio* methods, wherein simulations are aimed at solving the complex *many-body* Schrödinger equation of the atomic structure of a chemical system, using numerical algorithms [13]. For *many-body* systems (e.g., molecules), *ab initio* configuration-interaction methods, Møller–Plesset perturbation theory, and density functional theory (DFT) are used for many applications, including calculating the

---

* Deoxyribonucleic acid (DNA) is a biological unit (molecule) that encodes genetic information in living organisms and is composed of nucleic acids.

TABLE 5.1

Synopsis of Computational Molecular Modeling Techniques

Approaches	Methods	Pros	Cons	Applications
Ab initio	Based on QM and fundamental principles of physics	No prior data are required; very rigorous; useful for a wide range of chemical systems	Requires large computational computer resources and time; intractable for large systems	Calculation of electronic transition and exited states of small-sized systems (up to tens of atoms); systems requiring high degree of accuracy
Semiempirical	Based on QM and empirical parameters	Computationally less demanding than *ab initio* methods; useful for a wide range of chemical systems (e.g., energetic and biological materials)	Requires experimental or QM-derived data; less rigorous than *ab initio* methods	Calculation of electronic transition and exited states of medium-sized systems (up to a few hundreds of atoms); analysis and simulation of highly dynamic systems (e.g., fluids)
Molecular mechanics	Based on classical physical principles (purely empirical); uses force fields with empirically derived parameters	Computationally cheap; can be used for calculations that involve very large systems	Requires experimental or QM-derived data; less accurate than *ab initio* and semiempirical methods; applicable to a limited range of systems (mainly biologic systems, e.g., proteins)	Calculations of potential energy surfaces and spatial translations in large-sized systems (up to millions of atoms); chemical systems where breaking of chemical bonds is not involved

ground-state properties of many materials or chemical reactions. DFT is one of the most popular *ab initio* methods, particularly because it provides an effective compromise between computational tractability and accuracy, which is often within 0.1–1.0 kcal/mol. DFT models electron correlation as a functional (i.e., a function of a function) of the electron density, ρ, to describe the system of interacting electrons. The functional in current DFT methods partitions electronic energy (E) into a summation of four terms via Kohn–Sham equations [14,15]:

$$E = E^T + E^V + E^J + E^{XC},$$

where

E^T is the kinetic energy term (originate from motion of the electrons)

E^V is the potential energy term (includes nuclear–electron and electron–electron interactions)

E^J is the electron–electron repulsion term

E^{XC} is the electron correlation term

While the terms $E^T + E^V + E^J$ represent the classical energy of the electron distribution, the E^{XC} term represents the dynamic correlation energy due to the concerted motion of individual electrons and the quantum mechanical exchange energy, which accounts for electron spin [9].

The consistency and accuracy of the calculations provided by *ab initio* electronic structure methods are necessary when, for example, significant rearrangements of electrons in photonic devises occur or where chemical bonds are being broken or formed in chemical reactions. However, *ab initio* methods are extremely computationally expensive; they require extensive computer computational resources, and the time complexity of calculations is very large. Thus, application of *ab initio* methods has been mostly used to model the electronic structures of chemical systems comprising a limited number of atoms and restricted to short timescales. Computational methods, namely, semiempirical and molecular mechanics (MM) methods, have been developed to accelerate the calculations. The semiempirical methods are developed based on QM by replacing some explicit calculations with approximations (based upon experimental observations) or sometimes by omitting some parts of explicit calculations. MM methods, on the other hand, are entirely based on classical physics and therefore computationally fast. MM methods use force fields* for calculating the potential energy of chemical systems. However, since the explicit treatment of electronic structure is completely neglected in MM methods, such methods are limited in scope or sometimes unrealistic due to inaccuracies of the calculations. Nonetheless, MM methods currently provide the only feasible means to model very large and nonsymmetrical chemical systems (e.g., polymers, proteins, and DNA/RNA molecules).

5.2.2 Atomistic Level

Currently, empirical models and *ab initio* quantum mechanical methods are the two approaches used for modeling interactions between atoms. Both approaches can be used via the various computational methods available for simulating and modeling matter (e.g., lattice energy relaxation, lattice

---

* Force fields are characterized by a set of potential energy functions to describe chemical properties or forces, and are tuned against experimental data or quantum-level calculated data.

dynamics, molecular dynamics [MD], and Monte Carlo [MC] methods). The empirical model approach critically depends on mathematical equations reasonably describing dependency of interatomic interactions between atoms (e.g., space separation and local atomic coordination), and availability of some training data against which to *tune* parameters specified in the equation. One empirical model that has worked well for modeling ionic ceramic-based or mineral-based nanomaterials [8,11] is the Born–Mayer's [16] long-range Coulomb interactions with short-range repulsive interactions of the form

$$E(r) = A \exp\left(\frac{-r}{\rho}\right),$$

where

r is the interatomic spacing

A and ρ are parameters whose values are to be tuned against observed data

Other functions can be incorporated into this model, provided that there might be some physical or chemical justification. For example, the Born–Mayer–Buckingham form, which includes a term of the form $-Cr^{-6}$ (justified as representing known dispersive interactions), is well suited for calculating the partial density of states (PDOS) describing the properties of ionic crystals of ZnO NPs [17]. Over the years, empirical models have been expanded to include terms, such as bond-bending terms, distance-dependent terms, or multiatom terms, and are included in standard modeling software packages, like GULP [18] and DL_POLY [19].

Atomistic level methods such as MD and MC are popular methods that enable modeling systems of thousands or millions of atoms studied over time intervals of tens to hundreds of nanoseconds, depending on system's size and complexity [20,21]. In using MC or MD methods, details of the electronic structure of the system are lost; however, for many physical processes where electron perturbations are small, this is not important. The MC method is a stochastic technique to sample the infinite number of possible structure configurations of a system (e.g., molecule, molecule ensembles, and polymers) [7]. In the classic MC method implementation (known as Metropolis algorithm [22]), configurations are generated from a previous state using a transition probability, which depends on the energy difference between the initial and final states. This time-dependent behavior is described by the master formula

$$\frac{\partial P_n(t)}{\partial} = -\sum_{n \neq m} [P_n(t)W_{n \to m} - P_m(t)W_{m \to n}],$$

where

$P_n(t)$ is the probability of the system being in the state n at time t

$W_{n \to m}$ is the transition rate for $n \to m$

In equilibrium, $\partial P_n(t) = 0$ and both terms in the right-hand side of the previous equation must be equal. The probability of the nth state occurring in a classical system is given by the formula*

$$P_n(t) = \frac{e^{-E_n/K_BT}}{Z},$$

where

Z is a partition function[†]

the $\exp(-E_n/K_BT)$ represents a probability distribution (based on statistical mechanics and thermodynamics), where E is energy, T is temperature, and K_B is the Boltzmann constant

So if an n state is produced from an m state, the relative probability is the ratio of the individual probabilities, which results in only having to calculate the energy difference (ΔE) between the two states:

$$\Delta E = E_n - E_m.$$

Any change in the configuration or atomic ensemble of the system will give a change in energy. If the energy is negative (leading to a lowering of the systems' energy), the change is accepted. This procedure is repeated for a large number of steps leading to the evolution of the atomic ensemble through the multidimensional phase space. This approach ensures that the sampling procedure adheres to laws of thermodynamics and that most of the phase changes possible in the system are included for analysis.

The central idea of MD simulation methods is to exploit the fact that Newton's equations of motion link forces to acceleration [10,11]. Starting with the configuration of a large ensemble of atoms, empirical or QM methods are used to compute the force on each atom. Once the force is computed, it is converted to acceleration, which through a numerical time step algorithm can be combined with information on the current and previous atomic positions, velocities, and accelerations to predict the position of each atom a time step later. There are several algorithms commonly used to simulate the evolution of the positions and velocities of an ensemble of atoms, each providing different levels of stability and accuracy as required by the specific application. Common to all algorithms used in MD methods is the importance of the time step (Δt) concept. The motion of an atom ensemble (e.g., molecule) can become unstable due to large errors (inaccuracies) in calculations when time steps used are too large. Conversely, when time steps used are too small, computations may become inefficient due to the large computational times involved [23]. The typical time

---

* Detailed explanation and derivation of this formula can be found in Ref. [7], and other selected references outlined in Section 5.2.

† Equilibrium statistical mechanics is based upon the idea of a partition Z function.

step for MD simulations is in the order of 1 femtosecond (1 fs = 10^{-15} s) [24]. In the case of large quantum mechanical simulations, it is common to run time steps of 5 picoseconds (1 ps = 10^{-12} s) and up to several hundreds of picoseconds if empirical methods are used [8].

5.2.3 Mesoscale Level*

In the fully atomistic classical treatment of systems, the system (e.g., molecule) is modeled in atomistic detail and evolves to all possible regions of phase spaces, that is, the space of all possible positions and momenta for the atoms in the system [25]. At any instant, the system will be at some point in the multidimensional phase space and its evolution in time is represented by a trajectory in phase space, which is determined by inter- and intramolecular forces among the atoms and any external fields. Although this treatment is highly desirable, for applications involving very large systems (with millions or billions of atoms), it is not feasible. Similarly, atomistic classical treatment is not feasible for modeling processes that require large timescales (processes typically requiring over 1 ms to occur, e.g., self-assembly of proteins). For modeling such types of systems and processes, various mesoscale methods have been developed that are of particular interest in applications requiring modeling at multiple levels and timescales.

MM methods, also known as *force field* methods, bridge the gap between quantum and continuum mechanics and have been extensively used to study mesoscopic effects in several systems, including modeling energetic materials [9] and biological systems (e.g., protein structures). In MM methods, molecules are described by a ball-and-spring model [26], where atoms are held together through chemical bonds. This strategy allows overcoming the problem of calculating the electronic energy via the Schrödinger equation for a given nuclear configuration. In addition to bypassing the solution of the electronic Schrödinger equation, MM methods also neglect quantum aspects of nuclear motion, which means that the dynamics of the atoms is treated by classical mechanics (i.e., Newton's second law).

The principles and techniques of MM methods are fully described in the following selected Refs. [11,12,27,28], and a review of the general applications of the MM methods is discussed in Ref. [29]. The concept of MM was first introduced by F. H. Westheimer, who in 1956 published the only MM calculation done by hand to determine the transition state of a tetrasubstituted biphenyl molecule [30], and refined a decade later by K.B. Wiberg, who in 1965 developed the first general MM-type program with ability to compute

energy minimum [31]. The basic assumptions of typical MM methods can be summarized as follows:

- Each atom (its electrons and nuclei included) is represented as a particle with a characteristic mass and radius.
- A chemical bond is represented as a *spring*, with a characteristic force constant determined by the potential energy of interaction between the two chemically bonded atoms.
- Potential energy functions describe intramolecular phenomena (stretching, torsion, and bending of bonds) and/or intermolecular phenomena such as electrostatic interactions or van der Waals forces.
- Potential energy functions (force fields) rely on empirically derived parameters from experimental data or from other calculations (e.g., QM).

In MM method applications, mathematical models are used to predict the energy of a molecule as a function of its conformations, which allows predictions of equilibrium geometries and transition states, and relative energies between conformers or between different molecules. Energy (E) due to the stretching of bonds between atoms in a molecule is expressed as a Taylor series[†] about the equilibrium position R_e:

$$E(R) = k_2(R - R_e)^2 + k_3(R - R_e)^3 + \cdots,$$

where

R is a bond length

k_2, k_3, etc., are parameters representing the potential energy of interaction between chemically bonded atoms (force constants) and are derived from experimental data or quantum mechanical calculations

The central idea of MM methods is that these constants are transferable to other molecules. For instance, most carbon–hydrogen (C–H) bond lengths are 1.06–1.10 Å in just about any molecule, with stretching frequencies between 2900 and 3300 cm^{-1}. This strategy relies on using different *atom types*.

Current MM models are characterized by a set of potential energy functions, known as force fields, to describe the chemical forces within a system. These force fields depend upon (1) atomic displacements, that is, bond lengths; (2) the *atom type*, that is, the characteristics of an element within a specific chemical context (hybridization, formal charges on the atom immediate bonded neighbors); and (3) one or various parameter sets relating *atom types* and bond characteristics to empirical data. MM models express the total energy of a system as a sum of Taylor series expansions for *stretches* (E_{str}) for every pair of bonded atoms, and additional potential energy terms obtained from *bending energy* (E_{bend}), *torsional*

---

* Not rigidly specified but it is typically defined as an *intermediate* length scale ranging from 100 nm (upper limit of nanoparticles) to a few thousand nm. The mesophysics involved at this length scale is important in nanotechnology applications, like in processes for miniaturizing devices to microscopic sizes (micron size).

---

† A mathematical representation of a function as an infinite sum of terms.

energy (E_{tors}), van der Waals (E_{vdw}), electrostatics (E_{el}), and cross (E_{cross}) terms:

$$E = E_{str} + E_{bend} + E_{tors} + E_{vdw} + E_{el} + E_{cross}.$$

MM models separates out the van der Waals and electrostatic terms and attempts to make the remaining constants (bond bending, stretching, etc.) more transferable to various different types of molecules. It should be clear that MM models are representative models of the real quantum mechanical systems [12]. The total neglect of the electronic structure of individual atoms forces the user to define explicitly three inputs prior to the calculations: (1) the type of atoms present, (2) define chemically bonded neighbors of each atom, and (3) a start guess of the geometry. The main advantage of MM methods is the speed at which calculations can be performed, enabling the treatment of very large systems where chemical bonds are not broken or formed. For systems, for which good empirical parameters are available, it is possible to make very good predictions of the geometries, relative energies of large number of molecules and energetic barriers for the interconversion between different conformations. One of the major problems of MM methods is the need for good parameters. For classes of molecules where little or no data are available, the use of MM is very limited. In addition, it is not possible to assess the error within the method. The quality of the predictions can only be judged by comparisons with other calculations on similar types of molecules for which experimental data exist. Finally, force fields used in current software packages have been optimized primarily for biochemistry and pharmaceutical applications involving biological systems (e.g., design of small protein ligands and self-assembly of protein molecules). So, there is some concern about whether simulations can accurately reproduce the behavior of other materials, such as energetic materials (ionic crystals, ceramics, etc.) without force fields necessitating further modifications [9]. Examples of force fields used in current MM software packages are as follows:

- AMBER (Assisted Modelling Building with Energy Refinement), designed for modeling biological systems like proteins and nucleic acid molecules (DNA/RNA) [32–35].
- CHARMM (Chemistry at HARvard Molecular Mechanics), designed for biological and pharmaceutical studies (e.g., protein–protein interactions and receptor–ligand binding) [36]. The BIO+ force field is an implementation of the CHARMM force field used in the HyperChem package, developed for biological macromolecules.
- MM+ (various versions, e.g., MM2 and MM3), designed for structural and thermodynamics studies of organic molecules [37–39].
- OPLS (Optimized Potentials for Liquid Simulations), designed for modeling proteins and nucleic acid molecules (DNA/RNA), and for reproducing the physicochemical properties of biomolecules in solution [39–41].

These force field methods can be divided into two classes, based on the rigor of the calculations: *Class I* methods, like MM1–MM4, use higher-order terms and cross terms, provide higher accuracy, and are used for small- to medium-sized molecules. *Class II* methods, like AMBER and CHARMM, use only quadratic Taylor expansions and neglect *cross* terms, are computationally faster, and are used for very large molecules.

5.2.4 Hybrid Computational Approaches

MM methods are inherently unable to describe details of chemical reactions involving the breaking or formation of chemical bonds, since treatment of the extensive rearrangements of electrons is neglected in the MM model. Such level of detail is important for studying certain properties and functions of interest, such as protein-binding site(s) and metal centers in metalloproteins. For systems that are too large to be treated by electronic structure methods, two *hybrid* approaches have been developed (e.g., Morokumas' ONIOM [*our own n-layered integrated molecular orbital molecular mechanics method*] [42] and quantum mechanics–molecular mechanics [QM/MM] [43] methods). The central idea of these approaches is to partition the system into distinct parts. The *uninteresting* parts of the molecule are treated by MM (force fields) methods and the *interesting* parts by high-accuracy electronic structure methods. This strategy is useful for studying systems where part of the molecule is needed at high accuracy or for which no force field parameters exist. Treating systems in such a manner has been beneficial for many types of applications, for example, in pharmaceutical applications directed at improving the binding affinity and specificity of a drug by modeling the target protein's active site. For these types of applications, one approach for modeling the system consists in calculating the structure of the active site by electron structure methods (typically, semiempirical methods, low-level *ab initio* or DFT methods), while the backbone structure of the molecule is calculated by force field methods [43]. Such an approach is commonly denoted as QM/MM. One of the problems in the QM/MM method is in deciding how the two parts of the molecule connect to each other (i.e., delimiting the boundaries of parts). If QM/MM method cuts covalent bonds, the QM part of the calculation is terminated by adding so-called link atoms to the dangling bonds (usually hydrogen atoms). The second approach was introduced by Morokumas' method, ONIOM. The idea is to partition the molecule into multiple layers for calculations. For example, the central part of the molecule is treated by high-level *ab initio* methods, the intermediate layer by low-level electronic structure methods, and the outer layer by force field methods [42].

5.2.5 Problems of Computing Simulations

The continued advancement of computational methods, such as those described in the earlier sections, has allowed simulations technologies to become predictive in nature. Many novel concepts and designs have been first proposed based on modeling and simulations, and then followed by their realization or verification through workbench experiments. For example, Hagan Bayley's group proposed that specially designed arrangements of lipid networks could function as *synthetic mimics* of living cells [4]. Their idea was that by incorporating membrane proteins into the lipid bilayers of aqueous droplets and arranging droplets into networks with a specific 3D structure design, such droplet networks would form a cohesive material and display collective behaviors similar to those observed in living tissues (e.g., coordinated electrical communications among individual cells). They first performed a series of computational simulations and modeling to predict the size of droplets, as well as the 3D structure, electrical and mechanical properties that droplet networks would need to have to display the desired behavior. Then, they implemented the knowledge gained from these simulations into the design and conducted tests on nano 3D printed materials built from heterologous lipid bilayer droplets, successfully demonstrating the validity of their idea.

From the computational simulation perspective, the problem of modeling system behaviors is truly size and timescale in nature [44]. Over a decade ago, even the simplest MM method calculation required some waiting time in a desktop computer, MD and MC methods required high-performance computing, and researchers wanting to conduct *ab initio* calculations needed to reserve an account on specialized computational resources (e.g., supercomputers). Today, many of the once difficult to perform calculations (due to computational complexity) are carried out with ease thanks to the availability of high-performance and low-cost computer computational resources. Nevertheless, for many applications, problems inherent in both the computational methods and computational algorithms continue to thwart the rapid progression and development of novel concepts and designs. It is critical to realize that searching for solutions to these problems is tightly linked to advances in computer science. Lately, two types of computational resources have emerged that provide large amounts of computing power, namely, *high-performance** and *high-throughput*[†] computational resources. High-performance computational resources, like traditional supercomputers or the more recently marketed heterogeneous CPU–graphics processing unit (GPU) desktops (the so-called personal supercomputers), typically provide thousands or hundreds of processors with high speed and low-latency links between individual processors. Software simulation programs designed to exploit these types of computational resources need to be able to turn over parts of the overall task to individual processors to be carried out in parallel with the other parts and to have efficient

procedures that can facilitate data transfer and orchestration among the various tasks. Unfortunately, not all methods (e.g., *ab initio* methods) are amenable to algorithms that can scale well over thousands of processors. In addition, method codes rarely exploit GPU resources. Software programs, like Vijay Pande's Folding@Home and David Baker's Rosetta@Home that can exploit high-throughput computational resources (in both cases, a grid of desktop computers), have served as a testing framework for new methods like CAMEO-3D [45], some of which have been especially valuable for molecular engineering applications. Developing codes for running parallel computations in high-throughput mode may particularly be beneficial to *ab initio* methods. Overview and possible solutions to the problems that arise from increased computing power, that is, mining, management, storage, archiving, and integration of the *big data* emerging from simulations, were previously discussed [8,46].

The development of better computational algorithm codes that can efficiently combine different methods is also an important component toward efforts of improving execution of computational simulations. For example, the *interesting* part of a very large system targeted for direct simulations extends from the atomistic scale to the range of tens to hundreds of nanometers and then to mesoscale levels (micron or larger size). At the atomistic level, there are accurate semiempirical and QM methods that feed into the large-scale MD simulations of a system with millions or billions of atoms, which could then be coupled to mesoscopic descriptions of the system through continuum mechanics or finite element–based approaches. Nakano and collaborators, as well as Rudd and collaborators, have made initial attempts to develop a *grand-simulation code*, successfully cutting across different scale lengths and achieving a seamless integration across the interfaces [20,47]. Similar types of approaches are needed to bridge phenomenon along the timescale.

5.3 COMPUTATIONAL METHODS FOR PROTEIN DESIGN

Protein function is determined by the 3D structure of the protein. To fully understand the mechanism of how a protein functions, great efforts have been made toward elucidating their structures. Experimentally, the structure of a folded protein can be revealed by several methods, such as x-ray crystallography, NMR, cryo-electron microscopy, and small-angle x-ray scattering, each providing different levels of resolutions (accuracy). Structural biologists have brought forth a wealth of protein structure information and have made them available through public accessible databases such as the protein data bank (PDB) and Cambridge structural database. This flora of structural information provides a unique mechanistic vista that has greatly enhanced our understanding of biological functions of proteins and their relevance in a wide range of biological and cellular processes, like in signaling and metabolic pathways, molecule recognition, cell death, and in many other important biological activities.

---

[*] High-performance computing (HPC) refers to applications where one task (i.e., a part of the simulation) is shared over many processors; this is denoted as parallel processing.

[†] High-throughput computing (HTC) refers to applications where the overall task is parallelized, thus allowing the run of two or more separate simulations. There are two types of HTC computational resources, namely, purpose-built clusters and grid of desktop computers.

As the knowledge on the structure–function relations grows, researchers started to advance this field by working on the reversed problem, designing proteins with improved, adapted, or new functions. Notably, computational approaches and methods that can predict how conformational changes may affect the properties or functions of proteins have become an indispensable component in protein or small molecule design for pharmaceutical and medical applications. Recent development of efficient computational approaches and methods to predict protein structure has also provided relevant contributions toward such types of endeavors. Computational protein design generally involves two approaches: One is template-based design, where an existing protein with a known structure is used as a start reference of the spatial geometry to confer adapted/improved attributes on a protein (i.e., target) such as stability and specificity. The other approach is the *de novo* design of proteins, which is based on using fundamental physicochemistry principles that govern the interactions between protein residues. This approach is very useful for modeling proteins for which no experimental data exist and for engineering novel proteins using only knowledge on the desired shape and function to guide the design. The first approach has been more successful simply due to the vast amount of available information that can be used in the design. Over the years, various methods and algorithms have been developed and proven successful.

Generally, the first stage of a protein design is to construct a preliminary target structure, which contains the coordinates for the backbone atoms. Upon obtaining this target backbone conformation, a collection of candidate protein sequences is then generated by efficiently searching over the sequence space with specialized sampling method so as to prevent overwhelming available computational resources by computationally intensive task of enumerating all possibilities from the entire protein sequence and conformational space. The sequences are then evaluated by an energy function that assesses the sequence-structure compatibility in order to discard those with high energies and are unlikely to assume the target conformation. More rigorous but computationally expensive methods, such as MD or even quantum calculations (usually in the context of QM/MM), will ensue to evaluate and refine the structures adopted by remaining sequences. In the end, complementary experimental approaches can be carried out to validate or improve the computationally designed protein. Before we could go through each of the aforementioned steps, a few key concepts need to be introduced.

5.3.1 Rotamer Library

During the sequence and conformational search in protein design, certain level of flexibility on the target structure is often introduced to increase the number of sequences that can fold into the structure. In a simple model, the backbone of the target structure is fixed while the side chains are allowed to adopt multiple conformations. The conformation of protein side chains has continuous multiple degrees of freedom in

bond lengths, bond angles, and χ dihedral angles. To reduce the size of the conformational space, rotamer libraries are generally used in protein design. Rotamer library contains discrete side-chain conformations of only optimal values of bond lengths, bond angles, and a limited number of frequently observed χ dihedral angles.

5.3.2 Energy Function

In protein design, it is important to differentiate sequences that favorably assume the target structure from those that do not. An energy function that can accurately estimate the energy of the structures, as well as rank each sequence structure on the basis of comparisons with the target structure, is ideal for this purpose. Obviously, the most accurate estimation of energies is obtained by high-level quantum calculation. However, this is not an option for protein design simply because any practical system size surpasses the current available computing resources for quantum calculations. Therefore, developing energy functions that balance between accuracy and computational complexity is a key in protein design. Energy functions currently in use generally include physics-based energy functions that are often adapted from the force fields used in MD, knowledge-based functions derived structural databases, or hybrid functions that employ both of them [48]. Two of the most widely used protein design softwares, RosettaDesign [49,50] and ORBIT [51,52], both employ hybrid energy functions.

The energy functions (force fields) that were developed for MD of biomolecules include AMBER [32–35], CHARMM [36], and GROMOS (Groningen Molecular Simulation) [53]. They are generally developed by fitting parameters to the results of high-level quantum calculations and/or thermodynamic, crystallographic, and spectroscopic data from experimental data. A typical form of the energy functions used in MD is [35]:

$$U(r) = \sum_{bonds} K_r (r - r_{eq})^2 + \sum_{angles} K_0 (\theta - \theta_{eq})^2$$

$$+ \sum_{dihedrals} K_\phi [1 + \cos(n\phi - \delta)]$$

$$+ \sum_{impropers} K_\varphi (\varphi - \varphi_{eq})^2 + \sum_{electrostatic} \frac{q_i q_j}{\varepsilon_l r_{ij}}$$

$$+ \sum_{vdw} \left[\varepsilon_{ij} \left(\frac{R_{ij}^{min}}{r_{ij}} \right)^{12} - 2 \left[\frac{R_{ij}^{min}}{r_{ij}} \right]^6 \right].$$

The total energy of a system $U(r)$ is calculated by summing up the pairwise energy between each atom pair, since the potential energies are so defined that they are pairwise decomposable.

This is similar for protein design if the state of a sequence can be denoted by $(\alpha_1, r(\alpha_1); \alpha_2, r(\alpha_2); ...; \alpha_i, r(\alpha_i); \alpha_{i+1}, r(\alpha_{i+1}); ...; \alpha_N, r(\alpha_N))$, where α_i stands for a particular residue and $r(\alpha_i)$ for the rotamer state of it. The total energy of a particular sequence assuming a particular conformation can then be computed as [54]:

$$E(\alpha_1, r(\alpha_1); \alpha_2, r(\alpha_2); \ldots; \alpha_N, r(\alpha_N))$$

$$= \sum_{i=1}^{N} \varepsilon_i(\alpha_i, r(\alpha_i)) + \sum_{i=1}^{N} \sum_{j>1}^{N} \gamma_{ij}(\alpha_i, r(\alpha_i); \alpha_j, r(\alpha_j)).$$

The first term on the right-hand side of equation takes care of the interatomic interactions between side-chain atoms of protein residue α_i. The second term on the right-hand side deals with the interactions between side-chain atoms of residue α_i and protein residue α_j.

The interaction between solvent molecules and protein residues is a critical factor in protein folding and structure maintaining. Therefore, solvent molecules have to be taken into account in protein design. Since the number of solvent molecules is often huge, treating them explicitly brings large computational overhead for protein design. As a result, they are often treated implicitly by including hydrophobic effect, variable dielectric constant, and solvent exposure propensity [54].

5.3.3 Defining Backbone Conformation

Since the ultimate purpose of designing a protein is to gain its function, the target structure of the designed protein depends on the functional role the protein is meant to act. In the case of template-based design, the backbone of the target structure should be similar enough to the template so that the backbone coordinates of the template protein can be adopted as the target structure. While in the case of *de novo* design, where a template structure is not accessible, the target structure can be assembled from modular secondary structure elements or peptide fragments library from PDB database [55]. The backbone coordinates are often constrained during the calculations to reduce the computational cost and allow for certain level of flexibility to accommodate reasonably deviated structure and sequence.

5.3.4 Sequence and Conformation Space Search

As 20 amino acids are available for each position in a protein sequence, the size of the sequence space becomes 20^N, where N is the number of protein residues. The conformational space of the side chains of a given sequence has to be explored to find the optimal structure without knowing it in advance. This is a challenging task even if we only consider a limited number of conformations for each protein residue. Thermodynamically, the native structure of a given protein is considered to be the most stable conformation, that is, with the lowest global free energy. Any energy minimization algorithm with energy function derived from MD force field only guarantees a local minimum potential energy; therefore, the entire global energy landscape or all local low-energy

conformers have to be searched to find the global minimum state. Given an energy function, the problem of protein design becomes finding the conformations of the lowest free energy conformers of each sequence and comparing them to retrieve the energetically favorable ones. In template-based protein design, it is often computationally affordable to simply introduce variations/mutations to the template protein sequence and generate a manageable size sequence library for subsequent calculations, whereas in *de novo* protein design, it is necessary to apply constraints on the calculations so that the vast sequence space can be explored more efficiently. A variety of algorithms have been developed to search the sequence and conformation space; they mainly fall into three categories: stochastic, deterministic, and probabilistic approach.

5.3.4.1 Monte Carlo

MC methods are widely used in stochastic sampling [56]. In a simple form of MC sampling, a random starting sequence of the target structure is generated and the energy is calculated with a given energy function. In each of the following sampling steps, a randomly chosen rotamer of any protein residue is introduced and the energy of the system re-evaluated. The substitution is accepted or rejected based on a probability that is a function of the Boltzmann distribution, as it is shown in Equation 5.3, where K_B is the Boltzmann constant and T is the simulation temperature:

$$p = e^{-\frac{E_{(i+1)} - E_i}{K_B T}}.$$

MC sampling does not guarantee that a global energy minimum can be retrieved in the end, and the sampling could get trapped in local energy minima [54]. Therefore, multiple runs of MC sampling are often initiated with different starting sequences to cover the energy landscape as much as possible. In addition, the temperature T can be set to a higher value in the beginning and ramp down in following iterations so that some high energy barriers can be overcome and the sampling is not confined to some local minima; this is the so-called simulated annealing [49,57]. Other variants of MC methods such as MC with quenching [56] and biased MC [58] have been developed to improve the sampling efficiency and provide better estimation on the lowest energy sequences.

5.3.4.2 Genetic Algorithm

Genetic algorithm (GA) is also used for stochastic sampling. In a typical implementation of a GA in protein design [59], a genome S is defined as a N dimensional vector, $S_1 S_2 \ldots S_N$, where N corresponds to the length of the protein sequence that is being designed. A symbol in the genome stands for 1 protein residue out of the 20 possible ones. Before the sampling starts, a population of m sequences $S^i (i = 1, 2, \ldots, m)$ is generated with randomly chosen or predefined residues for each sequence. New generation of sequences is then derived from the initial population by operations such as mutation (change of a residue to a different residue), crossover, and selection. The number of mutations generated at each generation depends on an adjustable parameter, the mutation probability. Crossover operations exchange subsequences pairwisely among sequences that have certain level of sequence-structure compatibility (evaluated by a given energy function). For each generation, a subgroup of the existing population is selected to produce a new generation using a specific selection strategy [56]. The iterations generate population of sequence population with increased fitness (a parameter describing the overall sequence-structure compatibility) in the end.

5.3.4.3 Dead-End Elimination

Dead-end elimination (DEE) is a deterministic approach. The DEE algorithm reduces the sequence space that is being explored by iteratively pruning out residues and rotamers that not needed to define the global minimum energy conformation (GMEC) [60,61]. Generally, all possible pairs of rotamers at each residue position are compared with respect to the sum of the side chain–backbone energy and the minimum side chain–side chain energy that can be achieved with all other possible combinations of rotamers. The rotamer that has higher energy than another rotamer can be eliminated. Although DEE is guaranteed to converge to the GMEC, the functional exhaustiveness of it prevents it from being applied to relatively large protein. Modification to DEE has shown promise in applying DEE to larger proteins, using clusters of rotamers comparison instead of individual rotamers [62]. Other useful variants of DEE include the revised elimination and flagging criteria, extended DEE, and type-dependent DEE [54].

5.3.4.4 Graph Search

If the conformational space is represented as a tree structure with each internal nod designating a distinct rotamer state, the goal nodes (leaf nodes) being the target conformations, exploring the conformational space becomes a search tree problem. The A^* search is a popular graph search algorithm used in protein design [63]. It finds the optimal path from the root (being an empty sequence) to a goal node in a search tree. The evaluation function $f = g + h$ is used in the A^* algorithm for any node, where g is the lowest cost of reaching one particular node and h is the estimated cost to reach a goal node from this node. g and h are formally defined in the following equations [63], with l denoting the last assigned rotamers in the partial conformation:

$$g = \sum_{i=1}^{l}(E(r_i)) + \sum_{j=i+1}^{l}(E(r_i, r_j)),$$

$$h = \sum_{j=l+1}^{n}\left[\min_{r_j}\left(E(r_j) + \sum_{i=1}^{l}E(r_i, r_j) + \sum_{j=i+1}^{l}\min_{r_k}E(r_i, r_j)\right)\right].$$

This algorithm stores a list of nodes that are stored in an array by the algorithm, sorted according to their f values. At each iteration, the node with the lowest value of f is expanded to its successive nodes; the f values are then computed for its successive nodes and the new nodes are added to the list in the appropriate position. The iteration stops once it reaches to a goal node that has the minimum cost path from the root node.

5.3.4.5 Self-Consistent Mean Field

Self-consistent mean field (SCMF) is another popular deterministic sampling approach [64,65]. With SCMF approach, the energy of an individual rotamer is a function of the *mean-field* energy of the rotamers at other residue sites. In other words, the energy contribution from each rotamer of other residue site is proportional to its probability as shown in the following equation [54]:

$$\varepsilon_i(\alpha_i, r(\alpha_i)) = \sum_{r(\alpha_i)}\sum_{j,\alpha_j} w_j(\alpha_j, r(\alpha_j))\varepsilon_{ij}(\alpha_i, r(\alpha_i); \alpha_j, r(\alpha_j))$$
$$+ \varepsilon_i^0(\alpha_i, r(\alpha_i)),$$

where

$\varepsilon_i^0(\alpha_i, r(\alpha_i))$ is the side chain–backbone interactions energy of residue α_i

$\varepsilon_{ij}(\alpha_i, r(\alpha_i); \alpha_j, r(\alpha_j))$ is the interaction energy between residue α_i and α_j at two different residue sites

The $w_j(\alpha_j, r(\alpha_j))$ is the probability of the residue α_j with its rotamer state of $r(\alpha_j)$. This probability is usually converted from the energy state by using Boltzmann distribution as $w_j(\alpha_j, r(\alpha_j)) \propto e^{-\varepsilon_j(\alpha_j, r(\alpha_j))/(k_B T)}$. SCMF generally starts with a uniform probability for all the rotamers. By having the interaction energy for an individual rotamer computed, the probability interaction energy is then updated using the Boltzmann distribution. The method generally converges to a small number of high-probability rotamers at each position. The residues with the highest probability at each site are then selected to determine the sequence.

5.3.4.6 Probabilistic Approach

Different from directly identifying particular sequences in either stochastic or deterministic sampling approach, probabilistic approaches estimate the probabilities of site-specific residue for sequences assuming a target structure [66,67]. Since the energy functions are largely semiempirical and subject to inaccuracy, the directed sampling approaches may become problematic in identifying the global optima. For cases when only incomplete information about a problem is available, probabilistic approaches may become more appropriate. Probabilistic approaches are based on the concept of entropy maximization from statistical mechanics. The site-specific probabilities of the residues are determined by maximization of the total conformational entropy [54], which is defined as

$$S_c = -\sum_{i,\alpha_i, r(\alpha_i)} w_i(\alpha_i, r(\alpha_i)) \ln w_i(\alpha_i, r(\alpha_i)).$$

The notations are made consistent with the ones used in SCMF method. Desired energetic and functional constraints are needed in the maximization process to yield the site-specific probabilities. The probabilistic approaches also have the advantage of being able to address large systems that are computationally formidable to direct sampling, since the entire sequence space can be readily characterized using this approach [54].

5.3.5 Examples of Protein Design Applications

Novel protein designs started in 1970s, succeeded through assemble secondary structure modules together [68,69], or hydrophobic patterning [70], based on largely a qualitative knowledge on protein chemistry. The advance in computer power and force fields derived from the expanding protein structure databases has brought great momentum into computational protein design. Following will be introduced the landmarks in protein design.

5.3.5.1 Enzyme Design

The first fully automated *de novo* protein design algorithm was introduced by Mayo et al. in 1997 to screen a library of 1.9×10^{27} peptide sequences for designing a ββα protein motif based on the zinc finger domain [71,72]. In 2003, Baker's group combined *ab initio* prediction, atomic-level energy refinement, and sequence design in Rosetta to develop a general approach for designing novel protein folds that incorporates backbone flexibility into structure optimization based on rotamer libraries only. Using this approach, they designed a 93-residue protein with a novel sequence and topology, Top7 [57]. Efforts were also made on redesign or *de novo* design of enzymes and bio-catalysts. Jiang and colleagues developed a computational approach to design 32 enzymes showing unprecedented retro-aldolase activity [73]. Later on, they designed eight enzymes with different catalytic motifs to catalyze the Kemp elimination [74]. Similar success has also been achieved in designing novel enzyme for the Diels–Alder reaction by Siegel et al. [75]. More recently, Khare et al. redesigned a zinc-containing adenosine deaminase for organophosphate hydrolysis with the designed structure validated by crystal structure [76].

Designing proteins with increased or altered specificity, that is, differentially bind to structurally similar binding proteins, has also been a huge interest in this field. Maranas and colleagues successfully altered the *Candida boidinii* reductase (CbXR) cofactor specificity from NADPH to NADH with the iterative protein redesign and optimization framework [77], which is an iterative framework in which the optimization of substitute side chain is followed by relaxation of backbone atoms through local minimizations to better accommodate the new side chains [78]. Keating's group has developed a computational framework, CLASSY (cluster expansion and linear programming-based analysis of specificity and stability), to design protein-interaction specificity; they used the CLASSY to engineer the peptide-binding specificities of multiple members leucine zipper (bZIP) transcription factors [79].

5.3.5.2 Self-Assembling Proteins

Self-assembling peptides or proteins are important components in many biological activities; to understand and even be able to control the assembly process have great potential in using self-assembling proteins to develop biologically functional materials or molecular machineries. Saven's group has designed a protein crystal by redesigning the interfaces between the homotrimeric parallel coiled-coil templates [80]. Baker's group constructed 12- and 24-subunit nanocages by docking natural trimeric protein subunits to obtain an optimal packing framework, and then further refining the interfaces between the subunits to minimize the self-assembly energy [81,82]. The resolved crystal structures closely matched the designed model [81,82]. The approach of symmetric docking followed by protein–protein interface design has direct application in designing self-assembling protein nanomaterials for a variety of purposes like drug delivery, vaccine, and plasmonics [82]. Rufo et al. designed a series of amyloid-forming peptides that not only catalyze their own assembly but also act as Zn^{2+}-dependent esterases, which implied a promising future of designing self-assembling nanostructured catalysts [83]. Floudas's group recently designed three self-assembling tripeptides based on a known self-assembling tripeptide template using a two-stage design framework, with which they first generated low potential energy sequences and then calculated the fold specificity and/or approximate association affinity of the sequences to select the candidates for experimental validation [84]. Wang et al. designed self-assembling hexapeptide with similar motifs of amyloid-β (Aβ) peptides to inhibit the Aβ aggregation and reduce Aβ-induced toxicity using a hybrid-throughput computational method, which can be applied to design peptide inhibitors against various of amyloid diseases [85].

5.3.5.3 Therapeutic Peptide or Protein Design

Apart from the native interests in achieving new or adapted functions by protein design, the promise of designing new peptides or proteins for therapeutic application is also an important driving force to this field. As of 2010, over 200 peptides, proteins, and antibody therapeutics had been marketed [86]. Successful explorations have been made in designing peptides to interfere with a variety of diseases. Baker's group developed a general computational protocol for designing proteins of particular binding affinity to a target macromolecule surface [87]. They successfully engineered a protein binds to a conserved region on the stem of H1N1 influenza hemagglutinin, which is essential in the process of viral entry to host cells [87]. Floudas et al. designed HIV-1 entry inhibitors targeting gp41 using TINKER and Rosetta *ab initio* [88,89]. The approach they adopted in designing the inhibiting peptides has been developed into an interactive web interface, Protein WISDOM [90]. Hao et al. computationally optimized the binding free energy of the binding peptides, which they initially obtained through phage display, to CRIP1, a biomarker for early detection of cancers, and experimentally validated the optimization [91]. Istivan et al. implemented the resonant recognition model to design a peptide analogue to a myxoma virus protein that exhibits antitumor activity [92]. For a more complete list of recent publications on therapeutic peptide or protein design relevant to cancer, HIV, and Alzheimer's disease, readers are referred to a recent review paper published by Floudas and coworkers [84].

Antibody therapeutics has also been a topic of great interest in protein design, since they are extensively used in diagnostics and therapeutics. Many breakthroughs have been made in antibody CDR modeling, predicting V_L/V_H domain orientations, antibody–antigen recognition, modeling of somatic mutations (affinity maturation), stability improvement, and antigen design to elicit neutralizing antibodies (NAbs) [93]. A few examples covering some of the aforementioned aspect will be given to highlight the progresses. Sivasubramanian et al. developed RosettaAntibody for homology modeling of the variable regions of antibody [94]. The benchmark they did on 54 antibody crystal structures revealed *functional accuracy* of the resulting models [94]. Correia et al. designed epitope scaffolds in which contiguous structural epitope 4E10 is presented to improve conformational stability and higher affinity to monoclonal antibody, inhibits HIV neutralization by HIV+ sera [95]. Based on the chimeric protein they designed, they proposed two computational protocols for antigen optimization and protein engineering [96]. Miklos et al. managed to substantially improve the thermal inactivation resistance and affinity of an antibody by replacing multiple surface residues with charged residues using a supercharging protocol [97].

So far, the most widely used antibody modeling softwares include free RosettaAntibody [98], PIGS [99], WAM [100], and commercially available ones like Accelry's Discovery Studio and Chemical Computing Group's Molecular Operating Environment. Other protocols are also gaining popularity among researchers. Notably, Pantazes and colleagues introduced the OptCDR framework for designing antibodies to target any specific antigen epitope with affinity, and tested it with designing antibodies for a peptide from the capsid of hepatitis C, the hapten fluorescein, and the vascular endothelial growth factor with very promising results [101]. In 2013, they developed the Modular Antibody Parts (MAPs) database, which consists 929 analogous to the variable, diversity, and joining genes those were extracted from 1169 antibody structures from different species, for predicting antibody structures [102]. With this database, a target sequence would be able to find matching parts with the minimal mismatches. Using a test case consisting 260 target antibodies, the predicted structures showed an average all-atom root-mean-square deviation (RMSD) of only 1.9 Å [102].

5.4 ARTIFICIAL INTELLIGENCE APPLICATIONS IN SIMULATIONS AND MODELING

Nanotechnology is particularly conspicuous in the medical field (commonly referred to as nanomedicine). In this field, nanoscience is used in the manufacture of drug delivery vehicles (e.g., virus-like particles [VLPs] and NPs) [103], diagnostic devices (e.g., cantilevers) [104], contrast agents in therapeutics for treating cancer and cell imaging, etc. Manufacturing materials with desired properties engineered specifically for such intended applications is only possible through nanotechnology, which heavily relies on many mathematical theories and computational methods, including the methods overviewed in the previous sections of this chapter. At present, however, nanotechnology encounters physical limitations of its working scale, where the nanophysics is completely different from that of the macroscopic scale. This means that correctly interpreting the results obtained from the modeling and simulation of any system at the nanoscale is one of the issues that nanotechnology continues to face. It is in this context that AI paradigms, such as artificial neural networks (ANNs) and evolutionary algorithms, the so-called GAs, have been proposed as key tools for producing scientific results as well as for the development of nanotechnology applications [105,106]. Gomez and Varona discuss several examples of the applications and uses of AI methods in nanotechnology research [105]. For example, application of ANNs in simulation software has been useful in improving the quality of the simulations, where numerical methods (e.g., semiempirical or *ab initio* methods) are used to simulate the electrostatic potential around an *interesting* part and the ANN routine is used to minimize the error at the surface. Similarly, application of GAs in simulation software is proving helpful in *de novo* protein design, where numerical methods are used to compute pairwise potentials to ensure that gross overpacking of the protein core is avoided and the GA routine is used to optimize side-chain conformations [59].

As mentioned in Section 5.2, the use of force constants (derived from experimental data or QM calculations) to simulate and model biomolecule systems (e.g., proteins) originates from the premise that these constants are transferable to different types of molecules. It should also be clear that knowledge regarding the physical and chemical properties of the biomolecules comprising the system is very important, particularly, for the *de novo* design of proteins. From a molecular epidemiological perspective, accurate modeling and simulation of biological systems (e.g., protein and genes) provide powerful information on several aspects about disease-causing pathogens, including their evolution, transmission, and virulence. In the remaining parts of this chapter, discussion will be focused on the application of AI methods to predict the *interesting* parts in macromolecule systems (i.e., genes and proteins), more specifically the so-called biomarkers. Discussion will address the relevance of biomarkers and AI models for applications, such as in the design of novel tools purposed for the monitoring of infectious pathogens (i.e., surveillance tools) or for the detection of disease and virulence (i.e., diagnostic tools). In particular, ANN [107,108] and Bayesian network (BN) [109] applications will be overviewed.

5.4.1 QUANTITATIVE STRUCTURE RELATIONSHIP MODELS

A quantitative structure-activity relationship (QSAR) model is a mathematical formalism that is used to predict physicochemical properties or theoretical descriptors of molecules based on their 3D structures, such as electronic or thermodynamic properties, which are not available from experimental data (e.g., x-ray crystallographic data). QSAR modeling is based on the concept that molecules (e.g., proteins) with similar 3D structures have similar functions/activities, which is a concept known as structure–activity relationship. Formalisms of QSAR, in combination with quantum and/or MM methods, constitute important tools for designing novel therapeutic and diagnostic peptides or proteins, for predicting functions of new molecules based on comparisons of 3D structures between different molecules. The ANN model is a form of AI based on brain theory, which originated from neurosciences and mathematics over seven decades ago, and finds applications in diverse fields such as simulation and modeling, time series analysis, pattern recognition, and signal processing [107,108]. In simple terms, an ANN is a mathematical model of neuron operation that can, in principle, *learn* from input data and be used to compute any type of function (e.g., arithmetic and logical functions). It consists of two types of elements: processing or computation units (called *neurons*) and connections with adjustable *weights* or *strengths*. The ANN model has been successfully applied in QSAR modeling (ANN–QSAR models) to accurately predict the chemical or biological activities of many different types of molecules. Description of the ANN method and its use in QSAR model approaches and applications, for example, prediction of protein-binding interfaces, drug-binding regions, and other relevant activities that are important for designing improved therapeutic proteins, was previously reviewed in Ref. [110].

ANN–QSAR models, additionally being useful to accurately predict the activity of an area (region) in the macromolecule system, can also be used to accurately predict which parts in that area (i.e., specific set of molecules) are relevant for the activity. For example, an ANN–QSAR model of the antigenic property of the hepatitis C virus (HCV) NS3 protein was shown to predict regions of antigen-antibody binding, that is, conformational epitopes (100% agreement with a structure-based approach) as well as the antigenic activity of several NS3 variants with 90.0% accuracy [111]. In addition, relevant antigenic markers (determinants of antigenic activity) were predicted. Later, the predicted markers of antigenicity were shown to be useful parameters in classifier models, classifying NS3 variants on the basis of corresponding antigenic activity based on the physicochemical properties of 3D structures [112] with 100% accuracy. Although use of accurate descriptors, which are only obtainable from 3D structure data, contributes in large measure to the robustness of QSAR models, sequence information alone can also provide useful descriptors that describe the properties of such biologic systems. As described in the following, AI application has been successful in several studies solely based on the sequence data.

5.4.2 Discovery of Medically Relevant Biomarkers

With the recent advent of high-throughput sequencing technologies, a large body of genomic information has emerged, and is available in public databases, like NIH's genetic sequence database, commonly denoted as GenBank [113]. It is possible to calculate physicochemical properties of macromolecules, like proteins (e.g., electric potential energies and charges) from amino-acid sequence information, albeit, with low accuracy. In addition, there are thousands of numerical descriptors, experimentally measured or statistically derived, that are used to describe the physicochemical and biochemical properties of amino acids that comprise protein sequences [114]. Similarly, there is a large body of studies that report measured or statistically derived properties of nucleic acids comprised in DNA or RNA sequences. Finally, the 20 amino acids or 4 nucleic acids that make up proteins or DNA molecules, respectively, can be represented using an alphabetical term corresponding to each amino or nucleic acid type. The several different types of descriptors that are available for representing the properties or attributes of protein and DNA sequences have achieved popularity in several fields involving computational sequence analysis, ranging from the classic phylogenetic tree analysis to the more complex network sequence analysis.

AI models applied in computational sequence analysis have also relied on such descriptors to accurately predict, for instance, association between a specific virus strain and clinical progression of disease [115], the susceptibility of specific strains to a specific drug treatment [112,116,117], the distinct transmission pattern behaviors of strains [118], and the particular specificity of host adaptation of strains [112,118]. For these types of computational sequence–based applications, AI models based on probabilistic graphical techniques, like BNs [109], have improved the accuracy of predictions and have been very helpful for identifying genetic markers relevant for associating distinct strains of infectious agents to the phenomenon of interest. For example, Yury Khudyakov's group, using BN models, recently provided the first strong evidence of resistance to combined interferon ribavirin (IFN/RBV) treatments being an adaptive genetic trait of the HCV [111,112]. After building several BN models, which learned dependencies in amino-acid substitutions among coevolving sites, they show that BN models parameterized on specific genetic markers from two distinct proteins, namely, the NS5A and the E2 proteins, can predict HCV strains association to outcomes of IFN/RBV therapy with 85.0% and 98.0% accuracy, respectively. Furthermore, the study also demonstrates how BN models provide insights on how complex traits, such as drug resistance, can emerge in viral strains through broad epistatic connectivity among many coevolving sites (i.e., many genetic pathways).

Accuracies provided by AI models (ANN or BN models), in sequence-based applications, such as QSAR modeling or models of dependency in amino-acid/nucleic acid substitutions to identify predictive biomarkers (e.g., antigenic activity, drug resistance, and virulence), have several implications. From a nanotechnology perspective, such models could aid in accelerating the development of nanosensors specifically designed to diagnose disease or virulence of infecting strains as well as therapeutic agents against specific infecting strain(s). From the surveillance and medicoclinical points of view, the models themselves could provide new tools (e.g., *in silico* assays) for the rapid detection and accurate monitoring of infectious strains circulating in the population and for supporting clinicians in decision making.

5.5 CONCLUSIONS

Nanotechnology holds a great future for providing solutions in the area of medicine and biology, which in turn, could bring a new era in health-care and molecular epidemiology applications. To address the growing need for potent, inexpensive, and safe medicinal products, researchers are exploring the use of nanocrystals/polymer composites [119], nanoscale 3D printing devices [4], among other endeavors. For example, photoluminescence of dyes used for diagnostic imaging has been enhanced by the use of nanocrystals/polymer composites [119]. Also, advances in nano 3D printing technologies promise applicability to the manufacturing of VLP- and NP-based vaccines, which could potentially lead to the manufacturing of more efficient and cheaper vaccines for use against prevalent and emergent infectious diseases. Geneticist J. Craig Venter stated with respect to nano 3D printers that "such a devise could be used to instantly produce vaccines, medications or biological materials anywhere in the world simply through the transfer of a digital file" [120].

There are some major challenges facing nanotechnology. With regard to manufacturing medicines, the challenges in the meso- or nanoscopic realm are how to produce the medicine, how to deliver it to the tissue(s)/organ(s) where it is needed, and how safe the medicine is. The latter inquiry is especially pertinent to metal NPs used in medical and consumer products. For example, silver (Ag) or gold (Au) NPs are used as dyes in medical application such as imaging [121,122]. More recently, nanotechnology has made possible to greatly enhance its antibiotic properties [123] and broaden the number of products where it is used. A consumer report by the Wilson Center lists over 140 consumer products where silver nanoparticles (Ag NPs) have been incorporated in a wide variety of products, like furnishings, cosmetics, food packaging containers, and food-and-beverage supplements. However, Ag NPs, engineered to be as tiny as 5 nm, exhibit properties and behaviors that differ from the traditional silver. While silver is considered to be a low-toxicity metal and fairly inert in the body, new evidence indicates that the tiniest Ag NPs could pose serious health and environmental risks as they show increased reactivity with several types of biomolecules [124], significant cytotoxicity [125], and genotoxicity [126].

Computational protein design is a young field, many exciting breakthroughs in producing novel catalysts, protein-binding interfaces, protein inhibitors, and self-assembling peptides have been made since its birth. Regardless of the detailed procedure and algorithm used in a particular design, the choice of the candidate sequences is largely guided by the energy function being used to compare the sequence-structure compatibilities of different sequences and the sequence/conformation sampling method. Therefore, the accuracy of the energy functions and the efficiency of the sampling methods are determinants to the current success rate of computational protein design. So far, the success rate has not been very high—a few successes out of tens of designs [127]. Examining the unsuccessful designs revealed weaknesses in the energy functions derived from force fields, most evidently in computing the long-range electrostatic interaction and solvent model [128]. Using experimental approach to characterize and screen computationally designed peptides/protein, and feed the results back to subsequent iterative computational design has overcome some of the deficiencies in computational domain and yielded encouraging results. However, even with high-throughput experimental approaches, screening hundreds or sometimes even thousands of candidate proteins are usually nontrivial investments. Therefore, continuously improving the current force fields for more accurately treating electrostatic and solvent interaction is important to more effectively discriminate multiple candidate sequences with small energy gaps, for the ultimate purpose of reducing experimental obligations substantially, if not eliminating them.

So far, human intervention during computational protein design has been another obstacle to fully automate the processes. Instances like manually removing sequences/structures with unsatisfied hydrogen bonds and/or suboptimal packed domains upon visual inspection happen more often than not in various designs. Adjustments to energy functions are often carried out on a case-by-case basis to reach an optimal set of sequence selection using training sets. These human interventions are effective, but at the same time prevent the particular method that has been easily applied to new systems. A more profound understanding of the physicochemistry and biophysical principles underlying the interactions between biological system at molecular and atomistic level is critical for alleviating this situation. Moreover, *ab initio* and other simulation methods continue to be revised and improved, which in combination with innovations in computer science, that is, design of faster processors and more powerful GPUs, as well as advances in computational science, that is, algorithms specifically developed to take advantage of these new computer technologies, will hopefully help in eliminating many of the present computational challenges that prevent simulations of larger and more complex chemical systems.

Finally, innovations across platform technologies—nanotechnology, genomics, AI, robotics, and ubiquitous connectivity (i.e., broadband adoption, mobile Internet access, and mobile devices)—will be mutually reinforcing. For example, advances in nanotechnology will enable the enhanced computational power necessary for breakthroughs in AI and spawn new technologies for manipulating DNA, which will accelerate advancements in genomics. As research in such platform technologies moves from concepts and prototypes to find applications in industrial processes, they will generate efficiency improvements that will accelerate advancements in biotechnology and medicine. A report by the Bain Company estimates that the combined effect of developments in these technologies will grow the U.S. global gross domestic product by $1 trillion by 2020 [129].

REFERENCES

1. Policy, I.N.I.C.C.P., Nano statistics. 2014 (accessed on August 9, 2014); Available from: http://statnano.com/.
2. Health, U.S.N.I.o., ClinicalTrials.gov. 2014 (accessed on August 9, 2014); Available from: https://clinicaltrials.gov/.
3. Sun, Y. and C.-H. Kiang, DNA-based artificial nanostructures: Fabrication, properties, and applications, in *Handbook of Nanostructured Biomaterials and Their Applications in Nanotechnoloty*, Vol. 2, H.S. Nalwa, ed., 2005. Valencia, CA: American Scientific Publishers, pp. 224–226.
4. Villar, G., A.D. Graham, and H. Bayley, A tissue-like printed material. *Science*, 2013. **340**(6128): 48–52.
5. Foresman, J.B. and A. Frisch, *Exploring Chemistry with Electronic Structure Methods*, 1996. Pittsburgh, PA: Gaussian.
6. Gubbins, K.E. and J.D. Moore, Molecular modeling of matter: Impact and prospects in engineering. *Industrial and Engineering Chemistry Research*, 2010. **49**(7): 3026–3046.
7. Landau, D.P. and K. Binder, *A Guide to Monte Carlo Simulations in Statistical Physics*, 3rd edn., 2009. New York: Cambridge University Press. p. xv, 471pp.
8. Dove, M.T., An introduction to atomistic simulation methods. *Seminarios de la SEM*, 2008. **4**: 7–37.
9. Dorsett, H. and A. White, Overview of molecular modelling and ab initio molecular orbital methods suitable for use with energetic materials. Salisbury, South Australia, Australia: DSTO Aeronautical and Maritime Research Laboratory, 2000.
10. Allen, M.P. and D.J. Tildesley, *Computer Simulation of Liquids*, 1989. Oxford, U.K.: Oxford University Press. p. xix, 385pp.
11. Rapaport, D.C., *The Art of Molecular Dynamics Simulation*, 2nd edn., 2004. Cambridge, U.K.: Cambridge University Press. p. xiii, 549pp.
12. Jensen, F., *Introduction to Computational Chemistry*, 2007. Chichester, U.K.: John Wiley & Sons. p. 1 online resource (p. xx, 599pp.).
13. Payne, G.F. et al., Accessing biology's toolbox for the mesoscale biofabrication of soft matter. *Soft Matter*, 2013. **9**(26): 6019–6032.
14. Parr, R.G. and W. Yang, *Density-Functional Theory of Atoms and Molecules*. International Series of Monographs on Chemistry, 1989. Oxford, U.K.: Oxford University Press. p. ix, 333pp.
15. Joubert, D. ed., Density functionals: Theory and applications. *Proceedings of the 10th Chris Engelbrecht Summer School in Theoretical Physics* Meerensee, Cape Town South Africa, January 19–29, 1997. *Lecture Notes in Physics*, 1998. New York: Springer. p. xvi, 194pp.
16. Born, M. and J.E. Mayer, Zur gittertheorie der ionenkristalle. *Zeitschrift für Physik*, 1932. **75**(1–2): 1–18.
17. Zhang, S.L. ed., Theoretical fundamentals of Raman scattering in nanostructures, in *Raman Spectroscopy and Its Application in Nanostructures*. John Wiley & Sons, Ltd, Chichester, UK, pp. 249–308.
18. Gale, J.D. and A.L. Rohl, The general utility lattice program (GULP). *Molecular Simulation*, 2003. **29**(5): 291–341.
19. Todorov, I.T. and W. Smith, DL_POLY_3: The CCP5 national UK code for molecular-dynamics simulations. *Philosophical Transactions of the Royal Society of London Series A—Mathematical Physical and Engineering Sciences*, 2004. **362**(1822): 1835–1852.
20. Nakano, A. et al., Multiscale simulation of nanosystems. *Computing in Science and Engineering*, 2001. **3**(4): 56–66.

21. Pechenik, A., R.K. Kalia, and P. Vashishta, *Computer-Aided Design of High-Temperature Materials*, 1999. Oxford University Press, New York.
22. Metropolis, N. et al., Equation of state calculations by fast computing machines. *The Journal of Chemical Physics*, 1953. **21**(6): 1087–1092.
23. Ciccotti, G., W.G. Hoover, and Società italiana di fisica, Molecular-dynamics simulation of statistical-mechanical systems: Varenna on Lake Como, Villa Monastero, July 23–August 2, 1985. *Proceedings of the International School of Physics "Enrico Fermi"*, 1986. New York: North-Holland. p. xvii, 610pp., 1 leaf of plates.
24. Choe, J.-I. and B. Kim, Determination of proper time step for molecular dynamics simulation. *Bulletin of the Korean Chemical Society*, 2000. **21**(4): 419–424.
25. Musa, S.M., *Computational Finite Element Methods in Nanotechnology*, 2012. Boca Raton, FL: CRC Press. p. 1 online resource (p. xxi, 606pp.).
26. Dinur, U. and A.T. Hagler, New approaches to empirical force fields. *Reviews in Computational Chemistry*, 1991. **2**: 99–164.
27. Haile, J.M., *Molecular Dynamics Simulation: Elementary Methods*, 1992. John Wiley & Sons, Ltd, Chichester, UK.
28. Griffiths, D.J., *Introduction to Quantum Mechanics*, 2nd edn., 2005. Upper Saddle River, NJ: Pearson Prentice Hall. p. ix, 468pp.
29. Rappé, A.K. and C.J. Casewit, *Molecular Mechanics across Chemistry*, 1997. Sausalito, CA: University Science Books. p. xii, 444pp.
30. Taft, R. and M. Newman, *Steric Effects in Organic Chemistry*, 1956. New York: Wiley, 556pp.
31. Wiberg, K.B., A scheme for strain energy minimization. Application to the cycloalkanes. *Journal of the American Chemical Society*, 1965. **87**(5): 1070–1078.
32. Cornell, W.D. et al., A second generation force field for the simulation of proteins, nucleic acids, and organic molecules. *Journal of the American Chemical Society*, 1995. **117**: 5179; 1996. **118**(9): 2309.
33. Case, D.A. et al., The Amber biomolecular simulation programs. *Journal of Computational Chemistry*, 2005. **26**(16): 1668–1688.
34. Pearlman, D.A. et al., Amber, a package of computer-programs for applying molecular mechanics, normal-mode analysis, molecular-dynamics and free-energy calculations to simulate the structural and energetic properties of molecules. *Computer Physics Communications*, 1995. **91**(1–3): 1–41.
35. Weiner, S.J. et al., A new force field for molecular mechanical simulation of nucleic acids and proteins. *Journal of the American Chemical Society*, 1984. **106**(3): 765–784.
36. Brooks, B.R. et al., CHARMM: A program for macromolecular energy, minimization, and dynamics calculations. *Journal of Computational Chemistry*, 1983. **4**(2): 187–217.
37. Allinger, N.L., Conformational analysis. 130. MM2. A hydrocarbon force field utilizing V1 and V2 torsional terms. *Journal of the American Chemical Society*, 1977. **99**(25): 8127–8134.
38. Lii, J.H. and N.L. Allinger, The Mm3 force-field for amides, polypeptides and proteins. *Journal of Computational Chemistry*, 1991. **12**(2): 186–199.
39. Froimowitz, M., Hyperchem(Tm)—A software package for computational chemistry and molecular modeling. *Biotechniques*, 1993. **14**(6): 1010–1013.
40. Jorgensen, W.L. and J. Tiradorives, The OPLS potential functions for proteins—Energy minimizations for crystals of cyclic-peptides and crambin. *Journal of the American Chemical Society*, 1988. **110**(6): 1657–1666.

41. Jorgensen, W.L. and J. Tirado-Rives, Potential energy functions for atomic-level simulations of water and organic and biomolecular systems. *Proceedings of the National Academy of Sciences of the United States of America*, 2005. **102**(19): 6665–6670.

42. Svensson, M. et al., ONIOM: A multilayered integrated MO+MM method for geometry optimizations and single point energy predictions. A test for Diels–Alder reactions and Pt(P(t-Bu)(3))(2)+H-2 oxidative addition. *Journal of Physical Chemistry*, 1996. **100**(50): 19357–19363.

43. Field, M.J., P.A. Bash, and M. Karplus, A combined quantum-mechanical and molecular mechanical potential for molecular-dynamics simulations. *Journal of Computational Chemistry*, 1990. **11**(6): 700–733.

44. Srivastava, D. and S.N. Atluri, Computational nanotechnology: A current perspective. *CMES—Computer Modeling in Engineering and Sciences*, 2002. **3**(5): 531–538.

45. Haas, J. et al., The protein model portal—A comprehensive resource for protein structure and model information. *Database (Oxford)*, 2013. **2013**: bat031.

46. Walker, A.M. et al., Integrating computing, data and collaboration grids: The RMCS tool. *Philosophical Transactions of the Royal Society A—Mathematical Physical and Engineering Sciences*, 2009. **367**(1890): 1047–1050.

47. Rudd, R.E. and J.Q. Broughton, Concurrent coupling of length scales in solid state systems. *Physica Status Solidi B—Basic Research*, 2000. **217**(1): 251–291.

48. Boas, F.E. and P.B. Harbury, Potential energy functions for protein design. *Current Opinion in Structural Biology*, 2007. **17**(2): 199–204.

49. Dantas, G. et al., A large scale test of computational protein design: Folding and stability of nine completely redesigned globular proteins. *Journal of Molecular Biology*, 2003. **332**(2): 449–460.

50. Zanghellini, A. et al., New algorithms and an in silico benchmark for computational enzyme design. *Protein Science*, 2006. **15**(12): 2785–2794.

51. Bolon, D.N. and S.L. Mayo, Enzyme-like proteins by computational design. *Proceedings of the National Academy of Sciences of the United States of America*, 2001. **98**(25): 14274–14279.

52. Dahiyat, B.I. and S.L. Mayo, Protein design automation. *Protein Science*, 1996. **5**(5): 895–903.

53. Scott, W.R.P. et al., The GROMOS biomolecular simulation program package. *The Journal of Physical Chemistry A*, 1999. **103**(19): 3596–3607.

54. Samish, I. et al., Theoretical and computational protein design. *Annual Review of Physical Chemistry*, 2011. **62**: 129–149.

55. McAllister, K.A. et al., Using alpha-helical coiled-coils to design nanostructured metalloporphyrin arrays. *Journal of the American Chemical Society*, 2008. **130**(36): 11921–11927.

56. Voigt, C.A., D.B. Gordon, and S.L. Mayo, Trading accuracy for speed: A quantitative comparison of search algorithms in protein sequence design. *Journal of Molecular Biology*, 2000. **299**(3): 789–803.

57. Kuhlman, B. et al., Design of a novel globular protein fold with atomic-level accuracy. *Science*, 2003. **302**(5649): 1364–1368.

58. Cootes, A.P., P.M.G. Curmi, and A.E. Torda, Biased Monte Carlo optimization of protein sequences. *Journal of Chemical Physics*, 2000. **113**(6): 2489–2496.

59. Jones, D.T., De-novo protein design using pairwise potentials and a genetic algorithm. *Protein Science*, 1994. **3**(4): 567–574.

60. Desmet, J. et al., The dead-end elimination theorem and its use in protein side-chain positioning. *Nature*, 1992. **356**(6369): 539–542.

61. Goldstein, R.F., Efficient rotamer elimination applied to protein side-chains and related spin-glasses. *Biophysical Journal*, 1994. **66**(5): 1335–1340.

62. Looger, L.L. and H.W. Hellinga, Generalized dead-end elimination algorithms make large-scale protein side-chain structure prediction tractable: Implications for protein design and structural genomics. *Journal of Molecular Biology*, 2001. **307**(1): 429–445.

63. Leach, A.R. and A.P. Lemon, Exploring the conformational space of protein side chains using dead-end elimination and the A* algorithm. *Proteins: Structure Function and Genetics*, 1998. **33**(2): 227–239.

64. Koehl, P. and M. Delarue, Application of a self-consistent mean-field theory to predict protein side-chains conformation and estimate their conformational entropy. *Journal of Molecular Biology*, 1994. **239**(2): 249–275.

65. Lee, C., Predicting protein mutant energetics by self-consistent ensemble optimization. *Journal of Molecular Biology*, 1994. **236**(3): 918–939.

66. Kono, H. and J.G. Saven, Statistical theory for protein combinatorial libraries. Packing interactions, backbone flexibility, and the sequence variability of a main-chain structure. *Journal of Molecular Biology*, 2001. **306**(3): 607–628.

67. Calhoun, J.R. et al., Computational design and characterization of a monomeric helical dinuclear metalloprotein. *Journal of Molecular Biology*, 2003. **334**(5): 1101–1115.

68. Eisenberg, D. et al., The design, synthesis, and crystallization of an alpha-helical peptide. *Proteins*, 1986. **1**(1): 16–22.

69. Moser, R., R.M. Thomas, and B. Gutte, An artificial crystalline DDT-binding polypeptide. *FEBS Letters*, 1983. **157**(2): 247–251.

70. Kamtekar, S. et al., Protein design by binary patterning of polar and nonpolar amino-acids. *Science*, 1993. **262**(5140): 1680–1685.

71. Dahiyat, B.I., C.A. Sarisky, and S.L. Mayo, De novo protein design: Towards fully automated sequence selection. *Journal of Molecular Biology*, 1997. **273**(4): 789–796.

72. Dahiyat, B.I. and S.L. Mayo, De novo protein design: Fully automated sequence selection. *Science*, 1997. **278**(5335): 82–87.

73. Jiang, L. et al., De novo computational design of retro-aldol enzymes. *Science*, 2008. **319**(5868): 1387–1391.

74. Rothlisberger, D. et al., Kemp elimination catalysts by computational enzyme design. *Nature*, 2008. **453**(7192): 190–195.

75. Siegel, J.B. et al., Computational design of an enzyme catalyst for a stereoselective bimolecular Diels-Alder reaction. *Science*, 2010. **329**(5989): 309–313.

76. Khare, S.D. et al., Computational redesign of a mononuclear zinc metalloenzyme for organophosphate hydrolysis. *Nature Chemical Biology*, 2012. **8**(3): 294–300.

77. Khoury, G.A. et al., Computational design of *Candida boidinii* xylose reductase for altered cofactor specificity. *Protein Science*, 2009. **18**(10): 2125–2138.

78. Saraf, M.C. et al., IPRO: An iterative computational protein library redesign and optimization procedure. *Biophysical Journal*, 2006. **90**(11): 4167–4180.

79. Grigoryan, G., A.W. Reinke, and A.E. Keating, Design of protein-interaction specificity gives selective bZIP-binding peptides. *Nature*, 2009. **458**(7240): 859–864.

80. Lanci, C.J. et al., Computational design of a protein crystal. *Proceedings of the National Academy of Sciences of the United States of America*, 2012. **109**(19): 7304–7309.

81. King, N.P. et al., Computational design of self-assembling protein nanomaterials with atomic level accuracy. *Science*, 2012. **336**(6085): 1171–1174.

82. King, N.P. et al., Accurate design of co-assembling multi-component protein nanomaterials. *Nature*, 2014. **510**(7503): 103–108.

83. Rufo, C.M. et al., Short peptides self-assemble to produce catalytic amyloids. *Nature Chemistry*, 2014. **6**(4): 303–309.

84. Smadbeck, J. et al., De novo design and experimental characterization of ultrashort self-associating peptides. *PLoS Computational Biology*, 2014. **10**(7): e1003718.

85. Wang, Q. et al., De novo design of self-assembled hexa-peptides as beta-amyloid (Abeta) peptide inhibitors. *ACS Chemical Neuroscience*, 2014. **5**: 972–981.

86. Vlieghe, P. et al., Synthetic therapeutic peptides: Science and market. *Drug Discovery Today*, 2010. **15**(1–2): 40–56.

87. Fleishman, S.J. et al., Computational design of proteins targeting the conserved stem region of influenza hemagglutinin. *Science*, 2011. **332**(6031): 816–821.

88. Bellows, M.L. et al., Discovery of entry inhibitors for HIV-1 via a new de novo protein design framework. *Biophysical Journal*, 2010. **99**(10): 3445–3453.

89. Rajgaria, R., S.R. McAllister, and C.A. Floudas, Distance dependent centroid to centroid force fields using high resolution decoys. *Proteins*, 2008. **70**(3): 950–970.

90. Smadbeck, J. et al., Protein WISDOM: A workbench for in silico de novo design of biomolecules. *Journal of Visualized Experiments*, 2013. **77**: 1–25.

91. Hao, J. et al., Identification and rational redesign of peptide ligands to CRIP1, a novel biomarker for cancers. *PLoS Computational Biology*, 2008. **4**(8): e1000138.

92. Istivan, T.S. et al., Biological effects of a de novo designed myxoma virus peptide analogue: Evaluation of cytotoxicity on tumor cells. *PLoS ONE*, 2011. **6**(9): e24809.

93. Kuroda, D. et al., Computer-aided antibody design. *Protein Engineering, Design and Selection*, 2012. **25**(10): 507–521.

94. Sivasubramanian, A. et al., Toward high-resolution homology modeling of antibody Fv regions and application to antibody-antigen docking. *Proteins*, 2009. **74**(2): 497–514.

95. Correia, B.E. et al., Computational design of epitope-scaffolds allows induction of antibodies specific for a poorly immunogenic HIV vaccine epitope. *Structure*, 2010. **18**(9): 1116–1126.

96. Correia, B.E. et al., Computational protein design using flexible backbone remodeling and resurfacing: Case studies in structure-based antigen design. *Journal of Molecular Biology*, 2011. **405**(1): 284–297.

97. Miklos, A.E. et al., Structure-based design of supercharged, highly thermoresistant antibodies. *Chemistry and Biology*, 2012. **19**(4): 449–455.

98. Sircar, A., E.T. Kim, and J.J. Gray, RosettaAntibody: Antibody variable region homology modeling server. *Nucleic Acids Research*, 2009. **37**: W474–W479.

99. Marcatili, P., A. Rosi, and A. Tramontano, PIGS: Automatic prediction of antibody structures. *Bioinformatics*, 2008. **24**(17): 1953–1954.

100. Whitelegg, N.R.J. and A.R. Rees, WAM: An improved algorithm for modelling antibodies on the WEB. *Protein Engineering*, 2000. **13**(12): 819–824.

101. Pantazes, R.J. and C.D. Maranas, OptCDR: A general computational method for the design of antibody complementarity determining regions for targeted epitope binding. *Protein Engineering, Design and Selection*, 2010. **23**(11): 849–858.

102. Pantazes, R.J. and C.D. Maranas, MAPs: A database of modular antibody parts for predicting tertiary structures and designing affinity matured antibodies. *BMC Bioinformatics*, 2013. **14**(1): 168.

103. Martins, P. et al., Nanoparticle drug delivery systems: Recent patents and applications in nanomedicine. *Recent Patents on Nanomedicine*, 2013. **3**(2): 105–118.

104. Longo, G. et al., Rapid detection of bacterial resistance to antibiotics using AFM cantilevers as nanomechanical sensors. *Nature Nanotechnology*, 2013. **8**(7): 522–526.

105. Sacha, G.M. and P. Varona, Artificial intelligence in nanotechnology. *Nanotechnology*, 2013. **24**: 45.

106. Demming, A., An intelligent approach to nanotechnology. *Nanotechnology*, 2013. **24**: 45.

107. Arbib, M.A., *The Handbook of Brain Theory and Neural Networks*, 2nd edn., 2003. Cambridge, MA: MIT Press. p. xvii, 1290pp.

108. Kecman, V., *Learning and Soft Computing: Support Vector Machines, Neural Networks, and Fuzzy Logic Models*, 2001. Cambridge, MA: MIT Press.

109. Korb, K.B. and A.E. Nicholson, Introducing Bayesian networks, in *Bayesian Artificial Intelligence*, 2004. Series in *Computer Science and Data Analysis, Series*, J. Lafferty, D. Madigan, F. Murtagh, and P. Smith, eds., London, U.K.: Chapman & Hall, pp. 29–52.

110. Lara, J., Artificial neural networks for therapeutic protein engineering, in *Medicinal Protein Engineering*, Y. Khudyakov, ed., 2008. London, U.K.: CRC Press, pp. 23–56.

111. Lara, J. et al., Artificial neural network for prediction of antigenic activity for a major conformational epitope in the hepatitis C virus NS3 protein. *Bioinformatics*, 2008. **24**(17): 1858–1864.

112. Lara, J. and Y. Khudyakov, Association of antigenic properties to structure of the hepatitis C virus NS3 protein. *In Silico Biology*, 2011. **11**(5–6): 203–212.

113. Benson, D.A. et al., GenBank. *Nucleic Acids Research*, 2013. **41**(Database issue): D36–D42.

114. Kawashima, S. et al., AAindex: Amino acid index database, progress report 2008. *Nucleic Acids Research*, 2008. **36**(Database issue): D202–D205.

115. Lara, J. et al., Computational models of liver fibrosis progression for hepatitis C virus chronic infection. *BMC Bioinformatics*, 2014. **15**(Suppl. 8): S5.

116. Campo, D.S. et al., Coordinated evolution of the hepatitis B virus polymerase. *In Silico Biology*, 2011. **11**(5–6): 175–182.

117. Thai, H. et al., Convergence and coevolution of hepatitis B virus drug resistance. *Nature Communications*, 2012. **3**: 789.

118. Lara, J., M.A. Purdy, and Y.E. Khudyakov, Genetic host specificity of hepatitis E virus. *Infection, Genetics and Evolution*, 2014. **24**: 127–139.

119. Hessel, C.M. et al., Alkyl passivation and amphiphilic polymer coating of silicon nanocrystals for diagnostic imaging. *Small*, 2010. **6**(18): 2026–2034.

120. Bredenberg, A. DNA pioneer craig venter developing a 3D printer for vaccines. 2012 (accessed on September 11, 2014); Available from: http://inhabitat.com/dna-pioneer-craig-venter-announces-work-on-a-3d-printer-for-vaccines/.

121. Mohiti-Asli, M., B. Pourdeyhimi, and E.G. Loboa, Skin tissue engineering for the infected wound site: Biodegradable PLA nanofibers and a novel approach for silver ion release evaluated in a 3D coculture system of keratinocytes and Staphylococcus aureus. *Tissue Engineering Part C: Methods*, 2014. **20**: 790–797.

122. Liu, T. et al., Anti-TROP2 conjugated hollow gold nanospheres as a novel nanostructure for targeted photothermal destruction of cervical cancer cells. *Nanotechnology*, 2014. **25**(34): 345103.

123. Morones-Ramirez, J.R. et al., Silver enhances antibiotic activity against gram-negative bacteria. *Science Translational Medicine*, 2013. **5**(190): 190ra81.

124. Chernousova, S. and M. Epple, Silver as antibacterial agent: Ion, nanoparticle, and metal. *Angewandte Chemie— International Edition*, 2013. **52**(6): 1636–1653.

125. Jiang, X. et al., Fast intracellular dissolution and persistent cellular uptake of silver nanoparticles in CHO-K1 cells: Implication for cytotoxicity. *Nanotoxicology*, 2014. [Epub ahead of print]. doi:10.3109/17435390.2014.907457.

126. Gliga, A.R. et al., Size-dependent cytotoxicity of silver nanoparticles in human lung cells: The role of cellular uptake, agglomeration and Ag release. *Particle and Fibre Toxicology*, 2014. **11**: 11.

127. Khare, S.D. and S.J. Fleishman, Emerging themes in the computational design of novel enzymes and protein-protein interfaces. *FEBS Letters*, 2013. **587**(8): 1147–1154.

128. Lippow, S.M. and B. Tidor, Progress in computational protein design. *Current Opinion in Biotechnology*, 2007. **18**(4): 305–311.

129. Harris, K., A. Kim, and A. Scwedel, eds., The great eight: Trillion-dollar growth trends to 2020, in *Insights*. 2011. New York: Bain & Company, p.6; http://www.bain.com/Images/BAIN_BRIEF_8MacroTrends.pdf (accessed on August 25, 2014).

6 Viral Nanoparticles
Principles of Construction and Characterization

Andris Zeltins

CONTENTS

6.1 INTRODUCTION

Since the Nobel Prize laureate Richard Feynman in 1959 proposed the idea about manipulating the material structure at the atomic level, nanoscience and nanotechnology have been developed as special areas of research involving a broad spectrum of methodologies from physics, chemistry, and biology. In Feynman's historical lecture called "There's Plenty of Room at the Bottom," he postulated several principally new technologic ideas in fields such as information

processing and storage, mechanical construction of small devices, friction and lubrication problems, and microsurgery [1]. Notably, he pointed also on the importance of collaboration between scientists of different scientific areas. Talking about the progress in biology, Feynman indicated on several molecular problems that the biologists solved (or still continue to study!) only in next decades, for example, how proteins are synthesized or how the light is converted into chemical energy in the nature. He stressed also the important role of physicists for the progress in biology, which is not only the methodological input but also innovative solutions for the construction of better laboratory equipment for the analysis of biological objects. Especially important opinion, which is still crucial also in current molecular biology, is the *look at the thing* principle. Modern cryoelectron microscopes are able to provide the resolutions close to atomic level and can serve as a final proof *to look* on newly constructed materials for different nanotechnology applications. However, the real atomic resolution by electron microscopy (EM) technique is still in the development [2].

The term nanotechnology has been introduced by Norio Taniguchi in 1974 [3]. He suggested to describe the nanotechnology as a complex of measures for processing of materials by one atom or one molecule. As defined by Goodsell, bionanotechnology is a special case of nanotechnology that looks to natural structures or processes for the start of development of new materials at the atomic or molecular level [4].

Now, if we go closer to the topic of this chapter, we should return to the history of virology.

The idea that viruses are objects of nanometer scale was clear already at the end of the 1930s, when Nobel Prize laureate Stanley isolated a crystalline substance with tobacco mosaic virus (TMV) properties [5] and Kausche et al. identified rod-shaped particles of 20–40 nm in diameter if TMV was observed with EM [6]. In decades preceding the fundamental idea of Feynman, different physical and chemical experiments with viruses were already started, which we would today characterize as clearly *nanoscientific*, for example, disassembly/reassembly studies [7] or first attempts to chemically modify the plant virus TMV [8]. Additionally, the experiments with purified plant virus nucleic acids demonstrated that the RNA of TMV alone can be infectious [9]. This finding can be regarded as a starting point for recombinant virus technologies. Later, the general development of molecular techniques such as DNA cloning using restrictases, polymerase chain reaction, RNA copying into cDNA using reverse transcriptases, and DNA sequencing allowed to precisely manipulate with viral nucleic acid cDNAs and construct gene vectors that enabled the initiation of infection processes from artificially created nucleic acids [10]. This idea is highly important for the construction and isolation of recombinant infectious viruses and virus-like particles (VLPs) after the expression of viral structural genes in heterologous hosts.

The next important milestone leading to the development of viral nanotechnologies is bound to the progress in protein crystallography, providing visual structural models of different viruses. Since the first 3D virus structure at near-atomic resolution was determined in 1978 by Harrison and collaborators for an icosahedral plant virus (tomato bushy stunt virus [11]), approximately 400 3D viral structural protein models are built, using experimental techniques such as x-ray crystallography and cryo-EM (as seen July 17, 2014 from the VIPER database—http://viperdb.scripps.edu/ [249]).

Viruses are natural objects in nanometer scale and are highly structurally organized; therefore, they represent powerful native

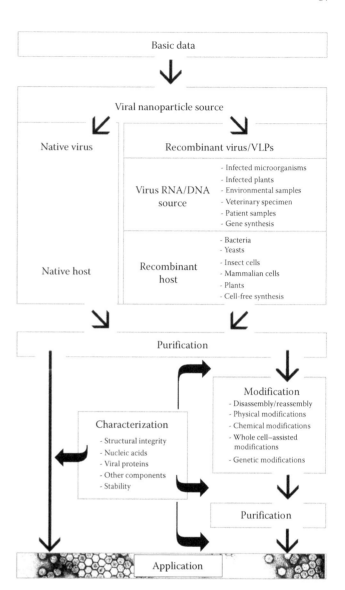

FIGURE 6.1 Overview of typical steps in the construction process of new virus-derived nanoparticles.

structural blocks for new human-made nanomaterials. This chapter is an attempt to summarize in a concise form published information and provide the reader with insight on how to obtain new viral nanoparticles and what options are known to adapt them to create new nanoparticles with defined properties and how these properties can be shown with experimental procedures. The graphical summary of this chapter is presented in Figure 6.1.

6.2 NATIVE VIRUSES AS NANOPARTICLE COMPONENTS

From the nanotechnology viewpoint, viruses can be regarded as structurally strongly organized organic multimolecular structures of nanometer scale representing nearly ideal natural building blocks for the creation of a wide variety of new materials for different purposes [12]. Viruses are built up from several hundreds to thousands of one to few types of structural proteins (coat or capsid proteins) and nucleic acid(s) coding for proteins necessary for the virus life cycle. Coat proteins (CPs) in native virions mostly are arranged according to icosahedral or helical symmetry in strongly repetitive fashion. If the data about biological, chemical, and physical properties including 3D structure of the virus are available, this repetitive organization allows the introduction of different functional molecules in the virus structure at precisely defined sites [13].

Native and attenuated viruses have been used already hundreds of years ago in early vaccination technologies (*variolation*) directly without isolation and purification. The filtration technique can be regarded as the first technology to purify the viral fraction, as suggested by Iwanowsky in 1892, who demonstrated that the causative agent of tobacco mosaic passed through the filter, which was effective in removing bacteria from infected plant extracts [14].

The next landmark in developing the necessary techniques for viral nanoparticle production is bound with the research of bacteriophages as potential therapeutic agents for veterinary and medicinal purposes since the beginning of the twentieth century. Experience gained at that time from the isolation and purification of enzymes allowed to obtain partially purified and concentrated phage preparations from cell lysates after several filtration, ammonium sulfate precipitation, and ultracentrifugation steps already in the 1930s. Chemical analysis suggested that bacteriophages are protein-containing objects and consist also of nucleic acids. In the 1940s, using the newly introduced EM technique, it became possible to visualize bacteriophages, clearly demonstrating their particulate nature and differences in morphology (for a review, see [15]). The advantages of these technological ideas are still broadly used in virus or VLP isolation and characterization experiments.

Native viruses originating from their native hosts can be used in the creation of new nanomaterials based on their intrinsic properties such as natural stability, binding to different molecules and cells, surface charge, presence of defined amino acid (AA) sequences in CPs, and their ability to bind nucleic acids and other negatively charged molecules. The first task to be solved in the creation of viral nanoparticles is to obtain a virus material in preparative amounts, allowing construction of new materials for potential nanotechnological applications. As seen from examples provided in Table 6.1, some native viruses can be purified also from their natural hosts without the need to clone and overexpress corresponding viral genes and introduced as new nanomaterials either without any chemical or genetic modification or after a few steps of simple adaptation. Mostly, well physically, biochemically, and structurally characterized bacterial and plant viruses are used for this purpose due to their availability in significant amounts and safety for mammalian organisms.

An important prerequisite for the creation of new nanomaterials derived from viral sources is the availability of biological, structural, chemical, and physical characteristics of the corresponding virus. Therefore, different basic studies of the virus are highly important, because these can reveal unexpected properties that can be used for subsequent technology development. Already mentioned plant virus TMV is probably the best characterized virus. Physically, TMV virions are highly stable 18 nm × 300 nm helical structures possessing exceptional stability at temperatures above 90°C as well as at high pressure or in vacuum. TMV is stable also in different buffer solutions at pH values between 2.8 and 8.5 as well as in organic solvents, providing the possibility to modify the virus in different chemical reactions. More detailed information about properties of TMV as well as its nanotechnological applications can be found in the review article [28].

Cowpea mosaic virus (CPMV), another plant virus with an icosahedral structure, is also widely used for different bionanotechnological applications. The CPMV particles are good candidates for use in the creation of new nanomaterials due to high outputs from infected plants and stability at elevated temperatures (60°C) in different solutions at pH 4–10, including also organic solvent–containing ones. According to the

TABLE 6.1
Examples of Native Viruses Used for Nanomaterial Preparation

Virus	Native Host	Application	Yield	Citations
Bacteriophages				
M13	*E. coli*	New tissue engineering materials	>500 mg/L	[16]
		Biomaterials with new optical properties		[17]
MS2	*E. coli*	Carrier for drug delivery	10^{10} PFU/mL	[18]
Pf1	*P. aeruginosa*	Liquid-crystalline medium for NMR	500 mg/L	[19]
		Biomaterials with tunable viscosity		[20]
Plant viruses				
CCMV	*Vigna ungiculata*	Gene delivery agents for mammalian cells	300 mg/kg	[21]
CPMV	*Vigna ungiculata*	New MRI contrasting agents	1500 mg/kg	[22]
		Template for new metal-containing nanomaterial		[23]
		Carrier for drug delivery		[24]
PVX	Potato	Tumor cell targeting	350 mg/kg	[25]
TMV	Tobacco	New MRI contrasting agents	3000 mg/kg	[26]
		Template for new metal		[27]

3D structural model data, native CPMV virions contain five surface-exposed lysine, nine carboxyl, and several tyrosine residues per asymmetric unit (L and S CPs), which allows to modify the virus in different chemical reactions [23]. Also other native plant viruses such as brome mosaic virus, cowpea chlorotic mottle virus (CCMV), red clover necrotic mosaic virus (RCNMV), potato virus X (PVX) and turnip yellow mosaic virus (TYMV) are demonstrated as potential nanomaterial candidates (for recent reviews, see [23,29–31]).

As shown in Table 6.1, several bacteriophages are broadly used for the construction of diverse nanomaterials. In some cases, purified filamentous bacteriophages (e.g., M13, Pf1) are used as advanced nanomaterials without any treatment. These can be produced in preparative amounts to the tune of a few grams and possess direction-dependent (anisotropic) physical properties. At high concentrations, phage particles can form ordered configurations (nematic-phase liquid crystals), which are anisotropically susceptible in magnetic field. This property is used in NMR structure determinations of different biological macromolecules. Bacteriophage Pf1 is an example when unmodified virus is commercially available as a product for NMR purposes, allowing to partially align proteins and nucleic acids for NMR structure determinations.

6.3　CREATING ARTIFICIAL VIRAL NANOPARTICLES

As already mentioned, native viruses without any changes can be used for nanoparticle construction only in limited number of cases. Mostly, virus-derived materials are obtained from cloned copies of viral genomes or structural protein genes. In this chapter, the main principles on how to obtain viral materials for subsequent nanoparticle construction with defined properties will be analyzed.

6.3.1　CONSTRUCTION OF NEW VIRUS-LIKE PARTICLES

The very first recombinant VLPs were obtained from cloned CP genes from the polyoma virus [32], HBV core antigen (HBcAg) [33], the HBV surface antigen (HBsAg) [34], and TMV [35] at the late 1970s and early 1980s. The first EM visualizations demonstrated that the polyoma and HBcAg VLPs were similar to viral structures that were isolated from infected hospital samples [32,36] and suggested the generally accepted principle that VLPs are identical or highly similar to their native progenitors.

As shown in Figure 6.1, several important stages can be identified in construction process of new VLPs. Further, these stages will be shortly analyzed step by step.

6.3.1.1　Basic Research Data

Construction of recombinant VLPs in most cases requires detailed information from basic studies of corresponding virus. First of all, the nucleotide sequence data are absolutely necessary. Taking into account the fact that over 2,000,000 nucleotide sequence records can be found after a search of the Life Science Search Engine (NCBI), the sequence(s) of

the viral structural protein(s) of interest can likely be found in the databases. Further, to identify viral structural genes, the information about genomic organization is highly helpful. Any data from basic research like virus life cycle; interactions with host cells; and physical, chemical, and structural information can suggest an idea for nanomaterial construction for certain applications.

6.3.1.2　Virus RNA/DNA Sources

The next stage of VLP construction is the cloning of the necessary structural genes. As the source of necessary genes, environmental samples like infected microorganisms or plants as well as clinical or veterinary specimens can be chosen for this purpose and can be amplified using different PCR techniques. Here also the careful analysis of existing information is highly important. The samples chosen for gene cloning should be at least tested at the preliminary stage for the presence of the virus under study (e.g., ELISA or PCR tests). If several virus isolate sequences are available in the databases, the sequence alignments should be carried out to find the conserved regions surrounding structural gene(s) for correct oligonucleotide design. Such an approach enhances the probability of the successful amplification of chosen genes.

As an efficient alternative in case of well-characterized viruses, gene synthesis can be used as a source of viral structural gene(s); this process is widely available also as a commercial service.

6.3.1.3　Infectious cDNAs/Cloned Structural Genes

As indicated previously, also native, fully infective viruses can serve as the source of assembly-competent protein structures for nanoparticle construction. Infectious viral cDNA is a very helpful tool allowing to obtain a virus material in the absence of viable viral stocks and providing the possibilities for different gene engineering manipulations for basic research and nanotechnological developments.

For bacteriophages like M13, phage single-stranded DNA during the infection is converted into a double-stranded supercoiled replicative form (RF), which can be isolated from infected bacteria and used for initiation of the infection in susceptible *Escherichia coli* cells [37]. This RF DNA can be also modified using standard molecular techniques to introduce the modifications in double-stranded phage DNA intermediate. The advantage of infectious phage DNA is widely used for phage display techniques as well as for construction of different M13-based nanomaterials [38]. Also for other bacteriophages, infectious cDNAs are available, for example, Qβ, T4, T7, and lambda (for a review, see [39]).

Infectious RNAs and cDNAs are widely used in plant virus research. cDNA construction and subsequent plant inoculations serve as a final proof to demonstrate the correctness of the sequence(s) of cloned plant viral nucleic acids. Simple systems to initiate the infections of plants with RNA-containing viruses require only RNA, obtained in vitro from the cDNA template under a strong bacterial promoter and corresponding RNA polymerase [40]. For plant DNA viruses, the bacterial plasmid with cloned full-length DNA, which is pretreated with an appropriate restriction enzyme to remove the plasmid vector part, can serve as the infectious cDNA, as shown in case of CaMV [41].

The principle of usage of infectious cDNAs is shown to be efficient also for different mammalian viruses. Since the first report on infectivity of the cloned poliovirus cDNA for mammalian cells [42], the specific DNA- or RNA-initiated infections are demonstrated for many viruses including members of the *Alphavirus*, *Flavovirus*, *Picornavirus*, and other taxonomical groups [43].

Regarding the cloned viral structural genes, there are a huge number of published cases. More than 100 examples are summarized in the recent review article when such cloning and expression in the appropriate host system resulted in VLP formation [44].

6.3.1.4 Expression Vector/Host Systems

For VLP production, depending on the intended nanoparticle application, one of the critical factors is the choice of an appropriate expression vector/host system. More than 170 recombinant expression systems are reported, demonstrating the successful synthesis of artificial viral particles in different host systems. Bacterial vectors and hosts are used in 28% of reported cases, mainly for the production of bacterial and plant VLPs. The insect cell systems also are widely used (28% of cases), especially for complicated mammalian VLPs, whereas yeast (20%), mammalian (15%), and plant hosts (9%) are predominantly used to obtain VLPs with specific properties for intended downstream applications [44].

6.3.1.4.1 Bacterial Systems

For VLP production in bacterial systems, mostly well-characterized commercial *E. coli* strains and expression vectors are used. Typically, viral structural protein genes are cloned in plasmid vectors under strong promoters (T7, *tac*, *trp*), allowing to achieve a high level of production of the recombinant proteins. However, proteins originating from eukaryotic cells often form insoluble inclusion bodies after expression in *E. coli* cells. This drawback can be efficiently exploited and viral structural proteins can be successfully produced in the form of insoluble inclusion bodies, purified under denaturing conditions, refolded, and self-assembled, as shown with the parvovirus B19 [45] and plant viruses CCMV [46] and CMV [47].

Alternatively, in some cases a simple change in cultivation conditions can lead to an increased production of soluble proteins, for example, low temperature *E. coli* cultures considerably stimulate the VLP formation in the cells, as demonstrated in cases of potyvirus PVY [48] or densovirus IHHNV [49]. In addition, factors such as the antibiotic resistance genes of the expression vectors and the components of the cultivation medium can affect the formation of VLPs in bacterial cells, as shown in the case of bacteriophage Qβ VLPs [50]. Moreover, simultaneous synthesis of specific RNA sequence called origin of assembly from the expression vector also can stimulate the self-assembly of VLPs in *E. coli* [51].

In most cases, efficient production of a recombinant viral structural protein using *E. coli* is possible in cases in which the target VLPs contain only one CP. However, *E. coli* cells can be used also in more complicated cases when VLPs under study require more than one type of structural protein for their formation. As demonstrated in experiments with avibirnavirus infectious bursal disease virus (IBDV), during simultaneous coexpression of VP2, VP3, and VP4 genes from the cloned polyprotein gene, the resulting polyprotein was cotranslationally autoprocessed by protease VP4, resulting in the formation of corresponding VLPs in *E. coli* cells [52]. Alternatively, simple coexpression of two CP genes from single bacterial expression vector can be sufficient to complete the VLP synthesis in bacterial cells [53].

Another approach to achieve higher expression levels and solubility of viral CPs is based on the usage of fusion protein systems. As an example, the papillomavirus HPV L1 protein was obtained from recombinant *E. coli* cells in the form of a glutathione-*S*-transferase (GST) fusion protein. The L1-VLP formation was achieved in vitro after the GST part was removed using thrombin. As a result, authors succeeded in the isolation of high-quality HPV L1 particles and were able to obtain L1 crystals for x-ray crystallographic analysis and solved the 3D structure of T = 1 icosahedral HPV L1 particles [54]. The fusion protein strategy was also successfully demonstrated in case of the picornavirus FMDV [55] and polyomavirus MuPyV [56] VLP formation.

Alternatively, also other recombinant prokaryotic organisms have been demonstrated as efficient VLP producers, for example, *Pseudomonas* was suggested to produce soluble VLPs after the expression of the plant virus CCMV structural gene [57] and HPV L1-producing *Lactobacillus* was proposed as an edible vaccine candidate [58].

However, several disadvantages have to be considered if bacteria-based approach is chosen for viral nanoparticle production. Typical for eukaryotic proteins, bacterial cells are unable to introduce posttranslational modifications, recombinant proteins do not have correct disulfide bridges in most cases, and the preparations often contain endotoxin contaminations. Besides, the *E. coli*–derived VLPs in some cases are heterogeneous in their shape, diameter, and other physical properties, and also incomplete particles are present in VLP preparations [59]. Despite these drawbacks, bacterial systems are well known, generally accepted technology for the production of different recombinant proteins, including viral structural proteins. Last developments of bacterial hosts allow to overcome at least a part of these disadvantages, for example, posttranslational modification can be completed in special *E. coli* strains, coexpressing necessary modifying enzymes [60,61]. Other ideas for recombinant protein as well as for VLP production in bacterial expression systems can be found in recent review articles [62–64].

6.3.1.4.2 Yeast Systems

Different yeast systems are broadly used for the expression of viral structural genes and the production of numerous VLPs of bacterial, yeast, plant, and mammalian origin, including commercial prophylactic vaccines against hepatitis B virus (HBV [34]) and human papilloma virus (HPV [65];

other examples can be found in Supplemental Tables of [44]). Most yeast expression vectors are based on multicopy plasmid 2 μm and are constructed as *E. coli*/yeast shuttle-vectors, containing selectable marker genes for both host cells. For efficient transcription of foreign gene(s), these vectors contain strong promoters and terminators. Useful yeast vectors were constructed by Sasnauskas and coworkers, which contain galactose-inducible promoter and form-aldehyde resistance gene, allowing to select *Saccharomyces cerevisiae* transformants using formaldehyde [66]. This vector system was efficient in producing different VLPs of bacterial and mammalian origin, including bacterio-phages Qβ and GA [67,68] and mammalian viruses, such as Schmallenberg virus [69], parvovirus [70], and human para-influenza virus [71]. Other extrachromosomal vector–based yeast systems have been used also to obtain VLPs derived from yeast [72] and plant viruses [73].

Commercial yeast expression systems based on *Pichia* and *Hansenula* strains allow to generate recombinant clones with multiple insertions of foreign gene in the yeast chro-mosome and ensuring high-level expressions. These systems have also been used for VLP production from bacteriophage [74], plant [75], or mammalian viral structural genes [76].

The aforementioned examples demonstrate the production of VLPs from the expression of a single viral structural gene. Yeast expression systems have been demonstrated as useful also in the creation of VLPs containing several structural components. The expression of rotavirus VP2, VP6, and VP7 genes in *S. cerevisiae* cells from a single plasmid vector initi-ated the formation of multilayered VLPs. The optimization of culture conditions revealed that fed-batch cultures resulted in maximum rotavirus VLP production in comparison to simple batch cultures [77]. Recent studies on HPV L1 VLPs demonstrated the influence of culture conditions not only on the output of VLPs but also on the structural stability, antige-nicity, and immunogenicity of vaccines. If higher concentra-tions of carbon source were used for *S. cerevisiae* cultivation, the resulting VLPs were able to induce higher neutralizing antibody titers against HPV as control cultures [78].

However, yeast systems are not always suitable for the synthesis of eukaryotic VLPs. So, attempts to obtain VLPs from enveloped viruses like HIV-2 from *S. cerevisiae* cells were unsuccessful, probably due to the absence of necessary host factors in yeast cells [79].

The yeast system can be used also to introduce the neces-sary mRNA in the structure of corresponding VLPs directly in the cultivation process without the need to disassemble and reassemble the particles. As shown for yeast-produced bacteriophage MS2 VLPs, simultaneous synthesis of the mRNAs coding for CP and that of the model protein from the same expression vector are necessary. Such a combination ensures the encapsidation of functional heterologous mRNA, provided that the MS2 packaging sequence is included in these mRNAs [80].

The yeast expression system is also useful for solving the solubility problems of the target viral proteins. In the case of CCMV VLPs, production in *Pichia* cells results in soluble

viral nanoparticles, whereas recombinant *E. coli* cells form insoluble CP containing inclusion bodies [75].

Yeast cells as expression hosts are suitable also for the introduction of necessary posttranslational modifications in VLPs, as shown with recombinant *Pichia pastoris* producing phosphorylated HBcAg VLPs [76] or *Hansenula polymor-pha* synthesizing partially glycosylated HBsAg particles [81].

Generally, yeast-based systems combine advantages of eukaryotic cells in posttranslational modifications and absence of bacterial endotoxins with bacterial system properties such as cultivations in the short periods of time in media of sim-ple composition and comparably easy genetic manipulations. These systems are regarded as safe and are broadly introduced for the production of different medical proteins in many cases. More detailed information about different yeast systems and their applications is provided in recent reviews [82,83].

6.3.1.4.3 Insect Cell Systems

Insect cells as expression hosts for recombinant proteins, produced from baculovirus vectors, are known since the 1980s [84]. Baculovirus vectors are genetically reconstructed rod-shaped insect viruses, where the naturally occurring polyhedrin gene is replaced with cDNA of a recombinant protein and put under strong polyhedrin or p10 promoters. Recombinant proteins from this system can be processed, posttranslationally modified, and targeted to different loca-tions in insect cells. Mostly, lepidopteran cell lines such as Sf9, Sf21, and Tn5 are used for the production of recombinant proteins. The insect cell line derived from *Trichoplusia ni* (Tn5) is used for the production of the commercial HPV VLP bivalent vaccine. In addition, recently regulatory authori-ties approved such isolated VLPs, from recombinant insect cells, as veterinary vaccines against classical swine fever and human seasonal influenza subunit vaccine [85].

Insect cell systems are especially useful for the construc-tion of complex, nonsymmetric VLPs derived from enveloped viruses [86]. Here, different baculovirus vector construction strategies can be exploited: coinfection strategy, where insect cells are infected with multiple different monocistronic recombinant baculoviruses, and coexpression strategy, where a single polycistronic baculovirus vector serves as a vector [87]. Using the coinfection for VLP production, up to six dif-ferent proteins have been demonstrated to be simultaneously synthesized and assembled into herpesvirus-like particles [88]. For influenza VLPs, the tricistronic baculovirus vector ensured the production of corresponding viral structures in insect cells [89].

Similar to bacterial and yeast systems, also recombinant baculovirus/insect cell system allows to obtain VLPs from viruses with polyprotein genome organization. After clon-ing of the whole polyprotein gene of alphavirus SAV [90] or plant virus CPMV [91] in the baculovirus vector, the cultiva-tion of recombinant cells resulted in efficient VLP production in both cases. Interestingly, CPMV particles obtained from insect cells were empty, without encapsidated RNA.

Establishing the optimal culture conditions is important also for recombinant protein production in insect cells.

As already mentioned, SAV VLPs were obtained only after reducing the cultivation temperature [90]; the effect can be bound with intrinsic properties of fish-borne SAV structural proteins to form correct folding at water environment temperatures. As shown in a recent study, changing the pH of the cultivation medium can considerably enhance the VLP production in insect cells. The specially adapted cell line SfBasic at higher pH values produced up to 11 times higher amounts of Chikungunya VLPs than parental Sf21 cells [92].

It should be mentioned that the described baculovirus vectors can also be used for whole insect larvae infections, allowing to produce VLPs containing up to six different structural proteins in their structure from a single polycistronic vector [93].

Baculovirus expression systems are extensively used for VLP production in basic research and on an industrial scale due to its several advantages, such as the ability of post-translational modification of recombinant proteins similar to mammalian cells, fast growth rates in animal product-free media, and the capacity for large-scale cultivations.

6.3.1.4.4 Plant Systems

If other recombinant protein production systems for industrial use are known since the 1980s, plant systems are comparably new; only in the last few years, initial developments have reached the industry level and approvals from regulatory authorities. Expression systems based on manipulated plant viral genomes are extensively studied since the late 1980s when the first demonstrations of infectivity of artificially synthesized nucleic acids in whole plants were demonstrated, as discussed earlier. The postulated ideas about edible vaccines and new agriculture supplying the market with recombinant medicinal proteins promoted intensive basic and applied studies with an aim to create efficient production tools from plants suitable for industrial use at cGMP conditions [94,95]. Most effective plant systems are very complicated contrary to that of other organisms and are based on the agrobacterial transfer of viral nucleic acid into plant cell nucleus without stable integration in the plant genome, where infective transcripts are synthesized and translocated to the cytosol and infection process is initiated. Probably, tobamovirus-derived *magnifection* technology is most efficient and is developed toward usage at industrial scale [96]. This system exploits several important principles:

- The vacuum infiltration technique can ensure the infection of most part of cells in whole plant leaves.
- Several *Agrobacterium* cells are able to infect the same plant cell simultaneously and transfer recombinant DNAs into the plant cell nucleus.
- Recombinant nucleic acids can be reduced in size by separation in several vector modules and subsequently joined by site-specific recombination.
- Viral transcripts can be made more *nucleus-like* by inserting intron sequences.

- Characteristic virus-induced posttranslational silencing in plants can be reduced by special transgene modules.
- CP genes can be removed from the vector and replaced by a foreign gene, if the system does not require the CPs for systemic movement function of viral transcripts through the whole plant.

Based on these principles also potato virus X–derived vector is constructed. Combination of these two vectors allows to produce hetero-oligomeric recombinant proteins simultaneously. The system is validated by the synthesis of more than 20 different monoclonal antibodies (mAbs), cytokines, interferon, different antigens, and other medicinal proteins. For these vectors, nonfood tobacco plant host *Nicotiana benthamiana*, its usage at industry-scale conditions, and principles of process biosafety for plant-made pharmaceuticals are suggested [96,97]. The *magnifection* technology was demonstrated to be efficient also in the production of VLPs for nanotechnological applications with a high output (>3 g/kg of plant biomass [98]).

Furthermore, based on agrobacterial transfer of viral transgene, also icosahedral plant viruses have been developed as efficient plant vectors. As an example, very effective transient expression system derived from plant virus CPMV was introduced several years ago. The main component of this system is a modified 5′-untranslated region (UTR) from CPMV RNA-2, which is able to considerably enhance the translation of foreign gene when inserted in the front of a transgene. Making use of CPMV UTR-based binary vectors, several VLPs were produced for different immunological and nanotechnological studies, as summarized in recent review [99].

Regarding the plant DNA-containing viruses, several ssDNA viruses including geminiviruses are used for the creation of expression vectors. Interested readers can find more detailed information about the history and newest developments of these expression systems in the recent review [100].

Plants are very promising as the source of recombinant nanoparticles, as recently developed systems allow to produce considerable amounts of them in a comparably short time at the industrially accepted, contained, environmentally save conditions; products can contain posttranslational modifications similar to that of human proteins. The potential of plant systems for industrial applications is supported by the fact that Norwalk VLPs are manufactured using *magnifection* technology under cGMP conditions and the VLP vaccine entered the Phase I clinical trial [101].

6.3.1.4.5 Mammalian Cell Systems

Mammalian hosts for the industrial production of different medicinal proteins are well known since the 1980s. As recombinant products from these hosts are structurally authentic and are correctly posttranslationally modified, more than a half of the recombinant proteins in pharmacological industry are produced using several mammalian cell lines [102,103], and approximately 70% of them are Chinese hamster ovary (CHO) cells [104]. The expression systems for mammalian cells can be constructed based on plasmid-type transient expression vectors, which contains active origin of replication, ensuring a high copy number in transfected cells, efficient promoter, and mRNA processing signals. Alternatively, also engineered mammalian virus–based vectors or stable cell lines are used for foreign protein productions [104,105]. Despite complicated and expensive construction and cultivation processes, mammalian systems are widely used in basic research and also in industry for the production of nanoparticles of complex enveloped viruses, mainly for vaccine candidate purposes. As recent examples of efficient expression systems, influenza [103] or Japanese encephalitis virus (JEV [106]) VLP production in mammalian cell lines should be mentioned here. Interestingly, as shown with JEV VLPs, also in case of mammalian cells, differences in cultivation conditions can influence the VLP production and targeted optimization can considerably enhance the outputs of recombinant antigens.

Furthermore, the advantages of viral polyprotein expression strategy are demonstrated also in case of mammalian nanoparticle productions. Using the lentiviral vector with alphavirus CHIKV polyprotein–encoded gene, transient expression resulted in VLP formation with a high output; VLPs were able to elicit strong virus neutralizing antibodies, suggesting a high structural similarity with native CHIKV virions [107].

The main benefit of mammalian cell systems for VLP production is the mentioned structural identity to their native viral counterparts, which is highly important for efficient vaccine construction. For other purposes, the usage of mammalian systems is probably too complicated and expensive due to low yields, complex media, long periods of sterile cultivations, and risks of virus contaminations.

6.3.1.4.6   Cell-Free Systems

Basic research on cell-free protein synthesis is known since the 1960s, when extracts from bacterial, wheat germ, insect cells, and rabbit reticulocytes were used in transcription and translation studies [108]. In last years, among of different applications, cell-free systems progressed also in field of viral nanoparticle production. Using such models as VLPs derived from HBcAg, bacteriophages MS2 and Qβ, it was shown that output cell-free synthesized viral nanoparticles can reach the level of 1000 mg/L. Moreover, the system allows to introduce disulfide bounds for VLP stabilization, incorporate unnatural AA, and even achieve ordered and oriented display of flagellin from *Salmonella* on the HBcAg surface [109]. Additional conceptual ideas and different aspects of cell-free technologies can be found in the recent review article [110].

6.3.2   Purification of Viral Nanoparticles

The next logical stage in the creation of new viral nanoparticles is elaboration of efficient purification protocols, allowing to obtain pure preparations of synthesized VLPs without loss of their structural and functional properties for further modifications and/or nanotechnological applications. Even more, chosen purification procedures may not only affect overall particulate structure of VLPs, but also influence physical properties of the surface and internal content of virus-derived assemblies [111]. Therefore, also at this stage, the data from basic research of corresponding virus have to be considered. For newly obtained, uncharacterized VLPs, initial conditions for purification can be chosen from those established in isolation of corresponding native viruses.

As in most cases, VLPs after synthesis are located in host cells; the very first important step in purification is cell lysis, to transfer target particles to the solution for further treatment. Some eukaryotic cell types require only mild treatment with detergent solutions to obtain cell-free extracts. However, in the case of bacterial, yeast, and plant cells, strong mechanic disintegration procedures such as French press, ultrasonication, and grinding with abrasives (e.g., aluminum oxide or carborundum) are applied. To facilitate cell disruption, additionally repeated freezing/thawing cycles and enzymatic treatments can be used. However, several mammalian and insect cells are able to secrete VLPs outside of the cells; for these systems mechanical treatment is not necessary [112]. The choice of cell lysis protocol is very important, because at conditions influencing the particle integrity, the VLPs can be lost at a very early stage, and wrong conclusions can be drawn about the suitability of the chosen vector/host system to synthesize and self-assemble desired viral particles. Therefore, it is necessary to test several mechanic treatment and extraction buffer combinations, in which supplements such as salts, protease inhibitors, detergents, low concentrations of urea, reducing agents, and chelating agents can be included.

In some cases, if the target VLPs are thermostable, the cell extracts can be treated at elevated temperatures. As shown with HBcAg from *E. coli*, incubation of the cell homogenate at 60°C before the removal of cell debris resulted in a higher yield and purity of VLPs [113].

As a next step, VLP-containing solutions are clarified by centrifugation or ultrafiltration and further concentrated and purified. To reduce the volume of the extract and amounts of host-derived impurities, for stable, salt-resistant VLPs, the precipitations with ammonium sulfate or polyethylene glycol (PEG) can be included in the purification protocol. This treatment effectively reduces the number of necessary chromatography and/or centrifugal steps. At small-scale VLP purifications, further successive rounds of low- and high-speed centrifugations, as well as different sucrose- or CsCl-gradient systems, can be applied to obtain sufficiently pure preparations for subsequent applications (e.g., [48,114]). However, ultracentrifugation is less compatible with large-scale preparations in industry, due to problematic scalability and labor intensity [112]. Therefore, ultracentrifugations are often substituted with different ultrafiltration and chromatographic processes, which at least in some cases can yield even several times higher outputs of VLPs than centrifugation techniques [115].

The choice of different size-exclusion, ion-exchange, hydrophobic interaction, and affinity chromatography techniques in purification largely depends on intrinsic properties of corresponding viral particles. Very efficient purification of VLPs can be achieved by broadly applied size-exclusion chromatography, in which molecules and their assemblies in solution are separated by their size. This technique is also widely used in industry. As shown, for example, with the RHDV VLP vaccine, produced in *P. pastoris* on a large scale, a long column packaged with sepharose CL4B was a highly effective approach in viral particle purification [116]. On the laboratory scale, a combination of different buffer solutions and precipitations with only one size-exclusion chromatography step can be sufficient to obtain satisfactory pure VLP preparations for further studies [117].

Next, ion-exchange chromatography is also often used in protein purification. The principle of separation is based on the ability of biological macromolecules to bind charged polymers and to release at elevated ionic strength. Therefore, the efficient purification on ion-exchange columns will be successful only in case if VLPs are stable in solutions with high salt concentrations. However, in some cases, the disassembly of VLPs during the ion-exchange chromatography step can be effectively used for purification of assembly-competent CP dimers, as shown with φCb5 VLPs [74]. Ion exchangers have been shown suitable also for the purification of filamentous viruses. In the recent study based on the usage of a monolith column material, it was shown that only one chromatography step was necessary to purify the plant virus PVY to apparent homogeneity after optimization of the conditions [118].

Mentioned monolith materials are reported to be more efficient in VLP purifications than conventional beaded resins. As demonstrated with HBcAg VLPs, the hydrophobic interaction chromatography using the monolith column is three times more efficient than conventional resin. However, the pretreatment step was necessary to remove most part of lipids from the yeast cell extract [119].

If corresponding binding partners are known from basic research for VLPs under study, special affinity supports are synthesized in order to purify the particles with high specificity. Moreover, passing the VLPs through such affinity column can even improve the structural and immunological properties, as demonstrated with L1-VLPs derived from HPV using heparin chromatography [120].

Viral nanoparticles after purification should be free from host-derived impurities. If the particles are intended for medical or veterinary use, chosen purification scheme have to ensure the separation the target product also from such contaminations as endotoxins. For this purpose, different specially designed chromatography materials are used, as well as extractions with detergents such as Triton X-114 [121,122].

Detailed analysis of different aspects of VLP purification can be found in recent review articles [123,124].

6.4 MODIFICATION OF VIRAL NANOPARTICLES

As discussed in previous chapters of this review, native and recombinant viral nanoparticles can be used directly for intended nanotechnological applications only in a limited number of cases, for example, in immunizations or in some physical applications. Mostly, obtained viral particles serve as a starting material for necessary modification processes to construct the nanoparticles with desired properties for further use. Here, the principles of physical, chemical, or genetic modifications will be discussed and most interesting examples will be provided.

6.4.1 PHYSICAL TECHNIQUES

In this chapter, different physical techniques that can be applied to new viral nanoparticle construction will be discussed. In many cases, to achieve the chosen physical interaction also modification techniques such as chemical or genetic modifications are necessary, which will be analyzed later.

6.4.1.1 Disassembly/Reassembly

In many occasions, to obtain new nanomaterials, the viral particles first have to be disassembled and subsequently reassembled to introduce the required functionalities. Generally, viruses are built up from one or several types of structural proteins and encapsidated nucleic acids, and the multimolecular viral structures are stabilized by protein–protein and protein–nucleic acid interactions. Depending on the virus species, additional stability is often achieved by the involvement of metal ions [125] and/or disulfide bridges [126], which provide supplementary bounds between CP subunits in the virion structure.

The simplest way to disassemble the viral structures is the application of typical protein denaturing conditions using buffer systems containing high concentrations of chaotropic agents like urea or guanidinium salts, as well as strong detergents in or without combination with reducing agents, as usual at sodium dodecyl sulfate polyacrylamide gel electrophoresis (SDS/PAGE) analysis. However, most viral structural proteins are irreversibly denatured at such conditions, and after such a treatment, it is not possible to achieve efficient reassembly in terms of the characteristic particle structure and yields. Therefore, it is necessary to use the disassembly protocols adapted to corresponding virus or VLP. For some objects such as norovirus VLPs, efficient disassembly can be completed by simply changing the pH of the solution to basic. The reassembly can be further accomplished by increasing the ionic strength; the resulting VLPs are empty, with heterogeneous isometric morphology [127]. A similar principle of pH-dependent disassembly/reassembly process was demonstrated in bacteriophage MS2: CP dimers formed after virus treatment with acetic acid, and viral structures can be reassembled again in neutral solutions in the presence of RNA [128].

For metal-containing viruses, the addition of chelating agents at high ionic strength is necessary to destroy the structure, as shown with plant virus SBMV. Then the reduction of ionic strength and adding of bivalent cations together with specific nucleic acids in the solution of disassembled SBMV CPs leads to the reassembly of T = 3 symmetry icosahedral particles [129].

In some cases, as shown for recombinant HBcAg, VLP disassembly can be achieved using enzymes that degrade encapsidated nucleic acids. Subsequent incubations with polyanions such as nucleic acids and polyglutamic and polyacrylic acids restore the VLP structure of different HBc VLP variants [130].

For enveloped viruses, for example, alphaviruses, the reassembly process is more complicated. First, alphavirus core-like particles are generated in vitro from recombinant CP in the presence of nucleic acids or other cargo molecules. Then these particles are introduced in mammalian cells that transiently express alphavirus glycoproteins. As a result, viral nanoparticles released from the cells contain a cargo-containing core, lipid envelope, and typical glycoproteins [131].

The assembly process, based on the intrinsic properties of viral proteins, is one of the most important steps in viral nanoparticle production. Moreover, the ability of proteins of enveloped viruses to incorporate in phospholipid membrane vesicles and self-assembly in vitro without involving the cell system is the basic principle exploited in the so-called virosome technology. Several developed virosomes are suggested as prophylactic or therapeutic vaccines, including against influenza [132].

For nanoparticle technology development, the inclusion of disassembly/reassembly into the production process can improve the properties of the target viral particles, as shown with enhanced stability and immunological properties of HPV VLPs [133].

Previously discussed examples demonstrate clearly that there are no generally applicable disassembly/reassembly conditions for all viruses and VLPs; they largely depend on the intrinsic properties of the virus under study. As seen from the data in Table 6.2, also here the most effective protocol has to be established individually.

6.4.1.2 Nucleic Acid Packaging

One of the most important characteristics of viral structural protein is the ability to bind nucleic acids. Currently, the assembly process of viruses often is interpreted as a sequence-independent nonspecific packaging of the genomic nucleic acid based only on electrostatic interactions with CPs. However, recent studies suggest that specific

nucleic acids play very important role in viral particle formation, especially at early stages [142,143]. Viral nucleic acid conformation is subjected to significant collapse by nearly stoichiometric binding of CP molecules to the multiple packaging signals present in viral genomes. After this first stage, additional CP molecules bind to the intermediate until the viral particle is completed. As shown in model experiments with bacteriophage MS2, unspecific RNA is not collapsed during the binding. As a result, unspecific host RNAs if they even support the assembly process can initiate the formation of irregular structures [143], which are often found when VLPs are produced in heterologous systems. For practical purposes, to package desired nucleic acids in viral nanoparticles in vitro, these should be adapted to the corresponding CP or vice versa—the necessary changes should be introduced in the structure of CPs [142]. Alternatively, unchanged binding partners are tested in order to assess their ability to form VLPs.

The encapsidation preference of specific RNA was recently demonstrated in next-generation sequencing experiments with insect virus FHV. It was found that native particles contain only 1% of host-derived RNAs, whereas recombinant FHV VLPs are able to encapsidate a large number of different insect cell host RNAs, including rRNA, transcripts from the baculovirus vector, and its own CP mRNA [144]. The encapsidation of different host RNAs, such as rRNAs, tRNAs, and mRNA of the transgene, was shown with different VLPs [145–148]. However, not always these most abundant host RNAs are found to be encapsidated in VLPs, as demonstrated with MRFV from *E. coli* or CCMV from *Pseudomonas* [53,57]. Interestingly, plant virus STNV genomic RNA contains at least eight different secondary structures that are built up from similar sequences and are able to strongly bind to immobilized CPs in the SELEX system. Further sequence analysis revealed that the STNV genome can contain up to 30 such secondary structure/sequence motifs, which can bind the CPs with different affinities [149].

The aforementioned data suggest that one of the important factors for successful VLP formation in heterologous host is

TABLE 6.2
Examples of Disassembly and Reassembly Conditions for Different Viruses or VLPs

Virus/VLP	Disassembly Conditions	Reassembly Conditions	Citations
M13	Detergents (SDS, cholate)	Phospholipid bilayers	[134]
MS2	66% acetic acid	50 mM Tris; 100 mM NaCl specific RNA	[135]
PP7	6 M urea, 10 mM DTT; storage: Na acetate, pH 4	Tris pH 8.5; RNA	[136]
CCMV	0.9 M NaCl, Tris pH 7.4	0.9 M NaCl Na acetate pH 4.8	[137]
PVX	2M LiCl, at −20°C	10 mM Tris pH 7.6 +20°C	[138]
TMV	0.1 M carbonate pH 10.5	30 mM acetate pH 6; nucleic acid	[139]
HBcAg	1.5 M guanidine/HCl, 0.5 M LiCl, 50 mM HEPES pH 7.5, 2 mM DTT	0.25 M NaCl, 50 mM HEPES pH 7.5, 2 mM DTT	[140]
HPV	20 mM Tris pH 8.2, 50 mM NaCl, 2 mM DTT	1 M NaCl, pH 6.2–8.2	[141]

the availability of RNAs in the transcriptome with necessary sequence/secondary structure properties to initiate and support VLP formation.

To encapsidate different nucleic acids in VLPs, first viral and/or host nucleic acids present in particles have to be removed. It can be achieved by the previously discussed disassembly process and then the necessary nucleic acid is introduced by subsequent reassembly. For example, if disassembled plant virus CCMV CPs are incubated with virus-specific in vitro synthesized RNA, they form VLPs highly similar to native virions [46]. Also bacteriophage φCb5 CP dimers are able to package the model RNA during in vitro VLP formation [74].

However, for some VLPs, the disassembly is not necessary to remove host nucleic acids. Instead, simple prolonged incubations at basic pH [122] or RNAse treatment can effectively eliminate the VLP RNA content. Subsequent incubations with oligonucleotides result in filled VLP, as known from experiments with Qβ and HBcAg VLPs [150].

Alternatively, chosen nucleic acids can be packaged in VLPs also during the cultivation of recombinant host. Such an approach was used to elucidate the size limit for polyomavirus JC VLPs that are able to encapsidate model plasmids up to 9 kbp directly, when CP is simultaneously expressed with model plasmid synthesis in *E. coli* cells [151]. Also the yeast expression system allows to package model mRNA in bacteriophage VLPs in vivo [152].

To facilitate the targeted packaging into VLPs, the necessary nucleic acids can be provided with packaging signals of corresponding virus. The well-known example is the translational operator (TR) sequence from bacteriophage MS2, which can be added to different functional RNAs to stimulate the efficient encapsidation in MS2 VLPs [153]. Additional aspects of interaction between viral structural proteins and nucleic acids are discussed in recent publications [154,155].

6.4.1.3 Protein Encapsidation

Most part of successful protein encapsidation systems is knowledge based. The first principle used for this purpose is the selection of a corresponding viral carrier that encapsidates additional proteins naturally. Possibly, the best developed carrier for packaging of different functional proteins based on this principle is constructed from bacteriophage P22. The VLPs derived from P22 are built up from two major proteins, 420 copies of CP and up to 330 copies of scaffolding protein (SP). The SPs are incorporated inside of VLP through the noncovalent binding. As only C-terminal part of SP is necessary for VLP formation, the N-terminal part can be used for genetic fusions of different proteins in order to achieve controlled encapsidation during the cultivation of recombinant *E. coli* cells. In case of alcohol dehydrogenase, up to 250 copies of enzyme were loaded into the interior of P22 VLPs [156].

Another principle, which can be exploited in encapsidation of different proteins inside of VLP carrier, is the inclusion of special modifications in viral CP, which stimulate the packaging of desired molecules. An elegant solution

for foreign protein packaging was demonstrated with plant virus CCMV using small leucine zipper-like AA motives (E-coil and K-coil) built in the structure of CP and the binding partner. Here, using genetic fusions, the N-terminal part of CP and the C-terminal part of the model protein were provided with coiled-coil motifs. In vitro binding of E-coil- and K-coil-derived proteins and encapsidation with the help of purified unmodified CPs from native virus resulted in encapsidation of 14 green fluorescent protein (GFP) molecules inside of CCMV VLPs [157].

Another modification that is reported to stimulate the encapsidation of proteins in the VLP interior is based on RNA–protein interaction. If simultaneously with bacteriophage Qβ CP gene, the gene containing the sequence of model protein, arginine-rich peptide and its RNA-binding partner along with Qβ packaging signal is coexpressed from two plasmid vectors in *E. coli*, the polynucleotide-mediated packaging system produces VLPs with up to 18 model enzyme molecules inside 1 particle. Encapsidation in this case is directed by RNA aptamer sequences that join the CP and a peptide tag fused to the chosen enzyme [158].

Interesting results in protein encapsidation were shown with plant RNA virus CMV. The VLPs from this virus are able to reassemble in the presence of nonspecific, 20–45 nt long ssDNA oligonucleotides or short dsDNA. If preincubated biotinylated oligonucleotides and streptavidin were introduced into the reassembly reaction, resulting VLPs contained streptavidin, one of the largest (53 kDa) proteins known to be introduced inside of VLPs [159].

In some cases, simple disassembly/reassembly technique can be used without any changes in viral CPs and protein to be encapsidated. It was demonstrated with the same CCMV as carrier and horseradish peroxidase, taken in excess in reassembly experiment. At these conditions, a low number of enzymes can be packaged inside of VLPs based on the statistical process, which is dependent on the initial concentrations of proteins [160].

6.4.1.4 Binding and Packaging of Different Functional Molecules

The idea that viral structural proteins can be used for the packaging of different materials other than nucleic acids is well known for many years. For encapsidation of small molecules, the property of several VLPs such as the ability to open or close the pores in viral shells at different conditions can be successfully exploited. If empty CCMV VLPs were incubated in comparatively highly concentrated solutions of paratungstate at conditions with open pores, then after lowering of pH and pore closure, the crystals of corresponding salts can be obtained. Such an approach—*virus mineralization*—allows to produce single crystals with defined dimensions [161].

Similar conditions with changing salt concentrations and pH can be also used for encapsidation of other low-molecular-weight compounds in VLPs, including hydrophobic therapeutic substances [162,163].

In special cases, when a ligand strongly binding to the VLPs is found in previous studies, it is possible to achieve the placing of functional molecule on the VLP surface. This principle is demonstrated for HBcAg, where the recombinant IL2 molecule with a selected binding peptide is complexed with HBcAg particles without chemical coupling or genetic fusion [164].

After corresponding modifications also metal nanoparticles can be coated with viral structural proteins, keeping the typical viral morphology. Gold nanoparticles are able to actively react with different mercapto group–containing compounds; this reaction can help to adapt the particles for encapsidation. If nanogold particles are coated via mercapto groups with long PEG-like oligomer mercaptononyl heptaoxatricosanoic acid, the resulting negative charge on gold surface stimulates the formation of typical bacteriophage VLPs with encapsidated gold [74]. Analogous protocol is efficient also for iron oxide particle encapsidation; however, in this case, a carboxy-terminated PEG-like oligomer with a hydrophobic tail is necessary to form micelles around the iron oxide nanoparticles. The final step of encapsidation is similar to that of previous example and is based on interaction between negative charges in polymer and positive charges in CPs of plant virus BMV [165]. Alternatively, simple metal nanoparticle binding on the virus surface can be achieved without chemical or genetic modifications for viruses, which are resistant to acidic solutions [27].

For certain nanotechnological applications, it is necessary to cover different surfaces with highly structurally ordered protein structures such as viruses or VLPs. So negatively charged CCMV particles can be assembled without chemical reactions into highly ordered configurations with the help of amphiphilic dendrimers, resulting in supermolecular structures with a clear cubic structure and long-range order [166]. Viral particles can be attached to polyelectrolyte layers also directly, as shown with plant virus TMV. If modified TMV, which contains cell-binding peptides on the virus exterior, is incubated on surfaces coated with several polyallylamine/polystyrene sulfonate layers, the resulting ordered surface is highly stimulating for bone cell differentiation [167]. Other examples of exploitation of physical binding principle in virus-based nanomaterial construction can be found elsewhere [13,38].

6.4.2 CHEMICAL TECHNIQUES

Generally, the editing of proteins is well known from processes observed in living cells, where different posttranslational modifications serve as tools for their adaptation to certain functions. Bioorganic chemistry offers a huge amount of artificial ways to chemically modify proteins and other biomolecules; however, not all of them are compatible with the preservation of the native structure and biological activity. The main requirements for successful chemical transformation of biological molecules are bound with aspects such as effectivity of the reaction in the water solvent with physiological pH, at temperatures acceptable for living organisms, and low concentrations of comparably nontoxic reagents [168]. In case of monomeric proteins, the conditions for the reaction

have to ensure the maintenance of biological activity. For viral structural proteins, additionally the typical 3D viral structure has to be preserved, if the potential downstream application foresees the usage of nanoparticles with a defined multimeric structure.

In this chapter, the principles of most popular protein modification methods will be analyzed and examples of chemically modified viral nanoparticles will be provided.

6.4.2.1 Modifications of Amino Groups

Freely accessible lysine residues on the surface of VLPs are probably the most popular targets for modifications and coupling of different functional molecules. Lysine residues contain primary amines, which can be modified at physiological pH values, despite their protonated state at these conditions.

The simplest way to modify virus nanoparticles at free amino groups is their reactions with commercially available *N*-hydroxysuccinimide (NHS) esters, which result in formation of bound stable amide (Figure 6.2a). This method is widely used for fluorescent labeling of different proteins including viral CPs, as shown with Alexa Fluor–labeled reoviruses [169] or plant CPMV VLPs [170]. NHS-modified PEGs are widely used reagents for PEGylation of different potentially valuable therapeutic proteins and VLPs to enhance the circulation time after injections in mammalian organisms [25]. However, often necessary NHS esters are not commercially available. Then, for construction of nanomaterials, new NHS esters have to be prepared from the free carboxylic group–containing ligands of choice in order to modify viral particles at free amine groups. Similarly, also isothiocyanates are well-known labeling reagents for viruses, for example, FITC [171].

Accessible amino groups on viral particles can be modified also using hydrazone chemistry in a two-step reaction (Figure 6.2c). First, using NHS chemistry, lysine residues on the VLP surface are modified with formylbenzamide; then the reaction with the aldehyde group of the peptide of choice results in the formation of a stable hydrazone bond [172].

One of the most used amino group modification techniques is the so-called *click* chemistry (Figure 6.2d). The principle of modification is based on primary amine conversion into organic azide and further reaction with alkyne-containing ligand in presence of Cu(I) to reduce the temperature necessary for cycloaddition [173]. It is applied in many cases also for virus and VLP modification for different downstream applications, for example, for VLP decorating with tumor-associated carbohydrate antigens [174], biotin, different dyes, and PEGs [175], as well as for introduction of multiple sulfated ligands [176]. However, before deciding to use this technique, some drawbacks should be considered, such as the ability of Cu(I) to generate reactive oxygen species, which can irreversibly affect the structural and functional integrities of viral coats. To overcome this disadvantage, a new Cu-free azide-alkyne cycloaddition method was suggested recently [177]. Extensive analysis of reaction mechanisms of *click* chemistry as well as applications in the construction of different nanomaterials can be found in a recent review article [178].

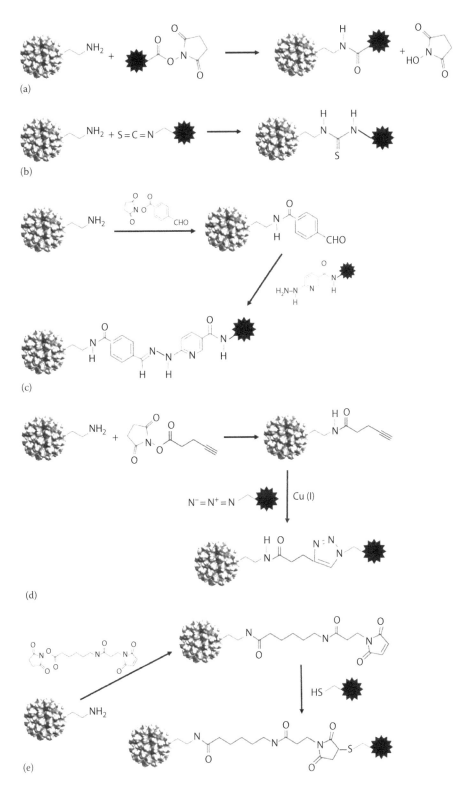

FIGURE 6.2 Modification of free amino groups in the VLP structure [168,178]. (a) Reaction of primary amines (e.g., Lys) with NHS esters forms a stable amide bond; (b) reaction of primary amines with isothiocyanates results in thiourea derivatives; (c) two-step hydrazone chemistry reaction; (d) principal scheme of *click* chemistry reactions; and (e) example of reaction of primary amine with a heterobifunctional linker reagent (succinimidyl-6-[(β-maleimidopropionamido)hexanoate], SMPH).

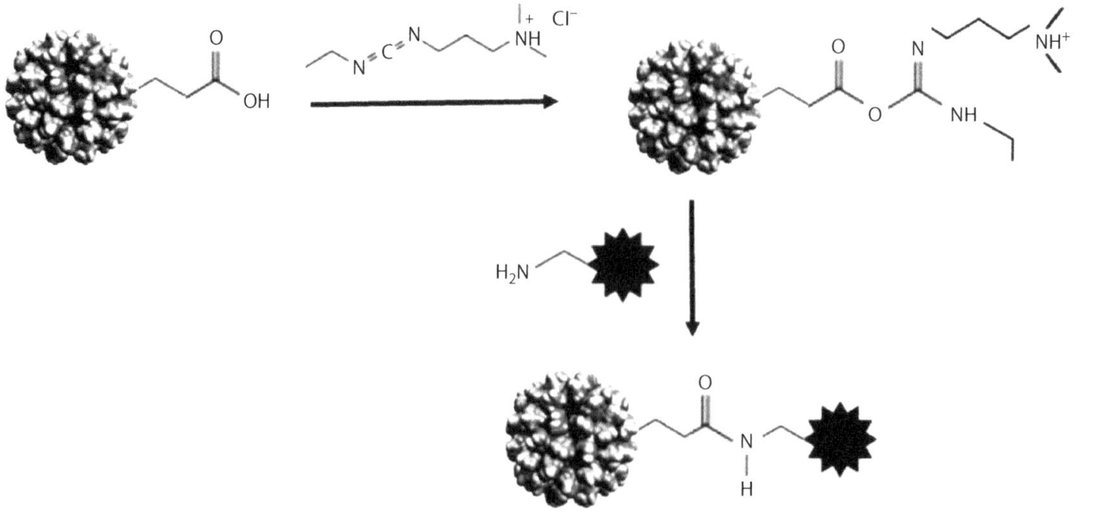

FIGURE 6.3 Modification of free carboxylic groups in the VLP structure [168]. Carboxylic groups (e.g., Asp, Glu, or protein *C*-termini) form a stable amide bound with free amino groups in a two-step reaction with a water-soluble carbodiimide.

Lysine residues located on surface can be activated with different bifunctional linker reagents, allowing them to react, for example, with sulfhydryl groups of chosen ligands (Figure 6.2e). The cross-linkers have an amine-reactive NHS ester at one end and a sulfhydryl-reactive group (e.g., maleimide) on the other end. Following this idea, Qβ-VLPs are successfully provided with multiple copies of model antigen, which contains a cysteine residue at C-terminal part of the peptide [150].

For nanomaterial construction purposes also different cross-linking agents are useful, for example, glutaraldehyde, which is able to cross-link adjacent amine groups with chemically stable covalent bonds [179].

6.4.2.2 Modification of Carboxylic Groups

Viruses and VLPs often contain also freely accessible carboxylic groups, which can be effectively used for ligand coupling. For that, usually the carbodiimide chemistry is applied (Figure 6.3). As an example, plant virus CPMV has 180 addressable carboxylates on particle surface, which can be activated with water-soluble carbodiimide EDC. Subsequent reaction with NHS results in succinimide-esterified CPMV VLPs, which further forms an amide bond with molecules containing a primary amine group, for example, the anti-cancer drug doxorubicin. This protocol allows producing VLPs with 80 covalently bound doxorubicin molecules on the surface of the CPMV [24]. Similarly, carbodiimide and NHS treatment of Asp and Glu residues localized on spikes of HBcAg results in coupling of Lys-containing HeLa cell-internalizing peptide to the VLPs with a high efficiency [180].

6.4.2.3 Modification of Sulfhydryl Groups

Free cysteine residues in VLP structures can be also used for ligand coupling for different purposes (Figure 6.4). As shown in experiments with Qβ VLPs, reduced and free sulfhydryl groups actively react with maleimide derivatives, forming stable thioether linkages [181]. Here also mentioned bifunctional linker reagents connecting amine groups and sulfhydryl

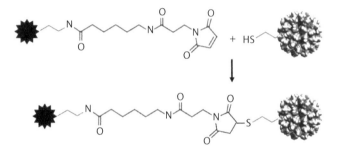

FIGURE 6.4 Modification of free thiol groups in the VLP structure [168]. The thiol group in the free Cys structure reacts selectively and stoichiometrically with maleimide derivatives.

groups are efficient tools for the coupling of desired ligands to VLPs with freely accessible sulfhydryl groups. Alternatively, the Cys residues located inside of VLPs also can be used for the coupling of desired functional molecules [182].

Interestingly, maleimide reactions are useful also for the construction of gold nanoparticle–containing VLPs. Water-soluble gold/p-mercaptobenzoic acid clusters after functionalization with maleimide linkers are able to react with free cysteines in enteroviruses, and surprisingly, all the chemical manipulations do not influence the infectivity of the viruses considerably [183].

If cysteine residues are located close to each other in the viral particle structure, the oxidation with an aim to obtain disulfide bonds between adjacent CP molecules often leads to an enhanced stability of VLPs. For Qβ VLPs, disulfide bond introduction between CPs in vitro after incubation in less concentrated hydrogen peroxide solution leads to significantly higher thermal stability (more than 40°C difference) than reduced VLPs [184].

6.4.2.4 Modification of Tyrosine Residues

If structural analysis of the virus reveals freely accessible tyrosine residues, it is possible to cross-link these residues in oxidative reaction using nickel(II) ions in the presence of a free radical source such as monoperoxyphthalate. As demonstrated with adeno-associated virus, the oxidation reaction results in the cross-linking of Tyr localized on neighboring CPs, yielding approximately 50% of viral proteins converted in the form of covalent dimers [185].

Exposed Tyr residues can be also modified using previously discussed copper-catalyzed azide-alkyne cycloaddition technology. It was successfully exploited to introduce the phosphate groups in TMV structure with an aim to achieve efficient calcium ion binding on TMV-coated substrates for improved cultivation of bone marrow stem cells [186].

6.4.2.5 Whole Cell–Assisted Modifications

In special cases, well-characterized cell systems are shown to be helpful for the introduction of necessary modifications in viral particles. If mammalian Vero cells are cultivated in medium that contains biotinylated dioleoyl-glycero-phosphoethanolamine, cell membranes become modified with biotin. Subsequent cultivations of pseudorabies viruses in these cells result in spontaneous biotin labeling of viral particles during the viral assembly process; the labeled viruses can be further used for different nanotechnological applications [187]. Similar whole-cell principle was demonstrated also for quantum dot packaging in mammalian virus VSV [188].

6.4.3 Genetic Techniques

Different genetic modifications are very powerful tools for creating viral nanoparticles for defined downstream applications. These techniques allow to change AA in viral coats and build in peptides and whole proteins into the virus structure at exactly distinct positions. Genetic techniques allow a better control of the amounts of introduced functional AA stretches, if compared with chemical ways, where often the couplings do not result in complete modifications at all possible positions on the viral coat.

6.4.3.1 Manipulations with Viral Structural Genes

For nanoparticle construction, frequently the manipulations with viral genes coding for structural proteins without changing the AA sequence of the viral carrier are necessary to achieve, for example, better production levels or improved particle properties. For different expression host systems, the adaptation of AA codons may be necessary to accomplish efficient synthesis of VLPs, because the codon usage in transgene can influence the target protein production considerably. However, rare tRNA synthesis in *E. coli* host cells can be supplemented from plasmid vectors coding for these tRNAs [189]; the vectors have to be introduced in the cells simultaneously with the expression vector. Moreover, our observations suggest that changes in transgene mRNA can reduce the output of VLPs, at least for VLPs possibly requiring the mRNA for VLP formation. This effect possibly is bound with potential packaging sequences/secondary structures located in the mRNA as discussed previously; the introduced codon adaptation changes can meddle with these structures. Therefore, the usage of unchanged viral genes and

plasmid vectors ensuring additional synthesis of rare tRNAs in *E. coli* cells should be tested before codon exchange.

A versatile system, based on HBcAg VLPs (SplitCore), is recently developed. If the HBcAg gene is split into two parts at the encoded immunodominant region and both the fragments are expressed in *E. coli* cells, the ability of VLP the formation is retained. Based on this, this system allows the display of medicinal proteins, which have sizes of more than 300 AA, on the VLP surface. This splitting principle is also applicable to other VLPs [190].

Recently, several VLP production systems are constructed, where the expression of a duplicated structural gene is used for VLP production, such as Qβ [181], PP7 [136], MS2 [191], and HBcAg (G. Lomonossoff, pers. commun. or Chapter 16 of this book). The duplicated gene system results in covalently linked CP dimer where the monomers are connected with the AA-linker. These VLPs are more stable and, importantly, can accommodate different peptides more efficiently than VLPs from single gene and can be used in peptide display experiments like filamentous bacteriophages [191].

6.4.3.2 Introduction of Desired Amino Acids

Genetic techniques are frequently used for changes or insertions of additional AAs to ensure the required properties of chosen nanoparticles. In special cases, to achieve efficient production of VLPs from heterologous host, it is necessary to change AA in the CP structure. After detailed analysis of the spatial structure of plant virus TMV, it was suggested that negatively charged Asp and Glu residues in the CP structure can negatively influence the helical rod assembly. The exchange of these AA against neutrally charged Asn and Gln resulted in high-level production of typical TMV particles in bacterial expression system [192].

If chosen VLPs do not possess the necessary AA at the defined position in the structure, it can be introduced by simple mutagenesis of the target gene, as shown with Lys insertion in plant virus TMV [193]. Different applications have been demonstrated with cysteine incorporation in the VLP structure. First, Cys residues if placed on adjacent stretches of two CP molecules can significantly improve the stability of the particles [181,194]. Next, if introduced on VLP surface, Cys-sulfhydryl groups are highly active against different metal ions and provide the chemical basis for electroless metal depositions on nanoparticle surface [195]. Surface-inserted Cys is frequently used for the coupling of different functional molecules, for example, as shown with infectious mutant of plant virus CPMV [196]. Furthermore, if the nanoparticle construction strategy foresees the packaging of the active substance inside of the particle, the coupling can be engineered also in VLP internal part *via* inside-introduced Cys residues [197].

For different purposes, also other AAs can be introduced in VLP structures. Interesting example with multiple changes of AA is demonstrated with plant virus CCMV VLPs: it is possible to exchange several positive charged AA inside the VLPs against acidic without influencing the ability to form typical particles. The CCMV particles with a negatively charged interior are suggested for usage as protein cages for iron nanoparticle preparations [198]. As known from basic studies, the exchange of the AA in VLPs may be helpful, for example, to mimic the serine phosphorylation in VLPs of bacterial origin by exchanging the Ser against Asp or Glu [199].

6.4.3.3 Introduction of Peptides

The introduction of foreign peptide sequences in viral structures is well known from experiments with an aim to create new vaccine candidates, based on bacterial, plant, insect, and mammalian virus carriers [86], as well as from phage display studies [39].

The idea to fuse an immunologically active peptide with a self-assembly-competent carrier molecule was suggested nearly 30 years ago, when the recombinant gene was constructed by gene synthesis, resulting in the TMV CP gene with eight AA long polio epitope. Recombinant protein was overexpressed in *E. coli*, TMV-like particles were purified, and the formation of neutralizing antibodies against poliovirus was demonstrated in rat immunization experiments [35]. The idea is still widely exploited in the construction of different immunologically active compounds. The main problem of these constructs in several cases is the fact that the peptide epitope placed on the surface of the VLP structurally represents a spatial conformation that differs from that in the original virus. As a result, formed antibodies do not recognize the native virus and the vaccine is inefficient. However, the spatial structure of the peptide exposed on the surface of the heterologous viral carrier can be optimized toward native conformation using existing 3D structures, careful modeling work, and experimental testing of different constructions, as shown with the influenza epitope on polyoma VLPs [200].

Introduced peptides can provide the nanoparticles with a wide variety of different functions for downstream applications. Using phage display methodology, a short peptide circularized by disulfide bonds is engineered on the surface of M13. The peptide successfully replaces the necessity of chemically coupling of biotin residues to the phage and bind streptavidin-conjugated growth factors for subsequent stimulation of different cell types during cultivation [201]. Additionally, VLPs with genetically inserted cell receptor–recognizing peptides efficiently target different tissues including cancer cells and are suggested as an alternative for chemical coupling [202,203].

The inclusion of hexahistidine peptide in structure of recombinant proteins is a usual measure to ensure simple one-step chromatographic purification. As demonstrated with TMV VLPs, His-tag can be efficiently exploited also for nanogold deposition on virus surface [204].

Interesting results were obtained in experiments with introduced elastin-like polypeptides at the N-terminal part of CCMV CP. Modified CPs are able to form smaller than typical VLPs in vitro, but the process is dependent on the temperature; at enhanced temperatures, the T = 1 form of VLP dominates, whereas CP dimers are mostly found at room temperature [205].

6.4.3.4 Introduction of Protein Domains

For vaccine purposes, the introduction of immunologically active protein domains or even whole proteins into the VLP structure seems to be very attractive, because their spatial structure is expected to be closer to that found in native pathogens. However, the introduction of large AA sequences in sizes comparable with viral structural proteins in VLP structure is a challenging task. First of all, whole proteins and their domains often have their own stable structure, which can considerably influence the structure of the CP part, resulting in the loss of typical viral morphology. Therefore, for a potential viral carrier, the capacity tests should be carried out. Examples such as successful accommodation of 238 AA long GFP [206], or 189 AA long OspC protein of *Borrelia burgdorferi* in internally located major immunodominant region of HBcAg [207], are exemptions, rather than a general rule. The insertions of larger protein domains are expected to be compatible with VLP formation in recombinant cells, if added to surface-located N- or C-terminus of corresponding CPs. As an example shown with plant virus PVY, 71 AA long rubredoxin-derived protein can be fused to the N-terminus of CP without loss of formation of filamentous VLPs [48].

The success of introduction of the protein domain in the VLP structure in the form of genetic fusion strongly depends not only on the VLP insertion capacity but also on the intrinsic properties of the chosen protein domain. For internal fusions, the structural property of the domain such as spatial localization of the protein termini plays an important role in successful VLP construction. Closely juxtaposed ends are stimulating for VLP formation, whereas the assembly can be influenced destructively, if the termini of foreign protein are located far away from each other. This problem is successfully solved in already discussed SplitCore system, at least for tested model proteins [190]. Next, when the domain is highly soluble as protein A IgG-binding derivatives from *S. aureus*, it promotes the solubility of the fusion protein and VLP assembly with exposed domain [50]. For domains, forming inclusion bodies in bacterial expression systems, like Dengue virus DIII fused to HBcAg, the purification under denaturing conditions and subsequent refolding is necessary to obtain nanoparticles for subsequent immunization experiments [208].

Highly efficient whole protein introduction can be achieved if the carrier VLPs intrinsically are built up from several proteins, as shown for bacteriophage P22 VLPs. As discussed previously, if SP is genetically fused with a model protein, it is possible to obtain P22 particles with T = 7 symmetry filled with approximately 250 molecules of SP-foreign protein fusion. These P22-based nanoparticles are suggested as nanoreactors for different applications including enzymatic transformations ([161] and Chapter 22 of this book—T. Douglas). Additional examples of introduction of enzymatically active proteins in virus shells are presented in the review article [209].

6.5 CHARACTERIZATION OF VIRAL NANOPARTICLES

The selection of procedures useful for the characterization of viral nanoparticles at different stages of development is highly important. It is necessary to choose efficient methods permitting the early identification of VLPs during the initial steps of construction as well as allowing the demonstration of suitability of nanoparticles for the intended downstream applications, including the tests for verifying the functionality of active molecules introduced in the structure of viral nanoparticles. As known from pharmaceutical industry, the whole complex of different analytical methods based on physical, chemical, and biological properties of virus-based vaccines is required to ensure the necessary monitoring of the manufacturing processes according to the established quality standards [124,210]. In this chapter, a short overview of the methods useful for viral nanoparticle characterization will be provided, emphasizing the simple, generally applicable and/or most interesting new ones.

6.5.1 INTEGRITY OF VIRAL NANOPARTICLES

After completing the cultivation process and cell disruption stage, it is highly important to identify the VLPs at initial stages of construction. The very first step in VLP characterization is the evaluation of production of the target protein. As usual in protein research, it can be done by denaturing polyacrylamide gel analysis (SDS/PAGE), which helps to evaluate the amount, purity, and approximate molecular mass of the target protein (Figure 6.5). At this stage, the solubility of viral structural proteins is determined in extracts after cell disruption by analyzing the soluble and insoluble fractions. If recombinant protein is soluble, the extract can be further subjected to analytical ultracentrifugation in sucrose gradient. Assembled proteins are typically found in fractions with higher sucrose concentrations as monomeric or oligomeric proteins from host cells. The sucrose gradient experiments provide the necessary information about the multimeric state and structural homogeneity of particles; for uniformly shaped

VLPs, the amounts of particles after separation are found in the gradient fractions following the Gaussian curve. Broad distribution of the target protein in the gradient fractions suggests the presence of partially disassembled and/or irregularly aggregated viral proteins in the cell extract. Moreover, fractions from analytical gradients can serve as a source of partially purified samples for the first analytical tests.

As a next tool for VLP early detection, the agarose gel tests can be used in parallel to the sucrose gradient analysis at the stage of crude lysates, if the dimensions of VLPs are compatible with pores in agarose gels. To reduce the amount of high-molecular-weight nucleic acids in the solution, it is advisable to treat the lysates with nucleases (e.g., Benzonase, Figure 6.6), which can facilitate the signal identification in the gels. Characteristic signals on ethidium bromide stained gel that overlap with signals from Coomassie-stained gel suggest the presence of ordered protein structures with protected

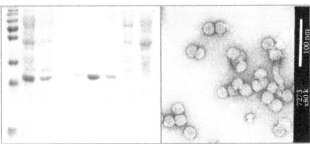

FIGURE 6.5 SDS/PAGE analysis of VLP production and purification. M, protein size marker; S, soluble *E. coli* cell proteins; P, insoluble *E. coli* cell proteins; 60%–0%, sucrose concentrations in protein-containing gradient fractions after separation in ultracentrifuge. Left, Coomassie-stained SDS/PAGE gel; right, EM analysis of 40% fraction.

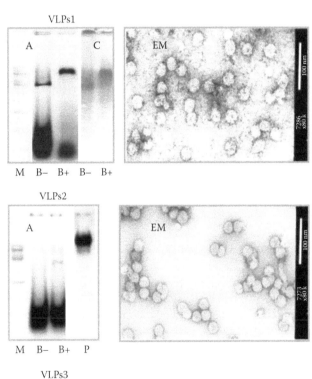

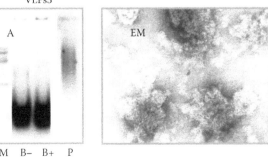

FIGURE 6.6 Early identification of viral structural protein/nucleic acid complexes in *E. coli* cell lysates. A, ethidium bromide–stained agarose gels; C, Coomassie-stained agarose gels, B−/B+, benzonase-untreated/benzonase-treated samples; P, PEG8000 precipitate; EM, EM analysis; M, DNA size marker.

nucleic acid. Such complexes are typical VLPs in most cases. If the VLPs are stable enough to preserve their typical structure after the PEG precipitation, the agarose gel analysis can provide more clear result with specific signals. In the absence of well-structured VLPs, the analysis of PEG precipitate results in a more diffused signal (VLPs3, Figure 6.6).

Additionally, agarose gel analysis allows to separate structurally different VLPs, as demonstrated with HBcAg particles of T = 3 and T = 4 symmetry. The method is useful not only for identification of VLP species; subsequent particle electroelution from excised gel pieces results in the isolation of small amounts of VLPs for further experiments [211].

The disassembled/reassembled state of VLPs can be monitored also using protein fluorescence spectroscopy. HPV L1 proteins in the form of VLPs demonstrate considerably higher fluorescence intensity than disassembled proteins. Additionally, if conformation-specific mAbs are available, corresponding ELISA test is a very sensitive approach to confirm the VLP integrity [212].

The next method in the analysis of VLP assembly, which can serve as an alternative to EM visualizations, is dynamic light scattering. It allows to determine hydrodynamic size of viral particles and estimate their electrokinetic properties in solutions used for purification or applications [114,213].

Recently, additional instrumental techniques such as capillary zone electrophoresis [214] and asymmetric flow field flow fractionation (AF4; [215]) are suggested as valuable tools for VLP characterization.

An indispensable tool in VLP research is the EM technique. It serves as a final visual proof of VLP integrity at all stages of construction and application, because other aforementioned methods may identify unstructured protein aggregates as VLPs. However, as samples in transmission EM (TEM) are negatively stained with different salts and analyzed in a dried state, the possible artifacts should be considered also for this technique. Therefore, several sample preparation protocols have to be tested in case of uncharacterized viral objects to find the optimal treatment for the TEM analysis. Alternatively, as cryo-EM and atomic force microscopy methodologies are bound with rapid freezing of virus suspensions, minimal possible alterations in the object morphologies are expected [216,217].

Viral nanoparticle construction principles require the structural data to exploit virus intrinsic properties or introduce functional molecules; therefore, a 3D model with a near-atomic resolution is one of the prerequisites for generating ideas for new viral nanoparticles. It is not surprising that all VLPs suggested for different nanotechnological applications are well characterized from the structural point of view, and introduction of new virus-derived structures as nanoparticles is often limited by the absence of such models. The principles and historical overview of virus structure elucidation are presented by Michael Rossmann in his excellent review article [218].

As 3D virus structure determination is a time- and labor-consuming process in many cases, as an alternative the high-resolution cryo-EM with subsequent modeling in silico is suggested, based on the 3D model of a similar virus, as shown, for example, in the case of the plant virus BRV [219].

VLP stability issues are highly important for nanoparticle construction and applications. Therefore, the methods allowing to efficiently obtain the necessary data should be emphasized among the already discussed approaches for particle integrity analysis. The influence of different environmental factors on VLP stability can be demonstrated by experimental techniques such as agarose gels [220], sucrose gradient centrifugations [221], size-exclusion chromatography [116], ELISA using mono- and polyclonal antibodies [222], fluorescence spectroscopy with SYPRO Orange at elevating temperatures [48], and other methods depending on the VLP properties [210].

6.5.2 NUCLEIC ACIDS

Encapsidated nucleic acids can be easily visualized on agarose gels after treatment with nucleases, as discussed previously. For the identification of known nucleic acids, typical methods of molecular biology such as reverse transcription/polymerase chain reaction (RT-PCR) and different nucleic acid hybridizations are frequently used. To detect unknown nucleic acid species inside of VLPs after cultivation in recombinant host, the next-generation sequencing technique is a very useful tool. It allows to identify not only host and expression vector-derived transcripts in VLPs but also encapsidated minor RNAs from native host cells inside of infectious RNA viruses [144].

To characterize the RNA secondary structures, an interesting approach, selective 2′-hydroxyl acylation and primer extension (SHAPE footprinting), was suggested several years ago [223] and recently applied to the analysis of genomic RNA from plant virus satellite tobacco mosaic virus (STMV). The method is based on RNA-selective hydroxyl acylation analyzed by primer extension and results in the identification of locally flexible, solvent accessible regions in the RNA structure. The SHAPE approach allows to identify long-distance interactions in viral secondary structures and surprisingly points on the possible weakening effect of CPs on the stability of some intramolecular RNA secondary structures inside of the capsid [224].

6.5.3 VIRAL STRUCTURAL PROTEINS

Generally applicable methods of protein analysis such as SDS/PAGE, immunoblots, and ELISA tests with virus- or peptide-specific antibodies were already discussed. Additionally, a wide variety of techniques are demonstrated as useful for the characterization of viral structural proteins. Whereas the SDS/PAGE analysis allows to clear up the approximate size of the protein, the mass spectrometric (MS) measurements provide accurate molecular mass, supporting the identification of viral proteins with introduced modifications [225] and allow the mapping of the products of proteolytic degradation [48] and posttranslational modifications [76] in recombinant VLPs. Moreover, whole-cell matrix-assisted laser desorption/ionization time-of-flight mass spectrometry (MALDI-TOF MS) fingerprints are demonstrated as efficient tools for monitoring the VLP production process and suggested as appropriate for the biotechnological industry [226].

The influence of different environmental factors on viral structural proteins can be demonstrated by circular dichroism spectroscopy. As shown with far-UV spectra of differentially treated HBsAg, prolonged incubations at elevated temperatures result in a significant loss of α-helices in the structure of surface proteins, and these changes are bound with lipid organization at the particle surface and substantial reduction of immunogenicity [227].

The next approach in analysis of viral proteins is bound with site-specific proteolytic degradation, which is connected to high-resolution mass spectrometry. First of all, the identification of characteristic signature peptides from viral proteins allows typing human viruses with an efficiency and reliability, comparable to RT-PCR [228]. Furthermore, partial proteolysis with trypsin and subsequent MS analysis suggest significant differences in physicochemical properties and posttranslational modifications of plant virus BMV, if obtained from diverse plant hosts [229]. Additionally, if freely accessible Lys residues in viral structural protein are modified in the amidination reaction, it is possible to identify the lysines involved in interactions with RNA or neighboring CP residues [230]. Topological features of viruses without known 3D structures can be discovered also by chemical cross-linking with subsequent proteolysis and MS analysis (PIR technology [231]).

6.5.4 QUANTIFICATION

The evaluation of virus or VLP concentrations in solution often is a challenging task. The estimation of a number of infectious particles in virus-containing solutions is well known from bacteriophage studies, where the ability of plaque formation on densely plated susceptible microorganisms is tested. Also for mammalian and plant viruses, tests of similar principle can be carried out, at least for a part of viruses [232,233]. However, for nanotechnological applications, it is important to know the concentration of all viral particles, not only infectious ones. As modified viruses or recombinant VLPs mostly are noninfectious, alternative methods are necessary for these purposes. Therefore, different particle counting approaches are shown as useful, such as quantitative capillary electrophoresis using fluorescently labeled viral particles or nucleic acids [234] or nanoparticle tracking analysis (NTA). However, the latter method is limited in the particle size; viral nanoparticles smaller than 30 nm cannot be effectively counted using NTA. Alternatively, also EM, atomic force microscopy, and quantitative RT-PCR analysis is suitable for viral particle counting after considering the advantages and disadvantages of the corresponding method [235].

Next, for different applications, molar concentration data are required to calculate the amounts of reacting components necessary for nanoparticle construction processes, as well as to evaluate the efficiency of the corresponding reaction.

For uncharacterized viruses, the approximate molecular weight (MW) of virions can be determined, based on sedimentation coefficients, which can be obtained by the ultracentrifugation technique [236]. In case of well-characterized viruses, the relation between protein and nucleic acid content is known and allows to calculate the MW of the assembled virus from sequence and structural data. For example, icosahedral plant viruses BMV with T = 3 symmetry is built up from 180 CP molecules with encapsidated viral RNAs of known sequence; calculation of theoretical mass of the virus summarizing MWs of all CPs and nucleic acids results in MW = 4.63 MDa [237].

In case of recombinant VLPs, the exact determination of MW can be more complicated. If native viruses essentially package their specific nucleic acids, then the efficiency of host nucleic acid packaging in VLPs obtained from heterologous expression systems is expected to be lower. As shown with Qβ VLPs, host cell RNAs in VLPs constitute only 25% of the total mass, whereas up to 50% of specific nucleic acids are found in native Qβ [238]. Reduced amounts of host nucleic acids are found also in different variants of HBcAg VLPs, if UV analysis data are compared with theoretical packaging capacities. Despite the presence of an unknown amount of uncharacterized nucleic acids in VLPs, UV spectroscopy allows to calculate the actual concentrations of highly purified VLPs, taking into account the fact that RNA absorbance is not considerably influenced after packaging and the light scattering effects can be included in calculations. The method is suggested for concentration estimations also for other VLPs [237].

To evaluate VLP concentrations, the densitometric analysis of signals on Coomassie-stained SDS/PAGE gels can provide the necessary information, for example, about the coupling efficiency of an antigen to the VLP carrier [239]. Alternatively, different protein staining protocols are frequently used in quantification of purified VLPs, such as bicinchoninic protein assay (BCA) [240] and Bradford test [241]. However, the influence of nucleic acids inside of VLPs on measurement results should be considered before the selection of an appropriate method [242].

For special cases such as vaccine production, the quantitative evaluation of the production process and of the final product is highly important. The coupling of AF4 with a multiangle light scattering device permits very capable analysis of VLPs in vaccine preparations, including MW calculations for single VLPs, as well as dimer, trimer, and oligomeric associations of particles [243]. For quantification of influenza vaccines, the whole complex of analytical technologies is suggested, including hemagglutination assay, single radial immunodiffusion assay, neuraminidase enzymatic activity measurements, Western blot, and EM [244]. If corresponding mAbs are available, different ELISA tests are also included in the quantification of VLPs during vaccine production [116].

6.6 CONCLUDING REMARKS

Knowledge from basic research, a continuously growing number of characterized viruses, and first successes in the commercial application of viral nanoparticles in vaccinology are major driving forces in developing different virus-based technologies. The application ideas of viral nanoparticles are based on the intrinsic properties of viruses, such as the ability to self-assemble in the presence or absence of nucleic acids, AA content of viral coats as targets for different chemical and genetic modifications, and multiple spatially repeated AA sequence motifs recognizing certain cell types and inducing strong immunological responses in mammalian organisms and serving as templates for the creation of new nanomaterials of inorganic and organic nature. Ideas derived from virus architecture principles are extensively exploited in the creation of nonviral nanomaterials.

As shown in several cases, native viruses can be used directly as components of different nanomaterials, after simple physical or chemical adaptation procedures. However, mostly the amounts of viruses produced in their native hosts are not sufficient to isolate them in preparative amounts; moreover, such isolation and applications are not compatible with biosafety rules. Therefore, the main sources of viral nanoparticles are different recombinant organisms, allowing to obtain VLPs in nearly unlimited amounts, especially from simple bacterial and yeast expression systems. For special purposes, such as vaccines, VLPs often are built up from several structural proteins and are posttranslationally modified, which usually require the usage of mammalian expression systems. Otherwise, also yeast, insect cell, and whole plant systems can serve as efficient alternatives in the synthesis of VLPs for vaccine purposes.

From a methodological point of view, the progress in basic science and in instrumental developments continuously increases the amount of available methods useful for viral nanoparticle construction and characterization. One of the most important factors influencing the efficiency of nanoparticle-based technologies, especially recombinant vaccines, is the fact that introduced functional AA stretches engineered into a viral carrier often structurally differ from that of the natural pathogen, which makes the intended application inefficient. Therefore, the development of capable methods, allowing rapid structural characterization of VLPs, is still needed. As one possible solution, computational simulations of whole viral nanoparticles are suggested [245]. Additionally, also new instrumental techniques are under development, such as biomolecular magic-angle spinning solid-state NMR, which potentially is able to provide the real viral and VLP structure models as alternatives to crystallography approaches [246].

Characterization process of viral nanoparticles can reveal also unexpected properties potentially important for technology development. Interesting examples, which should be mentioned here, are the ability of plant virus PVX to recognize and penetrate mammalian tumor tissues [247] and human antibodies against native plant virus TMV, possibly preventing smokers against Parkinson's disease due to epitope sequence similarities according to the molecular mimicry principle [248]. Therefore, also in the future, new ideas coming from different areas of basic studies as well as new instrumental developments will ensure the further progress in virus-based bionanotechnology and will result in the introduction of new nanomaterials for industrial applications.

ACKNOWLEDGMENTS

I thank Prof. Dr. P. Pumpens, Prof. Dr. M. Bachmann, Prof. Dr. K. Tars, and Dr. A. Kazaks for their helpful discussions and support during the preparation of this chapter. Dr. V. Ose is acknowledged for help in the preparation of EM pictures. Writing of the review was supported by Grant No. SP 672/14 provided by the Latvian Science Council.

REFERENCES

1. Feynman RP. There's plenty of room at the bottom. *Eng. Sci.* 1960; 23: 22–36.
2. Dukes MJ, Jacobs BW, Morgan DG, Hegde H, Kelly DF. Visualizing nanoparticle mobility in liquid at atomic resolution. *Chem. Commun. (Camb.).* 2013; 49: 3007–3009.
3. Taniguchi N. On the basic concept of 'nano-technology'. *Proceedings of the International Conference on Production Engineering,* Part II, Japan Society of Precision Engineering, Tokyo, Japan. 1974, pp. 5–10.
4. Goodsell DS. The quest for nanotechnology. In: *Bionanotechnology: Lessons from Nature,* DS Goodsell, ed. Wiley-Liss, Hoboken, New Jersey. 2004, pp. 1–9.
5. Stanley WM. Isolation of a crystalline protein assessing the properties of tobacco mosaic virus. *Science* 1935; 81: 644–645.
6. Kausche GA, Pfankuch E, Ruska H. Die Sichtbarmachung von pfanzlichem Viren im Übermikroskop. *Naturwissenschaften* 1939; 27: 292–299.
7. Lauffer MA, Ansevin AT, Cartwright TE, Brinton CC. Polymerization–depolymerizatic tobacco mosaic virus protein. *Nature* 1958; 181: 1338–1339.
8. Miller GL, Stanley WM. Derivatives of tobacco mosaic virus. II. Carbobenzoxy, p-chlorobenzoyl, and benzenesulfonyl virus. *J. Biol. Chem.* 1942; 146: 331–338.
9. Gierer A, Schramm G. Infectivity of ribonucleic acid from tobacco mosaic virus. *Nature* 1956; 177: 702–703.
10. Ahlquist P, French R, Janda M, Loesch-Fries LS. Multicomponent RNA plant virus infection derived from cloned viral cDNA. *Proc. Natl. Acad. Sci. USA* 1984; 81: 7066–7070.
11. Harrison SC, Olson AJ, Schutt CE, Winkler FK, Bricogne G. Tomato bushy stunt virus at 2.9 Å resolution. *Nature* 1978; 276: 368–373.
12. Fischlechner M, Donath E. Viruses as building blocks for materials and devices. *Angew. Chem. Int. Ed. Engl.* 2007; 46: 3184–3193.
13. Love AJ, Makarov V, Yaminsky I, Kalinina NO, Taliansky ME. The use of tobacco mosaic virus and cowpea mosaic virus for the production of novel metal nanomaterials. *Virology* 2014; 449: 133–139.
14. Iwanowsky D. Über die Mosaikkrankheit der Tabakspflanze. *Bulletin Scientifique publié par l'Académie Impériale des Sciences de Saint-Pétersbourg/Nouvelle Serie III.* 1892; 35: 67–70.
15. Summers WC. Bacteriophage research: Early history. In: *Bacteriophages: Biology and Applications,* E. Kutter, A. Sulakvelidze, eds. CRC Press, Boca Raton, FL. 2005, pp. 5–27.

16. Zhu H, Cao B, Zhen Z, Laxmi AA, Li D, Liu S, Mao C. Controlled growth and differentiation of MSCs on grooved films assembled from monodisperse biological nanofibers with genetically tunable surface chemistries. *Biomaterials* 2011; 32: 4744–4752.

17. Chung WJ, Oh JW, Kwak K, Lee BY, Meyer J, Wang E, Hexemer A, Lee SW. Biomimetic self-templating supramolecular structures. *Nature* 2011; 478: 364–368.

18. Wu W, Hsiao SC, Carrico ZM, Francis MB. Genome-free viral capsids as multivalent carriers for taxol delivery. *Angew. Chem. Int. Ed. Engl.* 2009; 48: 9493–9497.

19. Hansen MR, Mueller L, Pardi A. Tunable alignment of macromolecules by filamentous phage yields dipolar coupling interactions. *Nat. Struct. Biol.* 1998; 5: 1065–1074.

20. Huisman EM, Wen Q, Wang YH, Cruz K, Kitenbergs G, Erglis K, Zeltins A, Cebers A, Janmey PA. Gelation of semi-flexible polyelectrolytes by multivalent counterions. *Soft Matter* 2011; 7: 7257–7261.

21. Azizgolshani O, Garmann RF, Cadena-Nava R, Knobler CM, Gelbart WM. Reconstituted plant viral capsids can release genes to mammalian cells. *Virology* 2013; 441: 12–17.

22. Prasuhn DE, Yeh RM, Obenaus A, Manchester M, Finn MG. Viral MRI contrast agents: Coordination of Gd by native virions and attachment of Gd complexes by azide-alkyne cycloaddition. *Chem. Commun.* 2007; 12: 1269–1271.

23. Lomonossoff GP, Evans DJ. Applications of plant viruses in bionanotechnology. *Curr. Top. Microbiol. Immunol.* 2014; 375: 61–68.

24. Aljabali AA, Shukla S, Lomonossoff GP, Steinmetz NF, Evans DJ. CPMV-DOX delivers. *Mol. Pharm.* 2013; 10: 3–10.

25. Shukla S, Ablack AL, Wen AM, Lee KL, Lewis JD, Steinmetz NF. Increased tumor homing and tissue penetration of the filamentous plant viral nanoparticle potato virus X. *Mol. Pharm.* 2013; 10: 33–42.

26. Bruckman MA, Hern S, Jiang K, Flask CA, Yu X, Steinmetz NF. Tobacco mosaic virus rods and spheres as supramolecular high-relaxivity MRI contrast agents. *J. Mater. Chem. B Mater. Biol. Med.* 2013; 1: 1482–1490.

27. Khan AA, Fox EK, Górzny MŁ, Nikulina E, Brougham DF, Wege C, Bittner AM. pH control of the electrostatic binding of gold and iron oxide nanoparticles to tobacco mosaic virus. *Langmuir* 2013; 29: 2094–2098.

28. Alonso JM, Górzny MŁ, Bittner AM. The physics of tobacco mosaic virus and virus-based devices in biotechnology. *Trends Biotechnol.* 2013; 31: 530–538.

29. van Kan-Davelaar HE, van Hest JC, Cornelissen JJ, Koay MS. Using viruses as nanomedicines. *Br. J. Pharmacol.* 2014; 171: 4001–4009.

30. Glasgow J, Tullman-Ercek D. Production and applications of engineered viral capsids. *Appl. Microbiol. Biotechnol.* 2014; 98: 5847–5858.

31. Yildiz I, Shukla S, Steinmetz NF. Applications of viral nanoparticles in medicine. *Curr. Opin. Biotechnol.* 2011; 22: 901–908.

32. Brady JN, Consigli RA. Chromatographic separation of the polyoma virus proteins and renaturation of the isolated VP1 major capsid protein. *J. Virol.* 1978; 27: 436–442.

33. Burrell CJ, MacKay P, Greenaway PJ, Hofschneider PH, Murray K. Expression in *Escherichia coli* of hepatitis B virus DNA sequences cloned in plasmid pBR322. *Nature* 1979; 279: 43–47.

34. Valenzuela P, Medina A, Rutter WJ, Ammerer G, Hall BD. Synthesis and assembly of hepatitis B virus surface antigen particles in yeast. *Nature* 1982; 298: 347–350.

35. Haynes JR, Cunningham J, von Seefried A, Lennick M, Garvin RT, Shen SH. Development of a genetically-engineered, candidate polio vaccine employing the self-assembling properties of the tobacco mosaic virus coat protein. *Biotechnology* 1986; 4: 637–641.

36. Cohen BJ, Richmond JE. Electron microscopy of hepatitis B core antigen synthesized in *E. coli*. *Nature* 1982; 296: 677–678.

37. Messing J, Gronenborn B, Müller-Hill B, Hans Hopschneider P. Filamentous coliphage M13 as a cloning vehicle: Insertion of a HindII fragment of the lac regulatory region in M13 replicative form in vitro. *Proc. Natl. Acad. Sci. USA* 1977; 74: 3642–3646.

38. Yang SH, Chung WJ, McFarland S, Lee SW. Assembly of bacteriophage into functional materials. *Chem. Rec.* 2013; 13: 43–59.

39. Henry M, Debarbieux L. Tools from viruses: Bacteriophage successes and beyond. *Virology* 2012; 434: 151–161.

40. Dawson WO, Beck DL, Knorr DA, Grantham GL. cDNA cloning of the complete genome of tobacco mosaic virus and production of infectious transcripts. *Proc. Natl. Acad. Sci. USA* 1986; 83: 1832–1836.

41. Lebeurier G, Hirth L, Hohn T, Hohn B. Infectivities of native and cloned DNA of cauliflower mosaic virus. *Gene* 1980; 12: 139–146.

42. Racaniello VR, Baltimore D. Cloned poliovirus complementary DNA is infectious in mammalian cells. *Science* 1981; 214: 916–919.

43. Boyer JC, Haenni AL. Infectious transcripts and cDNA clones of RNA viruses. *Virology* 1994; 198: 415–426.

44. Zeltins A. Construction and characterization of virus-like particles: A review. *Mol. Biotechnol.* 2013; 53: 92–107.

45. Sánchez-Rodríguez SP, Münch-Anguiano L, Echeverría O, Vázquez-Nin G, Mora-Pale M, Dordick JS, Bustos-Jaimes I. Human parvovirus B19 virus-like particles: In vitro assembly and stability. *Biochimie* 2012; 94: 870–878.

46. Zhao X, Fox JM, Olson NH, Baker TS, Young MJ. In vitro assembly of cowpea chlorotic mottle virus from coat protein expressed in *Escherichia coli* and in vitro-transcribed viral cDNA. *Virology* 1995; 207: 486–494.

47. Xu Y, Ye J, Liu H, Cheng E, Yang Y, Wang W, Zhao M, Zhou D, Liu D, Fang R. DNA-templated CMV viral capsid proteins assemble into nanotubes. *Chem. Commun. (Camb.).* 2008; 1: 49–51.

48. Kalnciema I, Skrastina D, Ose V, Pumpens P, Zeltins A. Potato virus Y-like particles as a new carrier for the presentation of foreign protein stretches. *Mol. Biotechnol.* 2012; 52: 129–139.

49. Hou L, Wu H, Xu L, Yang, F. Expression and self-assembly of virus-like particles of infectious hypodermal and hematopoietic necrosis virus in *Escherichia coli*. *Arch. Virol.* 2009; 154: 547–553.

50. Brown SD, Fiedler JD, Finn MG. Assembly of hybrid bacteriophage Qbeta virus-like particles. *Biochemistry* 2009; 48: 11155–11157.

51. Hwang DJ, Roberts IM, Wilson TM. Expression of tobacco mosaic virus coat protein and assembly of pseudovirus particles in *Escherichia coli*. *Proc. Natl. Acad. Sci. USA* 1994; 91: 9067–9071.

52. Rogel A, Benvenisti L, Sela I, Edelbaum O, Tanne E, Shachar Y, Zanberg Y, Gontmakher T, Khayat E, Stram Y. Vaccination with *E. coli* recombinant empty viral particles of infectious bursal disease virus (IBDV) confer protection. *Virus Genes* 2003; 27: 169–175.

53. Hammond RW, Hammond J. Maize rayado fino virus capsid proteins assemble into virus-like particles in *Escherichia coli*. *Virus Res.* 2010; 147: 208–215.

54. Chen XS, Garcea RL, Goldberg I, Casini G, Harrison SC. Structure of small virus-like particles assembled from the L1 protein of human papillomavirus 16. *Mol. Cell* 2000; 5: 557–567.

55. Lee CD, Yan YP, Liang SM, Wang TF. Production of FMDV virus-like particles by a SUMO fusion protein approach in *Escherichia coli*. *J. Biomed. Sci.* 2009; 16: 69.

56. Middelberg AP, Rivera-Hernandez T, Wibowo N, Lua LH, Fan Y, Magor G, Chang C, Chuan YP, Good MF, Batzloff MR. A microbial platform for rapid and low-cost virus-like particle and capsomere vaccines. *Vaccine* 2011; 29: 7154–7162.

57. Phelps JP, Dao P, Jin H, Rasochova L. Expression and self-assembly of cowpea chlorotic mottle virus-like particles in *Pseudomonas fluorescens*. *J. Biotechnol.* 2007; 128: 290–296.

58. Cortes-Perez NG, Kharrat P, Langella P, Bermúdez-Humarán LG. Heterologous production of human papillomavirus type-16 L1 protein by a lactic acid bacterium. *BMC Res. Notes* 2009; 2: 167.

59. Zhao Q, Allen MJ, Wang Y, Wang B, Wang N, Shi L, Sitrin RD. Disassembly and reassembly improves morphology and thermal stability of human papillomavirus type 16 virus-like particles. *Nanomedicine* 2012; 8: 1182–1189.

60. Sugase K, Landes MA, Wright PE, Martinez-Yamout M. Overexpression of post-translationally modified peptides in *Escherichia coli* by co-expression with modifying enzymes. *Protein Expr. Purif.* 2008; 57: 108–115.

61. Neumann H, Peak-Chew SY, Chin JW. Genetically encoding N(epsilon)-acetyllysine in recombinant proteins. *Nat. Chem. Biol.* 2008; 4: 232–234.

62. Kushnir N, Streatfield SJ, Yusibov V. Virus-like particles as a highly efficient vaccine platform: Diversity of targets and production systems and advances in clinical development. *Vaccine* 2012; 31: 58–83.

63. Chen R. Bacterial expression systems for recombinant protein production: *E. coli* and beyond. *Biotechnol. Adv.* 2012; 30: 1102–1107.

64. Ramón A, Señorale-Pose M, Marín M. Inclusion bodies: Not that bad.... *Front. Microbiol.* 2014; 5: 56.

65. Hofmann KJ, Cook JC, Joyce JG, Brown DR, Schultz LD, George HA, Rosolowsky M, Fife KH, Jansen KU. Sequence determination of human papillomavirus type 6a and assembly of virus-like particles in *Saccharomyces cerevisiae*. *Virology* 1995; 209: 506–518.

66. Samuel D, Sasnauskas K, Jin L, Beard S, Zvirbliene A, Gedvilaite A, Cohen B. High level expression of recombinant mumps nucleoprotein in *Saccharomyces cerevisiae* and its evaluation in mumps IgM serology. *J. Med. Virol.* 2002; 66: 123–130.

67. Freivalds J, Dislers A, Ose V, Skrastina D, Cielens I, Pumpens P, Sasnauskas K, Kazaks A. Assembly of bacteriophage Qbeta virus-like particles in yeast *Saccharomyces cerevisiae* and *Pichia pastoris*. *J. Biotechnol.* 2006; 123: 297–303.

68. Freivalds J, Rumnieks J, Ose V, Renhofa R, Kazaks A. High-level expression and purification of bacteriophage GA virus-like particles from yeast *Saccharomyces cerevisiae* and *Pichia pastoris*. *Acta Univer. Latv.* 2008; 745: 75–85.

69. Lazutka J, Zvirbliene A, Dalgediene I, Petraityte-Burneikiene R, Spakova A, Sereika V, Lelesius R, Wernike K, Beer M, Sasnauskas K. Generation of recombinant schmallenberg virus nucleocapsid protein in yeast and development of virus-specific monoclonal antibodies. *J. Immunol. Res.* 2014; 2014: 160316.

70. Tamošiūnas PL, Simutis K, Kodžė I, Firantienė R, Emužytė R, Petraitytė-Burneikienė R, Zvirblienė A, Sasnauskas K. Production of human parvovirus 4 VP2 virus-like particles in yeast and their evaluation as an antigen for detection of virus-specific antibodies in human serum. *Intervirology* 2013; 56: 271–277.

71. Juozapaitis M, Zvirbliene A, Kucinskaite I, Sezaite I, Slibinskas R, Coiras M, de Ory Manchon F et al. Synthesis of recombinant human parainfluenza virus 1 and 3 nucleocapsid proteins in yeast *Saccharomyces cerevisiae*. *Virus Res.* 2008; 133: 178–186.

72. Powilleit F, Breinig T, Schmitt MJ. Exploiting the yeast L-A viral capsid for the in vivo assembly of chimeric VLPs as platform in vaccine development and foreign protein expression. *PLoS ONE* 2007; 2: e415.

73. Krol MA, Olson NH, Tate J, Johnson JE, Baker TS, Ahlquist P. RNA-controlled polymorphism in the in vivo assembly of 180-subunit and 120-subunit virions from a single capsid protein. *Proc. Natl. Acad. Sci. USA* 1999; 96: 13650–13655.

74. Freivalds J, Kotelovica S, Voronkova T, Ose V, Tars K, Kazaks A. Yeast-expressed bacteriophage-like particles for the packaging of nanomaterials. *Mol. Biotechnol.* 2014; 56: 102–110.

75. Brumfield S, Willits D, Tang L, Johnson JE, Douglas T, Young M. Heterologous expression of the modified coat protein of Cowpea chlorotic mottle bromovirus results in the assembly of protein cages with altered architectures and function. *J. Gen. Virol.* 2004; 85: 1049–1053.

76. Freivalds J, Dislers A, Ose V, Pumpens P, Tars K, Kazaks A. Highly efficient production of phosphorylated hepatitis B core particles in yeast *Pichia pastoris*. *Protein Expr. Purif.* 2011; 75: 218–224.

77. Rodríguez-Limas WA, Tyo KE, Nielsen J, Ramírez OT, Palomares LA. Molecular and process design for rotavirus-like particle production in *Saccharomyces cerevisiae*. *Microb. Cell Fact.* 2011; 10: 33.

78. Kim HJ, Jin Y, Kim HJ. The concentration of carbon source in the medium affects the quality of virus-like particles of human papillomavirus type 16 produced in *Saccharomyces cerevisiae*. *PLoS ONE* 2014; 9: e94467.

79. Morikawa Y, Goto T, Yasuoka D, Momose F, Matano T. Defect of human immunodeficiency virus type 2 Gag assembly in *Saccharomyces cerevisiae*. *J. Virol.* 2007; 81: 9911–9921.

80. Legendre D, Fastrez J. Production in *Saccharomyces cerevisiae* of MS2 virus-like particles packaging functional heterologous mRNAs. *J. Biotechnol.* 2005; 117: 183–194.

81. Janowicz ZA, Melber K, Merckelbach A, Jacobs E, Harford N, Comberbach M, Hollenberg CP. Simultaneous expression of the S and L surface antigens of hepatitis B, and formation of mixed particles in the methylotrophic yeast, *Hansenula polymorpha*. *Yeast* 1991; 7: 431–443.

82. Frenzel A, Hust M, Schirrmann T. Expression of recombinant antibodies. *Front. Immunol.* 2013; 4: 217.

83. Meehl MA, Stadheim TA. Biopharmaceutical discovery and production in yeast. *Curr. Opin. Biotechnol.* 2014; 30C: 120–127.

84. Luckow VA, Summers MD. Trends in the development of baculovirus expression vectors. *Nat. Biotechnol.* 1988; 6: 47–55.

85. Liu F, Wu X, Li L, Liu Z, Wang Z. Use of baculovirus expression system for generation of virus-like particles: Successes and challenges. *Protein Expr. Purif.* 2013; 90: 104–116.

86. Pushko P, Pumpens P, Grens E. Development of virus-like particle technology from small highly symmetric to large complex virus-like particle structures. *Intervirology* 2013; 56: 141–165.

87. Sokolenko S, George S, Wagner A, Tuladhar A, Andrich JM, Aucoin MG. Co-expression vs. co-infection using baculovirus expression vectors in insect cell culture: Benefits and drawbacks. *Biotechnol. Adv.* 2012; 30: 766–781.

88. Tatman JD, Preston VG, Nicholson P, Elliott RM, Rixon FJ. Assembly of herpes simplex virus type 1 capsids using a panel of recombinant baculoviruses. *J. Gen. Virol.* 1994; 75: 1101–1113.

89. Pushko P, Tumpey TM, Bu F, Knell J, Robinson R, Smith G. Influenza virus-like particles comprised of the HA, NA, and M1 proteins of H9N2 influenza virus induce protective immune responses in BALB/c mice. *Vaccine* 2005; 23: 5751–5759.

90. Metz SW, Feenstra F, Villoing S, van Hulten MC, van Lent JW, Koumans J, Vlak JM, Pijlman GP. Low temperature-dependent salmonid alphavirus glycoprotein processing and recombinant virus-like particle formation. *PLoS ONE* 2011; 6: e25816.

91. Saunders K, Sainsbury F, Lomonossoff GP. Efficient generation of cowpea mosaic virus empty virus-like particles by the proteolytic processing of precursors in insect cells and plants. *Virology* 2009; 393: 329–337.

92. Wagner JM, Pajerowski JD, Daniels CL, McHugh PM, Flynn JA, Balliet JW, Casimiro DR, Subramanian S. Enhanced production of Chikungunya virus-like particles using a high-pH adapted *Spodoptera frugiperda* insect cell line. *PLoS ONE* 2014; 9: e94401.

93. Yao L, Wang S, Su S, Yao N, He J, Peng L, Sun J. Construction of a baculovirus-silkworm multigene expression system and its application on producing virus-like particles. *PLoS ONE* 2012; 7: e32510.

94. Pogue GP, Lindbo JA, Garger SJ, Fitzmaurice WP. Making an ally from an enemy: Plant virology and the new agriculture. *Annu. Rev. Phytopathol.* 2002; 40: 45–74.

95. Zeltins, A. Plant virus biotechnology platforms for expression of medicinal proteins. In: *Medicinal Protein Engineering*, YE Khudyakov, ed. CRC Press, Boca Raton, FL. 2009, pp. 481–517.

96. Klimyuk V, Pogue G, Herz S, Butler J, Haydon H. Production of recombinant antigens and antibodies in *Nicotiana benthamiana* using 'magnifection' technology: GMP-compliant facilities for small- and large-scale manufacturing. *Curr. Top. Microbiol. Immunol.* 2014; 375: 127–154.

97. Gleba YY, Tusé D, Giritch A. Plant viral vectors for delivery by Agrobacterium. *Curr. Top. Microbiol. Immunol.* 2014; 375: 155–192.

98. Werner S, Marillonnet S, Hause G, Klimyuk V, Gleba Y. Immunoabsorbent nanoparticles based on a tobamovirus displaying protein A. *Proc. Natl. Acad. Sci. USA* 2006; 103: 17678–17683.

99. Saunders K, Lomonossoff GP. Exploiting plant virus-derived components to achieve in planta expression and for templates for synthetic biology applications. *New Phytol.* 2013; 200: 16–26.

100. Rybicki EP, Martin DP. Virus-derived ssDNA vectors for the expression of foreign proteins in plants. *Curr. Top. Microbiol. Immunol.* 2014; 375: 19–45.

101. Lai H, Chen Q. Bioprocessing of plant-derived virus-like particles of Norwalk virus capsid protein under current Good Manufacture Practice regulations. *Plant Cell Rep.* 2012; 31: 573–584.

102. Walsh, G. Biopharmaceutical benchmarks 2006. *Nat. Biotechnol.* 2006; 24: 769–776.

103. Wu CY, Yeh YC, Yang YC, Chou C, Liu MT, Wu HS, Chan JT, Hsiao PW. Mammalian expression of virus-like particles for advanced mimicry of authentic influenza virus. *PLoS ONE* 2010; 5: e9784.

104. Butler M, Spearman M. The choice of mammalian cell host and possibilities for glycosylation engineering. *Curr. Opin. Biotechnol.* 2014; 30C: 107–112.

105. Kaufman RJ. Overview of vector design for mammalian gene expression. *Mol. Biotechnol.* 2000; 16: 151–160.

106. Hua RH, Li YN, Chen ZS, Liu LK, Huo H, Wang XL, Guo LP, Shen N, Wang JF, Bu ZG. Generation and characterization of a new mammalian cell line continuously expressing virus-like particles of Japanese encephalitis virus for a subunit vaccine candidate. *BMC Biotechnol.* 2014; 14: 62.

107. Akahata W, Yang ZY, Andersen H, Sun S, Holdaway HA, Kong WP, Lewis MG, Higgs S, Rossmann MG, Rao S, Nabel GJ. A virus-like particle vaccine for epidemic Chikungunya virus protects nonhuman primates against infection. *Nat. Med.* 2010; 16: 334–338.

108. Carlson ED, Gan R, Hodgman CE, Jewett MC. Cell-free protein synthesis: Applications come of age. *Biotechnol. Adv.* 2012; 30: 1185–1194.

109. Lu Y, Welsh JP, Chan W, Swartz JR. *Escherichia coli*-based cell free production of flagellin and ordered flagellin display on virus-like particles. *Biotechnol. Bioeng.* 2013; 110: 2073–2085.

110. Smith MT, Wilding KM, Hunt JM, Bennett AM, Bundy BC. The emerging age of cell-free synthetic biology. *FEBS Lett.* 2014; 588: 2755–2761.

111. Dika C, Gantzer C, Perrin A, Duval JF. Impact of the virus purification protocol on aggregation and electrokinetics of MS2 phages and corresponding virus-like particles. *Phys. Chem. Chem. Phys.* 2013; 15: 5691–5700.

112. Vicente T, Roldão A, Peixoto C, Carrondo MJ, Alves PM. Large-scale production and purification of VLP-based vaccines. *J. Invertebr. Pathol.* 2011; 107(Suppl.): S42–S48.

113. Ng MY, Tan WS, Abdullah N, Ling TC, Tey BT. Heat treatment of unclarified *Escherichia coli* homogenate improved the recovery efficiency of recombinant hepatitis B core antigen. *J. Virol. Methods* 2006; 137: 134–139.

114. Skrastina D, Petrovskis I, Petraityte R, Sominskaya I, Ose V, Lieknina I, Bogans J, Sasnauskas K, Pumpens P. Chimeric derivatives of hepatitis B virus core particles carrying major epitopes of the rubella virus E1 glycoprotein. *Clin. Vaccine Immunol.* 2013; 20: 1719–1728.

115. Nishimura Y, Takeda K, Ishii J, Ogino C, Kondo A. An affinity chromatography method used to purify His-tag-displaying bio-nanocapsules. *J. Virol. Methods* 2013; 189: 393–396.

116. Fernández E, Toledo JR, Méndez L, González N, Parra F, Martín-Alonso JM, Limonta M et al. Conformational and thermal stability improvements for the large-scale production of yeast-derived rabbit hemorrhagic disease virus-like particles as multipurpose vaccine. *PLoS ONE* 2013; 8: e56417.

117. Skrastina D, Bulavaite A, Sominskaya I, Kovalevska L, Ose V, Priede D, Pumpens P, Sasnauskas K. High immunogenicity of a hydrophilic component of the hepatitis B virus preS1 sequence exposed on the surface of three virus-like particle carriers. *Vaccine* 2008; 26: 1972–1981.

118. Rupar M, Ravnikar M, Tušek-Žnidarič M, Kramberger P, Glais L, Gutiérrez-Aguirre I. Fast purification of the filamentous Potato virus Y using monolithic chromatographic supports. *J. Chromatogr. A* 2013; 1272: 33–40.

119. Burden CS, Jin J, Podgornik A, Bracewell DG. A monolith purification process for virus-like particles from yeast homogenate. *J. Chromatogr. B Analyt. Technol. Biomed. Life Sci.* 2012; 880: 82–89.

120. Kim HJ, Lim SJ, Kwag HL, Kim HJ. The choice of resin-bound ligand affects the structure and immunogenicity of column-purified human papillomavirus type 16 virus-like particles. *PLoS ONE* 2012; 7: e35893.

121. Guan Q, Weiss CR, Qing G, Ma Y, Peng Z. An IL-17 peptide-based and virus-like particle vaccine enhances the bioactivity of IL-17 in vitro and in vivo. *Immunotherapy* 2012; 4: 1799–1807.

122. Tumban E, Peabody J, Peabody DS, Chackerian B. A universal virus-like particle-based vaccine for human papillomavirus: Longevity of protection and role of endogenous and exogenous adjuvants. *Vaccine* 2013; 31: 4647–4654.

123. Vicente T, Mota JP, Peixoto C, Alves PM, Carrondo MJ. Rational design and optimization of downstream processes of virus particles for biopharmaceutical applications: Current advances. *Biotechnol. Adv.* 2011; 29: 869–878.

124. Josefsberg JO, Buckland B. Vaccine process technology. *Biotechnol. Bioeng.* 2012; 109: 1443–1460.

125. Johnson JE. Functional implications of protein–protein interactions in icosahedral viruses. *Proc. Natl. Acad. Sci. USA* 1996; 93: 27–33.

126. Simon C, Klose T, Herbst S, Han BG, Sinz A, Glaeser RM, Stubbs MT, Lilie H. Disulfide linkage and structure of highly stable yeast-derived virus-like particles of murine polyomavirus. *J. Biol. Chem.* 2014; 289: 10411–10418.

127. Tresset G, Decouche V, Bryche JF, Charpilienne A, Le Cœur C, Barbier C, Squires G, Zeghal M, Poncet D, Bressanelli S. Unusual self-assembly properties of Norovirus Newbury2 virus-like particles. *Arch. Biochem. Biophys.* 2013; 537: 144–152.

128. Lago H, Parrott AM, Moss T, Stonehouse NJ, Stockley PG. Probing the kinetics of formation of the bacteriophage MS2 translational operator complex: Identification of a protein conformer unable to bind RNA. *J. Mol. Biol.* 2001; 305: 1131–1144.

129. Savithri HS, Erickson JW. The self-assembly of the cowpea strain of southern bean mosaic virus: Formation of T = 1 and T = 3 nucleoprotein particles. *Virology* 1983; 126: 328–335.

130. Newman M, Chua PK, Tang FM, Su PY, Shih C. Testing an electrostatic interaction hypothesis of hepatitis B virus capsid stability by using an in vitro capsid disassembly/reassembly system. *J. Virol.* 2009; 83: 10616–10626.

131. Cheng F, Tsvetkova IB, Khuong YL, Moore AW, Arnold RJ, Goicochea NL, Dragnea B, Mukhopadhyay S. The packaging of different cargo into enveloped viral nanoparticles. *Mol. Pharm.* 2013; 10: 51–58.

132. Moser C, Müller M, Kaeser MD, Weydemann U, Amacker M. Influenza virosomes as vaccine adjuvant and carrier system. *Expert Rev. Vaccines* 2013; 12: 779–791.

133. Zhao Q, Modis Y, High K, Towne V, Meng Y, Wang Y, Alexandroff J et al. Disassembly and reassembly of human papillomavirus virus-like particles produces more virion-like antibody reactivity. *Virol. J.* 2012; 9: 52.

134. Spruijt RB, Hemminga MA. The in situ aggregational and conformational state of the major coat protein of bacteriophage M13 in phospholipid bilayers mimicking the inner membrane of host *Escherichia coli*. *Biochemistry* 1991; 30: 11147–11154.

135. Glasgow JE, Capehart SL, Francis MB, Tullman-Ercek D. Osmolyte-mediated encapsulation of proteins inside MS2 viral capsids. *ACS Nano* 2012; 6: 8658–8664.

136. Caldeira JC, Peabody DS. Stability and assembly in vitro of bacteriophage PP7 virus-like particles. *J. Nanobiotechnol.* 2007; 5: 10.

137. Lavelle L, Michel JP, Gingery M. The disassembly, reassembly and stability of CCMV protein capsids. *J. Virol. Methods* 2007; 146: 311–316.

138. Karpova OV, Zayakina OV, Arkhipenko MV, Sheval EV, Kiselyova OI, Poljakov VY, Yaminsky IV, Rodionova NP, Atabekov JG. Potato virus X RNA-mediated assembly of single-tailed ternary 'coat protein-RNA-movement protein' complexes. *J. Gen. Virol.* 2006; 87: 2731–2740.

139. Fraenkel-Conrat H, Williams RC. Reconstitution of active tobacco mosaic virus from its inactive protein and nucleic acid components. *Proc. Natl. Acad. Sci. USA* 1955; 41: 690–698.

140. Porterfield JZ, Dhason MS, Loeb DD, Nassal M, Stray SJ, Zlotnick A. Full-length hepatitis B virus core protein packages viral and heterologous RNA with similarly high levels of cooperativity. *J. Virol.* 2010; 84: 7174–7184.

141. Mukherjee S, Thorsteinsson MV, Johnston LB, DePhillips PA, Zlotnick A. A quantitative description of in vitro assembly of human papillomavirus 16 virus-like particles. *J. Mol. Biol.* 2008; 381: 229–237.

142. Zlotnick A, Porterfield JZ, Wang JC. To build a virus on a nucleic acid substrate. *Biophys. J.* 2013; 104: 1595–1604.

143. Borodavka A, Tuma R, Stockley PG. Evidence that viral RNAs have evolved for efficient, two-stage packaging. *Proc. Natl. Acad. Sci. USA* 2012; 109: 15769–15774.

144. Routh A, Domitrovic T, Johnson JE. Host RNAs, including transposons, are encapsidated by a eukaryotic single-stranded RNA virus. *Proc. Natl. Acad. Sci. USA* 2012; 109: 1907–1912.

145. Branco LM, Grove JN, Geske FJ, Boisen ML, Muncy IJ, Magliato SA, Henderson LA, Schoepp RJ, Cashman KA, Hensley LE, Garry RF. Lassa virus-like particles displaying all major immunological determinants as a vaccine candidate for Lassa hemorrhagic fever. *Virol. J.* 2010; 7: 279.

146. Bragard C, Duncan GH, Wesley SV, Naidu RA, Mayo MA. Virus-like particles assemble in plants and bacteria expressing the coat protein gene of Indian peanut clump virus. *J. Gen. Virol.* 2000; 81: 267–272.

147. Lokesh GL, Gowri TD, Satheshkumar PS, Murthy MR, Savithri HS. A molecular switch in the capsid protein controls the particle polymorphism in an icosahedral virus. *Virology* 2002; 292: 211–223.

148. Hema M, Subba Reddy ChV, Savithri HS, Sreenivasulu P. Assembly of recombinant coat protein of sugarcane streak mosaic virus into potyvirus-like particles. *Indian J. Exp. Biol.* 2008; 46: 793–796.

149. Bunka DH, Lane SW, Lane CL, Dykeman EC, Ford RJ, Barker AM, Twarock R, Phillips SE, Stockley PG. Degenerate RNA packaging signals in the genome of satellite tobacco necrosis virus: Implications for the assembly of a T = 1 capsid. *J. Mol. Biol.* 2011; 413: 51–65.

150. Storni T, Ruedl C, Schwarz K, Schwendener RA, Renner WA, Bachmann MF. Nonmethylated CG motifs packaged into virus-like particles induce protective cytotoxic T cell responses in the absence of systemic side effects. *J. Immunol.* 2004; 172: 1777–1785.

151. Fang CY, Lin PY, Ou WC, Chen PL, Shen CH, Chang D, Wang M. Analysis of the size of DNA packaged by the human JC virus-like particle. *J. Virol. Methods* 2012; 182: 87–92.

152. Rūmnieks J, Freivalds J, Cielēns I, Renhofa R. Specificity of packaging mRNAs in bacteriophage GA virus-like particles in yeast *Saccharomyces cerevisiae*. *Acta Univ. Latv.* 2008; 745: 145–154.

153. Galaway FA, Stockley PG. MS2 viruslike particles: A robust, semisynthetic targeted drug delivery platform. *Mol. Pharm.* 2013; 10: 59–68.

154. Zlotnick A, Mukhopadhyay S. Virus assembly, allostery and antivirals. *Trends Microbiol.* 2011; 19: 14–23.

155. Cuervo A, Daudén MI, Carrascosa JL. Nucleic acid packaging in viruses. *Subcell. Biochem.* 2013; 68: 361–394.

156. Patterson DP, Prevelige PE, Douglas T. Nanoreactors by programmed enzyme encapsulation inside the capsid of the bacteriophage P22. *ACS Nano* 2012; 6: 5000–5009.

157. Minten IJ, Hendriks LJ, Nolte RJ, Cornelissen JJ. Controlled encapsulation of multiple proteins in virus capsids. *J. Am. Chem. Soc.* 2009; 131: 17771–17773.

158. Fiedler JD, Brown SD, Lau JL, Finn MG. RNA-directed packaging of enzymes within virus-like particles. *Angew. Chem. Int. Ed. Engl.* 2010; 49: 9648–9651.

159. Lu X, Thompson JR, Perry KL. Encapsidation of DNA, a protein and a fluorophore into virus-like particles by the capsid protein of cucumber mosaic virus. *J. Gen. Virol.* 2012; 93: 1120–1126.

160. Comellas-Aragonès M, Engelkamp H, Claessen VI, Sommerdijk NA, Rowan AE, Christianen PC, Maan JC, Verduin BJ, Cornelissen JJ, Nolte RJ. A virus-based single-enzyme nanoreactor. *Nat. Nanotechnol.* 2007; 2: 635–639.

161. Douglas T, Young M. Host–guest encapsulation of materials by assembled virus protein cages. *Nature* 1998; 393: 152–155.

162. Ma Y, Nolte RJ, Cornelissen JJ. Virus-based nanocarriers for drug delivery. *Adv. Drug. Deliv. Rev.* 2012; 64: 811–825.

163. Arcangeli C, Circelli P, Donini M, Aljabali AA, Benvenuto E, Lomonossoff GP, Marusic C. Structure-based design and experimental engineering of a plant virus nanoparticle for the presentation of immunogenic epitopes and as a drug carrier. *J. Biomol. Struct. Dyn.* 2014; 32: 630–647.

164. Blokhina EA, Kupriyanov VV, Ravin NV, Skryabin KG. The method of noncovalent in vitro binding of target proteins to virus-like nanoparticles formed by core antigen of hepatitis B virus. *Dokl. Biochem. Biophys.* 2013; 448: 52–54.

165. Huang X, Bronstein LM, Retrum J, Dufort C, Tsvetkova I, Aniagyei S, Stein B et al. Self-assembled virus-like particles with magnetic cores. *Nano Lett.* 2007; 7: 2407–2416.

166. Mikkilä J, Rosilo H, Nummelin S, Seitsonen S, Ruokolainen J, Kostiainen MA. Janus dendrimer-mediated formation of crystalline virus assemblies. *ACS Macro Lett.* 2013; 2: 720–724.

167. Lee LA, Muhammad SM, Nguyen QL, Sitasuwan P, Horvath G, Wang Q. Multivalent ligand displayed on plant virus induces rapid onset of bone differentiation. *Mol. Pharm.* 2012; 9: 2121–2125.

168. Baslé E, Joubert N, Pucheault M. Protein chemical modification on endogenous amino acids. *Chem. Biol.* 2010; 17: 213–227.

169. Fecek RJ, Busch R, Lin H, Pal K, Cunningham CA, Cuff CF. Production of Alexa Fluor 488-labeled reovirus and characterization of target cell binding, competence, and immunogenicity of labeled virions. *J. Immunol. Methods* 2006; 314: 30–37.

170. Plummer EM, Manchester M. Endocytic uptake pathways utilized by CPMV nanoparticles. *Mol. Pharm.* 2013; 10: 26–32.

171. Detmer SE, Gramer MR, Goyal SM, Torremorell M. In vitro characterization of influenza A virus attachment in the upper and lower respiratory tracts of pigs. *Vet. Pathol.* 2013; 50: 648–658.

172. Wu Z, Chen K, Yildiz I, Dirksen A, Fischer R, Dawson PE, Steinmetz NF. Development of viral nanoparticles for efficient intracellular delivery. *Nanoscale* 2012; 4: 3567–3576.

173. Rostovtsev VV, Green LG, Fokin VV, Sharpless KB. A stepwise huisgen cycloaddition process: Copper(I)-catalyzed regioselective "ligation" of azides and terminal alkynes. *Angew. Chem. Int. Ed. Engl.* 2002; 41: 2596–2599.

174. Yin Z, Comellas-Aragones M, Chowdhury S, Bentley P, Kaczanowska K, Benmohamed L, Gildersleeve JC, Finn MG, Huang X. Boosting immunity to small tumor-associated carbohydrates with bacteriophage qβ capsids. *ACS Chem. Biol.* 2013; 8: 1253–1262.

175. Steinmetz NF, Mertens ME, Taurog RE, Johnson JE, Commandeur U, Fischer R, Manchester M. Potato virus X as a novel platform for potential biomedical applications. *Nano Lett.* 2010; 10: 305–312.

176. Mead G, Hiley M, Ng T, Fihn C, Hong K, Groner M, Miner W, Drugan D, Hollingsworth W, Udit AK. Directed polyvalent display of sulfated ligands on virus nanoparticles elicits heparin-like anticoagulant activity. *Bioconjug. Chem.* 2014; 25: 1444–1452.

177. Washington-Hughes CL, Cheng Y, Duan X, Cai L, Lee LA, Wang Q. In vivo virus-based macrofluorogenic probes target azide-labeled surface glycans in MCF-7 breast cancer cells. *Mol. Pharm.* 2013; 10: 43–50.

178. Lallana E, Sousa-Herves A, Fernandez-Trillo F, Riguera R, Fernandez-Megia E. Click chemistry for drug delivery nanosystems. *Pharm. Res.* 2012; 29: 1–34.

179. Chen PY, Dang X, Klug MT, Qi J, Dorval Courchesne NM, Burpo FJ, Fang N, Hammond PT, Belcher AM. Versatile three-dimensional virus-based template for dye-sensitized solar cells with improved electron transport and light harvesting. *ACS Nano* 2013; 7: 6563–6574.

180. Lee KW, Tey BT, Ho KL, Tejo BA, Tan WS. Nanoglue: An alternative way to display cell-internalizing peptide at the spikes of hepatitis B virus core nanoparticles for cell-targeting delivery. *Mol. Pharm.* 2012; 9: 2415–2423.

181. Fiedler JD, Higginson C, Hovlid ML, Kislukhin AA, Castillejos A, Manzenrieder F, Campbell MG, Voss NR, Potter CS, Carragher B, Finn MG. Engineered mutations change the structure and stability of a virus-like particle. *Biomacromolecules* 2012; 13: 2339–2348.

182. Niikura K, Sugimura N, Musashi Y, Mikuni S, Matsuo Y, Kobayashi S, Nagakawa K et al. Virus-like particles with removable cyclodextrins enable glutathione-triggered drug release in cells. *Mol. Biosyst.* 2013; 9: 501–507.

183. Marjomäki V, Lahtinen T, Martikainen M, Koivisto J, Malola S, Salorinne K, Pettersson M, Häkkinen H. Site-specific targeting of enterovirus capsid by functionalized monodisperse gold nanoclusters. *Proc. Natl. Acad. Sci. USA* 2014; 111: 1277–1281.

184. Bundy BC, Swartz JR. Efficient disulfide bond formation in virus-like particles. *J. Biotechnol.* 2011; 154: 230–239.

185. Horowitz ED, Finn MG, Asokan A. Tyrosine cross-linking reveals interfacial dynamics in adeno-associated viral capsids during infection. *ACS Chem. Biol.* 2012; 7: 1059–1066.

186. Kaur G, Wang C, Sun J, Wang Q. The synergistic effects of multivalent ligand display and nanotopography on osteogenic differentiation of rat bone marrow stem cells. *Biomaterials* 2010; 31: 5813–5824.

187. Huang BH, Lin Y, Zhang ZL, Zhuan F, Liu AA, Xie M, Tian ZQ, Zhang Z, Wang H, Pang DW. Surface labeling of enveloped viruses assisted by host cells. *ACS Chem. Biol.* 2012; 7: 683–688.

188. Zhang Y, Ke X, Zheng Z, Zhang C, Zhang Z, Zhang F, Hu Q, He Z, Wang H. Encapsulating quantum dots into enveloped virus in living cells for tracking virus infection. *ACS Nano* 2013; 7: 3896–3904.

189. Chuan YP, Lua LH, Middelberg AP. High-level expression of soluble viral structural protein in *Escherichia coli*. *J. Biotechnol.* 2008; 134: 64–71.

190. Walker A, Skamel C, Nassal M. SplitCore: An exceptionally versatile viral nanoparticle for native whole protein display regardless of 3D structure. *Sci. Rep.* 2011; 1: 5.

191. Chackerian B, Caldeira JC, Peabody J, Peabody DS. Peptide epitope identification by affinity selection on bacteriophage MS2 virus-like particles. *J. Mol. Biol.* 2011; 409: 225–237.

192. Brown AD, Naves L, Wang X, Ghodssi R, Culver JN. Carboxylate-directed in vivo assembly of virus-like nanorods and tubes for the display of functional peptides and residues. *Biomacromolecules* 2013; 14: 3123–3129.

193. Smith ML, Lindbo JA, Dillard-Telm S, Brosio PM, Lasnik AB, McCormick AA, Nguyen LV, Palmer KE. Modified tobacco mosaic virus particles as scaffolds for display of protein antigens for vaccine applications. *Virology* 2006; 348: 475–488.

194. Zhou K, Li F, Dai G, Meng C, Wang Q. Disulfide bond: Dramatically enhanced assembly capability and structural stability of tobacco mosaic virus nanorods. *Biomacromolecules* 2013; 14: 2593–2600.

195. Chiang CY, Epstein J, Brown A, Munday JN, Culver JN, Ehrman S. Biological templates for antireflective current collectors for photoelectrochemical cell applications. *Nano Lett.* 2012; 12: 6005–6011.

196. Wang Q, Lin T, Johnson JE, Finn MG. Natural supramolecular building blocks. Cysteine-added mutants of cowpea mosaic virus. *Chem. Biol.* 2002; 9: 813–819.

197. Farkas ME, Aanei IL, Behrens CR, Tong GJ, Murphy ST, O'Neil JP, Francis MB. PET imaging and biodistribution of chemically modified bacteriophage MS2. *Mol. Pharm.* 2013; 10: 69–76.

198. Douglas T, Strable E, Willits D, Aitouchen A, Libera M, Young M. Protein engineering of a viral cage for constrained nanomaterials synthesis. *Adv. Mater.* 2002; 14: 415–418.

199. Wang JC, Dhason MS, Zlotnick A. Structural organization of pregenomic RNA and the carboxy-terminal domain of the capsid protein of hepatitis B virus. *PLoS Pathog.* 2012; 8: e1002919.

200. Anggraeni MR, Connors NK, Wu Y, Chuan YP, Lua LH, Middelberg AP. Sensitivity of immune response quality to influenza helix 190 antigen structure displayed on a modular virus-like particle. *Vaccine* 2013; 31: 4428–4435.

201. Yoo SY, Merzlyak A, Lee S-W. Facile growth factor immobilization platform based on engineered phage matrices. *Soft Matter* 2011; 7: 1660.

202. Hovlid ML, Steinmetz NF, Laufer B, Lau JL, Kuzelka J, Wang Q, Hyypiä T, Nemerow GR, Kessler H, Manchester M, Finn MG. Guiding plant virus particles to integrin-displaying cells. *Nanoscale* 2012; 4: 3698–3705.

203. Ranka R, Petrovskis I, Sominskaya I, Bogans J, Bruvere R, Akopjana I, Ose V, Timofejeva I, Brangulis K, Pumpens P, Baumanis V. Fibronectin-binding nanoparticles for intracellular targeting addressed by *B. burgdorferi* BBK32 protein fragments. *Nanomedicine* 2013; 9: 65–73.

204. Wnęk M, Górzny ML, Ward MB, Wälti C, Davies AG, Brydson R, Evans SD, Stockley PG. Fabrication and characterization of gold nano-wires templated on virus-like arrays of tobacco mosaic virus coat proteins. *Nanotechnology* 2013; 24: 025605.

205. van Eldijk MB, Wang JC, Minten IJ, Li C, Zlotnick A, Nolte RJ, Cornelissen JJ, van Hest JC. Designing two self-assembly mechanisms into one viral capsid protein. *J. Am. Chem. Soc.* 2012; 134: 18506–18509.

206. Kratz PA, Böttcher B, Nassal M. Native display of complete foreign protein domains on the surface of hepatitis B virus capsids. *Proc. Natl. Acad. Sci. USA* 1999; 96: 1915–1920.

207. Nassal M, Skamel C, Vogel M, Kratz PA, Stehle T, Wallich R, Simon MM. Development of hepatitis B virus capsids into a whole-chain protein antigen display platform: New particulate Lyme disease vaccines. *Int. J. Med. Microbiol.* 2008; 298: 135–142.

208. Arora U, Tyagi P, Swaminathan S, Khanna N. Chimeric Hepatitis B core antigen virus-like particles displaying the envelope domain III of dengue virus type 2. *J. Nanobiotechnol.* 2012; 10: 30.

209. Cardinale D, Carette N, Michon T. Virus scaffolds as enzyme nano-carriers. *Trends Biotechnol.* 2012; 30: 369–376.

210. Zhao Q, Li S, Yu H, Xia N, Modis Y. Virus-like particle-based human vaccines: Quality assessment based on structural and functional properties. *Trends Biotechnol.* 2013; 31: 654–663.

211. Yoon KY, Tan WS, Tey BT, Lee KW, Ho KL. Native agarose gel electrophoresis and electroelution: A fast and cost-effective method to separate the small and large hepatitis B capsids. *Electrophoresis* 2013; 34: 244–253.

212. Rajendar B, Sivakumar V, Sriraman R, Raheem M, Lingala R, Matur RV. A simple and rapid method to monitor the disassembly and reassembly of virus-like particles. *Anal. Biochem.* 2013; 440: 15–17.

213. Vicente T, Peixoto C, Alves PM, Carrondo MJ. Modeling electrostatic interactions of baculovirus vectors for ion-exchange process development. *J. Chromatogr. A* 2010; 1217: 3754–3764.

214. Hahne T, Palaniyandi M, Kato T, Fleischmann P, Wätzig H, Park EY. Characterization of human papillomavirus 6b L1 virus-like particles isolated from silkworms using capillary zone electrophoresis. *J. Biosci. Bioeng.* 2014; 118: 311–314.

215. Chuan YP, Fan YY, Lua L, Middelberg AP. Quantitative analysis of virus-like particle size and distribution by field-flow fractionation. *Biotechnol. Bioeng.* 2008; 99: 1425–1433.

216. Goldsmith CS, Miller SE. Modern uses of electron microscopy for detection of viruses. *Clin. Microbiol. Rev.* 2009; 22: 552–563.

217. de Pablo PJ. Atomic force microscopy of viruses. *Subcell. Biochem.* 2013; 68: 247–271.

218. Rossmann MG. Structure of viruses: A short history. *Q. Rev. Biophys.* 2013; 46: 133–180.

219. Seitsonen JJ, Susi P, Lemmetty A, Butcher SJ. Structure of the mite-transmitted Blackcurrant reversion nepovirus using electron cryo-microscopy. *Virology* 2008; 378: 162–168.

220. Plevka P, Kazaks A, Voronkova T, Kotelovica S, Dishlers A, Liljas L, Tars K. The structure of bacteriophage phiCb5 reveals a role of the RNA genome and metal ions in particle stability and assembly. *J. Mol. Biol.* 2009; 391: 635–647.

221. Garmann RF, Comas-Garcia M, Gopal A, Knobler CM, Gelbart WM. The assembly pathway of an icosahedral single-stranded RNA virus depends on the strength of inter-subunit attractions. *J. Mol. Biol.* 2014; 426: 1050–1060.

222. Lin SY, Chung YC, Chiu HY, Chi WK, Chiang BL, Hu YC. Evaluation of the stability of enterovirus 71 virus-like particle. *J. Biosci. Bioeng.* 2014; 117: 366–371.

223. Merino EJ, Wilkinson KA, Coughlan JL, Weeks KM. RNA structure analysis at single nucleotide resolution by selective 2′-hydroxyl acylation and primer extension (SHAPE). *J. Am. Chem. Soc.* 2005; 127: 4223–4231.

224. Archer EJ, Simpson MA, Watts NJ, O'Kane R, Wang B, Erie DA, McPherson A, Weeks KM. Long-range architecture in a viral RNA genome. *Biochemistry* 2013; 52: 3182–3190.

225. Bruckman MA, Randolph LN, VanMeter A, Hern S, Shoffstall AJ, Taurog RE, Steinmetz NF. Biodistribution, pharmacokinetics, and blood compatibility of native and PEGylated tobacco mosaic virus nano-rods and -spheres in mice. *Virology* 2014; 449: 163–173.

226. Franco CF, Mellado MC, Alves PM, Coelho AV. Monitoring virus-like particle and viral protein production by intact cell MALDI-TOF mass spectrometry. *Talanta* 2010; 80: 1561–1568.

227. Greiner VJ, Manin C, Larquet E, Ikhelef N, Gréco F, Naville S, Milhiet PE, Ronzon F, Klymchenko A, Mély Y. Characterization of the structural modifications accompanying the loss of HBsAg particle immunogenicity. *Vaccine* 2014; 32: 1049–1054.

228. Nguyen AP, Downard KM. Proteotyping of the parainfluenza virus with high-resolution mass spectrometry. *Anal. Chem.* 2013; 85: 1097–1105.

229. Ni P, Vaughan RC, Tragesser B, Hoover H, Kao CC. The plant host can affect the encapsidation of brome mosaic virus (BMV) RNA: BMV virions are surprisingly heterogeneous. *J. Mol. Biol.* 2014; 426: 1061–1076.

230. Vaughan R, Running WE, Qi R, Kao C. Mapping protein–RNA interactions. *Virus Adapt. Treat.* 2012; 4: 29–41.

231. Chavez JD, Cilia M, Weisbrod CR, Ju H-J, Eng JK, Gray SM, Bruce JE. Cross-linking measurements of the potato leafroll virus reveal protein interaction topologies required for virion stability, aphid transmission, and virus-plant interactions. *J. Proteome Res.* 2012; 11: 2968–2981.

232. Maeda A, Maeda J. Review of diagnostic plaque reduction neutralization tests for flavivirus infection. *Vet. J.* 2013; 195: 33–40.

233. Scholthof KB. Making a virus visible: Francis O. Holmes and a biological assay for tobacco mosaic virus. *J. Hist. Biol.* 2014; 47: 107–145.

234. Azizi A, Mironov GG, Muharemagic D, Wehbe M, Bell JC, Berezovski MV. Viral quantitative capillary electrophoresis for counting and quality control of RNA viruses. *Anal. Chem.* 2012; 84: 9585–9591.

235. Kramberger P, Ciringer M, Štrancar A, Peterka M. Evaluation of nanoparticle tracking analysis for total virus particle determination. *Virol. J.* 2012; 9: 265.

236. Francki RI, Randles JW, Chambers TC, Wilson SB. Some properties of purified cucumber mosaic virus (Q strain). *Virology* 1966; 28: 729–741.

237. Porterfield JZ, Zlotnick A. A simple and general method for determining the protein and nucleic acid content of viruses by UV absorbance. *Virology* 2010; 407: 281–288.

238. Maurer P, Bachmann MF. Therapeutic vaccines for nicotine dependence. *Curr. Opin. Mol. Ther.* 2006; 8: 11–16.

239. Röhn TA, Jennings GT, Hernandez M, Grest P, Beck M, Zou Y, Kopf M, Bachmann MF. Vaccination against IL-17 suppresses autoimmune arthritis and encephalomyelitis. *Eur. J. Immunol.* 2006; 36: 2857–2867.

240. Joyce JG, Tung JS, Przysiecki CT, Cook JC, Lehman ED, Sands JA, Jansen KU, Keller PM. The L1 major capsid protein of human papillomavirus type 11 recombinant virus-like particles interacts with heparin and cell-surface glycosaminoglycans on human keratinocytes. *J. Biol. Chem.* 1999; 274: 5810–5822.

241. Tissot AC, Renhofa R, Schmitz N, Cielens I, Meijerink E, Ose V, Jennings GT, Saudan P, Pumpens P, Bachmann MF. Versatile virus-like particle carrier for epitope based vaccines. *PLoS ONE* 2010; 5: e9809.

242. Wenrich BR, Trumbo TA. Interaction of nucleic acids with Coomassie Blue G-250 in the Bradford assay. *Anal. Biochem.* 2012; 428: 93–95.

243. Lang R, Winter G, Vogt L, Zurcher A, Dorigo B, Schimmele B. Rational design of a stable, freeze-dried virus-like particle-based vaccine formulation. *Drug Dev. Ind. Pharm.* 2009; 35: 83–97.

244. Thompson CM, Petiot E, Lennaertz A, Henry O, Kamen AA. Analytical technologies for influenza virus-like particle candidate vaccines: Challenges and emerging approaches. *Virol. J.* 2013; 10: 141.

245. Lua LH, Connors NK, Sainsbury F, Chuan YP, Wibowo N, Middelberg AP. Bioengineering virus-like particles as vaccines. *Biotechnol. Bioeng.* 2014; 111: 425–440.

246. Goldbourt A. Biomolecular magic-angle spinning solid-state NMR: Recent methods and applications. *Curr. Opin. Biotechnol.* 2013; 24: 705–715.

247. Wen AM, Rambhia PH, French RH, Steinmetz NF. Design rules for nanomedical engineering: From physical virology to the applications of virus-based materials in medicine. *J. Biol. Phys.* 2013; 39: 301–325.

248. Liu R, Vaishnav RA, Roberts AM, Friedland RP. Humans have antibodies against a plant virus: Evidence from tobacco mosaic virus. *PLoS ONE* 2013; 8: e60621.

249. Carrillo-Tripp M, Shepherd CM, Borelli IA, Venkataraman S, Lander G, Natarajan P, Johnson JE, Brooks CL 3rd, Reddy VS. VIPERdb2: An enhanced and web API enabled relational database for structural virology. *Nucleic Acids Res.* 2009; 37(Database issue): D436–D442.

7 Immunology of Virus-Like Particles

Martin F. Bachmann and Franziska Zabel

CONTENTS

7.1 INTRODUCTION

The human body and in particular the human immune system have evolved a large number of mechanisms to control viral infections. Multiple mechanical and immunological barriers exist that inhibit and restrict viral entry, infections as well as replication. A first barrier is the surface of the body, which usually does not allow penetration by viral particles and blocks viral infection from the very start. If viruses successfully overcome this first barrier, a number of innate sensing mechanisms exist in practically every cell type (e.g., RIG-I, Mda5, LGP2, STING, AIM2 [Imaizumi et al. 2002; Gitlin et al. 2006; Ishikawa et al. 2009; Schmidt et al. 2009; Satoh et al. 2010; Barber 2011]). These intracellular molecules sense viral infection, causing the production of type I IFN, which globally inhibits viral replication by inducing degradation of viral nucleic acids or inhibition of viral replication (Bowie and Unterholzner 2008; Yan and Chen 2012). Plasmacytoid dendritic cells (pDCs) of the innate immune system have evolved a particularly powerful sensing system for viral infection and are the dominant producers of antiviral type I interferon (Colonna et al. 2004; McKenna et al. 2005). In addition to recognition of viral nucleic acids, both the innate and adaptive immune systems have evolved various ways to recognize and inactivate exogenous viral particles as well as infected cells (Zinkernagel 1996). Both arms of the immune system collaborate to achieve maximal specific immune responses.

Virus-like particles (VLPs), in particular those carrying nucleic acids, induce immune responses very similar to those induced by life viruses. An important difference between VLPs and viruses is that the former induce a better definable response, as they do not replicate. This is a major advantage, as vaccines based on VLPs do not have to account for the complexity of viral replication, which may differ from host to host and is affected by multiple factors, including host genetics and concomitant infections with unrelated pathogens (D'Argenio and Wilson 2010; Pulendran et al. 2010).

The immune system recognizes viral particles largely based on two characteristics (Bachmann and Jennings 2010): (1) they have a unique size and geometry as they have highly organized and repetitive surface and (2) they deliver nucleic acids to the endosomal compartment of antigen-presenting cells (APCs) and B cells, activating toll-like receptors 7/8 or 9. These features alone are sufficient to explain why most viruses and VLPs induce potent antibody responses (Zinkernagel 1996; Zinkernagel et al. 2001), why the dominant antibody isotypes are IgG2a in the mouse (Coutelier et al. 1987) and IgG1/3 in the human (Beck 1981; Litwinska et al. 1993; Cavacini et al. 2003), why strong mucosal and systemic IgA responses are generated (Macpherson et al. 2008; Cerutti et al. 2011), and why induced antibody responses are long lived (Wrammert and Ahmed 2008). These considerations also explain why most antiviral vaccines are based on virus-shaped structures and direct a way to the development of novel, improved vaccines.

7.2 ROLE OF VIRAL SIZE

Vaccines including VLPs need to reach lymphoid organs for the initiation of protective T and B cell responses (Batista and Harwood 2009). Antigens additionally have to be displayed in a native state, and relevant epitopes have to be in proper conformation if protective antibody responses are to be induced. It is, therefore, a highly relevant question how antigens reach the lymphoid organs from the periphery. One important factor is their size (Swartz et al. 2008; Gonzalez et al. 2011). In essence, there are three relevant size classes:

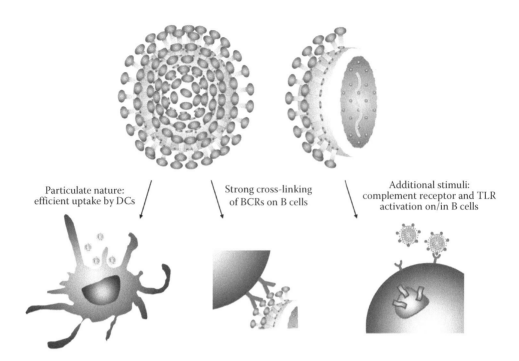

FIGURE 7.1 VLPs efficiently activate the immune system. VLPs usually occur with a size between 20 and 200 nm as particulate structure, which is easily recognized by DCs and can efficiently be taken up. The repetitive structure displayed on the surface of the VLPs induces a strong activation of B cells by cross-linking their BCRs. Moreover, carrying TLR ligands inside the VLPs supplies an additional stimulation for DCs or B cells. VLPs decorated with soluble complement factors bind to complement receptors on B cells.

(1) soluble and small protein complexes that are smaller than 10 nm. This is the regular self-protein and therefore the dominant size of host proteins and protein complexes in the extracellular space. (2) Particles of a size between 20 and 200 nm (Figure 7.1). Almost all viruses exhibit this size and such particles are not commonly found in the extracellular space in the absence of infection. (3) Particles that are larger than 200–500 nm. Those particles are usually of bacterial origin or derived from eukaryotic cells (fungal, parasitic, or host origin).

Both soluble proteins and particles <200 nm can freely drain and efficiently enter the lymphatic system by directly crossing lymphatic vessel walls through pores (Manolova et al. 2008). Indeed, lymphoid vessel walls are fenestrated at their peripheral endings and exhibit pores with a size of about 200 nm, only allowing the entry of particles <200 nm. Soluble proteins (<70 kDa, <5 nm) are subsequently transported into lymph nodes by specialized conduits (Sixt et al. 2005; Pape et al. 2007; Roozendaal et al. 2009), which virtually represent a *highway* for small antigen trafficking. However, it is likely that an important function of these conduits is the facilitation of intercellular communication rather than induction of immune responses. Proteins or particles >70 kDa or >5 nm are too large to enter conduits and need to be actively transported from the subcapsular sinus (SCS) to B cell follicles within the lymph node. To this end, various cell types, including myeloid cells such as SCS macrophages and B cells, bind such antigens on their surface and shuttle them into B cell follicles of lymph nodes and the spleen (see the following) (Qi et al. 2006;

Carrasco and Batista 2007; Cinamon et al. 2008; Phan et al. 2009; Gonzalez et al. 2010a,b).

Macrophages mediate trafficking of intact and lymph borne antigen from the SCS to B cell follicles within LNs as well as medullary DCs, which bind viral particles to their surface via SIGN-R1 (also known as CD209b) or DC-SIGN (Gonzalez et al. 2010a). Alternatively, it has also been shown that antigens can be stored in nondegrading intracellular compartments before being recycled to the cell surface (Bergtold et al. 2005). Nevertheless, such DC-mediated transportation has only been described within lymphoid organs and not from the periphery to B cell follicles. There is some evidence that DC-SIGN on DCs binds live HIV virions and transports them from mucosal surfaces to LNs and allows shuttling of the particles to T regions for infection of CD4+ T cells (Geijtenbeek et al. 2000). Little is known, however, about mechanisms that allow efficient interaction with B cells for large particulate antigens for the induction of protective antibody responses.

In contrast to soluble proteins and relatively small particles, larger particles of a size >200–500 nm are transported to lymphoid organs by an entirely different mechanism. First of all, they fail to enter the lymphatic system in free form since the pores in vessel walls are too small for passing. Thus, large particles (>200 nm) require cell-mediated transport by macrophages and peripheral DCs already from the site of entry or injection into the lymphatic system before they can be transported to lymph nodes in cell-associated form (Pflicke and Sixt 2009). A complicating factor is that DCs usually digest antigens they have taken up and therefore do not preserve their native antigenic structure. Hence as mentioned above, it is currently not entirely clear how such larger particles are able to trigger B cells in lymphoid organs. Taken together, viral particles and in particular VLPs exhibit the optimal size to reach B cell follicles of LN. Transportation happens in a two-step process. First, VLPs enter lymphatic vessels in free form by diffusing through pores in lymphatic vessel walls and subsequently passively drain to the subcapsular regions of LNs (Manolova et al. 2008). In a second step, VLPs are transported on the cell surface of macrophages, DCs, and, as discussed in the next section, B cells (Phan et al. 2009; Gonzalez et al. 2010a; Bessa et al. 2012; Link et al. 2012) into B cell follicles for the activation of B cells and induction of antibody responses.

7.3 ROLE OF VIRAL SURFACE GEOMETRY

Most viruses, in particular RNA viruses, have a small genome. These viruses are, therefore, unable to have complex surfaces but are forced to build up their surface using few different building blocks. As a consequence, most viruses have a highly repetitive and quasicrystalline surface (Bachmann and Zinkernagel 1997; Bachmann and Jennings 2010). A prominent viral structure is the icosahedron, consisting of 180 identical protein subunits assembled into a football-like form. This highly repetitive appearance of viral surfaces is different from almost all macromolecular structures in the host and can easily be identified as foreign by the innate as well as adaptive immune system. Consequently, such highly organized and repetitive structures may be considered a geometric pathogen-associated molecular pattern (PAMP) (Bachmann and Zinkernagel 1997). Indeed, repetitiveness is well recognized by both the innate and adaptive immune systems, which is a rather unique feature for a PAMP.

7.3.1 ROLE OF VIRAL SURFACE GEOMETRY FOR THE INNATE IMMUNE SYSTEM

Complement, natural antibodies, and pentraxins are important components of the innate humoral immune system (Bottazzi et al. 2010). An important feature of these molecules is that they are multimeric: C1q, mannan-binding protein, and pentraxins are pentameric, while natural IgM is even decameric (consisting of five covalently linked bivalent antibodies). Such multimeric molecules are ideal for recognition of repetitive structures, as they can interact and multiply with the repetitive viral surface. Hence, even if the affinity of the interaction between a single component of the multimeric molecules and a single viral capsid subunit is very low, the avidity of a pentavalent or decavalent interaction of the same molecules may be of considerable strength, clearly favoring such interactions. Consistent with these considerations, we have recently shown that VLPs bind natural IgM and fix C1q in vivo, leading to their efficient deposition on follicular dendritic cells (FDCs). In contrast, soluble dimeric

coat protein of 28 kDa derived from the same VLPs, which exhibit the same antigenic epitopes as the VLPs, failed to be recognized by natural IgM and/or C1q and was not efficiently deposited on FDCs in vivo. As a direct consequence, germinal center (GC) responses were strongly reduced when soluble protein rather than viral particles was used for immunization (Link et al. 2012) (see the following).

7.3.2 ROLE OF VIRAL SURFACE GEOMETRY FOR THE ADAPTIVE IMMUNE SYSTEM

Monomeric, soluble antigens are ignored by B cells in the majority of circumstances, since they fail to cause cross-linking of BCRs. In marked contrast, the highly organized and repetitive structure of viral particles induces strong cross-linking of BCRs, directly stimulating B cell activation even in the absence of T cell help (Bachmann and Zinkernagel 1997). For this reason, organization and repetitiveness is a key stimulus for B cells and may even be able to override B cell tolerance under some circumstances, resulting in the activation of anergic or *unresponsive* B cells (Bachmann et al. 1993; Chackerian et al. 2008). The importance of possible activation of self-specific B cells by highly repetitive antigens is underscored by the observation that some of the most important autoantigens involved in antibody-mediated autoimmune diseases are highly repetitive. As an example, myasthenia gravis is caused by antibodies specific for acetylcholine receptors (Naparstek and Plotz 1993) and Factor VIII is the most frequent target of autoantibodies in autoantibody-mediated hemophilia (Whelan et al. 2013). Strikingly, acetylcholine receptors, which are localized within the neuromuscular endplate, are as densely packed as viral surfaces and individual receptors are spaced by a distance of about 10 nm. This distance is very similar to the spacing of individual coat proteins on the viral surface and optimal for B cell activation (Fambrough et al. 1973; Bachmann et al. 1995). Factor VIII is a soluble molecule itself but binds to the von-Willebrand factor in vivo, which is a large and repetitive complex, consisting of up to 80 and more identical subunits (Hassan et al. 2012). Hence, Factor VIII may, therefore, also be considered as a highly repetitive protein in vivo.

The fact that the repetitive surface of VLPs recruits components of the innate humoral immune system has important consequences for B cell responses, in particular the GC reaction in secondary lymphoid organs, which is the hallmark of the adaptive humoral immune response and is required for induction of long-term host protection. One important first step for the induction of GCs is the transport and retention of intact Ag on FDCs to select specific B cells into the GC and for the generation of BCRs with an increased affinity in a process called Ag affinity maturation. As discussed earlier, soluble proteins reach B cell follicles by conduits. They are, however, normally not deposited on FDCs as they fail to activate complement and are not recognized by natural antibodies. This conduit system is probably more important for the transport of signaling substances (chemokines, cytokines, etc.) than Ags for induction of immune responses. Soluble proteins are only deposited on FDCs if they form ICs with specific antibodies in a complement receptor–dependent manner (Brown et al. 1973; Heinen et al. 1986; Ferguson et al. 2004). In line with this, it has been shown that specific Abs can have a strong influence on humoral responses (Heyman 2000; Hjelm et al. 2006). Indeed, immune complexes formed by the protein Ag and IgM as well as IgG may promote a stronger Ab response than free protein, a phenomenon which is most likely due to cross-linking of the BCRs, complement activation, and consequent efficient deposition on FDCs. Excess amounts of antibodies, however, can also suppress antibody responses, most likely by binding and neutralizing the Ag, preventing Ag access to B cells. It has been reported that there is an interesting difference between IgM^+ and IgG^+ memory B cells as the latter are able to efficiently respond to antigenic stimulation in the presence of specific IgG (Pape et al. 2011). Particulate Ags are finally transported into B cell follicles in a cell-associated way by cognate or noncognate B cells or DCs (Manolova et al. 2008; Gonzalez et al. 2010a; Bessa et al. 2012; Link et al. 2012). However, the role of natural and specific Abs for FDC deposition of particulate Ags is different compared to soluble protein Ags. Our lab recently published data on the transport of the same antigenic protein into B cell follicles, either as soluble protein consisting of 2 Qβ-monomers (28 kDa protein) or as Qβ-VLP which forms an icosahedral structure of 180 subunits (2520 kDa) (Link et al. 2012). The particulate and repetitive VLP was efficiently transported into B cell follicles and deposited on the surface of FDCs, a process which was dependent on natural IgM antibodies as well as components of the classical pathway of complement activation (C1q, C3) and complement receptors (CR1, CR2). The influence of specific antibodies was further investigated. Whereas specific IgM accelerated deposition of VLPs on FDCs, specific IgG Abs dramatically reduced this process. In contrast, the soluble form of the protein (Qβ-dimer) was not found on FDCs in a primary immune response. However, when mice were passively immunized with specific IgM or IgG antibodies 24 h prior challenging, efficient deposition on FDCs could be observed. These findings demonstrate a strong influence of antigen size and repetitiveness on deposition on FDCs for the formation of GCs. They also demonstrate that VLPs effectively recruit the innate humoral immune system for deposition on VLPs, while soluble antigens only efficiently deposit on FDCs in the presence of specific IgM or IgG. This may explain why soluble proteins fail to efficiently induce GCs during primary immune responses in the absence of strong adjuvants.

As discussed earlier, organization and repetitiveness of viral particles enhances complement fixation and activation, which facilitates their deposition on FDCs. In addition to a role in antigen trafficking, complement also drives more efficient antibody responses by direct interaction with B cells. Complement receptor CD21 on B cells forms a complex with the BCR and CD19, which is known to enhance BCR signaling (Carter and Fearon 1992). Consistent with this observation, soluble proteins fused to C3d have been shown to induce up to 1000-fold stronger antibody responses for model proteins (Dempsey et al. 1996). Furthermore, viral particles decorated with complement fragment have been shown to require a reduced number of BCRs to be cross-linked for efficient B cell activation (Jegerlehner et al. 2002). In addition, triggering of CD21 enhances the expression of the master-transcription factors XBP-1 and BLIMP-1, enabling the differentiation of GC-B cells into antigen-independent long-lived plasma cells (Gatto et al. 2005). In conclusion, antibody responses against VLPs are enhanced by components of the complement system at multiple levels, including (1) enhanced transport to and deposition on FDCs by, for example, shuttling B cells, (2) increased B cell activation by reducing numbers of BCRs that need to be engaged, and (3) increased induction of long-lived plasma cells by inducing transcription factors such as XBP-1 and BLIMP-1.

Quite surprisingly, highly organized and repetitive viral particles not only efficiently recruit the innate humoral immune system by multimeric interactions but are also able to bind to low-affinity BCRs of B cells by high-avidity interactions. We have recently shown that this process is relevant for the generation of antibody responses after intranasal immunization, as it allows binding of viral particles to low-affinity BCRs of B cells in the lung. This facilitates subsequent transport of the viral particles to B cell follicles of the spleen for deposition on FDCs followed by efficient induction of GC responses (Bessa et al. 2012). The observation that viral particles are bound by BCRs of B cells for antigen transport rather than their activation raises interesting considerations about the term *specificity* of the B cells involved. Specifically, the affinity or *strength* of the interaction between B cells and viral particles is apparently sufficient for binding to BCRs and cellular transport but insufficient for proper B cell activation. In line with these considerations, it is not the transporting B cells which will mount the antibody responses but spleen resident B cells instead.

7.4 TOLL-LIKE RECEPTOR STIMULATION

Viruses carry genetic information consisting of DNA and RNA. Interestingly, there are a number of differences between nucleic acids from viruses versus eukaryotic cells. Those include presence of viral 5′-triphosphate mRNA as well as of double-stranded RNA and double-stranded DNA (often intermediates) in the cytoplasm (Ronald and Beutler 2010; Yan and Chen 2012). In addition, RNA and DNA are both found in intracellular compartments under physiological conditions, and endosomal localization is unusual and recognized by the immune system as an indicator of infection. For this reason, there are specific toll-like receptors (TLRs) localized within endosomes of myeloid cells (mainly DCs) as well as B cells and, upon activation, also other cells that recognize nucleic acids. TLR3 is activated by double-stranded RNA, TLR7/8 by single-stranded RNA, and TLR9 by DNA. TLR9 is not activated by all forms of DNA, but only by DNA rich in nonmethylated CG motifs

(CpGs), which is mainly found in bacterial and viral DNA (Barton et al. 2006).

It has been suggested that B cell responses cannot be induced in the absence of innate signals mediated by TLR signaling (Pasare and Medzhitov 2005). A wealth of previous and subsequent studies has, however, shown that this is not the case in general (Gavin et al. 2006). It has been known for decades that highly pure but aggregated proteins mount strong B cell response. Furthermore, mice deficient in TLR signaling have been shown to mount robust IgG responses upon viral infection (Heer et al. 2007; Meyer-Bahlburg et al. 2007). We have generated VLPs loaded with RNA, CpGs, or inert polyglutamate that is not engaging TLRs. Using such particles and/or mice with specific ablation of TLR signaling in B cells and/or DCs, it has been shown that TLR signaling in B cells rather than DCs is important for IgG class-switching (Jegerlehner et al. 2007; Bessa et al. 2009; Hou et al. 2011). Hence, TLR signaling affects B cells directly and not via the induction of T_H1 cells. Specifically, wild-type mice immunized with RNA- or CpG-loaded VLPs mounted strong IgG responses dominated by the isotype IgG2a. In contrast, the same mice immunized with VLPs devoid of nucleic acids or mice selectively lacking TLR-signaling in B cells mounted substantial but reduced IgG responses, which were dominated by IgG1 isotypes. B cells recognizing viral particles carrying RNA or DNA are therefore directly stimulated to undergo isotype switching to IgG2a in mice and correspondingly to IgG1 in humans.

Packaging of RNA or CpGs into VLPs not only enhanced IgG responses overall but biased the response toward more protective IgG2a antibodies. These insights may translate to the development of more efficacious vaccines, and inclusion of TLR7/8/9 ligands into vaccine formulations may be a general way to enhance the protective potential of antibody responses. Indeed, many classical vaccines are *naturally formulated* in such a way as live viral vaccines naturally carry RNA or DNA. However, this may be different for chemically inactivated vaccines; specifically, formaldehyde treatment may not only eliminate the replicative potential of viruses but may also reduce the ability of the nucleic acids to engage TLRs. Recombinant viral vaccines based on VLPs are currently devoid of TLR ligands, which may limit their efficacy. In line with this, formulation of hepatitis B surface antigen (HBsAg) for vaccination against hepatitis B virus with CpGs enhances antibody responses in humans (Barry and Cooper 2007). It is interesting to note in this context that packaging TLR ligands into VLPs is superior to simple mixing, as it not only enhances their adjuvants potential but also strongly reduces side effects (Storni et al. 2004; Barry and Cooper 2007). Hence, TLR ligands packaged into VLPs may effectively target endosomal compartments of B and myeloid cells for the activation of specific B cells as well as cells presenting VLP-derived antigens.

Interestingly, induction of IgG2a antibodies did not need TLR expression in DCs nor induction of T_H1 responses (Jegerlehner et al. 2007; Bessa et al. 2009; Hou et al. 2011). This indicates that T_H1 responses and IgG2a antibodies are not necessarily causally related and that the correlation between both of them simply reflects the fact that both are caused by TLR signaling.

IgG responses against VLPs are to a high extent dependent on T_H cells (Pulendran and Ahmed 2006). Dependence of IgG responses on follicular T_H cells, which are thought to be the key *helper cells* for B cells, is however less strict as TLR signaling is able to overcome dependence on follicular T_H cells. Specifically, while VLPs lacking packaged nucleic acids cause strongly reduced IgG responses in the absence of IL-21 and follicular T_H cells, VLPs loaded with RNA induce essentially normal IgG responses in these mice. Thus, TLR ligands in VLPs reduce the dependence of the IgG response on T_H cells, in particular follicular T_H cells. The observation that the presence of TLR ligands inside VLPs lessens T_H cell dependence of antibody responses is even more pronounced for IgA. Indeed, systemic IgA responses are completely T cell independent upon immunization with VLPs loaded with RNA or CpGs. Interestingly, IgA responses are regulated in a different way than IgG responses (Fagarasan et al. 2010). IgA responses induced by systemic exposure to VLPs loaded with TLR ligands are dependent on TLR signaling in B cells but occur independently of all types of T_H cells or TLR signaling in DCs (Bergqvist et al. 2006, 2010; Bessa et al. 2008, 2009). In marked contrast, mucosal IgA responses are dependent on TLR signaling in DCs and alveolar macrophages (causing upregulation of transforming growth factor [TGF]-β and transmembrane activator and calcium modulator and cyclophilin ligand interactor [TACI]) but independent of TLR signaling in B cells (Bessa et al. 2009). Hence, IgG2a and systemic IgA responses are dependent on TLR stimulation in B cells, while mucosal IgA responses are driven by TLR signaling in DCs.

7.5 CONCLUSION

Viral particles and VLPs exhibit key features rendering them highly immunogenic for B cells. This offers an explanation why vaccines based on viral particles or VLPs have been so successful in the past. From a translational point of view, these insights may be used to optimize B cell responses against antigens of choice, which are not naturally present on viral particles. By displaying antigens covalently attached to this type of particles, they become equally immunogenic as the underlying VLP and VLPs may, therefore, be used as a general scaffold to increase the immunogenicity of any antigen. Such VLP-based conjugate vaccines have shown promise in mice and humans and deserve further investigation.

ACKNOWLEDGMENTS

We thank Alexander Link and Thomas Kündig for their helpful discussions and for critically reading the manuscript.

CONFLICT OF INTEREST

All authors are involved in the development of VLP-based vaccines.

REFERENCES

Bachmann, M. F., H. Hengartner, and R. M. Zinkernagel (1995). T helper cell-independent neutralizing B cell response against vesicular stomatitis virus: Role of antigen patterns in B cell induction? *Eur J Immunol* **25**(12): 3445–3451.

Bachmann, M. F. and G. T. Jennings (2010). Vaccine delivery: A matter of size, geometry, kinetics and molecular patterns. *Nat Rev Immunol* **10**(11): 787–796.

Bachmann, M. F., U. H. Rohrer, T. M. Kundig, K. Burki, H. Hengartner, and R. M. Zinkernagel (1993). The influence of antigen organization on B cell responsiveness. *Science* **262**(5138): 1448–1451.

Bachmann, M. F. and R. M. Zinkernagel (1997). Neutralizing antiviral B cell responses. *Annu Rev Immunol* **15**: 235–270.

Barber, G. N. (2011). Innate immune DNA sensing pathways: STING, AIM11 and the regulation of interferon production and inflammatory responses. *Curr Opin Immunol* **23**(1): 10–20.

Barry, M. and C. Cooper (2007). Review of hepatitis B surface antigen-1018 ISS adjuvant-containing vaccine safety and efficacy. *Expert Opin Biol Ther* **7**(11): 1731–1737.

Barton, G. M., J. C. Kagan, and R. Medzhitov (2006). Intracellular localization of Toll-like receptor 9 prevents recognition of self DNA but facilitates access to viral DNA. *Nat Immunol* **7**(1): 49–56.

Batista, F. D. and N. E. Harwood (2009). The who, how and where of antigen presentation to B cells. *Nat Rev Immunol* **9**(1): 15–27.

Beck, O. E. (1981). Distribution of virus antibody activity among human IgG subclasses. *Clin Exp Immunol* **43**(3): 626–632.

Bergqvist, P., E. Gardby, A. Stensson, M. Bemark, and N. Y. Lycke (2006). Gut IgA class switch recombination in the absence of CD40 does not occur in the lamina propria and is independent of germinal centers. *J Immunol* **177**(11): 7772–7783.

Bergqvist, P., A. Stensson, N. Y. Lycke, and M. Bemark (2010). T cell-independent IgA class switch recombination is restricted to the GALT and occurs prior to manifest germinal center formation. *J Immunol* **184**(7): 3545–3553.

Bergtold, A., D. D. Desai, A. Gavhane, and R. Clynes (2005). Cell surface recycling of internalized antigen permits dendritic cell priming of B cells. *Immunity* **23**(5): 503–514.

Bessa, J., A. Jegerlehner, H. J. Hinton, P. Pumpens, P. Saudan, P. Schneider, and M. F. Bachmann (2009). Alveolar macrophages and lung dendritic cells sense RNA and drive mucosal IgA responses. *J Immunol* **183**(6): 3788–3799.

Bessa, J., N. Schmitz, H. J. Hinton, K. Schwarz, A. Jegerlehner, and M. F. Bachmann (2008). Efficient induction of mucosal and systemic immune responses by virus-like particles administered intranasally: Implications for vaccine design. *Eur J Immunol* **38**(1): 114–126.

Bessa, J., F. Zabel, A. Link, A. Jegerlehner, H. J. Hinton, N. Schmitz, M. Bauer, T. M. Kundig, P. Saudan, and M. F. Bachmann (2012). Low-affinity B cells transport viral particles from the lung to the spleen to initiate antibody responses. *Proc Natl Acad Sci U S A* **109**(50): 20566–20571.

Bottazzi, B., A. Doni, C. Garlanda, and A. Mantovani (2010). An integrated view of humoral innate immunity: Pentraxins as a paradigm. *Ann Rev Immunol* **28**(1): 157–183.

Bowie, A. G. and L. Unterholzner (2008). Viral evasion and subversion of pattern-recognition receptor signalling. *Nat Rev Immunol* **8**(12): 911–922.

Brown, J. C., G. Harris, M. Papamichail, V. S. Sljivic, and E. J. Holborow (1973). The localization of aggregated human-globulin in the spleens of normal mice. *Immunology* **24**(6): 955–968.

Carrasco, Y. R. and F. D. Batista (2007). B cells acquire particulate antigen in a macrophage-rich area at the boundary between the follicle and the subcapsular sinus of the lymph node. *Immunity* **27**(1): 160–171.

Carter, R. H. and D. T. Fearon (1992). CD19: Lowering the threshold for antigen receptor stimulation of B lymphocytes. *Science* **256**(5053): 105–107.

Cavacini, L. A., D. Kuhrt, M. Duval, K. Mayer, and M. R. Posner (2003). Binding and neutralization activity of human IgG1 and IgG3 from serum of HIV-infected individuals. *AIDS Res Hum Retroviruses* **19**(9): 785–792.

Cerutti, A., K. Chen, and A. Chorny (2011). Immunoglobulin responses at the mucosal interface. *Ann Rev Immunol* **29**(1): 273–293.

Chackerian, B., M. R. Durfee, and J. T. Schiller (2008). Viruslike display of a neo-self antigen reverses B cell anergy in a B cell receptor transgenic mouse model. *J Immunol* **180**(9): 5816–5825.

Cinamon, G., M. A. Zachariah, O. M. Lam, F. W. Foss, and J. G. Cyster (2008). Follicular shuttling of marginal zone B cells facilitates antigen transport. *Nat Immunol* **9**(1): 54–62.

Colonna, M., G. Trinchieri, and Y. J. Liu (2004). Plasmacytoid dendritic cells in immunity. *Nat Immunol* **5**(12): 1219–1226.

Coutelier, J. P., J. T. van der Logt, F. W. Heessen, G. Warnier, and J. Van Snick (1987). IgG2a restriction of murine antibodies elicited by viral infections. *J Exp Med* **165**(1): 64–69.

D'Argenio, D. A. and C. B. Wilson (2010). A decade of vaccines: Integrating immunology and vaccinology for rational vaccine design. *Immunity* **33**(4): 437–440.

Dempsey, P. W., M. E. Allison, S. Akkaraju, C. C. Goodnow, and D. T. Fearon (1996). C3d of complement as a molecular adjuvant: Bridging innate and acquired immunity. *Science* **271**(5247): 348–350.

Fagarasan, S., S. Kawamoto, O. Kanagawa, and K. Suzuki (2010). Adaptive immune regulation in the gut: T cell-dependent and T cell-independent IgA synthesis. *Ann Rev Immunol* **28**(1): 243–273.

Fambrough, D. M., D. B. Drachman, and S. Satyamurti (1973). Neuromuscular junction in myasthenia gravis: Decreased acetylcholine receptors. *Science* **182**(4109): 293–295.

Ferguson, A. R., M. E. Youd, and R. B. Corley (2004). Marginal zone B cells transport and deposit IgM-containing immune complexes onto follicular dendritic cells. *Int Immunol* **16**(10): 1411–1422.

Gatto, D., T. Pfister, A. Jegerlehner, S. W. Martin, M. Kopf, and M. F. Bachmann (2005). Complement receptors regulate differentiation of bone marrow plasma cell precursors expressing transcription factors Blimp-1 and XBP-1. *J Exp Med* **201**: 993–1105.

Gavin, A. L., K. Hoebe, B. Duong, T. Ota, C. Martin, B. Beutler, and D. Nemazee (2006). Adjuvant-enhanced antibody responses in the absence of Toll-like receptor signaling. *Science* **314**(5807): 1936–1938.

Geijtenbeek, T. B. H., D. S. Kwon, R. Torensma, S. J. van Vliet, G. C. F. van Duijnhoven, J. Middel, I. L. M. H. A. Cornelissen et al. (2000). DC-SIGN, a dendritic cell–specific HIV-1-binding protein that enhances trans-infection of T cells. *Cell* **100**(5): 587–597.

Gitlin, L., W. Barchet, S. Gilfillan, M. Cella, B. Beutler, R. A. Flavell, M. S. Diamond, and M. Colonna (2006). Essential role of mda-5 in type I IFN responses to polyriboinosinic:polyribocytidylic acid and encephalomyocarditis picornavirus. *Proc Natl Acad Sci U S A* **103**(22): 8459–8464.

Gonzalez, S. F., S. E. Degn, L. A. Pitcher, M. Woodruff, B. A. Heesters, and M. C. Carroll (2011). Trafficking of B cell antigen in lymph nodes. *Ann Rev Immunol* **29**(1): 215–233.

Gonzalez, S. F., V. Lukacs-Kornek, M. P. Kuligowski, L. A. Pitcher, S. E. Degn, Y. A. Kim, M. J. Cloninger, L. Martinez-Pomares, S. Gordon, S. J. Turley, and M. C. Carroll (2010a). Capture of influenza by medullary dendritic cells via SIGN-R1 is essential for humoral immunity in draining lymph nodes. *Nat Immunol* **11**(5): 427–434.

Gonzalez, S. F., V. Lukacs-Kornek, M. P. Kuligowski, L. A. Pitcher, S. E. Degn, S. J. Turley, and M. C. Carroll (2010b). Complement-dependent transport of antigen into B cell follicles. *J Immunol* **185**(5): 2659–2664.

Hassan, M. I., A. Saxena, and F. Ahmad (2012). Structure and function of von Willebrand factor. *Blood Coagulat Fibrinol* **23**(1): 11–22. doi: 10.1097/MBC.1090b1013e32834cb32835d.

Heer, A. K., A. Shamshiev, A. Donda, S. Uematsu, S. Akira, M. Kopf, and B. J. Marsland (2007). TLR signaling fine-tunes anti-influenza B cell responses without regulating effector T cell responses. *J Immunol* **178**(4): 2182–2191.

Heinen, E., M. Braun, P. G. Coulie, J. Van Snick, M. Moeremans, N. Cormann, C. Kinet-Denoel, and L. J. Simar (1986). Transfer of immune complexes from lymphocytes to follicular dendritic cells. *Eur J Immunol* **16**(2): 167–172.

Heyman, B. (2000). Regulation of antibody responses via antibodies, complement, and Fc receptors. *Annu Rev Immunol* **18**: 709–737.

Hjelm, F., F. Carlsson, A. Getahun, and B. Heyman (2006). Antibody-mediated regulation of the immune response. *Scand J Immunol* **64**(3): 177–184.

Hou, B., P. Saudan, G. Ott, M. L. Wheeler, M. Ji, L. Kuzmich, L. M. Lee, R. L. Coffman, M. F. Bachmann, and A. L. DeFranco (2011). Selective utilization of Toll-like receptor and MyD88 signaling in B cells for enhancement of the antiviral germinal center response. *Immunity* **34**(3): 375–384.

Imaizumi, T., S. Aratani, T. Nakajima, M. Carlson, T. Matsumiya, K. Tanji, K. Ookawa et al. (2002). Retinoic acid-inducible gene-I is induced in endothelial cells by LPS and regulates expression of COX-2. *Biochem Biophys Res Commun* **292**(1): 274–279.

Ishikawa, H., Z. Ma, and G. N. Barber (2009). STING regulates intracellular DNA-mediated, type I interferon-dependent innate immunity. *Nature* **461**(7265): 788–792.

Jegerlehner, A., P. Maurer, J. Bessa, H. J. Hinton, M. Kopf, and M. F. Bachmann (2007). TLR9 signaling in B cells determines class switch recombination to IgG2a. *J Immunol* **178**(4): 2415–2420.

Jegerlehner, A., T. Storni, G. Lipowsky, M. Schmid, P. Pumpens, and M. F. Bachmann (2002). Regulation of IgG antibody responses by epitope density and CD21-mediated costimulation. *Eur J Immunol* **32**(11): 3305–3314.

Link, A., F. Zabel, Y. Schnetzler, A. Titz, F. Brombacher, and M. F. Bachmann (2012). Innate immunity mediates follicular transport of particulate but not soluble protein antigen. *J Immunol* **188**(8): 3724–3733.

Litwinska, B., B. Bucholc, A. Biesiadecka, and M. Kantoch (1993). Antiviral activity of IgG subclasses of immunoglobulin preparations for intravenous use. *Med Dosw Mikrobiol* **45**(4): 523–528.

Macpherson, A. J., K. D. McCoy, F. E. Johansen, and P. Brandtzaeg (2008). The immune geography of IgA induction and function. *Mucosal Immunol* **1**(1): 11–22.

Manolova, V., A. Flace, M. Bauer, K. Schwarz, P. Saudan, and M. F. Bachmann (2008). Nanoparticles target distinct dendritic cell populations according to their size. *Eur J Immunol* **38**(5): 1404–1413.

McKenna, K., A. S. Beignon, and N. Bhardwaj (2005). Plasmacytoid dendritic cells: Linking innate and adaptive immunity. *J Virol* **79**(1): 17–27.

Meyer-Bahlburg, A., S. Khim, and D. J. Rawlings (2007). B cell intrinsic TLR signals amplify but are not required for humoral immunity. *J Exp Med* **204**(13): 3095–3101.

Naparstek, Y. and P. H. Plotz (1993). The role of autoantibodies in autoimmune disease. *Ann Rev Immunol* **11**(1): 79–104.

Pape, K. A., D. M. Catron, A. A. Itano, and M. K. Jenkins (2007). The humoral immune response is initiated in lymph nodes by B cells that acquire soluble antigen directly in the follicles. *Immunity* **26**(4): 491–502.

Pape, K. A., J. J. Taylor, R. W. Maul, P. J. Gearhart, and M. K. Jenkins (2011). Different B cell populations mediate early and late memory during an endogenous immune response. *Science* **331**(6021): 1203–1207.

Pasare, C. and R. Medzhitov (2005). Control of B-cell responses by Toll-like receptors. *Nature* **438**(7066): 364–368.

Pflicke, H. and M. Sixt (2009). Preformed portals facilitate dendritic cell entry into afferent lymphatic vessels. *J Exp Med* **206**(13): 2925–2935.

Phan, T. G., J. A. Green, E. E. Gray, Y. Xu, and J. G. Cyster (2009). Immune complex relay by subcapsular sinus macrophages and noncognate B cells drives antibody affinity maturation. *Nat Immunol* **10**(7): 786–793.

Pulendran, B. and R. Ahmed (2006). Translating innate immunity into immunological memory: Implications for vaccine development. *Cell* **124**(4): 849–863.

Pulendran, B., S. Li, and H. I. Nakaya (2010). Systems vaccinology. *Immunity* **33**(4): 516–529.

Qi, H., J. G. Egen, A. Y. C. Huang, and R. N. Germain (2006). Extrafollicular activation of lymph node B cells by antigen-bearing dendritic cells. *Science* **312**(5780): 1672–1676.

Ronald, P. C. and B. Beutler (2010). Plant and animal sensors of conserved microbial signatures. *Science* **330**(6007): 1061–1064.

Roozendaal, R., T. R. Mempel, L. A. Pitcher, S. F. Gonzalez, A. Verschoor, R. E. Mebius, U. H. von Andrian, and M. C. Carroll (2009). Conduits mediate transport of low-molecular-weight antigen to lymph node follicles. *Immunity* **30**(2): 264–276.

Satoh, T., H. Kato, Y. Kumagai, M. Yoneyama, S. Sato, K. Matsushita, T. Tsujimura, T. Fujita, S. Akira, and O. Takeuchi (2010). LGP2 is a positive regulator of RIG-I– and MDA5-mediated antiviral responses. *Proc Natl Acad Sci* **107**(4): 1512–1517.

Schmidt, A., T. Schwerd, W. Hamm, J. C. Hellmuth, S. Cui, M. Wenzel, F. S. Hoffmann et al. (2009). 5′-triphosphate RNA requires base-paired structures to activate antiviral signaling via RIG-I. *Proc Natl Acad Sci U S A* **106**(29): 12067–12072.

Sixt, M., N. Kanazawa, M. Selg, T. Samson, G. Roos, D. P. Reinhardt, R. Pabst, M. B. Lutz, and L. Sorokin (2005). The conduit system transports soluble antigens from the afferent lymph to resident dendritic cells in the T cell area of the lymph node. *Immunity* **22**(1): 19–29.

Storni, T., C. Ruedl, K. Schwarz, R. A. Schwendener, W. A. Renner, and M. F. Bachmann (2004). Nonmethylated CG motifs packaged into virus-like particles induce protective cytotoxic T cell responses in the absence of systemic side effects. *J Immunol* **172**(3): 1777–1785.

Swartz, M. A., J. A. Hubbell, and S. T. Reddy (2008). Lymphatic drainage function and its immunological implications: From dendritic cell homing to vaccine design. *Seminars in Immunology* **20**(2): 147–156.

Whelan, S. F. J., C. J. Hofbauer, F. M. Horling, P. Allacher, M. J. Wolfsegger, J. Oldenburg, C. Male et al. (2013). Distinct characteristics of antibody responses against factor VIII in healthy individuals and in different cohorts of hemophilia A patients. *Blood* **121**: 1039–1048.

Wrammert, J. and R. Ahmed (2008). Maintenance of serological memory. *Biol Chem* **389**(5): 537–539.

Yan, N. and Z. J. Chen (2012). Intrinsic antiviral immunity. *Nat Immunol* **13**(3): 214–222.

Zinkernagel, R. M. (1996). Immunology taught by viruses. *Science* **271**(5246): 173–178.

Zinkernagel, R. M., A. LaMarre, A. Ciurea, L. Hunziker, A. F. Ochsenbein, K. D. McCoy, T. Fehr, M. F. Bachmann, U. Kalinke, and H. Hengartner (2001). Neutralizing antiviral antibody responses. *Adv Immunol* **79**: 1–53.

8 Nanomedicine
General Considerations and Examples

Kenza Snoussi and Michael Kann

CONTENTS

8.1 INTRODUCTION

8.1.1 WHAT DO WE TALK ABOUT?

Nanosize materials in medicine raise hope and chances for new and more efficient treatments. The term "nanomedicine" is, however, far away from being well defined. The field of medicine is wide, and the word "nano" is not necessarily taken literal. Thus, even European and U.S. scientific organizations use different definitions shown in Table 8.1. In this chapter, we will try to focus on those aspects, which are covered more or less by both definitions, and we will restrict our outline to those tools that are advanced in development. Further, the multiplicity of diseases and the numerous tools make the number of potential applications astronomical and detailed description would exceed the volume of an entire book. The numerous possible combinations of tools make it likely that we forgot important aspects for which we apologize. In the following paragraphs, we would like to reflect some general aspects of nanomedicine, giving some examples for existing therapies or clinical trials.

8.1.2 NANO

When using the word "nano" to the nanometer scale from 1 to 100 as the National Institutes of Health (NIH) for nanomaterial does, we have to include small peptides as treatment options, but we exclude slightly larger structures such as herpes simplex virus capsids. However, peptides do not differ too much from classical drugs or peptide hormones for necessarily justifying the inclusion, and the arbitrary threshold of 100 nm excludes complex structures slightly larger.

8.1.3 MEDICINE

Medicine comprises prevention, diagnostics, and treatment. Planning nanomedical applications requires to primarily define the area of the application as it comprises different safety standards. Prevention needs higher safety standards than treatment; treatment of a terminally ill person tolerates higher secondary effects than common diseases. It is, thus, not surprising that most nanodrugs target cancers, in particular after first-line classical treatment failed.

8.2 WHAT DO WE NEED TO CONSIDER?

All scientists in biology or medicine know that nothing happens to 100%. This experience is contradictory to the wish designing the perfect (nano-) drug. However, one should take into account that many diseases do not need this 100% perfection so that the technique can be chosen on the basis of the required efficiency and specificity. Both depend, among other parameters, on the form of application (local or systemic).

TABLE 8.1
Definitions of Nanomedicine

European Science Foundation (ESF)—European Medical Research Council (EMRC)	National Institute of Health (NIH)
The field of *nanomedicine* is the science and technology of diagnosing, treating, and preventing disease and traumatic injury, of relieving pain, and of preserving and improving human health, using molecular tools and molecular knowledge of the human body. It was perceived as embracing five main subdisciplines that in many ways are overlapping and underpinned by the following common technical issues [1]: • Analytical tools • Nanoimaging • Nanomaterials and nanodevices • Novel therapeutics and drug delivery systems • Clinical, regulatory, and toxicological issues	Nanomedicine, an offshoot of nanotechnology, refers to highly specific medical intervention at the molecular scale for curing disease or repairing damaged tissues, such as bone, muscle, or nerve. A nanometer is one-billionth of a meter, too small to be seen with a conventional laboratory microscope. It is at this size scale—about 100 nm or less—that biological molecules and structures operate in living cells [2].

8.2.1 ON WHICH ANATOMICAL LEVEL DO WE HAVE TO TREAT?

The different fields—prevention, diagnostics, and treatment—can be further subdivided, depending upon the anatomical level. For instance, does the intervention target special cell types, such as metastatic cancer cells, that can be diffuse in the organism, or is the process localized in defined environments as, for instance, solid tumors, emboli, and ischemic areas? Evidently, the first scenario asks for systemic application, while the last also allows local application, for instance, by a catheter. The latter can help in minimizing systemic side effects, but it is less applicable for redundant applications. Also on the local tissue level would be a repopulation of a compartment with transduced or otherwise modified cells. Finally, a mixture of both can be scheduled, for instance, when addressing the whole organism, for example, by using locally transduced cells for hormone production.

8.2.2 SPECIFICITY

Important to know is the needed specificity. Targeting specific cells or environments might be achievable when significant differences exist on cell surface as it is the case for different cell types and also for some cancers. Evidently, specificity is highly required in diagnostics to avoid false-positive results. As diagnostics frequently relies on different approaches, compensation can be obtained by applying different techniques so that nanomedical approaches can complement established methods.

One example for increased specificity upon systemic application is Abraxane, which consists of the anticancer agent paclitaxel (a microtubule polymerization inhibitor) coupled to albumin. While paclitaxel inhibits generally cell division, the albumin component causes enrichment in some tumors (see Table 8.2). Physically, Abraxane is 130 nm particles, sizes that are normally not freely diffusible. The albumin part, however, increases transcytosis, thus allowing its application for the treatment of breast cancer, non–small cell lung cancer, metastatic pancreatic cancer, and bladder cancer.

Another example is tissue plasminogen activator (tPA)-coated nanoparticles, which consist of a tPA coated to nanoparticles for the treatment of blood clots. Their specificity—as shown in mice—is caused by their activation in areas with increased shear stress. The advantage of the nanoapplication compared to plasminogen activator administration, which causes bleedings, is that only 1/50th has to be administered.

One example for diagnostic purposes is the X-shaped nanoparticle. Their specificity comes from an aptamer specifically binding to cancer cells. The presence of arms on the RNA allows the addition of siRNA, microRNA, or ribozyme functions, when used for treatment.

Apart from designing more specific nanobodies, local application increases specificity at least as long as most of the nanobodies are absorbed/act closed to the administration site. Such local administration can involve catheters using, for example, the tPA-coated nanoparticles. When the compartment is externally accessible, easier applications are possible, for instance, through inhalation to cure respiratory tract infections. This is exemplified by the nanostructured cationic lipid nanocarrier system (NLCS); RP1 or Bcl2 mRNAs-targeting siRNAs were used, which inhibit pumps or nonpump gene products related to cellular drug resistance.

Yet another opportunity to increase specificity is the modification of cells outside the body. Evidently, this technique requires the possibility to select specific cells.

8.2.3 EFFICIENCY

Apart from specificity, this is a crucial point not only for treatment but also for prevention. While some applications require enormous efficiency to obtain significant benefit to a person—as it is the case in cancer treatment—other applications need little transduction as a replacement for hormone-producing cells. Yet another example for limited but sufficient effect is the sun blocker for which nobody would ask for a 100% protection against UV light. An intermediate is cystic fibrosis where a significant number of cells must be transduced for full replacement.

TABLE 8.2
List of Approaches Exemplifying Tools, Applications, and Nature

Drug	Nature	Application	Targets	Principle/Comments	References
Abraxane	Paclitaxel albumin-bound particles with a diameter of 130 nm	Injection	Breast cancer, non-small cell lung cancer, pancreatic cancer (late-stage [metastatic] pancreatic cancer), and bladder cancer after failure of combination chemotherapy for metastatic disease or relapse within 6 months of adjuvant chemotherapy	Utilizes the natural albumin pathways including gp60 and caveolae-mediated transcytosis and increased intratumoral accumulation, through association with tumor-derived SPARC protein, extensive extravascular distribution, and/or tissue binding of paclitaxel.	[3–7]
Doxil	Anthracycline topoisomerase inhibitor in STEALTH® liposomes (PEGylated liposomes) with a diameter of 100 nm	Multiple intravenous injection	HIV-related Kaposi's sarcoma, ovarian cancer, multiple myeloma	Adverse reactions: asthenia, fatigue, fever, anorexia, nausea, vomiting, stomatitis, diarrhea, constipation, hand and foot syndrome, rash, neutropenia, thrombocytopenia, and anemia.	[8, 9]
Magnetic iron-oxide nanoparticles	Three magnetic iron-oxide nanospheres, chemically linked to one doxorubicin-loaded liposome	Intravenous injection, followed by activation using vibrating radiofrequency that disperses the released drug	Breast cancer	Compared with cancerous rats receiving normal doses of doxorubicin, those treated with the nanochains experienced half the tumor growth and a higher rate of cancer cell death. In those rats that received two doses of the nanotreatment, tumor growth was reduced to 1/10th that of those treated traditionally. The twice treated rats also survived for 31 more days than the traditionally treated ones. The nanochains contain only about 5%–10% of the doxorubicin used in standard chemo, reducing the chances of toxic drugs harming healthy cells.	[10]
PEG nanoparticles	PEG (poly(D,L-lactic-co-glycolic acid)-b-poly (L-histidine)-b-poly(ethylene glycol) (PLGA-PLH-PEG)) loaded with vancomycin	Injection	Treatment of bacterial infection	Strong binding to bacteria in acidic environment. Switch of charge releases drug.	[11]

(Continued)

TABLE 8.2 (Continued)
List of Approaches Exemplifying Tools, Applications, and Nature

Drug	Nature	Application	Targets	Principle/Comments	References
tPA-coated nanoparticles (NP)	tPA-coated nanoparticles	Injection or by catheter	Selective activation in regions of high shear stress	Dissolving blood clots, e.g., upon apoplexy. Use 1/50th of normal therapeutic dose → less side effects, better safety.	[12]
X-shaped RNA nanoparticles	Reengineered RNA fragment carrying four therapeutic and diagnostic modules/arms comprising an aptamer to target cancer cells, and each further arms carrying siRNA, microRNA, or a ribozyme	Intravenous application	For therapeutic and diagnostic functions	Highly stable (chemically and thermodynamically), polyvalent nature, which allows simultaneous delivery of multiple functional molecules to achieve synergistic effects; modular design, which enables controlled self-assembly with defined structure.	[13]
Leuko-like vectors (LLVs)	Synthetic nanoparticles functionalized with biomimetic leukocyte membranes possess cell-like functions. Loaded with doxorubicin	Intravenous injection		Particles behave like leukocytes; particle opsonization and the consequent specific clearance are inhibited by the membrane coating.	[14]
Nanostructured lipid nanocarrier-based system (NLCS)	Nanostructured PEGylated lipid carriers (NLCs) associated with doxorubicin or paclitaxel or siRNAs targeted to MRP1 or BCL2 mRNAs as a suppressor of pump or nonpump cellular resistance	Inhalation	Lung cancer	NLCS penetrates lung cells after inhalation and accumulates in the cytoplasm. NLCS targeted specifically to cancer cells accumulated predominantly in the areas of lungs contained tumor cells, keeping nontumorous lung tissues intact. Tumor size in animals treated with LHRH–NLC–TAX–siRNAs (MRP1 and BCL2) shrank to 2.6 ± 3.0 mm^3 ($P < 0.05$ when compared with other treatments and initial tumor volume).	[15]
Nanodiamond	Diamond particles with trapped drug (doxorubicin)	Injection into tumor	Chemotherapy-resistant liver and breast cancers	Slow release of the payload. Stable and biocompatible. Overcomes drug efflux and significantly increased apoptosis beyond conventional doxorubicin treatment in both murine liver tumor and mammary carcinoma model. Decreased toxicity in vivo compared to standard doxorubicin treatment.	[16]

A common technique for increasing efficiency is the generation of particles. If high specificity is given, higher doses of a drug are accumulated. Further particles may allow an increased half-life of a drug in the circulation. When an increased specificity is given, the longer circulation time further decreases the amount of drug need for treatment. To this class Doxil can be counted. Doxil is the first nanodrug approved by the FDA, which is used for ovarian cancer, multiple myeloma, and HIV-related Karposi's sarcoma after first-line treatment failed. Doxil consists of STEALTH liposomes filled with anthracycline from *Streptomyces peucetius var. caesius* as daunorubicin. Anthracycline is a topoisomerase inhibitor, affecting cell division. The STEALTH liposomes increase in the circulation with a half-life of 55 h, during which the drug is 90% attached. The increase of specificity for tumors is assumed to be based on a better infiltration of the highly vascularized tumors.

Another well-established reagent to generate particles is polyethylene glycol (PEG). PEG is used in numerous applications including PEG interferons for the treatment of chronic viral infections, PEG vancomycin is used for (mostly) Gram-positive bacterial infection, and PEG paclitaxel is used for breast, lung, and ovarian cancers. The vancomycin loaded exhibits further specificity as it needs the acidic pH present in bacterial infection sites for drug delivery.

8.2.4 DURATION

Evidently, one has to distinguish among a unique event, multiple events, and the need for continuous changes. The latter two scenarios likely need redundant applications with possible immune reactions, for instance, the antibody response after viral vector application.

Unique events include all types of occlusion diseases that might be treated with, for example, vascular endothelial growth factor (VEGF). However, as these diseases occur several times during a person's life, vector application can cause inefficiency when being highly immunogenic as nonenveloped viral vectors. Although a unique event, cancer treatment must also be considered to need multiple applications due to the required high efficiency. This is also the case for the anticancer nanodrugs mentioned earlier.

In the best case, chronic diseases require a single treatment. This is the case when cells of the body can be transduced in a permanent manner, sometimes requiring regulation. Evidently, this is restricted to cases when the cells/organism can produce the effector molecule, normally a protein or a hormone. However, even this seemingly simple scenario means that the genetic information must be incorporated into host cell nuclei, that it must be kept upon cell division generally requiring integration and that expression should not be silenced. If the therapeutic concept does not allow such transduction, multiple applications are required practically ruling out all kinds of virus-based approaches due to the humoral immune response. Exceptions are injections into immune-protected environments, the eye, for instance, for the treatment of macular degenerations using adeno-associated viral vectors. These vectors can express VEGF inhibitors, including the monoclonal antibody fragment ranibizumab for the treatment of wet age-related macular degeneration.

8.2.5 SUPPRESSION VERSUS SUBSTITUTION OR REPLACEMENT

Obviously, a substitution in particular for a limited time is the easiest goal as it may mimic already existing therapies based on local or systemic application. Elimination seems also achievable as long as no 100% efficiency or 100% specificity must be obtained and as long as the target cells or structures can be detected from outside the cell.

The ultimate problem is the replacement on the cellular or subcellular level. There are already sophisticated techniques such as the site-specific CRE/Lox and the FLP–FRT recombination system, but the efficiency is variable and depends on different factors.

8.2.6 DIRECT VERSUS INDIRECT TREATMENT

Practically all of the examples listed earlier are direct treatments. Indirect effects were mentioned in the context of secondary effects, in particular the immune response of the host. However, a significant number of approaches use the immune system for treatment, in particular when the target cells are diffusely distributed in the body. As tools we have the entire spectrum including cell type–specific T cells used in cancer treatment and antibodies used for targeting specific cells or structures either for their elimination or for drug delivery. It further comprises innate immune response, in particular by causing local secretion of interferons. Other indirect phenomena are, for instance, the bystander effect, affecting cells adjacent to the target cell.

REFERENCES

1. European Science Foundation (ESF). 2004. Nanomedicine—An ESF–European Medical Research Councils (EMRC) Forward Look Report. Strasbourg cedex, France. http://www.nanowerk.com/nanotechnology/reports/reportpdf/report53.pdf (accessed October, 2013).
2. National Institutes of Health (NIH). 2006. National Institute of Health Roadmap for Medical Research: Nanomedicine. https://commonfund.nih.gov/nanomedicine/overview.aspx (accessed May 15, 2006).
3. U.S. Food and Drug Administration. 2013 FDA approves Abraxane for late-stage pancreatic cancer. http://www.fda.gov/NewsEvents/Newsroom/PressAnnouncements/ucm367442.htm (accessed January 2015).
4. U.S. Food and Drug Administration. 2011. Warning letters and notice of violation letters to pharmaceutical companies. http://www.fda.gov/downloads/drugs/guidancecompliance regulatoryinformation/enforcementactivitiesbyfda/warning lettersandnoticeofviolationletterstopharmaceuticalcompanies/ucm289192.pdf (accessed January 2015).
5. European Medicines Agency. 2014. Summary Of Product Characteristics. http://www.ema.europa.eu/docs/fr_FR/document_library/EPAR__Product_Information/human/000778/WC500020435.pdf (accessed January 2015).

6. Abaza, Y.M. and Alemany, C. 2014. Nanoparticle albumin-bound-paclitaxel in the treatment of metastatic urethral adenocarcinoma: The significance of molecular profiling and targeted therapy. Case Rep Urol 2014. http://www.ncbi.nlm.nih.gov.gate1.inist.fr/pmc/articles/PMC4151539/ (accessed January 2015).

7. Hoy, S.M. 2014. Albumin-bound paclitaxel: A review of its use for the first-line combination treatment of metastatic pancreatic cancer. *Drugs* 74, 1757–1768.

8. Doxil.com. 2014. Highlights of prescribing information. http://www.doxil.com/shared/product/doxil/prescribing-information.pdf (accessed January 2015).

9. Barenholz, Y. (Chezy). 2012. Doxil®—The first FDA-approved nano-drug: Lessons learned. *J. Control. Release* 160, 117–134.

10. Peiris, P.M., Bauer, L., Toy, R., Tran, E., Pansky, J., Doolittle, E., Schmidt, E. et al. 2012. Enhanced delivery of chemotherapy to tumors using a multicomponent nanochain with radio-frequency-tunable drug release. *ACS Nano* 6, 4157–4168.

11. Radovic-Moreno, A.F., Lu, T.K., Puscasu, V.A., Yoon, C.J., Langer, R., and Farokhzad, O.C. 2012. Surface charge-switching polymeric nanoparticles for bacterial cell wall-targeted delivery of antibiotics. *ACS Nano* 6, 4279–4287.

12. WYSS Institute. 2012. Harvard's Wyss Institute develops novel nanotherapeutic that delivers clot-busting drugs directly to obstructed blood vessels. http://wyss.harvard.edu/viewpressrelease/87/ (accessed January 2015).

13. Haque, F., Shu, D., Shu, Y., Shlyakhtenko, L.S., Rychahou, P.G., Mark Evers, B., and Guo, P. 2012. Ultrastable synergistic tetravalent RNA nanoparticles for targeting to cancers. *Nano Today* 7, 245–257.

14. Parodi, A., Quattrocchi, N., van de Ven, A.L., Chiappini, C., Evangelopoulos, M., Martinez, J.O., Brown, B.S. et al. 2013. Synthetic nanoparticles functionalized with biomimetic leukocyte membranes possess cell-like functions. *Nat Nanotechnol* 8, 61–68.

15. Taratula, O., Kuzmov, A., Shah, M., Garbuzenko, O.B., and Minko, T. 2013. Nanostructured lipid carriers as multifunctional nanomedicine platform for pulmonary co-delivery of anticancer drugs and siRNA. *J. Control. Release* 171, 349–357.

16. Xi, G., Robinson, E., Mania-Farnell, B., Vanin, E.F., Shim, K.-W., Takao, T., Allender et al. 2014. Convection-enhanced delivery of nanodiamond drug delivery platforms for intracranial tumor treatment. *Nanomedicine* 10, 381–391.

Section II

Applications

9 Virus-Like Particles
A Versatile Tool for Basic and Applied Research on Emerging and Reemerging Viruses

Sandra Diederich, Alma Gedvilaite, Aurelija Zvirbliene, Andris Kazaks, Kestutis Sasnauskas, Nicholas Johnson, and Rainer G. Ulrich

CONTENTS

9.1 NOVEL, EMERGING, AND REEMERGING VIRUSES: AN INTRODUCTION

Large and devastating disease outbreaks and rapid epidemic or even pandemic pathogen spread have been of pivotal importance for human and animal health worldwide and have sparked enormous public interest. These outbreaks or clusters of human or animal diseases might be caused by known endemic pathogens or alternatively by emerging or reemerging pathogens. Emerging pathogens are either novel agents that have never been detected before due to a lack of diagnostic tools or result from their introduction into the human population from an animal source, or are agents that have modified their virulence or their geographic range (Johnson 2014). The emergence of infectious diseases (EIDs) has risen over time with a peak incidence in the 1980s (Jones et al. 2008). These EID events are dominated by zoonotic pathogens, with the majority originating from wildlife. The emergence of novel pathogens is driven by socioeconomic, environmental, and ecological factors. Multiple factors related to the pathogen itself, the reservoir, and human and livestock populations are influencing the spatial and temporal patterns of pathogen emergence (Johnson 2014).

The past 20 years illustrate that different regions of the world can be affected by the emergence of novel human and animal viruses (Figure 9.1). Many of these emerging and reemerging viruses belong to different taxonomic families with the majority possessing RNA genomes (Table 9.1). A number of emerging viruses were previously identified after clusters of human or animal disease cases, but frequently leaving the question of the true natural reservoir, at least initially, unanswered (Muller et al. 2007; Reusken et al. 2013). In addition, the application of newly developed techniques, such as next-generation sequencing (NGS) or rolling circle amplification (RCA) and also application of generic RT-PCR/PCR assays, has resulted in the identification of novel viruses within tissue specimen of diseased persons or animals (Hoffmann et al. 2012). The role of these new and/or emerging pathogens for disease in humans or animals is usually confirmed by epidemiological evidence. In contrast, recent ongoing pathogen hunting activities resulted in the identification of numerous unknown viruses, even of novel higher taxa, without clear association with disease in human or domestic animals (Phan et al. 2011; Bodewes et al. 2013; Drexler et al. 2013a,b; Ehlers and Wieland 2013; Smits et al. 2013; Sachsenroder et al. 2014). However, several of these agents with known hosts may allow the development of interesting animal models, such as rat-borne hepatitis E virus (HEV) or rodent-borne hepaciviruses (Johne et al. 2010; Drexler et al. 2013a).

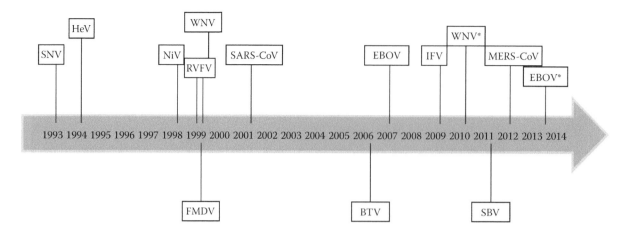

FIGURE 9.1 Timeline of emerging or reemerging viruses over the past 20 years. The boxes above the timeline show the emergence of zoonotic viruses including: Sin Nombre virus (SNV) in the Four Corners region of the United States, Hendra virus (HeV) in Australia, Nipah virus (NiV) in Malaysia, Rift Valley fever virus (RVFV) along the Arabian Peninsula, West Nile virus (WNV) lineage 1 in eastern North America (WNV), Ebola virus (EBOV) in Uganda, Severe Acute Respiratory Syndrome Coronavirus (SARS-CoV) in China, H1N1 Influenza virus (IFV) in Mexico, West Nile virus (WNV*) lineage 2 in Greece, Middle East Respiratory Syndrome Coronavirus (MERS-CoV) in Saudi Arabia, and EBOV* in West Africa. The boxes below the timeline display the emergence of animal-only viruses including foot and mouth disease virus (FMDV) in the United Kingdom, Bluetongue virus (BTV), and Schmallenberg virus (SBV) in Western Europe.

Many emerging and reemerging viruses are classified as either biosafety level 3 or biosafety level 4 (BSL3/4) agents due to their association or likelihood to cause serious or lethal human disease, respectively (Table 9.1). Preventive or therapeutic interventions may be available for BSL3 agents but are usually absent for BSL4 agents. Due to the risk associated with these pathogens, studies with infectious virus must be conducted in BSL3 or BSL4 high-containment laboratories that are rare and cost intensive. Thus, the ability to work on these agents at a lower containment level is highly desirable. VLPs that resemble the intact virions in many aspects but lack the potential to induce disease can be handled in lower containment facilities and therefore offer a major benefit in conducting research on emerging viruses. Within this manuscript, the use of VLPs as a tool for basic and applied research to develop diagnostic tools and potential vaccines for emerging viruses is reviewed.

9.2 VLP APPROACHES: A GENERAL OVERVIEW

The term *VLPs* has been used for different entities: One of these is based on electron microscopical findings of structures similar to viruses within cells and has been designated *VLPs*. Here, we use the term *VLPs* for structures that are generated specifically for selected applications. In general, three major strategies were followed to generate those virus surrogates: (1) noninfectious, replication-incompetent VLPs that are free of viral nucleic acid (niVLPs), (2) transcription- and replication-competent VLPs harboring a minigenome construct (trVLPs), and (3) infectious pseudoviruses (pseudotypes) consisting of autologous and heterologous surface proteins and the entire carrier virus genome (Figure 9.2).

niVLPs can be produced by heterologous expression of individual viral structural proteins, such as the envelope, the matrix, or capsid proteins of the original virus. These structural

proteins may have the intrinsic ability to spontaneously self-assemble into highly organized particles (*autologous* VLPs; Figure 9.2a), which resemble the virus itself morphologically in size and shape, but also in some functional and immunological respects. The simplest autologous VLPs are composed of one single protein, for example, a capsid protein. Such VLPs, frequently termed *capsid-like* particles (CLPs), were successfully generated for viruses belonging to different families. Examples include hepatitis B virus (HBV) core protein (for review, see Pumpens et al. 2008), papilloma virus L1 protein (for review, see Kazaks and Voronkova 2008), retrovirus Gag polyproteins (for review, see Ludwig and Wagner 2007), polyomavirus VP1 (Sasnauskas et al. 2002), norovirus capsid protein (for review, see Ramani et al. 2014), and different bacteriophages (Freivalds et al. 2014). Alternatively, VLPs have been generated using envelope proteins or matrix proteins alone, for example, for HBV surface antigen (HBsAg; for review, see Pumpens et al. 2008). More complex VLPs are composed of different protein layers, such as those generated for bluetongue virus (for review, see Roy 2004), rotaviruses (for review, see Li et al. 2014), or lentiviruses (for review, see Ludwig and Wagner 2007).

For several viruses, the expression of structural proteins or the formation of autologous VLPs failed or was found to be inefficient for subsequent applications. Therefore, alternative approaches have been developed: genetic fusion or chemical coupling of a protein segment to a self-assembly competent carrier protein was successfully exploited for the generation of *chimeric* VLPs (Figure 9.2b). Coexpression of the nonfused carrier moiety itself and a carrier-fusion or carrier-read-through protein has been demonstrated to allow the formation of *mosaic* particles (Figure 9.2c). Assembly-competent authentic structural proteins have been coexpressed with fusion proteins of another authentic structural protein, resulting in *pseudotype* VLPs (Figure 9.2d). In polyomavirus-derived pseudotype VLPs,

TABLE 9.1

Pathogenic Viruses Including Emerging and Reemerging Viruses

Virus Family[a] and Genome	Genus[a]	Virus
Arenaviridae (ambisense ssRNA)	*Arenavirus*	Bear Canyon virus
		Flexal virus
		Guanarito virus[b]
		Junin virus[b]
		Lassa virus[b]
		Lujo virus[b]
		Machupo virus[b]
		Sabia virus[b]
		Lymphocytic choriomeningitis virus (neurotropic)
		Whitewater Arroyo virus
Asfarviridae (ds DNA)	*Asfivirus*	African swine fever virus
Bunyaviridae (negative ssRNA)	*Hantavirus*	**Andes virus[c]**
		Bayou virus
		Black Creek Canal virus
		Cano Delgadito virus
		Dobrava-Belgrade virus[c]
		El Moro Canyon virus
		Hantaan virus[c]
		Khabarovsk virus
		Laguna Negra virus
		Muleshoe virus
		New York virus
		Puumala virus[c,d]
		Rio Segundo virus
		Seoul virus[c]
		Sin Nombre virus[c]
		Topografov virus
	Nairovirus	Dugbe virus
		Nairobi Sheep Disease virus
		Crimean Congo hemorrhagic fever virus[b]
		Khasan virus
	Orthobunyavirus	Akabane virus-JaGAr39
		Oropouche virus[c]
		Utive virus
		Douglas virus
		Termeil virus
		Schmallenberg virus
	Phlebovirus	**Rift Valley fever virus[c]**
Caliciviridae (positive ssRNA)	*Vesivirus*	Vesicular exanthema of swine virus
Coronaviridae (positive ssRNA)	*Betacoronavirus*	**Severe acute respiratory syndrome-related coronavirus (SARS-CoV)[c]**
		Middle East respiratory syndrome coronavirus (MERS-CoV)[c]
Filoviridae (negative ssRNA)	*Ebolavirus*	**Ebolavirus Bundibugyo[b]**
		Ebolavirus Cote d'Ivoire[b]
		Ebolavirus Sudan[b]
		Ebolavirus Zaire[b]
		Ebolavirus Reston[b]
	Marburgvirus	Lake Victoria Marburg virus[b]

(Continued)

TABLE 9.1 (*Continued*)
Pathogenic Viruses Including Emerging and Reemerging Viruses

Virus Family[a] and Genome	Genus[a]	Virus
Flaviviridae	*Flavivirus*	Cacipacorevirus
(positive ssRNA)		**Dengue virus 1–4[c]**
		Yellow fever virus[a]
		Israel turkey meningoencephalitis virus
		Japanese encephalitis virus[c]
		Koutango virus
		Kyasanur Forest disease virus[b]
		Louping III virus[c]
		Murray Valley encephalitis virus[c]
		Powassan virus
		Rociovirus
		St Louis encephalitis virus[c]
		Omsk hemorrhagic fever virus[b]
		Wesselsbron virus[c]
		West Nile virus[c]
		Yokose virus
		Tick-borne encephalitis virus
		Zika virus[c]
	Hepacivirus	Hepatitis C virus[c]
	Pestivirus	Classical swine fever virus
Hepeviridae	*Hepevirus*	**Hepatitis E virus[c]**
(positive ssRNA)		
Hepadnaviridae	*Orthohepadnavirus*	Hepatitis B virus[c]
(dsDNA-RT)		
Herpesviridae	*Simplexvirus*	Herpesvirus simiae[b]
(dsDNA)	*Varicellovirus*	Pseudorabies virus
Orthomyxoviridae	*Influenza A virus*	**e.g., highly pathogenic avian influenza viruses**
(negative ssRNA)	*Isavirus*	Infectious salmon anemia virus
Paramyxoviridae	*Henipavirus*	**Hendra virus[b]**
(negative ssRNA)		**Nipah virus[b]**
	Morbillivirus	Rinderpest virus
		Peste-des-petits-ruminants virus
Picornaviridae	*Aphtovirus*	**Foot and mouth disease virus**
(positive ssRNA)	*Enterovirus*	Porcine enterovirus B, serotype 9
Poxviridae	*Orthopoxvirus*	**Monkeypox virus[c]**
(dsDNA)		Variola virus[b]
Reoviridae	*Orbivirus*	African-Horse-Sickness virus
(dsRNA)		**Bluetongue virus**
		Orungovirus
		Epizootic hemorrhagic disease virus
Retroviridae	*Deltaretrovirus*	Human T-lymphotropic virus 1/2[c]
(ssRNA-RT)		
	Lentivirus	**Human immunodeficiency virus 1/2[c]**
Rhabdoviridae	*Lyssavirus*	Lyssavirus genotypes 2–7[c]
(negative ssRNA)		**Rabies virus (lyssavirus genotype 1)[c]**
	Vesiculovirus	Piry virus[c]

(*Continued*)

TABLE 9.1 (*Continued*)

Pathogenic Viruses Including Emerging and Reemerging Viruses

Virus Family[a] and Genome	Genus[a]	Virus
Togaviridae	*Alphavirus*	Cabassou virus
(positive ssRNA)		**Chikungunya virus[c]**
		Everglades virus[c]
		Getah virus[c]
		Mayaro virus[c]
		Mucambo virus[c]
		Ndumu virus[c]
		Eastern Equine encephalitis virus[c]
		Sindbis virus (Kyzylagach virus)
		Tonate virus[c]
		Venezuelan Equine encephalitis virus[c]
		Western Equine encephalitis virus[c]

Source: BGRCI (Berufsgenossenschaft Rohstoffe und Chemische Industrie), Viren: Einstufung biologischer Arbeitsstoffe. Merkblatt B 004, 2011.

Notes: Excerpt of list with human and animal pathogens. Emerging and reemerging pathogens are in bold letters.

[a] Taxonomy according to the International Committee on Taxonomy of Viruses (King et al. 2011).

[b] Human pathogen hazard group 4.

[c] Human pathogen hazard group 3.

[d] Puumala virus was included as it causes large hantavirus outbreaks in different parts of Europe (Heyman et al. 2011).

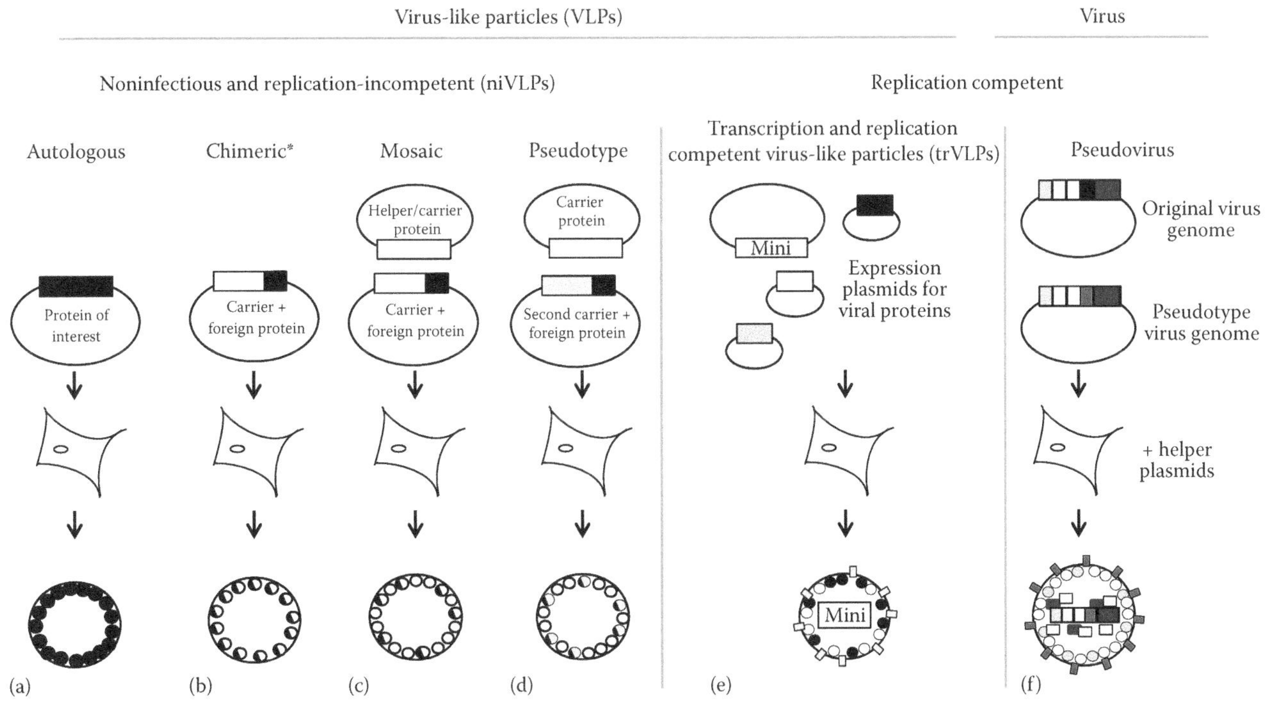

FIGURE 9.2 Potential ways to generate noninfectious virus-like particles (niVLPs), transcription- and replication-competent VLPs (trVLPs) or pseudoviruses. niVLPs can be produced by heterologous expression of a single or multiple structural proteins of a virus (autologous VLPs, a), by fusion (*or by chemical coupling) of the protein or protein segment of interest to a self-assembly-competent carrier (chimeric VLPs, b), by coexpression of the carrier itself with a carrier-insert fusion (mosaic VLPs, c) or a second protein with the insert that interacts with the carrier (pseudotype VLPs, d). trVLPs are generated by assembly-competent viral structural proteins that encapsidate a minigenome construct (e). Pseudoviruses (pseudotype viruses) are generated by incorporation of a surface protein from the virus of interest within a virus particle (f). Usually, vesicular stomatitis virus (VSV) or retroviruses have been used to generate these.

an intact carrier VP1 protein is functioning as a *helper*, mediating VLP formation of both VP1 and the modified VP2 protein molecule. In contrast to polyomavirus-derived chimeric VLPs, this approach allows the presentation of very large-sized foreign sequences as demonstrated for the entire human tumor-associated antigen Her2 (Tegerstedt et al. 2005), cellular marker p16[INK4A] (Lasickiene et al. 2012), and a Fc-engineered antibody molecule (Pleckaityte et al. 2011). This approach has not been applied for emerging virus-derived proteins so far.

Frequently used assembly-competent carriers for these approaches are the yeast Ty protein (for review, see Burns et al. 1994), HBV core (HBc) and surface protein (for review, see Pumpens et al. 2008), bacteriophage-derived proteins (for references, see Freivalds et al. 2006), polyomavirus major capsid protein VP1 (for review, see Kazaks and Voronkova 2008), and human immunodeficiency virus (HIV) Gag protein precursors (for review, see Ludwig and Wagner 2007).

All these aforementioned VLP types do not contain (intact) viral nucleic acid and are therefore noninfectious and replication incompetent. However, to be able to study additional aspects of the viral replication cycle such as transcription or replication under lower biosafety conditions and thus replacing labor- and cost-intensive screening experiments under high-containment conditions, so-called transcription- and replication-competent VLPs (trVLPs) have been developed that contain a minigenome instead of the viral genome encapsidated in the ribonucleoprotein (RNP) complex (Figure 9.2e; see Neumann et al. 2000; Hoenen et al. 2011 and references therein). Upon *infection* of target cells or better, transfer of the trVLPs onto target cells, trVLPs can enter and deliver their minigenome and mediate a single replication cycle in these cells. However, trVLPs cannot establish a multicycle infection and are thus noninfectious, making them a perfect tool to study highly pathogenic emerging viruses outside a BSL3 or BSL4 containment laboratory. Alternatively, pseudoviruses have been generated using vesicular stomatitis virus (VSV) or various retroviruses. Here, the genome of a low pathogenic virus is modified to express a surface protein of the virus of interest, resulting in an infectious pseudovirus, which can be investigated in lower containment (Figure 9.2f; for review see King and Daly 2014).

9.3 NONINFECTIOUS AUTOLOGOUS VLPs/CLPs OF EMERGING VIRUSES

The formation of autologous VLPs has been demonstrated for a broad range of emerging and reemerging viruses using different expression systems (Table 9.2). Thus, for several emerging paramyxoviruses, the formation of CLPs was demonstrated by heterologous expression of their entire nucleoproteins within yeast *Saccharomyces cerevisiae* (Figure 9.3). Autologous VLPs were shown to represent useful surrogates for complete virions to address a variety of basic research questions. Firstly, VLPs can serve as an excellent model system to study virus structure, assembly, and budding processes (Table 9.2). Initial structural information on HEV capsids was obtained by cryo-electron microscopy and x-ray analysis of capsid

protein-derived CLPs (for review, see Mori and Matsuura 2011). Using VLP technology, the viral structural elements participating in assembly, packaging, and the essential major driving forces in budding can be identified. In addition, biochemical methods can reveal details of protein domains or single amino acid residues essential for VLP formation, as evidenced for the HIV Gag precursor (for review, see Kattenbeck et al. 1996). Similarly, investigations of the VLP formation of truncated variants of the HEV capsid protein in an insect cell system confirmed that its proximal parts are dispensable for particle formation, as expected from analysis of patient stool-derived virions (Li et al. 2005).

Another prominent example for the study of assembly and budding of an emerging virus using VLP technology is Marburg virus (MARV) (Brauburger et al. 2012 and references therein). MARV is a hemorrhagic fever virus first isolated following human infection resulting from contact with African monkeys imported into Europe (Brauburger et al. 2012). It was shown that MARV VLP budding is mainly driven by the major matrix protein VP40 and that VP40 is also responsible for inducing the formation and release of filamentous VLPs. The coexpression of VP40 and GP surface protein results in the release of filamentous VLPs into the supernatant (Figure 9.4). However, the additional expression of the nucleoprotein NP, glycoprotein GP, and VP24 enhanced VP40-induced VLP release. Furthermore, cellular proteins linked to the endosomal sorting complexes required for transport (ESCRT) machinery, that is, Tsg101, Vps4A/B, and Nedd4.1., that promote particle release, have been demonstrated to be involved in MARV budding. Additionally, MARV VLP budding at filopodia was shown to depend on actin and not being sensitive to depolymerization of microtubules. For the closely related Ebola virus (EBOV), coexpression of VP40 with nucleoprotein NP also resulted in enhanced release of VLPs, again suggesting that cooperation between the nucleoprotein NP and VP40 is important for efficient budding (Licata et al. 2004). In agreement with the finding for MARV VLPs, EBOV VLPs consisting of EBOV VP40 and glycoprotein GP induced rearrangement of the actin cytoskeleton in endothelial cells (Wahl-Jensen et al. 2005). To determine the minimal requirements for influenza VLP formation and release, different production strategies have been addressed with variations in viral strains, the type of gene delivery utilized, and the nature of the host-cell expression system. Influenza VLPs have been constructed using one or both of the two viral glycoproteins, the hemagglutinin HA and the neuraminidase NA (Haynes 2009), confirming that they are the driving force for VLP budding. In some cases, influenza VLPs have been produced with one or both of the two influenza matrix proteins (M1 or M2); however, it was shown that the matrix protein only exhibits limited contribution to particle formation (Pushko et al. 2005; Chen et al. 2007; Lai et al. 2010; Wu et al. 2010).

In contrast to MARV that expresses a typical matrix protein, hantaviruses and other bunyaviruses do not have a matrix protein that mediates assembly and budding. For these viruses, the cytoplasmic tails of the glycoproteins are

TABLE 9.2

Generation of Autologous VLPs of Emerging Viruses and Their Application

Virus Family Genus	Viral Protein(s)	Expression System	Application	References
Arenaviridae				
Arenavirus				
Lassa virus (LASV)	NP, GP, Z	Mammalian	Immunological studies	Branco et al. (2010)
	Z	Mammalian	Budding	Sakuma et al. (2009), Strecker et al. (2003)
Junin virus (JUNV)	Z	Mammalian	Antiviral testing	Lu et al. (2014)
Bunyuaviridae				
Phlebovirus				
Rift Valley fever virus (RVFV)	Gn, Gc, N	Insect	Proof of principle	Liu et al. (2008)
	Gn, Gc, N	Mammalian, insect	Vaccine studies Proof of principle	Mandell et al. (2010a,b)
Hantavirus				
Hantaanvirus (HTNV)	N, Gn, Gc	Mammalian	Immunological studies	Li et al. (2010)
	N, G1, G2	Mammalian	Proof of principle	Betenbaugh et al. (1995)
Andes virus (ANDV)	Gn, Gc	Mammalian	Assembly and budding studies	Acuna et al. (2013)
Puumula virus (PUUV)	Gn, Gc	Mammalian	Assembly and budding studies	Acuna et al. (2013)
Coronaviridae				
Betacoronavirus				
Severe acute respiratory syndrome-related coronavirus (SARS-CoV)	S, E, M	Insect	Immunological studies	Lu et al. (2007)
	E, M/S, E, M	Insect	Proof of principle	Ho et al. (2004)
	N, M/S, N, M	Mammalian	Proof of principle, assembly	Huang et al. (2004)
	N, M, S, E	Mammalian	Packaging, assembly	Hsieh et al. (2005)
	M, N	Mammalian	Assembly	Tseng et al. (2013)
Filoviridae				
Ebolavirus				
Ebolavirus (EBOV)	VP40, GP	Mammalian	Vaccine candidate[a]	Warfield et al. (2003)
	VP40, GP	Insect	Vaccine candidate[a]	Sun et al. (2009)
Marburgvirus				
Marburg virus (MARV)	VP40, GP	Mammalian	Vaccine candidate[a]	Warfield et al. (2004)
	VP40, GP	Mammalian	Proof of principle	Swenson et al. (2004)
	NP, GP, VP40	Mammalian	Budding	Urata et al. (2007)
Flaviviridae				
Flavivirus				
Dengue virus (DENV)	capsid	Bacterial	Proof of principle	Lopez et al. (2009)
	prM, E	Mammalian	Proof of principle Immunological studies Vaccination studies	Konishi and Fujii (2002), Purdy and Chang (2005), Wang et al. (2009), Zhang et al. (2011)
	prM, E	Insect	Immunological studies	Kuwahara and Konishi (2010)
	prM, E	Yeast	Proof of principle Immunological studies	Liu et al. (2010)
Japanese encephalitis virus (JEV)	prM, E	Insect	Immunological studies	Kuwahara and Konishi (2010)
	prM, E	Mammalian	Vaccine candidate[a]	Hua et al. (2014)
West Nile virus (WNV)	M, E	Mammalian	Vaccine candidate[a]	Ohtaki et al. (2010)
Hepacivirus				
Hepatitis C virus (HCV)	C, E1, E2, p7	Insect	Proof of principle	Baumert et al. (1998)
Hepeviridae				
Hepevirus				
Hepatitis E virus (HEV)	CP	Insect	Antigenicity studies Structural studies	Li et al. (1997)
HEV, genotype 4	CP	Insect	Structural studies	

(Continued)

TABLE 9.2 (*Continued*)
Generation of Autologous VLPs of Emerging Viruses and Their Application

Virus Family Genus	Viral Protein(s)	Expression System	Application	References
Rat hepatitis E virus	CP	Insect	Proof of principle Immunological studies	Li et al. (2011)
Orthomyxoviridae				
Influenza A virus				
A/Victoria/3/75 (H3N2)	PB1, PB2, PA, NP, HA, NA, NS1, NS2, M1, M2	Mammalian	Assembly, budding	Mena et al. (1996), Gomez-Puertas et al. (2000)
A/Udorn/72 (H3N2)	M1, HA	Insect	Vaccine candidate[a]	Galarza et al. (2005)
A/Indonesia/05/2005 (H5N1)	HA, NA, M1	Insect	Vaccine candidate[a]	Mahmood et al. (2008)
A/Cambodia/ JP52a/2005(H5N1)	NA	Mammalian	Proof of principle	Lai et al. (2010)
A/Hong Kong/1073/99 (H9N2)	HA, NA, M1	Insect	Vaccine candidate[a]	Pushko et al. (2005)
A/Indonesia/05/2005 (H5N1)	HA, M1	Plant	Vaccine candidate[a]	D'Aoust et al. (2008)
A/New Caledonia/20/99 (H1N1)	HA, M1	Plant	Vaccine candidate[a]	D'Aoust et al. (2008)
A/Taiwan/083/2006 (H3N2)	HA, NA, M1, M2	Mammalian	Proof of principle Vaccine candidate[a]	Wu et al. (2010, 2012)
A/Hanoi/30408/2005 (H5N1)	HA, NA, M1, M2	Mammalian	Proof of principle Vaccine candidate[a]	Wu et al. (2010, 2012)
Paramyxoviridae				
Rubulavirus				
Menangle virus (MeV)	N	Yeast	Diagnostic application Generation of monoclonal antibodies	Juozapaitis et al. (2007a), Zvirbliene et al. (2010)
Tioman virus	N	Yeast	Proof of principle	Petraityte et al. (2009)
Henipavirus				
Nipah virus (NiV)	F, G, M, NP[b]	Mammalian	Assembly, budding	Patch et al. (2007)
	G, F and M	Mammalian	Structural/immunological studies	Walpita et al. (2011)
	N	Yeast	Proof of principle	Juozapaitis et al. (2007b)
Hendra virus (HeV)	N	Yeast	Proof of principle	Juozapaitis et al. (2007b)
Picornaviridae				
Aphtovirus				
Foot and mouth disease virus (FMDV)	VP0, VP1, VP3	Bacterial	Proof of principle, assembly	Lee et al. (2009)
	VP0, VP3, VP1–2	Insect	Proof of principle	Cao et al. (2010)
	P1–2A, 3C	Insect	Vaccine candidate[a]	Bhat et al. (2013)
Togaviridae				
Alphavirus				
Chikungunya virus (CHIKV)	C-E3-E2-6K-E1	Mammalian	Vaccine candidate[a]	Akahata et al. (2010)
	C-E3-E2-6K-E1	Insect	Vaccine studies	Metz et al. (2013)
	C-E3-E2-6K-E1	Mammalian and insect	Proof of principle Immunological studies	Wagner et al. (2014)
Reoviridae				
Orbivirus				
Bluetongue Virus (BTV)	VP3, VP7, VP2, VP5; VP3, VP7	Insect	Proof of principle	French et al. (1990), French and Roy (1990)
	VP2; VP2, VP3, VP5, VP7	Insect	Vaccine candidate[a]	Roy et al. (1994), Stewart et al. (2013)
	VP2, VP6, VP7, NS1	Insect	Proof of principle	Belyaev and Roy (1993)
Retroviridae				
Lentivirus				
Simian immunodeficiency virus (SIV)	Gag polyprotein	Insect	Proof of principle	Delchambre et al. (1989)
	Pr56 Gag and Env	Insect	Immunological studies	Notka et al. (1999)

(Continued)

TABLE 9.2 (*Continued*)

Generation of Autologous VLPs of Emerging Viruses and Their Application

Virus Family Genus	Viral Protein(s)	Expression System	Application	References
Human immunodeficiency virus (HIV)	Gag polyprotein	Insect	Assembly Immunological studies	Gheysen et al. (1989), Speth et al. (2008)
		Mammalian	Assembly	Cen et al. (2004)
		Yeast	Immunological studies	Tsunetsugu-Yokota et al. (2003)
	Pr55 Gag gp120	Insect cell	Immunological studies	Wagner et al. (1998)
			Proof of principle	Buonaguro et al. (2001)
Rhabdoviridae				
Lyssavirus				
Rabies virus	G	Mammalian	Vaccine studies	Fontana et al. (2014)

Abbreviations for viral proteins used in table: Glycoproteins: GP, G1, G2, Gn, Gc, F, G, HA, NA, S, E, E1, E2, Env; Matrix protein: Z, M, VP40; Nucleocapsid or capsid proteins: N, NP, C, CP, Gag; Nonstructural proteins: NS1, NS2; Polymerase: PB1, PB2, PA; Viral proteins: VP.

[a] Protective immunity shown in animal model.

[b] Various combinations of proteins.

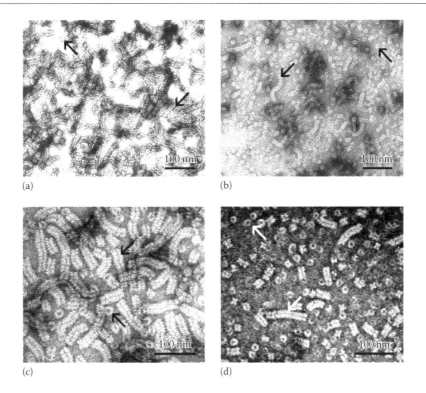

FIGURE 9.3 Negative stain electron microscopic images of VLPs formed by yeast-expressed nucleocapsid proteins of different emerging paramyxoviruses. The entire nucleocapsid proteins of Hendra virus (a), Nipah virus (b), Menangle virus (c) or Tioman virus (d) synthesized in yeast *S. cerevisiae* formed spontaneously helical nucleocapsid-like particles. Arrows indicate nucleocapsid-like rods and ring-like structures.

proposed to stabilize the virus structure (Hepojoki et al. 2010; Wang et al. 2010; Strandin et al. 2013). Initially, it was reported that hantavirus VLPs are produced, when glycoproteins Gn and Gc and nucleocapsid proteins are coexpressed, but with low efficiency (Betenbaugh et al. 1995). However, for other members of the *Bunyaviridae* family such as phleboviruses, it was demonstrated that the glycoproteins are the only viral components required for the

formation of VLPs (Overby et al. 2006; Mandell et al. 2010b). Recently, Acuna et al. confirmed that Andes virus (ANDV) and Puumala virus (PUUV) VLP formation depends solely on plasmid-driven glycoprotein expression and thus, the release of virus-like structures does not require the participation of the viral nucleocapsid protein (Acuna et al. 2013).

Secondly, VLPs have been an important tool in identifying potential cellular receptors, identifying viral proteins and their

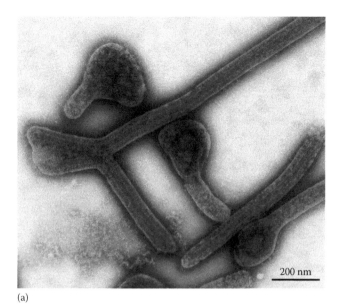

200 nm

(a)

100 nm

(b)

FIGURE 9.4 Electron microscopic images of Marburg virus (a) and virus-like particles released into the supernatant from cells expressing Marburg virus matrix protein VP40 and viral surface protein GP (b). (Images courtesy of Dr. Larissa Kolesnikova, Institute of Virology, Philipps-University Marburg, Marburg, Germany.)

functional domains that are involved in receptor recognition, binding, and entry pathways (Table 9.2). As an example of receptor interaction, Buonaguro et al. evaluated the binding of HIV-VLPs to cellular chemokine coreceptors using carboxyfluorescein succinimidyl ester labeling and cellular uptake (Buonaguro et al. 2013). Hepatitis C virus (HCV)-derived VLPs containing the structural proteins of HCV H77 strain were taken to study the early events in virus–cell interaction, namely, attachment and entry processes (Triyatni et al. 2002). Virus uptake of the BSL4 agent Nipah virus (NiV) into host cells in the presence of endocytosis inhibitors has been analyzed by using VLPs consisting of the NiV glycoprotein G (Diederich et al. 2008). Furthermore, using a β-lactamase-NiV matrix fusion protein, β-lactamase-matrix VLPs that incorporated NiV or HeV fusion F and attachment G proteins were produced (Wolf et al. 2009). Upon VLP entry, an enzymatic and fluorescent conversion of the preloaded β-lactamase substrate (CCF2-AM) followed cytosolic delivery of β-lactamase and thereby, entry characteristics and entry inhibitors can be analyzed. Similarly, Martinez et al. used β-lactamase-tagged EBOV VLPs to answer questions to various aspects of EBOV entry into cells. For example, they were able to show that cathepsin L only plays a minimal role in EBOV infection of human dendritic cells, whereas entry depends on cathepsin B

(Martinez et al. 2010). Furthermore, in another study, they demonstrated that amino acid substitutions in EBOV glycoprotein GP can modulate EBOV entry in a host-species-specific and cell-type-specific manner (Martinez et al. 2013).

Besides the application of VLPs for basic research, VLPs have been widely accepted in applied science. Since VLPs mimic the original virions outstandingly well with regard to antigenic epitopes displayed, VLP technology offers many advantages in the development of new diagnostic tools to determine the corresponding antibodies in virus-infected human and animals, that is, replacing the need for infectious virus in the production process of antigens for diagnostic applications. Thus, there is no need to propagate the infectious virus, there is no risk of virus transmission or infection, production levels are higher, and production itself is cheaper. Holmes et al. established an ELISA for the detection of virus-specific antibodies in serum panels from patients with recent Dengue virus (DENV), St. Louis encephalitis virus (SLEV), and West Nile virus (WNV) infection using a VLP mixture of DENV serotypes 1–4, Japanese encephalitis virus (JEV), WNV, and SLEV and demonstrated that specificity was higher using VLPs compared to the traditionally used virus-infected suckling mouse brain-derived antigens (Holmes et al. 2005). Also, the generation of VLPs has overcome the lack of an efficient cell culture system for HEV. Here, capsid protein–derived VLPs have been applied for the detection of antibodies against human pathogenic HEV (Rose et al. 2010) and recently discovered HEV strains from rat (Li et al. 2011) and ferret (Yang et al. 2013).

Additionally, VLPs have been used extensively as vaccine candidates (Table 9.2). One of the early drivers was the search for a vaccine against HIV. In general, lentiviruses have been used widely to generate VLPs (see later section) and some of the earliest VLPs using the Gag protein of HIV (Adams et al. 1987) were considered as a potential vaccine candidate (Weber et al. 1995). Unfortunately, the hopes for an effective HIV vaccine have not been realized, but the use of VLPs as potential vaccines has expanded dramatically to include many virus species (see Roy and Noad 2008 and references therein). In addition, emerging virus-derived CLPs/VLPs have been applied for immunological studies and the generation of specific antisera or monoclonal antibodies.

9.4 NONINFECTIOUS CHIMERIC AND MOSAIC VLPs/CLPs

Initial failure of efficient CLP or VLP formation of emerging virus-derived heterologously expressed proteins resulted in the generation of chimeric Ty- or HBc-derived VLPs presenting epitopes or protein segments of HIV (Adams et al. 1987; Borisova et al. 1989). Frequently, proof of principle studies were performed to prove the assembly of chimeric or mosaic VLPs, identifying the optimal insertion site within the carrier and estimating empirically the insertion capacity of the carrier moiety for foreign proteins (Table 9.3). Thus, HBc- and hamster polyomavirus (HaPyV)-derived VLPs

TABLE 9.3

Generation of Chimeric and Mosaic VLPs of Emerging Viruses and Their Application

Virus Family Genus	Viral Protein	Expression System	Chimeric (C) Mosaic (M)	Carrier	Application	References
Bunyaviridae						
Phlebonirus						
Rift Valley fever virus (RVFV)	Gn, Gc, N	Mammalian	C	MoMLV	Vaccine studies	Mandell et al. (2010a)
Hantavirus						
Puumula virus (PUUV)	N	Bacterial	C	HBcAg	Proof of principle Immunological studies Structural investigations	Ulrich et al. (1999), Koletzki et al. (1999, 2000), Geldmacher et al. (2004, 2005)
	N	Bacterial	C	HBcAg	Vaccine candidate[a]	Koletzki et al. (2000), Ulrich et al. (1998)
	N	Bacterial	M	HBcAg	Proof of principle Immunological studies	Ulrich et al. (1999), Koletzki et al. (1997, 1999, 2000), Kazaks et al. (2002)
	N	Yeast	C	HaPyV-VP1	Immunological studies Production of monoclonal antibodies	Gedvilaite et al. (2004), Zvirbliene et al. (2006)
	Gc	Yeast	C	HaPyV-VP1	Production of monoclonal antibodies	Zvirbliene et al. (2014)
Hantaan virus (HTNV), Dobrava-Belgrade virus (DOBV)	N	Bacterial	C	HBcAg	Immunological studies	Geldmacher et al. (2004, 2005)
Flaviviridae						
Flavivirus						
Dengue virus (DENV)	E	Yeast	C	HBsAg	Proof of principle	Bisht et al. (2001)
West Nile virus (WNV)	E	Bacterial	M	RNA phage AP205	Immunological studies	Cielens et al. (2014)
Hepacivirus						
Hepatitis C virus (HCV)	E1 or E2	Mammalian	M	HBsAg	Proof of principle	Patient et al. (2009)
	E2	Mammalian	C	HBsAg	Immunological studies Vaccine studies	Netter et al. (2001) Vietheer et al. (2007)
	E2	Bacterial	C	PapMV C	Immunological studies	Denis et al. (2007)
	C	Bacterial	C	HBcAg	Proof of principle	Yoshikawa et al. (1993)
Hepeviridae						
Hepatitis E virus (HEV)	C	Yeast	C	HBsAg	Proof of principle Immunological studies	Li et al. (2004)
Orthomyxoviridae						
Influenza A virus						
Influenza A/PR/8/34 (H1N1)	M2	Bacterial	C	HBcAg	Vaccine candidate[a]	Neirynck et al. (1999), De Filette et al. (2005)
			C[b]		Vaccine studies	Jegerlehner et al. (2002)
	HA		C[b]	Bacteriophage Qβ	Vaccine candidate[a]	Jegerlehner et al. (2013)
Influenza A/California/07/2009	HA	Bacterial	C[b]	Bacteriophage Qβ	Vaccine candidate[a]	Low et al. (2014)
	HA			MuPyV	Vaccine study	Anggraeni et al. (2013)
Influenza A/Aichi/72 (H3N2)	M2	Yeast	C	HBcAg	Vaccine studies	Fu et al. (2009)

(Continued)

TABLE 9.3 (*Continued*)

Generation of Chimeric and Mosaic VLPs of Emerging Viruses and Their Application

Virus Family Genus	Viral Protein	Expression System	Chimeric (C) Mosaic (M)	Carrier	Application	References
Picornaviridae						
Aphtovirus						
Foot and mouth disease virus (FMDV)	3A	Bacterial	C	RHDV VP60	Immunological studies	Crisci et al. (2012)
	VP1	Bacterial	C	HBcAg	Immunological studies Receptor binding	Clarke et al. (1987)
	VP1	Plant	C	HBcAg	Vaccine candidate[a]	Chambers et al. (1996)
	VP1	Yeast	C	HBcAg	Immunological studies	Beesley et al. (1990)
Togaviridae						
Alphavirus						
Venezuelan Equine encephalitis virus (VEEV)	E2	Bacterial	M	HBcAg	Immunological studies	Loktev et al. (1996)
Retroviridae						
Lentivirus						
Human immunodeficiency virus (HIV)	gp120	Yeast	C	Ty p1	Immunological studies	Layton et al. (1993)
	gp41	Bacterial	C	HBcAg	Proof of principle	Borisova et al. (1989)
	p17/p24	Bacterial	C	HBcAg	Proof of principle and immunological studies	Ulrich et al. (1992)
	gp41, p34, p17	Bacterial	C	HBcAg	Proof of principle	Isaguliants et al. (1996)
	gp120	Yeast	C	Ty p1	Vaccine studies	Griffiths et al. (1991)
	gp120	Mammalian	M	HBsAg	Vaccine studies	Schlienger et al. (1992), Michel et al. (1988, 1990)

Abbreviations for viral proteins used in table: Capsid protein: VP; Glycoproteins: gp, Gc, E, E1, E2; Matrix protein: M2; Nucleocapsid or capsid protein: N, C; Nonstructural proteins: 3A.

Abbreviations for carrier: HaPyV-VP1: Hamster polyomavirus VP1; HBcAg: Hepatitis B virus core antigen; HBsAg: Hepatitis B virus surface antigen; MoMLV: Moloney murine leukemia virus; MuPyV: murine polyomavirus; PapMV C: Papaya mosaic virus capsid protein; RHDV VP60: Rabbit hemorrhagic disease virus viral protein 60; Ty p1: particle-forming p1 protein of the yeast retro-transposon Ty; VSV: Vesicular stomatitis virus.

[a] Protective immunity shown in animal model.

[b] Chemical coupling.

were demonstrated to allow the surface exposure of hantavirus nucleocapsid protein segments of up to 120 amino acids in length (Koletzki et al. 1999; Gedvilaite et al. 2004) (see Figures 9.5 and 9.6). In addition, HBc constructs were used to generate mosaic particles consisting of the carrier itself and a carrier-hantavirus protein fusion synthesized by a suppressor tRNA-mediated read-through mechanism (Koletzki et al. 1997; Kazaks et al. 2002).

Application of this VLP technology was based on findings that chimeric VLPs can induce a strong immune response against the emerging virus-derived protein segment. Therefore, immunological studies were frequently performed to identify and characterize immunodominant and protective regions within the viral proteins presented on the VLPs, for example, hantavirus nucleocapsid protein segments presented on HBc particles (Ulrich et al. 1998; Lundkvist et al. 2002). In addition, chimeric VLPs were used

to generate hantavirus-specific monoclonal antibodies. Here, the HaPyV-VP1 derived VLPs harboring an N-terminal segment of the PUUV nucleocapsid protein resulted in the generation of monoclonal antibodies that are able to recognize the authentic viral protein in infected cells (Kucinskaite-Kodze et al. 2011; Zvirbliene et al. 2014). In addition, an epitope region of PUUV Gc presented on this carrier allowed the generation of a PUUV Gc-specific monoclonal antibody (Zvirbliene et al. 2014). Although this antibody is non-neutralizing, it recognizes a highly conserved Gc protein sequence identical among Gc proteins of different hantaviruses. Therefore, this broadly reactive antibody represents a promising universal tool for hantavirus detection in infected cells (Zvirbliene et al. 2014).

Moreover, chimeric and mosaic VLPs harboring HIV, hantavirus, and influenza virus insertions were tested as vaccine candidates. Thus, HIV-derived VLPs built up by

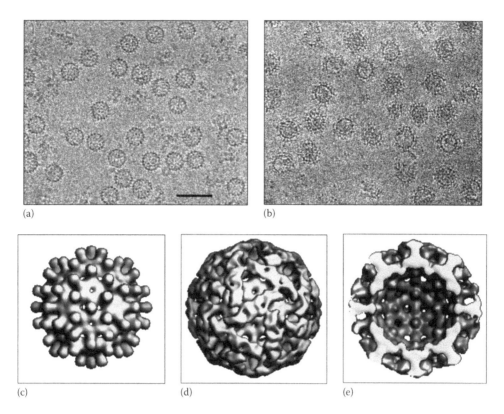

FIGURE 9.5 Electron cryo-microscopy of core particles formed by a truncated hepatitis B virus core protein without insertion (a, c) or with an internal insertion of 120 amino acids of the Puumala virus nucleocapsid protein segment (b, d, and e). (Reprinted from *Virology*, 323, Geldmacher, A., Skrastina, D., Petrovskis, I., Borisova, G., Berriman, J.A., Roseman, A.M., Crowther, R.A., Fischer, J., Musema, S., Gelderblom, H.R. et al., An amino-terminal segment of hantavirus nucleocapsid protein presented on hepatitis B virus core particles induces a strong and highly cross-reactive antibody response in mice, 108–119, Copyright (2004), with permission from Elsevier.) Scale bar in a (for a and b), 50 nm.

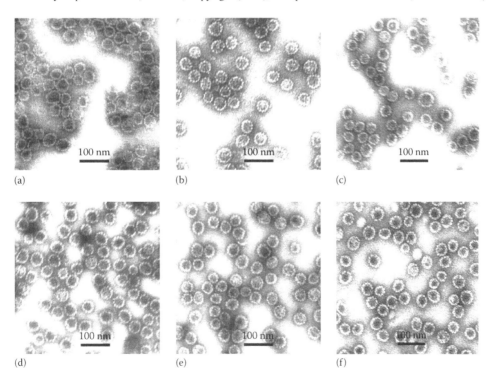

FIGURE 9.6 Negative stain electron microscopic images of chimeric hamster polyomavirus (HaPyV) VP1-derived VLPs harboring Puumala virus nucleocapsid protein segments (a through e) and unmodified yeast-expressed HaPyV-VP1 VLPs (f). Puumala virus nucleocapsid protein segments spanning amino acid (aa) residues 1–120 were inserted at site 1 between aa residues 80–89 (a) or at site 4 between aa residues 288–295 (d), aa residues 1–45 at site 1 (b) or at site 4 (e), and aa residues 1–80 at site 1 (c).

Gag-Env fusions were demonstrated to induce strong antibody and T-cell responses in animal models (Wagner et al. 1998). Chimeric HBc particles were also found to induce a protective immunity against PUUV challenge in a bank vole model (Ulrich et al. 1998). Interestingly, the level of protection was strongly influenced by the insertion site and the PUUV strain used (Koletzki et al. 2000). Moreover, chimeric core particles seem to outperform mosaic particles in terms of protective immunity (Koletzki et al. 2000). A strong antibody response against hantavirus nucleocapsid segment presented on either HBc or HaPyV-VP1 VLPs was reported after administration without additional adjuvant (Gedvilaite et al. 2004; Geldmacher et al. 2005). HBc particles were also used to prove the potential influence of a preexisting anti-HBc immunity on the hantavirus-specific antibody response after boost with chimeric HBc-hantavirus particles (Geldmacher et al. 2004). Similarly, the alternating vaccination with HaPyV-VP1 VLPs harboring the hantavirus insert at two different positions within the carrier might reduce the induction of anti-VP1 antibodies during multiple booster immunizations as evidenced by analyzing the hybridoma repertoire with no hybridoma clones specific to the VP1-carrier being found (Zvirbliene et al. 2006).

In recent years, a lot of attention has been paid to a highly conserved extracellular domain of the influenza virus M2 protein, namely, M2e, being proposed as a candidate for a universal influenza A virus vaccine. A number of attempts were performed to improve immunogenicity of this 23-amino-acid-long peptide via incorporation into VLPs. At first, chimeric VLPs harboring M2e sequence incorporated at the N-terminus of HBc were found to induce a protective immune response in mice (Neirynck et al. 1999). Later on, protectivity of this type of VLPs was improved by linking of three consecutive M2e copies (De Filette et al. 2005). Another variant of chimeric VLPs carrying M2e insertion into major immunodominant region of HBc was found to be protective in mice but not in rhesus monkeys (Fu et al. 2009). Besides genetic fusion, chimeric VLPs harboring M2e epitope were also generated by chemical coupling techniques using HBc and human papillomavirus capsid protein as carriers (Jegerlehner et al. 2002; Ionescu et al. 2006). Although nearly all M2e-containing chimeric VLPs ensured high-level protection against influenza A virus challenge in mice, it is generally agreed that other conserved influenza virus epitopes derived from the nucleoprotein or the hemagglutinin need to be added to design an universal influenza vaccine for humans. Using this approach, three tandem copies of M2e along with the nucleoprotein epitopes were incorporated within HBc VLPs, which produced robust M2e-specific antibodies and cellular immune responses in mice (Gao et al. 2013). Although the concept of a universal influenza vaccine seems very attractive, an alternative strategy is devoted to prepare a set of subtype-specific VLP vaccines to be rapidly generated from previously produced components upon necessity. Thus, bacterially expressed globular head domain of the influenza virus hemagglutinin, comprising most of the protein's neutralizing epitopes, was covalently

conjugated to bacteriophage Qβ-derived VLPs (Jegerlehner et al. 2013). Very recently, safety and immunogenicity of this H1N1 influenza virus VLP vaccine in Phase I clinical trial in healthy Asian volunteers has been reported (Low et al. 2014). Alternatively, helix 190 from hemagglutinin was exposed on murine polyomavirus VLPs in its native conformation, which indeed led to an improved immune response in a mouse model (Anggraeni et al. 2013).

Finally, chimeric HBc particles harboring a segment of the foot and mouth disease virus (FMDV) VP1 protein were applied for receptor binding studies (Chambers et al. 1996; Sharma et al. 1997). As demonstrated for hantavirus and influenza virus, there were several attempts to expose FMDV peptides on VLPs to generate vaccine candidates. Thus, immunogenicity of a 20–21-amino-acid stretch of the VP1 protein of FMDV was greatly enhanced within chimeric HBc VLPs obtained both from bacterial and plant cells (Clarke et al. 1987; Huang et al. 2005). Subsequently, more complex HBc VLPs carrying FMDV multiepitopes have resulted in enhanced humoral and cellular immune responses in mice (Zhang et al. 2007).

9.5 INFECTIOUS VLPs AND PSEUDOVIRUSES

To expand the area of possible applications of VLPs mentioned earlier to modeling of further aspects of the viral replication cycle such as transcription or genome replication, generation of VLPs was combined with the minigenome system, resulting in the production of transcription and replication competent VLPs (trVLPs) (see Section 9.2; Figure 9.2e). trVLPs do not replicate and are thus noninfectious (Hoenen et al. 2011). However, these VLPs are able to infect target cells with respect to entering the cell and releasing the RNP complex (minigenome) into the cytoplasm. Transcription and replication of the delivered RNP-packed minigenomes is driven by a RNP complex already expressed in the target cell either by transfection of expression plasmids or by helper virus infection. A plasmid-based trVLP system with such pretransfected target cells was established to investigate the role of the minor matrix protein VP24 in EBOV morphogenesis and budding (Watanabe et al. 2004). In recent studies, EBOV trVLPs have been utilized not only to investigate entry pathways in different host cells but also to study the genus-specific recruitment of the RNP complex into the budding particle (Aleksandrowicz et al. 2011; Spiegelberg et al. 2011). This trVLP system was further developed by Hoenen et al. using naïve target cells, demonstrating the fundamental role of EBOV VP24 for primary transcription and/or formation of RNP complexes capable of supporting primary transcription (Hoenen et al. 2006). Similarly, trVLP systems have been established for MARV and MARV trVLPs have been used to screen for MARV neutralizing antibodies and thus, replacing experiments that require BSL4-conditions (Krahling et al. 2010; Wenigenrath et al. 2010). Several trVLP systems have been established for bunyaviruses. The role of the nonstructural protein NSm in Bunyamwera virus assembly and morphogenesis as well as the contribution of the glycoprotein to RNP packaging and budding for Uukuniemi virus (UUKV) was investigated by infecting pretransfected target cells with trVLPs (Shi et al. 2006; Overby et al. 2007a,b). Using trVLPs on naïve target cells, Habjan et al. (2009) demonstrated that the cellular MxA protein can impair both primary and secondary transcription of Rift Valley fever virus (RVFV). With the advantage of modeling, many aspects of the viral replication cycle outside a BSL3 or BSL4 containment, trVLPs are a promising tool in future (high-throughput) screening for antiviral substances.

An alternative to niVLPs and trVLPs with regard to studying virus assembly, receptor tropism, antiviral drug discovery, diagnostic assay application, such as for the detection of neutralizing antibodies, or the generation of potential vaccine candidates is the pseudotyping of low or apathogenic viruses with the envelope glycoproteins of the virus of interest. This again avoids the need to work under high-level biosafety containment. One such example is the VSV which is known for its genetic flexibility. Recombinant VSV (rVSV) has been generated where the VSV glycoprotein G is replaced by foreign viral glycoproteins that are incorporated in a rather unselective manner (Schnell et al. 1996). Another advantage of VSV pseudotypes is the potential for propagation on many conventional mammalian cell lines and their growth to high infectious titers. Insertional mutagenesis, an inherent risk associated with retroviral vectors (see the following), can be excluded, since VSV replicates exclusively in the cytosol.

VSV pseudotyped with the surface glycoproteins hemagglutinin and neuraminidase from a highly pathogenic avian influenza virus can be deployed to detect neutralizing antibodies and to study different hemagglutinin and neuraminidase reassortants (Cheresiz et al. 2014; Zimmer et al. 2014). In agreement, VSV-derived pseudoviruses have also been generated for prototype Hantaan virus and other highly pathogenic hantaviruses such as Seoul virus, ANDV, and PUUV, allowing the performance of neutralization tests under low containment conditions (Ogino et al. 2003; Ray et al. 2010; Higa et al. 2012). To replace the plaque reduction neutralization test for NiV, which needs to be performed under BSL4 conditions, a neutralization assay was generated using VSV particles pseudotyped with NiV fusion protein (F) and glycoprotein (G) (Tamin et al. 2009). Furthermore, hantavirus pseudoviruses as well as NiV pseudoviruses were used to investigate virus–host cell interactions with regard to virus entry and receptor interaction (Tamin et al. 2009; Ray et al. 2010; Higa et al. 2012).

The use of rVSV in preventing virus infection has been demonstrated for a variety of viruses. Live attenuated rVSV expressing HIV Env and Gag proteins protected rhesus monkeys from AIDS following challenge with a pathogenic HIV (Rose et al. 2001). In recent studies, VSV-based vaccines for both MARV and EBOV have been established and have been shown to lead to nearly complete or complete protection against filovirus hemorrhagic fever in nonhuman primates (reviewed in Warfield and Olinger 2011 and Geisbert

and Feldmann 2011). Furthermore, these live rVSV vectors expressing either MARV or EBOV glycoprotein GP have been shown to offer promising results in postexposure treatment in monkeys and also in a human case of emergency vaccination (reviewed in Geisbert and Feldmann 2011; Gunther et al. 2011). However, safety is a remaining concern with these VSV-based vaccines, since the VSV vector is replication competent.

Generation of *de novo* pseudotypes based on a lentivirus such as HIV involves transfection of target cells with three plasmids containing a promoter with (1) the *gag/pol* genes of HIV, (2) the envelope gene of choice, and (3) a reporter gene, such as luciferase or green fluorescent protein, flanked by the HIV long-terminal repeat (LTR) sequences (Figure 9.7). Successful expression of the *gag*, *pol*, and *env* genes leads to the release of pseudotyped virus particles from the cell surface and subsequent infection of a second cell-line. This infection with pseudovirus leads to integration and expression of the reporter gene, enabling the infection to be visualized (Figure 9.7). However, in contrast to VSV pseudotypes, no further virus replication can take place following this single-round infection cycle.

As described for VSV pseudotypes, retroviral pseudotypes have been used for various applications. A retrovirus-based virus pseudotyped with the precursor membrane and envelope glycoproteins of four different serotypes of DENV has been used not only to characterize different field isolates on a molecular level but also to investigate entry of the virus into host cells (Hu et al. 2007). Also, retroviral pseudotypes have been used for the detection of neutralizing antibodies against highly pathogenic influenza A viruses (Temperton et al. 2007; Wang et al. 2008; Alberini et al. 2009; Tsai et al. 2009) and more recently to measure neutralization of the emergent novel human coronavirus from the Middle East, MERS-CoV (Zhao et al. 2013). Additionally, influenza virus pseudotypes have been used for high-throughput screening of new influenza virus inhibitors (Ao et al. 2008).

A major application for this system has been the production of safe, surrogate pseudovirions for use in neutralization assays involving lyssaviruses (Wright et al. 2008). This virus group has expanded in recent years to include 14 virus species, the majority isolated from bat reservoirs (Banyard et al. 2011). Rabies virus (RABV) has caused numerous outbreaks around the world and is considered a neglected reemerging disease, while many of the other virus species within the lyssavirus genus have caused fatal spill-over infections in humans (Johnson et al. 2010). Antibody neutralization of lyssaviruses occurs through binding to antigenic sites on the virus glycoprotein (Figure 9.7). A key biological property of the glycoprotein that has aided development of lentiviral-based pseudoviruses is the presence of a strong membrane targeting signal at the N-terminus of the protein (Anilionis et al. 1982). All lyssaviruses have this 19 residue targeting sequence that contains a high proportion of hydrophobic residues but shows little sequence similarity between species in contrast to the downstream sequence. This targeting sequence efficiently

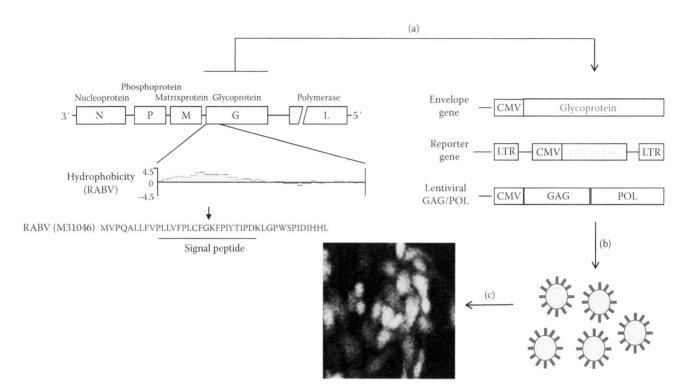

FIGURE 9.7 Schematic of the lyssavirus genome, generation of pseudoviruses, and expression of the reporter gene (green fluorescent protein) following cellular infection. The schematic shows the general structure of the lyssavirus genome with focus on the N-terminal signal sequence encoding part of the rabies virus (RABV—GenBank M31046). The glycoprotein coding sequence is cloned into a plasmid under the control of the human cytomegalovirus (HCMV) promoter (a). This plasmid is cotransfected into a cell-line allowing the expression of genes under the HCMV promoter along with two other plasmids containing a reporter gene between two lentiviral long-terminal repeats (LTRs) and lentiviral structural proteins to generate pseudoviruses (b). De novo generated pseudoviruses can infect permissive cells and express the reporter gene (c). (Photograph courtesy of Dr. Edward Wright, University of Westminster, London, U.K.)

directs nascent protein to insert into membranes where the targeting sequence is cleaved at a conserved cleavage/proteolytic site, allowing the mature glycoprotein to form trimeric structures on the membrane surface of the infected cell (Gaudin et al. 1992). Serology for lyssaviruses, especially RABV, is usually performed either by the rapid fluorescent focus inhibition test or the fluorescent antibody virus neutralization test (Louie et al. 1975; Cliquet et al. 1998). Both tests require live RABV and must be conducted under strict biological containment. Pseudotyped lentiviral particles have proven a robust, low containment alternative for measuring rabies virus neutralizing titers (Wright et al. 2008). This has been extended to the measurement of serological responses to infection with European bat lyssaviruses and a number of rare lyssaviruses found in Africa (Wright et al. 2008, 2009).

This technology has also been used to characterize the neutralizing properties of monoclonal antibodies being considered for use as immunotherapy for postexposure treatment of rabies. A panel of 10 pseudoviruses, including 3 divergent RABV strains and 6 other lyssavirus species, has been used to quantify the neutralizing properties of a recombinant monoclonal antibody 62-71-3 against a diverse range of lyssaviruses (Both et al. 2013). With supplies of RABV-specific immunoglobulin decreasing, it is hoped that alternatives such as recombinant monoclonal

antibodies may provide a more reliable alternative to current treatment options.

9.6 CONCLUSIONS

The emergence and reemergence of viruses is multifactorial and includes the properties of the virus itself, such as host receptor preference, changes in the vectors that transmit the disease, and the reservoirs that maintain the virus. Beyond this there are environmental factors that could promote virus transmission and probably the most important factors related to human behavior and activities, including the management of domestic animal populations. Disease surveillance offers an approach to the early detection of new and emerging diseases but must be accompanied by the establishment of platforms for rapid diagnostic and vaccine development. VLPs allow not only the generation of noninfectious tools for basic research studies on emerging viruses but also the generation of diagnostic tools and potential vaccines. For several emerging viruses autologous, chimeric or mosaic VLPs have been generated, allowing characterization of the viruses, generation of virus-specific monoclonal antibodies, or development of virus neutralization assays at low biosafety conditions.

A further point to consider is the acceptance of VLPs for clinical use. To some extent, the successful development of vaccines such as that for papilloma viruses has overcome such concerns and the inability of niVLPs to replicate provides a level of assurance for the potential outcomes of their application. Cost can also be an issue in the development and production of new biologicals using this technology. Despite these issues, future investigations on emerging viruses will profit from the broad spectrum of VLP-based tools that can be applied for rapid virus characterization and development of diagnostic tools.

The emergence of novel pathogens requires interdisciplinary and international approaches often described as a One Health approach that crosses barriers between human and veterinary medicine. Besides international surveillance for EID events and monitoring wildlife populations, preparation for an EID event is urgently needed and the application of VLP technology forms part of this response.

ACKNOWLEDGMENTS

The investigations in the laboratory of RGU are supported by the German Center for Infection Research (DZIF). The research of AG and AZ was funded by the European Social Fund under the Global Grant measure (Grant No. VPI-3.1-SMM-07-K-02-039). NJ is supported by EU FP7 project Anticipating The Global Onset of Novel Epidemics (ANTIGONE) project number 278076. The research of AK was supported by ERDF grant 2DP/2.1.1.1.0/14/APIA/VIAA/013. The authors kindly acknowledge Larissa Kolesnikova, Institute of Virology, Philipps-University Marburg, Germany, and Edward Wright, University of Westminster, United Kingdom, for providing Figures 9.4 and 9.7, respectively. RGU would like to thank Paul Pumpens for a long-lasting and productive collaboration.

REFERENCES

Acuna, R., N. Cifuentes-Munoz, C. Marquez, M. Bulling, J. Klingstrom, R. Mancini, P. Y. Lozach, and N. D. Tischler. 2013. Hantavirus Gn and Gc glycoproteins self-assemble into virus-like particles. *J Virol* 88 (4):2344–2348. doi: 10.1128/JVI.03118-13.

Adams, S. E., K. M. Dawson, K. Gull, S. M. Kingsman, and A. J. Kingsman. 1987. The expression of hybrid HIV: Ty virus-like particles in yeast. *Nature* 329 (6134):68–70. doi: 10.1038/329068a0.

Akahata, W., Z. Y. Yang, H. Andersen, S. Sun, H. A. Holdaway, W. P. Kong, M. G. Lewis, S. Higgs, M. G. Rossmann, S. Rao et al. 2010. A virus-like particle vaccine for epidemic Chikungunya virus protects nonhuman primates against infection. *Nat Med* 16 (3):334–338. doi: 10.1038/nm.2105.

Alberini, I., E. Del Tordello, A. Fasolo, N. J. Temperton, G. Galli, C. Gentile, E. Montomoli, A. K. Hilbert, A. Banzhoff, G. Del Giudice et al. 2009. Pseudoparticle neutralization is a reliable assay to measure immunity and cross-reactivity to H5N1 influenza viruses. *Vaccine* 27 (43):5998–6003. doi: 10.1016/j.vaccine.2009.07.079.

Aleksandrowicz, P., A. Marzi, N. Biedenkopf, N. Beimforde, S. Becker, T. Hoenen, H. Feldmann, and H. J. Schnittler. 2011. Ebola virus enters host cells by macropinocytosis and clathrin-mediated endocytosis. *J Infect Dis* 204 (Suppl. 3):S957–S967. doi: 10.1093/infdis/jir326.

Anggraeni, M. R., N. K. Connors, Y. Wu, Y. P. Chuan, L. H. Lua, and A. P. Middelberg. 2013. Sensitivity of immune response quality to influenza helix 190 antigen structure displayed on a modular virus-like particle. *Vaccine* 31 (40):4428–4435. doi: 10.1016/j.vaccine.2013.06.087.

Anilionis, A., W. H. Wunner, and P. J. Curtis. 1982. Amino acid sequence of the rabies virus glycoprotein deduced from its cloned gene. *Comp Immunol Microbiol Infect Dis* 5 (1–3):27–32.

Ao, Z., A. Patel, K. Tran, X. He, K. Fowke, K. Coombs, D. Kobasa, G. Kobinger, and X. Yao. 2008. Characterization of a trypsin-dependent avian influenza H5N1-pseudotyped HIV vector system for high throughput screening of inhibitory molecules. *Antiviral Res* 79 (1):12–18. doi: 10.1016/j.antiviral.2008.02.001.

Banyard, A. C., D. Hayman, N. Johnson, L. McElhinney, and A. R. Fooks. 2011. Bats and lyssaviruses. *Adv Virus Res* 79:239–289. doi: 10.1016/B978-0-12-387040-7.00012-3.

Baumert, T. F., S. Ito, D. T. Wong, and T. J. Liang. 1998. Hepatitis C virus structural proteins assemble into viruslike particles in insect cells. *J Virol* 72 (5):3827–3836.

Beesley, K. M., M. J. Francis, B. E. Clarke, J. E. Beesley, P. J. Dopping-Hepenstal, J. J. Clare, F. Brown, and M. A. Romanos. 1990. Expression in yeast of amino-terminal peptide fusions to hepatitis B core antigen and their immunological properties. *Biotechnology (N Y)* 8 (7):644–649.

Belyaev, A. S. and P. Roy. 1993. Development of baculovirus triple and quadruple expression vectors: Co-expression of three or four bluetongue virus proteins and the synthesis of bluetongue virus-like particles in insect cells. *Nucleic Acids Res* 21 (5):1219–1223.

Betenbaugh, M., M. Yu, K. Kuehl, J. White, D. Pennock, K. Spik, and C. Schmaljohn. 1995. Nucleocapsid- and virus-like particles assemble in cells infected with recombinant baculoviruses or vaccinia viruses expressing the M and the S segments of Hantaan virus. *Virus Res* 38 (2–3):111–124.

BGRCI (Berufsgenossenschaft Rohstoffe und Chemische Industrie). 2011. Viren: Einstufung biologischer Arbeitsstoffe. Merkblatt B 004.

Bhat, S. A., P. Saravanan, M. Hosamani, S. H. Basagoudanavar, B. P. Sreenivasa, R. P. Tamilselvan, and R. Venkataramanan. 2013. Novel immunogenic baculovirus expressed virus-like particles of foot-and-mouth disease (FMD) virus protect guinea pigs against challenge. *Res Vet Sci* 95 (3):1217–1223. doi: 10.1016/j.rvsc.2013.07.007.

Bisht, H., D. A. Chugh, S. Swaminathan, and N. Khanna. 2001. Expression and purification of Dengue virus type 2 envelope protein as a fusion with hepatitis B surface antigen in *Pichia pastoris*. *Protein Expr Purif* 23 (1):84–96. doi: 10.1006/prep.2001.1474.

Bodewes, R., J. van der Giessen, B. L. Haagmans, A. D. Osterhaus, and S. L. Smits. 2013. Identification of multiple novel viruses, including a parvovirus and a hepevirus, in feces of red foxes. *J Virol* 87 (13):7758–7764. doi: 10.1128/JVI.00568-13.

Borisova, G. P., I. Berzins, P. M. Pushko, P. Pumpen, E. J. Gren, V. V. Tsibinogin, V. Loseva, V. Ose, R. Ulrich, H. Siakkou et al. 1989. Recombinant core particles of hepatitis B virus exposing foreign antigenic determinants on their surface. *FEBS Lett* 259 (1):121–124.

Both, L., C. van Dolleweerd, E. Wright, A. C. Banyard, B. Bulmer-Thomas, D. Selden, F. Altmann, A. R. Fooks, and J. K. Ma. 2013. Production, characterization, and antigen specificity of recombinant 62-71-3, a candidate monoclonal antibody for rabies prophylaxis in humans. *FASEB J* 27 (5):2055–2065. doi: 10.1096/fj.12-219964.

Branco, L. M., J. N. Grove, F. J. Geske, M. L. Boisen, I. J. Muncy, S. A. Magliato, L. A. Henderson, R. J. Schoepp, K. A. Cashman, L. E. Hensley et al. 2010. Lassa virus-like particles displaying all major immunological determinants as a vaccine candidate for Lassa hemorrhagic fever. *Virol J* 7:279. doi: 10.1186/1743-422X-7-279.

Brauburger, K., A. J. Hume, E. Muhlberger, and J. Olejnik. 2012. Forty-five years of Marburg virus research. *Viruses* 4 (10):1878–1927. doi: 10.3390/v4101878.

Buonaguro, L., F. M. Buonaguro, M. L. Tornesello, D. Mantas, E. Beth-Giraldo, R. Wagner, S. Michelson, M. C. Prevost, H. Wolf, G. Giraldo et al. 2001. High efficient production of Pr55(gag) virus-like particles expressing multiple HIV-1 epitopes, including a gp120 protein derived from an Ugandan HIV-1 isolate of subtype A. *Antiviral Res* 49 (1):35–47.

Buonaguro, L., M. Tagliamonte, and M. L. Visciano. 2013. Chemokine receptor interactions with virus-like particles. *Methods Mol Biol* 1013:57–66. doi: 10.1007/978-1-62703-426-5_5.

Burns, N. R., J. E. Gilmour, S. M. Kingsman, A. J. Kingsman, and S. E. Adams. 1994. Production and purification of hybrid Ty-VLPs. *Mol Biotechnol* 1 (2):137–145.

Cao, Y., P. Sun, Y. Fu, X. Bai, F. Tian, X. Liu, Z. Lu, and Z. Liu. 2010. Formation of virus-like particles from O-type foot-and-mouth disease virus in insect cells using codon-optimized synthetic genes. *Biotechnol Lett* 32 (9):1223–1229. doi: 10.1007/s10529-010-0295-8.

Cen, S., M. Niu, J. Saadatmand, F. Guo, Y. Huang, G. J. Nabel, and L. Kleiman. 2004. Incorporation of pol into human immunodeficiency virus type 1 Gag virus-like particles occurs independently of the upstream Gag domain in Gag-pol. *J Virol* 78 (2):1042–1049.

Chambers, M. A., G. Dougan, J. Newman, F. Brown, J. Crowther, A. P. Mould, M. J. Humphries, M. J. Francis, B. Clarke, A. L. Brown et al. 1996. Chimeric hepatitis B virus core particles as probes for studying peptide-integrin interactions. *J Virol* 70 (6):4045–4052.

Chen, B. J., G. P. Leser, E. Morita, and R. A. Lamb. 2007. Influenza virus hemagglutinin and neuraminidase, but not the matrix protein, are required for assembly and budding of plasmid-derived virus-like particles. *J Virol* 81 (13):7111–7123. doi: 10.1128/JVI.00361-07.

Cheresiz, S. V., A. A. Kononova, Y. V. Razumova, T. S. Dubich, A. A. Chepurnov, A. A. Kushch, R. Davey, and A. G. Pokrovsky. 2014. A vesicular stomatitis pseudovirus expressing the surface glycoproteins of influenza A virus. *Arch Virol* 159 (10):2651–2658. doi: 10.1007/s00705-014-2127-y.

Cielens, I., L. Jackevica, A. Strods, A. Kazaks, V. Ose, J. Bogans, P. Pumpens, and R. Renhofa. 2014. Mosaic RNA phage VLPs carrying domain III of the West Nile virus E protein. *Mol Biotechnol* 56 (5):459–469. doi: 10.1007/s12033-014-9743-3.

Clarke, B. E., S. E. Newton, A. R. Carroll, M. J. Francis, G. Appleyard, A. D. Syred, P. E. Highfield, D. J. Rowlands, and F. Brown. 1987. Improved immunogenicity of a peptide epitope after fusion to hepatitis B core protein. *Nature* 330 (6146):381–384. doi: 10.1038/330381a0.

Cliquet, F., M. Aubert, and L. Sagne. 1998. Development of a fluorescent antibody virus neutralisation test (FAVN test) for the quantitation of rabies-neutralising antibody. *J Immunol Methods* 212 (1):79–87.

Crisci, E., L. Fraile, N. Moreno, E. Blanco, R. Cabezon, C. Costa, T. Mussa, M. Baratelli, P. Martinez-Orellana, L. Ganges et al. 2012. Chimeric calicivirus-like particles elicit specific immune responses in pigs. *Vaccine* 30 (14):2427–2439. doi: 10.1016/j.vaccine.2012.01.069.

D'Aoust, M. A., P. O. Lavoie, M. M. Couture, S. Trepanier, J. M. Guay, M. Dargis, S. Mongrand, N. Landry, B. J. Ward, and L. P. Vezina. 2008. Influenza virus-like particles produced by transient expression in *Nicotiana benthamiana* induce a protective immune response against a lethal viral challenge in mice. *Plant Biotechnol J* 6 (9):930–940. doi: 10.1111/j.1467-7652.2008.00384.x.

De Filette, M., W. Min Jou, A. Birkett, K. Lyons, B. Schultz, A. Tonkyro, S. Resch, and W. Fiers. 2005. Universal influenza A vaccine: Optimization of M2-based constructs. *Virology* 337 (1):149–161. doi: 10.1016/j.virol.2005.04.004.

Delchambre, M., D. Gheysen, D. Thines, C. Thiriart, E. Jacobs, E. Verdin, M. Horth, A. Burny, and F. Bex. 1989. The GAG precursor of simian immunodeficiency virus assembles into virus-like particles. *EMBO J* 8 (9):2653–2660.

Denis, J., N. Majeau, E. Acosta-Ramirez, C. Savard, M. C. Bedard, S. Simard, K. Lecours, M. Bolduc, C. Pare, B. Willems et al. 2007. Immunogenicity of papaya mosaic virus-like particles fused to a hepatitis C virus epitope: Evidence for the critical function of multimerization. *Virology* 363 (1):59–68. doi: 10.1016/j.virol.2007.01.011.

Diederich, S., L. Thiel, and A. Maisner. 2008. Role of endocytosis and cathepsin-mediated activation in Nipah virus entry. *Virology* 375 (2):391–400. doi: 10.1016/j.virol.2008.02.019.

Drexler, J. F., V. M. Corman, M. A. Muller, A. N. Lukashev, A. Gmyl, B. Coutard, A. Adam, D. Ritz, L. M. Leijten, D. van Riel et al. 2013a. Evidence for novel hepaciviruses in rodents. *PLoS Pathog* 9 (6):e1003438. doi: 10.1371/journal.ppat.1003438.

Drexler, J. F., A. Geipel, A. Konig, V. M. Corman, D. van Riel, L. M. Leijten, C. M. Bremer, A. Rasche, V. M. Cottontail, G. D. Maganga et al. 2013b. Bats carry pathogenic hepadnaviruses antigenically related to hepatitis B virus and capable of infecting human hepatocytes. *Proc Natl Acad Sci USA* 110 (40):16151–16156. doi: 10.1073/pnas.1308049110.

Ehlers, B. and U. Wieland. 2013. The novel human polyomaviruses HPyV6, 7, 9 and beyond. *APMIS* 121 (8):783–795. doi: 10.1111/apm.12104.

Fontana, D., R. Kratje, M. Etcheverrigaray, and C. Prieto. 2014. Rabies virus-like particles expressed in HEK293 cells. *Vaccine* 32 (24):2799–2804. doi: 10.1016/j.vaccine.2014.02.031.

Freivalds, J., A. Dislers, V. Ose, D. Skrastina, I. Cielens, P. Pumpens, K. Sasnauskas, and A. Kazaks. 2006. Assembly of bacteriophage Qbeta virus-like particles in yeast *Saccharomyces cerevisiae* and *Pichia pastoris*. *J Biotechnol* 123 (3):297–303. doi: 10.1016/j.jbiotec.2005.11.013.

Freivalds, J., S. Kotelovica, T. Voronkova, V. Ose, K. Tars, and A. Kazaks. 2014. Yeast-expressed bacteriophage-like particles for the packaging of nanomaterials. *Mol Biotechnol* 56 (2):102–110. doi: 10.1007/s12033-013-9686-0.

French, T. J., J. J. Marshall, and P. Roy. 1990. Assembly of double-shelled, viruslike particles of bluetongue virus by the simultaneous expression of four structural proteins. *J Virol* 64 (12):5695–5700.

French, T. J. and P. Roy. 1990. Synthesis of bluetongue virus (BTV) corelike particles by a recombinant baculovirus expressing the two major structural core proteins of BTV. *J Virol* 64 (4):1530–1536.

Fu, T. M., K. M. Grimm, M. P. Citron, D. C. Freed, J. Fan, P. M. Keller, J. W. Shiver, X. Liang, and J. G. Joyce. 2009. Comparative immunogenicity evaluations of influenza A virus M2 peptide as recombinant virus like particle or conjugate vaccines in mice and monkeys. *Vaccine* 27 (9):1440–1447. doi: 10.1016/j.vaccine.2008.12.034.

Galarza, J. M., T. Latham, and A. Cupo. 2005. Virus-like particle (VLP) vaccine conferred complete protection against a lethal influenza virus challenge. *Viral Immunol* 18 (1):244–251. doi: 10.1089/vim.2005.18.244.

Gao, X., W. Wang, Y. Li, S. Zhang, Y. Duan, L. Xing, Z. Zhao, P. Zhang, Z. Li, R. Li et al. 2013. Enhanced influenza VLP vaccines comprising matrix-2 ectodomain and nucleoprotein epitopes protects mice from lethal challenge. *Antiviral Res* 98 (1):4–11. doi: 10.1016/j.antiviral.2013.01.010.

Gaudin, Y., R. W. Ruigrok, C. Tuffereau, M. Knossow, and A. Flamand. 1992. Rabies virus glycoprotein is a trimer. *Virology* 187 (2):627–632.

Gedvilaite, A., A. Zvirbliene, J. Staniulis, K. Sasnauskas, D. H. Kruger, and R. Ulrich. 2004. Segments of Puumala hantavirus nucleocapsid protein inserted into chimeric polyomavirus-derived virus-like particles induce a strong immune response in mice. *Viral Immunol* 17 (1):51–68. doi: 10.1089/088282404322875458.

Geisbert, T. W. and H. Feldmann. 2011. Recombinant vesicular stomatitis virus-based vaccines against Ebola and Marburg virus infections. *J Infect Dis* 204 (Suppl. 3):S1075–S1081. doi: 10.1093/infdis/jir349.

Geldmacher, A., D. Skrastina, G. Borisova, I. Petrovskis, D. H. Kruger, P. Pumpens, and R. Ulrich. 2005. A hantavirus nucleocapsid protein segment exposed on hepatitis B virus core particles is highly immunogenic in mice when applied without adjuvants or in the presence of pre-existing anti-core antibodies. *Vaccine* 23 (30):3973–3983. doi: 10.1016/j.vaccine.2005.02.025.

Geldmacher, A., D. Skrastina, I. Petrovskis, G. Borisova, J. A. Berriman, A. M. Roseman, R. A. Crowther, J. Fischer, S. Musema, H. R. Gelderblom et al. 2004. An amino-terminal segment of hantavirus nucleocapsid protein presented on hepatitis B virus core particles induces a strong and highly cross-reactive antibody response in mice. *Virology* 323 (1):108–119. doi: 10.1016/j.virol.2004.02.022.

Gheysen, D., E. Jacobs, F. de Foresta, C. Thiriart, M. Francotte, D. Thines, and M. De Wilde. 1989. Assembly and release of HIV-1 precursor Pr55gag virus-like particles from recombinant baculovirus-infected insect cells. *Cell* 59 (1):103–112.

Gomez-Puertas, P., C. Albo, E. Perez-Pastrana, A. Vivo, and A. Portela. 2000. Influenza virus matrix protein is the major driving force in virus budding. *J Virol* 74 (24):11538–11547.

Griffiths, J. C., E. L. Berrie, L. N. Holdsworth, J. P. Moore, S. J. Harris, J. M. Senior, S. M. Kingsman, A. J. Kingsman, and S. E. Adams. 1991. Induction of high-titer neutralizing antibodies, using hybrid human immunodeficiency virus V3-Ty viruslike particles in a clinically relevant adjuvant. *J Virol* 65 (1):450–456.

Gunther, S., H. Feldmann, T. W. Geisbert, L. E. Hensley, P. E. Rollin, S. T. Nichol, U. Stroher, H. Artsob, C. J. Peters, T. G. Ksiazek et al. 2011. Management of accidental exposure to Ebola virus in the biosafety level 4 laboratory, Hamburg, Germany. *J Infect Dis* 204 (Suppl. 3):S785–S790. doi: 10.1093/infdis/jir298.

Habjan, M., N. Penski, V. Wagner, M. Spiegel, A. K. Overby, G. Kochs, J. T. Huiskonen, and F. Weber. 2009. Efficient production of Rift Valley fever virus-like particles: The antiviral protein MxA can inhibit primary transcription of bunyaviruses. *Virology* 385 (2):400–408. doi: 10.1016/j.virol.2008.12.011.

Haynes, J. R. 2009. Influenza virus-like particle vaccines. *Expert Rev Vaccines* 8 (4):435–445. doi: 10.1586/erv.09.8.

Hepojoki, J., T. Strandin, H. Wang, O. Vapalahti, A. Vaheri, and H. Lankinen. 2010. Cytoplasmic tails of hantavirus glycoproteins interact with the nucleocapsid protein. *J Gen Virol* 91 (Pt. 9):2341–2350. doi: 10.1099/vir.0.021006–0.

Heyman, P., C. S. Ceianu, I. Christova, N. Tordo, M. Beersma, M. Joao Alves, A. Lundkvist, M. Hukic, A. Papa, A. Tenorio et al. 2011. A five-year perspective on the situation of haemorrhagic fever with renal syndrome and status of the hantavirus reservoirs in Europe, 2005–2010. *Euro Surveill* 16 (36).

Higa, M. M., J. Petersen, J. Hooper, and R. W. Doms. 2012. Efficient production of Hantaan and Puumala pseudovirions for viral tropism and neutralization studies. *Virology* 423 (2):134–142. doi: 10.1016/j.virol.2011.08.012.

Ho, Y., P. H. Lin, C. Y. Liu, S. P. Lee, and Y. C. Chao. 2004. Assembly of human severe acute respiratory syndrome coronavirus-like particles. *Biochem Biophys Res Commun* 318 (4):833–838. doi: 10.1016/j.bbrc.2004.04.111.

Hoenen, T., A. Groseth, F. de Kok-Mercado, J. H. Kuhn, and V. Wahl-Jensen. 2011. Minigenomes, transcription and replication competent virus-like particles and beyond: Reverse genetics systems for filoviruses and other negative stranded hemorrhagic fever viruses. *Antiviral Res* 91 (2):195–208. doi: 10.1016/j.antiviral.2011.06.003.

Hoenen, T., A. Groseth, L. Kolesnikova, S. Theriault, H. Ebihara, B. Hartlieb, S. Bamberg, H. Feldmann, U. Stroher, and S. Becker. 2006. Infection of naive target cells with virus-like particles: Implications for the function of Ebola virus VP24. *J Virol* 80 (14):7260–7264. doi: 10.1128/JVI.00051-06.

Hoffmann, B., M. Scheuch, D. Hoper, R. Jungblut, M. Holsteg, H. Schirrmeier, M. Eschbaumer, K. V. Goller, K. Wernike, M. Fischer et al. 2012. Novel orthobunyavirus in cattle, Europe, 2011. *Emerg Infect Dis* 18 (3):469–472. doi: 10.3201/eid1803.111905.

Holmes, D. A., D. E. Purdy, D. Y. Chao, A. J. Noga, and G. J. Chang. 2005. Comparative analysis of immunoglobulin M (IgM) capture enzyme-linked immunosorbent assay using virus-like particles or virus-infected mouse brain antigens to detect IgM antibody in sera from patients with evident flaviviral infections. *J Clin Microbiol* 43 (7):3227–3236. doi: 10.1128/JCM.43.7.3227-3236.2005.

Hsieh, P. K., S. C. Chang, C. C. Huang, T. T. Lee, C. W. Hsiao, Y. H. Kou, I. Y. Chen, C. K. Chang, T. H. Huang, and M. F. Chang. 2005. Assembly of severe acute respiratory syndrome coronavirus RNA packaging signal into virus-like particles is nucleocapsid dependent. *J Virol* 79 (22):13848–13855. doi: 10.1128/JVI.79.22.13848-13855.2005.

Hu, H. P., S. C. Hsieh, C. C. King, and W. K. Wang. 2007. Characterization of retrovirus-based reporter viruses pseudotyped with the precursor membrane and envelope glycoproteins of four serotypes of dengue viruses. *Virology* 368 (2):376–387. doi: 10.1016/j.virol.2007.06.026.

Hua, R. H., Y. N. Li, Z. S. Chen, L. K. Liu, H. Huo, X. L. Wang, L. P. Guo, N. Shen, J. F. Wang, and Z. G. Bu. 2014. Generation and characterization of a new mammalian cell line continuously expressing virus-like particles of Japanese encephalitis virus for a subunit vaccine candidate. *BMC Biotechnol* 14 (1):62. doi: 10.1186/1472-6750-14-62.

Huang, Y., W. Liang, Y. Wang, Z. Zhou, A. Pan, X. Yang, C. Huang, J. Chen, and D. Zhang. 2005. Immunogenicity of the epitope of the foot-and-mouth disease virus fused with a hepatitis B core protein as expressed in transgenic tobacco. *Viral Immunol* 18 (4):668–677. doi: 10.1089/vim.2005.18.668.

Huang, Y., Z. Y. Yang, W. P. Kong, and G. J. Nabel. 2004. Generation of synthetic severe acute respiratory syndrome coronavirus pseudoparticles: Implications for assembly and vaccine production. *J Virol* 78 (22):12557–12565. doi: 10.1128/JVI.78.22.12557-12565.2004.

Ionescu, R. M., C. T. Przysiecki, X. Liang, V. M. Garsky, J. Fan, B. Wang, R. Troutman, Y. Rippeon, E. Flanagan, J. Shiver et al. 2006. Pharmaceutical and immunological evaluation of human papillomavirus viruslike particle as an antigen carrier. *J Pharm Sci* 95 (1):70–79. doi: 10.1002/jps.20493.

Isaguliants, M. G., P. Kadoshnikov Iu, T. I. Kalinina, E. Khuliakov Iu, A. Semiletov Iu, V. D. Smirnov, and B. Wahren. 1996. Expression of HIV-1 epitopes included in particles formed by human hepatitis B virus nucleocapsid protein. *Biokhimiia* 61 (3):532–545.

Jegerlehner, A., A. Tissot, F. Lechner, P. Sebbel, I. Erdmann, T. Kundig, T. Bachi, T. Storni, G. Jennings, P. Pumpens et al. 2002. A molecular assembly system that renders antigens of choice highly repetitive for induction of protective B cell responses. *Vaccine* 20 (25–26):3104–3112.

Jegerlehner, A., F. Zabel, A. Langer, K. Dietmeier, G. T. Jennings, P. Saudan, and M. F. Bachmann. 2013. Bacterially produced recombinant influenza vaccines based on virus-like particles. *PLoS ONE* 8 (11):e78947. doi: 10.1371/journal.pone.0078947.

Johne, R., A. Plenge-Bonig, M. Hess, R. G. Ulrich, J. Reetz, and A. Schielke. 2010. Detection of a novel hepatitis E-like virus in faeces of wild rats using a nested broad-spectrum RT-PCR. *J Gen Virol* 91 (Pt. 3):750–758. doi: 10.1099/vir.0.016584-0.

Johnson, N. (ed.). 2014. *The Role of Animals in Emerging Viral Diseases*. San Diego, CA: Academic Press.

Johnson, N., A. Vos, C. Freuling, N. Tordo, A. R. Fooks, and T. Muller. 2010. Human rabies due to lyssavirus infection of bat origin. *Vet Microbiol* 142 (3–4):151–159. doi: 10.1016/j.vetmic.2010.02.001.

Jones, K. E., N. G. Patel, M. A. Levy, A. Storeygard, D. Balk, J. L. Gittleman, and P. Daszak. 2008. Global trends in emerging infectious diseases. *Nature* 451 (7181):990–993. doi: 10.1038/nature06536.

Juozapaitis, M., A. Serva, I. Kucinskaite, A. Zvirbliene, R. Slibinskas, J. Staniulis, K. Sasnauskas, B. J. Shiell, T. R. Bowden, and W. P. Michalski. 2007a. Generation of menangle virus nucleocapsid-like particles in yeast *Saccharomyces cerevisiae*. *J Biotechnol* 130 (4):441–447. doi: 10.1016/j.jbiotec.2007.05.013.

Juozapaitis, M., A. Serva, A. Zvirbliene, R. Slibinskas, J. Staniulis, K. Sasnauskas, B. J. Shiell, L. F. Wang, and W. P. Michalski. 2007b. Generation of henipavirus nucleocapsid proteins in yeast *Saccharomyces cerevisiae*. *Virus Res* 124 (1–2):95–102. doi: 10.1016/j.virusres.2006.10.008.

Kattenbeck, B., A. Rohrhofer, M. Niedrig, H. Wolf, and S. Modrow. 1996. Defined amino acids in the gag proteins of human immunodeficiency virus type 1 are functionally active during virus assembly. *Intervirology* 39 (1–2):32–39.

Kazaks, A., S. Lachmann, D. Koletzki, I. Petrovskis, A. Dislers, V. Ose, D. Skrastina, H. R. Gelderblom, A. Lundkvist, H. Meisel et al. 2002. Stop codon insertion restores the particle formation ability of hepatitis B virus core-hantavirus nucleocapsid protein fusions. *Intervirology* 45 (4–6):340–349.

Kazaks, A. and Voronkova, T. 2008. Papillomavirus-derived virus-like particles. In *Medicinal Protein Engineering*, Y. Khudyakov (ed.), pp. 277–298. Boca Raton FL: Taylor & Francis.

King, A. M. Q., M. J. Adams, E. B. Carstens, and E. J. Lefkowitz (eds.). 2011. *Virus Taxonomy: Classification and Nomenclature of Viruses; Ninth Report of the International Committee on Taxonomy of Viruses*. London, U.K.: Elsevier/Academic Press.

King, B. and J. Daly. 2014. Pseudotypes: Your flexible friends. *Future Microbiol* 9 (2):135–137. doi: 10.2217/fmb.13.156.

Koletzki, D., S. S. Biel, H. Meisel, E. Nugel, H. R. Gelderblom, D. H. Kruger, and R. Ulrich. 1999. HBV core particles allow the insertion and surface exposure of the entire potentially protective region of Puumala hantavirus nucleocapsid protein. *Biol Chem* 380 (3):325–333. doi: 10.1515/BC.1999.044.

Koletzki, D., A. Lundkvist, K. B. Sjolander, H. R. Gelderblom, M. Niedrig, H. Meisel, D. H. Kruger, and R. Ulrich. 2000. Puumala (PUU) hantavirus strain differences and insertion positions in the hepatitis B virus core antigen influence B-cell immunogenicity and protective potential of core-derived particles. *Virology* 276 (2):364–375. doi: 10.1006/viro.2000.0540.

Koletzki, D., A. Zankl, H. R. Gelderblom, H. Meisel, A. Dislers, G. Borisova, P. Pumpens, D. H. Kruger, and R. Ulrich. 1997. Mosaic hepatitis B virus core particles allow insertion of extended foreign protein segments. *J Gen Virol* 78 (Pt 8):2049–2053.

Konishi, E. and A. Fujii. 2002. Dengue type 2 virus subviral extracellular particles produced by a stably transfected mammalian cell line and their evaluation for a subunit vaccine. *Vaccine* 20 (7–8):1058–1067.

Krahling, V., O. Dolnik, L. Kolesnikova, J. Schmidt-Chanasit, I. Jordan, V. Sandig, S. Gunther, and S. Becker. 2010. Establishment of fruit bat cells (*Rousettus aegyptiacus*) as a model system for the investigation of filoviral infection. *PLoS Negl Trop Dis* 4 (8):e802. doi: 10.1371/journal.pntd.0000802.

Kucinskaite-Kodze, I., R. Petraityte-Burneikiene, A. Zvirbliene, B. Hjelle, R. A. Medina, A. Gedvilaite, A. Razanskiene, J. Schmidt-Chanasit, M. Mertens, P. Padula et al. 2011. Characterization of monoclonal antibodies against hantavirus nucleocapsid protein and their use for immunohistochemistry on rodent and human samples. *Arch Virol* 156 (3):443–456. doi: 10.1007/s00705-010-0879-6.

Kuwahara, M. and E. Konishi. 2010. Evaluation of extracellular subviral particles of dengue virus type 2 and Japanese encephalitis virus produced by *Spodoptera frugiperda* cells for use as vaccine and diagnostic antigens. *Clin Vaccine Immunol* 17 (10):1560–1566. doi: 10.1128/CVI.00087–10.

Lai, J. C., W. W. Chan, F. Kien, J. M. Nicholls, J. S. Peiris, and J. M. Garcia. 2010. Formation of virus-like particles from human cell lines exclusively expressing influenza neuraminidase. *J Gen Virol* 91 (Pt. 9):2322–2330. doi: 10.1099/vir.0.019935-0.

Lasickiene, R., A. Gedvilaite, M. Norkiene, V. Simanaviciene, I. Sezaite, D. Dekaminaviciute, E. Shikova, and A. Zvirbliene. 2012. The use of recombinant pseudotype virus-like particles harbouring inserted target antigen to generate antibodies against cellular marker p16INK4A. *Sci World J* 2012:1–8. doi: 10.1100/2012/263737.

Layton, G. T., S. J. Harris, A. J. Gearing, M. Hill-Perkins, J. S. Cole, J. C. Griffiths, N. R. Burns, A. J. Kingsman, and S. E. Adams. 1993. Induction of HIV-specific cytotoxic T lymphocytes in vivo with hybrid HIV-1 V3: Ty-virus-like particles. *J Immunol* 151 (2):1097–1107.

Lee, C. D., Y. P. Yan, S. M. Liang, and T. F. Wang. 2009. Production of FMDV virus-like particles by a SUMO fusion protein approach in *Escherichia coli*. *J Biomed Sci* 16:69. doi: 10.1186/1423-0127-16-69.

Li, C., F. Liu, M. Liang, Q. Zhang, X. Wang, T. Wang, J. Li, and D. Li. 2010. Hantavirus-like particles generated in CHO cells induce specific immune responses in C57BL/6 mice. *Vaccine* 28 (26):4294–4300. doi: 10.1016/j.vaccine.2010.04.025.

Li, H. Z., H. Y. Gang, Q. M. Sun, X. Liu, Y. B. Ma, M. S. Sun, and C. B. Dai. 2004. Production in *Pichia pastoris* and characterization of genetic engineered chimeric HBV/HEV virus-like particles. *Chin Med Sci J* 19 (2):78–83.

Li, T., H. Lin, Y. Zhang, M. Li, D. Wang, Y. Che, Y. Zhu, S. Li, J. Zhang, S. Ge et al. 2014. Improved characteristics and protective efficacy in an animal model of *E. coli*-derived recombinant double-layered rotavirus virus-like particles. *Vaccine* 32 (17):1921–1931. doi: 10.1016/j.vaccine.2014.01.093.

Li, T. C., N. Takeda, T. Miyamura, Y. Matsuura, J. C. Wang, H. Engvall, L. Hammar, L. Xing, and R. H. Cheng. 2005. Essential elements of the capsid protein for self-assembly into empty virus-like particles of hepatitis E virus. *J Virol* 79 (20):12999–13006. doi: 10.1128/JVI.79.20.12999–13006.2005.

Li, T. C., Y. Yamakawa, K. Suzuki, M. Tatsumi, M. A. Razak, T. Uchida, N. Takeda, and T. Miyamura. 1997. Expression and self-assembly of empty virus-like particles of hepatitis E virus. *J Virol* 71 (10):7207–7213.

Li, T. C., K. Yoshimatsu, S. P. Yasuda, J. Arikawa, T. Koma, M. Kataoka, Y. Ami, Y. Suzaki, T. Q. Mai le, N. T. Hoa et al. 2011. Characterization of self-assembled virus-like particles of rat hepatitis E virus generated by recombinant baculoviruses. *J Gen Virol* 92 (Pt. 12):2830–2837. doi: 10.1099/vir.0.034835-0.

Licata, J. M., R. F. Johnson, Z. Han, and R. N. Harty. 2004. Contribution of Ebola virus glycoprotein, nucleoprotein, and VP24 to budding of VP40 virus-like particles. *J Virol* 78 (14):7344–7351. doi: 10.1128/JVI.78.14.7344-7351.2004.

Liu, L., C. C. Celma, and P. Roy. 2008. Rift Valley fever virus structural proteins: Expression, characterization and assembly of recombinant proteins. *Virol J* 5:82. doi: 10.1186/1743-422X-5-82.

Liu, W., H. Jiang, J. Zhou, X. Yang, Y. Tang, D. Fang, and L. Jiang. 2010. Recombinant dengue virus-like particles from *Pichia pastoris*: Efficient production and immunological properties. *Virus Genes* 40 (1):53–59. doi: 10.1007/s11262-009-0418-2.

Loktev, V. B., A. A. Ilyichev, A. M. Eroshkin, L. I. Karpenko, A. G. Pokrovsky, A. V. Pereboev, V. A. Svyatchenko, G. M. Ignat'ev, M. I. Smolina, N. V. Melamed et al. 1996. Design of immunogens as components of a new generation of molecular vaccines. *J Biotechnol* 44 (1–3):129–137. doi: 10.1016/0168-1656(95)00089-5.

Lopez, C., L. Gil, L. Lazo, I. Menendez, E. Marcos, J. Sanchez, I. Valdes, V. Falcon, M. C. de la Rosa, G. Marquez et al. 2009. In vitro assembly of nucleocapsid-like particles from purified recombinant capsid protein of dengue-2 virus. *Arch Virol* 154 (4):695–698. doi: 10.1007/s00705-009-0350-8.

Louie, R. E., M. B. Dobkin, P. Meyer, B. Chin, R. E. Roby, A. H. Hammar, and V. J. Cabasso. 1975. Measurement of rabies antibody: Comparison of the mouse neutralization test (MNT) with the rapid fluorescent focus inhibition test (RFFIT). *J Biol Stand* 3 (4):365–373.

Low, J. G., L. S. Lee, E. E. Ooi, K. Ethirajulu, P. Yeo, A. Matter, J. E. Connolly, D. A. Skibinski, P. Saudan, M. Bachmann et al. 2014. Safety and immunogenicity of a virus-like particle pandemic influenza A (H1N1) 2009 vaccine: Results from a double-blinded, randomized Phase I clinical trial in healthy Asian volunteers. *Vaccine* 32 (39):5041–5048. doi: 10.1016/j.vaccine.2014.07.011.

Lu, J., Z. Han, Y. Liu, W. Liu, M. S. Lee, M. A. Olson, G. Ruthel, B. D. Freedman, and R. N. Harty. 2014. A host-oriented inhibitor of Junin Argentine hemorrhagic fever virus egress. *J Virol* 88 (9):4736–4743. doi: 10.1128/JVI.03757-13.

Lu, X., Y. Chen, B. Bai, H. Hu, L. Tao, J. Yang, J. Chen, Z. Chen, Z. Hu, and H. Wang. 2007. Immune responses against severe acute respiratory syndrome coronavirus induced by virus-like particles in mice. *Immunology* 122 (4):496–502. doi: 10.1111/j.1365-2567.2007.02676.x.

Ludwig, C. and R. Wagner. 2007. Virus-like particles-universal molecular toolboxes. *Curr Opin Biotechnol* 18 (6):537–545. doi: 10.1016/j.copbio.2007.10.013.

Lundkvist, A., H. Meisel, D. Koletzki, H. Lankinen, F. Cifire, A. Geldmacher, C. Sibold, P. Gott, A. Vaheri, D. H. Kruger et al. 2002. Mapping of B-cell epitopes in the nucleocapsid protein of Puumala hantavirus. *Viral Immunol* 15 (1):177–192. doi: 10.1089/088282402317340323.

Mahmood, K., R. A. Bright, N. Mytle, D. M. Carter, C. J. Crevar, J. E. Achenbach, P. M. Heaton, T. M. Tumpey, and T. M. Ross. 2008. H5N1 VLP vaccine induced protection in ferrets against lethal challenge with highly pathogenic H5N1 influenza viruses. *Vaccine* 26 (42):5393–5399. doi: 10.1016/j.vaccine.2008.07.084.

Mandell, R. B., R. Koukuntla, L. J. Mogler, A. K. Carzoli, A. N. Freiberg, M. R. Holbrook, B. K. Martin, W. R. Staplin, N. N. Vahanian, C. J. Link et al. 2010a. A replication-incompetent Rift Valley fever vaccine: Chimeric virus-like particles protect mice and rats against lethal challenge. *Virology* 397 (1):187–198. doi: 10.1016/j.virol.2009.11.001.

Mandell, R. B., R. Koukuntla, L. J. Mogler, A. K. Carzoli, M. R. Holbrook, B. K. Martin, N. Vahanian, C. J. Link, and R. Flick. 2010b. Novel suspension cell-based vaccine production systems for Rift Valley fever virus-like particles. *J Virol Methods* 169 (2):259–268. doi: 10.1016/j.jviromet.2010.07.015.

Martinez, O., J. Johnson, B. Manicassamy, L. Rong, G. G. Olinger, L. E. Hensley, and C. F. Basler. 2010. Zaire Ebola virus entry into human dendritic cells is insensitive to cathepsin L inhibition. *Cell Microbiol* 12 (2):148–157. doi: 10.1111/j.1462-5822.2009.01385.x.

Martinez, O., E. Ndungo, L. Tantral, E. H. Miller, L. W. Leung, K. Chandran, and C. F. Basler. 2013. A mutation in the Ebola virus envelope glycoprotein restricts viral entry in a host species- and cell-type-specific manner. *J Virol* 87 (6):3324–3334. doi: 10.1128/JVI.01598-12.

Mena, I., A. Vivo, E. Perez, and A. Portela. 1996. Rescue of a synthetic chloramphenicol acetyltransferase RNA into influenza virus-like particles obtained from recombinant plasmids. *J Virol* 70 (8):5016–5024.

Metz, S. W., J. Gardner, C. Geertsema, T. T. Le, L. Goh, J. M. Vlak, A. Suhrbier, and G. P. Pijlman. 2013. Effective Chikungunya virus-like particle vaccine produced in insect cells. *PLoS Negl Trop Dis* 7 (3):e2124. doi: 10.1371/journal.pntd.0002124.

Michel, M. L., M. Mancini, Y. Riviere, D. Dormont, and P. Tiollais. 1990. T- and B-lymphocyte responses to human immunodeficiency virus (HIV) type 1 in macaques immunized with hybrid HIV/hepatitis B surface antigen particles. *J Virol* 64 (5):2452–2455.

Michel, M. L., M. Mancini, E. Sobczak, V. Favier, D. Guetard, E. M. Bahraoui, and P. Tiollais. 1988. Induction of anti-human immunodeficiency virus (HIV) neutralizing antibodies in rabbits immunized with recombinant HIV-hepatitis B surface antigen particles. *Proc Natl Acad Sci USA* 85 (21):7957–7961.

Mori, Y. and Y. Matsuura. 2011. Structure of hepatitis E viral particle. *Virus Res* 161 (1):59–64. doi: 10.1016/j.virusres.2011.03.015.

Muller, M. A., J. T. Paweska, P. A. Leman, C. Drosten, K. Grywna, A. Kemp, L. Braack, K. Sonnenberg, M. Niedrig, and R. Swanepoel. 2007. Coronavirus antibodies in African bat species. *Emerg Infect Dis* 13 (9):1367–1370. doi: 10.3201/eid1309.070342.

Neirynck, S., T. Deroo, X. Saelens, P. Vanlandschoot, W. M. Jou, and W. Fiers. 1999. A universal influenza A vaccine based on the extracellular domain of the M2 protein. *Nat Med* 5 (10):1157–1163. doi: 10.1038/13484.

Netter, H. J., T. B. Macnaughton, W. P. Woo, R. Tindle, and E. J. Gowans. 2001. Antigenicity and immunogenicity of novel chimeric hepatitis B surface antigen particles with exposed hepatitis C virus epitopes. *J Virol* 75 (5):2130–2141. doi: 10.1128/JVI.75.5.2130-2141.2001.

Neumann, G., T. Watanabe, and Y. Kawaoka. 2000. Plasmid-driven formation of influenza virus-like particles. *J Virol* 74 (1):547–551.

Notka, F., C. Stahl-Hennig, U. Dittmer, H. Wolf, and R. Wagner. 1999. Accelerated clearance of SHIV in rhesus monkeys by virus-like particle vaccines is dependent on induction of neutralizing antibodies. *Vaccine* 18 (3–4):291–301.

Ogino, M., H. Ebihara, B. H. Lee, K. Araki, A. Lundkvist, Y. Kawaoka, K. Yoshimatsu, and J. Arikawa. 2003. Use of vesicular stomatitis virus pseudotypes bearing Hantaan or Seoul virus envelope proteins in a rapid and safe neutralization test. *Clin Diagn Lab Immunol* 10 (1):154–160.

Ohtaki, N., H. Takahashi, K. Kaneko, Y. Gomi, T. Ishikawa, Y. Higashi, T. Kurata, T. Sata, and A. Kojima. 2010. Immunogenicity and efficacy of two types of West Nile virus-like particles different in size and maturation as a second-generation vaccine candidate. *Vaccine* 28 (40):6588–6596. doi: 10.1016/j.vaccine.2010.07.055.

Overby, A. K., R. F. Pettersson, and E. P. Neve. 2007a. The glycoprotein cytoplasmic tail of Uukuniemi virus (Bunyaviridae) interacts with ribonucleoproteins and is critical for genome packaging. *J Virol* 81 (7):3198–3205. doi: 10.1128/JVI.02655-06.

Overby, A. K., V. Popov, E. P. Neve, and R. F. Pettersson. 2006. Generation and analysis of infectious virus-like particles of Uukuniemi virus (Bunyaviridae): A useful system for studying bunyaviral packaging and budding. *J Virol* 80 (21):10428–10435. doi: 10.1128/JVI.01362–06.

Overby, A. K., V. L. Popov, R. F. Pettersson, and E. P. Neve. 2007b. The cytoplasmic tails of Uukuniemi Virus (Bunyaviridae) G(N) and G(C) glycoproteins are important for intracellular targeting and the budding of virus-like particles. *J Virol* 81 (20):11381–11391. doi: 10.1128/JVI.00767-07.

Patch, J. R., G. Crameri, L. F. Wang, B. T. Eaton, and C. C. Broder. 2007. Quantitative analysis of Nipah virus proteins released as virus-like particles reveals central role for the matrix protein. *Virol J* 4:1. doi: 10.1186/1743-422X-4-1.

Patient, R., C. Hourioux, P. Vaudin, J. C. Pages, and P. Roingeard. 2009. Chimeric hepatitis B and C viruses envelope proteins can form subviral particles: Implications for the design of new vaccine strategies. *N Biotechnol* 25 (4):226–234. doi: 10.1016/j.nbt.2009.01.001.

Petraityte, R., P. L. Tamosiunas, M. Juozapaitis, A. Zvirbliene, K. Sasnauskas, B. Shiell, G. Russell, J. Bingham, and W. P. Michalski. 2009. Generation of Tioman virus nucleocapsid-like particles in yeast *Saccharomyces cerevisiae*. *Virus Res* 145 (1):92–96. doi: 10.1016/j.virusres.2009.06.013.

Phan, T. G., B. Kapusinszky, C. Wang, R. K. Rose, H. L. Lipton, and E. L. Delwart. 2011. The fecal viral flora of wild rodents. *PLoS Pathog* 7 (9):e1002218. doi: 10.1371/journal.ppat.1002218.

Pleckaityte, M., A. Zvirbliene, I. Sezaite, and A. Gedvilaite. 2011. Production in yeast of pseudotype virus-like particles harboring functionally active antibody fragments neutralizing the cytolytic activity of vaginolysin. *Microb Cell Fact* 10:109. doi: 10.1186/1475-2859-10-109.

Pumpens, P., R. Ulrich, K. Sasnauskas, A. Kazaks, V. Ose, and E. Grens. 2008. Construction of novel vaccines on the basis of virus-like particles: Hepatitis B virus proteins as vaccine carriers. In *Medicinal Protein Engineering*, Y. Khudyakov (ed.), pp. 204–248. Boca Raton, FL: Taylor & Francis.

Purdy, D. E. and G. J. Chang. 2005. Secretion of noninfectious dengue virus-like particles and identification of amino acids in the stem region involved in intracellular retention of envelope protein. *Virology* 333 (2):239–250. doi: 10.1016/j.virol.2004.12.036.

Pushko, P., T. M. Tumpey, F. Bu, J. Knell, R. Robinson, and G. Smith. 2005. Influenza virus-like particles comprised of the HA, NA, and M1 proteins of H9N2 influenza virus induce protective immune responses in BALB/c mice. *Vaccine* 23 (50):5751–5759. doi: 10.1016/j.vaccine.2005.07.098.

Ramani, S., R. L. Atmar, and M. K. Estes. 2014. Epidemiology of human noroviruses and updates on vaccine development. *Curr Opin Gastroenterol* 30 (1):25–33.

Ray, N., J. Whidby, S. Stewart, J. W. Hooper, and A. Bertolotti-Ciarlet. 2010. Study of Andes virus entry and neutralization using a pseudovirion system. *J Virol Methods* 163 (2):416–423. doi: 10.1016/j.jviromet.2009.11.004.

Reusken, C. B., M. Ababneh, V. S. Raj, B. Meyer, A. Eljarah, S. Abutarbush, G. J. Godeke, T. M. Bestebroer, I. Zutt, M. A. Muller et al. 2013. Middle East Respiratory Syndrome coronavirus (MERS-CoV) serology in major livestock species in an affected region in Jordan, June to September 2013. *Euro Surveill* 18 (50):20662.

Rose, N., A. Boutrouille, C. Fablet, F. Madec, M. Eloit, and N. Pavio. 2010. The use of Bayesian methods for evaluating the performance of a virus-like particles-based ELISA for serology of hepatitis E virus infection in swine. *J Virol Methods* 163 (2):329–335. doi: 10.1016/j.jviromet.2009.10.019.

Rose, N. F., P. A. Marx, A. Luckay, D. F. Nixon, W. J. Moretto, S. M. Donahoe, D. Montefiori, A. Roberts, L. Buonocore, and J. K. Rose. 2001. An effective AIDS vaccine based on live attenuated vesicular stomatitis virus recombinants. *Cell* 106 (5):539–549.

Roy, P. 2004. Genetically engineered structure-based vaccine for bluetongue disease. *Vet Ital* 40 (4):594–600.

Roy, P., D. H. Bishop, H. LeBlois, and B. J. Erasmus. 1994. Long-lasting protection of sheep against bluetongue challenge after vaccination with virus-like particles: Evidence for homologous and partial heterologous protection. *Vaccine* 12 (9):805–811.

Roy, P. and R. Noad. 2008. Virus-like particles as a vaccine delivery system: Myths and facts. *Hum Vaccin* 4 (1):5–12.

Sachsenroder, J., A. Braun, P. Machnowska, T. F. Ng, X. Deng, S. Guenther, S. Bernstein, R. G. Ulrich, E. Delwart, and R. Johne. 2014. Metagenomic identification of novel enteric viruses in urban wild rats and genome characterization of a group A rotavirus. *J Gen Virol* 95 (Pt 12):2734–2747. doi: 10.1099/vir.0.070029-0.

Sakuma, T., T. Noda, S. Urata, Y. Kawaoka, and J. Yasuda. 2009. Inhibition of Lassa and Marburg virus production by tetherin. *J Virol* 83 (5):2382–2385. doi: 10.1128/JVI.01607–08.

Sasnauskas, K., A. Bulavaite, A. Hale, L. Jin, W. A. Knowles, A. Gedvilaite, A. Dargeviciute, D. Bartkeviciute, A. Zvirbliene, J. Staniulis et al. 2002. Generation of recombinant virus-like particles of human and non-human polyomaviruses in yeast *Saccharomyces cerevisiae*. *Intervirology* 45 (4–6):308–317.

Schlienger, K., M. Mancini, Y. Riviere, D. Dormont, P. Tiollais, and M. L. Michel. 1992. Human immunodeficiency virus type 1 major neutralizing determinant exposed on hepatitis B surface antigen particles is highly immunogenic in primates. *J Virol* 66 (4):2570–2576.

Schnell, M. J., L. Buonocore, E. Kretzschmar, E. Johnson, and J. K. Rose. 1996. Foreign glycoproteins expressed from recombinant vesicular stomatitis viruses are incorporated efficiently into virus particles. *Proc Natl Acad Sci USA* 93 (21):11359–11365.

Sharma, A., Z. Rao, E. Fry, T. Booth, E. Y. Jones, D. J. Rowlands, D. L. Simmons, and D. I. Stuart. 1997. Specific interactions between human integrin alpha v beta 3 and chimeric hepatitis B virus core particles bearing the receptor-binding epitope of foot-and-mouth disease virus. *Virology* 239 (1):150–157. doi: 10.1006/viro.1997.8833.

Shi, X., A. Kohl, V. H. Leonard, P. Li, A. McLees, and R. M. Elliott. 2006. Requirement of the N-terminal region of orthobunyavirus nonstructural protein NSm for virus assembly and morphogenesis. *J Virol* 80 (16):8089–8099. doi: 10.1128/JVI.00579–06.

Smits, S. L., V. S. Raj, M. D. Oduber, C. M. Schapendonk, R. Bodewes, L. Provacia, K. J. Stittelaar, A. D. Osterhaus, and B. L. Haagmans. 2013. Metagenomic analysis of the ferret fecal viral flora. *PLoS ONE* 8 (8):e71595. doi: 10.1371/journal.pone.0071595.

Speth, C., S. Bredl, M. Hagleitner, J. Wild, M. Dierich, H. Wolf, J. Schroeder, R. Wagner, and L. Deml. 2008. Human immunodeficiency virus type-1 (HIV-1) Pr55gag virus-like particles are potent activators of human monocytes. *Virology* 382 (1):46–58. doi: 10.1016/j.virol.2008.08.043.

Spiegelberg, L., V. Wahl-Jensen, L. Kolesnikova, H. Feldmann, S. Becker, and T. Hoenen. 2011. Genus-specific recruitment of filovirus ribonucleoprotein complexes into budding particles. *J Gen Virol* 92 (Pt. 12):2900–2905. doi: 10.1099/vir.0.036863-0.

Stewart, M., E. Dubois, C. Sailleau, E. Breard, C. Viarouge, A. Desprat, R. Thiery, S. Zientara, and P. Roy. 2013. Bluetongue virus serotype 8 virus-like particles protect sheep against virulent virus infection as a single or multi-serotype cocktail immunogen. *Vaccine* 31 (3):553–558. doi: 10.1016/j.vaccine.2012.11.016.

Strandin, T., J. Hepojoki, and A. Vaheri. 2013. Cytoplasmic tails of bunyavirus Gn glycoproteins-could they act as matrix protein surrogates? *Virology* 437 (2):73–80. doi: 10.1016/j.virol.2013.01.001.

Strecker, T., R. Eichler, J. ter Meulen, W. Weissenhorn, H. Dieter Klenk, W. Garten, and O. Lenz. 2003. Lassa virus Z protein is a matrix protein and sufficient for the release of virus-like particles [corrected]. *J Virol* 77 (19):10700–10705.

Sun, Y., R. Carrion, Jr., L. Ye, Z. Wen, Y. T. Ro, K. Brasky, A. E. Ticer, E. E. Schwegler, J. L. Patterson, R. W. Compans et al. 2009. Protection against lethal challenge by Ebola virus-like particles produced in insect cells. *Virology* 383 (1):12–21. doi: 10.1016/j.virol.2008.09.020.

Swenson, D. L., K. L. Warfield, K. Kuehl, T. Larsen, M. C. Hevey, A. Schmaljohn, S. Bavari, and M. J. Aman. 2004. Generation of Marburg virus-like particles by co-expression of glycoprotein and matrix protein. *FEMS Immunol Med Microbiol* 40 (1):27–31.

Tamin, A., B. H. Harcourt, M. K. Lo, J. A. Roth, M. C. Wolf, B. Lee, H. Weingartl, J. C. Audonnet, W. J. Bellini, and P. A. Rota. 2009. Development of a neutralization assay for Nipah virus using pseudotype particles. *J Virol Methods* 160 (1–2):1–6. doi: 10.1016/j.jviromet.2009.02.025.

Tegerstedt, K., J. A. Lindencrona, C. Curcio, K. Andreasson, C. Tullus, G. Forni, T. Dalianis, R. Kiessling, and T. Ramqvist. 2005. A single vaccination with polyomavirus VP1/VP2Her2 virus-like particles prevents outgrowth of HER-2/neu-expressing tumors. *Cancer Res* 65 (13):5953–5957. doi: 10.1158/0008-5472.CAN-05-0335.

Temperton, N. J., K. Hoschler, D. Major, C. Nicolson, R. Manvell, V. M. Hien, Q. Ha do, M. de Jong, M. Zambon, Y. Takeuchi et al. 2007. A sensitive retroviral pseudotype assay for influenza H5N1-neutralizing antibodies. *Influenza Other Respir Viruses* 1 (3):105–112. doi: 10.1111/j.1750–2659.2007.00016.x.

Triyatni, M., B. Saunier, P. Maruvada, A. R. Davis, L. Ulianich, T. Heller, A. Patel, L. D. Kohn, and T. J. Liang. 2002. Interaction of hepatitis C virus-like particles and cells: A model system for studying viral binding and entry. *J Virol* 76 (18):9335–9344.

Tsai, C., C. Caillet, H. Hu, F. Zhou, H. Ding, G. Zhang, B. Zhou, S. Wang, S. Lu, P. Buchy et al. 2009. Measurement of neutralizing antibody responses against H5N1 clades in immunized mice and ferrets using pseudotypes expressing influenza hemagglutinin and neuraminidase. *Vaccine* 27 (48):6777–6790. doi: 10.1016/j.vaccine.2009.08.056.

Tseng, Y. T., C. H. Chang, S. M. Wang, K. J. Huang, and C. T. Wang. 2013. Identifying SARS-CoV membrane protein amino acid residues linked to virus-like particle assembly. *PLoS ONE* 8 (5):e64013. doi: 10.1371/journal.pone.0064013.

Tsunetsugu-Yokota, Y., Y. Morikawa, M. Isogai, A. Kawana-Tachikawa, T. Odawara, T. Nakamura, F. Grassi, B. Autran, and A. Iwamoto. 2003. Yeast-derived human immunodeficiency virus type 1 p55(gag) virus-like particles activate dendritic cells (DCs) and induce perforin expression in Gag-specific CD8(+) T cells by cross-presentation of DCs. *J Virol* 77 (19):10250–10259.

Ulrich, R., G. P. Borisova, E. Gren, I. Berzin, P. Pumpen, R. Eckert, V. Ose, H. Siakkou, E. J. Gren, R. von Baehr et al. 1992. Immunogenicity of recombinant core particles of hepatitis B virus containing epitopes of human immunodeficiency virus 1 core antigen. *Arch Virol* 126 (1–4):321–328.

Ulrich, R., D. Koletzki, S. Lachmann, A. Lundkvist, A. Zankl, A. Kazaks, A. Kurth, H. R. Gelderblom, G. Borisova, H. Meisel, and D. H. Kruger. 1999. New chimaeric hepatitis B virus core particles carrying hantavirus (serotype Puumala) epitopes: Immunogenicity and protection against virus challenge. *J Biotechnol* 73 (2–3):141–153.

Ulrich, R., A. Lundkvist, H. Meisel, D. Koletzki, K. B. Sjolander, H. R. Gelderblom, G. Borisova, P. Schnitzler, G. Darai, and D. H. Kruger. 1998. Chimaeric HBV core particles carrying a defined segment of Puumala hantavirus nucleocapsid protein evoke protective immunity in an animal model. *Vaccine* 16 (2–3):272–280.

Urata, S., T. Noda, Y. Kawaoka, S. Morikawa, H. Yokosawa, and J. Yasuda. 2007. Interaction of Tsg101 with Marburg virus VP40 depends on the PPPY motif, but not the PT/SAP motif as in the case of Ebola virus, and Tsg101 plays a critical role in the budding of Marburg virus-like particles induced by VP40, NP, and GP. *J Virol* 81 (9):4895–4899. doi: 10.1128/JVI.02829–06.

Vietheer, P. T., I. Boo, H. E. Drummer, and H. J. Netter. 2007. Immunizations with chimeric hepatitis B virus-like particles to induce potential anti-hepatitis C virus neutralizing antibodies. *Antivir Ther* 12 (4):477–487.

Wagner, J. M., J. D. Pajerowski, C. L. Daniels, P. M. McHugh, J. A. Flynn, J. W. Balliet, D. R. Casimiro, and S. Subramanian. 2014. Enhanced production of Chikungunya virus-like particles using a high-pH adapted Spodoptera frugiperda insect cell line. *PLoS ONE* 9 (4):e94401. doi: 10.1371/journal.pone.0094401.

Wagner, R., V. J. Teeuwsen, L. Deml, F. Notka, A. G. Haaksma, S. S. Jhagjhoorsingh, H. Niphuis, H. Wolf, and J. L. Heeney. 1998. Cytotoxic T cells and neutralizing antibodies induced in rhesus monkeys by virus-like particle HIV vaccines in the absence of protection from SHIV infection. *Virology* 245 (1):65–74. doi: 10.1006/viro.1998.9104.

Wahl-Jensen, V. M., T. A. Afanasieva, J. Seebach, U. Stroher, H. Feldmann, and H. J. Schnittler. 2005. Effects of Ebola virus glycoproteins on endothelial cell activation and barrier function. *J Virol* 79 (16):10442–10450. doi: 10.1128/JVI.79.16.10442–10450.2005.

Walpita, P., J. Barr, M. Sherman, C. F. Basler, and L. Wang. 2011. Vaccine potential of Nipah virus-like particles. *PLoS ONE* 6 (4):e18437. doi: 10.1371/journal.pone.0018437.

Wang, H., A. Alminaite, A. Vaheri, and A. Plyusnin. 2010. Interaction between hantaviral nucleocapsid protein and the cytoplasmic tail of surface glycoprotein Gn. *Virus Res* 151 (2):205–212. doi: 10.1016/j.virusres.2010.05.008.

Wang, P. G., M. Kudelko, J. Lo, L. Y. Siu, K. T. Kwok, M. Sachse, J. M. Nicholls, R. Bruzzone, R. M. Altmeyer, and B. Nal. 2009. Efficient assembly and secretion of recombinant subviral particles of the four dengue serotypes using native prM and E proteins. *PLoS ONE* 4 (12):e8325. doi: 10.1371/journal.pone.0008325.

Wang, W., E. N. Butler, V. Veguilla, R. Vassell, J. T. Thomas, M. Moos, Jr., Z. Ye, K. Hancock, and C. D. Weiss. 2008. Establishment of retroviral pseudotypes with influenza hemagglutinins from H1, H3, and H5 subtypes for sensitive and specific detection of neutralizing antibodies. *J Virol Methods* 153 (2):111–119. doi: 10.1016/j.jviromet.2008.07.015.

Warfield, K. L., C. M. Bosio, B. C. Welcher, E. M. Deal, M. Mohamadzadeh, A. Schmaljohn, M. J. Aman, and S. Bavari. 2003. Ebola virus-like particles protect from lethal Ebola virus infection. *Proc Natl Acad Sci USA* 100 (26):15889–15894. doi: 10.1073/pnas.2237038100.

Warfield, K. L. and G. G. Olinger. 2011. Protective role of cytotoxic T lymphocytes in filovirus hemorrhagic fever. *J Biomed Biotechnol* 2011:1–13. doi: 10.1155/2011/984241.

Warfield, K. L., D. L. Swenson, D. L. Negley, A. L. Schmaljohn, M. J. Aman, and S. Bavari. 2004. Marburg virus-like particles protect guinea pigs from lethal Marburg virus infection. *Vaccine* 22 (25–26):3495–3502. doi: 10.1016/j.vaccine.2004.01.063.

Watanabe, S., T. Watanabe, T. Noda, A. Takada, H. Feldmann, L. D. Jasenosky, and Y. Kawaoka. 2004. Production of novel Ebola virus-like particles from cDNAs: An alternative to Ebola virus generation by reverse genetics. *J Virol* 78 (2):999–1005.

Weber, J., R. Cheinsong-Popov, D. Callow, S. Adams, G. Patou, K. Hodgkin, S. Martin, F. Gotch, and A. Kingsman. 1995. Immunogenicity of the yeast recombinant p17/p24: Ty virus-like particles (p24-VLP) in healthy volunteers. *Vaccine* 13 (9):831–834.

Wenigenrath, J., L. Kolesnikova, T. Hoenen, E. Mittler, and S. Becker. 2010. Establishment and application of an infectious virus-like particle system for Marburg virus. *J Gen Virol* 91 (Pt. 5):1325–1334. doi: 10.1099/vir.0.018226–0.

Wolf, M. C., Y. Wang, A. N. Freiberg, H. C. Aguilar, M. R. Holbrook, and B. Lee. 2009. A catalytically and genetically optimized beta-lactamase-matrix based assay for sensitive, specific, and higher throughput analysis of native henipavirus entry characteristics. *Virol J* 6:119. doi: 10.1186/1743-422X-6-119.

Wright, E., S. McNabb, T. Goddard, D. L. Horton, T. Lembo, L. H. Nel, R. A. Weiss, S. Cleaveland, and A. R. Fooks. 2009. A robust lentiviral pseudotype neutralisation assay for in-field serosurveillance of rabies and lyssaviruses in Africa. *Vaccine* 27 (51):7178–7186. doi: 10.1016/j.vaccine.2009.09.024.

Wright, E., N. J. Temperton, D. A. Marston, L. M. McElhinney, A. R. Fooks, and R. A. Weiss. 2008. Investigating antibody neutralization of lyssaviruses using lentiviral pseudotypes: A cross-species comparison. *J Gen Virol* 89 (Pt. 9):2204–2213. doi: 10.1099/vir.0.2008/000349–0.

Wu, C. Y., Y. C. Yeh, J. T. Chan, Y. C. Yang, J. R. Yang, M. T. Liu, H. S. Wu, and P. W. Hsiao. 2012. A VLP vaccine induces broad-spectrum cross-protective antibody immunity against H5N1 and H1N1 subtypes of influenza A virus. *PLoS ONE* 7 (8):e42363. doi: 10.1371/journal.pone.0042363.

Wu, C. Y., Y. C. Yeh, Y. C. Yang, C. Chou, M. T. Liu, H. S. Wu, J. T. Chan, and P. W. Hsiao. 2010. Mammalian expression of virus-like particles for advanced mimicry of authentic influenza virus. *PLoS ONE* 5 (3):e9784. doi: 10.1371/journal.pone.0009784.

Yang, T., M. Kataoka, Y. Ami, Y. Suzaki, N. Kishida, M. Shirakura, M. Imai, H. Asanuma, N. Takeda, T. Wakita, and T. C. Li. 2013. Characterization of self-assembled virus-like particles of ferret hepatitis E virus generated by recombinant baculoviruses. *J Gen Virol* 94 (Pt. 12):2647–2656. doi: 10.1099/vir.0.056671–0.

Yoshikawa, A., T. Tanaka, Y. Hoshi, N. Kato, K. Tachibana, H. Iizuka, A. Machida, H. Okamoto, M. Yamasaki, Y. Miyakawa et al. 1993. Chimeric hepatitis B virus core particles with parts or copies of the hepatitis C virus core protein. *J Virol* 67 (10):6064–6070.

Zhang, S., M. Liang, W. Gu, C. Li, F. Miao, X. Wang, C. Jin, L. Zhang, F. Zhang, Q. Zhang et al. 2011. Vaccination with dengue virus-like particles induces humoral and cellular immune responses in mice. *Virol J* 8:333. doi: 10.1186/1743-422X-8-333.

Zhang, Y. L., Y. J. Guo, K. Y. Wang, K. Lu, K. Li, Y. Zhu, and S. H. Sun. 2007. Enhanced immunogenicity of modified hepatitis B virus core particle fused with multiepitopes of foot-and-mouth disease virus. *Scand J Immunol* 65 (4):320–328. doi: 10.1111/j.1365-3083.2007.01900.x.

Zhao, G., L. Du, C. Ma, Y. Li, L. Li, V. K. Poon, L. Wang, F. Yu, B. J. Zheng, S. Jiang et al. 2013. A safe and convenient pseudovirus-based inhibition assay to detect neutralizing antibodies and screen for viral entry inhibitors against the novel human coronavirus MERS-CoV. *Virol J* 10:266. doi: 10.1186/1743-422X-10-266.

Zimmer, G., S. Locher, M. Berger Rentsch, and S. J. Halbherr. 2014. Pseudotyping of vesicular stomatitis virus with the envelope glycoproteins of highly pathogenic avian influenza viruses. *J Gen Virol* 95 (Pt. 8):1634–1639. doi: 10.1099/vir.0.065201-0.

Zvirbliene, A., I. Kucinskaite-Kodze, M. Juozapaitis, R. Lasickiene, D. Gritenaite, G. Russell, J. Bingham, W. P. Michalski, and K. Sasnauskas. 2010. Novel monoclonal antibodies against Menangle virus nucleocapsid protein. *Arch Virol* 155 (1):13–18. doi: 10.1007/s00705-009-0543-1.

Zvirbliene, A., I. Kucinskaite-Kodze, A. Razanskiene, R. Petraityte-Burneikiene, B. Klempa, R. G. Ulrich, and A. Gedvilaite. 2014. The use of chimeric virus-like particles harbouring a segment of hantavirus Gc glycoprotein to generate a broadly-reactive hantavirus-specific monoclonal antibody. *Viruses* 6 (2):640–660. doi: 10.3390/v6020640.

Zvirbliene, A., L. Samonskyte, A. Gedvilaite, T. Voronkova, R. Ulrich, and K. Sasnauskas. 2006. Generation of monoclonal antibodies of desired specificity using chimeric polyomavirus-derived virus-like particles. *J Immunol Methods* 311 (1–2):57–70. doi: 10.1016/j.jim.2006.01.007.

10 Virus-Like Particles Derived from Hepatitis Viruses

Wolfram H. Gerlich

CONTENTS

10.1 INTRODUCTION

In the late 1960s, the agents of viral hepatitis had not yet been identified and diagnosis was exclusively based on clinical observations and some biochemical markers of liver damage or dysfunction. The detection and study of hepatitis viruses was possible only by epidemiological studies or experimental transmission to humans. In this highly unsatisfactory situation, the accidental discovery of the virus-like particles (VLPs) associated with hepatitis B virus (HBV) infections was the starting point for the entire research on viral hepatitis (for review, see Ref. [1]). VLPs are particularly important for hepatitis B, but their potential significance for the diagnosis and prevention of the other hepatitis viruses will be briefly discussed as well.

10.2 VLPs EXPRESSED BY HEPATITIS B VIRUS: HEPATITIS B SURFACE ANTIGEN (HBsAg)

10.2.1 DISCOVERY OF SUBVIRAL SURFACE ANTIGEN PARTICLES OF HBV

The geneticist Baruch S. Blumberg searched in the 1960s for markers associated with diseases of presumably genetic origin. Since the tools of molecular biology were not yet available, he used an immunological approach. Blumberg postulated that people, who had received blood products from many donors, would develop antibodies against proteins that show small individual genetic differences. Blumberg's coworker, Harvey Alter, discovered indeed a new antigen in samples of a worldwide serum collection, quite often in Australian aborigines, for whom the Australia Antigen (AuAg) was named. At first, Blumberg believed that AuAg was indeed a genetic marker, but soon it was found that it might have something to do with hepatitis. Subsequently, Alfred Prince found that AuAg was indeed a marker for acute or chronic hepatitis B and that there were apparently healthy AuAg carriers. Under the electron microscope (EM), purified AuAg appeared as small round particles which, unlike real viruses, were variable in sizes between 17 and 25 nm. Most importantly, Blumberg and his team found that these particles consisted of lipoprotein but did not contain any nucleic acid. David S. Dane, however, discovered that AuAg appeared not only on these small pleomorphic particles but also on larger, virus-like objects 42 nm in size. The identity of these *Dane* particles with the infectious agent of hepatitis B was strongly suggested by the discovery of DNA within the cores of the Dane particles by William S. Robinson. Three groups of molecular biologists led by Pierre Tiollais, Kenneth Murray, and William Rutter showed by cloning and sequencing that this DNA was indeed the HBV genome. The results of these pioneering studies revealed that AuAg was formed by the surface (or envelope) proteins of HBV, and consequently, it was named hepatitis B surface antigen or HBsAg (reviewed in Ref. [1]). The discovery of HBsAg facilitated the dependable diagnosis of hepatitis B and the distinction of HBV from other hepatitis viruses.

10.2.2 BIOLOGICAL SIGNIFICANCE OF HBsAg VLPs

The biology of HBV and its relatives in the virus family Hepadnaviridae is characterized by a narrow host range and a strict hepatotropism. It builds on the ability to establish viral persistence and high-level viremia. Only this particular strategy allows for transmission from the infected liver to the not-yet infected liver of the next host. A precondition for this survival strategy is the ability to evade the host immune system on both the innate and adaptive levels. The VLPs expressed by HBV-infected cells (also called subviral particles [SVP]) are an important component of the viral evasion strategy. They are secreted very efficiently and are able to circulate in the blood for long time [1].

The amount of HBsAg protein on VLPs compared to that on complete virions is highly variable during the course of acute or chronic HBV infection. However, in the so-called immunotolerant phase (characterized by low or absent inflammation and very high HBV replication), a relatively constant ratio of VLPs to virions is found irrespective of the HBV genotype. In an international reference panel of 15 plasmas (containing HBV genotypes A–G, see Section 10.2.8), VLPs contained 2700 ± 1300 times more HBsAg than the complete virions (calculated from data published in Ref. [2]). The exact function of this huge excess is not clear, but it appears that the VLPs are dummies, which distract neutralizing antibodies from the infectious virus. Furthermore, the large amount of HBsAg particles is probably able to exhaust the HBsAg-specific T cells necessary for initiation of antibody production and cytotoxic reactions directed against HBV-infected cells. This putative immunomodulatory function adds to the tolerogenic location of the virus in hepatocytes. Thus, even in healthy immunocompetent adults, HBV can replicate for weeks or months up to very high levels of 10^9 virions per mL blood before the immune response is finally activated. As soon as protective immune reactions develop, the ratio of VLPs to virions increases transiently even more. The half-life of HBsAg in plasma during the virus elimination phase of *acute* hepatitis B is 8 days versus 1.6 days of the complete virus [3]. In *chronic* HBV infection, the immunotolerant phase may last for decades, but it proceeds very often to a variable and sometimes fluctuating *immune elimination* phase, which is connected with chronic hepatic inflammation and progressing fibrosis. If T cell immunity is finally effective in this phase, HBV replication may decrease to low levels $<10^4$ mL^{-1}, resulting in the so-called *inactive HBsAg carrier* state. While viremia is low and sometimes even undetectable (<10 mL^{-1}) in these carriers, HBsAg may be still relatively high up to 30 μg/mL [1].

10.2.3 DIAGNOSTIC SIGNIFICANCE OF QUALITATIVE HBsAg DETECTION

HBsAg reaches in extreme cases concentrations of 1 mg/mL. More typical are 1–100 μg/mL in cases of acute [3] or chronic [2] HBV infection. Due to these large amounts,

HBsAg could be discovered 50 years ago by the very insensitive technique of agar gel double diffusion according to Ouchterlony, which requires several µg protein per mL [1]. Since the introduction of modern solid-phase immune assays using labeled antibodies against HBsAg (anti-HBs), sensitivity has improved over the last 40 years dramatically by a factor up to 10^6 [4]. Today, virtually all cases of active HBV infection can be detected, provided the antibodies used in the assay are able to react with all relevant variants of HBsAg. For several years, assays were on the market, which failed to detect the HBsAg of HBV genotype F, which is most divergent from the other genotypes. Another gap (which is almost closed meanwhile) was the failure to detect escape mutants of HBV, which are selected when the virus replicates in the presence of anti-HBs [5]. Even with the best assays using most advanced labeling and signal generation remain two gaps where an HBV infection may be undetectable by an HBsAg assay:

1. The early window phase occurs when the HBsAg is still below the detection limit of the most sensitive assays (ca. 10 pg/mL) although infectious virions are already present in low number. This residual problem is relevant only for blood and liver donation and very small; the residual risk of HBV transmission is 1:>100,000 in countries with low HBV endemicity [6] but quite relevant in regions with a high incidence of new HBV infections [7]. The problem can partially (but not completely) be solved by testing the donor plasma samples for HBV DNA using highly sensitive nucleic acid amplification tests (NAT) [8,9].

2. The occult HBV infection (OBI) is defined as HBsAg negative *after* the early window phase. The additional criterion *HBV DNA positive* (in plasma) is often mentioned, but this is misleading because occult means in most cases that HBV is hidden in the liver and does not appear in detectable numbers in plasma. The clinical relevance of OBI is debatable. A known problem is the reactivation of OBI to a highly replicative infection under immunosuppression of B cells [10,11]. Another well-documented problem is the transmission of HBV from blood with OBI that is infrequent but not negligible [12,13]. Deferral of donors with antibodies to HBV core antigen (anti-HBc) is one way to prevent this; testing donations for HBV DNA by NAT is the other way. Contrary to the early window phase, NAT is able to detect virtually all infectious donors with OBI because only a small proportion of the HBV particles is infectious [13]. In contrast, transplantation of a liver with OBI results virtually always in transmission and reactivation of HBV [14]. It remains open whether supersensitive tests for HBsAg would be able to close these gaps.

10.2.4 Quantitation of HBsAg

The concentration of HBsAg can be determined in weight units/mL [15], but different authors used different methods to calibrate their quantitative immune assays (discussed in Ref. [16]). Thus, World Health Organization (WHO) has decided to use arbitrary International Units (IU) for HBsAg reactivity in International Standard (IS) preparations irrespective of their HBs protein content. Using native HBsAg (without heat or chemical treatment) and optimal HBsAg assays, 1 IU HBsAg corresponds to 0.88 ± 0.20 ng HBs protein with little HBV genotype or HBsAg subtype related differences [4]. One problem is that neither the first nor the second IS contained native HBsAg (although derived from HBV carrier plasma) because these samples were heat treated to remove HBV infectivity [16]. This remains true for the prospective third IS, which is made from a plasma-derived hepatitis B vaccine produced in Vietnam (D. Wilkinson, NIBSC, U.K., personal communication). In an international trial of the WHO, the majority of test systems reacted equally well with the first and second IS and the 15 HBsAg samples of the HBV genotype panel mentioned earlier. However, some tests reacted better with the IS than with the native samples, other samples reacted worse with the IS than with the native sample of the same genotype. Some tests had a sensitivity problem with certain genotypes. However, the well-established tests from the major producers were able to quantitate native HBsAg from all HBV genotypes correctly [4].

For long time, quantitation of HBsAg was neglected in the diagnosis and monitoring of hepatitis B. It was just considered as a side aspect of quality control assuring the detection limit of an assay. Due to a lack of correct quantification, no clear relationship between HBsAg concentration and infectivity or disease activity in chronic HBV infection could be found. HBV DNA in plasma was and is considered the optimal parameter for following replication and resolution [17]. However, HBsAg concentration was already in the 1970s found to be an early predictor for resolution of acute hepatitis B [18]. Immunotolerant HBV carriers with HBV e antigen (HBeAg) have typically >30 µg/mL and do not reach a sustained response to interferon therapy [19,20]. More recently, the kinetics of HBsAg has found attention for monitoring the interferon therapy. A sustained response after 1 year of strenuous therapy can only be expected if HBV DNA and HBsAg decrease within the first 2 months significantly. This observation provides a stopping rule due to an unfavorable prognosis [21].

The HBsAg quantity is an indirect marker for the number and expression activity of HBV genomes in the form of covalently closed circular (ccc) DNA [22]. Thus, it is a useful parameter for monitoring antiviral therapy with inhibitors of the HBV reverse transcriptase. While the envelopment and exocytosis of HBV cores depend on the reverse transcription of the encapsidated pregenomic RNA, the transcription and translation of the mRNAs still go on allowing unaltered secretion of HBsAg [23]. Stopping antiviral therapy does require not only a stable negative result for HBV DNA in serum but also a low HBsAg concentration [24].

10.2.5 Biochemical Structure of HBsAg VLPs

The vast majority of HBsAg VLPs appear in EM as hollow spherical particles of variable diameter between 17 and 25 nm (Figure 10.1, right part). Their density is ca. 1.16 g/mL after ultracentrifugation into sucrose density gradients [25] and 1.20 g/mL into cesium chloride gradients [15], confirming the lipid content of HBsAg [1]. The lipid content is ER derived and lower than in enveloped viruses generated by budding from the plasma membrane [26]. The essential structural element of HBsAg VLPs is a 226 amino acid (aa) long hydrophobic small (S) HBs polypeptide (P24) [27], which is translated at the ER. The HBsAg containing membrane patches bud at the ER–Golgi intermediate compartment (ERGIC) and are secreted via the Golgi apparatus. The SHBs sequence contains at least two transmembrane alpha helices, TMI and TMII, which act as translocation signals and three hydrophobic alpha helices TMIII–TMV with a variable membrane association. This topology generates in the VLPs an internal loop between TMI and TMII and an external antigenic loop (AGL) exposing the HBsAg determinants between TMII and TMIII [28] (Figure 10.2). In native HBsAg, the AGL is highly cross-linked by disulfide bridges and forms rigid conformational epitopes that are inactivated by reductive cleavage [29]. In vertebrate cell cultures, ca. 40% of the S protein are cotranslationally glycosylated at Asn 146. Thus, P24 is usually accompanied by glycoprotein GP27 [27]. It has been reported that HBsAg VLPs would have a symmetric structure [30], but this finding is difficult to reconcile with the pleomorphic appearance of the natural VLPs under the EM.

SHBs is essential for the envelopment and release of the virions and sufficient for the formation of the VLPs. It binds specifically to heparan sulfate proteoglycans via the conformational AGL and is required as one of at least two surface components for viral attachment [31].

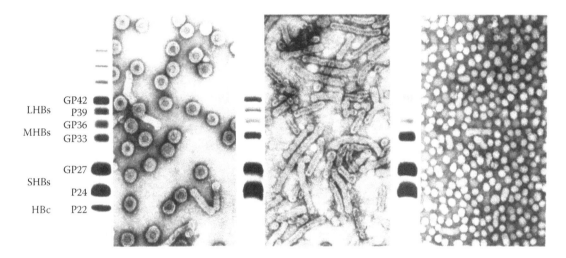

FIGURE 10.1 Morphology and protein composition of HBV (left) and its associated filamentous (center) or spherical (right) VLPs consisting of excessive surface antigen (HBsAg). The EM pictures show particles highly purified from the plasma of an HBeAg-positive HBV carrier and negative stained with phosphotungstic acid. The protein composition was analyzed after denaturation with SDS and dithiothreitol by polyacrylamide gel electrophoresis and subsequent silver staining as described in Ref. [29].

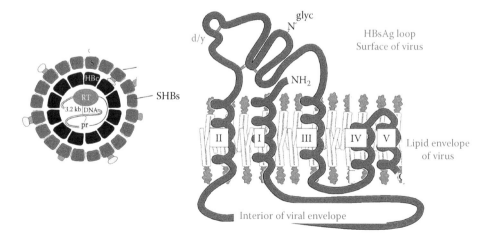

FIGURE 10.2 Model of the HBV particle (left) and 2D model of the membrane topology of SHBs P24 protein. The roman numerals indicate the membrane-spanning alpha helices, the short lines putative disulfide bridges. Glyc, N-linked glycan present in GP27; RT, reverse transcriptase; pr, primer domain.

Considerable heterogeneity of the VLPs is created by the fact that the open reading frame (ORF) for HBsAg may encode three cocarboxyterminal polypeptides: small SHBs, middle MHBs, and large LHBs protein. This multifunctionality is generated and regulated at the transcriptional and translational levels (see Figure 10.3). Furthermore, the three proteins come as two glycosylation variants. Natural HBsAg VLPs show a typical protein pattern in SDS gel electrophoresis with P24 or GP 27 as main SHBs components, GP33 or G36 as moderately prevalent MHBs components, and P39 or GP42 as minor LHBs components [29] (Figure 10.1, right). The exact proportion of the S, M, or L protein is variable from patient to patient and not easy to determine by normal protein stains like coomassie blue or colloidal silver. HBeAg-positive

immunotolerant HBV carriers show up to 16% L and 22% M protein when analyzed by a quantitative fluorescent Western blot using a monoclonal antibody (mab) against denatured S protein [32]. The VLPs contain less GP27 than P24, and less GP36 than GP33, but more GP42 than P39 (Figure 10.1). The composition of two examples is shown in Table 10.1. The strange protein pattern is generated by three mechanisms:

1. The existence of two mRNAs for the HBs proteins (Figure 10.3). The longer one is regulated by the preS1 promoter and encodes LHBs. The shorter has fuzzy 5′ ends and may code for MHBs or SHBs, depending on the presence or absence of the translational start codon of preS2.

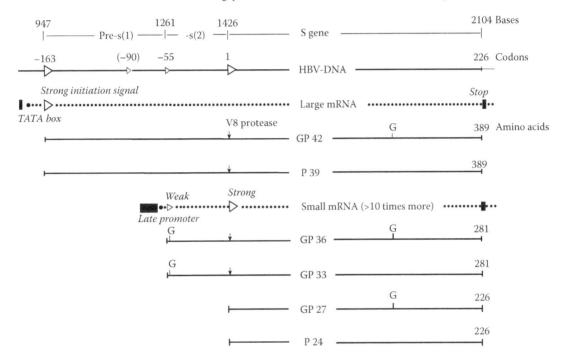

FIGURE 10.3 Organization and potential polypeptide products of the S ORF. (Reprinted from Heermann, K.H. et al., *J. Virol.*, 52(2), 396, 1984.) The upper line shows the DNA sequence of the entire ORF with the preS-(1) or -s(2) and S domains with the numeration beginning at the 5′ end of the pregenomic RNA. The second line shows the codons of the S ORF beginning with 1 at the start codon of SHBs. The dotted lines show the two mRNAs with the translational start codons as large or small arrows. G, N-linked glycan at Asn 4 of preS2 and Asn 146 of the S domain.

TABLE 10.1

Relative Protein Composition in % of Small Spherical HBsAg VLPs Purified from the Plasma of HBV Carriers with HBV Genotypes A2 or D

Protein	Size (kD)	Genotype A2	Genotype D
LHBs	GP42	7	4
	P39	7	3
MHBs	GP36	4	7
	GP33	9	14
SHBs	GP27	21	19
	P24	36	31

Notes: Proteins were quantified by Western blot using mab H1, which detects a sequential epitope in the S domain. Protein bands were visualized and quantitated with the Odyssee system (from Ref. [32]). The percentages do not add up to 100, because the system registered also dimers.

2. The fuzzy initiation of translation furthermore favors the production of SHBs because the start codon of MHBs is in a weak *Kozak* context, whereas the start codon of SHBs is in a strong initiation context.

3. The membrane topology of the three hydrophobic sequences in SHBs governs the glycosylation pattern in a complex way. TMI inserts during translation into the ER membrane in a way that it supports translocation of sequences N-terminal of it like preS2 [28]. PreS2 contains a conserved N-glycosylation motif, which is well recognized within the lumen of the ER. Thus, MHBs is always glycosylated at Asn4 of preS2, yielding GP33 and GP36. Long aminoterminal sequences are no longer translocated by TMI and remain cytosolic during translation and consequently, the entire preS domain, that is, preS1 and preS2, is not glycosylated although it has several N-glycosylation motifs [29,33,34]. Thus, LHBs may appear as unglycosylated P39. All three HBs proteins may be glycosylated at Asn146 in the S domain, but the efficiency of glycosylation is partially governed by the preS domains. It is decreased by the preS2 domain in MHBs but enhanced by the entire preS domain in LHBs.

10.2.6 ROLE OF PRES1

LHBs is a minor component in the main fraction of small spherical HBsAg VLPs, but its proportion is significantly larger in complete virions (Figure 10.1, right) and in the filamentous form of VLPs (Figure 10.1, center) [29]. It appears that the preS1 domain alters the intracellular pathway and the structure of HBV-associated particles in several ways:

1. A too high proportion of LHBs exceeding 30% of all HBs proteins prohibits secretion of HBsAg VLPs [35]. Storage of LHBs may cause ER stress and contribute to liver fibrosis and hepatocellular carcinoma [36].
2. The morphology of the VLPs changes from small spheres to filaments with a diameter of ca. 20–25 nm and variable length [29,37].
3. The intracellular membranes with much LHBs interact with mature (i.e., HBV DNA containing) HBV cores and provide the envelope for HBV [38]. This occurs with the participation of the endosomal sorting complex required for transport (ESCRT) and leads to the exocytosis of virions via multivesicular bodies [39].

There are three functions of *LHBs* in the viral life cycle:

1. The preS domain contains a transcription activating signal, which enhances the viral gene expression if it is located in the cytosol [40].
2. The preS domain links as a kind of matrix protein the viral nucleocapsid (often named core) with the nascent envelope [41].

3. The aminoterminal part of preS1 functions as attachment site to the species-specific high-affinity receptor of HBV: sodium-dependent (in German *N*atrium-abhängiges) taurocholate cotransporter polypeptide (NTCP) [42]. This receptor is also recognized by related primate or bat hepadnaviruses [43].

The first two functions require a cytosolic location or an internal position of preS in the virion. The latter function needs surface exposure. Surprisingly, LHBs can provide both topologies. A not-yet identified mechanism allows for a posttranslational translocation of the preS domain from the interior side of virion envelopes or the VLPs to the surface [33].

10.2.7 ROLE OF PRES2

MHBs seems to be nonessential for the viral life cycle [38] although it is conserved in all members of the mammalian hepadnaviruses. The preS2 domain has gained transiently some interest for the characterization of HBV infections (reviewed in Ref. [44]). The preS2 domain was initially detected by its ability to bind *polymerized* human serum albumin (pHSA). Since pHSA binds also to hepatocyte membranes, it was speculated in the early 1970s that pHSA would be a *receptor* for HBV [45]. This view was corroborated by the fact that only pHSA from humans or primates but not from other species was bound, reflecting the then known host range of HBV [46]. The pHSA was artificially generated by cross-linking with glutaraldehyde, but a very small proportion of natural monomeric HSA has also the ability to bind to preS2 [47], suggesting a not yet known role in the biology of primate HBV species. While preS2 is not essential for the viral life cycle, certain monoclonal antibodies (mabs) against the pHSA binding site do neutralize the infectivity of HBV [48] and preS2 antigens can induce protective immunity in experimental animals (reviewed in Ref. [44]).

10.2.8 GENETIC HETEROGENEITY OF HBV AND HBSAG

The genetic heterogeneity of HBV strains was recognized first in the 1970s in the form of the HBsAg subtype determinants *a*, *d*, or *y*, and *w1–4* or *r* (reviewed in Ref. [1]). Later HBV strains were grouped according to their genome sequence into genotypes A–I differing by >7.5% and at least 29 subgenotypes A1–I2 differing by >3.5%. Furthermore, recombinants between the (sub)genotypes exist [49]. Variability is restricted by the multiple use of the DNA sequence for coding overlapping ORFs and for regulatory elements. A major target for selection by neutralizing antibodies is the HBs antigen loop in the SHBs protein [50] reaching from amino acid (aa) 100 to ca. 170 aa with up to 23% variability [51]. The same region codes for the reverse transcriptase domain including some sites associated with antiviral drug resistance [52]. The other parts of the 389–400 codons spanning ORF are much less variable (<3%, [51]). Provided optimal HBsAg assays are used (e.g., as described in Ref. [53]), the variability of the S gene does not play a role in the diagnosis of HBV infections, but it has a tremendous impact on the efficacy of S gene–derived hepatitis B vaccines (reviewed in Ref. [44]).

10.2.9 NOMENCLATURE OF HBS PROTEINS

The additional aminoterminal parts of L- and MHBs have been termed for historical preS regions (or antigen or domain), simply to describe the position of the coding region in the genome upstream of gene S encoding HBsAg. This nomenclature is a bit unfortunate because it suggests to persons not so familiar with HBV that preS may be a part of the precursor protein for HBsAg, which is not the case. Furthermore, the use of the term HBsAg is ambiguous. Originally, it was meant as the entity of epitopes present on the surface of HBV particles. Since it was believed for several years that SHBs protein would be the only protein in the HBV envelope, the term HBsAg was synonymous for P24/GP27 and most people use it this way still today. The author would, however, discourage the use of preS1 protein for LHBs or preS2 protein for MHBs, because LHBs contains preS2 and the S domain as well, and MHBs contains in addition to preS2 the S domain.

Another point is the use of terminology HBV *envelope* protein for HBsAg. While *envelope* correctly describes the function of HBsAg, it may again be misleading because there is the officially approved but unexplained terminology (e = enigma?) for HBeAg, which has occasionally been mistaken as abbreviation for *envelope* or *early*. The term HBsAg was introduced by a consensus group in the early 1970s, it is widely used in medicine, and there is no need to add another term to describe the HBV surface antigen(s) or proteins. Very often, experts take the term HBsAg synonymous for the antigenicity or the particle or the protein.

10.2.10 HBSAG VLPS AS VACCINES

The use of HBsAg VLPs from HBV carriers as vaccine against hepatitis B was already suggested and patented by Blumberg in 1970, and these plans became reality in the early 1980s. The first-generation vaccine worked very well but questionable biosafety and foreseeability limited availability of this natural material from chronically infected people were reasons to attempt expression of HBsAg with the newly developed methods of gene technology [1]. Expression of the entire HBs ORF in mammalian cell lines under the control of the two natural promoters leads to secreted VLPs with a protein composition quite similar to the small spherical VLPs in the blood of HBV carriers [54]. This could have been the logical approach to replace the plasma-derived vaccine by a safer and more sustainable product. However, the evolution of the current hepatitis B vaccines took a different way (more extensively reviewed in Ref. [44]). Expression of native HBsAg was possible in mammalian cell cultures, but the yield was disappointingly low. HBV carriers have often 10–100 µg/mL of HBsAg in the plasma, but cell culture media contain rarely >10 ng/mL and do not exceed 100 ng/mL and such cell cultures are expensive. A much more efficient production system were cell cultures of the

yeast *Saccharomyces cerevisiae* transformed with the gene for SHBs, which produce up to 800 mg of SHBs protein per liter culture medium. This SHBs protein was not secreted, and it was not glycosylated. It was, however, possible to extract the SHBs protein in a particular form that exposed at least partially the conformational AGL. This type of second-generation vaccine proved to be efficacious for prevention of mother to child transmission and in other epidemiological settings. Thus, it became the standard vaccine recommended for worldwide application in neonates or infants and in high-risk groups. During the last three decades, it has facilitated the reduction of HBV prevalence and incidence by a factor of 3–10 in the vaccinated age groups, depending on the vaccine coverage and its optimal usage.

10.2.11 WEAKNESSES OF THE SECOND-GENERATION VACCINES

While the overall success of the second-generation vaccine is undebatable, some weaknesses ought to be mentioned.

Slow protection. The standard regimen for immunization of adults against hepatitis B requires three or even four doses of 10–20 µg HBsAg given i.m. together with aluminum compounds (e.g., hydroxide) as adjuvant at 0, 1, 6, or 12 months. Most vaccinees develop protective levels of >10 mIU/mL anti-HBs only after the last dose. There are rapid immunization schedules, but they are less efficient. Rapid protection is relevant for persons encountering a risk within short notice, which may happen to health care workers (HCWs) or before travel to highly endemic regions. Most important is a rapid protection in neonates born to HBV-positive mothers.

Nonresponders. While young healthy people respond very well to three doses of 10–20 µg HBsAg, persons with certain risk factors like older age, male gender, high body mass index, smoking, or moderately impaired immune reactivity (e.g., due to diabetes, hemodialysis, or asymptomatic HIV infection) often do not develop a protective immune response. It is attempted to overcome this by using higher and/or additional HBsAg doses and by using more potent adjuvants instead of aluminum compounds [55].

Incomplete protection. It has been observed that successfully vaccinated persons may develop a transient, asymptomatic HBV infection detectable by anti-HBc seroconversion or an anti-HBs titer increase without additional vaccine doses [56]. Furthermore, many vaccinated and HBV-exposed people show a cellular immune response to HBV cores or polymerase but no anti-HBc [57]. In several cases, transient viremia lasting for several months has been observed. This nonsterilizing, incomplete immunity may create a certain problem for the safety of blood donations [58]. More importantly, it may result in persistent OBI, which can reactivate under severe immunosuppression [59]. It appears that complete protection against HBV strains with heterologous HBsAg subtypes not present in the vaccine is less effective than against homologous subtypes and requires a ten times higher anti-HBs titer [44,58].

Failure to prevent perinatal transmission. The most serious weakness is the failure of vaccination in neonates from mothers with very high viremia of $>10^7$ mL^{-1}. These children become virtually always chronic HBV carriers. The simultaneous passive immunization does not overcome this problem, but an additional antiviral therapy of the mother in the last trimester can reduce the viremia to levels sufficiently low for protection of the neonate by postexposure vaccination [60]. Besides, frank breakthrough OBI is found very frequently in vaccinated neonates from HBV-positive mothers [61].

Selection of escape mutants. Incomplete protection leads to replication in presence of neutralizing anti-HBs and to selection of escape mutants with mutations in the AGL. Escape mutants are detected in ca. 25% of children in whom neonatal vaccination had failed [62] and in liver transplant recipients who received passive immunization with anti-HBs. They are, however, also found in asymptomatic persons with OBI who had been vaccinated early in childhood [63] or later [59].

10.2.12   THIRD-GENERATION VACCINES

These vaccines have been developed to mitigate the shortcomings mentioned earlier. Two well-studied products are Sci-B-Vac from an Israelian company [64] and Hepacare that is no longer available. Both vaccines were expressed in mammalian cells and contain besides SHBs, MHBs and preS1 antigen, Sci-B-Vac as entire LHBs, Hepacare as aminoterminal part linked to SHBs. Both vaccines induced a protective anti-HBs response more rapidly than the second generation, and they were able to induce a response in a part of nonresponders to the second-generation vaccines. In a preliminary study, Sci-B-Vac was found to protect neonates of HBV-positive mothers more efficiently than the second-generation vaccine [65]. Studies on complete (sterilizing) immunity induced by third-generation vaccines are not yet done, but a stronger protective capacity is likely because the preS domains contain additional neutralizing B cell epitopes and T helper cell epitopes, which may overcome genetic unresponsiveness to SHBs. Complete immunity would also prevent the selection of escape mutants by vaccination and probably decrease the risk of breakthrough caused by naturally occurring escape mutants from OBI cases.

There are still many options for further improvements without switching to radically new concepts like DNA vaccination or using vectors for antigen delivery. A more adequate coverage of the worldwide HBV genotypes would be desirable but is not provided by the major second-generation vaccines or Sci-B-Vac. The amount and immunogenicity of preS1 is rather low in Sci-B-Vac and similar VLPs. It could be enhanced if preS1 is presented on more immunogenic VLPs than SHBs particles, for example, HBV core particles [66]. The topics of Sections 10.2.10 through 10.2.12 have been reviewed more extensively in Refs. [44] and [64].

10.2.13   HBsAg VLPs AS DIAGNOSTIC REAGENTS

Besides being an important diagnostic marker for an active HBV infection and a successful prophylactic vaccine, HBsAg is an important reagent for the detection of anti-HBs and for the development or quality control of HBsAg assays.

Sandwich immune assays for anti-HBs. Originally, anti-HBs was detected by the Ouchterlony technique, which was, however, too insensitive in most cases. Ca. 1000-fold more sensitive modern *sandwich* immune assays for anti-HBs require HBsAg bound to a solid-phase form and as labeled signal-generating reagent. The two binding sites of one IgG molecule with anti-HBs specificity (or the five binding sites of IgM) cross-link the HBsAg at the solid phase with the labeled HBsAg VLPs in the liquid phase [1].

Diagnostic significance of anti-HBs. Anti-HBs appears weeks or even months *after* resolution of acute hepatitis B and disappearance of HBsAg and is usually absent in chronic HBV infection. Its production can be induced by vaccination with HBsAg. A positive specific anti-HBs result indicates protection against clinical disease or chronicity caused by a new HBV infection, but it does not dependably exclude the possibility of inapparent transient HBV infection if the anti-HBs titer is low [44,58].

Coexistence of HBsAg and anti-HBs. This seemingly paradoxical serological pattern is found in some percent of chronic HBV infections, but in this case, anti-HBs is not a favorable marker. The anti-HBs is in these cases virtually always directed against HBsAg subtypes different from the coexisting HBsAg [67]. The anti-HBs may be induced by the polyvalent nature of the B cell response, by double infection with more than one HBV genotype, or by (innocuous but useless) vaccination of an HBV carrier. If the associated viremia is low, HBV mutants may be selected by the anti-HBs, but usually viremia is high and no such selection occurs (reviewed in Ref. [67]). Immune complexes consisting of HBsAg and anti-HBs are frequent in chronic HBV carriers, but usually the HBsAg is present in large excess and no free anti-HBs is detectable [68].

Anti-HBs titers. The anti-HBs titer is particularly important if passive immunization with hepatitis B immunoglobulin (HBIG) is administered to nonimmune recipients after exposure to HBV.

Typical applications are protection of HCWs after work accidents with HBV-infected patients, HBV-positive recipients after liver transplantation [69] or most importantly neonates to HBV-positive mothers [70]. The amount of anti-HBs required for passive immunization is relatively high and blood donors with a titer of anti-HBs sufficient for HBIG production rare. Thus, WHO has attempted to standardize the anti-HBs contents of HBIG by introducing an IS for HBIG and the definition of an arbitrary anti-HBs IU. One milliliter of the IS for HBIG contains 100 IU anti-HBs/mL, and all HBIG preparations on the market have to exceed this concentration [71]. While for short-term postexposure prophylaxis only one dose is necessary, liver transplant recipients need to have 100 mIU/mL in their blood in order to prevent reappearance of their original HBV strain [69]. One IU of anti-HBs is able to bind 0.9 μg native HBsAg from carrier plasma [72].

The quantity of anti-HBs is also important for the follow-up of active immunization. Protection against a clinically relevant HBV infection is only to be expected if the vaccinee has >10 mIU/mL anti-HBs 1–4 weeks after the last dose. This somewhat arbitrary limit has prompted many providers or users of anti-HBs tests to consider results <10 mIU/mL by definition as negative, which is not really justified. For persons with an elevated risk >100 mIU/mL are recommended and an additional dose should be given if the standard regimen did not induce that level of anti-HBs. Unfortunately, different test kits yield quite often different quantitative results [73].

Heterogeneity of anti-HBs and HBsAg. A discussion of this problem is virtually absent in the literature but would be urgently needed. As mentioned in Sections 10.2.8 and 10.2.9, the meaning of the term *HBsAg* is not clearly defined. Most producers of anti-HBs tests leave it open (or keep it secret) whether the HBsAg VLPs in their test come from HBV carriers or from cell culture, whether they contain preS1 and preS2 or not, which HBV subgenotypes are included, how purification (and virus inactivation if necessary) may have altered the VLP structure, and how labeling may have affected the (cysteine and/or lysine containing) HBs epitopes. The best response (if any) from producers to such questions is that the HBsAg contains *HBsAg subtype ad and ay* which, however, reflects the state of knowledge in 1972. It is very likely that different producers use different HBsAg preparations for their anti-HBs assays and this would be the explanation that anti-HBs titers in one and the same sample vary considerably between different tests. In this situation, it is safe to use the lowest value, because the higher value may be due to subtype-specific antibodies. As mentioned earlier, HBsAg subtype-specific antibodies may be more important for HBV neutralization than the group-specific antibodies. The correlation between anti-HBs titers in immune assays and neutralization titer has never been accurately determined and may be overestimated.

Anti-preS. The potential detection of anti-preS antibodies is also relevant although preS antigen is usually not present in the hepatitis B vaccines. The anti-HBs test is used not only for control of the response to the vaccine but also for monitoring of HBV infection and the detection of naturally acquired immunity. For this application, anti-preS antibodies may be even more important than the anti-SHBs antibodies because they appear earlier after recovery [74] and may be more important for neutralization (see earlier text).

Development and quality control of HBsAg assays. These assays need to detect rare mutants, which are often available only in minute amounts. Since the sequence of these mutants can be determined in trace amounts of blood, expression of the corresponding HBsAg is possible and this antigen may be used for the generation of diagnostic antibodies or for the quality control of existing assays. It is of course important to express these HBsAg samples in mammalian cells to obtain all conformational epitopes and the correct glycosylation pattern.

10.3 HBV CORE PARTICLES

The core particle of HBV (HBcAg) was the first VLP generated and discovered by gene technology [75]. Surprisingly, bacteria transformed with HBV DNA produced HBcAg without construction of a real expression vector containing bacterial promoters and initiation signals, and surprisingly, these particles assembled to core particles that could not be distinguished from natural core particles in the EM [76]. The core protein assembles spontaneously to dimers and then to icosahedral particles with 180 or 240 subunits with T3 or T4 symmetry, respectively [77]. The core particle has an N-terminal assembly domain and a C-terminal RNA-binding domain. HBcAg VLPs derived from transformed *Escherichia coli* encapsidate nonspecifically RNA or oligonucleotides. The structure of the core protein and the use of these VLPs as epitope carrier are described elsewhere in this book.

10.3.1 Functions of Core Particles

10.3.1.1 Role in Replication

Core particles were recognized by EM as interior shell of Dane particles. If released from HBV particles, they were found to contain the viral DNA genome with such unusual features as being relaxed circular, covalently bound to a protein and partially single stranded, partially double stranded. At this stage in the viral life cycle, the core particle has the task to deliver the viral genome to the nuclear pore by the aid of its nuclear translocation signals where it is released into the nucleoplasm. After the conversion of the viral genome to a completely double-stranded cccDNA, expression of the HBV genome via the cellular RNA polymerase II is possible, leading to the pregenomic (pg) RNA and other viral mRNAs. The pgRNA is translated to the core protein and the viral polymerase. These three components assemble in the cytosol with the support of cellular chaperones to form the immature core particle that encapsidates the pgRNA. The polymerase transcribes the pgRNA to DNA using its N-terminal domain as primer for the first-strand synthesis. Mature core particles gain an affinity to the cytosolic preS1 domain of the LHBs protein, become enveloped, and are released as complete infectious viruses via exocytosis. The viruses are able to enter the target cell via endocytosis and to start a new round of replication (for more details, see Ref. [1] or other reviews). The incoming core protein ends in the nucleoplasm where it could be harmful due to its nucleic acid–binding domain. However, it reassembles in the nucleoplasm to core particles and encapsidates nonspecific RNA as in bacteria [78]. As a consequence, hepatocytes of heavily infected liver contain HBcAg both in the cytosol where it may acquire the envelope and in the nucleus where it is stored as a by-product of the viral life cycle.

10.3.1.2 Immunogenicity

HBcAg is a very strong B cell antigen and is able to induce T helper cell–independent anti-HBc production [79]. The major B cell epitope of HBcAg is conformation dependent and located on the tip of spikes formed by two alpha helices in the central part of the core protein (see Chapters 2 and 11 through 13). The encounter of the HBcAg with B cells is facilitated by a surprising exocytosis mechanism of nonenveloped core particles using the factor Alix [80]. It appears that HBV supports by this removal of excessive core particles the encounter of HBcAg with the immune system outside of the hepatocytes, leading to the formation of high-titered anti-HBc, which, however, cannot react with the infectious enveloped HBV. HBcAg carries also important T cell epitopes, which are believed to support resolution of acute disease and to contribute to chronic disease if the cytotoxic T cell response is present but too weak [81]. The secreted and soluble version of HBcAg, the HBeAg, seems to counteract this cellular defense by unknown mechanisms. HBeAg-positive HBV carriers have usually very high titers of HBV DNA, HBsAg, HBeAg, and hepatic HBcAg but no inflammatory liver disease.

10.3.2 Diagnostic Significance of HBcAg

10.3.2.1 HBcAg

The detection of HBcAg in HBV-infected liver biopsies (using immune staining with anti-HBc) had for a while some diagnostic significance but is today practically obsolete. Free HBcAg in serum is not detectable even in immunotolerant HBV carriers without anti-HBc, but it can be released from the enveloped HBV particle with mild detergents [82]. In most cases, the released HBcAg is immediately covered by

the large excess of coexisting anti-HBc, but it is possible to denature the immune complexes and to detect the unfolded core antigen with conformation-independent antibodies in an immune assay. Coexisting HBeAg is also detected by this method due to its largely overlapping protein sequence. Thus, the antigen detected by this diagnostic test is called HBc-related antigen or HBcrAg [83]. The assay is less sensitive than detection of HBV DNA by nucleic acid amplification and rarely used.

10.3.3 Diagnostic Significance of Anti-HBc

10.3.3.1 Specificity and Sensitivity

Anti-HBc is considered the most universal serological marker of resolved or ongoing HBV infection. A negative anti-HBc result virtually excludes HBV as cause of an acute hepatitis. After resolution, anti-HBc remains positive for long time, in many cases lifelong. The reasons for this persistence are not only the high immunogenicity of HBcAg but also the fact that HBV usually persists after resolution in occult form in the liver where it replicates at low level under the control of the immune system. The highest anti-HBc titers are found during chronic HBV infection irrespective of the inflammatory activity. Rare exceptions are immunotolerant HBV carriers who had a quasi-ablated immune system at the time of infection [82] or with inherited agammaglobulinemia. Newborns in which perinatal vaccination failed remain also in rare cases anti-HBc negative in spite of persistent HBV infection [84].

More problematic is the detection of occult infections with a very low level of HBcAg production and a protective T cell response against HBV. In these frequently occurring cases, anti-HBc may be borderline positive or false negative. In normal life, this problem is irrelevant, but in blood or organ donors, it may lead to transmission of HBV [14,58]. Most critical is the failure to detect an OBI before bone marrow transplantation or lymphoma therapy. In the absence of functional T and B cells, HBV reactivates its replication to very high levels. This remains unnoticed until after immune reconstitution the surviving T memory cells are activated and attack the HBV-infected hepatocytes, which may lead to lethal fulminant hepatitis [10,69].

10.3.3.2 HBcAg: A Problematic Reagent for Anti-HBc Assays

The first anti-HBc assays were done with the classical complement fixation test, soon after with various formats of solid-phase radio or enzyme immune assays [84–86]. Using the anti-μ capture format, it was also possible to detect selectively anti-HBc of the IgM class [87]. The antigen was either extracted from Dane particles with mild detergent or post mortem from infected liver. The amount of Dane particle–derived HBcAg in plasma was, however, much lower than that of HBsAg. The infected livers contained much more HBcAg in its free form, but human livers are not readily available, and in most cases, the HBcAg was complexed during extraction with the excessive anti-HBc. Thus, it was essential for the general introduction of anti-HBc assays to replace the natural HBcAg by the HBcAg expressed in transformed *E. coli* clones. The *recombinant* HBcAg (rHBcAg) was relatively easy to extract from the bacteria because it was already assembled (and contained even some RNA as the natural HBcAg from hepatocyte nuclei), but the results of the anti-HBc assays using rHBcAg were not as clear-cut as those using natural HBcAg. The number of borderline or nonreproducible results was much larger and the sensitivity lower [84,86,88]. This created many problems in the testing of blood or organ donors and of persons at risk. In order to avoid unnecessary loss of valuable blood or organ donations, cutoff levels were raised to point where the rate of positives seemed to be acceptable to the blood donation services.

10.3.3.3 Confirmatory Assays for Anti-HBc

For confirmation of questionable results, it is officially recommended to repeat tests with a kit from another producer; divergent results are considered false positive. Unfortunately, divergences were very frequent with low positive results [88]. The sources of these divergences are not understood. In contrast to the HBsAg, there is virtually no genetic or biochemical heterogeneity in HBcAg.

An effort was undertaken to distinguish true- from false-positive results by preincubating the serum sample with recombinant HBcAg and repeating subsequently the assay [89]. Using this inhibition assay about 50% of questionable results are either false positive or false negative. The existence of many false-negative results is indicated by the finding of T cell immunity against HBcAg in anti-HBc-negative vaccinated HCWs [57], and the anti-HBs titer rises in vaccinated persons without preceding booster dose [56]. The results corroborate the need for more sensitive and specific anti-HBc assays. The improvement of rHBcAg for diagnostic purposes is probably the most important step. Different sources of rHBcAg may also be considered including eukaryotic cell lines.

10.4 HEPATITIS DELTA VIRUS: A VLP WITH AN HBV ENVELOPE

10.4.1 Unique Nature of HDV

The first hint on the existence of hepatitis delta virus (HDV) was a new antigen, δ-antigen, discovered in liver biopsies from patients with chronic hepatitis B [90]. Subsequent studies showed that δ-antigen was part of a small defective virus, requiring the envelope of HBV for its morphogenesis, release, attachment, and entry. The genome of HDV resembles viroids, ultrasmall plant pathogens consisting only of a short circular RNA with ribozyme properties but without any protein. HDV differs from viroids in that it encodes in addition to its viroid part a coaminoterminal pair of HD proteins SHDAg and LHDAg. These two proteins govern

the replication, intracellular localization and transport of the HDV ribonucleoprotein (reviewed in Ref. [91]). With ca. 15 million chronic patients, HDV is relatively rare compared to HBV alone with ca. 240 million chronic carriers, but its medical importance is not negligible because the coinfection seems to take a worse course than chronic hepatitis B alone.

10.4.2 HDV as Research Tool

HDV is in many studies a useful substitute for authentic HBV. Transfected hepatoma cell lines produce infectious HBV in relatively low amounts, and susceptibility of hepatic cells for these HBV particles is even lower. Interestingly, HDV particles with an HBV envelope of choice are produced and released in larger amounts than the corresponding HBV and these HDV pseudotypes infect more readily the target cell. One potential reason for this is the high adaption of HBV to differentiated hepatocytes, while the HDV genome replicates in virtually every mammalian cell line. HDV particles are relatively easy to generate by cotransfecting a vector expressing the HDV replicon together with a vector for the entire ORF encoding the three HBs proteins. Using this and other approaches, the high-affinity NTCP receptor for HBV and HDV was identified [42].

When three new species of bat HBV were identified by amplifying, cloning, and sequencing their genomes, it was very difficult to obtain enough virus particles from the minute blood or tissue samples or from cells transfected with the viral genomes. However, it was important to find out whether these bat viruses could infect human hepatocytes. HDV ribonucleoprotein VLPs coated with the surface proteins of the bat viruses could readily be expressed. The VLPs with the bat HBV envelope proteins from tent-making bats (living in Central America) were able to infect cell cultures exposing the human NTCP, while the envelopes from the other two bat HBV species were not infectious for these target cells. This points to a zoonotic potential of this new bat virus, TBHBV [43]. Given the potential risk of TBHBV for humans, it was also interesting to know whether anti-HBs induced by the second-generation vaccine would neutralize its infectivity, which was not the case. This was not surprising because the antigen loop of TBHBV corresponding to the HBsAg loop shows 26 amino acid exchanges. It should be noted that human escape mutants of HBV may also have many exchanges in the HBsAg loop. In one case 16 exchanges were found and this heavily mutated HBsAg was not detected by the best available immune assays. Application of pseudotyped HDV particles with highly mutated HBs proteins would facilitate studies on the improvement of current hepatitis B vaccines and attachment inhibitors [92]. Attachment inhibitors are particularly important for the therapy of chronic hepatitis D because the current HBV replication inhibitors are not active against HDV or the expression of the HBs proteins.

10.5 HEPATITIS E VIRUS

10.5.1 Nature of HEV

Hepatitis E virus (HEV) was recognized indirectly after the identification of HBV and hepatitis A virus (HAV) by the observation that many cases of acute hepatitis were caused neither by HAV nor by HBV. For a while, the infectious agents of these diseases were called *nonA, nonB hepatitis viruses*. Epidemiological evidence suggested that there were at least two types of nonA, nonB viruses. One type was associated with post-transfusion hepatitis suggesting parenteral transmission (now hepatitis C virus [HCV]) and the other caused enterically transmitted hepatitis, today called HEV. Both virus types were identified by cloning and sequencing their genomes. HEV is a nonenveloped small virus with a plus strand RNA genome in its own virus family *Hepeviridae*. The closest related virus family are the enveloped *Togaviridae*. This seems to be surprising, but it is assumed that the life cycle of HEV includes transient envelopment as well. In contrast to the other human hepatitis viruses or togaviruses, HEV cannot dependably be propagated in cell culture. Thus, expression of VLPs using gene technology is particularly important.

10.5.2 Prevalence of HEV

HEV was believed for long time to be a kind of tropical disease linked to transmission by contaminated water. Later it was recognized that the majority of patients in seemingly nonendemic regions did not report a travel to endemic regions before the onset of the disease; the infections were *autochthonous* [93]. The initial error was partly due to the unsatisfactory diagnostic methods. Soon after sequencing the first HEV genome immune assays were developed, which used partial peptides to detect antibodies against HEV. These assays produced inacceptable results in quality control trials [94] but were nevertheless used for many years. VLPs composed of the HEV capsid protein expressed in suitable host cells are necessary to detect anti-HEV antibodies, but they are only recently used in viral diagnosis in some laboratories.

With these improved assays, it became apparent that the HEV infection is much more prevalent than believed. Sequencing showed that there were four genotypes of HEV with remarkable differences in their geographical and host distribution. Genotypes 1 and 2 are adapted to humans, while genotypes 3 and 4 are prevalent worldwide in many animal species, particularly in swine. Genotype 3 or 4 may be transmitted to other species including man and cause the autochthonous cases. These are meanwhile in many regions more frequent than acute hepatitis A but usually they are not diagnosed because the knowledge of this infection is still underdeveloped, which in turn is caused by insufficient laboratory tests. Ca. 30% of the European or North American population have encountered HEV, but only 1 in 1000 infections with genotypes 3 or 4 leads to clinically apparent hepatitis. The disease is usually self-limiting, but in immunosuppressed patients it may take a chronic course [95].

10.5.3 HEV VLPs

As mentioned, VLPs are essential for sensitive and specific anti-HEV tests. Fortunately, the large genetic heterogeneity does not lead to problems for the serologic test because there is only one serotype. This means antibodies against one genotype protect against all genotypes of human HEV.

HEV VLPs have been expressed in various cell types including insect cells but also bacteria. Proteolytic processing of the full-length capsid protein in eukaryotic cells leads to icosahedral VLPs with a T3 or T4 morphology similar to the natural HEV particles with the typical protrusions and clefts [96]. These particles can be used for vaccination against HEV and were reported to be very efficient [97]. China has introduced vaccination against HEV genotype 1 with an *E. coli*–derived vaccine [98].

10.6 HEPATITIS A VIRUS

HAV as a member of the *Picornaviridae* can be reasonably well propagated in cell cultures after some adaption and could be studied without the need for VLPs. The antigens necessary both for diagnostic test and for vaccines are obtained from infected cell cultures [99]. The complicated interplay of structural and nonstructural proteins and the strictly regulated sequence of proteolytic processing make it difficult to generate VLPs with authentic antigenicity of HAV.

10.7 HEPATITIS C VIRUS

HCV was initially recognized as agent of the parenterally transmitted form of nonA, nonB hepatitis and remained elusive for many years before its genome could be identified. The genome sequence clearly identifies HCV as member of the *Flaviviridae* where it forms with its seven genotypes an own genus hepacivirus [100]. HCV could not be replicated in cell cultures for many years. As an intermediate step, replicons were generated, which could replicate that part of the genome that encodes the nonstructural (NS). Later, genomes of selected HCV strains were transduced to special hepatoma cell cultures, which were then able to generate infectious HCV particles [101]. While HCV is meanwhile one of the most studied viruses, VLPs of HCV have never gained importance. One reason may be that HCV itself has no clearly defined spatial structure like other flaviviruses. Its unstructured ribonucleoprotein is initially embedded in intracellular lipid droplets, which later interact with the membrane-bound envelope glycoproteins E1 and E2. The lipid droplets are exported as very-low-density lipoprotein (VLDL) particles of heterogeneous size with the VLDL protein and E1 and E2 at their surface. E1 and E2 containing VLPs without HCV core protein or RNA have been described, but they are not so numerous [102]. The other reason is that conformational epitopes play apparently no role in the immune reaction against HCV. The serodiagnosis is mainly based on the partial peptides of the HCV core and *E. coli*–derived NS3/NS4 antigens. The neutralizing epitopes of E2 are hypervariable and have not yet contributed to a promising vaccine. It is, however, possible to insert these epitopes into the tip of *split* HBV core particles and to induce neutralizing antibodies with potentially broader specificity (see Chapter 12 and Ref. [103]).

10.8 CONCLUSION

In spite of the long history of VLPs in the research on hepatitis viruses, their potential for diagnosis, prevention, and possibly immunotherapy has not yet been fully exploited. The medical applications of HBsAg VLPs in diagnosis and prevention should be improved, by including both the preS domains and the variety of predominant HBV genotypes. Using optimized hepatitis B vaccines with the major neutralizing epitopes, it appears possible to eradicate HBV. HBcAg VLPs are excellent epitope carriers, but their reactivity in diagnosis of OBI is still unsatisfactory. Wider use of HEV VLPs in the detection of antibodies against this underestimated virus is desirable. The necessity and performance of HEV vaccine has to be seen.

ACKNOWLEDGMENTS

The work of the author during 40 years has been supported by numerous grants from the German Research Council (DFG), the Federal Ministry of Research, the European Union, WHO, Paul Ehrlich-Institute, Robert Koch Institute, and the Behring Röntgen Foundation. He thanks all colleagues who have contributed to that work, in particular Gerhard May (Frankfurt), Reiner Thomssen, Werner Stibbe, Klaus-Hinrich Heermann, Angela Uy, Volker Bruss (all Göttingen), William Robinson (Stanford), Michael Kann, Stephan Schaefer, Gregor Caspari, Dieter Glebe, Ulrike Wend, Wulf Willems, Christian Schüttler (all Giessen), and Paul Pumpens (Riga) whom he additionally thanks for editing this chapter.

REFERENCES

1. Gerlich, W.H., Medical virology of hepatitis B: How it began and where we are now. *Virol J*, 2013. **10**: 239.
2. Chudy, M. et al., First WHO International Reference Panel containing hepatitis B virus genotypes A-G for assays of the viral DNA. *J Clin Virol*, 2012. **55**(4): 303–309.
3. Chulanov, V.P. et al., Kinetics of HBV DNA and HBsAg in acute hepatitis B patients with and without coinfection by other hepatitis viruses. *J Med Virol*, 2003. **69**(3): 313–323.
4. Chudy, M. et al., Performance of hepatitis B surface antigen tests with the first WHO international hepatitis B virus genotype reference panel. *J Clin Virol*, 2013. **58**(1): 47–53.
5. Gerlich, W.H., Diagnostic problems caused by HBsAg mutants—A consensus report of an expert meeting. *Intervirology*, 2004. **47**(6): 310–313.
6. Hourfar, M.K. et al., Experience of German Red Cross blood donor services with nucleic acid testing: Results of screening more than 30 million blood donations for human immunodeficiency virus-1, hepatitis C virus, and hepatitis B virus. *Transfusion*, 2008. **48**(8): 1558–1566.
7. Vermeulen, M. et al., Hepatitis B virus transmission by blood transfusion during 4 years of individual-donation nucleic acid testing in South Africa: Estimated and observed window period risk. *Transfusion*, 2012. **52**(4): 880–892.
8. Gerlich, W.H. et al., HBsAg non-reactive HBV infection in blood donors: Transmission and pathogenicity. *J Med Virol*, 2007. **79**(S1): S32–S36.
9. Weusten, J. et al., Refinement of a viral transmission risk model for blood donations in seroconversion window phase screened by nucleic acid testing in different pool sizes and repeat test algorithms. *Transfusion*, 2011. **51**(1): 203–215.
10. Westhoff, T.H. et al., Fatal hepatitis B virus reactivation by an escape mutant following rituximab therapy. *Blood*, 2003. **102**(5): 1930.
11. Raimondo, G., R. Filomia, and S. Maimone, Therapy of occult hepatitis B virus infection and prevention of reactivation. *Intervirology*, 2014. **57**(3–4): 189–195.
12. Gerlich, W.H. et al., Occult hepatitis B virus infection: Detection and significance. *Dig Dis*, 2010. **28**(1): 116–125.
13. Allain, J.P. et al., Infectivity of blood products from donors with occult hepatitis B virus infection. *Transfusion*, 2013. **53**(7): 1405–1415.
14. Blaich, A. et al., Reactivation of hepatitis B virus with mutated hepatitis B surface antigen in a liver transplant recipient receiving a graft from an antibody to hepatitis B surface antigen- and antibody to hepatitis B core antigen-positive donor. *Transfusion*, 2012. **52**(9): 1999–2006.
15. Gerlich, W. and R. Thomssen, Standardized detection of hepatitis B surface antigen: Determination of its serum concentration in weight units per volume. *Dev Biol Stand*, 1975. **30**: 78–87.
16. Schüttler, C.G. et al., Antigenic and physicochemical characterization of the 2nd International Standard for hepatitis B virus surface antigen (HBsAg). *J Clin Virol*, 2010. **47**(3): 238–242.
17. European Association for the Study of the Liver. EASL clinical practice guidelines: Management of chronic hepatitis B virus infection. *J Hepatol*, 2012. **57**(1): 167–185.
18. Gerlich, W., B. Stamm, and R. Thomssen, Prognostic significance of quantitative HBsAg determination in acute hepatitis B. Partial report of a cooperative clinical study of the DFG-focus of "virus hepatitis". *Verh Dtsch Ges Inn Med*, 1977. **83**: 554–557.
19. Burczynska, B. et al., The value of quantitative measurement of HBeAg and HBsAg before interferon-alpha treatment of chronic hepatitis B in children. *J Hepatol*, 1994. **21**(6): 1097–1102.
20. Erhardt, A. et al., Mutations of the core promoter and response to interferon treatment in chronic replicative hepatitis B. *Hepatology*, 2000. **31**(3): 716–725.
21. Brunetto, M.R. and F. Bonino, Interferon therapy of chronic hepatitis B. *Intervirology*, 2014. **57**(3–4): 163–170.
22. Werle-Lapostolle, B. et al., Persistence of cccDNA during the natural history of chronic hepatitis B and decline during adefovir dipivoxil therapy. *Gastroenterology*, 2004. **126**(7): 1750–1758.
23. van Bömmel, F. and T. Berg, Antiviral therapy of chronic hepatitis B. *Intervirology*, 2014. **57**(3–4): 171–180.
24. Perez-Cameo, C., M. Pons, and R. Esteban, New therapeutic perspectives in HBV: When to stop NAs. *Liver Int*, 2014. **34**(Suppl. 1): 146–153.
25. Glebe, D. and W.H. Gerlich, Study of the endocytosis and intracellular localization of subviral particles of hepatitis B virus in primary hepatocytes. *Methods Mol Med*, 2004. **96**: 143–151.

26. Gavilanes, F., J.M. Gonzalez-Ros, and D.L. Peterson, Structure of hepatitis B surface antigen. Characterization of the lipid components and their association with the viral proteins. *J Biol Chem*, 1982. **257**(13): 7770–7777.

27. Peterson, D.L., I.M. Roberts, and G.N. Vyas, Partial amino acid sequence of two major component polypeptides of hepatitis B surface antigen. *Proc Natl Acad Sci U S A*, 1977. **74**(4): 1530–1534.

28. Eble, B.E., V.R. Lingappa, and D. Ganem, The N-terminal (pre-S2) domain of a hepatitis B virus surface glycoprotein is translocated across membranes by downstream signal sequences. *J Virol*, 1990. **64**(3): 1414–1419.

29. Heermann, K.H. et al., Large surface proteins of hepatitis B virus containing the pre-s sequence. *J Virol*, 1984. **52**(2): 396–402.

30. Gilbert, R.J. et al., Hepatitis B small surface antigen particles are octahedral. *Proc Natl Acad Sci U S A*, 2005. **102**(41): 14783–14788.

31. Sureau, C. and J. Salisse, A conformational heparan sulfate binding site essential to infectivity overlaps with the conserved hepatitis B virus a-determinant. *Hepatology*, 2013. **57**(3): 985–994.

32. Grün-Bernhard, S., Molekulare Determinanten der Infektiosität von Hepatitis B Virus Partikeln (Molecular determinants of the infectivity of hepatitis B virus). PhD thesis, 2008. University of Giessen, Giessen, Germany. urn:nbn:de:hebis:26-opus-67254.

33. Bruss, V. et al., Post-translational alterations in transmembrane topology of the hepatitis B virus large envelope protein. *EMBO J*, 1994. **13**(10): 2273–2279.

34. Bruss, V. and D. Ganem, Mutational analysis of hepatitis B surface antigen particle assembly and secretion. *J Virol*, 1991. **65**(7): 3813–3820.

35. Marquardt, O. et al., Cell type specific expression of preS1 antigen and secretion of hepatitis B virus surface antigen. *Arch Virol*, 1987. **96**(3–4): 249–256.

36. Churin, Y. et al., Pathological impact of hepatitis B virus surface proteins on the liver is associated with the host genetic background. *PLoS ONE*, 2014. **9**(3): e90608.

37. Heermann, K.H. et al., Immunogenicity of the gene S and Pre-S domains in hepatitis B virions and HBsAg filaments. *Intervirology*, 1987. **28**(1): 14–25.

38. Bruss, V. and D. Ganem, The role of envelope proteins in hepatitis B virus assembly. *Proc Natl Acad Sci U S A*, 1991. **88**(3): 1059–1063.

39. Prange, R., Host factors involved in hepatitis B virus maturation, assembly, and egress. *Med Microbiol Immunol*, 2012. **201**(4): 449–461.

40. Hildt, E. et al., The PreS2 activator MHBs(t) of hepatitis B virus activates c-raf-1/Erk2 signaling in transgenic mice. *EMBO J*, 2002. **21**(4): 525–535.

41. Bruss, V. and K. Vieluf, Functions of the internal pre-S domain of the large surface protein in hepatitis B virus particle morphogenesis. *J Virol*, 1995. **69**(11): 6652–6657.

42. Yan, H. et al., Sodium taurocholate cotransporting polypeptide is a functional receptor for human hepatitis B and D virus. *Elife*, 2012. **1**: e00049.

43. Drexler, J.F. et al., Bats carry pathogenic hepadnaviruses antigenically related to hepatitis B virus and capable of infecting human hepatocytes. *Proc Natl Acad Sci U S A*, 2013. **110**(40): 16151–16156.

44. Gerlich, W.H., Prophylactic vaccination against hepatitis B— Achievements, challenges and perspectives. *Med Microbiol Immunol*, 2015. **204**: 39–56.

45. Trevisan, A. et al., Demonstration of albumin receptors on isolated human hepatocytes by light and scanning electron microscopy. *Hepatology*, 1982. **2**(6): 832–835.

46. Machida, A. et al., A hepatitis B surface antigen polypeptide (P31) with the receptor for polymerized human as well as chimpanzee albumins. *Gastroenterology*, 1983. **85**(2): 268–274.

47. Krone, B. et al., Interaction between hepatitis B surface proteins and monomeric human serum albumin. *Hepatology*, 1990. **11**(6): 1050–1056.

48. Glebe, D. et al., Pre-s1 antigen-dependent infection of Tupaia hepatocyte cultures with human hepatitis B virus. *J Virol*, 2003. **77**(17): 9511–9521.

49. Kramvis, A., Genotypes and genetic variability of hepatitis B virus. *Intervirology*, 2014. **57**(3–4): 141–150.

50. Weinberger, K.M. et al., High genetic variability of the group-specific a-determinant of hepatitis B virus surface antigen (HBsAg) and the corresponding fragment of the viral polymerase in chronic virus carriers lacking detectable HBsAg in serum. *J Gen Virol*, 2000. **81**(Pt 5): 1165–1174.

51. Saniewski, M., Struktur und Funktion der Oberflächenproteine von ungewöhnlichen Hepatitis B Virus-Varianten. (Structure and function of surface proteins from unusual hepatitis B variants). PhD thesis, 2009: University of Giessen, Giessen, Germany. urn:nbn:de:hebis:26-opus-73574.

52. Torresi, J. et al., Reduced antigenicity of the hepatitis B virus HBsAg protein arising as a consequence of sequence changes in the overlapping polymerase gene that are selected by lamivudine therapy. *Virology*, 2002. **293**(2): 305–313.

53. Avellon, A. et al., European collaborative evaluation of the Enzygnost HBsAg 6.0 assay: Performance on hepatitis B virus surface antigen variants. *J Med Virol*, 2011. **83**(1): 95–100.

54. Shouval, D. et al., Improved immunogenicity in mice of a mammalian cell-derived recombinant hepatitis B vaccine containing pre-S1 and pre-S2 antigens as compared with conventional yeast-derived vaccines. *Vaccine*, 1994. **12**(15): 1453–1459.

55. Leroux-Roels, G., Old and new adjuvants for hepatitis B vaccines. *Med Microbiol Immunol*, 2015. **204**: 69–78.

56. Poovorawan, Y. et al., Evidence of protection against clinical and chronic hepatitis B infection 20 years after infant vaccination in a high endemicity region. *J Viral Hepat*, 2011. **18**(5): 369–375.

57. Werner, J.M. et al., The hepatitis B vaccine protects re-exposed health care workers, but does not provide sterilizing immunity. *Gastroenterology*, 2013. **145**(5): 1026–1034.

58. Stramer, S.L. et al., Nucleic acid testing to detect HBV infection in blood donors. *N Engl J Med*, 2011. **364**(3): 236–247.

59. Feeney, S.A. et al., Reactivation of occult hepatitis B virus infection following cytotoxic lymphoma therapy in an anti-HBc negative patient. *J Med Virol*, 2013. **85**(4): 597–601.

60. Deng, M. et al., The effects of telbivudine in late pregnancy to prevent intrauterine transmission of the hepatitis B virus: A systematic review and meta-analysis. *Virol J*, 2012. **9**: 185.

61. Pande, C. et al., Hepatitis B vaccination with or without hepatitis B immunoglobulin at birth to babies born of HBsAg-positive mothers prevents overt HBV transmission but may not prevent occult HBV infection in babies: A randomized controlled trial. *J Viral Hepat*, 2013. **20**(11): 801–810.

62. Xu, L. et al., Occult HBV infection in anti-HBs-positive young adults after neonatal HB vaccination. *Vaccine*, 2010. **28**(37): 5986–5992.

63. Lai, M.W. et al., Increased seroprevalence of HBV DNA with mutations in the s gene among individuals greater than 18 years old after complete vaccination. *Gastroenterology*, 2012. **143**(2): 400–407.

64. Shouval, D., H. Roggendorf, and M. Roggendorf, Enhanced immune response to hepatitis B vaccination through immunization with a Pre-S1/Pre-S2/S vaccine. *Med Microbiol Immunol*, 2015. **204**: 57–68.

65. Saed, N. et al., *49th European Association for the Study of the Liver International Liver Congress (EASL* 2014). London, U.K., April 9–13, 2014. Abstract O121. *J Hepatol*, 2014. **60**(S1): S50.

66. Bremer, C.M. et al., N-terminal myristoylation-dependent masking of neutralizing epitopes in the preS1 attachment site of hepatitis B virus. *J Hepatol*, 2011. **55**(1): 29–37.

67. Gerlich, W.H., The enigma of concurrent hepatitis B surface antigen (HBsAg) and antibodies to HBsAg. *Clin Infect Dis*, 2007. **44**(9): 1170–1172.

68. Madalinski, K. et al., Analysis of viral proteins in circulating immune complexes from chronic carriers of hepatitis B virus. *Clin Exp Immunol*, 1991. **84**(3): 493–500.

69. Roche, B. and D. Samuel, Prevention of hepatitis B virus reinfection in liver transplant recipients. *Intervirology*, 2014. **57**(3–4): 196–201.

70. Gerlich, W.H., Reduction of infectivity in chronic hepatitis B virus carriers among healthcare providers and pregnant women by antiviral therapy. *Intervirology*, 2014. **57**(3–4): 202–211.

71. Ferguson, M., M.W. Yu, and A. Heath, Calibration of the second International Standard for hepatitis B immunoglobulin in an international collaborative study. *Vox Sang*, 2010. **99**(1): 77–84.

72. Stamm, B., W. Gerlich, and R. Thomssen, Quantitative determination of antibody against hepatitis B surface antigen: Measurement of its binding capacity. *J Biol Stand*, 1980. **8**(1): 59–68.

73. Huzly, D. et al., Comparison of nine commercially available assays for quantification of antibody response to hepatitis B virus surface antigen. *J Clin Microbiol*, 2008. **46**(4): 1298–1306.

74. Gerken, G. et al., Pre-S encoded surface proteins in relation to the major viral surface antigen in acute hepatitis B virus infection. *Gastroenterology*, 1987. **92**(6): 1864–1868.

75. Pasek, M. et al., Hepatitis B virus genes and their expression in *E. coli. Nature*, 1979. **282**(5739): 575–579.

76. Cohen, B.J. and J.E. Richmond, Electron microscopy of hepatitis B core antigen synthesized in *E. coli. Nature*, 1982. **296**(5858): 677–679.

77. Crowther, R.A. et al., Three-dimensional structure of hepatitis B virus core particles determined by electron cryomicroscopy. *Cell*, 1994. **77**(6): 943–950.

78. Rabe, B. et al., Nuclear entry of hepatitis B virus capsids involves disintegration to protein dimers followed by nuclear reassociation to capsids. *PLoS Pathog*, 2009. **5**(8): e1000563.

79. Milich, D.R. and A. McLachlan, The nucleocapsid of hepatitis B virus is both a T-cell-independent and a T-cell-dependent antigen. *Science*, 1986. **234**(4782): 1398–1401.

80. Bardens, A. et al., Alix regulates egress of hepatitis B virus naked capsid particles in an ESCRT-independent manner. *Cell Microbiol*, 2011. **13**(4): 602–619.

81. Chisari, F.V., M. Isogawa, and S.F. Wieland, Pathogenesis of hepatitis B virus infection. *Pathol Biol*, 2010. **58**(4): 258–266.

82. Possehl, C. et al., Absence of free core antigen in anti-HBc negative viremic hepatitis B carriers. *Arch Virol Suppl*, 1992. **4**: 39–41.

83. Suzuki, F. et al., Correlation between serum hepatitis B virus core-related antigen and intrahepatic covalently closed circular DNA in chronic hepatitis B patients. *J Med Virol*, 2009. **81**(1): 27–33.

84. Kantelhardt, V.C. et al., Re-evaluation of anti-HBc nonreactive serum samples from patients with persistent hepatitis B infection by immune precipitation with labelled HBV core antigen. *J Clin Virol*, 2009. **46**(2): 124–128.

85. Wolff, W. and W.H. Gerlich, Direct radioimmunoassay of antibody against hepatitis B core antigen using 32P-labelled core particles. *Eur J Clin Microbiol*, 1984. **3**(1): 25–29.

86. Gerlich, W.H., W. Lüer, and R. Thomssen, Diagnosis of acute and inapparent hepatitis B virus infections by measurement of IgM antibody to hepatitis B core antigen. *J Infect Dis*, 1980. **142**(1): 95–101.

87. Gerlich, W.H. et al., Cutoff levels of immunoglobulin M antibody against viral core antigen for differentiation of acute, chronic, and past hepatitis B virus infections. *J Clin Microbiol*, 1986. **24**(2): 288–293.

88. Caspari, G. et al., Unsatisfactory specificities and sensitivities of six enzyme immunoassays for antibodies to hepatitis B core antigen. *J Clin Microbiol*, 1989. **27**(9): 2067–2072.

89. Huzly, D. et al., Simple confirmatory assay for anti-HBc reactivity. *J Clin Virol*, 2011. **51**(4): 283–284.

90. Rizzetto, M. et al., Immunofluorescence detection of new antigen–antibody system (delta/anti-delta) associated to hepatitis B virus in liver and in serum of HBsAg carriers. *Gut*, 1977. **18**(12): 997–1003.

91. Taylor, J.M., Virology of hepatitis D virus. *Semin Liver Dis*, 2012. **32**(3): 195–200.

92. Urban, S. et al., Strategies to inhibit entry of HBV and HDV into hepatocytes. *Gastroenterology*, 2014. **147**(1): 48–64.

93. Purcell, R.H. and S.U. Emerson, Hepatitis E: An emerging awareness of an old disease. *J Hepatol*, 2008. **48**(3): 494–503.

94. Mast, E.E. et al., Evaluation of assays for antibody to hepatitis E virus by a serum panel. Hepatitis E Virus Antibody Serum Panel Evaluation Group. *Hepatology*, 1998. **27**(3): 857–861.

95. Hoofnagle, J.H., K.E. Nelson, and R.H. Purcell, Hepatitis E. *N Engl J Med*, 2012. **367**(13): 1237–1244.

96. Guu, T.S. et al., Structure of the hepatitis E virus-like particle suggests mechanisms for virus assembly and receptor binding. *Proc Natl Acad Sci U S A*, 2009. **106**(31): 12992–12997.

97. Shrestha, M.P. et al., Safety and efficacy of a recombinant hepatitis E vaccine. *N Engl J Med*, 2007. **356**(9): 895–903.

98. Zhang, J. et al., Development of the hepatitis E vaccine: From bench to field. *Semin Liver Dis*, 2013. **33**(1): 79–88.

99. Vaughan, G. et al., Hepatitis A virus: Host interactions, molecular epidemiology and evolution. *Infect Genet Evol*, 2014. **21**: 227–243.

100. Smith, D.B. et al., Expanded classification of hepatitis C virus into 7 genotypes and 67 subtypes: Updated criteria and genotype assignment web resource. *Hepatology*, 2014. **59**(1): 318–327.

101. Bartenschlager, R. et al., Assembly of infectious hepatitis C virus particles. *Trends Microbiol*, 2011. **19**(2): 95–103.

102. Scholtes, C. et al., High plasma level of nucleocapsid-free envelope glycoprotein-positive lipoproteins in hepatitis C patients. *Hepatology*, 2012. **56**(1): 39–48.

103. Lange, M. et al., Hepatitis C virus hypervariable region 1 variants presented on hepatitis B virus capsid-like particles induce cross-neutralizing antibodies. *PLoS ONE*, 2014. **9**(7): e102235.

11 History and Potential of Hepatitis B Virus Core as a VLP Vaccine Platform

Hadrien Peyret, Sam L. Stephen, Nicola J. Stonehouse, and David J. Rowlands

CONTENTS

11.1 INTRODUCTION

Virus-like particles (VLPs) comprising viral antigenic proteins capable of inducing protective immunity are attractive candidates as alternative vaccines for a number of reasons (Roy and Noad, 2008; Kushnir et al., 2012). These include their inherent safety profile when produced by recombinant expression technology; since they contain no viral genetic material, they are incapable of accidentally transmitting infection. They can also present a means of producing practicable amounts of material, which is often difficult or impossible to obtain by culturing the target pathogen *in vitro*. In addition, proteins that can assemble into relatively large and regularly organized structures (VLPs) are inherently far more immunogenic than unassembled monomeric proteins (Bachmann and Jennings, 2010). These basic advantages have spurred research into VLP vaccines for more than a quarter of a century, but there are still only fully licensed VLP vaccines against two target pathogens, hepatitis B virus (HBV; Crovari et al., 1987; Krugman and Davidson, 1987) and human papillomavirus (Shank-Retzlaff et al., 2006; Monie et al., 2008), although a number of clinical trials have been conducted (Plummer and Manchester, 2010; Kushnir et al., 2012).

It is interesting that two quite distinct VLP structures that have been the subject of a great deal of work since the early 1980s are produced from different structural protein genes of the same virus, HBV. HBV infection is characterized by the massive overproduction of two viral structure-related proteins: s-antigen (sAg) and e-antigen (eAg). Both of these viral products are present at extraordinarily high levels in the blood of patients persistently infected with HBV and are thought to be involved in immune evasion by the virus, in essence acting as immunological *smoke screens*.

The sAg is present as 22 nm diameter spheres and tubes and appears to induce immune tolerance at the levels present in HBV carriers. The same protein is the major component of the outer lipoprotein layer of the Dane particle, the infectious HB virion, and antibodies directed against it can efficiently protect against HBV infection. Early vaccines against hepatitis B were produced from chemically inactivated sAg derived from blood donated by chronic HBV carriers (Ellis and Gerety, 1989). Although this vaccine was efficacious, it was clearly limited in supply and carried the inherent dangers associated with inactivation of live virus–containing material. In 1982, it was shown that sAg could be efficiently expressed in yeast and had the ability to assemble into authentic sAg particles in the absence of other viral components (Valenzuela et al., 1982). The yeast-expressed vaccine, which was introduced in 1986, has been used to immunize millions of people and has become a major component of childhood vaccines in many parts of the world (Hilleman, 2011).

The eAg is present in the blood of hepatitis B carriers as a soluble protein and is a modified form of the HBcAg protein, which forms the viral nucleocapsid. The eAg is translated from a translation initiation site upstream of that from which the regular core protein is translated. Translation from this upstream initiation site produces a core protein with an N-terminal extension, the precore region, which contains a secretory signal. As a consequence of this difference, the precore extended protein is secreted via the ER where it is processed at both its N- and C-termini and folded into a form that cannot self-assemble into particulate structures. It is this soluble form of the protein that expresses the distinct "e" antigenicity. The switch from eAg to anti-eAg is one of the serological markers of resolution of chronic HBV infection.

The core (HBcAg) protein is translated from an initiation site downstream of the precore initiation codon and lacks a secretory signal. The HBcAg protein remains intracellular and adopts a predominantly α-helical folding, which is quite distinct from that of eAg. HBcAg protein rapidly dimerizes to form structural subunits that go on to assemble in the presence of the viral reverse transcriptase and pregenomic RNA (pgRNA) into icosahedral nucleocapsid HBc particles. The HBcAg protein can also self-assemble into similar particles in the absence of viral pgRNA. These HBc VLPs are highly immunogenic and have been explored extensively as a platform for the development of novel recombinant vaccines (Pumpens and Grens, 2001).

11.2   ROLE OF HBc IN THE VIRAL REPLICATION CYCLE

The infectious virion of HBV is known as the Dane particle and comprises an icosahedral nucleocapsid core enveloped by a lipid membrane incorporating the viral glycoproteins, S, M, and L, which are a nested set of proteins with a common C-terminal region. HBcAg is predominately α-helical but has a less structured region containing a large number of arginine residues toward its C-terminus (Figure 11.1). This highly positively charged *protamine-like* domain has a high affinity for nucleic acids and is largely responsible for the encapsidation of the viral pgRNA. However, specificity of encapsidation of pgRNA over other intracellular RNAs is determined by the association of the viral polymerase (reverse transcriptase) with pgRNA, which is its translational template, to reveal preferred sites for interaction with HBcAg and thus encapsidation. The pgRNA is reverse transcribed within the core particle; a process that is modulated by the phosphorylation status of serine resides within the C-terminal domain, to ultimately produce the DNA form of the viral genome. The nucleocapsid core particle is enveloped by a membrane incorporating the S, M, and L multiple

membrane-spanning viral glycoproteins of which the smallest (S) is numerically the most abundant. A domain formed at the tips of the prominent spikes decorating the core particle has an affinity for a sequence present in a luminal portion of the S protein, and the association of these two regions is important during the viral assembly and maturation process (Dyson and Murray, 1995). The exact details of mature Dane particle assembly are unclear and the core particle envelopment appears to depend on the status of DNA transcription within the particle.

11.3   STRUCTURE DETERMINATION AND ASSEMBLY OF HBc

The 183–185 amino acid protein, HBcAg, self-assembles both *in vivo* and *in vitro* to form capsids with icosahedral symmetry, also referred to as cores. The capsids are unusual in that they exist in two forms, having either T = 3 or T = 4 symmetry. These are composed of 180 or 240 copies of the HBcAg protein, respectively, arranged as dimers. The monomeric protein consists of two domains separated by a short linker region, an assembly domain (aa 1–140) necessary and sufficient for capsid assembly, and the C-terminal domain (aa 150–183) (Birnbaum and Nassal, 1990; Nassal, 1992). The arginine-rich C-terminus contains a nucleic acid–binding domain that interacts with the pgRNA in association with the viral polymerase in a highly specific manner, selecting this for packaging into the viral capsid. The greater the truncation from aa149, the higher the proportion of T = 4 vs. T = 3 capsid result. The reasons for this are not fully understood, although it has been suggested to be due to destabilization of the T = 3 form rather than T = 4 stabilization.

The first information on the structure of the capsid came from single-particle reconstruction of T = 4 assembly domain capsids (aa 1–149) (Böttcher et al., 1997), and the crystal structure of the assembly domain was solved in 1999 (Wynne et al., 1999). From this structure, the capsid measured 175 Å in diameter with a luminal diameter of 150 Å. The dimer can be considered an upturned T, composed of 10 helices (5 α-helices per monomer), where helices 3 and 4 of each monomer form a 4-helix bundle comprising the vertical part (i.e., spikes on the capsid surface) and the remainder of the protein forms the horizontal, that is, the structural lattice of the icosahedral capsid (Figure 11.1). The packing of the helices results in the burial of ~2000 Å of solvent-accessible surface area of which a large proportion is hydrophobic. This extensive surface contact area and the potential exposure of hydrophobic residues in the monomer are likely to account for why the protein favors dimerization.

At the tips of the dimers, two adjacent loops are present, each linking helices 3 and 4 of the monomeric components. These loops constitute the major immunodominant region (MIR) of the protein. Together, the intra- and interdimer interactions result in a highly stable capsid, which can tolerate high pH and temperature. Cys-61 of helix 3 can form a disulfide bond in the dimer. Although this bond is not

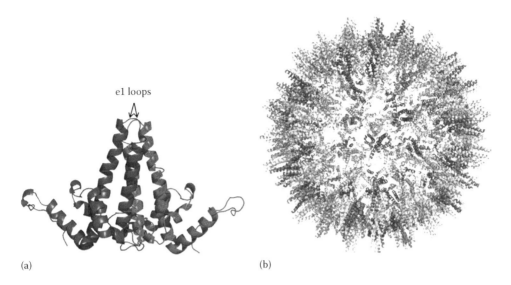

FIGURE 11.1 The structure of the hepatitis B core antigen in the context of the T = 4 capsid. PDB:1QGT. (a) The structure of the dimer taken from the T = 4 capsid and (b) the structure of the T = 4 capsid at 3.9 Å resolution. The four quasi-equivalent monomers are colored in magenta, green, yellow, and blue.

necessary for capsid assembly, it has been shown to contribute to the stability of resulting capsids. Although this disulfide bond was not observed in samples analyzed by cryo-EM (Yu et al., 2013), comparisons between wild-type (WT) capsids and a Cys61Ala mutant by mass spectrometry showed that when collision voltage is increased, the mutant capsids eject a single monomer, whereas the WT protein ejects a dimer (Uetrecht et al., 2011).

Although the *in vivo* capsid assembly process proceeds to form a nucleocapsid via the formation of a nucleus of pgRNA and reverse transcriptase (Bartenschlager and Schaller, 1992), *in vitro* assembly of HBc particles readily occurs in the absence of viral components other than the HBc protein. This has provided an interesting model system for structural studies (as explained earlier) and also for the study of capsid assembly. There have been extensive studies into the *in vitro* assembly behavior of the assembly domain primarily by Zlotnick et al., using high ionic strength conditions to initiate assembly of a solution of dimeric protein into capsids (Zlotnick et al., 1996; Ceres and Zlotnick, 2002). Zlotnick et al. combined experimental data with theoretical models to propose a mechanism for the capsid assembly process (Zlotnick et al., 1999). The experimental data fitted a kinetically limited assembly path, requiring the formation of a nucleus before polymerization can occur. The data suggested a trimeric nucleus, a trimer of dimers, followed by the sequential addition of dimers (Stray et al., 2004). Further evidence for the significance of the trimeric nucleus came from ion mobility–mass spectrometry studies by the group of Albert Heck (Uetrecht et al., 2011), suggesting that these were competent to go on to form capsids.

Anti-HBV small molecules called heteroaryldihydropyrimidines (HAPs) (Deres et al., 2003) have been shown to have an accelerating and misdirecting effect on assembly (Stray et al., 2005). In low concentrations, the compounds are thought to accelerate the assembly process by binding to the C-terminal part of the assembly domain and increasing nucleation and hence the overall rate of assembly. At high concentrations, the compounds may promote the formation of kinetic traps by promoting the formation of nuclei at the expense of there being any free dimers available to complete the assembly process.

11.3.1 CONFORMATIONAL DIVERSITY

The ability to explore a range of conformations is central to the function of many proteins. For HBcAg, the protein needs to be able to exist both free in a dimeric form and as part of both T = 3 and T = 4 forms of the viral capsid.

In the crystal structure, the C-terminus is present on the inner surface of the capsid. Kinetic hydrolysis studies by Hilmer et al. (2008) demonstrated that the C-terminus was able to protrude from the capsid outer surface and be protease accessible. Two forms, open (when C-terminus extended) and closed (when folded back toward the body of the protein), were also demonstrated in the free dimer and were proposed to be in equilibrium in solution (Hilmer et al., 2008). The C-terminus has also been shown by cryo-EM to become surface accessible, consistent with a loss of electron density from the capsid interior. C-terminus dynamics have also been implied by NMR studies. C-terminal residues displayed a significant extent of mobility on the picosecond timescale (Freund et al., 2008). We have demonstrated that the presence of a HAP compound affected the dynamics of the dimer in solution and resulted in oligomers different in shape, as characterized by ion mobility–mass spectrometry (Shepherd et al., 2013).

11.4   RECOMBINANT EXPRESSION OF HBc

The region of the HBV genome that encodes the HBcAg protein was first identified in 1979 by sequencing of the viral DNA inserts in plasmids that conferred on *Escherichia coli* the ability to express antigen that was both reactive with HBV positive serum and able to induce HBc-reactive antibodies when injected into rabbits (Pasek et al., 1979). At that time, it was noted that the C-terminal portion of the protein is extraordinarily rich in positively charged residues (arginine), and it was speculated that this was the nucleic acid–binding domain and to be important for particle assembly (Edman et al., 1981). In these early studies, it was noted that a portion of the anti-HBc-reactive protein was present as high-molecular-weight aggregates—the first indications that the protein was able to self-assemble into particulate structures. The faithful assembly of HBc protein expressed in *E. coli* into regular 27 nm particles resembling the core particles derived from virus-infected liver was later established by biophysical characterization and electron microscopy (Cohen and Richmond, 1982; Stahl et al., 1982). It was subsequently shown that HBcAg could also be expressed to form 27 nm VLPs in a wide variety of systems including yeast, mammalian, insect, and plant cells. Moreover, the protamine-like highly charged C-terminal of the protein could be deleted without eliminating assembly but resulting in considerably reduced nucleic acid incorporation.

11.5   ANTIGENIC/IMMUNOGENIC PROPERTIES OF HBc PARTICLES PRESENTING FOREIGN SEQUENCES

The first practical applications of the recombinant expression of HBcAg were the generation of immunological reagents for hepatitis B diagnostic purposes, and during these developments, it was noted that the HBc particles were highly immunogenic in a range of mammalian species. At that time, in the early 1980s, there was considerable interest in the possibilities offered by synthetic peptide epitopes as potential vaccine candidates. Newly developed molecular techniques had facilitated the identification of important antigenic sequences of a wide range of infectious agents, especially viruses, and a lot of effort was expended into converting these *minimal epitopes* into practical vaccines. Although these studies provided a wealth of understanding of the nature of antigenic epitopes, none stood the test of time to become practical vaccines. This was largely due to the fact that the great majority of natural epitopes are *conformational*, that is, they are normally only present in their native conformation when present within the complex structure of the protein/particle from which their sequence was derived. In addition, most epitopes comprise component parts derived from multiple separate regions of the protein/particle that only appear as a complete epitope when present in the native protein structure. These characteristics are typical of the great majority of natural epitopes capable of eliciting humoral immunity and are extremely difficult if not impossible to reproduce in

a synthetic peptide mimic. A notable exception to this general rule is the VP1 G–H loop sequence of foot-and-mouth disease virus (FMDV; Bittle et al., 1982). A variety of immunological and biochemical approaches had demonstrated that this sequence contains an important epitope, and moreover, it was capable of eliciting virus-neutralizing antibodies when used as a synthetic immunogen. The resolution of the molecular structure of the virus by x-ray crystallography helped to explain this apparent anomaly since it was found that the VP1 G–H loop was present on the surface of the particle with little interaction with other components of the virus particle. Thus, this critical sequence protruded from the virus surface as an isolated and unconstrained feature with many of the characteristics of a synthetic peptide.

This was an exciting finding and stimulated a great deal of work dedicated to improving the immunogenicity of the peptide sequence. Covalent coupling of the synthetic peptide to keyhole limpet hemocyanin (KLH), a common method of improving immunogenicity, improved immune responses to the peptide, as did extension of the synthetic sequence to include defined helper T-cell epitopes. In addition, because of its proven high immunogenicity, the sequence was linked to the N-terminus of HBcAg and expressed as a fusion protein in mammalian cells (Clarke et al., 1987). The expressed fusion protein assembled into regular 27 nm particles that had acquired the antigenic characteristics of the FMDV VP1 G–H loop. Immunization of small laboratory animals (guinea pigs) with the recombinant particles induced high levels of virus-neutralizing antibodies, and quantitatively their immunogenicity was comparable to FMDV particles. Unfortunately, practical peptide epitope-based FMDV vaccines have never been developed from these approaches for two main reasons. First, the induction of protective immunity was found to be quite species specific and the efficacy observed in guinea pigs was considerably less in pigs and cattle. Second, reliance of a vaccine on induction of immune responses to a single epitope, as compared to the multivalent responses induced against the whole virus particle, encourages the selection of antigenic variant viruses that can evade immunity. This was in fact observed in a peptide immunization trial in cattle.

This experience highlights the urgency to develop recombinant VLP platforms that can present larger, multiepitope immunogens, for example, whole proteins. This may overcome the problems associated with induction of too narrow an immune spectrum, as seen with single-epitope vaccines, and boost the level of response due to the *adjuvanting* effect of multimeric presentation on a VLP backbone.

11.6   INSERTION SITES IN HBc PROTEIN

The first recombinant fusion constructs linking foreign epitope sequences to HBc protein were made by genetically linking the sequence of antigens of interest to the N-terminus of the nucleocapsid protein (Clarke et al., 1987). Inclusion of a short portion of the precore sequence between the foreign epitope and HBcAg was necessary to allow full exposure of the antigen, but recombinant N-terminal fusion HBc particles were shown to be excellent immunogens, orders of magnitude superior to synthetic peptides alone and approximately tenfold better than peptides chemically linked to a large carrier protein, KLH.

Surprisingly, despite its internal localization in the HBc particle, foreign sequences can also be fused to the C-terminal portion of HBcAg and still assemble into particles presenting the added sequence at their surfaces (Stahl and Murray, 1989; Schödel et al., 1992; Koletzki et al., 2000). An explanation of this is that the C-terminal protamine-like domain appears to be transiently exposed at the surface of the natural HBc particle, as discussed earlier, and the arginine-rich regions are thought to direct its transport into the nucleus. Fusion of foreign sequences is tolerated by a range of C-terminally truncated HBcAg proteins, and the capacity of the site to accept large inserts while maintaining the ability to form particles is remarkable. However, C-terminal fusions are not as immunogenic as particles containing inserts at other positions in the molecule (Schödel et al., 1992; Lachmann et al., 1999).

Epitope mapping of HBcAg together with the resolution of structure of the particle suggested that an internal region of the HBcAg protein sequence might be the optimal site for insertion of foreign antigens. This is located between amino acids 74–87 and is at the tips of the four-helix bundle spikes that are prominently exposed at the surface of the assembled particle. This is the location of the e1 loop epitope characteristic of HBc particles and the great majority of the HBc-reactive antibodies in sera are directed to this site. As a consequence, the e1 loop containing this prominent antigenic site has been termed the MIR. It soon became apparent that the immunodominance of the MIR was transferred to antigens/epitopes inserted at this site (Brown et al., 1991; Chambers et al., 1996; Ulrich et al., 1998). Furthermore, with appropriate sequences, the native conformation of the antigen was more favorably preserved when presented at this site. In addition, insertions made in this region effectively destroyed the ability of the chimeric particle to induce HBcAg-reactive antibodies. Surprisingly large inserts can be accommodated at the MIR while maintaining their native conformation. The green fluorescent protein (GFP) was inserted, via flexible linker sequences, and the assembled chimeric HBc particles were both fluorescent and immunogenic (Kratz et al., 1999).

11.7 ALTERNATIVE MODES OF ANTIGEN PRESENTATION ON HBc PARTICLES

11.7.1 MOSAIC PARTICLES

Numerous groups have developed alternatives to straightforward genetic fusions in order to develop easier antigen presentation strategies or to overcome the limits to insertions described earlier. The most common strategy used to force HBcAg particles to carry large inserts at the C-terminus is to produce mosaics, particles containing a mix of modified and unmodified HBcAg proteins. Koletzki et al. (1999) used this strategy to fuse 120 amino acids of hantavirus protein at

the C-terminus of a truncated Δ144 HBcAg. A stop codon read-through mechanism was exploited to produce modified and unmodified HBcAg protein from the same construct. This allowed particle formation with this insert, whereas a straightforward C-terminal fusion did not. This was seen again by Ulrich et al. (1999), who produced mosaic particles with Δ144 HBcAg containing 144 extra amino acids at the C-terminus. This dual expression strategy was studied in more detail by Kazaks et al. (2002), who found that it permitted the formation of particles with 94, 114, or 213, but not 433 amino acids fused to the C-terminus of Δ144 HBcAg, when Δ144 HBcAg fusion proteins alone failed to allow the formation of regular particles (or gave prohibitively low yields). The authors conclude that the mosaic strategy can be of use to incorporate *problematic* (i.e., long or hydrophobic) sequences into HBcAg particles at the C-terminus. The mosaic technique clearly provides another tool for the insertion of peptides into the C-terminus of HBcAg protein, but it should be noted that there is an inherent lack of control of the assembly process so that particles could contain very few copies of the insert, and some particles may not contain any insert at all.

11.7.2 CHEMICAL CONJUGATION

An alternative strategy for display in the MIR is by chemical conjugation. It is possible to genetically modify HBcAg protein by inserting a lysine residue in the MIR and removing all cysteines from the sequence. A peptide insert with a cysteine on one terminus can then be chemically conjugated to the MIR lysine via its cysteine (Jegerlehner et al., 2002). This allows display of peptides chemically conjugated to the MIR, but it should be noted that the efficiency of conjugation decreases as the size of the insert increases. Indeed, a small peptide like the FLAG tag was coupled with about 50% efficiency, M2e (23 amino acids) coupling was about 40% efficient, and coupling of a 66 amino acid–long epitope from *Toxoplasma gondii* protein GRA2 was only about 30% efficient. The efficiency of coupling of a 134 amino acid–long PLA2 glycoprotein from bee venom was estimated to be even lower. Chemical coupling may therefore be an option if genetic fusion is unsuccessful or if the insert must present a free terminus.

11.7.3 AFFINITY BINDING

Another strategy is noncovalent interaction of the insert and the HBc particle carrier through a peptide tag that binds to the MIR. The natural interaction between HBcAg and HBsAg (Dyson and Murray, 1995) was exploited to design an oligopeptide (GSLLGRMKGA), which binds the MIR of HBcAg (Blokhina et al., 2013). This peptide tag, when fused to M2e, allowed display of the epitope on the HBcAg particles. The strength and stability of the interaction deserves further study, however, as sucrose gradient ultracentrifugation disrupted the interaction between the particles and the tagged epitope.

11.7.4 SPLIT CORE

The HBcAg protein can tolerate large inserts that may still form regular icosahedral particles, especially when containing domains with adjacent termini (Kratz et al., 1999). The split core strategy was designed to overcome conformational restrictions associated with the insertion of antigenic sequences in which the natural termini are well separated and exploits the high-affinity noncovalent association within the HBcAg particles following posttranslational cleavage of the insert in the MIR (Walker et al., 2008). In the first application of this principle, the insert was designed to contain the cleavage site for tobacco etch virus protease at its N-terminus. The modified HBcAg protein construct was then expressed, together with the protease, which cleaved the insert at its N-terminus without preventing capsid assembly. This technique allowed regular particle assembly when a straightforward fusion would not, as in the case of the *Borrelia burgdorferi* protein OspA, a 28 kDa protein, in which the N- and C-termini are very far apart. Although the particles formed when the OspA insert alone was introduced were not morphologically comparable to normal HBc particle, their immunogenicity was comparable to that of a validated vaccine (Simon et al., 1991) bearing the same epitope (Nassal et al., 2005). However, inclusion of the TEV protease recognition site and coexpression with the protease resulted in regular icosahedral HBc-like particles (Walker et al., 2008). This approach can therefore release the tension present when the natural termini of the inserted sequence are far apart, thus distorting or preventing HBc particle assembly.

The strategy was simplified to give SplitCore technology in which the HBcAg sequence is separated within the MIR between amino acids 79 and 80 into two distinct open reading frames, coreN and coreC, and expressed together as a bicistronic operon in *E. coli* (Walker et al., 2011). Thus, the N- and C-terminal portions of the protein are expressed separately within the same cell but associate through hydrophobic

interactions to reconstitute functional dimers and complete particles. This allows large inserts to be fused either at the C-terminus of the upstream (N-terminal) half monomer or at the N-terminus of the downstream (C-terminal) half monomer (Figure 11.2). The generation of irregular particles was abolished when the small 27 aa hepatitis C virus hypervariable region I (HVRI) or the large 255 aa OspA protein was fused to either of the half cores and the particles were immunogenic in mice (Lange et al., 2014). An even larger (319 aa) circumsporozoite protein of the malaria parasite *Plasmodium falciparum* expressed from a SplitCore produced regular VLPs, whereas the same protein, when expressed via a standard HBc protein, was insoluble, and in a murine model, the particles afforded more protection than two other malaria vaccine candidates. The N- and C-terminal halves of a self-assembling split GFP combined to generate fluorescent particles when fused to the coreN and coreC fragments. Several other protein inserts resulted in assembly of regular VLPs. These include the 56 aa B1 immunoglobulin-binding domain from protein G (GB1), the ~210 aa outer surface protein C (OspC), or the even larger ~240 aa attenuated enterotoxin B from *Staphylococcus aureus* (SEB), and the OspA domain which was successfully displayed from either or simultaneously as two copies from both of the split GFP domains. The principle was also successfully applied to woodchuck HBcAg to circumnavigate potential complications due to preexisting human HBV immunity (Walker et al., 2011).

11.7.5 TANDEM CORE

The MIR is present at the tips of the four-helix bundle spikes formed by the dimeric assembly of HBcAg protein, and therefore, two identical preferred insertion sites are inevitably closely adjacent in space. This is likely to restrict the insertion of large globular proteins due to potential conformational clashes between the inserted sequences. Tandem core is a strategy designed to overcome these restrictions by covalently linking two HBcAg proteins via a flexible linker

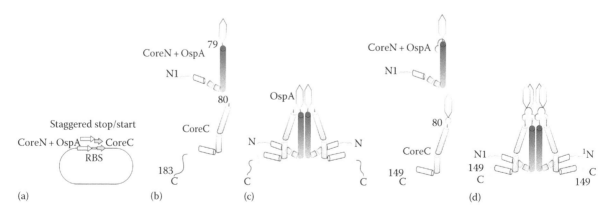

FIGURE 11.2 Development of the split core. (a) The HBc core is split into the N-terminal region containing coreN and the C-terminal region containing coreC and is expressed either as a bicistronic mRNA with the two parts of the core flanking a second ribosome binding site or by the use of a staggered stop codon of the coreN leading into the start codon of coreC in a plasmid. The individual domains (b) assemble into SplitCore dimers (c). Self-assembling domains (e.g., of GFP) can be inserted separately in the MIRs (d). RBS, ribosome binding site; MIR, major immunodominant region of the HBV core protein; α1–5, the alpha chains of the HBV core protein; C-term domain, nucleic acid–binding C-terminal domain of the HBV core protein.

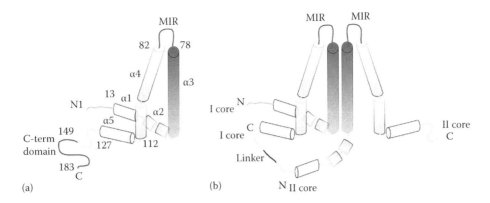

FIGURE 11.3 Development of the tandem core. (a) Monomer of the HBV core protein. (b) In the tandem core, a linker connects the C-terminus of the first core protein to the N-terminus of the second core protein. MIR, major immunodominant region of the HBV core protein; α1–5, the alpha chains of the HBV core protein; C-term domain, nucleic acid–binding C-terminal domain of the HBV core protein.

that defines the dimeric association of the up- and downstream proteins in the structural dimer (Figure 11.3). The linker replaces the C-terminal protamine-like domain of the upstream HBcAg protein and is fused to the N-terminus of the downstream copy (Peyret et al., 2015). The downstream copy may include the C-terminal domain, or this may be deleted with consequences for nucleic acid binding and particle stability. GFP as a test protein can be readily expressed in these tandem-fused constructs to produce regular icosahedral particles, especially when expressed in plants. This strategy also provides the potential to insert different antigens simultaneously into HBc particles.

11.7.6 Nanobody Tandem Core

The tandem core concept has been used to produce the nanobody tandem core. In this application, the *proof of principle* GFP insert in tandem core was replaced by the terminal antigen-recognizing domain of a camelid single-chain heavy-chain antibody (the so-called variable heavy-chain domain of a heavy chain antibody (VHH) or nanobody). A nanobody specific for GFP (Kirchhofer et al., 2010) was inserted into one of the MIRs of the tandem core construct and the resulting expressed protein assembled to form regular HBc particles displaying the nanobody at their surface. Moreover, the nanobody retained biological function and bound avidly to GFP. These results show that in principle HBc particles decorated with an appropriate nanobody could be used as generic adapters to bind proteins of interest bearing the cognate epitope. Interestingly, expression of nanobody fused into the MIR of monomeric HBcAg failed to produce assembled particles.

11.8 CONCLUSIONS

The HBV core particle is a highly versatile foundation platform for the construction of novel VLPs capable of presenting a wide range of epitopes and proteins at its surface. The protein structure and assembly is extremely robust and can be expressed in a wide range of systems. It is intrinsically highly immunogenic due to its regular repetitive epitope

presentation, and the presence of strong Th epitopes and encapsidated nucleic acid can have a strong adjuvanting effect. Several ingenious modifications to the basic system have facilitated the presentation of extremely large full protein inserts at the particle surface. The HBcAg particle is expressed and assembled within the cytosol and is not, per se, suitable for the presentation of glycosylated proteins, such as viral envelope proteins. However, chemical coupling via judiciously placed lysine residues or via VHH-mediated capture may address this limitation. Although most work has been done with the HBcAg protein of human HBV, the equivalent protein from woodchuck hepatitis virus may have additional advantages due to extra stability and the lack of epitopes of the human virus.

REFERENCES

Bachmann, M.F. and Jennings, G.T. (2010). Vaccine delivery: A matter of size, geometry, kinetics and molecular patterns. *Nat Rev Immunol* **10**, 787–796.

Bartenschlager, R. and Schaller, H. (1992). Hepadnaviral assembly is initiated by polymerase binding to the encapsidation signal in the viral RNA genome. *EMBO J* **11**, 3413–3420.

Birnbaum, F. and Nassal, M. (1990). Hepatitis B virus nucleocapsid assembly—Primary structure requirements in the core protein. *J Virol* **64**, 3319–3330.

Bittle, J.L., Houghten, R.A., Alexander, H., Shinnick, T.M., Sutcliffe, J.G., Lerner, R.A., Rowlands, D.J., and Brown, F. (1982). Protection against foot-and-mouth disease by immunization with a chemically synthesized peptide predicted from the viral nucleotide sequence. *Nature* **298**, 30–33.

Blokhina, E.A., Kuprianov, V.V., Stepanova, L.A., Tsybalova, L.M., Kiselev, O.I., Ravin, N.V., and Skryabin, K.G. (2013). A molecular assembly system for presentation of antigens on the surface of HBc virus-like particles. *Virology* **435**, 293–300.

Böttcher, B., Wynne, S.A., and Crowther, R.A. (1997). Determination of the fold of the core protein of hepatitis B virus by electron cryomicroscopy. *Nature* **386**, 88–91.

Brown, A.L., Francis, M.J., Hastings, G.Z., Parry, N.R., Barnett, P.V., Rowlands, D.J., and Clarke, B.E. (1991). Foreign epitopes in immunodominant regions of hepatitis B core particles are highly immunogenic and conformationally restricted. *Vaccine* **9**, 595–601.

Ceres, P. and Zlotnick, A. (2002). Weak protein–protein interactions are sufficient to drive assembly of hepatitis B virus capsids. *Biochemistry* **41**, 11525–11531.

Chambers, M.A., Dougan, G., Newman, J., Brown, F., Crowther, J., Mould, A.P., Humphries, M.J., Francis, M.J., Clarke, B., Brown, A.L., and Rowlands, D. (1996). Chimeric hepatitis B virus core particles as probes for studying peptide–integrin interactions. *J Virol* **70**, 4045–4052.

Clarke, B.E., Newton, S.E., Carroll, A.R., Francis, M.J., Appleyard, G., Syred, A.D., Highfield, P.E., Rowlands, D.J., and Brown, F. (1987). Improved immunogenicity of a peptide epitope after fusion to hepatitis B core protein. *Nature* **330**, 381–384.

Cohen, B.J. and Richmond, J.E. (1982). Electron-microscopy of hepatitis B core antigen synthesized in *Escherichia coli*. *Nature* **296**, 677–678.

Crovari, P., Crovari, P.C., Petrilli, R.C., Icardi, G.C., and Bonanni, P. (1987). Immunogenicity of a yeast-derived hepatitis B vaccine (Engerix-B) in healthy young adults. *Postgrad Med J* **63**(Suppl. 2), 161–164.

Deres, K., Schroder, C.H., Paessens, A., Goldmann, S., Hacker, H.J., Weber, O., Kramer, T. et al. (2003). Inhibition of hepatitis B virus replication by drug-induced depletion of nucleocapsids. *Science* **299**, 893–896.

Dyson, M.R. and Murray, K. (1995). Selection of peptide inhibitors of interactions involved in complex protein assemblies: Association of the core and surface antigens of hepatitis B virus. *Proc Natl Acad Sci USA* **92**, 2194–2198.

Edman, J.C., Hallewell, R.A., Valenzuela, P., Goodman, H.M., and Rutter, W.J. (1981). Synthesis of hepatitis B surface and core antigens in *E. coli*. *Nature* **291**, 503–506.

Ellis, R.W. and Gerety, R.J. (1989). Plasma-derived and yeast-derived hepatitis B vaccines. *Am J Infect Control* **17**, 181–189.

Freund, S.M., Johnson, C.M., Jaulent, A.M., and Ferguson, N. (2008). Moving towards high-resolution descriptions of the molecular interactions and structural rearrangements of the human hepatitis B core protein. *J Mol Biol* **384**, 1301–1313.

Hilleman, M. (2011). Three decades of hepatitis vaccinology in historic perspective. A paradigm of successful pursuits. In *History of Vaccine Development*, S.A. Plotkin (ed.), Springer, New York, pp. 233–246.

Hilmer, J.K., Zlotnick, A., and Bothner, B. (2008). Conformational equilibria and rates of localized motion within hepatitis B virus capsids. *J Mol Biol* **375**, 581–594.

Jegerlehner, A., Tissot, A., Lechner, F., Sebbel, P., Erdmann, I., Kundig, T., Bachi, T. et al. (2002). A molecular assembly system that renders antigens of choice highly repetitive for induction of protective B cell responses. *Vaccine* **20**, 3104–3112.

Kazaks, A., Lachmann, S., Koletzki, D., Petrovskis, I., Dislers, A., Ose, V., Skrastina, D. et al. (2002). Stop codon insertion restores the particle formation ability of hepatitis B virus core-hantavirus nucleocapsid protein fusions. *Intervirology* **45**, 340–349.

Kirchhofer, A., Helma, J., Schmidthals, K., Frauer, C., Cui, S., Karcher, A., Pellis, M. et al. (2010). Modulation of protein properties in living cells using nanobodies. *Nat Struct Mol Biol* **17**, 133–138.

Koletzki, D., Biel, S.S., Meisel, H., Nugel, E., Gelderblom, H.R., Kruger, D.H., and Ulrich, R. (1999). HBV core particles allow the insertion and surface exposure of the entire potentially protective region of Puumala hantavirus nucleocapsid protein. *Biol Chem* **380**, 325–333.

Koletzki, D., Lundkvist, A., Sjolander, K.B., Gelderblom, H.R., Niedrig, M., Meisel, H., Kruger, D.H., and Ulrich, R. (2000). Puumala (PUU) hantavirus strain differences and insertion positions in the hepatitis B virus core antigen influence B-cell immunogenicity and protective potential of core-derived particles. *Virology* **276**, 364–375.

Kratz, P.A., Bottcher, B., and Nassal, M. (1999). Native display of complete foreign protein domains on the surface of hepatitis B virus capsids. *Proc Natl Acad Sci USA* **96**, 1915–1920.

Krugman, S. and Davidson, M. (1987). Hepatitis B vaccine: Prospects for duration of immunity. *Yale J Biol Med* **60**, 333–339.

Kushnir, N., Streatfield, S.J., and Yusibov, V. (2012). Virus-like particles as a highly efficient vaccine platform: Diversity of targets and production systems and advances in clinical development. *Vaccine* **31**, 58–83.

Lachmann, S., Meisel, H., Muselmann, C., Koletzki, D., Gelderblom, H.R., Borisova, G., Kruger, D.H., Pumpens, P., and Ulrich, R. (1999). Characterization of potential insertion sites in the core antigen of hepatitis B virus by the use of a short-sized model epitope. *Intervirology* **42**, 51–56.

Lange, M., Fiedler, M., Bankwitz, D., Osburn, W., Viazov, S., Brovko, O., Zekri, A.R. et al. (2014). Hepatitis C virus hypervariable region 1 variants presented on hepatitis B virus capsid-like particles induce cross-neutralizing antibodies. *PLoS ONE* **9**, e102235.

Monie, A., Hung, C.F., Roden, R., and Wu, T.C. (2008). Cervarix: A vaccine for the prevention of HPV 16, 18-associated cervical cancer. *Biol: Targ Ther* **2**, 97–105.

Nassal, M. (1992). The arginine-rich domain of the hepatitis B virus core protein is required for pregenome encapsidation and productive viral positive-strand DNA synthesis but not for virus assembly. *J Virol* **66**, 4107–4116.

Nassal, M., Skamel, C., Kratz, P.A., Wallich, R., Stehle, T., and Simon, M.M. (2005). A fusion product of the complete *Borrelia burgdorferi* outer surface protein A (OspA) and the hepatitis B virus capsid protein is highly immunogenic and induces protective immunity similar to that seen with an effective lipidated OspA vaccine formula. *Eur J Immunol* **35**, 655–665.

Pasek, M., Goto, T., Gilbert, W., Zink, B., Schaller, H., Mackay, P., Leadbetter, G., and Murray, K. (1979). Hepatitis B virus genes and their expression in *Escherichia coli*. *Nature* **282**, 575–579.

Peyret, H., Gehin, A., Thuenemann, E.C., Blond, D., El Turabi, A., Beales, L., Clarke, D., Gilbert, R.J.C., Fry, E.E., Stuart, D.I., Holmes, K., Stonehouse, N.J., Whelan, M., Rosenberg, W., Lomonossoff, G.P., and Rowlands, D.J. (2015). Tandem fusion of hepatitis B core antigen allows assembly of virus-like particles in bacteria and plants with enhanced capacity to accommodate foreign proteins. *PLoS ONE*. In Press.

Plummer, E.M. and Manchester, M. (2010). Viral nanoparticles and virus-like particles: Platforms for contemporary vaccine design. *Wiley Interdiscip Rev: Nanomed Nanobiotechnol* **3**, 174–196.

Pumpens, P. and Grens, E. (2001). HBV core particles as a carrier for B cell/T cell epitopes. *Intervirology* **44**, 98–114.

Roy, P. and Noad, R. (2008). Virus-like particles as a vaccine delivery system: Myths and facts. *Hum Vaccin* **4**, 5–12.

Schödel, F., Moriarty, A.M., Peterson, D.L., Zheng, J., Hughes, J.L., Will, H., Leturcq, D.J., Mcgee, J.S., and Milich, D.R. (1992). The position of heterologous epitopes inserted in hepatitis B virus core particles determines their immunogenicity. *J Virol* **66**, 106–114.

Shank-Retzlaff, M.L., Zhao, Q., Anderson, C., Hamm, M., High, K., Nguyen, M., Wang, F. et al. (2006). Evaluation of the thermal stability of Gardasil. *Hum Vaccin* **2**, 147–154.

Shepherd, D.A., Holmes, K., Rowlands, D.J., Stonehouse, N.J., and Ashcroft, A.E. (2013). Using ion mobility spectrometry–mass spectrometry to decipher the conformational and assembly characteristics of the hepatitis B capsid protein. *Biophys J* **105**, 1258–1267.

Simon, M.M., Schaible, U.E., Kramer, M.D., Eckerskorn, C., Museteanu, C., Muller-Hermelink, H.K., and Wallich, R. (1991). Recombinant outer surface protein a from *Borrelia burgdorferi* induces antibodies protective against spirochetal infection in mice. *J Infect Dis* **164**, 123–132.

Stahl, S., MacKay, P., Magazin, M., Bruce, S.A., and Murray, K. (1982). Hepatitis B virus core antigen: Synthesis in *Escherichia coli* and application in diagnosis. *Proc Natl Acad Sci USA* **79**, 1606–1610.

Stahl, S.J. and Murray, K. (1989). Immunogenicity of peptide fusions to hepatitis B virus core antigen. *Proc Natl Acad Sci USA* **86**, 6283–6287.

Stray, S.J., Bourne, C.R., Punna, S., Lewis, W.G., Finn, M.G., and Zlotnick, A. (2005). A heteroaryldihydropyrimidine activates and can misdirect hepatitis B virus capsid assembly. *Proc Natl Acad Sci USA* **102**, 8138–8143.

Stray, S.J., Ceres, P., and Zlotnick, A. (2004). Zinc ions trigger conformational change and oligomerization of hepatitis B virus capsid protein. *Biochemistry* **43**, 9989–9998.

Uetrecht, C., Barbu, I.M., Shoemaker, G.K., van Duijn, E., and Heck, A.J. (2011). Interrogating viral capsid assembly with ion mobility-mass spectrometry. *Nat Chem* **3**, 126–132.

Ulrich, R., Koletzki, D., Lachmann, S., Lundkvist, A., Zankl, A., Kazaks, A., Kurth, A., Gelderblom, H.R., Borisova, G., Meisel, H., and Kruger, D.H. (1999). New chimeric hepatitis B virus core particles carrying hantavirus (serotype Puumala) epitopes: Immunogenicity and protection against virus challenge. *J Biotechnol* **73**, 141–153.

Ulrich, R., Nassal, M., Meisel, H., and Kruger, D.H. (1998). Core particles of hepatitis B virus as carrier for foreign epitopes. *Adv Virus Res* **50**, 141–182.

Valenzuela, P., Medina, A., Rutter, W.J., Ammerer, G., and Hall, B.D. (1982). Synthesis and assembly of hepatitis B virus surface antigen particles in yeast. *Nature* **298**, 347–350.

Walker, A., Skamel, C., and Nassal, M. (2011). SplitCore: An exceptionally versatile viral nanoparticle for native whole protein display regardless of 3D structure. *Sci Rep* **1**, 5.

Walker, A., Skamel, C., Vorreiter, J., and Nassal, M. (2008). Internal core protein cleavage leaves the hepatitis B virus capsid intact and enhances its capacity for surface display of heterologous whole chain proteins. *J Biol Chem* **283**, 33508–33515.

Wynne, S.A., Crowther, R.A., and Leslie, A.G.W. (1999). The crystal structure of the human hepatitis B virus capsid. *Mol Cell* **3**, 771–780.

Yu, X., Jin, L., Jih, J., Shih, C., and Zhou, H. (2013). 3.5Å cryoEM structure of hepatitis B virus core assembled from full-length core protein. *PLoS ONE* **8**, e69729.

Zlotnick, A., Cheng, N., Conway, J.F., Booy, F.P., Steven, A.C., Stahl, S.J., and Wingfield, P.T. (1996). Dimorphism of hepatitis B virus capsids is strongly influenced by the C-terminus of the capsid protein. *Biochemistry* **35**, 7412–7421.

Zlotnick, A., Palmer, I., Kaufman, J.D., Stahl, S.J., Steven, A.C., and Wingfield, P.T. (1999). Separation and crystallization of T = 3 and T = 4 icosahedral complexes of the hepatitis B virus core protein. *Acta Crystallogr D: Biol Crystallogr* **55**, 717–720.

12 SplitCore

Advanced Nanoparticulate Molecular Presentation Platform Based on the Hepatitis B Virus Capsid

Philipp Kolb, Thi Thai An Nguyen, Andreas Walker, and Michael Nassal

CONTENTS

12.1 INTRODUCTION

The icosahedral capsid of hepatitis B virus (HBV), serologically defined as hepatitis B core antigen (HBcAg), has become one of the most attractive viral nanoparticles for surface display of heterologous molecules, largely due to its exceptional immunogenicity and the easy accessibility of recombinant wild-type (wt) and modified versions of the constituent core protein (HBc) in the form of capsid-like particles (CLPs); we will use the term *CLP* rather than virus-like particle (VLP) in this article because in HBV, the envelope around the capsid defines the appearance of virions. While various aspects of HBc as a particulate carrier are covered in separate chapters of this book (Chapters 2, 11, and 13), the focus here is on a newly developed HBc-based display system termed *SplitCore*. Its fundamentally distinct feature is the ability of two separately expressed fragments of the HBc to efficiently associate into a 3D structure that retains the assembly competence of the wt protein; however, the interruption in the central part of the primary sequence forming the most surface-exposed region of the particle structure provides for an unprecedented flexibility in the choice of molecules that can be presented [1]. In the following, we outline the various rationales that led to the SplitCore concept and provide examples for SplitCore CLPs presenting foreign proteins that are incompatible with the structural constraints of the conventional contiguous chain HBc system. We also discuss options to engineer multilayered particles and the potential advantages of the open-ended versus clamped-in display of smaller peptides. Finally, we examine current limitations of the SplitCore system and how they may be overcome. We hope this will also spur the interest of those working on other proteinaceous nanoparticles to explore the extended capabilities the split protein approach can offer.

12.2 NATURAL HBcAg: THE NUCLEOCAPSID OF HEPATITIS B VIRUS

HBV is a small, hepatotropic DNA virus (hepadnavirus) that replicates by protein-primed reverse transcription [2]; related viruses occur in some mammals (*orthohepadnaviruses*) including in woodchucks (WHV), ground squirrels (GSHV), and bats [3] and in several bird species (*avihepadnaviruses*), for example, in ducks [4,5]. In virions, the nucleocapsid forms the inner-core particle [6], which is surrounded by a lipid envelope in which the large (L), middle (M; absent from *avihepadnaviruses*), and small (S) surface proteins, collectively termed *HBsAg*, are embedded [7]. In its lumen, the core particle carries the viral genome plus the viral polymerase. For virion morphogenesis, initially one of the viral RNAs, the pregenomic (pg) RNA, is specifically copackaged with the P protein into a shell of 120 (and, to a lesser extent, 90) dimers of a single core protein of ~180 (~260 for *avihepadnaviruses*) amino acids (aa) in length [8,9]; in a multistep process, the RNA is then converted into DNA (for review, see [10]). While formation of authentic nucleocapsids is thus highly complex, it was found early on that expression of just the core open reading frame (ORF) in *Escherichia coli* [11], yet also in other organisms (for review, see [12]), is sufficient to generate HBc CLPs (see Chapter 11). Hence, the information to assemble into ordered particles is intrinsically encoded in the core protein primary sequence; this also holds for the core proteins of WHV and GSHV (see Chapter 13) and HBV (DHBV) [13]. These heterologous expression systems were instrumental for structural studies. Deletion analyses revealed a two-domain structure for HBc [14] in which the N-terminal ~140 residues are sufficient to form the capsid shell; the Arg-rich C-terminal domain (CTD) serves as a nucleic acid–binding module that, in HBV replication, has an active role in pgRNA packaging and reverse transcription [15]. During heterologous expression, the CTD causes the CLPs to nonsequence specifically package host RNAs [14,16]; conversely, CLPs from C-terminally truncated HBc, for instance, the widely used variant HBc1-149, contain much less RNA. The particularly high expression levels achieved with this construct allowed for high-resolution structural analyses by electron cryomicroscopy ([17,18] and Chapter 3; see also Figure 12.1a) and x-ray crystallography (see [19] and Chapter 4). As shown in

Figure 12.1, they revealed a characteristic all α-helical fold for the assembly domain (Figure 12.1b and c), dominated by the long antiparallel central α3-helices and α4-helices; this α-helical hairpin forms a large interaction surface for a second monomer, yielding a highly stable symmetric dimer. The C proximal α5-helices and further downstream residues project sideward from the dimer and provide the interfaces for interdimer contacts, eventually resulting in formation of icosahedral lattices comprising 120 dimers (triangulation number T = 4) or 90 dimers (T = 3); the 4-helix bundles at the intradimer interfaces protrude as prominent spikes from the particle surface. Consistent with strong intradimer and weaker interdimer interactions [20], the CLPs can be dissociated under modestly denaturing conditions into individual dimers that remain folded and competent for reassembly once the denaturant is removed [21,22]. This highly stable fold, combined with substantial plasticity [23], is key to the suitability of HBc as well as SplitCore CLPs as a nanoparticulate display platform.

12.3 EXCEPTIONAL IMMUNOGENICITY OF HBcAg

A hallmark of chronic hepatitis B, caused by persistent HBV infection, is the lack of a sufficient immune response to clear the virus. Clearance requires T cell–mediated elimination of virus-infected cells [24], yet also antibodies to neutralize circulating virus and prevent reinfection. The prime target for such antibodies is the HBsAg in the viral envelope, and the failure to mount anti-HBsAg antibodies is a key feature of chronic hepatitis B. Much in contrast, virtually all HBV-infected individuals rapidly develop high-titered, long-lasting antibodies against HBcAg. Due to the

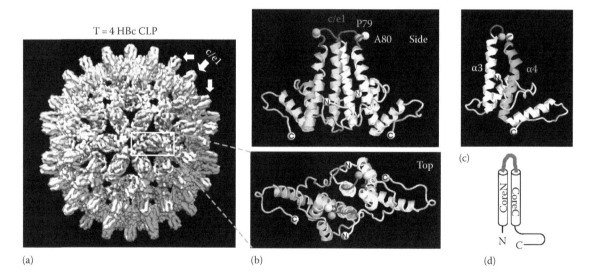

FIGURE 12.1 Basic structural features of HBc CLPs. (a) 3D reconstruction of T = 4 CLPs at 7.5 Å resolution. (M. Nassal and B. Böttcher, unpublished data.) (b) Side and top view of the HBc dimer. The graphic representations are based on PDB accession number: 1QGT. The constituent monomers are shown in light blue and green; the loop between helices α3 and α4 comprising the c/e1 epitope is depicted in red. The orange and yellow spheres represent residues Pro79 (P79) and Ala80 (A80), a preferred location for heterologous insertions and the splitting site in SplitCore CLPs. (c) The HBc monomer. The central helices α3 and α4 are shown in white and light blue, respectively. (d) Schematic representation of the HBc monomer fold, highlighting only the central helices and the surface-exposed c/e1 epitope.

virion-internal localization of HBcAg, such antibodies are not neutralizing; however, their nearly universal induction clearly demonstrates the exceptional immunogenicity of HBcAg. Notably, recombinant HBV CLPs have very similar properties. Studies mainly in mice have shown that the ordered multimeric nature of HBcAg is key to this exceptional immunogenicity; dissociated subunits or completely denatured core protein is drastically less immunogenic [25]. While this property is shared with other particulate antigens (see Chapter 7), a distinctive feature of HBc particles is their ability to directly activate B cells via B cell receptor (BCR) clustering, that is, without the usually required T cell help; hence, HBcAg can act as T cell–independent antigen that even may induce IgM to IgG class switching [25,26]. HBcAg has also been shown to interact with naive B1 cells, a conserved splenic B cell subset considered to be part of the innate immune system, which may act as particularly potent antigen-presenting cells for HBc-specific CD4+ T cells [27]. Not the least, potent T cell epitopes within HBc can as well stimulate the classical T cell–dependent arms of the immune response ([28]; see also Chapter 13).

12.4   CONVENTIONAL USES OF HBc AS AN IMMUNE-ENHANCING PARTICULATE CARRIER FOR FOREIGN ANTIGENS

The extraordinary immunological properties of HBcAg and the availability of efficient heterologous expression systems spurred the interest in exploiting HBc as a particulate carrier for foreign antigens even before its structure was known ([29,30]; see Chapter 11). As for most fusion proteins, the simplest way for connecting a heterologous sequence is to add it to one of the termini of the *carrier protein* as this provides the highest probability that the two parts can independently fold. Various peptides fused to HBc in this fashion did indeed not compromise particle formation; however, the corresponding CLPs, in particular those with a C-terminally fused partner, often failed to induce the expected strong antipeptide antibody responses when used as immunization antigens; instead, the dominant response was directed against the carrier [31]. Empirical searches then identified a site in the center of the primary sequence of the core protein's assembly domain (around 80 aa; see Figure 12.1) as tolerant toward (at least small) insertions of foreign sequences [29,32]; moreover, the corresponding CLPs induced indeed high-titered antibodies against the insert accompanied by reduced responses against HBc [31]. Notably, the insertion-tolerant region corresponded to the known, immunodominant "c/e1" epitope on HBcAg [33] against which >90% of the anti-HBcAg antibodies in patients are directed. The emerging structural data explained these observations. In the particle, the N termini of the HBc subunits are located in the canyons surrounding the spikes (Figure 12.1); hence, depending on the length of the added foreign sequence, it will barely be accessible on the CLP surface to the bulky BCRs. For C-terminal fusions, the situation is even worse

in that the C-terminal HBc residues are located in the particle interior, and so are heterologous sequences as long as their size does not exceed the available luminal space. This feature can in fact be exploited to package up to 240 small heterologous protein domains into the CLPs, for example, the ~16 kDa *Staphylococcus aureus* nuclease [34], though not the ~28 kDa green fluorescent protein (GFP). When coexpressed with HBV in eukaryotic cells, the HBc-nuclease fusion protein can coassemble with the wt HBc subunits and digest the packaged viral genome [35]. Though interesting as an antiviral strategy, it is obvious that the internal disposition of a heterologous sequence makes it very poorly immunogenic on the B cell level. In contrast, the sequence comprising the c/e1 epitope was found to reside at the tip of the capsid spikes, that is, at their most surface-exposed region on the particle (Figure 12.1), explaining its immunodominant nature. Most subsequent attempts to generate HBc CLPs presenting foreign antigens therefore focused on insertions into this site.

12.5   TOLERANCE TOWARD c/e1 INSERTIONS IS DEFINED BY INSERT 3D STRUCTURE, NOT MERE SEQUENCE LENGTH

Numerous heterologous peptides have successfully been inserted into the c/e1 epitope without compromising particle formation, reflecting the enormous stability of HBc's helical framework. For many such CLPs, the induction of strong antipeptide antibody responses has been demonstrated (for reviews, see [36,37]), such that HBc became one of the most attractive CLP carriers [28]. Not surprisingly given their sequence similarity, the core proteins of the woodchuck (WHV) and ground squirrel (GSHV) hepatitis B viruses can be used as display platforms as well [38].

However, the use of peptide antigens for vaccination faces several drawbacks, all related to their short length. The limited epitope number compromises the breadth of the antibody response, such that already existing genetic variants of a pathogen, for example, different strains, may be missed, and it facilitates the emergence of escape variants because a few amino acid exchanges suffice to avoid antibody recognition (low genetic barrier). Second, a flexible peptide may not adopt an identical conformation as in the native protein or, especially when clamped over both of its termini into the heterologous framework of the carrier, may even be forced into an improper, nonnatural conformation [39]. As BCRs see the native antigen, this may lead to antibodies that, even if high titered, are irrelevant for recognizing the native antigen (see Figure 12.14). Furthermore, due to MHC polymorphisms, peptide-specific T cell help may be expected in only a fraction of individuals from an outbred vaccine population.

These arguments make a strong case for using the whole-protein antigen rather than short peptides for immunization [40]. However, genetically grafting complete proteins on HBc CLPs (or other VLPs) poses its own challenges.

Initial attempts to insert ever larger peptides into the c/e1 epitope suggested that the longer the sequence, the more likely would it compromise particle formation. This led to the concept that the c/e1 epitope has a limited insertion capacity, as measured by the number of amino acids in the insert [37]. As is obvious in retrospect, such a linear view of protein *size* is only a very rough approximation; rather, the fold of a protein defines its volume. A highly relevant feature of this fold with respect to fusion protein formation is the accessibility of the protein's termini, in particular when that protein is intended to act as an *internal* insert where both of its ends are fixed to the carrier. Based on this structural concept, we indeed demonstrated that GFP, which adopts a barrel-like structure [41] from which both termini protrude from one side (Figure 12.2), can efficiently be inserted into the c/e1 epitope, giving rise to ordered particles that display fluorescent, that is, properly folded, GFP on their surface [42]. Moreover, these GFP CLPs induced high-titered GFP-specific antibodies, suggesting that the immune-enhancing effect of the HBc CLP carrier also holds for complete proteins presented on its surface.

Beyond 3D structure, also insert quaternary structure can play a decisive role in CLP formation, as revealed by color variants of GFP. While all GFP-related proteins adopt a similar barrel architecture, many form stable antiparallel dimers

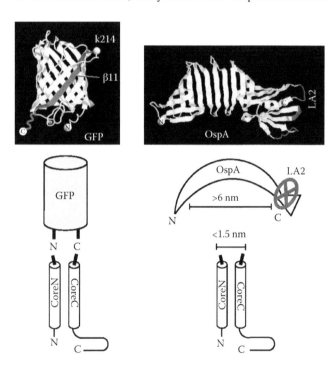

FIGURE 12.2 3D structure determines compatibility with c/e1 insertions in HBc CLPs. The molecular 3D representations of GFP and *B. burgdorferi* OspA on the top are based on x-ray crystallographic data (PDB accession numbers: GFP, 1YFP; OspA, 2OYB). White spheres labeled N and C refer to the protein termini; note that these are close in GFP but far apart in OspA. In GFP, the highlighted residue K214 indicates the splitting site in splitGFP constructs separating β1 to β10 from β11 (red); the wiggly extension symbolizes the disordered C-terminal residues not visible in the x-ray structure. In OspA, the neutralizing LA2 epitope is indicated in red. The following cartoon representations emphasize the natural fit of the GFP termini, but not the OspA termini, into the HBc c/e1 acceptor sites.

or tetramers. HBc fusions carrying these proteins in c/e1 were well expressed and the fluorescent protein parts developed the typical colors; however, the proteins were largely if not completely insoluble. Formation of soluble CLPs was rescued by instead using engineered monomeric fluorescent protein variants [43]. This indicated that interfluorescent protein interactions competed, or occurred simultaneously, with the desired intercore protein interactions, causing aggregate formation. However, not every higher-order structure is detrimental to CLP formation, as demonstrated by the efficient assembly of CLPs carrying the dimeric outer surface protein C (OspC) from *Borrelia burgdorferi* (the Lyme disease agent) on their surface [44]. Different from the dimeric GFP variants, the two subunits of the OspC dimer are arranged in a parallel fashion [45]. All four termini of the dimer protrude in close proximity from the mushroomlike structure and thus neatly fit to the four acceptor sites at the tip of the core protein dimer. Also these CLPs evoked strong and neutralizing antibody responses against *B. burgdorferi*.

Hence, major criteria for successful presentation of whole-protein antigens on HBc CLPs are an appropriate tertiary structure of the inserted protein with closely juxtaposed termini and, with few exceptions such as OspC, the absence of inter-insert interactions. However, such features are not abundant among all proteins that may represent relevant targets for immunization.

An illuminating example is the outer surface protein A (OspA) from *Borrelia*, a lipoprotein in which the tripalmitoyl-*S*-glyceryl-cysteine (Pam3Cys) moiety acts as a built-in strong toll-like receptor (TLR) agonist [46]. As lipidated OspA was the active ingredient in the first approved Lyme disease vaccine [47], it provided a gold standard against which the immunogenic potential of OspA-bearing HBc CLPs could be compared.

OspA is monomeric but folds into an extended structure with the N and C termini at opposite ends (Figure 12.2); their distance of ~6.3 nm drastically exceeds that between the ends of the c/e1 insertion sites in the core protein (<1.5 nm). Trying to span this distance by long flexible connecting linkers, the best we achieved was a fusion protein that formed heterogeneous, though soluble, aggregates but few or no regular particles (Figures 12.3 and 12.7a). In line with recent data for other aggregated proteins [48], the aggregates evoked a strong and even protective anti-OspA response in mice [49]. However, their heterogeneous nature and the presence of the extra linker sequences with an unknown immunogenic potential would preclude use of such preparations in human vaccines. HBc fusions with various other insert proteins showed similar undesired properties, prompting us to search for a more universal solution.

12.6 SPLITCORE PRINCIPLE

As shown in Figure 12.3, the key issue for any insertion into the core protein (or any other carrier protein) is the two-sided fixation of the insert. Structurally flexible sequences, such as many short peptides, can adapt into the given geometry of the acceptor sites; the same holds for a folded protein as long

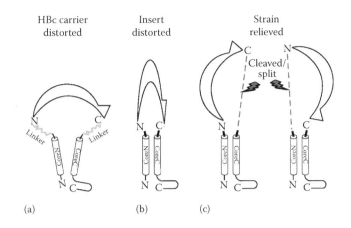

FIGURE 12.3 Possible outcomes of inserting OspA (or other incompatible molecules) into c/e1. (a) Folding of OspA into its native conformation splays the central HBc helices, preventing proper monomer folding, dimerization, and CLP assembly. Spanning the large distances by long flexible linkers (wiggly gray lines) provides only a partial remedy. (b) Folding of HBc into an assembly-competent conformation distorts OspA. (c) Disconnecting one or the other insert linkage to the HBc carrier relieves the steric strain, allowing both protein parts to adopt their native conformation.

as it has an insertion site-compatible structure, such as GFP. However, for proteins not meeting this criterion, such as OspA, there are two possible outcomes, neither of which is desirable.

Insert proteins with a very stable fold may splay the central core protein helices beyond a point where they can properly interact with each other and the helical hairpin of the sister sub-unit; this will prevent dimer formation and subsequent assembly into CLPs. Conversely, less stably folded inserts may be distorted or altogether be prevented from folding. This may

result in aggregation or, at the least, in the loss of important conformational epitopes. A potential remedy is to cut one of the two connections between carrier and insert (Figure 12.3). If the capsid structure is retained, the *insert* would still remain fixed to the CLPs, yet without the steric strain caused by the two-sided fixation. As a proof of principle, we introduced a protease recognition site into the c/e1 epitope and showed that the preformed CLPs remained intact after cleavage. When applied to the core-OspA fusion protein, formation of regular CLPs was greatly enhanced [50]. In search for a simpler system, we then tested whether the N-terminal segment (coreN; comprising the N-terminal core protein sequence to upstream insertion site, usually around 79 aa; see Figure 12.1) and the C-terminal segment (coreC; from the start of the downstream insertion site to the core protein C terminus) would be able to associate when *a priori* expressed as separate entities. In fact, the concept worked surprisingly well, especially when the initial two-plasmid system was replaced by a single vector encoding a bicistronic RNA (Figure 12.4); translation initiation at the downstream cistron could be mediated by overlapping stop codon/start codon arrangements (Figure 12.4b) or, in our experience more broadly applicable, by a separate second ribosome binding site (RBS2; Figure 12.4c). Figure 12.5 shows a representative SDS-PAGE analysis of fractions from a sucrose gradient where the cosedimentation of the two small fragments into HBV CLP–typical fractions already suggested efficient particle formation. This was confirmed by native agarose gel electrophoresis (NAGE) and directly by electron microscopy (EM), which revealed abundant particles with an appearance indistinguishable from contiguous chain HBc CLPs even in 3D reconstructions from cryoelectron micrographs (see Chapter 3). Furthermore, SplitCore shows the same CTD-dependent

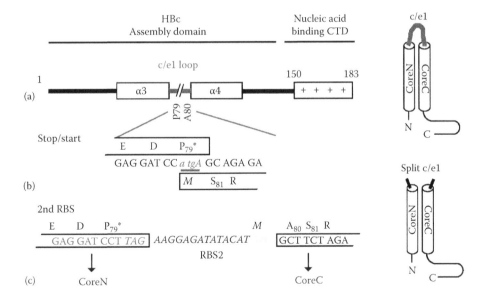

FIGURE 12.4 Basic constructs for SplitCore expression. (a) Domain structure of HBc. Key to the SplitCore system is the expression of the HBc primary sequence in two separate parts, comprising the coreN segment from the N terminus to the c/e1 epitope and the coreC segment from c/e1 to the C terminus of the assembly domain, plus, if desired, parts or the entire nucleic acid–binding CTD. For expression in *E. coli*, this is most efficiently achieved using bicistronic constructs. Translation of the downstream segment can be directed (b) by overlapping stop/start arrangements, for example, by the third nucleotide of the last upstream ORF codon, plus the following stop codon forming a start ATG in the −1 frame for the downstream ORF or (c) by a RBS2 preceding the downstream cistron.

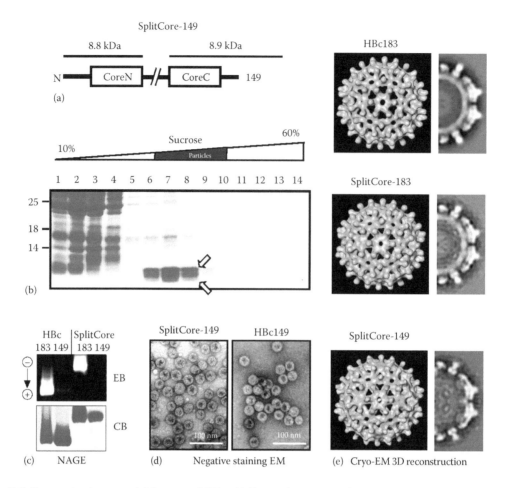

FIGURE 12.5 SplitCore maintains essential features of HBc. (a) Expression cassette for SplitCore-149. (b) Cosedimentation of coreN and coreC fragments in a sucrose gradient. The bicistronic SplitCore-149 construct was expressed in *E. coli*, and cleared lysates were run on a 10%–60% (w/v) sucrose gradient. Proteins in individual fractions were monitored by SDS-PAGE and CB staining. The nearly identically sized coreN and coreC fragments cosedimented into the particle-typical gradient fractions. (c) CTD-dependent RNA packaging in conventional versus SplitCore CLPs. Contiguous chain HBc183 and HBc149 (lacking the CTD) and their corresponding split versions were subjected to NAGE where ordered CLPs run as distinct bands. EB strongly stains the packaged RNA in both types of CLPs from full-length but not CTD-less HBc. Similar protein content was confirmed by staining the same gel with CB. (d) Negative staining EM. CLPs from contiguous HBc149 and SplitCore-149 show a similar appearance. All HBc149-based constructs contained a C-terminal His6 tag that is not shown. (e) Cryo-EM-based 3D reconstructions. The reconstructions confirm a comparable CLP architecture for conventional versus SplitCore CLPs. The left panels (gold) show surface representations at about 15 Å resolution; the panels on the right represent equatorial sections. The inner ring of density in the CTD containing yet absent from the truncated SplitCore-149 CLPs originates from packaged RNA. Cryo-EM and image reconstructions were performed by Britta Gerlach and Bettina Böttcher.

nonspecific RNA packaging phenotype, that is, constructs based on coreC fragments encompassing the CTD (e.g., HBc183) package much more RNA than CTD-lacking truncated derivates such as HBc149 [14]. This was revealed by ethidium bromide (EB) versus Coomassie blue (CB) staining after NAGE (Figure 12.5c) and by the presence versus absence of capsid internal density in 3D reconstructions (Figure 12.5e). Accumulating evidence indicates that the packaged bacterial RNA can act as TLR agonist, biasing the antibody response from a Th2/IgG1 to Th1/IgG2a type [27,51–53]. Hence, SplitCore can offer the same flexibility for cargo-mediated modulation of immune responses as the conventional HBc system, as is supported by our results with different protein-displaying SplitCore CLPs (A. Walker, R. Wallich, M. Simon, D. Milich, and M. Nassal, unpublished data).

12.7 SplitCore ENABLES CLP SURFACE DISPLAY OF PROTEIN REFRACTORY TO PRESENTATION ON CONVENTIONAL HBc CLPs

GFP provided a simple test system to validate the general suitability of the SplitCore system as protein presentation platform. Fusion of GFP to coreN as well as to coreC yielded similar amounts of green fluorescent CLPs as the contiguous chain construct (Figure 12.6), indicating that the split core protein framework had no major negative impact on assembly efficiency. This was further corroborated by successful CLP formation from a construct in which OspC was fused to coreN [1]. Most remarkably, however, assembly competence was also retained when the

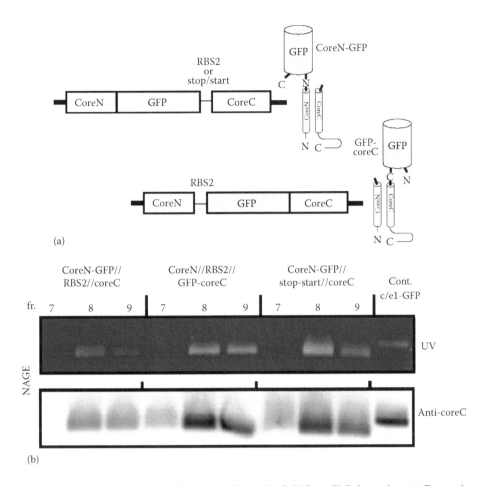

(a)

(b)

FIGURE 12.6 Fusion of GFP to coreN and to coreC is compatible with SplitCore CLP formation. (a) Expression cassettes used and expected products. (b) Confirmation of formation of CLPs displaying natively folded GFP by NAGE. Cleared lysates of bacteria transformed with the constructs from (a) were separated by sucrose gradient sedimentation as in Figure 12.5b. Material from the center gradient fractions was subjected to NAGE. GFP containing material was visualized by UV illumination (top panel); the presence of the coreC fragment in the same band was shown by immunoblotting.

previously refractory OspA protein was fused to either the coreN or the coreC segment (Figure 12.7), fully confirming the concept that the two-sided insert fixation is the key obstacle for CLP formation in the conventional c/e1 insertion approach [1].

An intrinsic difference between fusions to coreN and coreC is the orientation of the insert; fusion to the C terminus of coreN occurs through the foreign protein's N terminus, fusion to coreC through its C terminus. Especially for elongated protein structures like that of OspA, this causes a strong polarity in the regions of the protein that are solvent exposed in one versus the other orientation. Evidence for the biological relevance of such polarity came from immunization studies in mice with the two types of OspA CLPs. To this end, it is important that the major neutralizing epitope within OspA, LA2, is located in the C-terminal part of the protein [54] (Figures 12.2 and 12.7a).

In terms of total OspA-specific antibody titer, both SplitCore CLPs outperformed lipidated OspA, confirming that the immune-enhancing effect is maintained in the split version of the HBc CLP carrier. However, while total antibody titers against OspA fused to coreC exceeded those induced by the fusion to coreN, their neutralizing potency was much lower, as demonstrated by protection of only one out of six versus six out of six mice from the *B. burgdorferi* challenge [1]. These data are fully consistent with exposure versus shielding of the neutralizing LA2 epitope. An immediate inference is that immunization with a mixture of CLPs displaying a foreign protein in both orientations will yield the broadest possible antibody response and thus circumvent potential problems caused by the location of important epitopes close to the CLP surface and thus inaccessible to BCR recognition.

A further example for the superior ability of the SplitCore system to display large foreign proteins is the circumsporozoite protein (CSP; Figure 12.7b) of the malaria agent *Plasmodium falciparum* (see Chapter 13). CSP encompasses >300 aa with an immunodominant central region containing a variable number of repeats of the sequence motifs NANP and NVDP. The C-terminal half of CSP fused to the HBV S protein is the active ingredient in the most advanced, yet only modestly effective, preerythrocytic antimalaria vaccine RTS,S [55]. Several of the repeat

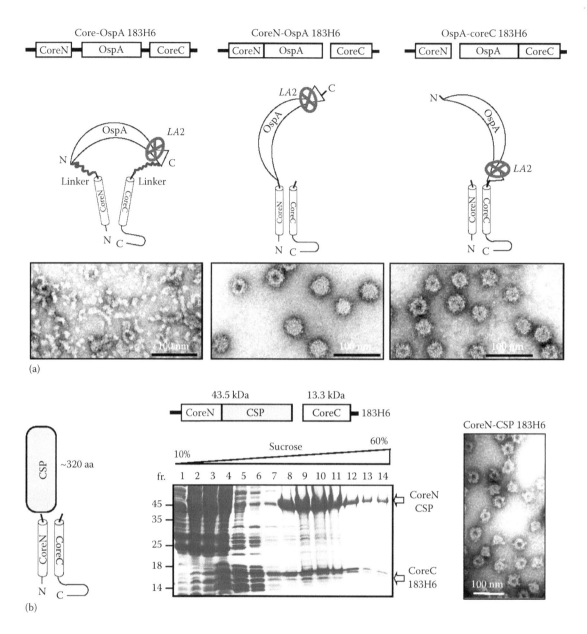

FIGURE 12.7 Efficient formation of SplitCore CLPs presenting whole proteins incompatible with contiguous chain c/e1 insertion. (a) *B. burgdorferi* OspA. From the top, the panels show the expression constructs, expected products, and negative staining electron micrographs of the purified preparations. The neutralizing LA2 epitope is indicated in red. Note that LA2 is solvent exposed when OspA is fused to coreN but buried when OspA is fused to coreC. (b) *P. falciparum* CSP. The panels show, from left to right, the expected CSP-SplitCore product, the expression construct used, sucrose gradient separation, and negative staining EM of the material in the center gradient fractions.

motifs inserted into the c/e1 epitope plus the universal CSP T cell epitope fused to the C terminus of contiguous chain HBc form the basis of another experimental vaccine, ICC-1132 [56]. RTS,S lacks the strong immune enhancement provided by the HBc carrier plus potentially important N proximal CSP epitopes, whereas ICC-1132 solely displays the CSP central repeats. These limitations are overcome by presenting the entire CSP. As for OspA, a contiguous chain construct carrying the entire 319 aa CS27IVC CSP derivative [57] failed to assemble, whereas a corresponding SplitCore version with the CSP fused to coreN formed abundant, ninja-shaped CLPs (Figure 12.7b). In mice,

these SplitCore-CSP CLPs induced similarly high-titered antibodies against the NANP repeat-motif peptide as ICC-1132, yet against CSP as test antigen, the SplitCore-induced antibody titers were even higher, consistent with the induction of additional antibodies against nonrepeat epitopes; conversely, the anti-HBc carrier response was about tenfold lower [1]. Because *P. falciparum* does not infect rodents, an important next step is to evaluate the protective potential of the SplitCore-induced anti-CSP antibodies in chimeric rodent-specific plasmodia bearing *P. falciparum* CSP sequences [58] and/or in humanized mouse models infectable by human-specific plasmodia [59,60].

12.8   SplitCore CLPs DISPLAYING SPLIT INSERTS: ACCESS TO MULTILAYER CLPs

The results described earlier showed that heterologous proteins can be attached to either the coreN or the coreC fragment. To evaluate the system's ability to simultaneously present sequences fused to both segments, we used a split version of GFP in which a fragment encompassing β-strands 1–10 of the GFP β-barrel structure was linked to coreN and the remaining β-strand 11 to coreC (Figure 12.2). As expected from the efficient association of the two segments, fluorescent SplitCore–splitGFP (SCSG) CLPs were generated in good yield. Moreover, the new termini at the solvent-exposed splitting site within GFP (at K214; see Figure 12.2) now provided an opportunity for the fusion of yet another sequence, thus creating triple-layer CLPs. As a first test case, the small (~50 aa) and highly soluble immunoglobulin-binding GB1 domain from streptococcal protein G was linked to the small GFP segment (Figure 12.8a) and efficiently yielded fluorescent CLPs (Figure 12.8b and c). The GB1 domains retained their immunoglobulin-binding capacity, as shown by their ability to directly detect antibodies in immunofluorescence (Figure 12.8d) and Western blot formats [1]. Comparable results were obtained by fusing GB1 to the large GFP fragment. Hence, these CLPs represent prototypic genetically accessible nanoparticles that combine an ordered multimeric structure (contributed by the HBc parts), an easily detectable reporter (GFP), and a targeting domain (GB1). Other reporter-targeting domain combinations can easily be envisaged, as indicated by the successful generation of fluorescent SCSG CLPs in which OspC, OspA, or an attenuated version of the *S. aureus* enterotoxin B was fused to the split GFP part [1]. If the outermost positioned protein contains an accessible terminus or other derivatizable group, engineering particles with additional layers are easily imagined.

As an example for CLPs whose surface can enzymatically be modified, we sought to generate SCSG CLPs displaying N-terminal PreS sequences of the large surface proteins of HBV and DHBV, which in virions are N-myristoylated (see Section 12.10). Unexpectedly, bicistronic expression

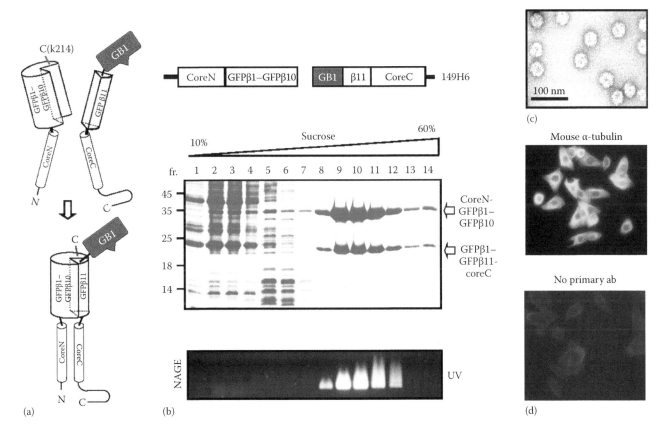

FIGURE 12.8   GB1-SplitCore-split GFP (GB1-SCSG), a prototypic triple-layer CLP. (a) Rationale. Complementation between the split HBc and split GFP segments should yield fluorescent CLPs. As the splitting site in GFP is solvent exposed, this should also hold for a foreign molecule, here the GB1 domain of protein G, fused to the C terminus of the large or the N terminus of the small GFP fragment. (b) Expression construct used, sucrose gradient separation, and NAGE analysis of the gradient fractions for native GFP containing material by UV illumination. (c) Negative staining EM of material from fraction 8. (d) The GB1 domains on GB1-SCSG CLPs is immunoglobulin-binding competent. Permeabilized HeLa cells were incubated with a mouse monoclonal antibody against tubulin (top), or not (bottom); thereafter, the cells were incubated with GB1-SCSG CLPs and analyzed by fluorescence microscopy. The strong fluorescence in the top but not the bottom panel indicates specific binding of the CLPs to the primary antitubulin antibody.

vectors analogous to those yielding high levels of GB1-SCSG failed to do so for PreS-SCSG. Reversing the cistron order turned out to be a solution that, based on similar results with other to-be-displayed proteins, appears generally useful when the conventional cistron order constructs do not give the expected results.

12.9 REVERSING THE CISTRON ORDER ON BICISTRONIC SPLITCORE CONSTRUCTS (SPLITCORE-REV) CAN BOOST CLP FORMATION

A key requirement for the SplitCore system to work is that the N- and C-terminal segments be expressed at nearly equal molar rates. In our hands, this was often achieved using the bicistronic constructs shown in Figure 12.4 where the coreN encoding cistron is 5′ of the coreC encoding cistron. However, for some fusion proteins, one segment was clearly more efficiently expressed than the other, as is shown in Figure 12.9a for an SCSG fusion protein carrying the HBV PreS1 sequence 1–48. Whereas the large fragment was well expressed, the smaller PreS1-containing GFPβ11-coreC segment was hardly detectable. In contrast, reversing the cistron order on the bicistronic construct led to expression of both segments in approximately equal amounts (Figure 12.9b). A likely explanation is the differential impact of the different sequence contexts of the initiator codons on their accessibility to the translation machinery. As proper secondary structure prediction for long RNAs is still demanding, a practical conclusion is that whenever a particular SplitCore fusion is poorly expressed from one type of construct, it is worthwhile to also try the other.

A further example is provided by the salivary proteins Salp15 and Iric-1 from North-American *Ixodes scapularis* and Eurasian *I. ricinus* ticks. These ticks are vectors for various human pathogens, including several Lyme disease–causing *Borrelia* species [61] and tick-borne encephalitis virus [62]. Salp15 and probably also its ortholog Iric-1 are immunosuppressive components of tick saliva [63,64], which facilitate the tick's blood meal and enhance transmission of tick-borne pathogens [65,66]. Antitick vaccines inducing neutralizing antibodies against these (and/or other) tick proteins may therefore represent a new strategy to reduce tick infestation per se as well as pathogen transmission [67,68]. As neutralizing epitopes on the tick saliva proteins are not yet known, presentation of the whole proteins on HBc CLPs appears as an attractive way to counteract their immunosuppressive activity. Both Salp15 and Iric-1 are only modestly sized (~115 aa after cleavage of the N-terminal signal sequence), yet beyond their natural occurrence as secreted glycosylated proteins, their content of seven Cys residues poses special challenges to recombinant expression in bacterial systems; this has thus far also prevented any structure determinations. By comparing a large number

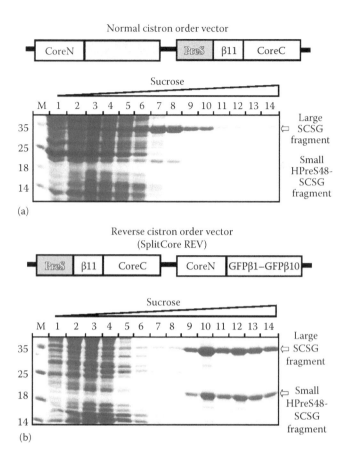

FIGURE 12.9 Reverse cistron order constructs can rescue imbalanced coreN versus coreC expression. (a) Lack of expression of the small fragment from a normal cistron order vector for HBV PreS-SCSG. *E. coli* cells transformed with the expression vector shown on the top produced large amounts of the coreN-GFPβ1-10 fragment (~35 kDa) but not the smaller HBV PreS1-48 (PreS48)-GFPβ11-coreC fragment (~22 kDa). (b) Rescue by reversing the cistron order. In this and other reverse cistron order (REV) constructs, the coding sequence for the C-terminal fragment is present as first cistron on the bicistronic RNA. Both fragments were efficiently expressed in assembly-competent form, as indicated by their cosedimentation.

of solubility-enhancing partner proteins, we eventually obtained appreciable amounts of both proteins and Cys-free variants in soluble form from *E. coli* as fusions with DsbA [69]; however, neither protein allowed CLP formation upon conventional c/e1 insertion into HBc. In contrast, the Cys-free variants efficiently formed soluble CLPs when expressed from a reverse order bicistronic SplitCore construct (Figure 12.10a and b). Notably, mice immunized with these CLPs developed antibodies that cross-react with the Cys-containing DsbA fusions and, more importantly, with glycosylated Salp15 from eukaryotic cells (Figure 12.10c) [70], whether these antibodies protect from tick infestation and *Borrelia* transmission is under investigation. Protein-chemically, however, the data already lend further support to the superior capacity of the SplitCore system to display whole-protein antigens.

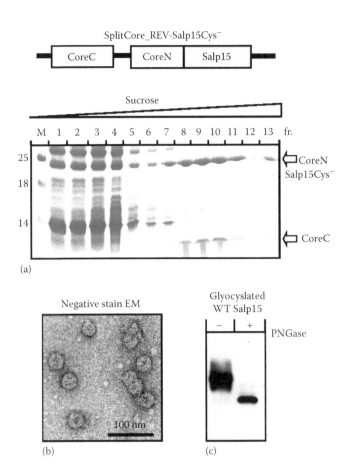

SplitCore_REV-Salp15Cys⁻

(a)

Negative stain EM

(b)

Glyocyslated WT Salp15

(c)

FIGURE 12.10 SplitCore CLPs presenting the tick saliva protein Salp15 from reverse cistron order constructs. Natural Salp15 contains seven cysteines and interfered with CLP formation in the conventional and in the SplitCore system. However, a Cys-free Salp15 variant formed CLPs exclusively in the SplitCore system and only from a REV construct. (a) Gradient purification of Salp15Cys⁻-SplitCore CLPs. (b) Negative staining EM of the material from fraction 9. (c) Antibodies induced by Salp15Cys⁻-SplitCore CLPs recognize glycosylated wt Salp15. Salp15 from supernatant of HEK293 transfected with a vector for secreted Salp15 was used for immunoblotting either directly (–PNGase) or after deglycosylation with peptide-*N*-glycosidase F (+PNGase). Both forms were equally well detected by antisera from mice immunized with Salp15Cys⁻-SplitCore CLPs.

12.10 ENZYMATIC MYRISTOYLATION OF PreS-SplitCore-SPLIT GFP PROTEINS

Using the preceding described reverse cistron order constructs, SCSG CLPs presenting a variety of N-terminal peptides from the PreS1 and PreS domain, respectively, of the large envelope proteins of HBV and DHBV (collectively termed *PreS*) fused to the small GFP segment became accessible (Figure 12.11). Naturally, the N-terminal Gly residues in PreS remaining after removal of the starting Met residue are N-myristoylated and this modification is essential for infectivity. In the SCSG CLPs, the PreS N termini are likewise accessible to myristoylation (Figure 12.11a). This was achieved either in vitro by incubation of purified CLPs

with isolated N-myristoyl transferase (NMT) and myristoyl coenzyme A or in situ by coexpression of the PreS-SCSG proteins with NMT in media supplemented with myristic acid [71]. Successful in situ myristoylation was suggested by a redistribution of the green fluorescence from cell-internal foci to the cell periphery, very similar to that seen with a GFP-tagged version of the genuine *E. coli* membrane protein Nuo J (Figure 12.11c). Furthermore, replacement of myristic acid by its ω-azido analog (Figure 12.11a) in either system allowed conjugation with an alkyne-bearing Texas Red derivative by click chemistry [72], leading to specific fluorescence labeling of the PreS-bearing small SplitCore fragment (Figure 12.11e) and thus demonstrating the potential for chemical modification of the CLP surface. Not unexpectedly, myristoylation drastically decreased water solubility of the CLPs, as is evident from partitioning of the PreS-SCSG protein coexpressed with NMT and myristic acid into the detergent-rich phase upon extraction with Triton X114, expressed as such the protein remained in the low detergent phase (Figure 12.11f). Because a non-assembling SCSG fusion protein based on a truncated HBc ending with 124 aa remained soluble upon myristoylation (Figure 12.11f), the likely reason for the insolubility of the assembling variants is that the entire CLP surface becomes covered with the hydrophobic fatty acid moieties. Their poor water solubility limits the usefulness of myristoylated PreS-SCSG CLPs as *virion mimics* for interaction studies with cells. A possible solution is mosaic CLPs in which only some subunits carry PreS sequences. However, already the current data provide proof of principle that the SplitCore-specific feature of exposing additional termini at the splitting site can be exploited to further manipulate the CLP surface. Moreover, the PreS-SCSG CLPs induced strong PreS-specific antibody responses in rabbits that, for DHBV PreS CLPs, were shown to protect against DHBV infection [71]. Hence, the split GFP-carrying CLPs seem to maintain the immune-enhancing effects of simpler HBc CLPs.

12.11 SplitCore DISPLAY OF PROTEIN SEGMENTS AND PEPTIDES

The tolerance of the SplitCore carrier toward the insert structure allows a much larger range of whole proteins to be presented on the CLP surface than on the contiguous chain HBc carrier. However, the system also provides for maximal surface exposure of shorter protein segments or peptides. As many such sequences have already successfully been displayed on contiguous HBc CLPs [36,37], what then distinguishes the two systems for this type of application?

A key issue is again one-sided versus two-sided fixation (Figure 12.12). Even relatively short peptides may adopt stable conformations that are incompatible with insertion into contiguous chain HBc, for instance, α-helices in which N and C termini may be even further apart than in a folded whole protein. A negative example regarding the conventional HBc

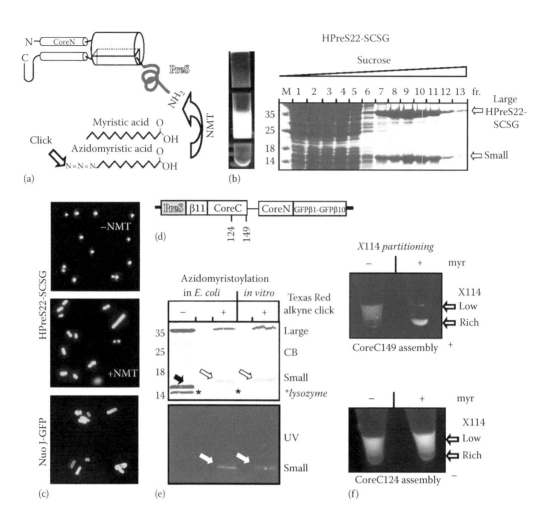

FIGURE 12.11 Enzymatic surface modification of SCSG CLPs presenting hepadnaviral PreS sequences. (a) Rationale. Various N-terminal sequences from the HBV and DHBV PreS1 and PreS domains, respectively, were fused to the N terminus of the β11 GFP fragment in SCSG. Via an N-myristoyltransferase (NMT) plus activated myristic acid or a homolog such as ω-azidomyristic acid, the N-terminal PreS Gly residues remaining after cleavage of the initiator Met should be myristoylated as in virions. (b) Sucrose gradient purification of SCSG presenting the HBV PreS1 sequence 1–22 (HPreS22-SCSG). (c) Coexpression of HPreS22-SCSG with NMT plus myristic acid causes membrane association. *E. coli* cells from induced cultures, either without (top) or with NMT coexpression and myristic acid supplementation (middle), were analyzed by fluorescence microscopy. Note the redistribution of GFP fluorescence to the cell periphery, similar to that in cells expressing a GFP-tagged version of the *E. coli* membrane protein Nuo J (bottom). (d) Expression construct used. (e) Chemical proof for successful CLP myristoylation. Either HPreS22-SCSG was coexpressed with NMT plus azidomyristic acid supplementation in *E. coli* or purified CLPs were incubated with NMT plus azidomyristoyl coenzyme A (in vitro). Subsequent click reaction with an alkyne-bearing Texas Red caused a retardation of the small HPreS-containing SCSG fragment and its specific fluorescent labeling. (f) Myristoylation drastically reduces water solubility of HPreS-SCSG CLPs but not individual subunits. Fusions containing an assembly competent coreC fragment ending after 149 aa, or an assembly deficient coreC ending after 124 aa, were expressed as such or under myristoylation conditions. Upon Triton X114 phase partitioning, both nonmyristoylated proteins remained in the low detergent (upper) phase; upon myristoylation, the assembly competent protein partitioned into the detergent-rich (lower) phase.

system is the *Acid* and *Base* peptide pair of leucine zippers [73], which were designed to form heterodimeric coiled coils (Figure 12.12). Insertion of the *Acid* sequence into the c/e1 epitope allowed CLP formation, yet addition of the complementary *Base* peptide disintegrated the particles [23]. Likely, the inserted *Acid* sequence on its own is sufficiently flexible to adapt into the c/e1 acceptor sites, whereas the stable heterodimeric coiled-coil structure forming upon *Base* addition interfered with maintenance of an assembly-competent structure. Though not yet experimentally proven, such problems would likely be prevented by releasing one end of the helix from the carrier.

Another situation where c/e1 display on SplitCore rather than conventional HBc CLPs promises advantages is peptides derived from *terminal* sequences of an antigen. Only in the SplitCore system is the naturally free terminus maintained. The feasibility of this concept is demonstrated by SplitCore CLPs displaying an N-terminal peptide from the influenza virus M2 protein. M2, essential for the viral life cycle, forms proton channels in the viral envelope and in membranes of virus-infected cells (see Chapter 25). Its ~100 aa sequence is organized into an N-terminal ectodomain of ~24 aa (termed *M2e*), a transmembrane domain, and a C-terminal cytoplasmic domain [74].

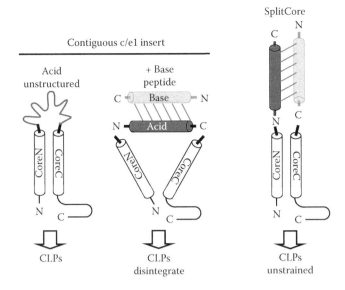

FIGURE 12.12 Distinct properties of SplitCore versus conventional HBc for presentation of short but structured peptides or protein segments. *Acid* and *Base* are ~35 aa long peptides designed to form antiparallel coiled coils. *Acid* alone conventionally integrated into c/e1 did not interfere with CLP formation, but addition of the complementary *Base* peptide caused CLP disintegration, likely due to formation of a stable coiled-coil structure. The openended fixation to the carrier in SplitCore CLPs is expected to prevent such interference. Note that all short sequences adopting a stable α-helical fold on their own would only be compatible with the SplitCore system.

Due to its high conservation, M2e has attracted interest as target for a universal influenza vaccine, including the presentation on HBc CLPs to enhance M2e's intrinsically low immunogenicity [75,76].

A well-studied derivative contains a triplicated consensus M2e sequence fused to the N terminus of HBc (Figure 12.13a). Triplication of the M2e peptide probably helps to alleviate its suboptimal solvent exposure caused by the location of the fusion site in the canyons around the CLP spikes. Such M2e-HBc CLPs induce high-titered M2-specific antibodies yet also an even stronger (~80-fold) anti-HBc response [77]; this is expected as the immunodominant c/e1 epitope remains intact in these constructs. Still, immunization provided protection in several animal models including ferrets against challenge with different HA subtypes of influenza virus [75].

For comparison, we fused the consensus M2e sequence from human influenza virus and that from swine-originated influenza virus (SOIV) to coreC in the SplitCore system (Figure 12.13b) and obtained, again from reverse cistron order constructs, substantial amounts of soluble regular CLPs (Figure 12.13). In mice, these CLPs induced similarly high titers of anti-M2 antibodies as CLPs carrying triplicated M2e at the HBc N terminus; however, the response against HBc was fivefold lower than that against M2, a 400-fold difference in relative antibody response compared to M2e fused to the HBc N terminus (A. Walker and U. Blohm, unpublished data). This highlights the dramatic impact of the attachment site on the ratio of the desired anti-insert versus undesired

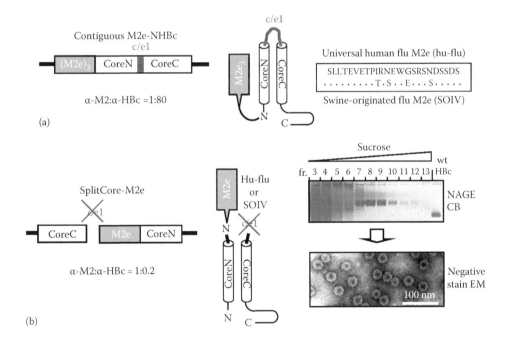

FIGURE 12.13 SplitCore allows free-ended display of terminal protein sequences. M2e refers to the N-terminal ectodomain of the influenza virus M2 protein. (a) A candidate *universal* influenza vaccine mimics this arrangement by fusion of a triplicated consensus M2e sequence from human influenza viruses to the N terminus of HBc, leaving the c/e1 epitope intact. Although these CLPs evoke a strong anti-M2 response, the anti-HBc response is about 80 times stronger. (b) SplitCore-M2e. M2e sequences corresponding to the universal human influenza virus sequence, or one from the SOIV, were fused to coreN and expressed from a reverse cistron order construct. CLP formation was verified by NAGE analysis after gradient separation and by negative staining EM (shown for the SOIV construct). In mice, both CLPs induced a comparably strong anti-M2 response as CLPs from (a), yet the anti-HBc response was fivefold weaker than that against M2.

anticarrier response. Even though the protective potency of these SplitCore-based M2 vaccines is yet to be determined, the current data already suggest such SplitCore derivatives as a potentially superior alternative to N-terminal HBc fusions.

Flexible or less stably structured protein-internal peptides represent a different case, with two possible extreme outcomes in the contiguous chain system that are highly relevant for vaccine applications (Figure 12.14a). Many neutralizing epitopes are conformational; hence, only antibodies recognizing this native conformation are relevant for protection. In the best case, clamping the peptide into the c/e1 loop of HBc (or a surface-exposed loop of any other scaffold) will stabilize the native conformation of the peptide and promote the induction of neutralizing antibodies. In the worst case, the peptide will be forced into a nonnative conformation (Figure 12.14b), inducing exclusively irrelevant antibodies.

In contrast, in an open-end attachment as in SplitCore, the heterologous sequence is not subjected to such forces

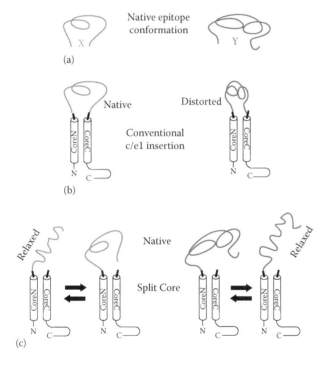

(a)

(b)

(c)

FIGURE 12.14 Distinct properties of SplitCore versus conventional HBc for presentation of short protein segments without strong folding potential. (a) X and Y symbolize two peptides that adopt a defined conformation in their natural protein context but have a low propensity to maintain this native fold in isolation. (b) Conventional c/e1 insertion. If X fits naturally into c/e1, its native fold will be stabilized by the helical framework of the HBc carrier, and upon immunization, exclusively relevant antibodies would be induced. In contrast, the nonfitting sequence Y will be distorted upon c/e1 insertion and induce only irrelevant antibodies. (c) SplitCore fusion. Due to open-ended fixation to the carrier, both X and Y may be present in an ensemble of different (and interchanging) structures that also include the native conformation. For most natural protein segments, the situation represented by Y will apply; hence, chances for inducing relevant antibodies are higher in the SplitCore than in the conventional c/e1 insertion system.

and, depending on its specific folding propensity, may be present in an ensemble of different conformations that interchange over time and may include the native structure (Figure 12.14c). On the positive side, this will increase chances that, compared to the worst-case scenario for two-sided fixation, at least a fraction of the antibodies induced will recognize the native antigen. On the negative side, at least a fraction of the antibodies are likely to be irrelevant. These considerations emphasize the importance of assessing, beyond immunogenicity, the protective potential of new vaccine candidates.

An example for the successful application of the SplitCore system to antigen-internal peptide sequences is provided by SplitCore CLPs displaying peptides from hepatitis C virus (HCV). HCV is an RNA virus causing acute and, more importantly, chronic hepatitis C. While effective drugs for virus elimination are becoming available [78], they are exceedingly expensive and provide no protection from reinfection. Hence, a prophylactic vaccine would be a highly desirable and affordable alternative; however, the high sequence variability of the virus has thus far prevented the generation of vaccines eliciting a broadly cross-neutralizing anti-HCV response [79].

Based on the strong association of an early antibody response with virus elimination [80] and the likely key role of the hypervariable region I (HVRI) of the E2 envelope protein as target for neutralizing antibodies [81], we generated SplitCore CLPs in which one of four different HCV genotype 1–based 27-mer HVRI sequences was fused to coreN so as to deliberately expose the more conserved C-terminal HVRI parts (Figure 12.15). All fusions were efficiently expressed in *E. coli* as soluble, regularly shaped CLPs. When used individually to immunize mice, all CLPs induced strong anti-HVRI antibody responses that exerted a modest but well detectable HCV genotype-specific cross-reactivity with a range of HVRI peptides and cross-neutralization in cell culture infection models. Importantly, immunization with a mixture of all four CLPs strongly broadened the cross-reactivity and cross-neutralization potency, resulting in protection from infection by a substantial proportion (>20%) of genotype (gt) 1a and gt 1b isolates. Given the easy access to different HVRI SplitCore CLPs demonstrated in this study, the approach could feasibly be broadened to include HCV gt 2 and 3 HVRI sequences to achieve intergenotypic protection.

Together, these studies indicate that SplitCore, beyond its superior ability to accept folded whole proteins, is as well suited to present short protein segments in an immune-enhancing fashion. For lack of experimental data, a quantitative comparison with the neutralizing potency of the same HCV peptides inserted into the conventional HBc carrier system is currently not possible, but, given the conceptional considerations outlined earlier, the odds for such CLPs displaying these sequences in exactly the same (unknown) conformation as in the authentic HCV E2 protein appear rather slim. However, if protein-chemically feasible, the most comprehensive approach is to validate both carrier systems for a given heterologous sequence.

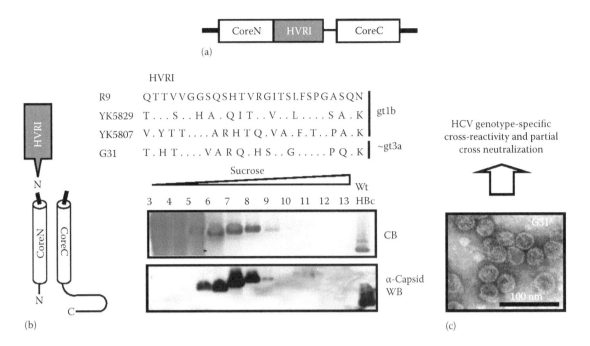

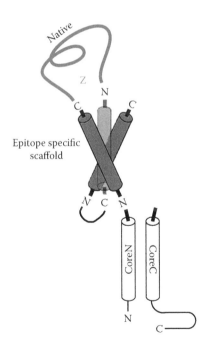

FIGURE 12.15 Partially cross-reactive and cross-neutralizing antibodies from SplitCore CLPs presenting HVRI from HCV E2 protein. (a) Expression cassettes used. Note that fusion to coreN served to deliberately expose the less variable C-terminal part of the HVRI. (b) Expected products, specific HVRI sequences used, and confirmation of CLP formation by NAGE and immunoblotting. (c) Negative staining EM.

FIGURE 12.16 SplitCore CLPs for epitope-focused vaccines. Key to this new concept is the design of an artificial protein scaffold that stably maintains a given epitope (Z) in exactly the same conformation as in its natural protein context (sequence X in Figure 12.14 represents a special case where the α-helical hairpin of HBc would act as such as scaffold). Thus far described scaffolds are based on three-helix bundle proteins. Due to their topology, such structures would be compatible with the SplitCore carrier but not conventional c/e1 insertion.

Notably, a novel systematic (but highly laborious) approach toward presenting peptide epitopes in their native conformation so as to elicit exclusively relevant antibodies has recently been presented as *epitope-focused vaccine design* [82]. In essence, a relevant pathogen epitope is identified, for example, via protective antibodies from patients, and then a protein scaffold, for example, a three-helix bundle, is designed specifically around this epitope so as to warrant formation and stabilization of its native conformation in the new context. Such scaffold-presented epitopes from respiratory syncytial virus (RSV) were able to induce RSV-neutralizing antibodies in macaques; notably, their protective potency was markedly higher when the epitope-presenting scaffolds were chemically conjugated to HBc CLPs.

The topology of three-helix bundle scaffolded epitopes is incompatible with c/e1 insertion into contiguous chain HBc, yet SplitCore should provide a suitable and simple, purely genetic platform for their presentation (Figure 12.16). Once additional specific scaffold–epitope combinations, including for highly variable viruses like HCV and HIV-1 as proposed [82], have been identified, this appears as a highly worthwhile approach.

12.12 NONHUMAN HEPADNAVIRUS-BASED SPLITCORE CLPs

The estimated number of ~250 million chronic HBV carriers [83] has raised issues regarding the efficacy of HBc CLP–based human vaccines. For instance, preexisting anti-HBc antibodies might blunt immunogenicity of the incoming vaccine CLPs. Alternatively, CLP opsonization may even enhance Fc receptor-mediated uptake by antigen-presenting cells and boost their immunogenicity. There is evidence favoring the former [84] as well as the latter notion [85,86]. As the vast majority of HBV infection-based anti-HBc antibodies are directed against c/e1 [33], this issue is probably much more relevant for HBc CLPs with an intact c/e1 epitope, for example, the earlier described M2e fusions to the HBc N terminus, than for c/e1 insertions or SplitCore CLPs in which the c/e1 epitope is, in addition, physically disrupted. The loss of c/e1 should also reduce interference with diagnostic discrimination between previous HBV infection (presence of anti-HBsAg plus anti-HBcAg antibodies) and HBV immunization, commonly done with purified HBsAg, thus exclusively evoking an anti-HBsAg response. For wt SplitCore CLPs, we found a drastically lower reactivity with patient anti-HBcAg than for contiguous chain HBc CLPs and a complete absence of anti-c/e1 antibodies [1]. Hence, specific testing for anti-c/e1 antibodies would provide a means to distinguish between infection and HBc CLP–based vaccination.

Perhaps more important is a potential lack of immune enhancement by carrier-dependent T cell responses in HBc-tolerant chronic carriers, which may be remedied by using the core proteins of WHV or GSHV ([87] and Chapter 13). Despite a similar CLP architecture, these proteins show less

than 60% sequence identity to human HBc, and they are largely non-cross-reactive with HBc on the B cell and T cell levels [38].

We therefore also evaluated the SplitCore approach for the WHV core protein (WHc), with comparable results [1]. Splitting WHc in the sequence that structurally corresponds to the c/e1 loop of HBc allowed for efficient assembly of CLPs from separately expressed segments (here termed *WHcN* and *WHcC*; Figure 12.17a). Regular CLPs were also formed upon fusing a 20-mer CSP repeat peptide, or the entire CSP sequence to the WHcN fragment (Figure 12.17b). While both CLPs induced strong anti-CSP repeat peptide responses, the titers were markedly higher for the full-length CSP-CLP-induced immune sera, which contained only a low level of anti-WHc antibodies. The reverse was seen with the repeat peptide CLP-induced antisera (Figure 12.17c). This suggests that steric shielding of the inner CLP shell by large fusion partners contributes to reduced anticarrier responses. Altogether, SplitWHc appears similarly suited as antigen carrier as HBc-derived SplitCore. Notably, CLP formation was also observed between coreN and WHcC, suggesting that most of the sequence diversity between the two core proteins relates to positions that are not essential for intracore protein subunit, intradimer, and interdimer contacts. This also indicates the possibility to

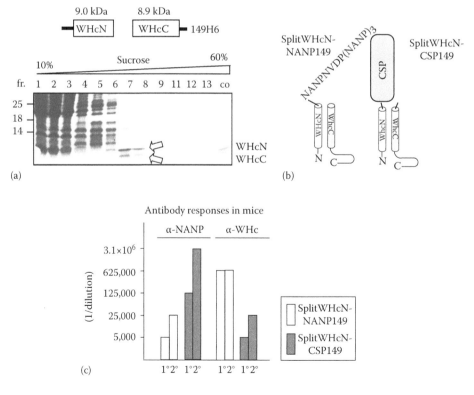

FIGURE 12.17 The SplitCore approach is applicable to nonhuman HBV core proteins. (a) SplitCore system based on WHV core protein (WHc). The expression cassette used and cosedimentation of the expected two WHV core protein segments WHcN and WHcC upon sucrose gradient sedimentation are shown. (b) WHc SplitCore CLPs presenting *P. falciparum* antigens. Either a sequence from the repeat region within CSP or the entire CSP sequence was fused to WHcN. CLP formation was confirmed by sedimentation and NAGE analysis (not shown). (c) Immune sera from mice immunized once (1°) or twice (2°) with either type of CLP were analyzed by ELISA for reactivity with the repeat peptide (α-NANP) or recombinant wt WHc. Titers are given as the maximal dilution giving a signal three times above background. Note the exceptionally high anti-NANP and low anti-WHc titers induced by the full-length CSP presenting CLPs.

mutationally modulate the T cell reactivity of human HBc as carrier without affecting its assembly competence.

12.13 CONCLUSIONS AND CHALLENGES

The preceding data presented establish the SplitCore system as a novel molecular presentation platform that maintains all of the known advantageous properties of HBc CLPs as a carrier platform while greatly expanding its potential applications. Foremost is its ability to support efficient CLP formation upon fusion with whole proteins whose structures are incompatible with the contiguous chain HBc system. This opens new avenues for a whole spectrum of protein-antigen-based CLP vaccines, with intrinsic assets such as native epitope presentation and almost complete B cell epitope coverage [40]. The restriction refers to the reduced accessibility of residues of the displayed protein that are close to the inner HBc shell, as shown by the distinct antibody profiles elicited when OspA was fused to either coreN or coreC. Conversely, the tolerance toward attachment in either orientation also offers a simple means to direct the antibody response toward specific regions in a protein or to combine the two orientations to widen the antibody repertoire.

Moreover, also peptide-based vaccine approaches can benefit from the increased structural tolerance of the SplitCore carrier, on one hand by the easy access it provides to multiple CLPs displaying different peptides [88] and on the other as a genetic alternative to chemical conjugation for the presentation of epitope-focused scaffolds [82].

Another SplitCore-intrinsic novel property is the exposure of new termini on the CLP surface that enables additional modifications, not necessarily restricted to vaccine applications. As an example, an 18-mer biotin acceptor peptide [89] fused to coreN was sufficiently flexible to serve as substrate for in vivo biotinylation by the *E. coli* biotin–protein ligase BirA, as shown by the specific association of the corresponding CLPs with streptavidin [1]. This might be extended to streptavidin and avidin that are themselves linked to other functional moieties. Our data with the SCSG- and SCSG-GB1 CLPs demonstrate that the principle of surface-accessible new termini can be extended to the displayed protein itself, thus allowing the successive addition of further layers around the inner shells, as shown by GB1 as an immunoglobulin targeting device and HBV PreS sequences as substrates for fatty acid modification of the CLP surface. Numerous other combinations of functional moieties can easily be imagined, especially considering the possibility of *packaging* additional sequences into the CLP lumen via their fusion to the C-terminal end of the HBc assembly domain [34].

Despite the new opportunities offered by SplitCore, there are still limitations as to which molecules, in particular proteins, can be presented without interference with regular CLP formation. One is size, because the available surface area of the CLPs (split or not) is geometrically confined. Another is nonproductive interactions in the fusion proteins that could compete with the desired interactions required for regular CLP assembly. To this end, the folding pathways of the carrier parts and the to-be-displayed protein need to remain independent, which is more likely to occur when all partners have a strong intrinsic folding propensity. HBc itself meets this criterion well, as implied by its highly ordered and stable structure; efficient SplitCore assembly indicates that this also holds for the separate coreN and coreC modules. However, the folding propensities of candidate display proteins may widely differ, as is evident from the ease with which some proteins (including HBc itself) can be solubly expressed in *E. coli* while others are notorious in forming insoluble aggregates. As a rule of thumb, proteins that are already difficult to express by themselves are unlikely to yield soluble SplitCore (or other) CLPs; in fact, the enormous local concentration on the assembling CLP may even promote contacts between immature folding intermediates and thus aggregation. A notable exception is CSP. Free CSP expressed in *E. coli* accumulated largely in insoluble form, and in contrast to the SplitCore-CSP CLPs, the small fraction of soluble protein started to precipitate soon after purification [90]. Hence in special cases, arraying a protein on the CLP surface may be beneficial but such behavior is as yet unpredictable. Various other mechanisms of sequence-dependent interference with soluble CLP formation can be imagined. For instance, even a short peptide, or a protein segment in a folding intermediate, might expose (perhaps transiently) interfaces that could compete with the desired HBc carrier interactions required for dimerization and particle assembly (between the two segments within one functional HBc subunit, between the two subunits within the dimer, or between the dimers). Hence, SplitCore overcomes the fundamental steric constriction imposed on the contiguous chain HBc system by the defined geometry of the two acceptor sites, but it does not, or only rarely (see CSP), alleviate folding problems that are intrinsic to the protein to be displayed. As is true for all proteinaceous carriers, it is the sum of all potential interactions that eventually determines whether carrier and insert fold independently or not.

Hence additional means will be required to further expand the range of HBc displayable molecules, which may also benefit from the specific features of SplitCore.

12.14 EXPRESSION HOSTS WITH HIGHER PROTEIN FOLDING CAPACITY

E. coli provides the simplest system for large-scale production of many recombinant proteins [91], yet proper folding and soluble expression of many others can only be achieved in higher organisms [92,93]. Hence expressing SplitCore proteins in eukaryotes may support the formation of CLPs presenting heterologous sequences that prevent ordered assembly in *E. coli*. One widely used system is insect cells transduced with recombinant baculoviruses [94]. We therefore used the splitGFP presenting SplitCore SCSG protein as a model to obtain proof of principle for the feasibility of SplitCore CLP formation in eukaryotic cells. As eukaryotic mRNAs are usually monocistronic, internal ribosome entry sites (IRESs) from eukaryotic viruses may be used to generate bicistronic

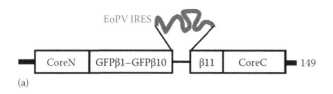

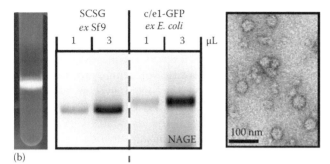

FIGURE 12.18 SplitCore CLPs can be expressed in eukaryotic cells. (a) Expression cassette used. The construct is analogous to that for SCSG expression in *E. coli*, except the prokaryotic RBS2 sequence has been replaced by the IRES from the insect virus EoPV. For expression in insect cells, the modified SCSG expression cassette was used to generate recombinant baculoviruses. (b) Evidence for successful SCSG CLP formation in insect cells. Sucrose gradient sedimentation of lysate from Sf9 cells transduced with the recombinant SCSG baculovirus led to the accumulation of a brightly fluorescent band in the gradient center. Aliquots thereof produced a distinct green fluorescent band during NAGE analysis with similar mobility and appearance as *E. coli* derived CLPs bearing GFP inside the c/e1 epitope. Bands were detected via GFP fluorescence using a Typhoon laser scanner. EM revealed the presence of ordered particles.

constructs for the simultaneous expression of the two protein segments from one mRNA (Figure 12.18a). A pilot study in Sf9 cells transduced with a recombinant baculovirus encoding a bicistronic expression cassette with the IRES element from *Ectropis obliqua* picorna-like virus (EoPV; [95]) indeed demonstrated successful formation of SCSG CLPs (Figure 12.18b). During sucrose gradient centrifugation, a visibly fluorescent band accumulated in the center fractions, the material from these fractions migrated as distinct fluorescent band in native agarose gels, and negative staining EM revealed the presence of ordered CLPs. Hence with further optimization, including the use of alternative IRES elements such as from white spot syndrome virus (WSSV) [96] or other mechanisms for downstream cistron translation, insect and possibly higher eukaryotic cells may be used for SplitCore CLPs that cannot be produced in *E. coli*.

12.15 GLYCOPROTEIN AND CARBOHYDRATE ANTIGEN DISPLAY

A large number of medically relevant antigens are glycoproteins and carbohydrates whose immunogenicity should benefit from presentation on CLPs. At present, this cannot be achieved by purely genetic approaches. Even though

eukaryotic expression hosts would allow fusion protein glycosylation, the corresponding machinery is confined to the secretory compartments that assembled CLPs are unlikely to reach. A possible solution is postassembly conjugation, that is, the coupling of the antigen to preassembled CLPs. One alternative to conventional conjugation chemistries [97] is enzymatic coupling, for example, by sortases. These transpeptidases from Gram-positive bacteria [98] can covalently join the N terminus of oligoglycine chains to the C terminus of a residue inside a sorting signal motif, such as the Thr residue in the LPXTG motif, where X refers to any amino acid [99]. The SplitCore system lends itself to this approach by allowing the simple addition of the LPXTG sequence to the C terminus of the coreN segment (Figure 12.19a). However, in preliminary experiments, in vitro sortase coupling to GFP carrying a Gly_5 sequence at its N terminus was inefficient, and this also held for coexpression of the two substrate proteins and the enzyme (P. Kolb, A. Walker, and M. Nassal, unpublished data). While improvements may be possible, a principal restriction is that the sortase enzyme (~25 kDa) must be able to access the conjugation site, which in the final structure will be located between the inner-core shell and the outer layer of the protein to be conjugated. Hence with increasing occupancy, conjugating further molecules to the CLP surface will become increasingly difficult. A more promising option is therefore to employ advanced *chemical* methods such as click chemistry or one of its various modifications [100,101]. For instance, combined with genetic approaches for incorporation of nonnatural amino acids [102], an alkyne-containing amino acid may be introduced into an exposed site, such as the c/e1 epitope (Figure 12.19b), on the core protein and then bioorthogonally be coupled to azide-modified glycoproteins or carbohydrates [103]. Again, employing the SplitCore carrier and incorporating a suitable nonnatural amino acid at the exposed ends of the coreN or coreC segment, preferentially combined with a flexible linker, should greatly enhance accessibility of the reactive group and thus promote successful conjugation even with bulky partners. In this way, the advantages of the SplitCore system may also be exploited for developing vaccines against the large number of medically relevant glycoprotein and carbohydrate antigens [104], which by themselves are unable to induce immunological memory [105].

12.16 PERSPECTIVES

Given its recent development, the number of applications of the SplitCore system is still small compared to those reported for the conventional HBc system. However, the current examples already document the highly increased flexibility of this new approach even though several of the data sets have yet to be fully explored. As emphasized in this and other chapters in this book, HBc provides distinct and highly useful features, especially for vaccine applications, that are fully maintained in the SplitCore system. However, depending on

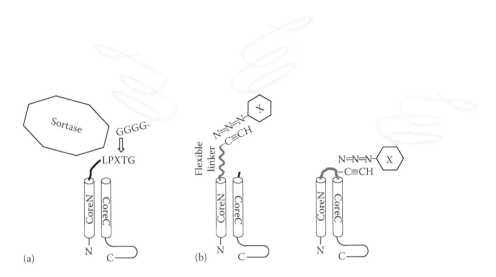

FIGURE 12.19   SplitCore CLPs for postassembly modification. (a) Sortase-mediated conjugation. Sortase links N-terminal Gly residues covalently to appropriate target sequences, for example, the Thr residue in the motif LPXTG (X = any aa). Fusing this target sequence to coreN allowed sortase-mediated conjugation to sequences bearing an N-terminal Gly5 motif. However, efficacy was limited likely because the bulky sortase enzyme encounters increasing steric hindrance when the CLP surface becomes increasingly covered. This is prevented by chemical conjugation methods. (b) SplitCore (*left*) versus conventional HBc CLPs (*right*) for click-chemistry-mediated conjugation. Nonnatural amino acids providing side chains for click chemistry (e.g., an alkyne) can be genetically incorporated into contiguous chain HBc; however, display on SplitCore CLPs should enhance accessibility, especially when combined with a flexible linker sequence. While any molecule X bearing a suitable click-chemistry partner moiety, such as here an azide, should be conjugatable, accessibility of the CLP-borne alkyne will be particularly important when X bears a large moiety Y. As azide labeling of sugars is now routinely possible, this or similar bioorthogonal approaches hold promise for the CLP presentation of glycoproteins and carbohydrates.

the application, other viral nanoparticles may as well be suitable as carriers. As the SplitCore concept is based on general structure principles, it can be expected to be equally applicable to such alternative carrier proteins.

ACKNOWLEDGMENTS

The authors thank numerous colleagues who contributed to this work, including Jolanta Vorreiter, Bettina Böttcher, Britta Gerlach, Ulrike Blohm, Markus Simon, Reinhard Wallich, and David Milich. This work was supported in part by the Deutsche Forschungsgemeinschaft, most recently through grant NA154/11-1, and a fellowship to TTAN by the German Academic Exchange Service (DAAD).

REFERENCES

1. Walker, A., C. Skamel et al., SplitCore: An exceptionally versatile viral nanoparticle for native whole protein display regardless of 3D structure. *Sci Rep*, 2011. **1**: 5.
2. Beck, J. and M. Nassal, Hepatitis B virus replication. *World J Gastroenterol*, 2007. **13**(1): 48–64.
3. Drexler, J. F., A. Geipel et al., Bats carry pathogenic hepadnaviruses antigenically related to hepatitis B virus and capable of infecting human hepatocytes. *Proc Natl Acad Sci USA*, 2013. **110**(40): 16151–16156.
4. Schultz, U., E. Grgacic et al., Duck hepatitis B virus: An invaluable model system for HBV infection. *Adv Virus Res*, 2004. **63**: 1–70.
5. Dallmeier, K. and M. Nassal. 2008. Hepadnaviruses have a narrow host range—Do they?, pp. 303–339. In O. Weber and U. Protzer (eds.), *Comparative Hepatitis*, Birkhäuser, Basel, Switzerland.
6. Dane, D. S., C. H. Cameron et al., Virus-like particles in serum of patients with Australia-antigen-associated hepatitis. *Lancet*, 1970. **1**(7649): 695–698.
7. Bruss, V., Hepatitis B virus morphogenesis. *World J Gastroenterol*, 2007. **13**(1): 65–73.
8. Kenney, J. M., C. H. von Bonsdorff et al., Evolutionary conservation in the hepatitis B virus core structure: Comparison of human and duck cores. *Structure*, 1995. **3**(10): 1009–1019.
9. Roseman, A. M., J. A. Berriman et al., A structural model for maturation of the hepatitis B virus core. *Proc Natl Acad Sci USA*, 2005. **102**(44): 15821–15826.
10. Nassal, M., Hepatitis B viruses: Reverse transcription a different way. *Virus Res*, 2008. **134**(1–2): 235–249.
11. Cohen, B. J. and J. E. Richmond, Electron microscopy of hepatitis B core antigen synthesized in *E. coli. Nature*, 1982. **296**(5858): 677–679.
12. Pumpens, P., G. P. Borisova et al., Hepatitis B virus core particles as epitope carriers. *Intervirology*, 1995. **38**(1–2): 63–74.
13. Nassal, M., I. Leifer et al., A structural model for duck hepatitis B virus core protein derived by extensive mutagenesis. *J Virol*, 2007. **81**(23): 13218–13229.
14. Birnbaum, F. and M. Nassal, Hepatitis B virus nucleocapsid assembly: Primary structure requirements in the core protein. *J Virol*, 1990. **64**(7): 3319–3330.
15. Nassal, M., The arginine-rich domain of the hepatitis B virus core protein is required for pregenome encapsidation and productive viral positive-strand DNA synthesis but not for virus assembly. *J Virol*, 1992. **66**(7): 4107–4116.

16. Porterfield, J. Z., M. S. Dhason et al., Full-length hepatitis B virus core protein packages viral and heterologous RNA with similarly high levels of cooperativity. *J Virol*, 2010. **84**(14): 7174–7184.

17. Conway, J. F., N. Cheng et al., Visualization of a 4-helix bundle in the hepatitis B virus capsid by cryo-electron microscopy. *Nature*, 1997. **386**(6620): 91–94.

18. Böttcher, B., S. A. Wynne et al., Determination of the fold of the core protein of hepatitis B virus by electron cryomicroscopy. *Nature*, 1997. **386**(6620): 88–91.

19. Wynne, S. A., R. A. Crowther et al., The crystal structure of the human hepatitis B virus capsid. *Mol Cell*, 1999. **3**(6): 771–780.

20. Ceres, P. and A. Zlotnick, Weak protein–protein interactions are sufficient to drive assembly of hepatitis B virus capsids. *Biochemistry*, 2002. **41**(39): 11525–11531.

21. Ceres, P., S. J. Stray et al., Hepatitis B virus capsid assembly is enhanced by naturally occurring mutation F97L. *J Virol*, 2004. **78**(17): 9538–9543.

22. Vogel, M., M. Diez et al., In vitro assembly of mosaic hepatitis B virus capsid-like particles (CLPs): Rescue into CLPs of assembly-deficient core protein fusions and FRET-suited CLPs. *FEBS Lett*, 2005. **579**(23): 5211–5216.

23. Böttcher, B., M. Vogel et al., High plasticity of the hepatitis B virus capsid revealed by conformational stress. *J Mol Biol*, 2006. **356**(3): 812–822.

24. Schuch, A., A. Hoh et al., The role of natural killer cells and CD8(+) T cells in hepatitis B virus infection. *Front Immunol*, 2014. **5**: 258.

25. Milich, D. R. and A. McLachlan, The nucleocapsid of hepatitis B virus is both a T-cell-independent and a T-cell-dependent antigen. *Science*, 1986. **234**(4782): 1398–1401.

26. Fehr, T., D. Skrastina et al., T cell-independent type I antibody response against B cell epitopes expressed repetitively on recombinant virus particles. *Proc Natl Acad Sci USA*, 1998. **95**(16): 9477–9481.

27. Lee, B. O., A. Tucker et al., Interaction of the hepatitis B core antigen and the innate immune system. *J Immunol*, 2009. **182**(11): 6670–6681.

28. Whitacre, D. C., B. O. Lee et al., Use of hepadnavirus core proteins as vaccine platforms. *Expert Rev Vaccines*, 2009. **8**(11): 1565–1573.

29. Clarke, B. E., S. E. Newton et al., Improved immunogenicity of a peptide epitope after fusion to hepatitis B core protein. *Nature*, 1987. **330**(6146): 381–384.

30. Stahl, S. J. and K. Murray, Immunogenicity of peptide fusions to hepatitis B virus core antigen. *Proc Natl Acad Sci USA*, 1989. **86**(16): 6283–6287.

31. Schödel, F., A. M. Moriarty et al., The position of heterologous epitopes inserted in hepatitis B virus core particles determines their immunogenicity. *J Virol*, 1992. **66**(1): 106–114.

32. Nassal, M., Total chemical synthesis of a gene for hepatitis B virus core protein and its functional characterization. *Gene*, 1988. **66**(2): 279–294.

33. Salfeld, J., E. Pfaff et al., Antigenic determinants and functional domains in core antigen and e antigen from hepatitis B virus. *J Virol*, 1989. **63**(2): 798–808.

34. Beterams, G., B. Bottcher et al., Packaging of up to 240 subunits of a 17 kDa nuclease into the interior of recombinant hepatitis B virus capsids. *FEBS Lett*, 2000. **481**(2): 169–176.

35. Beterams, G. and M. Nassal, Significant interference with hepatitis B virus replication by a core-nuclease fusion protein. *J Biol Chem*, 2001. **276**(12): 8875–8883.

36. Ulrich, R., M. Nassal et al., Core particles of hepatitis B virus as carrier for foreign epitopes. *Adv Virus Res*, 1998. **50**: 141–182.

37. Pumpens, P. and E. Grens, HBV core particles as a carrier for B cell/T cell epitopes. *Intervirology*, 2001. **44**(2–3): 98–114.

38. Billaud, J. N., D. Peterson et al., Comparative antigenicity and immunogenicity of hepadnavirus core proteins. *J Virol*, 2005. **79**(21): 13641–13655.

39. Taylor, K. M., T. Lin et al., Influence of three-dimensional structure on the immunogenicity of a peptide expressed on the surface of a plant virus. *J Mol Recognit*, 2000. **13**(2): 71–82.

40. Nassal, M., C. Skamel et al., Development of hepatitis B virus capsids into a whole-chain protein antigen display platform: New particulate Lyme disease vaccines. *Int J Med Microbiol*, 2008. **298**(1–2): 135–142.

41. Ormo, M., A. B. Cubitt et al., Crystal structure of the *Aequorea victoria* green fluorescent protein. *Science*, 1996. **273**(5280): 1392–1395.

42. Kratz, P. A., B. Bottcher et al., Native display of complete foreign protein domains on the surface of hepatitis B virus capsids. *Proc Natl Acad Sci USA*, 1999. **96**(5): 1915–1920.

43. Vogel, M., J. Vorreiter et al., Quaternary structure is critical for protein display on capsid-like particles (CLPs): Efficient generation of hepatitis B virus CLPs presenting monomeric but not dimeric and tetrameric fluorescent proteins. *Proteins*, 2005. **58**(2): 478–488.

44. Skamel, C., M. Ploss et al., Hepatitis B virus capsid-like particles can display the complete, dimeric outer surface protein C and stimulate production of protective antibody responses against *Borrelia burgdorferi* infection. *J Biol Chem*, 2006. **281**(25): 17474–17481.

45. Kumaran, D., S. Eswaramoorthy et al., Crystal structure of outer surface protein C (OspC) from the Lyme disease spirochete, *Borrelia burgdorferi*. *EMBO J*, 2001. **20**(5): 971–978.

46. Basto, A. P. and A. Leitao, Targeting TLR2 for vaccine development. *J Immunol Res*, 2014. **2014**: 619410.

47. Wallich, R., M. D. Kramer et al., The recombinant outer surface protein A (lipOspA) of *Borrelia burgdorferi*: A Lyme disease vaccine. *Infection*, 1996. **24**(5): 396–397.

48. Ratanji, K. D., J. P. Derrick et al., Immunogenicity of therapeutic proteins: Influence of aggregation. *J Immunotoxicol*, 2014. **11**(2): 99–109.

49. Nassal, M., C. Skamel et al., A fusion product of the complete *Borrelia burgdorferi* outer surface protein A (OspA) and the hepatitis B virus capsid protein is highly immunogenic and induces protective immunity similar to that seen with an effective lipidated OspA vaccine formula. *Eur J Immunol*, 2005. **35**(2): 655–665.

50. Walker, A., C. Skamel et al., Internal core protein cleavage leaves the hepatitis B virus capsid intact and enhances its capacity for surface display of heterologous whole chain proteins. *J Biol Chem*, 2008. **283**(48): 33508–33515.

51. Riedl, P., D. Stober et al., Priming Th1 immunity to viral core particles is facilitated by trace amounts of RNA bound to its arginine-rich domain. *J Immunol*, 2002. **168**(10): 4951–4959.

52. Sominskaya, I., D. Skrastina et al., A VLP library of C-terminally truncated Hepatitis B core proteins: Correlation of RNA encapsidation with a Th1/Th2 switch in the immune responses of mice. *PLoS ONE*, 2013. **8**(9): e75938.

53. Ibanez, L. I., K. Roose et al., M2e-displaying virus-like particles with associated RNA promote T helper 1 type adaptive immunity against influenza A. *PLoS ONE*, 2013. **8**(3): e59081.

54. Ding, W., X. Huang et al., Structural identification of a key protective B-cell epitope in Lyme disease antigen OspA. *J Mol Biol*, 2000. **302**(5): 1153–1164.

55. Olotu, A., G. Fegan et al., Four-year efficacy of RTS,S/AS01E and its interaction with malaria exposure. *N Engl J Med*, 2013. **368**(12): 1111–1120.

56. Gregson, A. L., G. Oliveira et al., Phase I trial of an alhydrogel adjuvanted hepatitis B core virus-like particle containing epitopes of *Plasmodium falciparum* circumsporozoite protein. *PLoS ONE*, 2008. **3**(2): e1556.

57. Cerami, C., U. Frevert et al., The basolateral domain of the hepatocyte plasma membrane bears receptors for the circumsporozoite protein of *Plasmodium falciparum* sporozoites. *Cell*, 1992. **70**(6): 1021–1033.

58. Persson, C., G. A. Oliveira et al., Cutting edge: A new tool to evaluate human pre-erythrocytic malaria vaccines: Rodent parasites bearing a hybrid *Plasmodium falciparum* circumsporozoite protein. *J Immunol*, 2002. **169**(12): 6681–6685.

59. Vaughan, A. M., S. H. Kappe et al., Development of humanized mouse models to study human malaria parasite infection. *Future Microbiol*, 2012. **7**(5): 657–665.

60. Espinosa, D. A., A. Yadava et al., Development of a chimeric *Plasmodium berghei* strain expressing the repeat region of the *P. vivax* circumsporozoite protein for in vivo evaluation of vaccine efficacy. *Infect Immun*, 2013. **81**(8): 2882–2887.

61. Estrada-Pena, A. and J. de la Fuente, The ecology of ticks and epidemiology of tick-borne viral diseases. *Antiviral Res*, 2014. **108C**: 104–128.

62. Mansfield, K. L., N. Johnson et al., Tick-borne encephalitis virus—A review of an emerging zoonosis. *J Gen Virol*, 2009. **90**(Pt 8): 1781–1794.

63. Juncadella, I. J. and J. Anguita, The immunosuppresive tick salivary protein, Salp15. *Adv Exp Med Biol*, 2009. **666**: 121–131.

64. Hovius, J. W., T. J. Schuijt et al., Preferential protection of *Borrelia burgdorferi* sensu stricto by a Salp15 homologue in *Ixodes ricinus* saliva. *J Infect Dis*, 2008. **198**(8): 1189–1197.

65. Ramamoorthi, N., S. Narasimhan et al., The Lyme disease agent exploits a tick protein to infect the mammalian host. *Nature*, 2005. **436**(7050): 573–577.

66. Hovius, J. W., A. P. van Dam et al., Tick-host-pathogen interactions in *Lyme borreliosis*. *Trends Parasitol*, 2007. **23**(9): 434–438.

67. Brossard, M. and S. K. Wikel, Tick immunobiology. *Parasitology*, 2004. **129**(Suppl.): S161–S176.

68. Merino, O., P. Alberdi et al., Tick vaccines and the control of tick-borne pathogens. *Front Cell Infect Microbiol*, 2013. **3**: 30.

69. Kolb, P., J. Vorreiter et al., Soluble cysteine-rich tick saliva proteins Salp15 and Iric-1 from E. coli. *FEBS Open Bio*, 2015. **5**: 42–55.

70. Kolb, P., Novel anti-tick vaccine candidates based on HBV capsid-like particles, PhD thesis, University of Freiburg, 2015.

71. Nguyen, T. T. A., Recombinant fluorescent PreS-containing hepatitis B virus capsid-like particles, PhD thesis, University of Freiburg, 2012.

72. Heal, W. P., S. R. Wickramasinghe et al., N-myristoyl transferase-mediated protein labelling in vivo. *Org Biomol Chem*, 2008. **6**(13): 2308–2315.

73. O'Shea, E. K., K. J. Lumb et al., Peptide 'Velcro': Design of a heterodimeric coiled coil. *Curr Biol*, 1993. **3**(10): 658–667.

74. Cross, T. A., H. Dong et al., M2 protein from influenza A: From multiple structures to biophysical and functional insights. *Curr Opin Virol*, 2012. **2**(2): 128–133.

75. Schotsaert, M., M. De Filette et al., Universal M2 ectodomain-based influenza A vaccines: Preclinical and clinical developments. *Expert Rev Vaccines*, 2009. **8**(4): 499–508.

76. Schmitz, N., R. R. Beerli et al., Universal vaccine against influenza virus: Linking TLR signaling to anti-viral protection. *Eur J Immunol*, 2012. **42**(4): 863–869.

77. Neirynck, S., T. Deroo et al., A universal influenza A vaccine based on the extracellular domain of the M2 protein. *Nat Med*, 1999. **5**(10): 1157–1163.

78. Liang, T. J. and M. G. Ghany, Current and future therapies for hepatitis C virus infection. *N Engl J Med*, 2013. **368**(20): 1907–1917.

79. Drummer, H. E., Challenges to the development of vaccines to hepatitis C virus that elicit neutralizing antibodies. *Front Microbiol*, 2014. **5**: 329.

80. Osburn, W. O., B. E. Fisher et al., Spontaneous control of primary hepatitis C virus infection and immunity against persistent reinfection. *Gastroenterology*, 2010. **138**(1): 315–324.

81. Ray, R., K. Meyer et al., Characterization of antibodies induced by vaccination with hepatitis C virus envelope glycoproteins. *J Infect Dis*, 2010. **202**(6): 862–866.

82. Correia, B. E., J. T. Bates et al., Proof of principle for epitope-focused vaccine design. *Nature*, 2014. **507**(7491): 201–206.

83. Ott, J. J., G. A. Stevens et al., Global epidemiology of hepatitis B virus infection: New estimates of age-specific HBsAg seroprevalence and endemicity. *Vaccine*, 2012. **30**(12): 2212–2219.

84. Jegerlehner, A., M. Wiesel et al., Carrier induced epitopic suppression of antibody responses induced by virus-like particles is a dynamic phenomenon caused by carrier-specific antibodies. *Vaccine*, 2010. **28**(33): 5503–5512.

85. Geldmacher, A., D. Skrastina et al., A hantavirus nucleocapsid protein segment exposed on hepatitis B virus core particles is highly immunogenic in mice when applied without adjuvants or in the presence of pre-existing anti-core antibodies. *Vaccine*, 2005. **23**(30): 3973–3983.

86. De Filette, M., W. Martens et al., Universal influenza A M2e-HBc vaccine protects against disease even in the presence of pre-existing anti-HBc antibodies. *Vaccine*, 2008. **26**(51): 6503–6507.

87. Billaud, J. N., D. Peterson et al., Advantages to the use of rodent hepadnavirus core proteins as vaccine platforms. *Vaccine*, 2007. **25**(9): 1593–1606.

88. Lange, M., M. Fiedler et al., Hepatitis C virus hypervariable region 1 variants presented on hepatitis B virus capsid-like particles induce cross-neutralizing antibodies. *PLoS ONE*, 2014. **9**(7): e102235.

89. Beckett, D., E. Kovaleva et al., A minimal peptide substrate in biotin holoenzyme synthetase-catalyzed biotinylation. *Protein Sci*, 1999. **8**(4): 921–929.

90. Walker, A., Recombinant hepatitis B virus capsid-like particels from split core proteins for native display of heterologous protein antigens, PhD thesis, University of Freiburg, 2009.

91. Rosano, G. L., and E. A. Ceccarelli, Recombinant protein expression in microbial systems. *Front Microbiol*, 2014. **5**: 341.

92. Sodoyer, R., Expression systems for the production of recombinant pharmaceuticals. *BioDrugs*, 2004. **18**(1): 51–62.

93. Demain, A. L. and P. Vaishnav, Production of recombinant proteins by microbes and higher organisms. *Biotechnol Adv*, 2009. **27**(3): 297–306.

94. van Oers, M. M., Opportunities and challenges for the baculovirus expression system. *J Invertebr Pathol*, 2011. **107**(Suppl.): S3–S15.

95. Lu, J., Y. Hu et al., *Ectropis obliqua* picorna-like virus IRES-driven internal initiation of translation in cell systems derived from different origins. *J Gen Virol*, 2007. **88**(Pt 10): 2834–2838.

96. Kang, S. T., H. C. Wang et al., The DNA virus white spot syndrome virus uses an internal ribosome entry site for translation of the highly expressed nonstructural protein ICP35. *J Virol*, 2013. **87**(24): 13263–13278.

97. Jegerlehner, A., A. Tissot et al., A molecular assembly system that renders antigens of choice highly repetitive for induction of protective B cell responses. *Vaccine*, 2002. **20**(25–26): 3104–3112.

98. Spirig, T., E. M. Weiner et al., Sortase enzymes in Gram-positive bacteria. *Mol Microbiol*, 2011. **82**(5): 1044–1059.

99. Popp, M. W., S. K. Dougan et al., Sortase-catalyzed transformations that improve the properties of cytokines. *Proc Natl Acad Sci USA*, 2011. **108**(8): 3169–3174.

100. Becer, C. R., R. Hoogenboom et al., Click chemistry beyond metal-catalyzed cycloaddition. *Angew Chem Int Ed Eng*, 2009. **48**(27): 4900–4908.

101. Grammel, M. and H. C. Hang, Chemical reporters for biological discovery. *Nat Chem Biol*, 2013. **9**(8): 475–484.

102. Chin, J. W., Expanding and reprogramming the genetic code of cells and animals. *Annu Rev Biochem*, 2014. **83**: 379–408.

103. Zhang, X. and Y. Zhang, Applications of azide-based bioorthogonal click chemistry in glycobiology. *Molecules*, 2013. **18**(6): 7145–7159.

104. Anish, C., B. Schumann et al., Chemical biology approaches to designing defined carbohydrate vaccines. *Chem Biol*, 2014. **21**(1): 38–50.

105. Mazmanian, S. K., and D. L. Kasper, The love–hate relationship between bacterial polysaccharides and the host immune system. *Nat Rev Immunol*, 2006. **6**(11): 849–858.

13 Use of VLPs in the Design of Malaria Vaccines

David Whitacre, Diego Espinosa, Darrell Peterson, Fidel Zavala, and David R. Milich

CONTENTS

13.1 RATIONALE FOR A PREERYTHROCYTIC/ LIVER STAGE MALARIA VACCINE

Malaria is the world's most important tropical parasitic disease that kills more people than any other communicable disease with the exception of tuberculosis. The causative agents in humans are four species of *Plasmodium* protozoa: *P. falciparum*, *P. vivax*, *P. ovale*, and *P. malariae*. Of these, *P. falciparum* accounts for the majority of infections and is the most lethal. Malaria is a public health problem today in more than 106 countries, inhabited by a total of 3.4 billion people—50% of the world's population. Worldwide prevalence of the disease is estimated to be on the order of 135–287 million clinical cases each year. Mortality due to malaria is estimated to be in the range of 473,000–789,000 each year.

The vast majority of deaths occur among young children in Africa, especially in remote rural areas with poor access to health services. Therefore, *P. falciparum* vaccine development has become a high research priority. The malaria parasite has 14 chromosomes, an estimated 5300 genes (many of which vary extensively between strains) and a complex 4-stage life cycle as it passes from a mosquito vector to humans and back again. Furthermore, the natural *P. falciparum* infection does not result in immunity, and partial immunity occurs only after years of recurring infections and illnesses. Therefore, a vaccine must outperform the immune response to the natural infection. This complexity has impeded vaccine development.

Although a number of antigens from the various life cycle stages are being pursued as vaccine candidates, the most progress has been made toward the development of a pre-erythrocytic stage vaccine [1]. Sporozoites, which represent the infective stage, are injected into the host by the bite of the mosquito and within 30 min leave the circulation and enter hepatocytes. The relatively low antigen load (<100 sporozoites per bite) and brief circulation time may explain the lack of protective immunity toward this stage after a natural infection [2]. Consistent with this interpretation, protection in mice injected intravenously with gamma-irradiated, attenuated *Plasmodium berghei* sporozoites is dose dependent [3,4]. Sporozoite-induced protection has also been achieved in rhesus monkeys [5] and humans [6].

Studies in the 1980s demonstrated that the circumsporozoite coat protein (CS) was the dominant target of protective antibodies [7–9]. This has been documented by an elegant study in CS-transgenic mice that are tolerant to CS epitopes [10]. In the absence of T-cell-dependent immune responses to CS protein, protection induced by immunization with two doses of irradiated sporozoites was greatly reduced [10]. Further, the dominant antibody epitope was represented by the CS central repeat sequences ($NANP_n$ in *P. falciparum*) [11,12]. Several studies have shown that antibodies specific for the CS-repeat sequences are protective: in vitro, CS-specific monoclonal antibodies can inhibit sporozoite motility, prevent sporozoite invasion of host hepatocytes, inhibit development of the intrahepatic parasite, and eliminate *P. berghei*–, *P. yoelii*–, or *P. falciparum*–infected hepatocytes from culture. In vivo, CS-specific antibody-mediated protective immunity against *P. berghei*, *P. yoelii*, *P. vivax*, or *P. knowlesi* sporozoite challenge has been demonstrated by passive immunization studies in mice and monkeys. To be effective, antisporozoite antibodies must be present in circulation at high titers and exert their activity within minutes of infection [13], consistent with the conclusion that sporozoites are thought to invade the hepatocytes within 2–30 min of inoculation [14]. A more recent study suggests that sporozoites may remain in the skin for several hours prior to trafficking to the liver, which may increase the time for CS-specific antibodies to neutralize liver infection [15]. Overall, data in animal models suggest that sporozoite-specific antibodies can confer protection against challenge in the absence of other parasite-specific immune responses. However, antisporozoite antibody responses are not considered the only effector mechanism of the protective immunity induced by immunization with radiation-attenuated sporozoites. Therefore, the design of CS-based subunit vaccines should include both T-cell and B-cell-protective epitopes. An attractive attribute of the CS protein is that it contains protective B and T cell epitopes and it is highly conserved among *P. falciparum* strains, especially within the CS-repeat domain.

13.2 RATIONALE FOR A VLP-BASED CS SUBUNIT MALARIA VACCINE

The studies demonstrating that CS-repeat-specific antibodies were protective prompted human clinical trials using recombinant [16] and synthetic [17,18] forms of the CS. These antigens formulated in alum were poorly immunogenic in terms of anti-NANP antibody titers determined by direct enzyme-linked immunosorbent assay (ELISA) or immunofluorescence antibody assay (IFA) tests on sporozoites (i.e., IFA titers of 10^2–10^3). Predictably, the weak immunogenicity was accompanied by limited protection. Similarly, in endemic areas, low levels of antibodies to $(NANP)_n$ do not appear to protect [19]. In an attempt to increase the immunogenicity of the CS-repeat sequences, a virus-like particle (VLP) consisting of 16 NANP repeats from the CS of *P. falciparum* fused to the N-terminus of the hepatitis surface antigen (HBsAg) was developed and tested in a phase I clinical trial. The immunogenicity was suboptimal [20]. A second vaccine based on the use of the HBsAg carrier known as RTS,S consisting of 19 NANP repeats plus the majority of the C-terminus of the CS (from aa 207 to 395 of *P. falciparum* [3D7]) fused to the N-terminus of the HBsAg was tested in a human challenge study [21]. The RTS,S vaccine formulated in alum was not protective, but the addition of 3-deacylated-monophosphoryl lipid A (MPL) to the adjuvant elicited protection in two of eight vaccinees with the highest antibody levels [21]. The use of a more potent adjuvant termed AS02A (MPL plus a saponin derivative QS-21 in an oil-in-water emulsion) protected six of seven vaccinees challenged 3 weeks after the third dose of vaccine [22]. This important study demonstrated that a single sporozoite antigen can produce full protection in a majority of human recipients. However, in a rechallenge experiment, only one of five of the originally protected vaccinees remained protected approximately 6 months after the initial challenge [23]. Similarly, in a field trial of RTS,S/AS02A in a malaria-endemic region, high levels of protection were achieved but again appeared transient [24]. Analysis of parasite genotypes in breakthrough infections of vaccinated individuals showed no evidence of an increased frequency of parasites with T cell epitopes in the CS protein that were not present in the vaccine strain [25], supporting a main role for antibodies specific for the conserved $(NANP)_n$ repeat in the partial protection against malaria that was observed. Furthermore, in a phase IIb clinical trial of RTS,S/AS02A in 1–4-year-old children from Mozambique, a 29.8% reduction in overall malaria incidence rate was observed. However, the clinical malaria incidence rates after 7 weeks of follow-up seemed to be similar in vaccinees and controls, indicating that efficacy might have waned rapidly, as in the other trials [26]. A phase I/IIb trial of RTS,S/AS02D in infants reported an adjusted vaccine efficacy against severe disease of 65.9% during a 3-month follow-up although protection against parasite infection remained within the 20%–30% range [27]. More recent phase II/III trials using a liposome-based adjuvant (AS01 series), which appears more immunogenic, demonstrate similar efficacies. In general, efficacy appears to vary as a function of malaria transmission intensity, time since vaccination, adjuvant, and age at vaccination [28]. Although the variability and transient nature of protection elicited by the CS-HBsAg-based RTS,S vaccine is problematic and needs to be addressed, comparison of its efficacy with the multitude of other vaccine candidates illustrates the superiority of a particulate platform containing epitopes from a single CS protein.

13.3   RATIONALE FOR AN HBcAg-BASED AS OPPOSED TO AN HBsAg-BASED CARRIER PLATFORM FOR MALARIA EPITOPES

Although RTS,S is the most protective malaria vaccine yet tested, the requirement for potent adjuvants and the rather transient protection period [23,24,26] prompted the developers of this vaccine to state that "further optimization … will be required to induce longer-lasting protective immunity" [23]. An alternative advantageous approach is the use of the HBcAg, which is significantly more immunogenic (100–1000-fold) than the HBsAg in mice and during natural infection in man [29,30]; furthermore, in a recent study comparing the immunogenicity of purified, soluble HBcAg and HBsAg given intranasally to humans, the HBcAg was significantly more immunogenic [31]; HBcAg can be produced in bacterial expression systems unlike HBsAg; no nonresponder MHC genotypes have been identified for the HBcAg in mice or humans, which is not true for the HBsAg [32–34]; and the HBcAg has been shown to be a highly versatile and efficient carrier platform for a number of pathogen-specific epitopes [33,35]. Specifically, we have demonstrated that immunization with hybrid-HBcAg particles containing CS-repeat sequences from the *P. berghei* and *P. yoelii* rodent malaria species elicited high-titer anti-CS-repeat antibodies and protected 90%–100% of vaccinated mice against these murine malaria strains [36,37]. Because no malaria-specific Th cell epitopes were inserted into the HBcAg platform, the 90%–100% protection was mediated solely by anti-CS antibodies. Based on the protective efficacy achieved against murine malaria, we developed a preerythrocytic *P. falciparum* subunit vaccine candidate composed of CS repeats and a CS-specific Th cell epitope inserted onto the HBcAg [36,38]. A start-up biotech company, Apovia, attempted (without the participation of the majority of the inventors of the technology) to develop the CS-HBcAg particles (designated V12. PF3.1 [38] and later designated ICC-1132 [39]) into a viable malaria vaccine and conducted a number of phase I and II clinical trials [40–42]. Although preclinical animal studies in mice and nonhuman primates looked very promising [38,39,43], the performance of the ICC-1132 vaccine in human phase IIa studies proved disappointing [42]. There were a number of procedural factors that may explain the negative clinical results. Briefly, a very low dose of ICC-1132 (50 μg corresponding to 5.0 μg of CS-repeat sequence) was given in a single injection without a booster injection. It would be difficult to imagine any nonreplicating, much less subunit, vaccine succeeding using this protocol. In fact, the RTS,S vaccine failed to adequately protect in two separate trials after three doses in nonstimulatory adjuvant systems [20,21]. Only after RTS,S was formulated with MPL and QS-21 in an oil emulsion and given in three injections did it achieve partial protection in humans [22,23]. Another biotech firm, Acambis, recently published a press release that an HBcAg-based vaccine candidate carrying the M2 influenza A protective epitope was immunogenic in humans in a phase I trial. However, two doses in a QS-21 adjuvant system were used. These early results obtained using VLP-hybrid recombinant vaccine candidates illustrate the importance of determining optimal vaccine dose, route, number of injections, and formulation prior to clinical trial.

13.4   RATIONALE FOR THE USE OF THE WHcAg AS OPPOSED TO THE HBcAg AS A CARRIER PLATFORM FOR MALARIA EPITOPES

Although HBcAg is a highly immunogenic particulate antigen, the existing HBcAg-based platform technology has problems that may limit its full potential as a vaccine carrier for the human population [44]. The main limitation of the current HBcAg platform technology can best be described as the *preexisting immunity* problem. Because HBcAg is derived from a human pathogen, preexisting anti-HBc antibodies present in individuals previously exposed to HBV infection may adversely affect immunogenicity. Further, the anti-HBc antibodies elicited by an HBcAg-based vaccine will compromise the usefulness of the anti-HBc assay currently employed as a diagnostic for current or recent HBV infection [44]. Most importantly, T cell immune tolerance toward the HBcAg and the HBsAg is present in individuals chronically infected with HBV (400 million globally) [45]. This is especially relevant to a malaria vaccine because chronic HBV infection is relatively common (i.e., 10%–20%) in many areas of the world endemic for malaria infection. It is important to note that these same *preexisting immunity* problems are as true for the HBsAg (used in the RTS,S vaccine) as for the HBcAg. As a means of addressing the *preexisting immunity* problem, we have developed the core proteins from the rodent hepadnaviruses (i.e., WHcAg) as carrier platforms [44–47]. A number of advantages to the use of the rodent hepadnaviral core proteins as opposed to the HBcAg for vaccine design have been defined including the following: (1) the rodent core proteins are equally or more immunogenic than the HBcAg at the T and B cell levels [47]; (2) the rodent core proteins will not substantially compromise the use of the anti-HBc diagnostic assay because they are not significantly cross-reactive with HBcAg at the antibody level [44]; (3) preexisting anti-HBc antibodies in HBV chronically infected patients or in previously infected and recovered persons may limit the efficacy of the HBcAg platform, whereas the hybrid-WHcAg platform does not bind preexisting anti-HBc antibodies [44]; and (4) the HBcAg-specific as well as HBsAg-specific immune tolerance present in HBV chronic carriers can be circumvented by the use of the WHcAg platform because the HBcAg is only partially cross-reactive at the T cell level with WHcAg [44,47].

We have utilized an HBV-Tg mouse model of HBV chronic infection to investigate this hypothesis. HBeAg-Tg mice produce the secreted form of the HBcAg and HBeAg-Tg mice on a (B10.S × Balb/c)$_{F1}$ background are tolerant to the HBcAg at the Th cell level, which mimics the immune status of HBV chronic carriers. When HBeAg-Tg mice are immunized with an HBcAg-based CS malaria vaccine (i.e., ICC-1132) minimal anti-HBc and anti-NANP antibodies are produced [44]. In contrast, when HBeAg-Tg mice are immunized with a WHcAg-based CS malaria vaccine (i.e., WHc-Mal-78-UTC), the HBeAg-Tg mice produce high levels of anti-WHc and anti-NANP antibodies equivalent to wild-type mice. Therefore, the negative effects of immune tolerance to the HBcAg can be circumvented by using the WHcAg as a vaccine platform for the malaria CS epitopes.

As previously discussed, the disappointing results using the HBcAg-based ICC-1132 candidate vaccine in human phase IIa trials suggest that modifications in addition to the switch to the WHcAg platform may be required to optimize a hybrid malaria-core vaccine construct. For this reason, we proposed to insert a greater diversity of malaria-specific T cell epitopes into the WHcAg platform than were used in the previous ICC-1132 vaccine candidate and utilized a full-length WHcAg to accommodate the inclusion of ssRNA within the core particle to act as a TLR7/8 ligand [48]. The ssRNA within full-length CS-WHcAg VLPs enhances anti-CS antibody responses and CD4+ T cell priming mediated through TLR7 [48]. This is consistent with the recent finding that a third signal in addition to BCR and Th signals is required to fully activate human B cells and TLR activation can serve as the third signal [49].

13.5   SELECTION OF A BASIC MALARIA VACCINE CANDIDATE

During the course of evaluating the WHcAg combinatorial technology, we used the *P. falciparum* major CS-repeat region as a source of model epitope inserts and produced a large panel of over 30 hybrid CS-WHcAg particles. From this large panel of hybrid CS-WHcAg constructs, we selected WHc-Mal-78-UTC as a primary candidate. This particle is composed of two well-defined neutralizing B cell epitopes derived from the CS protein (NANP and NVDP) [11] combined in the sequence NANP NVDP(NANP)$_3$ inserted into position 78 within the exposed loop region of the full-length WHcAg. To the C-terminus of the WHcAg, a well-defined human Th cell recognition region, designated UTC (universal T cell epitope) aa 333–352 [50], was added. An additional human Th cell recognition region (CST3 378–392) [51] has recently been added. These two Th cell regions contain many defined Th cell sites recognized in the context of numerous human MHC class II molecules [50,51]. Furthermore, a number of MHC class I–restricted T cell recognition sites for human CD8+ T cells are also present within these two regions [52]. When we added the CST3 372–392 T cell site, it was necessary to alter the insert position of the B cell repeats to residue 74 to achieve efficient assembly. The hybrid particle is designated WHc-Mal-74-TH (Figure 13.1). Because this is a newer construct, most of the preliminary data that follow were generated using the WHc-Mal-78-UTC hybrid particle. However, comparative immunogenicity studies demonstrate that the WHc-Mal-78-UTC and WHc-Mal-74-TH particles elicit similar anti-WHc, anti-NANP, and anti-NVDP antibody

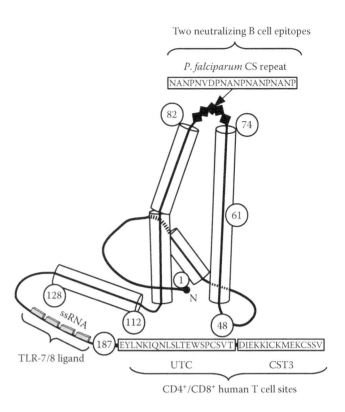

FIGURE 13.1 Vaccine candidate WHc-Mal-74-TH. Schematic representation of the preliminary vaccine candidate: WHc-Mal-74-TH (the particle is composed of 240 copies of this subunit).

responses in (B10 × B10.S)$_{F1}$ mice (Figure 13.2). The criteria for selecting candidate WHcAg-CS VLPs are summarized in Figure 13.3. The NANP and NVDP CS-repeat B cell epitopes and UTC and CST3 Th cell epitopes were selected due to their known protective efficacy and human Th cell recognition, respectively. From many tested orientations, the NANPNVDP(NANP)$_3$ sequence was chosen due to its superior antigenicity in terms of binding NANP- and NVDP-specific Mabs. Insertion of the NANPNVDP(NANP)$_3$ sequence into many positions on WHcAg with a number of C-terminal modifications were tested by ELISAs for protein expression levels, particle assembly, and insert antigenicity in the lysates of transformed *Escherichia coli* and for the yield of purified CS-WHcAg-hybrid particles.

13.6   BIOCHEMICAL/STRUCTURAL CHARACTERIZATION

Purified WHc-Mal-78-UTC hybrid particles were compared to WT-WHcAg (noninserted) particles in terms of several biochemical/biophysical parameters (Figure 13.4). The WHc-Mal-78-UTC particle is similar to WT-WHcAg in its secondary structure as measured by CD spectra (high alpha-helix content) and thermal stability (74°C) but has a larger diameter (48.5 vs. 21.3 nm). This is consistent with the CS epitopes being added as surface features of the particle that do not grossly affect the WHcAg particle structure. We were also interested in determining if WHc-Mal-78-UTC could tolerate lyophilization and glutaraldehyde treatments and

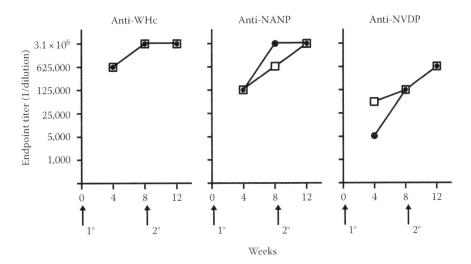

FIGURE 13.2 FLw-Mal-74-TH and FLw-Mal-78-UTC are equivalently immunogenic. Mice were immunized with 20 µg of the indicated WHc-CS particles and boosted with 10 µg in IFA. At the indicated time points, serum was collected and anti-WHc, anti-NANP, and anti-NVDP endpoint titers were determined by ELISA.

Task	Selection Process	End-Point
1. Select B cell epitopes	(NANP)$_4$, (NVDP), CS-non-repeats	α-Peptide binds CS protein and sporozoited
1a. Epitope optimization	NANP-NVDP-(NANP)$_3$	++Antigenicity/immunogenicty
Select T cell epitopes	UTC; TH.3R; CS(CST3)	Th cell function in humans
Select molecular adjuvants	ssRNA	++ Immunogenicity
2. Produce construct	Combinatorial technology 17 insertion sites/21 C-termini	
3. Transform *E. coli* with constructs (0.5 L culture volume)	Screen lysates (ELISAs) Protein expression Particle assembly Epitope antigenicity	2–3 units 3–4 units 3–4 units
4. CS-WHcAg purification (5 L culture volume)	Assembly (1% agarose, CL-4B) Yield	Particulate ≥80 mg/L
5. Biochemical/structural characterization	Particle size Electron microscopy CD specra Thermal stability	30–50 nm Formation of particle structures WHcAg like (high alpha helix) >70°C
6. Immunogenicity in mice	*In vivo* Ab Production	α-Insert titer of 10^6–10^7 IgG isotypes, Ab persistence
	In vitro/in vivo Th cell activity (1 prime/1 boost)	+ IL-2, + IFN$_\gamma$
7. *In vitro* immunofluoresence assay (IFA) on sporozoites		IFA titer of 10^4–10^5
8. *In vivo* protection liver/blood	a. Direct immunization/challenge b. Passive transfer of antisera	CD4+/antibody % protection only antibody % protection

FIGURE 13.3 Selection criteria for improvements to the preliminary FLw-Mal-74-TH vaccine candidate and new vaccine particles.

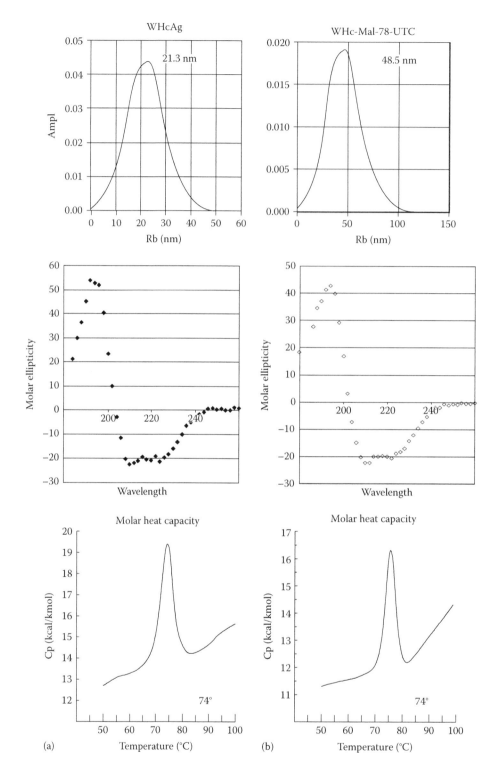

FIGURE 13.4 Comparison of biochemical and biophysical characteristics of the wild-type WHcAg (a) versus the WHc-Mal-78-UTC (b) candidate.

retain its physical–chemical characteristics. Lyophilization and rehydration had no measurable detrimental effects on the physical–chemical parameters of WHc-Mal-78-UTC (not shown). Glutaraldehyde treatment with a relatively high concentration (50 mM) did not alter the secondary structure (CD) or the diameter of WHc-Mal-78-UTC (i.e.,

no interparticle cross-linking), and the only observable change was a 3.5°C increase in thermal stability. The fact that WHc-Mal-78-UTC can tolerate both lyophilization and glutaraldehyde cross-linking has especially important implications for a malaria vaccine. Lyophilization and/or glutaraldehyde treatment will increase long-term stability even if stored unrefrigerated, which in malaria-endemic regions is advantageous because maintaining a cold chain can be problematic.

13.7 IMMUNOGENICITY IN MICE

Ultimately, the most important characteristic of a malaria vaccine candidate is its immunogenicity. The initial screen for immunogenicity has been performed in mice. A single injection of 20 µg of WHc-Mal-78-UTC elicited an anti-NANP end point titer of 1:3 × 10⁶ after 10 weeks (Figure 13.5). In addition, two versions of the CS-WHcAg-hybrid particle were compared in mice, WHc-Mal-78-UTC and an identical construct with the exception that the C-terminus of WHcAg was truncated at residue 149 (149-Mal-78-UTC). Full-length WHcAg/HBcAg binds *E. coli*-derived ssRNA (5.20 ng RNA/µg protein) in a size range between 30 and 3000 nucleotides [53]. The absence of the C-terminal 38 residues of WHcAg$_{149}$ eliminates the nucleic acid binding sites and the particles are ssRNA⁻. As shown in Figure 13.5, the presence of ssRNA within

the WHc-Mal-78-UTC particle enhances anti-NANP and anti-NVDP antibody production as measured by ELISA determined on the respective peptides (two left panels) and as measured by the IFA assay on *P. falciparum* sporozoites (right panel). The presence of ssRNA appears to enhance the primary antibody response to a greater degree than the boosted response. In addition to enhancing total IgG anti-CS-repeat antibody production, the presence of ssRNA within the CS-WHcAg-hybrid particle also affected the IgG isotype produced (Table 13.1). In general, 149-Mal-78-UTC lacking ssRNA elicited significantly more IgG$_1$ and less IgG$_{2a}$ to the WHcAg carrier and to the NANP and NVDP CS repeats than did WHc-Mal-78-UTC/ssRNA⁺, which elicited very high levels of IgG$_{2a}$ to all the epitopes and relatively low levels of IgG$_1$ specific for the WHcAg and the NVDP CS repeat. The WHc-Mal-78-UTC/ssRNA⁺ particle also elicited greater levels of IgG$_{2b}$ to the CS-repeat epitopes than did the 149-Mal-78-UTC/ssRNA⁻ particle. Because the IgG$_{2a}$ isotype is regulated by IFNγ in mice, it appears that the ssRNA⁺-containing particle elicits a greater IFNγ response than the ssRNA⁻ particle.

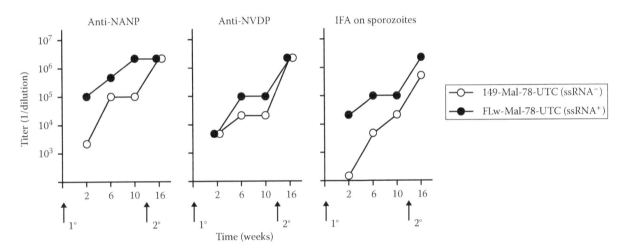

FIGURE 13.5 Comparison of WHc-Mal-78-UTC VLPs with and without encapsidated ssRNA. Mice were immunized with 20 µg of the indicated CS-WHc particles in IFA and boosted with 10 µg. Sera were tested for anti-NANP and anti-NVDP binding by ELISA and for binding to sporozoites by an IFA.

TABLE 13.1
IgG Isotype Distribution of Primary Antisera Raised against WHc-Mal Hybrid Particles Either Containing or Lacking ssRNA

Immunogen	Anti-WHc Titer				Anti-NANP Titer				Anti-NVDP Titer			
	IgG$_1$	IgG$_{2a}$	IgG$_{2b}$	IgG$_3$	IgG$_1$	IgG$_{2a}$	IgG$_{2b}$	IgG$_3$	IgG$_1$	IgG$_{2a}$	IgG$_{2b}$	IgG$_3$
149-Mal-78-UTC (ssRNA⁻)	250,000	25,000	3 × 10⁶	50,000	625,000	5,000	125,000	250,000	25,000	1,000	50,000	50,000
FLw-Mal-78-UTC (ssRNA⁺)	25,000	625,000	3 × 10⁶	125,000	625,000	3 × 10⁶	3 × 10⁶	250,000	1,000	625,000	625,000	50,000

Note: Mice were immunized with 20 µg of indicated immunogens emulsified in IFA. Six weeks after the single immunization, sera were evaluated by IgG isotype-specific ELISA.

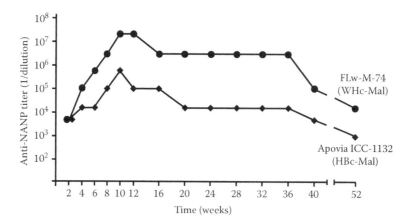

FIGURE 13.6 Comparative immunogenicity of HBcAg-based (HBc-Mal) and WHcAg-based (HBc-Mal) CS malaria immunogens carrying the same *P. falciparum* CS-repeat epitopes.

13.8 DIRECT COMPARISON BETWEEN HBcAg-BASED AND WHcAg-BASED CS-VLP CONSTRUCTS

To directly compare the WHcAg and the HBcAg as carrier platforms, two hybrid core VLPs (i.e., WHc-M/WHc-Mal-74 and HBc-M/ICC-1132) were produced, which carried the same *P. falciparum* malaria CS-repeat epitopes [M, NANPNVDP(NANP)$_3$]. The WHc-M-74 construct is similar to WHc-Mal-74-TH except it does not contain any malaria-specific Th regions. A single 20 µg dose of WHc-M in IFA elicited higher levels of anti-NANP antibodies with a better persistence profile than the same dose of HBc-M (Figure 13.6). These data indicate that the WHcAg is certainly a feasible alternative to the HBcAg as a vaccine carrier platform for CS epitopes. It is also important to note that the NANP-specific antibodies produced by both core particle carriers also bind *P. falciparum* sporozoites with IFA titers similar to the ELISA titers (data not shown). It is also notable that HBc-M does not contain ssRNA.

13.9 ROLE OF TLR-7 IN THE IgG$_{2a}$ RESPONSE TO WHc-Mal-78-UTC

To determine if a known receptor for ssRNA (i.e., TLR-7) was responsible for mediating the effects of ssRNA on the WHc-Mal-78-UTC response, wild-type B10 (+/+) and B10 TLR-7^{KO} mice were immunized with WHc-Mal-78-UTC, which contains ssRNA or 149-Mal-78-UTC that is ssRNA negative. Six weeks after a single immunization, anti-WHc, anti-NANP, and anti-NVDP endpoint titers were measured by ELISA (Figure 13.7). Wild-type and TLR-7^{KO} mice respond equally to 149-Mal-78-UTC particles, which do not contain ssRNA (Figure 13.7, left panel). However, TLR-7^{KO} mice respond significantly less well than +/+ mice to WHc-Mal-78-UTC particles, which contain ssRNA. Note also that the WHcAg-, NANP-, and NVDP-specific antibody responses are all reduced in TLR-7^{KO} mice. These data indicate that all three antibody responses are regulated through TLR-7 binding to

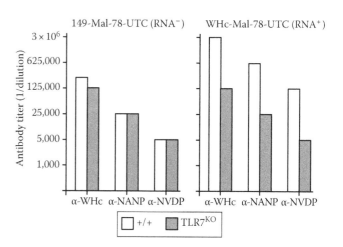

FIGURE 13.7 The encapsidated ssRNA acts through TLR7. B6-wt and TLR7^{KO} mice were immunized with 10 µg of 149-Mal-78-UTC (ssRNA$^-$) or WHc-Mal-78-UTC (ssRNA$^+$) in IFA and 6-week sera were analyzed for anti-WHc, anti-NANP, and anti-NVDP by ELISA. (Adapted from Lee, B.O. et al., *J. Immunol.*, 182, 6670, 2009.)

ssRNA contained within the particle because ssRNA is behaving as an endogenous and localized adjuvant for WHcAg particles. The ssRNA is protected by the core particle and is delivered into the cytosol of WHcAg-specific B cells and/or other APCs where the TLR-7 resides [48]. The *E. coli* ssRNA within CS-WHcAg particles is heterogeneous but the exact ssRNA species that binds human TLR 7/8 is not known; therefore, it may be superior to retain the bacterial ssRNA.

13.10 ABSENCE OF GENETIC NONRESPONDER MURINE STRAINS TO A WHcAg-MALARIA HYBRID PARTICLE

Efforts to produce *P. falciparum* vaccine candidates based on the CS-repeat sequences have been plagued by low immunogenicity and severe genetic restriction characterized by low responders in human clinical trials and low or nonresponder murine MHC genotypes in mice. Therefore, for malaria B

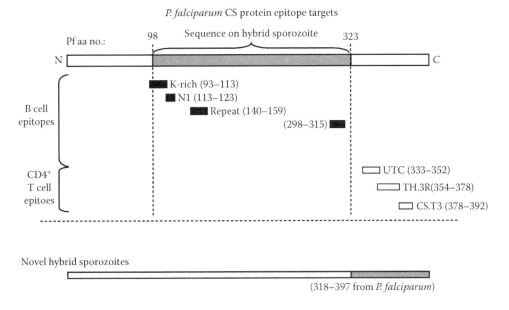

FIGURE 13.8 Candidate T and B cell epitopes on CSP. Schematic representation of the existing (upper panel) and novel (lower panel) hybrid sporozoites. Native *P. berghei* sequence is indicated by light shading and transgenic *P. falciparum* sequence by dark shading. Neutralizing or presumptive neutralizing B cell epitopes are denoted by black bars and human and murine CD4+ T cell epitopes by white bars.

cell epitopes, it is imperative that the carrier platform provides sufficient T cell helper function in the context of a wide variety of MHC haplotypes to guarantee the absence of antibody nonresponders. To directly examine the issue of MHC-linked restriction of the antibody response to a WHcAg-based immunogen, B10 H-2 congenic murine strains expressing eight different H-2 haplotypes were immunized with a malaria CS-WHcAg-hybrid particle in IFA and anti-WHc and anti-NANP serum IgG antibody titers were determined. First and importantly, all H-2 haplotypes responded and produced both anti-WHc and anti-NANP antibodies after a primary immunization and nonresponder H-2 haplotypes were not identified. Second, all strains at all time points produced an equal or greater antibody response to the insert (anti-NANP) as compared to anti-WHc [44]. The lack of genetic nonresponders to this experimental WHcAg-based immunogen is consistent with the absence of nonresponders to the WHcAg platform itself at the antibody and T cell levels [47]. This represents an important difference between the WHcAg carrier and the HBsAg carrier used in the current RTS,S malaria vaccine because genetic nonresponder MHC haplotypes to the HBsAg have been identified in both humans [32,54] and mice [34].

13.11 RODENT PARASITES CARRYING HYBRID *Plasmodium falciparum* CSP

Because *P. falciparum* sporozoites do not infect rodents, unlike the rodent malaria *P. berghei*, a recombinant parasite bearing the *P. falciparum* CSP extended repeat region from aa 98 to 323 was generated on an otherwise *P. berghei* sporozoite by Persson et al. in 2002 [55]. The hybrid sporozoites are fully infectious in vitro and in vivo in rodents and the

P. berghei/P. falciparum hybrid CSP allows the efficacy of human *P. falciparum* CS-based vaccines to be examined in mice. In the same year, another group developed transgenic sporozoites expressing the full-length *P. falciparum* CSP on *P. berghei* sporozoites [56]. We used the hybrid *P. berghei/P. falciparum* sporozoites bearing the extended repeat region because of greater infectivity in mice. However, the extended repeat region does not contain the dominant *P. falciparum* T cell domains. Therefore, Dr. Zavala constructed a second hybrid sporozoite containing aa 318–397 from the C-terminus of *P. falciparum* CSP substituted for the C-terminus of *P. berghei* (Figure 13.8) [68].

13.12 CONSIDERATION OF NONREPEAT OR FLANKING REGIONS OF THE CS PROTEIN FOR VACCINE DESIGN

A number of interesting candidate epitopes outside the CS-repeat domain have been described. For example, nonrepeat B cell epitopes that have been shown to elicit in vitro neutralizing antibodies include 93–113 (lysine-rich region), 113–122 (conserved N1), and 298–315 [57–59]. Similarly, a high percentage of adults and lesser numbers of children living in malaria-endemic areas possess antibodies specific for CS C-terminal sequences that represent CD4+ and CD8+ recognition sites for human and murine T cells (i.e., UTC, TH3.R and CS.T3 regions) [60]. For several reasons, the consideration of these nonrepeat, CS B cell epitopes for vaccine design has been marginalized. First, the immunodominance of the NANP and NVDP repeats and the established neutralizing efficacy of anti-CS-repeat antibodies have reduced interest in nonrepeat B cell epitopes somewhat. Second, the induction of high-titer CS-specific antibodies to nonrepeat epitopes has been difficult with most immunogens. Third, even when anti-CS, nonrepeat antibodies are induced, the neutralizing potential of these antibodies has not been determined in vivo because of the absence of an infectious *P. falciparum* model.

The combination of the WHcAg platform technology, which allows us to insert virtually any CS sequence onto WHcAg and obtain high-titer antibody even if the CS sequence may be cryptic on the native CS protein, and the hybrid *P. berghei/P. falciparum* sporozoite technology (see the earlier text), which allows us to determine the protective efficacy of any CS sequence by virtue of inserting the *P. falciparum* B and T cell candidate epitopes in the CS protein of *P. berghei* sporozoites, permits us to overcome the problems that have prevented the analysis of CS-nonrepeat T and B cell sites in the past. For example, antibody to the conserved N1 region, believed to be involved in a proteolytic processing step required for sporozoite–hepatocyte invasion, has been shown to neutralize sporozoite infection in vitro, yet this anti-CS, nonrepeat specificity is not readily induced by immunization with native CS protein or sporozoites (i.e., cryptic). Furthermore, the protective efficacy of anti-N1 antibodies has not been tested in an in vivo *P. falciparum* challenge model. We inserted only the *P. falciparum* N1 region into the loop of WHcAg and used the WHcAg-N1 hybrid as the immunogen and hybrid *P. falciparum/P. berghei* sporozoites, which contain the *P. falciparum* N1 region, for the sporozoite challenge. Therefore, the immunogenicity of the N1 region on the WHcAg platform and the protective efficacy of the *P. falciparum* N1 region can be assessed in an infectious system. We also tested other *P. falciparum* nonrepeat B and T cell epitopes in a similar manner.

In order to examine the protective efficacy of *P. falciparum* T cell epitopes, the same approach can be used. Individual *P. falciparum* CD4+ or CD8+ T cell epitopes recognized by murine T cells can be inserted at the C-terminus of WHcAg and used to immunize mice, which can be challenged with hybrid sporozoites containing the homologous *P. falciparum*

T cell site as the immunizing WHcAg-hybrid particle enabling the protective efficacy of the isolated T cell epitope to be tested. The protective role of malaria-specific CD4+ T cells has been previously demonstrated [61]. Of course, the ultimate goal is to combine all neutralizing and protective *P. falciparum* T and B cell sites from both the repeat and flanking regions of the CS protein. Due to its multivalency, the WHcAg platform technology can accommodate multiple T and B cell epitopes on the same hybrid-WHcAg particle or hybrid-WHcAg particles containing different inserts can be mixed prior to immunization.

13.13　COMPARATIVE IMMUNOGENICITY OF WHc-HYBRID VLPs CARRYING TWO REPEATS (NANP, NVDP) VERSUS THREE NONREPEAT B CELL EPITOPES FROM THE CS PROTEIN

We produced, characterized, and examined the immunogenicity of hybrid-WHcAg VLPs carrying NANP/NVDP repeat epitopes and the three selected nonrepeat CS-specific B cell epitopes: the N1 region (aa 113–122), the lysine (K-rich) region (aa 93–113), and the aa 298–315 region (see Figure 13.8). These B cell epitopes have been previously shown to elicit antibodies with neutralizing potential in vitro but have not been vigorously tested in vivo for protective efficacy or have been proven to be poorly immunogenic or even cryptic on the CS protein. Consistent with the cryptic nature of the nonrepeat CS B cell epitopes, immunization with the full-length rCS protein elicited no antibody to the $CS_{298-315}$ region, extremely low antibody production in the N1 region (i.e., 1:1,000 titer), and relatively low antibody production to the K-rich region (i.e., 1:125,000 titer) after a primary and secondary immunization (Table 13.2). Immunization with rCS protein elicited very high antibody production to the two repeat epitopes, NANP and NVDP. In contrast, the nonrepeat

TABLE 13.2
Characterization of WHc-CS VLPs Carrying Repeat and Nonrepeat B Cell Epitopes

WHc-CS-VLP (Insert)	Endpoint Dilution Titers									Reduction in Liver Stage In Vivo (%)
	α-WHc	α-NANP	α-NVDP	α-N1	α-K Rich	α-CS (298–315)	α-CSP Solid Phase	IFA Dry Hybrid spzt	IFA Viable Hybrid spzt	
NANP/NVDP	6 × 10⁶	15 × 10⁶	6 × 10⁶	—	—	—	>15 × 10⁶	16,000	++ >300	98.7
N1 (113–122)	3 × 10⁶	—	—	3 × 10⁶	—	—	6 × 10⁶	1,800	—	18
K rich (93–113)	625,000	—	—	—	3 × 10⁶	—	6 × 10⁶	600	—	0
CS (298–315)	6 × 10⁶	—	—	—	—	3 × 10⁶	>15 × 10⁶	16,000	—	44
CS protein Full length	0	6 × 10⁶	625,000	1,000	125,000	0	6 × 10⁶	N.D.	N.D.	N.D.

Source: Whitacre, D.C. et al., *PLoS ONE*, 10(5), e0124856.

Notes: The listed WHcAg-hybrid VLPs and full-length rCSP protein were used to immunize mice (two doses, 20 and 10 μg in IFA). Secondary antisera were serially diluted and analyzed by ELISA for binding to solid-phase WHcAg; repeat peptides NANP and NVDP; nonrepeat peptides N1, K-rich, and CSP 298–315; and rCSP. Endpoint titers are shown. Antisera were also evaluated by IFA on dry or viable sporozoites. The protective efficiency after in vivo challenge with 10,000 hybrid sporozoites of mice immunized with the listed WHc-CS VLPs is also shown.

B cell regions *excised* from the CS protein and inserted onto hybrid-WHcAg VLPs elicited high levels of anti-insert antibodies (i.e., 1:3 × 10⁶ titers) (Table 13.2). Furthermore, the repeat and nonrepeat anti-insert antibodies bound rCS protein in ELISAs and also bound dry, hybrid *P. berghei/P. falciparum* sporozoites to varying degrees detected by IFA (Table 13.2). Interestingly, only the repeat-specific anti-insert antibodies (i.e., NANP/NVDP) bound live sporozoites. These observations suggest that the three nonrepeat B cell epitopes on the CS protein are cryptic on intact, live, or viable sporozoites [68].

13.14 PROTECTIVE EFFICACY OF CS-SPECIFIC B AND T CELL EPITOPES IN THE MURINE CHALLENGE MODEL USING HYBRID *Plasmodium berghei/Plasmodium falciparum* SPOROZOITES

We performed immunization/challenge experiments to determine the protective efficacy of hybrid-WHcAg VLPs carrying the two repeat B cell epitopes (NANP/NVDP) and the three nonrepeat B cell epitopes described previously. As shown in Figure 13.9, immunization (two doses of 20 and 10 μg) with 2 versions of VLPs carrying the repeat B cell epitopes protected mice challenged with 10,000 sporozoites at levels of 94.2% and 98.7% in terms of 18S rRNA copies detected in liver compared to mice immunized with a control hybrid-WHcAg VLP carrying an irrelevant insert from the hepatitis B virus (HBV). In contrast, immunization (three doses of 20, 10, and 10 μg) with the hybrid-WHcAg VLPs carrying each of the three nonrepeat B cell epitopes provided little to no protection against sporozoite challenge despite the fact that high levels of anti-insert antibodies were present in the immunized mice (Figure 13.9 and Table 13.2). These results may be explained by the cryptic nature of the nonrepeat B cell epitopes on viable sporozoites in vivo [68]. The results suggest that it may not be productive to include these three nonrepeat B cell epitopes in a CS-VLP vaccine candidate. A caveat to this interpretation is that the nonrepeat B cell epitopes in the context of the VLPs may not represent the epitope structures present within the native CS protein.

13.15 CONFIRMATION THAT CS-REPEAT ANTIBODIES ARE PREDOMINANT IN PROVIDING PROTECTION

As an alternate approach to addressing the question of the importance of repeat versus nonrepeat CS-specific antibodies, we performed an experiment using rCSP as the immunogen rather than hybrid VLPs (Figure 13.10). Mice were immunized with two doses of rCSP (20 μg/10 μg in IFA), and sera that were pooled and preincubated with the 10,000 sporozoites used for the challenge provided significant protection compared to sporozoites preincubated in normal mouse sera (NMS). However, if the anti-rCSP antisera were adsorbed with repeat-containing VLPs

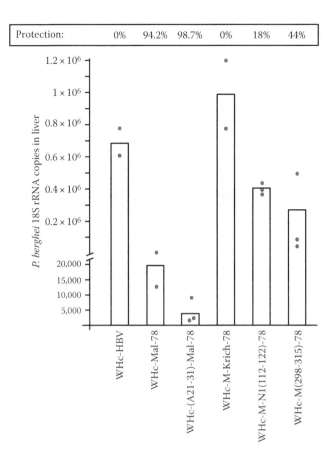

FIGURE 13.9 Comparison of protective efficacy of WHc-CS VLPs. Groups of mice were immunized with WHc-HBV negative control and CS-repeat (WHc-Mal-78 and WHc(A21–31)-Mal-78) VLPs, two doses of 20 and 10 μg in IFA; CS-nonrepeat VLPs, three doses of 20, 10, and 10 μg with the indicated WHc-hybrid VLPs. From 2 to 3.5 months after the last immunization dose, all mice were challenged with 10,000 hybrid sporozoites. *P. berghei* 18S rRNA copy number in the liver was determined by qPCR 40 h after infection. (Adapted from Whitacre, D.C. et al., *PLoS ONE*, 10(5), e0124856.)

(ΔNANP, NVDP) prior to being added to the 10,000 sporozoites, the protective efficacy was largely lost [68].

13.16 ANTI-NANP ANTIBODIES DEMONSTRATE GREATER PROTECTIVE EFFICACY THAN ANTI-NVDP ANTIBODIES

Having demonstrated the protective efficacy of hybrid-WHcAg VLPs carrying both CS-repeat B cell epitopes, it was of interest to determine the relative contribution of each specificity separately. For this purpose, rabbits were immunized (one dose, 200 μg) with a hybrid-WHcAg carrying both NANP and NVDP epitopes (WHc-Mal-78-UTC), and primary (4 week) diluted antisera were tested by preincubation with sporozoites prior to infection. One rabbit serum was absorbed with hybrid-HBcAg VLPs carrying only the NANP repeat epitope rendering the antisera highly NVDP specific, and the other rabbit serum was unabsorbed and contained both anti-NANP and anti-NVDP antibodies. The NANP-specific Mab 2A10 served as a positive control and naïve rabbit serum

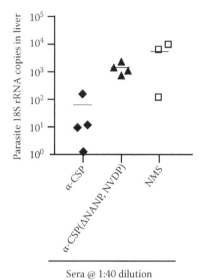

	Anti-CSP titers (1/dilution)			
	CSP	NANP	NVDP	CSP(93–113)
α-CSP	3×10^6	50 K	1 K	20 K
α-CSP (ΔNANP, NVDP)	625 K	1 K	0	20 K

In vivo protection

Sera @ 1:40 dilution
prior to passive transfer

FIGURE 13.10 Only anti-CSP repeat antibodies protect against a sporozoite challenge. Mice were immunized with rCSP and sera either unadsorbed or adsorbed with NANP/NVDP-containing VLPs (ΔNANP, NVDP) were incubated with 10,000 sporozoites prior to challenge. NMS, normal mouse sera. (From Whitacre, D.C. et al., *PLoS ONE*, 10(5), e0124856.)

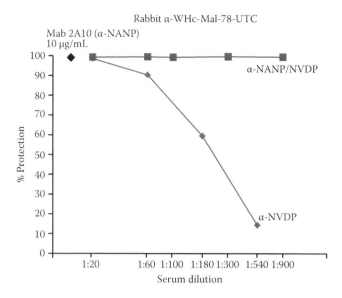

FIGURE 13.11 Comparison of protective efficacy of anti-NANP versus anti-NANP/NVDP antibodies. A rabbit was injected with 200 μg of WHc-Mal-78-UTC emulsified in IFA and 4 weeks after these primary immunization sera were collected. The serum was divided and one serum was absorbed with HBcAg-NANP VLPs to remove anti-NANP antibodies. The other was unabsorbed and contained both anti-NANP and anti-NVDP antibodies. The sera were serially diluted and preincubated with 10,000 hybrid sporozoites prior to injection into at least three mice for each dilution. Forty hours later, liver samples were assayed for *P. berghei* 18S rRNA and the percent protection calculated by comparison to dilutions of naïve rabbit sera (negative control). Mab 2A10 (10 μg/mL, NANP specific) served as a positive control.

served as a negative control. As shown in Figure 13.11, the NANP-absorbed antiserum was significantly less protective as shown by dilution analysis. For example, all dilutions of the unabsorbed rabbit serum provided almost total protection, whereas anti-NVDP antibodies in the absorbed serum provided significantly less protection.

13.17 OPTIMIZATION OF THE WHcAg CARRIER BY DELETING AN ENDOGENOUS WHcAg-SPECIFIC B CELL EPITOPE

To minimize production of anticarrier antibodies, we have initiated studies to mutate the WHcAg in order to delete as many endogenous WHc-specific B cell sites as possible while still permitting efficient assembly of hybrid-WHcAg VLPs. One such mutation involves aa 21–31 on WHcAg, which represents part of a WHcAg-specific B cell site. Partial substitution within this site with alanines *knocks out* B cell recognition by a Mab specific for this site. In order to determine if deletion of an endogenous WHcAg-specific B cell site

would enhance an anti-insert response, we compared wild-type WHcAg (WHc-Mal-78) versus A_{21-31}-mutated WHcAg (WHc(A_{21-31})-Mal-78) both carrying the same malaria-specific NANP/NVDP CS-repeat sequence (Figure 13.12). Both the primary and secondary anti-NANP and anti-NVDP antibody responses were superior when the A_{21-31}-mutated WHcAg-hybrid VLP was used for immunization. It was also notable that the anti-WHc carrier antibody response was decreased after (A_{21-31})-WHcAg immunization. Further, as shown in Figure 13.9, the A_{21-31} mutant elicited stronger protective efficacy (98.7% vs. 94.2%).

13.18 WHc-MAL-78-UTC ELICITS PROTECTIVE ABS IN RABBITS

To examine immunogenicity and protective efficacy in a second species, two rabbits were immunized with WHc-Mal-78-UTC (Figure 13.13) and sera were passively transferred into naïve murine recipients. The recipients of anti-VLP sera were either challenged (i.v. with 10,000 hybrid sporozoites) and parasite burden in the liver determined (Figure 13.14) or challenged by the bites of infected mosquitoes and blood-stage parasitemia monitored over a 10–14-day period (Figure 13.15). As shown in Figure 13.13, both rabbits (#73 and #74) produced high-titer anti-NANP, anti-NVDP, and anti-CSP Abs detected by ELISA and by IFA on hybrid

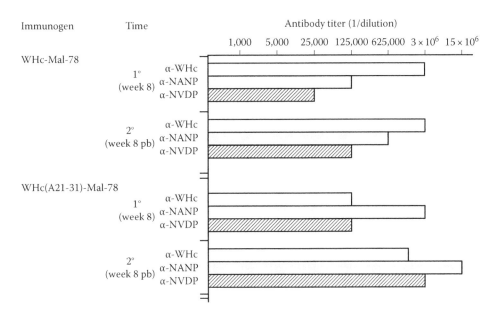

FIGURE 13.12 Immunogenic effect of deleting a carrier B cell epitope from WHc-Mal-78. Mice were immunized (two doses, 20 and 10 μg in IFA) with the indicated WHcAg-hybrid VLPs. Primary (1°) and secondary (2°) sera were serially diluted and analyzed by ELISA for binding to solid-phase WHcAg, NANP, and NVDP. Endpoint titers of pooled sera are shown.

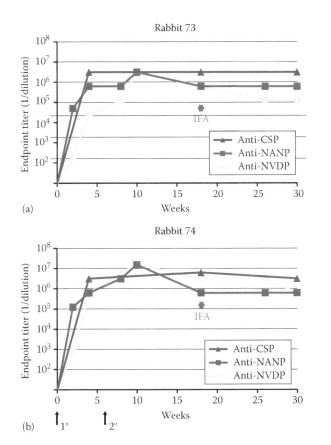

FIGURE 13.13 Immunogenicity of WHc-Mal-78-UTC in rabbits. Animals were primed with 200 μg of WHc-Mal-78-UTC emulsified in IFA and boosted at week 6 with (a) 100 μg emulsified in IFA (rabbit 73) or (b) 200 μg in saline (rabbit 74). Serum was collected at the indicated time points and endpoint titers against NANP, NVDP, and rCSP determined by ELISA. The sporozoite-specific IFA assay was performed on 18-week antisera (*). (Adapted from Whitacre, D.C. et al., *PLoS ONE*, 10(5), e0124856.)

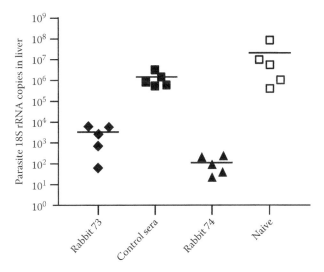

FIGURE 13.14 Protection against liver stage *P. falciparum/ P. berghei* infection. Mice were injected with 500 μL of indicated rabbit antisera and challenged with 10,000 sporozoites i.v. shortly after receiving the antisera. Liver burden was determined by qPCR 40 h after challenge. Control, NMS. (Adapted from Whitacre, D.C. et al., *PLoS ONE*, 10(5), e0124856.)

sporozoites [68]. Antisera (0.5 mL) from both rabbits were passively transferred (i.v.) to naïve mice and the mice were immediately challenged with 10,000 sporozoites (i.v.), and 40 h later, the parasite liver burdens were determined. Passively transferred anti-VLP sera from both rabbits significantly reduced the parasite liver burden as compared to control rabbit sera, although rabbit #74 sera were most effective (Figure 13.14) [68]. Rabbit #74 sera were chosen to passively transfer (0.2 mL) to murine recipients, which were challenged with the bites of from 3 to 12 infected mosquitoes over a 5 min time frame. Blood-stage parasitemia was

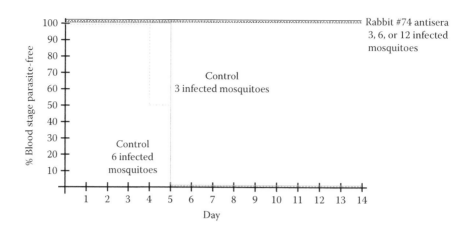

FIGURE 13.15 Protection against blood-stage *P. falciparum/P. berghei* infection. Two-hundred microliters of sera from rabbit #74 (see Figure 13.13) or from a control naïve rabbit were passively transferred to groups of seven or four mice, respectively, by i.v. injection. Mice were then challenged by allowing 3, 5, 6, or 12 mosquitoes infected with *P. falciparum/P. berghei* hybrid sporozoites to feed on the mice for 5 min. Mice were bled daily starting on day 4 postchallenge and blood-stage infection assessed by microscopy on stained blood smears. (Adapted from Whitacre, D.C. et al., *PLoS ONE*, 10(5), e0124856.)

monitored for the next 10–14 days. All 21 mice receiving the anti-WHc-Mal-78-UTC antisera were totally protected from blood-stage parasitemia, whereas the 8 control mice demonstrated infection by day 4 or 5 (Figure 13.15) [68].

13.19 CONSTRUCTION OF HYBRID-WHcAg VLPs CARRYING *Plasmodium falciparum*-CS-SPECIFIC T CELL DOMAINS

Another important goal is to add *P. falciparum*-CS-specific T cell sites to the final vaccine candidates in order to prime CS-specific CD4+/CD8+ T cells as well as elicit CS-specific

neutralizing antibodies. For this purpose, we have added one, two, or all three (i.e., UTC, TH.3R, and CS.T3) well-characterized human T cell domains to a standard hybrid-WHcAg VLP carrying the two CS-specific repeats (i.e., WHc-Mal-78). The T cell domains were added to the C-terminus of the hybrid-WHcAg VLPs and all three hybrid VLPs were successfully produced and were shown to be approximately equally immunogenic in terms of anti-NANP and anti-NVDP antibody production (Figure 13.16) [68]. In order to determine the contribution of *P. falciparum*–specific T cells to the protective efficacy of candidate VLP vaccines, the established protective efficacy of anti-NANP/NVDP antibodies must be excluded.

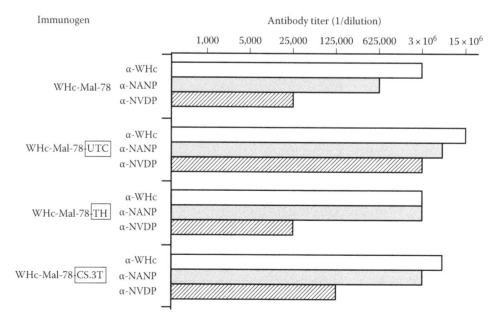

FIGURE 13.16 Comparison of WHc-CS VLPs containing malaria-specific T cell epitopes. Groups of three mice were immunized (two doses, 20 and 10 μg in IFA) with the indicated WHcAg-hybrid VLPs. Secondary antisera were serially diluted and analyzed by ELISA for binding to solid-phase WHcAg, NANP, and NVDP. Endpoint titers of pooled sera are shown. (From Whitacre, D.C. et al., *PLoS ONE*, 10(5), e0124856.)

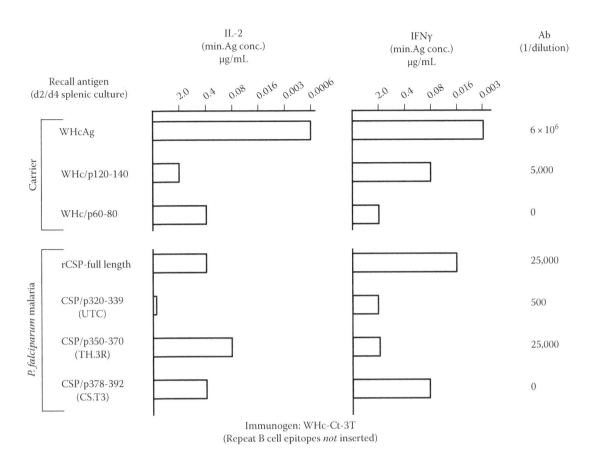

FIGURE 13.17 WHc-Ct-3T primes malaria-specific as well as WHcAg-specific CD4+ T cells. To assess T cell priming, mice were immunized with WHc-Ct-3T (a single 20 µg dose in IFA), and 10 days later, spleen cells were harvested and cultured with varying concentrations of the indicated recall antigens. Culture supernatants were collected and pooled at day 2 for determination of IL-2 and day 4 for determination of IFNγ. The minimum concentration of each antigen necessary to yield detectable cytokine is shown. For antibody production, mice were immunized (20 µg IFA) and boosted (10 µg IFA). (From Whitacre, D.C. et al., *PLoS ONE*, 10(5), e0124856.)

For that purpose, we constructed a hybrid-WHcAg VLP carrying only the 3T cell regions and devoid of the neutralizing CS-repeat B cell epitopes designated WHc-Ct-3T. As shown in Figure 13.17, immunization with WHc-Ct-3T primed both WHcAg-specific T cells and CS protein–specific T cells as determined by cytokine production elicited by splenic T cells cultured with a panel of WHcAg and CS protein–specific proteins and peptides [68]. Also note that WHc-Ct-3T immunization elicited low level Ab production to rCSP and the TH.3R site, which is also a B cell epitope in addition to a CD4+ T cell epitope.

Because the *P. berghei/P. falciparum* hybrid sporozoites used in the previous studies do not contain the *P. falciparum* T cell domains, Dr. Zavala produced a new transgenic sporozoite containing the complete C-terminus (i.e., aa 318–397) from *P. falciparum* CSP [68]. Therefore, we were able to perform an immunization/challenge experiment with WHc-Ct-3T VLPs and observed that although this hybrid VLP was immunogenic for both CS-specific B and T cell epitopes (see Figure 13.17), no protection against a 10,000 sporozoite challenge occurred (data not shown) [68].

13.20 MODIFICATION OF THE C-TERMINUS OF THE WHcAg PLATFORM MAY BROADEN APPLICATION TO LARGER INSERTS

Historically, the HBc/WHcAg platforms can tolerate C-terminal insertions of large size (i.e., >100 aa); however, the inserts are largely internalized and not accessible on the VLP surface, which renders them poorly or nonimmunogenic in terms of antibody production. Therefore, we have previously used the C-terminus exclusively for adding CD4+ T cell epitopes, which do not need to be accessible on the VLP surface. Recently, we have found that substituting the arginine (Arg) motifs at the C-terminus with alanines (Ala) may induce a conformational change that allows C-terminal insertions to become more exposed on the surface of VLPs. Therefore, we inserted the CS-repeat epitopes on the C-terminus of WT-WHcAg (WHc-Ct-Mal) or on mutated WHcAg [WHc(Ala)-Ct-Mal] and immunized mice (Figure 13.18). Immunization with two doses of unmutated WHcAg (WHc-Ct-Mal) elicited no anti-NANP/NVDP antibodies as predicted; however, the mutated WHcAg carrier (WHc(Ala)-Ct-Mal) elicited significant anti-NANP/NVDP antibodies.

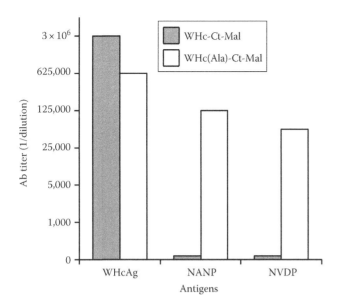

FIGURE 13.18 Substituting Ala for Arg motifs renders C-terminal insertions immunogenic. Mice were primed and boosted (20 µg/10 µg) with WHc-Ct-Mal and mutant WHc(Ala)-Ct-Mal, which contains Ala substitutions for the Arg motifs. Sera were collected 4 weeks after the boost and tested by ELISA for anti-WHc, anti-NANP, and anti-NVDP antibodies.

This is an important result because the C-terminus can tolerate very large insertions including entire proteins, and it may be possible to insert very large sequences at the C-terminus of the WHc(Ala) mutant such as other *P. falciparum* protective proteins (i.e., LSA-3 or CelTOS) or domains that may broaden the protective efficacy of hybrid VLPs.

13.21 COMPARISON OF PROTECTIVE EFFICACY OF VLPs CONTAINING ONLY *Plasmodium falciparum* NANP REPEAT B CELL EPITOPES VERSUS VLPs CONTAINING REPEAT B CELL EPITOPES PLUS THREE T CELL DOMAINS OF CSP

We compared the protective efficacy of a standard VLP (WHc-Mal5-78) containing four NANP repeats, which was previously shown to elicit significant protection against a hybrid sporozoite challenge, and a VLP containing an NANPNVDP(NANP)$_3$ B cell insert in the loop of WHcAg and three T cell domains (WHc-Mal-78-3T) inserted at the C-terminus of WHcAg. Groups of six mice were immunized with WHc-Mal5-78 or WHc-Mal-78-3T formulated either in saline only (200 µg VLPs), alum (100 µg VLPs), or Montanide ISA720 (50 µg VLPs) and given one booster injection in the same formulations (Figure 13.19). Both VLPs elicited significant reduction in parasite liver burden (at least 90% reduction in 18S rRNA copies in liver) in all three formulations compared to naïve control challenged mice (Figure 13.19a). However, the 3T cell domain-containing VLP (WHc-Mal-78-3T) elicited superior protection in saline (99.1% vs. 95% protection) and in alum (99.2% vs. 91.7% protection) compared to the (NANP)$_4$ B cell only–containing VLP

(WHc-Mal5-78). Both VLPs were equally protective when formulated in Montanide ISA720 (Figure 13.19a) [68].

We measured anti-CSP, anti-NANP, and anti-NVDP antibodies and performed IFAs to determine if differential antibody levels would explain the superior protective efficacy of the WHc-Mal-78-3T VLP (Figure 13.19b). No significant serological differences were noted between the two VLPs. IgG isotype testing also revealed no significant differences. This suggests that malaria-specific CD4$^+$ T cells primed by immunization with the WHc-Mal-78-3T VLP may have contributed to the greater efficacy either indirectly by providing an additional source of T helper cell function or, more likely, by directly exerting a negative effect on liver stage development via cytokine production [68].

Although the hybrid sporozoites used for challenge did not contain the *P. falciparum* T cell domains engineered into the WHc-Mal-78-3T VLPs, the 3T cell domains of *P. falciparum* and *P. berghei* share a significant degree of homology as shown in Figure 13.20 [68]. In any event, the superior performance of the WHc-Mal-78-3T VLP elevated this VLP to a primary vaccine candidate.

13.22 EFFECT OF FORMULATION ON PROTECTIVE EFFICACY

A primary VLP candidate, WHc-Mal-78-3T, was formulated in saline, alum, Montanide ISA720, or incomplete Freund's adjuvant (IFAd) for the prime and single boost or formulated in Montanide ISA 720 for the prime and saline for the boost. The Montanide ISA720 prime/saline boost was chosen to address the reactogenicity problems often associated with a Montanide ISA720 prime/boost. To the same end, Montanide ISA720 was used at 40% oil/antigen ratio instead of the standard 70% oil/antigen ratio recommended by the manufacturer (Figure 13.21). Although all formulations demonstrated significant levels of protection (from 99.1% to 99.98%), the order of efficacy was IFA/IFA = Montanide/Montanide = Montanide/saline > alum/alum > saline/saline. The performance of the Montanide ISA720 prime/saline boost was impressive and suggests that reactogenicity problems associated with the use of Montanide ISA720 could be mitigated by using less oil and boosting with WHc-Mal-78-3T in saline. The performances of WHc-Mal-78-3T formulated in alum and saline were also encouraging and suggest that additions of acceptable immunostimulants (e.g., QS-21, MPL) to either saline or alum may be sufficient to elicit sterile immunity as apparently achieved with the oil-containing adjuvants.

13.23 SELECTION OF A CS-WHcAg VLP CANDIDATE THAT CAN ELICIT STERILE IMMUNITY TO BLOOD-STAGE MALARIA USING ALUM ADJUVANTS

WHc-Mal-78-3T performed well in terms of reducing parasite load in the liver after a 10,000 sporozoite challenge (up to 99.98% reduction; Figure 13.21); however,

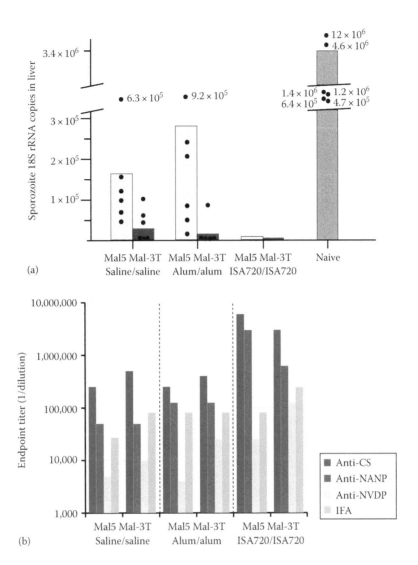

FIGURE 13.19 Comparison of protective efficacy and serology for WHc-Mal5-78 versus WHc-Mal-78-3T. Groups of six mice each were immunized with WHc-Mal5-78 or WHc-Mal-78-3T formulated in either saline, alum, or Montanide ISA-720 and given a single booster injection. After the boost, mice were challenged with 10,000 hybrid sporozoites. (a) Liver burden was assessed by determining sporozoite 18S rRNA copies in the liver. (b) Postboost antibody levels were determined by ELISA using rCSP, NANP, or NVDP peptides as solid-phase ligands. IFA titers were performed on dry sporozoites. (From Whitacre, D.C. et al., *PLoS ONE*, 10(5), e0124856.)

```
Sequence _____UTC_____  _____TH.3R_____  _____CS.T3_____
P. berghei  EFVKQIRDSITEEWSQCNVT C GSGIRVRKRKGSNKKAEDLTLEDI DTE--ICKMDKCSSIFN
            |..  |  |.. |||  |  ||   |  || ||  . ||  |   |    . |  |  ||||.||||.
on VLP      EYLNKIQNSLSTEEWSPCSVT S GNGIQVRIKPGSANKPKDELDYEN DIEKKICKMEKCSSV--
```

| Identical aa's: | 9/20 (45%) | 9/24 (38%) | 10/15 (67%) |
| Similar aa's: | 13/20 (65%) | 11/24 (46%) | 12/15 (80%) |

FIGURE 13.20 Conservation of *P. falciparum* CS T cell epitopes on *P. berghei* CS. Alignment of the *P. berghei* UTC, TH.3R, and CS.T3 T cell domains with the *P. falciparum* T cell domains incorporated into the WHc-Mal-78-3T VLP. The percentage represents homologies between the two sequences. (From Whitacre, D.C. et al., *PLoS ONE*, 10(5), e0124856.)

to determine if this level of reduction in liver burden is sufficient to yield full protection from blood-stage parasitemia, an immunization/challenge experiment monitoring blood-stage parasitemia as the final endpoint is required because a single surviving sporozoite infecting the liver can result in a blood-stage infection [62]. For this experiment we modified WHc-Mal-78-3T by a point mutation

(C61S) in the WHcAg, which eliminated the intermolecular disulfide bond at residue 61. The C61S mutation on WHcAg-hybrid VLPs can reduce anti-WHc (carrier-specific) antibody production and/or increase anti-insert antibody production as shown in Table 13.3 for WHc-Mal-78-3T versus WHc(C61S)-Mal-78-3T and for several other hybrid VLPs.

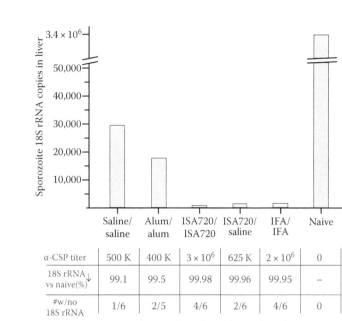

	Saline/ saline	Alum/ alum	ISA720/ ISA720	ISA720/ saline	IFA/ IFA	Naive
α-CSP titer	500 K	400 K	3×10^6	625 K	2×10^6	0
18S rRNA↓ vs naive(%)	99.1	99.5	99.98	99.96	99.95	–
#w/no 18S rRNA	1/6	2/5	4/6	2/6	4/6	0

WHc-Mal-78-3T formulation (prime/boost)

FIGURE 13.21 Effect of VLP formulation on protective efficacy. WHc-Mal-78-3T was formulated in the adjuvants shown. Groups of six mice each were immunized and boosted once with WHc-Mal-78-3T in the indicated formulations. After the boost, mice were challenged with 10,000 hybrid sporozoites and liver burden determined by 18S rRNA copy number in the liver. Also shown are the anti-CSP antibody titers, 18S rRNA reduction (%), and the number (#) of mice with no detectable 18S rRNA in the livers (i.e., sterile immunity).

TABLE 13.3

Effect of C61S Mutation on Anti-WHc/Anti-Insert Ab Production

	Titer (1/Dilution)	
WHcAg Hybrid	**Anti-WHc**	**Anti-Insert**
WHc-Mal-78-3T	625,000	50,000
WHc(C61S)-Mal-78-3T	625,000	**625,000**
WHc-RSV1-78	15×10^6	125,000
WHc(C61S)-RSV1-78	**625,000**	125,000
WHc-Mal5-78	1.2×10^6	625,000
WHc(C61S)-Mal-78	**250,000**	625,000
WHc-HBV1.3-78	625,000	125,000
Whc(C61S)-HBV1.3(+)-78	**125,000**	3×10^6
WHc-HBV1.6-78	125,000	125,000
WHc(C61S)-HBV1.6-78	125,000	**625,000**

Note: Bold represents at least fivefold difference in antibody titer.

Groups of 10 mice each were immunized and boosted with WHc(C61S)-Mal-78-3T either formulated in aluminum hydroxide (alum), alum + QS-21, or primed with an emulsion of Montanide ISA 720 (40%) and boosted in alum. The control group was primed with WHcAg (no insert), emulsified in Montanide ISA 720, and boosted in alum (Figure 13.22). Six weeks after the boost, mice were challenged by exposure to the bites of 12 infected mosquitoes for 5 min. This method of challenge was chosen because it represents

a more physiologically relevant route of infection as compared to i.v. injection of sporozoites. Blood was sampled over the next 14 days and examined for parasitemia. As shown in Figure 13.22, 10 of 10 WHcAg-immunized control mice became positive for blood-stage malaria within a mean of 4.4 days. In contrast, 0 of 9 mice immunized with WHc(C61S)-Mal-78-3T formulated in alum + QS-21 became infected, 1 of 10 mice immunized with Montanide/ alum became infected, and 2 of 10 mice immunized with alum only became infected [68]. The 3 of 29 mice in the experimental groups that did become infected demonstrated delayed parasitemia (mean of 6.0 days), suggesting a possible elimination of 99% of sporozoites given that 90% elimination is required to obtain a 1-day delay in developing a patent blood-stage infection. The serology of each group prechallenge and of the survivors 3 months postchallenge is shown in Table 13.4. Although anti-CSP Ab titers decreased over time, anti-CSP Abs were still in excess of 1×10^6 endpoint titer 3 months postchallenge [68].

The results indicate that immunization with an epitope-based VLP containing selected B and T cell epitopes from the *P. falciparum* CSP formulated in adjuvants licensed for human use can elicit sterile immunity against blood-stage malaria if sufficient anti-CSP protective Abs are produced. It would be useful to compare CSP-based vaccine candidates, including the industry standard RTS,S, in this standardized challenge model. A number of CSP-based vaccines have been developed recently [63–66]. Typically, protective efficacy has been determined using different challenge methods and different transgenic rodent parasites; therefore, comparative efficacy has not been possible to determine.

13.24 COMPARATIVE CHARACTERISTICS OF VLP-BASED CSP VACCINES

In the absence of direct head-to-head comparative studies, Table 13.5 lists comparative characteristics of the RTS, S vaccine, a WHc-CS-VLP (WHc(C61S)-Mal-78-3T), and an HBc-CS-VLP (ICC-1132). The WHc(C61S)-Mal-78-3T candidate embodies a number of unique characteristics that may be advantageous, especially for immunization in countries where malaria and HBV infection are both endemic. For example, the WHcAg carrier is derived from a nonhuman pathogen, whereas immune tolerance to both the HBsAg and HBcAg carriers is problematic in people chronically infected with HBV. Although we believe that there are distinct advantages to the use of full-length WHcAg as a carrier for malaria CSP epitopes (see Table 13.5), the disappointing results obtained with the HBcAg-based ICC-1132 candidate in a phase IIa trial were unexpected given the strong preclinical results in animals.

In preclinical tests of the ICC-1132 vaccine candidate, a variety of adjuvant systems were tested in mice and various alum formulations were relatively poor adjuvants [38]. In the first clinical trial in humans, three 50 µg doses of ICC-1132 absorbed on AL-PO$_4$ were used [40]. Alum tends to elicit Th2-like responses and the HBc/WHcAgs elicit

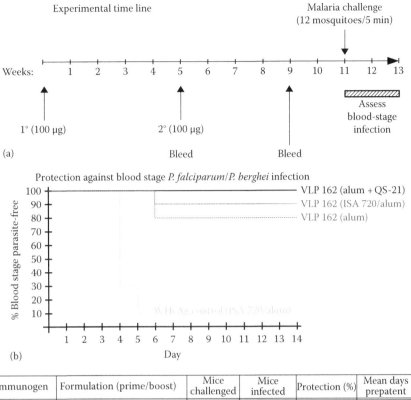

FIGURE 13.22 In vivo evaluation of VLP162 against *P. falciparum/P. berghei* malaria infection. Groups of 10 mice were primed and boosted with 100 μg of VLP-162 formulated in alum (250 μg/dose) and alum + QS-21 (20 μg/dose) or emulsified in Montanide ISA720 (50%, vol/vol) as indicated. (a) Timeline showing the schedule of prime, boost, and challenge with eight mosquitoes infected with *P. falciparum/P. berghei* hybrid sporozoites allowed to feed on the mice for 5 min. (b) Graphic representation of the percentage of mice remaining protected (i.e., free of blood-stage parasites) during the 14-day monitoring period. (c) Tabular summary of results. One mouse from the alum + QS-21 group died before the challenge, resulting in only nine mice in that challenge group. (From Whitacre, D.C. et al., *PLoS ONE*, 10(5), e0124856.)

TABLE 13.4
IgG Ab Titers from Primary Immunization with WHc(C61S)-Mal-78-3T to 3 Months Postchallenge

Immunogen	Formulation		Endpoint Titer (1/Dilution)			
			α-rCSP	α-NANP	α-NVDP	α-WHcAg
WHc(C61S)-	Alum	1°	82.5 K	43 K	5.5 K	125 K
Mal-78-3T	Alum	2°	4.7×10^6	4×10^6	900 K	3.5×10^6
		3 months postchallenge	2.2×10^6	1.6×10^6	540 K	875 K
WHc(C61S)-	Alum + QS-21	1°	103 K	68 K	13.5 K	125 K
Mal-78-3T	Alum + QS-21	2°	4.5×10^6	4.8×10^6	1.5×10^6	3×10^6
		3 months postchallenge	1.25×10^6	925 K	242 K	1.7×10^6
WHc(C61S)-	ISA-720	1°	183 K	113 K	17.8 K	475 K
Mal-78-3T	Alum	2°	4.3×10^6	4.3×10^6	1.3×10^6	2.6×10^6
		3 months postchallenge	1.25×10^6	1.1×10^6	160 K	708 K
WHcAg	ISA-720	1°	0	0	0	625 K
	Alum	2°	0	0	0	11.7×10^6
		3 months postchallenge	na	na	na	na

Source: Whitacre, D.C. et al., *PLoS ONE*, 10(5), e0124856.

TABLE 13.5

Comparison of WHc-CS Malaria VLP Vaccine Features to Previous-Generation Malaria VLP Vaccines RST,S and ICC-1132

		RTS,S	WHc-CS	Apovia (ICC-1132)
Malaria	B epitopes	CS repeats: (NANP)$_{16}$ only	CS repeats (NANP and NVDP)	CS repeats (NANP and NVDP)
	T epitopes	CS, C-terminus 302–395	CS 318–337 (**UTC**) CS 339–363 (**TH.3R**) CS 363–377 (**CS.T3**)	CS 318–337
Carrier		HBsAg (human pathogen)	WHcAg (full length) (**nonhuman pathogen**)	HBcAg (truncated) (human pathogen)
		T cell dependent	T cell dependent or independent	T cell dependent or independent
		100–1000-fold less immunogenic in mice	100–1000-fold more immunogenic in mice	100–1000-fold more immunogenic in mice
		Soluble HBsAg less immunogenic in humans	N.D.	Soluble HBcAg more immunogenic in humans
		MHC nonresponder genotypes (mice and human)	**No MHC nonresponders identified**	**No MHC nonresponders identified**
		Preexisting anti-HBs from HBV infection	**Not relevant**	Preexisting anti-HBc from HBV infection
		Immune tolerance in HBV chronics	**No immune tolerance in HBV chronics**	Immune tolerance in HBV chronics
		N.D.	**Δ carrier B cell sites**	N.D.
		N.D.	**Replace spikes**	N.D.
		N.D.	**Modify C-terminus**	N.D.
		Requires coexpression of HBsAg for HBsAg-CS assembly?	Self-assembly of WHc-CS	Self-assembly of HBc-CS
		Cannot express in bacteria	Bacterial expression	Bacterial expression
		N.D.?	**Sterile immunity to blood stage in Tg sporozoite model**	N.D.
Endogenous molecular adjuvants		None	**ssRNA–TLR7/8 ligands**	None
Stability to fixation		Lyophilization	Lyophilization	N.D.
Cold chain		Not required	Not required	N.D.
Formulation/adjuvants		Absolute requirement for immunostimulatory adjuvant (AS01B, MPL/QS-21/oil/liposome)	Saline, alum, Montanide ISA-720	Montanide ISA-720, alum
Phase II clinical results		1–3 dose, 0 protection	N.D.	1 dose, 0 protection
		3 dose + AS02A, 30%–50%	N.D.	N.D.

Note: Bold represents characteristics unique to WHc(C61S)-Mal-78-3T.

Th1-like responses so that the combination may cause cross-regulation and result in poor immunogenicity. However, adding an immunostimulant to alum (i.e., MPL or QS-21) may allow alum to be used more effectively. For example, inclusion of QS-21 in alum increased the protective efficacy of WHc(C61S)-Mal-78-3T to 100% from 80% in alum only (Figure 13.22). In the second and third clinical trials, a single 50 μg dose emulsified in the Montanide ISA 720 water-in-oil adjuvant was used at a 70:30 ratio with antigen [41,42]. A reactogenicity problem was encountered using the ISA 720 adjuvant with ICC-1132 in rhesus macaques [43]. The cause of the sterile abscesses that formed at the site of injection was determined to be a DTH reaction to the HBcAg

and not the other components of the vaccine. This indicates that strong HBcAg-specific T cell priming occurred in the macaques. In fact, cutaneous DTH reactions have been found to be a useful indicator of functional T cell immunity for malaria vaccines [67]. The authors of the phase IIa malaria challenge trial stated that the reactogenicity problem is the reason they used only a single 50 μg dose prior to malaria challenge. In murine studies, a 30:70 ratio of ISA 720/ICC-1132 was nearly as effective as a 70:30 ratio [38]. In the macaque trial, a 70:30 ratio of ISA 720/ICC-1132 was used and a reduction in the amount of the oil may have attenuated the risk of abscess formation. Furthermore, no sterile abscesses were observed in another species of macaque [43] nor in the clinical trials when Montanide ISA 720 was used [41,42]. Second, a single vaccine injection of ICC-1132 was used in the phase IIa trial, which is unprecedented for a subunit protein given to naïve recipients even in animal immunization protocols. At the very least, a booster injection in saline or alum should have been given, especially since the serum anti-NANP titers after the single injection were well below the levels known to be necessary to protect against a malaria challenge. These issues aside, possibly the more important issue was the ICC-1132 VLP dose given. The 50 μg human dose of ICC-1132 is only approximately 1.6 times the total dose (i.e., 30 μg) given to mice in IFA to elicit a protective Ab response and almost 4 times less than the optimal dose given to rhesus macaques [43]. When we used alum as an adjuvant for WHc(C61S)-Mal-78-3T in mice, we adjusted the dose up to 100 μg prime/100 μg boost, which represents a fourfold higher total mouse dose than was used for the ICC-1132 vaccine in the phase IIa clinical trial. A single 0.4 μg dose of ICC-1132 in IFA is the limiting dose at which 50% of BALB/c mice produce anti-NANP antibodies [38]. Interestingly, the limiting factor is NANP-specific B cells and not HBc-specific Th cells [38]. It is also important to consider that the inserted B cell epitope represents only 10% (by weight) of the total hybrid particle; therefore, at the mouse limiting dose, only 40 ng of the insert sequence is injected. The limiting dose for a human normalized for weight would be 0.8 mg of a VLP representing 80 μg of CS B cell epitopes. Other particulate human vaccines such as the HBsAg or HPV vaccines are given in three to four injections at 40–100 μg/dose usually in alum. However, the antibody response is directed to the entirety of the particle and not to an inserted sequence representing 10% (by weight) of the vaccine. For example, the limiting dose of the native HBcAg in mice is much less than 0.4 μg and is on the order of 0.004 μg (in IFA). Human dose escalation studies are an imperative for any CS-VLP vaccine candidate to be tested in the future. Some observers suggested that the failed phase IIa trial of ICC-1132 should end interest in the CS-VLP approach or suggested it be taken as a warning that preclinical results in animals are not predictive of human vaccine responses. The more realistic conclusions are to optimize dose and formulation adjusted for the size of the animal/human to be vaccinated and to not subvert basic tenets of immunology by dispensing with a booster injection.

REFERENCES

1. Good, M.F., Kaslow, D.C., and Miller, L.H. (1998) Pathways and strategies for developing a malaria blood-stage vaccine. *Annu Rev Immunol* 16: 57–87.

2. Nussenzweig, V. and Nussenzweig, R.S. (1989) Rationale for the development of an engineered sporozoite malaria vaccine. *Adv Immunol* 45: 283–334.

3. Nussenzweig, R.S., Vanderberg, J., Most, H., and Orton, C. (1967) Protective immunity produced by the injection of x-irradiated sporozoites of *Plasmodium berghei*. *Nature* 216: 160–162.

4. Nussenzweig, R.S., Vanderberg, J., Spitalny, G.L., Rivera, C.I., Orton, C., and Most, H. (1972) Sporozoite-induced immunity in mammalian malaria: A review. *Am J Trop Med Hyg* 21: 722–728.

5. Collins, W.E. and Contacos, P.G. (1972) Immunization of monkeys against *Plasmodium cynomolgi* by X-irradiated sporozoites. *Nat New Biol* 236: 176–177.

6. Clyde, D.F. (1975) Immunization of man against falciparum and vivax malaria by use of attenuated sporozoites. *Am J Trop Med Hyg* 24: 397–401.

7. Potocnjak, P., Yoshida, N., Nussenzweig, R.S., and Nussenzweig, V. (1980) Monovalent fragments (Fab) of monoclonal antibodies to a sporozoite surface antigen (Pb44) protect mice against malarial infection. *J Exp Med* 151: 1504–1513.

8. Yoshida, N., Nussenzweig, R.S., Potocnjak, P., Nussenzweig, V., and Aikawa, M. (1980) Hybridoma produces protective antibodies directed against the sporozoite stage of malaria parasite. *Science* 207: 71–73.

9. Zavala, F., Tam, J.P., Barr, P.J., Romero, P.J., Ley, V., Nussenzweig, R.S., and Nussenzweig, V. (1987) Synthetic peptide vaccine confers protection against murine malaria. *J Exp Med* 166: 1591–1596.

10. Kumar, K.A., Sano, G., Boscardin, S., Nussenzweig, R.S., Nussenzweig, M.C., Zavala, F., and Nussenzweig, V. (2006) The circumsporozoite protein is an immunodominant protective antigen in irradiated sporozoites. *Nature* 444: 937–940.

11. Zavala, F., Cochrane, A.H., Nardin, E.H., Nussenzweig, R.S., and Nussenzweig, V. (1983) Circumsporozoite proteins of malaria parasites contain a single immunodominant region with two or more identical epitopes. *J Exp Med* 157: 1947–1957.

12. Zavala, F., Tam, J.P., Hollingdale, M.R., Cochrane, A.H., Quakyi, I., Nussenzweig, R.S., and Nussenzweig, V. (1985) Rationale for development of a synthetic vaccine against *Plasmodium falciparum* malaria. *Science* 228: 1436–1440.

13. Charoenvit, Y., Mellouk, S., Cole, C., Bechara, R., Leef, M.F., Sedegah, M., Yuan, L.F., Robey, F.A., Beaudoin, R.L., and Hoffman, S.L. (1991) Monoclonal, but not polyclonal, antibodies protect against *Plasmodium yoelii* sporozoites. *J Immunol* 146: 1020–1025.

14. Fairley, N.H. (1949) Malaria; with special reference to certain experimental, clinical, and chemotherapeutic investigations; chemotherapy. *Br Med J* 2: 891–897.

15. Yamauchi, L.M., Coppi, A., Snounou, G., and Sinnis, P. (2007) Plasmodium sporozoites trickle out of the injection site. *Cell Microbiol* 9: 1215–1222.

16. Ballou, W.R., Hoffman, S.L., Sherwood, J.A., Hollingdale, M.R., Neva, F.A., Hockmeyer, W.T., Gordon, D.M. et al. (1987) Safety and efficacy of a recombinant DNA *Plasmodium falciparum* sporozoite vaccine. *Lancet* 1: 1277–1281.

17. Etlinger, H.M., Felix, A.M., Gillessen, D., Heimer, E.P., Just, M., Pink, J.R., Sinigaglia, F. et al. (1988) Assessment in humans of a synthetic peptide-based vaccine against the sporozoite stage of the human malaria parasite, *Plasmodium falciparum*. *J Immunol* 140: 626–633.

18. Herrington, D.A., Clyde, D.F., Losonsky, G., Cortesia, M., Murphy, J.R., Davis, J., Baqar, S. et al. (1987) Safety and immunogenicity in man of a synthetic peptide malaria vaccine against *Plasmodium falciparum* sporozoites. *Nature* 328: 257–259.

19. Hoffman, S.L., Oster, C.N., Plowe, C.V., Woollett, G.R., Beier, J.C., Chulay, J.D., Wirtz, R.A., Hollingdale, M.R., and Mugambi, M. (1987) Naturally acquired antibodies to sporozoites do not prevent malaria: Vaccine development implications. *Science* 237: 639–642.

20. Vreden, S.G., Verhave, J.P., Oettinger, T., Sauerwein, R.W., and Meuwissen, J.H. (1991) Phase I clinical trial of a recombinant malaria vaccine consisting of the circumsporozoite repeat region of *Plasmodium falciparum* coupled to hepatitis B surface antigen. *Am J Trop Med Hyg* 45: 533–538.

21. Gordon, D.M., McGovern, T.W., Krzych, U., Cohen, J.C., Schneider, I., LaChance, R., Heppner, D.G. et al. (1995) Safety, immunogenicity, and efficacy of a recombinantly produced *Plasmodium falciparum* circumsporozoite protein-hepatitis B surface antigen subunit vaccine. *J Infect Dis* 171: 1576–1585.

22. Stoute, J.A., Slaoui, M., Heppner, D.G., Momin, P., Kester, K.E., Desmons, P., Wellde, B.T., Garçon, N., Krzych, U., and Marchand, M. (1997) A preliminary evaluation of a recombinant circumsporozoite protein vaccine against *Plasmodium falciparum* malaria. RTS,S Malaria Vaccine Evaluation Group. *N Engl J Med* 336: 86–91.

23. Stoute, J.A., Kester, K.E., Krzych, U., Wellde, B.T., Hall, T., White, K., Glenn, G. et al. (1998) Long-term efficacy and immune responses following immunization with the RTS,S malaria vaccine. *J Infect Dis* 178: 1139–1144.

24. Bojang, K.A., Milligan, P.J., Pinder, M., Vigneron, L., Alloueche, A., Kester, K.E., Ballou, W.R. et al. (2001) Efficacy of RTS,S/AS02 malaria vaccine against *Plasmodium falciparum* infection in semi-immune adult men in The Gambia: A randomised trial. *Lancet* 358: 1927–1934.

25. Alloueche, A., Milligan, P., Conway, D.J., Pinder, M., Bojang, K., Doherty, T., Tornieporth, N., Cohen, J., and Greenwood, B.M. (2003) Protective efficacy of the RTS,S/AS02 *Plasmodium falciparum* malaria vaccine is not strain specific. *Am J Trop Med Hyg* 68: 97–101.

26. Alonso, P.L., Sacarlal, J., Aponte, J.J., Leach, A., Macete, E., Milman, J., Mandomando, I. et al. (2004) Efficacy of the RTS,S/AS02A vaccine against *Plasmodium falciparum* infection and disease in young African children: Randomised controlled trial. *Lancet* 364: 1411–1420.

27. Aponte, J.J., Aide, P., Renom, M., Mandomando, I., Bassat, Q., Sacarlal, J., Manaca, M.N. et al. (2007) Safety of the RTS,S/AS02D candidate malaria vaccine in infants living in a highly endemic area of Mozambique: A double blind randomised controlled phase I/IIb trial. *Lancet* 370: 1543–1551.

28. Bejon, P., White, M.T., Olotu, A., Bojang, K., Lusingu, J.P.A., Salim, N., Otsyula, N.N. et al. (2013) Efficacy of RTS,S malaria vaccines: Individual-participant pooled analysis of phase 2 data. *Lancet Infect Dis* 13: 319–327.

29. Hoofnagle, J.H., Gerety, R.J., and Barker, L.F. (1973) Antibody to hepatitis-B-virus core in man. *Lancet* 2: 869–873.

30. Milich, D.R. and McLachlan, A. (1986) The nucleocapsid of hepatitis B virus is both a T-cell-independent and a T-cell-dependent antigen. *Science* 234: 1398–1401.

31. Betancourt, A.A., Delgado, C.A.G., Estévez, Z.C., Martínez, J.C., Ríos, G.V., Aureoles-Roselló, S.R.M., Zaldívar, R.A. et al. (2007) Phase I clinical trial in healthy adults of a nasal vaccine candidate containing recombinant hepatitis B surface and core antigens. *Int J Infect Dis* 11: 394–401.

32. Alper, C.A. (1995) The human immune response to hepatitis B surface antigen. *Exp Clin Immunogenet* 12: 171–181.

33. Milich, D.R., Peterson, D.L., Zheng, J., Hughes, J.L., Wirtz, R., and Schödel, F. (1995) The hepatitis nucleocapsid as a vaccine carrier moiety. *Ann N Y Acad Sci* 754: 187–201.

34. Milich, D.R. and Chisari, F.V. (1982) Genetic regulation of the immune response to hepatitis B surface antigen (HBsAg). I. H-2 restriction of the murine humoral immune response to the a and d determinants of HBsAg. *J Immunol* 129: 320–325.

35. Pumpens, P. and Grens, E. (2001) HBV core particles as a carrier for B cell/T cell epitopes. *Intervirology* 44: 98–114.

36. Schödel, F., Wirtz, R., Peterson, D., Hughes, J., Warren, R., Sadoff, J., and Milich, D. (1994) Immunity to malaria elicited by hybrid hepatitis B virus core particles carrying circumsporozoite protein epitopes. *J Exp Med* 180: 1037–1046.

37. Schödel, F., Peterson, D., Milich, D.R., Charoenvit, Y., Sadoff, J., and Wirtz, R. (1997) Immunization with hybrid hepatitis B virus core particles carrying circumsporozoite antigen epitopes protects mice against *Plasmodium yoelii* challenge. *Behring Inst Mitt* 98: 114–119.

38. Milich, D.R., Hughes, J., Jones, J., Sällberg, M., and Phillips, T.R. (2001) Conversion of poorly immunogenic malaria repeat sequences into a highly immunogenic vaccine candidate. *Vaccine* 20: 771–788.

39. Birkett, A., Lyons, K., Schmidt, A., Boyd, D., Oliveira, G.A., Siddique, A., Nussenzweig, R., Calvo-Calle, J.M., and Nardin, E. (2002) A modified hepatitis B virus core particle containing multiple epitopes of the *Plasmodium falciparum* circumsporozoite protein provides a highly immunogenic malaria vaccine in preclinical analyses in rodent and primate hosts. *Infect Immun* 70: 6860–6870.

40. Nardin, E.H., Oliveira, G.A., Calvo-Calle, J.M., Wetzel, K., Maier, C., Birkett, A.J., Sarpotdar, P., Corado, M.L., Thornton, G.B., and Schmidt, A. (2004) Phase I testing of a malaria vaccine composed of hepatitis B virus core particles expressing *Plasmodium falciparum* circumsporozoite epitopes. *Infect Immun* 72: 6519–6527.

41. Oliveira, G.A., Wetzel, K., Calvo-Calle, J.M., Nussenzweig, R., Schmidt, A., Birkett, A., Dubovsky, F. et al. (2005) Safety and enhanced immunogenicity of a hepatitis B core particle *Plasmodium falciparum* malaria vaccine formulated in adjuvant Montanide ISA 720 in a phase I trial. *Infect Immun* 73: 3587–3597.

42. Walther, M., Dunachie, S., Keating, S., Vuola, J.M., Berthoud, T., Schmidt, A., Maier, C. et al. (2005) Safety, immunogenicity and efficacy of a pre-erythrocytic malaria candidate vaccine, ICC-1132 formulated in Seppic ISA 720. *Vaccine* 23: 857–864.

43. Langermans, J.A.M., Schmidt, A., Vervenne, R.A.W., Birkett, A.J., Calvo-Calle, J.M., Hensmann, M., Thornton, G.B., Dubovsky, F., Weiler, H., Nardin, E., and Thomas, A.W. (2005) Effect of adjuvant on reactogenicity and long-term immunogenicity of the malaria Vaccine ICC-1132 in macaques. *Vaccine* 23: 4935–4943.

44. Billaud, J., Peterson, D., Lee, B.O., Maruyama, T., Chen, A., Sallberg, M., Garduño, F., Goldstein, P., Hughes, J., Jones, J., and Milich, D. (2007) Advantages to the use of rodent hepadnavirus core proteins as vaccine platforms. *Vaccine* 25: 1593–1606.

45. Ferrari, C., Penna, A., Bertoletti, A., Valli, A., Antoni, A.D., Giuberti, T., Cavalli, A., Petit, M.A., and Fiaccadori, F. (1990) Cellular immune response to hepatitis B virus-encoded antigens in acute and chronic hepatitis B virus infection. *J Immunol* 145: 3442–3449.

46. Billaud, J., Peterson, D., Barr, M., Chen, A., Sallberg, M., Garduno, F., Goldstein, P., McDowell, W., Hughes, J., Jones, J., and Milich, D. (2005) Combinatorial approach to hepadnavirus-like particle vaccine design. *J Virol* 79: 13656–13666.

47. Billaud, J., Peterson, D., Schödel, F., Chen, A., Sallberg, M., Garduno, F., Goldstein, P., McDowell, W., Hughes, J., Jones, J., and Milich, D. (2005) Comparative antigenicity and immunogenicity of hepadnavirus core proteins. *J Virol* 79: 13641–13655.

48. Lee, B.O., Tucker, A., Frelin, L., Sallberg, M., Jones, J., Peters, C., Hughes, J., Whitacre, D., Darsow, B., Peterson, D.L., and Milich, D.R. (2009) Interaction of the hepatitis B core antigen and the innate immune system. *J Immunol* 182: 6670–6681.

49. Ruprecht, C.R. and Lanzavecchia, A. (2006) Toll-like receptor stimulation as a third signal required for activation of human naive B cells. *Eur J Immunol* 36: 810–816.

50. Calvo-Calle, J.M., Hammer, J., Sinigaglia, F., Clavijo, P., Moya-Castro, Z.R., and Nardin, E.H. (1997) Binding of malaria T cell epitopes to DR and DQ molecules in vitro correlates with immunogenicity in vivo: Identification of a universal T cell epitope in the *Plasmodium falciparum* circumsporozoite protein. *J Immunol* 159: 1362–1373.

51. Reece, W.H.H., Pinder, M., Gothard, P.K., Milligan, P., Bojang, K., Doherty, T., Plebanski, M. et al. (2004) A CD4(+) T-cell immune response to a conserved epitope in the circumsporozoite protein correlates with protection from natural *Plasmodium falciparum* infection and disease. *Nat Med* 10: 406–410.

52. Wang, R., Doolan, D.L., Le, T.P., Hedstrom, R.C., Coonan, K.M., Charoenvit, Y., Jones, T.R. et al. (1998) Induction of antigen-specific cytotoxic T lymphocytes in humans by a malaria DNA vaccine. *Science* 282: 476–480.

53. Riedl, P., Stober, D., Oehninger, C., Melber, K., Reimann, J., and Schirmbeck, R. (2002) Priming Th1 immunity to viral core particles is facilitated by trace amounts of RNA bound to its arginine-rich domain. *J Immunol* 168: 4951–4959.

54. Craven, D.E., Awdeh, Z.L., Kunches, L.M., Yunis, E.J., Dienstag, J.L., Werner, B.G., Polk, B.F. et al. (1986) Nonresponsiveness to hepatitis B vaccine in health care workers. Results of revaccination and genetic typings. *Ann Intern Med* 105: 356–360.

55. Persson, C., Oliveira, G.A., Sultan, A.A., Bhanot, P., Nussenzweig, V., and Nardin, E. (2002) Cutting edge: A new tool to evaluate human pre-erythrocytic malaria vaccines: Rodent parasites bearing a hybrid *Plasmodium falciparum* circumsporozoite protein. *J Immunol* 169: 6681–6685.

56. Tewari, R., Spaccapelo, R., Bistoni, F., Holder, A.A., and Crisanti, A. (2002) Function of region I and II adhesive motifs of *Plasmodium falciparum* circumsporozoite protein in sporozoite motility and infectivity. *J Biol Chem* 277: 47613–47618.

57. Aley, S.B., Bates, M.D., Tam, J.P., and Hollingdale, M.R. (1986) Synthetic peptides from the circumsporozoite proteins of *Plasmodium falciparum* and *Plasmodium knowlesi* recognize the human hepatoma cell line HepG2-A16 in vitro. *J Exp Med* 164: 1915–1922.

58. Rathore, D., Nagarkatti, R., Jani, D., Chattopadhyay, R., de la Vega, P., Kumar, S., and McCutchan, T.F. (2005) An immunologically cryptic epitope of *Plasmodium falciparum* circumsporozoite protein facilitates liver cell recognition and induces protective antibodies that block liver cell invasion. *J Biol Chem* 280: 20524–20529.

59. White, K., Krzych, U., Gordon, D.M., Porter, T.G., Richards, R.L., Alving, C.R., Deal, C.D. et al. (1993) Induction of cytolytic and antibody responses using *Plasmodium falciparum* repeatless circumsporozoite protein encapsulated in liposomes. *Vaccine* 11: 1341–1346.

60. Calle, J.M., Nardin, E.H., Clavijo, P., Boudin, C., Stüber, D., Takacs, B., Nussenzweig, R.S., and Cochrane, A.H. (1992) Recognition of different domains of the *Plasmodium falciparum* CS protein by the sera of naturally infected individuals compared with those of sporozoite-immunized volunteers. *J Immunol* 149: 2695–2701.

61. Takita-Sonoda, Y., Tsuji, M., Kamboj, K., Nussenzweig, R.S., Clavijo, P., and Zavala, F. (1996) *Plasmodium yoelii*: Peptide immunization induces protective CD4+ T cells against a previously unrecognized cryptic epitope of the circumsporozoite protein. *Exp Parasitol* 84: 223–230.

62. White, M.T., Bejon, P., Olotu, A., Griffin, J.T., Riley, E.M., Kester, K.E., Ockenhouse, C.F., and Ghani, A.C. (2013) The relationship between RTS,S vaccine-induced antibodies, CD4+ T cell responses and protection against *Plasmodium falciparum* infection. *PLoS ONE* 8: e61395.

63. Kaba, S.A., McCoy, M.E., Doll, T.A.P.F., Brando, C., Guo, Q., Dasgupta, D., Yang, Y. et al. (2012) Protective antibody and CD8+ T-cell responses to the *Plasmodium falciparum* circumsporozoite protein induced by a nanoparticle vaccine. *PLoS ONE* 7: e48304.

64. Kastenmüller, K., Espinosa, D.A., Trager, L., Stoyanov, C., Salazar, A.M., Pokalwar, S., Singh, S., Dutta, S., Ockenhouse, C.F., Zavala, F., and Seder, R.A. (2013) Full-length *Plasmodium falciparum* circumsporozoite protein administered with long-chain poly(I·C) or the Toll-like receptor 4 agonist glucopyranosyl lipid adjuvant-stable emulsion elicits potent antibody and CD4+ T cell immunity and protection in mice. *Infect Immun* 81: 789–800.

65. Porter, M.D., Nicki, J., Pool, C.D., DeBot, M., Illam, R.M., Brando, C., Bozick, B. et al. (2013) Transgenic parasites stably expressing full-length *Plasmodium falciparum* circumsporozoite protein as a model for vaccine down-selection in mice using sterile protection as an endpoint. *Clin Vaccine Immunol* 20: 803–810.

66. Przysiecki, C., Lucas, B., Mitchell, R., Carapau, D., Wen, Z., Xu, H., Wang, X. et al. (2012) Sporozoite neutralizing antibodies elicited in mice and rhesus macaques immunized with a *Plasmodium falciparum* repeat peptide conjugated to meningococcal outer membrane protein complex. *Front Cell Infect Microbiol* 2: 146.

67. Stewart, V.A., Walsh, D.S., McGrath, S.M., Kester, K.E., Cummings, J.F., Voss, G., Delchambre, M., Garçon, N., Cohen, J.D., and Heppner, D.G.J. (2006) Cutaneous delayed-type hypersensitivity (DTH) in a multi-formulation comparator trial of the anti-falciparum malaria vaccine candidate RTS,S in rhesus macaques. *Vaccine* 24: 6493–6502.

68. Whitacre, D.C., Espinosa, D.A., Peters, C.J., Jones, J.E., Tucker, A.E., Peterson, D.L., Zavala, F.P., and Milich, D.R. (2015) *P. falciparum* and *P. vivax* epitope-focused VLPs elicit sterile immunity to blood stage infections. *PLoS ONE*, 10(5), e0124856.

14 Use of HBcAg as an Adjuvant in DNA-Based Vaccines

An Unexpected Journey

Matti Sällberg

CONTENTS

14.1 HBcAg AND THE HOST IMMUNE SYSTEM

The way that HBcAg interacts with the host immune system has resulted in a significant number of papers. It was early recognized that patients infected by the hepatitis B virus (HBV) developed high levels of antibodies to HBcAg (anti-HBc) [1]. Early in the acute phase, high levels of anti-HBc IgM appear and have become the gold standard for diagnosing acute HBV [2]. The anti-HBc levels are very high, being one of the first signs of the immunogenicity of HBcAg, early in the infection, and then wane within a 6-month period [3]. However, in the chronic infection, extremely high levels of anti-HBc IgG appear, again supporting the immunogenicity of HBcAg. One of the early explanations of the high immunogenicity of HBcAg was suggested by Milich and McLachlan in 1986, where they showed that HBcAg was both T-cell-dependent and T-cell-independent antigens [4]. Albeit the role of the T cell independence of HBcAg in the immunogenicity of HBcAg is unclear, this led to two additional discoveries. First, extracellular HBcAg uses the B cell as the primary APC [5]. Second, this was later explained by the fact that HBcAg has the ability to bind naive human [6] and murine [7] B cells through the B cell receptor (BCR) (Figure 14.1). The structure of the latter interaction was recently solved [8]. Importantly, one of the key adjuvant effects seems to be that the HBcAg particles, or proteins, bind single-stranded RNA, which in turn activates the innate Toll-like receptor TLR-7 signaling pathways [9,10]. Thus, it is likely that the HBcAg particles will both cross-link BCRs to induce an activated state in the B cells, and after uptake, the TLR-7 pathway is induced by encapsidated RNA (Figure 14.1). The origin of the encapsidated RNA in recombinantly produced HBcAg particles seems to be of the origin from the cell where the capsids were expressed. Thus, the capsids seem to contain heterologous RNA [10].

The central role of HBcAg in the immune response and control of the HBV infection was first illustrated by that strong T cell response to HBcAg appeared in the early acute HBV infection [11]. Here, cytotoxic T lymphocyte (CTL) epitopes were identified in humans, and the appearance of CTLs to these epitopes was found to correlate with clearance [12,13]. Thus, HBcAg produced in hepatocytes has the ability to be recognized by CTLs. However, whether the priming of the HBcAg-specific CTLs occurs in the liver or requires transport by dendritic cells to a specialized lymphoid tissue is not known.

Data suggest that HBcAg is poorly presented to T cells by dendritic cells [5]. A key experiment in this chapter showed that HBcAg and the hepatitis B e antigen (HBeAg) were equally well presented to a T cell hybridoma by non–B cell APCs. In contrast, when B cells were used as APCs, HBcAg much was more efficiently presented to the T cell hybridomas than HBeAg [5]. Finally, HBcAg and the HBeAg share 150 amino acids. Hence, a T cell may well cross-recognize these two antigens, evidenced by the ability of HBeAg to induce T cell tolerance to both HBeAg and HBcAg [14]. However, it cannot be excluded that the fine specificity of the T cells may differ between HBcAg and HBeAg, depending on the exact sequence of the antigen and if any heterologous sequence has been added. Overall, HBcAg behaves in an unexpected way when processed by the host immune system.

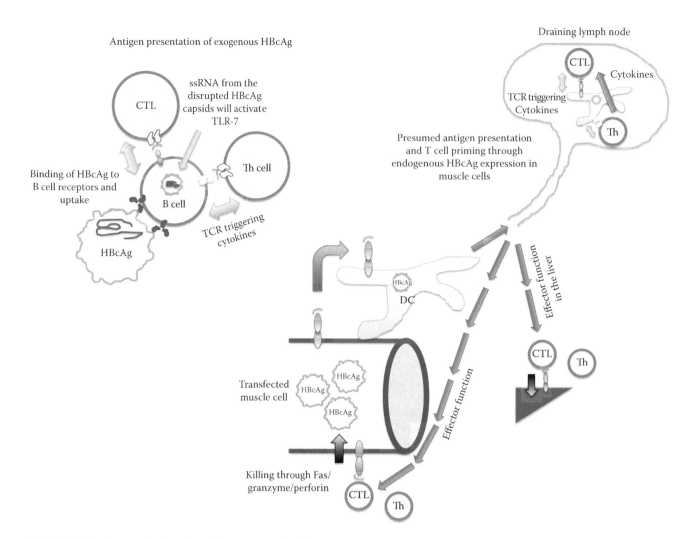

FIGURE 14.1 Carton showing the antigen presentation of exogenous/extracellular HBcAg (left) and the antigen presentation and T cell priming of HBcAg expressed by a transfected muscle tissue.

14.2 HBcAg AS A PROTEIN-BASED IMMUNOGEN

HBcAg has been shown to be highly immunogenic, regardless of which species the HBcAg is derived from [15]. This has translated into that HBcAg can act as a carrier for foreign sequences [16–21]. The ability of HBcAg to accept inserts of rather impressive sizes has made this platform widely explored [22]. However, the inserts may affect the stability of the particle, which in turn may affect the immunogenicity and the suitability for long-term storage. In the rodent HBcAg system, it has been shown that the flanking residues that surround an insert at the tip of the HBcAg spike have a strong influence of both stability and immunogenicity [16,17]. Thus, there is still much to be learned regarding how HBcAg is best used and produced as an antigen for foreign epitopes.

In animal models, HBcAg has been found to act as a highly potent carrier for foreign epitopes. The results in humans have so far been disappointing. In a first clinical trial where a chimeric HBcAg particle expresses the malaria circumsporozoite (CS) repeat epitope, three 50 μg doses of the vaccine in aluminum hydroxide induced anti-CS mean titers of up

to 1:400 [23]. In a second study, a better adjuvant was used and the anti-CS titers reached 1:1000 after a single dose [24]. However, these responses did unfortunately not result in protection against experimental malaria challenge [25]. It is not known whether a higher dose or more injections would have resulted in protection. Higher dose may, depending on the adjuvant used, constitute a problem due to side effects seen in nonhuman primates [26]. Thus, the potential for HBcAg as a carrier for foreign epitopes in humans certainly needs to be further explored. The initial test should be the ability of the particles to raise high titer functional antibodies in humans. In addition, one may consider delivering the chimeric particles as a genetic immunogen. However, the use of HBcAg as a protein-based particulate immunogen has been reviewed extensively in other chapters in this book.

14.3 HBcAg HAS ALSO BEEN TESTED AS A GENETIC IMMUNOGEN

This is where the unexpected journey of HBcAg begins. It would be predicted based on all evidence from HBV-infected humans and from the immunogenicity of particulate protein-based (exogenous) HBcAg that also endogenously expressed HBcAg is highly immunogenic. The answer to this question is it depends. And it depends on how, or where, the endogenous HBcAg is expressed and what the measurement of the immunogenicity is (Figure 14.1).

With respect to the site of HBcAg gene delivery and expression, some discrepant results have been obtained. Among the first studies of HBcAg and HBeAg DNA immunogens revealed that CTLs could be induced to novel epitopes [27,28]. We could, around the same time, show that an HBcAg-neomycin phosphoryl transferase fusion gene had the ability to prime HBcAg-specific CTLs in H-2^k mice [29,30]. However, this was a quite difficult task and the ability was dependent on several factors. Overall, the responses were comparatively weak. The vaccine was then evaluated in three chimpanzees with chronic HBV infection, and one out of these seroconverted from HBeAg to anti-HBe, simultaneous with transient increases in alanine aminotransferase levels and a drop in the HBV viral load [31]. Thus, this was a very early proof of concept for the therapeutic vaccination in chronic HBV using a genetic vaccine.

We could later find that the immunogenicity of endogenous HBcAg was, unexpectedly, affected by how the HBcAg was delivered. The delivery of a DNA plasmid containing a codon-optimized HBcAg gene to regular C57BL/6 (H-2^b) and Balb/C (H-2^d) mice transdermally using the gene gun (BioRad) resulted in a poor priming of HBcAg-specific CTLs [32]. We evaluated whether a preexisting CD4+ T helper cell population could improve CTL priming but it did not. In contrast, HBcAg-specific CTLs were more efficiently induced by administering the HBcAg-encoding plasmid by an intramuscular injection [32]. More importantly, the immunogenicity was even further improved when combined with in vivo electroporation. Thus, unexpectedly, the immunogenicity of endogenously produced HBcAg was strongly influenced by where the HBcAg plasmid was expressed. During a dinner conversation at a vaccine meeting, one of the dinner guests mentioned that we should reevaluate our findings with a different helium pressure. Although this may alter the results in that the depth of the injection in the tissue may change, this cannot alter our conclusions. This I base on the fact that when we deliver a hepatitis C virus nonstructural 3/4A–expressing plasmid using the same pressure and at the same or even lower doses, we see an excellent priming of NS3-specific CTLs [32]. Thus, there is certainly a difference in how endogenously HBcAg is processed and presented to the host immune system as compared to, for example, NS3/4A. This is further supported by the observation that NS3/4A-DNA primes CTLs at approximately 100-fold lower doses than HBcAg DNA, even when delivered i.m. using in vivo electroporation (EP) [32,33].

14.4   SO HOW DO YOU MAKE ENDOGENOUS HBcAg HIGHLY IMMUNOGENIC?

There are a number of tricks that one could bring out from the toolbox to improve the immunogenicity of an endogenously expressed HBcAg gene. First, codon optimize the gene, which we found improved immunogenicity by at least 10-fold. Next, improve the uptake of the gene, which can be achieved by many ways. One may use various viral vectors such as retroviral, adenoviral, or various poxvirus vectors [29–31,34]. Although these are simple to deliver and generally has no safety concerns, a major limitation is that they can usually be delivered only once with the maximum effect due to a potent antivector immunity. Thus, if viral vectors are used, they are then given in various prime-boost combinations with different vectors used for the prime and the boost. This is where one advantage of plasmid DNA, or naked RNA, becomes evident, the lack of antivector responses.

DNA and RNA can be delivered multiple times with a clear boost effect [33]. However, since the DNA uptake in larger animals is extremely poor, various delivery devices are required for a potent immunogenicity. One such that has proven to be effective is in vivo EP [35]. Here, electrical pulses are introduced over the tissue volume where the DNA has been injected to destabilize the cell membranes to enable DNA uptake by the cell [35]. We could show that the priming of HBcAg-specific CTLs by DNA was less dependent on B cells, in contrast to extracellular HBcAg [36]. With respect to HBcAg DNA, we could show that in vivo EP improved the immunogenicity by 10–100 times as compared to a standard i.m. injection [32]. It is highly likely that the tissue damage caused by the in vivo EP also contributes to the improvement of the immunogenicity. We could also show that this technology can be safely used in humans with chronic viral hepatitis [37,38].

Thus, the first steps to greatly improve the immunogenicity of HBcAg DNA is to codon optimize the gene and deliver by a technique that ensures a high transfection rate of cells in vivo. As additional techniques may add various genes expressing, for example, cytokines that improve the priming milieu at the site where the HBcAg is endogenously produced.

An approach to improve immunogenicity that we have tested, but unfortunately without success, is to generate fusion proteins between HBcAg from human HBV and HBV from various animal species. With respect to this concept, we have noted an unexpected competitive effect that may even eliminate the HBcAg-specific CTL priming (unpublished data). The reason for this needs further investigation. One possibility is that the HBcAg-specific epitopes presented by mouse MHC are low avidity binders and therefore are easily outcompeted.

14.5   ENDOGENOUS HBcAg CAN ACT AS AN ADJUVANT FOR FOREIGN SEQUENCES

We found in our studies on HBcAg and NS3/4A that endogenously expressed HBcAg effectively adjuvants the CTL priming by NS3/4A [39]. However, unexpectedly, this does not seem to work the other way around. We generated several fusions between NS3/4A and human HBcAg, both particulate and cleaved versions of HBcAg. Very unexpectedly, the most potent adjuvant effect was seen from the cleaved versions of HBcAg. Neither an uncleaved NS3–4A–HBcAg fusion nor with NS3/4A and HBcAg as free proteins were as effective as when HBcAg was cleaved by either introduced NS3/4A-cleavage sequences or autoproteolytic P2A sequences [39]. The mechanism behind this is still not known but may relate to how HBcAg is processed by APCs (Figure 14.2). We now have new data that this is true for several foreign antigens and that cleaved avian HBcAg may be even more effective than human HBcAg. In fact, there are many reasons for not using human HBcAg in an adjuvant setting. In many areas where these new vaccines may be used, there is an underlying high frequency of HBV

infection and even carriers. Thus, if human HBcAg was used as a carrier or an adjuvant sequence, this ability may be impaired by the imprint made by the HBV infection on the host with a possible tolerant HBcAg-specific T cells. By using an avian HBcAg sequence, this disadvantage can be completely circumvented.

The theoretical basis for the adjuvant effect of HBcAg when used as a genetic adjuvant is not fully understood. However, it is likely that the coexpressed and presented HBcAg will recruit healthy naive T cells that will improve the local priming milieu, regardless dysfunctional T cells or if a regulatory response is targeted (Figure 14.2).

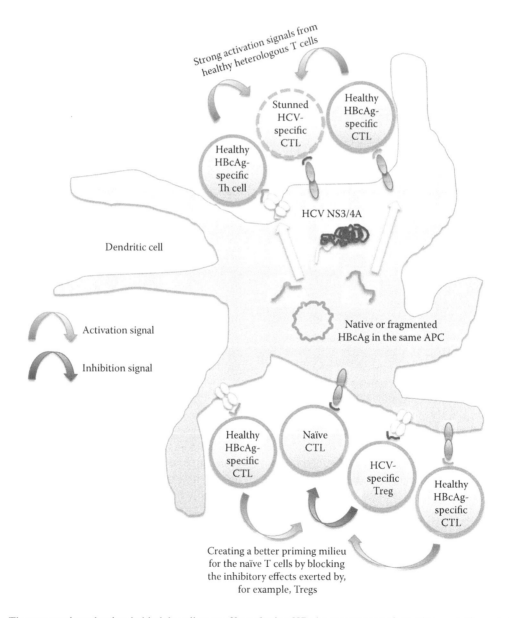

FIGURE 14.2 The proposed mechanism behind the adjuvant effect of using HBcAg sequences as heterologous antigens to improve immunogenicity. On top of the dendritic cell, an example showing how a tolerized, or stunned, dysfunctional HCV-specific T cell can be activated by a healthy heterologous HBcAg-specific T cell providing the appropriate cytokine environment. On the bottom of the DC, an example is shown when the heterologous HBcAg-specific T cells are counteracting the effects of Tregs, by again creating a better environment for T cell priming.

14.6 CONCLUSIONS

It is clear that HBcAg has unique properties that often result in an unexpected behavior. HBcAg is highly immunogenic in humans in inducing both antibodies and T cells in the context of the replicating virus. Less is known about the ability of exploiting this immunogenicity in human vaccines. If animal models are predictive of this, then HBcAg should be a potent adjuvant in both protein-based and genetic vaccines. Care should be taken to select the most immunogenic version of HBcAg in combination with a particular epitope or antigen since this may vary greatly.

ACKNOWLEDGMENTS

Mr H.D. Sällberg is thankfully acknowledged for invaluable help and inspiration throughout the preparation of this chapter.

REFERENCES

1. Vyas, G.N. and I.M. Roberts, Radioimmunoassay of hepatitis B core antigen and antibody with autologous reagents. *Vox Sang*, 1977. **33**(6): 369–372.
2. Cohen, B.J., The IgM antibody responses to the core antigen of hepatitis B virus. *J Med Virol*, 1978. **3**(2): 141–149.
3. Gerlich, W.H. and W. Luer, Selective detection of IgM-antibody against core antigen of the hepatitis B virus by a modified enzyme immune assay. *J Med Virol*, 1979. **4**(3): 227–238.
4. Milich, D.R. and A. McLachlan, The nucleocapsid of hepatitis B virus is both a T-cell-independent and a T-cell-dependent antigen. *Science*, 1986. **234**(4782): 1398–1401.
5. Milich, D.R. et al., Role of B cells in antigen presentation of the hepatitis B core. *Proc Natl Acad Sci U S A*, 1997. **94**(26): 14648–14653.
6. Cao, T. et al., Hepatitis B virus core antigen binds and activates naive human B cells in vivo: Studies with a human PBL-NOD/SCID mouse model. *J Virol*, 2001. **75**(14): 6359–6366.
7. Lazdina, U. et al., Molecular basis for the interaction of the hepatitis B virus core antigen with the surface immunoglobulin receptor on naive B cells. *J Virol*, 2001. **75**(14): 6367–6374.
8. Watts, N.R. et al., Non-canonical binding of an antibody resembling a naive B cell receptor immunoglobulin to hepatitis B virus capsids. *J Mol Biol*, 2008. **379**(5): 1119–1129.
9. Lee, B.L. et al., Interaction of the hepatitis B core antigen and the innate immune system. *J Immunol*, 2009. **182**(11): 6670–6681.
10. Porterfield, J.Z. et al., Full-length hepatitis B virus core protein packages viral and heterologous RNA with similarly high levels of cooperativity. *J Virol*, 2010. **84**(14): 7174–7184.
11. Ferrari, C. et al., Cellular immune response to hepatitis B virus-encoded antigens in acute and chronic hepatitis B virus infection. *J Immunol*, 1990. **145**(10): 3442–3449.
12. Bertoletti, A. et al., HLA class I-restricted human cytotoxic T cells recognize endogenously synthesized hepatitis B virus nucleocapsid antigen. *Proc Natl Acad Sci U S A*, 1991. **88**(23): 10445–10449.
13. Ferrari, C. et al., Identification of immunodominant T cell epitopes of the hepatitis B virus nucleocapsid antigen. *J Clin Invest*, 1991. **88**(1): 214–222.
14. Milich, D.R. et al., Is a function of the secreted hepatitis B e antigen to induce immunologic tolerance in utero? *Proc Natl Acad Sci U S A*, 1990. **87**(17): 6599–6603.
15. Billaud, J.N. et al., Comparative antigenicity and immunogenicity of hepadnavirus core proteins. *J Virol*, 2005. **79**(21): 13641–13655.
16. Billaud, J.N. et al., Combinatorial approach to hepadnavirus-like particle vaccine design. *J Virol*, 2005. **79**(21): 13656–13666.
17. Billaud, J.N. et al., Advantages to the use of rodent hepadnavirus core proteins as vaccine platforms. *Vaccine*, 2007. **25**(9): 1593–1606.
18. Malik, I.R. et al., A bi-functional hepatitis B virus core antigen (HBcAg) chimera activates HBcAg-specific T cells and preS1-specific antibodies. *Scand J Infect Dis*, 2012. **44**(1): 55–59.
19. Pushko, P., P. Pumpens, and E. Grens, Development of virus-like particle technology from small highly symmetric to large complex virus-like particle structures. *Intervirology*, 2013. **56**(3): 141–165.
20. Sallberg, M. et al., A malaria vaccine candidate based on a hepatitis B virus core platform. *Intervirology*, 2002. **45**(4–6): 350–361.
21. Schodel, F. et al., Immunity to malaria elicited by hybrid hepatitis B virus core particles carrying circumsporozoite protein epitopes. *J Exp Med*, 1994. **180**(3): 1037–1046.
22. Kratz, P.A., B. Bottcher, and M. Nassal, Native display of complete foreign protein domains on the surface of hepatitis B virus capsids. *Proc Natl Acad Sci U S A*, 1999. **96**(5): 1915–1920.
23. Nardin, E.H. et al., Phase I testing of a malaria vaccine composed of hepatitis B virus core particles expressing *Plasmodium falciparum* circumsporozoite epitopes. *Infect Immun*, 2004. **72**(11): 6519–6527.
24. Oliveira, G.A. et al., Safety and enhanced immunogenicity of a hepatitis B core particle *Plasmodium falciparum* malaria vaccine formulated in adjuvant Montanide ISA 720 in a phase I trial. *Infect Immun*, 2005. **73**(6): 3587–3597.
25. Walther, M. et al., Safety, immunogenicity and efficacy of a pre-erythrocytic malaria candidate vaccine, ICC-1132 formulated in Seppic ISA 720. *Vaccine*, 2005. **23**(7): 857–864.
26. Langermans, J.A. et al., Effect of adjuvant on reactogenicity and long-term immunogenicity of the malaria Vaccine ICC-1132 in macaques. *Vaccine*, 2005. **23**(41): 4935–4943.
27. Kuhrober, A. et al., DNA immunization induces antibody and cytotoxic T cell responses to hepatitis B core antigen in H-2b mice. *J Immunol*, 1996. **156**(10): 3687–3695.
28. Kuhrober, A. et al., DNA vaccination with plasmids encoding the intracellular (HBcAg) or secreted (HBeAg) form of the core protein of hepatitis B virus primes T cell responses to two overlapping Kb- and Kd-restricted epitopes. *Int Immunol*, 1997. **9**(8): 1203–1212.
29. Sallberg, M. et al., Characterization of humoral and CD4+ cellular responses after genetic immunization with retroviral vectors expressing different forms of the hepatitis B virus core and e antigens. *J Virol*, 1997. **71**(7): 5295–5303.
30. Townsend, K. et al., Characterization of CD8+ cytotoxic T-lymphocyte responses after genetic immunization with retrovirus vectors expressing different forms of the hepatitis B virus core and e antigens. *J Virol*, 1997. **71**(5): 3365–3374.
31. Sallberg, M. et al., Genetic immunization of chimpanzees chronically infected with the hepatitis B virus, using a recombinant retroviral vector encoding the hepatitis B virus core antigen. *Hum Gene Ther*, 1998. **9**(12): 1719–1729.

32. Nystrom, J. et al., Improving on the ability of endogenous hepatitis B core antigen to prime cytotoxic T lymphocytes. *J Infect Dis*, 2010. **201**(12): 1867–1879.

33. Ahlen, G. et al., In vivo electroporation enhances the immunogenicity of hepatitis C virus nonstructural 3/4A DNA by increased local DNA uptake, protein expression, inflammation, and infiltration of CD3+ cells. *J Immunol*, 2007. **179**(7): 4741–4753.

34. Boukhebza, H. et al., Comparative analysis of immunization schedules using a novel adenovirus-based immunotherapeutic targeting hepatitis B in naive and tolerant mouse models. *Vaccine*, 2014. **32**(26): 3256–3263.

35. Mathiesen, I., Electropermeabilization of skeletal muscle enhances gene transfer in vivo. *Gene Ther*, 1999. **6**(4): 508–514.

36. Lazdina, U. et al., Priming of cytotoxic T cell responses to exogenous hepatitis B virus core antigen is B cell dependent. *J Gen Virol*, 2003. **84**(Pt 1): 139–146.

37. Ahlen, G. et al., Containing "The Great Houdini" of viruses: Combining direct acting antivirals with the host immune response for the treatment of chronic hepatitis C. *Drug Resist Updat*, 2013. **16**(3–5): 60–67.

38. Weiland, O. et al., Therapeutic DNA vaccination using in vivo electroporation followed by standard of care therapy in patients with genotype 1 chronic hepatitis C. *Mol Ther*, 2013. **21**(9): 1796–1805.

39. Chen, A. et al., Heterologous T cells can help restore function in dysfunctional hepatitis C virus nonstructural 3/4A-specific T cells during therapeutic vaccination. *J Immunol*, 2011. **186**(9): 5107–5118.

15 Bacteriophage Virus-Like Particles as a Platform for Vaccine Discovery

Bryce Chackerian and David S. Peabody

CONTENTS

15.1 VACCINES BASED ON VIRUS-LIKE PARTICLES

Vaccines are among the most successful and cost-effective public health interventions ever devised. Up until the 1980s, most viral vaccines were based on attenuated or inactivated pathogens. While these traditional vaccine technologies enabled the production of vaccines for many diseases, both approaches have limitations that are significant barriers to vaccine development and implementation in the current regulatory climate. Attenuated vaccines have the potential to revert to more virulent forms and are also contraindicated in immunodeficient individuals. The process of pathogen inactivation, particularly when using chemical agents, may alter the structure of the virus, affecting the quality of the immune response, as exemplified by the inactivated RSV vaccine candidate introduced in the 1960s that made vaccinated individuals more susceptible to disease-related pathology [1]. The advent of recombinant techniques allowed the preparation of subunit vaccines that consist of isolated antigens. Subunit vaccines generally have good safety profiles, but vaccines based on individual proteins or peptides are far less effective than whole virus preparations. In particular, subunit vaccines generally do not have the highly multivalent, repetitive structure that is characteristic of highly immunogenic antigens. Thus, subunit vaccines are typically poorly immunogenic; they require more frequent and larger doses of antigen in combination with potent adjuvants. In addition, the immune responses elicited by subunit vaccines are typically short lived, meaning that booster immunizations are often required in order to maintain optimal protection.

VLPs combine many of the best features of whole virus and subunit-based vaccines. VLPs structurally and antigenically resemble the viruses from which they are derived, meaning that they can elicit strong and specific antiviral cellular and humoral immune responses. Like subunit vaccines, VLPs can be produced by recombinant technologies, using expression systems (such as bacteria, yeast, or insect cells) that can generate large amounts of recombinant protein without relying on the ability of the parental virus to replicate. Because of their unique particulate structure, VLPs can be purified easily by various standard techniques. VLPs lack viral genomic nucleic acid, cannot replicate, and are therefore intrinsically safer than attenuated virus. VLPs make excellent vaccines against the virus from which they were derived. Hepatitis B virus (HBV) and human papillomavirus (HPV) vaccines are two examples of clinically approved VLP-based vaccines. Both of these vaccines safely and consistently induce high-titer, durable antibody responses in humans [2,3]. VLP-based vaccines for many other human viruses are currently in clinical and preclinical development [4].

VLPs make effective vaccines because their particulate nature and multivalent structure provoke strong immune responses. These structural features are particularly important in enhancing interactions with B cells, allowing VLPs to elicit strong antibody responses. Antibody production is initiated by interactions between antigen and its cognate B cell receptor (BCR) on the surface of naïve B cells. The magnitude of B cell responsiveness to antigenic stimulation can vary dramatically depending on the nature of the antigen [5]. Antigens that have highly dense, multivalent structures, such as VLPs, can activate B cells and induce antibody responses at much lower concentrations than monomeric antigens and without the requirement for exogenous adjuvants [6–9]. Seminal studies by Howard Dintzis, Rolf Zinkernagel, and Martin Bachmann (and colleagues) demonstrated that antigens that contain repeated epitopes on their surfaces with a spacing of 50–100 Å, such as most virus particles, optimally induce B cell responses [10–12]. These highly multivalent antigens provoke extensive cross-linking of the BCR leading to the formation of stable lipid raft microdomains that are associated with enhanced signaling to the B cell [13]. This signaling stimulates B cell proliferation, migration, and upregulation of the expression of molecules that permit subsequent interactions with T helper cells [9], the generation of memory B cells, and the long-lived plasma cells that produce high-titer antibody responses for years to decades after vaccination [14].

As mentioned earlier, most recombinant subunit vaccines induce low-titer antibody responses after a single vaccination. At least one (and more often two) booster immunizations are usually required to achieve maximum antibody titers and optimal efficacy. However, VLP-based vaccines may be an exception to this requirement. An NCI-sponsored trial of the HPV VLP vaccine Cervarix in Costa Rica has shown that two doses of Cervarix are as protective as three doses [15]. In a more recent study, the same authors showed that even a single immunization with Cervarix provides 100% protection from persistent HPV16/18 infection over a 4-year period [16]. Although one dose of Cervarix elicits antibody titers that are lower than two or three doses, it is notable that a single dose is sufficient to elicit long-lasting, stable antibody titers, with no reduction in antibody titers during a 4-year follow-up period.

15.2 USING RECOMBINANT VLPs TO INDUCE ANTIBODY RESPONSES AGAINST HETEROLOGOUS ANTIGENS

VLPs can be used to produce vaccines against the viruses from which they are derived, but they also show great promise for presentation of epitopes from other sources, even when those targets are normally poorly immunogenic. Thus, the very features that make VLPs such effective stand-alone vaccines have been exploited to develop vaccines in which VLPs are used as platforms to display heterologous target antigens in a multivalent format that virtually guarantees strong immunogenicity. Linking target antigens, either genetically or chemically, to the surfaces of VLPs causes them to be displayed at high density. This high-density display, in turn, dramatically enhances the ability of linked antigens to induce antibody responses. An impressive accumulation of data demonstrates the effectiveness of VLP presentation as a method for boosting antibody responses to diverse molecules, including epitopes derived from pathogens, chemical agents, and even self-antigens [17,18]. A number of VLP-based vaccines have

shown efficacy in animal models and have progressed to human clinical trials. These include, but are not limited to, VLP-based vaccines for malaria, allergy, and smoking addiction that target circumsporozoite protein [19], dust mite allergen [20], and nicotine [21], respectively.

Highlighting the potent immunogenicity of VLP display, a number of groups, including our own, have shown that VLP display can be used to overcome the mechanisms of B cell tolerance to induce antibody responses against self-antigens. Self-antigens displayed on VLPs are inherently immunogenic at low doses and without exogenous adjuvants. Using our own data as an example, we showed that display of an epitope derived from a self-antigen (TNF-alpha) on VLPs elicited IgG titers that were 1000 times higher than when the epitope was simply linked to a foreign T helper epitope [22]. Moreover, when displayed on the VLP, TNF-alpha was as immunogenic as a foreign antigen presented in the same context. The magnitude of the anti-self IgG responses correlated with the density at which the self-antigen was displayed on the VLP surface, with maximum antibody responses induced when antigens were displayed with a spacing of 50–100 Å [10,11]. These results indicate that B cell recognition of *foreign-like* multivalent structural elements overwhelms the mechanisms that normally maintain B cell tolerance. VLP display of self-antigens has been successfully used to target molecules that are involved in the pathogenesis of a variety of chronic diseases, including Alzheimer's, hypertension, and certain cancers. Many of these vaccines have shown clinical efficacy in animal models, and several have been tested in human clinical trials [17].

15.3 EPITOPE-BASED VACCINES

The conventional view is that vaccines should elicit broad polyclonal responses against pathogens rather than targeting individual neutralizing epitopes. In part, this view was due to the poor immunogenicity of peptide-based vaccines. Until the recent development of platform-based technologies, it was nearly impossible to elicit peptide-specific responses of significant magnitude and longevity to have prophylactic effects. In addition, most pathogens have developed strategies to evade responses by presenting epitopes to the immune system that can readily undergo antigen variation, while hiding highly conserved sites that are essential for protein function. Until recently, it was unclear whether it was even possible to target these conserved domains, but a number of groups have used sensitive techniques to isolate rare human monoclonal antibodies that target conserved epitopes and have broadly neutralizing activity. The existence of these broadly neutralizing antibodies against HIV, influenza, hepatitis C, and other pathogens [23,24] indicate that it may be possible to identify immunogens capable of eliciting similar antibody responses. Unfortunately, there has been little success in translating the knowledge of these epitopes into useful vaccines, primarily because these epitopes are poorly immunogenic in their native context. Nevertheless, presentation of such epitopes in a sufficiently immunogenic form could yield effective and broadly protective vaccines.

15.4 BACTERIOPHAGE VLPs AS A VACCINE PLATFORM

VLPs derived from diverse virus types can serve as effective platforms for antigen display. Our laboratories have focused on utilizing VLPs derived from a family of related RNA bacteriophages in the Leviviridae family, including MS2, PP7, and Qβ. These are small icosahedral viruses with a short (<4 kb) single-stranded (+)-sense RNA genome encoding four proteins, coat, replicase, maturase, and lysis. The RNA phage virion is assembled from 180 copies of coat protein (with T = 3 icosahedral symmetry), one molecule of the maturation protein, and one copy of the RNA genome. Capsid assembly apparently initiates when coat protein associates with its specific recognition target (or *pac* site) in the viral genome, an RNA hairpin just 3′ of the coat protein coding sequence. The same RNA-binding event represses translation of the viral replicase cistron, and this hairpin is frequently referred to as the replicase cistron's translational operator.

The infectious phage is not suited to peptide display, because its complexly folded and structurally dynamic RNA genome is intolerant of arbitrary genetic manipulations. However, coat protein expressed from a plasmid readily self-assembles into a noninfectious VLP. For example, MS2 VLPs can assemble spontaneously when coat protein is expressed from a plasmid in *Escherichia coli*. MS2 VLPs can be expressed from plasmids in *E. coli* as about half of total soluble protein and are rapidly purified in a single chromatographic step [25]. The existence of detailed structural information facilitates genetic engineering of the coat protein and, as we detail later, underlies the suitability of MS2 VLPs as platform for the display of heterologous antigens.

15.5 ENGINEERING THE MS2 COAT PROTEIN FOR ANTIGENIC DISPLAY

Chimeric VLPs can be constructed by genetic insertion of target epitopes into viral structural proteins. In order for a site of insertion to be useful for antibody induction, it must be present on the surface of the VLP, and the insertion must not interfere with protein folding and VLP assembly. In many cases, proper folding and presentation can be accomplished by replacing exposed immunodominant viral epitopes (these are predominantly loop structures) with the target epitope. Because loop structures are frequently where relevant epitopes are found, this is often a natural location for peptide display. Unfortunately, the effects of peptide insertions into viral structural proteins are notoriously difficult to predict and all too often result in protein folding failures. Peptide length, hydrophobicity, charge, and structure can all influence success rates. As a consequence, the generation of chimeric VLPs in most systems described to date is a largely empirical process of trial and error. However, we have engineered the coat proteins of MS2 and PP7 so that they broadly tolerate peptide insertions.

The structure of the MS2 coat protein dimer is shown in Figure 15.1. One feature that is prominently displayed on

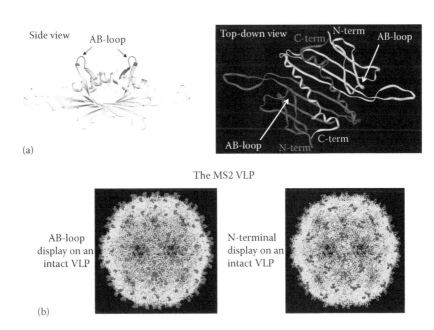

The MS2 coat protein dimer

Side view   AB-loop

(a)

Top-down view   C-term   N-term   AB-loop

AB-loop   N-term   C-term

The MS2 VLP

AB-loop display on an intact VLP

N-terminal display on an intact VLP

(b)

FIGURE 15.1 The structure of the MS2 coat protein. (a) Two views of the MS2 coat protein dimer. The left panel shows an edge-on view of coat protein. The two polypeptide chains are colored blue and green, with the three amino acids of each of the AB-loops shown in red. The panel on the right shows a top-down view of the dimer. The locations of the AB-loop, N-terminus, and C-terminus are indicated. Note the proximity of N-terminus and C-terminus. (b) The structure of the VLP. In the left panel, the location of the AB-loop on the intact VLP is shown in purple. The right panel shows the location of the N-terminus (in red). Note that the AB-loop is dispersed evenly on the surface of the particle, whereas the N-termini are clustered at the threefold axis of symmetry. Display of peptides at either site results in repetitive, multivalent display on the surface of the MS2 VLPs.

the surface of the MS2 VLP is the so-called AB-loop. The AB-loop is a 3-residue β-turn connecting coat protein's A and B β-strands. The surface accessibility and regular geometric spacing of the AB-loop in the MS2 VLP make it an attractive site for the display of foreign peptides. Peptides inserted here are predicted to be highly accessible and, because they are tethered at both ends, conformationally constrained. However, prior efforts to produce active coat proteins with AB-loop insertions have met with mixed success [26]. In some rare cases, insertions were tolerated, but when they were not, the protein failed to fold correctly and either aggregated in inclusion bodies or was proteolytically degraded. Fortunately, we have found a straightforward solution to this problem.

Coat protein is a symmetric dimer of identical polypeptide chains with the N-terminus of one monomer lying in close physical proximity to the C-terminus of the other monomer. Fusing two copies of coat protein into one long reading frame results in a functional protein, in which the coat protein dimer's two halves are synthesized as a single polypeptide chain. This so-called single-chain dimer has all the functions of normal coat protein; it folds correctly, it represses translation by specifically binding the translational operator of the replicase cistron, and it assembles normally into a VLP. The single-chain dimer is substantially more thermodynamically stable than its parent; it is considerably more resistant to thermal and chemical denaturation and, importantly, is dramatically more tolerant of various mutational perturbations [27,28]. For example, we inserted random 6, 8, and 10 amino acid sequences into the AB-loop of the MS2 coat protein monomer and the downstream copy of the MS2 coat protein single-chain dimer and then used a high-throughput assay based on the translational repressor activity of the correctly folded coat protein dimer to measure the percentage of insertions that were compatible with protein folding. While only a small minority (<2%) of six amino acid insertions into the AB-loop of the coat protein monomer were compatible with dimer formation, random 6-mer sequence insertions into the downstream copy of coat protein in the single-chain dimer were largely (96%) tolerated (Figure 15.2). Similarly, 94% of 8-mer and 92% of 10-mer insertions into the single-chain dimer gave functional translational repressors. A significant percentage of the small number of repressor-defective clones in each library was not the results of failure to tolerate peptide insertions, but had other defects, typically oligonucleotide synthesis errors, indicating that the peptide insertion tolerance was actually higher. Importantly, the vast majority (~90%) of repressor clones that were picked from the 6-, 8-, and 10-mer libraries also formed VLPs. A single-chain dimer version of the PP7 coat protein was similarly tolerant of 6–10 amino acid insertions into its AB-loop [29].

In addition to the AB-loop, both the N-terminus and C-terminus of coat protein are displayed on the surface of the VLP, near the threefold axis of symmetry. We have also begun an effort to systematically develop the coat protein N-terminus as an alternate display site, and we have constructed VLPs

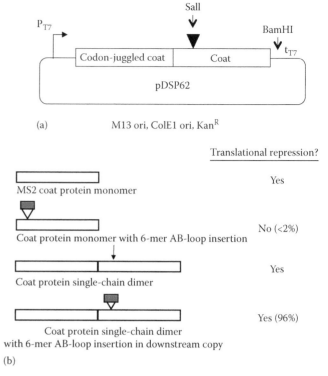

FIGURE 15.2 Display of heterologous peptides on the MS2 coat protein is dramatically enhanced by the use of a coat protein single-chain dimer. (a) The plasmid pDSP62. This plasmid contains a single-chain dimer sequence expressed under control of the T7 transcription initiation and termination signals. This plasmid also contains a ColE1-type replication origin and expresses resistance to kanamycin. The pDSP62 vector was produced to facilitate the cloning of random peptide sequence insertions in the AB-loop sequence of the downstream half of the coat protein single-chain dimer by simply replacing the Sal I–Bam HI fragment. VLPs synthesized from pDSP62 display foreign peptides at 90 per particle. We also introduced a M13 origin of replication into pDSP62 for production of single-strand phagemid DNAs. Diverse libraries of VLPs containing random peptide insertions can be constructed by producing single-stranded phagemid DNA and then annealing a mutagenic primer to the single-stranded circular template, which is then converted to a covalently closed double-stranded circle by the action in vitro of DNA polymerase and DNA ligase. To ensure that annealing of the mutagenic primer is directed to only one-half of the single-chain dimer, we replaced the upstream half of the single-chain dimer with a synthetic *codon-juggled* sequence containing the maximum possible number of silent mutations. (b) To assess the compatibility of random insertions with coat protein dimerization, we created a library plasmids displaying of random 6–amino acid insertions in the AB-loop of MS2 coat protein and then assessed functionality in a translational repression assay. Use of the single-chain dimer largely rescues the folding defects introduced by inserting random 6-mer sequences into the AB-loop of coat protein.

displaying peptides at the N-terminus of MS2 and PP7 [30]. Similar to our experience with the AB-loop, use of the single-chain dimer dramatically increases the percentage of insertions at the N-terminus that are compatible with VLP assembly. Unlike the AB-loop, where peptides are displayed in a conformationally constrained manner, display at either termini is likely to allow peptides to adopt a greater range of conformations.

The ability to display diverse peptide sequences on the surface of bacteriophage VLPs has allowed us to produce VLP-based vaccines displaying epitopes of interest by two different routes. First, we can simply rationally engineer the display of a specific, previously identified epitope on the surface of VLPs by genetic insertion into the viral coat protein at the AB-loop or the N-terminus of the single-chain dimer. The multivalent presentation of the epitope on the VLP surface renders it potently immunogenic. Second, we can use the MS2 VLP platform to identify vaccines from a vast library of VLPs displaying random or targeted peptides by affinity selection using antibodies [31]. This is a technique that is analogous to filamentous phage display, but unlike phage display, VLPs display peptides in a highly immunogenic context, meaning that we can then use the affinity selected VLPs directly as vaccines. In the next sections, we will describe our use of these complementary techniques to develop novel vaccines.

15.6 RATIONAL DESIGN OF BACTERIOPHAGE VLP VACCINES

As described earlier, we have developed a VLP platform that allows nearly universal display of short peptide sequences. Our experience is that peptides less than 15–20 amino acids are best tolerated, although this has not been systematically tested. These size limitations are a general feature of genetic display of targets on chimeric VLPs; the size of peptide insertions is usually limited to 20–30 amino acids or fewer, although there are some exceptions to this rule [32,33]. These size limitations seemingly complicate vaccine design, particularly when targeting pathogens that undergo antigenic variation. Antigenic variation typically limits the effectiveness of vaccines that target single epitopes—a pathogen could simply evade immune responses by undergoing mutagenesis such that the targeted epitopes is no longer recognized by the induced antibody response. Nevertheless, there is growing awareness that many pathogens harbor highly conserved, broadly neutralizing epitopes. These epitopes have been identified through the isolation of rare monoclonal antibodies that have broad viral neutralizing activity [34]. These sorts of antibodies are typically not induced upon infection or vaccination because the epitopes are poorly immunogenic in their normal contexts. Nevertheless, these cryptic epitopes could form the basis of an effective vaccine if they were rendered sufficiently immunogenic. We have targeted one such cryptic epitope from HPV and have data suggesting that a VLP displaying this epitope may serve as a broadly effective, second-generation HPV vaccine.

The current HPV vaccines (Gardasil and Cervarix) are comprised of VLPs derived from the HPV major capsid protein, L1. Both vaccines are highly immunogenic and strongly protect immunized individuals against infection with the high-risk (carcinogenic) HPV types (16 and 18) included in the vaccines. However, while HPV16 and HPV18 cause approximately 70% of cervical cancer cases [35,36], these vaccines largely do not provide cross-protection against other high-risk HPV types (because neutralizing antibodies against L1 are generally type specific) [37–41]. Thus, vaccinated individuals are at a decreased risk for the development of cancer, but are still vulnerable to infection by the other 12 or so high-risk HPV types that cumulatively cause ~30% of cervical cancers.

As an alternative, we have targeted a highly conserved broadly neutralizing epitope from the HPV minor capsid protein, L2. Although L2 is not required for the formation of HPV VLPs, it plays essential roles in viral entry and assembly. During natural HPV infection, L2 is poorly immunogenic, probably reflecting the fact that L2 is only transiently exposed on the surface of the virus particle during the infectious process [42]. Immunization with L2 can raise antibodies that have the ability to broadly neutralize diverse HPV types [43–45], but recombinant L2 protein is poorly immunogenic, and immunization with recombinant L2 elicits low-titer neutralizing antibodies. To present L2 in a more immunogenic form, we introduced a 17–amino acid peptide from a well-conserved region within the N-terminal domain of L2 into the AB-loop of the PP7 VLP and the N-terminus of MS2 coat protein [29,30,46–48]. Animals immunized with either L2-VLP mounted high-titer anti-L2 responses and were protected against infection by HPV pseudovirions with highly divergent serotypes. Although both VLP types provoked broadly neutralizing antibodies, the MS2 N-terminal fusions elicited significantly higher cross-neutralizing titers. In this case, we think that the greater conformational freedom of the peptide displayed at the N-terminus allows it to elicit a broader spectrum of antibodies.

Our lead vaccine candidate is a VLP in which HPV16 L2 amino acids 17–31 are displayed at the N-terminus of the MS2 coat protein (16L2-MS2). Immunization with 16L2-MS2 VLPs induces antibodies that strongly neutralize all eleven of the HPV types that we have tested in an in vivo genital HPV pseudovirus mouse challenge model [30]. L2-VLP vaccines also elicit long-lasting protective antibody responses in mice. In a study in which we followed mice for over 18 months after immunization, we observed only minimal reductions in the anti-L2 titer in these animals. Importantly, when challenged with three HPV PsV types 12–18.5 months after vaccination, the mice were strongly protected from infection [48]. These data support our hypothesis that VLPs can elicit long-lived antibody responses. Strikingly, even a single immunization with low dose of vaccine (2 µg) is sufficient to elicit high-titer antibody responses in mice. Since ~85% of cervical cancer cases occur in the developing world, a vaccine that is effective after a single immunization would have obvious practical advantages.

This is just one example in which we've used the RNA bacteriophage VLP platform for display of a previously identified target peptide. It should be noted that many other VLP platforms have been adapted to display specific peptide epitopes. For example, VLPs derived from HPV, AAV, and TMV have also been used as scaffolds to display epitopes derived from HPV L2 [49–51]. Our bias is that RNA bacteriophage platform has some advantages over other VLP platform technologies for these rational design applications. Bacteriophage VLPs are composed of a single protein that can be expressed in bacteria at around half of total cellular protein. Bacteriophage VLPs are easily purified away from bacterial contaminants (including LPS) by standard chromatographic techniques. We can make ~10–20 mg of recombinant VLPs from a liter of culture medium. Bacteriophage VLPs also encapsidate their own endogenous adjuvant (ssRNA), and, as described earlier, the coat protein can be readily modified without interfering with VLP assembly. However, we believe that the major advantage of the MS2 is described in the next section. We can use the platform to create large, complex libraries of VLPs displaying random peptide sequences and then to use affinity selection to identify vaccines from that population.

15.7 VLP PLATFORM FOR AFFINITY SELECTION OF VACCINES

One of the major hurdles in any effort to develop new vaccines able to elicit desired antibody responses is the difficulty in both identifying the relevant target epitopes and then in presenting those epitopes in a highly immunogenic context. In the past, VLP technology has not been adapted for use in epitope identification because recombinant VLPs are not well suited for the construction of diverse peptide libraries. As described earlier, peptide insertions into viral structural proteins often result in protein folding failures that interfere with VLP assembly. In contrast, phage display using filamentous phages, such as M13, is a robust technology for epitope identification. The screening of M13 phage libraries using mAbs and polyclonal sera has facilitated the identification of numerous specific epitopes and mimotopes [52]. Nevertheless, a major disadvantage of filamentous phage display is that, in general, filamentous phages are poor immunogens because they do not readily present foreign peptides at high densities required for potent immunogenicity [53]. For example, the most common display technique utilizes fusion to pIII, a protein found in only a few copies at one end of the M13 particle. As a consequence, the use of M13-identified peptides as vaccines requires that they be produced synthetically and then linked to a more immunogenic carrier protein. The epitope is therefore presented in a structural context unrelated to the one in which it was selected. Under these conditions, peptides seldom maintain the target affinities they showed when present on the phage particle, and they frequently lose the ability to induce antibodies with activities mimicking those of the selecting antibody [54]. Therefore, an ability to identify epitopes on the same structural platform to be used later in their presentation as a vaccine would have important advantages for

vaccine discovery. The MS2 VLP provides such an integrated platform, suitable both for epitope identification and for immunogenic presentation of epitopes to the immune system as a vaccine.

The use of the MS2 VLP for affinity selection applications depends on three key features: (1) a surface-exposed site in coat protein that tolerates insertions without disruption of coat protein folding or VLP assembly; (2) the encapsidation of nucleic acid that encodes the coat protein and any guest peptide it displays, allowing recovery of the genetic material postselection; and (3) the ability to create very large diverse libraries of VLPs. To satisfy the first criterion, we engineered a single-chain dimer version of coat protein that is much more stable thermodynamically and dramatically more tolerant of foreign AB-loop insertions, as described earlier. The second requirement is satisfied by the ability of recombinant VLPs to encapsidate the mRNA that encodes coat protein and any guest peptide it carries [29,55]. It has long been assumed that the same RNA hairpin bound by coat protein to repress replicase translation is also responsible for specific genome encapsidation, presumably by nucleating capsid assembly on the RNA. We find, however, that coat protein expressed from a plasmid assembles into VLPs that contain large amounts of coat-specific RNA without the need for the packaging hairpin [55]. Packaging of coat-specific RNA establishes the genotype/phenotype linkage that makes affinity selection possible. The third requirement, library construction, has been facilitated by creation of a plasmid vector that has allowed us to construct highly complex libraries of VLPs displaying random peptides. This plasmid vector, called pDSP62 (Figure 15.2), embodies the following features: (1) a kanamycin resistance cassette, (2) an M13 origin of replication, and (3) a synthetic *codon-juggled* coat gene. The latter two features are critical and described in more detail in the following.

Our first random sequence peptide libraries were constructed by use of polymerase chain reaction (PCR) to introduce random sequences in the coat protein sequence, which was then cloned into the appropriate plasmid vector. But methods that involve ligation are inconvenient to scale up to levels that result in very high library complexities. Therefore, we have made use of an old, but very efficient method for site-directed mutagenesis described by Kunkel [56]. It relies on the ability of plasmids containing an M13 replication origin to produce single-stranded circular DNA in the presence of an M13 helper phage. When grown in a dut-, ung-host, the DNA incorporates some dUTP in place of dTTP. A mismatched oligonucleotide primer is annealed to this DNA template, is elongated using DNA polymerase, and is then ligated to produce closed circular DNA. Subsequent transformation of any ung+ strain results in strongly preferential propagation of the mutant strand. The primer extension mutagenesis reaction can be conducted on relatively large quantities of DNA (e.g., 20 µg), enough to readily generate on the order of 10^{11} individual recombinants by electroporation. To facilitate the application of this method, pDSP62 contains an M13 origin of replication. Since the usual helper phages (e.g., M13KO7) confer resistance to kanamycin, they are unsuited to the propagation of pDSP62. We therefore also constructed a chloramphenicol-resistant M13 helper phage we call M13CM1. Using this system, we have readily produced random sequence for $[NNS]_6$, $[NNS]_7$, $[NNS]_8$, and $[NNS]_{10}$ peptide libraries (i.e., displaying random 6-, 7-, 8-, and 10-mer peptides) with complexities in excess of 10^{10} individual members. (Note that the nucleotide sequence NNS [N = any base, S = G or C] allows us to encode all 20 amino acids and minimizes the likelihood of stop codons.) Our method relies on the specific insertion of peptides in one of the two AB-loops of the coat protein single-chain dimer. Library construction using the primer extension method described earlier requires the ability to direct primer annealing specifically to only one-half of the single-chain dimer sequence. For this reason, pDSP62 the upstream copy of the coat sequence consists of a *codon-juggled* version that introduces the maximum possible number of silent nucleotide substitutions, making it possible to direct primer annealing specifically to the desired half of the single-chain dimer, while preserving the wild-type amino acid sequence.

Thus, these features allow us to generate complex libraries of VLPs displaying random peptide sequences at the level of plasmid DNA [31]. When the DNA is expressed in bacteria, a corresponding library of VLPs is produced, which can then be subjected to affinity selection (shown schematically in Figures 15.3 and 15.4), typically using a monoclonal antibody (mAb). Selected sequences can be recovered by reverse transcription (RT)-PCR, used to regenerate a library of selectant plasmids, and then this process can be repeated iteratively. An added feature of this process is that we can monitor the status of selection by deep sequencing the RT-PCR product after each round of selection. Using bioinformatics techniques can thus obtain a comprehensive view of the peptides obtained through the selection process.

We assume that selection of peptides having the highest affinity for a given mAb will provide the best molecular mimics of the native antigen and that these are the most likely to induce a relevant antibody response. Ideally, one conducts the first round of selection using multivalent display, thus obtaining a relatively complex population including all peptides having some minimal affinity for the target. Reducing the display valency in subsequent rounds increases the stringency of affinity selection. To achieve lower densities, we constructed a version of the coat protein expression cassette that places a nonsense codon on the junction between the two halves of the dimer. This means that ribosomes translating the coat sequence will normally terminate after synthesis of the first half of the single-chain dimer, that is, after synthesis of the wild-type

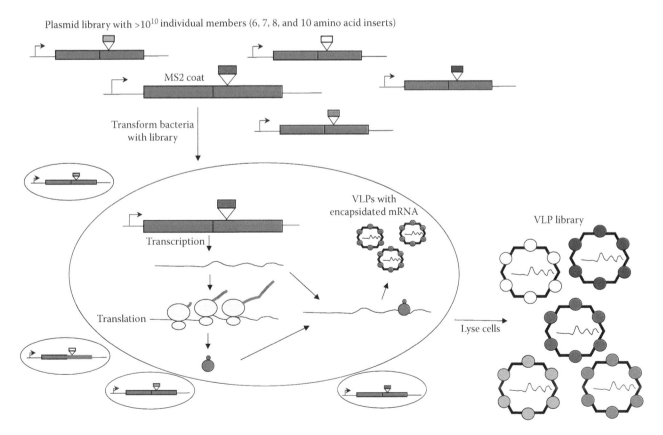

FIGURE 15.3 VLP library construction. Plasmid libraries displaying random 6–, 7–, 8–, and 10–amino acid sequence in the downstream copy of coat protein in the single-chain dimer are introduced into a T7 expression strain of *E. coli* by high-efficiency transformation. Upon induction, cells express recombinant coat protein, which self-assembles into VLPs and encapsidates its encoding mRNA. We have used this technique to create large (with greater than 10^{10} individual members) libraries of VLPs.

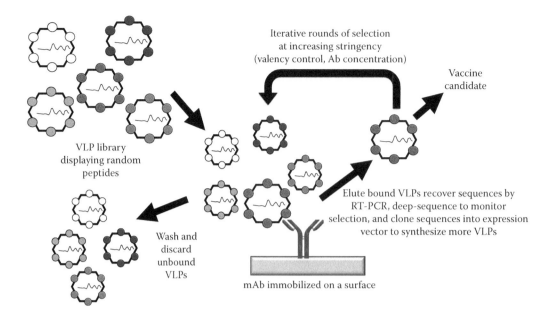

FIGURE 15.4 VLP affinity selection. Affinity selection is achieved by immobilizing a monoclonal antibody and allowing it to incubate with a VLP library. Typically, we combine libraries displaying peptides of varying lengths because different monoclonal antibodies often have preferences for specific peptide lengths. After washing, antibody-bound VLPs are eluted by an acid wash, and then the encapsidated nucleic acid is recovered by reverse transcription followed by PCR. At this point, we have the option of monitoring the status of the selection by deep sequencing the amplified DNA. After PCR, selected sequences are cloned back into the expression vector and used to generate a library of selectant VLPs. The selection can then be repeated iteratively. To increase the stringency of selection, we can decrease the valency of display by utilizing a version of the coat protein expression plasmid that contains a suppressible stop codon at the junction between the two halves of the dimer. Use of this plasmid allows us to decrease peptide valency to about three copies per VLP. We can also decrease the amount of mAb used during the affinity selection to allow for enhanced competition between VLPs for antibody binding. Selected VLPs can be used directly as vaccine candidates.

form of coat protein. It, of course, will proceed to assemble into VLPs. In the presence of the appropriate nonsense suppressor tRNA, however, some percentage of ribosomes will translate through the stop codon to produce the single-chain dimer, which coassembles with the wild-type protein to form mosaic VLPs. Since the peptide insertion resides only in the downstream half of the sequence, only the single-chain dimer displays it. Therefore, the average display valency of the VLPs is determined by the efficiency of nonsense suppression, which depends in turn on the expression level of the suppressor. We constructed a synthetic alanine-inserting suppressor tRNA and expressed it from a second plasmid. We estimate that VLPs present about three peptides per particle on average with the suppressor we normally use.

The ultimate goal of this system is to affinity select VLPs that can, in turn, elicit potent antibody responses with activities that mimic the selecting mAb. We have previously published the use of the MS2 affinity selection system to identify epitopes for several previously characterized mAbs that recognize linear epitopes (against the FLAG epitope and an epitope from anthrax protective antigen) [31]. When used to immunize mice, the selected VLPs elicited high-titer antibodies that bound to the FLAG peptide and protective antigen [31]. In unpublished data, we have used this approach to identify vaccines for diverse pathogens. Using monoclonal antibodies (mAbs) against

Nipah virus (NiV), the malaria blood stage antigen RH5, and the *Staphylococcus aureus* quorum sensing peptide AIP4 [57], we have used our affinity selection technique to identify VLPs that elicit antibodies with neutralizing activity. For NiV, we performed selections using a neutralizing antibody and identified a VLP displaying a peptide that extensively matched a sequence within the G-protein itself (illustrating the value of the MS2 platform for the mapping of linear epitopes). Serum from mice immunized with a selectant strongly neutralizes a luciferase-producing NiV-G pseudotyped vesicular stomatitis virus. For the blood-stage malaria antigen RH5, we performed selections using an anti-Rh5 mAb with high activity in *P. falciparum* growth inhibition assays (GIAs). These selections resulted in the identification of a peptide with homology to a linear amino acid sequence in the Rh5 protein [58]. The selected VLPs elicited antibodies with extremely high GIA activity. Lastly, we used a mAb that blocks the activity of AIP4, a cyclic thiolactone ring–containing 10–amino acid peptide that mediates *S. aureus* quorum sensing and the consequent expression of a host of *Staph* virulence genes. We identified two VLPs displaying peptides that have no sequence homology to the auto-inducing peptide (AIP) peptide that they mimic [59]. Mice immunized with selected VLPs had decreased ulcer size upon intradermal challenge with an AIP4-expressing *S. aureus* strain. Co-vaccination with the two VLPs in combination was even more protective.

15.8 STRUCTURAL CONSIDERATIONS: IDENTIFICATION OF LINEAR EPITOPES AND MIMOTOPES

Filamentous phage display has often been used to identify linear epitopes, but it has been possible in favorable cases to utilize affinity selection to isolate so-called mimotopes, molecular mimics of epitopes. Sometimes, it has been possible to mimic the structures of complex, conformational epitopes, and occasionally, the peptides thus identified are even able to elicit antibodies that react with the original epitope in its native environment. However, in many cases, these antibodies react well with the immunizing synthetic peptide, but do not recognize the epitope in its native environment. When inserted into the coat protein AB-loop, RNA phage VLPs present peptides in a conformationally constrained manner. Perhaps more importantly, the MS2 VLP platform allows both affinity selection and immunization to be carried out on a single structural framework, without the necessity of transferring a peptide optimized in one structural environment to a different environment on a more immunogenic platform for vaccination. The use of RNA phage VLPs may, therefore, increase the frequency with which mimotopes able to induce a desired response can be identified. Our experience performing selections using the mAb against AIP4 serves as a case in point. These selections resulted in the identification of five VLPs displaying peptide sequences that were, for the most part, unrelated to one another. In addition, none of the selected peptides had any sequence similarity to the AIP4 peptide, suggesting that the selectants were mimotopes of AIP4. As mentioned earlier, two of the selectants elicited antibody responses that blocked AIP4-mediated pathology. However, the other three selectants did not neutralize AIP4 activity. Thus, these selectants likely represent peptides that can bind to the paratope, but do not represent true immunologic epitope mimics.

We suspect that even linear epitopes can exist in a conformationally optimized context. For, example, after two rounds of selection with the anti-RH5 antibody we obtained a virtually homogeneous population of selectants, each of which displayed the 8-mer sequence, SAIKKPVT. This peptide contains a 4–amino acid identity to a sequence near the Rh5 N-terminus (AIKK), suggesting this site represents the epitope recognized by this mAb. Interestingly, deep sequence analysis of round 1 selectants identified a family of selectants that contained an AIKK (or AIKR) tetrapeptide, but with different flanking sequences. Many of them show additional similarities to one another outside this core motif. It is interesting that not only is the core sequence conserved, but also its relative position in the peptide. The AIK(K/R) always occupies amino acids 2–5 (from the N-end). We suspect its position and the amino acids surrounding the core epitope function to present the epitope to the antibody in a favorable conformation for binding. After the second round of selection, the pool of VLPs consisted essentially of a single sequence. The fact that affinity selection so clearly favored one sequence over the other members of the sequence family suggests that amino acid residues outside the AIKK identity may serve to most effectively present the core epitope to the antibody in the context of the coat protein AB-loop. Thus, even linear sequences may be conformationally optimized through the affinity selection process.

15.9 ADDITIONAL APPLICATIONS OF THE VLP PLATFORM

The identification of epitopes of mAbs is relatively straightforward, requiring iterative rounds of biopanning to enrich VLP libraries for binders and typically resulting in a relatively simple population of selectants or selectant families, depending on the mAb used. We have been interested in also using the VLP approach to characterize antibody responses to infection by performing selections using serum. Obviously, a polyclonal serum likely has both pathogen relevant and nonrelevant antibodies present. Identification of epitopes targeted by antibodies in polyclonal serum therefore requires both bioinformatics tools to analyze the data and specific considerations to enrich the selectant population for binders while maintaining enough diversity in the selectant population such that the full repertoire of epitopes recognized is maintained. One approach that we have taken is to generate libraries of VLPs that display overlapping peptides derived from either a single antigen or even a pathogen's entire *peptidome*.

We have used this approach to map the human antibody response to dengue virus (DENV) infection (Figure 15.5). First, we created an antigen fragment VLP library based on the amino acid sequence of the DENV polyprotein. We took advantage of newly available massively parallel microchip-based synthesis methods that are capable of making many thousands of specific oligonucleotides. The oligonucleotides were designed by virtually dividing the DENV proteome into 10–amino acid fragments, each with a 9–amino acid overlap with the adjacent fragment. In other words, the 10-mers scan through the sequence in single–amino acid steps, meaning that theoretically the library contains all possible DENV 10-mer peptides. After silent mutation of any codons rarely found in *E. coli*, the peptide-encoding sequences were flanked with 18-nucleotide sequences that mediate annealing of the primers to the site of insertion in the MS2 coat protein's AB-loop. Since the synthesis-on-chip method typically yields only around 10 pmol of the oligonucleotide mixture, the primers were amplified by PCR and then used to prime DNA synthesis on a single-stranded pDSP62 template in order to antigen fragment plasmid libraries. The relatively small size of the DENV genome, about 10 kb encoding a single polyprotein of about 3400 amino acids, means that scanning the entire genome with 10-mers at 1–amino acid increments (i.e., with 9–amino acid overlaps) required only about 3400 oligonucleotides, a number that was well within the capabilities of the microchip-based synthesis method. From these plasmid libraries, we created a DENV antigen fragment VLP library with greater than 10^9 individual

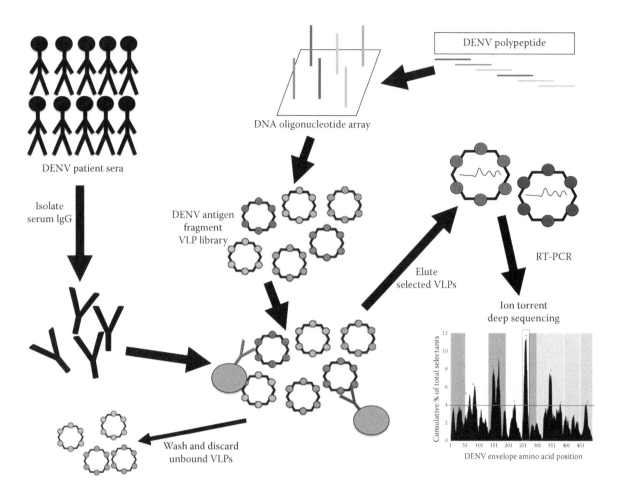

FIGURE 15.5 Affinity selection of antigen fragment libraries using polyclonal sera. We can create libraries of VLPs that display overlapping peptides from a target polypeptide (such as the DENV polypeptide) by virtually dividing the target proteome into 10–amino acid fragments, designing oligonucleotides that encode these peptides, and then utilizing massively parallel microchip-based synthesis methods to generate a library of primers. Antigen fragment VLP libraries can be screened using IgG purified from convalescent patients. Nucleic acid from selected VLPs can be recovered and then deep sequenced. Selected epitopes can be mapped against the target polypeptide (here DENV E protein) in order to create a detailed epitope map for each patient.

transformants, exceeding the needed complexity by many orders of magnitude and meaning that each DENV peptide is represented redundantly in the library. To validate library construction, we deep sequenced the resultant plasmid and VLP libraries to confirm adequate representation of the DENV proteome 10-mers.

To create an *antigenic fingerprint* for DENV, the DENV antigen fragment VLP library was exposed to protein A/G magnetic bead-immobilized IgG from DENV-infected patients. Nucleic acid from selected VLPs was recovered by RT-PCR, and then this product was used as template for Ion Torrent deep sequencing. Ion Torrent sequencing data were analyzed using custom MATLAB™ scripts and a variety of open-source bioinformatics tools. Our analysis consisted of quality control of the sequences included in the analysis, identification of the peptide sequences selected in the biopanning experiment, alignment against the DENV envelope (E) protein, and calculation of the enrichment of each sequence compared to the initial starting library. Epitopes within the DENV E protein recognized by patients were then identified by the presence of peaks when the normalized fold enrichment was plotted against the DENV E

amino acid position to create a detailed epitope map for each patient (Figure 15.4). Epitopes of interest, specifically those shared among a majority of patients, were then located on the quaternary structure of the DENV E protein and examined for variation among the four serotypes of DENV to identify potential vaccine targets. For example, the peak designated by the pink arrow in Figure 15.4 is highly conserved among DENV serotypes and is displayed on the surface of the E protein.

Of course, a limitation of antigen fragment libraries is that they are restricted to the presentation of linear epitopes. A certain proportion of an antigen's neutralizing epitopes are usually nonlinear, and because they may involve the conjunction in space of elements distant from one another in primary structure, they will certainly not be encountered in a 10-mer antigen-fragment library. As an alternative, it may be possible to screen random-sequence libraries using polyclonal sera. This technique would confirm the importance of many of the linear epitopes found in the antigen fragment library and provide additional information about the specific sequence requirements of those epitopes and may also be able to identify mimics of discontinuous epitopes.

15.10 FINAL THOUGHTS

Traditionally, vaccines have relied on one strategy—to mimic as closely as possible the immunological consequences of natural infection. This approach usually relies on immunization with a killed or attenuated version of the pathogenic agent, and for many diseases, it has been spectacularly successful. Nevertheless, whole pathogen–based vaccines consistently fail in cases where natural infection itself does not provoke the desired immunity. Infectious agents coevolve with their hosts, and many have elaborated strategies that rely on immunodominant presentation of highly variable epitopes while hiding from view the potentially broadly neutralizing epitopes they cannot afford to change. HIV, for example, eludes the immune system, not by failing to elicit antibodies at all, but rather by constantly mutating its dominant epitopes to stay one step ahead of the immune response and hiding critical invariant epitopes from immune surveillance. Other pathogens, such as respiratory syncytial virus, malaria, *Chlamydia*, and many others, also have mechanisms for dampening immune responses, meaning that long-lasting immunity is rarely sustained after infection is resolved. Thus, in these cases, antibodies to conserved epitopes, although able to effectively neutralize pathogens, are only rarely produced in sufficient amounts to influence the course of disease.

The MS2 VLP display system we describe has a number of advantages for vaccine development. The VLP has a simple chemical composition, is stable, and is easy to produce and purify in large amounts. It allows affinity selection of peptides in a highly immunogenic format, thus integrating the epitope discovery and vaccine functions into a single particle. Its amenability to construction of complex random-sequence and antigen-fragment libraries facilitates its application to affinity selection on both mono- and polyclonal antibodies. By targeting specific epitopes, it can elicit specific high-titer responses against vulnerable conserved domains of pathogens and can also avoid complications associated with nonneutralizing and infection-enhancing antibody responses. We have already presented evidence that it can open a path to vaccines based on linear epitopes as well as mimotopes of protein conformational epitopes. The empirical approaches described here have the potential to substantially accelerate the development of new vaccines against diverse targets.

REFERENCES

1. Anderson, L. J., Dormitzer, P. R., Nokes, D. J., Rappuoli, R., Roca, A., and Graham, B. S. (2013) *Vaccine* **31**(Suppl 2), B209–B215.
2. Poovorawan, Y., Chongsrisawat, V., Theamboonlers, A., Leroux-Roels, G., Kuriyakose, S., Leyssen, M., and Jacquet, J. M. (2011) *J Viral Hepat* **8**, 369–375.
3. Romanowski, B., de Borba, P. C., Naud, P. S., Roteli-Martins, C. M., De Carvalho, N. S., Teixeira, J. C., Aoki, F. et al. (2009) *Lancet* **374**, 1975–1985.
4. Pushko, P., Pumpens, P., and Grens, E. (2013) *Intervirology* **56**, 141–165.
5. Dintzis, H. M., Dintzis, R. Z., and Vogelstein, B. (1976) *Proc Natl Acad Sci U S A* **73**, 3671–3675.
6. Brunswick, M., Finkelman, F. D., Highet, P. F., Inman, J. K., Dintzis, H. M., and Mond, J. J. (1988) *J Immunol* **140**, 3364–3372.
7. Dintzis, R. Z., Middleton, M. H., and Dintzis, H. M. (1985) *J Immunol* **135**, 423–427.
8. Milich, D. R., Chen, M., Schodel, F., Peterson, D. L., Jones, J. E., and Hughes, J. L. (1997) *Proc Natl Acad Sci U S A* **94**, 14648–14653.
9. Chackerian, B., Durfee, M. R., and Schiller, J. T. (2008) *J Immunol* **180**, 5816–5825.
10. Jegerlehner, A., Storni, T., Lipowsky, G., Schmid, M., Pumpens, P., and Bachmann, M. F. (2002) *Eur J Immunol* **32**, 3305–3314.
11. Chackerian, B., Lenz, P., Lowy, D. R., and Schiller, J. T. (2002) *J Immunol* **169**, 6120–6126.
12. Dintzis, R. Z., Vogelstein, B., and Dintzis, H. M. (1982) *Proc Natl Acad Sci U S A* **79**, 884–888.
13. Thyagarajan, R., Arunkumar, N., and Song, W. (2003) *J Immunol* **170**, 6099–6106.
14. Amanna, I. J. and Slifka, M. K. (2010) *Immunol Rev* **236**, 125–138.
15. Kreimer, A. R., Rodriguez, A. C., Hildesheim, A., Herrero, R., Porras, C., Schiffman, M., Gonzalez, P. et al. (2011) *J Natl Cancer Inst* **103**, 1444–1451.
16. Safaeian, M., Porras, C., Pan, Y., Kreimer, A., Schiller, J. T., Gonzalez, P., Lowy, D. R. et al. (2013) *Cancer Prev Res (Phila)* **6**, 1242–1250.
17. Chackerian, B. (2007) *Expert Rev Vaccines* **6**, 381–390.
18. Jennings, G. T. and Bachmann, M. F. (2008) *Biol Chem.* **389**, 521–536.
19. Birkett, A., Lyons, K., Schmidt, A., Boyd, D., Oliveira, G. A., Siddique, A., Nussenzweig, R., Calvo-Calle, J. M., and Nardin, E. (2002) *Infect Immun* **70**, 6860–6870.
20. Senti, G., Johansen, P., Haug, S., Bull, C., Gottschaller, C., Muller, P., Pfister, T., Maurer, P., Bachmann, M. F., Graf, N., and Kundig, T. M. (2009) *Clin Exp Allergy* **39**, 562–570.
21. Cornuz, J., Zwahlen, S., Jungi, W. F., Osterwalder, J., Klingler, K., van Melle, G., Bangala, Y. et al. (2008) *PLoS ONE* **3**, e2547.
22. Chackerian, B., Lowy, D. R., and Schiller, J. T. (2001) *J Clin Invest* **108**, 415–423.
23. Koff, W. C., Burton, D. R., Johnson, P. R., Walker, B. D., King, C. R., Nabel, G. J., Ahmed, R., Bhan, M. K., and Plotkin, S. A. (2013) *Science* **340**, 1232910.
24. Burton, D. R., Poignard, P., Stanfield, R. L., and Wilson, I. A. (2012) *Science* **337**, 183–186.
25. Peabody, D. S. (1990) *J Biol Chem* **265**, 5684–5689.
26. Stockley, P. G. and Mastico, R. A. (2000) *Methods Enzymol* **326**, 551–569.
27. Peabody, D. S. and Chackerian, A. (1999) *J Biol Chem* **274**, 25403–25410.
28. Peabody, D. S. (1997) *Arch Biochem Biophys* **347**, 85–92.
29. Caldeira Jdo, C., Medford, A., Kines, R. C., Lino, C. A., Schiller, J. T., Chackerian, B., and Peabody, D. S. (2010) *Vaccine* **28**, 4384–4393.
30. Tumban, E., Peabody, J., Tyler, M., Peabody, D. S., and Chackerian, B. (2012) *PLoS ONE* **7**, e49751.
31. Chackerian, B., Caldeira Jdo, C., Peabody, J., and Peabody, D. S. (2011) *J Mol Biol* **409**, 225–237.
32. Kratz, P. A., Bottcher, B., and Nassal, M. (1999) *Proc Natl Acad Sci U S A* **96**, 1915–1920.
33. Tissot, A. C., Renhofa, R., Schmitz, N., Cielens, I., Meijerink, E., Ose, V., Jennings, G. T., Saudan, P., Pumpens, P., and Bachmann, M. F. (2010) *PLoS ONE* **5**, e9809.

34. Corti, D. and Lanzavecchia, A. (2013) *Annu Rev Immunol* **31**, 705–742.

35. Burd, E. M. (2003) *Clin Microbiol Rev* **16**, 1–17.

36. Munoz, N., Bosch, F. X., de Sanjose, S., Herrero, R., Castellsague, X., Shah, K. V., Snijders, P. J., and Meijer, C. J. (2003) *N Engl J Med* **348**, 518–527.

37. Villa, L. L., Costa, R. L., Petta, C. A., Andrade, R. P., Paavonen, J., Iversen, O. E., Olsson, S. E. et al. (2006) *Br J Cancer* **95**, 1459–1466.

38. Harper, D. M., Franco, E. L., Wheeler, C., Ferris, D. G., Jenkins, D., Schuind, A., Zahaf, T. et al. (2004) *Lancet* **364**, 1757–1765.

39. Paavonen, J., Jenkins, D., Bosch, F. X., Naud, P., Salmeron, J., Wheeler, C. M., Chow, S. N. et al. (2007) *Lancet* **369**, 2161–2170.

40. Brown, D. R., Kjaer, S. K., Sigurdsson, K., Iversen, O. E., Hernandez-Avila, M., Wheeler, C. M., Perez, G. et al. (2009) *J Infect Dis* **199**, 926–935.

41. Wheeler, C. M., Kjaer, S. K., Sigurdsson, K., Iversen, O. E., Hernandez-Avila, M., Perez, G., Brown, D. R. et al. (2009) *J Infect Dis* **199**, 936–944.

42. Buck, C. B., Cheng, N., Thompson, C. D., Lowy, D. R., Steven, A. C., Schiller, J. T., and Trus, B. L. (2008) *J Virol* **82**, 5190–5197.

43. Alphs, H. H., Gambhira, R., Karanam, B., Roberts, J. N., Jagu, S., Schiller, J. T., Zeng, W., Jackson, D. C., and Roden, R. B. (2008) *Proc Natl Acad Sci U S A* **105**, 5850–5855.

44. Christensen, N. D., Kreider, J. W., Kan, N. C., and DiAngelo, S. L. (1991) *Virology* **181**, 572–579.

45. Gambhira, R., Karanam, B., Jagu, S., Roberts, J. N., Buck, C. B., Bossis, I., Alphs, H., Culp, T., Christensen, N. D., and Roden, R. B. (2007) *J Virol* **81**, 13927–13931.

46. Hunter, Z., Tumban, E., Dziduszko, A., and Chackerian, B. (2011) *Vaccine* **29**, 4584–4592.

47. Tumban, E., Peabody, J., Peabody, D. S., and Chackerian, B. (2011) *PLoS ONE* **6**, e23310.

48. Tumban, E., Peabody, J., Peabody, D. S., and Chackerian, B. (2013) *Vaccine* **31**, 4647–4654.

49. Palmer, K. E., Benko, A., Doucette, S. A., Cameron, T. I., Foster, T., Hanley, K. M., McCormick, A. A., McCulloch, M., Pogue, G. P., Smith, M. L., and Christensen, N. D. (2006) *Vaccine* **24**, 5516–5525.

50. Schellenbacher, C., Roden, R., and Kirnbauer, R. (2009) *J Virol* **83**, 10085–10095.

51. Nieto, K., Weghofer, M., Sehr, P., Ritter, M., Sedlmeier, S., Karanam, B., Seitz, H. et al. (2012) *PLoS ONE* **7**, e39741.

52. Riemer, A. B. and Jensen-Jarolim, E. (2007) *Immunol Lett* **113**, 1–5.

53. Van Houten, N. E. and Scott, J. K. (2005) Phage libraries for developing antibody-targeted diagnostics and vaccines. In: *Phage Display in Biotechnology and Drug Discovery* (Sidhu, S. ed.), Taylor & Francis Group, Boca Raton, FL. pp. 165–254.

54. Irving, M. B., Craig, L., Menendez, A., Gangadhar, B. P., Montero, M., van Houten, N. E., and Scott, J. K. (2010) *Mol Immunol* **47**, 1137–1148.

55. Peabody, D. S., Manifold-Wheeler, B., Medford, A., Jordan, S. K., do Carmo Caldeira, J., and Chackerian, B. (2008) *J Mol Biol* **380**, 252–263.

56. Kunkel, T. A. (1985) *Proc Natl Acad Sci U S A* **82**, 488–492.

57. Park, J., Jagasia, R., Kaufmann, G. F., Mathison, J. C., Ruiz, D. I., Moss, J. A., Meijler, M. M., Ulevitch, R. J., and Janda, K. D. (2007) *Chem Biol* **14**, 1119–1127.

58. O'Rourke, J. P., Daly, S. M., Triplett, K. D., Peabody, D., Chackerian, B., and Hall, P. R. (2014) *PLoS ONE* **9**, e111198.

59. Ord, R. L., Caldeira, J. C., Rodriguez, M., Noe, A., Chackerian, B., Peabody, D. S., Gutierrez, G., and Lobo, C. A. (2014) *Malar J* **18**, 326.

16 Production of Virus-Like Particles in Plants

Pooja Saxena and George P. Lomonossoff

CONTENTS

16.1 INTRODUCTION

In the present-day scenario, where viral nanoparticles (VNPs) are being deployed in diverse fields ranging from material science to medicine, the need to develop versatile expression systems for the efficient production of such nanoparticles is essential. Plants represent an attractive platform for the production of VNPs since they offer several advantages over established bacterial, yeast, insect cell, mammalian cell, and cell-free production systems. These include high biomass, ease of scalability, cost-effectiveness and a low risk of contamination with endotoxins or human pathogens [1–3]. In addition, unlike prokaryotic expression systems, plants are capable of introducing eukaryotic posttranslational modifications such as glycosylation and hence can be used for the production of VNPs that assemble from complex eukaryotic proteins. Further, for VNPs that are being developed as vaccine candidates, their expression in plants offers the possibility of oral administration of the VNP [4].

The earliest examples of VNPs produced in plants were those produced during infection with a naturally occurring plant virus. Because of their ease of propagation, high yield, and stability, plant virus particles were the first VNPs

whose structures were determined to atomic resolution [5,6]. Subsequently, methods were developed for the genetic and/or chemical modification of infectious plant virus particles. This subject has been reviewed extensively in recent years [7–10] and is also discussed in Chapter 17. Therefore, this chapter focuses exclusively on noninfectious virus-like particles (VLPs), from whatever origin, which have been produced in plants. The pipeline of VLP production in plants is discussed, and the diversity of VLPs produced in plants, both in terms of their properties and potential applications, is reviewed.

16.2 PIPELINE FOR PRODUCTION OF VLPs IN PLANTS

Initial attempts of obtaining noninfectious particles from plants involved infection of plants with the virus, isolation of the virus particles, and subsequent inactivation of the viral genome using methods such as chemical inactivation [11,12] and ultraviolet irradiation [13,14]. While this approach generated noninfectious particles, there were a number of concerns involved with their use. First, particles derived from active virus risk retaining residual infectivity thereby raising biocontainment concerns. Second, the presence of genomic material within the virus, albeit inactivated, makes it challenging to get regulatory approval for use of these viral particles in clinical applications. Finally, the presence of genomic material within the viral particle limits the space available for loading with a desired cargo for such purposes as drug delivery. In the light of the limitations of nanoparticles derived from infectious viruses, modern approaches involve a generation of noninfectious VLPs that do not contain the viral genome by the expression of the viral proteins (VPs) necessary for capsid formation in plant hosts.

Protein expression in plants can be achieved using one of two approaches: (1) stable transformation of either the nuclear or plastid genome of the plant and (2) transient transformation of plant tissue. In the first approach, the desired gene is integrated into the plant nuclear or plastid genome making the transgene heritable through succeeding generations. Stable transformation of the plant genome has the advantage that once a transgenic line has been created, transgenic seeds can be stored for a long time and can be used for large-scale production of the transgenic protein with minimal effort. However, the process of the creation of transgenic lines is time-consuming, making this approach unsuitable for rapid screening. Conversely, the second approach of transient transformation allows for screening and protein expression in a matter of days. In this approach, plant tissue is transiently transformed using the soil bacterium *Agrobacterium tumefaciens* harboring the gene for the desired protein, either incorporated into a replication-competent viral genome or flanked by a promoter and terminator, within its tumor-inducing (Ti) plasmid. *A. tumefaciens* has the ability to transfer genetic material from its Ti plasmid to the plant host [15]. Once within the plant cell, this transfer DNA (T-DNA) is transported to the nucleus where it integrates into the host genome. This allows the T-DNA to be transcribed by the plant cell as if it were a part of the normal complement of plant genes. The resulting mRNA is then translated. If a replication-competent viral genome has been integrated, the integrated sequence can also be replicated to increase expression levels (Section 16.2.1). Expression from the T-DNA continues until cell senescence. In most cases, the Ti plasmid is modified to encode, in addition to the gene for the protein of interest, specific sequences from plant viruses to boost expression levels in the transformed tissue. Transformation is carried out by infiltration of leaves with agrobacteria using either positive (syringe infiltration, Figure 16.1) or negative pressure (vacuum infiltration).

While a number of plant species have been exploited for expression of different VLPs, *Nicotiana benthamiana* has long been the preferred host plant since it is well-studied and is particularly amenable to agroinfiltration with negligible damage being caused to the inoculated tissue during the process [16]. In addition, *N. benthamiana* can support the replication of a number of plant viruses that makes it a versatile host for virus-derived expression vectors. Figure 16.1 shows the generic pipeline of VLP production in *N. benthamiana* plants using *Agrobacterium*-mediated transient expression methods. The design of expression vectors, time of harvest, composition of extraction buffers, and choice of methods for purification are tailored specifically for optimal production of each kind of VLP, based on its physical and chemical properties.

16.2.1 Upstream Processes

The first step in the expression of VLPs in plants is the generation of an appropriate plasmid for expression of the protein(s) required for its assembly. These will include the proteins that comprise the capsid or shell of the VLP and any proteins, such as proteases, necessary for the generation of these structural proteins. For stable genetic transformation, the sequence(s) are inserted into an appropriate vector and used for either nuclear or plasmid transformation using standard transformation techniques [17,18]. Plants are then regenerated, self-fertilized, and used to further regenerate true-breeding lines. The main advantage of plastid transformation over nuclear transformation is that plastid-encoded transgenes have a very low risk of escaping into the environment since they are inherited maternally and not transmitted through pollen. Therefore, transplastomic plants raise fewer biocontainment concerns. In addition, each cell contains multiple copies of the plastid genome, leading to increased levels of protein production.

| Preparation of agrobacteria harboring plasmids encoding genes for VLP expression | Infiltration of a 3-week old *N. benthamiana* plant with an agrobacterial suspension | Homogenization of infiltrated plant tissue and filtration to obtain crude plant extract | Further purification using techniques such as density gradient centrifugation |

3–10 days

FIGURE 16.1 A representation of the pipeline of production of VLPs from plants using transient expression methods.

For transient expression of VLPs, there are two methods: the use of replicating plant viral vectors and direct expression via agroinfiltration. A number of plant viruses have been developed into expression vectors [19]. *Full virus* vectors based on cowpea mosaic virus (CPMV) [20] and potato virus X (PVX) [21] as well as *deconstructed* vectors based on tobacco mosaic virus (TMV) [22,23] and bean yellow dwarf virus [24] have been developed and used to produce high levels of proteins in plants by exploiting their ability to replicate and spread in the plant. These days, the viral genomes are usually initially introduced into plants via agroinfiltration. The advantage of replicating viral vectors is that extremely high levels of the target protein can be obtained due to both amplification of the target gene(s) and the ability of the replicating construct to move from cell to cell. However, there are disadvantages to this approach: during replication, the insert can undergo mutation or deletion, particularly if a vector based on an RNA virus is used, and it is difficult to express multiple polypeptides in the same cell due to the phenomenon of virus exclusion.

An alternative to the use of replicating vectors is to rely on the translation of mRNA molecules transcribed from Ti plasmids of agrobacteria to achieve expression after agroinfiltration. In this case, expression is achieved only in the infiltrated regions of a leaf meaning that it is essential to maximize expression levels in these regions. This can be achieved through the deployment of a suppressor of RNA silencing to stabilize the mRNAs thereby prolonging transgene expression [25] and the use of translational enhancers, such as viral 5′ and 3′ untranslated regions (UTRs). The "pEAQ" series of binary vectors for transient expression of heterologous proteins in plants without the need for viral replication makes use of these elements [26–28]. The gene to be expressed is inserted between modified sequences from CPMV RNA-2 or RNA-1 UTRs, to boost and control translational levels, and the vectors also encode P19, the suppressor of silencing from tomato bushy stunt virus (TBSV) [29,30]. One significant advantage of the pEAQ vectors is that they allow the simultaneous expression of multiple polypeptides within the same cell [31]. This becomes essential for expression of VLPs containing multiple polypeptides and has been discussed in Section 16.3.2. Images of some VLPs expressed in plants using pEAQ vectors are shown in Figure 16.2.

16.2.2 Downstream Processes

Purification of VLPs from plants involves homogenization of plant tissue, removal of plant cell debris, and enrichment of the plant extract for expressed VLPs. For maximum yields, it is essential to shear cellulose that comprises the rigid plant cell wall and remove any cellulose fibrils from the plant extract. This may be done by mechanical homogenization (Figure 16.1) or by enzymatic degradation. The use of enzymes for cellulose degradation is preferred in the case of enveloped and other stress-labile VLPs as mechanical shearing may tear them apart.

Further purification of VLPs from plant extracts is routinely done using methods such as ultracentrifugation, tangential flow filtration, and chromatography based on ion exchange, affinity, and size exclusion. Chromatography is often used for VLP production because of its high selectivity, high recovery rate, and ease of scalability. However, traditional chromatographic methods that use resins as matrices are designed for small proteins, and hence their diffusion rate is slow for VLPs. The use of matrices made from membranes instead of beads (membrane chromatography) is being developed for VLP purification since it takes into account their large size and has been successfully developed for downstream purification of adenovirus particles of around 100 nm in diameter [32]. Purification of enveloped VLPs, such as those from human immunodeficiency virus (HIV), presents a challenge as such VLPs are sensitive to osmotic shock. The lipid envelope of such particles may shear under pressure, and hence, the use of traditional chromatography is not desirable. The use of hollow-fiber cartridge membranes has been shown to lead to efficient purification of VLPs of HIV-1 with 95% recovery [33].

16.2.3 Scale-Up of Production

Once optimum conditions for VLP expression in plants have been established, production can be scaled-up from the benchtop to an industrial scale. Bulk production of VLPs from stable transgenic plants can be done by sowing transgenic seeds and cultivating regenerated plants over a large biocontained area. Large-scale transient production of VLPs can also be achieved by deploying industrial fermenters for growth of liters of agrobacterial culture, by adopting the

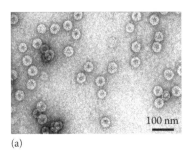
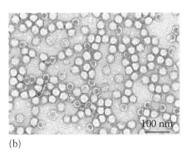
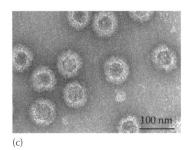

(a)   (b)   (c)

FIGURE 16.2 VLPs produced in plants using pEAQ vectors visualized using transmission electron microscopy of negatively stained samples. (a) Hepatitis B VLPs (35–40 nm). (b) Cowpea mosaic virus VLPs (28–30 nm). (c) Bluetongue virus VLPs (85–90 nm).

alternative approach of vacuum-mediated infiltration instead of the labor-intensive process of syringe infiltration and by utilizing industrial protocols for downstream protein purification. In vacuum-mediated infiltration, negative pressure is used to draw air out of intercellular spaces in leaves, making agroinfiltration easy and quick. By using this system, a whole tray of 20–30 plants (hundreds of grams of plant tissue) can be infiltrated in under 2 min. This ease of scalability makes plants a preferred system for VLP expression.

16.3 PLANT-PRODUCED VLPs

The relative ease of plant production of VLPs depends on the stability and number of kinds of structural proteins constituting the VLP and the requirement of processing for capsid formation. VLPs that are comprised of several copies of a single kind of protein can be produced in plants by high-level expression of the specific protein, relying on its subsequent self-assembly. Production of VLPs that involve self-assembly of more than one kind of protein in plants has the added complexity that the two or more proteins need to be available in the right stoichiometric ratios within the same cell for correct assembly. In addition, capsid formation may require other viral or host factors for the correct processing and folding of the capsid proteins prior to their assembly into icosahedral shells, rods, or other structures within a cell.

16.3.1 SINGLE-PROTEIN VLPs

A number of important diseases are caused by viruses whose capsids are of relatively simple structure, being nonenveloped and consisting of only a single type of capsid protein. Because of this simplicity, much of the early work on plant-based expression of VLPs was carried out on these viruses. This research established the principle that plants could be used to produce VLPs with properties comparable to VLPs produced in other systems, thereby paving the way for subsequent work on more complex structures.

16.3.1.1 Noroviruses

VLPs based on the prototype norovirus, Norwalk virus (NV), were the first simple, single-protein VLPs to be expressed in plants [34]. NV is a member of the Caliciviridae family and causes acute gastroenteritis in humans. The capsids consist of 180 copies of single protein of 58 kDa, which had been shown to be capable of self-assembly into 38 nm diameter empty capsids when expressed in insect cells [35]. The interest in making NV VLPs stems from the fact that there is currently no method of propagating the virus in cell culture, and hence traditional killed or live-attenuated vaccines have not been developed. Mason et al. [34] produced lines of tobacco and potato transgenic for the 58 kDa capsid protein and showed that the plant-produced protein was able to assemble into VLPs. Furthermore, VLPs partially purified from tobacco leaves or potato tubers expressing the protein could stimulate oral immunity in mice. When potato tubers expressing NV VLPs were fed to human volunteers, 19 out of 20 developed

some kind of immune response [36]. Though clearly pointing the way for the future development of a plant-based oral NV VLP vaccine, the levels of expression were relatively low (up to 0.23% of total soluble protein [TSP] in tobacco leaves and up to 0.37% in potato tubers representing 34 μg/g of tuber weight), and the tissues to be consumed (tobacco leaves or raw potato tubers) were not palatable. To increase expression levels and increase palatability, Zhang et al. [37] created a codon-optimized NV coat protein gene that was used to transform both tomato and potato. Dried powder from both tomato fruit and potato tubers was able to stimulate an immune response in mice, with the tomato material being more effective.

The disadvantages of using stable transformation to produce NV VLPs include the relatively long time needed to produce the transgenic lines and the often modest levels of expression obtained. As a result, transient expression has been investigated as a means of producing significant quantities of NV VLPs in plants in a short period. Using a TMV-derived transient expression system, Santi et al. [38] produced assembled, orally immunogenic NV VLPs at levels corresponding 0.8 mg/g in *N. benthamiana* leaves. The TMV system has been used to produce immunogenic VLPs of the related norovirus, Narita 104 virus, though in this case the yield was somewhat less due to the tendency of the construct to cause necrosis in the leaves [39]. A geminivirus-based replicon system was also found to be effective at producing NV VLPs in *N. benthamiana* leaves, though the yield at 0.34 mg/g leaf fresh weight was less than half that obtained with the TMV vector [40].

16.3.1.2 Papillomaviruses

Papillomaviruses (PVs) are nonenveloped viruses about 55 nm in diameter with a double-stranded circular genome of approximately 8 kb encapsidated by the major (L1) and minor (L2) capsid proteins; they form a large family with more than 100 species, the majority of which infect humans. Infection by human PVs (HPVs) can be associated with benign and malignant epithelial proliferations of skin and internal squamous mucosae. The L1 major capsid protein of HPV self-assembles into highly immunogenic VLPs or into capsomers made of five L1 monomers when expressed in different cell types [41,42]. HPV VLPs can exist either as T = 7 icosahedral particles consisting of 72 L1 capsomers that are morphologically identical to the native virions or as T = 1 particles consisting of 12 L1 capsomers [43,44]. Due to the strong immunogenicity of HPV capsomers and VLPs, they are also suitable as immune-enhancer carriers for the presentation of heterologous antigens.

Two types of HPV vaccines are currently marketed, targeting the mucosal HPV-16, HPV-18, and, in one formulation, also HPV-6 and HPV-11 that cause genital warts. By contrast, no vaccines against cutaneous HPVs are available. The aforementioned prophylactic vaccines consist of L1-based VLPs made in insect or yeast cells. Despite their efficacy, they suffer from high production costs, relatively low protection levels, the necessity of being administered via intramuscular injection, and the requirement of a cold chain. As a result, there has been considerable interest in using plant expression systems to overcome some of these problems, in particular regarding production costs. The first successful expression of L1 HPV-11 and HPV-16 was reported in transgenic tobacco and potato plants [45,46], and the protein was shown to assemble into 55 nm particles, similar in appearance to native HPV particles. The plant-produced VLPs were shown to be immunogenic when potato tubers were fed to animals. Despite this success, the expression levels obtained in transgenic plants were quite low, representing less than 1% TSP. Several modifications of the transgenic approach have been used to address this issue. By codon-optimizing the L1 sequence from HPV-16 for plant expression and directing the expressed protein to the chloroplasts of transgenic tobacco, Maclean et al. [47] achieved expression levels representing 11% TSP. Transplastomic tobacco plants have also been investigated as a source of HPV-16 VLPs [48,49]. In each case, the level of L1 expression was about 1.5% TSP, and there was evidence of assembly of the protein into capsomers and VLPs.

As an alternative to stable genetic transformation, transient expression, using both replicating and nonreplicating vectors, has been used to express HPV in plants. Varsani et al. [50] made use of a TMV-based vector to express HPV L1 in *N. benthamiana*. Although evidence of VLP assembly was obtained, the yield was low (20–37 μg/kg of fresh leaf material) probably because a native HPV-16 L1 gene was used; transient expression of a codon-optimized version of HPV-16 L1 using a nonreplicating vector gave up to 17% TSP [47]. Successful expression of a native (non-codon-optimized) version of HPV-8 L1 was achieved using either a replicating TMV-based vector or the nonreplicating pEAQ vector system [51], the latter approach giving 15-fold higher yield. In these studies, a truncated version of HPV8-L1, lacking the C-terminal 22 amino acids, accumulated at higher yields compared to the native full-length protein. Overall, yields ranged from 17 to 240 mg/kg of fresh weight of leaf material, according to the vectors and the protein length. The plant-expressed HPV8-L1 formed capsomers or VLPs of T = 1 or T = 7 symmetry.

In addition to the successful expression of VLPs based on native HPV L1, plant-based expression has also been used to express chimeric VLPs. To obtain HPV-16 VLPs suitable for presenting foreign epitopes, a synthetic, codon-optimized L1 gene was designed to contain unique restriction sites flanking sequences that encode amino acid regions predicted to be exposed on the VLP surface. Two putatively exposed regions were substituted by two versions of a conserved epitope of the ectodomain of the M2 protein of the influenza A virus [52]. The chimeric proteins were expressed using the pEAQ vector system at yields ranging from 30 to 120 mg/kg of fresh weight of leaf material and were all recognized by linear and conformation-specific anti-HPV-16 L1 monoclonal antibodies (MAbs); one of the chimeras carrying the M2e epitope also reacted with an anti-M2 MAb. Electron microscopy showed that the foreign epitopes did not disrupt the formation of T = 1 or T = 7 VLPs or capsomers. Using a similar approach, Pineo et al. [53] were able to express HPV-16 L1-based chimeras, containing cross-protective epitopes from the HPV L2 minor capsid protein in *N. benthamiana* using a variety of replicative and nonreplicative vectors. Consistent with the results of Maclean et al. [47], the highest yield was obtained when the modified L1 protein was directed to chloroplasts. The chimeras assembled differently to the native L1 protein, giving predominantly capsomers or T = 1 VLPs indicating that the length and nature of the L2 epitope can affect VLP assembly. The chimera containing L2 amino acids 108–120 was the most successful candidate vaccine, eliciting anti-L1 and anti-L2 responses in mice.

In addition to HPV, PVs causing veterinary diseases have also been expressed in plants. Kohl et al. [54] expressed a native cottontail rabbit papillomavirus (CRPV) L1 either transgenically in *Nicotiana tabacum* or transiently via a TMV vector in *N. benthamiana*. Though in both instances the L1 protein was detected in concentrated plant extracts, the protein appeared to assemble into capsomers rather than VLPs. Nevertheless, rabbits immunized with the plant-produced L1 were protected against wart development on subsequent challenge with live virus.

Bovine papillomavirus (BPV) is endemic throughout the world and causes a variety of economically important tumorigenic pathologies in horses and cattle. L1 capsomers and higher-order structures produced in yeast and insect cells are highly immunogenic and have been shown to confer protection against the BPV types from which they are derived. Consequently, L1 VLPs and their precursors are regarded as suitable prophylactic vaccines if they could be produced at sufficiently low cost. Love et al. [55] demonstrated that a codon-optimized version of BPV L1 can be expressed transiently to a high level in *N. benthamiana* using the pEAQ vector system. Electron microscopy revealed that the plant-expressed L1 self-assembled into T = 1 VLPs that were ~30 nm in diameter and had yields of 183 mg/kg fresh weight of leaf material of highly pure, structurally stable VLPs was obtained. When rabbits were immunized with purified L1 VLPs, the antisera generated were strongly cross-reactive with BPV VLPs produced in insect cells [55].

16.3.1.3 Hepatitis B Core Antigen

Although hepatitis B virus (HBV) is an enveloped virus (see Section 16.3.3), the core antigen (HBcAg) can form non-enveloped particles in the absence of infection. These core particles have attracted considerable attention as a potential source of vaccines. HBcAg has been shown to be capable of self-assembly into particles when expressed in a number of heterologous systems, including plants. The heterologously expressed particles have antigenic structures identical to those of native HBcAg particles and have attracted attention as potential carriers of foreign antigenic sequences. However, as HBcAg particles are the subject of the accompanying Chapter 11, they will not be discussed further here.

16.3.1.4 Alfalfa Mosaic Virus

The coat protein of the plant virus alfalfa mosaic virus (AlMV) forms particles of different sizes (20–60 nm) and shapes. The N-terminus of the protein is located on the surface of the assembled particles and can tolerate the addition of heterologous amino acids. Yusibov et al. [56] developed a system in which a modified version of the AlMV coat protein bearing epitopes from rabies virus and HIV-1 was expressed in tobacco using a TMV-based vector. In an infected leaf tissue, the modified AlMV CP subunits assembled into ellipsoid particles that expressed multiple copies of the antigenic insert. When purified and injected into mice, these particles elicited the production of appropriate virus-neutralizing antibodies. Particles presenting the rabies virus epitopes expressed in spinach were subsequently shown to protect mice against a normally lethal challenge with the virus when supplied either intraperitoneally or orally [57].

16.3.2 Multiple-Protein VLPs

Many nonenveloped viruses have capsids that consist of multiple copies of more than one polypeptide. The different polypeptides can originate either by translation from different, independent open reading frames or can arise as a result of processing of a precursor polyprotein. Complex VLPs of both types have been successfully produced in plants.

16.3.2.1 Reoviridae

Reoviruses are nonenveloped double-stranded RNA viruses whose particles consist of three concentric spheres containing four different VPs expressed from separate genome segments. To date, plant-based expression of reovirus VLPs has concentrated on rotavirus and the important veterinary pathogen, bluetongue virus (BTV).

Rotavirus particles consist of a core of VP2, a middle layer of VP6 and an outer layer of VP7 with spikes of VP4. In insect cells, coexpression of VP2 and VP6 results in the production of double-layered particles (2/6-VLPs), whereas coexpression of VP2, VP6, and VP7, with or without VP4, results in the production of triple-layered particles (2/6/ 7-VLPs or 2/4/6/7-VLPs), respectively, which are capable of stimulating an immune response in experimental animals [58–60]. The first attempt to express a rotavirus VLP in plants involved the expression of a murine VP6 from a PVX-based vector [61]. The VP6 assembled into a range of tubular and sheet forms within the plant cells and was shown to be able to assemble on to insect cell-expressed VP2 core particles. When the VP6 had the sequence of a 2A catalytic peptide at its *C*-terminus, it appeared to be able to form particles in the absence of VP2. When expressed as a transgene in potato, VP6 also appeared capable of forming higher-order structures that were immunogenic [62], and VP6 expressed in transgenic alfalfa was able to protect mice against challenge [63]. Coexpression of VP2 and VP6 in transgenic tomatoes resulted in the production of double-layered particles in the fruit [64], while the coexpression of VP2, VP6, and VP7,

also in transgenic tomato, resulted in the appearance of triple-layer particles [65]. In both cases, the assembled particles were shown to be capable of stimulating an immune response. One problem with much of this work, however, is the low levels of the rotavirus proteins produced in the plant tissue. To address this issue, Inka-Borchers et al. [66] created lines of transplastomic tobacco that accumulated VP6 at levels up to 15% TSP. The expressed VP6 was able to form trimers, but no evidence of assembly into higher-order structures or immunogenicity was presented.

BTV causes an insect-transmitted disease of domestic and wild ruminants. The structure of the particles resembles that of rotavirus, with an inner layer (subcore) consisting of VP3 on to which VP7 assembles to form core-like particles (CLPs). The outer shell consists of VP2 and VP5 that do not form a higher-order structure in the absence of VP3 and VP7. To achieve maximum immunogenicity, all four proteins should be present at the correct stoichiometry in a VLP. This means it is necessary to express 120 copies of VP3 to form a subcore upon which 780 copies of VP7 assemble to create CLPs. To create VLPs, it is necessary to additionally assemble VP5 (360 copies) and VP2 (180 copies) on to the CLPs; these outer proteins are the antigenic determinants.

Initial attempts at transient expression of BTV VLPs in plants involved expressing native sequences from BTV serotype 10 (BTV-10). Although coexpression of VP3 and VP7 from separate pEAQ constructs gave rise to CLPs, the yield was low. Furthermore, attempts to produce VLPs by coexpression with VP2 and VP5 were inconclusive [67]. To overcome the problem of low levels of expression, the sequences of all four structural proteins from BTV-8 were codon-optimized and expressed in *N. benthamiana* using the pEAQ vector system [68]. Coexpression of all four genes at high levels led to the formation of a mixture of subcores, CLPs and VLPs. It was rationalized that it should be possible to increase the proportion of fully assembled VLPs by decreasing the level of VP3 (which forms the inner subcores) relative to the outer proteins. This proved to be the case with the resulting preparations consisting mainly of VLPs containing all four structural proteins in the correct ratio. The total particulate BTV-8 protein yield was estimated to be more than 200 mg/kg leaf wet weight in these experiments. After purification, the plant-expressed BTV-8 VLPs were shown to elicit a strong antibody response in sheep. Furthermore, they provided protective immunity against a challenge with BTV-8. The results demonstrated that transient expression can be used to produce immunologically relevant complex heteromultimeric structures in plants in a matter of days.

16.3.2.2 Foot-and-Mouth Disease Virus

Viruses from several families require the processing of structural proteins from a precursor in order to form VLPs. Particularly prominent among these is the order Picornavirales, whose members include mammalian pathogens, such as foot-and-mouth disease virus (FMDV). To achieve assembly, it is necessary to coexpress both the coat precursor and its cognate proteinase.

Because of its importance as a veterinary disease, there have been a number of attempts to produce FMDV. Dus Santos et al. [69] and Pan et al. [70] coexpressed the P1 coat protein precursor of the structural proteins (VP0, VP1, and VP3) and the proteinase (3C) necessary for its processing in transgenic alfalfa and tomato, respectively. In both cases, the plants produced material that was able to protect mice or guinea pigs against challenge with FMDV. Although Pan et al. [70] showed processing of the FMDV P1 into VP0, VP1, and VP3, no direct evidence for the formation of FMDV VLPs has yet been presented. Indeed, the results of Wang et al. [71], who expressed FMDV P1 in transgenic rice in the absence of 3C, suggest that an immune response is not indicative of VLP formation.

16.3.2.3 Cowpea Mosaic Virus

Successful VLP formation has been reported for the plant member of the Picornavirales, CPMV. In this case, the aim was not to produce a candidate vaccine but to produce noninfectious VLPs for use in bionanotechnology. Such RNA-free particles of CPMV have been generated by the transient coexpression of the precursor of its coat proteins, VP60, and the viral proteinase, 24 K in *N. benthamiana* using the pEAQ expression system [72]. Cleavage of VP60 by the 24 K proteinase resulted in generation of large (L) and small (S) coat proteins, which efficiently assembled into empty VLPs or eVLPs. By optimizing the infiltration and purification procedures, the yield of eVLPs could reach levels comparable with the levels of native particles achieved via infection [31,73]. The eVLPs thus produced are being developed for various applications in bionanotechnology, including magnetic hyperthermia, epitope display, and cell-specific targeting of drugs [74,75].

16.3.3 Enveloped VLPs

Enveloped viruses have capsids that consist not only of virus-encoded polypeptides but also of lipid that is host derived. To produce these in plants, it is necessary to demonstrate that plant-derived lipids are able to substitute for the lipids derived from the original host. Fortunately, this has generally proved to be the case, and there are several reports of the successful production in plants of VLPs from enveloped viruses.

16.3.3.1 Hepatitis B Virus

Particles of HBV are doubled shelled. The outer shell consists of a lipid envelope into which three different versions (small, medium, and large) of the surface antigen (HBsAg) are inserted. The three versions of HBsAg share a common C-terminal domain and differ only in the length of the N-terminal extension. As well as providing the outer shell of virus particles, HBsAg can also self-assemble with lipids into noninfectious subviral particles. Within this outer shell, the core antigen (HBcAg) forms core particles (see Section 16.3.1) that encapsulate the viral genome. Because of its significance as a worldwide pathogen, there has been considerable interest in expressing both HBsAg and HBcAg in a variety of heterologous systems including plants.

The first report of the expression of HBsAg in plants concerned the expression of the small form of the protein (S-HBsAg) in transgenic tobacco [76]. The expressed protein formed particles of the expected size for subviral particles (average diameter 22 nm). The buoyant density of the plant-expressed HBsAg particles in CsCl was 1.16 g/mL, similar to that of particles isolated from human serum (1.20 g/mL) and consistent with them containing both lipid and protein. The particles produced in transgenic tobacco were shown to be immunogenic when injected into mice [77]. Though this work represented a promising start, the maximal levels of HBsAg expression obtained was only 0.01% TSP that was deemed to be an inadequate level for the efficient use of plants as a source of HBsAg for vaccine use. As a result, much subsequent work concentrated on developing methods for the higher-level expression of S-HBsAg using both stable transformation and transient expression in a variety of host species. The highest level of expression of HBsAg VLPs has been reported to be 295 μg/g leaf fresh weight obtained using a replicating TMV system [78], while HBsAg produced in transgenic potatoes was shown to be immunogenic when supplied orally to human volunteers [79]. The various approaches used and the immunological properties of the expressed VLPs have recently been reviewed by Pniewski [80].

In addition to S-HBsAg, the middle and large form of HBsAg (M-HBsAg and L-HBsAg) have also been produced in plants, using both the transgenic [81,82] and transient [83,84] approach. In general, the accumulation of M-/L-HBsAg was lower than that of S-HBsAg.

16.3.3.2 Influenza Virus

Influenza represents a serious and ongoing threat to human populations, and its prevention is a major public health challenge worldwide. Current vaccines are based on inactivated virus particles produced in embryonated eggs. However, the rapid emergence of new strains of the virus means that a faster method of production, using heterologous expression systems, is highly desirable. Such recombinant vaccines are generally based on the highly immunogenic virion surface proteins—hemagglutinin (HA) and neuraminidase (NA). In the case of plants, HA has been successfully expressed using a number of transient expression systems. In most cases, the HA has been expressed in a soluble form, generally as a monomer (e.g., [85–88]), but also as a trimer [89]. Plant-expressed soluble HA has proved immunogenic in a number of animal species and has recently been used in a Phase I clinical trial [90].

As an alternative to expressing soluble HA, D'Aoust et al. [91] developed a method for producing VLPs containing HA in plants using transient expression. The VLPs were shown to contain both HA and plant-derived lipids and resembled native influenza virus particles when examined by electron microscopy. Within *N. benthamiana* cells expressing the HA, the VLPs accumulated in apoplastic indentations of the plasma membrane. The VLPs were immunogenic in mice and conferred protective immunity. The expression and purification technologies for the influenza VLPs were adapted for the large-scale production of GMP-grade material; preclinical studies demonstrated the plant-produced VLPs induced a strong immune response in mice and ferrets [92]. A particular feature of the technology used was its speed of production: a preparation of purified VLPs can be obtained within 3 weeks of receiving a novel HA sequence [92]. Landry et al. [93] conducted a Phase I clinical study of an H5 VLP vaccine produced as described earlier. The material showed good safety and promising immunogenicity in humans, indicating that plant-based VLP vaccines have a promising future as a method for influenza vaccine, particularly in instances where speed is of the essence such as in prepandemic or pandemic situations where an effective flu vaccine is needed within weeks of detection of a new strain so that it can halt its spread. A commercial-scale plant-production facility named Medicago USA, Inc. has been designed to have the ability to produce 10 million doses of a VLP-based vaccine per month (www.medicago.com).

16.4   CHALLENGES AND HURDLES

VLPs produced in plants are a promising vaccine candidate due to their known similarity to infectious virions and their ability to be manipulated for use for antigenic display. They also present a novel and cost-effective approach for oral delivery of vaccines. While the success of several plant-derived pharmaceuticals in clinical trials makes plant expression look promising, there is no such product on the market yet. In addition, the use of VLPs, particularly from *N. benthamiana*, as pharmaceuticals is hindered by potential technical and regulatory difficulties owing to the presence of high levels of phenolics and toxic alkaloids in these plant species. Research has to now be targeted toward the development of better downstream purification processes and the expression of VLPs in various edible plants such as lettuce [94].

The success of any plant-derived pharmaceutical is absolutely essential for boosting further research in the field and addressing regulatory concerns with the use of such products. Recent developments such as the FDA approval granted to Protalix BioTherapeutics, Inc. for their drug *Elelyso* manufactured in carrot cells to treat Type I Gaucher disease in humans [95] and the success of *zMAPP*, a plant-produced immunotherapy candidate against Ebola virus, in trials in *Rhesus macaque* monkeys [96] suggest that this might not be a long way into the future.

KEY POINTS

1. Noninfectious VLPs based on viruses of diverse origins (animal, plant, and bacterial) and structures (rod shaped, icosahedral, enveloped, and multiple layer) have been successfully produced in plants.
2. Production of VLPs in plants is achieved either by stable transformation methods or via *Agrobacterium*-mediated transient transformation.
3. Large-scale production of VLPs can be achieved in plants due to their high biomass and ease of scalability.

4. Several VLPs produced in plants are being developed as vaccine candidates. Plant-made influenza VLPs are currently in Phase II clinical trials.
5. Plant expression systems are highly promising for the expression of VLPs in circumstances where speed is of the essence, such as in of the case of emerging pandemics.

STATEMENT OF DISCLOSURE

The authors are not aware of any affiliations, memberships, funding, or financial holdings that might be perceived as affecting the objectivity of this review.

ACKNOWLEDGMENTS

The authors thank Eva Thuenemann and Hadrien Peyret at John Innes Centre for providing images for parts of Figure 16.2. This work was supported by the U.K. Biotechnological and Biological Sciences Research Council (BBSRC) institute strategic program grant "Understanding and Exploiting Plant and Microbial Secondary Metabolism" (BB/J004596/1) and the John Innes Foundation.

REFERENCES

1. Fischer, R., Stoger, E., Schillberg, S., Christou, P., and Twyman, R. M. (2004). Plant-based production of biopharmaceuticals. *Curr Opin Plant Biol* 7: 152–158.
2. Ma, J. K., Drake, P. M., and Christou, P. (2003). The production of recombinant pharmaceutical proteins in plants. *Nature Rev Genet* 4: 794–805.
3. Twyman, R. M., Stoger, E., Schillberg, S., Christou, P., and Fischer, R. (2003). Molecular farming in plants: Host systems and expression technology. *Trends Biotechnol* 21: 570–578.
4. Sala, F., Manuela Rigano, M., Barbante, A., Basso, B., Walmsley, A. M., and Castiglione, S. (2003). Vaccine antigen production in transgenic plants: Strategies, gene constructs and perspectives. *Vaccine* 21: 803–808.
5. Harrison, S. C., Olson, A. J., Schutt, C. E., Winkler, F. K., and Bricogne, G. (1978). Tomato bushy stunt virus at 2.9 Å resolution. *Nature* 276: 368–373.
6. Abad-Zapatero, C., Abdel-Meguid, S. S., Johnson, J. E., Leslie, A. G., Rayment, I., Rossmann, M. G., Suck, D., and Tsukihara, T. (1980). Structure of southern bean mosaic virus at 2.8 Å resolution. *Nature* 286: 33–39.
7. Steinmetz, N. F., Lin, T., Lomonossoff, G. P., and Johnson, J. E. (2009). Structure-based engineering of an icosahedral virus for nanomedicine and nanotechnology. *Curr Top Microbiol Immunol* 327: 23–58.
8. Lomonossoff, G. P. (2010). Virus particles and the uses of such particles in bio- and nanotechnology. In: *Recent Advances in Plant Virology* (Caranta, C., Lopez-Moya, J. J., Aranda, M., and Tepfer, M., eds.). Norfolk, U.K.: Caister Academic Press, pp. 363–385.
9. Yildiz, I., Shukla, S., and Steinmetz, N. F. (2011). Applications of viral nanoparticles in medicine. *Curr Opin Biotechnol* 6: 901–908.
10. Lomonossoff, G. P. and Evans, D. J. (2014). Applications of plant viruses in bionanotechnology. *Curr Top Microbiol Immunol* 375: 61–87.

11. Ochoa, W. F., Chatterji, A., Lin, T., and Johnson, J. E. (2006). Generation and structural analysis of reactive empty particles derived from an icosahedral virus. *Chem Biol* 13: 771–778.

12. Phelps, J. P., Dang, N., and Rasochova, L. (2007). Inactivation and purification of cowpea mosaic virus-like particles displaying peptide antigens from *Bacillus anthracis*. *J Virol Methods* 141: 146–153.

13. Langeveld, J. P., Brennan, F. R., Martinez-Torrecuadrada, J. L., Jones, T. D., Boshuizen, R. S., Vela, C., Casal, J. I. et al. (2001). Inactivated recombinant plant virus protects dogs from a lethal challenge with canine parvovirus. *Vaccine* 19: 3661–3670.

14. Rae, C., Koudelka, K. J., Destito, G., Estrada, M. N., Gonzalez, M. J., and Manchester, M. (2008). Chemical addressability of ultraviolet-inactivated viral nanoparticles (VNPs). *PLoS One* 3: e3315.

15. Gelvin, S. B. (2005). Agricultural biotechnology: Gene exchange by design. *Nature* 433: 583–584.

16. Goodin, M. M., Zaitlin, D., Naidu, R. A., and Lommel, S. A. (2008). *Nicotiana benthamiana*: Its history and future as a model for plant-pathogen interactions. *Mol Plant Microbe Interact* 21: 1015–1026.

17. Horsch, R. B. and Klee, H. J. (1986). Rapid assay of foreign gene expression in leaf discs transformed by *Agrobacterium tumefaciens*: Role of T-DNA borders in the transfer process. *Proc Natl Acad Sci U S A* 83: 4428–4432.

18. Scotti, N., Rigano, M. M., and Cardi, T. (2012). Production of foreign proteins using plastid transformation. *Biotechnol Adv* 30: 387–397.

19. Gleba, Y. Y., Tusé, D., and Giritch, A. (2014). Plant viral vectors for delivery by *Agrobacterium*. *Curr Top Microbiol Immunol* 375: 155–192.

20. Canizares, M. C., Liu, L., Perrin, Y., Tsakiris, E., and Lomonossoff, G. P. (2006). A bipartite system for the constitutive and inducible expression of high levels of foreign proteins in plants. *Plant Biotechnol J* 4: 183–193.

21. Marusic, C., Rizza, P., Lattanzi, L., Mancini, C., Spada, M., Belardelli, F., Benvenuto, E., and Capone, I. (2001). Chimeric plant virus particles as immunogens for inducing murine and human immune responses against human immunodeficiency virus type 1. *J Virol* 75: 8434–8439.

22. Lindbo, J. A. (2007). TRBO: A high-efficiency tobacco mosaic virus RNA-based overexpression vector. *Plant Physiol* 145: 1232–1240.

23. Giritch, A., Marillonnet, S., Engler, C., van Eldik, G., Botterman, J., Klimyuk, V., and Gleba, Y. (2006). Rapid high-yield expression of full-size IgG antibodies in plants coinfected with noncompeting viral vectors. *Proc Natl Acad Sci U S A* 103: 14701–14706.

24. Mor, T. S., Moon, Y. S., Palmer, K. E., and Mason, H. S. (2003). Geminivirus vectors for high-level expression of foreign proteins in plant cells. *Biotechnol Bioeng* 81: 430–437.

25. Voinnet, O., Rivas, S., Mestre, P., and Baulcombe, D. (2003). An enhanced transient expression system in plants based on suppression of gene silencing by the p19 protein of tomato bushy stunt virus. *Plant J* 33: 949–956.

26. Sainsbury, F. and Lomonossoff, G. P. (2008). Extremely high-level and rapid transient protein production in plants without the use of viral replication. *Plant Physiol* 148: 1212–1218.

27. Sainsbury, F., Thuenemann, E. C., and Lomonossoff, G. P. (2009). pEAQ: Versatile expression vectors for easy and quick transient expression of heterologous proteins in plants. *Plant Biotechnol J* 7: 682–693.

28. Meshcheriakova, Y. A., Saxena, P., and Lomonossoff, G. P. (2014). Fine-tuning levels of heterologous gene expression in plants by orthogonal variation of the untranslated regions of a non-replicating transient expression system. *Plant Biotechnol J* 12: 718–727.

29. Peyret, H. and Lomonossoff, G. P. (2013). The pEAQ vector series: The easy and quick way to produce recombinant proteins in plants. *Plant Mol Biol* 83: 51–58.

30. Sainsbury, F. and Lomonossoff, G. P. (2014). Transient expressions of synthetic biology in plants. *Curr Opin Plant Biol* 19: 1–7.

31. Montague, N. P., Thuenemann, E. C., Saxena, P., Saunders, K., Lenzi, P., and Lomonossoff, G. P. (2011). Recent advances of cowpea mosaic virus-based particle technology. *Hum Vaccin* 7: 383–390.

32. Peixoto, C., Ferriera, T. B., Sousa, M. F., Carrondo, M. J., and Alves, P. M. (2008). Towards purification of adenoviral vectors based on membrane technology. *Biotechnol Prog* 24: 1290–1296.

33. Hammonds, J., Chen, X., Zhang, X., Lee, F., and Spearman, P. (2007). Advances in methods for the production, purification and characterization of HIV-1 Gag-Env pseudovirion vaccines. *Vaccine* 25: 8036–8048.

34. Mason, H. S., Ball, J. M., Shi, J. J., Jiang, X., Estes, M. K., and Arntzen, C. J. (1996). Expression of Norwalk virus capsid protein in transgenic tobacco and potato and its oral immunogenicity in mice. *Proc Natl Acad Sci U S A* 93: 5335–5340.

35. Jiang, X., Wang, M., Graham, D. Y., and Estes, M. K. (1992). Expression, self-assembly, and antigenicity of the Norwalk virus capsid protein. *J Virol* 66: 6527–6532.

36. Tacket, C. O., Mason, H. S., Losonsky, G., Estes, M. K., Levine, M. M., and Arntzen, C. J. (2000). Human immune responses to a novel Norwalk virus vaccine delivered in transgenic potatoes. *J Infect Dis* 182: 302–305.

37. Zhang, X., Buehner, N. A., Hutson, A. M., Estes, M. K., and Mason, H. S. (2006). Tomato is a highly effective vehicle for expression and oral immunization with Norwalk virus capsid protein. *Plant Biotechnol J* 4: 419–432.

38. Santi, L., Batchelor, L., Huang, Z., Hjelm, B., Kilbourne, J., Arntzen, C. J., Chen, Q., and Mason, H. S. (2008). An efficient plant viral expression system generating orally immunogenic Norwalk virus-like particles. *Vaccine* 26: 1846–1854.

39. Mathew, L. G., Herbst-Kralovetz, M. M., and Mason, H. S. (2014). Norovirus Narita 104 virus-like particles expressed in *Nicotiana benthamiana* induce serum and mucosal immune responses. *Biomed Res Int* 2014: 807539.

40. Huang, Z., Chen, Q., Hjelm, B., Arntzen, C., and Mason, H. (2009). A DNA replicon system for rapid high-level production of virus-like particles in plants. *Biotechnol Bioeng* 103: 706–714.

41. Kirnbauer, R., Booy, F., Cheng, N., Lowy, D. R., and Schiller, J. T. (1992). Papillomavirus L1 major capsid protein self-assembles into virus-like particles that are highly immunogenic. *Proc Natl Acad Sci U S A* 89: 12180–12184.

42. Modis, Y., Trus, B. L., and Harrison, S. C. (2002). Atomic model of the papillomavirus capsid. *EMBO J* 21: 4754–4762.

43. Baker, T. S., Newcomb, W. W., Olson, N. H., Cowsert, L. M., Olson, C., and Brown, J. C. (1991). Structures of bovine and human papillomaviruses: Analysis by cryoelectron microscopy and three-dimensional image reconstruction. *Biophys J* 60: 1445–1456.

44. Chen, X. S., Garcea, R., Goldberg, I., Casini, G., and Harrison, S. C. (2000). Structure of small virus-like particles assembled from the L1 protein of human papillomavirus 16. *Mol Cell* 5: 557–567.

45. Biemelt, S., Sonnewald, U., Galmbacher, P., Willmitzer, L., and Mueller, M. (2003). Production of Human Papillomavirus type 16 virus-like particles in transgenic plants. *J Virol* 77: 9211–9220.

46. Warzecha, H., Mason, H. S., Lane, C., Tryggvesson, A., Rybicki, E., Williamson, A. L., Clements, J. D., and Rose, R. C. (2003). Oral immunogenicity of human papillomavirus-like particles expressed in potato. *J Virol* 77: 8702–8711.

47. Maclean, J., Koekemoer, M., Olivier, A. J., Stewart, D., Hitzeroth, I. I., Rademacher, T., Fischer, R., Williamson, A. L., and Rybicki, E. P. (2007). Optimization of Human Papillomavirus type 16 (HPV-16) L1 expression in plants: Comparison of the suitability of different HPV-16 L1 gene variants and different cell-compartment localization. *J Gen Virol* 88: 1460–1469.

48. Lenzi, P., Scotti, N., Alagna, F., Tornesello, M. L., Pompa, A., Vitale, A., de Stradis, A. et al. (2008). Translational fusion of chloroplast-expressed Human Papillomavirus type 16 L1 capsid protein enhances antigen accumulation in transplastomic tobacco. *Transgenic Res* 17: 1091–1102.

49. Waheed, M. T., Thoenes, N., Mueller, M., Hassan, S. W., Razavi, N. M., Loessl, E., Kaul, H. P., and Loessl, A. E. (2011). Transplastomic expression of a modified Human Papillomavirus L1 protein leading to the assembly of capsomeres in tobacco: A step towards cost-effective second-generation vaccines. *Transgenic Res* 20: 271–282.

50. Varsani, A., Williamson, A. L., Stewart, D., and Rybicki, E. P. (2006). Transient expression of Human papillomavirus type 16 L1 protein in *Nicotiana benthamiana* using an infectious tobamovirus vector. *Virus Res* 120: 91–96.

51. Matić, S., Masenga, V., Poli, A., Rinaldi, R., Milne, R. G., Vecchiati, M., and Noris, E. (2012). Comparative analysis of recombinant Human Papillomavirus 8 L1 production in plants by a variety of expression systems and purification methods. *Plant Biotechnol J* 10: 410–421.

52. Matić, S., Rinaldi, R., Masenga, V., and Noris, E. (2011). Efficient production of chimeric Human Papillomavirus 16 L1 protein bearing the M2e influenza epitope in *Nicotiana benthamiana* plants. *BMC Biotechnol* 15: 106.

53. Pineo, C. B., Hitzeroth, I. I., and Rybicki, E. P. (2013). Immunogenic assessment of plant-produced human papillomavirus type 16 L1/L2 chimaeras. *Plant Biotechnol J* 11: 964–975.

54. Kohl, T., Hitzeroth, I. I., Stewart, D., Varsani, A., Govan, V. A., Christensen, N. D., Williamson, A. L., and Rybicki, E. P. (2006). Plant-produced cottontail rabbit papillomavirus L1 protein protects against tumor challenge: A proof-of-concept study. *Clin Vaccine Immunol* 13: 845–853.

55. Love, A. J., Chapman, S. N., Matic, S., Noris, E., Lomonossoff, G. P., and Taliansky, M. (2012). In planta production of a candidate vaccine against bovine papillomavirus type 1. *Planta* 236: 1305–1313.

56. Yusibov, V., Modelska, A., Steplewski, K., Agadjanyan, M., Weiner, D., Hooper, D. C., and Koprowski, H. (1997). Antigens produced in plants by infection with chimeric plant viruses immunize against rabies virus and HIV-1. *Proc Natl Acad Sci U S A* 94: 5784–5788.

57. Modelska, A., Dietzschold, B., Sleysh, N., Fu, Z. F., Steplewski, K., Hooper, D. C., Koprowski, H., and Yusibov, V. (1998). Immunization against rabies with plant-derived antigen. *Proc Natl Acad Sci U S A* 95: 2481–2485.

58. Labbé, M., Charpilienne, A., Crawford, S. E., Estes, M. K., and Cohen, J. (1991). Expression of rotavirus VP2 produces empty core-like particles. *J Virol* 65: 2946–2952.

59. Crawford, S. E., Cohen, M. L. J., Burroghs, M. H., Zhou, Y. J., and Estes, M. K. (1994). Characterization of virus-like particles produced by the expression of rotavirus capsid proteins in insect cells. *J Virol* 68: 5945–5952.

60. Bertolotti-Ciarlet, A., Ciarlet, M., Crawford, S. E., Connor, M. E., and Estes, M. K. (2003). Immunogenicity and protective efficacy of rotavirus 2/6-virus-like particles produced by a dual baculovirus expression vector and administered intramuscularly, intranasally, or orally to mice. *Vaccine* 21: 3885–3900.

61. O'Brien, G. J., Bryant, C. J., Voogd, C., Greenberg, H. B., Gardner, R. C., and Bellamy, A. R. (2000). Rotavirus VP6 expressed by PVX vectors in *Nicotiana benthamiana* coats PVX rods and also assembles into virus-like particles. *Virology* 270: 444–453.

62. Yu, J. and Langridge, W. H. R. (2003). Expression of rotavirus capsid protein VP6 in transgenic potato and its oral immunogenicity in mice. *Transgen Res* 12: 163–169.

63. Dong, J. L., Liang, B. G., Jin, Y. S., Zhang, W. J., and Wang, T. (2005). Oral immunization with pBsVP6-transgenic alfalfa protects mice against rotavirus infection. *Virology* 339: 153–163.

64. Saldaña, S., Guadarrama, F. E., Flores, T. D. J. O., Arias, N., López, S., Arias, C., Ruiz-Medrano, R. et al. (2006). Production of rotavirus-like particles in tomato (*Lycopersicon esculentum* L.) fruit by expression of capsid proteins VP2 and VP6 and immunological studies. *Viral Immunol* 19: 42–53.

65. Yang, Y., Li, X., Yang, H., Qian, Y., Zhang, Y., Fang, R., and Chen, X. (2011). Immunogenicity and virus-like particle formation of rotavirus capsid proteins produced in transgenic plants. *Sci China Life Sci* 54: 82–89.

66. Inka-Borchers, A. M., Gonzalez-Rabade, N., and Gray, J. C. (2012). Increased accumulation and stability of rotavirus VP6 protein in tobacco chloroplasts following changes to the 5′ untranslated region and the 5′ end of the coding region. *Plant Biotechnol J* 10: 422–434.

67. Thuenemann, E. C., Lenzi, P., Love, A. J., Taliansky, M., Bécares, M., Zuñiga, S., Enjuanes, L. et al. (2013). The use of transient expression systems for the rapid production of virus-like particles in plants. *Curr Pharm Des* 19: 5564–5573.

68. Thuenemann, E. C., Meyers, A. E., Verwey, J., Rybicki, E. P., and Lomonossoff, G. P. (2013). A method for rapid production of heteromultimeric protein complexes in plants: Assembly of protective bluetongue virus-like particles. *Plant Biotechnol J* 11: 839–846.

69. Dus Santos, M. J., Carrillo, C., Ardila, F., Ríos, R. D., Franzone, P., Piccone, M. E., Wigdorovitz, A., and Borca, M. V. (2005). Development of transgenic alfalfa plants containing the foot and mouth disease virus structural polyprotein gene P1 and its utilization as an experimental immunogen. *Vaccine* 23: 1838–1843.

70. Pan, L., Zhang, Y., Wang, Y., Wang, B., Wang, W., Fang, Y., Jiang, S., Lv, J., Wang, W., Sun, Y., and Xie, Q. (2008). Foliar extracts from transgenic tomato plants expressing the structural polyprotein, P1-2A, and protease, 3C, from foot-and-mouth disease virus elicit a protective response in guinea pigs. *Vet Immunol Immunopathol* 121: 83–90.

71. Wang, Y., Shen, Q., Jiang, Y., Song, Y., Fang, L., Xiao, S., and Chen, H. (2012). Immunogenicity of foot-and-mouth disease virus structural polyprotein P1 expressed in transgenic rice. *J Virol Methods* 181: 12–17.

72. Saunders, K., Sainsbury, F., and Lomonossoff, G. P. (2009). Efficient generation of cowpea mosaic virus empty virus-like particles by the proteolytic processing of precursors in insect cells and plants. *Virology* 393: 329–337.

73. Sainsbury, F., Saxena, P., Aljabali, A. A., Saunders, K., Evans, D. J., and Lomonossoff, G. P. (2014). Genetic engineering and characterization of Cowpea mosaic virus empty virus-like particles. *Methods Mol Biol* 1108: 139–153.

74. Sainsbury, F., Saunders, K., Aljabali, A. A., Evans, D. J., and Lomonossoff, G. P. (2011). Peptide-controlled access to the interior surface of empty virus nanoparticles. *ChemBioChem* 12: 2435–2440.

75. Wen, A. M., Shukla, S., Saxena, P., Aljabali, A. A., Yildiz, I., Dey, S., Mealy, J. E., Yang, A. C., Evans, D. J., Lomonossoff, G. P., and Steinmetz, N. F. (2012). Interior engineering of a viral nanoparticle and its tumor homing properties. *Biomacromolecules* 13: 3990–4001.

76. Mason, H. S., Lam, D. M. K., and Arntzen, C. J. (1992). Expression of hepatitis B surface antigen in transgenic plants. *Proc Natl Acad Sci USA* 89: 11745–11749.

77. Thanavala, Y., Yang, Y. F., Lyons, P., Mason, H. S., and Arntzen, C. J. (1995). Immunogenicity of transgenic plant-derived hepatitis B surface antigen. *Proc Natl Acad Sci U S A* 92: 3358–3361.

78. Huang, Z., LePore, K., Elkin, G., Thanavala, Y., and Mason, H. S. (2008). High-yield rapid production of hepatitis B surface antigen in plant leaf by a viral expression system. *Plant Biotechnol J* 6: 202–209.

79. Thanavala, Y., Mahoney, M., Pal, S., Scott, A., Richter, L., Natarajan, N., Goodwin, P., Arntzen, C. J., and Mason, H. S. (2005). Immunogenicity in humans of an edible vaccine for hepatitis B. *Proc Natl Acad Sci U S A* 102: 3378–3382.

80. Pniewski, T. (2013). The twenty-year story of a plant-based vaccine against hepatitis B: Stagnation or promising prospects? *Int J Mol Sci* 14: 1978–1998.

81. Joung, Y. H., Youm, J. W., Jeon, J. H., Lee, B. C., Ryu, C. J., Hong, H. J., Kim, H. C., Joung, H., and Kim, H. S. (2004). Expression of the hepatitis B surface S and preS2 antigens in tubers of *Solanum tuberosum*. *Plant Cell Rep* 22: 925–930.

82. Pniewski, T., Kapusta, J., Bociąg, P., Kostrzak, A., Fedorowicz-Strońska, O., Czyż, M., Gdula, M., Krajewski, P., Wolko, B., and Płucienniczak, A. (2012). Plant expression, lyophilisation and storage of HBV medium and large surface antigens for a prototype oral vaccine formulation. *Plant Cell Rep* 31: 585–595.

83. Huang, Z. and Mason, H. S. (2004). Conformational analysis of hepatitis B surface antigen fusions in an *Agrobacterium*-mediated transient expression system. *Plant Biotechnol J* 2: 241–249.

84. Huang, Z., Elkin, G., Maloney, B. J., Buehner, N., Arntzen, C. J., Thanavala, Y., and Mason, H. S. (2005). Virus-like particles expression and assembly in plants: Hepatitis B and Norwalk viruses. *Vaccine* 23: 1851–1858.

85. Kalthoff, D., Giritch, A., Geisler, K., Bettmann, U., Klimyuk, V., Hehnen, H. R., Gleba, Y., and Beer, M. (2010). Immunization with plant-expressed hemagglutinin protects chickens from lethal highly pathogenic avian influenza virus H5N1 challenge infection. *J Virol* 84: 12002–12010.

86. Shoji, Y., Farrance, C. E., Bautista, J., Bi, H., Musiychuk, K., Horsey, A., Park, H. et al. (2012). A plant-based system for rapid production of influenza vaccine antigens. *Influenza Other Respir Viruses* 6: 204–210.

87. Mortimer, E., Maclean, J. M., Mbewana, S., Buys, A., Williamson, A. L., Hitzeroth, I. I., and Rybicki. E. P. (2012). Setting up a platform for plant-based influenza virus vaccine production in South Africa. *BMC Biotechnol* 12: 14.

88. Kanagarajan, S., Tolf, C., Lundgren, A., Waldenström, J., and Brodelius, P. E. (2012). Transient expression of hemagglutinin antigen from low pathogenic avian influenza A (H7N7) in *Nicotiana benthamiana*. *PLoS One* 7: e33010.

89. Shoji, Y., Jones, R. M., Mett, V., Chichester, J. A., Musiychuk, K., Sun, X., Tumpey, T. M. et al. (2013). A plant-produced H1N1 trimeric hemagglutinin protects mice from a lethal influenza virus challenge. *Hum Vaccin Immunother* 9: 553–560.

90. Cummings, J. F., Guerrero, M. L., Moon, J. E., Waterman, P., Nielsen, R. K., Jefferson, S., Gross, F. L., Hancock, K., Katz, J. M., and Yusibov, V. (2014). Safety and immunogenicity of a plant-produced recombinant monomer hemagglutinin-based influenza vaccine derived from influenza A (H1N1) pdm09 virus: A Phase 1 dose-escalation study in healthy adults. *Vaccine* 32: 2251–2259.

91. D'Aoust, M. A., Lavoie, P. O., Couture, M. M., Trépanier, S., Guay, J. M., Dargis, M., Mongrand, S., Landry, N., Ward, B. J., and Vézina, L. P. (2008). Influenza virus-like particles produced by transient expression in *Nicotiana benthamiana* induce a protective immune response against a lethal viral challenge in mice. *Plant Biotechnol J* 6: 930–940.

92. D'Aoust, M. A., Couture, M. M., Charland, N., Trépanier, S., Landry, N., Ors, F., and Vézina, L. P. (2010). The production of hemagglutinin-based virus-like particles in plants: A rapid, efficient and safe response to pandemic influenza. *Plant Biotechnol J* 8: 607–619.

93. Landry, N., Ward, B. J., Trépanier, S., Montomoli, E., Dargis, M., Lapini, G., and Vézina, L. P. (2010). Preclinical and clinical development of plant-made virus-like particle vaccine against avian H5N1 influenza. *PLoS One* 5: e15559.

94. Lai, H., He, J., Engle, M., Diamond, M. S., and Chen, Q. (2012). Robust production of virus-like particles and monoclonal antibodies with geminiviral replicon vectors in lettuce. *Plant Biotechnol J* 10: 95–104.

95. Fox, J. L. (2012). First plant-made biologic approved. *Nat Biotechnol* 30: 472.

96. Qiu, X., Wong, G., Audet, J., Bello, A., Fernando, L., Alimonti, J. B., Fausther-Bovendo, H. et al. (2014). Reversion of advanced Ebola virus disease in nonhuman primates with ZMapp. *Nature*. 514: 47–53.

17 Bionanomaterials from Plant Viruses

Alaa A.A. Aljabali and David J. Evans

CONTENTS

17.1 INTRODUCTION

Bionanoscience, a subsection of nanoscience, involves the exploitation of biomaterials, devices, or methodologies on the nanoscale. It is multidisciplinary as it sits at the interface of chemistry, biology, physics, medicine, engineering, and materials science. It is the combination of biology and nanotechnology that takes advantage of natural nanoparticles to create materials and devices on the nanoscale. Nanosystems may be produced by either microfabrication, making big structures smaller and/or embedding smaller features into macroscopic materials (top-down), or using the techniques of component assembly and/or supramolecular chemistry to make small molecules bigger (bottom-up). Biomaterials such as DNA, RNA, proteins, and viruses can be used as building templates for nanomaterials. Viruses, in particular are of great interest to the nanotechnology field because of their highly monodisperse structures, small size, and the ability to self-assemble with high precession.

17.1.1 VIRUSES

The word "virus" is Latin and means poison. Viruses are obligate intracellular parasites, which only replicate within a host cell. Particles of simple (nonenveloped) virus typically consist of two components: one or more capsid proteins (shells) that assemble into precise 3D structures and genetic material. The genetic material is located in the interior of the capsid as circular, single-stranded, or double-stranded fragments. Enveloped viruses possess a bilipid layer on the exterior that provides targeting specificity to the virus. Plant and bacteria viruses (bacteriophages or phages) are not pathogenic to animals, so they are safe for use in materials science. The genome encodes the proteins necessary for the replication of the virus and the capsid protein(s); the capsid is mainly for the protection of the genetic material. Virus capsids also have roles in host cell targeting and cell entry [1]. Virus particles typically represent very stable structures that have evolved to withstand a broad range of environments yet are sensitive enough to release their nucleic acid when they infect a susceptible cell [2].

17.1.2 WHY VIRUS NANOPARTICLES

Viruses are considered as protein cages, scaffolds, and templates for the production of novel nanostructured materials on/in which organic and inorganic moieties can be incorporated in a very precise and controlled fashion. Genetic engineering (chimeric technology) enables the insertion or replacement of selected amino acids on virus capsids for uses from bioconjugation to mineralization [3,4]. Many virus coat proteins will assemble in vitro, naturally or after genetic manipulation, into noninfectious containers called virus-like particles (VLPs). Viruses can be found in a variety of distinct shapes, most commonly icosahedrons (sphere-like) and rod shaped (Figure 17.1), though, especially in the case of bacteriophages, more complicated structures can be formed.

Plant viruses are noninfectious toward other organisms, and they do not present a biological hazard. The production of the virus particles is simple and quick. When produced in the natural host, high expression yields can be achieved. For example, some plant virus particles can be obtained in gram scales from 1 kg of infected leaf material within 2–4 weeks [5]. Heterologous expression systems can also give rise to high yields of VLPs [6]. Viruses have unique properties, ease of functionalization, ability to self-assemble, monodispersity, and structural symmetry, and the majority of viruses are stable over a wide range of pH and temperatures. Further, they display a remarkable plasticity in their capsid structure and dynamics (coordinated assembly, disassembly, and uniformity).

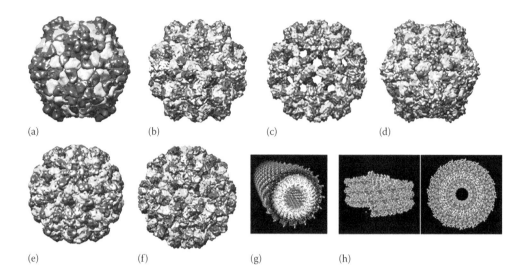

FIGURE 17.1 Space filling representation of some plant viruses. (a) RCNMV, (b) TYMV, (c) CCMV, (d) CPMV, (e) BMV, (f) CarMV, (g) TMV, and (h) TMV cross section.

Their amenability to genetic and chemical modification and diversity of sizes and shapes make them ideal workhorses with a variety of possible applications in nanotechnology.

Many plant viruses, due to these properties, have been investigated for their potential use for material fabrication on the nano-/microscale, including the icosahedral, sphere-like, viruses cowpea chlorotic mottle virus (CCMV), cowpea mosaic virus (CPMV), brome mosaic virus (BMV), cucumber mosaic virus (CMV), red clover necrotic mosaic virus (RCNMV), carnation mottle virus (CarMV) and turnip yellow mosaic virus (TYMV), and the rod-shaped tobacco mosaic virus (TMV). Viruses offer three different surfaces that can be exploited: the exterior, the interior, and the interface between subunits (Figure 17.2). The inner

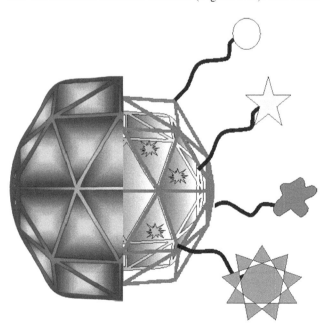

FIGURE 17.2 Schematic of potential chemical modification sites of an icosahedral plant virus capsid: external surface, internal surface, internal loading, external coating.

cavity of the virus capsid is accessible to small molecules and impermeable to larger ones, which allows the use of the interior space of VLPs as nanotemplates and nanoreactors. Furthermore, the exterior surface is widely considered as a robust platform and is used for both chemical and genetic modification allowing multivalent ligand display. The virus architectural interface is crucial for assembly and offers another route for the manipulation of the capsid architecture.

17.1.3 EMPTY VIRUS-LIKE PARTICLES

VLPs are supramolecular structures that have the form of rods or icosahedrons with diameters in the range of 25–100 nm [7]. They are composed of multiple copies of one or more recombinantly expressed virus structural proteins that spontaneously assemble into particles upon expression. In addition to being virus-like in structure, they are often antigenically indistinguishable from the virus from which they were derived. VLPs resemble viruses but are noninfectious because they do not contain any virus genetic material.

In vitro assembly and reassembly methods have been exploited for VLPs. Generally, particles are exposed to a disassembly buffer followed by dialysis against an assembly buffer. These disassembly approaches allow the release of nucleic acid and are used to generate VLPs for encapsulating materials of interest on reassembly. VLPs of TMV and CCMV have been demonstrated to efficiently self-assemble in vitro [8]. Alternatively, an alkaline hydrolysis method was developed to extract nucleic acid from CPMV [9] to generate empty particles. Recently, a method for the production of large quantities of CPMV VLPs in plants has been developed [10]. The mature large (L) and small (S) proteins of the capsid structure are produced by the cleavage of the precursor (VP60) by the action of a virus-encoded proteinase (24 K proteinase). CPMV VLP production takes advantage of the highly efficient plant transient expression

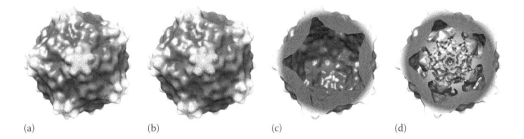

FIGURE 17.3 Cryoelectron microscopy reconstructions of CPMV particles. Purified particles were flash-frozen in vitreous ice and then subjected to cryoelectron microscopy. (a) reconstruction of wild-type CPMV exterior, (b) reconstruction of CPMV VLPs, (c) internal view of CPMV VLPs showing empty cavity, and (d) wild-type CPMV with encapsidated RNA in blue. (Images courtesy of Dr. Kyle Dent and Dr. Neil Ranson, University of Leeds, Leeds, U.K.)

system, pEAQ–HT [11,12]. The pEAQ–HT system is used to simultaneously express the VP60 coat protein precursor and the 24 K proteinase in plants via agroinfiltration. Efficient processing of VP60 to the L and S proteins occurs, leading to the formation of VLPs [10,13]. CPMV VLPs possess the same structure as wild-type CPMV but lack the genetic material within the capsid (Figure 17.3).

Virus platforms including CCMV [14], TMV [15], MS2 bacteriophage [16], and polyoma VLPs (capsoids) [17] have been used as constrained reaction vessels for packaging and nucleation sites for mineralization [14,18]. In addition, intact enzymes have been encapsulated within the CCMV capsid [19]. Further examples of the use of VLPs for the production on bionanomaterials will be presented later.

For RCNMV, the decoration of preformed synthetic nanoparticles with an origin of assembly site (OAS) that initiates coat protein monomer binding leads to self-assembled VLPs with synthetic nanoparticles within. Thiolated artificial OAS was attached to gold nanoparticles, and the OAS–gold nanoparticles were mixed with RCNMV coat protein monomers. This templates the in vitro self-assembly of VLPs around the gold nanoparticle (Figure 17.4) [20]. This approach was expanded to include encapsulation of quantum dots (QDs) with sizes of 5, 10, and 15 nm [20,21].

Alternatively, a nanoparticle is coated with negatively charged polymers to mimic RNA, an approach that was pioneered by Dragnea's group. Although citrate-stabilized gold nanoparticles resulted in VLPs encapsulating preformed gold nanoparticles, the process was not efficient [22]. However,

covalent functionalization of nanoparticles with carboxyl-terminated polyethylene glycol (PEG) [23,24] was used to mimic RNA nucleation and promote efficient in vitro self-assembly (Figure 17.5).

17.2 TOBACCO MOSAIC VIRUS

TMV is a rodlike plant virus consisting of 2130 asymmetric subunits arranged helically around a single-stranded RNA genome. Native TMV is 300 nm in length and 18 nm in diameter with a 4 nm cylindrical central cavity [25,26], which can be purified from infected tobacco plants in large quantities. The surface properties of TMV have been chemically and genetically manipulated without affecting the virus integrity or morphology of the capsid. The highly polar outer and inner surfaces of TMV contain hydroxyl and deprotonated carboxylate groups [27]. TMV has been exploited as a template to grow metal or metal oxide nanoparticles such as iron oxides, iron oxyhydroxides, cadmium sulfide, lead sulfide, gold, nickel, cobalt, silver, copper, CoPt, FePt, and silica [28–32].

In order to achieve successful coating based on electrostatic interactions, the deposition conditions should be varied in order to match the interaction between the virion surface and the deposition precursor. In the case of silica coating, carrying out the reaction at a pH less than three results in a positively charged TMV surface that will have strong interactions with the anionic silicate sols formed by hydrolysis of tetraethoxysilane. In contrast, cadmium sulfide, lead sulfide,

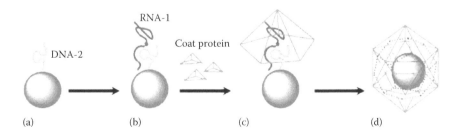

FIGURE 17.4 Assembly of RCNMV VLPs around a gold nanoparticle via OAS templating. (a) Conjugation of nanoparticle with DNA-2; (b) addition of RNA-1, which interacts with DNA-2 to form the functional OAS; (c) the artificial OAS templates the assembly of coat protein; and (d) formation of VLP with nanoparticle encapsidated. (Reprinted with permission from Loo, L. et al., Controlled encapsidation of gold nanoparticles by a viral protein shell, *J. Am. Chem. Soc.*, 128, 4502. Copyright 2006 American Chemical Society.)

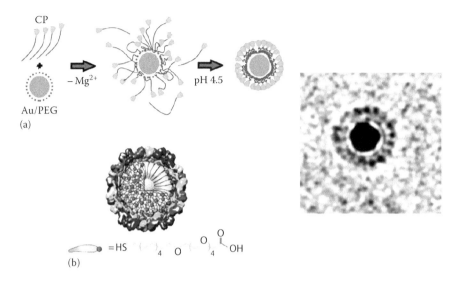

FIGURE 17.5 Proposed mechanism of VLP assembly from coat proteins. (a) Electrostatic interaction leads to the formation of disordered protein–gold nanoparticle complexes. The second step is a crystallization phase in which the protein–protein interactions lead to the formation of a regular capsid. (b) Schematic depiction of the encapsidated nanoparticle functionalized with carboxyl-terminated PEG chains. Right, cryoelectron micrograph of a single VLP shows the protein structure coating the 12 nm diameter gold nanoparticle (black disk). (Reprinted with permission from Chen, C. et al., Nanoparticle-templated assembly of viral protein cages, *Nano Lett.*, 6, 611. Copyright 2006 American Chemical Society.)

and iron oxides can be successfully coated on the outer surface at near neutral pH by specific metal ion binding with glutamate and aspartate residues [33]. As for metal deposition, in some cases, a suitable activation agent is needed in order to realize successful coating; Pd(II) and Pt(II) are two typical activation agents. The metal deposition can occur either inside the inner channel or at the outer surface of TMV [34].

Genetically engineered TMV can show enhanced deposition of metal onto its surface [35,36]. Native TMV can be genetically altered to display multiple metal binding sites through the insertion of two cysteine residues within the amino-terminus of the virus coat protein. In situ chemical reductions successfully deposited silver, gold, and palladium clusters onto the genetically modified TMV without any activation agent [35]. Furthermore, mixing TMV with aniline and a mild oxidant resulted in highly uniform, micron-length nanofibers measuring approximately 2 nm in diameter, likely due to the noncovalent interactions between aniline and TMV [36].

The exterior and interior of TMV have been utilized for inorganic material synthesis via metal electroless deposition (ELD). The interior surface of TMV is negatively charged under physiological conditions, whereas the exterior surface exhibits positive charge [37]. The exterior of TMV was coated with cadmium sulfide, lead sulfide, amorphous iron oxide, silica [15], and platinum and gold [38,39]. Silver and copper can be incorporated inside the central channel of TMV, the differential nucleation being electrostatically driven as a consequence of the difference in surface charges [40,41]. In addition, TMV particles were engineered to encode unique cysteine residues (TMV1cys); this allowed the assembly of TMV1cys onto a gold patterned surface,

the particles assembled in a vertically oriented fashion. The subsequent ELD resulted in uniform metal coatings up to 40 nm thick [42].

17.3 COWPEA CHLOROTIC MOTTLE VIRUS

CCMV is a member of the bromovirus group of the Bromoviridae family. The genome consists of three single-stranded, positive-sense RNAs that are encapsidated separately. The capsid is composed of 180 asymmetric (identical) protein subunits (20.3 kDa each). CCMV particles are ~28 nm in diameter formed as an icosahedral shell. Virus can be obtained in high yields of 1–2 g of purified particles per 1 kg of infected plant tissue. In addition, yeast-based expression systems of the coat proteins have been developed that can assemble VLPs. For example, expression of CCMV in *Pichia pastoris* yields up to 0.5 mg/g of wet cell mass [6]. Heterologous expression in yeast allows large-scale production of wild-type and genetically modified capsids or VLPs. The main advantage of heterologous expression is that it allows production of VLPs that would be unlikely to assemble and accumulate in the natural host cells.

The structure of the wild-type and swollen forms of CCMV has been determined to 3.2 Å [43]. The exterior CCMV capsid electrostatic interactions are well studied [44]. These interactions provide a reversible method to control the assembled capsid structure. The native form of CCMV swells at a pH > 6.5, resulting in 60 pores of 2 nm in size (Figure 17.6) [2,44,45]. The native genetic material can escape via the open pores, leaving an empty nanocontainer for the mineralization of polyoxometalate species, encapsulation of anionic polymers [2,14], and the synthesis of gold nanoparticle cores [46].

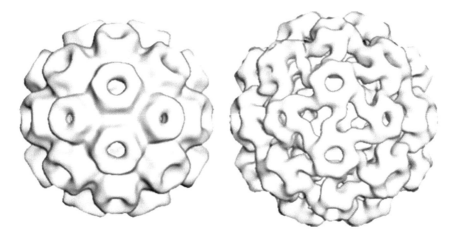

FIGURE 17.6 Cryoelectron microscopy and image reconstruction of CCMV. In an unswollen condition induced by low pH (on the left) and in a swollen condition induced by high pH (on the right). Swelling at the pseudo-threefold axis results in the formation of sixty 2 nm pores. (Reprinted by permission from Macmillan Publishers Ltd., *Nature*, Douglas, T. and Young, M., 1998, 393(6681), 152, copyright 1998).

The self-assembly of a virus is initiated by electrostatic interactions between RNA and the protein subunits. In the case of CCMV, mixing the coat protein monomers with differently sized polystyrenesulfonate (PSS) polymer to mimic the RNA resulted in two differently sized morphologies: 22 nm-sized particles generated by using PSS with molecular weight below 2 MDa and 27 nm-sized particles with PSS of 2 MDa or higher [47]. CCMV can also be used for the synthesis of biohybrid nanostructures. It can be self-assembled with a PEG functionalized exterior surface and can then internalize PSS, expanding the possibilities for new types of hybrid virus nanomaterials [48]. In related work, hibiscus chlorotic ring spot virus (HCRSV) particles have been used to encapsulate negatively charged polymers including PSS and polyacrylic acid. Reassembly yielded VLPs encapsulating the polymers. However, attempts to encapsulate neutral molecules such as fluorescein-labeled dextran failed, suggesting the encapsulation is based on electrostatic interactions [49].

Material encapsulation into assembled CCMV particles has been an active area of research. For example, the encapsulation of anionic polyanetholesulfonic acid (PASA) is controlled by pH-dependent gating of the CCMV pores [14]. Further, the entrapment of lanthanide ions [14] has been described. A similar approach has been adopted to infuse fluorescent dyes and drugs within RCNMV [50].

The mineralization reaction within CCMV is electrostatically driven. The interior is highly positively charged and therefore provides an interface for inorganic nucleation. The negatively charged tungstate, WO^{4-}, and vanadate, VO^{3-}, anions interact with the interior capsid surface via electrostatic interactions [14]. The same principle was used to generate internalized titanium oxide (β-TiO_2) [51] and Prussian blue nanoparticles [52]. Furthermore, the substitution of all the basic residues on the N-terminus of the coat protein with negatively charged glutamic acid [6] resulted in the mutant favoring interaction with ferrous

and ferric ions leading to the formation of magnetite (γ-Fe_2O_3) within the capsid [2].

17.4 COWPEA MOSAIC VIRUS

CPMV is a nonenveloped plant mosaic virus that is the type member of the comovirus genus of the family Comovirinae [53]. The virus is normally propagated in *Vigna unguiculata*, commonly known as cowpeas or black-eyed peas or beans. In the laboratory, the virus is introduced into the plant by means of mechanical inoculation, which results in mosaic and yellowing symptoms of the leaves. CPMV is considered one of the best investigated viruses and is widely used in bionanotechnology [54,55]. CPMV capsids have pseudo T = 3 icosahedral symmetry and a diameter of approximately 28 nm. The CPMV capsid structure comprises 60 small coat proteins (S, domain A) that fold into 1 jelly roll β-sandwich arranged as 12 pentamers at the fivefold axis and 60 large coat proteins (L, domains B and C) arranged as trimers at the threefold axis that fold into 2 jelly roll β-sandwich domains. The three domains form the asymmetric unit. The crystal structure is known to 2.8 Å resolution (Figure 17.7) [56]. CPMV is a single-stranded positive-sense bipartite RNA virus. Both of the separately encapsidated RNA segments are required for infection [57–59]. However, it was later demonstrated that RNA-1 is independently capable of replicating CPMV particles in protoplasts [60]. RNA-1 encodes the virus replication machinery as well as the virus proteinase and the virus protein genome-linked (VPg), which plays an essential role in initiating RNA synthesis. On the other hand, the smaller RNA-2 encodes the movement protein and the two capsid proteins.

CPMV particles produced during the infection process can be separated into three components on caesium chloride density gradients. These have identical protein compositions, but differ in their RNA contents [57]. The three centrifugal components are termed top (T), middle (M), and bottom (B)

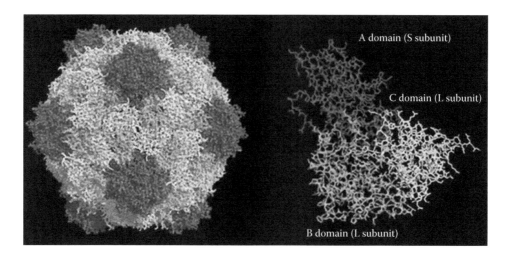

FIGURE 17.7 The structure of CPMV capsid and the asymmetric unit. CPMV capsid is comprised of small (S) and large (L) subunits. The A domain (in blue), B domain (in green), and C domain (in yellow). (From Steinmetz, N.F. and Evans, D.J., Utilisation of plant viruses in bionanotechnology, *Org. Biomol. Chem.*, 5(18), 2891–2902, 2007. Reproduced by permission of The Royal Society of Chemistry.)

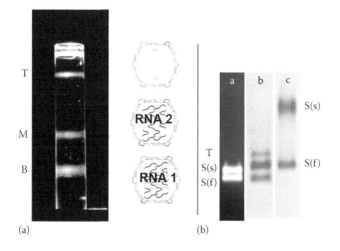

FIGURE 17.8 Separation of the different components of CPMV. (a) A caesium chloride gradient by ultracentrifugation. (b) Electrophoretic travel of particles containing the different forms of the S subunit, the slow (s) and the fast (f). (Images in (a) courtesy of Dr. Keith Saunders, John Innes Centre, Norwich, U.K.)

and contain no RNA, RNA-2, and RNA-1, respectively (Figure 17.8) [57,58,61]. The three components generated during CPMV infection also exist as two electrophoretic forms, slow and fast (Figure 17.8) [62,63]. The slow form can be converted to the fast form by proteolytic cleavage at the carboxyl (C)-terminus of the S protein [64,65]. The conversion occurs naturally during infection, with the fast (C-terminally processed) form of the S protein predominating at later times [64]. Cleavage of the S protein was shown to occur after leucine 189 resulting in the loss of the carboxyl-terminal 24 amino acids [10,66].

Wild-type CPMV was first mineralized after appropriate genetic modification that involved the use of infectious cDNA clones to modify the capsid, thereby allowing the presentation of foreign peptides on the external CPMV surface [67–69]. It had been established that additional amino acids can be inserted into the highly surface-exposed βB–βC

loop of the small subunit of CPMV. CPMV-based chimeric virus technology was then adapted for the environmentally benign synthesis of virus-templated monodisperse silica [3] and amorphous iron–platinum nanoparticles [4]. To avoid the lengthy process of generating chimeric CPMV particles and to avoid some of the disadvantages of the genetic engineering route (stability, reversion to wild type, and time taken), the possibility of chemically attaching peptides that promote specific mineralization to the surface of CPMV particles was investigated. Peptides previously identified by using phage display that specifically direct mineralization by CoPt (Cbz-CNAGDHANC), FePt (Cbz-HNKHLPSTQPLA), and zinc sulfide (Cbz-CNNPMHQNC) were selected [70,71], and chemically coupled to the surface-exposed lysines, the [Peptide]CPMV was subsequently incubated with the corresponding metal salt and reductant to generate CoPt–CPMV, FePt–CPMV, and ZnS–CPMV [72].

The ELD method was developed as an alternative approach to generate metal-coated virus particles based on the interaction of the CPMV surface amine groups with palladium ions acting as activator sites, which then formed palladium clusters after treatment with a reducing agent (dimethylamine borane). The clusters acted as nucleation sites for the subsequent adsorption of metal ions of interest to generate metal (nickel, cobalt, platinum, iron, nickel–iron, cobalt–platinum)-coated CPMV particles. This allowed production of metallized CPMV nanoparticles with a size of 35 nm in diameter, although deposition time could be extended to produce larger metallized spheres [73].

Furthermore, and importantly, it was subsequently shown that there is no requirement for the use of mineral-/metal-specific peptides or the ELD process for the CPMV-templated formation of monodisperse mineralized nanoparticles; it is sufficient to simply increase the virus surface negative charge. This is done by chemical modification of CPMV surface lysines with succinic acid to increase the overall negative charge on the virus capsid. The charge-modified CPMV ([Succinamate]CPMV) was coated with cobalt or iron oxide to give monodisperse nanoparticles of about 32 nm diameter. The iron oxide–CPMV nanoparticle surface can be functionalized further with, for example, oligosaccharides [74].

When a cationic polyelectrolyte, poly(allylamine hydrochloride) (PAH) was electrostatically adsorbed onto the external surface of the CPMV capsid, the polyelectrolyte promotes the adsorption of anionic gold complexes, which were then easily reduced, under mild conditions, to form a metallic gold layer. The process is simple and environmentally friendly, as only aqueous solvent and ambient temperature and pressure are required. This route offers a way to produce Au–CPMV with a narrow size distribution that can be further modified with thiol-containing moieties [75]. In contrast, reaction of polyelectrolyte-modified CPMV (^{PA}CPMV) with preformed gold nanoparticles resulted in the self-assembly of large, hexagonally packed, tessellated spheres (Figure 17.9) [75].

Small lanthanide molecules have been infused and entrapped inside the wild-type CPMV cavity by taking advantage of the negatively charge RNA interaction with gadolinium(III), Gd^{3+}, and terbium(III), Tb^{3+}, cations; approximately 80 Gd^{3+} or Tb^{3+} ions can be encapsulated [76]. However, as the importance of VLPs became apparent, some attempts to inactivate or eliminate the virus RNA within CPMV capsids were reported, including UV irradiation (254 nm) using a dosage of 2.0–2.5 J/cm². The inactivated particles were deemed as noninfectious to plants. However, particles remained intact and maintained chemical reactivity [77]. Moreover, the RNA content remains encapsidated so that the internal cavity of the particles is not accessible for material loading. Alternatively, alkaline hydrolysis was used to generate artificial empty CPMV particles [9]. The majority of the particles remain intact during the disassembly step. In addition, these processes risk altering the structural properties of the particles and generally do not actually remove RNA from the particles. The recent development of

a method for the production of large quantities of CPMV VLPs in plants, as described earlier [10], was the motivation behind the exploration of loading the generated CPMV VLPs with metal and metal oxide [78], dyes, and drugs [79]. It was noticed that the efficiency of loading with metal proved to be variable and depended on the particular VLP preparation used. The role of the carboxyl (C) terminus of the small coat (S) protein in controlling access to the interior of the VLP has been investigated through the determination of the efficiency of internal mineralization with cobalt [63]. The presence of a C-terminal 24–amino acid peptide of the S protein was found to inhibit internal loading, an effect that could be eliminated by enzymatic removal of this region with chymotrypsin. The amenability of the C-terminus to genetic modification has also been demonstrated. Substitution with six histidine residues generated stable particles and facilitated external mineralization by cobalt (Figure 17.10) [63].

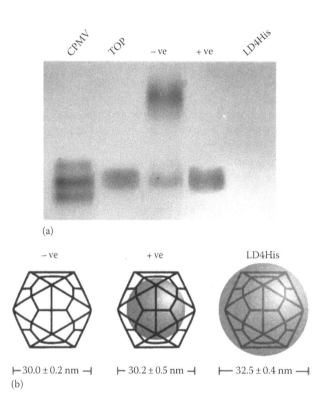

(a)

(b)

├─30.0 ± 0.2 nm ─┤   ├─30.2 ± 0.5 nm ─┤   ├─ 32.5 ± 0.4 nm ─┤

FIGURE 17.10 Loading of CPMV VLPs with cobalt. (a) Agarose gel of wild-type and LD4His VLPs subjected to the cobalt loading reaction. Wild-type VLPs were subjected to cobalt loading with (+ve) and without (−ve) pretreatment with chymotrypsin. CPMV virions (CPMV) and top components (top) were electrophoresed as controls. The gel was stained with Coomassie brilliant blue. (b) Representation of the pattern of mineralization of VLP variants showing dynamic light scattering measurements taken after the cobalt loading reaction: −ve shows no mineralization, +ve is internally mineralized, and LD4His is externally mineralized. (From Sainsbury, F., Saunders, K., Aljabali, A.A.A., Evans, D.J., and Lomonossoff, G.P.: Peptide-controlled access to the interior surface of empty virus nanoparticles. *ChemBioChem.* 2011. 12. 2435–2440. Copyright Wiley-VCH Verlag GmbH & Co. Reproduced with permission.)

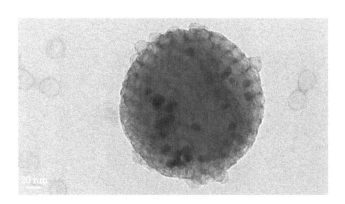

FIGURE 17.9 Stained TEM image of a single tessellated sphere showing the CPMV particles as hexagonally packed mosaic pieces in a spherical arrangement together with gold nanoparticles that appear as black circles. (Reprinted with permission from Aljabali, A.A.A. et al., CPMV-polyelectrolyte-templated gold nanoparticles, *Biomacromolecules*, 12(7), 2723. Copyright 2011 American Chemical Society.)

17.5 MODIFYING THE VIRUS SURFACE

The virus capsid surface can be addressed with a range of biological and chemical moieties. By doing so, a selection of functionalized nanobuilding blocks can be obtained. The external capsid surface of CCMV, CPMV, and TYMV possess a large number of selectively addressable amino acids allowing decoration with a large number of molecules. The native viruses display amine groups, from lysine, and carboxylate groups derived from aspartic and glutamic acids. These functional groups (–NH$_2$ or –COOH) on the surface offer a precise addressable site and are the most frequently used for chemical modification. This can be achieved using various techniques of covalent coupling [54,80–85] and by click chemistry, the copper(I)-catalyzed azide-alkyne [3 + 2] cycloaddition reaction [86–89]. Both wild-type and genetic variants of CPMV have been used as scaffolds for chemical modification; early investigations of the chemical reactivity of wild-type CPMV were aimed at lysine- and cysteine-selective derivatives. The X-ray structure and coordinates of CPMV indicated the presence of 5 solvent-exposed lysines per asymmetric unit, which equates to 300 exposed lysine side chains per CPMV particle (Figure 17.11) [90]. In a study on CPMV, single, double, triple, and quadruple lysine-minus mutants, in which the addressable lysines were sequentially replaced with arginines, were generated, and chemical labeling efficiency was measured. The studies indicated that all of the five lysines are available for functionalization; the degree of labeling efficiency varies between the different sites, and normally only four per subunit are modified [83,91]. The most reactive groups were found to be lysine 38 on the S protein and lysine 99 on the L protein [90].

Carboxylates derived from aspartic and glutamic acids have also been utilized. Chemical attachment of amine-containing compounds can be achieved by making use of the coupling reagents 1-ethyl-3-(3-dimethylaminopropyl) carbodiimide and *N*-hydroxysuccinimide. Up to 180 surface addressable carboxylates per CPMV virion have been modified with different moieties [83,92]. The native CPMV also displays two surface tyrosines per asymmetric unit that are available for chemical conjugation [80]. For CCMV, around 540 lysine residues and up to 560 carboxylates can be decorated [93]. In the case of TYMV, approximately 60 surface lysines and 90–120 carboxylate groups can be addressed [94].

Thiols derived from cysteine side chains are typically not found in a reactive form on the solvent-exposed surface of viruses, although cysteine side chains have been identified on the interior solvent-exposed surface of CPMV [86,95]. As thiols are nevertheless a useful group to use for conjugation, a range of cysteine-mutant viruses have been generated via genetic modification [8,95–101]. For CCMV$_{CYS}$ mutant, about 100 of the 300 introduced cysteinate thiols can be labeled [93], and for CPMV$_{CYS}$ mutant, 60 cysteinate thiols are introduced and addressed [95,102]. As an alternative route to genetic modification for the introduction of thiol groups at the capsid surface, a chemical method has been reported [103]. If required, histidines can be genetically engineered into the external surface [104].

The phenolic group of tyrosine side chains also provides a possible target for chemical modification. However, the phenolic group is only moderately reactive, and to achieve conjugation, the tyrosine has to be oxidized by one electron with, for example, peracid or persulfate reagents [105]. Commercial reagents are typically not available; therefore, the starting materials and coupling reagents must be chemically synthesized. The most common reaction utilizing tyrosine side chains on viruses is diazonium coupling, and this has been widely used with MS2 and TMV [106,107]. CPMV particles display two accessible tyrosine side chains located in the S subunit, which are available for chemical modification; fluorescein has been covalently attached [80].

CPMV has been utilized to construct enzyme-modified functional virus particles. The well-studied enzymes horseradish peroxidase (HRP) and glucose oxidase (GOX) were employed to prepare ^{Enzyme}CPMV conjugates by their conjugation to the external surface of wild-type CPMV. These

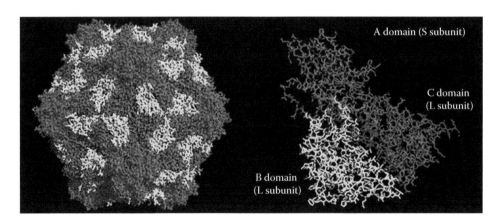

FIGURE 17.11 Schematic representation of the surface-exposed addressable amines and carboxylates of the CPMV capsid. (From Steinmetz, N.F. and Evans, D.J., Utilisation of plant viruses in bionanotechnology, *Org. Biomol. Chem.*, 5(18), 2891–2902, 2007. Reproduced by permission of The Royal Society of Chemistry.)

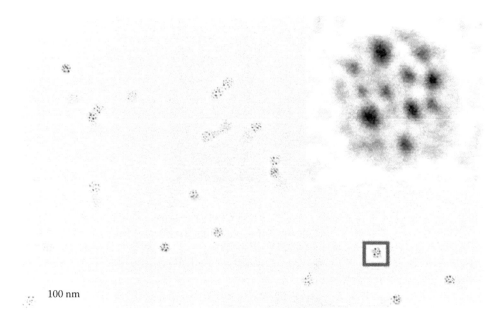

FIGURE 17.12 TEM image of [HRP-ADH]CPMV particles stained with 1% AgNO₃ solution that stains specifically the generated aldehyde groups as a result of the oxidation process confirming the presence of enzyme. Inset confirms the localization of the metallic silver on the virus capsid. (From Aljabali, A.A.A. et al., Controlled immobilisation of active enzymes on the Cowpea mosaic virus capsid, *Nanoscale*, 4(18), 5640–5645, 2012. Reproduced by permission of The Royal Society of Chemistry.)

relatively large biomacromolecules can be bound to the virus surface by simple coupling strategies without destroying their biological activity. The number of enzymes bound to the surface of the virus was determined to be approximately 11 HRP and 2–3 GOX, respectively. The [HRP]CPMV (Figure 17.12) and [GOX]CPMV particles have potential uses as building blocks for catalytic devices, diagnostic assays, or biosensors [108].

In this context, both carboxy- and amine-containing surface-exposed amino acids have been modified with electroactive moieties. The incorporation of ferrocene to CPMV ([Ferrocene]CPMV) provided nanostructures with well-defined sizes and functionalities [91]. Cyclic voltammetry established that [Ferrocene]CPMV is a redox-active nanoparticle. The ratio of the peak currents is close to unity showing the ferrocene/ferrocenium couple is electrochemically reversible. The peak currents were proportional to the square root of the scan rate indicating that oxidation and reduction were diffusion controlled [83,91]. Ferrocene dendrimers have shown similar properties due to fast rotation of the dendrimer compared to the electrochemical timescale, so that all the redox centers come close to the electrode within this timescale [109].

Over the past few years, a large number of biological, organic, and inorganic molecules have been attached to, especially, wild-type CPMV and mutant virions for different applications [110–112], demonstrating that these, and other virus nanoparticles (VNPs), can be regarded as robust and multi-addressable nanobuilding blocks. In a similar fashion to that described for the icosahedral viruses, TMV has been used as a scaffold for the selective attachment of fluorescent dyes and other small molecules. The native virions offer addressable exterior tyrosine residues and interior carboxylates derived from glutamic acid residues [113]. The tyrosine residues can be quantitatively functionalized with an alkyne residue by diazonium coupling; then subsequently, copper(I)-catalyzed azide-alkyne click chemistry facilitates the efficient conjugation of a wide range of compounds [114].

17.6 FILMS, LAYERS, AND ARRAYS

The formation of highly ordered nanostructured films, layers, and arrays are of great interest for the design of novel functional materials that might find applications in, for example, sensors, optoelectronics, nanoelectronics, and biomedical applications. VNPs have the capability to self-assemble into discrete particles but show also a propensity for self-organization.

Virus crystallization can lead to the formation of mesoscale self-organization in three dimensions. Icosahedral CPMV particles, for example, can crystallize into well-ordered arrays of 1×10^{13} particles in a typical 1 mm³ crystal [102]. CPMV will also self-assemble at the interface between two immiscible liquids. The virus particles segregate at a perfluorodecalin–water interface and stabilize the dispersion of the oil droplets. Cross-linking of the particles at the oil–water interface can be achieved with either glutaraldehyde or with biotin/avidin, which locks the assembly into place to give a well-defined, robust membrane; the integrity of the virus particles are not disrupted [115]. CPMV particles can also be assembled with polyamic acid in aqueous solution via the layer-by-layer technique. Then, upon thermal treatment, CPMV particles are removed, and polyamic acid is converted into polyimide in one step, resulting in a porous polyimide film [116].

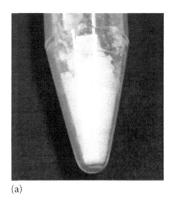
(a)

(b)

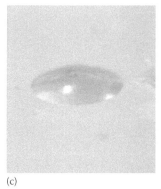
(c)

FIGURE 17.13 Optical photographs of sample tubes containing [Cat-CPMV][S] showing (a) a low-density white solid obtained after lyophilization and (b) a highly viscous translucent fluid obtained after annealing the freeze-dried material at 65°C for 10 min. (c) Optical image showing single liquid droplet of a solvent-free [Cat-CPMV][S] melt placed on a glass slide at 65°C. (From Patil, A.J., McGrath, N., Barclay, J.E. et al.: Liquid viruses by nanoscale engineering of capsid surfaces. *Adv. Mater.* 2012. 24(33). 4557–4563. Copyright Wiley-VCH Verlag GmbH & Co. Reproduced with permission.)

CPMV has been chemically engineered to generate a solvent-free liquid virus (Figure 17.13) [117]. The CPMV capsid asymmetric unit contains 51 acid residues (21 glutamic acid and 30 aspartic acid). The high anionic charge density was exploited to produce stoichiometric constructs capable of undergoing thermally induced melting in the absence of water. CPMV was cationized by covalent coupling of ethylenediamine. The cationic CPMV was further modified with anionic polymer–surfactant, poly(ethyleneglycol) 4-nonylphenyl-3-sulfopropyl ether. Extensive lyophilization of the single component polymer–surfactant/virus, [Cat-CPMV][S], produced a low-density white solid that on thermal annealing melted to a viscous translucent fluid that on cooling gave a soft, translucent solid. The ability to prepare solvent-free liquids comprising extremely high concentrations of structurally and functionally intact VNPs should have a significant impact in advancing the bioinspired design and processing of biologically derived nanostructures [117].

CCMV carries a negative net charge on its outer surfaces at neutral pH (isoelectric point, pI ≈ 3.8). The negative charge density on CCMV is not homogeneously distributed but is located in patches on the capsid surface. The presence of directional *sticky patches* can direct nanostructure formation, which is controlled by interactions between the patches [118]. 1-Pentanethiol-stabilized gold nanoparticles (AuNPs) were mixed with the virus in different pH conditions that resulted in various patterns of gold superlattices adsorbed onto the virus capsid [119]. TMV is adsorbed as a monolayer on gold surfaces or hydroxyl-containing surfaces, such as mica glass and silicon. In addition, covalent immobilization can be achieved utilizing acyl chloride that reacts with groups on the virus particles to form ester bonds [120]. Genetically modified CCMV$_{CYS}$ and CPMV$_{CYS}$ VNPs that present cysteine residues on the surface can be readily coupled to patterned gold templates using gold–thiol chemistry [96,121]. Combined with nanolithography techniques, this approach was utilized to control assembly of VNPs [121,122]. A similar method enables the deposition of a uniform monolayer of CPMV$_{CYS}$ mutant particles

onto a gold surface; this can be visualized by fluorescence microscopy when a suitable dye is conjugated to the virus particles [123]. Alternatively, biotinylated CPMV can be bound easily to streptavidin (SAv)-modified surfaces [123]. Further, CPMV$_{HIS}$ mutants, which have introduced histidines on the solvent-exposed surface, have been discretely immobilized on a NeutrAvidin-functionalized surface utilizing a biotin linker [124]. CPMV$_{HIS}$ can also be reversibly coupled to a nickel–nitrilotriacetic acid–terminated nanografted-patterned surface. The attachment/detachment of the CPMV$_{HIS}$ is tuneable by variation of conditions such as pH or addition of imidazole. This system has provided a means of defining the morphological and assembly kinetics of the entropic virus condensation and organization on the surface as a function of virus flux/mobility and intervirus interaction [125].

Both CCMV [126] and CPMV [123,127] 3D arrays of VNPs were assembled via a layer-by-layer approach using the high affinity interaction between biotin and SAv. In the case of CCMV, a monolayer of biotinylated CCMV was electrostatically conjugated to a functionalized silicon solid surface [44] and then further layers added [126]. For CPMV, the base layer was formed by immobilization by either direct binding of CPMV$_{CYS}$ particles onto gold or indirectly using CPMV–biotin mediated by a thiol-modified SAv. For ease of monitoring and tracking the assembly of the layers, one set of virus particles was labeled with the fluorescent dye AlexaFluor (AF) 488 and biotin and the another batch with AF568 and biotin. The first layer of building blocks was assembled with great control using the sulfur–gold interaction (SAv–thiol) and further layers by specific interactions between CPMV-bound biotin and SAv [123]. The successful assembly process was monitored by fluorescence microscopy and quartz crystal microbalance with dissipation monitoring [127].

Multilayer structures can also be fabricated by the use of electrostatic interactions. The electrostatic deposition of oppositely charged polyelectrolyte (polyallylamine and PSS) layers on spherical CarMV particles enabled the polyanion layer to be replaced by a monolayer of negatively charged

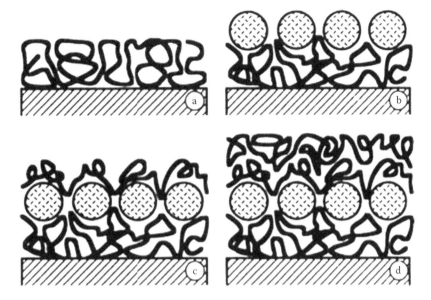

FIGURE 17.14 The deposition process of alternate layers of polyelectrolytes and CarMV. (a) Smooth polyelectrolyte precursor film. (b) Adsorbed virus particles partially intrude into the polymer layer. (c) The next polymer layer fills the empty space between the viruses and leads to a partial annealing of the surface. (d) Further polymer layers completely anneal the surface so as to give conventional roughness and layer thickness. (Reprinted with permission from Lvov et al., Successive deposition of alternate layers of polyelectrolytes and a charged virus, *Langmuir*, 10, 4232. Copyright 1994 American Chemical Society.)

CarMV particles [128]. The virus was electrostatically adsorbed onto the positively charged polymer and embedded within it. The next deposited layers, of several alternating polyanion and polycation layers, fill holes between the virus and restore the surface layer enabling another layer of virus to be deposited (Figure 17.14). The electrostatic interaction of CCMV with various surfaces was also investigated [44] and led to the construction of alternating polyelectrolyte (polylysine) and CCMV layers [126]. In this case, a cationic polymer layer was sufficient to promote the binding of successive layers of CCMV. The layer-by-layer approach has also been used to assemble CPMV and TMV into polyelectrolyte substrates [129]. Although CPMV and TMV have similar external surface charge and density, as a consequence of their shape, different layer architectures were obtained. The sphere-like particles are incorporated into the architecture, consistent with the properties of CarMV and CCMV described previously, to give a self-assembled, alternating structure of polyelectrolytes and VNPs. However, in stark contrast, the rod-shaped particles are excluded from the arrays and float on top of the polyionic layer architecture in an ordered state. This may provide a starting point for the design and construction of highly organized virus assemblies bound on polyelectrolyte films.

TMV can be organized and aligned by an alternative slow assembly technique, by pulling a meniscus containing the virus suspension over a solid surface, to give a film of controlled structure, thickness, and long-range virus orientation [130]. The propensity of TMV, at high concentration, to self-assemble into a nematic liquid crystal has been exploited to prepare mesostructured and mesoporous inverse silica replicas, by controlled solgel condensation of silicon dioxide on TMV liquid crystals followed by thermal degradation (calcination) [130,131]. Successful replication of the nematic phase requires balancing the rate of organosilane reagent hydrolysis and the time for realignment of the TMV after initial mixing. The highly ordered, micrometer-sized mesostructures that were obtained consist of a periodic array of coaligned, end-to-end-joined, TMV particles intercalated with a continuous filament of amorphous silica with a periodicity of about 20 nm. Calcination to remove the TMV liquid crystal template has a minimal effect on the periodicity of the inverse replica. At lower reagent concentrations, different nanoparticle forms are obtained that, after calcination, have a dense silica core of about 35 nm surrounded by a radial array of linear channels of about 50 nm length [131].

17.7 APPLICATIONS OF VIRUSES IN NANOTECHNOLOGY

The use of viruses in nanotechnology offers the prospect of rich and diverse technological advancements. For example, the deposition of platinum on the TMV capsid surface forms nanoparticles with an average diameter of 10 nm [30]. When Pt–TMV was placed between two aluminum electrodes, a unique memory effect was observed in which the RNA and the TMV capsid acted as charge donor and energy barrier, respectively. When a current was applied, the charge transferred from aromatic rings, such as guanine in the RNA, and was trapped in the platinum nanoparticles. The reversible charge transfer and charge trapped process have created a memory device. Similarly, another memory device was designed using CPMV with the exterior capsid modified with QDs [132]. The zinc sulfide capped layer of QD acted as charge storage, while the aromatic residues (tryptophan) on the CPMV capsid acted as a charge transporter. Ferromagnetic materials play an important role in the development of high-density storage devices, for example, a computer disk contains a 2D ferromagnetic thin film, where information is stored. Ferromagnetic nanoparticles, like FePt and CoPt [133], have emerged as potential high-density storage media. Therefore, the ability to generate highly monodisperse FePt– and CoPt–CPMV might be an alternative route to synthesize this type of particle [72], although their physical properties are yet to be established. Many VNPs are used to synthesize magnetic and metallic nanoparticles of various shapes. Besides, hollow metallic nanoparticles and nanoshells, which can differ in their properties, might hold great potential in nanomaterial development.

Furthermore, viruses are considered as naturally biocompatible and biodegradable, which is particularly important for medical applications. Plant viruses are less likely to interact with human receptors, and virus replication or gene expression is not supported in mammalian systems [85,134]. Nanoparticles and VNPs are currently under investigation in the field of nanomedicine (broadly defined as the application of nanotechnology to medicine) with the aim of control of drug delivery, targeting of diseased cells and as imaging agents. VNPs are less likely to interact specifically with the mammalian system and, therefore, are less likely to cause potential side effects [135]. For CPMV, it was shown that, even up to dosages of 10^{16} CPMV particles/kg body weight, no apparent toxic side effects were observed in mice [136]. It has been reported that CPMV and CCMV have a broad biodistribution and were detected in a wide variety of tissues throughout the body with no apparent toxic effects [137,138]. CPMV particles mostly accumulated in the liver and spleen [136]. To reduce their immunogenicity, VNPs can be PEGylated (modified with PEG) [139–141].

VNPs have been modified with fluorophores, QDs, and metallic nanoparticles, and gadolinium complexes have been developed for in vivo imaging agents [105,110,142,143]. FluorescentCPMV particles allowed high-resolution imaging of major blood vessels to a depth of up to 500 μm in comparison to fluorescent-labeled nanospheres [139]. Furthermore, it has been shown that Gd^{3+} can be covalently attached to the exterior or interior surface of viruses or electrostatically interacted with the encapsidated RNA molecules [76,144–146] indicating that VNPs could serve as excellent candidates for magnetic resonance imaging (MRI) contrast agents.

To conclude, in this chapter, it has been described how plant viruses can be utilized for the assembly of a wide range of bionanomaterials. The virus particles can be modified on their external surface, can incorporate material within their capsids, and can form films, layers, arrays, and liquid proteins. The potential for their application is extensive, ranging from new materials through novel nanodevices to medical imaging and therapy.

REFERENCES

1. Harrison, S. C. (1990). Principles of virus structure. In *Fundamental Virology*, Field, B. N. and Knipe, D. M., Eds., pp. 37–61, New York: Raven Press.
2. Liepold, L. O., J. Revis, M. Allen, L. Oltrogge, M. Young, T. Douglas (2005). Structural transitions in Cowpea chlorotic mottle virus (CCMV). *Physical Biology* 2(4): S166–S172.
3. Steinmetz, N. F., N. S. Shah, J. E. Barclay, G. Rallapalli, G. P. Lomonossoff, D. J. Evans (2009). Virus-templated silica nanoparticles. *Small* 5(7): 813–816.
4. Shah, S. N., N. F. Steinmetz, A. A. A. Aljabali, G. P. Lomonossoff, D. J. Evans (2009). Environmentally benign synthesis of virus-templated, monodisperse, iron-platinum nanoparticles. *Dalton Transactions* 40: 8479–8480.
5. Wellink, J. (1998). Comovirus isolation and RNA extraction. *Methods in Molecular Biology* 81: 205–209.
6. Brumfield, S., D. Willits, L. Tang, J. E. Johnson, T. Douglas, M. Young (2004). Heterologous expression of the modified coat protein of Cowpea chlorotic mottle bromovirus results in the assembly of protein cages with altered architectures and function. *Journal of Genetic Virology* 85(4): 1049–1053.
7. Johnson, J. E. and W. Chiu (2000). Structures of virus and virus-like particles. *Current Opinion in Structural Biology* 10(2): 229–235.
8. Miller, R. A., A. D. Presley, M. B. Francis (2007). Self-assembling light-harvesting systems from synthetically modified tobacco mosaic virus coat proteins. *Journal of the American Chemical Society* 129(11): 3104–3109.
9. Ochoa, W. F., A. Chatterji, T. W. Lin, J. E. Johnson (2006). Generation and structural analysis of reactive empty particles derived from an icosahedral virus. *Chemistry and Biology* 13(7): 771–778.
10. Saunders, K., F. Sainsbury, G. P. Lomonossoff (2009). Efficient generation of Cowpea mosaic virus empty virus-like particles by the proteolytic processing of precursors in insect cells and plants. *Virology* 393(2): 329–337.
11. Sainsbury, F. and G. P. Lomonossoff (2008). Extremely high-level and rapid transient protein production in plants without the use of viral replication. *Plant Physiology* 148(3): 1212–1218.
12. Sainsbury, F., E. C. Thuenemann, G. P. Lomonossoff (2009). pEAQ: Versatile expression vectors for easy and quick transient expression of heterologous proteins in plants. *Plant Biotechnology Journal* 7(7): 682–693.
13. Montague, N. P., E. C. Thuenemann, P. Saxena, K. Saunders, P. Lenzi, G. P. Lomonossoff (2011). Recent advances of Cowpea mosaic virus-based particle technology. *Human Vaccine* 7(3): 383–390.
14. Douglas, T. and M. Young (1998). Host-guest encapsulation of materials by assembled virus protein cages. *Nature* 393(6681): 152–155.
15. Shenton, W., T. Douglas, M. Young, G. Stubbs, S. Mann (1999). Inorganic–organic nanotube composites from template mineralization of tobacco mosaic virus. *Advanced Materials* 11: 253–256.
16. Anobom, C. D., S. C. Albuquerque, F. P. Albernaz et al. (2003). Structural studies of MS2 bacteriophage virus particle disassembly by nuclear magnetic resonance relaxation measurements. *Biophysical Journal* 84(6): 3894–3903.
17. Abbing, A., U. K. Blaschke, S. Grein et al. (2004). Efficient intracellular delivery of a protein and a low molecular weight substance via recombinant polyomavirus-like particles. *Journal of Biological Chemistry* 279(26): 27410–27421.

18. Douglas, T. and M. Young (1999). Virus particles as templates for materials synthesis. *Advanced Materials* 11(8): 679–681.

19. Comellas-Aragones, M., H. Engelkamp, V. I. Claessen et al. (2007). A virus-based single-enzyme nanoreactor. *Nature Nanotechnology* 2(10): 635–639.

20. Loo, L., R. H. Guenther, V. R. Basnayake, S. A. Lommel, S. Franzen (2006). Controlled encapsidation of gold nanoparticles by a viral protein shell. *Journal of the American Chemical Society* 128(14): 4502–4503.

21. Loo, L., R. H. Guenther, S. A. Lommel, S. Franzen (2007). Encapsidation of nanoparticles by Red clover necrotic mosaic virus. *Journal of the American Chemical Society* 129(36): 11111–11117.

22. Dragnea, B., C. Chen, E. S. Kwak, B. Stein, C. C. Kao (2003). Gold nanoparticles as spectroscopic enhancers for in vitro studies on single viruses. *Journal of the American Chemical Society* 125(21): 6374–6375.

23. Chen, C., M. C. Daniel, Z. T. Quinkert et al. (2006). Nanoparticle-templated assembly of viral protein cages. *Nano Letters* 6(4): 611–615.

24. Dixit, S. K., N. L. Goicochea, M. C. Daniel et al. (2006). Quantum dot encapsulation in viral capsids. *Nano Letters* 6(9): 1993–1999.

25. Pattanayek, R. and G. Stubbs (1992). Structure of the U2 strain of tobacco mosaic virus refined at 3.5 Å resolution using X-ray fiber diffraction. *Journal of Molecular Biology* 228(2): 516–528.

26. Stubbs, G. (1999). Tobacco mosaic virus particle structure and the initiation of disassembly. *Philosophical Transactions of the Royal Society of London Series B-Biological Sciences* 354(1383): 551–557.

27. Balci, S., A. M. Bittner, M. Schirra et al. (2009). Catalytic coating of virus particles with zinc oxide. *Electrochimica Acta* 54(22): 5149–5154.

28. Knez, M., A. Kadri, C. Wege, U. Gosele, H. Jeske, K. Nielsch (2006). Atomic layer deposition on biological macromolecules: Metal oxide coating of tobacco mosaic virus and ferritin. *Nano Letters* 6(6): 1172–1177.

29. Royston, E., S. Y. Lee, J. N. Culver, M. T. Harris (2006). Characterization of silica-coated tobacco mosaic virus. *Journal of Colloid and Interface Science* 298(2): 706–712.

30. Gorzny, M. L., A. S. Walton, M. Wnek, P. G. Stockley, S. D. Evans (2008). Four-probe electrical characterization of Pt-coated TMV-based nanostructures. *Nanotechnology* 19(16): 165704.

31. Liu, N., C. Wang, W. Zhang et al. (2012). Au nanocrystals grown on a better-defined one-dimensional tobacco mosaic virus coated protein template genetically modified by a hexahistidine tag. *Nanotechnology* 23(33): 335602.

32. Wnek, M., M. L. Gorzny, M. B. Ward et al. (2013). Fabrication and characterization of gold nano-wires templated on virus-like arrays of tobacco mosaic virus coat proteins. *Nanotechnology* 24(2): 025605.

33. Rong, J. H., F. Oberbeck, X. N. Wang et al. (2009). Tobacco mosaic virus templated synthesis of one dimensional inorganic-polymer hybrid fibres. *Journal of Materials Chemistry* 19(18): 2841–2845.

34. Balci, S., K. Hahn, P. Kopold et al. (2012). Electroless synthesis of 3 nm wide alloy nanowires inside tobacco mosaic virus. *Nanotechnology* 23(4): 045603.

35. Lee, S. Y., E. Royston, J. N. Culver, M. T. Harris (2005). Improved metal cluster deposition on a genetically engineered tobacco mosaic virus template. *Nanotechnology* 16(7): S435–S441.

36. Niu, Z., M. A. Bruckman, S. Q. Li et al. (2007). Assembly of tobacco mosaic virus into fibrous and macroscopic bundled arrays mediated by surface aniline polymerization. *Langmuir* 23(12): 6719–6724.

37. Namba, K. and G. Stubbs (1986). Structure of tobacco mosaic virus at 3.6 Å resolution: Implications for assembly. *Science* 231(4744): 1401–1406.

38. Dujardin, E., C. Peet, G. Stubbs, J. N. Culver, S. Mann (2003). Organisation of metallic nanoparticles using tobacco mosaic virus. *Nano Letters* 3: 413–417.

39. Bromley, K. M., A. J. Patil, A. W. Perriman, G. Stubbs, S. Mann (2008). Preparation of high quality nanowires by tobacco mosaic virus templating of gold nanoparticles. *Journal of Materials Chemistry* 18: 4796–4801.

40. Knez, M., M. Sumser, A. M. Bittner et al. (2004). Spatially selective nucleation of metal clusters on the tobacco mosaic virus. *Advanced Functional Materials* 14(2): 116–124.

41. Balci, S., A. M. Bittner, K. Hahn et al. (2006). Copper nanowires within the central channel of tobacco mosaic virus particles. *Electrochimica Acta* 51(28): 6251–6257.

42. Royston, E., A. Ghosh, P. Kofinas, M. T. Harris, J. N. Culver (2008). Self-assembly of virus-structured high surface area nanomaterials and their application as battery electrodes. *Langmuir* 24(3): 906–912.

43. Speir, J. A., S. Munshi, G. Wang, T. S. Baker, J. E. Johnson (1995). Structures of the native and swollen forms of Cowpea chlorotic mottle virus determined by X-ray crystallography and cryo-electron microscopy. *Structure* 3(1): 63–78.

44. Suci, P. A., M. T. Klem, F. T. Arce, T. Douglas, M. Young (2005). Influence of electrostatic interactions on the surface adsorption of a viral protein cage. *Langmuir* 21(19): 8686–8693.

45. Schneemann, A. and M. J. Young (2003). Viral assembly using heterologous expression systems and cell extracts. *Virus Structure* 64: 1–36.

46. Slocik, J. M., R. R. Naik, M. O. Stone, D. W. Wright (2005). Viral templates for gold nanoparticles synthesis. *Journal of Materials Chemistry* 15(7): 749–753.

47. Hu, Y. F., R. Zandi, A. Anavitarte, C. M. Knobler, W. M. Gelbart (2008). Packaging of a polymer by a viral capsid: The interplay between polymer length and capsid size. *Biophysical Journal* 94(4): 1428–1436.

48. Comellas-Aragones, M., A. de la Escosura, A. J. Dirks et al. (2009). Controlled integration of polymers into viral capsids. *Biomacromolecules* 10(11): 3141–3147.

49. Ren, Y. P., S. M. Wong, L. Y. Lim (2006). In vitro-reassembled plant virus-like particles for loading of polyacids. *Journal of General Virology* 87: 2749–2754.

50. Loo, L., R. H. Guenther, S. A. Lommel, S. Franzen (2008). Infusion of dye molecules into Red clover necrotic mosaic virus. *Chemical Communications* (1): 88–90.

51. Klem, M. T., M. Young, T. Douglas (2008). Biomimetic synthesis of β-TiO$_2$ inside a viral capsid. *Journal of Materials Chemistry* 18: 3821–3823.

52. de la Escosura, A., M. Verwegen, F. D. Sikkema et al. (2008). Viral capsids as templates for the production of monodisperse Prussian blue nanoparticles. *Chemical Communications* (13): 1542–1544.

53. Lin, T. and J. E. Johnson (2003). Structures of picorna-like plant viruses: Implications and applications. *Advances in Virus Research* 62: 167–239.

54. Evans, D. J. (2010). Bionanoscience at the plant virus-inorganic chemistry interface. *Inorganica Chimica Acta* 363(6): 1070–1076.

55. Sainsbury, F., M. C. Cañizares, G. P. Lomonossoff (2010). Cowpea mosaic virus: The plant virus-based biotechnology workhorse. *Annual Review of Phytopathology* 48(1): 437–455.

56. Lin, T. W., Z. G. Chen, R. Usha et al. (1999). The refined crystal structure of Cowpea mosaic virus at 2.8 Ångstrom resolution. *Virology* 265(1): 20–34.

57. Bruening, G. and H. O. Agrawal (1967). Infectivity of a mixture of Cowpea mosaic virus ribonucleoprotein components. *Virology* 32(2): 306–320.

58. van Kammen, A. (1967). Purification and properties of the components of Cowpea mosaic virus. *Virology* 31(4): 633–642.

59. Liu, L. and G. P. Lomonossoff (2002). Agroinfection as a rapid method for propagating Cowpea mosaic virus-based constructs. *Journal of Virological Methods* 105(2): 343–348.

60. Goldbach, R., G. Rezelman, A. Vankammen (1980). Independent replication and expression of B-component RNA of Cowpea mosaic virus. *Nature* 286(5770): 297–300.

61. Lomonossoff, G. P. and J. E. Johnson (1991). The synthesis and structure of Comovirus capsids. *Progress in Biophysics and Molecular Biology* 55(2): 107–137.

62. Semancik, J. S. (1966). Studies on electrophoretic heterogeneity in isometric plant viruses. *Virology* 30(4): 698–704.

63. Sainsbury, F., K. Saunders, A. A. A. Aljabali, D. J. Evans, G. P. Lomonossoff (2011). Peptide-controlled access to the interior surface of empty virus nanoparticles. *ChemBioChem* 12: 2435–2440.

64. Niblett, C. L. and J. S. Semancik (1969). Conversion of the electrophoretic forms of Cowpea mosaic virus in vivo and in vitro. *Virology* 38(4): 685–693.

65. Kridl, J. C. and G. Bruening (1983). Comparison of capsids and nucleocapsids from Cowpea mosaic virus-infected Cowpea protoplasts and seedlings. *Virology* 129(2): 369–380.

66. Taylor, K. M., V. E. Spall, P. J. G. Butler, G. P. Lomonossoff (1999). The cleavable carboxyl-terminus of the small coat protein of Cowpea mosaic virus is involved in RNA encapsidation. *Virology* 255(1): 129–137.

67. Dessens, J. T. and G. P. Lomonossoff (1993). Cauliflower mosaic virus 35S promoter-controlled DNA copies of Cowpea mosaic virus RNAs are infectious on plants. *Journal of Genetic Virology* 74(5): 889–892.

68. Lin, T., C. Porta, G. P. Lomonossoff, J. E. Johnson (1996). Structure-based design of peptide presentation on a viral surface: The crystal structure of a plant/animal virus chimera at 2.8 Å resolution. *Folding and Design* 1(3): 179–187.

69. Lomonossoff, G. P. and J. E. Johnson (1996). Use of macromolecular assemblies as expression systems for peptides and synthetic vaccines. *Current Opinion in Structural Biology* 6(2): 176–182.

70. Mao, C., D. J. Solis, B. D. Reiss et al. (2004). Virus-based toolkit for the directed synthesis of magnetic and semiconducting nanowires. *Science* 303(5655): 213–217.

71. Reiss, B. D., C. Mao, D. J. Solis, K. S. Ryan, T. Thomson, A. M. Belcher (2004). Biological routes to metal alloy ferromagnetic nanostructures. *Nano Letters* 4(6): 1127–1132.

72. Aljabali, A. A. A., S. N. Shah, R. Evans-Gowing, G. P. Lomonossoff, D. J. Evans (2011). Chemically-coupled-peptide-promoted virus nanoparticle templated mineralization. *Integrative Biology* 3(2): 119–125.

73. Aljabali, A. A. A., J. E. Barclay, G. P. Lomonossoff, D. J. Evans (2010). Virus templated metallic nanoparticles. *Nanoscale* 2: 2596–2600.

74. Aljabali, A. A. A., J. E. Barclay, O. Cespedes et al. (2011). Charge modified Cowpea mosaic virus particles for templated mineralization. *Advanced Functional Materials* 21(21): 4137–4142.

75. Aljabali, A. A. A., G. P. Lomonossoff, D. J. Evans (2011). CPMV-polyelectrolyte-templated gold nanoparticles. *Biomacromolecules* 12(7): 2723–2728.

76. Prasuhn, D. E., Jr., R. M. Yeh, A. Obenaus, M. Manchester, M. G. Finn (2007). Viral MRI contrast agents: Coordination of Gd by native virions and attachment of Gd complexes by azide-alkyne cycloaddition. *Chemical Communications* (12): 1269–1271.

77. Rae, C., K. J. Koudelka, G. Destito, M. N. Estrada, M. J. Gonzalez, M. Manchester (2008). Chemical addressability of ultraviolet-inactivated viral nanoparticles (VNPs). *PLoS One* 3(10): e3315.

78. Aljabali, A. A. A., F. Sainsbury, G. P. Lomonossoff, D. J. Evans (2010). Cowpea mosaic virus unmodified empty viruslike particles loaded with metal and metal oxide. *Small* 6(7): 818–821.

79. Wen, A. M., S. Shukla, P. Saxena et al. (2012). Interior engineering of a viral nanoparticle and its tumor homing properties. *Biomacromolecules* 13(12): 3990–4001.

80. Meunier, S., E. Strable, M. G. Finn (2004). Crosslinking of and coupling to viral capsid proteins by tyrosine oxidation. *Chemical Biology* 11(3): 319–326.

81. Hermanson, T. G. (2008). *Bioconjugate Techniques*. London, U.K.: Academic Press.

82. Laufer, B., N. F. Steinmetz, V. Hong, M. Manchester, H. Kessler, M. G. Finn (2009). Guiding VLP's the right way: Coating of virus like particles with peptidic integrin ligands. *Biopolymers* 92(4): 323–323.

83. Aljabali, A. A. A., J. E. Barclay, J. N. Butt, G. P. Lomonossoff, D. J. Evans (2010). Redox-active ferrocene-modified Cowpea mosaic virus nanoparticles. *Dalton Transactions* 39(32): 7569–7574.

84. Pokorski, J. K. and N. F. Steinmetz (2011). The art of engineering viral nanoparticles. *Molecular Pharmaceutics* 8(1): 29–43.

85. Steinmetz, N. F., C. F. Cho, A. Ablack, J. D. Lewis, M. Manchester (2011). Cowpea mosaic virus nanoparticles target surface vimentin on cancer cells. *Nanomedicine* 6(2): 351–364.

86. Wang, Q., T. R. Chan, R. Hilgraf, V. V. Fokin, K. B. Sharpless, M. G. Finn (2003). Bioconjugation by copper(I)-catalyzed azide-alkyne [3 + 2] cycloaddition. *Journal of the American Chemical Society* 125(11): 3192–3193.

87. Sen Gupta, S., J. Kuzelka, P. Singh, W. G. Lewis, M. Manchester, M. G. Finn (2005). Accelerated bioorthogonal conjugation: A practical method for the ligation of diverse functional molecules to a polyvalent virus scaffold. *Bioconjugate Chemistry* 16(6): 1572–1579.

88. Sen Gupta, S., K. S. Raja, E. Kaltgrad, E. Strable, M. G. Finn (2005). Virus-glycopolymer conjugates by copper(I) catalysis of atom transfer radical polymerization and azide-alkyne cycloaddition. *Chemical Communications* (34): 4315–4317.

89. Manchester, M. and P. Singh (2006). Virus-based nanoparticles (VNPs): Platform technologies for diagnostic imaging. *Advanced Drug Delivery Review* 58(14): 1505–1522.

90. Chatterji, A., W. Ochoa, M. Paine, B. R. Ratna, J. E. Johnson, T. W. Lin (2004). New addresses on an addressable virus nanoblock: Uniquely reactive Lys residues on Cowpea mosaic virus. *Chemical Biology* 11(6): 855–863.

91. Steinmetz, N. F., G. P. Lomonossoff, D. J. Evans (2006). Decoration of Cowpea mosaic virus with multiple, redox-active, organometallic complexes. *Small* 2(4): 530–533.

92. Steinmetz, N. F., G. P. Lomonossoff, D. J. Evans (2006). Cowpea mosaic virus for material fabrication: Addressable carboxylate groups on a programmable nanoscaffold. *Langmuir* 22(8): 3488–3490.

93. Gillitzer, E., D. Willits, M. Young, T. Douglas (2002). Chemical modification of a viral cage for multivalent presentation. *Chemical Communications* (20): 2390–2391.

94. Barnhill, H. N., R. Reuther, P. L. Ferguson, T. Dreher, Q. Wang (2007). Turnip yellow mosaic virus as a chemoaddressable bionanoparticle. *Bioconjugate Chemistry* 18(3): 852–859.

95. Wang, Q., T. Lin, J. E. Johnson, M. G. Finn (2002). Natural supramolecular building blocks. cysteine-added mutants of Cowpea mosaic virus. *Chemical Biology* 9(7): 813–819.

96. Klem, M. T., D. Willits, M. Young, T. Douglas (2003). 2-D array formation of genetically engineered viral cages on Au surfaces and imaging by atomic force microscopy. *Journal of the American Chemical Society* 125(36): 10806–10807.

97. Peabody, D. S. (2003). A viral platform for chemical modification and multivalent display. *Journal of Nanobiotechnology* 1(1): 5.

98. Kreppel, F., J. Gackowski, E. Schmidt, S. Kochanek (2005). Combined genetic and chemical capsid modifications enable flexible and efficient de- and retargeting of adenovirus vectors. *Molecular Therapy* 12(1): 107–117.

99. Khalil, A. S., J. M. Ferrer, R. R. Brau et al. (2007). Single M13 bacteriophage tethering and stretching. *Proceedings of the National Academy of Sciences of the United States of America* 104(12): 4892–4897.

100. Destito, G., A. Schneemann, M. Manchester et al. (2009). Biomedical nanotechnology using virus-based nanoparticles. *Current Topics of Microbiology and Immunology* 327: 95–122.

101. Blum, A. S., C. M. Soto, C. D. Wilson et al. (2004). Cowpea mosaic virus as a scaffold for 3-D patterning of gold nanoparticles. *Nano Letters* 4(5): 867–870.

102. Wang, Q., T. Lin, L. Tang, J. E. Johnson, M. G. Finn (2002). Icosahedral virus particles as addressable nanoscale building blocks. *Angewandte Chemie-International Edition* 41(3): 459–462.

103. Steinmetz, N. F., D. J. Evans, G. P. Lomonossoff (2007). Chemical introduction of reactive thiols into a viral nanoscaffold: A method that avoids virus aggregation. *ChemBioChem* 8(10): 1131–1136.

104. Chatterji, A., W. F. Ochoa, T. Ueno, T. W. Lin, J. E. Johnson (2005). A virus-based nanoblock with tunable electrostatic properties. *Nano Letters* 5(4): 597–602.

105. Manchester, M. and N. F. Steinmetz (2009). *Viruses and Nanotechnology* (Current Topics in Microbiology and Immunology). Berlin, Germany: Springer.

106. Kovacs, E. W., J. M. Hooker, D. W. Romanini, P. G. Holder, K. E. Berry, M. B. Francis (2007). Dual-surface-modified bacteriophage MS2 as an ideal scaffold for a viral capsid-based drug delivery system. *Bioconjugate Chemistry* 18(4): 1140–1147.

107. Schlick, T. L., Z. Ding, E. W. Kovacs, M. B. Francis (2005). Dual-surface modification of the tobacco mosaic virus. *Journal of the American Chemical Society* 127: 3718–3723.

108. Aljabali, A. A. A., J. E. Barclay, N. F. Steinmetz, G. P. Lomonossoff, D. J. Evans (2012). Controlled immobilisation of active enzymes on the Cowpea mosaic virus capsid. *Nanoscale* 4(18): 5640–5645.

109. Green, S. J., J. J. Pietron, J. J. Stokes et al. (1998). Three-dimensional monolayers: Voltammetry of alkanethiolate-stabilized gold cluster molecules. *Langmuir* 14(19): 5612–5619.

110. Steinmetz, N. F., D. J. Evans (2007). Utilisation of plant viruses in bionanotechnology. *Organic and Biomolecular Chemistry* 5(18): 2891–2902.

111. Evans, D. J. (2008). The bionanoscience of plant viruses: Templates and synthons for new materials. *Journal of Materials Chemistry* 18(32): 3746–3754.

112. Yildiz, I., S. Shukla, N. F. Steinmetz (2011). Applications of viral nanoparticles in medicine. *Current Opinion Biotechnology* 22(6): 901–908.

113. Young, M., D. Willits, M. Uchida, T. Douglas (2008). Plant viruses as biotemplates for materials and their use in nanotechnology. *Annual Review of Phytopathology* 46: 361–384.

114. Zhang, X., Y. Zhang (2013). Applications of azide-based bioorthogonal click chemistry in glycobiology. *Molecules* 18(6): 7145–7159.

115. Russell, J. T., Y. Lin, A. Boker et al. (2005). Self-assembly and cross-linking of bionanoparticles at liquid–liquid interfaces. *Angewandte Chemie-International Edition* 44(16): 2420–2426.

116. Peng, B., G. J. Wu, Y. Lin, Q. Wang, Z. H. Su (2011). Preparation of nanoporous polyimide thin films via layer-by-layer self-assembly of Cowpea mosaic virus and poly(amic acid). *Thin Solid Films* 519(22): 7712–7716.

117. Patil, A. J., N. McGrath, J. E. Barclay et al. (2012). Liquid viruses by nanoscale engineering of capsid surfaces. *Advanced Materials* 24(33): 4557–4563.

118. Glotzer, S. C., M. J. Solomon (2007). Anisotropy of building blocks and their assembly into complex structures. *Nature Materials* 6(8): 557–562.

119. Kostiainen, M. A., P. Hiekkataipale, A. Laiho et al. (2013). Electrostatic assembly of binary nanoparticle superlattices using protein cages. *Nature Nanotechnology* 8(1): 52–55.

120. Knez, M., M. P. Sumser, A. M. Bittner et al. (2004). Binding the tobacco mosaic virus to inorganic surfaces. *Langmuir* 20(2): 441–447.

121. Smith, J. C., K. B. Lee, Q. Wang et al. (2003). Nanopatterning the chemospecific immobilization of cowpea mosaic virus capsid. *Nano Letters* 3(7): 883–886.

122. Cheung, C. L., J. A. Camarero, B. W. Woods, T. W. Lin, J. E. Johnson, J. J. De Yoreo (2003). Fabrication of assembled virus nanostructures on templates of chemoselective linkers formed by scanning probe nanolithography. *Journal of the American Chemical Society* 125(23): 6848–6849.

123. Steinmetz, N. F., G. Calder, R. P. Richter, J. P. Spatz, G. P. Lomonossoff, D. J. Evans (2006). Plant viral capsids as nanobuilding blocks: Construction of arrays on solid supports. *Langmuir* 22(24): 10032–10037.

124. Medintz, I. L., K. E. Sapsford, J. H. Konnert et al. (2005). Decoration of discretely immobilized Cowpea mosaic virus with luminescent quantum dots. *Langmuir* 21(12): 5501–5510.

125. Cheung, C. L., S. W. Chung, A. Chatterji et al. (2006). Physical controls on directed virus assembly at nanoscale chemical templates. *Journal of the American Chemical Society* 128(33): 10801–10807.

126. Suci, P. A., M. T. Klem, F. T. Arce, T. Douglas, M. Young (2006). Assembly of multilayer films incorporating a viral protein cage architecture. *Langmuir* 22(21): 8891–8896.

127. Steinmetz, N. F., E. Bock, R. P. Richter, J. P. Spatz, G. P. Lomonossoff, D. J. Evans (2008). Assembly of multilayer arrays of viral nanoparticles via biospecific recognition: A quartz crystal microbalance with dissipation monitoring study. *Biomacromolecules* 9(2): 456–462.

128. Lvov, Y., H. Haas, G. Decher et al. (1994). Successive deposition of alternate layers of polyelectrolytes and a charged virus. *Langmuir* 10(11): 4232–4236.

129. Steinmetz, N. F., K. C. Findlay, T. R. Noel, R. Parker, G. P. Lomonossoff, D. J. Evans (2008). Layer-by-layer assembly of viral nanoparticles and polyelectrolytes: The film architecture is different for spheres versus rods. *ChemBioChem* 9(10): 1662–1670.

130. Kuncicky, D. M., R. R. Naik, O. D. Velev (2006). Rapid deposition and long-range alignment of nanocoatings and arrays of electrically conductive wires from tobacco mosaic virus. *Small* 2(12): 1462–1466.

131. Fowler, C. E., W. Shenton, G. Stubbs, S. Mann (2001). Tobacco mosaic virus liquid crystals as templates for the interior design of silica mesophases and nanoparticles. *Advanced Materials* 13(16): 1266–1269.

132. Portney, N. G., R. J. Tseng, G. Destito et al. (2007). Microscale memory characteristics of virus-quantum dot hybrids. *Applied Physics Letters* 90(21): 214104.

133. Sun, X. C., Y. H. Huang, D. E. Nikles (2004). FePt and CoPt magnetic nanoparticles film for future high density data storage media. *International Journal of Nanotechnology* 1(3): 328–346.

134. Koudelka, K. J., M. Manchester (2010). Chemically modified viruses: Principles and applications. *Current Opinion in Chemical Biology* 14(6): 810–817.

135. Koudelka, K. J., G. Destito, E. M. Plummer, S. A. Trauger, G. Siuzdak, M. Manchester (2009). Endothelial targeting of Cowpea mosaic virus (CPMV) via surface vimentin. *PLoS Pathogens* 5(5): e1000417.

136. Singh, P., D. Prasuhn, R. M. Yeh et al. (2007). Bio-distribution, toxicity and pathology of Cowpea mosaic virus nanoparticles in vivo. *Journal of Controlled Release* 120(1–2): 41–50.

137. Rae, C. S., I. W. Khor, Q. Wang et al. (2005). Systemic trafficking of plant virus nanoparticles in mice via the oral route. *Virology* 343(2): 224–235.

138. Kaiser, C. R., M. L. Flenniken, E. Gillitzer et al. (2007). Biodistribution studies of protein cage nanoparticles demonstrate broad tissue distribution and rapid clearance in vivo. *International Journal of Nanomedicine* 2(4): 715–733.

139. Lewis, J. D., G. Destito, A. Zijlstra et al. (2006). Viral nanoparticles as tools for intravital vascular imaging. *Nature Medicine* 12(3): 354–360.

140. Destito, G., R. Yeh, C. S. Rae et al. (2007). Folic acid-mediated targeting of Cowpea mosaic virus particles to tumor cells. *Chemistry and Biology* 14(10): 1152–1162.

141. Steinmetz, N. F., M. Manchester (2009). PEGylated viral nanoparticles for biomedicine: The impact of PEG chain length on VNP cell interactions in vitro and ex vivo. *Biomacromolecules* 10(4): 784–792.

142. Singh, P., M. J. Gonzalez, M. Manchester (2006). Viruses and their uses in nanotechnology. *Drug Development Research* 67: 23–41.

143. Steinmetz, N. F., T. Lin, G. P. Lomonossoff, J. E. Johnson (2009). Structure-based engineering of an icosahedral virus for nanomedicine and nanotechnology. *Current Topics of Microbiology and Immunology* 327: 23–58.

144. Allen, M., J. W. Bulte, L. Liepold et al. (2005). Paramagnetic viral nanoparticles as potential high-relaxivity magnetic resonance contrast agents. *Magnetic Resonance in Medicine* 54(4): 807–812.

145. Anderson, E. A., S. Isaacman, D. S. Peabody, E. Y. Wang, J. W. Canary, K. Kirshenbaum (2006). Viral nanoparticles donning a paramagnetic coat: Conjugation of MRI contrast agents to the MS2 capsid. *Nano Letters* 6(6): 1160–1164.

146. Hooker, J. M., A. Datta, M. Botta, K. N. Raymond, M. B. Francis (2007). Magnetic resonance contrast agents from viral capsid shells: A comparison of exterior and interior cargo strategies. *Nano Letters* 7(8): 2207–2210.

18 Assembly of a Bluetongue Virus-Like Particle

Multiprotein Complex and Its Use as Vaccine

Avnish Patel and Polly Roy

CONTENTS

18.1 INTRODUCTION

Viruses are obligate intracellular parasites that require the machinery of a host cell in order to undergo replication. This opportunistic lifestyle imposes particular biological constraints, to which the elegant solution by viruses governs the molecular characteristics of their particles.

Virions must be metastable. Particles may be disseminated into an extracellular environment in which the likelihood of an encounter with another host is low, and they must resist the relatively harsh physical demands of this setting in order to remain viable upon meeting such a host cell. Consequently, a degree of stability is required to achieve this, and viruses attain this by multiple sets of pseudo-equivalent interactions. However, a caveat must be observed as alluded to by the reencounter of a host cell, in that while possessing strength and resilience, the particle should be able to disassemble in an ordered fashion given the correct environment, in this case a productive host cell.

Moreover, given the reliance on host cell machinery for replication, viruses are not required to encode many factors considered essential for replication, for example, protein translational machinery, and as such they generally possess small genomes encoding a minimal set of genes. A closer consideration of this observation raises a key question: viral particles are relatively large structures consisting of hundreds or even thousands of protein molecules; however, their genomes only encode a minimal set of genes. For example, human immunodeficiency virus (HIV) has a virion of about 130 nanometers (nm) containing a capsid structure consisting of roughly 1500 proteins (Ganser-Pornillos et al. 2008); conversely, this virus only encodes 15 individual proteins from a genome of approximately 10 kilobase pairs (kbp) (Leitner et al. 2008). How is this structural size and complexity of viral particles accounted for by their minimal genomic content? Clearly, many identical copies of one or several proteins must be assembled into a scaffold that constitutes the virion. Indeed, this has shown to be the case for all viruses.

The overall structure of viruses falls into two major categories: enveloped, such as orthomyxoviruses and coronaviruses, and nonenveloped, such as reoviruses and rotaviruses, which is dictated by the utilization (enveloped) or lack of (nonenveloped) a virion outer membrane derived from the host cell. Under electron microscopy (EM), these two structural forms show a markedly different appearance (Figure 18.1). The outer membrane envelope leads to a pleomorphic appearance to enveloped viruses, as it is in itself inherently flexible (Figure 18.1a). In contrast, capsid viruses show a tight structure that is identical in presentation with clear symmetry axis seen (Figure 18.1b). This would lead to the conclusion that these virus types share little similarity in their structural composition; however, more detailed analysis shows this

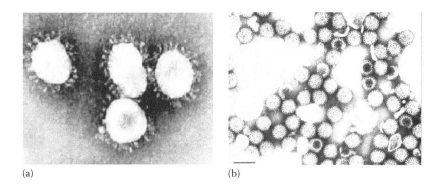

(a)   (b)

FIGURE 18.1   The contrasting morphology of enveloped and nonenveloped viruses. (a) Transmission EM of SARS corona virus particles, this enveloped virus displays a pleomorphic appearance with undefined symmetry. (b) Transmission EM of *Rotavirus* double-layered particles shows the regularity of structure for nonenveloped viruses; note each particle has an identical morphology.

to be artificial. The heterogeneous membrane of the envelope masks the underlying tight network of protein interactions consisting of matrix and nucleoproteins that assemble a tight ordered structure that packages the viral genome. Both protein coats, the capsid of nonenveloped viruses and the nucleocapsid of enveloped viruses, share similarity in their mechanism of assembly, both of which demonstrate multiple sets of pseudo-equivalent interactions mediated by simple molecular self-assembly of their protein constituents into higher-order structures.

The question of how viruses assemble these large complex structures from simple protein components was and still remains a highly insightful topic of research. In the 1980s, my group was the first to attempt to investigate this question using a nonenveloped virus, bluetongue virus (BTV), as a model system.

18.2   OVERVIEW OF BTV STRUCTURE

BTV is an arbovirus transmitted by the insect midge species *Culicoides*. It is the causative agent of bluetongue (BT) disease in ruminants, namely, cattle and sheep (Bowne 1971; Howell and Verwoerd 1971; Maclachlan 2011). BTV is a member of the *Orbivirus* genus within the family Reoviridae. Structurally, it consists of a complex icosahedral capsid approximately 82 nm in diameter comprised of three protein layers (Hewat et al. 1992b; Nason et al. 2004; Prasad et al. 1992) enclosing a genome of segmented double-stranded RNA (Figure 18.2). During an infection cycle, the virion particle exists as two forms (Martin and Zweerink 1972; Verwoerd and Huismans 1972), a triple-layered virion that attaches and enters the host cell and a double-layered core particle that arises when the virus enters the host cell cytosol (Eaton et al. 1990).

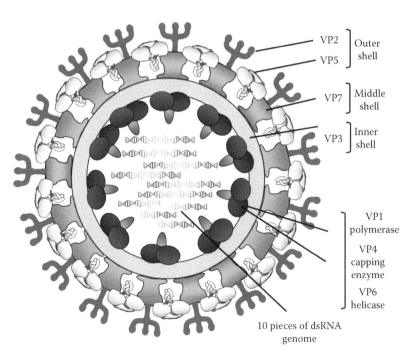

FIGURE 18.2   A cartoon depicting the structure of a BTV virion; note the triple-layered structure. The outer capsid is composed of trimers of VP2 (red forks) and VP5 (green globular clusters). The inner core is composed of VP7 (green knobs) and VP3 layers (blue circle), within which minor core proteins of the replication complex and segmented dsRNA genome are packaged.

18.2.1 Core Particle Structure

The core particle of BTV is an icosahedral particle of 70 nm in diameter, comprising five proteins; three minor proteins VP1, VP4, and VP6; and two major proteins VP3 and VP7, based on their abundance within particles (Huismans et al. 1987; Martin and Zweerink 1972). It exhibits a double-layered architecture displaying a knobbed surface (Figure 18.3a and b). Each knob is comprised of a trimer of VP7 (Grimes et al. 1995), and a total of 260 form the surface of the core (Prasad et al. 1992).

The VP7 protein itself is 38 kilodaltons (kDa) and is composed of two main domains: a beta sandwich domain, which projects outwardly from the core, and an alpha helical domain, which mediates attachment to the underlying inner core layer (Grimes et al. 1995). The VP7 trimers are formed via a cluster of hydrophobic methionine residues that line the intermolecular junction between VP7 monomers, providing hydrophobic interactions that drive the assembly of the highly stable timer (Grimes et al. 1995; Limn et al. 2000). These trimers are arranged pseudo-symmetrically on the core surface with 13 trimers present at each triangular face of the core icosahedron. Thus, the core VP7 outer layer displays T13 symmetry (Caspar and Klug 1962) encoded by 780 molecules of VP7.

Due to the symmetry mismatch between the underlying inner core layer, the T13 VP7 trimers occupy pseudo-equivalent positions that vary slightly in molecular interaction with the inner core (Grimes et al. 1997, 1998). These positions fall into a subset of trimer interactions, denoted as P to T, which dictate the mechanism of assembly of the VP7 layer (Limn et al. 2000; Limn and Roy 2003). This fundamental insight would not have been possible without the development of a virus-like particle (VLP) assembly system, which will be discussed in the section on VLPs.

The inner layer of the core particle comprises of a 103 kDa protein assigned as VP3. It forms a spherical structure consisting of 120 copies of VP3 (Grimes et al. 1998). The protein displays an overall teardrop structure with an extremely thin profile. Its structure can be subdivided into three distinct domains, assigned as apical, carapace, and dimerization. In all cases, the protein is positioned with the apical domain located at the icosahedral fivefold symmetry axis and dimerization domain facing the icosahedral twofold. The protomers form a T2 lattice, which is achieved by the use of two discrete structural isoforms denoted A and B. The B-form displays a rotation of its dimerization domain in relation to carapace, which causes it to be more curved, allowing for necessary distortion to fit within the gaps between A-form molecules (Grimes et al. 1998). This assembly forms decamers consisting of AB dimers; five A-form molecules create pentamers at the fivefold symmetry axis with five B-form molecules intercalated between them. A decamer is found at each of the 12 icosahedral fivefold axis, which together outlines the VP3 subcore sphere. These denotations are not arbitrary as studies with recombinant VP3 have shown; a deletion in the dimerization domain generates a decameric assembly intermediate, suggesting that the assembly pathway proceeds from AB dimers to decamers to sphere (Kar et al. 2004) (Figure 18.3a).

18.2.2 Structure of the Outer Capsid

As stated previously, core particles exhibit five proteins: VP1, VP3, VP4, VP6, and VP7. Whole virus particles, however, exhibit two extra proteins when analyzed by SDS–PAGE (Huismans et al. 1987; Martin and Zweerink 1972; Verwoerd et al. 1972), which differ noticeably in size and abundance. One is a smaller protein of 59 kDa, denoted as VP5 is observed, and the other is a larger protein, VP2, almost double in size with 110 kDa, which shows half the abundance of the smaller protein.

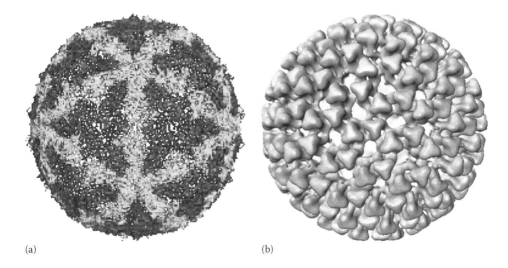

(a)   (b)

FIGURE 18.3 Structures of the core layers of BTV. (a) Structure of the VP3 subcore layer displaying 120 copies of VP3 that occurs in 2 isomers A (green) and B (red). (Modified from Grimes, J.M. et al., *Nature*, 395(6701), 470, 1998, doi:10.1038/26694. http://www.ncbi.nlm.nih.gov/pubmed/9774103.) (b) The structure of the core particle that exhibits 260 knob-like protrusions of VP7 trimers (blue) that sit atop the VP3 subcore layer (green). (Adapted from Grimes, J.M. et al., *Structure*, 5(7), 885, 1997, http://www.ncbi.nlm.nih.gov/pubmed/9261080.)

Complete BTV virions display a smoothed fuzzy appearance under EM (Figure 18.4a, right-hand panel). Closer analysis reveals the surface exhibits sail-like projections, of which there are 180 forming 60 triskelion trimers. Additionally, 120 globular structures are present, arranged underneath the triskelion

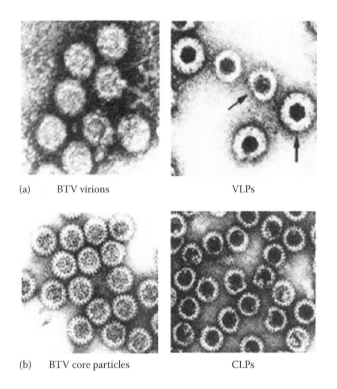

(a)   BTV virions   VLPs

(b)   BTV core particles   CLPs

FIGURE 18.4 A morphological comparison of purified virions with VLPs and core particles with CLPs. (a) Transmission EM of BTV virions (left) and BTV VLPs (right). (b) Transmission EM of BTV core particles (left) and CLPs (right). In both cases, VLPs and CLPs lack central density when compared to virions and core particles, indicating a lack of packaged genome.

layer (Hewat et al. 1992b; Nason et al. 2004). Taken together with SDS–PAGE analysis of whole virus particles, this indicates that VP2 comprises the triskelion outer surface and VP5 the underlying globular layer.

High-resolution cryo-EM has revealed an increased detail of the outer capsid VP2 and VP5 layer, which exhibits a symmetry arrangement unique to BTV, such that the outer capsid does not display the same T13 symmetry of the underlying inner capsid VP7 layer, as seen by other members of the Reoviridae family (Settembre et al. 2011; Zhang et al. 2010) (Figure 18.5a).

VP2 exhibits an elongated L-shaped profile with the base forming a hub domain of the triskelion trimers (Zhang et al. 2010). This region exhibits a β-sheet-rich galectin-like fold similar to that of mammalian sugar-binding galectin proteins (Barondess et al. 1994). The upward arm of the L represents a tip domain that projects from the surface of the virion and is highly solvent accessible (Zhang et al. 2010). The VP2 triskelions make contact with the underlying VP7 trimer layer through projections from the tip and hub domains.

The arrangement of VP5 is also unique, with trimers surrounding the VP2 triskelions and positioned underlying the gaps in the triskelion arms of VP2 (Figure 18.5b). The VP5 trimers possess a highly compact globular fold predominated by α-helices with a central coiled-coil motif facilitating trimerization, analogous to membrane viral fusion proteins such as the stalk of HIV gp41 (Chan et al. 1997; Zhang et al. 2010). Interactions between VP2 trimers and VP7 subshell are predicted to be weak owing to small-density "fingers" of contact; additionally, interactions between VP2 and VP5 are of a similarly weak density nature (Zhang et al. 2010). The paucity of intertrimer contacts of the outer capsid corresponds well with its observed fragility, owing to its ease of removal by high-salt treatment (Huismans et al. 1987).

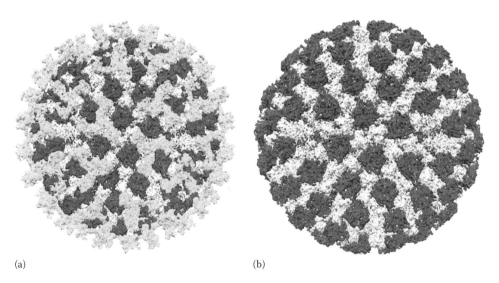

(a)   (b)

FIGURE 18.5 Structures of the outer capsid layers of BTV. (a) The structure of the BTV virion displaying 60 triskelion trimers of VP2 (emerald green) under which 120 globular trimers of VP5 (magenta) are localized. The underlying VP7 (light green) and VP3 (red) core structure is visible. (Image courtesy of Z. Hong Zhou.) (b) The structure of the virion with the outer VP2 layer removed, coloring is as (a). (Image courtesy of Z. Hong Zhou.)

In summary, the BTV virion is a nonenveloped icosahedral particle, which exhibits a triple-protein-layered arrangement: an almost spherical T2 VP3 inner core layer, a T13 surface core layer consisting of VP7, and an outer capsid of unique symmetry, composed of VP5 and VP2.

While much progress has been made on the structure of the virion, the process of assembly remains less thoroughly studied since any alteration in the capsid structure of a live virus is likely to be lethal. VLP technology overcomes this issue allowing for, amongst other applications, the study of the previously intractable process of virus assembly.

18.3 BACULOVIRUS EXPRESSION TECHNOLOGY AS A PROTEIN EXPRESSION SYSTEM FOR VLPs

In the 1980s, my lab set out to study the assembly process of BTV. As discussed in Section 18.2, BTV virions contain seven structural proteins (VP1-7), and these proteins vary markedly in molecular size from 36 kDa (VP6) at the smallest to 150 kDa (VP1) at the largest. At the time little was known about the role of each protein in the structure of the virion, this imposed great restrictions on the recombinant protein expression platform required to produce these proteins. A suitable system should accommodate the expression of the entire size range to allow the assembly process to be fully investigated. Five major recombinant protein systems exist: bacterial, yeast, baculovirus, mammalian, and cell-free, each of which shows advantages and disadvantages, on the whole leading to a trade-off between yields, speed of expression, and complexity of recombinant protein target.

Although bacterial and yeast expression systems allow rapid expression and ease of scaling of cultures, their disadvantage is tied to this very nature, in that the simplicity of their biology does not effectively support the expression of complex protein targets, such as those that are posttranslational modified and large complex protein folds (Andersen and Krummen 2002; Baneyx 1999; Chen 2014; Mattanovich et al. 2012; Terpe 2006). On the other side of the spectrum, mammalian and cell-free expression can accommodate such complex protein targets; however, they are not easily amenable to substantial scaling (Endo and Sawasaki 2006; Geisse et al. 1996; Wurm 2004). Baculovirus expression technology offers a balanced system in which complex protein targets can be effectively produced in substantial quantities by ease of scaling of expression culture, with only a minor trade-off in expression speed (Bishop 1990; Kost et al. 2005; Marek et al. 2011; Miller 1993). These properties made this system ideally suited to the expression of BTV structural proteins. More recently, advances in baculovirus expression vector systems (BEVSs) have further substantiated this as the expression system of choice for complex therapeutic proteins such as VLPs of numerous viruses.

18.3.1 BACULOVIRUS EXPRESSION TECHNOLOGY

The family Baculoviridae is comprised of large complex arthropod viruses with a circular dsDNA genome varying from 80 to 180 kbp in size (Van Oers and Vlak 2007). It is subdivided into four genera: alpha, beta, gamma, and delta, assigned by divergence of a core set of genes (Herniou and Jehle 2007; Jehle et al. 2006). The major species used for recombinant protein expression is that of *Autographa californica* multiple nucleopolyhedrosis virus (AcMNPV) and less commonly *Bombyx mori* nucleopolyhedrosis virus, which are both of the *Alphabaculovirus* genus. The natural hosts of these species are lepidopterans, infecting their larval stages (Blissard and Rohrmann 1990).

The virus exists in two forms: an unoccluded form in which virions are membrane bound and exist singularly and an occluded form in which multiple virions are present within a crystalline-like protein shell termed an inclusion body (Blissard and Rohrmann 1990). This protein shell leads to environmental persistence by allowing the virus to resist chemical and physical stresses. Larvae consume these inclusion bodies, whereby the alkaline environment of the insect midgut causes their dissociation, releasing unoccluded virions that infect intestinal epithelial cells (Engelhard et al. 1994; Volkman and Summers 1977). Once infected, the virus spread is systemic, leading to liquefaction of the host by viral proteases (Hawtin et al. 1997), thus allowing dissemination of inclusion bodies formed during infection into the environment in which they persist dominantly until uptake by another host.

Molecular biology and biochemical analysis of inclusion bodies show that their major constituent is polyhedron protein (Smith et al. 1983a, 1983c; Van der Beek et al. 1980), which accumulates to high levels during infection with almost one-third dry weight of infected caterpillars (Miller et al. 1983). A cell culture system for baculoviruses had been established in the 1970s (Goodwin et al. 1970; Faulkner and Henderson 1972); however, for the application of AcMNPV to recombinant protein expression, this extremely high level of polyhedrin was initially exploited. It was found that the polyhedrin gene is not required for growth of the virus in culture and that viruses harboring a deletion of this region could be selected based on variant plaque morphology (Smith et al. 1983a). This study paved the way for the use of AcMNPV as an expression vector through insertion of genes at the polyhedrin locus by homologous recombination between transfer vector and baculovirus genome, with human β-interferon being the first reported protein to be expressed in this manner (Smith et al. 1983b). Although operative, this system proved challenging and lengthy in isolation of recombinant virus from wild-type background. However, further engineering led to improvement by fusion of bacterial β-galactosidase to the polyhedrin gene allowing selective screening by 5-bromo-4-chloro-3-indolyl-D-galactopyranoside (IPTG) supplemented media (Pennock et al. 1984). Finally, further improvements were made to eliminate all background nonrecombinant viruses by deleting a portion of the essential *orf1629* gene, thus rendering the genome nonviable. By supplementation of this gene in the transfer plasmid used to introduce the recombinant expression cassette, only recombinant viruses would be complemented and rescued (Zhao et al. 2003). This breakthrough greatly increased the utility of the baculovirus system allowing high-throughput virus production by eliminating screening. Additionally, the recombination replaces the mini-F replicon and kanamycin resistance gene at the *orf1629* locus, potentially enhancing stability of recombinant insert (Pijlman 2003; Pijlman et al. 2004).

While expression of single proteins can be readily achieved, most viral capsids consist of more than one structural protein with some notable exceptions such as human papillomavirus (HPV) for which singly expressed L2 capsid protein has produced an effective immunogenic licenced vaccine (Harper et al. 2004, 2006; Kirnbauer et al. 1992; Koutsky et al. 2002; Suzich et al. 1995). Preliminary research into production of multiprotein complexes using BEVS relied on superinfection of cells with multiple viruses, each singly expressing one protein component of the multiprotein complex (French and Roy 1990; French et al. 1990). While multiprotein VLPs were attained, heterogeneity of purified samples was a major issue as not all cells are infected by the complete set of singly expressing viruses and by different ratios of virus. In order to address this issue, MultiBac systems were developed to allow expression of multiple genes from a single recombinant baculovirus genome. Initially, dual transfer vectors were engineered using inverted duplication of polyhedrin promoter *polh* (Emery and Bishop 1987; Matsuura et al. 1986; Takehara et al. 1988), allowing complexes of two proteins to be expressed. Although promising, baculoviruses express proteins that strongly promote homologous recombination (Crouch and Passarelli 2002; Martin and Weber 1997), which leads to instability of these vectors due to the back-to-back duplication of *polh* sequence. Further research identified the very late gene promoter of the *p10* locus as an additional strong promoter for expression of recombinant genes (Vlak et al. 1990). Utilizing both *p10* and *polh*, more stable dual transfer vectors could be achieved (Weyer and Possee 1991). Using a combinatorial approach, transfer vectors allowing triple and quadruple transfer vectors were created (Belyaev and Roy 1993; Belyaev et al. 1995); however, these are impractical to manipulate by conventional molecular biology techniques. Finally, recent advances in the recombineering system have allowed for the ease of manipulation of the baculovirus genome as a bacterial artificial chromosome within bacterial host strains (Copeland et al. 2001; Oppenheim et al. 2004). This has allowed the creation of viruses capable of expression of multiprotein complexes, with complexes containing eight or more subunits having been reported. Two approaches have been taken in this manner.

One is the use of the recET system (Datsenko and Wanner 2000) that first introduces a *loxP* site of the Cre recombinase system into the baculovirus genome knocking out the *v-cath* and *chiA* genes; this genome also contains a *Tn7* transposase site at the *polh* locus (Berger et al. 2004; Fitzgerald et al. 2007). These two sites, *loxP* and *Tn7*, allow for dual integration of transfer vectors that carry multiple expression cassettes; these are combinatorially assembled in separate bacterial hosts using the Cre-Lox recombinase system, from plasmids carrying donor and acceptor *loxP* sites (Liu et al. 1998). This methodology is fast and allows rapid assortment of subunits for the expression of multiprotein complexes, tailoring it to high-throughput structural biology (Fitzgerald et al. 2007). Some concerns over the long-term stability of viruses created by this methodology have been raised due to repetitive promoter and terminator sequences. Additionally the *Tn7* insertion system for expression has been shown to be prone to loss of recombinant insert after 20 passages, due to instability of the mini-F replicon sequence (Pijlman 2003; Pijlman et al. 2004; Simón et al. 2006); however, maintaining low passage number and multiplicity of infection upon passage, stable expression has been demonstrated (Fitzgerald et al. 2006).

The recET system has been utilized in a second manner to introduce multiple expression cassettes throughout the baculovirus genome (Noad et al. 2009) (Figure 18.6). In this system, transfer vectors containing expression cassettes paired with a selectable marker are flanked by targeting sequences homologous to regions within the baculovirus genome, at which the recombinant cassette is to be inserted. Linearized plasmids that contain an intact flanked cassette are electroporated into acceptor host strain containing bacmid and expressing the recET recombinase. A strong dual selection *Zeo^R LacZ* marker is used to screen recombinant BACs reducing false-positive rate. This selection marker is flanked by self-inactivating *loxP* sites allowing excision by Cre recombinase that leaves the recombinant gene cassette unaffected while allowing further insertion of expression cassettes using an iteration of the same process. This system allows context-dependent regulation of gene expression level as certain loci and orientations within loci were shown to exhibit differential gene expression levels (Noad et al. 2009). Using this system, high expression of protein complexes has been achieved at greater than 20 passages, suggesting enhanced stability of recombinant inserts (Roy and Noad, unpublished observations). In this system, viruses are recovered using recombination within host insect cells as developed by Zhao et al. (2003). This eliminates unstable mini-F replicon and kanamycin resistance genes, which, in combination with the wider genomic distribution of expression cassettes containing homologous DNA elements, may act synergistically to enhance insert stability of recombinant viruses.

18.4 BTV VLPs

My lab in the 1980s sought to investigate the process of BTV assembly that had remained ambiguous. Primarily, it was thought that while the core is structurally stable, the outer capsid may assemble a separate stable structure when the two protein components are expressed by BEVS. In addition, no clear evidence as to which viral protein within the core and

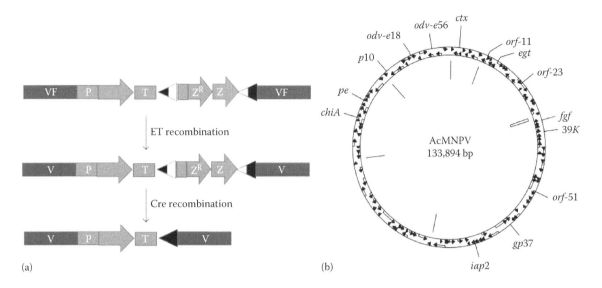

(a)   (b)

FIGURE 18.6 A schematic depicting the process of iterative modification of AcMNPV genome to insert multiple single-locus expression cassettes. (Reproduced from Roy, P. and Noad, R. *ISRN Microbiology*, 2012, article ID 628797, 11 pages. http://dx.doi.org/10.5402/2012/628797, 2012.) (a) The strategy used for repeated modification of the same bacmid genome to express multiple different recombinant proteins. Recombination of an expression cassette containing a promoter (P), target gene (olive arrow), and terminator (T) is targeted by viral flanking sequences to homologous sequences in the bacmid. Selection in *E. coli* is achieved using a bipartite marker consisting of a Zeocin resistance (Z^R) and LacZα (Z), flanked by lox71 (white triangle with black tip) and lox66 (black triangle with white tip) *loxP* sites. Following Cre-mediated recombination, this marker is removed, allowing a subsequent modification of the same DNA. (b) Loci within the AcMNPV were successfully modified using the iterative modification strategy.

virion composed specific structural features and a detailed view of their organization was not known.

We first sought to define the role of specific proteins within the architecture of a BTV virion. To this extent, from work on purified virions and cell-free translation of transcripts (Huismans et al. 1987; Mertens et al. 1984; Van Dijk and Huismans 1988; Verwoerd et al. 1972), proteins implicated in structural roles of the virion were cloned and expressed using BEVS.

Initially, individual expression of proteins allowed greater characterization of their function. Early baculovirus expression studies with BTV nonstructural proteins NS1 and NS3 allowed production of protein specific antibodies. These allowed greater analysis of events of BTV infection, and in addition, characterization of tubule formation in the case of NS1 (French et al. 1989; Urakawa and Roy 1988) demonstrated the utility of BEVS for the specific analysis of individual protein components by recombinant expression. Moreover, enzymatic function could be assigned as exemplified by studies of VP1, the largest core protein. Baculovirus technology facilitated sufficient expression of protein, allowing RNA polymerase activity to be demonstrated biochemically (Urakawa et al. 1989).

Taking these studies further, multiple expressions of structural proteins allowed delineation of their role in the formation of specific virion structures. Initially, core structure was examined, and dual expression of VP3 and VP7 demonstrated the production of core-like particles (CLPs) in insect cells that could be purified (French and Roy 1990) (Figure 18.4b, left-hand panel). These particles were similar to authentic BTV cores in terms of size, appearance, and the

stoichiometric arrangement of VP3 to VP7. CLP structures formed in the absence of BTV double-stranded viral RNA species or the associated minor core proteins, indicating an intrinsic self-assembly process that is independent of these components. In addition, the three BTV nonstructural proteins, NS1, NS2, and NS3, are not required to assist or direct the formation of empty CLPs from VP3 and VP7.

Structural analysis revealed CLPs were shown to structurally mimic authentic BTV cores when examined under cryo-EM; however, the VP7 P trimer found at the fivefold symmetry axis was absent, indicating weaker interlayer interactions at this region (Hewat et al. 1992a). This observation is not trivial and could be explained when the assembly process of the VP7 layer was analyzed using CLP expression (Limn and Roy 2003), which will be discussed further in the section on assembly.

Finally, further expression studies utilized coinfection of insect cells with two baculoviruses, one expressing CLP proteins VP3 and VP7 and the other expressing VP5 and VP2 outer capsid proteins, enabling production of double-shelled VLPs (French et al. 1990). When purified, these particles were found to have the same size and appearance as authentic BTV virions; however, VLPs demonstrated a lack of density in particle centers indicating a lack of incorporation of genomic RNA (Figure 18.4a, left-hand panel). VLPs exhibited high levels of hemagglutination activity as reported for BTV virions (Höbschle 1980), and additionally, antibodies raised to the expressed particles contained high titers of neutralizing activity against a homologous BTV serotype in cell culture (French et al. 1990). Again, cryo-EM reconstruction of these particles demonstrated that they structurally mimic authentic BTV virions, suggesting preservation of correct epitope presentation (Hewat et al. 1992b).

The assembly of these BT VLPs after the simultaneous expression of four separate proteins demonstrated the potential of this technology for the production of a new generation of viral vaccines. The correct structure of viral particles was preserved with no risk of reversion or incomplete inactivation as seen by traditional viral vaccines. Additionally, the exquisite control afforded by specific inclusion or exclusion of protein components required to form structures, in conjunction with established mutagenesis techniques, paved the way for insightful studies into the assembly process of viral capsids. Both vaccine and basic research outcomes of VLP technology for BTV will be addressed in the following sections.

18.5   USE OF VLPs TO STUDY BTV ASSEMBLY

VLP technology afforded the identification of the sole components required to form CLP and VLP structures without requiring viable virus. This allowed for the analysis of the assembly process of such structures, as invariably this could only be thoroughly studied using lethal mutations that disrupt the assembly process. Initially, the assembly of core structure was analyzed due to its simplicity and later the availability of an atomic resolution model.

Mutational analysis of VP3 and VP7 demonstrated key regions required to facilitate CLP assembly. To locate sites of interaction between VP3 and VP7 proteins, deletion analysis of conserved regions of VP3 was assessed for CLP assembly. This identified a key region, residues 499–508, which contains methionine 500 and an arginine 502, both

of which are required for CLP formation (Tanaka and Roy 1994). These residues lie in helix 9 central in the VP3 apical domain. Further analysis based on the atomic resolution structure of a BTV core (Grimes et al. 1998) allowed dissection of the assembly process of the VP3 inner core layer. The C-terminal dimerization domain of VP3, which was predicted to form intersubunit contacts at the icosahedral fivefold, was deleted. This led to the formation of VP3 decamers when expressed as a recombinant protein (Kar et al. 2004), indicating a trapped intermediate state of VP3 shell assembly. Interestingly, this VP3 decamer still bound replication complex proteins VP1 and VP4 but could not interact with double-stranded RNA (Kar et al. 2004). This finding allowed greater insight into the initial steps of BTV assembly allowing a hierarchical model of subcore assembly to be elucidated (Figure 18.7a). This model proposes VP3 first forms dimeric A and B forms (Grimes et al. 1998), which subsequently form decamers. These decamers then recruit replication complex proteins VP1 and VP4. Twelve of these decameric complexes then assemble to form a complete subcore. This assembled subcore, however, is relatively unstable as subcore-like particles are not seen when VP3 is expressed singly in insect cells (Inumaru et al. 1987) and can only be produced from stripping the VP7 outer layer from CLPs (Loudon and Roy 1991). This suggests a stabilizing effect of the VP7 layer upon the VP3 subcore shell.

In an extension of the analysis of VP3, the CLP expression system allowed dissection of specific motifs and process of VP7 coating of assembled subcores. Mutational analysis of VP7 demonstrated that stable core assembly depends on the strength of VP7 interaction, with mutants that form weak

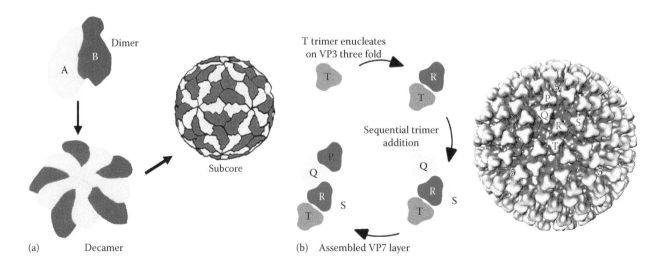

FIGURE 18.7   The assembly pathway of BTV CLPs. (a) The subcore is assembled from two structural isomers of VP3 A and B that initially assemble as dimers. Ten of these dimers then associate to form a pentameric decamer in which A-form VP3 molecules form a vertex. Twelve of these decamers then assemble to form a spherical subcore particle. (Adapted from Kar, A.K. et al., *Virology*, 324(2), 387, 2004, doi:10.1016/j.virol.2004.04.018, http://www.ncbi.nlm.nih.gov/pubmed/15207624.) (b) The coating of the VP7 layer follows an ordered assembly pathway in which trimers first bind strongly at the threefold symmetry axis denoted as the T position. This attachment then nucleates the coating of weaker interacting Q, S, and T position trimers that finally nucleate the coating of the weakest interacting P position trimer. These positions are indicated on the core structure. (Modified from Limn, C.-K. and Roy, P., *J. Virol.*, 77(20), 11114, 2003, doi:10.1128/JVI.77.20.11114.)

subcore interactions displaying reduced CLP yields (Limn et al. 2000). Extensive structure-based mutational analysis revealed that the VP7 coat is dependent on VP3 interactions with the trimer base through helix 2 as well as on lateral association of VP7 trimers mediated by helix 9 (Limn and Roy 2003). From cryo-EM studies, it was observed that CLPs lack the P trimer at the fivefold axis (Hewat et al. 1992a; Nason et al. 2004), suggesting that weak interactions occur at this localization (Limn and Roy 2003). Together, the data for VP7 mutagenesis indicate a model in which symmetry mismatch between the VP3 (T = 2) and VP7 layers (T = 13), causing localized nonequivalent interactions (Grimes et al. 1998), governs the order of assembly of VP7 trimers. A likely pathway of core assembly is that a number of strong VP7 trimer VP3 contacts act as multiple equivalent initiation sites that nucleate a second set of pseudo-equivalent weaker interactions that complete the outer layer of the core (sequentially T up to P). In such a model, VP7 T trimers (which have threefold symmetry) that sit atop the threefold axis of VP3 have the strongest interaction due to symmetry matching; thus, these act to nucleate assembly of the remaining VP7 coat of the core (Figure 18.7b).

In addition to the study of assembly, VLP technology can be applied to any fundamental research in which live virus would be rendered unviable. Receptor-binding studies of a putative integrin-binding RGD peptide (Mason et al. 1994) of BTV cores were made possible using CLPs (Tan et al. 2001). This motif is localized in the top jellyroll domain of VP7 on its outer surface and was mutated to analyze its role of BTV core entry of cells of the insect vector *Culicoides* sp. The results demonstrated that this tripeptide is important for insect cell attachment and may play a role in virus infection of the insect vector (Tan et al. 2001). Only VLP technology allows greater insight into essential functions such as receptor binding, as live virus would not be viable.

Taken together, the application of VLP technology to virus assembly is a powerful tool that enables the molecular assembly pathway of complex multiprotein capsids to be thoroughly dissected in a way independent of the need for viable virus. This insight is not only relevant at a basic research level but can guide the design of more effective VLP immunogens, potentially allowing these assembly rules to be applied to synthetic, specifically tailored VLP particles.

18.6 VLPs AS VACCINES FOR BLUETONGUE DISEASE

Traditionally, BTV vaccines have relied on live attenuated virus vaccines that have been used for over 40 years and are known to induce an effective and lasting immunity (Roy 1992, 2003; Roy et al. 2009). These vaccines are developed by serial passages in embryonated chicken eggs leading to attenuation of the virus over time. This process, while effective, has no guaranteed safety profile as the virus is still viable, that is, having a functional genome. In spite of their success in endemic areas, their use has

several drawbacks, and their teratological effects as a result of vaccination with attenuated BTV are well documented (Osburn et al. 1971; Schultz and Delay 1955). Other adverse reactions include depressed milk production in lactating sheep and abortion/embryonic death and teratogenesis in offspring from pregnant females that are vaccinated during the first half of gestation (Flanagan and Johnson 1995). Moreover, the viremia observed following vaccination both in laboratory experiments and in the field has been sufficient for vaccine strains to be transmitted to biting midges (Ferrari et al. 2005; Schultz and Delay 1955). This leads to a greater potential risk for spread by vectors, with eventual reversion to virulence and/or reassortment of vaccine virus genes with those of wild-type virus strains. The segmented nature of BTV genome allows for reassortment of genes between strains that coinfect the same animal (Stott et al. 1987). Indeed, this has been observed in the field in Italy in 2002, with BTV-16 strain circulating found to be a reassortment between BTV-2 and BTV-16 live attenuated virus vaccines (Batten et al. 2008). Attenuated vaccines for BTV may offer a route to control disease, but are not suitable for a program of disease eradication.

Due to their lack of genome and maintenance of correct capsid structure, VLP vaccines offer the technology that surmounts the drawbacks outlined for traditional vaccines, as has been demonstrated for BTV.

Initially, animal model experiments indicated that vaccination of guinea pigs with 75 µg of purified BTV-10 VLPs elicited strong neutralizing antibody titers when sera was tested in culture (French et al. 1990). This led the way for further clinical trials in which 1-year-old Merino sheep were vaccinated with a range of 10–200 µg of VLPs for BTV serotype 10 (Roy et al. 1992) (Figure 18.8a). All vaccinated animals developed demonstrable neutralizing antibodies and when challenged with virulent virus after 4 months of vaccination were completely protected from disease at doses as low as 10 µg VLP (Roy et al. 1992). In contrast, the nonvaccinated control group developed typical BT clinical symptoms. Owing to the success of monovalent vaccination, further trials to test combinatorial vaccination were performed. VLPs were made for five serotypes of BTV-1, BTV-2, BTV-10, BTV-13, and BTV-17. Purified particles were used either as single VLP types (BTV-10, BTV-17) or as a combination of all five serotypes, and sheep were challenged with homologous (BTV-10, BTV-13, BTV-17) or selected heterologous (BTV-4, BTV-11, BTV-16) viruses. In these animals, VLP vaccination provided complete protection against the vaccine serotypes and also partial protection from challenge with related nonvaccine serotypes with two doses of 10 µg of VLPs (Roy et al. 1994). The protective efficacy of vaccination in these trials extended over a long (14 months) period. This observation raises the possibility that a broad spectrum vaccine against all 24 BTV serotypes is achievable by combining VLPs from a relatively small number of serotypes.

Studies with BTV also demonstrate the advantages in efficacy of VLP vaccines over subunit vaccines, based on dissociated antigens or unassembled recombinant antigens.

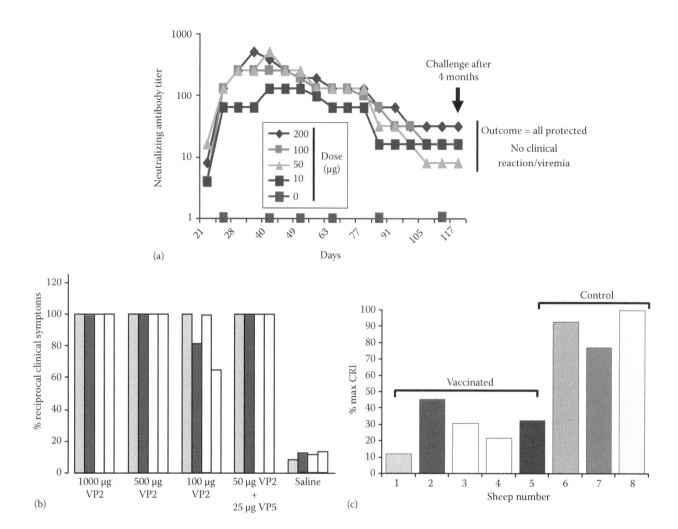

FIGURE 18.8 VLP and CLP vaccine trials. (a) Summary of neutralizing antibody response to VLP vaccination in Merino sheep. Sheep were vaccinated with two doses of VLPs with dose ranging from 0 to 200 μg as indicated. Neutralizing antibody titer was followed for 117 days, at which point the sheep were challenged with virulent BTV. (Adapted from Roy, P. et al., *Vaccine*, 10(1), 28, 1992, doi:10.1016/0264-410X(92)90415-G, http://www.sciencedirect.com/science/article/pii/0264410X9290415G.) (b) Vaccination with recombinant VP2 alone or VP2 with VP5 requires a significantly higher dose to elicit protection when compared with VLPs. (Reproduced from Roy, P. et al., *J. Virol.*, 64(5), 1998, 1990, http://jvi.asm.org/content/64/5/1998.short.) (c) Vaccination with CLPs alone reduces clinical symptoms of disease. Merino sheep were vaccinated with a dose of 50–100 μg of CLPs, or saline and clinical outcomes of disease were scored. Sheep vaccinated with CLPs displayed significantly less symptomatic response than the control group. (Adapted from Roy, P., *Dev. Biol.*, 114(January), 169, 2003, http://europepmc.org/abstract/MED/14677687.)

Vaccination with 100 μg BTV-10 VP2, the major serotype determining antigen, was only partially protective for a short duration (75 days) against virulent virus challenge (Roy et al. 1990). When vaccinated in conjunction with outer capsid protein 50 μg of VP2 combined with 25 μg, VP5 was protective (Roy et al. 1990) (Figure 18.8b). This indicates that when presented in a more authentic multiprotein complex, VP2 can elicit stronger immune stimulation. For VLPs, the minimum efficacious dose of 10 μg, containing only 1–2 μg VP2, afforded a better level of protection for a much longer duration (Roy et al. 1992). These studies demonstrate that assembly of antigens into VLPs results in a more effective immunogen due to the similarity of epitope presentation to authentic virus particles.

Another further advantage of VLPs is their flexibility in molecular engineering. BTV shows a multilayered capsid

with a highly conserved core and variable outer capsid proteins, which define serotype specificity (Mertens et al. 1989). Exploiting this, it was possible to coexpress the outer capsid proteins from different serotypes and coat them onto the conserved inner core (Liu et al. 1992). This opened an avenue to rapidly produce serotype-specific VLPs for BTV. VLPs for serotypes 2, 4, and 9 were produced in this manner, and serotype 2 was chosen to be validated, showing protection at a dose of 10 μg (Stewart et al. 2010) (Table 18.1).

Taking this approach further, the influence of evolutionary divergence (eastern and western topotypes) (Balasuriya et al. 2008; Carpi et al. 2010; Maan et al. 2012; Nomikou et al. 2009; Pritchard et al. 2004) on vaccine development was investigated. This study clearly highlighted that the development of serotype-specific neutralizing antibodies is more critical than variations in topotype of the circulating strain

TABLE 18.1

Summary of VLP Vaccination Trials in European Sheep Breeds

Vaccine Type	Serotype(s)	Neutralizing Antibody Titer (Prechallenge)	Challenge Virus	Viremia at Day 7 Postchallenge
Single	BTV-1	>128 (50–200 µg)	BTV-1	None
		64–128 (10 µg)	BTV-1	
Single	BTV-2	32–64 (10 µg)	BTV-2	None
		128 (20 µg)	BTV-2	None
Single	BTV-8	>128 (20 µg)	BTV-2	None
Cocktail	BTV-1	64–128 (20 µg)	BTV-1	None
	BTV-4	>16 (20 µg)	BTV-4	None
Cocktail	BTV-1			
	BTV-2	>128 (20 µg)		
	BTV-8	32–128 (20 µg)	BTV-8	None
		64–128 (20 µg)		

Note: Sheep were vaccinated with either BTV VLP vaccine consisting of a single serotype or a cocktail of multiple serotypes. Neutralization antibody titers for each serotype in the vaccine preparation were determined prechallenge with a virulent virus. At 7 days postchallenge, viremia was assayed.

(Stewart et al. 2012; Wilson and Mellor 2009). This suggests isolate variation or evolutionary distance of the strain is not a critical factor for BTV vaccines. Furthermore, production of VLPs in response to a viral outbreak is relatively rapid as demonstrated by the production of BTV serotype 8 VLPs from the highly pathogenic 2008 European outbreak strain (Stewart et al. 2013). This VLP was successfully incorporated into a cocktail of circulating European strains, as different serotypes did not affect the ability of each serotype to raise a specific neutralizing antibody response or interfere with the protective efficacy of the vaccine when animals were challenged (Stewart et al. 2013) (Table 18.1).

While humoral immunity has been readily demonstrated by VLP vaccination, interestingly, the vaccination of sheep with preparations of CLPs (lacking VP2 and VP5) showed a low level of protection against virulent virus challenges (Roy 2003; Stewart et al. 2012) (Figure 18.8c). This partial protection conferred by CLPs may be due, in part, to cell-mediated immune (CMI) response directed against antigenic sites on the inner core proteins. Indeed, each of the structural and nonstructural BTV proteins has been shown to induce a cytotoxic T lymphocyte response (Jones et al. 1996), and it is postulated that CMI response may reduce BTV viremia (Ellis et al. 1990; Jeggo et al. 1985; Jones et al. 1996), although the reduction in BTV viremia is unlikely to stop virus transmission (Stewart et al. 2012). More recently, both CD8+ and CD4+ T-cell epitopes have been mapped within VP7 sequence suggesting a role of this protein in the induction of CMI in vivo (Rojas et al. 2011).

In conclusion, of all studies with recombinantly derived vaccines, BTV VLPs, which display each protein in correct antigenic configuration, have been shown to be most promising. In each study, VLPs were proven to be excellent immunogens, both for preventing disease in sheep and for preventing detectable virus replication in small-scale experimental trials (summarized in Table 18.1). The success of BTV VLPs is likely due to its ability to elicit a strong neutralizing antibody response in

conjugation with the CMI, which also plays a role in recovery from infection and protection against reinfection. In addition to their efficacy, since they do not contain or express nonstructural proteins, it is possible to distinguish between vaccinated animals and those that are infected with the virus, thus addressing one of the major problems with current vaccines.

18.7 CONCLUSION

Through the avenue of basic research and advents in recombinant protein expression technology, it has been possible to produce BTV VLPs by solely expressing viral structural proteins. This technology has not only provided significant insights into the mechanism of viral capsid assembly but in doing so has provided the emergence of a new generation of vaccine types. These are not only highly immunogenic, as demonstrated by various successful vaccine trials, but have the highest safety profile as no pathogenic virus is utilized in their production. Additionally, their expression using BEVS allows for efficient scaling at industrial level, which is an essential property required for manufacturing VLPs as vaccine products.

While BTV has provided a model system for this technology, it has been applied to produce a wide variety of successful vaccine immunogens for numerous viral diseases (see review by Roy and Noad 2009). The success of these particles as immunogens has led to further research into their efficacy. These studies demonstrate that the high concentration of epitopes, due to their small size and repetitive structure, is key to their strong immune-stimulatory properties (Link et al. 2012; Manolova et al. 2008). As a result of the exquisite control and flexible nature of expression of individual VLP components, this strong immune stimulatory can now be readily exploited to produce designer immunogens, allowing this technology to be appropriated for nonviral targets by introduction of target epitopes (Grgacic and Anderson 2006; Plummer and Manchester 2010). In doing so, the design of such immunogens must be understood; this will require a thorough understanding of the molecular details of their fabrication. The CLP of BTV ideally suits such a purpose. Through basic research, the molecular details of assembly for both inner VP3 and outer VP7 layers have been well characterized, allowing for ease of manipulation without interference with this process. Additionally, it provides a large scaffold to which multiple epitopes can be incorporated into its VP7 outer layer. This, amongst other viral scaffolds, could provide an arsenal of compounds available for future vaccine designers.

In conclusion, the advent of VLP technology has developed a new age of vaccine design that is not limited in scope or form. In doing so, this technology may hold the promise to effectively treat all disease, both by pathogens and autoimmunity, through host-tailored immune-modulatory compounds. Only the future will see if this vast potential becomes a reality.

REFERENCES

Andersen, D C and L Krummen. 2002. Recombinant protein expression for therapeutic applications. *Current Opinion in Biotechnology* 13(2): 117–123. doi:10.1016/S0958-1669(02)00300-2. http://www.sciencedirect.com/science/article/pii/S0958166902003002.

Balasuriya, U B R, S A Nadler, W C Wilson, L I Pritchard, A B Smythe, G Savini, F Monaco et al. 2008. The NS3 proteins of global strains of bluetongue virus evolve into regional topotypes through negative (purifying) selection. *Veterinary Microbiology* 126(1–3): 91–100. doi:10.1016/j.vetmic.2007.07.006. http://www.sciencedirect.com/science/article/pii/S0378113507003343.

Baneyx, F. 1999. Recombinant protein expression in *Escherichia coli*. *Current Opinion in Biotechnology* 10(5): 411–421. doi:10.1016/S0958-1669(99)00003-8. http://www.sciencedirect.com/science/article/pii/S0958166999000038.

Barondess, S H, D N W Cooper, M A Gitts, and H Lefflerm. 1994. Galectins. Structure and function of a large family of animal lectins. *Journal of Biological Chemistry* 269(33): 20807–20810.

Batten, C A, S Maan, A E Shaw, N S Maan, and P P C Mertens. 2008. A European field strain of bluetongue virus derived from two parental vaccine strains by genome segment reassortment. *Virus Research* 137(1): 56–63. doi:10.1016/j.virusres.2008.05.016. http://www.sciencedirect.com/science/article/pii/S0168170208002268.

Belyaev, A S, R S Hails, and P Roy. 1995. High-level expression of five foreign genes by a single recombinant baculovirus. *Gene* 156(2): 229–233. http://www.ncbi.nlm.nih.gov/pubmed/7758961.

Belyaev, A S and P Roy. 1993. Development of baculovirus triple and quadruple expression vectors: Co-expression of three or four bluetongue virus proteins and the synthesis of bluetongue virus-like particles in insect cells. *Nucleic Acids Research* 21(5): 1219–1223. http://www.pubmedcentral.nih.gov/articlerender.fcgi?artid=309285&tool=pmcentrez&rendertype=abstract.

Berger, I, D J Fitzgerald, and T J Richmond. 2004. Baculovirus expression system for heterologous multiprotein complexes. *Nature Biotechnology* 22(12): 1583–1587. doi:10.1038/nbt1036. http://www.ncbi.nlm.nih.gov/pubmed/15568020.

Bishop, D H. 1990. Gene expression using insect cells and viruses. *Current Opinion in Biotechnology* 1(1): 62–67. http://www.ncbi.nlm.nih.gov/pubmed/1367918.

Blissard, G W and G F Rohrmann. 1990. Baculovirus diversity and molecular biology. *Annual Review of Entomology* 35(January): 127–155. doi:10.1146/annurev.en.35.010190.001015. http://www.ncbi.nlm.nih.gov/pubmed/2154158.

Bowne, J G. 1971. Bluetongue disease. *Advances in Veterinary Science and Comparative Medicine* 15(January): 1–46. http://www.ncbi.nlm.nih.gov/pubmed/4401394.

Carpi, G, E C Holmes, and A Kitchen. 2010. The evolutionary dynamics of bluetongue virus. *Journal of Molecular Evolution* 70(6): 583–592. doi:10.1007/s00239-010-9354-y. http://www.ncbi.nlm.nih.gov/pubmed/20526713.

Caspar, D L and A Klug. 1962. Physical principles in the construction of regular viruses. *Cold Spring Harbor Symposia on Quantitative Biology* 27(January): 1–24. http://www.ncbi.nlm.nih.gov/pubmed/14019094.

Chan, D C, Fass, D Berger, J M and Kim, P S 1997. Core structure of gp41 from the HIV envelope glycoprotein. *Cell*, 89(2): 263–273.

Chen, R. 2014. Bacterial expression systems for recombinant protein production: E. Coli and beyond. *Biotechnology Advances* 30(5): 1102–1107. doi:10.1016/j.biotechadv.2011.09.013. http://www.ncbi.nlm.nih.gov/pubmed/21968145.

Copeland, N G, N A Jenkins, and D L Court. 2001. Recombineering: A powerful new tool for mouse functional genomics. *Nature Reviews. Genetics* 2(10): 769–779. doi:10.1038/35093556. http://www.ncbi.nlm.nih.gov/pubmed/11584293.

Crouch, E A and A Lorena Passarelli. 2002. Genetic requirements for homologous recombination in *Autographa californica* nucleopolyhedrovirus. *Journal of Virology* 76(18): 9323–9334. http://www.pubmedcentral.nih.gov/articlerender.fcgi?artid=136457&tool=pmcentrez&rendertype=abstract.

Datsenko, K A and B L Wanner. 2000. One-step inactivation of chromosomal genes in *Escherichia coli* K-12 using PCR products. *Proceedings of the National Academy of Sciences of the United States of America* 97(12): 6640–6645. doi:10.1073/pnas.120163297. http://www.pnas.org/content/97/12/6640.long.

Eaton, B T, A D Hyatt, and S M Brookes. 1990. The replication of bluetongue virus. *Current Topics in Microbiology and Immunology* 162(January): 89–118. http://www.ncbi.nlm.nih.gov/pubmed/2166649.

Ellis, J A, A J Luedke, W C Davis, S J Wechsler, J O Mecham, D L Pratt, and J D Elliott. 1990. T lymphocyte subset alterations following bluetongue virus infection in sheep and cattle. *Veterinary Immunology and Immunopathology* 24(1): 49–67. doi:10.1016/0165-2427(90)90077-6. http://www.sciencedirect.com/science/article/pii/0165242790900776.

Emery, V C and D H Bishop. 1987. The development of multiple expression vectors for high level synthesis of eukaryotic proteins: Expression of LCMV-N and AcNPV polyhedrin protein by a recombinant baculovirus. *Protein Engineering* 1(4): 359–366. http://www.ncbi.nlm.nih.gov/pubmed/3334094.

Endo, Y and T Sawasaki. 2006. Cell-free expression systems for eukaryotic protein production. *Current Opinion in Biotechnology* 17(4): 373–380. doi:10.1016/j.copbio.2006.06.009. http://www.sciencedirect.com/science/article/pii/S0958166906000930.

Engelhard, E K, L N Kam-Morgan, J O Washburn, and L E Volkman. 1994. The insect tracheal system: A conduit for the systemic spread of *Autographa californica* M nuclear polyhedrosis virus. *Proceedings of the National Academy of Sciences of the United States of America* 91(8): 3224–3227. http://www.pubmedcentral.nih.gov/articlerender.fcgi?artid=43548&tool=pmcentrez&rendertype=abstract.

Faulkner, P and J F Henderson. 1972. Serial passage of a nuclear polyhedrosis disease virus of the cabbage looper (*Trichoplusia Ni*) in a continuous tissue culture cell line. *Virology* 50(3): 920–924. http://www.ncbi.nlm.nih.gov/pubmed/4629692.

Ferrari, G, C De Liberato, G Scavia, R Lorenzetti, M Zini, F Farina, A Magliano et al. 2005. Active circulation of bluetongue vaccine virus serotype-2 among unvaccinated cattle in central Italy. *Preventive Veterinary Medicine* 68(2–4): 103–113. doi:10.1016/j.prevetmed.2004.11.011. http://www.sciencedirect.com/science/article/pii/S0167587704002351.

Fitzgerald, D J, P Berger, C Schaffitzel, K Yamada, T J Richmond, and I Berger. 2006. Protein complex expression by using multigene baculoviral vectors. *Nature Methods* 3(12): 1021–1032. doi:10.1038/nmeth983. http://www.ncbi.nlm.nih.gov/pubmed/17117155.

Fitzgerald, D J, C Schaffitzel, P Berger, R Wellinger, C Bieniossek, T J Richmond, and I Berger. 2007. Multiprotein expression strategy for structural biology of eukaryotic complexes. *Structure (London, England : 1993)* 15(3): 275–279. doi:10.1016/j.str.2007.01.016. http://www.ncbi.nlm.nih.gov/pubmed/17355863.

Flanagan, M and S J Johnson. 1995. The effects of vaccination of merino ewes with an attenuated australian bluetongue virus serotype 23 at different stages of gestation. *Australian Veterinary Journal* 72(12): 455–457. doi:10.1111/j.1751-0813.1995.tb03488.x. http://doi.wiley.com/10.1111/j.1751-0813.1995.tb03488.x.

French, T J, S Inumaru, and P Roy. 1989. Expression of two related nonstructural proteins of bluetongue virus (BTV) type 10 in insect cells by a recombinant baculovirus: Production of polyclonal ascitic fluid and characterization of the gene product in BTV-infected BHK cells. *Journal of Virology* 63(8): 3270–3278.

French, T J, J J Marshall, and P Roy. 1990. Assembly of double-shelled, viruslike particles of bluetongue virus by the simultaneous expression of four structural proteins. *Journal of Virology* 64(12): 5695–5700. http://www.pubmedcentral.nih.gov/articlerender.fcgi?artid = 248707&tool = pmcentrez&rendertype = abstract.

French, T J and P Roy. 1990. Synthesis of bluetongue virus (BTV) corelike particles by a recombinant baculovirus expressing the two major structural core proteins of BTV. *Journal of Virology* 64(4): 1530–1536. http://jvi.asm.org/content/64/4/1530.short.

Ganser-Pornillos, B K, M Yeager, and W I Sundquist. 2008. The structural biology of HIV assembly. *Current Opinion in Structural Biology* 18(2): 203–217. doi:10.1016/j.sbi.2008.02.001. http://www.pubmedcentral.nih.gov/articlerender.fcgi?artid = 2819415&tool = pmcentrez&rendertype = abstract.

Geisse, S, H Gram, B Kleuser, and H P Kocher. 1996. Eukaryotic expression systems: A comparison. *Protein Expression and Purification* 8(3): 271–282. doi:10.1006/prep.1996.0101. http://www.sciencedirect.com/science/article/pii/S1046592896901011.

Goodwin, R H, J L Vaughn, J R Adams, and S J Louloudes. 1970. Replication of a nuclear polyhedrosis virus in an established insect cell line. *Journal of Invertebrate Pathology* 16(2): 284–288. http://www.ncbi.nlm.nih.gov/pubmed/5482779.

Grgacic, E V L and D A Anderson. 2006. Virus-like particles: Passport to immune recognition. *Methods (San Diego, Calif.)* 40(1): 60–65. doi:10.1016/j.ymeth.2006.07.018. http://www.sciencedirect.com/science/article/pii/S1046202306001629.

Grimes, J, A K Basak, P Roy, and D Stuart. 1995. The crystal structure of bluetongue virus VP7. *Nature* 373(6510): 167–170. doi:10.1038/373167a0. http://dx.doi.org/10.1038/373167a0.

Grimes, J M, J N Burroughs, P Gouet, J M Diprose, R Malby, S Ziéntara, P P Mertens, and D I Stuart. 1998. The atomic structure of the bluetongue virus core. *Nature* 395(6701): 470–478. doi:10.1038/26694. http://www.ncbi.nlm.nih.gov/pubmed/9774103.

Grimes, J M, J Jakana, M Ghosh, A K Basak, P Roy, W Chiu, D I Stuart, and B V Prasad. 1997. An atomic model of the outer layer of the bluetongue virus core derived from x-ray crystallography and electron cryomicroscopy. *Structure (London, England : 1993)* 5(7): 885–893. http://www.ncbi.nlm.nih.gov/pubmed/9261080.

Harper, D M, E L Franco, C Wheeler, D G Ferris, D Jenkins, A Schuind, T Zahaf et al. 2004. Efficacy of a bivalent L1 virus-like particle vaccine in prevention of infection with human papillomavirus types 16 and 18 in young women: A randomised controlled trial. *Lancet* 364(9447): 1757–1765. doi:10.1016/S0140-6736(04)17398-4. http://www.sciencedirect.com/science/article/pii/S0140673604173984.

Harper, D M, E L Franco, C M Wheeler, A-B Moscicki, B Romanowski, C M Roteli-Martins, D Jenkins, A Schuind, S A C Clemens, and G Dubin. 2006. Sustained efficacy up to 4.5 years of a bivalent L1 virus-like particle vaccine against human papillomavirus types 16 and 18: Follow-up from a randomised control trial. *Lancet* 367(9518): 1247–1255. doi:10.1016/S0140-6736(06)68439-0. http://www.sciencedirect.com/science/article/pii/S0140673606684390.

Hawtin, R E, T Zarkowska, K Arnold, C J Thomas, G W Gooday, L A King, J A Kuzio, and R D Possee. 1997. Liquefaction of *Autographa californica* nucleopolyhedrovirus-infected insects is dependent on the integrity of virus-encoded chitinase and cathepsin genes. *Virology* 238(2): 243–253. doi:10.1006/viro.1997.8816. http://www.ncbi.nlm.nih.gov/pubmed/9400597.

Hübschle, O J B. 1980. Bluetongue virus hemagglutination and its inhibition by specific sera. *Archives of Virology* 64(2): 133–140. doi:10.1007/BF01318017. http://link.springer.com/10.1007/BF01318017.

Herniou, E A and J A Jehle. 2007. Baculovirus phylogeny and evolution. *Current Drug Targets* 8(10): 1043–1050. http://www.ncbi.nlm.nih.gov/pubmed/17979664.

Hewat, E A, T F Booth, P T Loudon, and P Roy. 1992a. Three-dimensional reconstruction of baculovirus expressed bluetongue virus core-like particles by cryo-electron microscopy. *Virology* 189(1): 10–20. http://www.sciencedirect.com/science/article/pii/004268229290676G.

Hewat, E A, T F Booth, and P Roy. 1992b. Structure of bluetongue virus particles by cryoelectron microscopy. *Journal of Structural Biology* 109(1): 61–69. http://www.sciencedirect.com/science/article/pii/104784779290068L.

Howell, P G and D W Verwoerd. 1971. Bluetongue virus. *Virology Monographs. Die Virusforschung in Einzeldarstellungen* 9 (January): 35–74. http://www.ncbi.nlm.nih.gov/pubmed/4354734.

Huismans, H, A A van Dijk, and H J Els. 1987. Uncoating of parental bluetongue virus to core and subcore particles in infected L cells. *Virology* 157(1): 180–188. doi:10.1016/0042-6822(87)90327-8. http://www.ncbi.nlm.nih.gov/pubmed/3029957.

Inumaru, S, H Ghiasi, and P Roy. 1987. Expression of bluetongue virus group-specific antigen VP3 in insect cells by a baculovirus vector: Its use for the detection of bluetongue virus antibodies. *The Journal of General Virology* 68(Pt 6)(June): 1627–1635. http://www.ncbi.nlm.nih.gov/pubmed/3035063.

Jeggo, M H, R C Wardley, and J Brownlie. 1985. Importance of ovine cytotoxic T cells in protection against bluetongue virus infection. *Progress in Clinical and Biological Research* 178(January): 477–487. http://europepmc.org/abstract/MED/2989889.

Jehle, J A, G W Blissard, B C Bonning, J S Cory, E A Herniou, G F Rohrmann, D A Theilmann, S M Thiem, and J M Vlak. 2006. On the classification and nomenclature of baculoviruses: A proposal for revision. *Archives of Virology* 151(7): 1257–1266. doi:10.1007/s00705-006-0763-6. http://www.ncbi.nlm.nih.gov/pubmed/16648963.

Jones, L D, T Chuma, R Hails, T Williams, and P Roy. 1996. The non-structural proteins of bluetongue virus are a dominant source of cytotoxic T cell peptide determinants. *Journal of General Virology* 77(5): 997–1003. doi:10.1099/0022-1317-77-5-997. http://vir.sgmjournals.org/content/77/5/997.short.

Kar, A K, M Ghosh, and P Roy. 2004. Mapping the assembly pathway of bluetongue virus scaffolding protein VP3. *Virology* 324(2): 387–399. doi:10.1016/j.virol.2004.04.018. http://www.ncbi.nlm.nih.gov/pubmed/15207624.

Kirnbauer, R, F Booy, N Cheng, D R Lowy, and J T Schiller. 1992. Papillomavirus L1 major capsid protein self-assembles into virus-like particles that are highly immunogenic. *National Academy of Sciences of the United States of America* 89(24): 12180–12184. doi:10.1073/pnas.89.24.12180. http://www.pnas.org/content/89/24/12180.short.

Kost, T A, J P Condreay, and D L Jarvis. 2005. Baculovirus as versatile vectors for protein expression in insect and mammalian cells. *Nature Biotechnology* 23(5): 567–575. doi:10.1038/nbt1095. http://dx.doi.org/10.1038/nbt1095.

Koutsky, L A, K A Ault, C M Wheeler, D R Brown, E Barr, F B Alvarez, L M Chiacchierini, and K U Jansen. 2002. A controlled trial of a human papillomavirus type 16 vaccine—NEJM. *The New England Journal of Medicine*. 347: 1645–1651. doi:10.1056/NEJMoa020586. http://www.nejm.org/doi/full/10.1056/NEJMoa020586.

Leitner, T, B Hahn, Henry M Jackson Foundation, C Kuiken, B Foley, P Marx, S Wolinsky et al. 2008. HIV Sequence Compendium 2008 Editors.

Limn, C-K and P Roy. 2003. Intermolecular interactions in a two-layered viral capsid that requires a complex symmetry mismatch. *Journal of Virology* 77(20): 11114–11124. doi:10.1128/JVI.77.20.11114.

Limn, C-K, N Staeuber, P Gouet, P Roy, and K Monastyrskaya. 2000. Functional dissection of the major structural protein of bluetongue virus: Identification of key residues within VP7 essential for capsid assembly. *Journal of Virology* 78(18): 8658–8669. doi:10.1128/JVI.74.18.8658-8669.2000.Updated.

Link, A, F Zabel, Y Schnetzler, A Titz, F Brombacher, and M F Bachmann. 2012. Innate immunity mediates follicular transport of particulate but not soluble protein antigen. *Journal of Immunology* (*Baltimore, Md. : 1950*) 188(8): 3724–3733. doi:10.4049/jimmunol.1103312. http://www.ncbi.nlm.nih.gov/pubmed/22427639.

Liu, H M, T F Booth, and P Roy. 1992. Interactions between bluetongue virus core and capsid proteins translated in vitro. *The Journal of General Virology* 73(Pt 10)(October): 2577–2584. http://europepmc.org/abstract/MED/1328473.

Liu, Q, M Z Li, D Leibham, D Cortez, and S J Elledge. 1998. The univector plasmid-fusion system, a method for rapid construction of recombinant DNA without restriction enzymes. *Current Biology* 8(24): 1300–S1. doi:10.1016/S0960-9822(07)00560-X. http://www.sciencedirect.com/science/article/pii/S096098220700560X.

Loudon, P T and P Roy. 1991. Assembly of five bluetongue virus proteins expressed by recombinant baculoviruses: Inclusion of the largest protein VP1 in the core and virus-like particles. *Virology* 180(2): 798–802. doi:10.1016/0042-6822(91)90094-R. http://dx.doi.org/10.1016/0042-6822(91)90094-R.

Maan, S, N S Maan, G Pullinger, K Nomikou, E Morecroft, M Guimera, M N Belaganahalli, and P P C Mertens. 2012. The genome sequence of bluetongue virus type 10 from India: Evidence for circulation of a western topotype vaccine strain. *Journal of Virology* 86(10): 5971–5972. doi:10.1128/JVI.00596-12. http://jvi.asm.org/content/86/10/5971.short.

Maclachlan, N J. 2011. Bluetongue: History, global epidemiology, and pathogenesis. *Preventive Veterinary Medicine* 102(2): 107–111. doi:10.1016/j.prevetmed.2011.04.005. http://www.ncbi.nlm.nih.gov/pubmed/21570141.

Manolova, V, A Flace, M Bauer, K Schwarz, P Saudan, and M F Bachmann. 2008. Nanoparticles target distinct dendritic cell populations according to their size. *European Journal of Immunology* 38(5): 1404–1413. doi:10.1002/eji.200737984. http://www.ncbi.nlm.nih.gov/pubmed/18389478.

Marek, M, M M van Oers, F F Devaraj, J M Vlak, and O-W Merten. 2011. Engineering of baculovirus vectors for the manufacture of virion-free biopharmaceuticals. *Biotechnology and Bioengineering* 108(5): 1056–1067. doi:10.1002/bit.23028. http://www.ncbi.nlm.nih.gov/pubmed/21449023.

Martin, A and J Zweerink. 1972. Isolation and characterization virus of two particles of bluetongue. *Virology* 50(2): 495–506.

Martin, D W and P C Weber. 1997. DNA replication promotes high-frequency homologous recombination during *Autographa californica* multiple nuclear polyhedrosis virus infection. *Virology* 232(2): 300–309. doi:10.1006/viro.1997.8573. http://www.ncbi.nlm.nih.gov/pubmed/9191843.

Martin, S A and H J Zweerink. 1972. Isolation and characterization of two types of bluetongue virus particles. *Virology* 50(2): 495–506. doi:10.1016/0042-6822(72)90400-X. http://dx.doi.org/10.1016/0042-6822(72)90400-X.

Mason, P W, E Rieder, and B Baxt. 1994. RGD sequence of foot-and-mouth disease virus is essential for infecting cells via the natural receptor but can be bypassed by an antibody-dependent enhancement pathway. *Proceedings of the National Academy of Sciences of the United States of America* 91(5): 1932–1936. http://www.pubmedcentral.nih.gov/articlerender.fcgi?artid = 43278&tool = pmcentrez&rendertype = abstract.

Matsuura, Y, R D Possee, and D H Bishop. 1986. Expression of the S-coded genes of lymphocytic choriomeningitis arenavirus using a baculovirus vector. *The Journal of General Virology* 67(Pt 8)(August): 1515–1529. http://www.ncbi.nlm.nih.gov/pubmed/3525745.

Mattanovich, D, P Branduardi, L Dato, B Gasser, M Sauer, and D Porro. 2012. Recombinant protein production in yeasts. *Methods in Molecular Biology* (*Clifton, N.J.*) 824(January): 329–358. doi:10.1007/978-1-61779-433-9_17. http://www.ncbi.nlm.nih.gov/pubmed/22160907.

Mertens, P P, F Brown, and D V Sangar. 1984. Assignment of the genome segments of bluetongue virus type 1 to the proteins which they encode. *Virology* 135(1): 207–217. http://www.ncbi.nlm.nih.gov/pubmed/6328750.

Mertens, P P C, S Pedley, J Cowley, J N Burroughs, A H Corteyn, M H Jeggo, D M Jennings, and B M Gorman. 1989. Analysis of the roles of bluetongue virus outer capsid proteins VP2 and VP5 in determination of virus serotype. *Virology* 170(2): 561–565. doi:10.1016/0042-6822(89)90447-9. http://www.sciencedirect.com/science/article/pii/0042682289904479.

Miller, L K. 1993. Baculoviruses: High-level expression in insect cells. *Current Opinion in Genetics and Development* 3(1): 97–101. http://www.ncbi.nlm.nih.gov/pubmed/8453280.

Miller, L K, A J Lingg, and L A Bulla. 1983. Bacterial, viral, and fungal insecticides. *Science* (*New York, N.Y.*) 219(4585): 715–721. doi:10.1126/science.219.4585.715. http://www.sciencemag.org/content/219/4585/715.abstract.

Nason, E L, R Rothagel, S K Mukherjee, A K Kar, M Forzan, B V Venkataram Prasad, and P Roy. 2004. Interactions between the inner and outer capsids of bluetongue virus. *Journal of Virology* 78(15): 8059–8067. doi:10.1128/JVI.78.15.8059.

Noad, R J, M Stewart, M Boyce, C C Celma, K R Willison, and P Roy. 2009. Multigene expression of protein complexes by iterative modification of genomic bacmid DNA. *BMC Molecular Biology* 10(January): 87. doi:10.1186/1471-2199-10-87. http://www.pubmedcentral.nih.gov/articlerender.fcgi?artid = 2749033&tool = pmcentrez&rendertype = abstract.

Nomikou, K, C I Dovas, S Maan, S J Anthony, A R Samuel, M Papanastassopoulou, N S Maan, O Mangana, and P P C Mertens. 2009. Evolution and phylogenetic analysis of full-length VP3 genes of eastern mediterranean bluetongue virus isolates. *PLoS One* 4(7): e6437. doi:10.1371/journal.pone.0006437. http://www.plosone.org/article/info:doi/10.1371/journal.pone.0006437#pone-0006437-g004.

Oppenheim, A B, A J Rattray, M Bubunenko, L C Thomason, and D L Court. 2004. In vivo recombineering of bacteriophage lambda by PCR fragments and single-strand oligonucleotides. *Virology* 319(2): 185–189. doi:10.1016/j.virol.2003.11.007. http://www.ncbi.nlm.nih.gov/pubmed/14980479.

Osburn, B I, R T Johnson, A M Silverstein, R A Prendergast, M M Jochim, and S E Levy. 1971. Experimental viral-induced congenital encephalopathies. II. The pathogenesis of bluetongue vaccine virus infection in fetal lambs. *Laboratory Investigation: A Journal of Technical Methods and Pathology* 25(3): 206–210. http://europepmc.org/abstract/MED/4328760.

Pennock, G D, C Shoemaker, and L K Miller. 1984. Strong and regulated expression of *Escherichia coli* beta-galactosidase in insect cells with a baculovirus vector. *Molecular and Cellular Biology* 4(3):399–406. http://www.pubmedcentral.nih.gov/articlerender.fcgi?artid = 368716&tool = pmcentrez&rendertype = abstract.

Pijlman, G P. 2003. Spontaneous excision of BAC vector sequences from bacmid-derived baculovirus expression vectors upon passage in insect cells. *Journal of General Virology* 84(10): 2669–2678. doi:10.1099/vir.0.19438-0. http://vir.sgmjournals.org/content/84/10/2669.long.

Pijlman, G P, J de Vrij, F J van den End, J M Vlak, and D E Martens. 2004. Evaluation of baculovirus expression vectors with enhanced stability in continuous cascaded insect-cell bioreactors. *Biotechnology and Bioengineering* 87(6): 743–753. doi:10.1002/bit.20178. http://www.ncbi.nlm.nih.gov/pubmed/15329932.

Plummer, E M and M Manchester. 2010. Viral nanoparticles and virus-like particles: Platforms for contemporary vaccine design. *Wiley Interdisciplinary Reviews. Nanomedicine and Nanobiotechnology* 3(2): 174–196. doi:10.1002/wnan.119. http://www.ncbi.nlm.nih.gov/pubmed/20872839.

Prasad, B V, S Yamaguchi, and P Roy. 1992. Three-dimensional structure of single-shelled bluetongue virus. *Journal of Virology* 66(4): 2135–2142. http://www.pubmedcentral.nih.gov/articlerender.fcgi?artid = 289005&tool = pmcentrez&rendertype = abstract.

Pritchard, L I, I Sendow, R Lunt, S H Hassan, J Kattenbelt, A R Gould, P W Daniels, and B T Eaton. 2004. Genetic diversity of bluetongue viruses in south east Asia. *Virus Research* 101(2): 193–201. doi:10.1016/j.virusres.2004.01.004. http://www.sciencedirect.com/science/article/pii/S0168170204000243.

Rojas, J-M, T Rodríguez-Calvo, L Peña, and N Sevilla. 2011. T cell responses to bluetongue virus are directed against multiple and identical CD4+ and CD8+ T cell epitopes from the VP7 core protein in mouse and sheep. *Vaccine* 29(40): 6848–6857. doi:10.1016/j.vaccine.2011.07.061. http://www.sciencedirect.com/science/article/pii/S0264410X11011066.

Roy, P. 1992. From genes to complex structures of bluetongue virus and their efficacy as vaccines. *Veterinary Microbiology* 33(1–4): 155–168. doi:10.1016/0378-1135(92)90043-S. http://www.sciencedirect.com/science/article/pii/037811359290043S.

Roy, P. 2003. Nature and duration of protective immunity to bluetongue virus infection. *Developments in Biologicals* 114(January): 169–183. http://europepmc.org/abstract/MED/14677687.

Roy, P, D Bishop, H Leblois, and B Erasmus. 1994. Long-lasting protection of sheep against bluetongue challenge after vaccination with virus-like particles: Evidence for homologous and partial heterologous protection. *Vaccine* 12(9): 805–811. doi:10.1016/0264-410X(94)90289-5. http://www.sciencedirect.com/science/article/pii/0264410X94902895.

Roy, P, M Boyce, and R Noad. 2009. Prospects for improved bluetongue vaccines. *Nature Reviews. Microbiology* 7(2): 120–128. doi:10.1038/nrmicro2052. http://dx.doi.org/10.1038/nrmicro2052.

Roy, P, T French, and B J Erasmus. 1992. Protective efficacy of virus-like particles for bluetongue disease. *Vaccine* 10(1): 28–32. doi:10.1016/0264-410X(92)90415-G. http://www.sciencedirect.com/science/article/pii/0264410X9290415G.

Roy, P and R Noad. 2009. Virus-like particles as a vaccine delivery system: Myths and facts. *Advances in Experimental Medicine and Biology* 655(January): 145–158. doi:10.1007/978-1-4419-1132-2_11. http://www.ncbi.nlm.nih.gov/pubmed/20047040.

Roy, P and R Noad. 2012. Use of bacterial artificial chromosomes in baculovirus research and recombinant protein expression: current trends and future perspectives. *ISRN Microbiology* 2012, article ID 628797, 11 pages. http://dx.doi.org/10.5402/2012/628797.

Roy, P, T Urakawa, A A Van Dijk, and B J Erasmus. 1990. Recombinant virus vaccine for bluetongue disease in sheep. *Journal of Virology* 64(5): 1998–2003. http://jvi.asm.org/content/64/5/1998.short.

Schultz, G and P D Delay. 1955. Losses in newborn lambs associated with bluetongue vaccination of pregnancy ewes. *Journal of the American Veterinary Medical Association* 127(942): 224–226. http://europepmc.org/abstract/med/13251947.

Settembre, E C, J Z Chen, P R Dormitzer, N Grigorieff, and S C Harrison. 2011. Atomic model of an infectious rotavirus particle. *The EMBO Journal* 30(2): 408–416. doi:10.1038/emboj.2010.322. http://www.pubmedcentral.nih.gov/articlerender.fcgi?artid = 3025467&tool = pmcentrez&rendertype = abstract.

Simón, O, T Williams, P Caballero, and M López-Ferber. 2006. Dynamics of deletion genotypes in an experimental insect virus population. *Proceedings. Biological Sciences/The Royal Society* 273(1588): 783–790. doi:10.1098/rspb.2005.3394. http://rspb.royalsocietypublishing.org/content/273/1588/783.short.

Smith, G E, M J Fraser, and M D Summers. 1983a. Molecular engineering of the *Autographa californica* nuclear polyhedrosis virus genome: Deletion mutations within the polyhedrin gene. *Journal of Virology* 46(2): 584–593. http://jvi.asm.org/content/46/2/584.short.

Smith, G E, M D Summers, and M J Fraser. 1983b. Production of human beta interferon in insect cells infected with a baculovirus expression vector. *Molecular and Cellular Biology* 3(12): 2156–2165. http://www.pubmedcentral.nih.gov/articlerender.fcgi?artid = 370086&tool = pmcentrez&rendertype = abstract.

Smith, G E, J M Vlak, and M D Summers. 1983c. Physical analysis of *Autographa californica* nuclear polyhedrosis virus transcripts for polyhedrin and 10,000-molecular-weight protein. *Journal of Virology* 45(1): 215–225. http://www.pubmedcentral.nih.gov/articlerender.fcgi?artid = 256404&tool = pmcentrez&rendertype = abstract.

Stewart, M, Y Bhatia, T N Athmaran, R Noad, C Gastaldi, E Dubois, P Russo, et al. 2010. Validation of a novel approach for the rapid production of immunogenic virus-like particles for bluetongue virus. *Vaccine* 28(17): 3047–3054. doi:10.1016/j.vaccine.2009.10.072. http://www.sciencedirect.com/science/article/pii/S0264410X09016028.

Stewart, M, C I Dovas, E Chatzinasiou, T N Athmaram, M Papanastassopoulou, O Papadopoulos, and P Roy. 2012. Protective efficacy of bluetongue virus-like and subvirus-like particles in sheep: Presence of the serotype-specific VP2, independent of its geographic lineage, is essential for protection. *Vaccine* 30(12): 2131–2139. doi:10.1016/j.vaccine.2012.01.042. http://www.sciencedirect.com/science/article/pii/S0264410X12000680.

Stewart, M, E Dubois, C Sailleau, E Bréard, C Viarouge, A Desprat, R Thiéry, S Zientara, and P Roy. 2013. Bluetongue virus serotype 8 virus-like particles protect sheep against virulent virus infection as a single or multi-serotype cocktail immunogen. *Vaccine* 31(3): 553–558. doi:10.1016/j.vaccine.2012.11.016. http://www.sciencedirect.com/science/article/pii/S0264410X1201609X.

Stott, J L, R D Oberst, M B Channell, and B I Osburn. 1987. Genome segment reassortment between two serotypes of bluetongue virus in a natural host. *Journal of Virology* 61(9): 2670–2674. http://jvi.asm.org/content/61/9/2670.short.

Suzich, J A, S J Ghim, F J Palmer-Hill, W I White, J K Tamura, J A Bell, J A Newsome, A B Jenson, and R Schlegel. 1995. Systemic immunization with papillomavirus L1 protein completely prevents the development of viral mucosal papillomas. *Proceedings of the National Academy of Sciences of the United States of America* 92(25): 11553–11557. doi:10.1073/pnas.92.25.11553. http://www.pnas.org/content/92/25/11553.short.

Takehara, K, D Ireland, and D H Bishop. 1988. Co-expression of the hepatitis B surface and core antigens using baculovirus multiple expression vectors. *The Journal of General Virology* 69(Pt 11)(November): 2763–2777. http://www.ncbi.nlm.nih.gov/pubmed/3053987.

Tan, B-H, E Nason, N Staeuber, K Monastryrskaya, P Roy, and W Jiang. 2001. RGD tripeptide of bluetongue virus VP7 protein is responsible for core attachment to culicoides cells. *Journal of Virology* 75(8): 3937–3947. doi:10.1128/JVI.75.8.3937.

Tanaka, S and P Roy. 1994. Identification of domains in bluetongue virus VP3 molecules essential for the assembly of virus cores. *Journal of Virology* 68(5): 2795–2802. http://www.pubmedcentral.nih.gov/articlerender.fcgi?artid = 236767&tool = pmcentrez&rendertype = abstract.

Terpe, K. 2006. Overview of bacterial expression systems for heterologous protein production: From molecular and biochemical fundamentals to commercial systems. *Applied Microbiology and Biotechnology* 72(2): 211–222. doi:10.1007/s00253-006-0465-8. http://www.ncbi.nlm.nih.gov/pubmed/16791589.

Urakawa, T, D G Ritter, and P Roy. 1989. Expression of target RNA segment and synthesis of VP1 protein of bluetongue virus in insect cells by recombinant baculovirus: Association of VP1 protein with RNA polymerase activity. *Nucleic Acids Research* 17(18): 7395–7401. doi:10.1093/nar/17.18.7395. http://nar.oxfordjournals.org/content/17/18/7395.short.

Urakawa, T and P Roy. 1988. Bluetongue virus tubules made in insect cells by recombinant baculoviruses: Expression of the NS1 gene of bluetongue virus serotype 10. *Journal of Virology* 62(11): 3919–3127.

Van der Beek, C P, J D Saaijer-Riep, and J M Vlak. 1980. On the origin of the polyhedral protein of *Autographa californica* nuclear polyhedrosis virus isolation, characterization, and translation of viral messenger RNA. *Virology* 100(2): 326–333. doi:10.1016/0042-6822(80)90523-1. http://www.sciencedirect.com/science/article/pii/0042682280905231.

Van Dijk, A A and H Huismans. 1988. In vitro transcription and translation of bluetongue virus mRNA. *The Journal of General Virology* 69(Pt 3)(March): 573–581. http://www.ncbi.nlm.nih.gov/pubmed/2832524.

Van Oers, M M and J M Vlak. 2007. Baculovirus genomics. *Current Drug Targets* 8(10): 1051–1068. http://www.ncbi.nlm.nih.gov/pubmed/17979665.

Verwoerd, D W, H J Els, E-M De Villiers, and H Huismans. 1972. Structure of the bluetongue virus capsid. *Journal of Virology* 10(4): 783–794. http://jvi.asm.org/content/10/4/783.short.

Verwoerd, D W and H Huismans. 1972. Studies on the in vitro and the in vivo transcription of the bluetongue virus genome. *The Onderstepoort Journal of Veterinary Research* 39(4): 185–191. http://www.ncbi.nlm.nih.gov/pubmed/4352125.

Vlak, J M, A Schouten, M Usmany, G J Belsham, E C Klinge-Roode, A J Maule, J W Van Lent, and D Zuidema. 1990. Expression of cauliflower mosaic virus gene I using a baculovirus vector based upon the p10 gene and a novel selection method. *Virology* 179(1): 312–320. http://www.ncbi.nlm.nih.gov/pubmed/2219726.

Volkman, L E and M D Summers. 1977. *Autographa californica* nuclear polyhedrosis virus: Comparative infectivity of the occluded, alkali-liberated, and nonoccluded forms. *Journal of Invertebrate Pathology* 30(1): 102–103. http://www.ncbi.nlm.nih.gov/pubmed/336795.

Weyer, U and R D Possee. 1991. A baculovirus dual expression vector derived from the *Autographa californica* nuclear polyhedrosis virus polyhedrin and p10 promoters: Co-expression of two influenza virus genes in insect cells. *The Journal of General Virology* 72(Pt 12)(December): 2967–2974. http://www.ncbi.nlm.nih.gov/pubmed/1765769.

Wilson, A J and P S Mellor. 2009. Bluetongue in Europe: Past, present and future. *Philosophical Transactions of the Royal Society of London. Series B, Biological Sciences* 364(1530): 2669–2681. doi:10.1098/rstb.2009.0091. http://rstb.royalsocietypublishing.org/content/364/1530/2669.abstract.

Wurm, F M. 2004. Production of recombinant protein therapeutics in cultivated mammalian cells. *Nature Biotechnology* 22(11): 1393–1398. doi:10.1038/nbt1026. http://dx.doi.org/10.1038/nbt1026.

Zhang, X X, M Boyce, B Bhattacharya, S Schein, P Roy, and Z H Zhou. 2010. Bluetongue virus coat protein VP2 contains sialic acid-binding domains, and VP5 resembles enveloped virus fusion proteins. *Proceedings of the National Academy of Sciences of the United States of America* 107(14): 6292–6297. doi:10.1073/pnas.0913403107. http://www.pubmedcentral.nih.gov/articlerender.fcgi?artid = 2852009&tool = pmcentrez&rendertype = abstract.

Zhao, Y, D A Chapman, and I M Jones 2003. Improving baculovirus recombination. *Nucleic Acids Research* 31(2): 6e–6. doi:10.1093/nar/gng006. http://nar.oxfordjournals.org/content/31/2/e6.long.

19 Virus-Like Particles Based on Polyomaviruses and Human Papillomaviruses as Vectors for Vaccines, Preventive and Therapeutic Immunotherapy, and Gene Delivery

Tina Dalianis

CONTENTS

19.1 INTRODUCTION

To vaccinate against a viral infection or to conduct gene or immune therapy, many transfer systems based on viruses as vectors have been explored [1]. Earlier, some such vectors have had the disadvantage of maintaining viral genes that could potentially interfere with host genes or be abrogated by preexisting immunity to the virus [1]. This leads to the search for new vectors, and today, there are a plethora of different types of vectors of which some are based on virus-like particles (VLPs) [2,3]. Here, the focus is directed only on those based on polyomaviruses (PyVs) and human papillomaviruses (HPVs), which lack viral genes and are nonreplicative and nonpathogenic and where the former have been used for vaccination with very little side effects [2–8]. The fact that the capsids of these VLPs are easy to modify

and that they can be produced industrially for vaccine use has increased the awareness of the potential use of VLPs as vaccines, as well as vectors for immune therapy of cancer patients or patients with other diseases or conditions [2–8]. PyV and HPV VLPs consist of one or more viral structural proteins that can be produced in, for example, yeast cells, insect cells using baculoviruses, *Escherichia coli*, tobacco chloroplasts, or mammalian cells and spontaneously assemble into VLPs similar to the equivalent virions, but lacking viral genes [9–12]. Industrial production, as mentioned earlier, using yeast and baculovirus, respectively, has been used for the manufacturing of the two HPV vaccines Gardasil and Cervarix [5]. PyV and HPV VLPs can induce antibody responses against the corresponding virions and bind to cellular receptors, similar to that of the equivalent virions [2,3]. Nonspecies-specific VLPs can also be used to introduce DNA or proteins into cells in vivo, this way avoiding the disadvantage of preexisting antibodies that may abrogate the efficacy of the vectors [2–8,13–18]. However, recently, many more HPVs and human PyVs (HPyVs) have been discovered, potentially offering a broader spectrum of VLP choices [3]. In the following, some aspects of the use of PyV- and HPV-based VLPs as vaccines but also as vectors for immune therapy and gene therapy are presented.

19.2 PyVs AND PyV-BASED VLPs

Murine PyV (MPyV), the first PyV to be well characterized, was discovered in the 1950s as a transmissible agent causing many tumors when inoculated into newborn mice, thus its name "poly oma" in Greek referring to "many tumors" [19]. Since then, studies of MPyV have provided knowledge on transformation, replication, transcription, oncogenesis, cell cycle regulation, tumor immunity, tumor-specific antigens, and VLP assembly [19]. MPyVs, similar to other PyVs, have an around 5 kb genome consisting of double-stranded-circular DNA, arbitrarily divided into a noncoding regulatory region, an early and a late coding region, included with cellular histones in the 45 nm icosahedral virion [19]. Its early coding region encodes large T antigen, middle T antigen (lacking in most PyVs), and small T antigen, for example, regulatory proteins accountable for viral pathology and transformation, while its late coding region encodes the structural viral capsid proteins VP1-3 (Figures 19.1 and 19.2) [19]. PyV virions consist of 360 VP1 molecules, formed as 72 pentamers on the viral capsid, while VP2 and VP3 bind from the inside and are not necessary for VLP assembly in insect or yeast

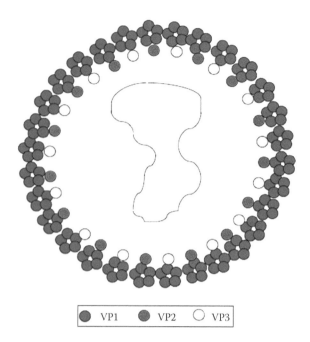

| ● VP1 | ● VP2 | ○ VP3 |

FIGURE 19.2 Schematic structure of a PyV particle. A single copy of a circular double-stranded DNA molecule is surrounded by a capsid. The capsid is in turn composed of 72 pentamers of the major structural protein VP1, and to each pentamer, either one VP2 or one VP3 molecule is bound.

cells but do enhance natural viral infectivity and viral production [19]. Notably, VP1 also binds MPyV DNA and is critical for viral uptake [19]. It is also important for binding to sialyloligosaccharide residues on gangliosides present on most cells [19].

Already in 1978, chromatographically separated and purified MPyV VP1 molecules were shown to have the capacity to form pentamers and capsomers, forming viral capsids [20]. Later, VLPs similar to native virions were produced by bacterial expression of recombinant VP1 or by VP1 produced in a baculovirus system or yeast [2–8,13–18,21].

Murine pneumotropic virus (MPtV), another MPyV that was isolated in 1953, was shown to cause fatal pneumonia in newborn, but not in adult, mice [22,23]. It encodes only two early proteins (LT and ST), similar to *Simian virus 40 (SV40)*, the first primate virus to be described in 1960 [23,24].

19.2.1 HPyVs

The first *HPyVs* BK virus (BKV) and JC virus (JCV), both possible to passage in tissue culture, were described in 1971 [25,26]. BKV was shown to cause hemorrhagic cystitis in hematopoietic stem cell–transplanted patients and PyV-associated nephritis in renal transplant patients [19]. JCV potentially caused progressive multifocal leukoencephalopathy in immunosuppressed patients, for example, those with HIV or more in recently multiple sclerosis patients treated with natalizumab [19]. It was not until 2007, by molecular means, that additional HPyVs were detected, with KIPyV, WUPyV, and MCV being the first three to be discovered [27–29]. Today, there are 12 or more known HPyVs, where

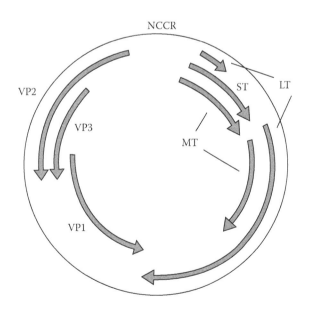

FIGURE 19.1 Structure of the genome of MPyV. Transcription occurs bidirectionally from the NCCR. Generally, most PyVs do not express middle T antigen, while some viruses express an additional protein, the agnoprotein, from the late region from an ORF located 5′ to the start codon of VP2. NCCR, noncoding control region; ST, small T antigen; MT, middle T antigen; LT, large T antigen.

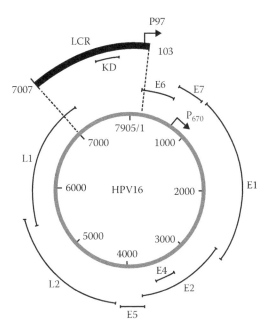

Viral protein	Functions and features
E1	It forms a heterodimer complex with E2 and controls viral replication.
E2	It regulates early gene promoter and together with E1 viral DNA replication.
E4	It may mediate the viral particle realease by destabilizing cytokeratin network.
E5	It stimulates mitogenic signals of growth factors.
E6	It inactivates many cellular proteins and is one of the major viral oncoproteins.
E7	It inactivates many cellular proteins and is one of the major viral oncoproteins.
L1	It is the major capsid protein and is the component of the HPV prophylactic vaccine.
L2	It is the minor capsid protein.

FIGURE 19.3 The double-stranded DNA HPV16 genome is represented by a gray circle annotated with the nucleotide numbers. The positions of the long control region and the early genes (E1–E7) and late genes (L1 and L2) are also shown. The early and late promoters, P97 and P670, respectively, are indicated by arrows. The main functions and features of the early and late gene products are listed in the table. (Reprinted from *Seminars in Cancer Biology*, 26C, Tommasino, M., The human papillomavirus family and its role in carcinogenesis, 13–21, Copyright (2014), with permission from Elsevier.)

notably MCV is linked to Merkel cell carcinoma and TSV associated to trichodysplasia spinulosa [30,31].

VLPs can be, to our knowledge, potentially produced from all PyVs and bind to many mammalian cells. Notably, however in contrast to MPyV, MPtV, SV40, BKV, and JCV are neuraminidase resistance indicating other ways of binding to cells [19]. Furthermore, many PyVs do not cross-react serologically, suggesting that they could be potentially be interchanged for prime boost VLP therapy [2,3,19]. VLPs from HPyVs have mainly been used for serology, showing that most HPyVs are relatively common in humans and that antibodies to these viruses are attained often in early childhood [19,32]. Nevertheless, the fact that there are several HPyVs allows for a broader repertoire of potential VLPs.

19.3 HPV AND HPV-BASED VLPs

There are >170 HPV types, some are cutaneous and some mucosal, where a separate type is defined if its L1 amino acid sequence differs >10% from another type [33,34]. In addition, some types can be associated with benign lesions, while others with malignancies, with the association to cervical, anogenital, and oropharyngeal cancers being well known [33,34].

HPVs are similar to HPyVs and double-stranded-circular DNA viruses with a genome of 7.2–8 kbp, arbitrarily divided into a regulatory noncoding control region and an early and a late coding region, enclosed in within a viral capsid of roughly 55 nm in diameter [33,34]. The early region encodes regulatory proteins E1, E2, and E4–E7 important for viral

pathogenesis, while the late region encodes the two structural capsid proteins L1 and L2 (Figure 19.3) [33,34]. Similar to HPyV, the major capsid protein L1 can self-assemble and around 360 L1 molecules make the 72 pentameric capsomers that create the viral capsid [33]. Trypsin treatment often inhibits HPV binding to cells implying that the HPV receptor is a protein and the so far best-studied and suggested candidate receptors are heparan sulfate proteoglycans that are expressed on many mammalian cells [35–40].

19.4 HPV AND HPyV VLPs AS PREVENTIVE VACCINES

19.4.1 HPV VLPs as Vaccines against Viral Infection

The best-known examples of the function of HPV VLPs as vaccines in humans are today's VLP-based HPV vaccines [4]. Both Gardasil (Merck), a vaccine against HPV types 6, 11, 16, and 18, and Cervarix (GlaxoSmithKline), against HPV16 and HPV18, induce high titers of neutralizing antibodies, which protect against infection with the equivalent HPV types, and in addition, also T-cell responses are obtained [3,4]. More importantly, both Gardasil and Cervarix protect against HPV16- or HPV18-induced cervical intraepithelial neoplasia accounting for around 70% of all cases, emphasizing the potent use of these vaccines in preventing the development of cervical cancer [3,4,41,42]. Today, more multivalent HPV vaccines are being produced and even more cases of cervical cancer will in this way be avoided [43,44]. The simple structure of HPV capsids, with L1 being so dominant and sufficient for formation of VLPs and the improvement of VLP production, was important for this success [43,44].

Alternative approaches have thus been discussed, of which one was to produce L1/L2 containing VLPs, since antibodies against L2 are more cross-reactive between HPV types and would cover more HPV types, as well as make antibodies that potentially could inhibit viral entry [42,43]. However, one drawback of this approach was that the antibody responses to L2 are generally weaker, so instead inserting part of L2 into L1 was suggested [42–47]. Another alternative, also a cheaper one, was to use a glutathione-S-transferase L1 fusion protein (GST L1) for vaccination [3,42–47]. In fact numerous methodologies are concerted, allowing for several options in the future. Nevertheless, once infected with HPV or when bearing an HPV-induced tumors, preventive vaccines are no longer useful [3,42–47].

19.4.2 PyV VLPs as Vaccines against Viral Infection in Normal and T-Cell Dysfunctional Mice and for Potential Use in Humans

MPyV VLP vaccination protects normal and CD4$^{-/-}$8$^{-/-}$ T-cell-deficient mice against MPyV infection [48,49]. VLP antibody titers were, however, lower in T-cell deficient than in normal mice, possibly due to less efficient IgG switching in the absence of efficient T-cell function [48]. MPyV VLPs and using a glutathione-S-transferase VP1 fusion protein (GST-VP1) for vaccination were compared, since the latter is easily produced in *E. coli* and cheaper as the aforementioned [48]. However, one must recall that GST-VP1 consists of dimers and pentamers and not VLPs. GST-VP1 immunized normal mice against MPyV infection, but could not prevent MPyV infection in all CD4$^{-/-}$8$^{-/-}$ mice, and the antibody responses were much weaker than those obtained after MPyV VLP vaccination [48]. The repetitive structures and folding of MPyV VLP are likely more potent as antigens than the more linear GST-VP1 proteins for triggering antibody responses in T-cell-deficient mice [48].

The data imply that GST-based vaccines be useful in humans with normal immunity, while VLP-based HPV vaccines of HPyV vaccines would be better in humans with a moderately T-cell dysfunctional immune system. This could be of special importance if considering, for example, producing preventive BKV or JCV vaccines for immunosuppressed individuals.

19.5 CHIMERIC VLPs FOR VACCINATION AGAINST VIRUS-INDUCED TUMORS

19.5.1 Chimeric VLPs

To induce immune responses against protein epitopes, these can be inserted into different loops or at the end of the major capsid proteins VP1 or L1 and thus be introduced onto the VLP surface but then have the limitation of disrupting VLP formation [2,3,19,50–54]. This approach is often preferred when wanting to elicit an antibody response.

An alternative approach is to fuse a protein or part of it to a minor capsid protein, for example, L2 or VP2/VP3, and coexpress the fusion protein with L1/VP in, for example, a baculovirus system, so that chimeric VLPs containing the fusion protein can assemble [2,50–54]. This alternative is often preferred for eliciting a cellular response, for example, for immunotherapy against cancer [2]. Moreover, an epitope situated within the VLPs will allow for the incorporation of a larger protein antigen without VLP disruption and less likely interfere with receptor VLP uptake, which could be essential for inducing a T-cell response [2,3,19,50–54].

19.5.2 Chimeric HPV VLPs for Vaccination against HPV-Induced Tumors

HPV E6 and HPV E7 are oncogenes, expressed in most HPV-associated malignancies [55]. To conduct immune therapy against HPV-associated malignancies or persisting HPV infections, numbers of different HPV16-based VLPs including full length or parts of HPV16 E6 and/or HPV16 E7 have been produced [51–54]. HPV16 VLPs, with L1 fused to an HPV16 E7 peptide, and likewise HPV16 L1/L2E7 VLPs, obtained by fusing HPV16 E7 to HPV16 L2 coexpressed with HPV16 L1, protect mice against outgrowth of an HPV16 positive tumor cell line, for example, by inducing E7-specific cytotoxic T cells [53,54]. Moreover, HPV16 L1E7 chimeric VLPs have been used to vaccinate against high-grade cervical intraepithelial neoplasia (CIN2/3), and both antibodies and cellular responses to L1 and E7 were obtained as well as a possible (nonstatistically significant) tendency for histological improvement [53,54]. It could, however, be potentially possible in the future to improve this type of approach by adding different adjuvants or dendritic cells (DCs) to the chimeric VLPs (see also Section 19.7).

19.5.3 MPyV VLPs as Vaccine against Induced MPyV Tumors

MPyV T antigens are responsible for transformation and expressed in MPyV-induced tumors, and T antigen–derived peptides have been shown to induce an immune response that can prevent MPyV tumor outgrowth [2,3,56]. T antigen peptide epitopes are therefore proposed to be presented by antigens of the major histocompatibility complex to the immune system [2,3,56]. However, VP1 is also expressed in some MPyV tumors, and here, VLP vaccination would be of benefit [57,58]. Likewise, HPV L1 expression by immunohistochemistry is occasionally observed in some HPV-associated malignancies, for example, oropharyngeal cancer [59]. To test the hypothesis, MPyV VLP–immunized mice were inoculated with three different MPyV-induced tumors. Complete and partial protection was obtained against two of the tumors, but not against the third tumor [57]. Thus, in some cases, the feasibility of using VP1 as target for immune rejection is possible, even though in general it is probably very low, since neither VP1 nor HPV L1 is required for tumor growth and is dispensable. Moreover, L1 is not expressed in basal epithelial cells where infection is maintained. Indeed, in a clinical trial including HPV16-/HPV18-infected women, no clearance of HPV DNA was obtained upon HPV VLP vaccination [60].

19.6 PyV CHIMERIC VLPs FOR PREVENTIVE AND THERAPEUTIC IMMUNE THERAPY OF NONVIRAL CANCER

MPyV chimeric VLPs have also been used for immune therapy nonviral cancer with the aim to break tolerance and elicit an immune response against a self-antigen [2,3]. The proto-oncogene Her2/*neu* and the prostate surface antigen (PSA) have been used as self-antigens in some studies (Figure 19.4) [5,61]. The proto-oncogene Her2/*neu* can be overexpressed in, for example, breast cancer and other types of tumors, and the transformed cell is often dependent on the proto-oncogene [62,63]. Also, in patients, for example, with Her2/neu-positive breast cancer, immune responses to Her2/*neu* can be found, suggesting that tolerance can be disrupted [63].

To examine the use of a Her2-VLP vaccine, chimeric Her2MPyV VLPs were made by fusing the human Her2/neu protein 683 amino acid extracellular and transmembrane domains to the MPyV VP2 protein and coexpressed with VP1 in a baculovirus/insect cell system [5]. Vaccination of mice with Her2MPyV VLPs resulted in the abrogation of outgrowth of a Her2-positive tumor [5]. Moreover, when combining Her2MPyV VLPs with CpG, therapeutic immunity was obtained up to 6 days after tumor inoculation [64]. Finally, when BALB-neuT transgenic mice, with a mutated rat Her2/neu gene developing neuT-induced mammary gland tumors of 15 weeks of age, were vaccinated with Her2MPyV

VLPs of 6 weeks of age, no tumors developed [5]. Her2-specific T-cell immunity, but no antibodies, was observed [5].

Vaccination with VP1 VLPs expressing a CD8+ T-cell ovalbumin epitope can also prevent outgrowth of a murine melanoma expressing ovalbumin and could thus also be used to potentially obtain therapeutic immunity [14]. Vaccination resulted in T-cell immunity with the induction of CD8+ T-cells specific for the equivalent ovalbumin epitope.

19.7 LOADING OF DENDRITIC CELLS WITH PyV VLPs AND HPV VLPs AND USE OF ADJUVANT

DCs incubated with VLPs in vitro prior to vaccination can improve T-cell responses and VLP vaccination efficacy. PyV VLPs and HPV VLPs are readily taken up by DCs, and inoculation of VLPs together with DCs usually enhances immunity irrespective of whether DC maturation is induced or not [61,64–71]. HPV16 and HPV18 VLPs induce maturation of human DCs, while MPtV, BKV, and JCV VLPs do not, and for MPyV VLPs the results vary [61,64–69]. In the preceding Her2 tumor model, it was possible to decrease the Her2-VLP dose 10-fold [69]. In a PSA model, it was possible to obtain tumor rejection only after immunization with DCs incubated with PSA VLPs together with the adjuvant CpG [61]. Moreover, HPV16 L1/L2-E7 VLPs loaded onto human DCs were useful for efficient induction of primary E7-specific T-cell responses in vitro [70,71].

19.8 PyV VLPs AS VECTORS FOR GENE THERAPY AND TO AUGMENT DNA IMMUNIZATION

PyVs generally lack of DNA, but they do have the ability to bind DNA [2,3]. Already in 1983, MPyV VLPs were shown to package viral DNA and transduce it into cells in vitro, resulting in viral gene expression, and later MPyV VLPs were shown to transduce plasmid DNA in vitro and in vivo, indicating that VLPs could potentially be used for gene therapy (Figure 19.5) [2,3,15,72–76]. Of special interest would be to use PyVs of nonhuman origin as vectors for gene therapy in humans. However, it must be mentioned that although large DNA fragments could be transduced, treatment of VLP–plasmid DNA complexes with DNase-degraded DNA larger than ~2.5 kb, indicating a limited length of the DNA protected within the VLPs [73]. Several studies have since then reported ways of how to optimize DNA packaging and transduction, and in a baculovirus system, SV40 VLPs incubated with nuclear extract from insect cells containing SV40 VP1 with plasmid DNA in the presence of ATP allowed packaging of plasmids ≤17.6 kb [2,15,72–78]. Moreover, SV40 VLPs, packaged with a Pseudomonas exotoxin gene inoculated into a tumor, caused tumor reduction, also showing the use VLPs for gene delivery in vivo [78].

FIGURE 19.4 Schematic figure of a cVLP based on PyV where the foreign protein is "hidden" within the cVLP. A foreign protein (yellow) is fused to VP2 (red). VP1 in blue.

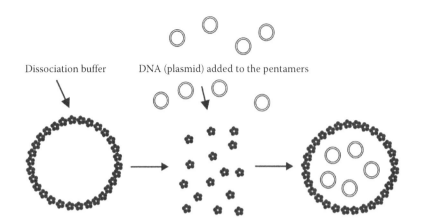

FIGURE 19.5 Formation of DNA/VLP complexes by disassembly/reassembly. Dissociation buffer is added to the VLPs (left), leading to dissociation into VP1 pentamers (middle). This is followed by mixing of the pentamers with DNA, and finally, the pentamers are reassociated into VLPs and the DNA is, hopefully, packaged within the VLP (right).

Finally, it was possible to increase the antibody titers 10-fold of a DNA plasmid encoding p37 (p24 and p17) nucleocapsid proteins of HIV1, when incubating the DNA with VLPs before vaccination [79].

19.9 MUTATED HPV VLPs AS VECTORS FOR GENE THERAPY

In the preceding discussions, examples of nonspecies-specific VLPs are discussed. Another approach is to mutate, for example, HPV-based VLPs as, for example, exemplified by the work of Fleury et al. [80]. Here, a few point mutations in the FG loop of the HPV16 L1 capsid protein resulted in a VLP that could escape actual or future vaccination and could potentially be of use for gene therapy in humans [80].

19.10 CONCLUSION

PyV VLPs and HPV VLPs are useful for vaccination against the corresponding viral infection not only in healthy individuals but potentially also in immunosuppressed individuals with some T-cell deficiency. Chimeric PyV or HPV VLPs containing viral or other tumor antigens can also potentially be extremely effective as vaccines against virus induced other types of cancer. Finally, VLPs can introduce DNA into cells and be likely be used for gene therapy as well as augment DNA vaccination.

ACKNOWLEDGMENTS

Dr. Kalle Andreasson is acknowledged for Figures 19.1, 19.2, 19.4, and 19.5, originating primarily from the frame of his academic PhD thesis. This work was partly supported by grants from the Swedish Cancer Foundation, King Gustav 5th Jubilee Society, Stockholm Cancer Society, Cancer and Allergy Foundation, Henning and Ida Perssons Research Foundation, Stockholm City Council, and Karolinska Institutet.

REFERENCES

1. Young, L.S., P.F. Searle, D. Onion, and V. Mautner, Viral gene therapy strategies: From basic science to clinical application. *J Pathol*, 2006. **208**: 299–318.
2. Ramqvist, T., K. Andreasson, and T. Dalianis, Vaccination, immune and gene therapy based on virus-like particles against viral infections. *Expert Opin Biol Ther*, 2007. **7**(7): 997–1007.
3. Ramqvist, T. and T. Dalianis, Immunotherapeutic polyoma and human papilloma virus-like particles. *Immunotherapy*, 2009. **1**(2): 303–312.
4. Schmiedeskamp, M.R. and D.R. Kockler, Human papillomavirus vaccines. *Ann Pharmacother*, 2006. **40**(7–8): 1344–1352.
5. Tegerstedt, K., J.A. Lindencrona, C. Curcio et al., A single vaccination with polyomavirus VP1/VP2Her2 virus-like particles prevents outgrowth of HER-2/neu-expressing tumors. *Cancer Res*, 2005. **65**(13): 5953–5957.
6. Kimchi-Sarfaty, C. and M.M. Gottesman, SV40 pseudovirions as highly efficient vectors for gene transfer and their potential application in cancer therapy. *Curr Pharm Biotechnol*, 2004. **5**(5): 451–458.
7. Chackerian, B., D.R. Lowy, and J.T. Schiller, Conjugation of a self-antigen to papillomavirus-like particles allows for efficient induction of protective autoantibodies. *J Clin Invest*, 2001. **108**(3): 415–423.
8. Maurer, P., G.T. Jennings, J. Willers et al., A therapeutic vaccine for nicotine dependence: Preclinical efficacy, and Phase I safety and immunogenicity. *Eur J Immunol*, 2005. **35**(7): 2031–2040.
9. Palkova, Z., T. Adamec, D. Liebl, J. Stokrova, and J. Forstova, Production of polyomavirus structural protein VP1 in yeast cells and its interaction with cell structures. *FEBS Lett*, 2000. **478**(3): 281–289.
10. Forstová, J., N. Krauzewicz, S. Wallace et al., Cooperation of structural proteins during late events in the life cycle of polyomavirus. *J Virol*, 1993. **67**(3): 1405–1413.
11. Salunke, D.M., D.I. Caspar, and R.L. Garcea, Self-assembly of purified polyomavirus capsid protein VP1. *Cell*, 1986. **46**(6): 895–904.
12. Chen, X.S., G. Casini, S.C. Harrison, and R.L. Garcea, Papillomavirus capsid protein expression in *Escherichia coli*: Purification and assembly of HPV11 and HPV16 L1. *J Mol Biol*, 2001. **307**(1): 173–182.

13. Fernández-San Millán, A., S.M. Ortigosa, S. Hervás-Stubbs et al., Human papillomavirus L1 protein expressed in tobacco chloroplasts self-assembles into virus-like particles that are highly immunogenic. *Plant Biotechnol J*, 2008. **6**: 427–441.

14. Brinkman, M., J. Walter, S. Grein et al., Beneficial therapeutic effects with different particulate structures of murine polyomavirus VP1-coat protein carrying self or non-self CD8 T cell epitopes against murine melanoma. *Cancer Immunol Immunother*, 2005. **54**(6): 611–622.

15. Heidari, S., N. Krauzewicz, M. Kalantari, A. Vlastos, B.E. Griffin, and T. Dalianis, Persistence and tissue distribution of DNA in normal and immunodeficient mice inoculated with polyomavirus VP1 pseudocapsid complexes or polyomavirus. *J Virol*, 2000. **74**(24): 11963–11965.

16. Tegerstedt, K., K. Andreasson, A. Vlastos, K.O. Hedlund, T. Dalianis, and T. Ramqvist, Murine pneumotropic virus VP1 virus-like particles (VLPs) bind to several cell types independent of sialic acid residues and do not serologically cross react with murine polyomavirus VP1 VLPs. *J Gen Virol*, 2003. **84**: 3443–3452.

17. Greenstone, H.L., J.D. Nieland, K.E. de Visser et al., Chimeric papillomavirus virus-like particles elicit antitumor immunity against the E7 oncoprotein in an HPV16 tumor model. *Proc Natl Acad Sci USA*, 1998. **95**: 1800–1805.

18. Tegerstedt, K., A.V. Franzén, K. Andreasson et al., Murine polyomavirus "virus-like" particles as vectors for gene and immune therapy and as vaccines. *Anticancer Res*, 2005. **25**(4): 2601–2608.

19. Dalianis, T. and H. Hirsch, Human polyomaviruses in disease and cancer. *Virology*, 2013. **432**: 63–72.

20. Brady, J.N. and R.A. Consigli, Chromatographic separation of the polyoma virus proteins and renaturation of the isolated VP1 major capsid protein. *J Virol*, 1978. **27**: 436–442.

21. Sasnauskas, K., A. Bulavaite, A. Hale et al., Generation of recombinant virus-like particles of human and non-human polyomaviruses in yeast *Saccharomyces cerevisiae*. *Intervirology*, 2002. **45**(4–6): 308–317.

22. Kilham, L. and H.W. Murphy, A pneumotropic virus isolated from C3H mice carrying the Bittner Milk Agent. *Proc Soc Exp Biol Med*, 1953. **82**: 133–137.

23. Greenlee, J.E., S.H. Clawson, R.C. Phelps, and W.G. Stroop, Distribution of K-papovavirus in infected Newborn mice. *J Comp Pathol*, 1994. **111**(3): 259–268.

24. Sweet, B.H. and M.R. Hilleman, The vacuolating virus S.V.40. *Proc Soc Exp Biol Med*, 1960. **105**: 420–427.

25. Gardner, S.D., A.M. Field, D.V. Coleman, and B. Hulme, New human papovavirus (B.K.) isolated from urine after renal transplantation. *Lancet*, 1971. **1**(7712): 1253–1257.

26. Padgett, B.L., D.L. Walker, G.M. ZuRhein, R.J. Eckroade, and B.H. Dessel, Cultivation of papova-like virus from human brain with progressive multifocal leucoencephalopathy. *Lancet*, 1971. **1**(7712): 1257–1260.

27. Allander, T., K. Andreasson, S. Gupta et al., Identification of a third human polyomavirus. *J Virol*, 2007. **81**: 4130–4136.

28. Gaynor, A.M., M.D. Nissen, D.M. Whiley et al., Identification of a novel polyomavirus from patients with acute respiratory tract infections. *PLoS Pathog*, 2007. **3**(5): e64.

29. Feng, H., M. Chuda, Y. Chang, and P.S. Moore, Clonal integration of a polyomavirus in human Merkel cell carcinoma. *Science*, 2008. **319**(5866): 1096–1100.

30. van der Meijden, E., R.W. Janssens, C. Lauber, J.N. Bouwes Bavinck, A.E. Gorbalenya, and M.C. Feltkamp, Discovery of a new human polyomavirus associated with trichodysplasia spinulosa in an immunocompromized patient. *PLoS Pathog*, 2010. **6**(7): e1001024.

31. Dalianis, T., Immunotherapy for polyomaviruses: Opportunities and challenges. *Immunotherapy*, 2012. **4**(6): 617–628.

32. Moens, U., M. VanGhelue, and B. Ehlers, Are human polyomaviruses co-factors for cancers induced by other oncoviruses? *Rev Med Virol*, 2014. **24**(5): 343–360.

33. zur Hausen, H., Papillomaviruses and cancer: From basic studies to clinical application. *Nat Rev Cancer*, 2002. **2**: 342–350.

34. Tommasino, M., The human papillomavirus family and its role in carcinogenesis. *Semin Cancer Biol*, 2014. **26C**: 13–21.

35. Qi, Y.M., S.W. Peng, K. Hengst et al., Epithelial cells display separate receptors for papillomavirus VLPs for soluble L1 protein. *Virology*, 1996. **216**: 35–45.

36. Volpers, C., F. Unckel, P. Schirmacher, R. Streek, and M. Sapp, Binding and internalization of human papillomavirus type 33 virus like particles by eukaryotic cells. *J Virol*, 1995. **69**: 3258–3264.

37. Lindahl, U., M. Kusche-Gullberg, and L. Kjellen, Regulated diversity of heparan sulfate. *J Biol Chem*, 1998. **273**: 24979–24982.

38. de Witte, L., Y. Zoughlami, B. Aengeneyndt et al., Binding of human papilloma virus L1 virus-like particles to dendritic cells is mediated through heparan sulfates and induces immune activation. *Immunobiology*, 2007. **212**(9–10): 679–691.

39. Giroglou, T., L. Florin, F. Schäfer, R.E. Streeck, and M. Sapp, Human papillomavirus infection requires cell surface heparan sulfate. *J Virol*, 2001. **75**(3): 1565–1570.

40. Richards, K.F., M. Bienkowska-Haba, J. Dasgupta, X.S. Chen, and M. Sapp, Multiple heparan sulfate binding site engagements are required for the infectious entry of human papillomavirus type 16. *J Virol*, 2013. **87**(21): 11426–11437.

41. Harper, D.M., E.L. Franco, C.M. Wheeler et al.; HPV Vaccine Study group, Sustained efficacy up to 4.5 years of a bivalent L1 virus-like particle vaccine against human papillomavirus types 16 and 18: Follow-up from a randomised control trial. *Lancet*, 2006. **367**(9518): 1247–1255.

42. Cadman, L., The future of cervical cancer prevention: Human papillomavirus vaccines. *J Fam Health Care*, 2008. **18**(4): 131–132.

43. Ma, B., B. Maraj, N.M. Tran et al., Emerging human papillomavirus vaccines. *Expert Opin Emerg Drugs*, 2012. **17**(4): 469–491.

44. Jagu, S., K. Kwak, and J.T. Schiller, Phylogenetic consideration in designing broadly protective multimeric L2 vaccine. *J Virol*, 2013. **87**(11): 6127–6136.

45. Roden, R.B., W.H. Yutzy IV, R. Fallon, S. Inglis, D.R. Lowy, and J.T. Schiller, Minor capsid protein of human genital papillomaviruses contains subdominant, cross-neutralizing epitopes. *Virology*, 2000. **270**(2): 254–257.

46. Kondo, K., H. Ochi, T. Matsumoto, H. Yoshikawa, and T. Kanda, Modification of human papillomavirus-like particle vaccine by insertion of the cross-reactive L2-epitopes. *J Med Virol*, 2008. **80**: 841–846.

47. Slupetzky, K., R. Gambhira, T.D. Culp et al., A papillomavirus-like particle (VLP) vaccine displaying HPV16 L2 epitopes induces cross-neutralizing antibodies to HPV11. *Vaccine*, 2007. **25**(11): 2001–2010.

48. Heidari, S., A. Vlastos, T. Ramqvist et al., Immunization of T-cell deficient mice against polyomavirus infection using viral pseudocapsids or temperature sensitive mutants. *Vaccine*, 2002. **20**: 1571–1578.

49. Vlastos, A., K. Andreasson, K. Tegerstedt et al., VP1 pseudocapsids, but not a glutathione-S-transferase VP1 fusion protein, prevent polyomavirus infection in a T-cell immune deficient experimental mouse model. *J Med Virol*, 2003. **70**: 293–300.

50. Gedvilaite, A., C. Frömmel, K. Sasnauskas et al., Formation of immunogenic virus-like particles by inserting epitopes into surface-exposed regions of hamster polyomavirus major capsid protein. *Virology*, 2000. **273**(1): 21–35.

51. Wakabayashi, M.T., D.M. Da Silva, R.K. Potkul, and W.M. Kast, Comparison of human papillomavirus type 16 L1 chimeric virus-like particles versus L1/L2 chimeric virus-like particles in tumor prevention. *Intervirology*, 2002. **45**(4–6): 300–307.

52. Jochmus, I., K. Schafer, S. Faath, M. Muller, and L. Gissmann, Chimeric virus-like particles of the human papillomavirus type 16 (HPV 16) as a prophylactic and therapeutic vaccine. *Arch Med Res*, 1999. **30**(4): 269–274.

53. Kaufmann, A.M., J. Nieland, M. Schinz et al., HPV16 L1E7 chimeric virus-like particles induce specific HLA-restricted T cells in humans after in vitro vaccination. *Int J Cancer*, 2001. **92**(2): 285–293.

54. Kaufmann, A.M., J.D. Nieland, I. Jochmus et al., Vaccination trial with HPV16 L1E7 chimeric virus-like particles in women suffering from high grade intraepithelial neoplasia (CIN 2/3). *Int J Cancer*, 2007. **121**: 2794–2800.

55. Munger, K. and P.M. Howley, Human papillomavirus immortalization and transformation functions. *Virus Res*, 2002. **89**(2): 213–228.

56. Ramqvist, T., G. Reinholdsson, M. Carlquist, T. Bergman, and T. Dalianis, A single peptide derived from the sequence common to polyoma small and middle T-antigen induces immunity against polyoma tumors. *Virology*, 1989. **172**(1): 359–362.

57. Vlastos Franzén, A., K. Tegerstedt, D. Holländerová, J. Forstová, T. Ramqvist, and T. Dalianis, Murine polyomavirus VP1 virus-like particles immunize against some polyomavirus induced tumors. *In Vivo*, 2005. **19**(2): 323–326.

58. Chen, L. and M. Fluck, Kinetic analysis of the steps of the polyomavirus lytic cycle. *J Virol*, 2001. **75**: 8368–8379.

59. Dahlstrand, H., L. Björnestähl, D. Lindquist, T. Dalianis, E. Munck-Wikland, and G. Elmberger, A high grade of p16^{ink4a} immunohistochemistry staining predicts response to radiotherapy and is similar to human papilloma virus (HPV) detection by PCR a favourable prognostic factor in tonsillar carcinoma. *Anticancer Res*, 2005. **25**(6C): 4375–4383.

60. Hildesheim, A., R. Herrero, S. Wacholder et al.; Costa Rican HPV Vaccine Trial Group, Effect of human papillomavirus 16/18 L1 viruslike particle vaccine among young women with preexisting infection: A randomized trial. *JAMA*, 2007. **298**(7): 743–753.

61. Eriksson, M., K. Andreasson, J. Weidmann et al., Murine polyomavirus virus-like particles carrying full-length human PSA protect BALB/cmice from outgrowth of a PSA expressing tumor. *PLoS One*, 2011. **6**(8): e23828.

62. Slamon, D.J., G.M. Clark, S.G. Wong, W.J. Levin, A. Ullrich, and W.L. McGuire, Human breast cancer: Correlation of relapse and survival with amplification of the HER-2/neu oncogene. *Science*, 1987. **235**(4785): 177–182.

63. Kiessling, R., W.Z. Wei, F. Herrmann et al., Cellular immunity to the Her-2/neu protooncogene. *Adv Cancer Res*, 2002. **85**: 101–144.

64. Andreasson, K., K. Tegerstedt, M. Eriksson et al., Murine pneumotropic virus chimeric Her2/neu virus-like particles as prophylactic and therapeutic vaccines against Her2/neu expressing tumours. *Int J Cancer*, 2009. **124**(1): 150–156.

65. Lenz, P., P.M. Day, Y.Y. Pang et al., Papillomavirus-like particles induce acute activation of dendritic cells. *J Immunol*, 2001. **166**(9): 5346–5355.

66. Lenz, P., D.R. Lowy, and J.T. Schiller, Papillomavirus virus-like particles induce cytokines characteristic of innate immune responses in plasmacytoid dendritic cells. *Eur J Immunol*, 2005. **35**(5): 1548–1556.

67. Gedvilaite, A., D.C. Dorn, K. Sasnauskas et al., Virus-like particles derived from major capsid protein VP1 of different polyomaviruses differ in their ability to induce maturation in human dendritic cells. *Virology*, 2006. **354**(2): 252–260.

68. Bickert, T., G. Wohlleben, M. Brinkman et al., Murine polyomavirus-like particles induce maturation of bone marrow-derived dendritic cells and proliferation of T cells. *Med Microbiol Immunol (Berl)*, 2007. **196**(1): 31–39.

69. Tegerstedt, K., A. Franzén, T. Ramqvist, and T. Dalianis, Dendritic cells loaded with polyomavirus VP1/VP2Her2 virus-like particles (VLPs) efficiently prevent outgrowth of a Her2/neu expressing tumor without inducing high anti-VLP serum titers. *Cancer Immunol Immunother*, 2007. **56**(9): 1335–1344.

70. Lenz, P., C.D. Thompson, P.M. Day, S.M. Bacot, D.R. Lowy, and J.T. Schiller, Interaction of papillomavirus virus-like particles with human myeloid antigen-presenting cells. *Clin Immunol*, 2003. **106**(3): 231–237.

71. Fausch, S.C., D.M. Da Silva, and W.M. Kast, Differential uptake and cross-presentation of human papillomavirus virus-like particles by dendritic cells and Langerhans cells. *Cancer Res*, 2003. **63**(13): 3478–3482.

72. Slilaty, S.N. and H.V. Aposhian, Gene transfer by polyoma-like particles assembled in a cell-free system. *Science*, 1983. **220**(4598): 725–727.

73. Goldmann, C., N. Stolte, T. Nisslein, G. Hunsmann, W. Lüke, and H. Petry, Packaging of small molecules into VP1-virus-like particles of the human polyomavirus JC virus. *J Virol Methods*, 2000. **90**(1): 85–90.

74. Krauzewicz, N., J. Stokrová, C. Jenkins, M. Elliott, C.F. Higgins, and B.E. Griffin, Virus-like gene transfer into cells mediated by polyoma virus pseudocapsids. *Gene Ther*, 2000. **7**(24): 2122–2131.

75. Stokrová, J., Z. Palková, L. Fischer et al., Interactions of heterologous DNA with polyomavirus major structural protein, VP1. *FEBS Lett*, 1999. **445**(1): 119–125.

76. Kimchi-Sarfaty, C., M. Arora, Z. Sandalon, A. Oppenheim, and M.M. Gottesman, High cloning capacity of in vitro packaged SV40 vectors with no SV40 virus sequences. *Hum Gene Ther*, 2003. **14**(2): 167–177.

77. Arad, U., E. Zeira, M.A. El-Latif et al., Liver-targeted gene therapy by SV40-based vectors using the hydrodynamic injection method. *Hum Gene Ther*, 2005. **16**(3): 361–371.

78. Kimchi-Sarfaty, C., W.D. Vieira, D. Dodds et al., SV40 Pseudovirion gene delivery of a toxin to treat human adenocarcinomas in mice. *Cancer Gene Ther*, 2006. **13**(7): 648–657.

79. Rollman, E.I., T. Ramqvist, B. Zuber et al., Genetic immunization is augmented by murine polyomavirus VP1 pseudocapsids. *Vaccine*, 2003. **21**(19–20): 2263–2267.

80. Fleury MJ, A. Touzé, and P. Coursaget, Human papillomavirus type 16 pseudovirions with few point mutations in L1 major capsid protein FG loop could escape actual or future vaccination for potential use in gene therapy. *Mol Biotechnol*, 2014. **56**: 479–486.

20 Applications of Viral Nanoparticles Based on Polyomavirus and Papillomavirus Structures

Jiřina Suchanová, Hana Španielová, and Jitka Forstová

CONTENTS

20.1 INTRODUCTION

Polyomaviruses (PyVs) and papillomaviruses (PVs) are small nonenveloped tumorigenic viruses. They share many common morphological features (Klug 1965) that led to their original coclassification as part of *Papovaviridae* family. In 1998, the detailed understanding of their distinct biology and genome organization resulted in the splitting of these viruses into two separate families: *Polyomaviridae* (containing the single genus Polyomavirus) and *Papillomaviridae* (containing the single genus Papillomavirus) (Van Regenmortel et al. 1999). Since their discovery, PyVs have served as research tools for revealing basic principles of viral capsid structure (Klug 1965; Anderer et al. 1967; Finch 1974; Rayment et al. 1982) and many important molecular processes in living cells. Moreover, since the early 1970s, the potential of empty polyoma viral particles as carriers of genes into mammalian cells has been recognized (Osterman et al. 1970; Qasba and Aposhian 1971; Aposhian et al. 1975). Research into PVs has lagged behind due to the difficulty of their cultivation. DNA recombinant technology has helped to regain research interest in PVs, especially with the discovery of the presence of two types of human PVs (HPVs), HPV16 and HPV18, in human cervical tumors (Dürst et al. 1983). Consequently, several studies have demonstrated that more than a dozen HPV types are important etiological agents in human cancer. These findings accelerated PV research and led to the development of virus-like particle (VLP) technology, mainly for vaccine production. On the contrary, well-known human PyVs, BK virus (BKPyV) and JC polyomavirus (JCPyV) discovered in the 1970s (Zurhein and Chou 1965; Gardner et al. 1971), were never recognized as oncogenic in humans (Abend et al. 2009; Maginnis and Atwood 2009), and research focused on nonhuman PyVs as vectors for gene and immune therapy in PyV-unrelated cancers (Krauzewicz and Griffin 2000; Tegerstedt et al. 2005a). However, the Merkel cell PyV (MCPyV), the newly discovered PyV found in biopsies of the rare and aggressive human neuroendocrine skin cancer Merkel cell carcinoma (MCC) (Feng et al. 2008), changed the view of the oncogenic potential of PyVs in humans, and

protective vaccines may soon become an important area of research. Conversely, the availability of two different protective vaccines against high-risk HPVs may result not only in progress toward therapeutic vaccines but also in the development of PV-based nanocarriers of drugs, diagnostic probes, or therapeutic genes, similar to those of PyV-based vectors. This chapter summarizes the important developments in PyV- and PV-based nanotechnology. It compares both systems for different applications when the biology and structure of the virus, interactions between the virus and its host cell, and vector production facilities are taken into account. The suitability of these viral nanotechnology tools as gene, protein, drug, or other compound nanocarriers as well as vaccines is discussed for both virus families together with the advantages and disadvantages connected to specific uses.

20.2 BIOLOGY OF PAPILLOMAVIRUSES AND POLYOMAVIRUSES

20.2.1 GENERAL CHARACTERISTICS

PyVs and PVs share many common features in their morphology and biology, but they differ in some aspects. Table 20.1 shows some of these features.

TABLE 20.1
Properties of PyVs and PVs

Characteristics	PyV	PV
Virion		
Capsid symmetry	Icosahedral	Icosahedral
Diameter	45 nm	55 nm
Composition	VP1, VP2, VP3,[a] genome with cellular histones	L1, L2, genome with cellular histones
Genome		
Type (size)	Circular dsDNA (5 kbp)	Circular dsDNA (8 kbp)
ORFs	6–7 (encoded by both DNA strands)	8–10 (encoded by the same DNA strand)
Infection		
Hosts	Mammals, birds	Mammals, birds, reptiles
Tissue tropism	Various	Skin and mucosa (epithelia)
Result of acute infection	Unapparent	Microlesions, benign warts
Persistent/latent infection	Yes	Yes
Oncogenic potential		
Tumors in immunocompetent host	No	Yes (high-risk types)[b]
Tumors in immunocompromised or nonpermissive host	Yes	Yes
In vitro cell transformation	Yes	Rarely
Individual members		
Infect humans (representative members)	12 HPyV species (BKPyV, JCPyV, MCPyV)	170 HPV types (HPV16, HPV18—high-risk types)[b]
Representative animal isolates	MPyV (mouse), SV40 (monkey), HaPyV (hamster)	BPV (bovine), CRPV (rabbit), COPV(canine)
Summary	NCBI taxonomy database[c] (Browser ID: 151340)	NCBI taxonomy database[c] (Browser ID: 151341)

[a] MCPyV capsid does not contain VP3 protein (Schowalter and Buck 2013).

[b] The International Agency for Research on Cancer (IARC) classified 12 different HPV types as carcinogenic to humans: types 16, 18,31, 33, 35, 39, 45, 51, 52, 56, 58, and 59, with HPV16 and HPV18 types most frequently found in cervical cancers (Bouvard et al. 2009); several animal PVs (bovine, bat, feline) are etiological agents in carcinoma (Rector and Van Ranst 2013).

[c] Benson et al. (2009).

At present, the NCBI taxonomy database (Benson et al. 2009) counts almost 100 PyV species in the Polyomaviridae family, so far comprising 12 human PyVs (Ehlers and Wieland 2013). PyVs infect only mammals and birds. Most mammalian PyVs have not been directly linked to acute disease after natural infection of an immunocompetent host. Primary infection usually results in lifelong persistence, and PyVs are probably widespread benign members of the extensive flora of viruses that are associated with the body. The major sites of persistence for human PyVs are the skin, the kidney, the central nervous system, and the hematopoietic system. Under immunosuppression, however, reactivation of the viruses can occur, leading to several disease patterns (Dalianis and Hirsch 2013). PyVs are believed to have a narrow host range. Even though virus replication in permissive cells is connected with the production of tumor antigens (T antigens), the subsequent virion production leads to virus-induced cell lysis, thus preventing cell transformation. However, most mammalian PyVs are able to induce malignant tumors after inoculation of nonpermissive hosts or exhibit transforming properties in cell culture. In contrast, PyVs of birds, which are highly pathogenic especially for young animals, do not exhibit tumorigenic properties at all (Johne and Müller 2007; zur Hausen 2008).

PV isolates are traditionally described as *types*, and the taxonomy of the PVs is rapidly evolving (De Villiers 2013). For human PVs, higher-order clusters based on the sequence identity of L1 open reading frame have been established for classification in genera and species (De Villiers et al. 2004). At present, 170 HPV types are known, compared to only 12 human PyVs (De Villiers 2013). Similar to PyVs, the nonHPVs have been recovered from a vast array of mammalian species, including the mouse (Ingle et al. 2010), whose PV was not known of for a long time. PVs have been found in birds and, unlike PyVs, also in three species of reptiles (Rector and Van Ranst 2013). Depending on the type of the tissue of origin, PV types are commonly grouped as either cutaneous or mucosal. The mucosal-type HPVs differ in their carcinogenic potential, and according to this, they are sorted into three groups: high risk, intermediate risk, and low risk (Bouvard et al. 2009). The low-risk group produces benign skin lesions and includes HPV6, HPV11, HPV42, HPV43, and HPV44. HPV types from this group are not associated with carcinomas, but HPV6 and HPV11 can cause genital warts. The intermediate group is composed of HPV31, HPV33, HPV35, HPV51, HPV52, and HPV58, which are detected in benign skin lesions as well as in cancer cells. The high-risk HPVs are preferentially detected in carcinomas and encompass, for example, HPV16, HPV18, HPV45, and HPV56 (Furumoto and Irahara 2002). Because the intermediate-risk group is also present in cancer cells, it is very difficult to distinguish HPV types between these two groups, and sometimes the intermediate-risk HPVs are presented as part of the high-risk group.

Productive infection takes place only in differentiating keratinocytes, and virions are frequently found in the skin swabs of healthy humans or animals (Antonsson and Hansson 2002; Rector and Van Ranst 2013). In some cases, PVs cause benign tumors (warts, papillomas) found in the skin and mucosal epithelia in their natural host and occasionally in related species (reviewed in zur Hausen 2001). Some papillomatous proliferations induced by specific types of PVs bear a high risk for malignant progression (reviewed in zur Hausen 2002). PVs seem to coexist with their host preferentially in a latent infection over long periods of time. Similar to PyVs, immune suppression can lead to reactivation or increased susceptibility to reinfection, and immunodeficiency may also predispose humans and animals to develop papillomas and carcinomas (reviewed in Sundberg et al. [2000] and Denny et al. [2012]).

20.2.2 VIRION STRUCTURE

The capsid proteins of PyVs and PVs exhibit only weak sequence homologies (Belnap et al. 1996), but the structures of monomeric VP1 and L1 are remarkably similar; they have a typical jelly roll structure comprised of an eight-stranded antiparallel barrel. Electron microscopy and image analysis of negatively stained PyVs and PVs (Klug 1965) have helped to determine that the capsids of both viruses have 72 pentameric capsomeres composed of major capsid proteins arranged with T = 7d icosahedral lattice symmetry. The atomic structures of simian virus 40 (SV40) (Liddington et al. 1991; Stehle et al. 1996), murine PyV (MPyV) (Rayment et al. 1982; Stehle et al. 1994), HPV16 (Chen et al. 2000), three other HPVs (HPV11, HPV18, and HPV35) (Bishop et al. 2007), and bovine papillomavirus 1 (BPV1) (Wolf et al. 2010) showed that capsomere arrangement and intercapsomere contacts are slightly different between the two families (Figure 20.1). The C-terminal arms of the major capsid protein always mediate interpentamer contacts. For PyVs, the C-terminal assembly domain (approximately 60 residues) invades the neighboring pentamer and terminates within the target subunit. In SV40, the C-terminal arm is anchored to the invaded pentamer by an interpentamer disulfide bond, and in MPyV, the invading arm is locked in place by an intrapentamer disulfide bond (Stehle et al. 1996). In contrast, the long C-terminal ends (approximately 90 aa residues) of BPV1 form elaborate loops to create the interpentamer contacts and reinsert into the core of the pentamer from which they emerge (Wolf et al. 2010). The formation of disulfide bonds is important for stable virion assembly in both viral families. In MPyV, disulfide bonds enable complete particle assembly and prevent capsid disassembly, but are not essential for the formation of VLPs (Schmidt et al. 2000). In SV40, transient disulfide bonding occurs during the intracellular folding and pentamerization of the major capsid protein VP1 (Li et al. 2002), and disulfide bonds that stabilize the capsid structure (Ishizu et al. 2001) are observed between SV40 VP1 pentamers (Stehle et al. 1996). For PVs, the extent of disulfide bonding can slightly differ between species and serotypes, as HPV16 L1 has been observed to dimerize and trimerize (Ishii et al. 2003) and BPV1 has more extensive cross-linking (Buck et al. 2004; Wolf et al. 2010). Moreover, the cellular DNA in recombinant VLPs increases the disulfide cross-linking of L1, indicating

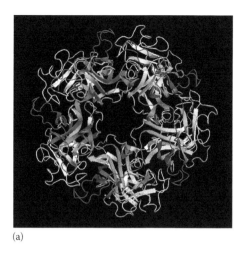
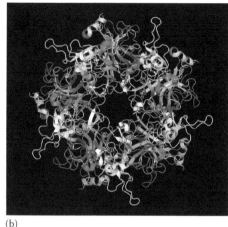

(a)   (b)

FIGURE 20.1 Structure of the recombinant VP1 pentamer (a) (PDB ID, 1VPN, Stehle and Harrison [1996]) and L1 pentamer (b) (PDB ID, 2R5I, Bishop et al. [2007]). Ribbon drawing of the pentamer with elaborate loop domains located on the exterior surface of the assembled pentamer. Each loop is highlighted in a different color: BC loop in red, DE loop in blue, EF loop in yellow, FG loop in magenta, and HI loop in green. Note that L1 loops are more elongated than VP1 loops and that the FG loop is not formed in a PyV pentamer. The C-termini of both, VP1 and L1, are disordered and likely extended into the interior space of the pentamer in the particle. Pentamers were visualized and colored using the PyMOL molecular graphics system. (From DeLano, W.L., *The PyMOL Molecular Graphics System*, version: v0.99, Schrödinger LCC, Cambridge, MA, 2006.)

that nucleic acid in the virion likely induces a capsid conformation that is structurally distinct from that of the VLP (Fligge et al. 2001). The recent model suggests that assembled virions of PVs undergo a slow process of disulfide bond formation and shuffling (i.e., maturation) in order to stabilize the virion by correct formation of inter-L1 disulfide bonds (Buck and Trus 2012). Interestingly, treatment of purified BPV1 with reduction agent dithiothreitol (DTT) is associated with a conformational change resulting in expansion of the capsids by approximately 10% in diameter (Li et al. 1998). This expansion allows the penetration of proteases and nucleases to the interior, which can then result in virion disruption. This structural change may correspond to the *open* capsids seen by cryoelectron microscopy in different PV species but not found in the PyVs (Belnap et al. 1996). This suggests that PyVs use additional factors for virion stabilization. Indeed, calcium ions, which are not used to stabilize the PV virions, are important for the assembly of PyV capsids. The calcium-binding sites consist mainly of acidic amino acids, which are a conserved feature in VP1 sequences across PyVs. Forming calcium salt bridges has been shown to be important for SV40 virion formation (Li et al. 2003) and MPyV capsid assembly (Haynes et al. 1993; Schmidt et al. 2000; Chuan et al. 2010).

The VP1 and L1 monomers adopt a very similar structure. The eight antiparallel strands are ordered in two β-sheets, which stick against one another in each monomer, forming the hydrophobic core of each protein. Additional β-strands align with the β-sandwiches from neighboring capsid protein molecules to form pentamers. Within the core domain, four or three predominant loops are located on the exterior surface in the assembled pentamer and particle in PyVs (Figure 20.1a; BC, DE, EF, HI loops) or PVs (Figure 20.1b; DE, EF, FG loops), respectively. L1 loops are more elongated than VP1 loops and mediate additional interpentameric contacts: the HI loop of one monomer intertwines with the FG and EF loop of the anticlockwise neighbors (Garcea and Chen 2007). The capsid protein sequences exposed on the surface of pentamers and virions represent the most variable regions among PV serotypes (Chen et al. 2000), BKPyV virus variants (Jin et al. 1993; Luo et al. 2012; Pastrana et al. 2013), and PyV species in general (Fang et al. 2010; Neu et al. 2011). This variability probably represents an immunologically driven evolution of serotypes and tissue specificity adaptation. Epitopes identified for neutralizing monoclonal antibodies for HPV16 and HPV11 can be mapped directly to surface loop domains on the capsomere (Chen et al. 2000). Similarly, it has been found that polymorphism located close to the receptor-binding site in the BC loop of VP1 in BKPyV genotypes can permit the escape from antibody-mediated neutralization and determine cellular tropisms and pathogenic potentials (Pastrana et al. 2013). Different strains of MPyV also exhibit different tropisms and pathogenic potentials depending on mutation in the surface loop of the VP1 protein (Mezes and Amati 1994; Bauer et al. 1995).

In PyV virions, the VP1 capsomeres associate with two minor capsid proteins, VP2 and VP3, which are not needed for capsid assembly but play important roles during the early steps of virus infection (Section 20.2.3). Crystallographic studies of MPyV have shown that minor proteins insert into the inward-facing cavity along the fivefold axis of a VP1 pentamer (Griffith et al. 1992). The C-terminus of VP2/VP3 inserts in an unusual, hairpin-like manner into the axial cavity of the VP1 pentamer, where it is anchored strongly by hydrophobic interactions. The sequence alignment of VP2 from eight different PyV species detected conserved amino acids in the region covering the contact structure between VP1 and VP2 (residues 269–296), thus suggesting that interaction between the VP2 and VP1 pentamers involves a similar structure in all PyVs. The N-terminal part of the minor protein appears to be flexible and not tightly folded. Therefore, it can, under the appropriate circumstances (e.g., during cell entry), emerge from the inside of the virion through the 12.5 Å capsomere openings (Chen et al. 1998). PyV virions contain an average of one minor capsid protein (either VP2 or VP3) per capsomere (Imperiale and Major 2007), but this issue has been revised by Schowalter and Buck (2013), and the possibility that two minor capsid proteins associate with one pentameric capsomere cannot be excluded.

Analogously, biochemical analysis of HPV16 capsid preparations showed that up to 72 molecules of L2 can be incorporated per capsid, and cryoelectron microscopy and image reconstruction analysis of these capsids have revealed an icosahedrally ordered L2-specific density beneath the axial lumen of each L1 capsomere (Buck et al. 2008). L1–L2 contacts are mediated by a well-characterized hydrophobic interaction domain in the C-terminal part of L2 and probably also by the interaction domain in the N-terminal part of L2 (Finnen et al. 2003). Both termini of L2 molecules seem to be closely apposed within the capsid (Buck et al. 2008), but a small portion of the N-terminal region of L2 is thought to be exposed on the surface of mature capsids (Hagensee et al. 1993; Liu et al. 1997; Kondo et al. 2007).

20.2.3 Interactions between Virus Proteins and Host Cell

This section will focus on the interactions of PyV and PV particles or their constituents during their attachment on the surface of host cells, the internalization process, and their trafficking toward the cell nucleus where gene expression and virus replication take place.

20.2.3.1 Interactions of Virions with Receptors on Plasma Membrane

Both PyVs and PVs enter cells by receptor-mediated endocytosis. The major structural protein VP1 of PyVs or L1 of PVs is responsible for attachment of virions to the cell surface.

20.2.3.1.1 Receptors of Papillomaviruses

PVs are highly species- and tissue-specific viruses. They infect skin and mucosa epithelial cells exclusively. Differentiation of these cells is vital for the completion of PV replication. This fact impedes PV propagation in vitro. Progress was made after developing and propagation of VLPs composed of the major structural protein, L1, or both PV structural proteins, L1 and L2 (Zhou et al. 1991; Kirnbauer et al. 1992; Hagensee et al. 1993; Kirnbauer et al. 1993; Rose et al. 1993; Touzé et al. 1996; Hildesheim et al. 2007). However, particles lacking L2 and/or viral DNA complexed with host cell histones exhibit slight conformation changes that more or less affect their interactions. This problem was at least in part overcome by the development of particles more resembling native virions, known as pseudovirions (PsVs), containing reporter plasmids in complex with cell histones (Buck et al.

2004, 2005a) (Section 20.4.2). The development of in vitro organotypic epithelial *raft* cultures, permitting full differentiation of keratinocytes, provided the successful propagation of some HPVs for control experiments (Meyers et al. 1992). Entry of PVs into in vitro cultured epithelial cells is initiated by attachment of virions to the cell surface mediated by receptors. Interestingly, in vivo experiments in a mouse model revealed that prior to transfer to the basal keratinocyte cell surface, the epithelial basement membrane underlying basal keratinocytes is the primary site of HPV PsV binding during infection of the genital tract (Roberts et al. 2007; Kines et al. 2009).

Since the 1990s, several proteins of host cells have been described as receptors or molecules interacting with the virions of PVs during their internalization by cells. First, integrin-α6 in combination with β4 or β1 was found to bind VLPs in vitro (Evander et al. 1997; McMillan et al. 1999). Later, several other receptor or possible coreceptor candidates were suggested as heparan sulfate proteoglycans (HSPGs) (Joyce et al. 1999); laminin-332, formerly named laminin-5 (Culp et al. 2006a,b); tetraspanins (Spoden et al. 2008; Scheffer et al. 2013); growth factor receptors (Surviladze et al. 2012); or annexin A2 (Woodham et al. 2012).

Now it is commonly accepted that HSPGs mediate the first interaction of PVs with the cell surface (Joyce et al. 1999; Combita et al. 2001; Giroglou et al. 2001; Richards et al. 2013). HSPGs are heterogeneous population of molecules, composed of cell surface or matrix proteins with covalently bound glycosaminoglycans, modified by sulfation and acetylation, especially with heparan sulfate (HS). They are present on the surface or extracellular matrix of most cells. Interaction of HSPG was demonstrated for VLPs or PsVs of HPV5, HPV11, HPV16, HPV18, HPV31, and HPV33 and also for BPV1 (as reviewed in Raff et al. 2013). Experiments with VLPs composed of L1 only, and structural studies, proved that the PV major structural protein is responsible for the primary interaction with HSPGs (Joyce et al. 1999; Knappe et al. 2007; Dasgupta et al. 2011). The process of virus attachment and internalization was intensively studied with HPV16 PsVs. After interaction of virions with HSPG, they undergo conformation changes, affecting both L1 and L2 and leading to the exposure of the N-terminus of the minor structural protein, L2. Cyclophilin B (peptidyl-prolyl *cis/trans* isomerase) was suggested to facilitate L2 exposure (Bienkowska-Haba et al. 2009). The same group showed later that cyclophilins are also employed at an additional, postinternalization step for dissociation of L1 from a complex of L2 and genome DNA prior to egress from endosomes (Bienkowska-Haba et al. 2012). The exposed L2 N-terminus is cleaved by furin or by the related proprotein convertase 5/6. The consensus sequence for furin cleavage is highly conserved among PVs, and cleavage is essential for infection (Richards et al. 2006). Antibodies against the exposed L2 sequences were found to neutralize the virus (Gambhira et al. 2007; Day et al. 2008), and recently, a vaccine based on these sequences was prepared and induced protective immunity in mice (Chen et al. 2014). A neutralizing epitope of L2 was described to interact with the annexin

A2 heterotetramer and inhibition of an endogenous annexin A2 with antibody against annexin A2 reduced HPV16 infection (Woodham et al. 2012). PsVs precleaved by furin (unlike normal mature PsVs or native virions) were able to infect HSPG negative cells (Day et al. 2008). It has been hypothesized that conformation changes of virions after HSPG binding and furin cleavage expose a secondary binding site on the virus particle for a putative secondary receptor (Schiller et al. 2010). Whether it is α4-integrin, or one of other aforementioned candidates, or an as-of-yet detected cell surface molecule remains to be explored.

Single-particle tracking of fluorescently labeled HPV16 PsVs reveals that they bind preferentially to filopodia and afterward move rapidly on the surface of the cultured cells toward the cell body (Schelhaas et al. 2008). This *surfing* depends on the actin cytoskeleton. Particles then accumulate in discrete membrane areas prior to internalization (Schiller et al. 2010). It is not clear whether the movement from the filopodia is in connection with the HSPG receptor. As the movement of the virus particles on filopodia resembles the movement of epidermal growth factor receptor (EGFR) and EGFR signaling disruption reduces the infectivity of the HPV16 PsVs, participation of EGFR in a complex mediating HPV16 PsV surfing is hypothesized (Raff et al. 2013).

Binding and/or internalization of PVs is apparently connected to the transient activation of signaling pathway(s). Signaling mediated by integrins has been described, supporting a role of integrins in virus internalization. It was observed that HPV16 PsVs induce after their adsorption to HSPG receptor activation of focal adhesion kinase (FAK), necessary for virus entry into early endosomes (EEs) (Abban and Meneses 2010). The authors suggest the role of α4-integrin in FAK induction. Furthermore, by α4β6-activated phosphatidylinositol 3-kinase (PI3 kinase) pathway, early postadsorption of VLPs of HPV types 6b, 18, 31, 35, and BPV1 was observed (Fothergill and McMillan 2006). Other possible secondary receptor candidates of growth factor receptors were found to be rapidly phosphorylated and downstream effectors activated after HPV16 VLP binding (Surviladze et al. 2012). A recent study revealed that cellular entry of HPV16 PsVs into cells involves activation of the PI-3/Akt/mTOR pathway, leading to autophagy inhibition (Surviladze et al. 2013).

20.2.3.1.2 Receptors of Polyomaviruses

After years of intensive but unsuccessful searching for the protein receptor recognized by SV40 or MPyV, Tsai et al. (2003) demonstrated that hydrophilic sialylated oligosaccharide moieties connected with hydrophobic ceramide gangliosides serve as cellular receptors for these viruses, GM1 for SV40, and GD1a and GT1b for MPyV. The surface loop of the major capsid protein, VP1, interacts with the oligosaccharide parts of the gangliosides. Oligosaccharide (glycan) moieties of gangliosides were shown to be recognized also by other studied PyVs; GD1b and GT1b gangliosides are utilized by human BKPyV (Low et al. 2006), and GT1b ganglioside was described as the receptor for MCPyV (Erickson et al.

2009). Sialic acid (5-N-acetyl neuraminic acid; Neu5Ac) is crucial for the interaction of gangliosides with VP1 proteins of MPyV, BKPyV, and MCPyV. Therefore, MPyV, BKPyV, and MCPyV are able to hemagglutinate guinea pig, human, and sheep red blood cells, respectively. SV40 does not hemagglutinate red blood cells. Hemagglutination assays suggested, and structural studies confirmed, that the methods of interaction of different PyVs with glycans differ (Erickson et al. 2009; Neu et al. 2012). Infection by simian B-lymphotropic PyV (LPyV) also depends on the sialic acid on the surface of host cells. Glycan array screening has revealed that LPyV specifically recognizes a linear carbohydrate motif terminating in α2,3-linked Neu5Ac (Neu et al. 2013). Its closest related human PyV 9 (HPyV9), with different tropism, preferentially binds a similar linear carbohydrate motif that, however, terminates in 5-N-glycolyl neuraminic acid (Neu5Gc) (Khan et al. 2014).

Later studies of interaction of MCPyV with surface molecules (Schowalter et al. 2011; Neu et al. 2012) revealed that the primary interaction of the MCPyV with the cell surface is not mediated by GT1b or other sialylated glycans but by glycosaminoglycans, such as HS. The authors suggested that HS is required for the initial interaction of MCPyV with host cells, and secondary interaction with a sialylated coreceptor (which might be the glycan of GT1b or of another glycolipid or glycoprotein) is then necessary for virus internalization.

Recent high-resolution x-ray structure analysis of the major capsid proteins, VP1, from human PyVs HPyV6 and HPyV7, revealed substantial differences in virion surfaces in comparison to all other known PyV structures. The VP1 groove employed in interaction with specific sialic acid–containing glycan receptors in other PyVs is blocked, and HPyV6 and HPyV7 VP1 apparently do not interact with sialylated compounds in solution or on cultured human cells (Ströh et al. 2014).

The search for a receptor for human neurotropic PyV, JCPyV, was rather complicated. Several studies collected evidence that glycoproteins or glycolipids (possibly gangliosides), terminated by sialylated oligosaccharides, can serve as JCPyV receptors (Liu et al. 1998; Komagome et al. 2002; Dugan et al. 2008). In 2004, the serotoninergic 5-HT2A receptor that belongs to the serotonin receptor family and is a G protein-coupled receptor was described as a cellular receptor for the human neurotropic PyV, JCPyV, on human glial cells (Elphick et al. 2004).

Later, by using a glycan array screen and structure analyses, the linear 2,6-linked pentameric oligosaccharide, lactoseries tetrasaccharide c (LSTc; sequence NeuNAc-α2, 6-Gal-β1,4-GlcNAc-β1,3-Gal-β1,4-Glc), the glycan which is not part of ganglioside was identified as the receptor motif for adsorption of human JCPyV on the surface of host cells (Neu et al. 2010). A recent study (Assetta et al. 2013) revealed that JCPyV infection requires both the LSTc and 5-HT2A receptors. While LSTc-VP1 interaction mediates the initial attachment of the virus to cells, the 5-HT2A receptor contributes to JCPyV infection by an as-yet unclear mechanism as a coreceptor, facilitating entry of virions into host cells.

Besides JCPyV and MCPyV, the concept of receptors and coreceptors might also be applicable to other PyVs (O'Hara et al. 2014). In earlier studies, SV40 binding to cells was shown to be blocked by antibodies directed against class major histocompatibility proteins (Breau et al. 1992), and, for the MPyV, $\alpha 4\beta 1$-integrin was suggested as one of the possible coreceptors acting at the postattachment level (Caruso et al. 2003).

There are several aspects of PyV infection (and VLP pseudoinfection) that should be taken into consideration:

1. Although the majority of the PyVs studied interact with glycan moieties containing sialic acids, their interactions can be significantly different, owing to differences in VP1 surface loop sequences and virion surface conformation (Jin et al. 1993; Stehle and Harrison 1996; Stehle et al. 1996; Chen et al. 2000; Luo et al. 2012; Pastrana et al. 2013).
2. Mutations in surface loops of VP1 can markedly change virus tropism (Mezes and Amati 1994; Bauer et al. 1995).
3. Binding some glycolipids or glycoproteins containing sialic acids (pseudoreceptors) can be counterproductive and may result in virus destruction instead of productive infection. In situ hybridization of viral genomes with fluorescently labeled MPyV genomic DNA proved that, indeed, the majority of virions internalized by cells never deliver carried genomes into the cell nucleus (Mannová and Forstová 2003). Using a ganglioside-deficient cell line, Quian and Tsai showed that GD1a is the functional entry receptor for MPyV, binding to the virus on the plasma membrane, forming part of a complex that is internalized and further transported for productive infection. They also observed that, in contrast, glycoproteins acted as *decoy receptors*, restricting the productive infection of MPyV (Qian and Tsai 2010).

20.2.3.2 Internalization of Virions by Cells and Virus Trafficking toward the Nucleus

20.2.3.2.1 Papillomaviruses

Internalization of PVs by cells and their trafficking to the cell nucleus has been intensively studied. However, studies have often resulted in diverse, controversial findings, and limited consensus has been found. Differences have been ascribed to the different natures of the virus particles used (L1 VLPs versus L1/L2 VLPs, mature versus immature PsVs, or native virions), to cell types and experimental conditions, and to various PV genotypes used, although discrepancies have appeared even in experiments performed with the same PV genotype.

Endocytosis via clathrin-coated pits was described for BPV1 and HPV16 and HPV58 (Bousarghin et al. 2003; Day et al. 2003). For HPV31, caveola-mediated uptake was described (Bousarghin et al. 2003; Smith et al. 2008), while in another study clathrin-dependent endocytosis was suggested (Hindmarsh and Laimins 2007). A novel clathrin- and caveolin-independent entry of HPVs was described first for HPV16 (Spoden et al. 2008). Authors showed that the inhibition of clathrin-, caveolin-, and membrane raft–dependent endocytic pathways by dominant-negative mutants and small interfering RNA (siRNA)–mediated knockdown, as well as inhibition of dynamin function, did not impair infection. Moreover, they suggested that HPV16 associates with tetraspanin proteins on the plasma membrane and that tetraspanin-enriched membrane microdomains might act as entry platforms for HPV16. Thorough study of HPV16 entry into epithelial cells (Schelhaas et al. 2012), exploiting biochemical and various microscopy methods, combined with green fluorescent protein (GFP) expression after plasmid delivery by PsVs, confirmed clathrin-, caveolin-, cholesterol-, and dynamin-independent HPV16 entry into HeLa and HaCaT cells. The pathway exhibited some features of macropinosis. Similar methods of entry were confirmed for HPV18 and HPV31 (Spoden et al. 2013).

Internalization of PVs is, in comparison with other nonenveloped viruses, very slow and asynchronous, lasting several hours (Giroglou et al. 2001; Culp and Christensen 2004; Schelhaas et al. 2008). Half times of the internalization of HPV16 ranged from 4 to 12 h p.i. for the fastest and average particles (Spoden et al. 2013), and expression of reporter gene of the pseudogenome could not be detected until 24–48 h postinfection (Schelhaas et al. 2012).

Further trafficking of PVs also remains obscure. Relatively good consensus exists in respect to the requirement of acidic endosomal pH for productive infection. Infection can be blocked by selective inhibitors of endosomal acidification (Selinka et al. 2002; Day et al. 2003; Dabydeen and Meneses 2009; Schelhaas et al. 2012). Analysis of the average time that HPV16 requires for acid activation revealed that the half time of activation was 6 h. The virus required several hours to be exposed to low pH or to an enzyme requiring low pH (Schelhaas et al. 2012).

Internalized papillomaviral particles enter early compartments, EEs, or macropinosomes. Accordingly, PV infection depends on GTPase Rab5. Colocalization with EE has been observed for BPV1 (Day et al. 2003; Laniosz et al. 2008) and for HPV31 (Smith et al. 2008). No significant colocalization with an EEA1 marker of EEs was found for HPV16 (Schelhaas et al. 2012). However, the authors of the study detected a brief comigration of the viral particles with Rab5-positive compartments. The virus then appeared in multivesicular bodies, in late endosomes (LEs), and in the endolysosomal compartment. The combination of endosomal acidic pH and action of endosomal proteases may result in an uncoating process.

Uncoating can be detected during passage through the endosomal compartment. Previous furin cleavage of L2 and cyclophilin B–mediated separation of L2 together with the viral genome from the major capsid protein, L1, are necessary for escape from the LE (Bienkowska-Haba et al. 2012; Day et al. 2013).

Despite the appearance of virus particles in the endolysosomal compartment, surprisingly, Rab7 GTPase was found to be dispensable for HPV16 or HPV31 infection (Smith et al. 2008; Schelhaas et al. 2012; Day and Schelhaas 2014).

Recently, it was observed that an uncoated viral pseudogenome in complex with L2 travels from LEs to the *trans*-Golgi network (TGN), while the major structural protein, L1, is retained mostly within the LE and appears to become degraded (Day et al. 2013). This traveling is dependent upon furin cleavage of L2. Infection in the presence of a furin inhibitor or with particles containing L2 furin cleavage mutants results in the accumulation of uncoated capsids, L2, and DNA in a late endosomal compartment. The traveling of PVs to TGN can also be prevented with inhibitors of anterograde and retrograde Golgi trafficking, brefeldin A, or golgicide A. GTPases Rab9a and Rab7b were determined to mediate this transit. Expression of dominant-negative versions of these GTPases (but not of Rab7a) inhibited HPV16 pseudovirus infection (Day et al. 2013). Genome-wide siRNA screening identified many retrograde transport factors required for efficient PV infection, including multiple subunits of the retromer, which were described to initiate retrograde transport from the EE (Lipovsky et al. 2013). PVs therefore seem to travel to the TGN from both EEs and LEs. Another sorting protein was found to interact with PVs on their way to the cell nucleus: nexin 17 (SNX17) was identified as an interacting partner of L2 protein, and its depletion was connected with the lysosomal degradation of L2 (Bergant Marušič et al. 2012; Bergant and Banks 2013). Thus, L2 protein is important for the escape of the genome from the LEs. Moreover, its conserved C-terminal peptide (the last 23 aa) was shown to be able to interact with and disrupt membranes (Kämper et al. 2006). The precise method of escape of the L2/genome complex from endosomes is not known.

Finally, PV genomes and L2 can be detected in the cell nucleus, predominantly localizing in promyelocytic leukemia (PML) bodies (Day et al. 2004). However, the way in which they reach the nucleus is not understood. One model suggests that an L2–genome complex released from the endosome is transported along microtubules and delivered to the nucleus via nucleopores. The hypothesis has been supported by the fact that nocodazole (tubulin-disrupting agent) inhibits PV infection (Day et al. 2003; Schelhaas et al. 2012) and, more importantly, by finding that L2 interacts directly with dynein light chains (Florin et al. 2006; Schneider et al. 2011). L2 protein also possesses a nuclear localization signal for karyopherin-mediated transport to the nucleus (Darshan et al. 2004).

Another model emerged from evidence that for the establishment of HPV infection and genome expression, cell division is required (Pyeon et al. 2009). Authors of this study suggest the possibility that nuclear envelope breakdown is necessary for the HPV genome to enter the nucleus. This model was very recently strongly supported by a systematic RNA interference (RNAi) silencing approach for the identification of host cell proteins required during HPV16 infection (Aydin et al. 2014). The screening uncovered a crucial role for mitosis in HPV16 nuclear entry and the HPV16 pseudogenome requirement of changes in nuclear envelope permeability facilitated by nuclear envelope breakdown.

20.2.3.2.2 Polyomaviruses

The entry of PyVs into host cells and trafficking from the cell membrane toward the cell nucleus has been intensively investigated on SV40 and MPyV model viruses. Unlike trafficking studies of PVs, research into PyV trafficking can be performed using native virions. However, the interpretation of observations has been complicated by the fact that only a minority of adsorbed and internalized virions successfully deliver their genomes into the cell nucleus, and it has not been easy to distinguish the productive pathway from that leading to virus destruction.

The most intensively studied PyV, SV40, was first described to exploit a unique endocytic pathway: after binding the GM1 ganglioside receptor, virions of SV40 become internalized by caveolae and then fuse with a large caveolin-rich endocytic compartment named by the authors as a *caveosome*. From this newly described nonacidic organelle, the virus was transported by unidentified vesicles that did not contain caveolin, along microtubules to the endoplasmic reticulum (ER) (Pelkmans et al. 2001). On the other hand, human JCPyV was found to enter glial cells by receptor-mediated clathrin-dependent endocytosis (Pho et al. 2000). At the same time, MPyV internalization was described as caveola and clathrin independent (Gilbert and Benjamin 2000). MPyV enters epithelial and fibroblast cells in smooth, tightly fitted monopinocytotic vesicles (Mackay and Consigli 1976; Richterová et al. 2001). However, caveolin-1 was detected in some monopinocytotic vesicles carrying MPyV virions by immunoelectron microscopy (Richterová et al. 2001), and Gilbert and Benjamin revealed that after addition of the GD1a receptor to rat glioma C6 cells deficient in complex ganglioside production, virus particles were internalized in caveolin-1-positive vesicles (Gilbert and Benjamin 2004). Furthermore, BKPyV was found to be associated with caveolin-1-positive vesicles in Vero cells or in human renal proximal tubular epithelial cells (Eash et al. 2004; Moriyama et al. 2007). Nevertheless, MPyV virions were efficiently internalized by Jurkat cells, which do not express caveolin-1, and lack of caveolae and overexpression of a caveolin-1 dominant-negative mutant in mouse epithelial cells did not prevent their productive infection. More recently, the caveosome concept described for SV40 has been disclaimed by authors (Mercer et al. 2010; Engel et al. 2011), and caveosomes have been described as artifacts of overexpression of caveolin fused with enhanced green fluorescent protein (EGFP) or as LEs. Now, it has been established that SV40, MPyV, and BKPyV become internalized by lipid raft microdomains (which may or may not contain caveolin-1) as infectivity of all three viruses is sensitive to cholesterol depletion (Pelkmans et al. 2001; Richterová et al. 2001; Eash et al. 2004; Gilbert and Benjamin 2004). Virion internalization is associated with tyrosine kinase signaling (Pelkmans et al. 2005; Swimm et al. 2010; Ewers and Helenius 2011) and transient actin disorganization (Richterová et al. 2001; Pelkmans et al. 2002).

Both MPyV and SV40 virions colocalize with an EEA1 marker of EEs and Rab5 GTPase (Mannová and Forstová

2003; Pelkmans et al. 2004), and infectivity of MPyV and SV40 was negatively affected by expression of a Rab5 dominant-negative mutant (Liebl et al. 2006; Engel et al. 2011). Then, the viruses appear in an endolysosomal compartment, LEs, and multivesicular bodies. Their infectivity is affected by overexpression of a dominant-negative Rab7 GTPase mutant (Qian et al. 2009; Engel et al. 2011; Zila et al. 2014). Qian and coworkers suggested that in endolysosome compartments, the GD1a receptor stimulates MPyV sorting from LEs and/or lysosomes to the ER. This suggestion is supported by previous observations that the addition of GD1a to cells deficient for gangliosides had no effect on the overall level of virus binding but mediated the transit of MPyV to ER (Gilbert and Benjamin 2004). There is a common consensus that the productive pathway continues by the transit of PyVs to the ER. However, the precise mechanism controlling the transport of MPyV to the ER remains to be clarified. No colocalization of SV40 or MPyV or BKPyV with Golgi apparatus has been observed (Pelkmans et al. 2001; Mannová and Forstová 2003; Moriyama and Sorokin 2008; Engel et al. 2011). Transport of SV40 to the cell nucleus was found to be inhibited by Brefeldin A, acting through inhibition of the ARF1 GTPase, which is known to regulate assembly of COPI coat complexes on Golgi cisternae (Norkin et al. 2002). Furthermore, colocalization of the virus with a βCOP subunit of the COP1 coatomer was detected (Norkin and Kuksin 2005). Moreover, a retrograde trafficking inhibitor of ricin and Shiga-like toxins inhibited infection by SV40 and human BKPyV and JCPyV (Nelson et al. 2013). However, no significant colocalization of MPyV with βCOP, and only mild sensitivity to Brefeldin A, was detected (Mannová and Forstová 2003). A common agreement exists that intact microtubules and a dynein motor are vital for PyV transport into the ER (Pelkmans et al. 2001; Ashok and Atwood 2003; Gilbert et al. 2003; Zila et al. 2014). Further studies are needed to solve the mechanism of PyV trafficking to the ER.

Even less clear is the mechanism by which PyVs deliver their genomes into the cell nucleus. Based on electron microscopy analyses, early papers suggest that SV40 (Maul et al. 1978) or MPyV (Mackay and Consigli 1976) enter the cell nucleus by fusion of vesicles carrying virions directly with the nuclear envelope, so bypassing nuclear pores. At present, a hypothesis that virions, partially disassembled in the ER, translocate by an as-yet unknown mechanism to the cytosol and travel to the nucleus through nuclear pores is commonly accepted. Although never proven, several findings supporting the hypothesis have been published. Schelhaas and collaborators described the dependence of SV40 infection on ER folding and quality control. Downregulation of thiol-disulfide oxidoreductases, ER57 and PDI-protein disulfide isomerase, and two ER membrane proteins, Derline 1 and Sel1L, involved in the export of misfolded proteins from the ER to the cytosol for proteasomal degradation, significantly inhibited virus infection (Schelhaas et al. 2007). For the MPyV, it has been found that downregulation of PDI and Derline 2 decreases the level of MPyV infection (Gilbert et al. 2006;

Lilley et al. 2006). However, the nature and extent of virus disassembly in the ER and the means by which viral genomes are transported to the cytosol and subsequently to the nucleus are not known. The possibility cannot be excluded that the virus moves directly from the ER to the cell nucleus, bypassing nuclear pores.

One possibility of transit from the ER to the cytosol is the utilization of channels for the elimination of misfolded or unassembled proteins from the ER for proteasomal degradation (ER-associated degradation). The crystal structure of the ER translocon showed that the pore of the protein-conductive channel would allow the passage of a molecule with a diameter of 10–12 Å. In its open form, the diameter of the ER translocon was estimated to be 40–60 Å (Meusser et al. 2005). However, the diameter of PyV virions is approximately 45 nm, and the nucleocore without the capsid shell is about 30 nm in diameter.

In vitro studies on the minor structural proteins, VP2 and VP3, of SV40 (Daniels et al. 2006) and of MPyV (Rainey-Barger et al. 2007; Huerfano et al. 2010) showed that the proteins are able to bind, insert into, fuse, and even perforate cell membranes. Thus, the minor proteins VP2 and VP3 might be good candidates for helping the virus to deliver genomes to the cell nucleus.

Although significant progress in understanding PV and PyV trafficking from a host cell membrane to the cell nucleus has been achieved, several gaps remain to be filled, and many details must be clarified.

20.3 VIRUS PROTEIN SELF-ASSEMBLY AND VLP PRODUCTION

The major capsid proteins of PyVs and PVs rapidly oligomerize through a self-assembly process in vitro without the help of minor capsid proteins. In fact, intact purified proteins are never found in the monomeric form under physiological conditions. Both proteins have been successfully produced in various expression systems. Depending on the virus type, designed mutation, and actual conditions, recombinant proteins are purified as viral capsid protein assemblies: pentamers, polymorphic capsid structures, VLPs, or pseudovirion-like particles (PLPs) (for terminology see Table 20.6). Figure 20.2 gives an example of a highly polydispersed preparation of VLPs that can be further purified to near homogeneity. Recent reviews (Teunissen et al. 2013) give an excellent summary of PyV (Cho et al. 2011) and of PV production systems. Table 20.2 presents an overview of the expression system used for VLP production of both viral families.

Mammalian cells were historically the first system for PyV VLP production and isolation, when empty capsids (Crawford et al. 1962) and PsVs containing host cell DNA (Michel et al. 1967) were observed in routine viral preparations. Recently, it has been shown for SV40 that vector DNA with a size that does not exceed the size of a viral genome can be encapsidated if the viral capsid proteins are produced in the system *in trans*. For such PsV production, the

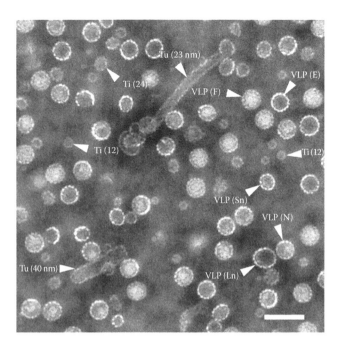

FIGURE 20.2 Electron micrograph of polymorphic structures found in MPyV VP1 VLP preparation from Sf9 cells. VLPs, which may be devoid of nucleic acid, appear as *empty* particles, VLPs (E), or may contain fragments of DNA and appear as *full* particles, VLPs (F). VLPs of various size and morphology can be formed: VLPs of normal size, VLPs (N); VLPs that are larger than normal, VLPs (Ln); VLPs that are smaller than normal, VLP (Sn); VLPs that can form tubular structures (Tu) of various diameters (in brackets) or can be composed of 12 or 24 capsomeres (labeled as *Tiny* Ti [12] or Ti [24]), respectively. Magnification 75,000×. Bar = 100 nm. Electron microscopy, Jiřina Suchanová.

wild-type virus serves as a helper (Oppenheim and Peleg 1989) to sustain high expression of capsid genes, which is difficult to achieve by expression from recombinant vectors. Due to the strictly differentiation-dependent expression of genes for capsid proteins, this approach was impossible to perform with PVs. Instead, it has been shown for PVs that several modifications of the coding sequence of capsid protein genes are needed to obtain reasonable production of capsid proteins from a heterogeneous vector (Zhou et al. 1999; Leder et al. 2001; Mossadegh et al. 2004). The genes should be optimized for codon usage and other properties known to aid protein expression, such as modification of mRNA secondary structures that might impede transcription or nuclear export. The same types of modification were proven useful for PyVs (Tolstov et al. 2009). Nowadays, the cotransfection of expression vectors encoding codon-optimized capsid protein genes with DNA that serves as a target for encapsidation is an established procedure for the production of PyV as well as PV PsVs for specialized applications (Section 20.4.2.2). For the production of VLPs devoid of target DNA, however, the mammalian system is not usually used, due to expensive transfection and cultivation conditions and a risk of potentially infectious contaminants. Interestingly, in mammalian cells, with a good transfection method and modified genes, the yield can be as high as 20 mg of the capsid protein, assembled into

PsVs, per liter of media (1 × 10⁹ cells) for PyVs (Pastrana et al. 2009; Tolstov et al. 2009) and 10 mg for PV VLPs (Buck and Thompson 2007; Buck 2012). Capsid proteins of PyVs as well as HPV PsVs have been also produced in mammalian cells with recombinant vaccinia viruses (Stamatos et al. 1987; Zhou et al. 1991; Unckell et al. 1997).

Insect cells, predominantly *Spodoptera frugiperda* (Sf9) and *Trichoplusia ni* (High Five™, H5) ovary cells and baculovirus expression systems, are a good choice for the production of most VLP types. The system is used for manufacturing one of the VLP-based vaccines against HPV Cervarix (GlaxoSmithKline). VLPs are usually assembled inside the cell nucleus. The quantity depends usually on the type or variant of the parental virus (Touzé et al. 1998) and capsid gene modification (e.g., codon optimization). The yield can be as high as 40 mg of VP1 protein assembled in VLPs per liter of cultivation media for a nonmodified VP1 gene from MPyV (our unpublished observation) and 10 mg of L1 protein per liter for PVs with optimally modified L1 genes (Xu et al. 2014). Moreover, it has been shown that a modified baculovirus-based (MultiBac) expression system can substantially improve VLP production by multiple folds (e.g., for HPV2 from 1 to 40 mg/L) for certain HPV types, whereas the conventional baculovirus expression system gives a low yield (Senger et al. 2009). VLPs can also be produced by baculoviruses in insect larvae with reasonable yield for HPV16 L1 recombinant protein produced in *T. ni* larvae; the yield was five times higher than that in cell culture and reached 21 mg/g of insect biomass (about four insect larvae) (Millán et al. 2010). For HPV6b, the VLPs were also reported to be successfully produced in larvae with a *Bombyx mori* nucleopolyhedrovirus bacmid expression system (Palaniyandi et al. 2012). The *Drosophila* inducible/secreted expression system was also used to produce VLPs. The yield of HPV16 L1 protein was 1.1 g/L of media (Zheng et al. 2008). The yield of MPyV VP1 tagged with a secretion signal for targeting to the extracellular medium was disappointingly low (2–4 mg VP1 per liter of media), and only a small fraction of the recombinant secreted protein assembled into VLP-like structures (Ng et al. 2007).

While production of VLPs from PyVs in plants has never been reported, the need for cost effective manufacturing of VLP-based vaccines against HPV led to exploration of the production of PV-based VLPs in several transgenic plant systems, using tobacco or potato plants (reviewed in Rybicki 2009). VLPs produced in these systems had correct morphology and elicited an immunological response after intravenous or oral administration, but the yields were very low. Varsani et al. (2003) reported a yield as low as 2–4 ng HPV16 L1 protein per gram of fresh tobacco leaf material, corresponding to 0.0003% (w/w) for expression from nonmodified capsid gene. Plant codon usage optimization (Warzecha et al. 2003) or usage of humanized L1 gene (Biemelt et al. 2003) has been shown to increase the yield of HPV11 L1 to 20 ng/g of fresh tuber in potato plants and to 14 μg of HPV16 L1 per gram of fresh tobacco leaves (0.5% of total soluble protein), respectively. The threshold for commercial

TABLE 20.2

Expression Systems for VLP Production

System	Virus (Protein/s)	References[a]	Structure[b]	Notes[c]
PV*s*				
Mammalian				
Vaccinia virus expression system	HPV1 (L1/L2)	Zhou et al. (1991)	VLPs	Incorrectly assembled, d = 35–40 nm
	HPV1 (L1/L2)	Hagensee et al. (1993)	VLPs	Expected morphology and size
	BPV1 (L1, L1/L2)	Zhou et al. (1993)	VLPs	L1/L2 VLPs encapsidated DNA
	HPV18 (L1, L1/L2)	Stauffer et al. (1998)	VLPs	L1/L2 VLPs encapsidated DNA
	HPV6b (L1)	Fang et al. (1999)	VLPs	
	HPV33 (L1, L1/L2)	Unckell et al. (1997)	VLPs	Empty and full capsids (PsVs), L2 not needed for encapsidation
Semliki forest virus–based expression	HPV16 (L1, L1/L2)	Heino et al. (1995)	VLPs	Correct morphology and size, no Tu
Fowlpox virus expression	HPV16 (L1)	Zanotto et al. (2011)	Prevalent Ti, some VLPs	Low expression of L1 protein, low yield of particles
Direct transfection of expression vectors	HPV16 (L1, L1/L2)	Leder et al. (2001)	VLPs	Codon-optimized L1 and L2 genes, VLPs formed abundantly in cell nucleus
	HPV11 (L1)	Mossadegh et al. (2004)	VLPs	Codon-optimized L1 gene, VLPs formed in the cell nucleus
	BPV1 (L1, L1/L2), HPV16 (L1/L2)	Buck et al. (2004)	VLPs	PsVs generated in 293TT cells
Insect				
Baculovirus expression system	BPV1 (L1), HPV16 (L1)	Kirnbauer et al. (1992)	VLPs	Correct morphology and size, also smaller particles
	HPV16 (L1/L2), HPV6 (L1), HPV11 (L1), CRPV (L1)	Kirnbauer et al. (1993)	VLPs	Better yield of VLPs from clinical sample variants and after L1/L2 coexpression
	HPV11 (L1)	Rose et al. (1993)	VLPs	VLPs formed in cell nucleus, but purification not efficient (in vitro assembly)
	HPV11 (L1)	Christensen et al. (1994)	VLPs	VLPs of variable size
	HPV11,16,18 (L1)	Rose et al. (1994)	VLPs	
	COPV(L1)	Suzich et al. (1995)	VLPs	Correct (d = 55 nm) size and morphology
	HPV33 (L1/L2)	Volpers et al. (1994)	VLPs, Tu	Spherical VLPs (d = 50–60 nm) and tubular structures (d = 25–30 nm or 50–60 nm)
	HPV6b (L1), HPV11 (L1, L1/L2), HPV16 (L1, L1/L2)	Muller et al. (1995)	VLPs	Regular VLPs
	HPV45 (L1)	Touzé et al. (1996)	VLPs	Low yield
	HPV6, HPV11, HPV16, HPV31, HPV33, HPV35, HPV18, HPV39, HPV45 (L1)	Giroglou et al. (2001)	VLPs	Regular VLPs
	HPV2, 3, 10, 27, 77B (L1) BPV5, BPV6 (L1)	Senger et al. (2009)	VLPs	MultiBac—high production system (8–40 times increase in yield)
	HPV16, HPV18 (L1)	Harper et al. (2004)	VLPs	Bivalent HPV vaccine Cervarix (GlaxoSmithKline)
	MusPV (L1, 4 variants)	Joh et al. (2014)	VLPs	
Drosophila expression system	HPV16 (L1)	Zheng et al. (2008)	VLPs	
Insect larvae/*T. ni*	HPV16 (L1)	Millán et al. (2010)	VLPs, Ti, Tu	Polymorphic structures formed
Insect larvae/*B. mori* nucleopolyhedrovirus expression system	HPV6b (L1)	Palaniyandi et al. (2012)	VLPs, Ti	Short-length and full-length L1 formed tiny or mixed population of VLP/Ti, respectively

(Continued)

TABLE 20.2 (*Continued*)
Expression Systems for VLP Production

System	Virus (Protein/s)	References[a]	Structure[b]	Notes[c]
Yeast				
S. cerevisiae	HPV6a (L1, L1/L2)	Hofmann et al. (1995)	VLPs	Smaller particles d = 40–50 nm, capsomeres and monomeric protein
	CRPV (L1, L1/L2)	Jansen et al. (1995)	VLPs	Spherical particles, d = 50 nm
	HPV11(L1), HPV11/6a hybrid	Neeper et al. (1996)	VLPs	VLPs detected in cell lysates d = 40–50 nm
	HPV16 (L1/L2)	Rossi et al. (2000)	VLPs	VLPs for gene transfer
	HPV11(L1)	Cook et al. (1999)	VLPs	VLPs of variable size d = 32–97 nm
	HPV16, HPV6 (L1/L2 coexpression 16/6)	Buonamassa et al. (2002)	VLPs	Coexpression of 4 proteins, chimeric VLPs
	HPV11, HPV6, HPV16 (L1)	Mach et al. (2006)	VLPs	Irregular shape, d = 30–60 nm
	HPV6, HPV11, HPV16, HPV18 (L1) (vaccine)	Markowitz et al. (2007)	VLPs	Quadrivalent HPV vaccine Gardasil® (Merck and Co., Inc., Whitehouse Station, New Jersey)
	HPV16 (L1)	Kim et al. (2007), Park et al. (2008)	VLPs	Particles d = 51 ± 15 nm, purification method described
S. pombe	HPV16 (L1, L1/L2), HPV16/HPV6 (L1/L2)	Sasagawa et al. (1995)	VLPs	L2 not incorporated in VLPs
Pichia pastoris	HPV16 (L1)	Liu et al. (2007)	VLPs	Variable size approx. 50 nm
	HPV16 (L1)	Bazan et al. (2009)	VLPs	L1 protein unstable, low yield (aggregation)
	HPV16, 18 (L1)	Rao et al. (2011)	VLPs	Variable size approx. 53 nm
Bacterial				
Escherichia coli	HPV11 (L1/L2)	Finnen et al. (2003)	Capsomeres	VLPs assembled in vitro
	HPV11 (L1, L1 mutants)	Li et al. (1997)	Capsomeres	VLPs assembled in vitro
Lactobacillus casei	HPV16 (L1)	Aires et al. (2006)	VLPs	VLPs produced intracellularly, d = 30–60 nm
Bacillus subtilis	HPV33 (L1)	Baek et al. (2012)	VLPs	Highly heterogeneous VLPs (d = 20–60 nm), problems with purification
Plant				
Tobacco and/or potato leaves	HPV11 (L1)	Warzecha et al. (2003)	VLPs	Uniform spherical particles, d = 55 nm
	HPV16 (L1)	Biemelt et al. (2003)	VLPs, capsomeres	Mainly capsomeres, VLPs, d = 55–65 nm and smaller
	HPV16 (L1)	Fernández-San Millán et al. (2008)	VLPs	VLPs assembled in the stroma of chloroplasts, high yield, d = 55–65 nm and smaller
	HPV8 (L1)	Matić et al. (2012)	VLPs, Ti	
PyVs				
Mammalian				
Direct transfection of expression vector	BKPyV, JCPyV, SV40, LPyV (VP1/VP2/VP3)	Nakanishi et al. (2008)	VLPs	VLPs in the form of PsVs
	SV40 (VP1/VP2/VP3)	Oppenheim and Peleg (1989)	VLPs	VLPs in the form of PsVs
	JCPyV (VP1/VP2/VP3)	Shishido et al. (1997)	VLPs	VLPs observed in the nucleus
	MPyV (VP1/VP2/VP3) MCPyV (VP1/VP2)	Tolstov et al. (2009)	VLPs	Uniform VLPs (d = 55–58 nm)
Avian				
Influenza virus expression system	APyV (VP1/VP2/VP3/VP4)	Johne and Müller (2004)	VLPs	VLP (d = 45 nm) purified

(Continued)

TABLE 20.2 (*Continued*)
Expression Systems for VLP Production

System	Virus (Protein/s)	References[a]	Structure[b]	Notes[c]
Insect				
Baculovirus expression system	HPyV6, HPyV7, TSPyV (VP1)	Nicol et al. (2013)	VLPs, Ti	HPyV6, predominantly Ti, some regular VLPs; HPyV7, regular VLPs; TSPyV, predominantly regular VLPs, some Ti
	MCPyV (VP1)	Touzé et al. (2010)	VLPs	Regular VLPs, d = 45 nm; different clinical isolate generates only protein aggregates
	HPyV9 (VP1)	Nicol et al. (2012)	Ti (VLPs)	Mainly tiny VLPs (d = 24 nm), few regular VLPs (d = 45 nm) in preparations
	APyV (VP1/VP2/VP3)	An et al. (1999)	Capsomeres	VLPs assembled in vitro
	JCPyV (VP1)	Chang et al. (1997)	VLPs	Regular VLPs, d = 45 nm
	MPyV (VP1)	Montross et al. (1991)	VLPs	Regular VLPs in nucleus, purified empty VLPs (d = 46 nm), smaller VLPs in minority
	MPyV (VP1/VP2/VP3)	Forstová et al. (1993)	VLPs	Regular VLPs
	LPyV (VP1)	Pawlita et al. (1996)	VLPs	VLPs of regular size (d = 45 nm) found in cell nucleus; VLPs contain DNA
	MPtV (VP1)	Tegerstedt et al. (2003)	VLPs	VLPs of regular size (d = 45 nm), antibody does not cross-react with MPyV
	SV40 (VP1, VP1/VP2/VP3)	Kosukegawa et al. (1996)	VLPs	VLPs of regular size (d = 45 nm)
	BKPyV (VP1)	Touzé et al. (2001)	VLPs	Regular empty VLPs, d = 45 nm
	TSPyV (VP1)	Chen et al. (2011)	VLPs	Regular d = 45 nm and smaller VLPs
	GHPyV (VP1)	Zielonka et al. (2006)	VLPs, capsomeres	Capsomeres prevalent in preparation, VLPs of regular size (d = 45 nm)
Drosophila expression system	HaPyV (VP1)	Voronkova et al. (2007)	VLPs	Regular VLPs, observed in cell nucleus
	MPyV (VP1)	Ng et al. (2007)	VLPs, aggregates	Secreted VP1, low yield, altered disulfide bonding, irregular deformed VLPs
Yeast				
S. cerevisiae	HaPyV (VP1)	Sasnauskas et al. (1999)	VLPs	Regular size, VLPs in nucleus and cytoplasm
	BKPyV, JCPyV, SV40, HaPyV, MPyV, BFPyV (VP1)	Sasnauskas et al. (2002)	VLPs, Ti	VLPs (except HaPyV and MPyV) heterogeneous in size, d = 45–50 nm; minor fraction of Ti VLPs, d = 20–25 nm
	BKPyV, JCPyV (VP1)	Hale et al. (2002)	VLPs, Ti	BKPyV VLPs, d = 45–50 nm and d = 20–25 nm particle. JCPyV VLPs, d = 50–55 nm
	GHPyV (VP1/VP2), APyV (VP1)	Zielonka et al. (2006)	VLPs, Ti, capsomeres	GHPyV forms exclusively VLPs, d = 20 nm, low stability of VLPs
	APyV, CPyV, FPyV, GHPyV (VP1)	Zielonka et al. (2012)	VLPs, Ti capsomeres	VLPs (d = 45 nm), Ti (d = 25 nm), variable yield of VLPs depending on virus type
	ChPyV (VP1)	Zielonka et al. (2011)	VLPs, Ti, capsomeres	Low efficiency of VLP (d = 45 nm, d = 25 nm)
	JCPyV (VP1)	Chen et al. (2001)	VLPs	Regular size of VLPs (d = 45 nm)
	MPyV (VP1)	Palková et al. (2000)	VLPs, Ti	VLPs (d = 45 nm) in the nucleus, minor fraction of Ti, VLPs containing naked DNA without histones

(Continued)

TABLE 20.2 (*Continued*)

Expression Systems for VLP Production

System	Virus (Protein/s)	References[a]	Structure[b]	Notes[c]
Bacterial				
E. coli	SV40 (VP1)	Wróbel et al. (2000)	VLPs	GroELS chaperone system and His-tag on N-terminus, VLPs detected in cell extract, no in vitro assembly reaction, empty capsids prevalent, some full capsids
	MPyV (VP1)	Salunke et al. (1986)	Capsomeres	In vitro assembly reaction for VLPs formation (see Table 20.3)
	JCPyV (VP1)	Ou et al. (1999)	VLPs	VLPs detected in cell extracts
	HaPyV (VP1)	Voronkova et al. (2007)	VLPs	VLPs detected in cells and cell extracts
	BFPyV (VP1)	Rodgers et al. (1994)	Capsomeres	VLPs assembled in vitro

[a] Only reports of the first production or conflicting reports are listed.
[b] Ti, tiny particles; Tu, tubular structures; see Figure 20.2.
[c] d, diameter.

production of recombinant protein in plants is considered to be 1% of total soluble protein (Fischer et al. 2004). Indeed, it has been shown for HPV16 VLP production that specific optimization of the transcriptional or translational context and chloroplast localization of expression can improve the yield manyfold (17% of total soluble protein) (Maclean et al. 2007; Fernández-San Millán et al. 2008; Lenzi et al. 2008; Matić et al. 2012), making the plant system an interesting alternative for vaccine production.

Both PyV- and PV-based VLPs can be successfully produced in yeast, and quadrivalent HPV6/HPV11/HPV16/HPV18 vaccine (Gardasil, Merck and Co., Inc.) produced in *Saccharomyces cerevisiae* has been successfully introduced into the market. The first VLPs consisting of either L1 alone or L1/L2 produced in yeast were derived from cottontail rabbit PV (CRPV) (Jansen et al. 1995). The study showed that the VLPs were morphologically indistinguishable from native virions and protected rabbits from CRPV-induced wart formation after immunization. Further reports of HPV VLP production followed (Hofmann et al. 1995; Sasagawa et al. 1995; Neeper et al. 1996; Cook et al. 1999). Some reports noted that HPV L1 VLPs produced in *S. cerevisiae* yeast display type-dependent properties of particles (Mach et al. 2006). Whereas HPV18 L1 protein forms uniformly assembled VLPs (60 nm in diameter), L1 proteins of HPV6, HPV11, and HPV16 tend to form more irregular particles of 30–50 nm in diameter, which has no effect on their immunogenic properties but limits particle stability. The authors introduced an efficient procedure of disassembling and reassembling the yeast-derived VLPs to achieve more uniform particle morphology (60 nm diameter spheres) and maximized stability. Disassembly and reassembly of particles seems nowadays to be an important step in manufacturing yeast-derived VLPs for better immunoreactivity (Zhao et al. 2012b) and morphology, decreased heterogeneity, and increased thermal stability (Zhao et al. 2012b). In fact, the limited stability and aggregation of the

recombinant HPV VLPs purified from yeast (Shi et al. 2005) may result in loss of HPV VLPs during purification procedures, and buffer conditions, such as high salt and nonionic surfactants, can substantially increase the yield (Kim et al. 2007; Park et al. 2008). The yields of VLPs from yeast can be high, and the yeast production system allows cultivation in large quantities with a relatively simple culture medium. Cook et al. (1999) reported production of HPV11 VLPs from 200 L, using a galactose-inducible *S. cerevisiae* expression system, where the yield of L1 came to approximately 15% of the total soluble protein of the yeast cell lysate and 6 mg/L of media of highly purified VLPs. Alternatively, the HPV VLPs can be produced in *Pichia pastoris* (Bazan et al. 2009) with similar (9.5 and 6.4 mg/L of HPV16 and HPV18 VLPs, respectively) (Rao et al. 2011) or even higher (20 mg/L of HPV58 VLPs) (Jiang et al. 2011) yields.

Interestingly, in contrast to PyVs, PVs have the inherent capacity to replicate in yeast (Angeletti et al. 2002), and packaging of actively replicating target DNA can lead to the production of PsVs containing full-length HPV genomes (Angeletti 2005) or plasmid DNA with a reporter gene (Rossi et al. 2000). The proof of the concept of PyV-based VLPs production in yeast was assessed using hamster PyV (HaPyV) (Sasnauskas et al. 1999). VP1 was expressed in *S. cerevisiae* and VLPs were abundantly formed in the nucleus as well as in the cytoplasmic compartment. Consequently, our group showed formation of VLPs from MPyV VP1 produced by a galactose-inducible *S. cerevisiae* yeast expression system (Palková et al. 2000). We also showed that a subpopulation of VLPs carried fragments of plasmid or linear chromosomal DNA and that newly synthesized VP1 can interact with mitotic microtubules, thus inhibiting yeast growth. In agreement with this, Sasnauskas et al. (2002) reported the importance of yeast strain selection for high expression of VP1 of PyVs from humans (JCPyV and BKPyV), rhesus monkeys (SV40), hamsters (HaPyV), mice (MPyV), and birds (budgerigar fledgling disease virus

[BFPyV]), but they showed the formation of VLPs devoid of nucleic acid. The reported yields of VLPs were high: 40 mg/L for mammalian PyVs and 5 mg/L for BFPyV. Generally, the yield in purified VP1 preparations may differ remarkably between viruses and can be quite low for goose hemorrhagic PyV (GHPyV) (1.2 mg/L) (Zielonka et al. 2006) or chimpanzee PyV (ChPyV) VP1 (0.3 mg/L) (Zielonka et al. 2011).

The expression of capsid proteins in a bacterial expression system often leads to the purification of protein in a pentameric form (Salunke et al. 1986; Li et al. 1997), but HaPyV or JCPyV VP1 expressed in *E. coli* assembles directly to VLPs (Ou et al. 1999; Voronkova et al. 2007). For HPV11 and HPV16, the purification of 3–5 mg of near-homogeneous L1 protein from 1 L of cell culture was reported (Chen et al. 2001). For MPyV, the effect of host, plasmid, and culture conditions on the expression of VP1 capsid protein in *E. coli* was examined, and the expression yield of 180 mg of soluble VP1 per liter of bacterial culture was obtained (Chuan et al. 2008). Several optimizing conditions enabled the same group to achieve even higher production rates ($\approx$0.3 g of glutathione S-transferase [GST]-VP1 protein per liter of culture) in laboratory shake-flask conditions (Lipin ct al. 2008). The extremely high production of MPyV VP1 (4.38 g of GST-VP1 protein per liter) in high-cell-density fed-batch cultivation in recombinant *E. coli* has been demonstrated (Liew et al. 2010). The disadvantage of the bacterial system is that recombinant protein preparations from bacteria always bear the risk of contaminating endotoxins, which are highly toxic in humans and therefore have to be eliminated from vaccine preparations and VLPs intended to be used in clinic.

Capsid proteins can also be produced in cell-free systems. VLP production, using this system with components derived either from bacteria or yeasts, has been so far reported only for PVs for PVs (Iyengar et al. 1996; Wang et al. 2008). The latter system yielded 50–70 μg of HPV58 L1 protein per milliliter of reaction volume after optimization (Wang et al. 2008).

Purified pentamers can be efficiently assembled into VLPs in vitro in high ionic strength with the addition of calcium (in case of PyVs) or oxidation of disulfide bonds (in case of PVs). Assembled VLPs can be further disassembled and reassembled into the desired structures by changing the buffer conditions and temperatures. Tables 20.3 and 20.4 show several reported disassembly/reassembly systems. These reaction systems are valuable tools for the study of virion assembly, which is a poorly understood phenomenon and also serves for the preparation of viral nanostructures for different biomedical applications (e.g., an increase in the stability of yeast-derived HPV vaccines).

20.4 UTILIZATION OF VLPs AS A CARGO DELIVERY SYSTEM FOR THERAPY AND DIAGNOSTICS

VLPs have been studied intensively as diagnostic and therapeutic compounds. Their structural stability, manipulation tolerance, and ability for molecule incorporation with fast and low-cost production make them an ideal tool for use in gene therapy, immunotherapy, and diagnostics. Here, VLPs serve as vehicles for the transport of therapeutic DNA, drugs, antigens, or contrast agents into target cells. In some cases, including PV and PyV, the application of VLPs can be limited by their nonspecific binding to various cell types. On the contrary, the application potential of VLPs can increase if the selectivity of VLPs for distinct cells is guaranteed. DNA technology enables the preparation of genetically modified capsids on demand, and purified VLPs can be used as the ideal polyvalent monodispersed protein nanoobjects for further chemical engineering. The exterior of the VLPs can be functionalized by the connection of targeting molecules, and the interior of the particles can encapsulate cargo molecules for cellular delivery.

20.4.1 RETARGETING OF VLPs

The concepts of vector targeting are well recognized throughout the gene therapy field (reviewed in Waehler et al. 2007) where current eukaryotic viral vectors can infect cells with high efficiency, but they have the disadvantage that their native tropism must be ablated to avoid the transduction of nontarget tissue. Similarly, PyV and PV VLPs, with their inherent capacity to bind to a wide array of cell types (Section 20.4.2), might need to be detargeted from primary receptor binding and retargeted to the new destination by attachment of the targeting moiety. Although in some instances the addition of the targeting ligand reduces the native tropism sufficiently, detargeting is usually achieved by the genetic mutation of several amino acids that are responsible for interaction with the primary receptor. To reprogram VLP cell binding and entry, ligands that direct targeting to specific cell types should be attached to the surface of VLPs. The rational design of VLP modification, therefore, requires knowledge of virion structure, which is fortunately known for species of PyV and PV (Section 20.2.2). The selection of the targeting moiety depends on the nature of the target cell and the actual biomedical application and usually consists of the polypeptide molecule (e.g., epidermal growth factor [EGF]) that naturally binds the receptors on target cells (EGF receptors are overrepresented in some cancer cells), monoclonal antibody (or single-chain antibody) against the target cell receptor, or small targeting peptide motifs. These small targeting peptides can be chosen either from library selection approaches or from naturally occurring motifs. For this purpose, the RGD motif (tripeptide Arg-Gly-Asp), which targets vectors to integrins overrepresented in tumors and vasculature, has most commonly been used. Furthermore, nonpeptide molecules, such as sugars, fatty acids, nonpeptide hormones, or small molecular compounds, can serve as targeting molecules (Waehler et al. 2007).

The major advantage of VLPs against other therapeutic carriers is their ability to expose a large number of targeting molecules whose number and orientation can be controlled. The attachment of a targeting moiety can be performed by genetic or chemical methods. Genetic approaches allow

TABLE 20.3

Main Conditions for In Vitro Reconstitution of VLPs from Capsomeres and the Outcome of Assembly Reaction after Dialysis

Protein(s)	System	References	Conc.	Time	T (°C)	pH	Salt	Ca	VLPs[a]	Tiny[b]	Tu[c]	Pentamers	Aggregates
VP1 (MPyV)	E. coli	Salunke et al. (1989)	0.5–1 mg/mL	2 days	RT	7.2	150 mM NaCl	0.5 mM	+++ N, Sn	+		+	
						8.5	150 mM NaCl	0.5 mM	Disrupted	+++		+	+
						5.0	150 mM NaCl	0.5 mM	+++ N, Ln	+		+	
						7.2	2 M (NH$_4$)$_2$SO$_4$	—	++ N, Sn	++	++ (15 nm)		
						8.5	2 M (NH$_4$)$_2$SO$_4$	—	++ N, Sn	++	+ (15 nm)	+	
						5.0	2 M (NH$_4$)$_2$SO$_4$	—	++++				
VP1 (SV40)	Insect	Kanesashi et al. (2003)	0.06 mg/mL	1 day	4	7.2	150 mM NaCl	2 mM					
					RT	5.0	150 mM NaCl	2 mM					++++
					RT (4)	7.2	1 M NaCl	2 mM	+++ N, Sn, Ln	+	++++ (30 nm)		
					4	7.2	1 M NaCl	—			++ (45 nm)		
					4	7.2	2 M (NH$_4$)$_2$SO$_4$	2 mM	++++ N, Sn	++++			
HPV11	E. coli	Li et al. (1997)	nd	[d]	RT	7.2	1 M NaCl	—	+++				+

Abbreviations: Conc., capsid protein concentration; Ca, calcium concentration; RT, room temperature; nd, not determined; number of "+" indicating quantity of each assembly form.

[a] VLPs of various size can be formed (N, normal size; Ln, larger than normal; Sn, smaller than normal)—see Figure 20.2.

[b] Particles composed of 12 or 24 capsomeres are labeled as *Tiny*.

[c] VLPs can form tubular structures (Tu) of various diameters (in brackets).

[d] VLPs assembled during elution from the phosphocellulose column in 1 M NaCl.

TABLE 20.4

Examples of Disassembly/Reassembly Procedures

Virus/VLPs (System)	References	Disassembly[a]	Reassembly	Notes
MPyV virion	Brady et al. (1977)	10 mM Tris–HCl (pH 8.5), 150 mM NaCl, 10 mM EGTA, 3 mM DTT, 30 min	Not done	Show stabilizing effect of high salt (1 M NaCl) and calcium ions (5 mM for disassembly); pH > 8.5 increases the disruption, high concentration (50–150 mM) of EGTA decreases disruption.
MPyV virion	Brady et al. (1979)	1 mM EGTA, 0.1 M ME, 0.15 M NaCl in 0.05 M Tris–HCl (pH 7.4), 30 min, RT	Dialyzing in 10%/DMSO, 0.01% Triton X-100 in PBS (pH 7.4) with 0.5 μM $CaCl_2$	Virions dissociated in the pH range of 7.4–7.8 have higher frequency of reassembly; dissociation at pH 8.0 is harmful to the reassembly; higher concentration of $CaCl_2$ is inhibitory to the assembly.
SV40 virion	Colomar et al. (1993)	50 mM Tris–HCl (pH 7.9), 150 mM NaCl, 1 mM EGTA, 20 mM DTT, 37°C, 1 h	Gradual addition of $CaCl_2$ (5 mM final)	Disassembly not complete—disassembly/reassembly verified by infectivity assay not EM.
SV40 VP1 VLPs (insect)	Kanesashi et al. (2003)	20 mM Tris–HCl (pH 7.9), 0.1% Nonidet P-40, 25 mM EGTA, 30 mM DTT, 1 h at 37°C plus gel filtration: 20 mM Tris–HCl (pH 7.9), 150 mM NaCl, 5 mM EGTA, 5 mM DTT	Dialyzing 9 μg of the purified pentamer preparation for 24 h against the various buffers or 2 M $(NH_4)_2SO_4$, 2 mM $CaCl_2$, pH 7.2, 4°C	Reports about aggregation in physiological salt condition (150–30 mM NaCl) and acidic pH (pH < 5.0), formation of VLPs in high concentrations of $(NH_4)_2SO_4$ did not require $CaCl_2$.
HPV16 L1/2 VLPs (insect)	Kawana et al. (1998)	1 mg of VLPs incubated in 1 mL of PBS containing ME (5%), 16 h at 4°C	VLPs mixed with 2 mg of plasmid and dialyzed against 4 L PBS, 0.5 M NaCl, 2 mM $CaCl_2$, 24 h at 4°C	Electron microscopy confirmed the presence of particles; 10 μg of PLPs from 1 mg of disassembled capsids.
HPV6, HPV11, HPV16 L1 VLPs (yeast)	Mach et al. (2006)	0.15 M NaCl, 35 mM sodium phosphate, 2 mM EDTA, 0.03% polysorbate 80, 100 mM Tris (pH 8.2), 10 mM DTT	Dialyzing against a solution of high salt concentration (0.5–1 M NaCl) at a lower pH (6–7)	Aggregation during the disassembly and initial reassembly in low salt solutions; disassembly performed under high salt conditions (0.5–1.2 M NaCl) with polysorbate 80 to eliminate the aggregation.
HPV11 L1 VLPs (insect)	McCarthy et al. (1998)	PBS, 150 mM NaCl, 5% ME, 16 h, 4°C	Capsomeres (0.5–5.0 mg) dialyzed versus 4 L of PBS with 0.5 M NaCl, 4°C, 24 h	The aggregated VLPs were resistant to disassembly; 0.5 M NaCl during reassembly designed to stabilize VLPs.
BKPyV VP1 VLPs (insect)	Touzé et al. (2001)	50 mM Tris–HCl (pH 7.5), 150 mM NaCl, 1 mM EGTA, 20 mM DTT, 30 min RT	Dilution in 50 mM Tris–HCl (pH 7.5), 150 mM NaCl, 2 mM $CaCl_2$, 1% DMSO; $CaCl_2$ molarity increased stepwise from 2 to 5 mM (1 mM/h) at 20°C to reach a final volume of 500 μL.	1 μg of plasmid DNA in 50 μL of 50 mM Tris–HCl buffer (pH 7.5), 150 mM NaCl added to 50 μL (10 μg) of disrupted VLPs and further diluted.

[a] ME, 2-mercaptoethanol; DTT, dithiothreitol.

ablation of natural tropism and introduce a targeting or adaptor molecule in one step but are usually limited by the size of the ligand that can be incorporated without compromising the assembly, stability, and yield of VLPs. In addition, the peptide affinities to the receptor may be influenced by the location in the capsid. More importantly, nonpeptide molecules cannot be used as targeting moieties, despite being high-affinity ligands for different targets (Waehler et al. 2007). On the contrary, the chemical modification of VLPs allows covalent attachment of specific ligands from a large collection of protein and nonprotein compounds by the classical techniques of protein alteration and cross-linking (reviewed in Wong [1991] and Strable and Finn [2009]). Moreover, the coupling of full-length proteins can be performed without negatively affecting the structural and biological integrity of VLPs. The techniques use acylation of the amino groups of lysine, alkylation of the sulfhydryl group of cysteine and activation of carboxylic acid residues (of aspartic and glutamic acids), and coupling with added amines. These amino acids, together with the aromatic groups of tyrosine and tryptophan, have distinct reactivity patterns and therefore predominantly serve for bioconjugation purposes (reviewed in Strable and Finn [2009]). On the other hand, the utilization of chemical approaches can be limited by the absence of these amino acids in the appropriate positions on the VLP surface or by the fact that reaction conditions promote disassembly or aggregation or severely reduce the yield of modified VLPs during purification steps. Conversely, chemical cross-linkers can, in some instances, significantly reduce the capacity of VLPs to disassemble intracellularly, which might be undesirable for specific applications.

The combination of genetic and chemical techniques for modifying the surface of viral particles seems to limit the disadvantages of both approaches. Genetic techniques can be used to introduce a specific chemical reactivity (amino acid[s]) at defined positions on the viral capsid surface. An addition of few amino acids usually has no effect on VLP stability, in contrast to extensive amino acid changes when genetic retargeting is applied. After production of the VLPs in a conventional expression system, the newly integrated amino acid(s) can be used to chemically couple ligands for targeting.

Generally, the single- or two-component systems can be used for the attachment of the targeting ligand to VLPs. The single-component system uses direct incorporation of a targeting moiety into or onto VLPs, whereas the second strategy uses the adaptor molecule to mediate the attachment of a targeting ligand to VLPs. Both of these strategies have been explored for retargeting PyV VLPs. Since the methods are essentially the same as for the preparation of VLPs designed to expose immunodominant epitopes for vaccination purposes (Section 20.5.2.2) or for peptide delivery (Section 20.4.3) application, PV VLPs are modified in the same way.

20.4.1.1   Targeting of VLPs by a Single-Component System

The single-component system, where the targeting moiety is directly attached to the VLPs, is more technically challenging and less versatile than the use of the two-component system, but might provide stable and functionally homogenous retargeted particles. In addition, this approach might simplify high-titer production since there is no need to create a separate adaptor or docking molecules. Both genetic and chemical methods are used for the creation of targeted VLPs.

Targeting by genetic modification requires the construction of genetic fusion between the capsid protein and the targeting ligand. This leads to the formation of a special variant of so-called chimeric VLPs (Table 20.6), where

the targeting sequence must be exposed on the surface of the VLP. Finding the suitable positions in the surface loop of capsid proteins that allow this type of modification without affecting the physical integrity of the particles is usually the main obstacle in the construction of these VLPs. It is also crucial that the displayed ligand maintains its bridging ability toward the target after fusion. Both PyVs and PVs have been used for genetic modification and the effect on VLPs stability analyzed, but only few reports of successful targeting exist.

Both phenomenon stability and targeting were systematically analyzed for SV40 VP1 VLPs by Takahashi et al. (2008). They inserted FLAG epitope (Asp-Tyr-Lys-Asp-Asp-Asp-Asp-Lys octapeptide) with short triglycine flexible linkers, which assist the molecule to achieve the most preferable conformation, in different positions of all surface loops of the VP1 protein (BC, DE, EF, and HI). For targeting, they inserted three consecutive RGD motifs. They demonstrated that only a small number of positions (one in DE and one in the HI loop) of VP1 can accommodate foreign peptides without affecting VLP formation and that at least three glycine residues flanking both sides of the foreign peptide were needed for VLP assembly. Moreover, the RGD motifs displayed on the VLPs were found to be directly involved in cell attachment, and interaction with cells was enhanced for VLPs displaying RGD in DE as well as HI loop compared to VLPs carrying the FLAG-tags at the same positions. The latter VLPs also associated with target cells only very weakly compared to wild-type VLPs, because both DE and HI loops are, together with the BC loop, involved in receptor binding (Neu et al. 2008).

The same targeting peptide has been used for the retargeting of VLPs derived from LPyV (Langner et al. 2004). The single RGD motif without a linker sequence was replaced in 11 different positions (in BC, DE, and HI loops) in the VP1 protein sequence, and only five mutant proteins (three in BC loop and one each in HI and DE loops) were found to yield VLPs. The modifications led to the loss of LPyV receptor binding of all VLPs. Specific binding to αvβ3-integrin, an important marker of angiogenesis in solid tumors, was shown for VLPs carrying RGD in the BC loop. Interestingly, no binding was observed with three other integrins, αvβ6, αIIbβ3, and αvβ5, which also recognize RGD with more restricted ligand recognition profiles than αvβ3.

MPyV VLPs are probably the VLPs most frequently modified by a genetic approach for different purposes; Table 20.5 summarizes these reports. The VP1 loops were modified to carry foreign epitopes (Sections 20.4.1 and 20.5.2.2) or adaptor sequences (Section 20.4.2.2) of various lengths, but only one study attempted to incorporate targeting molecule directly into VP1 protein (Shin and Folk 2003). In this study, VLPs were targeted to a urokinase-type plasminogen activator receptor (uPAR), which is a protein expressed by many cancer cells, where it correlates with metastasis and poor prognosis. To restrict the binding of VLPs to a natural receptor, the VP1 protein was modified by the FLAG in the HI loop before subsequent manipulation. The fragments of uPAR activator were inserted into all surface-exposed loops

TABLE 20.5

Overview of MPyV VP1 Protein Modification and Their Influence on the VLPs Assembly

Virus	Insert	Location	Utilization	Expression System	Assembled VLPs	References
MPyV	Pre-S1 phil—two hydrophilic fragments from HBV pre-S1 sequence (70 aa + 6 aa linkers)	HI loop	Immunization	*S. cerevisiae*	Yes	Skrastina et al. (2008)
MPyV	B-cell epitopes (12 and 14 aa)	BC loop	Immunization	*E. coli*	Yes	Neugebauer et al. (2006)
MPyV	Protein Z (antibody binding to a domain of protein A) (57 aa + 17 aa linkers)	HI loop	Retargeting	*E. coli*	Yes	Gleiter and Lilie (2001)
MPyV	WW domain (from murine FBP11) (38 aa)	DE loop HI loop	Retargeting	*E. coli*	No	Schmidt et al. (2001)
MPyV	Peptide with 8 glutamate and 1 cysteine residues	HI loop	Retargeting	*E. coli*	Yes	Stubenrauch et al. (2001)
MPyV	Peptide sequence binding uPAR (60 aa) or FLAG sequence (8 aa)	BC loop DE loop HI loop EF loop	Retargeting	Baculovirus (insect)	No Yes	Shin and Folk (2003)
MPyV	Peptide sequence from Bcr-Abl protein (25 aa)	HI loop	Immunization	Baculovirus (insect)	No	Španielová (unpublished results)

Abbreviations: aa, amino acids; FBP11, formin-binding protein 11; uPAR, urokinase-type plasminogen activator receptor.

(BC, DE, EF, HI) of a detargeted variant of VP1, but only the insertion of an N-terminal part of a 60 aa uPAR activator fragment into the EF loop had no detrimental effect on protein solubility and particle integrity. However, only VLPs with a diameter of 20 nm were formed. When these mutant VP1 proteins were coexpressed with a detargeted variant of VP1 protein, they formed heterotypic VLPs of a regular size. Compared to wild-type VLPs, these heterotypic VLPs did not bind cells with uninduced expression of uPAR. However, after induction of the expression of uPAR, the heterotypic VLPs bound specifically to uPAR on the surface of target cells. These data suggested that the EF loop, which is exposed laterally, far from the receptor-binding site, might still provide a targeting function and be more flexible in accommodating longer foreign sequences.

A combination of genetic and chemical techniques has been used for the preparation of SV40 VLPs targeted to the EGF receptor (Kitai et al. 2011). The reactive cysteine residue, which was introduced into the DE loop of the VP1 protein, was subsequently used for chemical conjugation of full-length human EGF to the surface of VLPs through a thiol group with heterobifunctional cross-linker SM(PEG)2, which possesses maleimide and succinimide moieties. The chemical reaction had a 44% yield of intact modified VLPs and resulted in the conjugation of approximately 40 molecules of human EGF per VLP. These EGF-VLPs were further shown to display enhanced selectivity for cells that overexpress the EGF receptor, and their internalization was 10-fold greater than that of unmodified VLPs. Moreover, the study provided the first evidence that VLPs targeted by single-component systems not only can increase the binding capacity of VLPs to specific target cells but also can enhance cellular uptake through the EGF receptor–mediated endocytosis.

20.4.1.2 Targeting of VLPs by a Two-Component System

The two-component system does not attach the targeting moiety directly to VLPs, but uses noncovalently bound adaptor molecules for this purpose. Adaptors are molecules with dual specificities: one end binds the viral protein, and the other binds the receptor on the target cell. The adaptor strategy possesses great flexibility, as different adaptors can readily be coupled to the same VLPs to allow for easy testing of several target receptors. On the other hand, the system might be limited by varying coupling efficiency and suboptimal stability of the VLP–adaptor complex, especially in vivo (Waehler et al. 2007). Nevertheless, this system was frequently used for the modification of PyV as well as PV VLPs for different applications.

In a two-component system, the VLPs and adaptor molecule can be designed in such a way that VLPs can be genetically or chemically modified to display the versatile adaptor-binding motif. This strategy has been explored primarily for MPyV VLPs, and different motifs have been used in several complementary studies.

In one study, the HI loop of VP1 protein was genetically engineered to display a 9 aa region of polyanionic peptide (Glu_8Cys, E8C) on the surface of VLPs (Stubenrauch et al. 2001). Polyionic fusion peptides are highly soluble, and their interaction does not depend on specific secondary structures. This manipulation therefore did not lead to a destabilization of VLPs (Stubenrauch et al. 2000). This region was subsequently used as a docking site for electrostatic association with a targeting molecule a tumor-specific antibody Fv fragment fused with a complementary polycationic (Arg_8Cys, R8C) tag. Finally, the cysteine residues of the polyionic peptides enabled covalent cross-linking under oxidizing conditions.

Approximately 30 antibody fragments were bound to engineered E8C VLPs and the system proved to be highly specific and efficient. The VLPs were packaged with a plasmid containing the reporter gene, and the selectivity of VLPs coupled with tumor-specific antibody for the target cells was determined from transduction assays. Transduction with targeted VLPs resulted in fivefold higher β-galactosidase activity in target cells than transduction with E8C VLPs not decorated with the antibody. However, the transduction efficiency was lower than the cell-type nonspecific binding transduction of wild-type VLPs and was generally rather low (3% and 5%, respectively) (Stubenrauch et al. 2001). The complementary study (May et al. 2002) verified the specific and efficient association of E8C VLPs with cellular targets mediated by the antibody fragment and suggested that their incapacity to surmount the endosomal membrane and escape lysosomal degradation is responsible for the lack of functional transduction of the respective cells (see Section 20.4.2).

In the other study, a 38 aa domain of protein Z was inserted via short serine–glycine linkers into the HI loop (Gleiter and Lilie 2001). Protein Z is a binding domain, derived from protein A of the bacteria *Staphylococcus aureus*, which is able to specifically bind antibody immunoglobulins. This insertion did not affect VLP stability or the functional integrity of the Z domain. A humanized monoclonal antibody bound the Z domain on the surface of VLPs with high affinity and a stoichiometry of around 0.8 antibody molecules per VP1-Z monomer. This specific targeting function was confirmed with Herceptin antibody. Herceptin binds selectively to the HER2 glycoprotein, a member of the EGF receptor family, which is present on several different human tumor cells. This antibody directed the respective particles specifically toward the cells with high expression of HER2. Without the antibody coupled to the VLPs or with the cell line, which does not express HER2, no targeting was observed.

Another study investigated a strategy that uses a genetic fusion of capsid protein with 28 aa of the WW domain of the mouse formin-binding protein 11 (Schmidt et al. 2001). The WW domains are very small protein domains that bind proline-rich ligands with high affinity. They are named after two conserved tryptophan residues that are essential for the maintenance of the native fold and specific binding ligands. The WW domain was flanked by serine–glycine linkers (5 aa) and inserted in either the DE or the HI loop of the VP1 sequence. Whereas the first variant of WW-VP1 fusion yielded completely formed particles after stabilization with disulfide bonds, the latter WW-VP1 fusion protein lost the capacity to form VLPs. WW VLPs were tested for their capacity to bind a proline-rich sequence with a PPLP (Pro-Pro-Leu-Pro) consensus motif by addition of GFP with the PPLP-tag. Approximately 25 ± 5 molecules of GFP could bind to the capsid surface. Unfortunately, the coupling efficiency as well as the stability of the complex was limited by the fast dissociation reaction, and the strategy was not used further for targeting experiments. The same system was, however, successfully used for VLP-mediated intracellular delivery of protein and peptides (Günther et al. 2001) (Section 20.4.3).

Besides targeting purposes, the similar two-component systems are often adapted for application, where VLPs serve as carriers of other substances. For example, BPV-based VLPs with a polyglutamic acid–cysteine sequence inserted into a surface-exposed region of the L1 major capsid protein were successfully used for the coupling of an antigen with an N-terminal polyarginine cysteine tag for immunization purposes (Pejawar-Gaddy et al. 2010). Interestingly, the E8C region influenced the stability of BPV when inserted to a BC or DE loop of L1, but HI loop modification yielded regular E8C VLPs that conjugated with the antigen with a higher efficiency than reported (Stubenrauch et al. 2001) for MPyV VLPs (14% and 8%, respectively). Even higher coupling efficiency with the two-component system based on strong interactions between biotin and streptavidin has been reported for BPV1 L1 VLPs (Chackerian et al. 2001), where biotinylated VLPs have bound to approximately 540 streptavidin tetramers.

20.4.1.3 Concluding Remarks

The numerous studies done mainly with PyV VLPs have shown that retargeting in respect to selective cellular binding is possible, but only few studies have demonstrated selective particle uptake. The low level of unspecific internalization of VLPs is evident in all of the studies. Technically, the chemical modification of particles seems to be more efficient and less challenging than the genetic approach. Most studies have conclusively suggested that the size of a foreign sequence genetically inserted into a capsid protein is a limiting factor for self-assembly into VLPs, but this is generally hard to predict. For MPyV VP1, sequences no longer than 40 aa together with flexible linkers (e.g., glycine–serine) can usually be inserted without compromising VLP integrity, but occasionally the incorporation of the whole 18 kDa enzyme can be successful (Gleiter et al. 1999). For PV, peptides of up to 60 aa can be fused to the truncated L1 without disrupting the assembly of VLPs (Müller et al. 1997). However, the integrity of the VLPs is probably influenced by the actual position and character of the foreign sequence introduced into the capsid protein, so various reports of unsuccessful manipulation in VLPs surface loops exist (see Table 20.5 for MPyV VLPs). Accordingly, the versatility of a two-component system that requires genetic modification of VLPs might be limited by particle stability, and chemical methods for adaptor coupling might be favorable for many applications.

20.4.2 Nucleic Acid Delivery

Nucleic acids are obvious cargo molecules for VLPs. Historically, the need for safe and efficient gene delivery vehicles in the flourishing field of gene therapy led to attempts to use VLPs exclusively for the delivery of DNA for gene expression. Today, the approaches for direct delivery of mRNA or silencing molecules (antisense DNA/RNA, siRNA) (Lund et al. 2010) are a more attractive direction for research. Moreover, VLPs can serve as shielding vehicles for the delivery of DNA vaccines (Section 20.5.1.2).

TABLE 20.6

PV and PyV Particle Terminology[a]

Virion	A complete virus particle composed of all capsid proteins and viral genome in the form of minichromosome; capable of initiating infection and expressing viral genes
Virus-like particle (VLP)	A noninfectious particle composed of major capsid protein with or without minor capsid proteins (should be specified); encapsidating no specific (but may contain unspecific) nucleic acid
Chimeric VLP (CVLP)	VLP formed from capsid protein(s) fused with foreign sequence(s)
Pseudocapsid	Equivalent to VLP, reflects differences in the composition of naturally occurring capsids and heterogeneously expressed particles; usually composed exclusively of major capsid protein; encapsidating no specific (but may contain unspecific) nucleic acid
Capsoid	Pseudocapsid-like structure formed after assembly of capsid protein, which was recombinantly expressed in *E. coli*
Pseudovirion (PsV)	A virus particle composed of major and minor protein(s) capable of transducing the reporter gene expression; plasmid encapsidated in the form of a minichromosome
Pseudovirion-like particle (PLP)	Similar to PsV, but the capsid protein composition differs from virion capsid or the nucleic acid as it is encapsidated as naked and not in the form of a minichromosome.
Quasivirion (QV)	An infectious particle assembled *artificially* in vivo (in 293TT cells) by supplying the capsid proteins and genome in *trans*; it is composed of all capsid proteins and viral genome in the form of a minichromosome; it is capable of initiating infection and expressing viral genes
Quasivirion-like particle (QVLP)	An infectious particle assembled *artificially* in vitro; it is composed of all capsid proteins and viral genome in the form of naked DNA (for PyVs the term polyoma-like particles was used)

[a] Compiled from Ozbun and Kivitz (2012) and literature cited in the text.

Terminologically, several papovaviral assemblies with nucleic acid can be recognized (Table 20.6). The general term *viruslike particles* is usually used for the description of noninfectious capsid-like particles (the actual protein composition in respect to the presence of minor proteins should be specified) produced in a heterologous expression system. VLPs made only from the major capsid protein are alternatively called pseudocapsids (Forstová et al. 1995). These particles may be completely devoid of a nucleic acid (appearing as *empty* particles under an electron microscope) or may contain fragments of DNA (appearing as *full* particles under electron microscope). These two populations of particles are separated into two distinct bands with slightly different buoyant densities during CsCl gradient ultracentrifugation. For MPyV VLPs, these particles are sometimes described as *light* ($\rho = 1.290$ g/cm^3) and *heavy* ($\rho > 1.300$ g/cm^3), respectively (Palková et al. 2000). It has been demonstrated for PyV VLPs produced in a baculovirus expression system that these DNA fragments originate either from the host cells or from the baculovirus genome and are complexed with cellular histones into pseudonucleocores (Pawlita et al. 1996; Gillock et al. 1997). In contrast, MPyV VLPs produced in *S. cerevisiae* did not assemble with cellular histones and *full* particles isolated from yeast cells contained linear DNA fragments, up to 3 kbp long, encapsidated as *naked* DNA by empty VP1 particles. VLPs purified from insect cells are heavily contaminated with RNases (Forstová, unpublished observation), but the VLPs produced in a yeast or bacterial system may be contaminated with RNA (Ou et al. 1999; Sasnauskas et al. 1999; Gedvilaite et al. 2000). VLPs can be complexed with a specific nucleic acid in vitro as well as in vivo. PsV-based technology established in a mammalian expression system (Section 20.4.2.1) allows the production of infectious particles that closely resemble a

native virion containing the major and minor proteins and specific DNA complexed with cellular histones in the form of a minichromosome. These particles are called PsVs. The same term, however, is sometimes used for DNA-loaded particles without minor capsid proteins, for particles purified from yeast expression system where DNA is not assembled with histones, as well as for particles prepared by an in vitro disassembly/reassembly approach, which contain only naked DNA. Although these types of particles can be infectious to some extent, we recommend the definition of these particles as PLPs. The same abbreviation was originally used for polyoma-like particles (Barr et al. 1979), the particles formed from purified empty capsids after incubation with supercoiled DNA. For this type of particle, we would today select the term *quasivirion-like particle (QVLP)* reflecting the term *quasivirion (QV)*, coined by the laboratory of Neil D. Christensen (Culp et al. 2006c) to describe virions with an authentic virus genome with a nucleocore produced in vivo in 293TT (Pyeon et al. 2005).

20.4.2.1 Preparation of VLP–Nucleic Acid Complexes

Several methods for the preparation of complexes made from VLPs and nucleic acids have been developed. VLPs loaded with nucleic acids can be prepared either in vivo or in a cell-free system (in vitro). The expression of the reporter gene is often used as the readout for the successful packaging and delivery of DNA into the cells.

20.4.2.1.1 *In Vivo Production of PsVs and PLPs*

PsVs are usually produced in the easy-to-transfect human embryonic kidney cells, 293TT, which express a high level of SV40 large T antigen (Buck et al. 2004), but other cell lines expressing T antigen can be used (e.g., Cos cells). The 293TT cell line is usually cotransfected with helper and target vectors. Both the helper vectors ensuring the expression of capsid proteins and the target vectors harboring the reporter gene carry the SV40 origin of replication (ori). These two SV40 regulatory elements (large T antigen and SV40 ori) are needed to achieve high-level production of capsid proteins as well as a high concentration of target DNA for encapsidation. Moreover, it has been shown for the SV40 virus that the ori sequence overlaps with the SV40 packaging signal (Oppenheim et al. 1992) but pseudovirions or QVs can be prepared in the absence of this signal (Pyeon et al. 2005) and the concentration of the target vector seems to be crucial for its encapsidation (Culp et al. 2006c; Španielová et al. 2014). The system is universal and allows the production of SV40 vectors with exchangeable capsids that exhibit differential efficiency of gene transduction to the target cells (Nakanishi et al. 2008). On the other hand, the system is also stochastic, promiscuous, and not very efficient. The efficiency of encapsidation of the reporter vector is estimated to be approximately 5% for BPV PsV (Buck et al. 2004), and the observed particle-to-infectivity ratios (expressed as reporter plasmid copies per infectious units) are highly variable for different HPV types (varies between 20 and 5000) (Handisurya et al. 2012). HPV16 and HPV18 L1/L2 PsV stocks contain substantial amounts of encapsidated cellular DNA (Buck et al. 2005b). This can complicate the subsequent PsV-based assays and generates safety issues connected with PsV production.

HPV33 PsVs were prepared with a vaccinia virus expression system with a packaging efficiency that was rather low (1 input plasmid per 25,000 particles), probably because cellular DNA fragments generated by the lytic infection of vaccinia virus competed with the target plasmid during encapsidation (Unckell et al. 1997). Based on the same principle, the BPV PsVs were prepared in insect Sf9 cells that were coinfected with L1/L2 recombinant and E2 recombinant baculoviruses and transfected with the target plasmid (Zhao et al. 2000). The packaging efficiency was estimated to be 0.01% (1 input plasmid per 10,000 particles).

PsVs and PLPs can be prepared in vivo in nonmammalian systems. Rossi et al. (2000) demonstrated the production of HPV16 PLPs containing L1 and L2 proteins and the GFP reporter plasmid in yeasts. These PLPs were successfully used for the delivery and expression of the reporter gene after in vitro infection of mammalian cells and after injection of PLPs into mice. Although promising, the approach exhibited substantial variations in yield, and finally, PLP production was lost entirely (Peiler 2004). The presence of a plasmid in VLPs derived from yeast has been demonstrated for PyVs (Palková et al. 2000), but the PyV-derived PLPs produced in yeast have never been used for gene delivery purposes. Interestingly, since JCPyV and HaPyV VLPs self-assemble into VLPs inside the bacterial cells, the PLPs composed of VP1 protein and reporter gene expression plasmid can be produced in vivo in E. coli. This method substantially increases the efficiency of subsequent gene transduction, from 2% recorded for JCPyV PLPs prepared in vitro to 80% for in vivo DNA packaged PLPs (Chen et al. 2010).

20.4.2.1.2 In Vitro Production of PsV and PLPs

Several procedures for in vitro complexation of VLPs with nucleic acid have been established.

During these procedures, nucleic acid is loaded to the VLPs by their exposure to osmotic shock, sequential disassembly and reassembly, chemical conjugation with nucleic acid, or simply direct mixing with nucleic acid. These cell-free methods require highly purified VLPs devoid of contaminating DNA (inside) and nuclease activity (outside). Preparation of VLPs is therefore crucial for successful encapsulation of the cargo molecule. VLPs formed during production in most expression systems (except for bacterial) contain fragments of cellular DNA, and VLPs must therefore be either purified to near homogeneity by density gradient ultracentrifugation to obtain just *empty* pseudocapsids or disassembled, exposed to nuclease treatment, and reassembled. According to our experience, the elimination of nuclease contamination requires at least one purification step by centrifugation in sucrose density gradient. Table 20.7 shows a selection of some experiments where VLPs complexed in vitro were used.

20.4.2.1.2.1 Osmotic Shock Procedure Nucleic acid loading by the passive osmotic shock procedure was originally developed for PyV PLPs (Barr et al. 1979). The loading process probably proceeds in several steps. First, a DNA/RNA molecule binds to an empty capsid to form what is designated as a DNA–capsid-binding complex. Forstová et al. (1995) showed that only empty or disrupted VLPs can form these complexes, whereas *full* virions do not interact with exogenous DNA. The second step consists of the lowering of ionic strength by the addition of distilled water, which facilitates the entrance of DNA into the capsid, as shown by Barr et al. (1979). In the third step, the added nuclease cleaves and removes the external DNA that is not able to enter the capsid. The encapsidated DNA is further protected from nuclease action. The fragment of DNA that is encapsidated does not consist of a specific sequence, and encapsidation is sequence independent. Either linear, circular, or supercoiled DNA, as well as single-stranded DNA, rRNA, and synthetic oligonucleotides, was originally used for PLP formation (Slilaty et al. 1982).

There is likely a size constraint on the amount of genetic information that might be packaged by this method. The limit appears to be between 1.8 and 2.5 kbp for dsDNA (Slilaty et al. 1982; Forstová et al. 1995) with circular DNA being packaged more efficiently than linear DNA (Forstová et al. 1995). The character of interactions between MPyV pseudocapsids and DNA during osmotic shock has been studied in detail by electron microscopy (Štokrová et al. 1999) (Figure 20.3). The study revealed that pseudocapsids form only weak interactions with internal parts of the circular form of DNA. In contrast, two pseudocapsids bound to each end of a linearized DNA molecule were found to form highly stable complexes where the DNA is partially encapsidated. The level of protection from nuclease activity was the same for linearized or circular DNA—approximately

TABLE 20.7

Selection of Reports Using In Vitro Methods of PLP Preparation for the Gene Transfer

Production System/Vector	Parental Virus (protein[s])	PLP/PsV Method	VLP/DNA Ratio	Detection System	Quantity of DNA (pg/cell)	Transduction Efficiency	References
Sf21/baculovirus	SV40 (VP1)	Nuclear extract (reassembly)	10/100	MDR, GFP (4.7 kbp)/various cell lines	1000	70%–100%	Kimchi-Sarfaty et al. (2004)
Sf21/baculovirus	HPV16 (L1)	Disassembly/reassembly	25/5	GFP (5 kbp) β-galactosidase (7.2 kbp), various cells	2	70% (GFP)	Touzé and Coursaget (1998)
Yeast	HPV11 and HPV16 (L1, L1/2)	Chemical conjugation	20/1.12	β-Lactamase (PCR fragment (1.8 kbp)/C33A	20	15%–40%	Yeager et al. (2000)
Sf21/baculovirus	BKPyV (VP1)	Direct mixing (plus other)	10/1	Luciferase (7.1 kbp), β-galactosidase/Cos-7	20	50% (β-gal)	Touzé et al. (2001)
E. coli	HaPyV (VP1)	Disassembly/reassembly	50/1	GFP(4 kbp)/Cos-7, CHO	1	Low (~20%)	Voronkova et al. (2007)
Sf158/baculovirus	JCPyV (VP1)	Disassembly/reassembly	3/1	β-Galactosidase (4.5 kbp)/Cos-7	0.25	20%	Goldmann et al. (1999)
Sf9/baculovirus	HPV6, HPV11, HPV16 (L1, L1/L2)	Chemical conjugation	25/10	β-Galactosidase (8.9 kbp)/various cell lines	100	1% 20% (with adenovirus)	Muller et al. (1995)
Sf9/baculovirus	MPyV (VP1)	Direct mixing	30/1	β-Galactosidase, GFP/Cos-7	20	Few cells (β-gal) 0.1%–0.5% (GFP)	Krauzewicz et al. (2000)
Sf21/baculovirus	MPtV (VP1)	Direct mixing	10–30/2	EGFP (4.7 kbp)/293 and Cos-1	1	0.03%	Tegerstedt et al. (2003)
Sf9/baculovirus	HPV16 (L1/L2)	Disassembly/reassembly	1/2	β-Galactosidase (6.8 kbp)/several cell lines	0.25	0.00425%	Kawana et al. (1998)

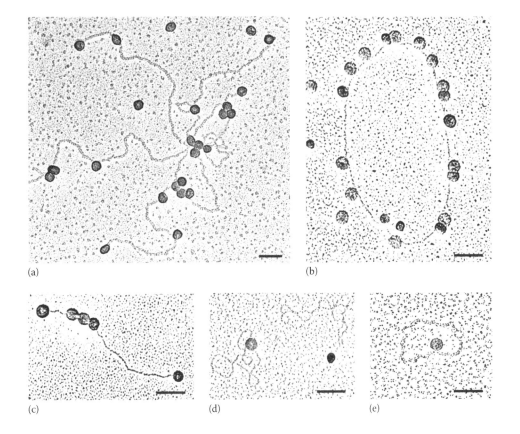

(a)

(b)

(c)

(d)

(e)

FIGURE 20.3 Electron micrographs of MPyV VLP interactions with linearized (panels a through c) or circular (panels d and e) bacterial plasmids. Aqueous spreading technique, bar = 100 nm. Electron microscopy, Jitka Štokrová.

2.5 kbp. The pentameric capsomeres exhibited high binding affinity for both linearized and circular DNAs, but the interaction did not lead to the protection of target DNA after nuclease treatment.

The procedure is widely used for the preparation of transduction-competent PyV-based PLPs (Forstová et al. 1995; Soeda et al. 1998; Henke et al. 2000; Krauzewicz et al. 2000b; Stubenrauch et al. 2001; Touzé et al. 2001), but HPV PLPs were also successfully prepared by this method (Combita et al. 2001). Empty capsids package DNA most efficiently when complexes are formed at a molar ratio of 5:1 for capsids/DNA (Aposhian et al. 1975), and for maximal transduction efficiency, the optimum loading ratio seems to be the same, despite the fact that a significant portion of DNA associated with pseudocapsids appears not to be packaged when observed by electron microscopy (Krauzewicz et al. 2000b). Oligonucleotides are incorporated into the VLPs at a higher ratio than plasmid DNA (72 oligonucleotides per VLP), and the process is pH dependent; the highest oligonucleotide encapsidation capacity occurred at pH 5 (Braun et al. 1999).

To improve the packaging capacity of VLPs, the polycationic amine, poly-L-lysine, was examined as a DNA-condensing agent. The addition of poly-L-lysine to the reaction before osmotic shock increased the size of nuclease-protected plasmid DNA (7.2 kbp) and enhanced transient, but not stable, expression of genes carried into cells by MPyV VP1 pseudocapsids (Soeda et al. 1998). Interestingly, the extent of protection did not correlate with transduction efficiency, and unprotected VLP/DNA complexes could sustain a high level of transduction in vitro and in vivo (Soeda et al. 1998).

20.4.2.1.2.2 Disassembly/Reassembly Procedure

Nucleic acids and other cargos can be encapsulated into VLPs during the formation of VLPs from capsomeres. The procedure was pioneered by Salunke et al. (1986). Depending on the conditions, several forms of viral assemblies can arise, and Table 20.3 gives an overview of the conditions used to form VLPs and/or other structures from capsomeres. The loading of nucleic acid is performed by its addition to capsomeres, which have been either purified from a bacterial system or prepared by a disassembly reaction from VLPs. The most widely used procedures for disassembly and reassembly are described in Table 20.4. Generally, the disassembly of VLPs requires a reducing reagent (DTT, β-mercaptoethanol) and chelating agent (ethylenediaminetetraacetic acid [EDTA], ethylene glycol-bis(2-aminoethylether)-N,N,N′,N′-tetraacetic acid [EGTA]) and a mildly basic pH (pH = 8.5). In our experience, the aggregation of material can be a complicating factor during disassembly and reassembly (see also Notes in Table 20.4), and the aggregate material should be removed either during the purification of capsomeres on column chromatography or during the centrifugation step, before the reassembly reaction is initiated. For reassembly, the capsomeres are usually dialyzed against buffers with an acidic to neutral pH, higher ionic strength, and, in the case of PyVs, the addition of calcium. Interestingly, the reassembly buffer consisting of 2 M $(NH_4)_2SO_4$ seems to give the best results even without the calcium ions (Table 20.3). Calcium ions are also not required

for capsid assembly when the reconstituted bacterial chaperones are used in the reaction (Chromy et al. 2003).

It has been shown that the presence of nucleic acids can enhance the assembly reaction of SV40 PLPs or MPyV PLPs (Braun et al. 1999; Mukherjee et al. 2010) and DNA can mediate the formation of 40 nm SV40 particles (Tsukamoto et al. 2007). The latter system requires DNA longer than 250 bp at a minimum concentration of 5 mg/L, and for plasmid DNA (4729 bp), optimal concentrations appear to be 10–20 mg/L (3.1–6.1 nM), corresponding to a molar ratio of 1:1 for capsids/DNA. Similar to osmotic shock procedure, nuclease treatment of these SV40 PLPs converted plasmid DNA to fragments of less than 2 kbp, suggesting that the naked DNA was only partially packaged into VP1 VLPs (Tsukamoto et al. 2007). Other reports, however, showed that the size limit for encapsidation of DNA by this method corresponds to the size of the parental virus genome and was 8 kbp for HPV16 PLPs (Touzé and Coursaget 1998). Occasionally, VLPs were reported to accommodate and transduce, albeit with low efficiency, plasmids of a bigger size than the viral genome, as shown for BKPyV (Touzé et al. 2001).

In theory, packaging efficiency can be enhanced by the addition of DNA-condensing agent before the assembly of capsomeres is initiated. This strategy has been used for the formation of MPyV PLPs (Henke et al. 2000). Here, the authors tried to condense DNA with histone sulfate or activated dendrimers. Dendrimers are able to condense DNA by the strong interactions of their positively charged amino groups with the negatively charged phosphate groups of the DNA. Although dendrimers formed aggregates that could be used for in vitro assembly reaction, electron microscopy revealed incomplete capsid assembly, and transduction experiments showed no expression of the reporter gene (Henke et al. 2000). DNA with histone sulfate formed large aggregates (500–6200 nm) that could not be used for assembly reaction. In contrast, compact SV40 PLPs were formed when reporter plasmid was in vitro associated with purified histones and used for encapsidation (Enomoto et al. 2011). The plasmid was fully protected from nuclease treatment, but the maximum length was not longer than the size of the SV40 genome.

Significantly longer target plasmid DNA can be encapsidated into SV40 PLPs by the procedure developed by the A. Oppenheim group (Sandalon et al. 1997). This method is based on the assumption that nuclear factors (e.g., chaperones) may increase the efficiency of encapsidation. The VLPs are produced by a baculovirus expression system in S. frugiperda cells, and disassembly, heterologous nucleic acid addition, and reassembly take place in the nuclease-treated nuclear extracts of these cells without VLP purification. The procedure was shown to yield 10 times more transduction-competent PLPs than the disassembly/reassembly method performed with purified VLPs. The improved protocol permitted the packaging of at least 17 kbp plasmid DNA without the requirement for any viral sequences (Kimchi-Sarfaty et al. 2003). It has been suggested (Mukherjee et al. 2007) that the absence of histones provides space and facilitates the packaging of significantly larger plasmids. The optimal VP1 capsid/DNA molar ratio was set at 1:1, which corresponds to a VP1/DNA ratio of 5:1 on a weight basis (for 5 kbp DNA molecule). The electron microscopy examination of the purified VP1/2/3-PLPs showed well-assembled particles of uniform size (45 nm) and shape. The PLPs were found to have an infectivity ratio 1:3 × 10^5 (1000-fold less than the SV40 virion) (Mukherjee et al. 2007). The same study indicated the minimal concentration of VP1 protein in nuclear extract to be 1 mg/mL (~5%–10% of nuclear extract stock). It is interesting to note that this method of PLP preparation uses a relatively large quantity of DNA, corresponding to 1 ng/cell during transduction (compared to 1–20 pg/cell used in other protocols; see Table 20.7).

20.4.2.1.2.3 Chemical Coupling Procedure It has been shown that targeted delivery of DNA can be enhanced by the coupling of DNA directly to the adenoviral capsids, which significantly reduces the number of viral particles used but still maintains high levels of gene expression (Wagner et al. 1992). A method where the viral capsid was chemically conjugated to poly(L-lysine) and bound ionically to DNA molecules (Cristiano et al. 1993) was used by Muller et al. (1995) to physically link the intact PV VLPs with the plasmid (8.9 kbp), harboring the reporter gene to study the uptake of particles into different cell lines. They showed that the VLP–DNA complexes were able to bind and penetrate into a broad range of cells, but for reporter gene expression, the coinfection with reporter constructed from a replication-defective adenovirus (facilitates lysis of lysosome membranes) was required. This suggests that (1) tropism of infection by different PV is controlled by events downstream of initial binding and uptake and that (2) this approach can be used for successful nucleic acid delivery if lysosome escape would be promoted (e.g., by arginine-rich cell-penetrating peptides) (El-Sayed et al. 2009). Another group showed even more promising results with a PCR-generated reporter gene (1.8 kbp) driven by a human cytomegalovirus promoter covalently cross-linked to the outside of the HPV VLPs. They suggested that the coupling ratio of the reporter construct to VLP might be crucial for success and reported that up to 40% of target cells can be consistently infected with these PLPs. Further, they demonstrated that L2 inclusion into the VLP dramatically improved infection efficiency for HPV16 and HPV11 PLPs (Yeager et al. 2000).

20.4.2.1.2.4 Direct Mixing Procedure Chemical coupling to the VLP exterior has clearly shown that encapsidation into the particle is unnecessary for successful delivery of nucleic acids into the cell. Some reports even suggested that physical linkage to VLPs is not needed and that VLPs can be mixed with DNA without any further treatment. Touzé et al. (2001), comparing osmotic shock, disassembly/reassembly, and direct mixing methods for BKPyV PLP preparation, found the later procedure to be the most efficient method in transduction assays. The same results were obtained for different types of HPV VLPs and indicated that a direct mixing procedure leads to the highest level of protection against nuclease (45% for osmotic shock, 31% for disassembly/reassembly, and 55% for a direct mixing

method in HPV16 PLP preparations) (Combita et al. 2001). Similarly, Krauzewicz et al. (2000a) showed that this method resulted in sustained ex vivo and in vivo transfers of a reporter gene by MPyV VLPs. Other studies, however, reported conflicting results. Clark et al. (2001) showed that the transfection efficiency of MPyV VLP–DNA complexes appeared to be the same or lower than that of DNA alone. Other studies also indicated that mixing DNA and VLPs results in poor short-term in vitro transfection (Touzé and Coursaget 1998; Ou et al. 1999). The reason could be that a significant proportion of DNA remains free in the preparations (Clark et al. 2001) and unpackaged DNA might trigger cellular defense mechanisms leading to loss or silencing of the transgenes and consequent inefficiency of the vector in transduction assays (Bishop et al. 2006).

20.4.2.2 Gene Transfer

Gene delivery vehicles should efficiently penetrate the cell and facilitate gene expression in the target cell. Despite the restricted tissue tropism of some PyV as well as PV, both viruses use for the primary cell bind widespread receptors' moieties (sialic acid and HS, respectively) and therefore enter many different cell lines. Muller et al. (1995) found that 15 out of 16 different cell lines that originated from different tissues and species were able to take up the HPV VLP–reporter plasmid complexes. Moreover, constructs composed of HPV16 L1 and L2 proteins delivered the DNA into cells as efficiently as the VLP–reporter plasmid complex made of L1 alone, suggesting that the L2 protein is not necessary for PV binding and penetration. In another study, the HPV33 L1 VLPs were also found to bind to all of the cell lines tested, including insect Sf9 cells (Volpers et al. 1995). Conversely, L1 VLPs derived from the nine different HPV types (16, 18, 31, 33, 39, 45, 58, 59, and 68) were able to transfer genes into Cos-7 cells, thus showing that most HPV types can be used for gene transfer (Combita et al. 2001). The binding and uptake of PyV VLPs by various cell lines was systematically investigated only for murine pneumotropic virus (MPtV) VLPs (Tegerstedt et al. 2003), but model PyVs (MPyV, SV40) are known to enter and efficiently transform heterologous cells in culture derived from various nonpermissive hosts (Pipas 2009). Moreover, numerous studies done with PyV PLPs and PsVs have confirmed that these are able to enter a variety of cell lines and even transduce reporter genes (e.g., see Table 20.7). Compared to PVs, which require 4–12 h for cellular uptake (Volpers et al. 1995), the internalization of PyVs and VLPs is rapid (Pelkmans et al. 2001; Richterová et al. 2001), and within 40 min approximately 50% of all membrane-bound capsids are internalized (Tegerstedt et al. 2003).

All these results indicate that polyoma- and papilloma-derived VLPs are efficient for penetration into a vast variety of cells. Although this can be good for application where the widespread distribution of a vector is needed, the selectivity is usually required for therapeutic purposes. To increase therapeutic potential, modification of capsid proteins can be performed to abrogate type-specific epitopes and induce retargeting of VLPs to specific cells (see Section 20.4.1).

Interaction of a virus capsid with the cell surface receptor is often an important, but not sole, determinant of success in delivering genetic information for gene expression. There is emerging evidence that events downstream of cell surface interactions such as endocytosis, virus-induced signaling, intracellular trafficking, and transcriptional regulators may also significantly contribute to the expression of genes carried by a virus-based vector. Numerous gene transduction experiments that have been performed with PyV and PV vectors over the last 20 years have helped to identify some of these factors. The experiments that substantially contributed to identify these factors or showed potential for further applications are summarized this section.

The pioneering work of Forstová et al. (1995) proved that PLPs made from MPyV VP1 pseudocapsids loaded with DNA by osmotic shock are able to stably transduce genes for expression. They used an interesting method to assess this delivery system: the linear fragment (1.6 kbp) of the PyV middle T antigen gene, the principal oncogene of MPyV, was transduced by PLPs into the immortalized rat-2 cells, and the transformed foci were observed 3 weeks later. Although the PLPs' loading with DNA was low (only 5%–10% of input DNA was protected from nuclease treatment), the efficiency of transduction was higher than that seen in the calcium phosphate or liposome transfection method when the same amount of input DNA was used. In the complementary experiment, the whole plasmid DNA containing either the reporter gene chloramphenicol acetyltransferase (CAT) (6.2 kbp) or the p43 gene was used for transduction into human liver CCL13 cells and into human embryonic lung fibroblast cells. The loading capacity of VP1 pseudocapsids was higher for the circular plasmid (30% of input DNA was protected against nucleases after osmotic shock) and the high expression of the reporter gene as well as the p43 gene was observed after 3 days. For the CAT gene, the transduction efficiency was clearly better than with the control lipofectin transfection. Follow-up studies (Krauzewicz et al. 2000a) demonstrated much lower transduction efficiency into Cos-7 (10 times lower than the calcium phosphate method) with the same MPyV VP1 pseudocapsids, but loaded with plasmid by the direct mixing method. These PLPs were, however, able to sustain ex vivo transfer into nondividing rabbit corneal explants and stable expression (after administration of 10^{13} PLPs per animal) of the reporter gene in several tissues of nude and immunocompetent mice. Complementary study (Heidari et al. 2000) examined the persistence and tissue distribution of PyV DNA in normal and immunodeficient mice inoculated in the form of MPyV PLPs, through plasmid DNA or as a natural virus. Mice inoculated with PLPs were found to carry 10–50-fold and 50–100-fold higher copy numbers than mice inoculated with plasmid alone in immunodeficient and normal mice, respectively. The number of DNA copies found in mice inoculated with PLPs was similar to that found in mice infected with PyV, but normal mice were found be more resistant to PLP/virus treatment (DNA detected in 7 out of 11 mice) than immunocompromised animals (DNA detected in 14 out of 15 animals). Importantly, when present in animals, DNA was widely distributed to almost all tissues up to 6 months p.i. The result confirmed previous results (Krauzewicz et al. 2000a) and indicated that the immune system may influence the persistence of viral DNA introduced by pseudocapsids but does not totally eliminate it.

In vivo gene transfer was also performed with MPtV PLPs prepared by direct mixing method with a small group (two animals) of normal mice. Three weeks after intraperitoneal inoculation of pure reporter plasmid (pEGFP-C1) or PLPs loaded with plasmid, the PCR detected pEGFP-C1 DNA in many organs of mice inoculated with PLPs, but not in mice inoculated with plasmid alone (Tegerstedt et al. 2003).

In general, these studies indicated that PLP-mediated DNA delivery favored long-term expression, whereas the initial expression was rather low. This discrepancy was partly explained by the fact that DNA has been readily integrated in host genetic information and by the observation that the DNA loading method (osmotic shock) yielded large aggregates of VLPs and DNA, which could reduce the effectiveness of transfer by sequestering the material. The efficiency of gene transfer correlates with the level of nuclease protection of DNA and is apparently an important factor during gene transfer (Combita et al. 2001). As demonstrated by Enomoto et al. (2011), naked DNA protruding from VLPs might inhibit cell attachment, whereas nucleosome arrangement enhances compact particle formation and cellular uptake. Unpackaged DNA might trigger cellular defense mechanisms leading to silencing of the transgenes, and consequent inefficiency of the vector in transduction assays (Bishop et al. 2006) and hyperacetylation of histones, as found in native SV40 virions (Chestier and Yaniv 1979; Coca-Prados et al. 1980), could significantly enhance reporter gene expression (Enomoto et al. 2011).

In vivo techniques for PLP and PsV production usually yield particles that completely protect encapsidated DNA from nuclease action. Studies with JCPyV PLPs that can be made in vivo in *E. coli* even indicate that nucleosomal arrangement, which is absent in this system, is not crucial for successful gene transfer. In fact, one of the few functional studies of the gene transfer was performed with JCPyV PLPs prepared by this system (Chen et al. 2010). In this study, the expression of plasmids harboring either the reporter gene (GFP) or a gene encoding a prodrug-converting enzyme (thymidine kinase [TK]) was packaged into the self-assembled VLPs. Purified PLPs that contained the full-length plasmid were used for the transduction of human carcinoma cells (COLO-320 HSR) since JCPyV seems to have a specific tropism for colon epithelial cells and is detected in carcinoma lesions (Coelho et al. 2010). GFP gene expression was achieved in 90% of cells. The same PLPs were selectively transduced in vivo to tumor nodules in nude mice bearing human COLO 320 HSR tumors, when intravenously injected into the mouse. Treatment with PLPs containing the TK gene resulted in total tumor growth inhibition after intraperitoneal injection of ganciclovir for 3 weeks, whereas there was no inhibition in the negative control groups (PLP-TK without ganciclovir or PBS with or without ganciclovir). The work demonstrated that JCPyV VP1 PLPs can efficiently protect genetic information and deliver it for gene expression in nude mice. The reason for high selectivity for tumor tissues is interesting, but not completely clear: the authors have suggested that human tissues may be more susceptible to JCPyV PLP infection than mouse cells.

The inconsistencies in gene transfer efficiencies during different experimental settings were also explained by the fact that two modes of entry of viral particles into the cells exist: the productive pathway, which leads to gene expression, and the nonproductive, default pathway for cargo, which enters the cell in an unspecific manner, for example, by phagocytosis. Differences in surface characteristics induced by DNA packaging or absence of VP2 and VP3 minor proteins were suggested to affect the interaction with the cell in a way that VLPs would be significantly less effective for DNA delivery than natural virions (Krauzewicz et al. 2000b).

For PV PsVs, numerous studies have clearly shown that PLPs consisting of L1 alone are infectious, but L2 enhances infectivity (Unckell et al. 1997). The same has been shown to be true for some PyV-derived PsVs. Specifically, a recent study (Schowalter and Buck 2013) indicates that infectivity of the BKPyV VP1–only PsVs is dramatically lower compared to the VP1 + VP2 + VP3 PsVs on all tested cell lines. In contrast, the effect of VP2 on MCPyV pseudovirus transduction efficiency differs dramatically between cell lines. Thus, infectious entry of the BKPyV pseudovirus appears to differ from the MCPyV pseudovirus with regard to its dependence on minor capsid proteins, indicating that differences can exist between virus types in general. Moreover, the study showed that not only the sole presence but also the ratio of minor proteins can dramatically accelerate PsV *infectivity*, which was never zero even for VP1-only particles. The same conclusion has been drawn from experiments performed with in vitro reconstituted SV40 QVs or PsVs. Although the minor capsid proteins VP2/3 strongly facilitated gene transducing activity, the gene expression level after transduction with PsVs was only 2% of that achieved by SV40 virions. The authors speculated that the slightly lower content of VP2/3 in VP1/2/3 PsVs affected postinternalization processes, such as virion disassembly, nuclear translocation, and DNA replication (Enomoto et al. 2011). This may explain why the repeated production of PsV stocks is more consistent for PV PsV than PyV PsVs. L1 and L2 genes are usually cotransfected on one plasmid, whereas for PyV PsVs, at least one capsid protein is usually encoded on a separate plasmid. The actual cotransfection efficiency then determines the yield of fully *infectious* PsV (Španielová et al. 2014). As demonstrated by the work of Nakanishi et al. (2008), the coexpression of all capsid proteins from one plasmid may be crucial for obtaining PsV stocks that can transduce permissive cells with efficiencies reaching 100% and similar to that of the virion. The same study also showed that capsid exchange could significantly alter the cell specificity of gene transfer.

The PsVs generated in mammalian cells, as well as in yeast systems, serve predominantly as a diagnostic tool. PsVs are used for the detection of neutralization antibodies in vaccinated or infected individuals. These neutralization assays depend on the ability of antibodies in a test serum to prevent infection of cells by a virus. Since infectious HPV virions are not readily available, the PsV-based technology is used to generate infectious particles (Buck et al. 2005a). PsVs containing an easily detectable reporter gene (alkaline phosphatase, luciferase, GFP) are purified on density gradients and are used to infect a detection cell line. To increase sensitivity, the detection cell line should express SV40 large T antigen to support replication of the reporter vector, which harbors the SV40 ori. The presence of antibodies to HPV in a test serum blocks infection of PsVs and reduces the signal from the reporter gene. These assays are more type specific than the enzyme-linked immunosorbent assays (ELISAs). This was demonstrated for HPV subtypes (Pastrana et al. 2004), as well as BKPyV subtypes, when serotypes not recognized by VLP-based ELISA could be discovered by neutralization assay (Pastrana et al. 2013). Neutralization can be used for characterizing candidate vaccines, quantification of seroresponsiveness, and diagnosis of viral subtypes.

For safety reasons, PsVs produced in mammalian cells have limited potential as vehicles for therapeutic genes. Nevertheless, some studies exist. One study using a gutless recombinant SV40 (rSV40) vector examined the feasibility of SV40 PsVs for gene therapy in cystic fibrosis (CF) (Mueller et al. 2010). The genetic defect in CF is caused by mutations in a cell membrane chloride channel, the cystic fibrosis transmembrane conductance regulator (CFTR). The human CFTR gene was cloned into a rSV40 vector under the control of the SV40 early promoter and was used as a target plasmid for production of PsVs in Cos-7. These cells carry an integrated copy of the wild-type SV40 genome that is defective at the ori. The authors noted that the SV40 capsid genes, under the control of the SV40 late promoter, are not expressed constitutively in Cos-7 cells, but that the presence of a replicating rSV40 vector that includes the late promoter is sufficient to activate Cos-7 transcription of the capsid genes in *trans* and the vector could be packaged. By this procedure, they were able to obtain high-titer stocks (approximately 10^9 infectious units [IU]/mL) of SV40 PsVs. They showed that rSV40–CFTR was able to induce the expression of CFTR protein, which localized to the plasma membrane and restored channel function to CFTR-deficient cells. When matched groups of 5 Cftr$^{-/-}$ mice were treated with 4.0×10^7 particles of either the rSV40–CFTR vector or the irrelevant rSV40–BUGT negative control via intratracheal injection, delivering rSV40–CFTR to the lungs of Cftr$^{-/-}$ mice resulted in a reduction of the pathology associated with intratracheal *Pseudomonas aeruginosa* infection. CFTR gene was stably expressed at least around 9 weeks post delivery and the authors reported that animals were free from virus revertants and no adverse effects or toxicity was evident.

Another study demonstrated targeting of SV40 PLPs to mouse liver by the use of the hydrodynamic tail-vein injection (Arad et al. 2005). The PLPs were prepared either in vivo or in vitro in nuclear extracts of Sf9 cells and contained the luciferase reporter gene. After injection, the luciferase activity in the mouse liver was monitored with a light detection camera. In vivo transduction of hepatocytes was efficient and persistent (lasted at least 107 days). Importantly, SV40-specific antibodies were observed in mice administered with SV40 PLPs, and antibody response was dose related. The authors observed a decline in luciferase activity during long-term observation and speculated that the immune response against the vector eliminates some transgene-expressing cells early after transduction.

In spite of the widely accepted notion that minor proteins are important for high PLP transfer, the approach of producing SV40 PLPs by the disassembly/reassembly method in the nuclear extract of Sf9 cells reproducibly achieved high efficiencies of transduction (close to 100% cells transduced) in a wide variety of cells without VP2/3 proteins (Kimchi-Sarfaty and Gottesman 2004; Kimchi-Sarfaty et al. 2004). Originally, minor proteins were included in the system (Oppenheim et al. 1986; Sandalon and Oppenheim 1997; Rund et al. 1998; Kimchi-Sarfaty et al. 2002) and suggested to improve transduction substantially (Sandalon et al. 1997). At the same time, the PLPs made solely from VP1 containing longer target DNAs (17 kbp) were observed to be larger (55 nm) than normal VLPs or virions (45 nm). The study speculated that the absence of minor proteins increased the space for encapsidation and decreased the rigidity of the capsid (Kimchi-Sarfaty et al. 2003). This corresponded with the previous observation that VP2/3 containing PLPs showed better protection against nuclease treatment (Sandalon et al. 1997).

The system with or without the minor proteins was, however, successfully used for gene transfer. In particular, SV40 PLPs were shown to be very efficient for gene delivery into human hematopoietic cells, and ex vivo transduction of hematopoietic stem cells with SV40 PLP permitted expression of, for example, multidrug resistance 1 (MDR1) gene or β-globin in several studies (Rund et al. 1998; Dalyot-Herman et al. 1999; Kimchi-Sarfaty et al. 2002; Kimchi-Sarfaty et al. 2004).

The potential of SV40 PLPs in anticancer therapy was also examined (Kimchi-Sarfaty et al. 2006). Using these SV40 PLPs, a truncated *Pseudomonas* exotoxin gene (PE38) was delivered into various human cells and dramatically reduced their viability, thus showing that the toxin has a similar toxicity in cells of various origins. Interestingly, this nontargeted route of PE38 delivery was found to be effective in the treatment of human adenocarcinomas growing in nude mice, when injected either intratumorally or systemically. Controls, including daily delivery of the same DNA in the naked form, as well as empty capsid proteins or GFP encapsidated using the same PsV system, all failed to inhibit tumor growth. The authors noted that treatment that started later, on larger tumors, was less effective than treatment that started only 2 days after tumor inoculation; thus, the effectiveness of the therapy is related to the ratio of PsVs to the number of tumor cells. Treatment also caused no abnormalities in mice. Three of the PE38-treated mice stayed tumor-free

after a year and a half. The authors speculate that the SV40 PsVs circulate in the blood and are taken up selectively by the tumors through the enhanced permeability and retention effect (Kimchi-Sarfaty et al. 2006).

20.4.2.3 Gene Silencing

RNAi is a sequence-specific, naturally occurring gene-silencing mechanism. A number of approaches have been developed in recent years that allow for exploitation of this process principally through the use either of (1) synthetic siRNAs or duplex RNA oligonucleotides or of (2) plasmid-expressed short hairpin RNAs (shRNAs) that must be endogenously processed to generate a siRNA. The in vivo use of RNAi therapy is limited by obstacles related to effective delivery into the cell, and PyV and PV VLPs may serve as efficient RNAi delivery vehicles.

20.4.2.3.1 RNA Transfer

SV40 PLPs produced in vitro in nuclear extracts of Sf9 cells were shown to transduce not only DNA but also siRNA (Kimchi-Sarfaty et al. 2005). The same system was used to deliver plasmid-expressed shRNAs and synthetic siRNAs into human cells. After transduction with SV40 PLPs loaded with siRNA corresponding to the GFP gene, the complete silencing of the GFP gene was observed in HeLa cells stably expressing GFP. The delivery of siRNA by SV40 PLPs was more efficient than the Lipofectamine method.

Packaging of short ssRNA (75–800 nucleotides) into SV40 PLPs was also performed by the disassembly/reassembly method (Kler et al. 2012). The encapsidation of RNA seemed to be very fast and the method yielded small particles (diameters of 24.5 nm) made from one molecule of 524 nucleotides RNA and 12 pentamers. Unfortunately, neither the packaging of longer ssRNA molecules nor the RNA transfer into cells was examined in the study. If efficient, the system can be an attractive means of ssRNA transduction for special applications.

20.4.2.3.2 Short Hairpin RNA Delivery

The utilization of PyV and PV VLPs for gene therapy application may be restricted by their packaging capacity, but they can be an efficient means of delivering RNAi effectors, such as DNA encoding the shRNA sequences.

The JCPyV PLPs were experimentally used for silencing of the IL-10 gene, which is overexpressed in patients with systemic lupus erythematosus. VP1 VLPs produced in yeasts were loaded (by osmotic shock) with a PCR fragment containing a sequence of IL-10 shRNA. After transduction into a murine macrophage cell line, IL-10 shRNA was found to reduce IL-10 expression by 85%–89%, as compared with unloaded VLPs. In BALB/c mice, IL-10 shRNA abolished 95% of IL-10 secretion (Chou et al. 2010).

More recently, JCPyV VP1–PLPs were successfully used for the silencing of BKPyV infection. PLPs produced in vivo in *E. coli* were used as a delivery system to transfer plasmid-encoding shRNA for BKPyV large T antigen into BKPyV-infected human kidney cells. PLP-mediated transduction with the plasmid decreased the proportions of BKPyV large T antigen and VP1-expressing cells by 73% and 82%, respectively (Lin et al. 2014).

The potential of shRNA transduction mediated by PV VLPs was explored for the inhibition of cervical cancer cell growth. HPV31 PLPs were loaded by the disassembly/reassembly method with plasmids encoding the shRNA for two main oncoproteins of high-risk HPV, E6, and E7 (Bousarghin et al. 2009). The silencing of both genes was achieved after transduction in HPV-positive cells lines. E6 silencing resulted in the accumulation of cellular p53 and reduced cell viability. Cell death was observed when E7 expression was suppressed. In mice, where murine TC1 cells expressing HPV16 E6 and E7 oncogenes induced fast-growing tumors, HPV PLPs coding only for an E7 shRNA were sufficient for dramatic inhibition of tumor growth.

20.4.2.3.3 Oligonucleotide Transfer

Antisense oligonucleotides are specific drugs to inhibit gene expression at the transcriptional level. MPyV VP1 VLPs were investigated as a means of oligonucleotide delivery system into the cells. Capsoids formed after expression in *E. coli* were loaded with the fluorescently labeled oligonucleotides directed against the N-methyl-D-aspartate (NMDA) receptor by an osmotic shock. These PLPs were used for the transduction of mouse fibroblasts that overexpress the NMDA receptor and allow a functional antisense oligonucleotide test system based on excitotoxicity (cell death). In comparison with several other delivery methods, MPyV VP1 VLPs showed very low uptake and only a moderate effect in a functional antisense oligonucleotide test (Weyermann et al. 2004).

In contrast, JCPyV VP1 VLPs showed better potential as an oligonucleotide delivery vehicle for human neurological disorders. In the model system of SV40-transformed human fetal glial cells, VLPs generated in yeast were used to package and deliver an antisense oligodeoxynucleotide against an SV40 large T antigen, the main SV40 oncoprotein. The oligonucleotide transfer resulted in the inhibition of large T antigen expression and subsequently led to cell death. As expected from JCPyV tropism, VLPs were able to deliver oligonucleotides into human astrocytoma, neuroblastoma, and glioblastoma cells with high efficiency, and in vivo delivery of oligonucleotides into a human neuroblastoma tumor nodule by VLP was also demonstrated (Wang et al. 2004).

An SV40 PLP system was also explored for the delivery of peptide nucleic acid (PNA) into the cells (Macadangdang et al. 2011). PNA is a synthetic DNA analog in which the sugar-phosphate backbone is replaced with a polyamide backbone. PNA can bind with high affinity and sequence specificity to complementary nucleic acid sequences and can be used to suppress gene expression (Corradini et al. 2007). PNA molecules designed to bind in the region of the major transcription initiation site of the MDR1 gene were loaded in vitro to SV40 VP1 VLPs in the nuclear extract of Sf9 cells. These PLPs could effectively decrease MDR1 mRNA levels in drug-resistant KB-8-5 cells. Compared to other delivery systems, SV40 VP1 PLP transduction times were very short (2.5 h versus 48–96 h), and treatment with micromolar concentrations of antisense PNA yielded almost a 30% reduction in MDR1 mRNA levels. Importantly, it also increased the sensitivity of cells to the chemotherapeutic agent Adriamycin. The study concluded that the combination of PNA with the SV40-based delivery system is a method for suppressing a gene of interest that could be broadly applied to numerous targets (Macadangdang et al. 2011).

20.4.2.4 Concluding Remarks

Numerous studies have used the polyoma- and papilloma-derived VLPs as gene delivery vehicles with various successes but have shown their potential for gene transfer applications. Due to differences in experimental settings, these studies are impossible to compare directly even if the same type of PLPs was used. The potential pitfalls were, however, identified and can be possibly solved by exploiting or, on the contrary, manipulating the inherent biological features of these viruses.

Two factors emerged as very important aspects for successful gene delivery: (1) efficient packaging of nucleic acid with no DNA exposed to the outside space and (2) the *virus-like* composition of particles with the correct ratio of minor proteins. The latter factor seems to be more important for some virus types (e.g., BKPyV) than for others (e.g., JCPyV). Nonviral production of such a capsid may be more difficult for PyV VLPs, which have two minor proteins, than for PV VLPs. The minor proteins probably facilitate endosome/lysosome escape, and inclusion of other penetrating peptides in the system can accelerate gene expression. Similarly, nucleosomal arrangement of genetic information with hyperacetylated histones can be beneficial but probably not necessary. The main challenges in gene (and some other cargos) delivery are, however, the unspecific binding and uptake of these particles and their immunogenicity. Both challenges can at least in part be overcome by the modification of capsid structures by chemical or genetic means or by selecting the virus with specific tropism for VLP production. The repeated administration of these VLPs for some applications will probably require sequential use of VLPs from different virus types. Luckily, PyVs and PVs offer a wide selection of materials for further exploration.

Some reports of gene transfer performed in animals have clearly shown long-term expression of transgenes introduced by PyV VLPs—vehicles. The immune system was found to partly restrict the persistence of a transgene but not totally eliminate it. The system was found to be suitable for ex vivo transduction of hematopoietic cells. Its potential for anticancer gene therapies has been also explored, but experiments performed with xenografted immunocompromised animals might be misleading in regard to actual efficacy and need further verification. The utilization of these VLPs for classical gene therapy applications is therefore still far from reach, but special applications, such a gene silencing, offer another interesting possibility for their use.

20.4.3 PROTEIN AND PEPTIDE DELIVERY

The use of many potent proteinaceous pharmaceutical agents might be complicated by their instability, side effects, and their need to be delivered intracellularly to exert their therapeutic action. VLPs of both PyVs and PVs are able to serve as nanocontainers for foreign peptides: the exterior can be used for labeling or targeting and the interior space for encapsulation of biologically active protein or peptide cargos. Encapsidation into a VLP has the advantage of hiding and protecting the proteins from external proteases and from recognition by the immune system. Peptide loading into the inner space of PV and PyV VLPs is usually achieved either by genetic fusion with a major capsid protein or by the utilization of minor protein in a single- or two-component attachment system (see Section 20.4.1 for description).

Although VLP-based vaccines do not usually require intracellular delivery of an antigen, its encapsulation may be important for the delivery of cytotoxic T-lymphocyte (CTL) epitopes. This strategy has been explored for peptide fusion to major capsid protein L1, since PV VLPs are widely used as transport vehicles of various antigens, including oncoproteins, for anticancer immunotherapy. An extensive study was conducted where the C-terminal 34 aa of HPV16 L1 protein were replaced with various segments of oncoprotein E7. Peptides up to 60 aa could be fused to the truncated L1 without disrupting the assembly of VLPs (Müller et al. 1997). Besides E7, many other epitopes or peptides have been subsequently added to the truncated C-terminal of L1 protein, for example, human immunodeficiency virus (HIV)-IIIB CTL epitope of gp160, gp120, reverse transcriptase, and protein Nef (Peng et al. 1998; Liu et al. 2000). Even though the C-terminal part of the PV L1 protein is preferable for peptide fusion, chimeric particles with GFP added to the C- and N-terminal part of the L1 protein prove that both ends are suitable for protein fusion without disrupting the assembly of VLPs, but the actual location of GFP molecules in VLPs has not been analyzed (Windram et al. 2008).

Genetic fusion with the major capsid protein of MPyV has been also explored for encapsidation of protein cargo—the GFP molecule (Günther et al. 2001). According to the theoretical calculation, 360 globular proteins with an average size of up to 17 kDa can be encapsidated into VP1 particles. PyV VP1 protein seems to be less flexible to peptide fusions than the L1 protein, and therefore the WW domain-based two-component system (see Section 20.4.1.2) was used for protein encapsidation. The WW domain was fused to the N-terminal part of MPyV VP1, and 260 polyproline-tagged GFP molecules were reported to have become encapsidated. The authors, however, did not attempt to verify the position of GFP in the particle but showed that these particles were able to deliver cargos to mouse fibroblasts NIH 3T3 (Günther et al. 2001).

An alternative approach to peptide delivery is conjugation of a foreign protein to a minor protein. The resulting fusion protein uses minor proteins as an anchor for noncovalent association with VP1 pentamers and, as such, should be hidden inside the particle. Even though the theoretical maximum amount of fused protein per VLP is 72, the actual number is usually lower. PV HPV16 L2 protein is mostly applied for the transfer of an immunogenic epitope of E7 oncoprotein. The C-terminal part of the L2 protein was successfully fused either with parts of E7-derived peptides (Rudolf et al. 2001; Wakabayashi et al. 2002) or with the whole oncoprotein (Greenstone et al. 1998). Another oncoprotein, E2, was also conjugated to L2 (Davidson 2003), and even together with E7 (Qian et al. 2006). The removal of the central amino acids of the L2 protein (70–390) allowed the simultaneous incorporation of three oncoproteins, E1, E2, and E7. Although this peptide sequence reached approximately 130 kDa, it did not disrupt VLP assembly and produced chimeric particles that were highly immunogenic (Tobery et al. 2003). Windram et al. (2008) proved that not only C-terminal but also N-terminal fusion of GFP to L2 protein led to chimeric VLPs without exerting an influence on their stability.

An electron density map of MPyV reveals that in VP2 amino acids between Val-269 and Tyr-296 are responsible for interaction with three VP1 monomers of the pentamer and that a 45 aa long C-terminal segment of VP2/VP3 protein comprising these residues should be sufficient for a tight association with the cavity of VP1 pentamers (Chen et al. 1998). Accordingly, the stretches of 49 aa of the C-terminal part of VP2 served as an anchor sequence for the loading of GFP. The VLPs produced were regularly shaped and were stable. The amount of encapsidated GFP did not reach the theoretical number of 72 but was still very high at 64 (Abbing et al. 2004). A similar approach was used for N-terminal fusion of EGFP to a truncated minor protein, VP3 (corresponding to VP2 residues in positions 225–324) (Boura et al. 2005). This construct served for the testing of various processes and immunity responses, for example, EGPF (Frič et al. 2008) and Bcr-Abl (Hrusková et al. 2009). The MPyV VP2 protein anchor was also used for sufficient transport of a 683 aa long fragment of HER2 (Tegerstedt et al. 2005b, 2007) and full-length human prostate-specific antigen (PSA) (Eriksson et al. 2011) into dendritic cells (DCs).

Analogously, the linkage of foreign protein to SV40 PyV minor capsid proteins was investigated using a series of EGFP fusions (Inoue et al. 2008). Based on the structural model of SV40, where VP2 C-terminal residues 275–302 interact with the inner surface of the conical cavity of the VP1 pentamer (Chen et al. 1998), the coding sequence for VP2 amino acids 222–352 was used as anchor for protein fusion. Interestingly, the N-terminal EGFP fusions either interfered with VLP formation or were not incorporated into the particles. The C-terminal fusions were efficiently incorporated into VLPs. The minimal C-terminal region sufficient for incorporation into VLPs was finally determined to be a 36 aa residue of VP2 protein (273–308). Furthermore, the study demonstrated that a full-length prodrug-converting enzyme, cytosine deaminase, could be incorporated through the VP2 anchor. The enzyme retained its activity (converted 5-fluorocytosine to 5-fluorouracil) in purified VLPs, thus showing that small molecules and ions can gain access to the interior of VLPs. VLPs were also shown to deliver cytosine deaminase activity to the cells.

In conclusion, the fusion of foreign protein to minor proteins seems to have—unlike fusion to the major capsid protein—a minor impact on VLP stability, but the amount of transported protein is significantly lower (72 versus 360 molecules). Moreover, the stoichiometry of minor protein-anchored cargo in VLPs might vary between batches. Unlike for other viruses (Wen et al. 2012), the chemical conjugation of the peptides into the interior space of PV or PyV VLPs has not been reported so far. Several studies have shown that PyV and PV particles can deliver peptides intracellularly. The potential commercial utilization was immediately identified by the authors of these transport systems, and their patenting shortly followed (US 6991795 B1; US 7011968 B1).

20.4.4 DELIVERY OF OTHER MOLECULAR COMPOUNDS

Highly sensitive noninvasive imaging techniques performed in the absence of ionizing radiation are urgently needed in clinics for early diagnoses of diseases (e.g., cancer) and frequent monitoring of therapeutic progress. One of these techniques is magnetic resonance imaging (MRI), which possesses high temporal and spatial resolution. The recently recognized potential of optical imaging, especially in invisible near-infrared (NIR) fluorescent light, has led to the commercial availability of NIR fluorescence imaging systems for diagnostics and image-guided surgery (Gioux et al. 2010). Contrast agents and imaging probes are a key part of these techniques, as they allow a higher resolution and greater sensitivity in diagnostic images. Nevertheless, for enhancement of diagnostic imaging capabilities, the sensitivity, biocompatibility, and biodistribution of various contrast materials need to be improved. VLPs, as proteinaceous biodegradable nanoscale platforms, offer the opportunity to modify the chemical and physical properties of contrast materials in order to overcome these concerns. The surface modification of VLPs with targeting ligands can highly improve the accuracy of imaging techniques and signal strength. Interior and exterior engineering allow the combination of different functional imaging modalities with therapeutic compounds into a single formulation of a *theranostic* tool (Rosen et al. 2011; Cheng et al. 2012). Both PyV and PV VLPs have been used experimentally as imaging agents or drug delivery agents.

Chemical conjugation of fluorophores to virion or VLP surfaces is a routinely used technique for studying virus–host cell interactions in vitro. In applied research, these techniques can be used for validating the interaction of VLPs with target cells or tissues during in vitro tests of targeted drug delivery. A useful system for imaging the intracellular uptake of VLPs was designed for PV VLPs. HPV6 L1 or HPV16 L1/L2 VLPs were chemically coupled with *silent* fluorochrome carboxyfluorescein diacetate succinimidyl ester (CFDA SE), which fluoresces only after exposure to intracellular esterases. The labeled VLPs had the same structure and appearance as unlabeled VLPs. Importantly, the internalization of labeled VLPs into the host cells was not altered (Bergsdorf et al. 2003; Drobni et al. 2003).

Utilization of the internal space of a particle for encapsulation of a fluorescent probe leaves the exterior free for interaction with the receptor and/or for the attachment of targeting ligands. This strategy has been reported for PyV VP1 VLPs. In one study, all naturally occurring cysteines in MPyV VP1 protein were replaced by serines, and the new cysteine was reintroduced to the GH loop exposed toward the interior of the VLP. Then, fluorescent dye (fluorescein; Texas Red) was specifically conjugated to the cysteine through a maleimide linker. These VLPs retained their receptor-binding ability and entered C2C12 mouse cells (Schmidt et al. 1999).

Diagnostic or therapeutic compounds could also possibly be incorporated by a simple encapsidation during VLP assembly. This method was investigated for JCPyV VP1 VLPs and enabled the packing of fluorescent dyes Cy3 (Qu et al. 2004) and propidium iodide (PI) (Goldmann et al. 2000). The fluorescence intensity of VLPs with encapsidated PI was dependent on the initial PI concentration. Importantly, only fully formed VLPs were stably associated with PI or Cy3 dye, which was not removed from particles during the long-term dialyzing step. The authors confirmed that dyes associated with VP1 noncovalently (Qu et al. 2004), but the absence of traces of DNA inside the VLPs before reassembly reaction was not verified. This could be of some importance, since VLPs are permeable for small molecules (Inoue et al. 2008; Kitai et al. 2011) and the principle of dye stabilization inside the particle is not clear.

Fluorescence is particularly suitable for in vitro imaging applications. Fluorescence imaging probes dedicated to in vivo imaging, however, need to be highly bright fluorophores absorbing and emitting in the NIR range and provide a high signal-to-background ratio. This requires the targeting abilities of the probes to accumulate specifically in cells to be labeled, while being cleared from surrounding tissues. In vivo imaging is limited by the spectral properties of body fluids. This defines a narrow *optical window* between 650 and 900 nm, where light is able to penetrate deeper (a few centimeters' depth) and tissue autofluorescence is also reduced (Mérian et al. 2012).

Quantum dots (QDs) are ideal candidates for in vivo imaging probes. QDs are nanocrystals made of semiconductor materials that display outstanding optical properties, such as a high absorption coefficient and fluorescence quantum yields (Li and Zhu 2013). The immediate optical feature of colloidal QDs is their color. The smaller the QD, the higher the energy of the light that is emitted (and so the bluer the fluorescence spectrum). QDs are also highly dense and therefore visible by electron microscopy without negative staining. Encapsulation of QD into VLPs can provide a platform for visualization of early steps of virus infection, QD targeting, and construction of clusters with different modalities. All this has been shown for SV40 VLPs. Packing of QD was achieved by the reassembly of SV40 VP1 pentamers in the presence of an assembly buffer with QDs and resulted in the formation of particles that were homogenous in size (approximately 24 nm), mainly T = 1 particles consisting of 12 VP1 pentamers. The ratio was one QD per VLP. These particles were highly stable under routine storage conditions. The internalization of these particles by target cells was unaltered and was similar to the early infection steps of the wild-type virus (Li et al. 2009). Recently, an interesting phenomenon of QD encapsidation was observed. Surprisingly, QDs enhanced assembly of SV40 VLPs from VP1 pentamers directly in the dissociation buffer, and a high affinity between the QD and SV40 VP1 protein was discovered (Gao et al. 2013). QDs were suggested to act as scaffolds, favoring the correct inter-pentamer contacts. Furthermore, it was established that the structural stability and integrity of these particles are absolutely dependent on disulfide bonds formed in VP1 during capsid assembly (Li et al. 2013). The same group also tested the encapsidation of QDs modified by different surface coatings. Their results indicated that positively, negatively, or neutrally charged QDs were encapsulated with comparable efficiencies, and all types of QDs–VLPs preserved their cell-entering ability (Li et al. 2010).

Another class of nanoparticles that possess—depending on their size and shape—interesting optical properties, such as strong absorption and scattering in the visible–NIR region, are gold nanoparticles (AuNPs). They have been used in various types of biomedical applications including photo-thermal therapy, biosensing, and gene delivery for anticancer therapy (see Khlebtsov et al. [2013] for review). AuNPs of different particle sizes and surface decoration were encapsulated within the SV40 capsids, where the encapsulation efficiency increased with the size of AuNPs modified with methoxypolyethylene glycol 750 (mPEG750) (from 10 to 30 nm). Encapsulation of AuNPs modified with negatively charged DNA ligands was also tested. The electrostatic interactions promoted the encapsidation efficiency of AuNPs with smaller diameters (10 and 15 nm). Moreover, the AuNPs encapsulated into SV40 VLPs were successfully delivered into living Vero cells, whereas the AuNPs alone were unable to enter these cells (Wang et al. 2011). Except for encapsidation, AuNPs could be conjugated to the surface of VLPs. This conjugation of AuNPs was published for the JCPyV. AuNPs of different sizes (5, 10, and 15 nm) were linked to JCPyV VLPs through sialic acid–linked lipid, which was conjugated to AuNPs by a ligand-exchange reaction. The most suitable size of AuNPs for optical detection was 15 nm (Niikura et al. 2009). This approach was combined with encapsidation of QDs or AuNPs inside the SV40 VLPs, and VNP-guided 3D hybrid nanoarchitectures were assembled (Li et al. 2011, 2012). First, the SV40 VP1 protein was genetically modified to display only one cysteine, so that five cysteines were exposed on the VP1 pentamer. These altered SV40 VLPs were used for encapsidation of QDs by self-assembly, and AuNPs were conjugated to the surface through the affinity of gold to thiol groups. As a result, 3D hybrid nanoarchitectures of SV40 VNPs with one QD per VNP and a certain number of AuNPs (from 1 to 12) were obtained (Li et al. 2011). Afterward, this group encapsidated two other types of QDs or AuNPs into SV40 VNPs. As previously, AuNPs were linked to the VNP surface, but this time the conjugation was based on electrostatic interaction between negatively charged

AuNPs and positively charged amino acids on the outer surface of SV40 VNPs, and the average AuNP number per VLPs was 27 (Li et al. 2012).

For MRI, various contrast agents exist. Magnetic iron oxide–based nanoparticles (MNPs) are often considered for theranostic applications. Their superparamagnetic nature enables MRI, and particles themselves can be used for therapy through techniques such as magnetic hyperthermia (Gupta and Gupta 2005). Association of VLPs with these particles can therefore help to extend the spectra of their application, for example, by addition of cancer-targeting moieties. This was achieved by Enomoto et al. (2013): the authors encapsulated citrate-coated MNP into SV40 VLPs and then chemically conjugated EGF to the surface of the SV40 VLPs. Linking EGF to these SV40 VLPs–MNPs enabled them to enter the cell lines of human epidermoid carcinoma and human colon adenocarcinoma in a selective manner. Tumor cells are more heat sensitive than normal cells, so these MNPs can serve not only as an imaging tool, but also as a heater element for hyperthermia (Gupta and Gupta 2005). These types of particles have great potential to be applied as excellent material for directed tumor diagnostics and therapy.

The proof of principle for loading of therapeutic molecules such as low-molecular-weight drugs into VLPs was reported for PyV VLPs. Utilizing the covalent linkage of MPyV to a VP2 anchor, 462 molecules of methotrexate were loaded per VLP. Methotrexate is a well-known antifolate used in tumor therapy. The drug was delivered to methotrexate-sensitive cells, CCRF-CEM, and clearly demonstrated time- and concentration-dependent cytotoxicity (Abbing et al. 2004). Other research groups used naturally presented cysteines of the JCPyV PyV for linking the drug molecule via disulfide bonds inside the particle. The hydrophobic drug paclitaxel (PTX) was conjugated to JCPyV VLPs through modified β-cyclodextrin with a thiol-reactive group. The authors were able to encapsulate up to 12.3 PTX molecules, and these particles had a clear cytotoxic effect on NIH 3T3 cells (Niikura et al. 2013).

The essential properties required for a particle drug delivery system are not only specific targeting and cell entry but also the retention and release of the drug molecule in the target cells. First, low-pH-controlled drug release from VLPs was established (Ohtake et al. 2010). This group used a hexa-histidine motif (His_6) tag fused to the N-terminus of a 225 aa VP2 anchor for noncovalent linking of fluorescent dye sulforhodamine 101 (SR). This dye was conjugated to nitri-lotriacetic acid (NTA), which targets the His_6-tag sequence. The hexahistidine motif is ideal for release triggered by changes in pH because the pK_a of this motif is approximately 6.5, and a lower pH eliminates the His_6-tag affinity. The NTA-SR entered JCPyV VLPs containing the His_6-tag through simple diffusion and bound specifically to the His_6-tag at the physiologic pH (7.4) and in the presence of cobalt ions. After adjustment of the pH to 5.0, the release saturation was achieved within 20 min. Second, the redox-responsive drug release was discovered (Niikura et al. 2013). It utilizes glutathione (GSH), which has been widely recognized as a ubiquitous stimulus for drug release in cells, and its concentration is three times higher inside than outside of the cell. JCPyV VLPs were conjugated with thiol β-cyclodextrin (CD-SSO$_3^-$) through disulfide bonds between cysteine residues and a thiol-reactive group. The release speed was highly dependent on GSH concentration. The plateau was achieved within 5 min with the 10 mM GSH, and with 10 μm GSH, it lasted several hours. The controlled release approaches are extremely important for increasing the possibility of using VLPs as intelligent nanocarriers for therapeutic compounds.

In conclusion, numerous studies have shown that PyV VLPs can potentially serve as versatile functional hybrid nanostructures for application in the field of targeted drug delivery and diagnostic imaging. The interior of VLPs can accommodate drugs as well as various imaging compounds and increase their biocompatibility. The utilization of minor proteins as noncovalently attached drug anchors offers a unique opportunity for controllable drug release. The exterior of VLPs can display multiple imaging molecules or addressable moieties, thus creating targeted multimodal nanoparticles. PV VLPs are less extensively studied than PyV for these types of applications, but their potential utilization as theranostic agents could be very similar to PyV VLPs.

20.5 UTILIZATION OF VIRUS-LIKE PARTICLES FOR IMMUNOTHERAPY

VLPs of both viral families are suitable candidates for immunotherapy due to the high immunogenicity of their major capsid protein. VLPs are characterized by their ability to induce a humoral as well as a cellular immune response. Their structure is similar to the wild-type virus but they lack genetic material, thus providing a safe alternative to attenuated or inactive vaccines. The most important characteristic of VLPs is their ability to induce the production of neutralizing antibodies, which are critical mediators of immunity against viral challenge. These antibodies could be produced by T-helper cell-dependent or T-helper cell-independent mechanisms. Through these mechanisms, VLPs are able to induce a rapid and fulminant humoral immune response without the need for an adjuvant (Jennings and Bachmann 2008). The T-helper cell-dependent immunogenicity of VLPs is entailed by the character of their interaction with DCs. At the beginning of an effective immune reaction is a process of activation of the immature antigen receiving DC into a mature antigen-presenting DCs. The presentation of antigen by DCs expressing costimulation molecules and cytokines is necessary for induction of an effective and strong T-cell-specific immune response followed by long-lasting memory. VLPs were tested in vitro for their ability to stimulate human DCs, and the results showed that both papillomaviral and polyomaviral VLPs could induce a variety of phenotypic and functional changes: enhancement of the expression of major histocompatibility complex (MHC) glycoproteins, costimulation molecules (CD40, CD80, and CD86), and production of cytokines (IL-12, IL-6) (Lenz et al. 2001; Rudolf et al. 2001; Gedvilaite et al. 2006; Bickert et al. 2007).

However, PyV-derived VLPs appeared not to be consistent in DC activation. In contrast to rodent PyVs, BKPyV, JCPyV, and SV40-derived VLPs demonstrated only weak DC maturation and, correspondingly, low secretion of IL-12. The difference in DC maturation might be caused by the quality of antigen preparation, by the level of VLP uptake, or by the more extensive trigger of pattern recognition receptors inducing human DC maturation (Gedvilaite et al. 2006).

T-helper cell-independent humoral immunity is induced by the repetitive and highly ordered structure of VLPs. These repetitive epitopes enable the cross-linking of specific immunoglobulins that comprise the B-cell receptor, which is a critical signal for B-cell activation (Bachmann et al. 1995; Thyagarajan et al. 2003). Generation of neutralizing antibodies by both PVs and PyVs has been shown to be largely T-helper cell independent (Szomolanyi-Tsuda et al. 2001; Yang et al. 2005).

VLPs can be utilized for vaccination in two different ways: against capsid proteins or against foreign proteins that have been attached to VLPs. The main usage is the application of VLPs for vaccination against the virus from which they derived. Some VLP-based vaccines are already available on the market, such as prophylactic vaccines against HPVs (e.g., Gardasil, Merck; Cervarix, GlaxoSmithKline) or against hepatitis B (e.g., Engerix, GlaxoSmithKline; Recombivax HB, Merck). Hepatitis B virus (HBV) vaccines were already confirmed to be highly effective, with the induction of lifelong immunity and protection against the development of hepatocellular carcinoma (FitzSimons et al. 2005). The long-term efficacy of HPV vaccines is under investigation. This section will summarize the current knowledge of the utilization of polyoma- and papilloma-based VLPs for the development of prophylactic and therapeutic vaccines.

20.5.1 PAPILLOMAVIRUSES

HPV infection plays a key role in the etiology of cervical carcinoma, which is the fourth most frequent malignancy for women worldwide. Globally, there are approximately 530,000 new cervical carcinoma cases per year, with a mortality of 270,000 women per year. Around 85% of all new cases come from the developing countries (WHO 2014), and the majority are diagnosed in women older than 40 years. HPV infection is also connected to other carcinomas such as those in the anogenital region, head and neck cancers, and also benign genital warts in both men and women. It is estimated that HPV causes around 80% of anal carcinomas and 40%–60% of vulvar, vaginal, and penile carcinomas. Nevertheless, these malignancies are relatively rare and preferentially affect adults older than 50 years. HPV16 is the most common culprit in these tumors (WHO 2009).

HPV infection of the genital tract in both men and women is nowadays known as the most frequent sexually transmitted disease. The entry gate of infection is enabled by the microtrauma of the mucous membranes of the skin. Transmission is possible not only by sexual intercourse itself but also during noncoital sexual activities. The percentage of high-risk

HPV infected women after the initiation of sexual life rises up to 30%, reached within adolescence (under 25 years old). The prevalence then gradually decreases to 5% because the majority of infected women undergo spontaneous clearance by their immune system. Nevertheless, in some cases, persistent infections occur and induce malignant cell transformation. It is impossible to predict whether spontaneous clearance or tumor growth will occur following HPV infection. While HPV infection is an etiological cause of cervical carcinoma development, there are a number of cofactors that support dysplastic changes, such as smoking, promiscuity, hormonal contraception usage, an early start to sexual life, or illnesses connected to immunosuppression (Muñoz et al. 2006).

The HPV life cycle is restricted to the mucous membranes: no viremia occurs. HPV infects the basal cells of the cervical epithelium but does not induce cell death or inflammatory changes; therefore, activation of antigen-presenting cells is insufficient and the immune reaction is low (Stanley et al. 2006). Approximately 50% of women infected by HPV will not experience induction of the antibody response. The other 50% will be shown to evolve low antibody titers, but the numbers are not sufficient to develop protection against repeated infections. The amount of neutralizing antibodies in cervicovaginal secretions plays a vital role in the development of HPV protection. These antibodies bind the viral particles and thus block infection of the basal cells. So, the sufficient level of neutralizing antibodies in the epithelial region is necessary for antiviral protection. Vaccination induces high and stable level of antibodies in the serum. The level of serum antibodies strongly correlates with antibody level in the cervicovaginal secretions, proving the penetration of antibodies from serum to cervicovaginal secretions (Schwarz et al. 2010).

VLP-based vaccine development programs began on the basis of the discovery by several academic groups that the papillomaviral major coat protein, L1, can self-assemble into VLPs when expressed as a recombinant protein in a heterologous eukaryotic system. VLP-based vaccines against PVs induce a specific immune response (Fausch et al. 2003) and are formed from the major capsid protein, L1 (Koutsky et al. 2002), or both capsid proteins, L1 and L2 (Lenz et al. 2001). Chimeric vaccines also exist, where the structural protein L1 or L2 is fused to the whole or part of an oncoprotein, E2, E6, or E7 (Greenstone et al. 1998; Rudolf et al. 2001; Tobery et al. 2003).

20.5.1.1 First-Generation Vaccines

PV-based commercial prophylactic vaccines, after their approval by the FDA, became well known worldwide. The development of these vaccines was a long, demanding, and expensive research process that began with the immunization of rabbits, cattle, and canines with appropriate papillomaviral VLPs that protected the animals against papillomaviral infection (Breitburd et al. 1995; Suzich et al. 1995; Kirnbauer et al. 1996). The positive effect of papillomaviral VLPs on animal models led to their application on human volunteers. First, L1 VLPs from HPV18 were tested. Forty women took part in the vaccination study, ten of which received a placebo.

It was demonstrated that the HPV18 L1 VLP vaccine was well tolerated and exhibited high immunogenicity (Ault et al. 2004). Then, HPV16 L1 VLPs were tested on 2000 women, some of whom received a placebo instead of the vaccine. The results showed that none of the vaccinated women were infected with HPV16 (Brown et al. 2004). Another study focused on L1 VLPs derived from HPV11 or HPV16 (Fife et al. 2004). This study also proved the high immunogenicity of papillomaviral VLPs, and every vaccinated woman developed antibodies against these VLPs. Both vaccine types were well tolerated, and the only adverse effect was a reaction in the site of injection.

Two marketed prophylactic vaccines contain VLPs, which consist of an L1 coat protein self-assembled into spheres mimicking the natural conformation of the virus. Both vaccines—bivalent (Cervarix, GSK) and quadrivalent (Gardasil, Merck)—are composed of two high-risk HPV16 and HPV18 VLPs. The quadrivalent vaccine has additional HPV6 and HPV11 VLPs that are responsible for nearly 90% of genital warts. The vaccines differ also in their production system and adjuvant composition. Gardasil vaccine is produced in yeast, and amorphous aluminum hydroxyphosphate sulfate is used as adjuvant. The bivalent vaccine Cervarix is produced in insect cells and adjuvanted with AS04, which combines aluminum hydroxide and monophosphoryl lipid-A, a modified endotoxin that is an antagonist of Toll-like receptor 4. Vaccines are administered intramuscularly, according to three dose protocols (0, 1, and 6 months for Cervarix; 0, 2, and 6 months for Gardasil), and the most common side effects are redness, fatigue, swelling, headache, and fever (Gonçalves et al. 2014). A comparison of observational studies revealed that Cervarix caused more side effects than Gardasil, probably due to the type of adjuvant system (Einstein et al. 2009). Vaccines protect against the HPV types included in the vaccination. Durable protection against the corresponding infection of HPV type has been observed for up to 9.4 years in the case of the bivalent vaccine (GlaxoSmithKline Vaccine HPV-007 Study Group et al. 2009; Lehtinen et al. 2012; Naud et al. 2014) and up to 5 years for the quadrivalent vaccine (Villa et al. 2006; The FUTURE I/II Study Group et al. 2010). The safety and efficacy of this vaccine has been confirmed for the vaccination of pregnant women (Goss et al. 2014).

The FDA licensed the Gardasil vaccine in 2006 for young females of 9–26 years old and the Cervarix vaccine for young females aged 10–25 years old. However, the vaccination could also be applied to women with evidence of prior HPV exposure. Although the therapeutic effect against anogenital cancer was not proven, these vaccines were able to prevent reinfection of HPV (Olsson et al. 2009; Miltz et al. 2014). The Gardasil vaccine was also, in 2009, approved by the FDA for the vaccination of 9–26-year-old males, and results from a phase III clinical trial proved a 90% reduction in genital lesions (Giuliano et al. 2011).

In a direct comparison, the quadrivalent vaccine produced lower levels of antibodies over 2 years than the bivalent vaccine, probably due to its adjuvant system being less efficient (Einstein et al. 2009). Another reason could be the production of VLPs of the bivalent vaccine in the baculovirus expression system: routine purification procedures might not precisely discriminate between VLPs and baculoviruses, and as a consequence, the VLP preparations could be contaminated by baculoviral particles that are able to enhance the adjuvant properties (Abe et al. 2003). Moreover, commercial vaccines differ also in their ability to induce cross protection against other related HPV types. The bivalent vaccine enabled wider cross protection, and the obtained vaccine efficacy was about 78% for HPV31, 45% for HPV33, and 77% for HPV45 (Paavonen et al. 2009; Lu et al. 2011; Wheeler et al. 2012). The vaccine efficacy of the quadrivalent vaccine was only around 46% for HPV31, 29% for HPV33, and 8% for HPV45 (Brown et al. 2009; Westra et al. 2013), and thus significant cross protection was obtained only against HPV31. For both vaccines, no statistically significant cross protection against persistent infection with HPV52 or HPV58 was detected (Kemp et al. 2011; Malagón et al. 2012). More data are needed to establish the duration of cross protection.

20.5.1.2 Second-Generation Vaccines

Current prophylactic vaccination programs should have a significant impact on the reduction of HPV-related malignancies, owing to their ability to provide long-term protection against cervical intraepithelial neoplasia (Cadman 2008). However, the vaccines currently registered have some limitations affecting their worldwide application: the main problems include their low stability, costly production, no therapeutic effect, and limited effective cross protection to other HPV types. Since 85% of all newly diagnosed HPV-related carcinomas come from developing countries, the pressure on the development of a second-generation HPV vaccine rises.

20.5.1.2.1 Increase in Stability and Cost Efficiency

The low-cost and noninvasive biopharmaceuticals enabling widespread immunization are desired mostly in developing countries. The application of vaccines by injection might transmit other infectious diseases, generally due to repeated needle usage. The vaccines currently available must be stored in a low-temperature environment of around 6°C, and their production also requires advanced techniques and facilities. Thus, repeated vaccination is very impractical for low-resource countries. Edible (oral) vaccines have gained positive attention in this context. The principle is based on the contact between the antigen and the surface of a mucosal membrane—ideally of the gastrointestinal tract or respiratory system. These vaccines could be applied in the form of capsules, aerosols, or gel, with a precise dosage regarding the required antigen concentration. However, oral applications need significantly higher doses of antigen (approximately a hundred times) to induce an effective immune response. The antigen dose could be lowered by combination with the appropriate adjuvants (Rose et al. 1999; Gerber et al. 2001). There are two patents of edible vaccine vehicles: the first one is for the HPV16 L1 protein produced in

the yeast *Schizosaccharomyces pombe* (US20090017063; 2009), which can induce a systemic immune response and production of HPV16 type-specific neutralizing and mucosal antibodies. The second one is for oncoprotein E7 of HPV16 carried by a lactic acid–based vehicle (WO2010079991; 2010), which induces an antigen-specific cellular immune response after oral administration in mice. In the last decade, the use of plants as a production system for vaccine antigen production has become more popular: plant expression systems offer a cheap, robust, and relatively fast alternative production system. The high potential of plant vaccines is in their utilization in third-world countries, but also in advanced countries where development is directed toward noninvasive applications that are more comfortable for patients. HPV VLPs self-assemble in the transgenic plant cell environment, and the robust plant cellulose membrane even protects VLPs against gastric low pH and allows their safe transport to the colon. The relatively high resistance of naked HPV VLPs to the gastrointestinal tract has been published (Rose et al. 1999). The production of an oral HPV vaccine is theoretically feasible; however, more research data are needed. On the other hand, the low-cost and simple application of these vaccines could potent their further research and help to introduce complex vaccine programs against HPV (Schiller and Nardelli-Haefliger 2006; Stanley et al. 2008).

20.5.1.2.2 Broadening of Cross Protection

Although immunization with current prophylactic vaccines is very efficient in inducing a protective antibody response against HPV6, HPV11, HPV16, and HPV18, such antibodies are predominantly type restricted. There is evidence of cross protection that could possibly be explained by phylogenetic similarities between the L1 genes of HPV types. The bivalent vaccine has shown cross protection against HPV31, HPV33, and HPV45, but the quadrivalent vaccine only against HPV31. However, the cross protective stimulation of antibody production represents <1% of the neutralizing activity induced by the dominant conformational epitopes. In order to prevent all HPV-related tumors, specific VLPs for each HPV type involved in cancer must be developed. Although it does not seem to be economically feasible, studies are ongoing to increase the number of HPV types to nine in the quadrivalent vaccines by adding the VLPs of HPV31, HPV33, HPV45, HPV52, and HPV58. This nonavalent vaccine is currently under development by Merck Research Laboratories, and a mathematical model predicts that this vaccine could raise protection from 70% to 90% of the infections responsible for invasive cervical cancer (Serrano et al. 2012). It is also possible that this vaccine will provide cross protection to other nonvaccinated HPV types, such as current commercial prophylactic vaccines. This nonavalent preventive vaccine is currently in a phase III clinical trial on women between 16 and 26 years old (National Cancer Institute 2014).

As an alternative approach to the increase in the number of various HPV L1 VLPs, the incorporation of L2 minor protein into VLPs has arisen as a possible means for cross protection improvement. In contrast to L1 protein, L2 protein is barely visible for the immune system during infection because L2 protein is not exposed until the virus binds to the basement membrane. Therefore, the neutralizing antibodies directed against L2 are not produced. The pressure to evolve L2 neutralizing epitopes is low, therefore L2 sequences (especially between 20 and 38 aa) are highly preserved among various high-risk HPV types (Karanam et al. 2009). A study based on the vaccination of sheep with HPV6, HPV16, or HPV18 L1/L2 VLPs proved that L2 protein contains a subdominant, cross neutralizing epitope. The sheep antiserum from each HPV type provided cross protection to the other tested HPV types (Roden et al. 2000). However, the immune response induced by HPV VLPs composed of both L1 and L2 was predominantly directed against the L1 protein. This low immunogenicity of L2 might be caused by a higher amount of L1 protein compared to L2 protein, by the distant spacing of L2 protein, or by the fact that L2 protein is hidden inside the L1 capsid (Karanam et al. 2009). In many studies, the L2 protein peptide sequences were inserted into the surface loops of the L1 protein; these studies revealed the most potent cross neutralizing antibody sequences to be on the N-terminus of the L2 protein (Kondo et al. 2007, 2008; McGrath et al. 2013). Although L2 protein vaccination evoked broad-spectrum immunity, the efficiency was low. However, HPV16 L1/L2 VLP vaccination of rabbits together with Freund's adjuvant resulted in a strong increase in the level of neutralizing antibody production (Schellenbacher et al. 2009). This was consistent with the previous data, where conjugation of modified adjuvant mLTK63 to C-terminus of HPV16 L2 protein induced higher titers of HPV16-specific, long-lasting neutralizing antibodies and splenocyte proliferation (Xu et al. 2008). The data demonstrate the requirement of repeated immunizations and adjuvant utilization for high anti-L2 immune response induction.

20.5.1.2.3 Development of Therapeutic VLP Vaccine

Papillomaviral VLPs are intensively studied as an immunology tool for the production of therapeutic vaccines against HPV-related cancer, vaccines against other viral infections, or even vaccines against nonviral diseases such as Alzheimer's disease or arthritis. First, the current possibilities for therapeutic vaccine development against diseases connected to HPV infection will be discussed.

Prophylactic HPV vaccines induce a potent immune response that results in the production of high titers of neutralizing antibodies sufficient to prevent infection (Day et al. 2010; Naud et al. 2014). However, for anticancer immunotherapy, the cellular immune response is more important than antibody production. Although HPV L1 VLPs have been shown to cause a potent cellular immune response (Woo et al. 2007), there is no evidence of their therapeutic effect (Hildesheim et al. 2007; Olsson et al. 2007; Miltz et al. 2014). Therefore, E6 and E7—the major viral oncoproteins of high-risk HPVs—are intensively studied as interesting T-cell response targets.

The E6 and E7 proteins are produced in the early phase of infection and inactivate two tumor suppressor proteins, p53

and pRb. Protein p53 is inactivated by E6, which forms a stable complex with p53, and, in turn, this complex undergoes a proteolysis (Scheffner et al. 1990). The protein pRb is bound by E7 and this interaction results in the release of the transcriptional factor E2F that promotes the transcription of genes required for cell DNA synthesis and cell cycle progression (Dyson et al. 1989). The viral oncogenes, E6 and E7, are thought to modify the cell cycle so as to retain the differentiating host keratinocyte in a state that is favorable to the amplification of viral genomes and consequent late gene expression (Münger and Howley 2002). It has also been shown that the E2 protein of high-risk HPVs can participate in cell transformation because its expression allows chromosomal instability and promotes integration of viral genome, which has been documented as one of the major steps leading to HPV-induced transformation (for review see Bellanger et al. 2011).

Recently, early antigens have represented a frequent target for therapeutic intervention in vaccine development. Various vaccine designs have been studied, including DNA vaccines, protein/peptide-based vaccines, or vaccines based on chimeric VLPs. Several series of studies of DNA vaccines, encompassing naked DNA and viral or bacterial-based DNA vaccines, have been published. These vaccines can be easily prepared and manipulated, are low cost, and are stable. On the other hand, naked DNA vaccines are limited by low transfection efficiency, and live vectors possess a potential risk of toxicity and activation of immunosuppressive factors in humans (for review see Hung et al. [2008] and Cho et al. [2011]). The protein-/peptide-based vaccines are based on the direct transfer of synthesized HPV early antigens, are poorly immunogenic, and need strong adjuvants for the induction of a protective immune response (for review see Bijker et al. [2007] and Hung et al. [2008]). Therefore, the next therapeutic vaccines are based on the transport of early proteins through HPV VLPs. These VLPs could potentially induce both a prophylactic and therapeutic immune response, resulting in the prevention of reinfection and/or control of reactivation.

The combination of early and late HPV proteins in chimeric VLPs is a strategy with which to meet this goal (Jochmus et al. 1999). In chimeric VLPs, foreign epitopes could be coupled to VLPs either by fusion to the major capsid protein, L1, or by fusion to the minor capsid protein, L2 (see Section 20.4.3). In an initial study of Müller et al. (1997), various parts of an E7 protein were conjugated to the C-terminally truncated HPV16 L1 protein for chimeric HPV16 L1/E7 VLP construction. Only chimeric VLPs bearing up to 55 N-terminal aa of the E7 protein gave a high yield of uniform particles (Müller et al. 1997). These HPV16 L1ΔCE7$_{1-55}$ chimeric VLPs induced an E7-specific T-cell response in vitro (Kaufmann et al. 2001) and were used for the vaccination of women with high-grade cervical intraepithelial neoplasia (Kaufmann et al. 2007). However, the difference between the immune responders and placebo recipients in terms of histological improvement was only 14%. Other research groups used the minor protein, L2, as an anchor for the coupling of early antigens. Chimeric HPV16 L1/L2E7 VLPs were obtained by conjugation of the full-length E7 protein to the C-terminus of the L2 protein and its coexpression with the L1 in a baculovirus expression system. The vaccination of mice with these VLPs induced protection against the outgrowth of an HPV E7–positive tumor (Greenstone et al. 1998). Later, the comparison of HPV16 L1 and L1/L2 chimeric particles conjugated with the first 57 aa of HPV16 E7 protein was described. The L1 chimeric particles induced a significantly higher E7-specific immune response than L1/L2 chimeric particles, where the E7 sequence was fused to the L2 protein (Wakabayashi et al. 2002). This is consistent with the amount of delivered epitope. Fusion to L1 could theoretically transfer 360 epitopes per particle, whereas fusion to an L2 protein would transfer only 72 epitopes. Afterward, another vaccine strategy for increasing the therapeutic potential by incorporating HPV16 E2 protein into VLPs was examined. Through the conjugation of full-length E7 and E2 in a row to the C-terminus of the L2 protein, chimeric HPV16 L1/L2E7E2 VLPs were prepared. Their morphology was similar to HPV16 L1 VLPs (Qian et al. 2006). Unfortunately, these chimeric particles were unable to induce a CTL response against E7 or E2 without immunomodulators. Moreover, the cost would be too high for their utilization as a commercial vaccine. The vaccination with chimeric particles could potentially induce both responses: the prophylactic one against the HPV16 virus and the therapeutic antitumor response against the cells transformed by HPV16. Nevertheless, if the patient is already infected or vaccinated and an immune response against the L1 protein is induced, the efficiency of this combined vaccine might be reduced by neutralizing anti-L1 antibodies (Da Silva et al. 2003).

HPV VLPs could be utilized also as a vaccination platform against other non-HPV-related diseases. The first option is the conjugation of immunodominant peptides to the capsid surface by genetic modification. Specific anti hepatitis B core antigen (HBc) antibody production was induced by the insertion of a hexameric DPASRE peptide from the HBV virus core antigen into different L1 surface loops, and only one (BC) loop failed to result in anti-HBc antibody induction in mice, suggesting that the epitope inserted into the BC loop was hidden inside the capsid (Sadeyen et al. 2003). BPV L1 protein was used as a platform for genetic incorporation of chemokine receptor type 5 (CCR5), which is the major coreceptor of the HIV. Inoculation of these chimeric L1-CCR5 VLPs into mice highly induced the production of anti-CCR5 autoantibodies, which bound the CCR5 receptor and inhibited the HIV infection in vitro (Chackerian et al. 1999). The other sequence inserted into BPV L1 was the first 9 aa of human amyloid-β protein, and stable chimeric VLPs were produced. Rabbits were chosen for the immunization because there exists a 100% identity match of the 9 aa sequence of amyloid-β between rabbits and humans. Chimeric VLPs generated specific autoantibodies against the amyloid-β peptide and thus broke the B-cell tolerance to this self-antigen (Zamora et al. 2006). Autoantibodies against amyloid-β were previously described as a method for

decreasing cognitive decline in Alzheimer's disease (Hock et al. 2003), and thus the Zamora research group used model transgenic mice of human Alzheimer's disease for immunization with these chimeric VLPs. The deposits of amyloid-β in the brain decreased, and more amyloid-β was present in the plasma, suggesting its release from the brain. It seems that these chimeric VLPs could serve as an efficient immunotherapeutic for human Alzheimer's disease (Zamora et al. 2006). The second option connects the protein epitopes to the PV VLP surface by chemical conjugation. The most frequently used technique is the conjugation of a streptavidin-fused epitope to biotinylated BPV VLPs. Peptides derived from tumor necrosis factor alpha (TNF-α; the proinflammatory cytokine), amyloid-β, or CCR5 were displayed on the surface and induced specific antibody production against the exposed peptides in mice (TNF-α, amyloid-β) or in macaques (CCR5) (Chackerian et al. 2001, 2004; Li et al. 2004). The VLPs with TNF-α were even successful in decreasing collagen-induced arthritis development in mice (Chackerian et al. 2001). The third option is based on the ability of the L1 protein to encapsidate DNA during self-assembly. For DNA transfer, VLPs lacking L2 and nucleocores are not suitable, as L2 protein and the correct conformation of particles are necessary for infection and efficient delivery of DNA into the cell nucleus. Therefore, for this purpose, PsVs produced in packaging cells, containing L1, L2, and DNA complexed with histones, have been used. HPV PsVs were tested as a platform for the delivery of a DNA vaccine against respiratory syncytial virus (Graham et al. 2010) or against HIV (using the simian immunodeficiency virus model) (Gordon et al. 2012). Both vaccines were delivered intravaginally and induced specific CD8+ T-cell immune responses in mice (respiratory syncytial virus) or in macaques (HIV). The vaginal application was necessary because HPV does not infect intact epithelial cells. Furthermore, an experimental DNA vaccine encoding the ovalbumin antigen was encapsidated into HPV PsVs. After subcutaneous application, a strong ovalbumin-specific CD8+ immune response was evoked (Peng et al. 1998). All PsV-DNA vaccinations were more efficient when compared to the naked DNA vaccines. Thus, HPV PsVs are considered as a promising platform for DNA-based therapeutic vaccination.

20.5.1.3 Concluding Remarks

Vaccination is the most effective way for prevention of infectious diseases. Despite the great promise for cervical and other HPV-related cancers, there are several characteristics of current HPV vaccines that need to be improved. The vaccines are unaffordable in many parts of the world, even though the quadrivalent vaccine has been offered to GAVI Alliance that increased the access to vaccine in poor countries, incredibly only for $5 per dose. Second-generation prophylactic HPV vaccines, currently in clinical trials, may overcome several limitations of the current commercially used vaccines. First, they could mediate the protection against additional oncogenic HPV types by broadening of cross protection through incorporation of L2 or multivalent vaccine production. Second, they should be less dependent on refrigerated conditions, needles, and low costs. The edible vaccines might solve all these deficiencies. Plants in particular are at the center of attention because they are cheap, can be orally administered, are grown easily in developing countries, and produce high amounts of recombinant protein. Third, the prophylactic together with the therapeutic effect is required. Therefore, the combination of late and early HPV proteins has been extensively studied. Early patents include chimeric vaccines containing HPV L1 with the E7 protein (US6649167, 2003; US7754430, 2010).

Unfortunately, only a few of the second-generation vaccine candidates discussed earlier are ready for clinical trials, their success is still uncertain, and the time for commercial use is very distant.

20.5.2 POLYOMAVIRUSES

In general, infection with mammalian PyVs is asymptomatic, and the vast majority of the population is seropositive. Representative examples of human PyVs that are intensively studied include BKPyV, JCPyV, and MCPyV. The BKPyV has the highest seropositivity: it increases with age and finally reaches almost 100%. Children up to 5 years of age are 63% seropositive, and children 10 years of age are nearly all seropositive (Stolt et al. 2003). BKPyV and JCPyV are present in cases of kidney transplantations, but only BKPyV is able to induce PyV-associated nephropathy, which is the most common reason for rejection of a renal transplant (Fishman 2002). JCPyV has significantly lower seropositivity: with 5-year-old children being 27% positive and adults reaching 72% seropositivity (Stolt et al. 2003). Reactivation of this virus needs an immunosuppressive state and induces a progressive multifocal leukoencephalopathy that is caused by the demyelination of neurons (Fishman 2002). The prevalence of MCPyV differs among the continents (Europe, 85% [Becker et al., 2009]; United States, 70%; and Australia, 25% [Garneski et al., 2009]) and gender, where women have a higher prevalence of MCPyV in MCC (aggressive neuroectodermal tumors) than men (Andres et al. 2010). MCPyV infects Merkel cells—cellular mechanoreceptors in skin—where the viral infection is one of the most leading factors for their malignant transformation (Andres et al. 2010). The MCCs are at least 80% positive for MCPyV presence, and the viral stimulator for tumor development could be the integration of the virus into the host cell genome (Feng et al. 2008).

Polyomaviral VLPs are highly immunogenic. They induce an antibody-mediated immune response, which prevents viral infection and also leads to a cell-mediated immune response that protects against the formation and evolution of virus-induced tumorigenesis. This immunogenic potential could be utilized in vaccine development against the native viral capsids as well as other illnesses, even of a nonviral origin. In this case, VLPs perform two functions: to carry the antigen and to serve as an adjuvant. As was described in Section 20.4.3, there are different ways in which antigens (obviously of protein origin) can be connected to the viral particle. The connection could result in the display of the antigen on the particle surface or lead to it being hidden inside the VLP core.

20.5.2.1 Native VLP Vaccines

The productive infection of PyVs is restricted to their natural host; therefore, the eligible and most widely used system for studying immunity and pathogenesis is the MPyV infection model. Heidari et al. (2002) reported that MPyV VLPs were successful against MPyV infection in the vaccination of normal and even T-cell-immunodeficient mice (CD4$^{-/-}$ CD8$^{-/-}$). After induced infection, approximately half of the mice from each group were MPyV DNA-free and protected. However, the titers of anti-VP1 antibodies were generally higher in the case of normal mice. This might be caused by a less-effective IgG switch in the absence of functional T cells (Heidari et al. 2002). In the following study, the authors tried to use different immunization protocols to improve the percentage of protected mice. The best results were obtained by subcutaneous rather than intraperitoneal application, irrespective of the presence of Freund's adjuvant (Vlastos et al. 2003). For comparison of immunogenicity, MPyV VLPs and VP1 protein fused to GST (in the form of pentamers) were administered as vaccines. While MPyV VLPs protected all mice, GST-VP1 protein protected 100% of normal but only 60% of T-cell-immunodeficient mice (CD4$^{-/-}$CD8$^{-/-}$) (Vlastos et al. 2003). The lower antibody production was probably caused by the disability of GST-VP1 fusion protein to assemble into VLPs. The repetitive structure of VLPs was proved to be necessary for inducing a strong T-cell-independent humoral response (Szomolanyi-Tsuda et al. 1998; Velupillai et al. 2006).

However, VLPs could be used not only for the prevention of viral infection but also for antitumor therapeutic vaccination. Peptides derived from MPyV T antigens were described as a target of the T-cell immune response. These peptides prevent tumor development and also mediate tumor rejection. They are presented by MHC molecules and called tumor-specific transplantation antigens (TSTAs) (reviewed in Ramqvist and Dalianis [2010]). As some MPyV tumors express VP1 protein, the hypothesis of using VP1-derived peptides as a TSTA was experimentally tested by Franzén et al. (2005). Vaccination with MPyV VLPs should prevent mice from outgrowth of three different MPyV tumors producing T antigens: hair follicle tumor (derived from ACA mouse strain), sarcoma and fibrosarcoma (derived from CBA mouse strain). ACA hair follicle tumor was completely rejected, while partial protection against the CBA fibrosarcoma and no protection against the CBA sarcoma outgrowth were achieved (Franzén et al. 2005). Although this experiment showed promising results for the application of VP1 VLPs as an antitumor agent, its widespread utilization is implausible, since VP1 production in MPyV- or MCPyV-induced tumors is minor (Talmage et al. 1992; Sanjuan et al. 2001; Holländerová et al. 2003; Haugg et al. 2014), is connected to episomal viral DNA presence (Talmage et al. 1992; Stubenrauch et al. 2001; Holländerová et al. 2003), and disappears with passaging of tumor cell line (Talmage et al. 1992; Holländerová et al. 2003).

The human PyVs BK and JC need immunosuppression for reactivation; therefore, vaccine development for that limited population might not be cost effective. On the other hand, accumulating evidence of the role of MCPyV in cancer induction justifies investigation into the vaccine development. Since MCC patients have very high antibody titers against MCPyV, it has been suggested that humoral immunity alone would not prevent the disease and that a cell-mediated immune mechanism might be involved in protection against MCPyV-induced malignancy (Pastrana et al. 2009). However, immunization of healthy individuals (seropositive or seronegative for MCPyV) by MCPyV VP1 VLPs led to helper T-cell responses, which were highly antigen specific and concentration dependent. IFN-γ was the most readily detectable cytokine (Kumar et al. 2011). This interferon has antiviral and tumor-suppressive functions, and its production is associated with a favorable prognosis of MCC (Paulson et al. 2011). Nowadays, the mechanism based on transfer of MCPyV-specific T cells as a source of reactive antitumor immunity was published (Chapuis et al. 2014). The therapy was applied to a 67-year-old man with metastatic MCPyV-expressing MCC. The vaccination was well tolerated and evoked a durable complete response in two of three metastatic lesions. The transferred CD8+ T cells preferentially accumulated in the tumor tissue.

Another field where vaccine production might be economically attractive is the poultry industry. One of the offending diseases is hemorrhagic nephritis and enteritis of geese, which is characterized by high morbidity and mortality in geese between 3 and 10 weeks of age. This illness is caused by the GHPyV. VLPs of this PyV were successfully produced in yeast or insect cells (Zielonka et al. 2006), and VLPs from insect cells were used for the vaccination of goslings. VLPs provided protection to goslings after their vaccination at 1 day of age and boosting after 17 days, and surprisingly even without boosting. The efficacy was not influenced by the dose of the antigen (Mató et al. 2009).

20.5.2.2 Chimeric VLPs as Vaccines against Foreign Epitopes

The ability to induce an immune response against either a self-antigen or an antigen of a pathogen is very desirable during therapeutic and prophylactic vaccine development. Therefore, effort was concentrated on the development of various agents that would exhibit this effect and combine it with other important features such as minimal side effects, safety for the organism, low-cost production, and relatively easy preparation and administration. VLPs meet all of the aforementioned requirements, and after their modification, they can be used as vectors for a multimeric presentation of foreign antigens in vaccine technology. One of the essential tasks is the development of an efficient tumor treatment, which is complicated by the very individual and heterogenic tumor character. The main goal of nonviral tumor immunotherapy is the induction of immune responses to autoantigens and the breaking of an established tumor tolerance. Utilization of VLPs in this

context should have a great future; however, there are still only a limited number of studies investigating this VLP potential.

20.5.2.2.1 Epitopes Displayed on the Surface of Polyomavirus-Based VLPs

In case of PyV-based VLPs, epitopes can be inserted into one of the surface loops of the major structural protein, VP1, generating VLPs with foreign epitopes exposed on the surface. This approach is limited by the inserted peptide length, as the longer peptides are able to influence the assembly and stability of VLPs.

As an alternative to the commercial HBV vaccine, chimeric VLPs carrying a hydrophilic component of the pre-S1 sequence of HBV (75 aa) in the HI loop of MPyV VP1 protein were constructed (Skrastina et al. 2008). The pre-S1 sequence is directly responsible for the binding of HBV to human hepatocytes and is not present on the 22 nm particles currently used for vaccination against HBV. The delivery of this sequence on the surface of chimeric VLPs is of the highest interest for multitarget HBV vaccine development. Chimeric VLPs were subcutaneously applied to mice and led to an induction of a strong antibody response against the inserted epitope, as well as to stimulation of IL-12 and IFN-γ production. Interestingly, the insertion of the epitope lowered the humoral response to MPyV VP1 protein (Skrastina et al. 2008). This phenomenon was not achieved in the previous study, where S1 epitopes derived from pre-S1 sequence were inserted into various loops of HaPyV VP1 protein. Assembled chimeric VLPs induced a strong anti-VP1 and anti-S1 antibody response in mice (Gedvilaite et al. 2000). As an alternative approach for influenza A vaccine production, VLPs derived from SV40 were tested. Chimeric SV40 VLPs were constructed by insertion of HLA-A*02:01 restricted CTL epitope corresponding to the influenza A virus matrix protein peptide 58–66 (FMP 58–66) into the DE or HI loop (Kawano et al. 2014). These chimeric VLPs effectively induced influenza-specific cytotoxic T cells and heterosubtypic protection against influenza A viruses without the need for an adjuvant.

HaPyV VLPs as a source for chimeric VLP production became the center of attention of the scientific group of R. G. Ulrich (Humboldt University, Berlin, Germany) and K. Sasnauskas (Institute of Biotechnology, Vilnius, Lithuania). Based on the crystal structure of SV40 VP1 protein (Chen et al. 1998), they predicted suitable insertion sites in HaPyV VP1 protein. First, a peptide (5 aa) derived from the S1 protein of HBV was inserted into four different regions of HaPyV. The positions corresponded to loops BC, EF, FG, and HI and were identified using numbers from one to four (Gedvilaite et al. 2000). In two additional constructs, the same peptide was inserted into two sites (positions 1 and 2 or 1 and 3). All recombinant VP1 proteins could self-assemble into chimeric VLPs and, together with Freund's complete adjuvant, were intraperitoneally administered to mice. The ability of chimeric VLPs to evoke specific antibodies (mainly IgG) against the S1-derived peptide was dependent on the place of insertion. The highest titers were obtained in position 1 (BC loop), and as expected, the combination of this site with another site for epitope presentation induced a stronger antigen-specific antibody response. Additionally, splenocytes derived from immunized animals exhibited the increased production of IL-12 and IFN-γ (Gedvilaite et al. 2000). In the next study, the same group inserted a T-cell-recognized epitope derived from carcinoembryonic antigen (CEA) to positions 1 or 4 or 1 and 4 with or without the flanking linker (double glycine-serine [GS] on both sites of the insert). Furthermore, a construct with insertion into all four positions with flanking linkers was prepared. Surprisingly, all recombinant proteins were able to form chimeric VLPs irrespective of the flanking linker presence, except that with insertions in all four sites, which did not assemble into VLPs. The highest antibody response was induced by chimeric VLPs with an inserted epitope in position 1, without any influence of linker presence on immunogenicity. The CEA-specific antibodies were detectable even 6 months after immunization (Lawatscheck et al. 2007). Subsequently, the insertion of a mucin-1 (MUC-1) CTL epitope into position 1 and/or 4, with or without glycine–serine linker, or again into all four sites with linkers, was tested (Zvirbliene et al. 2006; Dorn et al. 2008). In agreement with the previous study, only the VP1 with all four insertions was unable to self-assemble into VLPs. The chimeric VLPs carrying the MUC-1 with GS linkers in sites 1 and 4 were the most potent for specific anti-MUC-1 antibody production. These particles were also able to mature human DCs and evoke a specific CTL response in vitro (Dorn et al. 2008). In another study of CTL response, chimeric HaPyV VLPs with a GP33 CTL epitope derived from lymphocytic choriomeningitis virus (LCMV) (Pircher et al. 1990) incorporated into the BC or HI loop induced protective memory and a CTL response (Mazeike et al. 2012). T-cell proliferation was induced both in vitro and in vivo without adjuvant usage. After intravenous immunization of mice, 70% of them were fully and 30% were partially protected from LCMV infection.

Similar studies were performed with 45 aa, 80 aa, or 120 aa long segments from the N-terminal part of the nucleocapsid protein (NP) from the Puumala hantavirus (PUUV). These segments were inserted into all four positions (80 aa peptide only into positions 1 and 4) of HaPyV VP1 protein. Only positions 1 and 4 were able to tolerate insertion of the long foreign peptides without affecting VLP assembly (Gedvilaite et al. 2004). These chimeric VLPs were injected to BALB/c mice and, without any adjuvant, generated high titers of IgG antibodies against PUUV NP and stimulation of IL-2 and IFN-γ secretion. However, adjuvant usage induced a 10-fold higher antibody production. Both humoral and cellular immune responses were observed. The strongest immune response was observed after immunization with the longest 120 aa segment. Insertion of PUUV NP segments also reduced antigenicity of the HaPyV VP1 protein. The level of reduction was dependent on the size of the inserted protein; with an increased size of the inserted protein, the antigenicity of the VP1 protein decreased. These results were confirmed by a following study with chimeric HaPyV VLPs carrying a 120 aa PUUV NP segment, which promoted the generation

of five antigen-specific antibodies of IgG isotypes but no VP1-specific antibodies (Zvirbliene et al. 2006).

Further effort was directed toward the production of a low-cost vaccine against group A streptococcus (GAS), which causes severe infections in low-income nations. A highly conserved 20 aa peptide (p145) from the M protein of GAS has been shown to generate specific human antibodies that are able to opsonize multiple strains of GAS (Pruksakorn et al. 1994). Later, the minimal protective epitope within the p145 protein was defined precisely as 12 aa epitope called J8 (Hayman et al. 1997). In order to recognize a small difference in the peptide sequence, a strategy published by Relf et al. (1996), based on the insertion of p145 peptides into other peptides known to form α-helix, was applied. The flanking peptides for driving the constitution of a helical structure were taken from the GCN4 protein, a DNA-binding protein in yeast. The resulting 28 aa long peptide was constructed with GCN4-derived linkers (6 aa) and a J8 epitope in the middle (Hayman et al. 1997). This epitope GCN4-J8i-GCN4 was inserted into the HI loop of MPyV VP1 protein, and stable chimeric particles were produced. The subcutaneous delivery of these particles into mice induced high titers of J8i-specific antibodies with a bactericidal effect (Middelberg et al. 2011). Afterward, the efficacy of chimeric VLPs carrying two copies of the J8i antigenic element was examined and compared to those carrying only a single copy of J8i. IgG isotypes induced by both chimeric VLPs were similar, indicating a mixed T-helper cell response (Chuan et al. 2013; Rivera-Hernandez et al. 2013). Chuan et al. (2013) also showed that the chimeric VLPs displaying J8i successfully induced high titers of J8i-specific antibodies, even in mice that were previously immunized with chimeric MPyV VLPs and exhibited high anti-VP1 antibody titers.

In most studies the vaccines were administered subcutaneously or intraperitoneally. The chimeric VLPs previously mentioned, carrying J8i peptide, were also delivered intranasally without an adjuvant. This administration induced both IgG and antigen-specific mucosal IgA antibodies. Vaccinated mice showed improved survival; however, the statistically significant level of protection was two times lower than the positive control, which was vaccinated with GCN4-J8i-GCN4 conjugated to diphtheria toxoid (Rivera-Hernandez et al. 2013).

20.5.2.2.2 Epitopes Hidden Inside Polyomavirus-Based VLPs

Internalization of foreign sequences inside VLPs can be achieved by conjugation to VLPs through the major protein, VP1, or minor proteins VP2 or VP3. This approach enables longer peptide insertions, which is usually impossible in the case of surface-exposed loops. For vaccine development, the epitopes buried within the capsid core might induce an inefficient immune response. For the induction of a CTL response, however, it was shown that the display of epitopes on the capsid surface is not necessary. The immunodominant CD8+ T-cell epitope derived from ovalbumin was fused to the C-terminal part of the VP1 protein. Chimeric VLPs induced CD8+ and CD4+ T cells specific for the ovalbumin epitope (Bickert et al. 2007) and were able to protect mice from ovalbumin-expressing tumors (Brinkman et al. 2004). Moreover, mice vaccinated at 4 and 11 days after the melanoma tumor challenge were also protected against tumor outgrowth (Brinkman et al. 2005).

Vaccination by VLPs with antigen internalized inside the particle against tumors of nonviral origin is well described for the proto-oncogene HER2/neu model. This proto-oncogene is frequently overexpressed in breast, lung, ovarian, gastric, and pancreatic cancer. Transmembrane and extracellular domains of human HER2/neu protein were fused to the MPyV minor protein VP2, and chimeric VLPs (HER2$_{1-683}$PyVLPs) were produced by coexpression with MPyV VP1 protein in a baculovirus expression system. Two different in vivo models were used for vaccination experiments with these particles for testing the rejection of, and protection against, HER2/neu tumors. Protection was assessed in both models after a single vaccination. In the group of mice transfected with human HER2/neu, the vaccination protected mice and rejected their tumors, while in the group of BALB-neuT mice, developing spontaneous neuT-induced tumors in mammary glands, only protection against tumor growth was induced. Both models failed in the induction of HER2-specific antibodies. The protection elicited by these chimeric VLPs was provided by the cellular immune response. The presence of HER2-specific T cells was demonstrated using an enzyme-linked immunospot assay (Tegerstedt et al. 2005b). Similar results were also obtained for MPtV chimeric VLPs (Andreasson et al. 2009). The following study of chimeric MPtV VLPs revealed the stimulation of both CD4+ and CD8+ T-lymphocytes. According to the obtained results, CD4+ and CD8+ T-lymphocytes could act independently, in part, during tumor rejection after vaccination. In combination with CpG oligonucleotides, these chimeric particles induced a long-lasting immunologic memory persisting for at least 10 weeks (Andreasson et al. 2010).

The efficiency of T-cell stimulation could also be enhanced by vaccination with DCs loaded with VLPs in vitro. For the rejection of HER2/neu tumors, a dose of chimeric VP1/VP2-HER2 MPyV VLPs that was 10 times lower was sufficient, when the VLPs were not used directly but loaded onto DCs prior to administration. The vaccine efficiency was preserved, and 100% of treated animals were protected after a single immunization with DCs loaded with chimeric VP1/VP2-HER2 MPyV (Tegerstedt et al. 2005b). The same approach was useful during the vaccination against prostate cancer. Chimeric MPyV VLPs containing PSA fused to the minor structural VP2 protein protected the mice only marginally. However, vaccination with DCs loaded with VP1/VP2-PSA MPyV VLPs in the presence of CpG oligonucleotides induced the protection of mice against the outgrowth of a PSA-expressing tumor. The production of anti-VP1 antibodies was eight times lower compared to vaccination with VP1/VP2-PSA MPyV VLPs alone (Eriksson et al. 2011).

Model antigens for immunization were also conjugated to the minor protein, VP3. The C-terminal sequence (49 aa) of MPyV minor protein VP3 was used for the coupling of enhanced GFP protein. After intranasal administration, both VP1 and EGFP induced proliferation of specific T-helper cells and production of IL-12 and IFN-γ (Bouřa et al. 2005; Frič et al. 2008).

20.5.2.3 Concluding Remarks

The main advantages of PyV VLP–based vaccines can be summarized as follows: first, they are very tolerant to peptide insertions into the VP1 surface loops that generally cause the major problems of generating chimeric VLPs. Polyomaviral VLPs are capable of long peptide conjugations, as was proved by the fusion of the entire enzyme dihydrofolate reductase (Gleiter et al. 1999) or incorporation of a 120 aa long peptide from PUUV nucleoprotein (Gedvilaite et al. 2004). Second, they are highly immunogenic and stimulate the maturation of DCs irrespective of adjuvant presence. Third, they are very stable. They can tolerate relatively high temperatures (up to 70°C) or various pH levels (Nims and Plavsic 2013). The possibility of long-term storage without affecting the immunogenicity of VLPs was also confirmed (Caparrós-Wanderley et al. 2004). No reduced induction of immune responses in intranasally immunized mice was observed after storage of VLPs for 9 weeks at room temperature.

20.5.3 Conclusion

The success of prophylactic HPV vaccines has increased interest in artificial viral particles and has also led to an intensive development of new vaccine applications. Nowadays, a number of VLP-using approaches are being tested not only for vaccination purposes but also for the immunotherapeutic treatment of cancer (Kimchi-Sarfaty and Gottesman 2004; Tegerstedt et al. 2005), rheumatoid arthritis (Chackerian et al. 2001), or even smoking addiction (Maurer et al. 2005). The investigation of VLPs as carriers for protein and DNA delivery was also heightened. The immune response to a transgene carried by PsVs is stronger than the response to transgenes applied to the organism as naked DNA (Clark et al. 2001). Animal VLPs are superior for this kind of application compared to their human counterparts, which might inhibit the efficiency of the vector due to their preexisting immunity in the organism. The high immunogenicity of VLPs could also represent a problem for repeated applications; therefore, modifications of VLPs will be necessary to narrow the range of their target cells or lower VLP recognition by the immune system (Heidari et al. 2000). Recent results have shown that through the fusion of long epitopes, the immune response to the capsid protein is significantly decreased (Zvirbliene et al. 2006), suggesting the feasibility of such modifications. Although a great deal of data has been published, the generation of chimeric VLPs is largely empirical, and nowadays, it is almost impossible to predict whether the modifications will affect the assembly or whether the inserted epitopes will be immunogenic. Solving this uncertainty in the preparation of the desired VLPs will be a great challenge for the future.

ACKNOWLEDGMENTS

This work was supported by the Grant Agency of Charles University (Project GAUK/913613 (Jiřina Suchanová and Hana Španielová); the Grant Agency of the Czech Republic (Project P302/13-26115S) (Jitka Forstová and Hana Španielová); the Ministry of Education, Youth and Sports of the Czech Republic (Project SVV-2014-260081); and Charles University in Prague (Project UNCE 204013).

LIST OF ABBREVIATIONS

aa	amino acid
APyVs	avian polyomaviruses
AuNPs	gold nanoparticles
BFPyV	budgerigar fledgling disease virus, budgerigar fledgling polyomavirus
BKPyV	BK polyomavirus
BPV	bovine papillomavirus
CCR5	chemokine receptor type 5
CEA	carcinoembryonic antigen
CF	cystic fibrosis
CFDA SE	carboxyfluorescein diacetate succinimidyl ester
CFTR	cystic fibrosis transmembrane conductance regulator
ChPyV	chimpanzee polyomavirus
COPV	canine oral papillomavirus
CPyV	crow polyomavirus
CRPV	cottontail rabbit papillomavirus
CTL	cytotoxic T-lymphocyte
DC	dendritic cell
DTT	dithiothreitol
E. coli	*Escherichia coli*
E8C	polyglutamic acid–cysteine
EEs	early endosomes
EGF	epidermal growth factor
EGFP	enhanced green fluorescent protein
EGFR	epidermal growth factor receptor
ER	endoplasmic reticulum
FDA	Food and Drug Administration
FLAG	octapeptide Asp-Tyr-Lys-Asp-Asp-Asp-Asp-Lys
FPyV	finch polyomavirus
GAS	group A streptococcus
GFP	green fluorescent protein
GHPyV	goose hemorrhagic polyomavirus
GST	glutathione S-transferase
HaPyV	hamster polyomavirus
HBV	hepatitis B virus
HER2	human epidermal growth factor receptor 2
HPV	human papillomavirus
HPyV	human polyomavirus
HS	heparan sulfate
HSPGs	heparan sulfate proteoglycans
JCPyV	JC polyomavirus
LCMV	lymphocytic choriomeningitis virus
LEs	late endosomes
LPyV	simian B-lymphotropic polyomavirus

LSTc	lactoseries tetrasaccharide c
MCC	Merkel cell carcinoma
MCPyV	Merkel cell polyomavirus
MDR1	multidrug resistance 1
ME	2-mercaptoethanol
MHC	major histocompatibility complex
MNPs	magnetic nanoparticles
MPtV	murine pneumotropic virus
MPyV	murine polyomavirus
MRI	magnetic resonance imaging
MUC-1	mucin 1
MusPV	mouse papillomavirus
Neu5Ac	5-*N*-acetyl neuraminic acid, sialic acid
PLP	pseudovirion-like particle
PNA	peptide nucleic acid
PSA	prostate-specific antigen
PsV	pseudovirion
PUUV	Puumala hantavirus
PV	papillomavirus
PyV	polyomavirus
QD	quantum dot
QV	quasivirion
QVLP	quasivirion-like particle
R8C	polyarginine cysteine
RGD	tripeptide Arg-Gly-Asp
rSV40	recombinant SV40 vector
Sf9	cell line from *Spodoptera frugiperda*
shRNA	short hairpin RNA
siRNAs	small interfering RNAs
SV40	simian virus 40
T antigen	tumorigenic antigen
TGN	*trans*-Golgi network
TK	thymidine kinase
TNF-α	tumor necrosis factor alpha
TSPyV	Trichodysplasia spinulosa–associated polyomavirus
TSTA	tumor-specific transplantation antigen
uPAR	urokinase-type plasminogen activator receptor
VLP	virus-like particle
VNP	viral nanoparticle

REFERENCES

Abban, C.Y. and P.I. Meneses. 2010. Usage of heparan sulfate, integrins, and FAK in HPV16 infection. *Virology* 403(1) (July 20): 1–16.

Abbing, A., U.K. Blaschke, S. Grein, M. Kretschmar, C.M.B. Stark, M.J.W. Thies, J. Walter et al. 2004. Efficient intracellular delivery of a protein and a low molecular weight substance via recombinant polyomavirus-like particles. *The Journal of Biological Chemistry* 279(26) (June 25): 27410–27421.

Abe, T., H. Takahashi, H. Hamazaki, N. Miyano-Kurosaki, Y. Matsuura, and H. Takaku. 2003. Baculovirus induces an innate immune response and confers protection from lethal influenza virus infection in mice. *The Journal of Immunology* 171(3) (August 1): 1133–1139.

Abend, J.R., M. Jiang, and M.J. Imperiale. 2009. BK virus and human cancer: Innocent until proven guilty. *Seminars in Cancer Biology* 19(4) The Polyomaviruses (August): 252–260.

Aires, K.A., A.M. Cianciarullo, S.M. Carneiro, L.L. Villa, E. Boccardo, G. Pérez-Martinez, I. Perez-Arellano, M.L.S. Oliveira, and P.L. Ho. 2006. Production of human papillomavirus type 16 L1 virus-like particles by recombinant *Lactobacillus casei* cells. *Applied and Environmental Microbiology* 72(1) (January 1): 745–752.

An, K., S.A. Smiley, E.T. Gillock, W.M. Reeves, and R.A. Consigli. 1999. Avian polyomavirus major capsid protein VP1 interacts with the minor capsid proteins and is transported into the cell nucleus but does not assemble into capsid-like particles when expressed in the baculovirus system. *Virus Research* 64(2) (November): 173–185.

Anderer, F.A., H.D. Schlumberger, M.A. Koch, H. Frank, and H.J. Eggers. 1967. Structure of simian virus 40 II. Symmetry and components of the virus particle. *Virology* 32(3): 511–523.

Andreasson, K., M. Eriksson, K. Tegerstedt, T. Ramqvist, and T. Dalianis. 2010. CD4+ and CD8+ T cells can act separately in tumour rejection after immunization with murine pneumotropic virus Chimeric Her2/neu virus-like particles. ed. A. Gregson. *PLoS ONE* 5(7) (July 19): e11580.

Andreasson, K., K. Tegerstedt, M. Eriksson, C. Curcio, F. Cavallo, G. Forni, T. Dalianis, and T. Ramqvist. 2009. Murine pneumotropic virus chimeric Her2/*neu* virus-like particles as prophylactic and therapeutic vaccines against Her2/*neu* expressing tumors. *International Journal of Cancer* 124(1) (January 1): 150–156.

Andres, C., B. Belloni, U. Puchta, C.A. Sander, and M.J. Flaig. 2010. Prevalence of MCPyV in merkel cell carcinoma and non-mcc tumors. *Journal of Cutaneous Pathology* 37(1) (January): 28–34.

Angeletti, P.C. 2005. Replication and encapsidation of papillomaviruses in *Saccharomyces cerevisiae*. *Methods in Molecular Medicine* 119: 247–260.

Angeletti, P.C., K. Kim, F.J. Fernandes, and P.F. Lambert. 2002. Stable replication of papillomavirus genomes in *Saccharomyces cerevisiae*. *Journal of Virology* 76(7) (April): 3350–3358.

Antonsson, A. and B.G. Hansson. 2002. Healthy skin of many animal species harbors papillomaviruses which are closely related to their human counterparts. *Journal of Virology* 76(24) (December 15): 12537–12542.

Aposhian, H.V., R.E. Thayer, and P.K. Qasba. 1975. Formation of nucleoprotein complexes between polyoma empty capsides and DNA. *Journal of Virology* 15(3): 645–653.

Arad, U., E. Zeira, M.A. El-Latif, S. Mukherjee, L. Mitchell, O. Pappo, E. Galun, and A. Oppenheim. 2005. Liver-targeted gene therapy by SV40-based vectors using the hydrodynamic injection method. *Human Gene Therapy* 16(3) (March): 361–371.

Ashok, A. and W.J. Atwood. 2003. Contrasting roles of endosomal pH and the cytoskeleton in infection of human glial cells by JC virus and simian virus 40. *Journal of Virology* 77(2) (January): 1347–1356.

Assetta, B., M.S. Maginnis, I. Gracia Ahufinger, S.A. Haley, G.V. Gee, C.D.S. Nelson, B.A. O'Hara, S.A. Allen Ramdial, and W.J. Atwood. 2013. 5-HT2 receptors facilitate JC polyomavirus entry. *Journal of Virology* 87(24) (December): 13490–13498.

Ault, K.A., A.R. Giuliano, R.P. Edwards, G. Tamms, L.-L. Kim, J.F. Smith, K.U. Jansen et al. 2004. A Phase I study to evaluate a human papillomavirus (HPV) type 18 L1 VLP vaccine. *Vaccine* 22(23–24) (August): 3004–3007.

Aydin, I., S. Weber, B. Snijder, P. Samperio Ventayol, A. Kuhbacher, M. Becker, P.M. Day et al. 2014. Large scale RNAi reveals the requirement of nuclear envelope breakdown for nuclear import of human papillomaviruses. *PLoS Pathogens* 10(5) (May 29): e1004162. http://www.ncbi.nlm.nih.gov/pmc/articles/PMC4038628/. Accessed August 10, 2014.

Bachmann, M.F., H. Hengartner, and R.M. Zinkernagel. 1995. T Helper cell-independent neutralizing B cell response against vesicular stomatitis virus: Role of antigen patterns in B cell induction? *European Journal of Immunology* 25(12) (December 1): 3445–3451.

Baek, J.O., J.W. Seo, O. Kwon, S.M. Park, C.H. Kim, and I.H. Kim. 2012. Production of human papillomavirus type 33 L1 major capsid protein and virus-like particles from *Bacillus subtilis* to develop a prophylactic vaccine against cervical cancer. *Enzyme and Microbial Technology* 50(3) (March 10): 173–180.

Barr, S.M., K. Keck, and H.V. Aposhian. 1979. Cell-free assembly of a polyoma-like particle from empty capsids and DNA. *Virology* 96(2) (July 30): 656–659.

Bauer, P.H., R.T. Bronson, S.C. Fung, R. Freund, T. Stehle, S.C. Harrison, and T.L. Benjamin. 1995. Genetic and structural analysis of a virulence determinant in polyomavirus VP1. *Journal of Virology* 69(12) (January 12): 7925–7931.

Bazan, S.B., A. de A.M. Chaves, K.A. Aires, A.M. Cianciarullo, R.L. Garcea, and P.L. Ho. 2009. Expression and characterization of HPV-16 L1 capsid protein in *Pichia pastoris*. *Archives of Virology* 154(10) (October 1): 1609–1617.

Becker, J.C., R. Houben, S. Ugurel, U. Trefzer, C. Pföhler, and D. Schrama. 2009. MC polyomavirus is frequently present in merkel cell carcinoma of European patients. *The Journal of Investigative Dermatology* 129(1) (January): 248–250.

Bellanger, S., C.L. Tan, Y.Z. Xue, S. Teissier, and F. Thierry. 2011. Tumor suppressor or oncogene? A critical role of the human papillomavirus (HPV) E2 protein in cervical cancer progression. *American Journal of Cancer Research* 1(3): 373.

Belnap, D.M., N.H. Olson, N.M. Cladel, W.W. Newcomb, J.C. Brown, J.W. Kreider, N.D. Christensen, and T.S. Baker. 1996. Conserved features in papillomavirus and polyomavirus capsids. *Journal of Molecular Biology* 259(2): 249–263.

Benson, D.A., I. Karsch-Mizrachi, D.J. Lipman, J. Ostell, and E.W. Sayers. 2009. GenBank. *Nucleic Acids Research* 37(suppl 1) (January 1): D26–D31.

Bergant, M. and L. Banks. 2013. SNX17 facilitates infection with diverse papillomavirus types. *Journal of Virology* 87(2) (January): 1270–1273.

Bergant Marušič, M., M.A. Ozbun, S.K. Campos, M.P. Myers, and L. Banks. 2012. Human papillomavirus L2 facilitates viral escape from late endosomes via sorting nexin 17. *Traffic* (Copenhagen, Denmark) 13(3) (March): 455–467.

Bergsdorf, C., C. Beyer, V. Umansky, M. Werr, and M. Sapp. 2003. Highly efficient transport of carboxyfluorescein diacetate succinimidyl ester into COS7 cells using human papillomavirus-like particles. *FEBS Letters* 536(1–3) (February): 120–124.

Bickert, T., G. Wohlleben, M. Brinkman, C.M. Trujillo-Vargas, C. Ruehland, C.O.A. Reiser, J. Hess, and K.J. Erb. 2007. Murine polyomavirus-like particles induce maturation of bone marrow-derived dendritic cells and proliferation of T cells. *Medical Microbiology and Immunology* 196(1) (March): 31–39.

Biemelt, S., U. Sonnewald, P. Galmbacher, L. Willmitzer, and M. Müller. 2003. Production of human papillomavirus Type 16 virus-like particles in transgenic plants. *Journal of Virology* 77(17) (January 9): 9211–9220.

Bienkowska-Haba, M., H.D. Patel, and M. Sapp. 2009. Target cell cyclophilins facilitate human papillomavirus Type 16 infection. *PLoS Pathogens* 5(7) (July):e1000524. http://www.ncbi.nlm.nih.gov/pmc/articles/PMC2709439/. Accessed August 4, 2014.

Bienkowska-Haba, M., C. Williams, S.M. Kim, R.L. Garcea, and M. Sapp. 2012. Cyclophilins facilitate dissociation of the human papillomavirus type 16 capsid protein L1 from the L2/DNA complex following virus entry. *Journal of Virology* 86(18) (September): 9875–9887.

Bijker, M.S., C.J.M. Melief, R. Offringa, and S.H. van der Burg. 2007. Design and development of synthetic peptide vaccines: Past, present and future. *Expert Review of Vaccines* 6(4) (August): 591–603.

Bishop, B., J. Dasgupta, M. Klein, R.L. Garcea, N.D. Christensen, R. Zhao, and X.S. Chen. 2007. Crystal structures of four types of human papillomavirus L1 capsid proteins: Understanding the specificity of neutralizing monoclonal antibodies. *The Journal of Biological Chemistry* 282(43) (September 6): 31803–31811.

Bishop, C.L., M. Ramalho, N. Nadkarni, W. May Kong, C.F. Higgins, and N. Krauzewicz. 2006. Role for centromeric heterochromatin and PML nuclear bodies in the cellular response to foreign DNA. *Molecular and Cellular Biology* 26(7) (April): 2583–2594.

Bouřa, E., D. Liebl, R. Špíšek, J. Frič, M. Marek, J. Štokrová, V. Holáň, and J. Forstová. 2005. Polyomavirus EGFP-pseudocapsids: Analysis of model particles for introduction of proteins and peptides into mammalian cells. *FEBS Letters* 579(29) (December): 6549–6558.

Bousarghin, L., A. Touzé, G. Gaud, S. Iochmann, E. Alvarez, P. Reverdiau, J. Gaitan, M.-L. Jourdan, P.-Y. Sizaret, and P.L. Coursaget. 2009. Inhibition of cervical cancer cell growth by human papillomavirus virus-like particles packaged with human papillomavirus oncoprotein short hairpin RNAs. *Molecular Cancer Therapeutics* 8(2) (January 2): 357–365.

Bousarghin, L., A. Touzé, P.-Y. Sizaret, and P. Coursaget. 2003. Human papillomavirus types 16, 31, and 58 use different endocytosis pathways to enter cells. *Journal of Virology* 77(6) (March): 3846–3850.

Bouvard, V., R. Baan, K. Straif, Y. Grosse, B. Secretan, F.E. Ghissassi, L. Benbrahim-Tallaa, N. Guha, C. Freeman, and L. Galichet. 2009. A review of human carcinogens—Part B: Biological agents. *The Lancet Oncology* 10(4): 321–322.

Brady, J.N., J.D. Kendall, and R.A. Consigli. 1979. In vitro reassembly of infectious polyoma virions. *Journal of Virology* 32(2) (January 11): 640–647.

Brady, J.N., V.D. Winston, and R.A. Consigli. 1977. Dissociation of polyoma virus by the chelation of calcium ions found associated with purified virions. *Journal of Virology* 23(3) (January 9): 717–724.

Braun, H., K. Boller, J. Löwer, W.M. Bertling, and A. Zimmer. 1999. Oligonucleotide and plasmid DNA packaging into polyoma VP1 virus-like particles expressed in *Escherichia coli*. *Biotechnology and Applied Biochemistry* 29 (Pt 1) (February): 31–43.

Breau, W.C., W.J. Atwood, and L.C. Norkin. 1992. Class I major histocompatibility proteins are an essential component of the simian virus 40 receptor. *Journal of Virology* 66(4) (April): 2037–2045.

Breitburd, F., R. Kirnbauer, N.L. Hubbert, B. Nonnenmacher, C. Trin-Dinh-Desmarquet, G. Orth, J.T. Schiller, and D.R. Lowy. 1995. Immunization with viruslike particles from cottontail rabbit papillomavirus (CRPV) can protect against experimental CRPV infection. *Journal of Virology* 69(6): 3959–3963.

Brinkman, M., J. Walter, S. Grein, M.J.W. Thies, T.W. Schulz, M. Herrmann, C.O.A. Reiser, and J. Hess. 2005. Beneficial therapeutic effects with different particulate structures of murine polyomavirus VP1-coat protein carrying self or non-self CD8 T cell epitopes against murine melanoma. *Cancer Immunology, Immunotherapy* 54(6) (February 1): 611–622.

Brinkman, M., J. Walter, I. Jennes, M. Neugebauer, W.M. Bertling, S. Grein, M.J.W. Thies, M. Weigand, T. Beyer, and M. Herrmann. 2004. Recombinant murine polyoma virus-like-particles induce protective antitumour immunity. *Letters in Drug Design & Discovery* 1(2): 137–147.

Brown, D.R., K.H. Fife, C.M. Wheeler, L.A. Koutsky, L.M. Lupinacci, R. Railkar, G. Suhr et al. 2004. Early assessment of the efficacy of a human papillomavirus type 16 L1 virus-like particle vaccine. *Vaccine* 22(21–22) (July 29): 2936–2942.

Brown, D.R., S.K. Kjaer, K. Sigurdsson, O. Iversen, M. Hernandez-Avila, C.M. Wheeler, G. Perez et al. 2009. The impact of quadrivalent human papillomavirus (HPV; Types 6, 11, 16, and 18) L1 virus-like particle vaccine on infection and disease due to oncogenic nonvaccine HPV types in generally HPV-naive women aged 16–26 years. *The Journal of Infectious Diseases* 199(7) (April): 926–935.

Buck, C.B. 2012. Protocol for harvesting pseudovirus producer cells. http://home.ccr.cancer.gov/lco/pseudovirusproduction.htm. Accessed October 25, 2013.

Buck, C.B., N. Cheng, C.D. Thompson, D.R. Lowy, A.C. Steven, J.T. Schiller, and B.L. Trus. 2008. Arrangement of L2 within the papillomavirus capsid. *Journal of Virology* 82(11) (January 6): 5190–5197.

Buck, C.B., D.V. Pastrana, D.R. Lowy, and J.T. Schiller. 2004. Efficient intracellular assembly of papillomaviral vectors. *Journal of Virology* 78(2) (January): 751–757.

Buck, C.B., D.V. Pastrana, D.R. Lowy, and J.T. Schiller. 2005a. Generation of HPV pseudovirions using transfection and their use in neutralization assays. *Methods in Molecular Medicine* 119: 445–462.

Buck, C.B. and C.D. Thompson. 2007. Production of papillomavirus-based gene transfer vectors. *Current Protocols in Cell Biology*, Eds. J. S. Bonifacino, M. Dasso, J. B. Harford, J. Lippincott-Schwartz, and K. M. Yamada (Bethesda, MD) Chapter 26 (December): Unit 26.1.

Buck, C.B., C.D. Thompson, Y.-Y.S. Pang, D.R. Lowy, and J.T. Schiller. 2005b. Maturation of papillomavirus capsids. *Journal of Virology* 79(5) (March): 2839–2846.

Buck, C.B. and B.L. Trus. 2012. The papillomavirus virion: A machine built to hide molecular achilles' heels. In *Viral Molecular Machines*, ed. M.G. Rossmann and V.B. Rao, pp. 403–422. Advances in Experimental Medicine and Biology, Vol. 726. New York: Springer. http://link.springer.com/chapter/10.1007/978-1-4614-0980-9_18. Accessed January 2, 2014.

Buonamassa, D.T., C.E. Greer, S. Capo, T.S. Benedict Yen, C.L. Galeotti, and G. Bensi. 2002. Yeast coexpression of human papillomavirus types 6 and 16 capsid proteins. *Virology* 293(2): 335–344.

Cadman, L. 2008. The future of cervical cancer prevention: Human papillomavirus vaccines. *The Journal of Family Health Care* 18(4): 131–132.

Caparrós-Wanderley, W., B. Clark, and B.E. Griffin. 2004. Effect of dose and long-term storage on the immunogenicity of murine polyomavirus VP1 virus-like particles. *Vaccine* 22(3–4) (January): 352–361.

Caruso, M., L. Belloni, O. Sthandier, P. Amati, and M.-I. Garcia. 2003. Alpha4beta1 integrin acts as a cell receptor for murine polyomavirus at the postattachment level. *Journal of Virology* 77(7) (April): 3913–3921.

Chackerian, B., L. Briglio, P.S. Albert, D.R. Lowy, and J.T. Schiller. 2004. Induction of autoantibodies to CCR5 in macaques and subsequent effects upon challenge with an R5-tropic simian/human immunodeficiency virus. *Journal of Virology* 78(8) (April 15): 4037–4047.

Chackerian, B., D.R. Lowy, and J.T. Schiller. 1999. Induction of autoantibodies to mouse CCR5 with recombinant papillomavirus particles. *Proceedings of the National Academy of Sciences of the United States of America* 96(5): 2373–2378.

Chackerian, B., D.R. Lowy, and J.T. Schiller. 2001. Conjugation of a self-antigen to papillomavirus-like particles allows for efficient induction of protective autoantibodies. *The Journal of Clinical Investigation* 108(3) (August): 415–423.

Chang, D., C.Y. Fung, W.C. Ou, P.C. Chao, S.Y. Li, M. Wang, Y.L. Huang, T.Y. Tzeng, and R.T. Tsai. 1997. Self-assembly of the JC virus major capsid protein, VP1, expressed in insect cells. *Journal of General Virology* 78(6) (January 6): 1435–1439.

Chapuis, A.G., O.K. Afanasiev, J.G. Iyer, K.G. Paulson, U. Parvathaneni, J.H. Hwang, I. Lai et al. 2014. Regression of metastatic merkel cell carcinoma following transfer of polyomavirus-specific T cells and therapies capable of re-inducing HLA class-I. *Cancer Immunology Research* 2(1) (January 1): 27–36.

Chen, L.S., M. Wang, W.C. Ou, C.Y. Fung, P.L. Chen, C.F. Chang, W.S. Huang, J.Y. Wang, P.Y. Lin, and D. Chang. 2010. Efficient gene transfer using the human JC virus-like particle that inhibits human colon adenocarcinoma growth in a nude mouse model. *Gene Therapy* 17(8): 1033–1041.

Chen, P.-L., M. Wang, W.-C. Ou, C.-K. Lii, L.-S. Chen, and D. Chang. 2001. Disulfide bonds stabilize JC virus capsid-like structure by protecting calcium ions from chelation. *FEBS Letters* 500(3): 109–113.

Chen, T., P.S. Mattila, T. Jartti, O. Ruuskanen, M. Söderlund-Venermo, and K. Hedman. 2011. Seroepidemiology of the newly found trichodysplasia spinulosa-associated polyomavirus. *Journal of Infectious Diseases* 204(10) (November 15): 1523–1526.

Chen, X., H. Liu, T. Zhang, Y. Liu, X. Xie, Z. Wang, and X. Xu. 2014. A vaccine of L2 epitope repeats fused with a modified IgG1 Fc induced cross-neutralizing antibodies and protective immunity against divergent human papillomavirus types. *PLoS ONE* 9(5): e95448.

Chen, X.S., G. Casini, S.C. Harrison, and R.L. Garcea. 2001. Papillomavirus capsid protein expression in *Escherichia coli*: Purification and assembly of HPV11 and HPV16 L1. *Journal of Molecular Biology* 307(1) (March 16): 173–182.

Chen, X.S., R.L. Garcea, I. Goldberg, G. Casini, and S.C. Harrison. 2000. Structure of small virus-like particles assembled from the L1 protein of human papillomavirus 16. *Molecular Cell* 5(3): 557–567.

Chen, X.S., T. Stehle, and S.C. Harrison. 1998. Interaction of polyomavirus internal protein VP2 with the major capsid protein VP1 and implications for participation of VP2 in viral entry. *The EMBO Journal* 17(12): 3233–3240.

Cheng, Z., A.A. Zaki, J.Z. Hui, V.R. Muzykantov, and A. Tsourkas. 2012. Multifunctional nanoparticles: Cost versus benefit of adding targeting and imaging capabilities. *Science* 338(6109) (November 16): 903–910.

Chestier, A. and M. Yaniv. 1979. Rapid turnover of acetyl groups in the four core histones of simian virus 40 minichromosomes. *Proceedings of the National Academy of Sciences of the United States of America* 76(1) (January 1): 46–50.

Cho, H.-J., Y.-K. Oh, and Y.B. Kim. 2011. Advances in human papilloma virus vaccines: A patent review. *Expert Opinion on Therapeutic Patents* 21(3) (March): 295–309.

Chou, M.-I., Y.-F. Hsieh, M. Wang, J.T. Chang, D. Chang, M. Zouali, and G.J. Tsay. 2010. In vitro and in vivo targeted delivery of IL-10 interfering RNA by JC virus-like particles. *Journal of Biomedical Science* 17(1) (June 24): 51.

Christensen, N.D., R. Höpfl, S.L. DiAngelo, N.M. Cladel, S.D. Patrick, P.A. Welsh, L.R. Budgeon, C.A. Reed, and J.W. Kreider. 1994. Assembled baculovirus-expressed human papillomavirus type 11 L1 capsid protein virus-like particles are recognized by neutralizing monoclonal antibodies and induce high titres of neutralizing antibodies. *Journal of General Virology* 75(9) (January 9): 2271–2276.

Chromy, L.R., J.M. Pipas, and R.L. Garcea. 2003. Chaperone-mediated in vitro assembly of polyomavirus capsids. *Proceedings of the National Academy of Sciences of the United States of America* 100(18): 10477–10482.

Chuan, Y.P., Y.Y. Fan, L.H.L. Lua, and A.P.J. Middelberg. 2010. Virus assembly occurs following a pH- or Ca^{2+}-triggered switch in the thermodynamic attraction between structural protein capsomeres. *Journal of the Royal Society Interface* 7(44) (March 6): 409–421.

Chuan, Y.P., L.H.L. Lua, and A.P.J. Middelberg. 2008. High-level expression of soluble viral structural protein in *Escherichia coli*. *Journal of Biotechnology* 134(1–2) (March 20): 64–71.

Chuan, Y.P., T. Rivera-Hernandez, N. Wibowo, N.K. Connors, Y. Wu, F.K. Hughes, L.H.L. Lua, and A.P.J. Middelberg. 2013. Effects of pre-existing anti-carrier immunity and antigenic element multiplicity on efficacy of a modular virus-like particle vaccine. *Biotechnology and Bioengineering* 110(9) (September): 2343–2351.

Clark, B., W. Caparros-Wanderley, G. Musselwhite, M. Kotecha, and B.E. Griffin. 2001. Immunity against both polyomavirus VP1 and a transgene product induced following intranasal delivery of VP1 pseudocapsid–DNA complexes. *Journal of General Virology* 82(11): 2791–2797.

Coca-Prados, M., G. Vidali, and M.T. Hsu. 1980. Intracellular forms of simian virus 40 nucleoprotein complexes. III. Study of histone modifications. *Journal of Virology* 36(2): 353–360.

Coelho, T.R., L. Almeida, and P.A. Lazo. 2010. JC virus in the pathogenesis of colorectal cancer, an etiological agent or another component in a multistep process? *Virology Journal* 7: 42.

Colomar, M.C., C. Degoumois-Sahli, and P. Beard. 1993. Opening and refolding of simian virus 40 and in vitro packaging of foreign DNA. *Journal of Virology* 67(5) (May): 2779–2786.

Combita, A.L., A. Touzé, L. Bousarghin, P.Y. Sizaret, N. Muñoz, and P. Coursaget. 2001. Gene transfer using human papillomavirus pseudovirions varies according to virus genotype and requires cell surface heparan sulfate. *FEMS Microbiology Letters* 204(1) (October 16): 183–188.

Cook, J.C., J.G. Joyce, H.A. George, L.D. Schultz, W.M. Hurni, K.U. Jansen, R.W. Hepler et al. 1999. Purification of virus-like particles of recombinant human papillomavirus type 11 major capsid protein 11 from *Saccharomyces cerevisiae*. *Protein Expression and Purification* 17(3) (December): 477–484.

Corradini, R., S. Sforza, T. Tedeschi, F. Totsingan, and R. Marchelli. 2007. Peptide nucleic acids with a structurally biased backbone: Effects of conformational constraints and stereochemistry. *Current Topics in Medicinal Chemistry* 7(7): 681–694.

Crawford, L.V., E.M. Crawford, and D.H. Watson. 1962. The physical characteristics of polyoma virus. I. Two types of particle. *Virology* 18 (October): 170–176.

Cristiano, R.J., L.C. Smith, M.A. Kay, B.R. Brinkley, and S.L. Woo. 1993. Hepatic gene therapy: Efficient gene delivery and expression in primary hepatocytes utilizing a conjugated adenovirus–DNA complex. *Proceedings of the National Academy of Sciences of the United States of America* 90(24) (December 15): 11548–11552.

Culp, T.D., L.R. Budgeon, and N.D. Christensen. 2006a. Human papillomaviruses bind a basal extracellular matrix component secreted by keratinocytes which is distinct from a membrane-associated receptor. *Virology* 347(1) (March 30): 147–159.

Culp, T.D., L.R. Budgeon, M.P. Marinkovich, G. Meneguzzi, and N.D. Christensen. 2006b. Keratinocyte-secreted laminin 5 can function as a transient receptor for human papillomaviruses by binding virions and transferring them to adjacent cells. *Journal of Virology* 80(18) (September): 8940–8950.

Culp, T.D. and N.D. Christensen. 2004. Kinetics of in vitro adsorption and entry of papillomavirus virions. *Virology* 319(1) (February): 152–161.

Culp, T.D., N.M. Cladel, K.K. Balogh, L.R. Budgeon, A.F. Mejia, and N.D. Christensen. 2006c. Papillomavirus particles assembled in 293TT cells are infectious in vivo. *Journal of Virology* 80(22) (January 11): 11381–11384.

Dabydeen, S.A. and P.I. Meneses. 2009. The role of NH4Cl and cysteine proteases in human papillomavirus Type 16 infection. *Virology Journal* 6: 109.

Dalianis, T. and H.H. Hirsch. 2013. Human polyomaviruses in disease and cancer. *Virology* 437(2) (March 15): 63–72.

Dalyot-Herman, N., D. Rund, and A. Oppenheim. 1999. Expression of beta-globin in primary erythroid progenitors of beta-thalassemia patients using an SV40-based gene delivery system. *Journal of Hematotherapy & Stem Cell Research* 8(6) (December): 593–599.

Daniels, R., N.M. Rusan, P. Wadsworth, and D.N. Hebert. 2006. SV40 VP2 and VP3 insertion into ER membranes is controlled by the capsid protein VP1: Implications for DNA translocation out of the ER. *Molecular Cell* 24(6) (December 28): 955–966.

Darshan, M.S., J. Lucchi, E. Harding, and J. Moroianu. 2004. The L2 minor capsid protein of human papillomavirus type 16 interacts with a network of nuclear import receptors. *Journal of Virology* 78(22) (November): 12179–12188.

Dasgupta, J., M. Bienkowska-Haba, M.E. Ortega, H.D. Patel, S. Bodevin, D. Spillmann, B. Bishop, M. Sapp, and X.S. Chen. 2011. Structural basis of oligosaccharide receptor recognition by human papillomavirus. *The Journal of Biological Chemistry* 286(4) (January 28): 2617–2624.

Da Silva, D.M., J.T. Schiller, and W.M. Kast. 2003. Heterologous boosting increases immunogenicity of chimeric papillomavirus virus-like particle vaccines. *Vaccine* 21(23) (July): 3219–3227.

Davidson, E.J. 2003. Human papillomavirus type 16 E2- and L1-specific serological and T-Cell responses in women with vulval intraepithelial neoplasia. *Journal of General Virology* 84(8) (August 1): 2089–2097.

Day, P.M., C.C. Baker, D.R. Lowy, and J.T. Schiller. 2004. Establishment of papillomavirus infection is enhanced by promyelocytic leukemia protein (PML) expression. *Proceedings of the National Academy of Sciences of the United States of America* 101(39) (September 28): 14252–14257.

Day, P.M., R. Gambhira, R.B.S. Roden, D.R. Lowy, and J.T. Schiller. 2008. Mechanisms of human papillomavirus type 16 neutralization by l2 cross-neutralizing and l1 type-specific antibodies. *Journal of Virology* 82(9) (May): 4638–4646.

Day, P.M., R.C. Kines, C.D. Thompson, S. Jagu, R.B. Roden, D.R. Lowy, and J.T. Schiller. 2010. In vivo mechanisms of vaccine-induced protection against HPV infection. *Cell Host & Microbe* 8(3) (September): 260–270.

Day, P.M., D.R. Lowy, and J.T. Schiller. 2003. Papillomaviruses infect cells via a clathrin-dependent pathway. *Virology* 307(1) (March 1): 1–11.

Day, P.M., D.R. Lowy, and J.T. Schiller. 2008. Heparan sulfate-independent cell binding and infection with furin-precleaved papillomavirus capsids. *Journal of Virology* 82(24) (December): 12565–12568.

Day, P.M. and M. Schelhaas. 2014. Concepts of papillomavirus entry into host cells. *Current Opinion in Virology* 4 (February): 24–31.

Day, P.M., C.D. Thompson, R.M. Schowalter, D.R. Lowy, and J.T. Schiller. 2013. Identification of a role for the trans-golgi network in human papillomavirus 16 pseudovirus infection. *Journal of Virology* 87(7) (April): 3862–3870.

DeLano, W.L. 2006. *The PyMOL Molecular Graphics System*, version: v0.99. Cambridge, MA: Schrödinger LCC.

Denny, L.A., S. Franceschi, S. de Sanjosé, I. Heard, A.B. Moscicki, and J. Palefsky. 2012. Human papillomavirus, human immunodeficiency virus and immunosuppression. *Vaccine* 30, Supplement 5 (November 20): F168–F174.

De Villiers, E.-M. 2013. Cross-roads in the classification of papillomaviruses. *Virology* 445(1–2): 2–10.

De Villiers, E.-M., C. Fauquet, T.R. Broker, H.-U. Bernard, and H. zur Hausen. 2004. Classification of papillomaviruses. *Virology* 324(1): 17–27.

Dorn, D.C., R. Lawatscheck, A. Zvirbliene, E. Aleksaite, G. Pecher, K. Sasnauskas, M. Özel et al. 2008. Cellular and humoral immunogenicity of hamster polyomavirus-derived virus-like particles harboring a mucin 1 cytotoxic T-cell epitope. *Viral Immunology* 21(1) (March): 12–26.

Drobni, P., N. Mistry, N. McMillan, and M. Evander. 2003. Carboxy-fluorescein diacetate, succinimidyl ester labeled papillomavirus virus-like particles fluoresce after internalization and interact with heparan sulfate for binding and entry. *Virology* 310(1) (May): 163–172.

Dugan, A.S., M.L. Gasparovic, and W.J. Atwood. 2008. Direct correlation between sialic acid binding and infection of cells by two human polyomaviruses (JC Virus and BK Virus). *Journal of Virology* 82(5) (March): 2560–2564.

Dürst, M., L. Gissmann, H. Ikenberg, and H. zur Hausen. 1983. A papillomavirus DNA from a cervical carcinoma and its prevalence in cancer biopsy samples from different geographic regions. *Proceedings of the National Academy of Sciences of the United States of America* 80(12) (January 6): 3812–3815.

Dyson, N., P.M. Howley, K. Münger, and E. Harlow. 1989. The human papilloma virus-16 E7 oncoprotein is able to bind to the retinoblastoma gene product. *Science* (New York) 243(4893) (February 17): 934–937.

Eash, S., W. Querbes, and W.J. Atwood. 2004. Infection of vero cells by BK virus is dependent on caveolae. *Journal of Virology* 78(21) (November): 11583–11590.

Ehlers, B. and U. Wieland. 2013. The novel human polyomaviruses HPyV6, 7, 9 and beyond. *APMIS* 121(8): 783–795.

Einstein, M.H., M. Baron, M.J. Levin, A. Chatterjee, R.P. Edwards, F. Zepp, I. Carletti et al. 2009. Comparison of the immunogenicity and safety of cervarix and gardasil human papillomavirus (HPV) cervical cancer vaccines in healthy women aged 18–45 years. *Human Vaccines* 5(10) (October): 705–719.

Elphick, G.F., W. Querbes, J.A. Jordan, G.V. Gee, S. Eash, K. Manley, A. Dugan et al. 2004. The human polyomavirus, JCV, uses serotonin receptors to infect cells. *Science* (New York) 306(5700) (November 19): 1380–1383.

El-Sayed, A., S. Futaki, and H. Harashima. 2009. Delivery of macromolecules using arginine-rich cell-penetrating peptides: Ways to overcome endosomal entrapment. *The AAPS Journal* 11(1) (January 6): 13–22.

Engel, S., T. Heger, R. Mancini, F. Herzog, J. Kartenbeck, A. Hayer, and A. Helenius. 2011. Role of endosomes in simian virus 40 entry and infection. *Journal of Virology* 85(9) (May): 4198–4211.

Enomoto, T., M. Kawano, H. Fukuda, W. Sawada, T. Inoue, K.C. Haw, Y. Kita et al. 2013. Viral protein-coating of magnetic nanoparticles using simian virus 40 VP1. *Journal of Biotechnology* 167(1) (August 10): 8–15.

Enomoto, T., I. Kukimoto, M. Kawano, Y. Yamaguchi, A.J. Berk, and H. Handa. 2011. In vitro reconstitution of SV40 particles that are composed of VP1/2/3 capsid proteins and nucleosomal DNA and direct efficient gene transfer. *Virology* 420(1) (November 10): 1–9.

Erickson, K.D., R.L. Garcea, and B. Tsai. 2009. Ganglioside GT1b is a putative host cell receptor for the merkel cell polyomavirus. *Journal of Virology* 83(19) (October): 10275–10279.

Eriksson, M., K. Andreasson, J. Weidmann, K. Lundberg, K. Tegerstedt, T. Dalianis, and T. Ramqvist. 2011. Murine polyomavirus virus-like particles carrying full-length human PSA protect BALB/c mice from outgrowth of a PSA expressing tumor. ed. M.M. Rodrigues. *PLoS ONE* 6(8) (August 17): e23828.

Evander, M., I.H. Frazer, E. Payne, Y.M. Qi, K. Hengst, and N.A. McMillan. 1997. Identification of the alpha6 integrin as a candidate receptor for papillomaviruses. *Journal of Virology* 71(3) (March): 2449–2456.

Ewers, H., and A. Helenius. 2011. Lipid-mediated endocytosis. *Cold Spring Harbor Perspectives in Biology* 3(8) (August): a004721.

Fang, C.-Y., H.-Y. Chen, M. Wang, P.-L. Chen, C.-F. Chang, L.-S. Chen, C.-H. Shen et al. 2010. Global analysis of modifications of the human BK virus structural proteins by LC–MS/MS. *Virology* 402(1): 164–176.

Fang, N.X., I.H. Frazer, J. Zhou, and G.J. Fernando. 1999. Post translational modifications of recombinant human papillomavirus type 6b major capsid protein. *Virus Research* 60(2) (April): 113–121.

Fausch, S.C., D.M. Da Silva, and W.M. Kast. 2003. Differential uptake and cross-presentation of human papillomavirus virus-like particles by dendritic cells and langerhans cells. *Cancer Research* 63(13): 3478–3482.

Feng, H., M. Shuda, Y. Chang, and P.S. Moore. 2008. Clonal integration of a polyomavirus in human merkel cell carcinoma. *Science* (New York) 319(5866) (February 22): 1096–1100.

Fernández-San Millán, A., S.M. Ortigosa, S. Hervás-Stubbs, P. Corral-Martínez, J.M. Seguí-Simarro, J. Gaétan, P. Coursaget, and J. Veramendi. 2008. Human papillomavirus L1 protein expressed in tobacco chloroplasts self-assembles into virus-like particles that are highly immunogenic. *Plant Biotechnology Journal* 6(5) (June): 427–441.

Fife, K.H., C.M. Wheeler, L.A. Koutsky, E. Barr, D.R. Brown, M.A. Schiff, N.B. Kiviat et al. 2004. Dose-ranging studies of the safety and immunogenicity of human papillomavirus type 11 and type 16 virus-like particle candidate vaccines in young healthy women. *Vaccine* 22(21–22) (July 29): 2943–2952.

Finch, J.T. 1974. The surface structure of polyoma virus. *Journal of General Virology* 24(2) (January 8): 359–364.

Finnen, R.L., K.D. Erickson, X.S. Chen, and R.L. Garcea. 2003. Interactions between papillomavirus L1 and L2 capsid proteins. *Journal of Virology* 77(8) (April 15): 4818–4826.

Fischer, R., E. Stoger, S. Schillberg, P. Christou, and R.M. Twyman. 2004. Plant-based production of biopharmaceuticals. *Current Opinion in Plant Biology* 7(2) (April): 152–158.

Fishman, J.A. 2002. BK virus nephropathy—Polyomavirus adding insult to injury. *The New England Journal of Medicine* 347(7) (August 15): 527–530.

FitzSimons, D., G. François, A. Hall, B. McMahon, A. Meheus, A. Zanetti, B. Duval et al. 2005. Long-term efficacy of hepatitis B vaccine, booster policy, and impact of hepatitis B virus mutants. *Vaccine* 23(32) (July): 4158–4166.

Fligge, C., F. Schafer, H.-C. Selinka, C. Sapp, and M. Sapp. 2001. DNA-induced structural changes in the papillomavirus capsid. *Journal of Virology* 75(16) (August): 7727–7731.

Florin, L., K.A. Becker, C. Lambert, T. Nowak, C. Sapp, D. Strand, R.E. Streeck, and M. Sapp. 2006. Identification of a dynein interacting domain in the papillomavirus minor capsid protein l2. *Journal of Virology* 80(13) (July): 6691–6696.

Forstová, J., N. Krauzewicz, V. Sandig, J. Elliott, Z. Palková, M. Strauss, and B.E. Griffin. 1995. Polyoma virus pseudocapsids as efficient carriers of heterologous DNA into mammalian cells. *Human Gene Therapy* 6(3) (March): 297–306.

Forstová, J., N. Krauzewicz, S. Wallace, A.J. Street, S.M. Dilworth, S. Beard, and B.E. Griffin. 1993. Cooperation of structural proteins during late events in the life cycle of polyomavirus. *Journal of Virology* 67(3) (January 3): 1405–1413.

Fothergill, T. and N.A.J. McMillan. 2006. Papillomavirus virus-like particles activate the PI3-kinase pathway via alpha-6 beta-4 integrin upon binding. *Virology* 352(2) (September 1): 319–328.

Franzén, A.V., K. Tegerstedt, D. Holländerova, J. Forstová, T. Ramqvist, and T. Dalianis. 2005. Murine polyomavirus-VP1 virus-like particles immunize against some polyomavirus-induced tumours. *In Vivo* (Athens, Greece) 19(2) (April): 323–326.

Frič, J., M. Marek, V. Hrušková, V. Holáň, and J. Forstová. 2008. Cellular and humoral immune responses to chimeric EGFP-pseudocapsids derived from the mouse polyomavirus after their intranasal administration. *Vaccine* 26(26) (June): 3242–3251.

Furumoto, H. and M. Irahara. 2002. Human papilloma virus (HPV) and cervical cancer. *The Journal of Medical Investigation* 49(3–4) (August): 124–133.

Gambhira, R., B. Karanam, S. Jagu, J.N. Roberts, C.B. Buck, I. Bossis, H. Alphs, T. Culp, N.D. Christensen, and R.B.S. Roden. 2007. A protective and broadly cross-neutralizing epitope of human papillomavirus L2. *Journal of Virology* 81(24) (December): 13927–13931.

Gao, D., Z.-P. Zhang, F. Li, D. Men, J.-Y. Deng, H.-P. Wei, X.-E. Zhang, and Z.-Q. Cui. 2013. Quantum dot-induced viral capsid assembling in dissociation buffer. *International Journal of Nanomedicine* 8: 2119–2128.

Garcea, R.L. and X.S. Chen. 2007. Papillomavirus structure and assembly. In *The Papillomaviruses*. Eds. R.L. Garcea and D. DiMaio. New York: Springer.

Gardner, S.D., A.M. Field, D.V. Coleman, and B. Hulme. 1971. New human papovavirus (B.K.) isolated from urine after renal transplantation. *Lancet* 1(7712) (June 19): 1253–1257.

Garneski, K.M., A.H. Warcola, Q. Feng, N.B. Kiviat, J.H. Leonard, and P. Nghiem. 2009. Merkel cell polyomavirus is more frequently present in North American than Australian merkel cell carcinoma tumors. *Journal of Investigative Dermatology* 129(1) (January): 246–248.

Gedvilaite, A., D.C. Dorn, K. Sasnauskas, G. Pecher, A. Bulavaite, R. Lawatscheck, J. Staniulis et al. 2006. Virus-like particles derived from major capsid protein VP1 of different polyomaviruses differ in their ability to induce maturation in human dendritic cells. *Virology* 354(2) (October): 252–260.

Gedvilaite, A., C. Frömmel, K. Sasnauskas, B. Micheel, M. Özel, O. Behrsing, J. Staniulis, B. Jandrig, S. Scherneck, and R. Ulrich. 2000. Formation of immunogenic virus-like particles by inserting epitopes into surface-exposed regions of hamster polyomavirus major capsid protein. *Virology* 273(1) (July): 21–35.

Gedvilaite, A., A. Zvirbliene, J. Staniulis, K. Sasnauskas, D.H. Krüger, and R. Ulrich. 2004. Segments of puumala hantavirus nucleocapsid protein inserted into chimeric polyomavirus-derived virus-like particles induce a strong immune response in mice. *Viral Immunology* 17(1): 51–68.

Gerber, S., C. Lane, D.M. Brown, E. Lord, M. DiLorenzo, J.D. Clements, E. Rybicki, A.-L. Williamson, and R.C. Rose. 2001. Human papillomavirus virus-like particles are efficient oral immunogens when coadministered with *Escherichia coli* heat-labile enterotoxin mutant R192G or CpG DNA. *Journal of Virology* 75(10) (May 15): 4752–4760.

Gilbert, J. and T. Benjamin. 2004. Uptake pathway of polyomavirus via ganglioside GD1a. *Journal of Virology* 78(22) (November): 12259–12267.

Gilbert, J., W. Ou, J. Silver, and T. Benjamin. 2006. Downregulation of protein disulfide isomerase inhibits infection by the mouse polyomavirus. *Journal of Virology* 80(21) (November): 10868–10870.

Gilbert, J.M. and T.L. Benjamin. 2000. Early steps of polyomavirus entry into cells. *Journal of Virology* 74(18) (September): 8582–8588.

Gilbert, J.M., I.G. Goldberg, and T.L. Benjamin. 2003. Cell penetration and trafficking of polyomavirus. *Journal of Virology* 77(4) (February): 2615–2622.

Gillock, E.T., S. Rottinghaus, D. Chang, X. Cai, S.A. Smiley, K. An, and R.A. Consigli. 1997. Polyomavirus major capsid protein VP1 is capable of packaging cellular DNA when expressed in the baculovirus system. *Journal of Virology* 71(4): 2857–2865.

Gioux, S., H.S. Choi, and J.V. Frangioni. 2010. Image-guided surgery using invisible near-infrared light: Fundamentals of clinical translation. *Molecular Imaging* 9(5) (October): 237–255.

Giroglou, T., L. Florin, F. Schafer, R.E. Streeck, and M. Sapp. 2001. Human papillomavirus infection requires cell surface heparan sulfate. *Journal of Virology* 75(3) (February): 1565–1570.

Giroglou, T., M. Sapp, C. Lane, C. Fligge, N.D. Christensen, R.E. Streeck, and R.C. Rose. 2001. Immunological analyses of human papillomavirus capsids. *Vaccine* 19(13–14): 1783–1793.

Giuliano, A.R., J.M. Palefsky, S. Goldstone, E.D. Moreira, M.E. Penny, C. Aranda, E. Vardas et al. 2011. Efficacy of quadrivalent HPV vaccine against HPV infection and disease in males. *New England Journal of Medicine* 364(5) (February 3): 401–411.

GlaxoSmithKline Vaccine HPV-007 Study Group, B. Romanowski, P.C. de Borba, P.S. Naud, C.M. Roteli-Martins, N.S. De Carvalho, J.C. Teixeira et al. 2009. Sustained efficacy and immunogenicity of the human papillomavirus (HPV)-16/18 AS04-adjuvanted vaccine: Analysis of a randomised placebo-controlled trial up to 6.4 years. *Lancet* 374(9706) (December 12): 1975–1985.

Gleiter, S. and H. Lilie. 2001. Coupling of antibodies via protein Z on modified polyoma virus-like particles. *Protein Science* 10(2): 434–444.

Gleiter, S., K. Stubenrauch, and H. Lilie. 1999. Changing the surface of a virus shell fusion of an enzyme to polyoma VP1. *Protein Science* 8(12): 2562–2569.

Goldmann, C., H. Petry, S. Frye, O. Ast, S. Ebitsch, K.-D. Jentsch, F.-J. Kaup, F. Weber, C. Trebst, and T. Nisslein. 1999. Molecular Cloning and Expression of Major Structural Protein VP1 of the human polyomavirus JC virus: Formation of virus-like particles useful for immunological and therapeutic studies. *Journal of Virology* 73(5): 4465–4469.

Goldmann, C., N. Stolte, T. Nisslein, G. Hunsmann, W. Lüke, and H. Petry. 2000. Packaging of small molecules into VP1-virus-like particles of the human polyomavirus JC virus. *Journal of Virological Methods* 90(1): 85–90.

Gonçalves, A.K., R.N. Cobucci, H.M. Rodrigues, A.G. de Melo, and P.C. Giraldo. 2014. Safety, tolerability and side effects of human papillomavirus vaccines: A systematic quantitative review. *The Brazilian Journal of Infectious Diseases* 18(6) (April 27): 651–659. http://linkinghub.elsevier.com/retrieve/pii/S1413867014000695. Accessed July 14, 2014.

Gordon, S.N., R.C. Kines, G. Kutsyna, Z.-M. Ma, A. Hryniewicz, J.N. Roberts, C. Fenizia et al. 2012. Targeting the vaginal mucosa with human papillomavirus pseudovirion vaccines delivering simian immunodeficiency virus DNA. *The Journal of Immunology* 188(2) (January 15): 714–723.

Goss, M.A., F. Lievano, M.M. Seminack, and A. Dana. 2014. No adverse signals observed after exposure to human papillomavirus type 6/11/16/18 vaccine during pregnancy: 6-Year pregnancy registry data. *Obstetrics and Gynecology* 123 Suppl. 1 (May): 93S.

Graham, B.S., R. Kines, K.S. Corbett, J. Nicewonger, T.R. Johnson, M. Chen, D. LaVigne et al. 2010. Mucosal delivery of human papillomavirus pseudovirus-encapsidated plasmids improves potency of DNA vaccination. *Mucosal Immunology* 3(5) (September): 475–486.

Greenstone, H.L., J.D. Nieland, K.E. De Visser, M.L. De Bruijn, R. Kirnbauer, R.B. Roden, D.R. Lowy, W.M. Kast, and J.T. Schiller. 1998. Chimeric papillomavirus virus-like particles elicit antitumor immunity against the E7 oncoprotein in an HPV16 tumor model. *Proceedings of the National Academy of Sciences of the United States of America* 95(4): 1800–1805.

Griffith, J.P., D.L. Griffith, I. Rayment, W.T. Murakami, and D.L. Caspar. 1992. Inside polyomavirus at 25-A resolution. *Nature* 355(6361) (February 13): 652–654.

Günther, C., U. Schmidt, R. Rudolph, and G. Böhm. 2001. Protein and peptide delivery via engineered polyomavirus-like particles. *The FASEB Journal* 15(9) (May 9): 1646–1648. http://www.fasebj.org/content/early/2001/07/02/fj.00-0645fje.short. Accessed July 31, 2014.

Gupta, A.K. and M. Gupta. 2005. Synthesis and surface engineering of iron oxide nanoparticles for biomedical applications. *Biomaterials* 26(18) (June): 3995–4021.

Hagensee, M.E., N. Yaegashi, and D.A. Galloway. 1993. Self-assembly of human papillomavirus type 1 capsids by expression of the L1 protein alone or by coexpression of the L1 and L2 capsid proteins. *Journal of Virology* 67(1) (January): 315–322.

Hale, A.D., D. Bartkeviciūte, A. Dargeviciūte, L. Jin, W. Knowles, J. Staniulis, D.W.G. Brown, and K. Sasnauskas. 2002. Expression and antigenic characterization of the major capsid proteins of human polyomaviruses BK and JC in *Saccharomyces cerevisiae*. *Journal of Virological Methods* 104(1): 93–98.

Handisurya, A., P.M. Day, C.D. Thompson, C.B. Buck, K. Kwak, R.B.S. Roden, D.R. Lowy, and J.T. Schiller. 2012. Murine skin and vaginal mucosa are similarly susceptible to infection by pseudovirions of different papillomavirus classifications and species. *Virology* 433(2) (November 25): 385–394.

Harper, D.M., E.L. Franco, C. Wheeler, D.G. Ferris, D. Jenkins, A. Schuind, T. Zahaf et al. 2004. Efficacy of a bivalent L1 virus-like particle vaccine in prevention of infection with human papillomavirus types 16 and 18 in young women: A randomised controlled trial. *The Lancet* 364(9447): 1757–1765.

Haugg, A.M., D. Rennspiess, A. zur Hausen, E.-J.M. Speel, G. Cathomas, J.C. Becker, and D. Schrama. 2014. Fluorescence in situ hybridization and qPCR to detect merkel cell polyomavirus physical status and load in merkel cell carcinomas. *International Journal of Cancer* 135(12) (May 9): 2804–2815.

Hayman, W.A., E.R. Brandt, W.A. Relf, J. Cooper, A. Saul, and M.F. Good. 1997. Mapping the minimal murine T cell and B cell epitopes within a peptide vaccine candidate from the conserved region of the M protein of group A Streptococcus. *International Immunology* 9(11) (November): 1723–1733.

Haynes, J.I., D. Chang, and R.A. Consigli. 1993. Mutations in the putative calcium-binding domain of polyomavirus VP1 affect capsid assembly. *Journal of Virology* 67(5) (January 5): 2486–2495.

Heidari, S., N. Krauzewicz, M. Kalantari, A. Vlastos, B.E. Griffin, and T. Dalianis. 2000. Persistence and tissue distribution of DNA in normal and immunodeficient mice inoculated with polyomavirus VP1 pseudocapsid complexes or polyomavirus. *Journal of Virology* 74(24) (December): 11963–11965.

Heidari, S., A. Vlastos, T. Ramqvist, B. Clark, B.E. Griffin, M.-I. Garcia, M. Perez, P. Amati, and T. Dalianis. 2002. Immunization of T-cell deficient mice against polyomavirus infection using viral pseudocapsids or temperature sensitive mutants. *Vaccine* 20(11): 1571–1578.

Heino, P., J. Dillner, and S. Schwartz. 1995. Human papillomavirus type 16 capsid proteins produced from recombinant semliki forest virus assemble into virus-like particles. *Virology* 214(2) (December 20): 349–359.

Henke, S., A. Rohmann, W.M. Bertling, T. Dingermann, and A. Zimmer. 2000. Enhanced in vitro oligonucleotide and plasmid DNA transport by VP1 virus-like particles. *Pharmaceutical Research* 17(9) (September): 1062–1070.

Hildesheim, A., R. Herrero, S. Wacholder, A.C. Rodriguez, D. Solomon, M.C. Bratti, J.T. Schiller et al. 2007. Effect of human papillomavirus 16/18 L1 viruslike particle vaccine among young women with preexisting infection: A randomized trial. *The Journal of the American Medical Association* 298(7) (August 15): 743–753.

Hindmarsh, P.L. and L.A. Laimins. 2007. Mechanisms regulating expression of the HPV 31 L1 and L2 capsid proteins and pseudovirion entry. *Virology Journal* 4 (February 26): 19.

Hock, C., U. Konietzko, J.R. Streffer, J. Tracy, A. Signorell, B. Müller-Tillmanns, U. Lemke et al. 2003. Antibodies against B-amyloid slow cognitive decline in Alzheimer's disease. *Neuron* 38(4) (May 22): 547–554.

Hofmann, K.J., J.C. Cook, J.G. Joyce, D.R. Brown, L.D. Schultz, H.A. George, M. Rosolowsky, K.H. Fife, and K.U. Jansen. 1995. Sequence Determination of human papillomavirus type 6a and assembly of virus-like particles in *Saccharomyces cerevisiae*. *Virology* 209(2): 506–518.

Holländerová, D., H. Raslová, D. Blangy, J. Forstová, and M. Berebbi. 2003. Interference of mouse polyomavirus with the c-Myc gene and its product in mouse mammary adenocarcinomas. *International Journal of Oncology* 23(2) (August): 333–341.

Hrusková, V., A. Morávková, K. Babiarová, V. Ludvíková, J. Fric, V. Vonka, and J. Forstová. 2009. Bcr-Abl fusion sequences do not induce immune responses in mice when administered in mouse polyomavirus based virus-like particles. *International Journal of Oncology* 35(6) (December): 1247–1256.

Huerfano, S., V. Zíla, E. Boura, H. Spanielová, J. Stokrová, and J. Forstová. 2010. Minor capsid proteins of mouse polyomavirus are inducers of apoptosis when produced individually but are only moderate contributors to cell death during the late phase of viral infection. *The FEBS Journal* 277(5) (March): 1270–1283.

Hung, C.-F., B. Ma, A. Monie, S.-W. Tsen, and T.-C. Wu. 2008. Therapeutic human papillomavirus vaccines: Current clinical trials and future directions. *Expert Opinion on Biological Therapy* 8(4) (April): 421–439.

Imperiale, M.J. and E.O. Major. 2007. Polyomaviridae. In *Fields Virology*, eds. B.N. Fields, D.M. Knipe, and P.M. Howley, pp. 2263–2298. 5th edn. Philadelphia, PA: Wolters Kluwer Health/Lippincott Williams & Wilkins.

Ingle, A., S. Ghim, J. Joh, I. Chepkoech, A. Bennett Jenson, and J.P. Sundberg. 2010. Novel laboratory mouse papillomavirus (MusPV) infection. *Veterinary Pathology* 48(2) (August 4): 500–505.

Inoue, T., M. Kawano, R. Takahashi, H. Tsukamoto, T. Enomoto, T. Imai, K. Kataoka, and H. Handa. 2008. Engineering of SV40-based nano-capsules for delivery of heterologous proteins as fusions with the minor capsid proteins VP2/3. *Journal of Biotechnology* 134(1–2) (March 20): 181–192.

Ishii, Y., K. Tanaka, and T. Kanda. 2003. Mutational analysis of human papillomavirus type 16 major capsid protein L1: The cysteines affecting the intermolecular bonding and structure of L1-capsids. *Virology* 308(1) (March 30): 128–136.

Ishizu, K.-I., H. Watanabe, S.-I. Han, S.-N. Kanesashi, M. Hoque, H. Yajima, K. Kataoka, and H. Handa. 2001. Roles of disulfide linkage and calcium ion-mediated interactions in assembly and disassembly of virus-like particles composed of simian virus 40 VP1 capsid protein. *Journal of Virology* 75(1) (January 1): 61–72.

Iyengar, S., K.V. Shah, K.L. Kotloff, S.J. Ghim, and R.P. Viscidi. 1996. Self-assembly of in vitro-translated human papillomavirus type 16 L1 capsid protein into virus-like particles and antigenic reactivity of the protein. *Clinical and Diagnostic Laboratory Immunology* 3(6) (November): 733–739.

Jansen, K.U., M. Rosolowsky, L.D. Schultz, H.Z. Markus, J.C. Cook, J.J. Donnelly, D. Martinez, R.W. Ellis, and A.R. Shaw. 1995. Vaccination with yeast-expressed cottontail rabbit papillomavirus (CRPV) virus-like particles protects rabbits from CRPV-induced papilloma formation. *Vaccine* 13(16): 1509–1514.

Jennings, G.T. and M.F. Bachmann. 2008. The coming of age of virus-like particle vaccines. *Biological Chemistry* 389(5) (January 1): 521–536. http://www.degruyter.com/view/j/bchm.2008.389.issue-5/bc.2008.064/bc.2008.064.xml. Accessed July 30, 2014.

Jiang, Z., G. Tong, B. Cai, Y. Xu, and J. Lou. 2011. Purification and immunogenicity study of human papillomavirus 58 virus-like particles expressed in *Pichia pastoris*. *Protein Expression and Purification* 80(2) (December): 203–210.

Jin, L., P.E. Gibson, W.A. Knowles, and J.P. Clewley. 1993. BK virus antigenic variants: Sequence analysis within the capsid VP1 epitope. *Journal of Medical Virology* 39(1) (January): 50–56.

Jochmus, I., K. Schäfer, S. Faath, M. Müller, and L. Gissmann. 1999. Chimeric virus-like particles of the human papillomavirus type 16 (HPV 16) as a prophylactic and therapeutic vaccine. *Archives of Medical Research* 30(4): 269–274.

Joh, J., A.B. Jenson, A. Ingle, J.P. Sundberg, and S. Ghim. 2014. Searching for the initiating site of the major capsid protein to generate virus-like particles for a novel laboratory mouse papillomavirus. *Experimental and Molecular Pathology* 96(2) (April): 155–161. http://www.sciencedirect.com/science/article/pii/S001448001300155X. Accessed August 11, 2014.

Johne, R. and H. Müller. 2004. Nuclear localization of avian polyomavirus structural protein VP1 is a prerequisite for the formation of virus-like particles. *Journal of Virology* 78(2) (January 15): 930–937.

Johne, R. and H. Müller. 2007. Polyomaviruses of birds: Etiologic agents of inflammatory diseases in a tumor virus family. *Journal of Virology* 81(21) (January 11): 11554–11559.

Joyce, J.G., J.S. Tung, C.T. Przysiecki, J.C. Cook, E.D. Lehman, J.A. Sands, K.U. Jansen, and P.M. Keller. 1999. The L1 major capsid protein of human papillomavirus type 11 recombinant virus-like particles interacts with heparin and cell-surface glycosaminoglycans on human keratinocytes. *The Journal of Biological Chemistry* 274(9) (February 26): 5810–5822.

Kämper, N., P.M. Day, T. Nowak, H.-C. Selinka, L. Florin, J. Bolscher, L. Hilbig, J.T. Schiller, and M. Sapp. 2006. A membrane-destabilizing peptide in capsid protein L2 is required for egress of papillomavirus genomes from endosomes. *Journal of Virology* 80(2) (January): 759–768.

Kanesashi, S., K. Ishizu, M. Kawano, S. Han, S. Tomita, H. Watanabe, K. Kataoka, and H. Handa. 2003. Simian virus 40 VP1 capsid protein forms polymorphic assemblies in vitro. *Journal of General Virology* 84(7) (January 7): 1899–1905.

Karanam, B., S. Jagu, W.K. Huh, and R.B.S. Roden. 2009. Developing vaccines against minor capsid antigen L2 to prevent papillomavirus infection. *Immunology and Cell Biology* 87(4) (May): 287–299.

Kaufmann, A.M., J. Nieland, M. Schinz, M. Nonn, J. Gabelsberger, H. Meissner, R.T. Müller et al. 2001. HPV16 L1E7 chimeric virus-like particles induce specific HLA-restricted T cells in humans after in vitro vaccination. *International Journal of Cancer* 92(2): 285–293.

Kaufmann, A.M., J.D. Nieland, I. Jochmus, S. Baur, K. Friese, J. Gabelsberger, F. Gieseking et al. 2007. Vaccination trial with HPV16 L1E7 chimeric virus-like particles in women suffering from high grade cervical intraepithelial neoplasia (CIN 2/3). *International Journal of Cancer* 121(12): 2794–2800.

Kawana, K., H. Yoshikawa, Y. Taketani, K. Yoshiike, and T. Kanda. 1998. In vitro construction of pseudovirions of human papillomavirus type 16: Incorporation of plasmid DNA into reassembled L1/L2 capsids. *Journal of Virology* 72(12) (January 12): 10298–10300.

Kawano, M., K. Morikawa, T. Suda, N. Ohno, S. Matsushita, T. Akatsuka, H. Handa, and M. Matsui. 2014. Chimeric SV40 virus-like particles induce specific cytotoxicity and protective immunity against influenza A virus without the need of adjuvants. *Virology* 448 (January): 159–167.

Kemp, T.J., A. Hildesheim, M. Safaeian, J.G. Dauner, Y. Pan, C. Porras, J.T. Schiller, D.R. Lowy, R. Herrero, and L.A. Pinto. 2011. HPV16/18 L1 VLP vaccine induces cross-neutralizing antibodies that may mediate cross-protection. *Vaccine* 29(11) (March): 2011–2014.

Khan, Z.M., Y. Liu, U. Neu, M. Gilbert, B. Ehlers, T. Feizi, and T. Stehle. 2014. Crystallographic and glycan microarray analysis of human polyomavirus 9 VP1 identifies N-glycolyl neuraminic acid as a receptor candidate. *Journal of Virology* 88(11) (June): 6100–6111.

Khlebtsov, N., V. Bogatyrev, L. Dykman, B. Khlebtsov, S. Staroverov, A. Shirokov, L. Matora et al. 2013. Analytical and theranostic applications of gold nanoparticles and multifunctional nanocomposites. *Theranostics* 3(3) (February 20): 167–180.

Kim, S.N., H.S. Jeong, S.N. Park, and H.-J. Kim. 2007. Purification and immunogenicity study of human papillomavirus type 16 L1 protein in *Saccharomyces cerevisiae*. *Journal of Virological Methods* 139(1) (January): 24–30.

Kimchi-Sarfaty, C., N.S. Alexander, S. Brittain, S. Ali, and M.M. Gottesman. 2004. Transduction of multiple cell types using improved conditions for gene delivery and expression of SV40 pseudovirions packaged in vitro. *BioTechniques* 37(2) (August): 270–275.

Kimchi-Sarfaty, C., M. Arora, Z. Sandalon, A. Oppenheim, and M.M. Gottesman. 2003. High cloning capacity of in vitro packaged SV40 vectors with No SV40 virus sequences. *Human Gene Therapy* 14(2) (January 20): 167–177.

Kimchi-Sarfaty, C., S. Brittain, S. Garfield, N.J. Caplen, Q. Tang, and M.M. Gottesman. 2005. Efficient delivery of RNA interference effectors via in vitro-packaged SV40 pseudovirions. *Human Gene Therapy* 16(9) (September): 1110–1115.

Kimchi-Sarfaty, C. and M.M. Gottesman. 2004. SV40 pseudovirions as highly efficient vectors for gene transfer and their potential application in cancer therapy. *Current Pharmaceutical Biotechnology* 5(5) (October): 451–458.

Kimchi-Sarfaty, C., O. Ben-Nun-Shaul, D. Rund, A. Oppenheim, and M.M. Gottesman. 2002. In vitro-packaged SV40 pseudovirions as highly efficient vectors for gene transfer. *Human Gene Therapy* 13(2) (January 20): 299–310.

Kimchi-Sarfaty, C., W.D. Vieira, D. Dodds, A. Sherman, R.J. Kreitman, S. Shinar, and M.M. Gottesman. 2006. SV40 pseudovirion gene delivery of a toxin to treat human adenocarcinomas in mice. *Cancer Gene Therapy* 13(7) (July): 648–657.

Kines, R.C., C.D. Thompson, D.R. Lowy, J.T. Schiller, and P.M. Day. 2009. The initial steps leading to papillomavirus infection occur on the basement membrane prior to cell surface binding. *Proceedings of the National Academy of Sciences of the United States of America* 106(48) (December 1): 20458–20463.

Kirnbauer, R., F. Booy, N. Cheng, D.R. Lowy, and J.T. Schiller. 1992. Papillomavirus L1 major capsid protein self-assembles into virus-like particles that are highly immunogenic. *Proceedings of the National Academy of Sciences of the United States of America* 89(24) (December 15): 12180–12184.

Kirnbauer, R., L.M. Chandrachud, B.W. O'neil, E.R. Wagner, G.J. Grindlay, A. Armstrong, G.M. McGarvie, J.T. Schiller, D.R. Lowy, and M.S. Campo. 1996. Virus-like particles of bovine papillomavirus type 4 in prophylactic and therapeutic immunization. *Virology* 219(1): 37–44.

Kirnbauer, R., J. Taub, H. Greenstone, R. Roden, M. Dürst, L. Gissmann, D.R. Lowy, and J.T. Schiller. 1993. Efficient self-assembly of human papillomavirus type 16 L1 and L1-L2 into virus-like particles. *Journal of Virology* 67(12) (December): 6929–6936.

Kitai, Y., H. Fukuda, T. Enomoto, Y. Asakawa, T. Suzuki, S. Inouye, and H. Handa. 2011. Cell selective targeting of a simian virus 40 virus-like particle conjugated to epidermal growth factor. *Journal of Biotechnology* 155(2) (September 10): 251–256.

Kler, S., R. Asor, C. Li, A. Ginsburg, D. Harries, A. Oppenheim, A. Zlotnick, and U. Raviv. 2012. RNA encapsidation by SV40-derived nanoparticles follows a rapid two-state mechanism. *Journal of the American Chemical Society* 134(21) (May 30): 8823–8830.

Klug, A. 1965. Structure of viruses of the papilloma-polyoma type: II. Comments on other work. *Journal of Molecular Biology* 11(2): 424–431, IN45.

Knappe, M., S. Bodevin, H.-C. Selinka, D. Spillmann, R.E. Streeck, X.S. Chen, U. Lindahl, and M. Sapp. 2007. Surface-exposed amino acid residues of HPV16 L1 protein mediating interaction with cell surface heparan sulfate. *The Journal of Biological Chemistry* 282(38) (September 21): 27913–27922.

Komagome, R., H. Sawa, T. Suzuki, Y. Suzuki, S. Tanaka, W.J. Atwood, and K. Nagashima. 2002. Oligosaccharides as receptors for JC virus. *Journal of Virology* 76(24) (December): 12992–13000.

Kondo, K., Y. Ishii, H. Ochi, T. Matsumoto, H. Yoshikawa, and T. Kanda. 2007. Neutralization of HPV16, 18, 31, and 58 pseudovirions with antisera induced by immunizing rabbits with synthetic peptides representing segments of the HPV16 minor capsid protein L2 surface region. *Virology* 358(2) (February): 266–272.

Kondo, K., H. Ochi, T. Matsumoto, H. Yoshikawa, and T. Kanda. 2008. Modification of human papillomavirus-like particle vaccine by insertion of the cross-reactive L2-epitopes. *Journal of Medical Virology* 80(5) (May): 841–846.

Kosukegawa, A., F. Arisaka, M. Takayama, H. Yajima, A. Kaidow, and H. Handa. 1996. Purification and characterization of virus-like particles and pentamers produced by the expression of SV40 capsid proteins in insect cells. *Biochimica et Biophysica Acta (BBA)—General Subjects* 1290(1): 37–45.

Koutsky, L.A., K.A. Ault, C.M. Wheeler, D.R. Brown, E. Barr, F.B. Alvarez, L.M. Chiacchierini, and K.U. Jansen. 2002. A controlled trial of a human papillomavirus type 16 vaccine. *New England Journal of Medicine* 347(21): 1645–1651.

Krauzewicz, N., C. Cox, E. Soeda, B. Clark, S. Rayner, and B.E. Griffin. 2000a. Sustained ex vivo and in vivo transfer of a reporter gene using polyoma virus pseudocapsids. *Gene Therapy* 7(13) (July): 1094–1102.

Krauzewicz, N. and B.E. Griffin. 2000. Polyoma and papilloma virus vectors for cancer gene therapy. *Advances in Experimental Medicine and Biology* 465: 73–82.

Krauzewicz, N., J. Stokrová, C. Jenkins, M. Elliott, C.F. Higgins, and B.E. Griffin. 2000b. Virus-like gene transfer into cells mediated by polyoma virus pseudocapsids. *Gene Therapy* 7(24) (December): 2122–2131.

Kumar, A., T. Chen, S. Pakkanen, A. Kantele, M. Söderlund-Venermo, K. Hedman, and R. Franssila. 2011. T-Helper cell-mediated proliferation and cytokine responses against recombinant merkel cell polyomavirus-like particles. ed. H. Tse. *PLoS ONE* 6(10) (October 3): e25751.

Langner, J., B. Neumann, S.L. Goodman, and M. Pawlita. 2004. RGD-mutants of B-lymphotropic polyomavirus capsids specifically bind to αvβ3 integrin. *Archives of Virology* 149(10) (October 1): 1877–1896.

Laniosz, V., K.A. Holthusen, and P.I. Meneses. 2008. Bovine papillomavirus type 1: From clathrin to caveolin. *Journal of Virology* 82(13) (July): 6288–6298.

Lawatscheck, R., E. Aleksaite, J.A. Schenk, B. Micheel, B. Jandrig, G. Holland, K. Sasnauskas, A. Gedvilaite, and R.G. Ulrich. 2007. Chimeric polyomavirus-derived virus-like particles: The immunogenicity of an inserted peptide applied without adjuvant to mice depends on its insertion site and its flanking linker sequence. *Viral Immunology* 20(3) (September): 453–460.

Leder, C., J.A. Kleinschmidt, C. Wiethe, and M. Müller. 2001. Enhancement of capsid gene expression: Preparing the human papillomavirus type 16 major structural gene L1 for DNA vaccination purposes. *Journal of Virology* 75(19) (January 10): 9201–9209.

Lehtinen, M., J. Paavonen, C.M. Wheeler, U. Jaisamrarn, S.M. Garland, X. Castellsagué, S.R. Skinner, D. Apter, P. Naud, and J. Salmerón. 2012. Overall efficacy of HPV-16/18 AS04-adjuvanted vaccine against grade 3 or greater cervical intraepithelial neoplasia: 4-Year end-of-study analysis of the randomised, double-blind PATRICIA trial. *The Lancet Oncology* 13(1): 89–99.

Lenz, P., P.M. Day, Y.-Y.S. Pang, S.A. Frye, P.N. Jensen, D.R. Lowy, and J.T. Schiller. 2001. Papillomavirus-like particles induce acute activation of dendritic cells. *The Journal of Immunology* 166(9): 5346–5355.

Lenzi, P., N. Scotti, F. Alagna, M.L. Tornesello, A. Pompa, A. Vitale, A.D. Stradis et al. 2008. Translational fusion of chloroplast-expressed human papillomavirus type 16 L1 capsid protein enhances antigen accumulation in transplastomic tobacco. *Transgenic Research* 17(6) (December 1): 1091–1102.

Li, F., H. Chen, L. Ma, K. Zhou, Z.-P. Zhang, C. Meng, X.-E. Zhang, and Q. Wang. 2013. Insights into stabilization of a viral protein cage in templating complex nanoarchitectures: Roles of disulfide bonds. *Small* 10(3) (September 9): 536–543.

Li, F., H. Chen, Y. Zhang, Z. Chen, Z.-P. Zhang, X.-E. Zhang, and Q. Wang. 2012. Three-dimensional gold nanoparticle clusters with tunable cores templated by a viral protein scaffold. *Small* 8(24) (December 21): 3832–3838.

Li, F., D. Gao, X. Zhai, Y. Chen, T. Fu, D. Wu, Z.-P. Zhang, X.-E. Zhang, and Q. Wang. 2011. Tunable, discrete, three-dimensional hybrid nanoarchitectures. *Angewandte Chemie International Edition* 50(18) (April 26): 4202–4205.

Li, F., K. Li, Z.-Q. Cui, Z.-P. Zhang, H.-P. Wei, D. Gao, J.-Y. Deng, and X.-E. Zhang. 2010. Viral coat proteins as flexible nano-building-blocks for nanoparticle encapsulation. *Small* 6(20) (October 18): 2301–2308.

Li, F., Z.-P. Zhang, J. Peng, Z.-Q. Cui, D.-W. Pang, K. Li, H.-P. Wei, Y.-F. Zhou, J.-K. Wen, and X.-E. Zhang. 2009. Imaging viral behavior in mammalian cells with self-assembled capsid-quantum-dot hybrid particles. *Small* 5(6) (March 20): 718–726.

Li, J. and J.-J. Zhu. 2013. Quantum dots for fluorescent biosensing and bio-imaging applications. *Analyst* 138(9) (April 2): 2506–2515.

Li, M., P. Beard, P.A. Estes, M.K. Lyon, and R.L. Garcea. 1998. Intercapsomeric disulfide bonds in papillomavirus assembly and disassembly. *Journal of Virology* 72(3) (January 3): 2160–2167.

Li, M., T.P. Cripe, P.A. Estes, M.K. Lyon, R.C. Rose, and R.L. Garcea. 1997. Expression of the human papillomavirus type 11 L1 capsid protein in *Escherichia coli*: Characterization of protein domains involved in DNA binding and capsid assembly. *Journal of Virology* 71(4) (January 4): 2988–2995.

Li, P.P., A. Nakanishi, S.W. Clark, and H. Kasamatsu. 2002. Formation of transitory intrachain and interchain disulfide bonds accompanies the folding and oligomerization of simian virus 40 Vp1 in the cytoplasm. *Proceedings of the National Academy of Sciences of the United States of America* 99(3) (May 2): 1353–1358.

Li, P.P., A. Naknanishi, M.A. Tran, K.-I. Ishizu, M. Kawano, M. Phillips, H. Handa, R.C. Liddington, and H. Kasamatsu. 2003. Importance of Vp1 calcium-binding residues in assembly, cell entry, and nuclear entry of simian virus 40. *Journal of Virology* 77(13) (January 7): 7527–7538.

Li, Q., C. Cao, B. Chackerian, J. Schiller, M. Gordon, K.E. Ugen, and D. Morgan. 2004. Overcoming Antigen Masking of Anti-Amyloidbeta Antibodies Reveals Breaking of B Cell Tolerance by Virus-like Particles in Amyloidbeta Immunized Amyloid Precursor Protein Transgenic Mice. *BMC Neuroscience* 5(1): 21.

Liddington, R.C., Y. Yan, J. Moulai, R. Sahli, T.L. Benjamin, and S.C. Harrison. 1991. Structure of simian virus 40 at 3.8-Å Resolution. *Nature* 354(6351) (November 28): 278–284.

Liebl, D., F. Difato, L. Horníková, P. Mannová, J. Stokrová, and J. Forstová. 2006. Mouse Polyomavirus Enters Early Endosomes, Requires their acidic pH for productive infection, and meets transferrin Cargo in Rab11-positive endosomes. *Journal of Virology* 80(9) (May): 4610–4622.

Liew, M.W.O., A. Rajendran, and A.P.J. Middelberg. 2010. Microbial production of virus-like particle vaccine protein at gram-per-litre levels. *Journal of Biotechnology* 150(2): 224–231.

Lilley, B.N., J.M. Gilbert, H.L. Ploegh, and T.L. Benjamin. 2006. Murine polyomavirus requires the endoplasmic reticulum protein Derlin-2 to initiate infection. *Journal of Virology* 80(17) (September): 8739–8744.

Lin, M.-C., M. Wang, C.-Y. Fang, P.-L. Chen, C.-H. Shen, and D. Chang. 2014. Inhibition of BK virus replication in human kidney cells by BK virus large tumor antigen-specific shRNA delivered by JC virus-like particles. *Antiviral Research* 103 (March): 25–31.

Lipin, D.I., L.H.L. Lua, and A.P.J. Middelberg. 2008. Quaternary size distribution of soluble aggregates of glutathione-S-transferase-purified viral protein as determined by asymmetrical flow field flow fractionation and dynamic light scattering. *Journal of Chromatography A* 1190(1–2) (May 9): 204–214.

Lipovsky, A., A. Popa, G. Pimienta, M. Wyler, A. Bhan, L. Kuruvilla, M.-A. Guie et al. 2013. Genome-wide siRNA screen identifies the retromer as a cellular entry factor for human papillomavirus. *Proceedings of the National Academy of Sciences of the United States of America* 110(18) (April 30): 7452–7457.

Liu, C.K., G. Wei, and W.J. Atwood. 1998. Infection of glial cells by the human polyomavirus JC is mediated by an N-linked glycoprotein containing terminal alpha(2–6)-linked sialic acids. *Journal of Virology* 72(6) (June): 4643–4649.

Liu, D., Y. Zhang, X. Yu, C. Jiang, Y. Chen, Y. Wu, Y. Jin et al. 2007. Assembly and immunogenicity of human papillomavirus type 16 major capsid protein (HPV16 L1) in *Pichia pastoris*. *Chemical Research in Chinese Universities* 23(2) (March): 200–203.

Liu, W.J., L. Gissmann, X.Y. Sun, A. Kanjanahaluethai, M. Müller, J. Doorbar, and J. Zhou. 1997. Sequence close to the N-terminus of L2 protein is displayed on the surface of bovine papillomavirus type 1 virions. *Virology* 227(2) (January 20): 474–483.

Liu, W.J., X.S. Liu, K.N. Zhao, G.R. Leggatt, and I.H. Frazer. 2000. Papillomavirus virus-like particles for the delivery of multiple cytotoxic T cell epitopes. *Virology* 273(2) (August 1): 374–382.

Low, J.A., B. Magnuson, B. Tsai, and M.J. Imperiale. 2006. Identification of gangliosides GD1b and GT1b as receptors for BK virus. *Journal of Virology* 80(3) (February): 1361–1366.

Lu, B., A. Kumar, X. Castellsagué, and A. Giuliano. 2011. Efficacy and safety of prophylactic vaccines against cervical HPV infection and diseases among women: A systematic review & meta-analysis. *BMC Infectious Diseases* 11(1): 13.

Lund, P.E., R.C. Hunt, M.M. Gottesman, and C. Kimchi-Sarfaty. 2010. Pseudovirions as vehicles for the delivery of siRNA. *Pharmaceutical Research* 27(3) (March): 400–420.

Luo, C., H.H. Hirsch, J. Kant, and P. Randhawa. 2012. VP-1 quasispecies in human infection with polyomavirus BK. *Journal of Medical Virology* 84(1): 152–161.

Macadangdang, B., N. Zhang, P.E. Lund, A.H. Marple, M. Okabe, M.M. Gottesman, D.H. Appella, and C. Kimchi-Sarfaty. 2011. Inhibition of multidrug resistance by SV40 pseudovirion delivery of an antigene peptide nucleic acid (PNA) in cultured cells. *PLoS ONE* 6(3): e17981.

Mach, H., D.B. Volkin, R.D. Troutman, B. Wang, Z. Luo, K.U. Jansen, and L. Shi. 2006. Disassembly and reassembly of yeast-derived recombinant human papillomavirus virus-like particles (HPV VLPs). *Journal of Pharmaceutical Sciences* 95(10): 2195–2206.

Mackay, R.L. and R.A. Consigli. 1976. Early events in polyoma virus infection: Attachment, penetration, and nuclear entry. *Journal of Virology* 19(2) (August): 620–636.

Maclean, J., M. Koekemoer, A.J. Olivier, D. Stewart, I.I. Hitzeroth, T. Rademacher, R. Fischer, A.-L. Williamson, and E.P. Rybicki. 2007. Optimization of human papillomavirus type 16 (HPV-16) L1 expression in plants: Comparison of the suitability of different HPV-16 L1 gene variants and different cell-compartment localization. *Journal of General Virology* 88(5) (January 5): 1460–1469.

Maginnis, M.S. and W.J. Atwood. 2009. JC virus: An oncogenic virus in animals and humans? *Seminars in Cancer Biology* 19(4) The Polyomaviruses (August): 261–269.

Malagón, T., M. Drolet, M.-C. Boily, E.L. Franco, M. Jit, J. Brisson, and M. Brisson. 2012. Cross-protective efficacy of two human papillomavirus vaccines: A systematic review and meta-analysis. *The Lancet Infectious Diseases* 12(10) (October): 781–789.

Mannová, P. and J. Forstová. 2003. Mouse polyomavirus utilizes recycling endosomes for a traffic pathway independent of COPI vesicle transport. *Journal of Virology* 77(3) (February): 1672–1681.

Markowitz, L.E., E.F. Dunne, M. Saraiya, H.W. Lawson, H. Chesson, E.R. Unger, Centers for Disease Control and Prevention (CDC), and Advisory Committee on Immunization Practices (ACIP). 2007. Quadrivalent human papillomavirus vaccine: Recommendations of the advisory committee on immunization practices (ACIP). *Recommendations and Reports: Morbidity and Mortality Weekly Report. Recommendations and Reports/Centers for Disease Control* 56(RR-2) (March 23): 1–24.

Matić, S., V. Masenga, A. Poli, R. Rinaldi, R.G. Milne, M. Vecchiati, and E. Noris. 2012. Comparative analysis of recombinant human papillomavirus 8 L1 production in plants by a variety of expression systems and purification methods. *Plant Biotechnology Journal* 10(4): 410–421.

Mató, T., Z. Pénzes, P. Rueda, C. Vela, V. Kardi, A. Zolnai, F. Misák, and V. Palya. 2009. Recombinant subunit vaccine elicits protection against goose haemorrhagic nephritis and enteritis. *Avian Pathology* 38(3) (June): 233–237.

Maul, G.G., G. Rovera, A. Vorbrodt, and J. Abramczuk. 1978. Membrane fusion as a mechanism of simian virus 40 entry into different cellular compartments. *Journal of Virology* 28(3) (December): 936–944.

Maurer, P., G.T. Jennings, J. Willers, F. Rohner, Y. Lindman, K. Roubicek, W.A. Renner, P. Müller, and M.F. Bachmann. 2005. A therapeutic vaccine for nicotine dependence: Preclinical efficacy, and phase I safety and immunogenicity. *European Journal of Immunology* 35(7): 2031–2040.

May, T., S. Gleiter, and H. Lilie. 2002. Assessment of cell type specific gene transfer of polyoma virus like particles presenting a tumor specific antibody Fv fragment. *Journal of Virological Methods* 105(1) (August): 147–157.

Mazeike, E., A. Gedvilaite, and U. Blohm. 2012. Induction of insert-specific immune response in mice by hamster polyomavirus VP1 derived virus-like particles carrying LCMV GP33 CTL epitope. *Virus Research* 163(1) (January): 2–10.

McCarthy, M.P., W.I. White, F. Palmer-Hill, S. Koenig, and J.A. Suzich. 1998. Quantitative disassembly and reassembly of human papillomavirus type 11 viruslike particles in vitro. *Journal of Virology* 72(1) (January 1): 32–41.

McGrath, M., G.K. Villiers, E. Shephard, I.I. Hitzeroth, and E.P. Rybicki. 2013. Development of human papillomavirus chimeric L1/L2 candidate vaccines. *Archives of Virology* 158(10) (May 1): 2079–2088.

McMillan, N.A., E. Payne, I.H. Frazer, and M. Evander. 1999. Expression of the alpha6 integrin confers papillomavirus binding upon receptor-negative B-cells. *Virology* 261(2) (September 1): 271–279.

Mercer, J., M. Schelhaas, and A. Helenius. 2010. Virus entry by endocytosis. *Annual Review of Biochemistry* 79: 803–833.

Mérian, J., J. Gravier, F. Navarro, and I. Texier. 2012. Fluorescent nanoprobes dedicated to in vivo imaging: From preclinical validations to clinical translation. *Molecules* 17(5) (May 10): 5564–5591.

Meusser, B., C. Hirsch, E. Jarosch, and T. Sommer. 2005. ERAD: The long road to destruction. *Nature Cell Biology* 7(8) (August): 766–772.

Meyers, C., M.G. Frattini, J.B. Hudson, and L.A. Laimins. 1992. Biosynthesis of human papillomavirus from a continuous cell line upon epithelial differentiation. *Science* (New York) 257(5072) (August 14): 971–973.

Mezes, B. and P. Amati. 1994. Mutations of polyomavirus VP1 allow in vitro growth in undifferentiated cells and modify in vivo tissue replication specificity. *Journal of Virology* 68(2) (February): 1196–1199.

Michel, M.R., B. Hirt, and R. Weil. 1967. Mouse cellular DNA enclosed in polyoma viral capsids (pseudovirions). *Proceedings of the National Academy of Sciences of the United States of America* 58(4): 1381.

Middelberg, A.P.J., T. Rivera-Hernandez, N. Wibowo, L.H.L. Lua, Y. Fan, G. Magor, C. Chang, Y.P. Chuan, M.F. Good, and M.R. Batzloff. 2011. A microbial platform for rapid and low-cost virus-like particle and capsomere vaccines. *Vaccine* 29(41) (September 22): 7154–7162.

Millán, A.F.-S., S. Gómez-Sebastián, M.C. Nuñez, J. Veramendi, and J.M. Escribano. 2010. Human papillomavirus-like particles vaccine efficiently produced in a non-fermentative system based on insect larva. *Protein Expression and Purification* 74(1) (November): 1–8.

Miltz, A., H. Price, M. Shahmanesh, A. Copas, and R. Gilson. 2014. Systematic review and meta-analysis of L1-VLP-based human papillomavirus vaccine efficacy against anogenital pre-cancer in women with evidence of prior HPV exposure. ed. M. Scheurer. *PLoS ONE* 9(3) (March 3): e90348.

Montross, L., S. Watkins, R.B. Moreland, H. Mamon, D.L. Caspar, and R.L. Garcea. 1991. Nuclear assembly of polyomavirus capsids in insect cells expressing the major capsid protein VP1. *Journal of Virology* 65(9) (January 9): 4991–4998.

Moriyama, T., J.P. Marquez, T. Wakatsuki, and A. Sorokin. 2007. Caveolar endocytosis is critical for BK virus infection of human renal proximal tubular epithelial cells. *Journal of Virology* 81(16) (August): 8552–8562.

Moriyama, T. and A. Sorokin. 2008. Intracellular trafficking pathway of BK virus in human renal proximal tubular epithelial cells. *Virology* 371(2) (February 20): 336–349.

Mossadegh, N., L. Gissmann, M. Müller, H. Zentgraf, A. Alonso, and P. Tomakidi. 2004. Codon optimization of the human papillomavirus 11 (HPV 11) L1 gene leads to increased gene expression and formation of virus-like particles in mammalian epithelial cells. *Virology* 326(1) (August 15): 57–66.

Mueller, C., M.S. Strayer, J. Sirninger, S. Braag, F. Branco, J.-P. Louboutin, T.R. Flotte, and D.S. Strayer. 2010. In vitro and in vivo functional characterization of gutless recombinant SV40-derived CFTR vectors. *Gene Therapy* 17(2) (February): 227–237.

Mukherjee, S., M. Abd-El-Latif, M. Bronstein, O. Ben-nun -Shaul, S. Kler, and A. Oppenheim. 2007. High cooperativity of the SV40 major capsid protein VP1 in virus assembly. *PLoS ONE* 2(8) (August 22):e765 http://www.ncbi.nlm.nih.gov/pmc/articles/PMC1942081/. Accessed September 19, 2013.

Mukherjee, S., S. Kler, A. Oppenheim, and A. Zlotnick. 2010. Uncatalyzed assembly of spherical particles from SV40 VP1 pentamers and linear dsDNA incorporates both low and high cooperativity elements. *Virology* 397(1): 199–204.

Muller, M., L. Gissmann, R.J. Cristiano, X.Y. Sun, I.H. Frazer, A.B. Jenson, A. Alonso, H. Zentgraf, and J. Zhou. 1995. Papillomavirus capsid binding and uptake by cells from different tissues and species. *Journal of Virology* 69(2) (February): 948–954.

Müller, M., J. Zhou, T.D. Reed, C. Rittmüller, A. Burger, J. Gabelsberger, J. Braspenning, and L. Gissmann. 1997. Chimeric papillomavirus-like particles. *Virology* 234(1) (July 21): 93–111.

Münger, K. and P.M. Howley. 2002. Human papillomavirus immortalization and transformation functions. *Virus Research* 89(2) (November): 213–228.

Muñoz, N., X. Castellsagué, A.B. de González, and L. Gissmann. 2006. Chapter 1: HPV in the etiology of human cancer. *Vaccine* 24 (August): S1–S10.

Nakanishi, A., B. Chapellier, N. Maekawa, M. Hiramoto, T. Kuge, R. Takahashi, H. Handa, and T. Imai. 2008. SV40 vectors carrying minimal sequence of viral origin with exchangeable capsids. *Virology* 379(1) (September 15): 110–117.

National Cancer Institute. 2014. Broad spectrum HPV (human papillomavirus) vaccine study in 16- to 26-year-old women (V503-001) 2014 (January 16). http://www.cancer.gov/clinicaltrials/search/view. Accessed July 16, 2014.

Naud, P.S., C.M. Roteli-Martins, N.S. De Carvalho, J.C. Teixeira, P.C. de Borba, N. Sanchez, T. Zahaf, G. Catteau, B. Geeraerts, and D. Descamps. 2014. Sustained efficacy, immunogenicity, and safety of the HPV-16/18 AS04-adjuvanted vaccine: Final analysis of a long-term follow-up study up to 9.4 years post-vaccination. *Human Vaccines & Immunotherapeutics* 10(8) (June 19): 2147–2162

Neeper, M.P., K.J. Hofmann, and K.U. Jansen. 1996. Expression of the major capsid protein of human papillomavirus type 11 in *Saccharomyces cerevisiae*. *Gene* 180(1–2) (November 21): 1–6.

Nelson, C.D.S., D.W. Carney, A. Derdowski, A. Lipovsky, G.V. Gee, B. O'Hara, P. Williard, D. DiMaio, J.K. Sello, and W.J. Atwood. 2013. A retrograde trafficking inhibitor of ricin and shiga-like toxins inhibits infection of cells by human and monkey polyomaviruses. *mBio* 4(6): e00729–e00713.

Neu, U., H. Hengel, B.S. Blaum, R.M. Schowalter, D. Macejak, M. Gilbert, W.W. Wakarchuk et al. 2012. Structures of merkel cell polyomavirus VP1 complexes define a sialic acid binding site required for infection. *PLoS Pathogens* 8(7): e1002738.

Neu, U., Z.M. Khan, B. Schuch, A.S. Palma, Y. Liu, M. Pawlita, T. Feizi, and T. Stehle. 2013. Structures of B-lymphotropic polyomavirus VP1 in complex with oligosaccharide ligands. *PLoS Pathogens* 9(10) (October): e1003714.

Neu, U., M.S. Maginnis, A.S. Palma, L.J. Ströh, C.D.S. Nelson, T. Feizi, W.J. Atwood, and T. Stehle. 2010. Structure-function analysis of the human JC polyomavirus establishes the LSTc pentasaccharide as a functional receptor motif. *Cell Host & Microbe* 8(4) (October 21): 309–319.

Neu, U., J. Wang, D. Macejak, R.L. Garcea, and T. Stehle. 2011. Structures of the major capsid proteins of the Human Karolinska Institutet and Washington University Polyomaviruses. *Journal of Virology* 85(14) (July 15): 7384–7392.

Neu, U., K. Woellner, G. Gauglitz, and T. Stehle. 2008. Structural basis of GM1 ganglioside recognition by simian virus 40. *Proceedings of the National Academy of Sciences of the United States of America* 105(13) (April 1): 5219–5224.

Neugebauer, M., B. Walders, M. Brinkman, C. Ruehland, T. Schumacher, W.M. Bertling, E. Geuther et al. 2006. Development of a vaccine marker technology: Display of B cell epitopes on the surface of recombinant polyomavirus-like pentamers and capsids induces peptide-specific antibodies in piglets after vaccination. *Biotechnology Journal* 1(12) (December): 1435–1446.

Ng, J., O. Koechlin, M. Ramalho, D. Raman, and N. Krauzewicz. 2007. Extracellular self-assembly of virus-like particles from secreted recombinant polyoma virus major coat protein. *Protein Engineering, Design & Selection* 20(12) (December): 591–598.

Nicol, J.T.J., R. Robinot, A. Carpentier, G. Carandina, E. Mazzoni, M. Tognon, A. Touzé, and P. Coursaget. 2013. Age-specific seroprevalences of merkel cell polyomavirus, human polyomaviruses 6, 7, and 9, and trichodysplasia spinulosa-associated polyomavirus. *Clinical and Vaccine Immunology* 20(3) (January 3): 363–368.

Nicol, J.T.J., A. Touzé, R. Robinot, F. Arnold, E. Mazzoni, M. Tognon, and P. Coursaget. 2012. Seroprevalence and Cross-Reactivity of Human Polyomavirus 9. *Emerging Infectious Diseases* 18(8) (August): 1329–1332.

Niikura, K., K. Nagakawa, N. Ohtake, T. Suzuki, Y. Matsuo, H. Sawa, and K. Ijiro. 2009. Gold nanoparticle arrangement on viral particles through carbohydrate recognition: A non-cross-linking approach to optical virus detection. *Bioconjugate Chemistry* 20(10) (October 21): 1848–1852.

Niikura, K., N. Sugimura, Y. Musashi, S. Mikuni, Y. Matsuo, S. Kobayashi, K. Nagakawa et al. 2013. Virus-like Particles with Removable Cyclodextrins Enable Glutathione-Triggered Drug Release in Cells. *Molecular BioSystems* 9(3): 501.

Nims, R.W. and M. Plavsic. 2013. Polyomavirus inactivation—A review. *Biologicals: Journal of the International Association of Biological Standardization* 41(2) (March): 63–70.

Norkin, L.C., H.A. Anderson, S.A. Wolfrom, and A. Oppenheim. 2002. Caveolar endocytosis of simian virus 40 is followed by Brefeldin A-Sensitive transport to the endoplasmic reticulum, where the virus disassembles. *Journal of Virology* 76(10) (May): 5156–5166.

Norkin, L.C. and D. Kuksin. 2005. The Caveolae-mediated sv40 entry pathway bypasses the golgi complex En Route to the endoplasmic reticulum. *Virology Journal* 2: 38.

O'Hara, S.D., T. Stehle, and R. Garcea. 2014. Glycan receptors of the polyomaviridae: Structure, function, and pathogenesis. *Current Opinion in Virology* 7C (June 28): 73–78.

Ohtake, N., K. Niikura, T. Suzuki, K. Nagakawa, S. Mikuni, Y. Matsuo, M. Kinjo, H. Sawa, and K. Ijiro. 2010. Low pH-triggered model drug molecule release from virus-like particles. *ChemBioChem* 11(7) (May 3): 959–962.

Olsson, S.-E., S.K. Kjaer, K. Sigurdsson, O.-E. Iversen, M. Hernandez-Avila, C.M. Wheeler, G. Perez et al. 2009. Evaluation of quadrivalent HPV 6/11/16/18 vaccine efficacy against cervical and anogenital disease in subjects with serological evidence of prior vaccine type HPV infection. *Human Vaccines* 5(10) (October): 696–704.

Olsson, S.-E., L.L. Villa, R.L.R. Costa, C.A. Petta, R.P. Andrade, C. Malm, O.-E. Iversen et al. 2007. Induction of immune memory following administration of a prophylactic quadrivalent human papillomavirus (HPV) types 6/11/16/18 L1 virus-like particle (VLP) vaccine. *Vaccine* 25(26) (June): 4931–4939.

Oppenheim, A. and A. Peleg. 1989. Helpers for efficient encapsidation of SV40 pseudovirions. *Gene* 77(1) (April 15): 79–86.

Oppenheim, A., A. Peleg, E. Fibach, and E.A. Rachmilewitz. 1986. Efficient introduction of plasmid DNA into human hemopoietic cells by encapsidation in simian virus 40 pseudovirions. *Proceedings of the National Academy of Sciences of the United States of America* 83(18): 6925–6929.

Oppenheim, A., Z. Sandalon, A. Peleg, O. Shaul, S. Nicolis, and S. Ottolenghi. 1992. A cis-acting DNA signal for encapsidation of simian virus 40. *Journal of Virology* 66(9) (January 9): 5320–5328.

Osterman, J.V., A. Waddell, and H.V. Aposhian. 1970. DNA and gene therapy: Uncoating of polyoma pseudovirus in mouse embryo cells. *Proceedings of the National Academy of Sciences of the United States of America* 67(1): 37–40.

Ou, W.C., M. Wang, C.Y. Fung, R.T. Tsai, P.C. Chao, T.H. Hseu, and D. Chang. 1999. The major capsid protein, VP1, of human JC virus expressed in *Escherichia coli* is able to self-assemble into a capsid-like particle and deliver exogenous DNA into human kidney cells. *Journal of General Virology* 80(1) (January 1): 39–46.

Ozbun, M.A. and M.P. Kivitz. 2012. The art and science of obtaining virion stocks for experimental human papillomavirus infections. In *Small DNA Tumour Viruses*, ed. K. Gaston, pp. 19–35. Norfolk, U.K.: Horizon Scientific and Caister Academic Press.

Paavonen, J., P. Naud, J. Salmeron, C.M. Wheeler, S.N. Chow, D. Apter, H. Kitchener, X. Castellsague, J.C. Teixeira, and S.R. Skinner. 2009. Efficacy of human papillomavirus (HPV)-16/18 AS04-adjuvanted vaccine against cervical infection and precancer caused by oncogenic HPV types (PATRICIA): Final analysis of a double-blind, randomised study in young women. *The Lancet* 374(9686): 301–314.

Palaniyandi, M., T. Kato, and E.Y. Park. 2012. Expression of human papillomavirus 6b L1 protein in silkworm larvae and enhanced green fluorescent protein displaying on its virus-like particles. *SpringerPlus* 1 (October 4). http://www.ncbi.nlm.nih.gov/pmc/articles/PMC3725899/. Accessed July 14, 2014.

Palková, Z., T. Adamec, D. Liebl, J. Štokrová, and J. Forstová. 2000. Production of polyomavirus structural protein VP1 in yeast cells and its interaction with cell structures. *FEBS Letters* 478(3) (August 4): 281–289.

Park, M.-A., H.J. Kim, and H.-J. Kim. 2008. Optimum conditions for production and purification of human papillomavirus type 16 L1 protein from *Saccharomyces cerevisiae*. *Protein Expression and Purification* 59(1) (May): 175–181.

Pastrana, D.V., C.B. Buck, Y.-Y.S. Pang, C.D. Thompson, P.E. Castle, P.C. FitzGerald, S. Krüger Kjaer, D.R. Lowy, and J.T. Schiller. 2004. Reactivity of human sera in a sensitive, high-throughput pseudovirus-based papillomavirus neutralization assay for HPV16 and HPV18. *Virology* 321(2) (April 10): 205–216.

Pastrana, D.V., U. Ray, T.G. Magaldi, R.M. Schowalter, N. Çuburu, and C.B. Buck. 2013. BK polyomavirus genotypes represent distinct serotypes with distinct entry tropism. *Journal of Virology* 87(18) (September 15): 10105–10113.

Pastrana, D.V., Y.L. Tolstov, J.C. Becker, P.S. Moore, Y. Chang, and C.B. Buck. 2009. Quantitation of human seroresponsiveness to merkel cell polyomavirus. ed. R.L. Garcea. *PLoS Pathogens* 5(9) (September 11): e1000578.

Paulson, K.G., J.G. Iyer, A.R. Tegeder, R. Thibodeau, J. Schelter, S. Koba, D. Schrama et al. 2011. Transcriptome-wide studies of merkel cell carcinoma and validation of intratumoral CD8+ lymphocyte invasion as an independent predictor of survival. *Journal of Clinical Oncology* 29(12) (April 20): 1539–1546.

Pawlita, M., M. Müller, M. Oppenländer, H. Zentgraf, and M. Herrmann. 1996. DNA encapsidation by viruslike particles assembled in insect cells from the major capsid protein VP1 of B-lymphotropic papovavirus. *Journal of Virology* 70(11) (January 11): 7517–7526.

Peiler, T. 2004. Pseudovirionen zur Simulation von Infektionen mit humanpathogenen Papillomaviren. Dissertation. http://archiv.ub.uni-heidelberg.de/volltextserver/4831/. Accessed December 22, 2013.

Pejawar-Gaddy, S., Y. Rajawat, Z. Hilioti, J. Xue, D.F. Gaddy, O.J. Finn, R.P. Viscidi, and I. Bossis. 2010. Generation of a tumor vaccine candidate based on conjugation of a MUC1 peptide to polyionic papillomavirus virus-like particles. *Cancer Immunology, Immunotherapy* 59(11) (July 21): 1685–1696.

Pelkmans, L., T. Bürli, M. Zerial, and A. Helenius. 2004. Caveolin-stabilized membrane domains as multifunctional transport and sorting devices in endocytic membrane traffic. *Cell* 118(6) (September 17): 767–780.

Pelkmans, L., E. Fava, H. Grabner, M. Hannus, B. Habermann, E. Krausz, and M. Zerial. 2005. Genome-wide analysis of human kinases in clathrin- and caveolae/raft-mediated endocytosis. *Nature* 436(7047) (July 7): 78–86.

Pelkmans, L., J. Kartenbeck, and A. Helenius. 2001. Caveolar endocytosis of simian virus 40 reveals a new two-step vesicular-transport pathway to the ER. *Nature Cell Biology* 3(5) (May): 473–483.

Pelkmans, L., D. Püntener, and A. Helenius. 2002. Local actin polymerization and dynamin recruitment in SV40-induced internalization of caveolae. *Science* (New York) 296(5567) (April 19): 535–539.

Peng, S., I.H. Frazer, G.J. Fernando, and J. Zhou. 1998. Papillomavirus virus-like particles can deliver defined CTL epitopes to the MHC class I pathway. *Virology* 240(1) (January 5): 147–157.

Pho, M.T., A. Ashok, and W.J. Atwood. 2000. JC virus enters human glial cells by clathrin-dependent receptor-mediated endocytosis. *Journal of Virology* 74(5) (March): 2288–2292.

Pipas, J.M. 2009. SV40: Cell Transformation and Tumorigenesis. Small Viruses, Big Discoveries: The Interwoven Story of the Small DNA Tumor Viruses. *Virology* 384(2): 294–303.

Pircher, H., D. Moskophidis, U. Rohrer, K. Bürki, H. Hengartner, and R.M. Zinkernagel. 1990. Viral escape by selection of cytotoxic T cell-resistant virus variants in vivo. *Nature* 346(6285) (August 16): 629–633.

Pruksakorn, S., B. Currie, E. Brandt, D. Martin, A. Galbraith, C. Phornphutkul, S. Hunsakunachai, A. Manmontri, and M.F. Good. 1994. Towards a vaccine for rheumatic fever: Identification of a conserved target epitope on M protein of group A streptococci. *Lancet* 344(8923) (September 3): 639–642.

Pyeon, D., P.F. Lambert, and P. Ahlquist. 2005. Production of infectious human papillomavirus independently of viral replication and epithelial cell differentiation. *Proceedings of the National Academy of Sciences of the United States of America* 102(26) (June 28): 9311–9316.

Pyeon, D., S.M. Pearce, S.M. Lank, P. Ahlquist, and P.F. Lambert. 2009. Establishment of human papillomavirus infection requires cell cycle progression. *PLoS Pathogens* 5(2) (February): e1000318.

Qasba, P.K. and H.V. Aposhian. 1971. DNA and gene therapy: Transfer of mouse DNA to human and mouse embryonic cells by polyoma pseudovirions. *Proceedings of the National Academy of Sciences of the United States of America* 68(10) (October): 2345–2349.

Qian, J., Y. Dong, Y.-Y.S. Pang, R. Ibrahim, J.A. Berzofsky, J.T. Schiller, and S.N. Khleif. 2006. Combined prophylactic and therapeutic cancer vaccine: Enhancing CTL responses to HPV16 E2 using a chimeric VLP in HLA-A2 mice. *International Journal of Cancer* 118(12) (June 15): 3022–3029.

Qian, M., D. Cai, K.J. Verhey, and B. Tsai. 2009. A lipid receptor sorts polyomavirus from the endolysosome to the endoplasmic reticulum to cause infection. *PLoS Pathogens* 5(6) (June): e1000465.

Qian, M. and B. Tsai. 2010. Lipids and proteins act in opposing manners to regulate polyomavirus infection. *Journal of Virology* 84(19) (October): 9840–9852.

Qu, Q., H. Sawa, T. Suzuki, S. Semba, C. Henmi, Y. Okada, M. Tsuda, S. Tanaka, W.J. Atwood, and K. Nagashima. 2004. Nuclear entry mechanism of the human polyomavirus JC virus-like particle: Role of importins and the nuclear pore complex. *The Journal of Biological Chemistry* 279(26) (June 25): 27735–27742.

Raff, A.B., A.W. Woodham, L.M. Raff, J.G. Skeate, L. Yan, D.M. Da Silva, M. Schelhaas, and W.M. Kast. 2013. The evolving field of human papillomavirus receptor research: A review of binding and entry. *Journal of Virology* 87(11) (June): 6062–6072.

Rainey-Barger, E.K., B. Magnuson, and B. Tsai. 2007. A chaperone-activated nonenveloped virus perforates the physiologically relevant endoplasmic reticulum membrane. *Journal of Virology* 81(23) (December): 12996–13004.

Ramqvist, T. and T. Dalianis. 2010. Lessons from immune responses and vaccines against murine polyomavirus infection and polyomavirus-induced tumours potentially useful for studies on human polyomaviruses. *Anticancer Research* 30(2): 279–284.

Rao, N.H., P.B. Babu, L. Rajendra, R. Sriraman, Y.-Y.S. Pang, J.T. Schiller, and V.A. Srinivasan. 2011. Expression of codon optimized major capsid protein (L1) of human papillomavirus type 16 and 18 in *Pichia pastoris*; purification and characterization of the virus-like particles. *Vaccine* 29(43) (October 6): 7326–7334.

Rayment, I., T.S. Baker, D.L. Caspar, and W.T. Murakami. 1982. Polyoma virus capsid structure at 22.5 A resolution. *Nature* 295(5845) (January 14): 110–115.

Rector, A. and M. Van Ranst. 2013. Animal papillomaviruses. *Virology* 445(1–2): 213–223.

Relf, W.A., J. Cooper, E.R. Brandt, W.A. Hayman, R.F. Anders, S. Pruksakorn, B. Currie et al. 1996. Mapping a conserved conformational epitope from the M protein of group A streptococci. *Pept Res.* 9(1) (Jan–Feb): 12–20.

Richards, K.F., M. Bienkowska-Haba, J. Dasgupta, X.S. Chen, and M. Sapp. 2013. Multiple heparan sulfate binding site engagements are required for the infectious entry of human papillomavirus type 16. *Journal of Virology* 87(21) (November): 11426–11437.

Richards, R.M., D.R. Lowy, J.T. Schiller, and P.M. Day. 2006. Cleavage of the papillomavirus minor capsid protein, L2, at a furin consensus site is necessary for infection. *Proceedings of the National Academy of Sciences of the United States of America* 103(5) (January 31): 1522–1527.

Richterová, Z., D. Liebl, M. Horák, Z. Palková, J. Stokrová, P. Hozák, J. Korb, and J. Forstová. 2001. Caveolae are involved in the trafficking of mouse polyomavirus virions and artificial VP1 pseudocapsids toward cell nuclei. *Journal of Virology* 75(22) (November): 10880–10891.

Rivera-Hernandez, T., J. Hartas, Y. Wu, Y.P. Chuan, L.H.L. Lua, M. Good, M.R. Batzloff, and A.P.J. Middelberg. 2013. Self-adjuvanting modular virus-like particles for mucosal vaccination against group A streptococcus (GAS). *Vaccine* 31(15) (April 8): 1950–1955.

Roberts, J.N., C.B. Buck, C.D. Thompson, R. Kines, M. Bernardo, P.L. Choyke, D.R. Lowy, and J.T. Schiller. 2007. Genital transmission of HPV in a mouse model is potentiated by nonoxynol-9 and inhibited by carrageenan. *Nature Medicine* 13(7) (July): 857–861.

Roden, R.B.S., W.H. Yutzy, R. Fallon, S. Inglis, D.R. Lowy, and J.T. Schiller. 2000. Minor capsid protein of human genital papillomaviruses contains subdominant, cross-neutralizing epitopes. *Virology* 270(2) (May): 254–257.

Rodgers, R.E., D. Chang, X. Cai, and R.A. Consigli. 1994. Purification of recombinant budgerigar fledgling disease virus VP1 capsid protein and its ability for in vitro capsid assembly. *Journal of Virology* 68(5): 3386–3390.

Rose, R.C., W. Bonnez, C. Da Rin, D.J. McCance, and R.C. Reichman. 1994. Serological differentiation of human papillomavirus types 11, 16 and 18 using recombinant virus-like particles. *The Journal of General Virology* 75(Pt 9) (September): 2445–2449.

Rose, R.C., W. Bonnez, R.C. Reichman, and R.L. Garcea. 1993. Expression of human papillomavirus type 11 L1 protein in insect cells: In vivo and in vitro assembly of viruslike particles. *Journal of Virology* 67(4) (April): 1936–1944.

Rose, R.C., C. Lane, S. Wilson, J.A. Suzich, E. Rybicki, and A.L. Williamson. 1999. Oral vaccination of mice with human papillomavirus virus-like particles induces systemic virus-neutralizing antibodies. *Vaccine* 17(17) (April 23): 2129–2135.

Rosen, J.E., S. Yoffe, A. Meerasa, M. Verma, and F.X. Gu. 2011. Nanotechnology and diagnostic imaging: New advances in contrast agent technology. *Journal of Nanomedicine & Nanotechnology* 02(05) (October 21):e1000115. http://www.omicsonline.org/2157-7439/2157-7439-2-115.digital/2157-7439-2-115.html. Accessed August 4, 2014.

Rossi, J.L., L. Gissmann, K. Jansen, and M. Müller. 2000. Assembly of human papillomavirus type 16 pseudovirions in *Saccharomyces cerevisiae*. *Human Gene Therapy* 11(8) (May 20): 1165–1176.

Rudolf, M.P., S.C. Fausch, D.M. Da Silva, and W.M. Kast. 2001. Human dendritic cells are activated by chimeric human papillomavirus type-16 virus-like particles and induce epitope-specific human T cell responses in vitro. *The Journal of Immunology* 166(10): 5917–5924.

Rund, D., M. Dagan, N. Dalyot-Herman, C. Kimchi-Sarfaty, P.V. Schoenlein, M.M. Gottesman, and A. Oppenheim. 1998. Efficient transduction of human hematopoietic cells with the human multidrug resistance gene 1 via SV40 pseudovirions. *Human Gene Therapy* 9(5) (March 20): 649–657.

Rybicki, E.P. 2009. Plant-produced vaccines: Promise and reality. *Drug Discovery Today* 14(1–2) (January): 16–24.

Sadeyen, J.-R., S. Tourne, M. Shkreli, P.-Y. Sizaret, and P. Coursaget. 2003. Insertion of a foreign sequence on capsid surface loops of human papillomavirus type 16 virus-like particles reduces their capacity to induce neutralizing antibodies and delineates a conformational neutralizing epitope. *Virology* 309(1) (April 25): 32–40.

Salunke, D.M., D.L. Caspar, and R.L. Garcea. 1989. Polymorphism in the assembly of polyomavirus capsid protein VP1. *Biophysical Journal* 56(5) (November 1): 887–900.

Salunke, D.M., D.L.D. Caspar, and R.L. Garcea. 1986. Self-assembly of purified polyomavirus capsid protein VP1. *Cell* 46(6) (September): 895–904.

Sandalon, Z., N. Dalyot-Herman, A.B. Oppenheim, and A. Oppenheim. 1997. In vitro assembly of SV40 virions and pseudovirions: Vector development for gene therapy. *Human Gene Therapy* 8(7) (May 1): 843–849.

Sandalon, Z. and A. Oppenheim. 1997. Self-assembly and protein–protein interactions between the SV40 capsid proteins produced in insect cells. *Virology* 237(2): 414–421.

Sanjuan, N., A. Porras, J. Otero, and S. Perazzo. 2001. Expression of major capsid protein VP-1 in the absence of viral particles in thymomas induced by murine polyomavirus. *Journal of Virology* 75(6) (March 15): 2891–2899.

Sasagawa, T., P. Pushko, G. Steers, S.E. Gschmeissner, M.A. Nasser Hajibagheri, J. Finch, L. Crawford, and M. Tommasino. 1995. Synthesis and assembly of virus-like particles of human papillomaviruses type 6 and type 16 in fission yeast *Schizosaccharomyces pombe*. *Virology* 206(1) (January 10): 126–135.

Sasnauskas, K., A. Bulavaite, A. Hale, L. Jin, W.A. Knowles, A. Gedvilaite, A. Dargeviciūte et al. 2002. Generation of recombinant virus-like particles of human and non-human polyomaviruses in yeast *Saccharomyces cerevisiae*. *Intervirology* 45(4–6): 308–317.

Sasnauskas, K., O. Buzaite, F. Vogel, B. Jandrig, R. Razanskas, J. Staniulis, S. Scherneck, D.H. Krüger, and R. Ulrich. 1999. Yeast cells allow high-level expression and formation of polyomavirus-like particles. *Biological Chemistry* 380(3) (March): 381–386.

Scheffer, K.D., A. Gawlitza, G.A. Spoden, X.A. Zhang, C. Lambert, F. Berditchevski, and L. Florin. 2013. Tetraspanin CD151 mediates papillomavirus type 16 endocytosis. *Journal of Virology* 87(6) (March): 3435–3446.

Scheffner, M., B.A. Werness, J.M. Huibregtse, A.J. Levine, and P.M. Howley. 1990. The E6 oncoprotein encoded by human papillomavirus types 16 and 18 promotes the degradation of p53. *Cell* 63(6) (December 21): 1129–1136.

Schelhaas, M., H. Ewers, M.-L. Rajamäki, P.M. Day, J.T. Schiller, and A. Helenius. 2008. Human papillomavirus type 16 entry: Retrograde cell surface transport along actin-rich protrusions. *PLoS Pathogens* 4(9): e1000148.

Schelhaas, M., J. Malmström, L. Pelkmans, J. Haugstetter, L. Ellgaard, K. Grünewald, and A. Helenius. 2007. Simian virus 40 depends on ER protein folding and quality control factors for entry into host cells. *Cell* 131(3) (November 2): 516–529.

Schelhaas, M., B. Shah, M. Holzer, P. Blattmann, L. Kühling, P.M. Day, J.T. Schiller, and A. Helenius. 2012. Entry of human papillomavirus type 16 by actin-dependent, clathrin-and lipid raft-independent endocytosis. *PLoS Pathogens* 8(4): e1002657.

Schellenbacher, C., R. Roden, and R. Kirnbauer. 2009. Chimeric L1–L2 virus-like particles as potential broad-spectrum human papillomavirus vaccines. *Journal of Virology* 83(19) (July 29): 10085–10095.

Schiller, J.T., P.M. Day, and R.C. Kines. 2010. Current understanding of the mechanism of HPV infection. *Gynecologic Oncology* 118(1 Suppl.) (June): S12–S17.

Schiller, J.T. and D. Nardelli-Haefliger. 2006. Chapter 17: Second generation HPV vaccines to prevent cervical cancer. *Vaccine* 24 (August): S147–S153.

Schmidt, U., J. Kenklies, R. Rudolph, and G. Böhm. 1999. Site-specific fluorescence labelling of recombinant polyomavirus-like particles. *Biological Chemistry* 380(3) (March): 397–401.

Schmidt, U., R. Rudolph, and G. Böhm. 2000. Mechanism of assembly of recombinant murine polyomavirus-like particles. *Journal of Virology* 74(4) (February 15): 1658–1662.

Schmidt, U., R. Rudolph, and G. Böhm. 2001. Binding of external ligands onto an engineered virus capsid. *Protein Engineering* 14(10) (January 10): 769–774.

Schneider, M.A., G.A. Spoden, L. Florin, and C. Lambert. 2011. Identification of the dynein light chains required for human papillomavirus infection. *Cellular Microbiology* 13(1) (January): 32–46.

Schowalter, R.M. and C.B. Buck. 2013. The merkel cell polyomavirus minor capsid protein. *PLoS Pathogens* 9(8) (August 22): e1003558.

Schowalter, R.M., D.V. Pastrana, and C.B. Buck. 2011. Glycosaminoglycans and sialylated glycans sequentially facilitate merkel cell polyomavirus infectious entry. *PLoS Pathogens* 7(7) (July): e1002161.

Schwarz, T.F., M. Kocken, T. Petäjä, M.H. Einstein, M. Spaczynski, J.A. Louwers, C. Pedersen et al. 2010. Correlation between levels of human papillomavirus (HPV)-16 and 18 antibodies in serum and cervicovaginal secretions in girls and women vaccinated with the HPV-16/18 AS04-adjuvanted vaccine. *Human Vaccines* 6(12) (December 1): 1054–1061.

Selinka, H.-C., T. Giroglou, and M. Sapp. 2002. Analysis of the infectious entry pathway of human papillomavirus type 33 pseudovirions. *Virology* 299(2) (August 1): 279–287.

Senger, T., L. Schädlich, L. Gissmann, and M. Müller. 2009. Enhanced papillomavirus-like particle production in insect cells. *Virology* 388(2): 344–353.

Serrano, B., L. Alemany, S. Tous, L. Bruni, G.M. Clifford, T. Weiss, F.X. Bosch, and S. de Sanjosé. 2012. Potential impact of a nine-valent vaccine in human papillomavirus related cervical disease. *Infect Agent Cancer* 7(1): 38.

Shi, L., G. Sanyal, A. Ni, Z. Luo, S. Doshna, B. Wang, T.L. Graham, N. Wang, and D.B. Volkin. 2005. Stabilization of human papillomavirus virus-like particles by non-ionic surfactants. *Journal of Pharmaceutical Sciences* 94(7) (July): 1538–1551.

Shin, Y.C. and W.R. Folk. 2003. Formation of polyomavirus-like particles with different VP1 molecules that bind the urokinase plasminogen activator receptor. *Journal of Virology* 77(21) (October 13): 11491–11498.

Shishido, Y., S. Nukuzuma, J. Mukaigawa, S. Morikawa, K. Yasui, and K. Nagashima. 1997. Assembly of JC virus-like particles in COS7 cells. *Journal of Medical Virology* 51(4) (April): 265–272.

Skrastina, D., A. Bulavaite, I. Sominskaya, L. Kovalevska, V. Ose, D. Priede, P. Pumpens, and K. Sasnauskas. 2008. High immunogenicity of a hydrophilic component of the hepatitis B virus preS1 sequence exposed on the surface of three virus-like particle carriers. *Vaccine* 26(16) (April): 1972–1981.

Slilaty, S.N., K.I. Berns, and H.V. Aposhian. 1982. Polyoma-like particle: Characterization of the DNA encapsidated in vitro by polyoma empty capsids. *The Journal of Biological Chemistry* 257(11) (October 6): 6571–6575.

Smith, J.L., S.K. Campos, A. Wandinger-Ness, and M.A. Ozbun. 2008. Caveolin-1-dependent infectious entry of human papillomavirus type 31 in human keratinocytes proceeds to the endosomal pathway for pH-dependent uncoating. *Journal of Virology* 82(19) (October): 9505–9512.

Soeda, E., N. Krauzewicz, C. Cox, J. Stokrová, J. Forstová, and B.E. Griffin. 1998. Enhancement by polylysine of transient, but not stable, expression of genes carried into cells by polyoma VP1 pseudocapsids. *Gene Therapy* 5(10) (October): 1410–1419.

Španielová, H., M. Fraiberk, J. Suchanová, J. Soukup, and J. Forstová. 2014. The encapsidation of polyomavirus is not defined by a sequence-specific encapsidation signal. *Virology* 450–451: 122–131.

Spoden, G., K. Freitag, M. Husmann, K. Boller, M. Sapp, C. Lambert, and L. Florin. 2008. Clathrin- and caveolin-independent entry of human papillomavirus type 16—Involvement of tetraspanin-enriched microdomains (TEMs). *PLoS ONE* 3(10): e3313.

Spoden, G., L. Kühling, N. Cordes, B. Frenzel, M. Sapp, K. Boller, L. Florin, and M. Schelhaas. 2013. Human papillomavirus types 16, 18, and 31 share similar endocytic requirements for entry. *Journal of Virology* 87(13) (July): 7765–7773.

Stamatos, N.M., S. Chakrabarti, B. Moss, and J.D. Hare. 1987. Expression of polyomavirus virion proteins by a vaccinia virus vector: Association of VP1 and VP2 with the nuclear framework. *Journal of Virology* 61(2) (February): 516–525.

Stanley, M., L. Gissmann, and D. Nardelli-Haefliger. 2008. Immunobiology of human papillomavirus infection and vaccination—Implications for second generation vaccines. *Vaccine* 26 (August): K62–K67.

Stanley, M., D.R. Lowy, and I. Frazer. 2006. Chapter 12: Prophylactic HPV vaccines: Underlying mechanisms. *Vaccine* 24 (August): S106–S113.

Stauffer, Y., K. Raj, K. Masternak, and P. Beard. 1998. Infectious human papillomavirus type 18 pseudovirions. *Journal of Molecular Biology* 283(3): 529–536.

Stehle, T., S.J. Gamblin, Y. Yan, and S.C. Harrison. 1996. The structure of simian virus 40 refined at 3.1 Å resolution. *Structure* 4(2): 165–182.

Stehle, T. and S.C. Harrison. 1996. Crystal structures of murine polyomavirus in complex with straight-chain and branched-chain sialyloligosaccharide receptor fragments. *Structure* (London, England: 1993) 4(2) (February 15): 183–194.

Stehle, T., Y. Yan, T.L. Benjamin, and S.C. Harrison. 1994. Structure of murine polyomavirus complexed with an oligosaccharide receptor fragment. *Nature* 369(6476) (May 12): 160–163.

Štokrová, J., Z. Palková, L. Fischer, Z. Richterová, J. Korb, B.E. Griffin, and J. Forstová. 1999. Interactions of heterologous DNA with polyomavirus major structural protein, VP1. *FEBS Letters* 445(1): 119–125.

Stolt, A., K. Sasnauskas, P. Koskela, M. Lehtinen, and J. Dillner. 2003. Seroepidemiology of the human polyomaviruses. *The Journal of General Virology* 84(Pt 6) (June): 1499–1504.

Strable, E. and M.G. Finn. 2009. Chemical modification of viruses and virus-like particles. *Current Topics in Microbiology and Immunology* 327: 1–21.

Ströh, L.J., U. Neu, B.S. Blaum, M.H.C. Buch, R.L. Garcea, and T. Stehle. 2014. Structure analysis of the major capsid proteins of the human polyomavirus 6 and 7 reveals an obstructed sialic acid binding site. *Journal of Virology* 88(18) (July 9): 10831–10839.

Stubenrauch, K., A. Bachmann, R. Rudolph, and H. Lilie. 2000. Purification of a viral coat protein by an engineered polyionic sequence. *Journal of Chromatography B: Biomedical Sciences and Applications* 737(1–2) (January 14): 77–84.

Stubenrauch, K., S. Gleiter, U. Brinkmann, R. Rudolph, and H. Lilie. 2001. Conjugation of an antibody Fv fragment to a virus coat protein: Cell-specific targeting of recombinant polyoma-virus-like particles. *The Biochemical Journal* 356(Pt 3) (June 15): 867–873.

Sundberg, J.P., M.V. Ranst, R. Montali, B.L. Homer, W.H. Miller, P.H. Rowland, D.W. Scott et al. 2000. Feline papillomas and papillomaviruses. *Veterinary Pathology Online* 37(1) (January 1): 1–10.

Surviladze, Z., A. Dziduszko, and M.A. Ozbun. 2012. Essential roles for soluble virion-associated heparan sulfonated proteoglycans and growth factors in human papillomavirus infections. *PLoS Pathogens* 8(2): e1002519.

Surviladze, Z., R.T. Sterk, S.A. DeHaro, and M.A. Ozbun. 2013. Cellular entry of human papillomavirus type 16 involves activation of the phosphatidylinositol 3-Kinase/Akt/mTOR pathway and inhibition of autophagy. *Journal of Virology* 87(5) (March): 2508–2517.

Suzich, J.A., S.J. Ghim, F.J. Palmer-Hill, W.I. White, J.K. Tamura, J.A. Bell, J.A. Newsome, A.B. Jenson, and R. Schlegel. 1995. Systemic immunization with papillomavirus L1 protein completely prevents the development of viral mucosal papillomas. *Proceedings of the National Academy of Sciences of the United States of America* 92(25) (December 5): 11553–11557.

Swimm, A.I., W. Bornmann, M. Jiang, M.J. Imperiale, A.E. Lukacher, and D. Kalman. 2010. Abl family tyrosine kinases regulate sialylated ganglioside receptors for polyomavirus. *Journal of Virology* 84(9) (May): 4243–4251.

Szomolanyi-Tsuda, E., J.D. Brien, J.E. Dorgan, R.L. Garcea, R.T. Woodland, and R.M. Welsh. 2001. Antiviral T-cell-independent type 2 antibody responses induced in vivo in the absence of T and NK cells. *Virology* 280(2) (February): 160–168.

Szomolanyi-Tsuda, E., Q.P. Le, R.L. Garcea, and R.M. Welsh. 1998. T-cell-independent immunoglobulin G responses in vivo are elicited by live-virus infection but not by immunization with viral proteins or virus-like particles. *Journal of Virology* 72(8): 6665–6670.

Takahashi, R., S. Kanesashi, T. Inoue, T. Enomoto, M. Kawano, H. Tsukamoto, F. Takeshita et al. 2008. Presentation of functional foreign peptides on the surface of SV40 virus-like particles. *Journal of Biotechnology* 135(4) (July): 385–392.

Talmage, D.A., R. Freund, T. Dubensky, M. Salcedo, P. Gariglio, L.M. Rangel, C.J. Dawe, and T.L. Benjamin. 1992. Heterogeneity in state and expression of viral DNA in polyoma virus-induced tumors of the mouse. *Virology* 187(2) (April): 734–747.

Tegerstedt, K., K. Andreasson, A. Vlastos, K.O. Hedlund, T. Dalianis, and T. Ramqvist. 2003. Murine pneumotropic virus VP1 virus-like particles (VLPs) bind to several cell types independent of sialic acid residues and do not serologically cross react with murine polyomavirus VP1 VLPs. *Journal of General Virology* 84(12) (January 12): 3443–3452.

Tegerstedt, K., A. Franzén, T. Ramqvist, and T. Dalianis. 2007. Dendritic cells loaded with polyomavirus VP1/VP2Her2 virus-like particles efficiently prevent outgrowth of a Her2/neu expressing tumor. *Cancer Immunology, Immunotherapy* 56(9) (June 26): 1335–1344.

Tegerstedt, K., A.V. Franzén, K. Andreasson, J. Joneberg, S. Heidari, T. Ramqvist, and T. Dalianis. 2005a. Murine polyomavirus virus-like particles (VLPs) as vectors for gene and immune therapy and vaccines against viral infections and cancer. *Anticancer Research* 25(4) (January 7): 2601–2608.

Tegerstedt, K., J.A. Lindencrona, C. Curcio, K. Andreasson, C. Tullus, G. Forni, T. Dalianis, R. Kiessling, and T. Ramqvist. 2005b. A single vaccination with polyomavirus VP1/VP2Her2 virus-like particles prevents outgrowth of HER-2/neu-expressing tumors. *Cancer Research* 65(13): 5953–5957.

Teunissen, E.A., M. de Raad, and E. Mastrobattista. 2013. Production and biomedical applications of virus-like particles derived from polyomaviruses. *Journal of Controlled Release* 172(1) (November): 305–321.

The FUTURE I/II Study Group, J. Dillner, S.K. Kjaer, C.M. Wheeler, K. Sigurdsson, O.E. Iversen, M. Hernandez-Avila et al. 2010. Four year efficacy of prophylactic human papillomavirus quadrivalent vaccine against low grade cervical, vulvar, and vaginal intraepithelial neoplasia and anogenital warts: Randomised controlled trial. *BMJ* 341 (July 20): c3493.

Thyagarajan, R., N. Arunkumar, and W. Song. 2003. Polyvalent antigens stabilize B cell antigen receptor surface signaling microdomains. *The Journal of Immunology* 170(12) (June 15): 6099–6106.

Tobery, T.W., J.F. Smith, N. Kuklin, D. Skulsky, C. Ackerson, L. Huang, L. Chen, J.C. Cook, W.L. McClements, and K.U. Jansen. 2003. Effect of vaccine delivery system on the induction of HPV16L1-specific humoral and cell-mediated immune responses in immunized rhesus macaques. *Vaccine* 21(13–14) (March 28): 1539–1547.

Tolstov, Y.L., D.V. Pastrana, H. Feng, J.C. Becker, F.J. Jenkins, S. Moschos, Y. Chang, C.B. Buck, and P.S. Moore. 2009. Human merkel cell polyomavirus infection II. MCV is a common human infection that can be detected by conformational capsid epitope immunoassays. *International Journal of Cancer* (*Journal International Du Cancer*) 125(6) (September 15): 1250–1256.

Touzé, A., L. Bousarghin, C. Ster, A.-L. Combita, P. Roingeard, and P. Coursaget. 2001. Gene transfer using human polyomavirus BK virus-like particles expressed in insect cells. *Journal of General Virology* 82(12) (January 12): 3005–3009.

Touzé, A. and P. Coursaget. 1998. In vitro gene transfer using human papillomavirus-like particles. *Nucleic Acids Research* 26(5) (March 1): 1317–1323.

Touzé, A., C. Dupuy, M. Chabaud, P. Le Cann, and P. Coursaget. 1996. Production of human papillomavirus type 45 virus-like particles in insect cells using a recombinant baculovirus. *FEMS Microbiology Letters* 141(1) (July 15): 111–116.

Touzé, A., J. Gaitan, F. Arnold, R. Cazal, M.J. Fleury, N. Combelas, P.-Y. Sizaret et al. 2010. Generation of merkel cell polyomavirus (MCV)-like particles and their application to detection of MCV antibodies. *Journal of Clinical Microbiology* 48(5) (January 5): 1767–1770.

Touzé, A., S.E. Mehdaoui, P.-Y. Sizaret, C. Mougin, N. Muñoz, and P. Coursaget. 1998. The L1 major capsid protein of human papillomavirus type 16 variants affects yield of virus-like particles produced in an insect cell expression system. *Journal of Clinical Microbiology* 36(7) (January 7): 2046–2051.

Tsai, B., J.M. Gilbert, T. Stehle, W. Lencer, T.L. Benjamin, and T.A. Rapoport. 2003. Gangliosides are receptors for murine polyoma virus and SV40. *The EMBO Journal* 22(17) (September 1): 4346–4355.

Tsukamoto, H., M. Kawano, T. Inoue, T. Enomoto, R. Takahashi, N. Yokoyama, N. Yamamoto et al. 2007. Evidence that SV40 VP1–DNA interactions contribute to the assembly of 40-Nm spherical viral particles. *Genes to Cells* 12(11): 1267–1279.

Unckell, F., R.E. Streeck, and M. Sapp. 1997. Generation and neutralization of pseudovirions of human papillomavirus type 33. *Journal of Virology* 71(4) (January 4): 2934–2939.

Van Regenmortel, M.H.V., D.H.L. Bishop, and C.M. Fauquet. 1999. *Virus Taxonomy: Seventh Report of the International Committee on Taxonomy of Viruses*. San Diego, CA: Academic.

Varsani, A., A.-L. Williamson, R.C. Rose, M. Jaffer, and E.P. Rybicki. 2003. Expression of human papillomavirus type 16 major capsid protein in transgenic Nicotiana tabacum cv. Xanthi. *Archives of Virology* 148(9) (September 1): 1771–1786.

Velupillai, P., R.L. Garcea, and T.L. Benjamin. 2006. Polyoma virus-like particles elicit polarized cytokine responses in APCs from tumor-susceptible and-resistant mice. *The Journal of Immunology* 176(2): 1148–1153.

Villa, L.L., R.L.R. Costa, C.A. Petta, R.P. Andrade, J. Paavonen, O.-E. Iversen, S.-E. Olsson et al. 2006. High sustained efficacy of a prophylactic quadrivalent human papillomavirus types 6/11/16/18 L1 virus-like particle vaccine through 5 years of follow-up. *British Journal of Cancer* 95(11) (December 4): 1459–1466.

Vlastos, A., K. Andreasson, K. Tegerstedt, D. Holländerová, S. Heidari, J. Forstová, T. Ramqvist, and T. Dalianis. 2003. VP1 pseudocapsids, but not a glutathione-S-transferase VP1 fusion protein, prevent polyomavirus infection in a T-cell immune deficient experimental mouse model. *Journal of Medical Virology* 70(2) (June): 293–300.

Volpers, C., P. Schirmacher, R.E. Streeck, and M. Sapp. 1994. Assembly of the major and the minor capsid protein of human papillomavirus type 33 into Virus-like particles and tubular structures in insect cells. *Virology* 200(2) (May 1): 504–512.

Volpers, C., F. Unckell, P. Schirmacher, R.E. Streeck, and M. Sapp. 1995. Binding and internalization of human papillomavirus type 33 virus-like particles by eukaryotic cells. *Journal of Virology* 69(6) (January 6): 3258–3264.

Voronkova, T., A. Kazaks, V. Ose, M. Özel, S. Scherneck, P. Pumpens, and R. Ulrich. 2007. Hamster polyomavirus-derived virus-like particles are able to transfer in vitro encapsidated plasmid DNA to mammalian cells. *Virus Genes* 34(3) (June 1): 303–314.

Waehler, R., S.J. Russell, and D.T. Curiel. 2007. Engineering targeted viral vectors for gene therapy. *Nature Reviews Genetics* 8(8) (August): 573–587.

Wagner, E., K. Zatloukal, M. Cotten, H. Kirlappos, K. Mechtler, D.T. Curiel, and M.L. Birnstiel. 1992. Coupling of adenovirus to transferrin-polylysine/DNA complexes greatly enhances receptor-mediated gene delivery and expression of transfected genes. *Proceedings of the National Academy of Sciences of the United States of America* 89(13) (July 1): 6099–6103.

Wakabayashi, M.T., D.M. Da Silva, R.K. Potkul, and W.M. Kast. 2002. Comparison of human papillomavirus type 16 L1 chimeric virus-like particles versus L1/l2 chimeric virus-like particles in tumor prevention. *Intervirology* 45(4–6): 300–307.

Wang, M., T.-H. Tsou, L.-S. Chen, W.-C. Ou, P.-L. Chen, C.-F. Chang, C.-Y. Fung, and D. Chang. 2004. Inhibition of simian virus 40 large tumor antigen expression in human fetal glial cells by an antisense oligodeoxynucleotide delivered by the JC virus-like particle. *Human Gene Therapy* 15(11) (November 1): 1077–1090.

Wang, T., Z. Zhang, D. Gao, F. Li, H. Wei, X. Liang, Z. Cui, and X.-E. Zhang. 2011. Encapsulation of gold nanoparticles by simian virus 40 capsids. *Nanoscale* 3(10): 4275.

Wang, X., J. Liu, Y. Zheng, J. Li, H. Wang, Y. Zhou, M. Qi, H. Yu, W. Tang, and W.M. Zhao. 2008. An optimized yeast cell-free system: Sufficient for translation of human papillomavirus 58 L1 mRNA and assembly of virus-like particles. *Journal of Bioscience and Bioengineering* 106(1): 8–15.

Warzecha, H., H.S. Mason, C. Lane, A. Tryggvesson, E. Rybicki, A.-L. Williamson, J.D. Clements, and R.C. Rose. 2003. Oral immunogenicity of human papillomavirus-like particles expressed in potato. *Journal of Virology* 77(16) (August 15): 8702–8711.

Wen, A.M., S. Shukla, P. Saxena, A.A.A. Aljabali, I. Yildiz, S. Dey, J.E. Mealy et al. 2012. Interior engineering of a viral nanoparticle and its tumor homing properties. *Biomacromolecules* 13(12) (December 10): 3990–4001.

Westra, T.A., I. Stirbu-Wagner, S. Dorsman, E.D. Tutuhatunewa, E.L. de Vrij, H.W. Nijman, T. Daemen, J.C. Wilschut, and M.J. Postma. 2013. Inclusion of the benefits of enhanced cross-protection against cervical cancer and prevention of genital warts in the cost-effectiveness analysis of human papillomavirus vaccination in the Netherlands. *BMC Infectious Diseases* 13(1): 75.

Weyermann, J., D. Lochmann, and A. Zimmer. 2004. Comparison of antisense oligonucleotide drug delivery systems. *Journal of Controlled Release* 100(3) (December 10): 411–423.

Wheeler, C.M., X. Castellsagué, S.M. Garland, A. Szarewski, J. Paavonen, P. Naud, J. Salmerón, S.-N. Chow, D. Apter, and H. Kitchener. 2012. Cross-protective efficacy of HPV-16/18 AS04-adjuvanted vaccine against cervical infection and precancer caused by non-vaccine oncogenic HPV types: 4-Year end-of-study analysis of the randomised, double-blind PATRICIA trial. *The Lancet Oncology* 13(1): 100–110.

WHO. 2009. Human papillomavirus vaccines. WHO position paper. Weekly Epidemiological Record no. 84, 15, pp. 117–132. http://www.who.int/wer/2009/wer8415.pdf?ua = 1. Accessed July 14, 2014.

WHO. 2014. WHO | Human Papillomavirus (HPV). WHO. http://www.who.int/immunization/diseases/hpv/en/. Accessed July 14, 2014.

Windram, O.P., B. Weber, M.A. Jaffer, E.P. Rybicki, D.N. Shepherd, and A. Varsani. 2008. An investigation into the use of human papillomavirus type 16 virus-like particles as a delivery vector system for foreign proteins: N- and C-terminal fusion of GFP to the L1 and L2 capsid proteins. *Archives of Virology* 153(3) (January 4): 585–589.

Wolf, M., R.L. Garcea, N. Grigorieff, and S.C. Harrison. 2010. Subunit interactions in bovine papillomavirus. *Proceedings of the National Academy of Sciences of the United States of America* 107(14) (June 4): 6298–6303.

Wong, S.S. 1991. *Chemistry of Protein Conjugation and Cross-Linking.* Boca Raton, FL: CRC Press.

Woo, M.-K., S.-J. Hur, S. Park, and H.-J. Kim. 2007. Study of cell-mediated response in mice by HPV16 L1 virus-like particles expressed in *Saccharomyces cerevisiae*. *Journal of Microbiology and Biotechnology* 17(10) (October): 1738–1741.

Woodham, A.W., D.M. Da Silva, J.G. Skeate, A.B. Raff, M.R. Ambroso, H.E. Brand, J.M. Isas, R. Langen, and W.M. Kast. 2012. The S100A10 subunit of the annexin A2 heterotetramer facilitates L2-mediated human papillomavirus infection. *PLoS ONE* 7(8): e43519.

Wróbel, B., Y. Yosef, A.B. Oppenheim, and A. Oppenheim. 2000. Production and purification of SV40 major capsid protein (VP1) in *Escherichia coli* strains deficient for the GroELS chaperone machine. *Journal of Biotechnology* 84(3) (December 28): 285–289.

Xu, X., T. Zhang, Y. Xu, and D. Fan. 2014. Virus-like particles of capsid proteins from human papillomavirus type 16/58/18/6/11 and the method for preparation and the uses thereof. Patent no. CN101148661 B 2013-01-02.

Xu, Y., H. Zhang, and X. Xu. 2008. Enhancement of vaccine potency by fusing modified LTK63 into human papillomavirus type 16 chimeric virus-like particles. *FEMS Immunology & Medical Microbiology* 52(1) (January): 99–109.

Yang, R., F.M. Murillo, M.J. Delannoy, R.L. Blosser, W.H. Yutzy, S. Uematsu, K. Takeda, S. Akira, R.P. Viscidi, and R.B. Roden. 2005. B lymphocyte activation by human papillomavirus-like particles directly induces Ig class switch recombination via TLR4-MyD88. *The Journal of Immunology* 174(12): 7912–7919.

Yeager, M.D., M. Aste-Amezaga, D.R. Brown, M.M. Martin, M.J. Shah, J.C. Cook, N.D. Christensen et al. 2000. Neutralization of human papillomavirus (HPV) pseudovirions: A novel and efficient approach to detect and characterize HPV neutralizing antibodies. *Virology* 278(2) (December 20): 570–577.

Zamora, E., A. Handisurya, S. Shafti-Keramat, D. Borchelt, G. Rudow, K. Conant, C. Cox, J.C. Troncoso, and R. Kirnbauer. 2006. Papillomavirus-like particles are an effective platform for amyloid-B immunization in rabbits and transgenic mice. *The Journal of Immunology* 177(4): 2662–2670.

Zanotto, C., E. Pozzi, S. Pacchioni, M. Bissa, C.D.G. Morghen, and A. Radaelli. 2011. Construction and characterisation of a recombinant fowlpox virus that expresses the human papilloma virus L1 protein. *Journal of Translational Medicine* 9(1) (November 4): 190.

Zhao, K.-N., K. Hengst, W.-J. Liu, Y.H. Liu, X.S. Liu, N.A.J. McMillan, and I.H. Frazer. 2000. BPV1 E2 protein enhances packaging of full-length plasmid DNA in BPV1 pseudovirions. *Virology* 272(2): 382–393.

Zhao, Q., M.J. Allen, Y. Wang, B. Wang, N. Wang, L. Shi, and R.D. Sitrin. 2012a. Disassembly and reassembly improves morphology and thermal stability of human papillomavirus type 16 virus-like particles. *Nanomedicine: Nanotechnology, Biology and Medicine* 8(7): 1182–1189.

Zhao, Q., Y. Modis, K. High, V. Towne, Y. Meng, Y. Wang, J. Alexandroff et al. 2012b. Disassembly and reassembly of human papillomavirus virus-like particles produces more virion-like antibody reactivity. *Virology Journal* 9 (February 22): 52.

Zheng, J., X. Yang, Y. Sun, B. Lai, and Y. Wang. 2008. Stable high-level expression of truncated human papillomavirus type 16 L1 protein in Drosophila schneider-2 cells. *Acta Biochimica et Biophysica Sinica* 40(5) (January 5): 437–442.

Zhou, J., W.J. Liu, S.W. Peng, X.Y. Sun, and I. Frazer. 1999. Papillomavirus capsid protein expression level depends on the match between codon usage and tRNA availability. *Journal of Virology* 73(6) (January 6): 4972–4982.

Zhou, J., D.J. Stenzel, X.-Y. Sun, and I.H. Frazer. 1993. Synthesis and assembly of infectious bovine papillomavirus particles in vitro. *Journal of General Virology* 74(4) (January 4): 763–768.

Zhou, J., X.Y. Sun, D.J. Stenzel, and I.H. Frazer. 1991. Expression of vaccinia recombinant HPV 16 L1 and L2 ORF proteins in epithelial cells is sufficient for assembly of HPV virion-like particles. *Virology* 185(1) (November): 251–257.

Zielonka, A., A. Gedvilaite, J. Reetz, U. Rösler, H. Müller, and R. Johne. 2012. Serological cross-reactions between four polyomaviruses of birds using virus-like particles expressed in yeast. *Journal of General Virology* 93(Pt 12) (January 12): 2658–2667.

Zielonka, A., A. Gedvilaite, R. Ulrich, D. Lüschow, K. Sasnauskas, H. Müller, and R. Johne. 2006. Generation of virus-like particles consisting of the major capsid protein VP1 of goose hemorrhagic polyomavirus and their application in serological tests. *Virus Research* 120(1–2) (September): 128–137.

Zielonka, A., E.J. Verschoor, A. Gedvilaite, U. Roesler, H. Müller, and R. Johne. 2011. Detection of chimpanzee polyomavirus-specific antibodies in captive and wild-caught chimpanzees using yeast-expressed virus-like particles. *Virus Research* 155(2): 514–519.

Zila, V., F. Difato, L. Klimova, S. Huerfano, and J. Forstova. 2014. Involvement of microtubular network and its motors in productive endocytic trafficking of mouse polyomavirus. *PLoS ONE* 9(5): e96922.

zur Hausen, H. 2001. Proliferation-inducing viruses in non-permissive systems as possible causes of human cancers. *The Lancet* 357(9253): 381–384.

zur Hausen, H. 2002. Papillomaviruses and cancer: From basic studies to clinical application. *Nature Reviews Cancer* 2(5) (May): 342.

zur Hausen, H. 2008. Novel human polyomaviruses—Re-Emergence of a well known virus family as possible human carcinogens. *International Journal of Cancer* 123(2): 247–250.

Zurhein, G. and S.M. Chou. 1965. Particles resembling papova viruses in human cerebral demyelinating disease. *Science* (New York) 148(3676) (June 11): 1477–1479.

Zvirbliene, A., L. Samonskyte, A. Gedvilaite, T. Voronkova, R. Ulrich, and K. Sasnauskas. 2006. Generation of mono-clonal antibodies of desired specificity using chimeric polyomavirus-derived virus-like particles. *Journal of Immunological Methods* 311(1–2) (April): 57–70.

21 Nanoparticles of Norovirus

Ming Tan and Xi Jiang

CONTENTS

21.1 INTRODUCTION

Noroviruses (NoVs) are a group of small round-structured RNA viruses that constitute the *Norovirus* genus in the family Caliciviridae. The viruses are highly contagious, causing epidemic acute gastroenteritis that affect millions of people and claim over 200,000 lives each year [1]. NoVs are nonenveloped viruses that are composed of an outer protein capsid encapsulating the single-stranded, positive-sense RNA genomes. The NoV capsid is made by a single major structure protein (VP1) that is encoded by open reading frame 2 of the genome. The crystal structures of the recombinant NoV virus-like particle (VLP) revealed that the NoV capsid exhibits a T = 3 icosahedral symmetry containing 180 copies of VP1s organized into 90 dimeric capsomeres [2].

The NoV VP1 protein consists of two major domains, the shell (S) and the protruding (P) domains, linked by a short, flexible hinge [2]. The S domain is responsible for the icosahedral shell formation, while the P domain forms the arch-like dimers extending outward from the shell. The P domain is further divided into P1 and P2 subdomains, constituting the leg and the head of the arch-like P dimer, respectively [2]. The P2 subdomain forms the outermost surface of the viral capsid, including the highly conserved sites for host receptor interaction [3–6] and the highly variable regions responding to the host immune selection.

Like many other viral capsid proteins, NoV VP1 exhibits strong intermolecular interactions, which are the major driving force of NoV capsid formation during viral assembly. As a result, in vitro expression of NoV VP1 assembles into VLPs. In addition, productions of VP1 subdomains form different smaller subviral particles, including the S particles that are formed by the S domain [7,8] and the P particles and small P particles that are formed by the P domain [9–11]. Chimeric polymers containing P domains of NoVs and other viruses have also been made [12–14]. Due to the lack of an efficient cell culture system and a small animal model for human NoVs (huNoVs), these recombinant subviral particles have been used extensively as models and tools for huNoV research. Currently, there is no prophylactic and therapeutic approach for control and prevention of huNoVs and their associated diseases. These variable subviral particles and complexes also provided attractive model systems for vaccine and antiviral development.

21.2 SUBVIRAL PARTICLES OF NoVs

Variable NoV VP1–based nanoparticles have been made in the past two decades as results of studies of the structures and functions of NoV capsids, including VLPs, P particles, and other P domain–based polymers. These subviral complexes have been widely used in characterization of the virus–host interactions, immune responses, diagnosis, and vaccine development of NoVs. The application of these subviral complexes as candidate vaccines against NoVs is the major focus of this chapter.

21.2.1 NoV VLPs

The NoV VLPs (Figure 21.1) assemble spontaneously when the full-length NoV VP1 or VP1 plus VP2 is expressed in a eukaryotic expression system, such as insect cells through a recombinant baculovirus [15,16], mammalian cells through a Venezuelan equine encephalitis replicon [17], a vesicular stomatitis virus (VSV) vector [18], or a plasmid [19], yeast (*Pichia pastoris*) [20], and several transgenic plants including tomato, potato, and tobacco [21–26]. Most of the expression systems are highly efficient for production of NoV VLPs, in which the insect cell/baculovirus expression system has been successfully used in production of the first NoV vaccine candidate for clinical trials [27–31]. Unfortunately, the prokaryotic *Escherichia coli* system seems not supporting VLP assembly [32].

The crystal structures of a recombinant VLP of Norwalk virus (NV, GI.1), the prototype of huNoVs, have been elucidated, which is believed to be indistinguishable from the authentic NoVs. It is ~37 nm in diameter, formed by 180 copies of the VP1 in a T = 3 icosahedral symmetry similar to many small RNA viruses [2] (Figure 21.1). The internal shell of the NoV capsid is formed by an S domain of VP1, and the exterior protrusions are constituted by the P domains. The P domain has been demonstrated to be responsible for virus/host interaction and immune response of NoVs. While the formation of full-size VLPs with 180 copies of the capsid protein is highly efficient, low percentages of smaller VLPs

with 60 copies of VP1 in a potential T = 1 icosahedral symmetry were also observed [33]. NoV VLPs provide an excellent model and valuable sources of materials for the study of huNoVs.

21.2.2 S Particles

The NoV S particles (~27 nm; Figure 21.1) form when the S domain alone is expressed in insect cells via a recombinant baculovirus [7,8]. They reveal the typical structures of the interior shell of a NoV capsid in an icosahedral symmetry (T = 3). Due to the lack of the P domains that contain the majority of the neutralizing epitopes, the S particle may not be a good choice for vaccine development. However, the polyvalent feature of the S particle makes it a potential vaccine platform for antigen displays for vaccine development against other pathogens. In addition, the empty S particle could be a good cargo carrier for target-specific drug delivery.

21.2.3 P Particles

Although the expression of the P domain alone generally forms P dimers and P particles at low efficiency [8,11], high yields of stable P particles can be made through a minor modification at the ends of the P domain [11]. Two types of the P particles have been constructed, the 24 mer P particle (~20 nm; Figure 21.1) and the 12 mer small P particle (~14 nm; Figure 21.1). The P particles were made from the P domain with a cysteine-containing peptide (CCP) at either end [11], while the small P particles were made from P domain with a CCP and a further peptide at each of the two ends [10]. Intermolecular disulfide bonds among cysteines of the CCPs significantly stabilize the P particle formation.

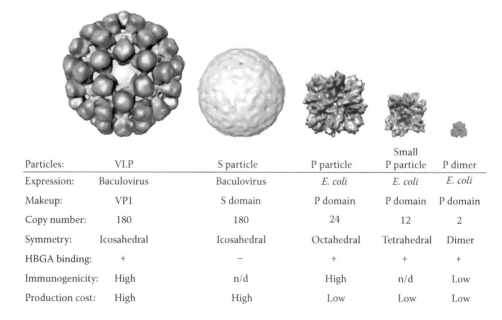

Particles:	VLP	S particle	P particle	Small P particle	P dimer
Expression:	Baculovirus	Baculovirus	*E. coli*	*E. coli*	*E. coli*
Makeup:	VP1	S domain	P domain	P domain	P domain
Copy number:	180	180	24	12	2
Symmetry:	Icosahedral	Icosahedral	Octahedral	Tetrahedral	Dimer
HBGA binding:	+	−	+	+	+
Immunogenicity:	High	n/d	High	n/d	Low
Production cost:	High	High	Low	Low	Low

FIGURE 21.1 Subviral particles of NoV. The crystal or cryo-EM structures of the NoV VLP, S particle, P and small P particles, and P dimer are shown in the top panel. Their general properties, including the in vitro expression systems, makeup, copy numbers, symmetry, host receptor binding function, immunogenicity, and production costs, are shown.

Unlike the VLP and S particles that assemble only in a eukaryotic system, both the P particles self-assemble in the prokaryotic system, allowing easy production of the P particles at high yields with low cost.

The structures of both P particles have been determined by electron cryomicroscopy (cryo-EM) [9,10]. The P particle exhibits an octahedral symmetry containing 24 copies of the P domains organized in 12 P dimers. The 12 P dimers interact with each other through their P1 subdomain forming the center of the P particle, in which the intermolecular disulfide bonds among the CCPs play a critical role in stabilizing these structures. There is an open cavity in the center of each P particle. The sizes of the cavity become larger when the hinge is present at the N-terminus of the P domain. The P2 subdomains of the 12 P dimers extend outward from the center forming 12 protrusions of the P particles [9]. As a result, as those in the VLPs and authentic virions, the P2 and most P1 regions are exposed. Therefore, the host interacting sites, such as the histo-blood group antigen (HBGA) binding sites and the major antigenic determinants, are preserved on the P particles, providing another vaccine candidate against huNoVs (see Section 21.3.2). It was noted that small populations of 18 and 36 mer P particles are present in the P particle assemblies [34,35], but their symmetry and structures remain to be defined.

The small P particles exhibit a tetrahedral symmetry containing 12 P monomers organized into 6 P dimers the following text Section 21.3.2). Similar to P particles, the 12 P1 subdomains of the small P particles interact with each other with the help of the intermolecular disulfide bonds among CCPs, forming the center with an open cavity of the small P particle [10]. The P2 subdomains are also exposed, being responsible for the interaction with the host HBGA receptors and the host immunity similar to those of P particles. Thus, the small P particle may also be used as a vaccine candidate against huNoVs.

21.2.4 CHIMERIC P DOMAIN POLYMERS

The fact that structural proteins of NoVs and other viruses exhibit strong homotypic interactions forming dimers and/or oligomers makes them attractive building blocks to construct chimeric polymers. This is achieved by the fusion of two or more such proteins into one molecule through DNA recombinant technology followed by in vitro expressions of the fusion proteins (Figure 21.2). The polymers self-assemble through intermolecular dimerization and/or oligomerization of the homologous protein components. Three types of polymers, the linear, network, and agglomerate polymers, containing the P domains of NoV, hepatitis E virus (HEV), and astrovirus have been made successfully [12–14]. These polymers can be of large sizes with several hundred copies of the protein components.

The linear polymers are made from a fusion of two dimeric proteins (Figure 21.2a and d). For example, the NoV P domain with a deletion of the C-terminal arginine cluster can be expressed in *E. coli* highly efficiently, forming stable dimers [13,36]. When two such P domains are fused together, stable linear polymers with many copies of the fusion proteins

assemble [13]. The two dimeric proteins in the polymers can be from any NoVs or even other pathogens, making the polymers potential bivalent vaccines against different antigenic types of NoVs as well as other pathogens.

Network polymers are usually formed when three dimeric proteins are fused together and expressed in bacteria (Figure 21.2b and e) [13]. The network polymers also have a large molecular weight with hundred copies of each protein component, providing a valuable opportunity for trivalent vaccine development.

When one of the dimeric proteins in the aforementioned fusion proteins is replaced by an oligomeric protein, agglomerate polymers form. For example, the dimeric glutathione S-transferase (GST) fused with a modified P domain of HEV formed such agglomerate polymers due to the ability of the modified HEV P domain to form tetramers [14]. Similarly, more sophisticated agglomerate polymers (Figure 21.2c and f) have also been made through fusion of the GST with the modified NoV P domain with a CCP that forms 24 mer P particles (see Section 21.2.3). The resulting agglomerate polymers are highly complex containing several hundred copies of each component.

21.3 NoV SUBVIRAL PARTICLES FOR VACCINE DEVELOPMENT

While NoV subviral particles and complexes may find different applications in biomedicine, the most evaluated one is vaccine development. After the introduction of the first two rotavirus (RV) vaccines in 2006, NoVs became the most important cause of nonbacterial acute gastroenteritis worldwide, and the development of an effective NoV vaccine has become a top priority. However, the inability of replication of huNoVs in a cell culture system prevents the use of traditional vaccine strategies of live attenuated and inactivated vaccines. Therefore, subunit vaccines based on recombinant NoV antigens must be developed.

21.3.1 NoV VLP AS VACCINE

As mentioned earlier, NoV VLPs are morphologically and antigenically similar to the authentic virions [2,15,16,30,37] and retain binding function to HBGA receptors [38–43]. It is generally assumed that VLPs preserve the virus-specific molecular patterns and major B and T cell epitopes to induce potent innate, humoral, and cellular immune responses, respectively [44,45]. Therefore, NoV VLPs are an excellent candidate for development of a subunit vaccine. After many preclinical studies in the past two decades, NoV VLP vaccines have been proven safe and effective recently through clinical trials.

21.3.1.1 Preclinical Studies of VLP Vaccines

NoV VLPs made by different eukaryotic expression systems have been extensively studied through animal models. For example, mice that were fed with NV VLPs in transgenic plants induced specific serum IgG and secretory IgA responses [21,24].

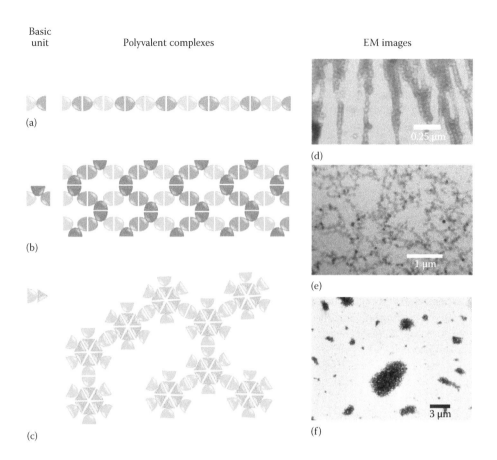

FIGURE 21.2 Schematic illustrations of polymer formations. (a) A linear polymer (right) assembles by a fusion of two dimeric proteins (left, green and blue half ovals) through intermolecular dimerizations (ovals) between two homologous dimeric proteins. (b) A network polymer (right) assembles by a fusion of three dimeric proteins (left, green, purple, and blue half ovals) via intermolecular dimerizations (ovals) between homologous dimeric protein components. (c) An agglomerate polymer (right) assembles by a fusion of a dimeric protein (left, green half oval) with an oligomeric protein (left, yellow triangle) through intermolecular dimerization (green ovals) and oligomerization (yellow hexagon) of the two protein components. In all three cases, only small portions of the large polymers are shown for clarity, while the actual polymers are much more complex. The oligomers in (c) are shown in a cross-sectional view, while they may be spherical in the actual agglomerate polymers. Electron microscopy (EM) examples of each polymer are shown in (d through f).

Similar immune responses were observed in mice that were orally immunized with the baculovirus-expressed NV VLPs [46]. Intranasal immunization of VLPs appeared more efficient, resulting in higher immune responses [47] than those induced by oral administration. Usage of adjuvants, such as LT(R192G), the mutant *E. coli* heat-labile toxin [47], further improved the immune responses of the VLP vaccine. Other immunization approaches of NoV VLP vaccines were also evaluated. For instance, inoculation of viral vector (rVSV-VP1) expressing NoV VLPs to mice induced significantly higher IgG and T cell responses compared with those stimulated by the VLP vaccination [18]. Similar results were also observed in mice after inoculation of alphavirus vectors expressing NV VLPs [17]. These preclinical studies built solid foundation for further development of VLP vaccines in humans.

21.3.1.2 Clinical Trials of VLP Vaccines

Phase I and phase II clinical trials of VLP vaccines have been performed, which demonstrated safety, immunogenicity, and protective efficacy of VLP vaccines in humans. In a recent

study, specific IgG and IgA antibodies were found to increase significantly in all subjects after intranasal immunization of the VLP vaccine with monophosphoryl lipid A and chitosan [48] adjuvants. Vaccinated subjects also developed IgA antibody-secreting cells (ASCs) that expressed molecules associated with homing to mucosal and peripheral lymphoid tissues. Another similar study showed B memory [B(M)] responses in humans and the B(M) cell frequencies correlated with serum antibody levels and mucosally primed ASC responses [49].

To evaluate the efficacy of the VLP vaccines, human subjects were vaccinated with GI.1 NV VLP intranasally followed by challenging orally with homologous NVs [50]. A fourfold or higher increase of specific IgA titers was observed in 70% of the vaccinees, and the vaccine protected subjects from NV infection and occurrence of gastroenteritis. In another human trial to assess a bivalent VLP vaccine (GI.1 NV/GII.4 consensus, [51]), immunized human subjects were challenged with a GII.4 2003 NoV (Cin-1) (See web site of Takeda Pharmaceutical Company, http://www.takeda.com/news/2013/20131007_6021.html) [62].

Comparing vaccination versus placebo groups, a significant decrease of vomiting and/or diarrhea was noted, while severity of disease reduced significantly among the vaccinated subjects with symptoms. Therefore, NoV VLP vaccines are safe and effective for protection against NoV infection and the associated diseases.

21.3.2 NoV P Particles as Vaccine Candidates

NoV P particles have also been evaluated as a candidate vaccine against huNoVs. Structural study strongly suggested that the P particles preserve the authentic structures of the P dimers [9], and therefore the antigenic epitopes of NoVs. The authentic binding function of the P particles to the HBGA receptors as their parental VLPs supports such assumption. In addition, the features of multivalence, high stability, high immunogenicity, and easy production through a prokaryotic expression system make the P particles a promising candidate of low-cost vaccines against huNoVs.

The P particles stimulated significantly higher antibody responses in mice than that induced by the free P dimers [52]. High titers of NoV-specific immunoglobulin (IgY) in chicken egg yolks were also induced via immunization of the P particles to hens, providing a strategy for large-scale production of NoV immunoglobulins for diagnosis and/or therapeutic purposes against NoV diseases [53,54]. In addition, the P particles induced substantial central memory CD4+ T cell phenotypes, which were inducible for production of IFN-γ by a specific CD4+ T cell epitope of the P domain [52]. Furthermore, immunization of the P particles to mice efficiently induced maturation of bone marrow–derived dendritic cells (DCs), and the mature DCs elicited proliferation of specific CD4+ T cells targeting NoV P domain [52]. These data indicated that the NoV P particles are a promising vaccine candidate to elicit both humoral and cellular immune responses against huNoVs.

21.3.3 Chimeric Polymers as Vaccine Candidates

Several polymers containing antigens of NoVs and other viral pathogens have been evaluated as vaccine candidates in mice. Generally, the polymers induce significantly higher humoral and cellular immune responses than those stimulated by the free antigens [12–14]. The linear and network polymers containing the P antigens of two genogroup II (GII) NoVs (VA387, GII.4 and VA207, GII.9) or a genogroup I (GI) NoV (VA115, GI.3) and a GII (VA387, GII.4) NoV induced specific antibodies against these genetically different NoVs. The hyperimmune mouse sera also blocked binding of these NoVs to corresponding HBGA receptors, strongly suggesting that the polymer vaccines may be effective against NoV infection. In addition, the polymer vaccines also induced NoV-specific CD4+ T cell responses, as shown by the significantly higher level of secretion of the CD4+ T cell cytokines IL-2, IFN-γ, and TNF-α [12–14]. Therefore, the P domain polymers are promising vaccine candidates against huNoVs.

21.4 NoV P PARTICLES AND P POLYMERS AS VACCINE PLATFORMS

The features of polyvalence, high stability, high immunogenicity, and easy production through a prokaryotic system of the P particles and P polymers make them excellent platforms for the display of foreign antigens for immune enhancement for vaccine development. Crystal structures of the P dimers indicate three surface loops on the distal end of each P domain, equivalent to the outermost surface of the P particles [55] and VLPs [2] (Figure 21.3). Therefore, inserting an exogenous antigen into such a loop will end up with 24 exposed antigens on the P particle, or hundreds of antigens on the P polymers. In addition, antigens can be fused to the ends of a protein component of the polymers. These will greatly increase the immune response of the inserted antigen. Several epitopes and antigens have been successfully displayed by the P particle and/or the P polymer platforms. Significantly, the high immune responses of the P particles and the P polymers retained, making these chimeric molecules dual vaccines against NoVs and other viral pathogens.

21.4.1 Dual Vaccine against NoV and Rotavirus

The major neutralization antigen VP8* (159 residues) of RV is inserted to a surface loop of the P particle, resulting in a chimeric VP8*-P particle [55,56] (Figure 21.3). The chimeric P particle vaccine induced significantly higher antibody response to RV VP8* with higher neutralizing titer against RV replication than that induced by the free VP8* alone [13,55]. A chimeric P particle vaccine containing a murine RV VP8* of EDIM (Epizootic diarrhea of infant mice) strain provided vaccinated mice a high level (~90%) of protection against a challenge of the EDIM murine RV [55]. As expected, the chimeric P particle also retains the ability to induce high titers of NoV-specific antibody that blocked the binding of NoV VLPs to HBGA receptors. This chimeric VP8*-P particle vaccine is under further development for a potential dual vaccine against both NoV and RV.

21.4.2 Dual Vaccine against NoV and Influenza Virus

This dual vaccine is a chimeric P particle that contains the ectodomain (M2e) of the M2 protein of influenza A viruses on its surface loop [57,58]. The M2e epitopes (23 residues) are highly conserved among most influenza A viruses [59,60] and thus are an attractive target for a broadly protective flu vaccine. Immunization of the chimeric M2e-P particle to mice induced significantly increased titers of M2e-specific antibody than those induced by free M2e peptides [61], and the chimeric M2e-P particle–vaccinated mice were fully protected (100% survival rate) from a lethal challenge of a mouse-adapted influenza virus (PR8 strain, H1N1). This dual vaccine candidate is currently studied in chicken and swine models (Jiang and Lee, unpublished data).

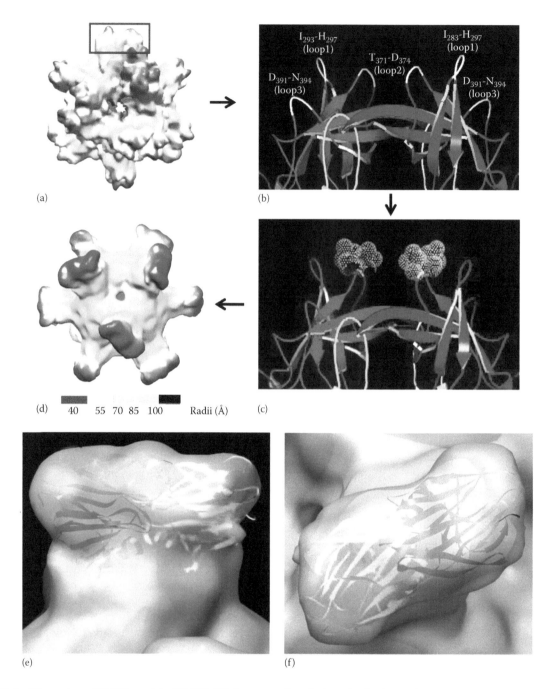

FIGURE 21.3 Illustration of NoV P particle (VA387, GII.4) as a vaccine platform for foreign antigen display. (a) The cryo-EM structure of a NoV P particle. (b) Crystal structure of the top region of a protrusion of NoV P particle with three surface loops. (c) Indication of the locations (amino acids in surface dot model) of loop 2 that a foreign antigen is inserted. (d) Cryo-EM structure of the chimeric P particle with an insertion of the RV surface spike protein VP8* in loop 2 of the P particle. (e and f) Fitting of the RV VP8* crystal structures into the protruding region of the chimeric VP8*-P particle in (e) side and (f) top views.

21.4.3 Dual Vaccine against NoV and Hepatitis E Virus

Two oligomeric complexes formed by the fusion proteins of the P domain antigens of NoV and HEV, with or without a GST, respectively, were developed as potential dual vaccines against NoV and HEV [12]. Both the complexes induced significantly higher antibody titers to NoV P and HEV P antigens, respectively, than those induced by a mixture of the NoV P and HEV P dimers. The complex-induced antisera also exhibited significantly higher neutralizing activity against HEV infection in cell cultures and higher blocking activity on NoV P particles binding to HBGA receptors than those of the dimer-induced antisera. Thus, the chimeric complexes may be promising bivalent vaccines against NoV and HEV.

21.5 CONCLUSION AND FUTURE PERSPECTIVE

A number of subviral particles and complexes, including the VLPs and the P particles of NoVs, as well as the chimeric particles and the polymers of NoV and other viral antigens, have been made through recombinant technology. These particles and complexes retain natural antigenic determinants of the viruses with a significantly enhanced immunogenicity of each viral antigen component, providing novel strategies for vaccine development against huNoVs, as well as other viral pathogens. In addition, the NoV VLPs and P particles have been widely used as tools or models for studying virus–host interaction, diagnosis, and immunology of huNoVs. Additional challenges in development of a useful NoV vaccine, however, remain, including the potential short life of protective immunity against huNoV after viral infection, the wide range of genetic and antigenic variations of huNoVs, and the rapid evolution and frequent emergence of new epidemic variants of huNoVs. Furthermore, the lack of a cell culture and an efficient small animal model for huNoVs also posts a hurdle in the vaccine development. Further studies to characterize these subviral particles and complexes, including designs of new particles and complexes facing to these challenges for highly efficient, low-cost, and broadly protective vaccines, are necessary.

REFERENCES

1. Patel MM, Widdowson MA, Glass RI, Akazawa K, Vinje J, Parashar UD. 2008. Systematic literature review of role of noroviruses in sporadic gastroenteritis. *Emerg Infect Dis* **14**:1224–1231.
2. Prasad BV, Hardy ME, Dokland T, Bella J, Rossmann MG, Estes MK. 1999. X-ray crystallographic structure of the Norwalk virus capsid. *Science* **286**:287–290.
3. Bu W, Mamedova A, Tan M, Xia M, Jiang X, Hegde RS. 2008. Structural basis for the receptor binding specificity of Norwalk virus. *J Virol* **82**:5340–5347.
4. Cao S, Lou Z, Tan M, Chen Y, Liu Y, Zhang Z, Zhang XC, Jiang X, Li X, Rao Z. 2007. Structural basis for the recognition of blood group trisaccharides by norovirus. *J Virol* **81**:5949–5957.
5. Chen Y, Tan M, Xia M, Hao N, Zhang XC, Huang P, Jiang X, Li X, Rao Z. 2011. Crystallography of a lewis-binding norovirus, elucidation of strain-specificity to the polymorphic human histo-blood group antigens. *PLoS Pathogens* **7**:e1002152.
6. Choi JM, Hutson AM, Estes MK, Prasad BV. 2008. Atomic resolution structural characterization of recognition of histo-blood group antigens by Norwalk virus. *Proc Natl Acad Sci U S A* **105**:9175–9180.
7. Bertolotti-Ciarlet A, White LJ, Chen R, Prasad BV, Estes MK. 2002. Structural requirements for the assembly of Norwalk virus-like particles. *J Virol* **76**:4044–4055.
8. Tan M, Hegde RS, Jiang X. 2004. The P domain of norovirus capsid protein forms dimer and binds to histo-blood group antigen receptors. *J Virol* **78**:6233–6242.
9. Tan M, Fang P, Chachiyo T, Xia M, Huang P, Fang Z, Jiang W, Jiang X. 2008. Noroviral P particle: Structure, function and applications in virus-host interaction. *Virology* **382**:115–123.
10. Tan M, Fang PA, Xia M, Chachiyo T, Jiang W, Jiang X. 2011. Terminal modifications of norovirus P domain resulted in a new type of subviral particles, the small P particles. *Virology* **410**:345–352.
11. Tan M, Jiang X. 2005. The p domain of norovirus capsid protein forms a subviral particle that binds to histo-blood group antigen receptors. *J Virol* **79**:14017–14030.
12. Wang L, Cao D, Wei C, Meng XJ, Jiang X, Tan M. 2014. A dual vaccine candidate against norovirus and hepatitis E virus. *Vaccine* **32**:445–452.
13. Wang L, Huang P, Fang H, Xia M, Zhong W, McNeal MM, Jiang X, Tan M. 2013. Polyvalent complexes for vaccine development. *Biomaterials* **34**:4480–4492.
14. Wang L, Xia M, Huang P, Fang H, Cao D, Meng X, McNeal M, Jiang X, Tan M. 2014. Branched-linear and agglomerate protein polymers as vaccine platforms. *Biomaterials* **35**:8427–8438.
15. Jiang X, Zhong WM, Farkas T, Huang PW, Wilton N, Barrett E, Fulton D, Morrow R, Matson DO. 2002. Baculovirus expression and antigenic characterization of the capsid proteins of three Norwalk-like viruses. *Arch Virol* **147**:119–130.
16. Jiang X, Wang M, Graham DY, Estes MK. 1992. Expression, self-assembly, and antigenicity of the Norwalk virus capsid protein. *J Virol* **66**:6527–6532.
17. Harrington PR, Yount B, Johnston RE, Davis N, Moe C, Baric RS. 2002. Systemic, mucosal, and heterotypic immune induction in mice inoculated with Venezuelan equine encephalitis replicons expressing Norwalk virus-like particles. *J Virol* **76**:730–742.
18. Ma Y, Li J. 2011. Vesicular stomatitis virus as a vector to deliver virus-like particles of human norovirus: A new vaccine candidate against an important noncultivable virus. *J Virol* **85**:2942–2952.
19. Taube S, Kurth A, Schreier E. 2005. Generation of recombinant norovirus-like particles (VLP) in the human endothelial kidney cell line 293T. *Arch Virol* **150**:1425–1431.
20. Xia M, Farkas T, Jiang X. 2007. Norovirus capsid protein expressed in yeast forms virus-like particles and stimulates systemic and mucosal immunity in mice following an oral administration of raw yeast extracts. *J Med Virol* **79**:74–83.
21. Mason HS, Ball JM, Shi JJ, Jiang X, Estes MK, Arntzen CJ. 1996. Expression of Norwalk virus capsid protein in transgenic tobacco and potato and its oral immunogenicity in mice. *Proc Natl Acad Sci U S A* **93**:5335–5340.
22. Tacket CO, Mason HS, Losonsky G, Estes MK, Levine MM, Arntzen CJ. 2000. Human immune responses to a novel Norwalk virus vaccine delivered in transgenic potatoes. *J Infect Dis* **182**:302–305.
23. Souza AC, Vasques RM, Inoue-Nagata AK, Lacorte C, Maldaner FR, Noronha EF, Nagata T. 2013. Expression and assembly of Norwalk virus-like particles in plants using a viral RNA silencing suppressor gene. *Appl Microbiol Biotechnol* **97**:9021–9027.
24. Zhang X, Buehner NA, Hutson AM, Estes MK, Mason HS. 2006. Tomato is a highly effective vehicle for expression and oral immunization with Norwalk virus capsid protein. *Plant Biotechnol J* **4**:419–432.
25. Santi L, Huang Z, Mason H. 2006. Virus-like particles production in green plants. *Methods* **40**:66–76.
26. Santi L, Batchelor L, Huang Z, Hjelm B, Kilbourne J, Arntzen CJ, Chen Q, Mason HS. 2008. An efficient plant viral expression system generating orally immunogenic Norwalk virus-like particles. *Vaccine* **26**:1846–1854.
27. Kissmann J, Ausar SF, Foubert TR, Brock J, Switzer MH, Detzi EJ, Vedvick TS, Middaugh CR. 2008. Physical stabilization of Norwalk virus-like particles. *J Pharmaceut Sci* **97**:4208–4218.
28. Koho T, Mantyla T, Laurinmaki P, Huhti L, Butcher SJ, Vesikari T, Kulomaa MS, Hytonen VP. 2012. Purification of norovirus-like particles (VLPs) by ion exchange chromatography. *J Virol Methods* **181**:6–11.

29. Koho T, Huhti L, Blazevic V, Nurminen K, Butcher SJ, Laurinmaki P, Kalkkinen N, Ronnholm G, Vesikari T, Hytonen VP, Kulomaa MS. 2012. Production and characterization of virus-like particles and the P domain protein of GII.4 norovirus. *J Virol Methods* **179**:1–7.

30. Herbst-Kralovetz M, Mason HS, Chen Q. 2010. Norwalk virus-like particles as vaccines. *Expert Rev Vaccines* **9**:299–307.

31. Willyard C. 2013. First vaccines targeting 'cruise ship virus' sail into clinical trials. *Nature Med* **19**:1076–1077.

32. Tan M, Zhong W, Song D, Thornton S, Jiang X. 2004. *E. coli*-expressed recombinant norovirus capsid proteins maintain authentic antigenicity and receptor binding capability. *J Med Virol* **74**:641–649.

33. White LJ, Hardy ME, Estes MK. 1997. Biochemical characterization of a smaller form of recombinant Norwalk virus capsids assembled in insect cells. *J Virol* **71**:8066–8072.

34. Bereszczak JZ, Barbu IM, Tan M, Xia M, Jiang X, van Duijn E, Heck AJ. 2012. Structure, stability and dynamics of norovirus P domain derived protein complexes studied by native mass spectrometry. *J Struct Biol* **177**:273–282.

35. Han L, Kitova EN, Tan M, Jiang X, Klassen JS. 2014. Identifying carbohydrate ligands of a norovirus P particle using a catch and release electrospray ionization mass spectrometry assay. *J Am Soc Mass Spectrom* **25**:111–119.

36. Tan M, Meller J, Jiang X. 2006. C-terminal arginine cluster is essential for receptor binding of norovirus capsid protein. *J Virol* **80**:7322–7331.

37. Jiang X, Matson DO, Ruiz-Palacios GM, Hu J, Treanor J, Pickering LK. 1995. Expression, self-assembly, and antigenicity of a snow mountain agent-like calicivirus capsid protein. *J Clin Microbiol* **33**:1452–1455.

38. Huang P, Farkas T, Marionneau S, Zhong W, Ruvoen-Clouet N, Morrow AL, Altaye M, Pickering LK, Newburg DS, LePendu J, Jiang X. 2003. Noroviruses bind to human ABO, Lewis, and secretor histo-blood group antigens: Identification of 4 distinct strain-specific patterns. *J Infect Dis* **188**:19–31.

39. Huang P, Farkas T, Zhong W, Tan M, Thornton S, Morrow AL, Jiang X. 2005. Norovirus and histo-blood group antigens: Demonstration of a wide spectrum of strain specificities and classification of two major binding groups among multiple binding patterns. *J Virol* **79**:6714–6722.

40. Tan M, Jiang X. 2005. Norovirus and its histo-blood group antigen receptors: An answer to a historical puzzle. *Trends Microbiol* **13**:285–293.

41. Tan M, Jiang X. 2007. Norovirus-host interaction: Implications for disease control and prevention. *Expert Rev Mol Med* **9**:1–22.

42. Tan M, Jiang X. 2010. Norovirus gastroenteritis, carbohydrate receptors, and animal models. *PLoS Pathogens* **6**:e1000983.

43. Tan M, Jiang X. 2010. Virus-host interaction and cellular receptors of caliciviruses, pp. 111–130. *In* Hansman G, Jiang X, Green K (eds.), *Caliciviruses*. Caister Academic Press, Norwich, U.K.

44. Plummer EM, Manchester M. 2010. Viral nanoparticles and virus-like particles: Platforms for contemporary vaccine design. *Wiley Interdiscip Rev Nanomed Nanobiotechnol.* **3**:174–196.

45. Zhao Q, Li S, Yu H, Xia N, Modis Y. 2013. Virus-like particle-based human vaccines: Quality assessment based on structural and functional properties. *Trends Biotechnol* **31**:654–663.

46. Ball JM, Hardy ME, Atmar RL, Conner ME, Estes MK. 1998. Oral immunization with recombinant Norwalk virus-like particles induces a systemic and mucosal immune response in mice. *J Virol* **72**:1345–1353.

47. Guerrero RA, Ball JM, Krater SS, Pacheco SE, Clements JD, Estes MK. 2001. Recombinant Norwalk virus-like particles administered intranasally to mice induce systemic and mucosal (fecal and vaginal) immune responses. *J Virol* **75**:9713–9722.

48. El-Kamary SS, Pasetti MF, Mendelman PM, Frey SE, Bernstein DI, Treanor JJ, Ferreira J et al. 2010. Adjuvanted intranasal Norwalk virus-like particle vaccine elicits antibodies and antibody-secreting cells that express homing receptors for mucosal and peripheral lymphoid tissues. *J Infect Dis* **202**:1649–1658.

49. Ramirez K, Wahid R, Richardson C, Bargatze RF, El-Kamary SS, Sztein MB, Pasetti MF. 2012. Intranasal vaccination with an adjuvanted Norwalk virus-like particle vaccine elicits antigen-specific B memory responses in human adult volunteers. *Clin Immunol* **144**:98–108.

50. Atmar RL, Bernstein DI, Harro CD, Al-Ibrahim MS, Chen WH, Ferreira J, Estes MK, Graham DY, Opekun AR, Richardson C, Mendelman PM. 2011. Norovirus vaccine against experimental human Norwalk virus illness. *N Engl J Med* **365**:2178–2187.

51. Parra GI, Bok K, Taylor R, Haynes JR, Sosnovtsev SV, Richardson C, Green KY. 2012. Immunogenicity and specificity of norovirus Consensus GII.4 virus-like particles in monovalent and bivalent vaccine formulations. *Vaccine* **30**:3580–3586.

52. Fang H, Tan M, Xia M, Wang L, Jiang X. 2013. Norovirus P particle efficiently elicits innate, humoral and cellular immunity. *PLoS One* **8**: e63269.

53. Dai YC, Wang YY, Zhang XF, Tan M, Xia M, Wu XB, Jiang X, Nie J. 2012. Evaluation of anti-norovirus IgY from egg yolk of chickens immunized with norovirus P particles. *J Virol Methods* **186**:126–131.

54. Dai YC, Zhang XF, Tan M, Huang P, Lei W, Fang H, Zhong W, Jiang X. 2012. A dual chicken IgY against rotavirus and norovirus. *Antiviral Res* **97**:293–300.

55. Tan M, Huang P, Xia M, Fang P-A, Zhong W, McNeal M, Wei C, Jiang W, Jiang X. 2011. Norovirus P particle, a novel platform for vaccine development and antibody production. *J Virol* **85**:753–764.

56. Tan M, Xia M, Huang P, Wang L, Zhong W, McNeal M, Wei C, Jiang X. 2011. Norovirus P particle as a platform for antigen presentation. *Procedia Vaccinol* **4**:19–26.

57. Zebedee SL, Richardson CD, Lamb RA. 1985. Characterization of the influenza virus M2 integral membrane protein and expression at the infected-cell surface from cloned cDNA. *J Virol* **56**:502–511.

58. Lamb RA, Zebedee SL, Richardson CD. 1985. Influenza virus M2 protein is an integral membrane protein expressed on the infected-cell surface. *Cell* **40**:627–633.

59. Ito T, Gorman OT, Kawaoka Y, Bean WJ, Webster RG. 1991. Evolutionary analysis of the influenza A virus M gene with comparison of the M1 and M2 proteins. *J Virol* **65**:5491–5498.

60. Liu W, Zou P, Ding J, Lu Y, Chen YH. 2005. Sequence comparison between the extracellular domain of M2 protein human and avian influenza A virus provides new information for bivalent influenza vaccine design. *Microbes Infect* **7**:171–177.

61. Xia M, Tan M, Wei C, Zhong W, Wang L, McNeal M, Jiang X. 2011. A candidate dual vaccine against influenza and noroviruses. *Vaccine* **29**:7670–7677.

62. Takada Pharmaceutical Company. 2013. Takeda highlights data from clinical trial of investigational norovirus vaccine candidate, Website: http://www.takeda.com/news/2013/20131007_6021.html, accessed October 7, 2013.

22 Virus-Like Particle Enzyme Encapsulation
Confined Catalysis and Metabolic Materials

Benjamin Schwarz, Dustin Patterson, and Trevor Douglas

CONTENTS

22.1 INTRODUCTION

22.1.1 COMPARTMENTALIZATION AND CATALYSIS

Compartmentalization is a common theme in biology. Barriers enforce chemical and physical separations defining *self* from the surrounding environment. In higher organisms, compartmentalization is hierarchical extending from the outer barriers that define the organism down to individual cells and subcellular structures such as organelles. Each level of compartmentalization enables essential functions of the organism. Cellular structural compartmentalization and organization of enzymes and metabolic pathways provide a control for the relative flux and efficiency of metabolism as well as protection and sequestration of enzymes and metabolites [1–3]. In addition, biological barriers enable macromolecular densities on the order of 300 mg/mL within the cell, which is vastly different from the conditions used for typical enzyme characterization [4,5]. Structural compartmentalization is a largely untapped resource for enzymatic and metabolic engineering as well as a key point of study for understanding metabolism at the cellular level. One factor inhibiting the study and utility of eukaryotic metabolic structural organization is the complexity and heterogeneity of these systems. Simpler candidates are needed to begin digging into enzyme compartmentalization as a tool and natural phenomenon.

22.1.2 BACTERIAL MICROCOMPARTMENTS

Protein-based structures, reminiscent of subcellular organelles, were observed in cyanobacteria as early as the late 1960s [6]. These protein-based structures have since been termed bacterial microcompartments (BMCs). BMCs, which self-assemble from multiple protein subunits, are closed shell capsid structures that encapsulate enzymes and enzymatic pathways. Some examples of BMCs include the carboxysome, propanediol utilization (Pdu) microcompartment, and ethanolamine utilization (Etu) microcompartment [7–9]. Encapsulation of enzymes within the BMCs is hypothesized to enhance the flux of metabolites through desired pathways, channeling substrates between the colocalized enzymes in the BMC and limiting alternative, unproductive, or detrimental processes. For instance, the carboxysome encapsulates carbonic anhydrase, which rapidly interconverts bicarbonate and carbon dioxide. This is thought to provide high local concentrations of substrate, enabling the kinetically slow ribulose bisphosphate carboxylase/oxygenase to effectively fix carbon dioxide with ribulose-1,5-bisphosphate by converting them to 3-phosphoglycerate [7,10]. Alternatively, the enzymes within the Pdu and Etu microcompartments produce volatile aldehyde intermediates that could easily diffuse or react with other cellular components if not sequestered and colocalized with downstream enzymes that are able to transform them into stable products [11–14]. While the number of BMC systems studied to date has been limited, analyses of prokaryotic genomes suggest that these systems are more prevalent than previously thought and may play a significant role in the catalysis and metabolism of these cells.

While further examination of new BMCs is of interest, a growing body of work has arisen that looks to synthetically mimic such structures by encapsulation of enzymes within nanocontainers. Within these controlled systems, researchers can isolate and study the effects of crowding and confinement on a specific enzyme or subset of enzymes, perhaps providing insights into intracellular behavior as well as the design of enzymatic systems for controlled catalysis. The striking similarity between some BMCs and the structures of viral capsids leads naturally to the use of well-characterized virus-like particles (VLPs) for application in making synthetic catalyst-encapsulated nanocompartments (Figure 22.1). Here, we examine the progress made to date in mimicking these structures by encapsulation of enzymes within protein cage structures and take a look at the future direction of research in the use of VLP enzyme encapsulation.

22.1.3 BIOINSPIRED ENCAPSULATION

While the discussion here focuses on encapsulation of enzymes within protein cages, we do not want to neglect peripherally related systems. There are many other approaches that have been examined for sequestration of enzymes in confining structures including systems such as polymerosomes, vesicles, and gel matrices, to name a few. However, what separates protein cages from these predominantly synthetic systems is the elegant self-assembly and homogeneous structures that form these biological systems. Protein cages self-assemble from a defined number of protein subunits to form symmetrical structures with a hollow interior by a genetically encoded preprogrammed assembly [15]. In general, they are composed of a specific number of protein subunits, with little variation in the range

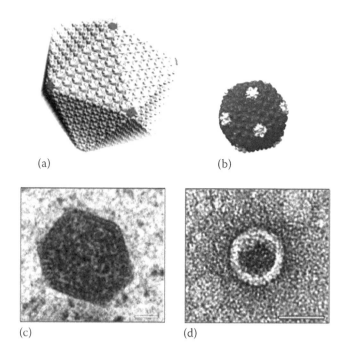

(a)   (b)

(c)   (d)

FIGURE 22.1 The structural similarity of BMCs and VLPs. (a) A model of the carboxysome based on subunit crystal structures and EM analysis of global structure shows a T = 75 icosahedron (740 hexamers, 12 pentamers) approximately 115 nm in diameter. (b) A cryo-EM reconstruction of the P22 VLP in the EX (PDB/2XYZ) shows a T = 7 icosahedron (60 hexamers, 12 pentamers) approximately 60 nm in diameter. Images are approximately to scale. (c and d) Transmission electron micrographs showing the carboxysome and the expanded P22 VLP, respectively. Scale bars are 50 nm. (Adapted from Tanaka, S. et al., *Science*, 319, 1083, 2008.)

of structures they assemble into. In addition, protein cage systems can be produced by heterologous expression of the protein subunits in bacterial hosts or seminative purification from plant hosts in high yield, making their synthesis and production extremely easy, efficient, and sustainable with no toxic solvents required [16–18]. As we will discuss later, the self-assembly and/or disassembly of the protein subunits that compose the cage wall can be exploited to allow incorporation of cargoes such as proteins and enzymes. The protein cage architectures discussed here are referred to generally as VLPs and can be divided into two types: viral VLPs and nonviral VLPs. Viral VLPs are derived from the coat protein (CP) or proteins that make up the capsid structure of a virus but typically lack any of the pathogenic components, such as the nucleic acid genome. Nonviral VLPs include multimeric protein architectures such as lumazine synthase, the E2 protein subunit of pyruvate dehydrogenase, heat shock and chaperone proteins, and the ferritin iron storage family of proteins. While there is some distinction, often, the term VLP is used interchangeably between the two groups, and for simplicity, we will use VLP to describe both for the remainder of this chapter. To begin to understand how VLPs are utilized for encapsulation, we will discuss methods for incorporation of protein cage cargoes.

22.2 METHODS OF VLP ENCAPSULATION

22.2.1 STATISTICAL VERSUS DIRECTED ENCAPSULATION

VLP encapsulation methodology can be divided into two main strategies: statistical and directed. A statistical approach assumes that if the capsid assembly process occurs in the presence of a high enough concentration of cargo, some will get encapsulated. While fundamentally wasteful, this process requires little engineering of either the capsid or the cargo and has been shown to be effective especially in the case of polymer- and gel-based encapsulation experiments. Directed encapsulation, on the other hand, utilizes a specific tag attached to the cargo to direct localization of the cargo to the capsid interior. This approach requires tagging of the cargo and/or manipulation of the capsid itself but creates a system that actively sequesters the target (Figure 22.2). Both of the strategies can potentially occur *in vitro*, the components are purified from cellular hosts prior to encapsulation, or *in vivo*, encapsulated constructs fully assemble prior to purification from cellular hosts, though no example of an *in vivo* statistical approach will be discussed here.

22.2.2 *IN VITRO* ENCAPSULATION

The first example of statistical *in vitro* enzyme encapsulation within a VLP used the controlled disassembly/assembly of the cowpea chlorotic mottle virus (CCMV). This VLP can be disassembled into subunit dimers at neutral pH (pH = 7.5) and subsequently reassembles into a T = 3 capsid upon lowering the pH (pH = 5.0). In their initial study, Cornelissen and coworkers aimed to encapsulate a single horseradish peroxidase per CCMV VLP [19]. This work was key toward demonstrating that, while the enzyme was entrapped within the VLP, the small-molecule substrates and products were free to diffuse across the viral capsid barrier. In later work adopting a partially directed *in vitro* strategy, the system was

modified by complementary peptides that formed a heterodimeric coiled-coil genetically fused to the CCMV CP and cargo protein [20]. Fusion proteins of the positively charged K-coil peptide were fused to the N-terminus of CCMV CP and complementary *E. coil* peptide to the C-terminus of enhanced green fluorescent protein (eGFP) allowing for the components to assemble via a specific noncovalent coiled-coil interaction prior to pH-mediated capsid assembly. This technique was later used to encapsulate *Pseudozyma antarctica* lipase B (PalB) [21]. While the *in vitro* coiled-coil strategy provided a level of control, the method required mixing unmodified CCMV CPs into the assembly system, together with the fusion proteins to avoid steric inhibition of assembly by the PalB cargo. While this method utilizes the specific coiled-coil interaction, the necessity of mixing with unmodified CCMV CP makes the encapsulation process random and thus the strategy remains statistical. Such trial and error strategies require significant amounts of each component and ultimately lead to a distribution of species, ranging from assembled capsids containing no protein or enzyme cargo to fully packed capsids. While statistical *in vitro* encapsulation has focused largely on the CCMV protein cage platform, directed *in vivo* encapsulation strategies have utilized a diverse array of protein cage platforms.

22.2.3 *IN VIVO* ENCAPSULATION

The general strategy for *in vivo* encapsulation relies on directed encapsulation utilizing a VLP interior-specific tag coupled to a target enzyme to direct encapsulation during the natural self-assembly process of the VLP. Thus, upon expression and assembly of the CP into the cage structure, the cargo is encapsulated within the cage. This is a genetically programmed assembly system that is not limited to a particular VLP or cargo and is inherently scalable for manufacturing applications.

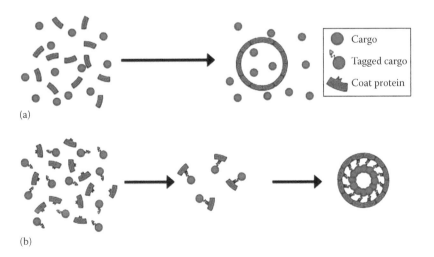

(a)

(b)

Cargo

Tagged cargo

Coat protein

FIGURE 22.2 Statistical versus directed encapsulation. In a statistical encapsulation process, (a) the capsid assembly is triggered in the presence of a cargo and some of the cargo ends up in the capsid as a function of the cargo and CP concentrations. In a directed encapsulation process, (b) the cargo contains a specific tag or directly fused to the CP. The cargo associates with the CP prior to assembly and, ideally, triggers or directs the encapsulation process resulting in a more controlled particle formation and higher-density packing.

In one such *in vivo* strategy utilized by Hilvert and coworkers, several exposed interior residues of the CP of lumazine synthase were mutated to glutamic acid to produce a negatively charged internal surface. A GFP cargo was successfully targeted for encapsulation through the genetic incorporation of a positively charged deca-arginine tag [22]. Later, directed evolution studies were performed on this lumazine synthase encapsulation system to optimize sequestration of HIV protease, wherein more successful sequestration was gauged by survival of cells producing the positively tagged HIV protease [23]. While toxicity was successfully negated through this encapsulation process, the encapsulated HIV protease was never shown to be active after sequestration leaving the actual mechanism of inactivity unclear. Another approach by Handa and coworkers exploited the minor CPs of SV40 virus to direct encapsulation of an enzyme cargo within the confines of a VLP [24]. Fusion of enzyme cargo to the VP2 minor CP, which is incorporated on the interior of SV40, and coexpressing it together with the minor CP VP3 and major CP VP1 yielded SV40 VLPs with active enzymes on the interior.

A unique strategy has been developed by Finn and coworkers utilizing the Qβ VLP platform that takes advantage of the natural nucleic acid packaging potential of viruses. In their approach, they utilize a chimeric single-stranded RNA composed of a hairpin that associates with the internal surface of the Qβ CP and an RNA aptamer sequence, developed to bind to an arginine-rich peptide, to direct the encapsulation of protein cargoes tagged with the arginine-rich peptide [25]. Conserving genetic space in the construct, the mRNA for the Qβ CP serves as the spacer between the capsid-binding hairpin and the cargo-binding aptamer. When the mRNA is transcribed, the same length of RNA can simultaneously serve as the CP transcript and, upon folding, as the encapsulation mediator. Coexpression of the CP/RNA mediator with the tagged protein cargo has been shown to be highly versatile, encapsulating a number of enzyme cargoes, as will be discussed in more detail later.

The P22 bacteriophage system has recently been shown to be a highly versatile platform for the directed *in vivo* encapsulation of a wide range of cargoes. The P22 bacteriophage VLP assembles from 420 CP subunits into a T = 7 icosahedral VLP with the aid of ~100–300 copies of a scaffolding protein (SP) [26]. The SPs template the capsid assembly and are incorporated on the interior of the P22 VLP. The 303-residue-long SP can be severely truncated to an essential C-terminal scaffolding domain [27]. This truncation is useful toward the goal of encapsulation because the SP serves as a space holder in the capsid after assembly. Thus, the truncation of the SP increases the amount of *free space* while simultaneously providing a minimal protein domain with the potential to direct encapsulation. In addition, the P22 capsid is thermally dynamic, undergoing an irreversible expansion, at 65°C, from the spherical procapsid (PC) to an expanded form (EX), which doubles interior volume and thus halves the internal density of the cargo [28]. In addition heating, at 75°C, removes the 12 pentons of the cage,

increasing accessibility by introducing 11 nm pores in a structure referred to as wiffle ball (WB) [29]. By coexpressing a genetic fusion of a protein cargo and the truncated SP together with CP, we have successfully encapsulated a host of protein and enzyme cargoes that maintain their functional activities, as will be discussed later in more detail [30].

22.2.4 BMC Nonnative Encapsulation

There is great potential for exploiting other self-assembling protein cage systems for the encapsulation of enzyme cargoes. BMCs are of particular interest, as they naturally encapsulate enzymes to purportedly improve the overall catalysis of metabolic reactions. While it has proven difficult to purify and express the BMC cages compared to systems discussed earlier, recently, it was found that cargo proteins are directed inside certain BMCs via short terminal peptide sequences. Yeates and coworkers showed that the fusion of the N-terminal 18 amino acids of propionaldehyde dehydrogenase to GFP, glutathione *S*-transferase, and maltose-binding protein effectively directed their encapsulation within 1,2-propandiol utilization microcompartment [31]. Additional studies by Ban and coworkers have identified C-terminal extensions that are responsible for the encapsulation of enzyme within encapsulins, bacterial protein shell-forming compartments [32]. In some cases, the enzyme does not have a C-terminal peptide extension that directs encapsulation, but is directly fused to the encapsulin CP. These studies lay a foundation for constructing new protein cage nanoreactors from diverse bacterial protein compartments, including hyperthermophilic organisms and other extremophiles with the potential to construct extremely robust and tolerant nanoreactors and catalytic nanomaterials. In addition, other nonvirus protein cage–derived particles, such as the E2 protein and vault proteins, are potential candidates for constructing new nanoreactors by encapsulation of enzymes on their interior.

22.3 SINGLE-ENZYME VLP ENCAPSULATION

22.3.1 Potential Results of Encapsulation

As mentioned in Section 22.1, naturally occurring BMCs and organelle sequestration of specific metabolic pathways are thought to lead to advantageous characteristics for the cell and organism. The pursuit of new and beneficial catalyst characteristics has been one of the main motivations for enzyme encapsulation in VLPs. In addition, synthetic VLP encapsulation systems have the potential to isolate a pure enzyme sample under cellular-like macromolecular densities, 300–400 mg/mL for *E. coli*, allowing for in-depth kinetic evaluation, which is not possible amidst the metabolically noisy background of a cellular environment. Studies at these densities have the potential to shed light on the behavior of enzymes in a simultaneously confined and crowded environment similar to a cell [4].

Several characteristics of a constrained VLP environment were anticipated to, and have since been shown to, have an effect on the behavior of enzymatic cargo. These include high local enzyme concentration, the presence of the capsid as a barrier, and high internal macromolecular density often termed crowding. As we have discussed with BMCs, the presence of a capsid wall could protect the encapsulated enzymes as well as control substrate transport across the barrier. The high local concentration of enzymes changes the distribution of reaction points in solution compared to free unencapsulated enzymes as well as the reaction potential of these centers. The crowded interior could limit the dynamics of encapsulated enzymes making them less susceptible to denaturation as well as either positively or negatively affecting kinetic behavior. In order to convincingly deconvolute these effects, it is ideal to generate VLP–enzyme systems with minimal heterogeneity and a high degree of control over the packing density. Because of the diverse factors potentially contributing to encapsulated enzyme kinetics, it is often more useful to refer to parameters such as the Michaelis–Menten constant (K_M) as an apparent value, K_M^{app}, to distinguish from the behavior of the free enzyme.

22.3.2 Fluorescent Protein Encapsulation

Prior to the systematic studies on enzyme encapsulation, many of the initial protein encapsulation studies first examined the encapsulation of fluorescent proteins on the interior of VLPs. Fluorescent proteins are robust, fold well, and provide immediate visual validation of the successful encapsulation of a properly folded protein. While much of the work utilizes fluorescent proteins as a proof of concept that properly folded proteins can be encapsulated within the protein cage, others have utilized fluorescent protein encapsulation to gauge the crowded environment within the VLP. Rome and coworkers examined quenching of a GFP variant encapsulated inside vault protein and observed super quenching by Congo red, indicative of the close adjacency of the fluorescent protein donors [33]. In addition, it was also found that encapsulation of the GFP variant inside vault protein provided protection against KCl quenching.

Utilizing the P22 system, coencapsulation of two different fluorescent proteins, GFP and mCherry, which serve as FRET partners, was demonstrated [34]. The internal concentration of these two fluorescent proteins was calculated to be ~10 mM, or ~380 mg/mL, which is well within the purported range of macromolecular density under cellular conditions. The high internal concentrations and localization of GFP and mCherry with one another resulted in a significantly enhanced FRET response as compared to free concatenated GFP–mCherry. In addition, the intracapsid density and corresponding FRET signal could be modulated by thermal expansion of the P22, resulting in a doubling of the internal capsid volume. These results suggest that encapsulation of enzymes inside the protein cages in high concentration may have significant effects due to crowding and confinement in the local environment, setting the stage for enzyme encapsulation studies that followed.

22.3.3 Enzyme Encapsulation

The first reported encapsulation of an enzyme was reported by Nassal and coworkers, who encapsulated the 17 kDa *Staphylococcus aureus* nuclease on the inside of the VLP derived from the hepatitis B virus (HBV) core protein [35]. To successfully encapsulate the DNA/RNA degrading nuclease (SN), the basic C-terminal domain of the HBV core protein was genetically removed and replaced with the SN gene to yield an HBV core–nuclease fusion protein upon expression. Recombinantly expressed HBV core–SN fusion proteins formed VLPs with sizes consistent with those of truncated HBV core VLPs, but containing a thickened wall indicative of the nuclease being localized on the interior of the VLP (Figure 22.3). Little activity was observed for HBV–SN until the VLP particles were denatured in urea, which allowed the DNA substrate to access the SN. In addition, the researchers performed coexpression of wild-type HBV core protein with the core–nuclease fusion, yielding mosaic particles via statistical incorporation of core protein variants.

Later, the Rome group showed that they could encapsulate luciferase (Luc), along with a GFP variant noted earlier, inside vault protein. Via genetic fusion to a vault-interaction domain (INT), Luc was successfully

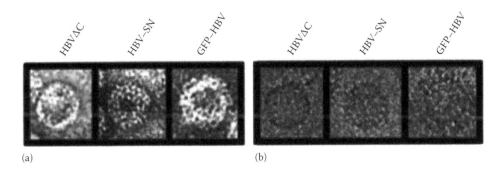

FIGURE 22.3 HBV core protein fusion particles shown by (a) TEM negative staining and (b) cryo-EM. Compared to the empty C-terminal truncation core particles (HBVΔC), fusion particles containing either the SN fused to the interior (HBV–SN) or GFP (GFP-HBV) displayed on the exterior show additional density corresponding to the cargo protein. (Adapted from Beterams, G. et al., *FEBS Lett.*, 481(2), 169, 2000.)

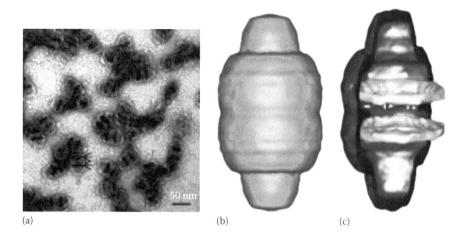

(a)   (b)   (c)

FIGURE 22.4 Vault protein encapsulating luciferase (Luc) via the vault interacting sequence (INT). (a) A negative-stained transmission electron micrograph showing vault with Luc-INT encapsulated. Cargo is evident in the dark stripes on either side of the particle equator. (b) The Luc-INT vault reconstruction with a cutaway view (c) showing the added interior density due to the Luc-INT. (Adapted from Kickhoefer, V.A. et al., *Proc. Natl. Acad. Sci. U.S.A.*, 102(12), 4348, 2005.)

encapsulated and remained active [33]. Vault particles exhibited morphologies identical to wild-type vault proteins by EM but showed two prominent darkly stained stripes across the central barrel of the vault structure where the INT domain is known to associate. Cryo-EM reconstruction was used to better resolve Luc density within the cavity (Figure 22.4). Vault-encapsulated Luc was found to retain activity; however, in order to obtain kinetic profiles similar to free Luc-INT, encapsulated Luc required preincubation with ATP, suggesting substrate diffusion limitations due to either substrate charge or size.

The first example of *in vitro* encapsulation targeted the enzyme horseradish peroxidase using the CCMV VLP. In their study, Cornelissen and coworkers kinetically examined the encapsulation of a single horseradish peroxidase within CCMV [19]. While detailed kinetics were not provided, this initial study was essential for showing that enzymes can remain active upon encapsulation and that substrates are permeable through the capsid shell. In addition, it was shown that modulation of the pore size in CCMV by changing the pH had a significant effect on substrate diffusion, with larger pores providing greater kinetic rates, presumably due to increased diffusion. Cornelissen and coworkers have expanded on this initial work by examining packaging of different copy numbers of enzymes within CCMV and the effects this had on overall activity. Using the coiled-coil partially directed packaging scheme discussed earlier, an average of 1–4 PalB enzymes was packaged per cage [21]. The initial rate of a singly encapsulated PalB was found to be nearly five times faster than the rate observed for non-encapsulated PalB. Interestingly, as the number of enzymes increased within the capsid, the relative kinetic rates dropped to become nearly the same as that of the free forms. While the exact cause of the kinetic change is not clear in this case, limitation of the enzyme dynamics or stabilization of an active state and a specific macromolecular density could be contributing factors.

Looking to exploit the SV40 VLP as a cellular delivery vehicle, Handa and coworkers looked at encapsulation of the yeast cytosine deaminase, an enzyme not found in mammalian cells, which is able to transform cytosine to uracil. In a display of the potential practical applications for enzyme encapsulation in protein cages for enzyme therapy, the cellular uptake propensity of SV40 VLP was used to incorporate cytosine deaminase into a target cell [24]. In the presence of cytosine deaminase, the prodrug 5-fluorocytosine (5-FC) is readily converted to the toxic 5-fluorouracil, an inhibitor of thymidylate synthase, inducing cell death. Incubation of CV-1 cells with VLPs encapsulating cytosine deaminase, followed by treatment of the cells with 5-FC, conferred complete sensitivity to the drug, which was not observed for treatment with unencapsulated cytosine deaminases, suggesting that the cytosine deaminase-VLPs were internalized and remained active. Interestingly, while the encapsulated cytosine deaminase was shown to be active, it was noted that, in contrast to the case of PalB discussed earlier, the activity decreased relative to the unencapsulated enzyme. This suggests that the specific effects of encapsulation on the target enzyme are not yet generalizable and vary from case to case. While these results demonstrate the potential of enzyme VLP systems as enzyme therapies, the inevitable immune response to these particles needs to be characterized and controlled before the technology is viable.

Another example of directed enzyme encapsulation has been reported by the Finn group utilizing the Qβ VLP RNA aptamer system. They have examined encapsulation of three enzymes, the 25 kDa N-terminal aspartate dipeptidase (PepE), the 62 kDa firefly luciferase (Luc), and a thermostable mutant firefly luciferase (tsLuc) variant, and performed full kinetic studies on each alongside their unencapsulated counterparts [25]. Encapsulation of 9 PepE per capsid was found to reduce the overall turnover of PepE by nearly twofold, but showed little effect on the K_M^{app} of the enzyme. Alternatively, a separate encapsulation of both Luc and tsLuc showed increased K_M^{app} for both enzymes, more than sixfold for encapsulation of two tsLuc. Increasing the number of either Luc or tsLuc encapsulated per capsid showed even greater increases in the K_M^{app} and also resulted in lower turnovers than observed for unencapsulated, which was unchanged for encapsulation of only two tsLuc. The fact that increasing the internal concentration of enzymes modulates the kinetic parameters suggests that such effects are due to macromolecular crowding. In addition, they were able to show that encapsulation provided protection against proteolysis, thermal denaturation, and prevented inactivation upon absorption on polystyrene beads, validating the practical application of enzyme encapsulation.

More recently we have examined the use of the bacteriophage P22 to encapsulate enzymes. To date, P22 is one of the largest protein cage systems (T = 7, 60 nm VLP) to be utilized to encapsulate protein cargoes and as such has shown the highest per particle loading of enzymes. Interestingly, this system has also yielded the highest internal densities of enzymes, potentially due to the robust directed *in vivo* encapsulation strategy developed using this capsid. The first enzyme encapsulated in P22 was the monomeric alcohol dehydrogenase D (AdhD) from the hyperthermophile *Pyrococcus furiosus*, which was encapsulated at an internal concentration of 7 mM in the PC form of P22, sevenfold higher than reported for any other VLP systems and equivalent to a cellularly relevant ~350 mg/mL protein density [17]. Taking advantage of the thermophilic nature of AdhD and the thermal expansion of the capsid to EX and WB forms the effects of internal enzyme concentration, and substrate accessibility on the kinetic parameters of encapsulated AdhD were examined. Transformation to EX and WB forms from PC effectively produced a twofold dilution of the internal concentration of enzyme. Our results showed that at 50°C, encapsulation produced a significant change in both the K_M and turnover, with a sevenfold decrease in turnover for all encapsulated variants (PC, EX, and WB) from the *free* AdhD. In addition to the change in turnover, the K_M^{app} showed a decrease of nearly sixfold in AdhD in the PC form, resulting in an apparent catalytic efficiency (k_{cat}/K_M^{app}) nearly identical to *free* AdhD, while the WB and EX forms showed K_M^{app} values intermediate of those of PC and *free* AdhD. As the temperature increased, lower relative turnover was observed as well as an increased relative K_M^{app} for the encapsulated constructs. The examination of temperature dependence in this study built on the crowding kinetic implications of previous studies by showing that the relative kinetics changes are temperature dependent further implicated limited dynamic potential and crowding as key factors in these kinetic effects.

We have examined the P22 encapsulation of a library of other enzymes, two of which we will examine in more detail here. To examine the effects of quaternary structure on P22 encapsulation, we encapsulated the tetrameric protein CelB, a thermostable beta-glycosidase from *P. furiosus* [36]. Our investigation showed no change in the apparent kinetic parameters of encapsulated CelB, indicating that the quaternary structure was likely not perturbed upon encapsulation. This was despite maintaining a 4 mM internal enzyme concentration or a protein density of ~250 mg/mL. Overall, CelB was found to be particularly robust, with almost no loss in activity for either *free* or encapsulated forms when treated with the protease trypsin, although encapsulation did provide protection to the enzyme when embedded in a gel. We have also exploited P22 to encapsulate the GalA alpha-galactosidase from *P. furiosus*, an industrially interesting enzyme that is inextricably expressed as an inclusion body in *E. coli*. By exploiting the robust, directed *in vivo* P22 encapsulation strategy, we were able to sequester and rescue GalA enzymes by encapsulation inside P22 [37]. Encapsulated GalA was readily purified from cell lysate in only a few simple steps and maintained a much greater activity at lower temperatures than reported for enzyme extracted from inclusion bodies. GalA demonstrates the utility of these VLP strategies toward broadening the available enzymes for engineered catalysis by increasing solubility and yield from heterologous expression.

22.4 KINETIC EFFECTS OF SINGLE-ENZYME ENCAPSULATION

When examining the kinetics of an enzyme, it would be optimal to maintain the authentic background environment of the enzyme. It is in this environment that the enzyme evolved and likely has been functionally optimized. For intracellular enzymes, it is difficult to replicate the exceptionally crowded and confined conditions of the cell [4]. One driver of VLP enzyme encapsulation is the concept that the crowded environment of the VLP interior may better mimic the cell and potentially produce enhanced kinetics by better matching the environment for which the enzyme has evolved.

22.4.1 TURNOVER OF SINGLE-ENZYME SYSTEMS

Unfortunately, the goal of producing enhanced enzymatic turnovers through single-enzyme encapsulation to date has largely been unfruitful. One exception performed by Cornelissen's group showed enhanced activities for encapsulated PalB in comparison to free PalB [21]. Interestingly, the activity of encapsulated PalB was found to have the highest turnover relative to free PalB when only one enzyme was encapsulated within the CCMV VLP. As the number of PalBs increased, and thus the concentration of enzyme within the cage, activities began to move toward those of free PalB. This finding seems to suggest that either there is a concentration optimum for PalB or there is an unanticipated specific interaction between PalB and CCMV that enhances kinetics. The later case can be eliminated by encapsulation of PalB at similar concentrations in an alternative VLP.

Most studies have shown either no change in the turnover or reduced rates upon encapsulation within protein cage/VLP structures. As noted earlier, encapsulation of yeast cytosine deaminase in SV40 was observed to have a slightly reduced activity, although the kinetic parameters were not examined in detail [24]. Finn and coworkers found that encapsulation produced a reduction in the activity of PepE and as did increasing internal concentrations of a thermostable luciferin from two to nine enzymes per capsid, similar to the trend observed in Cornelissen's study [25]. One of the most dramatic changes has been observed with encapsulation of thermophilic AdhD in P22, which showed a sevenfold decline in turnover compared to free AdhD [17]. No recovery in activity could be obtained by reducing the internal concentration of enzyme by expansion of the PC VLP form to EX or WB, and results from the WB, containing large 11 nm channels in the VLP structure, further suggested that substrate access was not responsible for lower activities. While limited in the number of studies performed, these initial findings suggest that crowding and confinement effects have a detrimental effect on enzyme behavior across internal enzyme concentrations ranging from 0.5 mM to as high as 7 mM. However, there could be alternative causes to these changes in activity, particularly in the case of enzymes that are encapsulated through *in vivo* encapsulation strategies. For example, reduced rates could be due to encapsulation of misfolded and inactive enzymes that are packaged alongside active enzymes. Encapsulation has been found to be rapid enough to rescue proteins from inclusion body formation, providing some support that proteins could be encapsulated before they are properly folded and are incorporated in a tightly constrained environment that doesn't allow ideal conditions for folding to the mature state. Determining whether encapsulated enzymes or proteins are properly folded and maintain their structure is experimentally difficult and is an area for future research to understand these systems better.

22.4.2 Assessing Encapsulation Effects with K_M^{app}

Whereas enzyme turnover is related to the percentage of active enzymes in a sample, the binding affinity, represented by the K_M, of an enzyme is independent of the percentage of active enzymes in a sample and is likely a better indicator of crowding and confinement effects separate from misfolding or inactivation problems. Changes in the K_M^{app} values have been observed for several enzymes upon encapsulation in a VLP. Luciferase encapsulated inside Qβ showed large changes in K_M^{app} for ATP and luciferin, with nearly a 10- and 20-fold increase, respectively, for the most crowded constructs, and showed a positive correlation with increasing number of encapsulated enzymes [25]. Interestingly, the most crowded luciferase constructs also showed reduced turnover, but not the construct containing only two luciferases per capsid, which showed increased K_M^{app} but unchanged turnover. Alternatively, encapsulation of the enzyme PepE in Qβ showed no change in K_M^{app} with a twofold lower turnover.

Results for encapsulation of AdhD in P22 showed a decrease in K_M^{app} values for its substrate, acetoin, with higher internal concentrations of enzymes despite a lower turnover [17]. Increasing the temperature of the activity assay for AdhD showed little change in the differences of the K_M^{app} values for the EX and WB morphologies, which maintained about a twofold lower K_M^{app} overall temperatures. Changes in K_M^{app} of the NADH cofactor utilized by AdhD showed an apparent increase of two- to threefold upon encapsulation, although these differences are barely resolvable statistically. While these examples show that significant differences arise from encapsulation, the question remains as to whether they are caused by crowding or other confinement effects. Indeed, crowding studies utilizing conventional methodology have seen a range of changes, increases and decreases in K_M and turnover values, as has been observed in the studies utilizing VLPs. However, crowding studies typically utilize techniques, such as circular dichroism, to monitor changes in enzyme structure that result from crowding reagents [38–41]. To date, there have been no reports of monitoring the structural arrangement of enzymes encapsulated within VLPs, likely due to the large background of protein secondary structure from the VLPs.

22.4.3 Kinetic Conclusions of Single-Enzyme Encapsulation to Date

The general behavior of enzymes in VLP encapsulation systems appears to vary from case to case, which is not entirely surprising considering the diversity of enzyme structures and mechanisms. Utilizing VLPs as a platform for examining potential effects caused by crowding and confinement experienced under conditions remains an area of great promise. Concentrations of macromolecules within cells have been estimated to be on the order of 300 mg/mL, which have been reached in several examples of enzyme encapsulation within P22. However, advancement of this area will require developing strategies and techniques to understand the changes that occur to the enzyme upon encapsulation. In addition, *in vivo* encapsulation methods need to be optimized to prevent enzyme misfolding. This can be done by simply employing a two-vector approach that allows expression and maturation of an enzyme of interest followed by expression of the CP, subsequently induced by an alternative inducer than the one controlling the enzyme expression.

Although there are few examples of enhanced activities upon encapsulation, other new properties that emerge as a consequence of encapsulation present a useful practical application for this technology. For instance, protection against proteases and increased thermostability are properties that have been imparted to enzymes encapsulated within VLPs. In addition, the capsid provides a sacrificial scaffold that can be used to attach enzymes to surfaces, which in the case of free enzyme often leads to inactivation, or the ability to build complex materials by connecting VLP-containing enzymes into hierarchically structured materials.

22.5   MULTIENZYME VLP ENCAPSULATION AND KINETICS

22.5.1   Two-Enzyme Encapsulation

The most recent advancement in enzyme encapsulation has been the ability to coencapsulate multiple different enzymes, with coupled activity, within the same protein cage, further mimicking BMCs. As noted in Section 22.1, natural BMCs, such as the carboxysome, are thought to provide enhanced activities within the pathway, preventing leakage of toxic intermediates and increasing local concentrations of substrates [7]. In this way, they function as metabolons, multienzyme metabolic complexes [3], with the added complexity of the capsid wall. It has been widely anticipated that simple colocalization of enzymes should result in enhanced kinetics due to the second enzyme in the pathway experiencing a higher local

concentration of intermediate as it nears the site of production at the first enzyme [2].

To date, there has been only one successful demonstration of multienzyme encapsulation within a protein cage or VLP cage, though multienzyme systems have been examined in polymerosomes but without the potential for precise kinetic examination [42–44]. Utilizing the P22 platform, a fusion protein of two or three enzymes connected through polyglycine peptide linkers fused to the SP was coexpressed with and encapsulated by the CP [45]. Expression of the enzymes as a concatenated single gene product guaranteed a 1:1 stoichiometry. For this study, we investigated the cascade reaction generated by coupling the previously mentioned CelB, to a downstream ADP-dependent glucokinase (GLUK), and an ATP-dependent galactokinase, all from *P. furiosus*, creating a synthetic encapsulated metabolon (Figure 22.5). The detailed kinetic

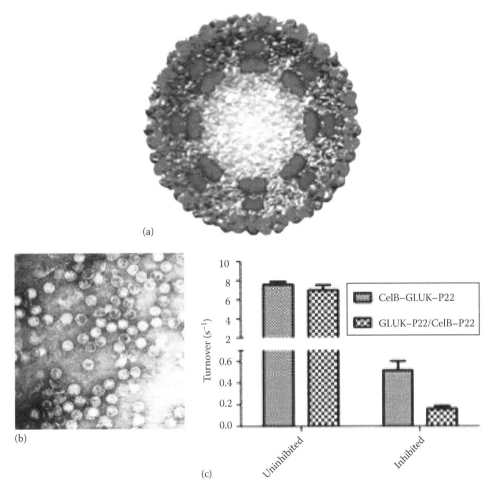

(a)

(b)

(c)

FIGURE 22.5   Multienzyme encapsulation in a single VLP using the P22 system. (a) An artistic representation of the CelB–GLUK–SP fusion protein encapsulated in P22 VLP. P22–CP (gray), CelB (red), GLUK (blue), and SP and linker regions (purple). (PDB: 2GP8, 1UA4, 3APG, 2XYY). (b) A negatively stained transmission electron micrograph showing P22 VLPs with the multienzyme fusion protein encapsulated. Scale bar 100 nm. (c) Maximum initial turnover of the coencapsulated enzyme construct (CelB–GLUK–P22) compared to a 1:1 stoichiometric mixture of the individually encapsulated enzymes (GLUK–P22/CelB–P22) under substrate-saturating conditions for enzyme 1 (CelB). No pathway advantage is observed under normal conditions, but an advantage can be induced by changing the kinetic balance. In this case, the balance is altered by selectively inhibiting CelB to the same degree in coencapsulated construct and the control. (Adapted from Patterson, D.P. et al., *ACS Chem. Biol.*, 9(2), 359, 2013.)

analysis of this study focused mainly on the two-enzyme system of CelB and GLUK and the conversion of lactose to glucose-6-phosphate. Contrary to the trend observed with many singly encapsulated constructs, neither of these enzymes was affected kinetically by the encapsulation process. Interestingly, it was found that colocalization of these specific enzymes in the same protein cage showed no enhancement in activity compared with a 1:1 mixture of singly encapsulated enzymes in separate P22 capsids. However, when the kinetic balance of the enzymes was altered, a kinetic advantage emerged for the coencapsulated construct (Figure 22.5). *In silico* examination of a two-enzyme system separated by diffusion of the intermediate substrate demonstrated that this kinetic-balance-dependent advantage was actually to be expected.

22.5.2 Simple Diffusional Channeling in Multienzyme Kinetics

Enhanced pathway kinetics can be thought of in several ways. Considering only a simple diffusional method of intermediate substrate channeling, that is, one that relies only on the second enzyme being spatially closer to the first enzyme, one can think about maximum initial pathway rate enhancement or steady-state enhancement. These two ways of thinking about a pathway advantage are not necessarily exclusive but are useful for discerning the results of current characterization of multienzyme systems. Steady-state analysis provides a more convincing validation of these systems for industrial applications. Initial maximum pathway rate analysis lends itself to comparison to the Michaelis–Menten characterization of the individual enzymes involved and thus has been more prevalently used in the literature in enzymology studies. Here, the maximum initial pathway rate is measured after a lag period characteristic of multistep reactions.

The CelB, GLUK study discussed here utilized maximum initial pathway rate analysis to compare turnover under saturating substrate conditions for the first enzyme CelB [45]. Under normal conditions, no pathway enhancement was observed. However, when the activity of the first enzyme was reduced relative to the second through small-molecule inhibition, a pathway advantage emerged. Inhibition of the first enzyme is the kinetic equivalent of utilizing a different enzymatic homolog or, only in the case of noncompetitive inhibition, it can be thought of as changing the stoichiometry of the enzymes. To look at the full effects of changing the kinetic balance of the enzymes in a simple diffusional system, we developed a simplified model consisting of two Michaelis–Menten enzymes separated by a spherical diffusion equation dependent on the intermediate substrate (Figure 22.6). By systematically changing distance, enzyme kinetic parameters, and stoichiometry, we showed that for a given enzyme pair, there is a necessary kinetic balance that is required to generate a maximum initial pathway rate advantage under saturating conditions for the first enzyme, namely, that the first enzyme should

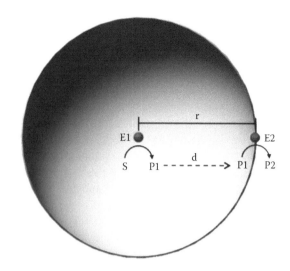

FIGURE 22.6 The two-enzyme simple diffusional channeling model displayed as a geometric schematic. The production of the intermediate substrate (P1) at enzyme 1 (E1) is treated as a Michaelis–Menten system in the initial phase of reaction with an approximately linear rate of P1 production. Diffusion of P1 to enzyme 2 (E2) is controlled by the intermediate diffusion coefficient (d) and the radial distance (r). A spherical boundary is used to define E2, and an average stoichiometry of 1:1 is used between E1 and E2. E2 is treated as a Michaelis–Menten system and responds to the available concentration of P1. (Adapted from Patterson, D.P. et al., *ACS Chem. Biol.*, 9(2), 359, 2013.)

be rate limiting. This balance is dependent on the diffusion behavior of the intermediate substrate. Even when the model does not demonstrate a rate advantage, the inter-enzyme distance can affect the length of the lag phase, though the implications of this result are not entirely clear at this point. Potentially, a shorter lag phase would mean that the entire pathway could respond more quickly to sudden changes in substrate concentration. This type of behavior would have advantageous implications for processes such as cellular signaling.

22.5.3 Complex Contributions to Pathway Activity

There are many possible deviations from the simple diffusional channeling case, and it remains unclear how this behavior will affect steady-state pathway catalysis. One example of alternative rate enhancement is seen in the carboxysome where it has been proposed that the nature of pores in the protein shell discourages carbon dioxide and oxygen diffusion but encourages bicarbonate diffusion resulting in a high local concentration of carbon dioxide [7]. While simple diffusional channeling accounts for the complete advantage in the two-enzyme P22 case earlier, it is feasible to design VLP systems that take advantage of specific, cage-mediated effects or individual enzyme enhancement. Discerning between these various contributions depends heavily on a detailed understanding of simple enzyme colocalization.

22.6 METABOLIC MATERIALS

22.6.1 METABOLIC PARTICLES

A very promising application of VLP-encapsulated enzymes, alluded to in our discussion of multienzyme encapsulation, is the development of metabolic particles and materials. As industrial catalysts, these particles and materials could be used to enhance enzyme lifetimes, pathway turnover, and catalyst separation from the reaction mixture [3]. In several cases, single-enzyme VLP encapsulation systems have been shown to enhance the toughness and immobilization potential of enzymes [25,36]. In addition, the aforementioned multienzyme encapsulation demonstrates that colocalization can lead to enhanced pathway kinetics with the correct engineering [45]. However, the maximum number of enzymes we were able to encapsulate within the P22 system was three before the structural integrity of the capsid was compromised. There are other candidate VLPs available that may be able to extend this number but likely at the expense of precisely controlling stoichiometry and particle population. In the context of many desirable metabolic conversions, three enzymes are a very small metabolic chunk.

22.6.2 HIGHER-ORDER METABOLIC MATERIALS

Advanced metabolic materials may be more easily constructed by connecting VLPs containing different enzymes into higher-order structures. One can envision connecting protein cages with different enzymes through either covalent or noncovalent mechanisms to produce extended networks of colocalized capsids. Examples of hierarchical assembly of VLPs have been described. Several examples of noncovalent hierarchical assembly have been described utilizing dendrons that associate with the VLPs through electrostatic interactions with the capsid exterior and self-associate with the other dendrons via hydrophobic interactions. One example reported by Corneliessen's group incorporated optically sensitive functional groups within the dendron that, upon exposure to light, caused photolysis of the dendron and release of the CCMV VLPs utilized in the assembly [46]. Other strategies that utilize chemical cross-linking or attachment of oligomerization domains to the exterior are also possible and can be looked to for constructing such higher-order assemblies in the future [47,48].

22.7 OUTLOOK AND CHALLENGES

The future of utilizing VLP systems for understanding biology and developing metabolic materials remains promising. Encapsulation of proteins and enzymes to explore fundamental problems in biochemistry, such as the effects of crowding and confinement on enzyme function by mimicking environmental conditions found in cells or utilizing them as models for natural BMCs, is compelling. Encapsulation of enzymes within VLPs has become routine in many systems, with the major limitation being the ability to examine enzyme

structural changes upon encapsulation that may lead to many of the observed kinetic effects. As noted earlier, several lines of evidence suggest that crowding and confinement effects are at play upon encapsulation, but new experiments must be developed to show this conclusively. Further development of these systems to construct nanoreactors and higher-order catalytic materials is another area of intense interest. A broad range of enzymes, with diverse quaternary structures from various host organisms, has been incorporated into VLPs. The incorporation of more complex heteromultimeric enzymes with cofactors may provide additional challenges but offer the ability to tap into unique and useful chemistries. The ability to coencapsulate enzymes within the same VLP provides new capabilities for coupling and designing new metabolic pathways within the VLPs. New mathematical models have been developed, with the potential to help in the design of optimized system, determining if coencapsulation will provide improved activities or whether longer localizations are just as advantageous. The ability of substrates to diffuse freely in and out of VLPs allows coupling of enzymes that are not coencapsulated, but which can be localized to one another through covalent or noncovalent methods, with hierarchical assembly strategies being particularly attractive. Future work looking at methods to produce highly ordered arrays and assemblies will be required to construct tailor-made catalytic materials composed of VLPs encapsulating enzymes.

REFERENCES

1. Minton, A.P., The influence of macromolecular crowding and macromolecular confinement on biochemical reactions in physiological media. *Journal of Biological Chemistry*, 2001. **276**(14): 10577–10580.
2. Lee, H., W.C. DeLoache, and J.E. Dueber, Spatial organization of enzymes for metabolic engineering. *Metabolic Engineering*, 2012. **14**(3): 242–251.
3. Schoffelen, S. and J.C.M. van Hest, Multi-enzyme systems: Bringing enzymes together in vitro. *Soft Matter*, 2012. **8**(6): 1736–1746.
4. Ellis, R.J., Macromolecular crowding: Obvious but underappreciated. *Trends in Biochemical Sciences*, 2001. **26**(10): 597–604.
5. Zimmerman, S.B. and S.O. Trach, Estimation of macromolecule concentrations and excluded volume effects for the cytoplasm of *Escherichia coli*. *Journal of Molecular Biology*, 1991. **222**(3): 599–620.
6. Gantt, E. and S.F. Conti, Ultrastructure of blue-green algae. *Journal of Bacteriology*, 1969. **97**(3): 1486–1493.
7. Cheng, S. et al., Bacterial microcompartments: Their properties and paradoxes. *Bioessays*, 2008. **30**(11–12): 1084–1095.
8. Kerfeld, C.A. et al., Protein structures forming the shell of primitive bacterial organelles. *Science*, 2005. **309**(5736): 936–938.
9. Yeates, T.O. et al., Protein-based organelles in bacteria: Carboxysomes and related microcompartments. *Nature Reviews Microbiology*, 2008. **6**(9): 681–691.
10. Tanaka, S. et al., Atomic-level models of the bacterial carboxysome shell. *Science*, 2008. **319**(5866): 1083–1086.

11. Havemann, G.D. and T.A. Bobik, Protein content of polyhedral organelles involved in coenzyme B12-dependent degradation of 1,2-propanediol in *Salmonella enterica* serovar Typhimurium LT2. *Journal of Bacteriology*, 2003. **185**(17): 5086–5095.

12. Havemann, G.D., E.M. Sampson, and T.A. Bobik, PduA is a shell protein of polyhedral organelles involved in coenzyme B12-dependent degradation of 1,2-propanediol in *Salmonella enterica* serovar Typhimurium LT2. *Journal of Bacteriology*, 2002. **184**(5): 1253–1261.

13. Sampson, E.M. and T.A. Bobik, Microcompartments for B12-dependent 1,2-propanediol degradation provide protection from DNA and cellular damage by a reactive metabolic intermediate. *Journal of Bacteriology*, 2008. **190**(8): 2966–2971.

14. Stojiljkovic, I., A.J. Bäumler, and F. Heffron, Ethanolamine utilization in *Salmonella typhimurium*: Nucleotide sequence, protein expression, and mutational analysis of the cchA cchB eutE eutJ eutG eutH gene cluster. *Journal of Bacteriology*, 1995. **177**(5): 1357–1366.

15. Douglas, T. and M. Young, Viruses: Making friends with old foes. *Science*, 2006. **312**(5775): 873–875.

16. Sainsbury, F., E.C. Thuenemann, and G.P. Lomonossoff, pEAQ: Versatile expression vectors for easy and quick transient expression of heterologous proteins in plants. *Plant Biotechnology Journal*, 2009. **7**(7): 682–693.

17. Patterson, D.P., P.E. Prevelige, and T. Douglas, Nanoreactors by programmed enzyme encapsulation inside the capsid of the bacteriophage P22. *ACS Nano*, 2012. **6**(6): 5000–5009.

18. Brumfield, S. et al., Heterologous expression of the modified coat protein of Cowpea chlorotic mottle bromovirus results in the assembly of protein cages with altered architectures and function. *Journal of General Virology*, 2004. **85**(4): 1049–1053.

19. Comellas-Aragones, M. et al., A virus-based single-enzyme nanoreactor. *Nature Nanotechnology*, 2007. **2**(10): 635–639.

20. Minten, I.J. et al., Controlled encapsulation of multiple proteins in virus capsids. *Journal of the American Chemical Society*, 2009. **131**(49): 17771–17773.

21. Minten, I.J. et al., Catalytic capsids: The art of confinement. *Chemical Science*, 2011. **2**(2): 358–362.

22. Seebeck, F.P. et al., A simple tagging system for protein encapsulation. *Journal of the American Chemical Society*, 2006. **128**(14): 4516–4517.

23. Wörsdörfer, B., K.J. Woycechowsky, and D. Hilvert, Directed evolution of a protein container. *Science*, 2011. **331**(6017): 589–592.

24. Inoue, T. et al., Engineering of SV40-based nano-capsules for delivery of heterologous proteins as fusions with the minor capsid proteins VP2/3. *Journal of Biotechnology*, 2008. **134**(1): 181–192.

25. Fiedler, J.D. et al., RNA-directed packaging of enzymes within virus-like particles. *Angewandte Chemie International Edition*, 2010. **49**(50): 9648–9651.

26. King, J., E.V. Lenk, and D. Botstein, Mechanism of head assembly and DNA encapsulation in Salmonella phage P22. II. Morphogenetic pathway. *Journal of Molecular Biology*, 1973. **80**(4): 697–731.

27. Parker, M.H., S. Casjens, and P.E. Prevelige Jr., Functional domains of bacteriophage P22 scaffolding protein. *Journal of Molecular Biology*, 1998. **281**(1): 69–79.

28. Parent, K.N. et al., P22 coat protein structures reveal a novel mechanism for capsid maturation: Stability without auxiliary proteins or chemical crosslinks. *Structure*, 2010. **18**(3): 390–401.

29. Teschke, C.M., A. McGough, and P.A. Thuman-Commike, Penton release from P22 heat-expanded capsids suggests importance of stabilizing penton–hexon interactions during capsid maturation. *Biophysical Journal*, 2003. **84**(4): 2585–2592.

30. O'Neil, A. et al., Genetically programmed in vivo packaging of protein cargo and its controlled release from bacteriophage P22. *Angewandte Chemie International Edition*, 2011. **50**(32): 7425–7428.

31. Fan, C. et al., Short N-terminal sequences package proteins into bacterial microcompartments. *Proceedings of the National Academy of Sciences of the United States of America*, 2010. **107**(16): 7509–7514.

32. Sutter, M. et al., Structural basis of enzyme encapsulation into a bacterial nanocompartment. *Nature Structural and Molecular Biology*, 2008. **15**(9): 939–947.

33. Kickhoefer, V.A. et al., Engineering of vault nanocapsules with enzymatic and fluorescent properties. *Proceedings of the National Academy of Sciences of the United States of America*, 2005. **102**(12): 4348–4352.

34. O'Neil, A. et al., Coconfinement of fluorescent proteins: Spatially enforced communication of GFP and mCherry encapsulated within the P22 capsid. *Biomacromolecules*, 2012. **13**(12): 3902–3907.

35. Beterams, G., B. Böttcher, and M. Nassal, Packaging of up to 240 subunits of a 17 kDa nuclease into the interior of recombinant hepatitis B virus capsids. *FEBS Letters*, 2000. **481**(2): 169–176.

36. Patterson, D.P. et al., Virus-like particle nanoreactors: Programmed encapsulation of the thermostable CelB glycosidase inside the P22 capsid. *Soft Matter*, 2012. **8**(39): 10158–10166.

37. Patterson, D.P., B. LaFrance, and T. Douglas, Rescuing recombinant proteins by sequestration into the P22 VLP. *Chemical Communications*, 2013. **49**(88): 10412–10414.

38. Tokuriki, N. et al., Protein folding by the effects of macromolecular crowding. *Protein Science*, 2004. **13**(1): 125–133.

39. Sasahara, K., P. McPhie, and A.P. Minton, Effect of dextran on protein stability and conformation attributed to macromolecular crowding. *Journal of Molecular Biology*, 2003. **326**(4): 1227–1237.

40. Eggers, D.K. and J.S. Valentine, Molecular confinement influences protein structure and enhances thermal protein stability. *Protein Science*, 2001. **10**(2): 250–261.

41. Jiang, M. and Z. Guo, Effects of macromolecular crowding on the intrinsic catalytic efficiency and structure of enterobactin-specific isochorismate synthase. *Journal of the American Chemical Society*, 2007. **129**(4): 730–731.

42. Baumler, H. and R. Georgieva, Coupled enzyme reactions in multicompartment microparticles. *Biomacromolecules*, 2010. **11**(6): 1480–1487.

43. Kuiper, S.M. et al., Enzymes containing porous polymersomes as nano reaction vessels for cascade reactions. *Organic & Biomolecular Chemistry*, 2008. **6**(23): 4315–4318.

44. van Dongen, S.F.M. et al., A three-enzyme cascade reaction through positional assembly of enzymes in a polymersome nanoreactor. *Chemistry—A European Journal*, 2009. **15**(5): 1107–1114.

45. Patterson, D.P. et al., Encapsulation of an enzyme cascade within the bacteriophage P22 virus-like particle. *ACS Chemical Biology*, 2013. **9**(2): 359–365.

46. Kostiainen, M.A. et al., Self-assembly and optically triggered disassembly of hierarchical dendron–virus complexes. *Nature Chemistry*, 2010. **2**(5): 394–399.

47. Broomell, C.C. et al., Protein cage nanoparticles as secondary building units for the synthesis of 3-dimensional coordination polymers. *Soft Matter*, 2010. **6**(14): 3167–3171.

48. Moon, H. et al., Fabrication of uniform layer-by-layer assemblies with complementary protein cage nanobuilding blocks via simple His-tag/metal recognition. *Journal of Materials Chemistry B*, 2013. **1**(35): 4504–4510.

23 Principles of Design of Virus Nanoparticles for Imaging Applications

Irina Tsvetkova and Bogdan Dragnea

CONTENTS

23.1 INTRODUCTION

Since viruses are nature's self-assembled biological nanoparticles, a renewed interest in them has come from their reprogramming for a wide spectrum of new potential applications ranging from electronics [1] to reconfigurable materials [2,3] and nanomedicine [4,5]. Here, we review one of the rapidly growing virus nanotechnology application areas: the use of virus-based imaging probes in nanomedicine [6]. Specific topics include virus-like particles (VLPs) as contrast agents for nonintrusive *in vivo* imaging in diagnostics, and for studies of the mechanisms of virus–host cell interaction, which are important for understanding and fighting pathogenesis.

Biocompatibility, uniformity, reconfigurability, diversity of sizes and morphologies, and the possibility of genetic engineering are advantages of VLPs over other types of nanoparticles. For many imaging and delivery applications, these unique properties are worth preserving and amending via addition of new contrast principles. VLP-enabled imaging probes provide a variety of imaging contrast possibilities resulting from the physical properties of the label moiety. This can be radioactivity (in positron-emission tomography [PET]), bioluminescence or fluorescence (for optical imaging), or magnetism (in magnetic resonance imaging [MRI]).

Viruses are formed of a coat made of proteins and sometimes lipids, which encapsulates the genome. The coat carries multiple roles, including genome protection and delivery, trafficking, and targeting. The regularly packed protein coats of isometric viruses are known as capsids. Most approaches in nanomedicine have focused on the modification of capsids, which can be engineered genetically or by chemical methods to provide new functions leading to improved contrast [7], targeting [8–10], and drug loading [11,12], while preserving some properties that are inherited from the virus itself.

Design of *in vivo* imaging probes with potential use in medicine is done taking into account the following constraints:

- Binding affinity and specificity to target. High affinity warrants high accumulation of the imaging probe in the targeted tissues with the benefits of increased contrast and reduced side effects [8,13].
- Sensitivity, which is a measure of the minimum detectable concentration of target moiety.
- Improved pharmacokinetic properties, which aim to ensure prolonged *in vivo* stability, low immunogenicity, and toxicity [14,15].
- Production and economical feasibility [16,17].

In this chapter, we review design principles and selected applications of modified virus particles for *in vivo* and *in vitro* imaging applications.

23.2 VLP DESIGN CONSIDERATIONS

Challenges associated with biomedical imaging technologies include minimal invasiveness, real-time monitoring capabilities, and access to the wide range of temporal and spatial scales over which biological systems operate. A wide variety of techniques have been developed, but most studies utilizing virus-based probes to date have focused on optical imaging, MRI, and PET [18]. To this end, a few approaches of engineering-desirable physical properties for imaging and targeting have sought to benefit from and preserve unique virus characteristics, such as precise stoichiometry down to molecular scale, responsivity to chemical cues, and rapid and efficient assembly [19]. Virus shell multivalency and versatility in chemical addressability are also significant advantages, which allow presentation of multiple ligands associated with a small probe volume—a simple, efficient way to increase sensitivity.

Designing VLPs for imaging has relied on two strategies: (1) the attachment of the imaging probe to one of the virus interfaces by binding to viral components, nucleic acid [20,21], coat protein [7], and the lipid envelope [22–24], or (2) encapsulation of the probe by assembly or direct synthesis inside the lumen of the virus capsid [25–28].

For instance, the choice of labeling of nucleic acid genome or encapsulating a bulky probe instead of it could be beneficial when the problem at hand requires intact external surface capsid properties or multiple labeling [29,30]. However, it is worth noting that, even if the added cargo or labels are buried inside the capsid and the surface morphology is preserved, this does not necessarily imply a preservation of the dynamic properties, which may depend on specific nucleic acid–capsid protein interactions [31].

Fluorescent labeling of viral RNA can be achieved through insertion of fluorescent protein–binding domains within the genomic molecule [30] and by *in vitro* chemical labeling of RNA [32]. Insertions of RNA aptamers, which can become fluorescent upon binding of small-molecular dyes, which are dark in dissociated state, are another promising technique that potentially ensures minimum disruption in RNA–protein interaction [33,34].

Virus capsids consist of hundreds to thousands of protein copies. In certain cases, capsids can be disassembled, purified, and reassembled *in vitro* [35] or made permeable to chemical reagents through morphological transitions in response to environmental changes [36]. This remarkable reconfigurability was used for direct synthesis of nanomaterials inside the capsid, some with optical or magnetic properties. For example, nanoparticles of polyoxometalate species, cobalt, Prussian blue, and titanium oxide have been synthesized inside virus capsid by biomineralization without capsid disassembly [36–39].

On the other hand, efficient self-assembly of protein capsid around functionalized nanoparticles was first used for gold particle encapsulation [40] and later for semiconductor quantum dots (QD) [25,41]. Spherical capsids of brome mosaic virus (BMV), simian virus 40, and rotavirus (capsid protein VP4) were used to encapsulate iron oxide nanoparticles for high-contrast MRI imaging [26,42,43]. The advantage of encapsulating nanoparticles by self-assembly is that the nanoparticles are obtained separately, by independent methods, which potentially provides optimal physical properties and nanoparticle surface chemistry and interfacial cargo–protein interactions that can be tailored to ensure the desired level of stability and regular protein packing. Remarkably, it was recently shown that encapsulation can be achieved during virus assembly in living cells by conjugation of QD to genomic RNA [44].

Small molecules can also be encapsulated in virus capsids when mixed with coat protein subunits during self-assembly triggered by a change in solution properties [12,45]. Infusion of a virus with dyes that electrostatically bind to RNA [11,46,47] has been shown to provide a simple, albeit nonspecific, modality of cargo loading.

Capsid modification with fluorescent proteins inserted via genetic engineering to loops at the virus surface [48] or on the inside of the virus capsid [49,50] was also achieved. The disadvantage of fluorescent protein insertion is its size, which can induce steric clashes and alteration of the biological properties of virus particles [51]. As an alternative, small tetracysteine motifs, which bind membrane-permeable Bi–As compounds, can be integrated in proteins with minimal alteration of their functions [52].

Capsid proteins also allow for the utilization of a variety of standard bioconjugation protocols, which target chemically reactive amino acid side chains. Most common reactions involve lysine, cysteine, and aspartic/glutamic acid residues [53,54].

Conjugation of chelate moieties to MS2 virus capsid surface through lysine modifications led to a remarkable surface density corresponding to ~500 of gadolinium ions for a 30 nm diameter capsid, a significant benefit for MRI applications requiring high contrast [55]. Comparison of modifications of outer or inner surfaces of MS2 viral capsid by conjugation of flexible lysines at the exterior and relatively rigid tyrosines in interior has shown that the interior of the capsid remains highly accessible to water. However, the rigidity of the linker is important for obtaining a high relaxivity enhancement [56–58]. Increase in the Gd payload by almost 30 times was obtained by synthesizing addressable polymer networks in the lumenal space of the P22 phage capsid [59,60]. Modified P22 particles exhibited an order-of-magnitude improvement in relaxivity over previous VLP systems. The capsids of other viruses, such as Qβ, cowpea mosaic virus (CPMV), and the rod-shaped TMV, have also been studied for Gd3+ ligand scaffolding, indicating that the approaches can be transferred between systems [61–63].

23.3   SELECTED IMAGING APPLICATIONS

Generating a spatial map of material properties or imaging can be achieved in two ways: by direct force application (e.g., profilometry, or atomic force microscopy) or by utilizing the interaction of radiation with matter. Here, we deal with the later approach, in particular with its optical microscopy, MRI, and PET embodiments.

23.3.1   OPTICAL IMAGING

Microscopic optical imaging has enabled biomedical discovery for more than three centuries. Its leadership position in the arsenal of biomedical methods is due to a set of features hard to encounter at the same time in other methods. Optical imaging is noninvasive because it utilizes nonionizing electromagnetic radiation, which interacts only weakly with cells and tissues. Its dynamic range covers the broad span of characteristic times for biological processes. Conventional optical spatial resolution and visible light penetration depth allow for observations from single molecules, to single cells and their organelles, to tissues. Moreover, intrinsic or extrinsic optical signal enhancement techniques have been widely exploited for visualizing chemical processes in cells, and even *in vivo*.

Limitations of optical imaging include increase of scattering and loss of contrast and spatial resolution for samples thicker than ~1–5 µm for wavelengths in the visible spectrum. A second fundamental challenge is diffraction-limited spatial resolution, which sets practical limits to ~300 nm in directions transversal to propagation and ~2 µm axially. However, the first challenge can be addressed by utilizing near-infrared (NIR) (800–900 nm) radiation, and NIR fluorescent tags have been developed to extend the advantages of fluorescence microscopy to deeper penetrating NIR radiation [64]. Concerning the second challenge, the past decade has witnessed the emergence of a variety of super-resolution methods, which have pushed the resolution limit of optical microscopes to nearly a 10th of the diffraction limit [65,66].

Fluorescently labeled plant virus particles, such as the CPMV, were utilized for intravital vascular imaging in mouse and chicken embryos [67,68]. It was found that CPMV was selectively internalized by endothelial cells, thus lining the vasculature and providing very high–resolution images of vasculature in living mice and chick embryos to a depth of ~500 µm for extended (>72 h) periods of time (Figure 23.1). Further improvement in sensitivity and imaging depth penetration of the CPMV system was achieved by implementation of commercially available NIR probes. Low concentrations of NIR (Alexa Fluor 647) labeled icosahedral CPMV and filamentous potato virus X VLPs have given good results in fluorescence tomographic imaging of prostate cancer in a chicken chorioallantoic membrane (CAM) tumor model [69,70].

Avian influenza H5N1 pseudotype virus (H5N1p) conjugated with NIR QDs by bioorthogonal reaction was used for *in vivo* monitoring of respiratory infection in mice [71]. QD-H5N1p in mouse lung tissues exhibited bright fluorescence and improved photostability in complex state. It was observed that the signal magnitude strongly correlated with the severity of viral infection. Moreover, the sensitivity of this system to the administration of antiviral drug oseltamivir carboxylate and mouse antiserum suggested potential usefulness in the evaluation of antiviral drugs in model systems.

Fluorescent labeling of VLPs has contributed significantly to fundamental studies of viruses and VLP interactions with cells and promises to provide basis for interfering with viral infection through new mechanistic insights in pathogenesis [72,73]. Single virus tracking with fluorescence microscopy is a powerful tool in studying the mechanisms of virus attachment, entry, and cargo release [74–76]. Furthermore,

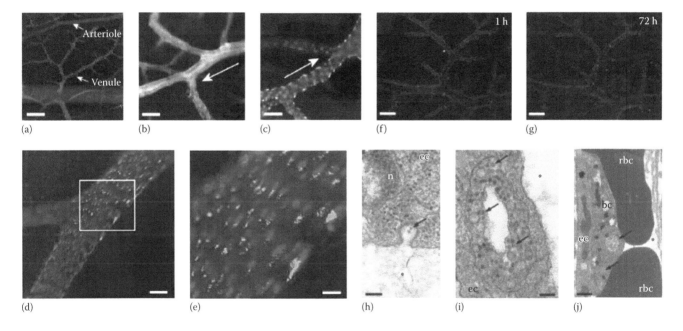

FIGURE 23.1 Intravital fluorescence imaging of chick CAM vasculature and subcellular localization of CPMV. (a) Fluorescence image looking down through surface of chick CAM showing multiple levels of vasculature. Scale bar, 100 µm. (b) Arteriole. Scale bar, 22 mm. (c) CAM venule (arrows in b and c denote blood flow direction). Scale bar, 22 µm. (d) Intravital image of large CAM vein, CPMV-A555 (orange), endothelial cell nuclei (blue) stained with Hoechst 33258. Box indicates area magnified in (e). Scale bar, 16 µm. (e) CPMV VLPs are restricted to perinuclear compartments in the vascular endothelial cells. Scale bar, 5.5 µm. (f and g) CPMV-A555 remains restricted to endothelial cells and allows intravital staining of the vasculature over long periods of time. Seventy-two hours after the initial injection of CPMV-A555, venular staining remains roughly equivalent to the initial staining. Scale bar, 160 µm. (h) Transmission electron micrograph of chick embryo CAM injected with CPMV-A555, showing CPMV particles being actively internalized into an endothelial cell (black arrows indicate CPMV particles). ec, endothelial cell; n, nucleus. Asterisk indicates vessel lumen. Scale bar, 100 nm. (i) Endothelial cell with CPMV-filled vesicle. Arrows indicate CPMV particles among many. Scale bar, 64 nm. (j) Macrophage (wbc) at the luminal periphery with large CPMV-containing vesicles (arrows). rbc, red blood cell. Scale bar, 693 nm. (Reproduced from Lewis, J.D. et al., *Nat. Med.*, 12(3), 354, 2006. With permission.)

ability of labeling virus components *in vivo* provides a platform for studies of virus assembly, release, and intercellular trafficking [77].

Genetically labeled HIV-1 particles were used in studies of assembly and virus release in HeLa cells [78]. By measuring the instantaneous velocity of the particle as a marker for extracellular release (Figure 23.2), the time from appearance of an assembly site to particle release of a complete HIV-1 particle from HeLa cells was estimated at ~30 min. The actual assembly of the Gag shell took place in 8–9 min. Analysis of dynamics of release by single-particle tracking allowed identification of the most likely rate-limiting steps. Such knowledge could prove essential in the development of antiretroviral drugs targeting assembly.

In addition, recent advances in computational image processing in combination with improved detectors for single-particle imaging allowed for quantitative analysis of trafficking kinetics and biodistribution of Cy5-labeled adeno-associated virus (AAV) in 3D animal cells and mouse tissues [79]. The study suggested that following intramuscular injection, AAVs spread across muscle tissues between myofibers instead of passing through target cells.

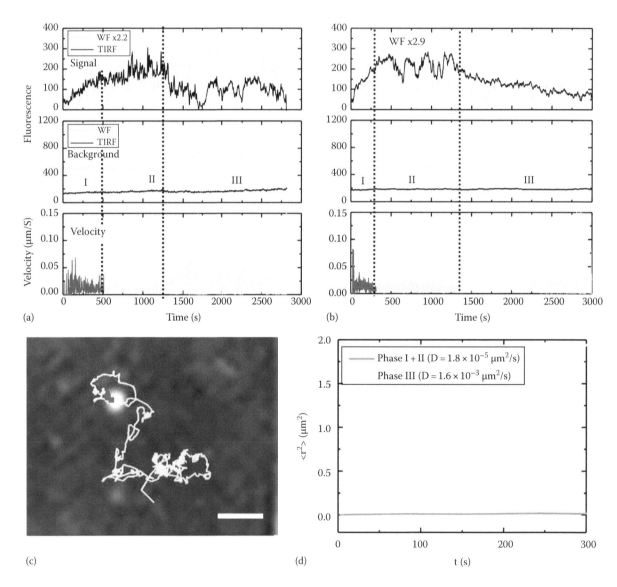

FIGURE 23.2 Three phases of HIV-1 assembly from single-particle tracking measurements. (a and b) Intensity traces of two individual Gag-eGFP clusters recorded through different optical contrast modalities are shown in the *top panels*. Three phases, separated by dashed lines, were observed: (phase I) a rapid rise in fluorescence intensity, (phase II) a plateau region with fluctuations in fluorescence intensity, and (phase III) a decay of the fluorescence signal. The *middle panels* show the intensity traces of the local background. The *bottom panel* displays a plot of the corresponding instantaneous velocities of the particles. (a) An abrupt increase in instantaneous velocity is observed concomitantly with the onset of phase III. (c) Trajectory of the individual Gag-eGFP cluster from panel A color-coded according to the three different phases. Scale bar: 1 μm. (d) Mean-square displacement analysis of the trajectory shown in panel C indicates the random character of motion in the different phases and shows a change by two orders of magnitude in diffusion coefficient between phases I + II and III. (Adapted from Ivanchenko, S. et al., *PLoS Pathog.*, 5(11), e1000652, 2009. With permission.)

23.3.2 MAGNETIC RESONANCE IMAGING

MRI is a nonintrusive, imaging tool suitable for whole organism studies, which owes its contrast to the alignment of protons in a strong magnetic field and relaxation to the ground state after the field is removed. MRI allows whole body imaging with submillimeter resolution, but at relatively low sensitivities, 10^{-3}–10^{-5} mol/L. MRI sensitivity depends on the length of proton relaxation times and is customarily increased by coupling contrast agents to macromolecular carriers. Relaxation pathways include spin–spin (T1) and spin–lattice (T2), each being characterized by a different correlation time. Most commonly used MRI contrast agents are chelates of gadolinium, which work through shortening of the T1 relaxation time of protons located nearby the metal ion. Superparamagnetic iron oxide (SPIO) reduces the T_2 relaxation time in absorbing tissues through local magnetic field gradient effect and thus provides negative or *dark* contrast.

Visualization of the mouse vascular system was demonstrated by administering Gd(III)–DTPA-conjugated P22 capsids to a BALB/C nude mouse [80]. Following injection of the Gd(III)–DTPA-conjugated P22 viral capsids, contrast enhancement resulted in a detailed map of blood vessels, including the carotid, mammary arteries, the jugular vein, and the superficial vessels of the head (Figure 23.3). The isotropic resolution was ~250 μm.

Gd-loaded CPMV particles (CPMV-Gd) were also studied as potential diagnostic agents for conditions of the central nervous system (CNS), for example, multiple sclerosis and experimental autoimmune encephalomyelitis [62]. Analysis of *in vivo* distribution of CPMV-Gd within the peripheral and CNS has shown probe accumulation in inflammatory lesions within the brain and spinal cord.

Bacteriophage M13, presenting phage display–developed targeting groups, coated with SPIO particles enhanced the *in vivo* imaging contrast sixfold on a prostate cancer tissue model [81].

Cubic iron oxide NPs encapsulated in BMV capsids exhibited at least five times higher relaxivities in MRI experiments with respect to unlabeled phantom samples. The target application in this case was the *in vivo* MRI tracking of systemic plant virus transport. Preliminary data from cell-to-cell and long-distance transport in *Nicotiana benthamiana* leaves have shown high penetration into tissue and long-distance transfer through the vasculature of SPIO–VLPs when compared to bare cubic iron oxide NPs [82]. Double-loaded iron oxide and dox-rotavirus VP4 VLPs were used for simultaneous imaging and killing of cancer cells with the same capsid-based vehicle [43].

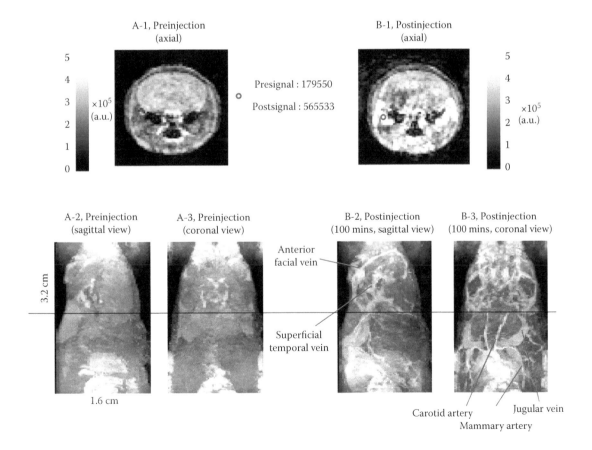

FIGURE 23.3 Axial preinjection and 100 min after injection 3D-FLASH images of BALB/C nude mouse (A-1 and B-1, respectively). Maximum intensity projections of preinjection images are shown along the sagittal (A-2) and coronal (A-3) directions. The corresponding postinjection images along sagittal (B-2) and coronal (B-3) directions are shown as well for direct comparison (total Gd injected dose, 25 μM Gd(III)/kg). (Adapted from Min, J. et al., *Biomacromolecules*, 14(7), 2332, 2013. With permission.)

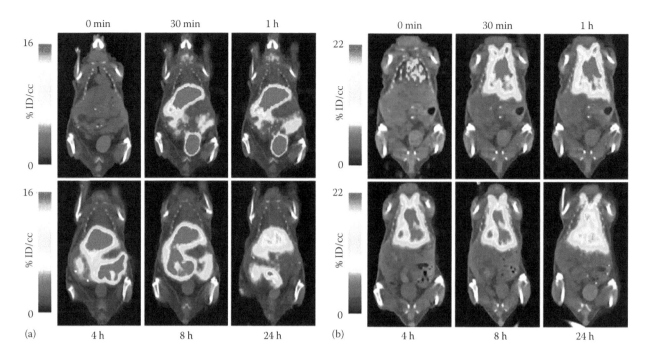

FIGURE 23.4 PET–CT images obtained from mouse injected with 64Cu show marked differences in biodistribution for VLPs. (a) No viral capsids and (b) injection with 64Cu-labeled DOTA-MS2. A dynamic scan was performed over the first 60 min, followed by scans obtained at 4, 8, and 24 h. The scale is reported as percent injected dose per milliliter (% ID/cc). (Adapted from Farkas, M.E. et al., *Mol. Pharm.,* 10(1), 69, 2013.)

23.3.3 POSITRON EMISSION TOMOGRAPHY

PET imaging relies on the detection of radiation emitted when certain atomic isotopes decay and is extremely sensitive, with detection levels of 10^{-11}–10^{-12} mol/L. The commonly used isotope is 18F, although 64Cu, 11C, 124I, 68Ga, 15O2, 13N, and 74As have also been used with good results. The main drawback of PET imaging is low spatial resolution.

Bacteriophage MS2 capsids labeled with both 18F and 64Cu PET [83,84] were explored in relation to the biodistribution of the untargeted particles and compared with nonlabeled MS2 distribution (Figure 23.4). It was shown that distribution, clearance rate, and pathways were not altered by cargo. Moreover, the lack of specificity for nonexcretory organs is beneficial ensuring low background activity in targeted delivery applications.

In a targeted PET imaging application, bacteriophage T7 capsid was modified with two different 64Cu chelators and RGD peptides, which targeted the integrins on tumor cell surfaces [85]. Despite significant liver uptake, particles were found to accumulate at sufficient levels in tumor tissue, and the choice of chelator showed significant difference at later times, which points to a possible stability difference between chelators.

In conclusion, VLPs are complementing more mature technologies based on the use of nanoparticles for imaging contrast enhancement, through additional properties that are inherent of viruses, but currently difficult to reproduce by any other nonbiological approach. The field of VLP-assisted imaging has seen significant excursions in the recent past in a variety of directions, in terms of both contrast mechanisms

and applications. With better understanding of virus assembly and dynamics, opportunities for future development are expected to arise.

ACKNOWLEDGMENT

Support from the U.S. Army Research Office under award W911NF-13-1-0490 is gratefully acknowledged.

REFERENCES

1. Lee, S.-Y., J.-S. Lim, and M.T. Harris, Synthesis and application of virus-based hybrid nanomaterials. *Biotechnology and Bioengineering,* 2012. **109**(1): 16–30.
2. Musick, M.A. et al., Reprogramming virus nanoparticles to bind metal ions upon activation with heat. *Biomacromolecules,* 2011. **12**(6): 2153–2158.
3. Li, F. and Q. Wang, Fabrication of nanoarchitectures templated by virus-based nanoparticles: Strategies and applications. *Small,* 2014. **10**(2): 230–245.
4. Yildiz, I., S. Shukla, and N.F. Steinmetz, Applications of viral nanoparticles in medicine. *Current Opinion in Biotechnology,* 2011. **22**(6): 901–908.
5. Lomonossoff, G.P., Virus particles and the uses of such particles in bio- and nanotechnology, in *Recent Advances in Plant Virology,* C. Caranta et al. (eds.), 2011, Caister Academic Press, Norfolk, England, pp. 363–385.
6. Destito, G., A. Schneemann, and M. Manchester, Biomedical nanotechnology using virus-based nanoparticles, in *Viruses and Nanotechnology,* M. Manchester and N. Steinmetz (eds.), 2009, Springer, Berlin, Germany, pp. 95–122.
7. Douglas, T. and M. Young, Viruses: Making friends with old foes. *Science,* 2006. **312**(5775): 873–875.

8. Manchester, M. and P. Singh, Virus-based nanoparticles (VNPs): Platform technologies for diagnostic imaging. *Advanced Drug Delivery Reviews*, 2006. **58**(14): 1505–1522.

9. Strable, E. and M.G. Finn, Chemical modification of viruses and virus-like particles, in *Viruses and Nanotechnology*, M. Manchester and N. Steinmetz (eds.), 2009, Springer, Berlin, Germany, pp. 1–21.

10. Rhee, J.-K. et al., Glycan-targeted virus-like nanoparticles for photodynamic therapy. *Biomacromolecules*, 2012. **13**(8): 2333–2338.

11. Loo, L. et al., Infusion of dye molecules into Red clover necrotic mosaic virus. *Chemical Communications*, 2008. (1): 88–90.

12. Brasch, M. et al., Encapsulation of phthalocyanine supramolecular stacks into virus-like particles. *Journal of the American Chemical Society*, 2011. **133**(18): 6878–6881.

13. Wu, Z.J. et al., Development of viral nanoparticles for efficient intracellular delivery. *Nanoscale*, 2012. **4**(11): 3567–3576.

14. Kaiser, C.R. et al., Biodistribution studies of protein cage nanoparticles demonstrate broad tissue distribution and rapid clearance in vivo. *International Journal of Nanomedicine*, 2007. **2**(4): 715–733.

15. Bruckman, M.A. et al., Biodistribution, pharmacokinetics, and blood compatibility of native and PEGylated tobacco mosaic virus nano-rods and -spheres in mice. *Virology*, 2014. **449**: 163–173.

16. Pattenden, L.K. et al., Towards the preparative and large-scale precision manufacture of virus-like particles. *Trends in Biotechnology*, 2005. **23**(10): 523–529.

17. Bundy, B.C., M.J. Franciszkowicz, and J.R. Swartz, *Escherichia coli*-based cell-free synthesis of virus-like particles. *Biotechnology and Bioengineering*, 2008. **100**(1): 28–37.

18. Fass, L., Imaging and cancer: A review. *Molecular Oncology*, 2008. **2**(2): 115–152.

19. Glasgow, J. and D. Tullman-Ercek, Production and applications of engineered viral capsids. *Applied Microbiology and Biotechnology*, 2014. **98**(13): 5847–5858.

20. Brandenburg, B. et al., Imaging poliovirus entry in live cells. *PLoS Biology*, 2007. **5**(7): e183.

21. Huang, L.-L. et al., A new stable and reliable method for labeling nucleic acids of fully replicative viruses. *Chemical Communications*, 2012. **48**(18): 2424–2426.

22. Ayala-Nuñez, N.V., J. Wilschut, and J.M. Smit, Monitoring virus entry into living cells using DiD-labeled dengue virus particles. *Methods*, 2011. **55**(2): 137–143.

23. Markosyan, R.M., F.S. Cohen, and G.B. Melikyan, Time-resolved imaging of HIV-1 Env-mediated lipid and content mixing between a single virion and cell membrane. *Molecular Biology of the Cell*, 2005. **16**(12): 5502–5513.

24. Floyd, D.L. et al., Single-particle kinetics of influenza virus membrane fusion. *Proceedings of the National Academy of Sciences of the United States of America*, 2008. **105**(40): 15382–15387.

25. Dixit, S.K. et al., Quantum dot encapsulation in viral capsids. *Nano Letters*, 2006. **6**(9): 1993–1999.

26. Huang, X. et al., Self-assembled virus-like particles with magnetic cores. *Nano Letters*, 2007. **7**(8): 2407–2416.

27. Goicochea, N.L. et al., Structure and stoichiometry of template-directed recombinant HIV-1 gag particles. *Journal of Molecular Biology*, 2011. **410**(4): 667–680.

28. Chen, C. et al., Nanoparticle-templated assembly of viral protein cages. *Nano Letters*, 2006. **6**(4): 611–615.

29. Liu, S.L. et al., High-efficiency dual labeling of influenza virus for single-virus imaging. *Biomaterials*, 2012. **33**(31): 7828–7833.

30. Jouvenet, N., S.M. Simon, and P.D. Bieniasz, Imaging the interaction of HIV-1 genomes and Gag during assembly of individual viral particles. *Proceedings of the National Academy of Sciences of the United States of America*, 2009. **106**(45): 19114–19119.

31. Vaughan, R. et al., The tripartite virions of the brome mosaic virus have distinct physical properties that affect the timing of the infection process. *Journal of Virology*, 2014. **88**(11): 6483–6491.

32. Borodavka, A., R. Tuma, and P.G. Stockley, Evidence that viral RNAs have evolved for efficient, two-stage packaging. *Proceedings of the National Academy of Sciences of the United States of America*, 2012. **109**(39): 15769–15774.

33. Paige, J.S., K.Y. Wu, and S.R. Jaffrey, RNA mimics of green fluorescent protein. *Science*, 2011. **333**(6042): 642–646.

34. Han, K.Y. et al., Understanding the photophysics of the Spinach–DFHBI RNA aptamer-fluorogen complex to improve live-cell RNA imaging. *Journal of the American Chemical Society*, 2013. **135**(50): 19033–19038.

35. Aniagyei, S.E. et al., Self-assembly approaches to nanomaterial encapsulation in viral protein cages. *Journal of Materials Chemistry*, 2008. **18**(32): 3763–3774.

36. Douglas, T. and M. Young, Host–guest encapsulation of materials by assembled virus protein cages. *Nature*, 1998. **393**(6681): 152–155.

37. Liu, C. et al., Magnetic viruses via nano-capsid templates. *Journal of Magnetism and Magnetic Materials*, 2006. **302**(1): 47–51.

38. Klem, M.T., M. Young, and T. Douglas, Biomimetic synthesis of beta-TiO(2) inside a viral capsid. *Journal of Materials Chemistry*, 2008. **18**(32): 3821–3823.

39. de la Escosura, A. et al., Viral capsids as templates for the production of monodisperse Prussian blue nanoparticles. *Chemical Communications*, 2008. (13): 1542–1544.

40. Chen, C. et al., Packaging of gold particles in viral capsids. *Journal of Nanoscience and Nanotechnology*, 2005. **5**(12): 2029–2033.

41. Li, F. et al., Imaging viral behavior in mammalian cells with self-assembled capsid-quantum-dot hybrid particles. *Small*, 2009. **5**(6): 718–726.

42. Enomoto, T. et al., Viral protein-coating of magnetic nanoparticles using simian virus 40 VP1. *Journal of Biotechnology*, 2013. **167**(1): 8–15.

43. Chen, W. et al., Rotavirus capsid surface protein VP4-coated Fe_3O_4 nanoparticles as a theranostic platform for cellular imaging and drug delivery. *Biomaterials*, 2012. **33**(31): 7895–7902.

44. Zhang, Y. et al., Encapsulating quantum dots into enveloped virus in living cells for tracking virus infection. *ACS Nano*, 2013. **7**(5): 3896–3904.

45. Jung, B., A.L.N. Rao, and B. Anvari, Optical nano-constructs composed of genome-depleted brome mosaic virus doped with a near infrared chromophore for potential biomedical applications. *ACS Nano*, 2011. **5**(2): 1243–1252.

46. Yildiz, I. et al., Infusion of imaging and therapeutic molecules into the plant virus-based carrier cowpea mosaic virus: Cargo-loading and delivery. *Journal of Controlled Release*, 2013. **172**(2): 568–578.

47. Kremser, L. et al., Labeling of capsid proteins and genomic RNA of human rhinovirus with two different fluorescent dyes for selective detection by capillary electrophoresis. *Analytical Chemistry*, 2004. **76**(24): 7360–7365.

48. Kratz, P.A., B. Böttcher, and M. Nassal, Native display of complete foreign protein domains on the surface of hepatitis B virus capsids. *Proceedings of the National Academy of Sciences of the United States of America*, 1999. **96**(5): 1915–1920.

49. O'Neil, A. et al., Coconfinement of fluorescent proteins: Spatially enforced communication of GFP and mCherry encapsulated within the P22 capsid. *Biomacromolecules*, 2012. **13**(12): 3902–3907.

50. Rhee, J.K. et al., Colorful virus-like particles: Fluorescent protein packaging by the Q beta capsid. *Biomacromolecules*, 2011. **12**(11): 3977–3981.

51. Tsvetkova, I.B. et al., Fusion of mApple and venus fluorescent proteins to the Sindbis virus E2 protein leads to different cell-binding properties. *Virus Research*, 2013. **177**(2): 138–146.

52. Whitt, M.A. and C.E. Mire, Utilization of fluorescently-labeled tetracysteine-tagged proteins to study virus entry by live cell microscopy. *Methods*, 2011. **55**(2): 127–136.

53. Steinmetz, N.F., Viral nanoparticles as platforms for next-generation therapeutics and imaging devices. *Nanomedicine: Nanotechnology, Biology and Medicine*, 2010. **6**(5): 634–641.

54. Young, M. et al., Plant viruses as biotemplates for materials and their use in nanotechnology. *Annual Review of Phytopathology*, 2008. **46**: 361–384 (Annual Reviews, Palo Alto, CA).

55. Anderson, E.A. et al., Viral nanoparticles donning a paramagnetic coat: Conjugation of MRI contrast agents to the MS2 capsid. *Nano Letters*, 2006. **6**(6): 1160–1164.

56. Datta, A. et al., High relaxivity gadolinium hydroxypyridonate-viral capsid conjugates: Nanosized MRI contrast agents. *Journal of the American Chemical Society*, 2008. **130**(8): 2546–2552.

57. Garimella, P.D. et al., Multivalent, high-relaxivity MRI contrast agents using rigid cysteine-reactive gadolinium complexes. *Journal of the American Chemical Society*, 2011. **133**(37): 14704–14709.

58. Hooker, J.M. et al., Magnetic resonance contrast agents from viral capsid shells: A comparison of exterior and interior cargo strategies. *Nano Letters*, 2007. **7**(8): 2207–2210.

59. Lucon, J. et al., Use of the interior cavity of the P22 capsid for site-specific initiation of atom-transfer radical polymerization with high-density cargo loading. *Nature Chemistry*, 2012. **4**(10): 781–788.

60. Qazi, S. et al., P22 viral capsids as nanocomposite high-relaxivity MRI contrast agents. *Molecular Pharmaceutics*, 2013. **10**(1): 11–17.

61. Prasuhn, D.E., Jr. et al., Viral MRI contrast agents: Coordination of Gd by native virions and attachment of Gd complexes by azide-alkyne cycloaddition. *Chemical Communications*, 2007. (12): 1269–1271.

62. Shriver, L.P. et al., Localization of gadolinium-loaded CPMV to sites of inflammation during central nervous system autoimmunity. *Journal of Materials Chemistry B*, 2013. **1**(39): 5256–5263.

63. Bruckman, M.A. et al., Tobacco mosaic virus rods and spheres as supramolecular high-relaxivity MRI contrast agents. *Journal of Materials Chemistry B*, 2013. **1**(10): 1482–1490.

64. Josephson, L. et al., Near-infrared fluorescent nanoparticles as combined MR/optical imaging probes. *Bioconjugate Chemistry*, 2002. **13**(3): 554–560.

65. Baddeley, D. et al., 4D super-resolution microscopy with conventional fluorophores and single wavelength excitation in optically thick cells and tissues. *PLoS ONE*, 2011. **6**(5): e20645.

66. Roy, R., Next-generation optical microscopy. *Current Science*, 2013. **105**(11): 1524–1536.

67. Lewis, J.D. et al., Viral nanoparticles as tools for intravital vascular imaging. *Nature Medicine*, 2006. **12**(3): 354–360.

68. Leong, H.S. et al., Intravital imaging of embryonic and tumor neovasculature using viral nanoparticles. *Nature Protocols*, 2010. **5**(8): 1406–1417.

69. Steinmetz, N.F. et al., Intravital imaging of human prostate cancer using viral nanoparticles targeted to gastrin-releasing peptide receptors. *Small*, 2011. **7**(12): 1664–1672.

70. Shukla, S. et al., Increased tumor homing and tissue penetration of the filamentous plant viral nanoparticle potato virus X. *Molecular Pharmaceutics*, 2013. **10**(1): 33–42.

71. Pan, H. et al., Noninvasive visualization of respiratory viral infection using bioorthogonal conjugated near-infrared-emitting quantum dots. *ACS Nano*, 2014. **8**(6): 5468–5477.

72. Barrow, E., A.V. Nicola, and J. Liu, Multiscale perspectives of virus entry via endocytosis. *Virology Journal*, 2013. **10**: 177.

73. Ruthardt, N., D.C. Lamb, and C. Brauchle, Single-particle tracking as a quantitative microscopy-based approach to unravel cell entry mechanisms of viruses and pharmaceutical nanoparticles. *Molecular Therapy*, 2011. **19**(7): 1199–1211.

74. Huang, L.-L. and H.-Y. Xie, Progress on the labeling and single-particle tracking technologies of viruses. *Analyst*, 2014. **139**(13): 3336–3346.

75. Brandenburg, B. and X. Zhuang, Virus trafficking—Learning from single-virus tracking. *Nature Reviews Microbiology*, 2007. **5**(3): 197–208.

76. Sun, E., J. He, and X.W. Zhuang, Live cell imaging of viral entry. *Current Opinion in Virology*, 2013. **3**(1): 34–43.

77. Zhou, P. et al., Multicolor labeling of living-virus particles in live cells. *Angewandte Chemie International Edition*, 2012. **51**(3): 670–674.

78. Ivanchenko, S. et al., Dynamics of HIV-1 assembly and release. *PLoS Pathogens*, 2009. **5**(11): e1000652.

79. Xiao, P.J. et al., Quantitative 3D tracing of gene-delivery viral vectors in human cells and animal tissues. *Molecular Therapy*, 2012. **20**(2): 317–328.

80. Min, J. et al., Implementation of P22 viral capsids as intravascular magnetic resonance T1 contrast conjugates via site-selective attachment of Gd(III)-chelating agents. *Biomacromolecules*, 2013. **14**(7): 2332–2339.

81. Ghosh, D. et al., M13-templated magnetic nanoparticles for targeted in vivo imaging of prostate cancer. *Nature Nanotechnology*, 2012. **7**(10): 677–682.

82. Huang, X.L. et al., Magnetic virus-like nanoparticles in *N. benthamiana* plants: A new paradigm for environmental and agronomic biotechnological research. *ACS Nano*, 2011. **5**(5): 4037–4045.

83. Hooker, J.M. et al., Genome-free viral capsids as carriers for positron emission tomography radiolabels. *Molecular Imaging and Biology*, 2008. **10**(4): 182–191.

84. Farkas, M.E. et al., PET imaging and biodistribution of chemically modified bacteriophage MS2. *Molecular Pharmaceutics*, 2013. **10**(1): 69–76.

85. Li, Z. et al., Trackable and targeted phage as positron emission tomography (PET) agent for cancer imaging. *Theranostics*, 2011. **1**: 371–80.

24 Enveloped Viruses with Single-Stranded Negative RNA Genome as Objects and Subjects of VLP Nanotechnology

Peter Pushko and Paul Pumpens

CONTENTS

24.1 INTRODUCTION

This chapter extends Chapter 9 and describes applications of VLPs for medically important viruses causing emerging and reemerging diseases. Here, we focus on viruses from the Baltimore's group of the negative single-stranded RNA (minus-ssRNA) viruses with antisense, or minus-strand, RNA as a genome. Although somewhat underrepresented among published VLP approaches and technologies, this group nevertheless includes a number of VLP structures and experimental vaccines. Some of these were used for the construction of chimeric VLPs or involved in the generation of novel bionanomaterials. This is hardly surprising since the members of the group generally represent highly pathogenic viruses requiring special high-level biocontainment facilities, which hampers research efforts. In addition, these viruses exhibit the least symmetrical morphology, possess only traces of icosahedral symmetry, and demonstrate at the same time rather complicated organization and self-assembly mechanisms that impede structural reconstructions and molecular manipulations. Nevertheless, this group is very important because of unprecedented medical and veterinary burden caused by its members and urgent interest in establishing novel VLP approaches to prepare safe and effective vaccines against the respective viruses.

We selected for this chapter the representatives of two families: (1) Arenaviridae and Bunyaviridae with segmented antisense RNA genomes and (2) the large Mononegavirales order consisting of five families with nonsegmented genomes. The remaining Orthomyxoviridae family that is in fact one of the most advanced in the sense of VLP applications among other members of the group is reviewed in Chapter 25 devoted exclusively to the efforts in the field of influenza VLPs. The genetic organization and structure of the respective viruses are schematically depicted in Figure 24.1, while published representative VLP structures are shown in Table 24.1 including VLPs from the orthomyxoviruses [1–5]. Regarding the order of narrative, we proceed from the predominantly spherical virions to the bullet-shaped and, at last, tubular and filamentous structures, which demonstrate many common features in the general principles of organization and self-assembly of their nucleocapsids and envelopes.

24.2 ARENAVIRUSES: LASSA VIRUS, LCMV, AND OTHERS

24.2.1 General Considerations

According to the latest issue of the official taxonomical International Committee on Taxonomy of Viruses (ICTV) documents [6], the arenaviruses are forming Arenaviridae family with a single *Arenavirus* genus represented by 25 species. The National Center for Biotechnology Information (NCBI) taxonomy [7] includes 55 arenaviral names and continues traditional separation of the arenaviruses into the Old World (Europe, Asia, and Africa) and the New World (America) virus groups with 2 group-independent species: Lujo and Luna viruses and 18 unclassified members of the family.

The arenaviruses infect rodents, and some of these viruses are responsible for severe human diseases that are the central focus of this review in the light of potential VLP applications for the arenavirus vaccines (for a recent review on the vaccination against arenavirus infection, see [8]). First of all, the most familiar reference strain Lassa virus (LASV) from the Old World virus group and Guanarito virus, Junin virus (JUNV), Machupo virus (MACV), Sabia virus (SABV), or Whitewater Arroyo virus from the New World virus group, as well as Lujo virus (LUJV) may cause severe hemorrhagic fever syndromes, which result in a significant mortality. The observed severe cases of disease have introduced LASV to the scientific community [9]. Second, the reference strain that is the most studied representative of the family in the virological and immunological sense, namely, lymphocytic choriomeningitis virus (LCMV), may cause not only influenza-like syndromes but also severe aseptic meningitis. LCMV exists in both geographic areas but is classified as an Old World virus [7].

The arenaviruses are enveloped pleomorphic, mostly spherical viruses (Figure 24.1) with a broad variation of sizes from 60 to 300 nm in diameter and segmented linear RNA genome encoding for four proteins: L-segment is about 7.5 kb and S-segment 3.5 kb [10]. According to the Baltimore classification [11], the arenaviruses are usually attributed to members of the Baltimore's group V: *minus-ssRNA viruses with minus-strand or antisense RNA*. However, their genome is ambisense, since both RNA directions are used to encode viral genes. The S-segment encodes the glycoprotein precursor (GPC), which is cleaved further into three parts: attachment glycoprotein GP1, fusion transmembrane glycoprotein GP2 and a stable signal peptide SSP, and, in ambisense, the nucleoprotein (NP); the L-segment encodes the matrix protein Z and, in negative sense, the multifunctional protein L [8].

Two RNA segments are embedded into the nucleocapsid that interacts with the Z protein under the plasma membrane and buds, releasing the virion covered with surface glycoprotein spikes. The Z protein forms homooligomers that represent structural scaffold of the virion. The life cycle of the arenaviruses is restricted to the cell cytoplasm. The arenavirus virions contain sand-looking particles that are responsible for the name of the family (*arena* means sand in Latin) and are nothing else as ribosomes acquired from the host cells.

24.2.2 Production of Arenaviral VLPs in Standard Expression Systems

Attempts to express the nearly full-length or fragmented structural LASV proteins NP [12–14] and GP1 and GP2 [14] in *Escherichia coli* led to assembly-defective products that were useful in the induction of the human T-cell proliferative response [13] and ELISA diagnostics [14]. In order to construct a putative live mucosal vaccine, the complete LASV NP sequence was cloned in a recombinant *aroA* attenuated *Salmonella typhimurium*, and the presence of the nonassembled NP in whole cell extracts was detected, with some cross protective efficiency of the recombinant *Salmonella* against LCMV infection demonstrated in mice [15,16]. Similarly to the bacterial system, baculovirus-driven expression of LASV NP [17,18] and GPC [19] in insect cells did not result in assembly-competent products, but warranted production of valuable diagnostic reagents.

Homologous eukaryotic vaccinia virus-driven expression of LASV NP resulted in the production of the corresponding nonassembled protein that was able to induce protection in guinea pigs against lethal challenge with LASV [20,21]. The same was true for the vaccinia virus-driven expression of LASV GPC that demonstrated in infected cells the presence of GPC precursor and GP1 and GP2 posttranslational cleavage products that did not show any signs of assembly, but nevertheless acted as an efficient vaccine against LASV lethal challenge in guinea pigs [21,22]. Moreover, vaccinia virus-driven expression of GPC protected rhesus monkeys against LASV challenge [23]. Simultaneous vaccinia virus-driven expression of NP and GPC resulted in the synthesis of authentic proteins in cells with respect to electrophoretic mobility, glycosylation, and posttranslational cleavage, but again, without signs of self-assembly [24]. In general, the early work on protective properties of

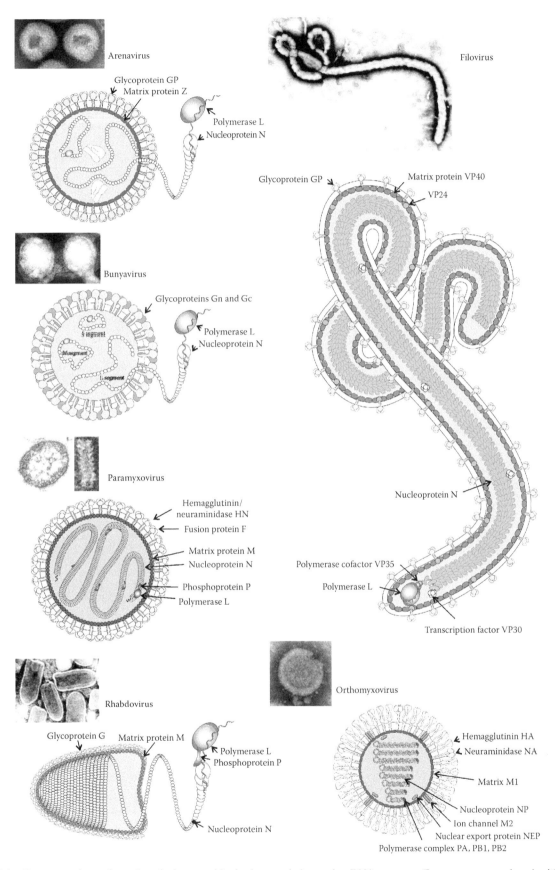

FIGURE 24.1 Representatives of enveloped viruses with single-stranded negative RNA genome. (Images are reproduced with a kind permission from the ViralZone, SIB [Swiss Institute of Bioinformatics], Hulo, C. et al., *Nucleic Acids Res.*, 39(Database issue), D576, 2011, http://viralzone.expasy.org/; The electron micrographs of the virions are from the CDC Public Health Image Library [PHIL].) In the case of paramyxoviruses, an image of nucleocapsid, or herringbone-like, structure is shown.

TABLE 24.1

VLP Formation Capability of the Indicated Structural Proteins Derived from the Enveloped Viruses with Single-Stranded Negative RNA Genome

Genus	Species	Expression System									
		Mammalian Cells		Insect Cells		Plants		Yeast		Bacteria	
		No	Yes	No	Yes	No	Yes	No	Yes	No	Yes
Order: not assigned; family: Arenaviridae; subfamily: not assigned											
Arenavirus	Junin virus		Z, NP, GP	Z						Z	
	Lassa virus	NP	Z	NP						NP	
		GPC	Z, GP	GPC						GPC	
		NP, GPC	Z, NP, GP								
	Lymphocytic choriomeningitis virus		Z								
	Mopeia virus		Z, NP								
	Tacaribe virus		Z, NP, GP								
Order: not assigned; family: Bunyaviridae; subfamily: not assigned											
Hantavirus	Andes virus		Gn, Gc								
	Dobrava-Belgrade virus							N			
	Hantaan virus	N Gn, Gc	N, Gn, Gc	N, Gn, Gc	N			Gc		N	
	Puumala virus		Gn, Gc			N		N			
Nairovirus	Crimean–Congo hemorrhagic fever virus	N			N						
Orthobunyavirus	California encephalitis virus (snowshoe hare bunyavirus)			N							
	Schmallenberg virus							N		N	
Phlebovirus	Rift Valley fever virus	Gn, Gc	N, Gn, Gc		N, Gc					N	
					N, Gn, Gc						
					Gn, Gc						
	Human severe fever with thrombocytopenia syndrome phlebovirus									N	
Tospovirus	Tomato spotted wilt virus					N				N	
Order: Mononegavirales; family: Paramyxoviridae; subfamily: Paramyxovirinae											
Avulavirus	NDV	N, F, HN	M		HN—yes?				M	N	
			M, N, F, HN								M[a]
Henipavirus	Nipah virus	N, F, G	M		N				N	N	
			M, N, F, G								M
			M, F, G								
	Hendra virus								N		
Morbillivirus	Cetacean morbillivirus (dolphin morbillivirus)				N						
	Measles virus		N		N				N		N
			M								
			M, F								
	Peste-des-petits-ruminants virus				M, N						N
					M, N, H						
					M, H, F						
	Rinderpest virus										N
Respirovirus	Human parainfluenza virus 1		M						N		
			M, N								
	Human parainfluenza virus 3								N		
	Sendai virus		M						N		
			M, N, F, HN								

(*Continued*)

TABLE 24.1 (*Continued*)

VLP Formation Capability of the Indicated Structural Proteins Derived from the Enveloped Viruses with Single-Stranded Negative RNA Genome

		Expression System								
		Mammalian Cells		Insect Cells		Plants		Yeast		Bacteria
Genus	Species	No	Yes	No	Yes	No	Yes	No	Yes	No	Yes
Rubulavirus	Mumps virus		M						N		N
			M, N, F								
	Tioman virus (unclassified rubulavirus)								N		

Order: Mononegavirales; family: Paramyxoviridae; subfamily: Pneumovirinae

Metapneumovirus	Human metapneumovirus	F	M, G, F		Influenza M1, G, F			N	M		
Pneumovirus	Bovine respiratory syncytial virus			F						N	
	Human respiratory syncytial virus	F	M, F		N			G	M		N
		M			F					F	
										G	

Order: Mononegavirales; family: Bornaviridae; subfamily: not assigned

Bornavirus	Borna disease virus										M
Ephemerovirus	Bovine ephemeral fever virus			G							
Lyssavirus	European bat lyssavirus type 1								N		
	European bat lyssavirus type 2								N		
	Lagos bat virus										M?
	Mokola bat virus										M?
	Thailand dog virus										M?
	Rabies virus		G		N	G	N?	N	N		N
					N, M						G

Order: Mononegavirales; family: Rhabdoviridae; subfamily: not assigned

Novirhabdovirus	Infectious hematopoietic necrosis virus									N	
										G	
Vesiculovirus	Chandipura virus	G									
	VSV			N						N	

Order: Mononegavirales; family: Filoviridae; subfamily: not assigned

Cuevavirus	Lloviu virus		NP, GP, VP40								
Ebolavirus	Zaire ebolavirus		GP, VP40		GP, VP40						GP
			NP, GP, VP40		NP, GP, VP40						
Marburgvirus	Marburg marburgvirus		GP, VP40		NP, GP, VP40						
			NP, GP, VP40								

Order: not assigned; family: Orthomyxoviridae; subfamily: not assigned

	Influenza A virus		HA, NA		HA, NA, M1		HA				
	Influenza B virus				HA, NA, M1						

Note: Capability of the protein or of the group of proteins to form VLPs is indicated by Yes, inability to form VLPs by No. In the case of the expression of NPs (N, NP) only, Yes means formation of nucleocapsid-like, or herringbone-like, structures. The question mark means a lack of information regarding VLP assembly. For references and comments, see text.

a After refolding in vitro.

the LASV and LCMV proteins was highly important in the sense of immunological theory, since this clearly demonstrated the importance of strong cellular immune response against arenavirus proteins to protect experimental animals from clinical disease, but not from infection, in the absence of measurable neutralizing antibodies (for more details, see a review [25]).

Expression of LASV NP and GPC and high protective efficiency of the generated vaccine constructs in guinea pigs was demonstrated by using highly efficient eukaryotic expression vector based on an alphavirus, namely, Venezuelan equine encephalitis (VEE) virus [26]. Furthermore, dual expression of glycoprotein genes of both Ebola and LASVs was achieved by using bivalent alphavirus particles [26]. However, the assembly of expressed LASV proteins was not studied.

Next, the yellow fever vaccine was used for the expression of LASV GPC or GP1 and GP2 [27] and demonstrated proper processing of GPC and protection against fatal LASV challenge in guinea pigs, but not sterilizing immunity, unlike immunization with live reassortant LASV vaccine ML29 [28].

Comprehension of general mechanisms of arenaviral self-assembly and the ways to involve arenaviruses in the VLP methodology evolved with understanding of the role played by the protein Z of LASV and LCMV and its ability to self-assemble in vitro [29] and in mammalian cells [30,31]. The small RING finger protein Z acted as a matrix protein that is a driving force for the viral assembly and budding. The protein Z self-assembled in the absence of any other viral proteins and was sufficient for the release of enveloped protein Z–based VLPs (for a recent review, see [32]). Although protein Z demonstrates strong variability among different arenaviral species, all arenaviruses encode and use it to initialize self-assembly.

24.2.3 High-Resolution Structure

Electron microscopy showed that the protein Z–induced VLPs did not significantly differ in their morphology and size from LASV particles [33]. The protein Z recruits NP to cellular membranes where virus assembly takes place, but mutation of two proline-rich domains PTAP and PPXY within the protein Z drastically reduces the release of VLPs [33]. The fine 3D structure of the LASV protein Z was determined by triple-resonance NMR techniques [34] and refined by the homology models and replica exchange molecular dynamics [35].

The crystal structure of the full-length LASV NP of Josiah [36] and AV [37] strains was solved at a resolution of 1.80 and 2.45 Å, respectively. Separately, crystal structures of functional LASV NP domains were determined. First, the structure of N-terminal domain in complex with single-stranded RNA was shown suggesting the likely assembly by which viral ribonucleoprotein complexes are organized [38]. Second, the structure of the immunosuppressive C-terminal portion that possesses exonuclease activity with strict specificity for double-stranded RNA substrates was demonstrated [39]. Recently, the crystal

structure of the homologous C-terminal NP portion was determined for the JUNV [40], LCMV [41], and Tacaribe arenavirus (TCRV) [42].

The crystal structures were presented also for the attachment glycoprotein GP1 of MACV at 1.7 Å resolution [43] and for the recombinant ectodomain of the LCMV transmembrane glycoprotein GP2 at 1.8 Å resolution [44]. Direct interaction of the protein Z with GPC, first of all, with SSC was shown for LASV and LCMV by confocal microscopy and coimmunoprecipitation assays [45].

The combination of all available structural information into an integrated arenaviral high-resolution structure with its publication in the VIPER-dB structural database [46] remains a problem of the nearest future.

24.2.4 VLP Applications: Arenaviruses as Objects and Subjects of VLP Nanotechnology

24.2.4.1 Arenaviruses as VLP Objects: Potential VLP Carriers

24.2.4.1.1 Generation of Arenaviral VLPs

Preparation and broad applications of arenavirus VLPs that are based on the protein Z self-assembly was hampered by the lack of assembled VLPs after the expression of arenavirus, namely, JUNV, protein Z in bacterial and baculovirus-driven insect expression systems, the two broadly used systems for VLP production [47]. General complexity of the arenaviral structural organization prevented therefore simple solutions with bacterial or baculovirus-driven expression. It is not excluded that the localization in membranes and strong need for correct myristoylation of the protein Z were the reason for this failure [48].

Nevertheless, a gradual progress in the elucidation of fine molecular mechanisms of self-assembly and budding of arenaviruses by the dissection of the whole processes into consecutive steps offered the prospect of the arenaviral VLP preparation at least in mammalian cells. A number of arenavirus VLPs have been prepared (Table 24.1). First, obstacles of homotypic LCMV and heterotypic (LCMV, LASV, and MACV) self-association of their NPs were documented [49]. It was shown that LASV NP forms trimers upon expression in mammalian cells [50]. Contribution of certain domains [51] and mapped amino acid residues [52] to self-assembly of TCRV NP was demonstrated.

Second, the ability of the protein Z to self-assemble into VLPs with the NP was shown by coexpression of the protein Z and NP of Mopeia virus, a close relative of the pathogenic LASV, when highly selective incorporation of the NP into protein Z–induced VLPs was observed [53]. The self-assembly of the NP and protein Z is complicated however by participation of a host protein, namely, an ESCRT-associated protein ALIX/AIP1 [54].

Third, the ability of the protein Z to self-assemble into VLPs with the GPs was demonstrated [55]. Fourth, all three structural arenaviral components, protein Z, NP, and GPs, were found to self-assemble into VLPs after dissected plasmid-based expression of the respective genes of JUNV and TCRV in mammalian cells [56]. Efficient multimilligram generation of LASV VLPs structurally and morphologically similar to native LASV virions, but lacking replicative functions, was achieved [57]. The LASV VLPs demonstrated typical pleomorphic distribution in size and shape by electron microscopy analysis and induced antibody response against individual viral proteins in mice in the absence of adjuvants [57].

The first application of the arenaviral chimeric VLPs was shown by the preparation of the JUNV Z–based fusion protein with eGFP [58]. Further, the ability of JUNV protein Z–eGFP fusion protein to generate VLPs in the absence of any other viral protein and the capacity of the Z protein to support fusions at its C-terminal without impairing budding activity in a standard mammalian cell line 293T were shown [59].

24.2.4.1.2 *Exchange of Arenaviral Envelopes, or Pseudotyped Arenaviruses*

Although the substitution of arenaviral envelopes with foreign ones does not fit precisely into the traditional concept of the VLPs, we decided to include these approaches into the section describing arenaviruses as potential VLP-based epitope carriers. First, a chimeric LCMV variant, where the glycoprotein G of vesicular stomatitis virus (VSV) was used to substitute the GP of LCMV, was constructed [60]. Therefore, a serious arenaviral *Achilles' heel* consisting in the ability to generate live attenuated arenavirus vaccines by reverse genetic engineering was found [61]. The rLCMV/VSVG chimeras provided strong immunological background for further development of GP exchange vaccines for combating arenaviral hemorrhagic fevers [62]. Further, placing of LASV GP on the backbone of LCMV resulted in a chimeric virus that displayed high tropism for dendritic cells following in vitro or in vivo infection of mice [63]. Introduction of point mutations and module exchanges into LASV GP on the pseudotyped rLCMV allowed further mapping of immunological and functional units of arenaviral GPs [64]. Another example of chimeric virus is the Mopeia–Lassa reassortant ML29 vaccine virus, which contained L and Z genes from Mopeia virus, while NP and GP genes were derived from LASV (reviewed in [65]).

24.2.4.2 Arenaviruses as VLP Subjects: Pseudotyping and a Source of Epitopes

24.2.4.2.1 *Pseudotyping of Non-Arenaviral VLPs by Arenaviral GPs*

The pseudotyping history started with the generation of replication-competent recombinant VSV (rVSV) expressing the glycoproteins of LASV and other viral hemorrhagic fever agents such as Ebola and Marburg viruses [66]. Further, pseudotyping of murine leukemia retroviral vectors by LASV, LCMV, JUNV, and MACV GPs not only allowed identification of arenaviral receptors on the cells but also explored the role played by the arenaviral GPs in viral entry in the absence of other arenavirus proteins [67]. An efficient model for the fast and quantifiable detection of neutralizing antibodies in human and animal sera was constructed on the basis of LASV GP–pseudotyped murine leukemia retroviral cores [68]. A lentiviral pseudotyping model was used for the studies on LCMV [69] and a nonpathogenic Pichinde virus (PICV) [70].

Very recently, pseudotyping of rVSV with a set of arenaviral GPs including LASV, JUNV, SABV, MACV, Chapare virus, and a novel highly pathogenic arenavirus LUJV with production of the respective pseudotyped viruses in 293T-cell line was performed [71].

Remarkably, pseudotyping of VSV by LCMV GP contributed to the clinical development of the VSV as an anticancer agent for the VSV-based oncolytic virotherapy by preventing neurotoxicity of the pseudotyped rVSV, when pseudotyped with the nonneurotropic LCMV GPs [72]. Furthermore, involvement of the rVSV pseudotyped with LCMV GPs solved the problem of antivector immunity induced by vaccine and/or gene therapy vector and preventing boosts with the same vector [73]. Implementation of LCMV-pseudotyped rVSV allowed the boost in mice receiving VSV vectors encoding ovalbumin as a model antigen [73].

24.2.4.2.2 *Arenaviruses as a Source of Epitopes*

A unique contribution of arenaviruses to the VLP methodology was the discovery and applications of widely used classic model epitopes. First of all, this is the cytotoxic T cell (CTL) epitope with a basic sequence KAVYNFATC derived originally from the leader peptide of LCMV GP1 [74,75] and known under the names gp33–41, or gp33, or p33, or finally as a *gold standard epitope* of vaccinology and gene therapy [76]. Immunodominant character and fine mechanisms of the gp33–41 presentation remained a subject of basic immunological investigations during more than 15 years after its discovery [77–82]. Crystal structure of the complex formed by the gp33–41 epitope with the murine MHC class I molecules was resolved [83,84].

The gp33–41 epitope was used as a model epitope in the following VLP models: hepatitis B virus core (HBc) [85–89], RNA phages Qβ [89–94] and AP205 [90], murine leukemia virus (MLV) [76,95], papaya mosaic virus [96], hamster polyomavirus (HaPyV) [97], and rabbit hemorrhagic disease virus [98].

Aside VLPs, but in the context of the potential vaccine applications, the gp33–41 epitope was used to label non-structural protein NS1 of influenza virus A/PR/8/34, which appeared to be a promising protective antigen for the design of novel modified live influenza virus vaccines [99]. Another application was the expression of the gp33–41 by the pseudogenome encapsidated into human papillomavirus VLPs as a potential gene therapy tool [100].

Another important LCMV CTL epitope covering amino acid residues 118–132 of the LCMV NP was inserted into the VP2 coat protein of the porcine parvovirus as a VLP model and assured protection of mice against lethal challenge with LCMV [101–105]. Further, the NP 118–132 epitope was used as an immunological marker for a very interesting and unusual VLP carrier, namely, tubule-forming nonstructural protein NS1 of bluetongue virus VLP model [106]. Chimeric NS1-based tubules protected the immunized mice against challenge with a lethal LCMV dose [106].

After these highly successful CTL epitopes, LASV, LCMV, and other arenaviruses continued to be a source of potential CTL epitopes for further use in the VLP studies (see, e.g., [107,108])

Finally, one study involved replacement of human immunodeficiency virus envelope protein domains, signal peptide, transmembrane, and cytoplasmic tail with the analogous elements of other viral envelopes including those of LASV GPs, in an attempt to enhance incorporation of chimeric products into VLPs [109].

24.3 BUNYAVIRUSES: HANTAAN VIRUS, CRIMEAN–CONGO HEMORRHAGIC FEVER VIRUS, BUNYAMWERA VIRUS, RIFT VALLEY FEVER VIRUS, TOMATO SPOTTED WILT VIRUS, AND OTHERS

24.3.1 GENERAL CONSIDERATIONS

The bunyaviruses demonstrate clear resemblance to the arenaviruses in their size, shape, general virion structure, and to some extent by genome organization (Figure 24.1) (for general comparison, see [8]). The Bunyaviridae family is, however, much more diverse than the Arenaviridae family and consists of 5 genera, *Hantavirus*, *Nairovirus*, *Orthobunyavirus*, *Phlebovirus*, and *Tospovirus*, represented in total by 97 species in accordance with the latest ICTV recommendations [6]. More than 700 species of the bunyaviruses including those unclassified to the aforementioned genera are listed in the NCBI taxonomy [7].

Similar to arenaviruses, most of the bunyaviruses infect rodents. Some of them including hantaviruses, Crimean–Congo hemorrhagic fever virus (CCHFV) from the *Nairovirus* genus, some subspecies of Bunyamwera virus from the *Orthobunyavirus* genus, and Rift Valley fever virus (RVFV) from the *Phlebovirus* genus can cause severe human hemorrhagic fever diseases (for a recent review, see [110]). With the exception of members of the *Hantavirus* genus that are transmitted by rodent feces, the bunyaviruses are vector-borne viruses and their transmission occurs via an arthropod vector: mosquitos, tick, or sand fly. A midge-transmitted orthobunyavirus isolated in 2011 and known under an informal name Schmallenberg virus is a persistent problem of animal husbandry in the northern hemisphere, since it causes congenital malformations and stillbirths in cattle, sheep, and goats (for a review, see [111]). Members of the *Tospovirus* genus infect plants, and tomato spotted wilt virus (TSWV) has become one of the limiting factors for vegetable crops such as tomato, pepper, and lettuce (for review, see [112]).

The bunyaviruses are enveloped spherical particles with a size from 80 to 120 nm in diameter, and their genome is comprised of three single-stranded RNA segments designated by size as L (6.8–12 kb), M (3.2–4.9 kb), and S (1–3 kb) RNAs,

which are encapsidated by the nucleocapsid (N) protein [10]. Regarding structural proteins, the M RNA encodes a single open reading frame that is processed into the viral glycoproteins Gn and Gc, which are embedded into the lipid bilayer envelope of the virion, but the S RNA segment encodes the protein N, which in addition to encapsidation of the genomic RNAs functions as an RNA chaperone [110]. Not all, but some of the bunyaviruses are using ambisense coding, like the arenaviruses. Unlike arenaviruses and other negative-sense RNA strand viruses, the bunyaviruses do not encode a matrix protein as a possible analogue of the protein Z that plays the central role in the assembly of arenaviral VLPs (see earlier text). It has been speculated that the function of matrix protein in the bunyaviruses is carried out by the cytoplasmic tail of glycoprotein Gn that interacts with the nucleocapsid protein and/or genomic RNA (for a recent review, see [113]).

24.3.2 PRODUCTION OF BUNYAVIRAL VLPS IN STANDARD EXPRESSION SYSTEMS

Early attempts to produce bunyaviral structural proteins N, Gn, and Gc in expression systems did not result in the generation of VLPs. First, it was observed by the vaccinia virus-driven expression of the glycoproteins Gn and Gc of RVFV [114] and Hantaan virus [115], as well as by the expression of the protein N of Hantaan virus [116] or CCHFV [117] in mammalian cells. Second, no VLPs were detected by the baculovirus-driven expression of the protein N of snowshoe hare bunyavirus [118] or CCHFV [119,120], glycoproteins Gn and Gc of RVFV [121,122], or N, Gn, and Gc of Hantaan virus [123,124] in insect cells. Third, no VLPs were detected after expression of the protein N of two hantaviruses [125], TSWV [126] or RVFV [127], in *E. coli*. Fourth, expression of the protein N of Puumala hantavirus in transgenic tobacco and potato plants [128] or homologous expression of the protein N of TSWV or other tospoviruses in plants [129] did not lead to any self-assembled products. Fifth, the same was true for the expression of the glycoprotein Gc of Hantaan virus in yeast *Pichia pastoris* [130] and of the protein N of Puumala [131], Dobrava [132], or other [133] hantavirus strains in *Saccharomyces cerevisiae*.

No data on production of VLPs were reported in the very recent efforts for the expression of the protein N of two emerging viruses: human severe fever with thrombocytopenia syndrome phlebovirus (SFTSV) in *E. coli* [134] and ruminant Schmallenberg orthobunyavirus in *E. coli* [135] or *S. cerevisiae* [136].

Expression of the bunyaviral nucleocapsid-like structures, similar to authentic ribonucleoproteins prepared by detergent disruption of virions, was reported for the first time by using expression of the protein N of Hantaan virus by the baculovirus-driven system in insect cells [137]. Furthermore, the first bunyaviral VLPs similar to the original Hantaan virions were observed during simultaneous expression of the protein N and envelope glycoproteins Gn and Gc by the vaccinia virus-driven system in mammalian cells, but not in insect cells [137]. The ability of the insect cells to produce RVFV

VLPs during baculovirus-driven expression of three structural proteins N, Gn, and Gc or two proteins N and Gc was clearly demonstrated by Roy and colleagues [138].

24.3.3 HIGH-RESOLUTION STRUCTURE

The presence of coiled-coil motifs and di- and trimerization with further multimerization was experimentally confirmed first for the protein N of hantavirus [139,140], TSWV [141], and RVFV [142]. Direct interaction of the protein N with the envelope glycoproteins Gn and Gc during intracellular trafficking was demonstrated, for example, in the case of TSWV [143,144], Uukuniemi phlebovirus [145], and hantaviruses [146,147].

X-ray crystallographic structure was determined first at 3.3 Å resolution for the Bunyamwera virus protein N expressed in *E. coli* [148]. The 1.6 Å crystal structure of the RVFV hexameric ring-shaped protein N forming a functional RNA binding site and expressed in *E. coli* was determined [149]. This structure explained the switch from an intra- to an intermolecular interaction mode of the N-terminal arm as a general principle that underlies multimerization and RNA encapsidation by bunyaviruses. In order to perform putative drug selection studies with the discovery of new antiviral reagents, the crystal structures of the protein N for other practically highly important bunyaviruses, CCHFV [150,151], SFTSV [152], Schmallenberg virus [153], and TSWV [154], were resolved recently. Notably, structural alignment of the CCHFV protein N with other bunyaviral N proteins revealed that the closest CCHFV relative is not other bunyavirus, but the Lassa arenavirus and suggested a potential revision of the current taxonomy of segmented negative-strand RNA viruses [155].

As to high-resolution structures of the bunyavirus envelope glycoproteins, the first one was the structure of the C-terminal cytoplasmic tail of the CCHFV glycoprotein Gn determined by NMR [156]. Then, the crystal structure of the RVFV glycoprotein Gc followed and demonstrated an important, but unanticipated evolutionary link between bunyavirus and flavivirus envelopes [157,158], namely, it was similar to that described for the alphavirus E1–E2 proteins.

The first complete bunyaviral 3D structures were resolved by electron cryotomography for Uukuniemi phlebovirus [159] and RVFV [160]. These structures suggested the presence of an icosahedral quasisymmetry with T = 12 triangulation for the glycoprotein protrusions, or spikes, in the case of the most regular particles. Intriguingly, this is the only known appearance of the T = 12 icosahedral symmetry in the viral world. Further electron cryomicroscopy and tomography studies showed that the structure of hantaviruses [161,162] demonstrates a unique paradigm and differs from the T = 12 symmetrical structure proposed for the phleboviruses. Recently, an electron cryotomography structure of the Bunyamwera orthobunyavirus was published [163].

A recent review on the bunyavirus structures [113] discussed the absence of matrix protein in bunyaviruses, which is present in arenaviruses as the protein Z that facilitates virus release from the host cells, acts as an anchor between the viral membrane and its genetic core, and plays the central role in the VLP preparation strategies (see earlier text). The authors suggest the potential role of the matrix function by the cytoplasmic tail of the bunyaviral glycoprotein Gn.

24.3.4 VLP APPLICATIONS: BUNYAVIRUSES AS OBJECTS AND SUBJECTS OF VLP NANOTECHNOLOGY

24.3.4.1 Bunyaviruses as VLP Objects: Potential VLP Carriers

24.3.4.1.1 Generation of Bunyaviral VLPs

A number of bunyaviral VLP structures have been prepared and described (Table 24.1). Authentic genome-free nucleocapsid-like particles formed by the protein N and similar to those generated for Hantaan virus and RVFV in mammalian and insect cells [137,138] were prepared for CCHFV by using a baculovirus–insect cell expression system [164]. Coexpression of the Hantaan protein N together with glycoproteins Gn and Gc in Chinese ovary hamster cells led to classic VLPs that stimulated highly specific antibody response against Hantaan virus N protein and glycoproteins, comparable to that induced by commercial inactivated bivalent hantaviruses vaccine, but a higher level of specific cellular response to N protein than that of the inactivated vaccine [165]. Efficient mammalian and insect cell expression systems were developed to produce high-quality RVFV VLPs consisting of the proteins N, Gn, and Gc [166]. Further, capability of the RVFV glycoproteins Gn and Gc alone, without the protein N, to self-assemble into VLPs was shown in a *Drosophila* insect cell expression system and used to generate an RVFV vaccine candidate that fully protected mice from a lethal challenge with RVFV [167]. Very recently, convincing data were presented that the expression of Andes and Puumala hantavirus Gn and Gc envelope glycoproteins in mammalian cells leads to their self-assembly into pleomorphic VLPs, which do not require the protein N and are released to cell supernatants [168]. Moreover, a Gc endodomain deletion mutant does not abrogate VLP formation [168].

The hypothesis that the glycoproteins Gn and Gc are the only viral components required for the formation of VLPs was described earlier on Uukuniemi phlebovirus in a so-called infectious VLP model [169]. It was shown in this model that the cytoplasmic tails of both glycoproteins Gn and Gc contain specific information necessary for efficient virus particle generation [170]. The same conclusions followed from the experiments on the role of the glycoproteins Gn and Gc from Bunyamwera orthobunyavirus [171,172].

In addition to the structural genes, the infectious Uukuniemi phlebovirus VLP model [169,170] included a minigenome encoding reporter genes within the VLPs and resembled more a gene therapy vector rather than the classic genome-free VLP approach. The same idea of a minigenome, or minireplicon according to authors' terminology, approach was realized in the case of RVFV, where transcriptionally active RVFV nucleocapsids were formed by expression of the recombinant polymerase L and the protein N from the minireplicon packaged into VLPs by additional expression of viral glycoproteins [173]. Such infectious VLPs resembled authentic RVFV virions, were able to infect new cells, and conferred protection against lethal RVFV challenge in mice after three subsequent immunizations [174]. Replacement of the protein N–encoding gene with the reporter gene within the RVFV minireplicon improved the safety and efficiency of the proposed vaccine against lethal RVFV challenge in mice [175].

The same minigenome, or minireplicon, idea was realized in the so-called nonspreading RVFV vaccine produced in a mammalian cell line [176]. The nonspreading vaccine that was improved by the expression of the glycoprotein Gn from the encapsidated minigenome demonstrated sterile protection against lethal RVFV challenge in lambs after a single intramuscular vaccination [177].

24.3.4.1.2 Generation of Chimeric Bunyaviral VLPs

Replacement of the N-terminal half of the glycoprotein Gc ectodomain of Bunyamwera virus with the sequences of fluorescent proteins, either eGFP or mCherry, led to the generation of chimeric infectious viruses that allowed visualization of different stages of the infection cycle [178]. Fusion of GFP to the Puumala hantavirus protein N expressed in *S. cerevisiae* did not result in the nucleocapsid-like particles [179]. The same was true for the expression of the TSWV protein N fusions with fluorescent proteins in mammalian cells [180] or plants [181].

24.3.4.1.3 Replacement of Bunyaviral Envelopes, or Pseudotyped Bunyaviruses

An example of this approach is represented by preparation of a recombinant virus from La Crosse and Jamestown Canyon viruses, both belonging to the California encephalitis virus serogroup of orthobunyaviruses [182]. As a result, the La Crosse virus expressing the attachment/fusion glycoproteins of Jamestown virus demonstrated protection in mice against lethal challenge with both viruses and against challenge with Jamestown Canyon virus in rhesus monkeys [182].

24.3.4.2 Bunyaviruses as VLP Subjects: Pseudotyping and a Source of Epitopes

24.3.4.2.1 Pseudotyping of Non-Bunyaviral Models by Bunyaviral GPs

First, Moloney MLV vector was used for the generation of particles pseudotyped by either La Crosse orthobunyavirus or Hantaan virus [183]. Second, the VSV pseudotype bearing hantavirus envelope glycoproteins was produced and used in a neutralization test as a substitute for native hantaviruses [184,185] and as an alternative vaccine in mice [186]. Third, the lentiviral gene therapy vectors were pseudotyped with hantavirus envelopes in order to improve the transduction efficiency into vascular smooth muscle and endothelial cells [187].

The recombinant pseudotyped lentivirus containing the Hantaan virus glycoproteins was used also as a potential vaccine candidate [188].

A specific approach similar to the pseudotyping concept was used by the generation of mosaic, or chimeric according to authors' terminology, RVFV VLPs that contained the gag protein of Moloney MLV together with the RVFV proteins N, Gn, and Gc and demonstrated efficient protection of mice and rats against lethal RVFV challenge [189].

24.3.4.2.2 Bunyaviruses as a Source of Epitopes

After determination of the N-terminal sequence of the Puumala hantavirus as a major antigenic domain of the whole virus [190,191], several studies were performed in order to expose the N-terminal hantavirus protein N domains on widely used HBc and HaPyV VLPs. The detailed description of these attempts that have been initiated by Rainer G. Ulrich in 1997 [192] and are active until now [193] is presented in Chapter 9. Until now, other bunyavirus proteins and/or their domains did not attract much attention of VLP researchers as a source of epitopes.

24.4 MONONEGAVIRALES ORDER

24.4.1 INTRODUCTORY REMARKS

The Mononegavirales order includes five families, Bornaviridae, Filoviridae, Nyamiviridae, Rhabdoviridae, and Paramyxoviridae, consisting of two subfamilies: Paramyxovirinae and Pneumovirinae. The most crucial structural features of the order members are the following: (1) the single-stranded RNA genome of negative polarity is not segmented, (2) the gene order 3'-UTR–core protein genes–envelope protein genes–RNA-dependent RNA polymerase gene–5'-UTR is characteristic, and (3) infectious helical ribonucleocapsids are enveloped into virions (Figure 24.1). The virions can be very similar to these of arenaviruses and bunyaviruses in the case of spherical paramyxoviruses, pneumoviruses, bornaviruses, and nyamiviruses, more distinct in the case of bullet-shaped rhabdoviruses, and markedly distinct in the case of long filamentous filoviruses. We shall follow this order of ribonucleocapsid shapes in the description of the VLP applications in the Mononegavirales order. The main emphasis will be made on paramyxoviruses and pneumoviruses, for which VLPs have been made and characterized in great detail. Two other families covering spherical viruses contain a small number of representatives with only a few studies: the Bornaviridae family is represented by a single genus with a sole Borna disease virus (BDV) species, while the Nyamiviridae family of three species is the youngest family in the ICTV list and was recognized as an independent family only in 2014 [6]. The bacilliform rhabdoviruses represent the largest family of the Mononegavirales order and are comprised of 11 genera. However, their impact to the VLP concept development was rather limited because of complexity of the rhabdoviral structure to manipulations.

Without any doubts, the filoviruses are the most urgent topic among members of the Mononegavirales order, but we will not describe this in detail, since comprehensive review of filovirus VLPs is presented in Chapter 9.

24.4.2   Paramyxovirinae Subfamily of the Paramyxoviridae Family: Paramyxoviruses

24.4.2.1   General Considerations

The paramyxoviruses represent enveloped spherical pleomorphic virions of about 150 nm in diameter (Figure 24.1). The structural proteins are the NP N, the matrix protein M, and the envelope proteins: fusion protein F and hemagglutinin–neuraminidase HN. With regard to the public health care, the most important and dangerous members of the paramyxoviruses are human parainfluenza viruses (PIVs) belonging to the *Respirovirus* and *Rubulavirus* genera, Nipah virus of the *Henipavirus* genus, measles virus of the *Morbillivirus* genus, and mumps virus of the *Rubulavirus* genus. The paramyxoviruses remain a serious burden of animal and poultry husbandry: Newcastle disease virus (NDV) of the *Avulavirus* genus causes highly contagious disease of birds, which is transmissible to humans, peste-des-petits-ruminants virus (PPRV) of the *Morbillivirus* genus causes disease affecting goats and sheep with up to 80% mortality rate in acute cases, and Atlantic salmon paramyxovirus of the *Aquaparamyxovirus* genus is one of the causes of proliferative gill inflammation of salmons that leads to considerable losses in fishery. Sendai virus of the *Respirovirus* genus is responsible for a highly transmissible respiratory tract infection in mice and occasionally in pigs. Importantly, the rinderpest, or cattle plague, caused by rinderpest virus (RPV) of the *Morbillivirus* genus was officially proclaimed by the UN Food and Agriculture Organization as fully eradicated, making it the second eliminated disease after smallpox in world history [194].

24.4.2.2   Production of Paramyxoviral VLPs in Standard Expression Systems

Chronologically, the first early report described the expression of the NDV HN protein in baculovirus-driven insect cell model and electron microscopic observation of NDV-like particles [195]. Production of VLPs was confirmed in many expression systems very soon for the protein N of paramyxoviruses (Table 24.1). First, production of nucleocapsid-like structures was shown by the vaccinia virus-driven expression of the measles virus protein N in avian and mammalian cells [196]. Second, high-level production of the nucleocapsid-like, or so-called herringbone-like, particles of approximately 20–22 nm in diameter and variable length from 50 to 100 nm was detected after baculovirus-driven expression in insect cells for the protein N of measles virus [197], Nipah virus [198], and dolphin morbillivirus [199]. Third, the same herringbone-like particles were formed also by the expression in *E. coli* in the case of the protein N of measles virus [200], NDV [201], RPV [202], PPRV [202],

Nipah virus [203], and mumps virus [204]. Fourth, the most impressing results on the expression of the protein N in the form of the herringbone-like structures were achieved in yeast. So, both *S. cerevisiae* and *P. pastoris* were used for the expression of the protein N in the case of mumps [205,206] and measles [207] viruses. The proteins N of Sendai virus [208], Hendra and Nipah henipaviruses [209], human PIVs 1 and 3 [210], and Tioman virus [211] were also expressed in yeast.

The first evidence on the determinative role of the protein M in the assembly of virus-like and not of the nucleocapsid-like structures stems from the experiments in mammalian cells on Sendai virus mutants [212] and resulted in an orderly mechanism of the protein M-dependent self-assembly of paramyxoviruses [213]. Expression of the Nipah virus protein M was shown to be sufficient for the production of VLPs in mammalian cells [214] and *E. coli* [215]. Recently, the protein M of NDV was expressed in its native form in *S. cerevisiae* and also in *E. coli*, after refolding in vitro [216].

24.4.2.3   High-Resolution Structure

Electron cryotomography at low resolution has revealed general features of the virions of Sendai virus [217], measles virus [218], and NDV [219]. In the latter case, the electron cryotomography was combined with x-ray crystallography and clearly revealed dimers of the matrix protein M that assemble into pseudotetrameric arrays generating the membrane curvature necessary for virus self-assembly, anchoring the glycoproteins in the gaps between the matrix proteins, and associating the helical nucleocapsids, but about 90% of virions lack matrix arrays [219].

The helical structure of the nucleocapsids formed by the protein N was revealed for measles virus at relatively low resolution, not higher than 12 Å [220,221]. Recently, NMR was employed to improve resolution of the nucleocapsid structure [222]. The high-resolution crystallographic structure of the fusion protein F was obtained for NDV [223], measles virus [224], and Nipa and Hendra viruses [225]. The crystal structure of the hemagglutinin–neuraminidase HN and its stalk region was resolved for NDV [226,227], as well as for the HN analogues in other paramyxoviruses: the hemagglutinin H of measles virus [228] and its complex with the CD46 receptor [229], the attachment glycoprotein G of Nipah and Hendra viruses [230–232]. Recently, the crystal structure of the Nipah virus multimeric phosphoprotein P that tethers the viral polymerase to the nucleocapsid was resolved [233].

24.4.2.4   VLP Applications: Paramyxoviruses as Objects and Subjects of VLP Nanotechnology

24.4.2.4.1   Paramyxoviruses as VLP Objects: Potential VLP Carriers

24.4.2.4.1.1   Generation of VLPs Coexpression of the proteins M and N of human PIV type 1 in mammalian cells showed that the M protein alone can induce the budding of VLPs and that the protein N can assemble into intracellular nucleocapsid-like structures [234]. Furthermore, coexpression of the proteins M, N, F, and HN of Sendai virus in mammalian cells resulted in the generation of VLPs that had morphology and density similar to those of authentic virus particles and allowed investigation of the roles played by individual proteins in the self-assembly and budding [235,236]. Similarly, coexpression of all possible combinations of N, M, F, and HN of NDV was performed in avian cells and resulted in the release of VLPs with densities and efficiencies of release similar to those of authentic virions [237]. Expression of the protein M alone, but not N, F, or HN protein individually, resulted in the efficient VLP release, while expression of all different combinations of proteins in the absence of the protein M did not result in VLP release [237]. The central role of the protein M in the self-assembly of N, F, and G (analogue of HN) was demonstrated also in the vaccinia virus-driven expression of Nipah virus proteins in mammalian cells [238]. The VLPs were detected after coexpression of the proteins M and F of measles virus [239] and the proteins M, N, and F of mumps virus [240]. The NDV VLPs directed the attachment to cell surfaces and induced immune response in mice that was comparable to the response to equivalent amounts of inactivated NDV vaccine virus [241]. The same vaccine potential was confirmed also in the case of Nipah VLPs composed of the proteins M, G, and F [242]. Recently, production of PPRV VLPs was achieved by baculovirus-driven coexpression of the proteins M and N [243]; M, N, and H [244]; and M, H, and F [245] in insect cells.

24.4.2.4.1.2 Chimeric VLPs as Epitope Carriers: Insertion of Foreign Proteins

The only examples of the insertion of epitopes into structural paramyxoviral monomers are related to the protein N as a scaffold of the nucleocapsid-like, or ringlike, or herringbone-like, particles. First, chimeric N proteins of NDV were constructed in which the antigenic regions of the NDV HN and F proteins, myc epitope, and a hexa-His tag were linked to the C-terminus of the N monomer, expressed in *E. coli* and self-assembled into ringlike structures [246]. Second, six fragments of the protein VP1 of enterovirus EV71 were fused to the C-terminal end of the full-length and truncated protein N of NDV with further expression of chimeric proteins in *E. coli* [247]. The chimeric proteins self-assembled into intact ringlike structures and induced a strong immune response against the complete EV71 VP1 in rabbits [247], mice [248], and hamsters [249]. Third, a fusion of the CS protein from *Plasmodium berghei* to the protein N of measles virus was expressed in *P. pastoris* [250]. The chimeric protein generated highly multimeric, but heterogenic NP bearing the CS protein on the surface that ensured significant reduction of parasitemia after immunization of mice with whole heat-inactivated yeast cells and following challenge with a high dose of parasites [250].

24.4.2.4.1.3 Chimeric VLPs Carrying Different Glycoproteins

Paramyxoviruses possess long and successful history of generation of viral vectors for the expression of foreign proteins, including foreign glycoproteins and/or chimeric glycoproteins consisting of foreign glycoprotein fragments inserted into authentic glycoproteins, and acting as vaccine and gene therapy candidates (for recent review, see [251,252]), as well as oncolytic viruses [253–255]. Chimeric NDV and PIV3 [256] or NDV and avian paramyxovirus [257] glycoproteins HN and mutually exchanged RPV and PPRV glycoproteins F and H [258] could be mentioned among the earliest examples of the vaccine and virus delivery approach based on the construction of *mixed* envelopes. Further, human PIV-vectored Ebola virus (EBOV) vaccine was constructed by the introduction of the EBOV glycoprotein GP [259] and demonstrated high protective efficiency in guinea pigs by intranasal inoculation [260]. NDV virus expressing the hemagglutinin H5 protected chickens against avian H5N1 influenza, as well as NDV challenge [261]. Very recently, promising bivalent human PIV5-based RSV [262] and chicken NDV-based infectious laryngotracheitis virus [263] vaccines were described.

Authentic VLP approach was realized by the construction of a chimeric RSV vaccine candidate based on the NDV VLPs formed by the proteins M and N and carrying a chimeric protein that contained the cytoplasmic and transmembrane domains of the NDV HN protein and the ectodomain of the human RSV protein G [264]. Further, the NDV-based RSV vaccine candidate was improved by addition of the ectodomains of both RSV F and G proteins fused to the transmembrane and cytoplasmic domains of NDV F and HN proteins, respectively [265]. The vaccine candidate was purified from avian cells and demonstrated complete protection of mice from RSV replication in lungs [265] and efficient stimulation of long-lived RSV-specific, T-cell-dependent secretion of neutralizing antibodies and RSV-specific memory response [266]. Next, an NDV-based influenza vaccine candidate was constructed by the expression of VLPs composed of the proteins M1 and HA from avian influenza virus (AIV) and a chimeric protein containing the cytoplasmic and transmembrane domains of AIV neuraminidase protein NA and the ectodomain of the NDV protein HN [267]. A single immunization of chickens with the chimeric VLPs induced both AIV H5– and NDV-specific antibodies and conferred complete protection against NDV challenge [267]. Recently, NDV VLPs containing the NDV protein F along with influenza virus matrix protein M1 were produced in insect cells and protected chickens against lethal NDV challenge [268].

24.4.2.4.2 Paramyxoviruses as VLP Subjects: Pseudotyping and a Source of Epitopes

24.4.2.4.2.1 Pseudotyping

The paramyxoviruses demonstrate numerous examples of pseudotyping on traditional pseudotype carriers: First, on Moloney MLV with the glycoprotein F of Sendai virus [269], the proteins H and F of measles virus [270]. Second, lentivirus vectors were pseudo-typed by the glycoproteins of Sendai virus [271], human PIV3 [272], measles virus [273], Nipah and Hendra henipaviruses [274], and tupaia paramyxovirus [275]. Third, pseudotyping of VSV vectors was performed by the glycoproteins of Nipah virus [276] and an advanced high-throughput serum neutralization assay for Nipah virus antibodies was constructed [277]. Fourth, the protein H of RPV was incorporated into extracellular baculovirus during expression in insect cells and ensured induction of neutralizing antibody response in cattle [278].

24.4.2.4.2.2 Paramyxoviruses as a Source of Epitopes   Five variants of the CTL epitope 325–332 from the protein N of Sendai virus were cloned within the Ty VLPs encoded by yeast retrotransposon Ty1 and used in parallel with other CTL epitopes as a model for the induction of specific CTL response by chimeric VLPs [279].

A candidate NDV vaccine on the basis of infectious cucumber mosaic virus (CMV) was generated by insertion of neutralizing B-cell epitopes: 17 aa long epitope of the protein F, 8 aa long epitope of the protein HN, or duplicated 8 aa HN epitope into the internal betaH–betaI loop (motif 5) within the coat protein of CMV and propagation of the chimeric viruses in plants [280]. Further, the appropriate genes were placed under the transcriptional control of the duplicated subgenomic coat protein promoter of a potato virus X (PVX)-based vector, and chimeric CMV-based VLPs carrying NDV epitopes were produced in the form of authentic VLPs [281]. The peptides corresponding to the NDV epitopes were cross-linked to chimeric maize rayado fino virus (MRFV) VLPs by chemical coupling to mutationally introduced cysteine anchors on the surface of MRFV VLPs produced in tobacco plants after expression by the PVX-based vector [282].

A stable infectious chimeric porcine circovirus type 2 (PCV2), which displayed on its surface a 14 aa V5 epitope tag from simian PIV type 5 that was added to the C-terminus of the PCV2 capsid protein, was constructed and demonstrated capability of the novel PCV2 platform to accept foreign insertions [283].

Finally, the long C-terminal domain 401–532 of the protein N from Nipah virus was inserted at the N-terminus and into the major immunodominant loop between aa residues 79 and 80 of HBc VLPs [284]. Unexpectedly, the long insertion was compatible with VLP formation at least partially, and some VLPs were detected by electron microscopy in both cases [284].

24.4.3 Pneumovirinae Subfamily of the Paramyxoviridae Family: Pneumoviruses

As the paramyxoviruses, the pneumoviruses possess enveloped spherical pleomorphic virions of about 150 nm in diameter. The structural proteins are the NP N, the matrix protein M, and the envelope proteins: fusion protein F and glycoprotein G. With regard to the public health care, human metapneumovirus (hMPV) of the *Metapneumovirus* genus and human respiratory syncytial virus (hRSV) of the *Pneumovirus* genus are the most noticeable infectious agent from the pneumoviruses. The hRSV and hMPV are among the leading causes of childhood hospitalization and a major health burden worldwide.

The protein N of hRSV formed typical herringbone-like structures by expression in insect cells [285], as well as the protein N of hMPV in yeast *S. cerevisiae* [286]. Expression of the hRSV protein N in *E. coli* led to formation of ringlike structures [287] that protected mice against hRSV challenge and provided cross protective immunity in calves against a viral challenge with bovine respiratory syncytial virus (bRSV) [288]. Attempts to prepare similar ringlike structures by expression of the protein N from bRSV were unsuccessful [288]. Further, the safety and efficacy of the mucosal vaccine based on the *E. coli*–derived hRSV protein N rings was demonstrated in mice neonates [289]. The *E. coli*–derived ringlike structures allowed generation of the first chimeric pneumoviral particles by the chimeric hRSV protein N carrying GFP added to its N-terminus [287].

Expression of the hRSV protein F in insect cells led to the production of the 40 nm nanoparticles composed of multiple hRSV F oligomers arranged in the form of rosettes that were able to induce neutralizing antibodies [290]. The Phase I trial of such hRSV F nanoparticle vaccine candidate in healthy adults demonstrated good toleration and induction of hRSV-neutralizing antibodies [291]. Another protein F–derived vaccine candidate was based on the trimers of a truncated secreted version of the hRSV protein F that was expressed in mammalian cells and demonstrated protective efficacy in mice after formulation with the CpG adjuvant [292].

The crystal structure of the hRSV nucleocapsid-like protein–RNA complex was determined at about 3 Å resolution [293,294], and general hRSV architecture, which indicated important structural differences between the Pneumovirinae and Paramyxovirinae subfamilies, was elucidated by electron cryotomography [295]. Recently, the crystal structure of the protein M of hMPV in its native dimeric state was resolved at 2.8 Å resolution [296].

The fine elucidation of functional hRSV self-assembly revealed absolute requirement for the protein M [297]. The coexpression of the hMPV proteins M, G, and F in mammalian cells led to formation of VLPs with a similar morphology to the filamentous virus morphology that was observed in hMPV-infected cells; unexpectedly, the protein G only was able to form VLPs in the absence of the other virus proteins [298]. Contrary to the idea of a dominant role of the protein G in the pneumoviral self-assembly, the proteins M and F were found sufficient to form hMPV VLPs by expression in mammalian cells [299]. Two doses of such VLPs conferred complete protection against hMPV replication in the lungs of mice [299].

Additional advance in the pneumoviral VLP approaches was achieved in insect cells by efficient baculovirus vector-driven coexpression of the hRSV proteins F and G together with the influenza matrix protein M1 [300]. Intramuscular vaccination of mice with such VLPs provided effective protection against hRSV infection [300], but the vaccine candidate was improved by addition of the protein F–encoding DNA [301,302]. Direct comparison of the combined VLP and DNA vaccines with the formalin-inactivated RSV vaccine in mice revealed clear advantages of the VLP vaccine [303].

As the paramyxoviruses, the pneumoviruses were used in a long line of the mixed infectious virus constructions with the envelope exchange. For example, the hRSV envelope proteins G and F were expressed on the surface of such chimeric viruses based on recombinant bovine/human PIV type 3 [304], Sendai virus [305–307], NDV [308], measles virus [309], and influenza virus [310–312].

Pseudotyping with the hRSV envelope proteins G and F was performed on VSV [313] and with the hMPV proteins G and F on MLV [314]. The major immunogenic domains of the hRSV proteins G and F were displayed on the surface of *Lactococcus lactis* by fusion to the appropriate bacterial anchors [315] and paved a way to a protective and safe vaccine candidate [316].

As a source of immunological epitopes, the hRSV provided first the protective B-cell epitope 173–187 from the glycoprotein G that was displayed on the surface of recombinant phage *fd* by the fusion to the coat protein pIII and provided the complete resistance of mice to hRSV infection [317]. Next, the same hRSV epitope was fused to the alfalfa mosaic virus (AlMV) coat protein, expressed in tobacco plants, purified as a part of recombinant alfalfa mosaic virions, and conferred protection of mice against challenge with hRSV [318].

The ectodomains of hRSV F and G proteins were fused to the transmembrane and cytoplasmic domains of NDV F and HN proteins, respectively, and provided complete protection of mice from hRSV replication in lungs [265,266] (see Section 24.4.2.4.1.3).

The CTL epitope from the hRSV protein M2 located at aa positions 82–90 was inserted into HBsAg and provided partial protection of mice against hRSV infection after immunization in the form of a DNA vaccine [319]. Another CTL epitope from the hRSV protein F located at aa positions 85–93 was inserted in the NA stalk of influenza virus and provided a significant reduction in the lung viral load upon a subsequent challenge with hRSV [320].

The neutralizing epitope of hRSV from the glycoprotein F played an important role for providing the proof of principle data for epitope-focused vaccine design [321]. This epitope was selected since it is targeted by the licensed, prophylactic neutralizing antibody palivizumab, also known as Synagis, or pali, and an affinity-matured variant, motavizumab, or mota (for references, see [321]).

Finally, the hRSV protein F played a pioneering role in the development of modern bionanomaterials as putative vaccines by display on a gold nanorod [322].

24.4.4 BORNAVIRUSES AND NYAMIVIRUSES

BDV, the only representative of the Bornaviridae family, possesses spherical virions of 70–130 nm in diameter [10]. BDV is the only animal RNA virus that establishes a persistent infection in the host cell nucleus. More, there are DNA sequences derived from the mRNAs of ancient bornaviruses in the genomes of vertebrates, including humans, designated as endogenous Borna-like elements, but BDV does not integrate into host's genome (for references, see [323]). However, the way of the bornavirus participation in the evolution of host genomes remains unclear. BDV is a neurotropic agent and infects a wide variety of mammalian species from rodents to birds and readily establishes a long-lasting, persistent noncytolytic infection in brain cells. BDV is involved in the pathogenesis of some human psychiatric disorders.

The Nyamiviridae family appeared in the ICTV list in 2014 and remains therefore one of the youngest accepted taxonomy units [324]. Two representatives of the *Nyavirus* genus of the Nyamiviridae family, Nyamanini virus (NYMV) and Midway virus, are unclassified tick-borne agents that infect land birds and seabirds, respectively [324]. A third agent, soybean cyst nematode virus 1 (previously named soybean cyst nematode nyavirus), was recently found to be a member of the family, but of the unassigned genus [324]. It is not surprising that the contribution of two families in question to the VLP methodology and applications is still rather minimal.

Expression of the BDV protein M in *E. coli* resulted in the production of tetramers with the tendency to assemble into high-molecular-mass lattice-like complexes [325]. The crystal structure of the BDV protein M tetramer exhibited structural similarity to the N-terminal domain of the EBOV protein M (VP40) [326]. More, it was found recently that two distinct proteins of NYMV serve a matrix protein function as previously described for members of the Filoviridae family [327]. The BDV protein N synthesized in *E. coli* revealed a planar tetrameric structure by crystallographic analysis at 1.76 Å resolution [328].

Pseudotyping by the BDV protein G was performed on the replication-competent VSV [329], but no epitopes from bornaviruses or nyamiviruses were used for the expression on VLP and/or non-VLP carriers yet.

The use of a reverse genetic approach allowed identifying the BDV proteins required for packaging of BDV RNA analogues into infectious VLPs and establish the BDV model as a vector for the cloning of foreign genes [330], first of all, for the specific expression in the central nervous system [331].

24.4.5 RHABDOVIRUSES: RABIES VIRUS, VESICULAR STOMATITIS VIRUS, IHNV, AND OTHERS

The rhabdoviruses are bullet shaped or bacilliform, usually 180 nm long and 75 nm wide [10]. The structural proteins of rhabdoviruses are the NP N, the matrix protein M, and the glycoprotein G. According to the latest issue of the official taxonomical ICTV documents [6], the broad Rhabdoviridae family involves 11 genera (the 12th genus is unassigned), with

2 new genera, *Sprivivirus* and *Tupavirus*, accepted in 2013. The total ICTV number of rhabdoviral species is 71. With regard to the public health care, rabies virus of the *Lyssavirus* genus deserves special consideration. VSV of the *Vesiculovirus* genus is one of the classical objects and subjects of modern virology, immunology, and gene therapy. Infectious hematopoietic necrosis virus (IHNV) of the *Novirhabdovirus* genus remains a global burden of fishery, since it causes the disease known as infectious hematopoietic necrosis in salmonid fish such as trout and salmon.

The *loosely coiled strands of varying length* formed by the protein N of rabies virus have been detected by electron microscopy at the very beginning of the genetic engineering era [332]. Although not always observed by electron microscopy, the structures similar to herringbone-like particles described earlier for other negative-stranded viruses might have been produced by the baculovirus-driven expression of the protein N from rabies virus [333–335] and VSV [336] in insect cells. Further evaluation confirmed the presence of the ringlike structures in the insect cell–derived preparations of the protein N from rabies virus [337,338]. Such preparations demonstrated immunogenicity [339] and conferred protective immunity in mice against challenge with lethal doses of rabies virus [340].

Expression of the rhabdoviral protein N in *E. coli* did not reveal any nucleocapsid-like structures in the case of IHNV [341], VSV [342], and rabies virus [343]. However, coexpression of the VSV protein N and phosphoprotein P resulted in the formation of the native N–P complex [344].

The typical nucleocapsid-like particles were obtained by the expression of the protein N from rabies virus and two other lyssaviruses: European bat lyssavirus 1 and European bat lyssavirus 2 in *S. cerevisiae* [345]. High-level expression of the protein N from rabies virus was achieved in transgenic tomato and tobacco plants and partial protection of mice with plant-derived material against viral challenge was presented, however, without any indications about the structural status of the produced protein N [346].

Similarly to the protein N, the rhabdoviral glycoprotein G has been expressed in mammalian cells very early in the case of rabies [347,348] and Chandipura [349] viruses. Next, the glycoprotein G of rabies virus [350], IHNV [351], and bovine ephemeral fever virus [352] was expressed in insect cells, but no data on its self-assembly were presented. The insect cell–derived glycoprotein G of rabies virus conferred protection against a lethal challenge with rabies virus to mice [350] and to raccoons by oral field vaccination [353], but the IHNV glycoprotein G provided very limited protection in rainbow trout [354]. Noticeably, according to two investigations, expression in mammalian cells of the glycoprotein G from VSV [355] and rabies virus [356] may lead to the appearance of some sort of virus-like particles.

Expression of the glycoprotein G of INHV [357] and rabies virus [358] in *E. coli* did not result in any self-assembled protein G products. The same was true for the expression of the glycoprotein G of rabies virus in yeast [359,360] and plants: tomatoes [361] and carrot [362].

The pivotal role of the matrix protein M in the rhabdoviral self-assembly was identified from the very beginning in the case of VSV [363] and rabies virus [364]. Bacterial expression of the protein M from three lyssaviruses, Lagos bat, Mokola, and Thailand dog, allowed high-quality purification for further crystallization and elucidation of their 3D structure [365].

The crystal structure was determined for the protein N from rabies virus [366,367] and VSV [368,369], for the glycoprotein G from VSV [370] and Chandipura virus [371], for the protein M from VSV [372,373] and lyssaviruses mentioned earlier [365,373]. Remarkably, the structures of the protein M from VSV and Lagos bat lyssavirus both share a common fold despite sharing no identifiable sequence homology [373].

Finally, an electron cryomicroscopy model of VSV was constructed and showed that each virion contains two nested, left-handed helices: an outer helix of matrix protein M and an inner helix of NP N and RNA [374].

The use of rhabdoviral structural proteins as putative carriers of foreign sequences includes fusion of GFP to the protein N that was efficiently expressed in mammalian cells and incorporated into infectious rabies virions [375]. Moreover, the fusion protein induced a strong humoral immune response against GFP in mice [375].

The rhabdoviruses are widely used objects and subjects of the pseudotyping technology, first of all VSV, which is one of the most common pseudotype objects functioning as a support for foreign envelope proteins, as well as subjects providing the protein G for the envelopment of retroviral and lentiviral cores and participating in many efficient gene therapy vectors. Numerous examples of VSV pseudotyping are given in the present chapter. In the absence of a recent specialized review on the VSV pseudotype potential, we would recommend a protocol book chapter [376] and restrict our consideration to some very recent examples of the rhabdoviral contribution to the pseudotyping approaches: retargeting of VSV-pseudotyped lentiviral vectors by polymer nanomaterials [377] and targeting of rabies virus glycoprotein-pseudotyped lentiviral vectors to assess nerve recovery in nerve injury models [378] and pseudotyping of VSV by rabies glycoprotein to study transport among neurons in vivo [379] and glycoproteins of highly pathogenic AIVs [380].

The first rhabdoviral sequence exposed on a VLP carrier was an epitope from the rabies virus glycoprotein G added to the surface of M13 bacteriophage 30 years ago [381]. Next, the immunogenic epitope from the rabies virus glycoprotein G was inserted into coat protein of tobacco mosaic virus (TMV) and displayed on the surface of infectious TMV virions [382]. A chimeric peptide carrying rabies virus epitopes from protein N (aa 404–418) and glycoprotein G (aa 253–275) was inserted into coat protein of AlMV, expressed as a chimeric virus in tobacco plants, and found protective against viral challenge in mice [383]. Moreover, human volunteers were involved and three of five volunteers who had previously been immunized against rabies virus with a conventional vaccine specifically responded against the peptide antigen after ingesting spinach leaves infected with the recombinant virus, but five of nine nonimmune individuals demonstrated significant antibody responses to either rabies virus or AlMV [383].

It is noteworthy that rhabdoviruses can be used as viral vectors for the delivery of VLPs, for example, of human norovirus VLPs [384]. A detailed protocol for the construction of rabies viral vectors is published recently [385].

24.4.6 FILOVIRUSES: EBOLAVIRUS AND MARBURGVIRUS

The Filoviridae family [6] consists of three genera: *Ebolavirus* with five species, *Marburgvirus* (MARV) with one species *Marburg marburgvirus*, and a novel, since 2014, genus *Cuevavirus* with one species *Lloviu cuevavirus* [386]. The work on the Filoviridae taxonomy is going on very actively [387]. Filoviruses are filamentous and reach the length of 790 nm in the case of *Marburgvirus* and 970 nm in the case of *Ebolavirus*, with about 80 nm in diameter (Figure 24.1) [10]. The filoviral structural proteins are the NP, the glycoprotein G1,2 (GP), the matrix protein M (VP40), and the protein VP24 that is not homologous to genes of other *Mononegavirales* representatives and participates in the nucleocapsid assembly, together with the NP and VP35 protein, a cofactor of the viral RNA polymerase complex. Regarding the filovirus evolutionary history, the evidence of their origin assuming EBOV and cuevavirus divergence from marburgviruses since the early Miocene was provided very recently [388]. The recent largest outbreak of EBOV disease spreading through Guinea, Liberia, Sierra Leone, and Nigeria in 2014 discovered novel virus variants [389] and stimulated search for novel vaccine candidates against filoviruses [390]. Notably, like bornaviruses, the filovirus genome fragments are found integrated into vertebrate genomes [391].

The first success in the construction of antifilovirus vaccines arose from the expression of MARV genes encoding NP, GP, VP40, VP35, or VP24 by RNA replicon based upon VEE virus, when NP or GP immunization induced protection of guinea pigs against viral challenge, while GP and simultaneous NP and GP immunization ensured full protection of nonhuman primates (cynomolgus macaques) [392]. Next, VEE virus vector-driven expression of EBOV genes encoding GP or both NP and GP resulted in full protection of mice and guinea pigs against lethal EBOV infection, while NP immunization protected mice, but not guinea pigs [393]. As stated before in Section 24.2.2, dual expression of bivalent alphavirus particles that carried glycoprotein genes of both EBOV and LASVs protected guinea pigs against challenges with Ebola and LASVs [26]. However, further evaluation led to conclusions that the disease observed in primates differed from that in rodents, suggesting that rodent models of EBOV may not predict the efficacy of candidate vaccines in primates and that protection of primates may require different mechanisms [394].

In parallel, the development of antifiloviral strategy focused on the VLP approach (Table 24.1). The EBOV and MARV VLPs composed of the glycoprotein GP and matrix protein VP40 and resembling the distinctively filamentous infectious virions were generated by the expression in mammalian cells [395]. Vaccination of rodents with EBOV [396] or MARV [397] VLPs conferred induction of the immune response and protected mice against EBOV [396] or guinea pigs against MARV [398] lethal challenge. Further, EBOV and MARV VLPs were shown to be more effective stimulators of human dendritic cells than the respective viruses [399]. A pan-filovirus hybrid VLP vaccine candidate on the basis of EBOV and MARV proteins GP and VP40 demonstrated that only GP was required and sufficient to protect against a homologous filovirus challenge [400]. Moreover, the EBOV and MARV VLPs produced in mammalian cells fully protected nonhuman primates (cynomolgus macaques) from the lethal viral challenge [401]. Finally, it was concluded that the expression of the matrix protein VP40 alone is sufficient to the VLP production, but addition of other filovirus proteins increases the efficiency of VLP production in mammalian cells and results in the case of the coexpression of GP and VP40 in the promising vaccine candidate (for full list of references, see review [402]). Very recently, VLPs consisting of NP, GP, and NP40 of lloviu cuevavirus were prepared [403]. The same GP and VP40-based VLPs as in the mammalian cells were produced by the baculovirus-driven expression in insect cells [404]. Next, the EBOV and MARV VLPs similar to those produced in mammalian cells were generated by the baculovirus-driven coexpression of the proteins GP, VP40, and NP, and their protective efficiency against a lethal viral challenge in mice was demonstrated [405]. The insect cell–produced EBOV GP–VP40 VLPs conferred full protection, but showed strong dose dependence by immunization of mice [406]. Very recently, efficiency of the insect cell–produced EBOV VLP vaccine [405] was confirmed by the first trial on captive chimpanzees [407]. Large volume of biophysical data on the conformational stability of insect cell–produced EBOV and MARV VLPs characterized by various spectroscopic techniques over a wide pH and temperature range was generated, in order to select optimized solution conditions for further vaccine formulation and long-term storage [408]. No attempts to produce the filoviral VLPs in bacteria or plants are known.

Regarding high-resolution elucidation of spatial structure of filoviral proteins, the crystal structures of the EBOV GP ectodomain [409,410], VP40 [411–413], GP [414], VP35 [415,416], and the MARV GP ectodomain [417] and VP24 [418] were published.

The crystal structure of the EBOV VP40 revealed its topological distinction from all other known viral matrix proteins, consisting of two domains with unique folds connected by a flexible linker [412]. Later, the crystal structure of a disk-shaped octameric form of VP40 comprised of the four antiparallel homodimers of the N-terminal domain was shown [413].

Successful attempts to view the complete picture of filoviral components and their morphogenesis [419–421], as well as the budding process [422], were undertaken by electron cryotomography.

A system of so-called infectious VLPs that carry a minigenome consisting of the negative-sense copy of the GFP [423] or luciferase [424] gene was constructed for EBOV and MARV, respectively. The reverse genetics approach is also fully adapted to filoviruses [425]. It is worth to mention the recombinant MARV with the EGFP gene that was inserted between the second and third genes, encoding VP35 and VP40, respectively, and conferred expression of the EGFP gene from an additional transcription unit [426].

The filoviruses also have a long and successful history of the use of their glycoproteins as pseudotyping agents. First, pseudotyped lentiviral vectors were constructed for the delivery of lentiviral vectors to the cells exposing receptors for the filoviruses, for example, folate receptor alpha, a glycosylphosphatidylinositol-linked surface protein on the apical surface of airway epithelia [427–429]. Another retroviral vector, namely, MLV vector, was used to construct cross protective antifiloviral vaccine candidates based on the retro-VLPs that were pseudotyped by the EBOV GP lacking mucin-like domain or on the appropriate DNA plasmids, or so-called plasmo-retroVLPs able to direct production of retro-VLPs [430].

Second, EBOV GP was used to pseudotype VSV vector and the VSV-based vaccine was shown to protect rhesus monkeys against an EBOV challenge [431]. Further, complete protection of cynomolgus macaques by Zaire EBOV or MARV VLPs against a challenge with the respective virus was demonstrated [432]. Furthermore, complete protection was reported also against a challenge with Sudan EBOV of rhesus monkeys immunized with VSV-based vaccine displaying the Sudan EBOV GP [433]. A single immunization with VSV-based vaccine expressing the Zaire EBOV GP, but not Côte d'Ivoire EBOV GP, provided some cross protection of cynomolgus macaques against a challenge with Bundibugyo EBOV that was found as a new EBOV species following an outbreak in Uganda in 2007 [434]. An advanced research on the VSV-based vaccines against both EBOV and MARV by improving cross protection [435,436] and avoiding neurovirulence in nonhuman primates [437] is now in progress. Very recently, the GP of the novel lloviu cuevavirus was used to pseudotype the VSV vector [403].

Third, the EBOV GP was used to pseudotype paramyxovirus vectors (see Section 24.4.2.4.2.1) of human PIV [260] and NDV [438].

Regarding the usage of filoviral epitopes, the only example is represented up to now by the insertion of the CTL epitope 43–54 of the NP from Zaire EBOV into mouse cytomegalovirus (MCMV) vector, namely, by the C-terminal addition of the EBOV epitope to the IE2, a nonessential MCMV protein, which conferred full protection of mice against a lethal challenge of animals by Zaire EBOV [439].

Switching from the filovirus VLPs to the development of novel bionanomaterials, an attempt to block the EBOV infection by the nested glycodendrimeric layers constructed on the basis of RNA phage Qβ is worth mentioning [440].

In conclusion, recombinant VLPs and nanoscale structures have been developed for many representatives of the group of enveloped viruses with single-stranded, negative-sense RNA genome. Currently, the majority of applications of these VLPs are in the vaccine field. Although more research is needed, VLPs appear to be safe and effective vaccines for many representatives of this group of viruses including, but not limited to, highly pathogenic arenaviruses, bunyaviruses, and filoviruses.

ACKNOWLEDGMENTS

We thank Professor Elmars Grens for continuous professional support over the years and Dr. Andris Dishlers and Dr. Laima Tihomirova for continuous interest to the VLP projects and valuable discussions and suggestions. We also thank the Centers for Disease Control and Prevention and Dr. Patrick Masson from ViralZone for permission to use images of viruses. The content is solely the responsibility of the authors and does not necessarily represent the official views of the funding agencies.

REFERENCES

1. Pushko, P. et al., Influenza virus-like particles comprised of the HA, NA, and M1 proteins of H9N2 influenza virus induce protective immune responses in BALB/c mice. *Vaccine*, 2005. **23**(50): 5751–5759.
2. Chen, B.J. et al., Influenza virus hemagglutinin and neuraminidase, but not the matrix protein, are required for assembly and budding of plasmid-derived virus-like particles. *J Virol*, 2007. **81**(13): 7111–7123.
3. Landry, N. et al., Preclinical and clinical development of plant-made virus-like particle vaccine against avian H5N1 influenza. *PLoS One*, 2010. **5**(12): e15559.
4. Pushko, P. et al., Influenza virus-like particle can accommodate multiple subtypes of hemagglutinin and protect from multiple influenza types and subtypes. *Vaccine*, 2011. **29**(35): 5911–5918.
5. Tretyakova, I. et al., Intranasal vaccination with H5, H7 and H9 hemagglutinins co-localized in a virus-like particle protects ferrets from multiple avian influenza viruses. *Virology*, 2013. **442**(1): 67–73.
6. *Virus Taxonomy: 2013 Release EC 45*, Edinburgh, U.K., July 2013 http://ictvonline.org/virusTaxonomy.asp. Accessed February 4, 2015.
7. *NCBI Taxonomy*, http://www.ncbi.nlm.nih.gov/taxonomy/, 2014. Accessed February 4, 2015.
8. Olschlager, S. and L. Flatz, Vaccination strategies against highly pathogenic arenaviruses: The next steps toward clinical trials. *PLoS Pathog*, 2013. **9**(4): e1003212.
9. Buckley, S.M., J. Casals, and W.G. Downs, Isolation and antigenic characterization of Lassa virus. *Nature*, 1970. **227**(5254): 174.
10. Hulo, C. et al., ViralZone: A knowledge resource to understand virus diversity. *Nucleic Acids Res*, 2011. **39**(Database issue): D576–D582.
11. Baltimore, D., The strategy of RNA viruses. *Harvey Lect*, 1974. **70**(Series): 57–74.
12. Barber, G.N., J.C. Clegg, and J. Chamberlain, Expression of Lassa virus nucleocapsid protein segments in bacteria: Purification of high-level expression products and their application in antibody detection. *Gene*, 1987. **56**(1): 137–144.
13. ter Meulen, J. et al., Characterization of human CD4(+) T-cell clones recognizing conserved and variable epitopes of the Lassa virus nucleoprotein. *J Virol*, 2000. **74**(5): 2186–2192.

14. Branco, L.M. et al., Bacterial-based systems for expression and purification of recombinant Lassa virus proteins of immunological relevance. *Virol J*, 2008. **5**: 74.

15. Djavani, M. et al., Murine immune responses to mucosally delivered *Salmonella* expressing Lassa fever virus nucleoprotein. *Vaccine*, 2000. **18**(15): 1543–1554.

16. Djavani, M. et al., Mucosal immunization with *Salmonella typhimurium* expressing Lassa virus nucleocapsid protein cross-protects mice from lethal challenge with lymphocytic choriomeningitis virus. *J Hum Virol*, 2001. **4**(2): 103–108.

17. Barber, G.N., J.C. Clegg, and G. Lloyd, Expression of the Lassa virus nucleocapsid protein in insect cells infected with a recombinant baculovirus: Application to diagnostic assays for Lassa virus infection. *J Gen Virol*, 1990. **71**(Pt 1): 19–28.

18. Saijo, M. et al., Development of recombinant nucleoprotein-based diagnostic systems for Lassa fever. *Clin Vaccin Immunol*, 2007. **14**(9): 1182–1189.

19. Hummel, K.B., M.L. Martin, and D.D. Auperin, Baculovirus expression of the glycoprotein gene of Lassa virus and characterization of the recombinant protein. *Virus Res*, 1992. **25**(1–2): 79–90.

20. Clegg, J.C. and G. Lloyd, Vaccinia recombinant expressing Lassa-virus internal nucleocapsid protein protects guinea pigs against Lassa fever. *Lancet*, 1987. **2**(8552): 186–188.

21. Morrison, H.G. et al., Protection of guinea pigs from Lassa fever by vaccinia virus recombinants expressing the nucleoprotein or the envelope glycoproteins of Lassa virus. *Virology*, 1989. **171**(1): 179–188.

22. Auperin, D.D. et al., Construction of a recombinant vaccinia virus expressing the Lassa virus glycoprotein gene and protection of guinea pigs from a lethal Lassa virus infection. *Virus Res*, 1988. **9**(2–3): 233–248.

23. Fisher-Hoch, S.P. et al., Protection of rhesus monkeys from fatal Lassa fever by vaccination with a recombinant vaccinia virus containing the Lassa virus glycoprotein gene. *Proc Natl Acad Sci U S A*, 1989. **86**(1): 317–321.

24. Morrison, H.G. et al., Simultaneous expression of the Lassa virus N and GPC genes from a single recombinant vaccinia virus. *Virus Res*, 1991. **18**(2–3): 231–241.

25. ter Meulen, J., Lassa fever: Implications of T-cell immunity for vaccine development. *J Biotechnol*, 1999. **73**(2–3): 207–212.

26. Pushko, P. et al., Individual and bivalent vaccines based on alphavirus replicons protect guinea pigs against infection with Lassa and Ebola viruses. *J Virol*, 2001. **75**(23): 11677–11685.

27. Bredenbeek, P.J. et al., A recombinant Yellow Fever 17D vaccine expressing Lassa virus glycoproteins. *Virology*, 2006. **345**(2): 299–304.

28. Jiang, X. et al., Yellow fever 17D-vectored vaccines expressing Lassa virus GP1 and GP2 glycoproteins provide protection against fatal disease in guinea pigs. *Vaccine*, 2011. **29**(6): 1248–1257.

29. Kentsis, A., R.E. Gordon, and K.L. Borden, Self-assembly properties of a model RING domain. *Proc Natl Acad Sci U S A*, 2002. **99**(2): 667–672.

30. Strecker, T. et al., Lassa virus Z protein is a matrix protein and sufficient for the release of virus-like particles [corrected]. *J Virol*, 2003. **77**(19): 10700–10705.

31. Perez, M., R.C. Craven, and J.C. de la Torre, The small RING finger protein Z drives arenavirus budding: Implications for antiviral strategies. *Proc Natl Acad Sci U S A*, 2003. **100**(22): 12978–12983.

32. Urata, S. and J. Yasuda, Molecular mechanism of arenavirus assembly and budding. *Viruses*, 2012. **4**(10): 2049–2079.

33. Eichler, R. et al., Characterization of the Lassa virus matrix protein Z: Electron microscopic study of virus-like particles and interaction with the nucleoprotein (NP). *Virus Res*, 2004. **100**(2): 249–255.

34. Volpon, L., M.J. Osborne, and K.L. Borden, NMR assignment of the arenaviral protein Z from Lassa fever virus. *Biomol NMR Assign*, 2008. **2**(1): 81–84.

35. May, E.R. et al., The flexible C-terminal arm of the Lassa arenavirus Z-protein mediates interactions with multiple binding partners. *Proteins*, 2010. **78**(10): 2251–2264.

36. Qi, X. et al., Cap binding and immune evasion revealed by Lassa nucleoprotein structure. *Nature*, 2010. **468**(7325): 779–783.

37. Brunotte, L. et al., Structure of the Lassa virus nucleoprotein revealed by X-ray crystallography, small-angle X-ray scattering, and electron microscopy. *J Biol Chem*, 2011. **286**(44): 38748–38756.

38. Hastie, K.M. et al., Crystal structure of the Lassa virus nucleoprotein-RNA complex reveals a gating mechanism for RNA binding. *Proc Natl Acad Sci U S A*, 2011. **108**(48): 19365–19370.

39. Hastie, K.M. et al., Structure of the Lassa virus nucleoprotein reveals a dsRNA-specific 3′ to 5′ exonuclease activity essential for immune suppression. *Proc Natl Acad Sci U S A*, 2011. **108**(6): 2396–2401.

40. Zhang, Y. et al., Crystal structure of Junin virus nucleoprotein. *J Gen Virol*, 2013. **94**(Pt 10): 2175–2183.

41. West, B.R., K.M. Hastie, and E.O. Saphire, Structure of the LCMV nucleoprotein provides a template for understanding arenavirus replication and immunosuppression. *Acta Crystallogr D Biol Crystallogr*, 2014. **70**(Pt 6): 1764–1769.

42. Jiang, X. et al., Structures of arenaviral nucleoproteins with triphosphate dsRNA reveal a unique mechanism of immune suppression. *J Biol Chem*, 2013. **288**(23): 16949–16959.

43. Bowden, T.A. et al., Unusual molecular architecture of the machupo virus attachment glycoprotein. *J Virol*, 2009. **83**(16): 8259–8265.

44. Igonet, S. et al., X-ray structure of the arenavirus glycoprotein GP2 in its postfusion hairpin conformation. *Proc Natl Acad Sci U S A*, 2011. **108**(50): 19967–19972.

45. Capul, A.A. et al., Arenavirus Z-glycoprotein association requires Z myristoylation but not functional RING or late domains. *J Virol*, 2007. **81**(17): 9451–9460.

46. *VIPERdb: VIrus Particle ExploreR*, http://viperdb.scripps.edu/index.php. Accessed February 4, 2015.

47. Goni, S.E. et al., Expression and purification of Z protein from Junin virus. *J Biomed Biotechnol*, 2010. **2010**: 970491.

48. Strecker, T. et al., The role of myristoylation in the membrane association of the Lassa virus matrix protein Z. *Virol J*, 2006. **3**: 93.

49. Ortiz-Riano, E. et al., Self-association of lymphocytic choriomeningitis virus nucleoprotein is mediated by its N-terminal region and is not required for its anti-interferon function. *J Virol*, 2012. **86**(6): 3307–3317.

50. Lennartz, F. et al., The role of oligomerization for the biological functions of the arenavirus nucleoprotein. *Arch Virol*, 2013. **158**(9): 1895–1905.

51. Levingston Macleod, J.M. et al., Identification of two functional domains within the arenavirus nucleoprotein. *J Virol*, 2011. **85**(5): 2012–2023.

52. D'Antuono, A. et al., Differential contributions of tacaribe arenavirus nucleoprotein N-terminal and C-terminal residues to nucleocapsid functional activity. *J Virol*, 2014. **88**(11): 6492–6505.

53. Shtanko, O. et al., A role for the C terminus of Mopeia virus nucleoprotein in its incorporation into Z protein-induced virus-like particles. *J Virol*, 2010. **84**(10): 5415–5422.

54. Shtanko, O. et al., ALIX/AIP1 is required for NP incorporation into Mopeia virus Z-induced virus-like particles. *J Virol*, 2011. **85**(7): 3631–3641.

55. Schlie, K. et al., Viral protein determinants of Lassa virus entry and release from polarized epithelial cells. *J Virol*, 2010. **84**(7): 3178–3188.

56. Casabona, J.C. et al., The RING domain and the L79 residue of Z protein are involved in both the rescue of nucleocapsids and the incorporation of glycoproteins into infectious chimeric arenavirus-like particles. *J Virol*, 2009. **83**(14): 7029–7039.

57. Branco, L.M. et al., Lassa virus-like particles displaying all major immunological determinants as a vaccine candidate for Lassa hemorrhagic fever. *Virol J*, 2010. **7**:279.

58. Garcia, C.C. et al., Characterization of Junin virus particles inactivated by a zinc finger-reactive compound. *Virus Res*, 2009. **143**(1): 106–113.

59. Borio, C.S. et al., Antigen vehiculization particles based on the Z protein of Junin virus. *BMC Biotechnol*, 2012. **12**: 80.

60. Pinschewer, D.D. et al., Recombinant lymphocytic choriomeningitis virus expressing vesicular stomatis virus glycoprotein. *Proc Natl Acad Sci U S A*, 2003. **100**(13): 7895–7900.

61. Bergthaler, A. et al., Envelope exchange for the generation of live-attenuated arenavirus vaccines. *PLoS Pathog*, 2006. **2**(6): e51.

62. Pinschewer, D.D. et al., Innate and adaptive immune control of genetically engineered live-attenuated arenavirus vaccine prototypes. *Int Immunol*, 2010. **22**(9): 749–756.

63. Lee, A.M. et al., Pathogenesis of Lassa fever virus infection: I. Susceptibility of mice to recombinant Lassa Gp/LCMV chimeric virus. *Virology*, 2013. **442**(2): 114–121.

64. Sommerstein, R. et al., Evolution of recombinant lymphocytic choriomeningitis virus/Lassa virus in vivo highlights the importance of the GPC cytosolic tail in viral fitness. *J Virol*, 2014. **88**(15): 8340–8348.

65. Carrion, R., Jr. et al., Vaccine platforms to control arenaviral hemorrhagic fevers. *J Vaccines Vaccin*, 2012. **3**(7): pii: 1000160.

66. Garbutt, M. et al., Properties of replication-competent vesicular stomatitis virus vectors expressing glycoproteins of filoviruses and arenaviruses. *J Virol*, 2004. **78**(10): 5458–5465.

67. Reignier, T. et al., Receptor use by pathogenic arenaviruses. *Virology*, 2006. **353**(1): 111–120.

68. Cosset, F.L. et al., Characterization of Lassa virus cell entry and neutralization with Lassa virus pseudoparticles. *J Virol*, 2009. **83**(7): 3228–3237.

69. Dylla, D.E. et al., Altering alpha-dystroglycan receptor affinity of LCMV pseudotyped lentivirus yields unique cell and tissue tropism. *Genet Vaccin Ther*, 2011. **9**: 8.

70. Kumar, N. et al., Characterization of virulence-associated determinants in the envelope glycoprotein of Pichinde virus. *Virology*, 2012. **433**(1): 97–103.

71. Tani, H. et al., Analysis of Lujo virus cell entry using pseudotype vesicular stomatitis virus. *J Virol*, 2014. **88**(13): 7317–7330.

72. Muik, A. et al., Pseudotyping vesicular stomatitis virus with lymphocytic choriomeningitis virus glycoproteins enhances infectivity for glioma cells and minimizes neurotropism. *J Virol*, 2011. **85**(11): 5679–5684.

73. Tober, R. et al., VSV-GP: A potent viral vaccine vector that boosts the immune response upon repeated applications. *J Virol*, 2014. **88**(9): 4897–4907.

74. Hombach, J. et al., Strictly transporter of antigen presentation (TAP)-dependent presentation of an immunodominant cytotoxic T lymphocyte epitope in the signal sequence of a virus protein. *J Exp Med*, 1995. **182**(5): 1615–1619.

75. Gairin, J.E. et al., Optimal lymphocytic choriomeningitis virus sequences restricted by H-2Db major histocompatibility complex class I molecules and presented to cytotoxic T lymphocytes. *J Virol*, 1995. **69**(4): 2297–2305.

76. Desjardins, D. et al., Recombinant retrovirus-like particle forming DNA vaccines in prime-boost immunization and their use for hepatitis C virus vaccine development. *J Gene Med*, 2009. **11**(4): 313–325.

77. Aichele, P. et al., Peptide antigen treatment of naive and virus-immune mice: Antigen-specific tolerance versus immunopathology. *Immunity*, 1997. **6**(5): 519–529.

78. Hudrisier, D., M.B. Oldstone, and J.E. Gairin, The signal sequence of lymphocytic choriomeningitis virus contains an immunodominant cytotoxic T cell epitope that is restricted by both H-2D(b) and H-2K(b) molecules. *Virology*, 1997. **234**(1): 62–73.

79. Ludewig, B. et al., Dendritic cells efficiently induce protective antiviral immunity. *J Virol*, 1998. **72**(5): 3812–3818.

80. Gallimore, A. et al., Induction and exhaustion of lymphocytic choriomeningitis virus-specific cytotoxic T lymphocytes visualized using soluble tetrameric major histocompatibility complex class I-peptide complexes. *J Exp Med*, 1998. **187**(9): 1383–1393.

81. Kotturi, M.F. et al., Naive precursor frequencies and MHC binding rather than the degree of epitope diversity shape CD8+ T cell immunodominance. *J Immunol*, 2008. **181**(3): 2124–2133.

82. Bunztman, A. et al., The LCMV gp33-specific memory T cell repertoire narrows with age. *Immun Ageing*, 2012. **9**(1): 17.

83. Achour, A. et al., A structural basis for LCMV immune evasion: Subversion of H-2D(b) and H-2K(b) presentation of gp33 revealed by comparative crystal structure analyses. *Immunity*, 2002. **17**(6): 757–768.

84. Sandalova, T. et al., Expression, refolding and crystallization of murine MHC class I H-2Db in complex with human beta2-microglobulin. *Acta Crystallogr Sect F Struct Biol Cryst Commun*, 2005. **61**(Pt 12): 1090–1093.

85. Storni, T. et al., Critical role for activation of antigen-presenting cells in priming of cytotoxic T cell responses after vaccination with virus-like particles. *J Immunol*, 2002. **168**(6): 2880–2886.

86. Ruedl, C. et al., Cross-presentation of virus-like particles by skin-derived CD8(−) dendritic cells: A dispensable role for TAP. *Eur J Immunol*, 2002. **32**(3): 818–825.

87. Marsland, B.J. et al., Innate signals compensate for the absence of PKC-{theta} during in vivo CD8(+) T cell effector and memory responses. *Proc Natl Acad Sci U S A*, 2005. **102**(40): 14374–14379.

88. Storni, T. and M.F. Bachmann, On the role of APC-activation for in vitro versus in vivo T cell priming. *Cell Immunol*, 2003. **225**(1): 1–11.

89. Storni, T. et al., Nonmethylated CG motifs packaged into virus-like particles induce protective cytotoxic T cell responses in the absence of systemic side effects. *J Immunol*, 2004. **172**(3): 1777–1785.

90. Schwarz, K. et al., Efficient homologous prime-boost strategies for T cell vaccination based on virus-like particles. *Eur J Immunol*, 2005. **35**(3): 816–821.

91. Bachmann, M.F. et al., Functional properties and lineage relationship of CD8+ T cell subsets identified by expression of IL-7 receptor alpha and CD62L. *J Immunol*, 2005. **175**(7): 4686–4696.

92. Bessa, J. et al., Efficient induction of mucosal and systemic immune responses by virus-like particles administered intranasally: Implications for vaccine design. *Eur J Immunol*, 2008. **38**(1): 114–126.

93. Agnellini, P. et al., Kinetic and mechanistic requirements for helping CD8 T cells. *J Immunol*, 2008. **180**(3): 1517–1525.

94. Keller, S.A. et al., Innate signaling regulates cross-priming at the level of DC licensing and not antigen presentation. *Eur J Immunol*, 2010. **40**(1): 103–112.

95. Derdak, S.V. et al., Direct stimulation of T lymphocytes by immunosomes: Virus-like particles decorated with T cell receptor/CD3 ligands plus costimulatory molecules. *Proc Natl Acad Sci U S A*, 2006. **103**(35): 13144–13149.

96. Lacasse, P. et al., Novel plant virus-based vaccine induces protective cytotoxic T-lymphocyte-mediated antiviral immunity through dendritic cell maturation. *J Virol*, 2008. **82**(2): 785–794.

97. Mazeike, E., A. Gedvilaite, and U. Blohm, Induction of insert-specific immune response in mice by hamster polyomavirus VP1 derived virus-like particles carrying LCMV GP33 CTL epitope. *Virus Res*, 2012. **163**(1): 2–10.

98. Li, K. et al., Antigen incorporated in virus-like particles is delivered to specific dendritic cell subsets that induce an effective antitumor immune response in vivo. *J Immunother*, 2013. **36**(1): 11–19.

99. Mueller, S.N. et al., Immunization with live attenuated influenza viruses that express altered NS1 proteins results in potent and protective memory CD8+ T-cell responses. *J Virol*, 2010. **84**(4): 1847–1855.

100. Shi, W. et al., Papillomavirus pseudovirus: A novel vaccine to induce mucosal and systemic cytotoxic T-lymphocyte responses. *J Virol*, 2001. **75**(21): 10139–10148.

101. Sedlik, C. et al., Recombinant parvovirus-like particles as an antigen carrier: A novel nonreplicative exogenous antigen to elicit protective antiviral cytotoxic T cells. *Proc Natl Acad Sci U S A*, 1997. **94**(14): 7503–7508.

102. Sedlik, C. et al., Intranasal delivery of recombinant parvovirus-like particles elicits cytotoxic T-cell and neutralizing antibody responses. *J Virol*, 1999. **73**(4): 2739–2744.

103. Casal, J.I., Use of parvovirus-like particles for vaccination and induction of multiple immune responses. *Biotechnol Appl Biochem*, 1999. **29**(Pt 2): 141–150.

104. Rueda, P. et al., Engineering parvovirus-like particles for the induction of B-cell, CD4(+) and CTL responses. *Vaccine*, 2000. **18**(3–4): 325–332.

105. Sedlik, C. et al., In vivo induction of a high-avidity, high-frequency cytotoxic T-lymphocyte response is associated with antiviral protective immunity. *J Virol*, 2000. **74**(13): 5769–5775.

106. Ghosh, M.K. et al., Induction of protective antiviral cytotoxic T cells by a tubular structure capable of carrying large foreign sequences. *Vaccine*, 2002. **20**(9–10): 1369–1377.

107. Botten, J. et al., Identification of protective Lassa virus epitopes that are restricted by HLA-A2. *J Virol*, 2006. **80**(17): 8351–8361.

108. Botten, J. et al., A multivalent vaccination strategy for the prevention of Old World arenavirus infection in humans. *J Virol*, 2010. **84**(19): 9947–9956.

109. Wang, B.Z. et al., Incorporation of high levels of chimeric human immunodeficiency virus envelope glycoproteins into virus-like particles. *J Virol*, 2007. **81**(20): 10869–10878.

110. Soldan, S.S. and F. Gonzalez-Scarano, The bunyaviridae. *Handb Clin Neurol*, 2014. **123**: 449–463.

111. Beer, M., F.J. Conraths, and W.H. van der Poel, 'Schmallenberg virus'—A novel orthobunyavirus emerging in Europe. *Epidemiol Infect*, 2013. **141**(1): 1–8.

112. Turina, M., L. Tavella, and M. Ciuffo, Tospoviruses in the Mediterranean area. *Adv Virus Res*, 2012. **84**: 403–437.

113. Strandin, T., J. Hepojoki, and A. Vaheri, Cytoplasmic tails of bunyavirus Gn glycoproteins—Could they act as matrix protein surrogates? *Virology*, 2013. **437**(2): 73–80.

114. Kakach, L.T., T.L. Wasmoen, and M.S. Collett, Rift Valley fever virus M segment: Use of recombinant vaccinia viruses to study Phlebovirus gene expression. *J Virol*, 1988. **62**(3): 826–833.

115. Pensiero, M.N. et al., Expression of the Hantaan virus M genome segment by using a vaccinia virus recombinant. *J Virol*, 1988. **62**(3): 696–702.

116. Schmaljohn, C.S. et al., Baculovirus expression of the small genome segment of Hantaan virus and potential use of the expressed nucleocapsid protein as a diagnostic antigen. *J Gen Virol*, 1988. **69**(Pt 4): 777–786.

117. Garcia, S. et al., Evaluation of a Crimean-Congo hemorrhagic fever virus recombinant antigen expressed by Semliki Forest suicide virus for IgM and IgG antibody detection in human and animal sera collected in Iran. *J Clin Virol*, 2006. **35**(2): 154–159.

118. Urakawa, T., D.A. Small, and D.H. Bishop, Expression of snowshoe hare bunyavirus S RNA coding proteins by recombinant baculoviruses. *Virus Res*, 1988. **11**(4): 303–317.

119. Saijo, M. et al., Recombinant nucleoprotein-based enzyme-linked immunosorbent assay for detection of immunoglobulin G antibodies to Crimean-Congo hemorrhagic fever virus. *J Clin Microbiol*, 2002. **40**(5): 1587–1591.

120. Dowall, S.D. et al., Development of an indirect ELISA method for the parallel measurement of IgG and IgM antibodies against Crimean-Congo haemorrhagic fever (CCHF) virus using recombinant nucleoprotein as antigen. *J Virol Methods*, 2012. **179**(2): 335–341.

121. Schmaljohn, C.S. et al., Baculovirus expression of the M genome segment of Rift Valley fever virus and examination of antigenic and immunogenic properties of the expressed proteins. *Virology*, 1989. **170**(1): 184–192.

122. Takehara, K., S. Morikawa, and D.H. Bishop, Characterization of baculovirus-expressed Rift Valley fever virus glycoproteins synthesized in insect cells. *Virus Res*, 1990. **17**(3): 173–190.

123. Schmaljohn, C.S. et al., Antigenic subunits of Hantaan virus expressed by baculovirus and vaccinia virus recombinants. *J Virol*, 1990. **64**(7): 3162–3170.

124. Yoshimatsu, K. et al., Protective immunity of Hantaan virus nucleocapsid and envelope protein studied using baculovirus-expressed proteins. *Arch Virol*, 1993. **130**(3–4): 365–376.

125. Gott, P. et al., Antigenicity of hantavirus nucleocapsid proteins expressed in *E. coli*. *Virus Res*, 1991. **19**(1): 1–15.

126. Richmond, K.E. et al., Characterization of the nucleic acid binding properties of tomato spotted wilt virus nucleocapsid protein. *Virology*, 1998. **248**(1): 6–11.

127. Jansen van Vuren, P. et al., Preparation and evaluation of a recombinant Rift Valley fever virus N protein for the detection of IgG and IgM antibodies in humans and animals by indirect ELISA. *J Virol Methods*, 2007. **140**(1–2): 106–114.

128. Kehm, R. et al., Expression of immunogenic Puumala virus nucleocapsid protein in transgenic tobacco and potato plants. *Virus Genes*, 2001. **22**(1): 73–83.

129. Chen, T.C. et al., Purification and serological analyses of tospoviral nucleocapsid proteins expressed by Zucchini yellow mosaic virus vector in squash. *J Virol Methods*, 2005. **129**(2): 113–124.

130. Ha, S.H. et al., Molecular cloning and high-level expression of G2 protein of hantaan (HTN) virus 76–118 strain in the yeast *Pichia pastoris* KM71. *Virus Genes*, 2001. **22**(2): 167–173.

131. Dargeviciute, A. et al., Yeast-expressed Puumala hantavirus nucleocapsid protein induces protection in a bank vole model. *Vaccine*, 2002. **20**(29–30): 3523–3531.

132. Geldmacher, A. et al., Yeast-expressed hantavirus Dobrava nucleocapsid protein induces a strong, long-lasting, and highly cross-reactive immune response in mice. *Viral Immunol*, 2004. **17**(1): 115–122.

133. Razanskiene, A. et al., High yields of stable and highly pure nucleocapsid proteins of different hantaviruses can be generated in the yeast *Saccharomyces cerevisiae*. *J Biotechnol*, 2004. **111**(3): 319–333.

134. Jiao, Y. et al., Preparation and evaluation of recombinant severe fever with thrombocytopenia syndrome virus nucleocapsid protein for detection of total antibodies in human and animal sera by double-antigen sandwich enzyme-linked immunosorbent assay. *J Clin Microbiol*, 2012. **50**(2): 372–377.

135. Zhang, Y. et al., Expression and purification of the nucleocapsid protein of Schmallenberg virus, and preparation and characterization of a monoclonal antibody against this protein. *Protein Expr Purif*, 2013. **92**(1): 1–8.

136. Lazutka, J. et al., Generation of recombinant schmallenberg virus nucleocapsid protein in yeast and development of virus-specific monoclonal antibodies. *J Immunol Res*, 2014. **2014**: 160316.

137. Betenbaugh, M. et al., Nucleocapsid- and virus-like particles assemble in cells infected with recombinant baculoviruses or vaccinia viruses expressing the M and the S segments of Hantaan virus. *Virus Res*, 1995. **38**(2–3): 111–124.

138. Liu, L., C.C. Celma, and P. Roy, Rift Valley fever virus structural proteins: Expression, characterization and assembly of recombinant proteins. *Virol J*, 2008. **5**: 82.

139. Alfadhli, A. et al., Hantavirus nucleocapsid protein oligomerization. *J Virol*, 2001. **75**(4): 2019–2023.

140. Kaukinen, P. et al., Interaction between molecules of hantavirus nucleocapsid protein. *J Gen Virol*, 2001. **82**(Pt 8): 1845–1853.

141. Uhrig, J.F. et al., Homotypic interaction and multimerization of nucleocapsid protein of tomato spotted wilt tospovirus: Identification and characterization of two interacting domains. *Proc Natl Acad Sci U S A*, 1999. **96**(1): 55–60.

142. Le May, N. et al., The N terminus of Rift Valley fever virus nucleoprotein is essential for dimerization. *J Virol*, 2005. **79**(18): 11974–11980.

143. Snippe, M. et al., Tomato spotted wilt virus Gc and N proteins interact in vivo. *Virology*, 2007. **357**(2): 115–123.

144. Ribeiro, D. et al., Tomato spotted wilt virus nucleocapsid protein interacts with both viral glycoproteins Gn and Gc in plants. *Virology*, 2009. **383**(1): 121–130.

145. Overby, A.K., R.F. Pettersson, and E.P. Neve, The glycoprotein cytoplasmic tail of Uukuniemi virus (Bunyaviridae) interacts with ribonucleoproteins and is critical for genome packaging. *J Virol*, 2007. **81**(7): 3198–3205.

146. Wang, H. et al., Interaction between hantaviral nucleocapsid protein and the cytoplasmic tail of surface glycoprotein Gn. *Virus Res*, 2010. **151**(2): 205–212.

147. Shimizu, K. et al., Role of nucleocapsid protein of hantaviruses in intracellular traffic of viral glycoproteins. *Virus Res*, 2013. **178**(2): 349–356.

148. Rodgers, J.W. et al., Purification, crystallization and preliminary X-ray crystallographic analysis of the nucleocapsid protein of Bunyamwera virus. *Acta Crystallogr Sect F Struct Biol Cryst Commun*, 2006. **62**(Pt 4): 361–364.

149. Ferron, F. et al., The hexamer structure of Rift Valley fever virus nucleoprotein suggests a mechanism for its assembly into ribonucleoprotein complexes. *PLoS Pathog*, 2011. **7**(5): e1002030.

150. Carter, S.D., J.N. Barr, and T.A. Edwards, Expression, purification and crystallization of the Crimean-Congo haemorrhagic fever virus nucleocapsid protein. *Acta Crystallogr Sect F Struct Biol Cryst Commun*, 2012. **68**(Pt 5): 569–573.

151. Wang, Y. et al., Structure of Crimean-Congo hemorrhagic fever virus nucleoprotein: Superhelical homo-oligomers and the role of caspase-3 cleavage. *J Virol*, 2012. **86**(22): 12294–12303.

152. Zhou, H. et al., The nucleoprotein of severe fever with thrombocytopenia syndrome virus processes a stable hexameric ring to facilitate RNA encapsidation. *Protein Cell*, 2013. **4**(6): 445–455.

153. Dong, H. et al., Crystal structure of Schmallenberg orthobunyavirus nucleoprotein-RNA complex reveals a novel RNA sequestration mechanism. *RNA*, 2013. **19**(8): 1129–1136.

154. Komoda, K. et al., Expression, purification, crystallization and preliminary X-ray crystallographic study of the nucleocapsid protein of tomato spotted wilt virus. *Acta Crystallogr Sect F Struct Biol Cryst Commun*, 2013. **69**(Pt 6): 700–703.

155. Carter, S.D. et al., Structure, function, and evolution of the Crimean-Congo hemorrhagic fever virus nucleocapsid protein. *J Virol*, 2012. **86**(20): 10914–10923.

156. Estrada, D.F. and R.N. De Guzman, Structural characterization of the Crimean-Congo hemorrhagic fever virus Gn tail provides insight into virus assembly. *J Biol Chem*, 2011. **286**(24): 21678–21686.

157. Dessau, M. and Y. Modis, Crystal structure of glycoprotein C from Rift Valley fever virus. *Proc Natl Acad Sci U S A*, 2013. **110**(5): 1696–1701.

158. Rusu, M. et al., An assembly model of rift valley Fever virus. *Front Microbiol*, 2012. **3**: 254.

159. Overby, A.K. et al., Insights into bunyavirus architecture from electron cryotomography of Uukuniemi virus. *Proc Natl Acad Sci U S A*, 2008. **105**(7): 2375–2379.

160. Freiberg, A.N. et al., Three-dimensional organization of Rift Valley fever virus revealed by cryoelectron tomography. *J Virol*, 2008. **82**(21): 10341–10348.

161. Huiskonen, J.T. et al., Electron cryotomography of Tula hantavirus suggests a unique assembly paradigm for enveloped viruses. *J Virol*, 2010. **84**(10): 4889–4897.

162. Battisti, A.J. et al., Structural studies of Hantaan virus. *J Virol*, 2011. **85**(2): 835–841.

163. Bowden, T.A. et al., Orthobunyavirus ultrastructure and the curious tripodal glycoprotein spike. *PLoS Pathog*, 2013. **9**(5): e1003374.

164. Zhou, Z.R. et al., Production of CCHF virus-like particle by a baculovirus-insect cell expression system. *Virol Sin*, 2011. **26**(5): 338–346.

165. Li, C. et al., Hantavirus-like particles generated in CHO cells induce specific immune responses in C57BL/6 mice. *Vaccine*, 2010. **28**(26): 4294–4300.

166. Mandell, R.B. et al., Novel suspension cell-based vaccine production systems for Rift Valley fever virus-like particles. *J Virol Methods*, 2010. **169**(2): 259–268.

167. de Boer, S.M. et al., Rift Valley fever virus subunit vaccines confer complete protection against a lethal virus challenge. *Vaccine*, 2010. **28**(11): 2330–2339.

168. Acuna, R. et al., Hantavirus Gn and Gc glycoproteins self-assemble into virus-like particles. *J Virol*, 2014. **88**(4): 2344–2348.

169. Overby, A.K. et al., Generation and analysis of infectious virus-like particles of uukuniemi virus (bunyaviridae): A useful system for studying bunyaviral packaging and budding. *J Virol*, 2006. **80**(21): 10428–10435.

170. Overby, A.K. et al., The cytoplasmic tails of Uukuniemi Virus (Bunyaviridae) G(N) and G(C) glycoproteins are important for intracellular targeting and the budding of virus-like particles. *J Virol*, 2007. **81**(20): 11381–11391.

171. Shi, X. et al., Role of the cytoplasmic tail domains of Bunyamwera orthobunyavirus glycoproteins Gn and Gc in virus assembly and morphogenesis. *J Virol*, 2007. **81**(18): 10151–10160.

172. Shi, X. et al., Functional analysis of the Bunyamwera orthobunyavirus Gc glycoprotein. *J Gen Virol*, 2009. **90**(Pt 10): 2483–2492.

173. Habjan, M. et al., Efficient production of Rift Valley fever virus-like particles: The antiviral protein MxA can inhibit primary transcription of bunyaviruses. *Virology*, 2009. **385**(2): 400–408.

174. Naslund, J. et al., Vaccination with virus-like particles protects mice from lethal infection of Rift Valley fever virus. *Virology*, 2009. **385**(2): 409–415.

175. Pichlmair, A. et al., Virus-like particles expressing the nucleocapsid gene as an efficient vaccine against Rift Valley fever virus. *Vector Borne Zoonotic Dis*, 2010. **10**(7): 701–703.

176. Kortekaas, J. et al., Creation of a nonspreading Rift Valley fever virus. *J Virol*, 2011. **85**(23): 12622–12630.

177. Oreshkova, N. et al., A single vaccination with an improved nonspreading Rift Valley fever virus vaccine provides sterile immunity in lambs. *PLoS One*, 2013. **8**(10): e77461.

178. Shi, X. et al., Visualizing the replication cycle of bunyamwera orthobunyavirus expressing fluorescent protein-tagged Gc glycoprotein. *J Virol*, 2010. **84**(17): 8460–8469.

179. Antoniukas, L., H. Grammel, and U. Reichl, Production of hantavirus Puumala nucleocapsid protein in *Saccharomyces cerevisiae* for vaccine and diagnostics. *J Biotechnol*, 2006. **124**(2): 347–362.

180. Snippe, M. et al., The use of fluorescence microscopy to visualise homotypic interactions of tomato spotted wilt virus nucleocapsid protein in living cells. *J Virol Methods*, 2005. **125**(1): 15–22.

181. Lacorte, C. et al., The nucleoprotein of tomato spotted wilt virus as protein tag for easy purification and enhanced production of recombinant proteins in plants. *Protein Expr Purif*, 2007. **55**(1): 17–22.

182. Bennett, R.S. et al., A recombinant chimeric La Crosse virus expressing the surface glycoproteins of Jamestown Canyon virus is immunogenic and protective against challenge with either parental virus in mice or monkeys. *J Virol*, 2012. **86**(1): 420–426.

183. Ma, M. et al., Murine leukemia virus pseudotypes of La Crosse and Hantaan Bunyaviruses: A system for analysis of cell tropism. *Virus Res*, 1999. **64**(1): 23–32.

184. Ogino, M. et al., Use of vesicular stomatitis virus pseudotypes bearing hantaan or seoul virus envelope proteins in a rapid and safe neutralization test. *Clin Diagn Lab Immunol*, 2003. **10**(1): 154–160.

185. Higa, M.M. et al., Efficient production of Hantaan and Puumala pseudovirions for viral tropism and neutralization studies. *Virology*, 2012. **423**(2): 134–142.

186. Lee, B.H. et al., A pseudotype vesicular stomatitis virus containing Hantaan virus envelope glycoproteins G1 and G2 as an alternative to hantavirus vaccine in mice. *Vaccine*, 2006. **24**(15): 2928–2934.

187. Qian, Z. et al., Targeting vascular injury using Hantavirus-pseudotyped lentiviral vectors. *Mol Ther*, 2006. **13**(4): 694–704.

188. Yu, L. et al., A recombinant pseudotyped lentivirus expressing the envelope glycoprotein of hantaan virus induced protective immunity in mice. *Virol J*, 2013. **10**: 301.

189. Mandell, R.B. et al., A replication-incompetent Rift Valley fever vaccine: Chimeric virus-like particles protect mice and rats against lethal challenge. *Virology*, 2010. **397**(1): 187–198.

190. Elgh, F. et al., A major antigenic domain for the human humoral response to Puumala virus nucleocapsid protein is located at the amino-terminus. *J Virol Methods*, 1996. **59**(1–2): 161–172.

191. Gott, P. et al., A major antigenic domain of hantaviruses is located on the aminoproximal site of the viral nucleocapsid protein. *Virus Genes*, 1997. **14**(1): 31–40.

192. Koletzki, D. et al., Mosaic hepatitis B virus core particles allow insertion of extended foreign protein segments. *J Gen Virol*, 1997. **78**(Pt 8): 2049–2053.

193. Zvirbliene, A. et al., The use of chimeric virus-like particles harbouring a segment of hantavirus Gc glycoprotein to generate a broadly-reactive hantavirus-specific monoclonal antibody. *Viruses*, 2014. **6**(2): 640–660.

194. A world without rinderpest: The FAO looks to the future. *Vet Rec*, 2014. **174**(13): 313.

195. Nagy, E. et al., Synthesis of Newcastle disease virus (NDV)-like envelopes in insect cells infected with a recombinant baculovirus expressing the haemagglutinin-neuraminidase of NDV. *J Gen Virol*, 1991. **72**(Pt 3): 753–756.

196. Spehner, D., A. Kirn, and R. Drillien, Assembly of nucleocapsidlike structures in animal cells infected with a vaccinia virus recombinant encoding the measles virus nucleoprotein. *J Virol*, 1991. **65**(11): 6296–6300.

197. Fooks, A.R. et al., Measles virus nucleocapsid protein expressed in insect cells assembles into nucleocapsid-like structures. *J Gen Virol*, 1993. **74**(Pt 7): 1439–1444.

198. Eshaghi, M. et al., Purification and characterization of Nipah virus nucleocapsid protein produced in insect cells. *J Clin Microbiol*, 2005. **43**(7): 3172–3177.

199. Grant, R.J. et al., Expression from baculovirus and serological reactivity of the nucleocapsid protein of dolphin morbillivirus. *Vet Microbiol*, 2010. **143**(2–4): 384–388.

200. Warnes, A. et al., Expression of the measles virus nucleoprotein gene in *Escherichia coli* and assembly of nucleocapsid-like structures. *Gene*, 1995. **160**(2): 173–178.

201. Errington, W. and P.T. Emmerson, Assembly of recombinant Newcastle disease virus nucleocapsid protein into nucleocapsid-like structures is inhibited by the phosphoprotein. *J Gen Virol*, 1997. **78**(Pt 9): 2335–2339.

202. Mitra-Kaushik, S., R. Nayak, and M.S. Shaila, Identification of a cytotoxic T-cell epitope on the recombinant nucleocapsid proteins of Rinderpest and Peste des petits ruminants viruses presented as assembled nucleocapsids. *Virology*, 2001. **279**(1): 210–220.

203. Tan, W.S. et al., Solubility, immunogenicity and physical properties of the nucleocapsid protein of Nipah virus produced in *Escherichia coli*. *J Med Virol*, 2004. **73**(1): 105–112.

204. Cox, R. et al., Characterization of a mumps virus nucleocapsidlike particle. *J Virol*, 2009. **83**(21): 11402–11406.

205. Samuel, D. et al., High level expression of recombinant mumps nucleoprotein in *Saccharomyces cerevisiae* and its evaluation in mumps IgM serology. *J Med Virol*, 2002. **66**(1): 123–130.

206. Slibinskas, R. et al., Synthesis of mumps virus nucleocapsid protein in yeast *Pichia pastoris*. *J Biotechnol*, 2003. **103**(1): 43–49.

207. Slibinskas, R. et al., Synthesis of the measles virus nucleoprotein in yeast *Pichia pastoris* and *Saccharomyces cerevisiae*. *J Biotechnol*, 2004. **107**(2): 115–124.

208. Juozapaitis, M. et al., Generation of Sendai virus nucleocapsid-like particles in yeast. *Virus Res*, 2005. **108**(1–2): 221–224.

209. Juozapaitis, M. et al., Generation of henipavirus nucleocapsid proteins in yeast *Saccharomyces cerevisiae*. *Virus Res*, 2007. **124**(1–2): 95–102.

210. Juozapaitis, M. et al., Synthesis of recombinant human parainfluenza virus 1 and 3 nucleocapsid proteins in yeast *Saccharomyces cerevisiae*. *Virus Res*, 2008. **133**(2): 178–186.

211. Petraityte, R. et al., Generation of Tioman virus nucleocapsid-like particles in yeast *Saccharomyces cerevisiae*. *Virus Res*, 2009. **145**(1): 92–96.

212. Stricker, R., G. Mottet, and L. Roux, The Sendai virus matrix protein appears to be recruited in the cytoplasm by the viral nucleocapsid to function in viral assembly and budding. *J Gen Virol*, 1994. **75**(Pt 5): 1031–1042.

213. Takimoto, T. and A. Portner, Molecular mechanism of paramyxovirus budding. *Virus Res*, 2004. **106**(2): 133–145.

214. Ciancanelli, M.J. and C.F. Basler, Mutation of YMYL in the Nipah virus matrix protein abrogates budding and alters subcellular localization. *J Virol*, 2006. **80**(24): 12070–12078.

215. Subramanian, S.K. et al., Production of the matrix protein of Nipah virus in *Escherichia coli*: Virus-like particles and possible application for diagnosis. *J Virol Methods*, 2009. **162**(1–2): 179–183.

216. Iram, N. et al., Heterologous expression, characterization and evaluation of the matrix protein from Newcastle disease virus as a target for antiviral therapies. *Appl Microbiol Biotechnol*, 2014. **98**(4): 1691–1701.

217. Loney, C. et al., Paramyxovirus ultrastructure and genome packaging: Cryo-electron tomography of sendai virus. *J Virol*, 2009. **83**(16): 8191–8197.

218. Liljeroos, L. et al., Electron cryotomography of measles virus reveals how matrix protein coats the ribonucleocapsid within intact virions. *Proc Natl Acad Sci U S A*, 2011. **108**(44): 18085–18090.

219. Battisti, A.J. et al., Structure and assembly of a paramyxovirus matrix protein. *Proc Natl Acad Sci U S A*, 2012. **109**(35): 13996–14000.

220. Schoehn, G. et al., The 12 Å structure of trypsin-treated measles virus N-RNA. *J Mol Biol*, 2004. **339**(2): 301–312.

221. Bhella, D., A. Ralph, and R.P. Yeo, Conformational flexibility in recombinant measles virus nucleocapsids visualised by cryo-negative stain electron microscopy and real-space helical reconstruction. *J Mol Biol*, 2004. **340**(2): 319–331.

222. Barbet-Massin, E. et al., Insights into the structure and dynamics of measles virus nucleocapsids by (1)H-detected solid-state NMR. *Biophys J*, 2014. **107**(4): 941–946.

223. Chen, L. et al., Cloning, expression, and crystallization of the fusion protein of Newcastle disease virus. *Virology*, 2001. **290**(2): 290–299.

224. Zhu, J. et al., Crystallization and preliminary X-ray crystallographic analysis of the trimer core from measles virus fusion protein. *Acta Crystallogr D Biol Crystallogr*, 2003. **59**(Pt 3): 587–590.

225. Lou, Z. et al., Crystal structures of Nipah and Hendra virus fusion core proteins. *FEBS J*, 2006. **273**(19): 4538–4547.

226. Crennell, S. et al., Crystal structure of the multifunctional paramyxovirus hemagglutinin-neuraminidase. *Nat Struct Biol*, 2000. **7**(11): 1068–1074.

227. Yuan, P. et al., Structure of the Newcastle disease virus hemagglutinin-neuraminidase (HN) ectodomain reveals a four-helix bundle stalk. *Proc Natl Acad Sci U S A*, 2011. **108**(36): 14920–14925.

228. Hashiguchi, T. et al., Crystal structure of measles virus hemagglutinin provides insight into effective vaccines. *Proc Natl Acad Sci U S A*, 2007. **104**(49): 19535–19540.

229. Santiago, C. et al., Structure of the measles virus hemagglutinin bound to the CD46 receptor. *Nat Struct Mol Biol*, 2010. **17**(1): 124–129.

230. Xu, K. et al., Host cell recognition by the henipaviruses: Crystal structures of the Nipah G attachment glycoprotein and its complex with ephrin-B3. *Proc Natl Acad Sci U S A*, 2008. **105**(29): 9953–9958.

231. Bowden, T.A. et al., Crystal structure and carbohydrate analysis of Nipah virus attachment glycoprotein: A template for antiviral and vaccine design. *J Virol*, 2008. **82**(23): 11628–11636.

232. Xu, K. et al., Crystal structure of the Hendra virus attachment G glycoprotein bound to a potent cross-reactive neutralizing human monoclonal antibody. *PLoS Pathog*, 2013. **9**(10): e1003684.

233. Bruhn, J.F. et al., Crystal structure of the nipah virus phosphoprotein tetramerization domain. *J Virol*, 2014. **88**(1): 758–762.

234. Coronel, E.C. et al., Human parainfluenza virus type 1 matrix and nucleoprotein genes transiently expressed in mammalian cells induce the release of virus-like particles containing nucleocapsid-like structures. *J Virol*, 1999. **73**(8): 7035–7038.

235. Sugahara, F. et al., Paramyxovirus Sendai virus-like particle formation by expression of multiple viral proteins and acceleration of its release by C protein. *Virology*, 2004. **325**(1): 1–10.

236. Gosselin-Grenet, A.S., G. Mottet-Osman, and L. Roux, Sendai virus particle production: Basic requirements and role of the SYWST motif present in HN cytoplasmic tail. *Virology*, 2010. **405**(2): 439–447.

237. Pantua, H.D. et al., Requirements for the assembly and release of Newcastle disease virus-like particles. *J Virol*, 2006. **80**(22): 11062–11073.

238. Patch, J.R. et al., Quantitative analysis of Nipah virus proteins released as virus-like particles reveals central role for the matrix protein. *Virol J*, 2007. **4**: 1.

239. Pohl, C. et al., Measles virus M and F proteins associate with detergent-resistant membrane fractions and promote formation of virus-like particles. *J Gen Virol*, 2007. **88**(Pt 4): 1243–1250.

240. Li, M. et al., Mumps virus matrix, fusion, and nucleocapsid proteins cooperate for efficient production of virus-like particles. *J Virol*, 2009. **83**(14): 7261–7272.

241. McGinnes, L.W. et al., Assembly and biological and immunological properties of Newcastle disease virus-like particles. *J Virol*, 2010. **84**(9): 4513–4523.

242. Walpita, P. et al., Vaccine potential of Nipah virus-like particles. *PLoS One*, 2011. **6**(4): e18437.

243. Liu, F. et al., Formation of peste des petits ruminants spikeless virus-like particles by co-expression of M and N proteins in insect cells. *Res Vet Sci*, 2014. **96**(1): 213–216.

244. Liu, F. et al., Budding of peste des petits ruminants virus-like particles from insect cell membrane based on intracellular co-expression of peste des petits ruminants virus M, H and N proteins by recombinant baculoviruses. *J Virol Methods*, 2014. **207**: 78–85.

245. Li, W. et al., Self-assembly and release of peste des petits ruminants virus-like particles in an insect cell-baculovirus system and their immunogenicity in mice and goats. *PLoS One*, 2014. **9**(8): e104791.

246. Rabu, A. et al., Chimeric Newcastle disease virus nucleocapsid with parts of viral hemagglutinin-neuraminidase and fusion proteins. *Acta Virol*, 2002. **46**(4): 211–217.

247. Sivasamugham, L.A. et al., Recombinant Newcastle Disease virus capsids displaying enterovirus 71 VP1 fragment induce a strong immune response in rabbits. *J Med Virol*, 2006. **78**(8): 1096–1104.

248. Ch'ng, W.C. et al., Immunogenicity of a truncated enterovirus 71 VP1 protein fused to a Newcastle disease virus nucleocapsid protein fragment in mice. *Acta Virol*, 2011. **55**(3): 227–233.

249. Ch'ng, W.C. et al., Immunization with recombinant enterovirus 71 viral capsid protein 1 fragment stimulated antibody responses in hamsters. *Virol J*, 2012. **9**: 155.

250. Jacob, D. et al., Whole *Pichia pastoris* yeast expressing measles virus nucleoprotein as a production and delivery system to multimerize *Plasmodium* antigens. *PLoS One*, 2014. **9**(1): e86658.

251. Nakanishi, M. and M. Otsu, Development of Sendai virus vectors and their potential applications in gene therapy and regenerative medicine. *Curr Gene Ther*, 2012. **12**(5): 410–416.

252. Le Bayon, J.C. et al., Recent developments with live-attenuated recombinant paramyxovirus vaccines. *Rev Med Virol*, 2013. **23**(1): 15–34.

253. Galanis, E., Therapeutic potential of oncolytic measles virus: Promises and challenges. *Clin Pharmacol Ther*, 2010. **88**(5): 620–625.

254. Zamarin, D. and P. Palese, Oncolytic Newcastle disease virus for cancer therapy: Old challenges and new directions. *Future Microbiol*, 2012. **7**(3): 347–367.

255. Hudacek, A.W., C.K. Navaratnarajah, and R. Cattaneo, Development of measles virus-based shielded oncolytic vectors: Suitability of other paramyxovirus glycoproteins. *Cancer Gene Ther*, 2013. **20**(2): 109–116.

256. Deng, R. et al., Functional chimeric HN glycoproteins derived from Newcastle disease virus and human parainfluenza virus-3. *Arch Virol Suppl*, 1997. **13**: 115–130.

257. Peeters, B.P. et al., Generation of a recombinant chimeric Newcastle disease virus vaccine that allows serological differentiation between vaccinated and infected animals. *Vaccine*, 2001. **19**(13–14): 1616–1627.

258. Das, S.C., M.D. Baron, and T. Barrett, Recovery and characterization of a chimeric rinderpest virus with the glycoproteins of peste-des-petits-ruminants virus: Homologous F and H proteins are required for virus viability. *J Virol*, 2000. **74**(19): 9039–9047.

259. Yang, L. et al., A paramyxovirus-vectored intranasal vaccine against Ebola virus is immunogenic in vector-immune animals. *Virology*, 2008. **377**(2): 255–264.

260. Bukreyev, A. et al., Chimeric human parainfluenza virus bearing the Ebola virus glycoprotein as the sole surface protein is immunogenic and highly protective against Ebola virus challenge. *Virology*, 2009. **383**(2): 348–361.

261. Nayak, B. et al., Immunization of chickens with Newcastle disease virus expressing H5 hemagglutinin protects against highly pathogenic H5N1 avian influenza viruses. *PLoS One*, 2009. **4**(8): e6509.

262. Phan, S.I. et al., A respiratory syncytial virus (RSV) vaccine based on parainfluenza virus 5 (PIV5). *Vaccine*, 2014. **32**(25): 3050–3057.

263. Zhao, W. et al., Newcastle disease virus (NDV) recombinants expressing infectious laryngotracheitis virus (ILTV) glycoproteins gB and gD protect chickens against ILTV and NDV challenges. *J Virol*, 2014. **88**(15): 8397–8406.

264. Murawski, M.R. et al., Newcastle disease virus-like particles containing respiratory syncytial virus G protein induced protection in BALB/c mice, with no evidence of immunopathology. *J Virol*, 2010. **84**(2): 1110–1123.

265. McGinnes, L.W. et al., Assembly and immunological properties of Newcastle disease virus-like particles containing the respiratory syncytial virus F and G proteins. *J Virol*, 2011. **85**(1): 366–377.

266. Schmidt, M.R. et al., Long-term and memory immune responses in mice against Newcastle disease virus-like particles containing respiratory syncytial virus glycoprotein ectodomains. *J Virol*, 2012. **86**(21): 11654–11662.

267. Shen, H. et al., Assembly and immunological properties of a bivalent virus-like particle (VLP) for avian influenza and Newcastle disease. *Virus Res*, 2013. **178**(2): 430–436.

268. Park, J.K. et al., Virus-like particle vaccine confers protection against a lethal newcastle disease virus challenge in chickens and allows a strategy of differentiating infected from vaccinated animals. *Clin Vaccin Immunol*, 2014. **21**(3): 360–365.

269. Spiegel, M. et al., Pseudotype formation of Moloney murine leukemia virus with Sendai virus glycoprotein F. *J Virol*, 1998. **72**(6): 5296–5302.

270. Voelkel, C. et al., Pseudotype-independent nonspecific uptake of gammaretroviral and lentiviral particles in human cells. *Hum Gene Ther*, 2012. **23**(3): 274–286.

271. Kobayashi, M. et al., Pseudotyped lentivirus vectors derived from simian immunodeficiency virus SIVagm with envelope glycoproteins from paramyxovirus. *J Virol*, 2003. **77**(4): 2607–2614.

272. Jung, C. et al., Lentiviral vectors pseudotyped with envelope glycoproteins derived from human parainfluenza virus type 3. *Biotechnol Prog*, 2004. **20**(6): 1810–1816.

273. Funke, S. et al., Pseudotyping lentiviral vectors with the wild-type measles virus glycoproteins improves titer and selectivity. *Gene Ther*, 2009. **16**(5): 700–705.

274. Khetawat, D. and C.C. Broder, A functional henipavirus envelope glycoprotein pseudotyped lentivirus assay system. *Virol J*, 2010. **7**: 312.

275. Enkirch, T. et al., Targeted lentiviral vectors pseudotyped with the Tupaia paramyxovirus glycoproteins. *Gene Ther*, 2013. **20**(1): 16–23.

276. Kaku, Y. et al., A neutralization test for specific detection of Nipah virus antibodies using pseudotyped vesicular stomatitis virus expressing green fluorescent protein. *J Virol Methods*, 2009. **160**(1–2): 7–13.

277. Kaku, Y. et al., Second generation of pseudotype-based serum neutralization assay for Nipah virus antibodies: Sensitive and high-throughput analysis utilizing secreted alkaline phosphatase. *J Virol Methods*, 2012. **179**(1): 226–232.

278. Sinnathamby, G. et al., Recombinant hemagglutinin protein of rinderpest virus expressed in insect cells induces humoral and cell mediated immune responses in cattle. *Vaccine*, 2001. **19**(28–29): 3870–3876.

279. Layton, G.T. et al., Induction of single and dual cytotoxic T-lymphocyte responses to viral proteins in mice using recombinant hybrid Ty-virus-like particles. *Immunology*, 1996. **87**(2): 171–178.

280. Zhao, Y. and R.W. Hammond, Development of a candidate vaccine for Newcastle disease virus by epitope display in the Cucumber mosaic virus capsid protein. *Biotechnol Lett*, 2005. **27**(6): 375–382.

281. Natilla, A., R.W. Hammond, and L.G. Nemchinov, Epitope presentation system based on cucumber mosaic virus coat protein expressed from a potato virus X-based vector. *Arch Virol*, 2006. **151**(7): 1373–1386.

282. Natilla, A. and R.W. Hammond, Maize rayado fino virus virus-like particles expressed in tobacco plants: A new platform for cysteine selective bioconjugation peptide display. *J Virol Methods*, 2011. **178**(1–2): 209–215.

283. Huang, L. et al., Construction and biological characterisation of recombinant porcine circovirus type 2 expressing the V5 epitope tag. *Virus Res*, 2011. **161**(2): 115–123.

284. Yap, W.B. et al., Display of the antigenic region of Nipah virus nucleocapsid protein on hepatitis B virus capsid. *J Biosci Bioeng*, 2012. **113**(1): 26–29.

285. Meric, C., D. Spehner, and V. Mazarin, Respiratory syncytial virus nucleocapsid protein (N) expressed in insect cells forms nucleocapsid-like structures. *Virus Res*, 1994. **31**(2): 187–201.

286. Petraityte-Burneikiene, R. et al., Generation of recombinant metapneumovirus nucleocapsid protein as nucleocapsid-like particles and development of virus-specific monoclonal antibodies. *Virus Res*, 2011. **161**(2): 131–139.

287. Roux, X. et al., Sub-nucleocapsid nanoparticles: A nasal vaccine against respiratory syncytial virus. *PLoS One*, 2008. **3**(3): e1766.

288. Riffault, S. et al., A new subunit vaccine based on nucleoprotein nanoparticles confers partial clinical and virological protection in calves against bovine respiratory syncytial virus. *Vaccine*, 2010. **28**(21): 3722–3734.

289. Remot, A. et al., Nucleoprotein nanostructures combined with adjuvants adapted to the neonatal immune context: A candidate mucosal RSV vaccine. *PLoS One*, 2012. **7**(5): e37722.

290. Smith, G. et al., Respiratory syncytial virus fusion glycoprotein expressed in insect cells form protein nanoparticles that induce protective immunity in cotton rats. *PLoS One*, 2012. **7**(11): e50852.

291. Glenn, G.M. et al., Safety and immunogenicity of a Sf9 insect cell-derived respiratory syncytial virus fusion protein nanoparticle vaccine. *Vaccine*, 2013. **31**(3): 524–532.

292. Garlapati, S. et al., Enhanced immune responses and protection by vaccination with respiratory syncytial virus fusion protein formulated with CpG oligodeoxynucleotide and innate defense regulator peptide in polyphosphazene microparticles. *Vaccine*, 2012. **30**(35): 5206–5214.

293. Tawar, R.G. et al., Crystal structure of a nucleocapsid-like nucleoprotein-RNA complex of respiratory syncytial virus. *Science*, 2009. **326**(5957): 1279–1283.

294. El Omari, K. et al., Structures of respiratory syncytial virus nucleocapsid protein from two crystal forms: Details of potential packing interactions in the native helical form. *Acta Crystallogr Sect F Struct Biol Cryst Commun*, 2011. **67**(Pt 10): 1179–1183.

295. Liljeroos, L. et al., Architecture of respiratory syncytial virus revealed by electron cryotomography. *Proc Natl Acad Sci U S A*, 2013. **110**(27): 11133–11138.

296. Leyrat, C. et al., Structure and self-assembly of the calcium binding matrix protein of human metapneumovirus. *Structure*, 2014. **22**(1): 136–148.

297. Mitra, R. et al., The human respiratory syncytial virus matrix protein is required for maturation of viral filaments. *J Virol*, 2012. **86**(8): 4432–4443.

298. Loo, L.H. et al., Evidence for the interaction of the human metapneumovirus G and F proteins during virus-like particle formation. *Virol J*, 2013. **10**: 294.

299. Cox, R.G. et al., Human metapneumovirus virus-like particles induce protective B and T cell responses in a mouse model. *J Virol*, 2014. **88**(11): 6368–6379.

300. Quan, F.S. et al., Viruslike particle vaccine induces protection against respiratory syncytial virus infection in mice. *J Infect Dis*, 2011. **204**(7): 987–995.

301. Ko, E.J. et al., Virus-like nanoparticle and DNA vaccination confers protection against respiratory syncytial virus by modulating innate and adaptive immune cells. *Nanomedicine*, 2015. **11**(1): 99–108.

302. Hwang, H.S. et al., Co-immunization with virus-like particle and DNA vaccines induces protection against respiratory syncytial virus infection and bronchiolitis. *Antiviral Res*, 2014. **110**: 115–123.

303. Lee, J.S. et al., Baculovirus-expressed virus-like particle vaccine in combination with DNA encoding the fusion protein confers protection against respiratory syncytial virus. *Vaccine*, 2014. **32**(44): 5866–5874.

304. Schmidt, A.C. et al., Recombinant bovine/human parainfluenza virus type 3 (B/HPIV3) expressing the respiratory syncytial virus (RSV) G and F proteins can be used to achieve simultaneous mucosal immunization against RSV and HPIV3. *J Virol*, 2001. **75**(10): 4594–4603.

305. Takimoto, T. et al., Recombinant Sendai virus expressing the G glycoprotein of respiratory syncytial virus (RSV) elicits immune protection against RSV. *J Virol*, 2004. **78**(11): 6043–6047.

306. Zimmer, G. et al., A chimeric respiratory syncytial virus fusion protein functionally replaces the F and HN glycoproteins in recombinant Sendai virus. *J Virol*, 2005. **79**(16): 10467–10477.

307. Zhan, X. et al., Respiratory syncytial virus (RSV) fusion protein expressed by recombinant Sendai virus elicits B-cell and T-cell responses in cotton rats and confers protection against RSV subtypes A and B. *Vaccine*, 2007. **25**(52): 8782–8793.

308. Martinez-Sobrido, L. et al., Protection against respiratory syncytial virus by a recombinant Newcastle disease virus vector. *J Virol*, 2006. **80**(3): 1130–1139.

309. Mok, H. et al., Evaluation of measles vaccine virus as a vector to deliver respiratory syncytial virus fusion protein or Epstein-Barr virus glycoprotein gp350. *Open Virol J*, 2012. **6**: 12–22.

310. Fonseca, W. et al., A recombinant influenza virus vaccine expressing the F protein of respiratory syncytial virus. *Arch Virol*, 2014. **159**(5): 1067–1077.

311. Bian, C. et al., Influenza virus vaccine expressing fusion and attachment protein epitopes of respiratory syncytial virus induces protective antibodies in BALB/c mice. *Antiviral Res*, 2014. **104**: 110–117.

312. Zhang, P. et al., Characterization of recombinant influenza A virus as a vector expressing respiratory syncytial virus fusion protein epitopes. *J Gen Virol*, 2014. **95**(Pt 9): 1886–1891.

313. Kahn, J.S. et al., Recombinant vesicular stomatitis virus expressing respiratory syncytial virus (RSV) glycoproteins: RSV fusion protein can mediate infection and cell fusion. *Virology*, 1999. **254**(1): 81–91.

314. Levy, C. et al., Virus-like particle vaccine induces crossprotection against human metapneumovirus infections in mice. *Vaccine*, 2013. **31**(25): 2778–2785.

315. Lim, S.H. et al., Surface display of respiratory syncytial virus glycoproteins in *Lactococcus lactis* NZ9000. *Lett Appl Microbiol*, 2010. **51**(6): 658–664.

316. Rigter, A. et al., A protective and safe intranasal RSV vaccine based on a recombinant prefusion-like form of the F protein bound to bacterium-like particles. *PLoS One*, 2013. **8**(8): e71072.

317. Bastien, N., M. Trudel, and C. Simard, Protective immune responses induced by the immunization of mice with a recombinant bacteriophage displaying an epitope of the human respiratory syncytial virus. *Virology*, 1997. **234**(1): 118–122.

318. Belanger, H. et al., Human respiratory syncytial virus vaccine antigen produced in plants. *FASEB J*, 2000. **14**(14): 2323–2328.

319. Woo, W.P. et al., Hepatitis B surface antigen vector delivers protective cytotoxic T-lymphocyte responses to disease-relevant foreign epitopes. *J Virol*, 2006. **80**(8): 3975–3984.

320. De Baets, S. et al., Recombinant influenza virus carrying the respiratory syncytial virus (RSV) F85–93 CTL epitope reduces RSV replication in mice. *J Virol*, 2013. **87**(6): 3314–3323.

321. Correia, B.E. et al., Proof of principle for epitope-focused vaccine design. *Nature*, 2014. **507**(7491): 201–206.

322. Stone, J.W. et al., Gold nanorod vaccine for respiratory syncytial virus. *Nanotechnology*, 2013. **24**(29): 295102.

323. Horie, M. et al., Comprehensive analysis of endogenous bornavirus-like elements in eukaryote genomes. *Philos Trans R Soc Lond B Biol Sci*, 2013. **368**(1626): 20120499.

324. Kuhn, J.H. et al., Nyamiviridae: Proposal for a new family in the order Mononegavirales. *Arch Virol*, 2013. **158**(10): 2209–2226.

325. Kraus, I. et al., Oligomerization and assembly of the matrix protein of Borna disease virus. *FEBS Lett*, 2005. **579**(12): 2686–2692.

326. Neumann, P. et al., Crystal structure of the Borna disease virus matrix protein (BDV-M) reveals ssRNA binding properties. *Proc Natl Acad Sci U S A*, 2009. **106**(10): 3710–3715.

327. Herrel, M. et al., Tick-borne Nyamanini virus replicates in the nucleus and exhibits unusual genome and matrix protein properties. *J Virol*, 2012. **86**(19): 10739–10747.

328. Rudolph, M.G. et al., Crystal structure of the borna disease virus nucleoprotein. *Structure*, 2003. **11**(10): 1219–1226.

329. Perez, M. et al., Generation and characterization of a recombinant vesicular stomatitis virus expressing the glycoprotein of Borna disease virus. *J Virol*, 2007. **81**(11): 5527–5536.

330. Perez, M. and J.C. de la Torre, Identification of the Borna disease virus (BDV) proteins required for the formation of BDV-like particles. *J Gen Virol*, 2005. **86**(Pt 7): 1891–1895.

331. Daito, T. et al., A novel borna disease virus vector system that stably expresses foreign proteins from an intercistronic noncoding region. *J Virol*, 2011. **85**(23): 12170–12178.

332. Schneider, L.G. et al., Rabies group-specific ribonucleoprotein antigen and a test system for grouping and typing of rhabdoviruses. *J Virol*, 1973. **11**(5): 748–755.

333. Reid-Sanden, F.L. et al., Rabies diagnostic reagents prepared from a rabies N gene recombinant expressed in baculovirus. *J Clin Microbiol*, 1990. **28**(5): 858–863.

334. Prehaud, C. et al., Expression, characterization, and purification of a phosphorylated rabies nucleoprotein synthesized in insect cells by baculovirus vectors. *Virology*, 1990. **178**(2): 486–497.

335. Fu, Z.F. et al., Rabies virus nucleoprotein expressed in and purified from insect cells is efficacious as a vaccine. *Proc Natl Acad Sci U S A*, 1991. **88**(5): 2001–2005.

336. Katz, J.B., A.L. Shafer, and K.A. Eernisse, Construction and insect larval expression of recombinant vesicular stomatitis nucleocapsid protein and its use in competitive ELISA. *J Virol Methods*, 1995. **54**(2–3): 145–157.

337. Pinto, R.M., A. Bosch, and D.H. Bishop, Structures associated with the expression of rabies virus structural genes in insect cells. *Virus Res*, 1994. **31**(1): 139–145.

338. Iseni, F. et al., Characterization of rabies virus nucleocapsids and recombinant nucleocapsid-like structures. *J Gen Virol*, 1998. **79**(Pt 12): 2909–2919.

339. Hooper, D.C. et al., Rabies ribonucleocapsid as an oral immunogen and immunological enhancer. *Proc Natl Acad Sci U S A*, 1994. **91**(23): 10908–10912.

340. Yin, X. et al., Rabies virus nucleoprotein expressed in silkworm pupae at high-levels and evaluation of immune responses in mice. *J Biotechnol*, 2013. **163**(3): 333–338.

341. Oberg, L.A. et al., Bacterially expressed nucleoprotein of infectious hematopoietic necrosis virus augments protective immunity induced by the glycoprotein vaccine in fish. *J Virol*, 1991. **65**(8): 4486–4489.

342. Das, T. and A.K. Banerjee, Expression of the vesicular stomatitis virus nucleocapsid protein gene in *Escherichia coli*: Analysis of its biological activity in vitro. *Virology*, 1993. **193**(1): 340–347.

343. Goto, H. et al., Expression of the nucleoprotein of rabies virus in *Escherichia coli* and mapping of antigenic sites. *Arch Virol*, 1995. **140**(6): 1061–1074.

344. Gupta, A.K. and A.K. Banerjee, Expression and purification of vesicular stomatitis virus N-P complex from *Escherichia coli*: Role in genome RNA transcription and replication in vitro. *J Virol*, 1997. **71**(6): 4264–4271.

345. Kucinskaite, I. et al., Antigenic characterisation of yeast-expressed lyssavirus nucleoproteins. *Virus Genes*, 2007. **35**(3): 521–529.

346. Perea Arango, I. et al., Expression of the rabies virus nucleoprotein in plants at high-levels and evaluation of immune responses in mice. *Plant Cell Rep*, 2008. **27**(4): 677–685.

347. Kieny, M.P. et al., Expression of rabies virus glycoprotein from a recombinant vaccinia virus. *Nature*, 1984. **312**(5990): 163–166.

348. Burger, S.R. et al., Stable expression of rabies virus glycoprotein in Chinese hamster ovary cells. *J Gen Virol*, 1991. **72**(Pt 2): 359–367.

349. Masters, P.S. et al., Structure and expression of the glycoprotein gene of Chandipura virus. *Virology*, 1989. **171**(1): 285–290.

350. Prehaud, C. et al., Immunogenic and protective properties of rabies virus glycoprotein expressed by baculovirus vectors. *Virology*, 1989. **173**(2): 390–399.

351. Koener, J.F. and J.A. Leong, Expression of the glycoprotein gene from a fish rhabdovirus by using baculovirus vectors. *J Virol*, 1990. **64**(1): 428–430.

352. Johal, J. et al., Antigenic characterization of bovine ephemeral fever rhabdovirus G and GNS glycoproteins expressed from recombinant baculoviruses. *Arch Virol*, 2008. **153**(9): 1657–1665.

353. Fu, Z.F. et al., Oral vaccination of raccoons (Procyon lotor) with baculovirus-expressed rabies virus glycoprotein. *Vaccine*, 1993. **11**(9): 925–928.

354. Cain, K.D. et al., Immunogenicity of a recombinant infectious hematopoietic necrosis virus glycoprotein produced in insect cells. *Dis Aquat Organ*, 1999. **36**(1): 67–72.

355. Rolls, M.M. et al., Novel infectious particles generated by expression of the vesicular stomatitis virus glycoprotein from a self-replicating RNA. *Cell*, 1994. **79**(3): 497–506.

356. Fontana, D. et al., Rabies virus-like particles expressed in HEK293 cells. *Vaccine*, 2014. **32**(24): 2799–2804.

357. Verjan, N. et al., A soluble nonglycosylated recombinant infectious hematopoietic necrosis virus (IHNV) G-protein induces IFNs in rainbow trout (Oncorhynchus mykiss). *Fish Shellfish Immunol*, 2008. **25**(1–2): 170–180.

358. Singh, A. et al., Enhanced expression of rabies virus surface G-protein in *Escherichia coli* using SUMO fusion. *Protein J*, 2012. **31**(1): 68–74.

359. Klepfer, S.R. et al., Characterization of rabies glycoprotein expressed in yeast. *Arch Virol*, 1993. **128**(3–4): 269–286.

360. Sakamoto, S. et al., Studies on the structures and antigenic properties of rabies virus glycoprotein analogues produced in yeast cells. *Vaccine*, 1999. **17**(3): 205–218.

361. McGarvey, P.B. et al., Expression of the rabies virus glycoprotein in transgenic tomatoes. *Biotechnology (N Y)*, 1995. **13**(13): 1484–1487.

362. Ashraf, S. et al., High level expression of surface glycoprotein of rabies virus in tobacco leaves and its immunoprotective activity in mice. *J Biotechnol*, 2005. **119**(1): 1–14.

363. Lyles, D.S. et al., Complementation of M gene mutants of vesicular stomatitis virus by plasmid-derived M protein converts spherical extracellular particles into native bullet shapes. *Virology*, 1996. **217**(1): 76–87.

364. Mebatsion, T., F. Weiland, and K.K. Conzelmann, Matrix protein of rabies virus is responsible for the assembly and budding of bullet-shaped particles and interacts with the transmembrane spike glycoprotein G. *J Virol*, 1999. **73**(1): 242–250.

365. Assenberg, R. et al., Expression, purification and crystallization of a lyssavirus matrix (M) protein. *Acta Crystallogr Sect F Struct Biol Cryst Commun*, 2008. **64**(Pt 4): 258–262.

366. Albertini, A.A. et al., Crystal structure of the rabies virus nucleoprotein-RNA complex. *Science*, 2006. **313**(5785): 360–363.

367. Albertini, A.A. et al., Isolation and crystallization of a unique size category of recombinant Rabies virus Nucleoprotein-RNA rings. *J Struct Biol*, 2007. **158**(1): 129–133.

368. Green, T.J. and M. Luo, Structure of the vesicular stomatitis virus nucleocapsid in complex with the nucleocapsid-binding domain of the small polymerase cofactor, P. *Proc Natl Acad Sci U S A*, 2009. **106**(28): 11713–11718.

369. Leyrat, C. et al., Structure of the vesicular stomatitis virus N(0)-P complex. *PLoS Pathog*, 2011. **7**(9): e1002248.

370. Roche, S. et al., Structure of the prefusion form of the vesicular stomatitis virus glycoprotein G. *Science*, 2007. **315**(5813): 843–848.

371. Baquero, E. et al., Crystallization and preliminary X-ray analysis of Chandipura virus glycoprotein G. *Acta Crystallogr Sect F Struct Biol Cryst Commun*, 2012. **68**(Pt 9): 1094–1097.

372. Gaudier, M., Y. Gaudin, and M. Knossow, Crystal structure of vesicular stomatitis virus matrix protein. *EMBO J*, 2002. **21**(12): 2886–2892.

373. Graham, S.C. et al., Rhabdovirus matrix protein structures reveal a novel mode of self-association. *PLoS Pathog*, 2008. **4**(12): e1000251.

374. Ge, P. et al., Cryo-EM model of the bullet-shaped vesicular stomatitis virus. *Science*, 2010. **327**(5966): 689–693.

375. Koser, M.L. et al., Rabies virus nucleoprotein as a carrier for foreign antigens. *Proc Natl Acad Sci U S A*, 2004. **101**(25): 9405–9410.

376. Lo, H.L. and J.K. Yee, Production of vesicular stomatitis virus G glycoprotein (VSV-G) pseudotyped retroviral vectors. *Curr Protoc Hum Genet*, 2007. Chapter 12: Unit 12.7.

377. Liang, M. et al., Retargeting vesicular stomatitis virus glycoprotein pseudotyped lentiviral vectors with enhanced stability by in situ synthesized polymer shell. *Hum Gene Ther Methods*, 2013. **24**(1): 11–18.

378. Wei, Y. et al., Lentiviral vectors enveloped with rabies virus glycoprotein can be used as a novel retrograde tracer to assess nerve recovery in rat sciatic nerve injury models. *Cell Tissue Res*, 2014. **355**(2): 255–266.

379. Beier, K.T. et al., Vesicular stomatitis virus with the rabies virus glycoprotein directs retrograde transsynaptic transport among neurons in vivo. *Front Neural Circuits*, 2013. **7**: 11.

380. Zimmer, G. et al., Pseudotyping of vesicular stomatitis virus with the envelope glycoproteins of highly pathogenic avian influenza viruses. *J Gen Virol*, 2014. **95**(Pt 8): 1634–1639.

381. Lathe, R.F. et al., M13 bacteriophage vectors for the expression of foreign proteins in *Escherichia coli*: The rabies glycoprotein. *J Mol Appl Genet*, 1984. **2**(4): 331–342.

382. Bendahmane, M. et al., Display of epitopes on the surface of tobacco mosaic virus: Impact of charge and isoelectric point of the epitope on virus-host interactions. *J Mol Biol*, 1999. **290**(1): 9–20.

383. Yusibov, V. et al., Expression in plants and immunogenicity of plant virus-based experimental rabies vaccine. *Vaccine*, 2002. **20**(25–26): 3155–3164.

384. Ma, Y. and J. Li, Vesicular stomatitis virus as a vector to deliver virus-like particles of human norovirus: A new vaccine candidate against an important noncultivable virus. *J Virol*, 2011. **85**(6): 2942–2952.

385. Osakada, F. and E.M. Callaway, Design and generation of recombinant rabies virus vectors. *Nat Protoc*, 2013. **8**(8): 1583–1601.

386. Kuhn, J.H. et al., Proposal for a revised taxonomy of the family Filoviridae: Classification, names of taxa and viruses, and virus abbreviations. *Arch Virol*, 2010. **155**(12): 2083–2103.

387. Kuhn, J.H. et al., Virus nomenclature below the species level: A standardized nomenclature for filovirus strains and variants rescued from cDNA. *Arch Virol*, 2014. **159**(5): 1229–1237.

388. Taylor, D.J. et al., Evidence that ebolaviruses and cuevaviruses have been diverging from marburgviruses since the Miocene. *Peer J*, 2014. **2**: e556.

389. Gire, S.K. et al., Genomic surveillance elucidates Ebola virus origin and transmission during the 2014 outbreak. *Science*, 2014. **345**(6202): 1369–1372.

390. Cohen, J., Infectious Disease. Ebola vaccines racing forward at record pace. *Science*, 2014. **345**(6202): 1228–1229.

391. Belyi, V.A., A.J. Levine, and A.M. Skalka, Unexpected inheritance: Multiple integrations of ancient bornavirus and ebolavirus/marburgvirus sequences in vertebrate genomes. *PLoS Pathog*, 2010. **6**(7): e1001030.

392. Hevey, M. et al., Marburg virus vaccines based upon alphavirus replicons protect guinea pigs and nonhuman primates. *Virology*, 1998. **251**(1): 28–37.

393. Pushko, P. et al., Recombinant RNA replicons derived from attenuated Venezuelan equine encephalitis virus protect guinea pigs and mice from Ebola hemorrhagic fever virus. *Vaccine*, 2000. **19**(1): 142–153.

394. Geisbert, T.W. et al., Evaluation in nonhuman primates of vaccines against Ebola virus. *Emerg Infect Dis*, 2002. **8**(5): 503–507.

395. Bavari, S. et al., Lipid raft microdomains: A gateway for compartmentalized trafficking of Ebola and Marburg viruses. *J Exp Med*, 2002. **195**(5): 593–602.

396. Warfield, K.L. et al., Ebola virus-like particles protect from lethal Ebola virus infection. *Proc Natl Acad Sci U S A*, 2003. **100**(26): 15889–15894.

397. Swenson, D.L. et al., Generation of Marburg virus-like particles by co-expression of glycoprotein and matrix protein. *FEMS Immunol Med Microbiol*, 2004. **40**(1): 27–31.

398. Warfield, K.L. et al., Marburg virus-like particles protect guinea pigs from lethal Marburg virus infection. *Vaccine*, 2004. **22**(25–26): 3495–3502.

399. Bosio, C.M. et al., Ebola and Marburg virus-like particles activate human myeloid dendritic cells. *Virology*, 2004. **326**(2): 280–287.

400. Swenson, D.L. et al., Virus-like particles exhibit potential as a pan-filovirus vaccine for both Ebola and Marburg viral infections. *Vaccine*, 2005. **23**(23): 3033–3042.

401. Warfield, K.L. et al., Ebola virus-like particle-based vaccine protects nonhuman primates against lethal Ebola virus challenge. *J Infect Dis*, 2007. **196**(Suppl 2): S430–S437.

402. Warfield, K.L. et al., Filovirus-like particles as vaccines and discovery tools. *Expert Rev Vaccin*, 2005. **4**(3): 429–440.

403. Maruyama, J. et al., Characterization of the envelope glycoprotein of a novel filovirus, lloviu virus. *J Virol*, 2014. **88**(1): 99–109.

404. Ye, L. et al., Ebola virus-like particles produced in insect cells exhibit dendritic cell stimulating activity and induce neutralizing antibodies. *Virology*, 2006. **351**(2): 260–270.

405. Warfield, K.L. et al., Filovirus-like particles produced in insect cells: Immunogenicity and protection in rodents. *J Infect Dis*, 2007. **196**(Suppl 2): S421–S429.

406. Sun, Y. et al., Protection against lethal challenge by Ebola virus-like particles produced in insect cells. *Virology*, 2009. **383**(1): 12–21.

407. Warfield, K.L. et al., Vaccinating captive chimpanzees to save wild chimpanzees. *Proc Natl Acad Sci U S A*, 2014. **111**(24): 8873–8876.

408. Hu, L. et al., Biophysical characterization and conformational stability of Ebola and Marburg virus-like particles. *J Pharm Sci*, 2011. **100**(12): 5156–5173.

409. Weissenhorn, W. et al., Crystal structure of the Ebola virus membrane fusion subunit, GP2, from the envelope glycoprotein ectodomain. *Mol Cell*, 1998. **2**(5): 605–616.

410. Malashkevich, V.N. et al., Core structure of the envelope glycoprotein GP2 from Ebola virus at 1.9-Å resolution. *Proc Natl Acad Sci U S A*, 1999. **96**(6): 2662–2667.

411. Dessen, A. et al., Crystallization and preliminary X-ray analysis of the matrix protein from Ebola virus. *Acta Crystallogr D Biol Crystallogr*, 2000. **56**(Pt 6): 758–760.

412. Dessen, A. et al., Crystal structure of the matrix protein VP40 from Ebola virus. *EMBO J*, 2000. **19**(16): 4228–4236.

413. Gomis-Ruth, F.X. et al., The matrix protein VP40 from Ebola virus octamerizes into pore-like structures with specific RNA binding properties. *Structure*, 2003. **11**(4): 423–433.

414. Lee, J.E. et al., Structure of the Ebola virus glycoprotein bound to an antibody from a human survivor. *Nature*, 2008. **454**(7201): 177–182.

415. Leung, D.W. et al., Expression, purification, crystallization and preliminary X-ray studies of the Ebola VP35 interferon inhibitory domain. *Acta Crystallogr Sect F Struct Biol Cryst Commun*, 2009. **65**(Pt 2): 163–165.

416. Leung, D.W. et al., Crystallization and preliminary X-ray analysis of Ebola VP35 interferon inhibitory domain mutant proteins. *Acta Crystallogr Sect F Struct Biol Cryst Commun*, 2010. **66**(Pt 6): 689–692.

417. Koellhoffer, J.F. et al., Crystal structure of the Marburg virus GP2 core domain in its postfusion conformation. *Biochemistry*, 2012. **51**(39): 7665–7675.

418. Zhang, A.P. et al., Crystal structure of Marburg virus VP24. *J Virol*, 2014. **88**(10): 5859–5863.

419. Noda, T. et al., Nucleocapsid-like structures of Ebola virus reconstructed using electron tomography. *J Vet Med Sci*, 2005. **67**(3): 325–328.

420. Bharat, T.A. et al., Cryo-electron tomography of Marburg virus particles and their morphogenesis within infected cells. *PLoS Biol*, 2011. **9**(11): e1001196.

421. Bharat, T.A. et al., Structural dissection of Ebola virus and its assembly determinants using cryo-electron tomography. *Proc Natl Acad Sci U S A*, 2012. **109**(11): 4275–4280.

422. Welsch, S. et al., Electron tomography reveals the steps in filovirus budding. *PLoS Pathog*, 2010. **6**(4): e1000875.

423. Watanabe, S. et al., Production of novel ebola virus-like particles from cDNAs: An alternative to ebola virus generation by reverse genetics. *J Virol*, 2004. **78**(2): 999–1005.

424. Wenigenrath, J. et al., Establishment and application of an infectious virus-like particle system for Marburg virus. *J Gen Virol*, 2010. **91**(Pt 5): 1325–1334.

425. Hoenen, T. et al., Minigenomes, transcription and replication competent virus-like particles and beyond: Reverse genetics systems for filoviruses and other negative stranded hemorrhagic fever viruses. *Antiviral Res*, 2011. **91**(2): 195–208.

426. Schmidt, K.M. et al., Recombinant Marburg virus expressing EGFP allows rapid screening of virus growth and real-time visualization of virus spread. *J Infect Dis*, 2011. **204**(Suppl 3): S861–S870.

427. Kobinger, G.P. et al., Filovirus-pseudotyped lentiviral vector can efficiently and stably transduce airway epithelia in vivo. *Nat Biotechnol*, 2001. **19**(3): 225–230.

428. Sinn, P.L. et al., Lentivirus vectors pseudotyped with filoviral envelope glycoproteins transduce airway epithelia from the apical surface independently of folate receptor alpha. *J Virol*, 2003. **77**(10): 5902–5910.

429. Medina, M.F. et al., Lentiviral vectors pseudotyped with minimal filovirus envelopes increased gene transfer in murine lung. *Mol Ther*, 2003. **8**(5): 777–789.

430. Ou, W. et al., Induction of ebolavirus cross-species immunity using retrovirus-like particles bearing the Ebola virus glycoprotein lacking the mucin-like domain. *Virol J*, 2012. **9**: 32.

431. Geisbert, T.W. et al., Vesicular stomatitis virus-based ebola vaccine is well-tolerated and protects immunocompromised nonhuman primates. *PLoS Pathog*, 2008. **4**(11): e1000225.

432. Geisbert, T.W. et al., Vesicular stomatitis virus-based vaccines protect nonhuman primates against aerosol challenge with Ebola and Marburg viruses. *Vaccine*, 2008. **26**(52): 6894–6900.

433. Geisbert, T.W. et al., Recombinant vesicular stomatitis virus vector mediates postexposure protection against Sudan Ebola hemorrhagic fever in nonhuman primates. *J Virol*, 2008. **82**(11): 5664–5668.

434. Falzarano, D. et al., Single immunization with a monovalent vesicular stomatitis virus-based vaccine protects nonhuman primates against heterologous challenge with Bundibugyo ebolavirus. *J Infect Dis*, 2011. **204**(Suppl 3): S1082–S1089.

435. Marzi, A. et al., Vesicular stomatitis virus-based Ebola vaccines with improved cross-protective efficacy. *J Infect Dis*, 2011. **204**(Suppl 3): S1066–S1074.

436. Geisbert, T.W. and H. Feldmann, Recombinant vesicular stomatitis virus-based vaccines against Ebola and Marburg virus infections. *J Infect Dis*, 2011. **204**(Suppl 3): S1075–S1081.

437. Mire, C.E. et al., Recombinant vesicular stomatitis virus vaccine vectors expressing filovirus glycoproteins lack neurovirulence in nonhuman primates. *PLoS Negl Trop Dis*, 2012. **6**(3): e1567.

438. Wen, Z. et al., Recombinant lentogenic Newcastle disease virus expressing Ebola virus GP infects cells independently of exogenous trypsin and uses macropinocytosis as the major pathway for cell entry. *Virol J*, 2013. **10**: 331.

439. Tsuda, Y. et al., A replicating cytomegalovirus-based vaccine encoding a single Ebola virus nucleoprotein CTL epitope confers protection against Ebola virus. *PLoS Negl Trop Dis*, 2011. **5**(8): e1275.

440. Ribeiro-Viana, R. et al., Virus-like glycodendrinanoparticles displaying quasi-equivalent nested polyvalency upon glycoprotein platforms potently block viral infection. *Nat Commun*, 2012. **3**: 1303.

25 Traditional and Novel Trends in Influenza Vaccines

Peter Pushko and Terrence M. Tumpey

CONTENTS

25.1 INTRODUCTION

25.1.1 INFLUENZA: BRIEF HISTORY

Influenza virus is a pathogen that causes significant human health impact and mortality worldwide. Influenza virus and humans are long-time companions. The disease may have been mentioned by Hippocrates as early as in the 412 BC as *the cough of Perinthus* in his work *Of the Epidemics*. The first clear description of influenza dates to 1100s [1,2]. The word *influenza* is originating from the fourteenth century Italy meaning *flow of liquid* or *influence*, the latter perhaps suggesting its use by the contemporary scholars to describe the impact of the stars on the appearance of influenza epidemics. In English, influenza is commonly abbreviated to *flu*. The synonym *grippe* is more often used in other languages, including French, German, Russian, and Spanish.

Influenza virus is the most frequent cause of acute respiratory illness requiring medical intervention. The causative agents were not isolated until after 1930; the first human influenza A virus was isolated in 1933 [3] and influenza B virus that had no antigenic similarity to influenza A was isolated in 1940 [4]. Two years later, in 1942, the first vaccine for influenza A and B viruses was introduced to the U.S. Armed Forces Epidemiological Board and licensed in 1945. Such a rapid introduction of the vaccine reflected the acute need for the preventive measures against the disease. After World War II, the vaccine was also used for civilians and has been continuously improved since then.

In spite of the long history of medical research and the fact that effective vaccines exist, influenza remains a public health problem of global proportions. Human influenza viruses efficiently infect people and spread from person to person, they affect all age groups, and they can recur in any individual [5,6]. According to the U.S. Centers for Disease Control and Prevention (CDC), from 1976 to 2007 in the United States, the estimated annual average of excess deaths resulting from respiratory and circulatory causes associated with seasonal influenza was about 23,600 (range 3,300–48,600) [7]. The World Health Organization (WHO) estimates that 250,000–500,000 people die annually from influenza worldwide [8], which makes influenza the cause of the greatest number of vaccine-preventable deaths.

Influenza viruses also present a pandemic threat, the worst of which has been the *Spanish* influenza pandemic, which circled the globe in 1918 and resulted in deaths estimated between 40 and 100 million worldwide, more than from any other virus on record [9–11]. The United States alone lost 675,000 people to the 1918 pandemic, which is more than casualties of World War I, World War II, the Korean War, and the Vietnam War combined. The 1918 pandemic virus has been reconstructed in order to study the pathogen and to find effective ways to respond to the pathogenic potential [12]. The second pandemic of the twentieth century, *Asian* H2N2 influenza in 1957–1958, resulted in an estimated one million deaths worldwide and 70,000 deaths in the United States. In 1968–1969, *Hong Kong* influenza pandemic caused by an H3N2 influenza virus

resulted in approximately 34,000 deaths in the United States. The relatively low death toll in the 1968 pandemic is thought to have been due to the residual cross-reactive immunity to viral hemagglutinin (HA) H3 and especially neuraminidase (NA) N2 proteins from the previous exposures [13].

The first pandemic of the twenty-first century was caused by the swine-origin 2009 H1N1 virus [14,15]. The 2009 H1N1 virus was identified as a pathogen of concern in March–April 2009 [16] and quickly spread from North America to all continents except Antarctica prompting the WHO to declare a pandemic on June 11, 2009 [17]. Other potentially pandemic influenza viruses also exist. In 2003, highly pathogenic H7N7 avian influenza virus caused an outbreak in poultry farms in the Netherlands, with 89 human cases and probable human-to-human transmission being reported in a family cluster [18]. Recently, an outbreak of human infection caused by a novel avian-origin influenza A H7N9 virus emerged in eastern China in the spring of 2013 [19]. In late March of 2013, the notification of three human cases of H7N9 in Shanghai and Anhui, which died after acute respiratory distress syndrome and multiorgan failure, raised public health concerns [19]. As of January 28, 2014, more than 200 laboratory-confirmed human cases of H7N9 virus infection have been reported, with a fatality rate of approximately 22%. In addition to H7N9 virus, avian-origin influenza viruses of the H5N1 subtype possess pandemic potential [20]. Since 2003 to January 2014, the number of laboratory-confirmed human cases of avian H5N1 influenza reached 650, from which 386 infections were fatal, representing a 59% mortality rate [21]. In addition to H7N9, H5N1, H9N2, and H6N1 viruses also represent human pathogens with a pandemic potential. For example, in 1999, the avian H9N2 influenza virus was isolated from two children who recovered from influenza-like illnesses in Hong Kong [22]. Since then, more human cases of H9N2 influenza have been reported including a confirmed case in December 2013. The first human case of H10N8 influenza has been also described [23].

As major human pathogens, influenza viruses remain among the most important targets for prophylactic and therapeutic interventions [13,24,25]. In this review, we summarize the traditional as well as novel approaches used in the development of vaccines. In more detail, we describe the development of VLPs as candidate vaccines for influenza, as well as application of protein engineering methods for the development of VLPs with predicted antigenic characteristics. Most of influenza vaccine research has been done with influenza A virus. However, in most cases, including the existing commercial vaccines, approaches developed for influenza A virus could also be successfully applied to vaccine development for influenza B.

25.1.2 INFLUENZA VIRUSES

All influenza viruses belong to the Orthomyxoviridae family of which there are three types, A, B, and C [26,27]. Influenza A viruses are responsible for all known influenza pandemics and annual epidemics. Serious outcomes of influenza infection can result in hospitalization or death, and as a result, influenza A virus is a primary vaccine target and an obligatory component of seasonal and pandemic vaccines. The natural reservoirs of influenza A are the wild aquatic birds of the world, and infection of domestic poultry is thought to occur due to contact with these aquatic birds. The virus has been also isolated from a number of other animal hosts including horses, whales, seals, mink, and swine [2]. The pathogenesis of influenza A virus in humans usually starts with infection of upper respiratory mucosal epithelium. The infection may progresses to an acute febrile illness, which is associated with myalgias, headache, cough, rhinitis, and otitis media. Influenza infection is usually self-limited but may progress to pneumonia. Other serious complications include encephalopathy, myocarditis, and myositis [28].

In addition to influenza A, influenza B viruses are also included in the annual vaccines. Influenza B virus is primarily a human pathogen, which is not associated with an animal reservoir. Influenza B causes similar symptoms and disease as influenza A and is often clinically indistinguishable from influenza A virus infections. However, the frequency of the severe cases of influenza B infections appears to be significantly lower than that of influenza A and children will usually show symptoms more frequently than adults infected with influenza B virus [29,30]. For influenza C, well-defined outbreaks have rarely been detected in humans, and the virus is rarely associated with severe syndromes, although outbreaks have been identified [31–33]. Influenza C virus has been also isolated from pigs [34]. Most people have antibody to influenza C by early adulthood. Administration of influenza C virus to volunteers induced only mild symptoms [35]. Due to these reasons, influenza C is not included in the current influenza vaccine formulations.

25.1.3 INFLUENZA PROTEINS

25.1.3.1 Envelope Proteins: Primary Targets for Vaccines and Antivirals

Influenza A and B virus particles are pleomorphic, mostly spherical in shape and 80–120 nm in diameter (Figure 25.1). Filamentous particles can also be found, with 80–120 nm diameter and up to 2000 nm in length. The virion of influenza A contains a negative-sense, single-stranded genome composed of eight RNA segments, numbered 1–8 on Figure 25.2a. The outer envelope of the virion particle is made of a lipid bilayer that contains glycoprotein spikes of two types, HA (~14 nm long trimer) and NA (~6 nm long tetramer) (Figure 25.2). The HA and NA represent the two major glycoprotein antigens on the surface of influenza virions. Depending on antigenic characteristics of the HA and NA, influenza A (but not B) viruses are divided into antigenic subtypes, or serotypes. For influenza A viruses, there are currently 16 different HA serotypes, H1–H16, and 9 different serotypes of NA (N1–N9).

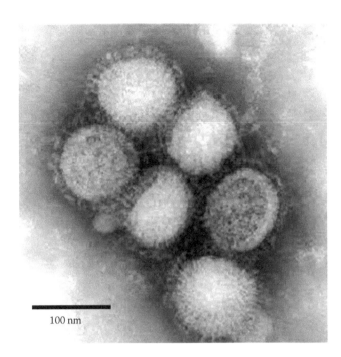

FIGURE 25.1 Influenza A virus particles, by negative staining transmission electron microscopy. (Image courtesy of the Centers for Disease Control and Prevention (CDC), Atlanta, GA/Cynthia S. Goldsmith.)

The 3D structures for the HA and NA have been determined with high resolution using x-ray crystallography (Figure 25.2b), which form the structural basis for rational design and engineering of vaccine-related proteins and antiviral drugs [36,37]. The HA protein is responsible for the attachment of the virus to cell receptors and subsequent fusion of virus envelope with host cell membrane. Human influenza viruses (H1–H3 subtypes) recognize cell surface glycoconjugates containing terminal alpha 2,6-linked sialyl-galactosyl (α2,6) sialic acid (SA) moieties that are found on the human respiratory tract epithelium. In contrast, avian influenza viruses preferentially bind SA linked to galactose by an alpha-2,3 linkage (α2,3 SA), which is found in high concentrations on the epithelial cells of the intestine of waterfowl and shorebirds. A genetic change in the avian virus HA protein resulting in increased α2,6 SA binding may only require one or two amino acid mutations in the HA protein [36]. HA protein is normally synthesized in the infected cells as HA0 precursor that is cleaved posttranslationally by cellular proteases into HA1 and HA2. Cleavage exposes hydrophobic N-terminus of HA2, which then mediates fusion between viral envelope and endosomal membrane. There is a direct link between cleavage and virulence of avian influenza viruses [38]. Highly pathogenic H5 and H7 viruses contain multiple basic amino acid cleavage site between HA1 and HA2, which can be recognized by furin and PC6 proteases in many host cells and organs that may lead to more efficient spread of the virus and more severe disease in humans and up to 100% mortality in birds.

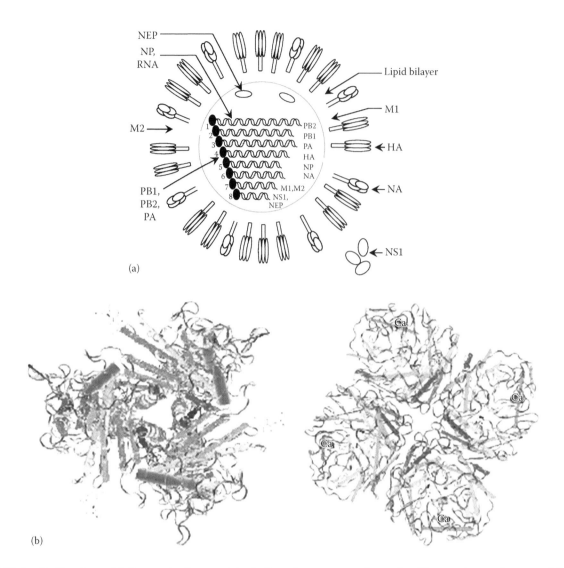

FIGURE 25.2 (a) Structure of influenza A virus particle. Indicated are structural proteins, lipoprotein membrane, and RNA segments. Numbers of RNA segments are also shown, as well as the RNA-encoded proteins. NEP, nuclear export protein (formerly NS2); NP, nucleoprotein; RNA, ribonucleic acid; M2, ion channel protein (translated from the same RNA as matrix protein M1); PB1, polymerase basic protein 1; PB2, polymerase basic protein 2; PA, polymerase acid protein; M1, matrix protein 1; HA, hemagglutinin; NA, neuraminidase; NS1, nonstructural protein 1. (b) Ribbon representations of 3D structures of the HA trimer (left) and NA tetramer (right). HA (H5 subtype) shown with the ligands as seen on Cn3D image of PDB Id 2IBX at NCBI [36]. NA (N1 subtype) monomers contain canonical six-bladed β-propeller structure; also shown are locations of bound ligands and Ca^{++} as seen on Cn3D image of PDB Id 2HTY at NCBI. (From Russell, R.J. et al., *Nature*, 443(7107), 45, 2006.)

The NA has a sialidase activity that removes SA from the HA, NA, and host cell surfaces thus facilitating the release of progeny virions from infected cells. In addition, NA may facilitate virus attachment to the epithelial cells by removal of SA from the mucin layer [39,40].

In addition to HA and NA, the outer lipoprotein envelope of the virus particle also contain several molecules of M2, an ion channel protein. The M2 protein is a minor component of the envelope that has been implicated in ion channel activity and efficient uncoating of incoming viruses during the infection cycle.

The outer envelope proteins are the attractive targets for vaccines as well as for therapeutic drugs including two types of currently available antivirals that are approved by the U.S. Food and Drug Administration (FDA). One type of

drugs, M2 blockers, such as amantadine and rimantadine, target specifically the M2 ion channel protein [41]. However, many influenza A isolates are resistant to M2 blockers [42] and these drugs are not active against influenza B virus. The second type of antivirals, NA inhibitors (such as oseltamivir, also known as Tamiflu), target the NA protein [43]. Drug resistance to the NA-specific antivirals has also been reported [44,45].

Because drug-resistant influenza viruses can be found in nature or generated in the process of treatment with antiviral drugs, prophylactic vaccination may be a more effective way to combat influenza. Induction of immune responses against surface envelope proteins is a preferred way, because virus-neutralizing responses to the viral envelope proteins would prevent early steps of viral infection. Due to significant variability of the HA and NA, two major envelope proteins, such responses are strain specific and not capable of efficient cross-neutralizing many strains of influenza. In contrast, M2 protein, a minor component of the envelope, is well-conserved among influenza viruses, and successful induction of a protective immune response to M2 potentially can result in a *universal* vaccine capable of protecting against many isolates of the virus. However, it is more difficult to induce effective protective response against a minor envelope protein. Recently, conserved stalk domain of HA has been considered as promising candidate for the development of universal influenza vaccines [46].

25.1.3.2 Inner Proteins of Influenza Virion

In addition to the lipoprotein envelope that includes viral HA, NA, and M2 proteins, influenza A virion particle contains an inner core composed of several internal structural and nonstructural proteins, which potentially can also serve as targets for the development of countermeasures against influenza (Figure 25.2a).

The inner side of the envelope is lined by the matrix protein (M1), which is the most abundant structural protein of the influenza A virion. The M1 and M2 proteins are encoded within the same RNA segment 7 (Figure 25.2a), from which M1 and M2 are generated by using alternatively spliced RNA.

The inner part of the virion contains eight genomic RNA segments complexed with the nucleoprotein (NP) into helical ribonucleoproteins. In addition to NP, three polymerase polypeptides (PB1, PB2, PA) are associated with each RNA segment. Viral RNA synthesis takes place in the nucleus of infected cell, where the virus subverts the cellular transcription machinery to express and replicate its own single-strand RNA genome. Influenza virus conserved inner proteins are being considered as candidates for universal influenza vaccines and therapies along with conserved epitopes within HA, NA, and M2 proteins [46–48].

The nuclear export protein (NEP), previously known as NS2, is also associated with the virion. Finally, NS1 is a cell-associated protein, which is normally not found in the virion particle. NS1 inhibits export of poly-A containing mRNA molecules from the nucleus, which gives preference to viral RNAs in transport and translation. NS1 has also been implicated in induction of apoptosis in influenza-infected cells [49] and in resistance to antiviral interferon (IFN) in highly pathogenic H5N1 influenza [50]. NS1 appears to participate in the blockade of the host innate immune response and is therefore being considered as a potential therapeutic target [51,52].

Influenza B virus has an additional protein, NB, which is generated by utilizing an alternative open reading frame from the same RNA segment that encodes the NA. The functions of NB are not completely understood. An NB-deficient live influenza virus could be generated, which was attenuated in mice [53].

Inner virus proteins are normally hidden within the particle and can only become accessible to the host immune effector mechanisms after initiation of infection, virus uncoating, or expression within the infected cells.

25.1.4 Immune Responses to Influenza

25.1.4.1 Innate Immunity

Understanding mechanisms of generating protective immunity is a primary requirement for the development of successful vaccines. The innate immune system is one of the first lines of defense against influenza [54]. Vaccine adjuvants, which act through the innate immune system, are known to enhance both humoral and T-cell-mediated responses to influenza vaccines. Adjuvant effects include enhanced antigen presentation, activation and maturation of dendritic cells, and production of inflammatory cytokines, which can affect the desired cell-mediated immune responses. The response of the innate immune system to influenza infection involves several mechanisms including virus interaction with toll-like receptors (TLRs), which recognize pathogen-associated molecular patterns and trigger resistance to infection through various pathways such as cytokine cascades. For example, TLR3 on respiratory epithelial cells can activate type 1 IFN production and release of cytokines in response to influenza A virus or double-stranded RNA [55]. TLR7, another TLR, which is expressed on dendritic cells and reacts with single-stranded RNA, appears also to be involved in IFN pathway and production of cytokines [56]. Induction of cytokine responses has been observed when volunteers were experimentally infected with influenza; furthermore, exacerbated cytokine responses have been detected in a limited number of H5N1-infected patients with severe disease [57,58]. IFN appears to contribute to recovery from influenza infection [59]. However, highly pathogenic H5N1 viruses appear to be resistant to the effects of type 1 IFN response when evaluated in a porcine epithelial cell monolayer [50].

Production of IFN-γ has been detected in peripheral blood natural killer (NK) cell subsets, as well as in influenza-specific memory CD8 T cells after exposure to influenza virus; IL-2 production by T cells was required for the IFN-γ response of NK cells, indicating that memory T cells enhance innate NK-mediated antiviral immunity [60].

Nonstructural protein 1 (NS1) of influenza A virus plays an important role in the blockade of the host innate immune response [51]. NS1 functions include targeting the tripartite motif-containing protein 25 (TRIM25), which is required to ubiquitinate and activate RIG-I leading to subsequent IFN production. TRIM25 needs to be oligomerized in order to perform its function, and NS1 inhibition abolishes this activity. NS1 also functions to impair the IFN-induced protein PKR. The latter can inhibit virus propagation via phosphorylation of the translation factor eIF2α [61].

Finally, the involvement of innate immunity in control of influenza infection has been also implicated, when intranasally (i.n.) administered baculovirus induced protection in mice to influenza challenge [62]. It has been shown that baculoviruses may induce maturation of dendritic cells in vivo and production of inflammatory cytokines and also promote humoral and CD8 T cell adaptive responses against coadministered antigens [63].

25.1.4.2 Adaptive Humoral and Cell-Mediated Immunity to Influenza

During human infection or vaccination with influenza A virus, serum antibodies to HA, NA, NP, and M1 are generated [64]. Although differences can be observed for distinct influenza strains, high antibody titers to HA and NA envelope glycoproteins typically correlate with a lower attack rate for infection and less severe influenza disease [26,65]. Neutralizing and receptor-blocking antibodies against HA are the primary means of protection from the spread of pandemic and seasonal strains. Anti-NA antibodies seem to play a secondary role by inhibiting virus release from infected cells, thus restricting spread of the virus [65]. Antibody to internal M1 and NP proteins are generally not associated with resistance to infection. Therefore, acquired or vaccine-induced immunity to influenza is typically measured as serum IgG antibodies to the HA and NA antigens of the circulating influenza A and B viruses. High levels of hemagglutination-inhibition (HI) antibody correlate with efficient neutralization of infectivity of the incoming virus before it infects host cells. Data from limited studies with H5N1 virus suggest that although initially robust, there is substantial waning of the serum antibody responses in survivors of H5N1 virus infection [66].

While HA-specific antibodies in the serum provide an excellent correlate of protection, the control of disease and clearance of the virus depend on more complex effector mechanisms provided by both humoral and cell-mediated immunity. Immune responses at both mucosal and systemic sites are presumed to be necessary for either natural or vaccine-induced protection. Both humoral and cell-mediated responses following influenza infection or vaccination have been studied [67–72]. Antibody and cytotoxic T cell epitopes have been found in all influenza proteins [73]. However, in spite of the fact that antigenic epitopes and immune responses to multiple influenza proteins have been found, their specific roles in protecting the human host against infection and reinfection still remains to be fully elucidated. Further, the immune mechanisms and relative importance of each component in the infected host are not completely understood. It has been shown that human T cell responses induced by infection with seasonal influenza viruses are directed to relatively conserved internal proteins and can cross-react with the incoming H5N1 virus. Furthermore, the role for T-cell-based heterosubtypic immunity against H5N1 viruses was suggested in animal studies as reviewed elsewhere [66]. It has been also shown that passive transfer of antibodies can clear influenza A virus infection in mice even in the absence of B or T cells [74].

25.2 TRADITIONAL INFLUENZA VACCINES

The majority of currently available prophylactic influenza vaccines licensed in the United States are based on viruses grown in embryonated chicken eggs. Recently, vaccines have also been developed in cell culture including Flucelvax

vaccine and the first recombinant FluBlok vaccine [75–77]. Seasonal vaccines address the issue of multiple virus strains and significant variability of influenza viruses by including two influenza A viruses and one or two influenza B viruses into the vaccine as well as by changing vaccine strains frequently to protect against the most current prevalent circulating viruses. The composition of seasonal influenza vaccines is reevaluated twice yearly by the WHO and is updated to reflect predominant circulating viruses [78]. A bivalent influenza type A/B vaccine was produced from 1942 after the discovery of influenza B. Until recently, influenza vaccines included two influenza A strains and one influenza B strain. However, for several decades, two antigenically distinct lineages of influenza B viruses, B/Yamagata and B/Victoria, have cocirculated globally. In many recent influenza seasons, the predominant circulating influenza B lineage was different from that contained in trivalent influenza vaccines. Therefore, in 2012, a quadrivalent inactivated (QIV) and live-attenuated influenza vaccine (Q/LAIV) have been introduced that each contained two A strains and two B strains [79,80] (Table 25.1). For example, Q/LAIV includes two B strains, one from each lineage, in order to provide broad protection against influenza B. Q/LAIV was recently approved for use in the United States in eligible individuals 2–49 years of age [80] (Table 25.1). In the United States and EU, criteria for vaccine immunogenicity, which are based on the induction of virus-neutralizing or HAI antibodies in the serum, have been implemented [81,82]. In 2000, approximately 240 million doses of influenza vaccine have been distributed worldwide [83]. A Federal Medicare rule in the United States requires all long-term care facilities to offer annual vaccination for influenza and to document the results [84].

25.2.1 INACTIVATED VACCINES

25.2.1.1 Whole-Virus Vaccines

Historically, the first influenza vaccines were preparations of the whole viruses grown in allantoic fluid of embryonated eggs and chemically inactivated, for example, by using formalin or beta-propiolactone [85]. Such whole-virus vaccines were often reactogenic because of contamination with egg-derived components. Introduction of zonal centrifugation [86] improved characteristics of the whole-virus vaccines. Generally, the whole-virus vaccines have high immunogenicity, exceeding immunogenicity of the currently used subvirion, or *split* virus, vaccines. However, with the introduction of split-virus vaccines, the use of the whole-virus vaccines has been essentially discontinued due to the higher rates of adverse reactions [87,88]. The reasons for improved immunogenicity of whole-virus vaccines remain to be fully elucidated. It has been hypothesized that presentation of influenza antigens in an intact lipid membrane in a whole-virus vaccine may result in better processing of antigen, resulting in advantageous epitope presentation and immune responses [89].

Due to pandemic threats and with introduction of egg-independent vaccine production methods, the whole-virus

TABLE 25.1

Examples of Influenza Vaccines: United States, 2013–2014 Influenza Season

Vaccine	Trade Name	Manufacturer	Age Indications	Route
1. Inactivated influenza vaccine, trivalent (IIV3), standard dose	Afluria	CSL Limited	≥9 years	i.m.
	Fluarix	GlaxoSmithKline	≥3 years	i.m.
	Flucelvax[a]	Novartis Vaccines	≥18 years	i.m.
	FluLaval	ID Biomedical	≥3 years	i.m.
	Fluvirin	Novartis Vaccines	≥4 years	i.m.
	Fluzone	Sanofi Pasteur	6–35 months	i.m.
			≥36 months	i.m.
			≥36 months	i.m.
			≥6 months	i.m.
	Fluzone Intradermal	Sanofi Pasteur	18–64 years	ID
2. Inactivated influenza vaccine, trivalent (IIV3), high dose	Fluzone High-Dose	Sanofi Pasteur	≥65 years	i.m.
3. Inactivated influenza vaccine, quadrivalent (IIV4), standard dose	Fluarix Quadrivalent	GlaxoSmithKline	≥3 years	i.m.
	FluLaval Quadrivalent	ID Biomedical	≥3 years	i.m.
	Fluzone Quadrivalent	Sanofi Pasteur	6–35 months	i.m.
			≥36 months	i.m.
			≥36 months	i.m.
4. Recombinant influenza vaccine, trivalent (RIV3)	FluBlok	Protein Sciences	18–49 years	i.m.
5. Live-attenuated influenza vaccine, quadrivalent (LAIV4)	FluMist Quadrivalent	MedImmune	2–49 years	i.n.

Source: Influenza Vaccines—United States. Centers for Disease Control and Prevention, Atlanta, GA (cdc.gov).

IIV, inactivated influenza vaccine; IIV3, inactivated influenza vaccine, trivalent; IIV4, inactivated influenza vaccine, quadrivalent; RIV, recombinant influenza vaccine; LAIV, live-attenuated influenza vaccine; i.m., intramuscular; ID, intradermal; i.n., intranasal.

[a] Produced in mammalian cell culture.

vaccines may return to the vaccine market, because higher immunogenicity in a naive population could reduce the vaccine dose needed to provide effective protective immunity [90]. Reduction of a vaccine dose may allow to vaccinate more people, which is important in a pandemic scenario [91]. Promising clinical results of safety and immunogenicity for cell-derived, inactivated whole-virus vaccine for H5N1 virus have been presented [92].

25.2.1.2 Split-Virus Trivalent and Quadrivalent Vaccines

A cornerstone of current prophylactic influenza vaccines is inactivated subvirion, or split-virus, vaccines. The examples of FDA-approved influenza vaccines for use in the United States are shown in Table 25.1. The standard split influenza vaccine is a trivalent inactivated virus preparation, given by intramuscular (i.m.) injection. Generally, one dose of trivalent inactivated vaccine for adults contains 45 μg of HA, that is, 15 μg for each strain. Recently approved QIV split-virus seasonal influenza vaccine Fluarix quadrivalent contains 15 μg of HA from each of the four influenza virus strains expected to circulate in the upcoming influenza season [79]. The methods for vaccine preparation and virus inactivation that results in a *split virus* were developed in the 1960s [86,93] and are used until today with several modifications. Before formulating vaccine, each of the influenza viruses is produced, inactivated, and purified separately. For example, in the process of vaccine manufacturing, virus-containing

allantoic fluids are harvested from eggs; each virus is then concentrated and purified by zonal centrifugation using a linear sucrose density gradient solution containing detergent to disrupt the viruses. After dilution, the virus preparation is further purified by diafiltration. Each influenza virus is chemically inactivated, for example, by the consecutive effects of sodium deoxycholate and formaldehyde leading to the production of a split virus. Inactivated vaccines usually represent injectable formulations indicated initially for adults 18 years of age and older [94]. QIV have been recently approved for active immunization of individuals aged 3 years or older [79].

On November 20, 2012, the U.S. FDA approved the use of the first trivalent inactivated influenza vaccine Flucelvax, which was manufactured using cell culture technology [95]. In place of fertilized chicken eggs, the cell-based vaccine manufacturing involved Madin–Darby Canine Kidney (MDCK) cells in liquid suspension cultures as a host for the growing influenza virus. This cell-based vaccine can be safely used for vaccinating people with allergic reactions to eggs. Furthermore, vaccine manufacturing is not dependent on an egg supply, which can be threatened in case of an outbreak of avian influenza or other agricultural disease that affects chicken flocks.

25.2.1.3 Pandemic Vaccines

The swine-origin influenza A virus, designated as A(H1N1) pdm09, emerged in the spring of 2009 and caused the first influenza pandemic in the twenty-first century. The 2009 pandemic placed unprecedented demand for vaccine production. Both split-virus and live-attenuated vaccines were produced by growing the virus in chicken eggs, with deliveries starting from November 2009. Vaccines proved to be efficacious; however, some adverse effects have been described, and the causative effects are still being elucidated [96].

With the increasing number of human infections with H5N1 avian influenza, pandemic vaccines have been developed against the H5N1 influenza virus [91,97]. Several monovalent H5N1 vaccines have been manufactured for emergency use in the case of H5N1 pandemic [98,99]. The first egg-based inactivated split vaccine for H5N1 influenza, which was based on influenza virus strain A/Vietnam/1203/2004 (H5N1, clade 1), has been licensed in the United States since 2007 [100] as a prepandemic vaccine for noncommercial use. Each 1 mL dose was formulated to contain 90 µg of HA. The vaccine was prepared from influenza virus harvested from embryonated chicken eggs and inactivated with formaldehyde. For the production of H5N1 pandemic vaccine, the virus was concentrated and purified in a linear sucrose density gradient using a continuous flow centrifuge. The virus was then chemically disrupted using polyethylene glycol p-isooctylphenyl ether (Triton X-100), a nonionic surfactant, producing a split virus, which was further purified by biochemical methods.

With emerging of new potentially pandemic influenza viruses, the demand for new vaccines is expected to continue. For example, an outbreak of H7N9 influenza in eastern China in the spring of 2013 showed significant human morbidity and mortality and raised demands for the H7N9 vaccine development. Furthermore, in December 2013, China reported the first human case of avian influenza A subtype H10N8 [23].

25.2.2 LIVE-ATTENUATED VACCINES

Since 2003, the i.n. administered LAIV, FluMist, has been licensed in the United States (Table 25.1). This live influenza A and B virus vaccine was initially indicated for healthy individuals aged 5–49 years as an alternative approach to influenza vaccination [101,102]. As of 2014, FluMist is indicated for vaccination of individuals 2–49 years of age (Table 25.1). In order to develop such a vaccine, influenza virus was passaged at 25°C in chicken kidney cells and in embryonated eggs that resulted in cold-adapted, highly attenuated virus [103,104]. FluMist is administered as nasal spray and does not require needles for administration. In comparison to inactivated vaccines, which induce good antibody response, live-attenuated vaccine appears to induce similar or even more balanced responses including mucosal and cellular immunity. In one study, a placebo-controlled study in 103 experimentally infected adult patients found that the protective efficacy of FluMist was 85%, compared to 71% with inactivated influenza vaccine [105]. Similarly to other live-attenuated virus vaccines, FluMist is generally not recommended for individuals with known or suspected immune deficiency diseases including thymic abnormalities, malignancies, or human immunodeficiency virus (HIV) infection, as well as to those who may be immunosuppressed because of radiation treatment or other immunosuppressive therapies.

Current FluMist vaccine that is a quadrivalent LAIV formulation that contains two A strains and two B strains has been developed [80]. Vaccine is approved for use in the United States in eligible individuals 2–49 years of age for 2013–2014 influenza season [80].

25.2.3 VETERINARY VACCINES

Outbreaks of avian influenza virus in poultry have greatly impacted the economy and international trade in affected countries. Successful control of influenza in poultry and other agricultural species is important for the eradication of the disease and preventing infections in humans. The number of influenza outbreaks in poultry has increased during the last years, and vaccination can be a powerful tool to prevent outbreaks along with other control measures [106]. Vaccination of poultry needs to be implemented as part of a comprehensive control strategy that also includes biosecurity, surveillance, education, and elimination of infected poultry [107]. Preventive measures should be considered not only for highly pathogenic avian influenza (HPAI) (H5 and H7) viruses with pandemic potential but also for subtypes of low pathogenic avian influenza (LPAI) viruses, because some LPAI viruses have the potential to evolve into HPAI viruses. Veterinary vaccines for influenza are available and have been used successfully in avian influenza control programs. Standard inactivated influenza vaccines are fully or conditionally licensed for parenteral administration and have been successful at providing protection against clinical signs and death among poultry. Avian influenza vaccines generated by reverse genetics [108] or recombinant genetically engineered fowlpox vector vaccines expressing HA have been also successfully tested [109].

25.3 CHALLENGES FOR VACCINE DEVELOPMENT

25.3.1 ANTIGENIC DRIFT AND ANTIGENIC SHIFT

The intrinsic genetic variability of influenza viruses presents the major obstacle for vaccination as well as for vaccine development. Inactivated and live-attenuated trivalent vaccines have generally been found very effective in preventing and limiting the spread of the disease. Lower efficiency of vaccination was observed during annual epidemics when components of a vaccine did not match well the circulating influenza strains. For example, in 2003, the H3N2 vaccine virus, A/Moscow/10/1999, did not antigenically match the circulating A/Fujian/411/2002-like viruses, resulting in reduced effectiveness against virus-caused illness [110]. However, egg grown A/Fujian/411/2002 virus was not available in time for vaccine production [25].

Influenza viruses have the ability to evolve into new variants that can overcome immunity and cause epidemics or pandemics through the genetic processes termed antigenic drift and antigenic shift. The emergence of new antigenic variants of influenza viruses through antigenic drift is a relatively frequent event, and this phenomenon is responsible for seasonal changes in circulating virus strains and for annual influenza epidemics.

In contrast to antigenic drift, antigenic shift is a relatively infrequent event. Unlike antigenic drift, which is caused by mutations within gene segments, antigenic shift is caused by the reassortment of gene segments between two influenza strains in permissive hosts. Influenza viruses belonging to any of the three different types (influenza A, B, or C) can undergo reassortment, but not between members of different types. For influenza A viruses, the potential for a pandemic arises when such an antigen shift occurs, because most people will not have immunity to such a novel reassortant virus. If such a novel reassortant virus can cause illness and is capable of efficient human-to-human transmission, a pandemic can occur. Examples of genetic shift include the 2009 swine-origin pandemic H1N1 virus, as well as the 2013 avian-origin H7N9 virus, in which multiple reassortant events have been suggested.

Broadly protective vaccines and immunotherapies based on the highly conserved epitopes are currently under development, which are based on M2, HA stalk, NA, as well as inner influenza virus proteins [46–48,51,111]. Potentially, novel vaccines could be developed into universal influenza virus vaccines that protect from infection with drifted seasonal as well as novel pandemic influenza virus strains, therefore obviating the need for annual vaccination and enhancing pandemic preparedness.

25.3.2 Low Immunogenicity in the Young and the Elderly

Even when vaccine matches circulating viruses, inactivated vaccines tend to have lower effectiveness in young children and especially in the elderly [101,112–114]. Unfortunately, the young and the elderly populations are also more vulnerable to the serious complications of influenza that might result in hospitalization or death [115]. The higher susceptibility of the host to serious influenza in the pediatric and geriatric populations likely reflects diminished capacities of both the innate and adaptive immune systems in the very young and the elderly [116,117]. Young children are often immunologically naive and may also have intrinsic limitations in immune cell functions. Aging of the immune system in the elderly results in weaker responses to both infection and vaccination, despite repeated priming of memory immunity. On December 23, 2009, the U.S. FDA approved a high-dose Fluzone formulation of the trivalent inactivated influenza vaccine, for prevention of influenza in people 65 years of age and older. Clinical studies have demonstrated that this vaccine, containing four times more HA than standard-dose inactivated influenza vaccines (Table 25.1), can produce an enhanced immunologic response in patients of 65 years of age and older, while maintaining a favorable safety profile [118].

25.3.3 Manufacturing in Eggs and Preparation of High-Yield Viruses

Preparation of influenza vaccines in eggs is a relatively lengthy process. In a typical chicken embryo operation, one egg is required to produce one 45 μg dose of vaccine protein. However, some influenza viruses do not grow in eggs at yields suitable for vaccine manufacturing. Reassortment with the high-yield influenza virus is done in order to generate a virus that combines the desired antigenic and growth characteristics. Manufacturers receive such seed strains through their government agencies, which generate high-yield viruses using classical reassortment [119] of seasonal strains with the laboratory strain, A/PR/8/34 for influenza A viruses. Such engineered reassortant viruses usually contain six genomic RNA segments from the laboratory strain and the remaining two segments (HA and NA) from the virus of interest. Reassortment involves multiple passaging of seed viruses to allow virus/egg adaptation. It should be noted that during repeated passage to improve yield, the HA and NA may acquire additional mutations, which may cause alteration in their functional and antigenic characteristics [120,121]. Recent introduction of reverse genetics methods that allow generation of influenza virus from DNA plasmids in cell culture can make production of reassortant seed viruses more efficient [122].

Reassortment is also used in the process of preparation of live-attenuated vaccines. Following reassortment, vaccine seed viruses are generated that contain two genomic segments from the circulating viruses, whereas the remaining six genomic segments are derived from the master seeds of cold-adapted attenuated A and B viruses.

For licensed vaccines, the entire vaccine manufacturing process including generation of seed viruses and all the required tests needs to be completed before annual epidemic begins. As previously mentioned, failure to isolate A/Fujian/411/2002 (H3N2) in eggs resulted in its absence from the 2003/2004 vaccine. Furthermore, because of the manufacturing constraints, there is a risk that demand may outstrip the supply of the vaccine. A significant shortage of influenza vaccine occurred in the United States in 2004. After routine testing required by FDA, one of the two suppliers of inactivated influenza vaccine for the United States found bacterial contamination in a limited number of vaccine lots. As a result, FDA announced that influenza vaccine manufactured by this supplier was not safe for use. The remaining vaccine doses were recommended for those at the highest risk of complications from influenza. As a result of the vaccine shortage, FDA had to identify additional sources of vaccine that could be made available under an FDA investigational new drug (IND) application [123]. These efforts resulted in FDA approving INDs that permitted the potential use of vaccines from additional suppliers such as GlaxoSmithKline and Berna Biotech. Dose-sparing studies have been carried out including intradermal injections and adjuvants in order to determine if the reduction of vaccine dose may allow both adequate protection and increase the number of available vaccine doses [124–126].

25.3.4 Effects of Adjuvants and the Route of Administration

Adjuvants offer alternative and complementary approaches to vaccines, particularly because of their ability to enhance immune responses and for their dose-sparing properties. In spite of the development of vaccine purification methods, influenza vaccines, especially those derived from eggs, may exhibit residual reactogenicity, especially in individuals with allergies to certain products. Therefore, vaccine dose reduction may reduce adverse reactions.

Therefore, attempts have been made to enhance the efficacy and safety of current inactivated vaccines by improving vaccine composition, optimizing the route of vaccine administration, and evaluating novel adjuvants [126,127]. For example, a split vaccine derived from live-attenuated recombinant H5N1 influenza virus was tested in mice i.n. with several adjuvants including cholera toxin B subunit containing a trace amount of holotoxin, synthetic double-stranded RNA, or chitin microparticles. Promising effects of these adjuvanted vaccines on IgA and IgG responses, expression of TLR3, and protection from live H5N1 infectious virus challenge have been observed [128]. Further, safety and immunogenicity of split trivalent influenza vaccine formulated with lipid/polysaccharide carrier has been evaluated in a clinical trial as nasal influenza vaccine [129]. Vaccine strains included in this study were A/Johannesburg/82/96 (H1N1), A/Nanchang/933/95 (H3N2), and B/Harbin/07/94. The authors concluded that the inactivated nasal influenza vaccine was well tolerated and immunogenic in healthy adults. In another study, split trivalent vaccines were evaluated in varying i.m. and i.n. dosages separately and combined [130]. The viruses used were trivalent inactivated vaccine for the 2001–2002 influenza season that included the following influenza virus strains: A/New Caledonia/20/99 (H1N1), A/Panama/2007/99 (H3N2), and B/Victoria/504/2000. Volunteers between the ages of 18 and 45 years received 15, 30, or 60 µg of vaccines by either i.n., i.m., or both routes, 120 µg of vaccine i.m., or placebo. All vaccine dosages and routes of vaccine administration were well-tolerated, safe, and induced serum as well as mucosal antibody responses. Overall, a high dose of i.m. vaccine with or without i.n. vaccine generated high levels of HAI- and virus-neutralizing serum antibody responses, suggesting that advantage for i.n. vaccine may be limited to induction of nasal secretion antibodies [130]. Finally, the 2009 H1N1 pandemic provided some the most recent insights into immunity against influenza and the role of adjuvants in immunity and protection. For example, the addition of oil-in-water emulsion adjuvants to inactivated vaccines provided enhanced functional antibody titers, greater breadth of antibody cross-reactivity, and antigen dose sparing. The MF59 adjuvant broadened the distribution of B cell epitopes recognized on HA and NA following immunization [65].

25.3.5 Traditional Vaccines for Pandemic Strains: More Challenges

The emergence of pandemic influenza viruses represents a major challenge for vaccine development [10,131,132]. It is unlikely that seasonal influenza vaccines will provide significant levels of protection against potential pandemic strains. This has been confirmed during the 2009 H1N1 pandemic. Antigenic and genetic diversity within a subtype are an additional challenge. For example, currently available H5N1-inactivated vaccine (Table 25.1) based on clade 1 A/Vietnam/1203/2004 strain may not protect against other H5N1 viruses, which in the case of a pandemic, a large number of vaccine doses may be required within a relatively short period of time. Such a scale-up may be difficult to achieve using egg-based influenza vaccine manufacturing technology. The pandemic virus may be highly lethal to birds including chickens, and the maintenance of a constant supply of embryonated eggs would be difficult, because virus can rapidly kill chicken embryos before virus can grow to the yields sufficient for effective production. The latter difficulty can be circumvented by reassortment of pandemic strain virus with the laboratory influenza strain(s). However, this may require additional biocontainment measures in order to prevent accidental generation of a new reassortant virus with unpredictable pathogenic and transmission characteristics. Protein engineering methods can be applied to reduce pathogenic potential of highly pathogenic reassortants. For example, deletion of the polybasic cleavage site within the H5 can be used to prevent generation of highly pathogenic virus. In any event, high levels of biosafety containment are needed for the production of pandemic vaccines using live pandemic influenza viruses in order to protect workers and to prevent the escape of the viruses into the environment.

Further, the risk of incomplete inactivation of a highly pathogenic pandemic virus such as H5N1 virus may be not fully acceptable for human vaccines. Live-attenuated vaccines for pandemic influenza viruses also cause concerns, because they may undergo reassortment with the wild-type influenza viruses to regenerate pathogenic or even a new pandemic virus. In addition, live-attenuated virus vaccines themselves may represent risks for some vaccine recipients, especially for those with immune disorders.

25.4 NOVEL TRENDS IN INFLUENZA VACCINES

25.4.1 Need for New Influenza Vaccines

Although traditional influenza vaccines have been used successfully for decades to reduce the impact of influenza, there is an unprecedented demand for the development of new influenza vaccines. It is clear that annual influenza epidemics and the threat of an influenza pandemic constitute major public health problems. This requires not only timely production of the seasonal influenza vaccines but also designing strategies for providing large numbers of doses of vaccines that in the case of pandemic can rapidly induce protective immunity in an immunologically naive population. The current technologies for producing influenza virus vaccines are time consuming. In order to overcome limitations of current vaccines, the development of egg-independent, cell culture–based influenza vaccines is one of the important priorities. This is being actively pursued with the goal of developing both seasonal and pandemic vaccines that are safe, effective, and simple in manufacturing [132]. Significant progress has been achieved. For example, live influenza vaccines have been approved during the last decade, and during the last year, the first cell-based vaccine, the first recombinant vaccine, and the first quadrivalent vaccine have been approved.

Other important priorities for improvement of influenza vaccines are the development and application of new technologies such as reverse genetics or protein engineering, which could be used to develop improved vaccines and to shorten the process of preparing current vaccines [25]. For example, application of reverse genetics technology instead of the traditional reassortment technique can facilitate and speed up the manufacturing process for the current inactivated as well as LAIVs. The development of reverse genetics for influenza viruses has greatly expanded knowledge about the virus and enabled researchers to generate influenza viruses with rationally designed genotypes [133]. However, the conventional 12- or 8-plasmid reverse genetics systems have relatively low transfection efficiency of such sets of plasmids, which may impede the rapid generation of vaccine seed viruses. In order to overcome this potential difficulty, the number of plasmids required to generate influenza virus by reverse genetics have been reduced to only four [134]. The improved system consists of (1) a plasmid that encodes the six gene segments for the internal proteins (PB2, PB1, PA, NP, M, and NS); (2) a second plasmid that encodes the HA and NA segments; (3) a third plasmid that expresses influenza NP protein; and (4) a fourth plasmid expressing PB2, PB1, and PA proteins. These four plasmids are used for generation of influenza virus using transfection of susceptible mammalian cell culture. This results in the process that is more efficient than the conventional process involving 12-plasmid systems.

Finally, novel molecular engineering approaches can be used for the engineering of alternative, conceptually novel, influenza vaccines. Such new influenza vaccines should address the issues that are not completely addressed by currently licensed vaccines, which otherwise have been proven effective and provide a standard for comparison with any new candidate vaccine. The critical parameters for assessment of new influenza vaccines are safety; efficacy in various populations including young, elderly, and the immunologically naive or compromised; cross-protective potential against virus variants; speed and ease of production; cost; and acceptance by regulatory agencies and public [25]. Vaccine approaches that have been developed in the past years included recombinant protein subunit vaccines, virosomes, naked DNA vaccines, virus vector vaccines, and other approaches.

Depending on the mechanism of inducing protective immunity, novel influenza vaccines can be divided into two large groups: genetic (or nucleic acid vector-based) vaccines and protein vaccines. In the genetic vaccines, influenza antigen is expressed in the tissues of vaccine recipient from an injected vector construct. In the protein vaccines, antigen is administered directly to vaccine recipient.

25.5 GENETIC VACCINES

25.5.1 DNA Vaccines

DNA vaccination is based upon inoculation of purified DNA expressing immunogen of interest in vivo. DNA vaccines may be especially attractive as *rapid response* vaccines, for example, in the case of a pandemic [135]. There are several methods of administration of DNA vaccines including direct injection into muscle cells, injection of DNA-covered gold particles (gene gun), topical application of DNA, microneedle delivery, as well as in vivo DNA electroporation. In any case, DNA inoculation results in the production of a protein by host cells that results in induction of immune response. The immune responses involve both T cell– and antibody-mediated immunity. Studies have also implicated stimulation by DNA of the innate immune system that creates favorable cytokine profiles for Th1 cell–mediated responses [136,137].

Several studies have demonstrated that inoculation of plasmid DNA encoding influenza antigens, particularly HA, elicited specific immune responses and provided protection against influenza virus in preclinical models including mice and ferrets [138–141]. For example, HA- and NA-based DNA vaccines protected mice from live virus challenge [142]. Conversely, M1-, NP-, or NS1-based DNA vaccine provided lower, if any, protection [138,142]. However, protection of mice with NP could be demonstrated using a higher DNA vaccine dose [143]. A DNA vaccine expressing full-length consensus M2 protein induced M2-specific antibody responses and protected mice against lethal influenza virus challenge [144].

Microneedle DNA vaccine delivery in the skin has been used to improve DNA vaccination. Recent studies have demonstrated that DNA vaccination in the skin using microneedles can induce higher humoral and cellular immune responses and improve protective immunity compared to conventional i.m. injection of HA-based DNA vaccine [145].

Successful DNA vaccination against influenza depends not only on the dose of DNA vaccine, method of DNA vaccine administration, and the animal model used but also on many other factors such as immunogen choice and design. Immunogenicity of HA-expressing DNA vaccines could be improved by using codon-optimized HA sequences, as shown for H1 or H3 serotypes of human influenza A virus [146]. The authors of this study also engineered two forms of HA antigen, a full-length HA and a secreted form of HA with transmembrane (TM) domain truncated. Both full-length and TM-truncated H3 induced high levels of HAI and neutralizing antibody responses. However, the full-length H1 induced significantly higher HI and virus-neutralizing antibody responses than did the TM-truncated HA. These data suggest that influenza antigens from different serotypes may have different requirements for the induction of optimal immune responses.

Promising preclinical results achieved with DNA vaccines have generated great interest in developing DNA constructs as a new generation of influenza vaccines with the ability to elicit effective humoral and cellular immune responses. However, DNA vaccines have yet to overcome technical and regulatory hurdles and proceed past phase I/II clinical trials, primarily due to a need to induce more potent immune responses in people [147–149]. Safety issues have been also raised with regard to DNA vaccines, such as development of autoimmunity to DNA or integration of vector into cellular DNA that potentially can lead to insertional mutagenesis, such as activation of oncogenes or inactivation of tumor suppressor genes. The United States and EU have developed specific advices in regard of safety testing for DNA vaccines [150,151].

The clinical trials of a DNA vaccine designed to protect against H5N1 influenza infection have been conducted [152,153]. It has been shown that influenza virus H5 DNA vaccine is immunogenic by i.m. and intradermal routes in people [153].

In order to improve efficacy of vaccination with DNA vaccines, a prime–boost approach has been often used, in which animals were first vaccinated with DNA and then received booster inoculation with protein. For example, this has been shown recently with DNA/protein prime–boost involving H7 and M1 proteins [154]. Prime–boost DNA/protein vaccination appeared to be more advantageous compared to DNA/DNA or protein/protein vaccinations [154]. Another strategy of improving efficacy of DNA vaccines may be primary vaccination with DNA and a boost with a virus vector vaccine. Thus, enhanced vaccine protection was afforded if a recombinant adenoviral boost immunization to NP was included as a DNA prime and boost regimen [155].

In addition to being a genetic vaccine, DNA can also be used as an adjuvant to enhance the efficacy of influenza vaccines. For example, it has been shown recently that a cationic lipid/DNA complex can improve the efficacy of the trivalent inactivated influenza vaccine Fluzone in elderly nonhuman primates [156].

25.5.2 Virus Vectors as Vaccines

Engineering of nanoparticle-scale viral vectors represent another strategy for the development of egg-independent, manufacturing-friendly influenza vaccines. Unlike DNA vaccines that involve direct inoculation of *naked* nucleic acid, virus vector vaccines are based upon inoculation of virus vectors [157]. The virus vector functions as an efficient vehicle to deliver the viral nucleic acid (DNA or RNA) into host cells. The viral nucleic acid is configured to express influenza vaccine–related immunogen of interest such as HA

in vivo. Similar to the DNA vaccines, inoculation of virus vector results in the production of a vaccine-relevant antigen by host cells in vivo, which results in the generation of both cellular and humoral immunity. Generally, virus vectors can be either propagation competent (live virus) or propagation incompetent. In the latter case, virus vectors are capable of infecting cells and expressing antigen of interest, but it cannot spread to other cells [158,159].

Poxviruses have been used as influenza vaccine vectors [160]. Propagation-competent, live recombinant fowlpox-based vaccines for avian influenza have been developed and used in chickens [109,161]. A fowlpox vaccine with an H5 gene insert protected chickens against clinical signs and death following challenge by nine different H5 avian influenza viruses, which were isolated from different continents over a 38 year period and had 87%–100% sequence similarity with the H5 within the vaccine [109]. Fowlpox-based vaccine expressing H5 gene was granted a license in the United States for emergency use in 1989 and full registration in Mexico, Guatemala, and El Salvador. This vaccine is administered to 1-day-old chickens and has been also found immunogenic in cats suggesting possibility for use in mammals [161].

Candidate vaccines expressing H5 gene from either A/Hong Kong/156/97 or A/Vietnam/1194/04 have also been developed from highly attenuated MVA strain of vaccinia virus. The vaccines were tested in C57BL/6J mice, and the data suggested that recombinant MVA expressing the H5 of influenza virus A/Vietnam/1194/04 is a promising candidate for the induction of protective immunity against various H5N1 influenza viruses [162].

Propagation-incompetent virus vectors have been also developed as influenza vaccine candidates. For example, alphavirus replicon vectors expressing HA from A/Puerto Rico/8/34 virus induced antibody response and protected BALB/c mice from challenge with homologous influenza virus [158]. A single dose of alphavirus replicon particles expressing HA from A/Hong Kong/156/97 (H5N1) completely protected two-week-old chickens from infection with lethal parent virus [163].

Adenoviruses are also used as vaccine vectors for influenza vaccine development [164]. For example, adenovirus types 4 and 7 (Ad4 and Ad7) have been shown to effectively express influenza H1 protein and induce anti-H1N1 immunity in mice against a heterologous challenge [164]. In another study, a propagation-incompetent, human adenoviral vector containing H5 influenza gene (HAd-H5), given to BALB/c mice, induced both humoral and cell-mediated immune responses against avian H5N1 influenza viruses. Vaccination of mice with HAd-H5 provided effective protection from H5N1 infection using antigenically distinct strains of H5N1 influenza viruses [159]. Another propagation-incompetent adenoviral vector expressing A/Vietnam/1203/2004 (H5N1) HA fully protected BALB/c mice, which were challenged with a lethal dose of homologous H5N1 virus. A single subcutaneous vaccination also protected chickens from a lethal i.n. challenge with live influenza virus [165].

Safety concerns are among the major issues that have been raised for the virus vector vaccines. Propagation-competent vectors, such as live poxviruses, may be not safe for the use in people, especially in immunocompromised individuals. Disseminated vaccinia was detected in a military recruit with HIV disease [166]. Propagation-incompetent vectors appear to be safer; however, in some cases, live virus may be rescued at low frequency as a result of recombination during virus vector preparation [158]. Additional limitation for the use of viral vectors is preexisting immunity, such as in the case of adenovirus vectors. Because a large part of the population has antibody to adenovirus, this may limit the ability of the vector to generate effective immune response against influenza. Inoculation with the virus vector may also induce strong antivector immunity, which can preclude effective booster vaccinations. In order to address such a limitation, a prime–boost approach is used, in which primary vaccination is carried out using DNA vector, whereas the booster injection is provided by virus vector. For example, mice primed with M2-DNA and then boosted with recombinant adenovirus expressing M2 (M2-Ad) had enhanced antibody responses that cross-reacted with human and avian M2 sequences and produced T cell responses [144]. This M2 prime–boost vaccination conferred broad protection against challenge with lethal influenza A, including an H5N1 strain.

25.6 PROTEIN VACCINES

25.6.1 PEPTIDE AND SUBUNIT VACCINES

Peptides and subunit proteins can be promising vaccine candidates because they are safe, not reactogenic, and they can be generated and purified in large quantities. Many antibody and T cell epitopes for influenza virus have been reported in the literature, including a number of protective epitopes, and the antigenic architecture of the HA for some influenza viruses have been determined at high resolution [73,167]. Methods for fine mapping of viral epitopes have been developed [168–173]. Potentially, information regarding protective epitopes can be used for the development of epitope-based peptide vaccines. However, despite high safety profile of synthetic peptides, attempts to develop influenza vaccines based on a limited number of peptides have encountered several problems including HLA polymorphism, limitations in the immunogenicity of peptide-based immunogens, and the high mutation rate of influenza viruses. Further, most epitopes are conformation dependent. The majority of identified influenza epitopes are located within the HA and NP proteins [73]. Peptides derived from the conserved parts of influenza proteins may offer some advantages as a basis for the development of potential peptide-based vaccines. Conserved epitopes of influenza virus include stem domain of the HA and the M2e domain. Through further development, peptide epitope vaccines based on the HA or M2e have the potential to protect vaccinated individuals against multiple influenza strains including unanticipated pandemic and epidemic influenza viruses. For example, M2 protein is highly conserved across

influenza A subtypes. The efficacy of M2-based peptide vaccine has been evaluated in mice [144]. Animals were vaccinated with M2 peptide of a widely shared consensus sequence. Vaccination induced serum antibodies that cross-reacted with divergent M2 peptide from an H5N1 subtype [144]. However, epitope vaccines appear to hold even more promise when they are administered not as soluble synthetic peptides but in conjunction with other immunologically active components such as adjuvants or engineered VLPs (See Section 25.6.3). For example, in M2 protein incorporated into liposomal formulations containing a lipid adjuvant, MPL were shown to enhance immune protection to M2 protein and elicit protection against lethal homologous challenge [174]. In a more recent study, three copies of M2e epitope were expressed from live-attenuated *Bordetella pertussis* vaccine, and immunogenicity of this recombinant vaccine was demonstrated in a prime–boost regimen [175]. In another example of peptide vaccine, conserved synthetic peptides derived from the HA protein induced broad humoral and T cell responses in a pig model [176].

Recent efforts for the development of broadly protective *universal* influenza vaccine have focused on the highly conserved HA stalk domain [46,177]. Synthetic peptide provided protection in mice against influenza viruses of the structurally divergent subtypes H3N2, H1N1, and H5N1 [178]. In another example, H3-stalk-based chimeric HA universal influenza virus vaccines were used to protect against H7N9 virus challenge in mice. Chimeric HA constructs protected from viral challenge in the context of different administration routes and a generic oil-in-water adjuvant [177].

In addition to peptide antigens, subunit vaccines comprising purified influenza proteins are also well tolerated clinically [179]. Recombinant subunit vaccine FluBlok recently has been approved by FDA as the first recombinant human influenza vaccine (Table 25.1). A trivalent, baculovirus-generated, recombinant subunit HA0 (rHA0) vaccine was safe and immunogenic in a healthy adult population, and inclusion of a NA protein did not appear to be required for protection [180]. Participants of the clinical study received (1) a single injection of saline placebo or (2) 75 µg of an rHA0 vaccine containing 15 µg of HA from influenza A/New Caledonia/20/99 (H1N1) and influenza B/Jiangsu/10/03 virus and 45 µg of HA from influenza A/Wyoming/3/03(H3N2) virus or (3) 135 µg of rHA0 containing 45 µg of HA each from all 3 components. HI antibody responses to the H1 component were seen in 51% of 75 µg vaccine and 67% of 135 µg vaccine recipients, while responses to B were seen in 65% of 75 µg vaccine and 92% of 135 µg vaccine recipients [180].

Commercial trivalent subunit vaccines have been used for vaccination of people, such as Influvac in the Netherlands that contains 15 µg of HA for each strain [181]. The safety and immunogenicity of such commercial trivalent subunit influenza vaccine was compared to an experimental virosome-formulated influenza vaccine in elderly patients [182]. The virosome vaccine was generated by incorporating egg-purified HA into the membrane of phosphatidylcholine liposomes. Both vaccines elicited anti-HA antibody titer to all three vaccine components within 1 month after immunization. However, significantly more patients vaccinated with the virosome vaccine mounted a more than fourfold higher response to influenza as compared with those who received subunit vaccine. Approximately 68% of patients immunized with the virosome vaccine attained protective levels of antibody to all three vaccine components versus 38% for the subunit vaccine [182].

In general, most of the peptide and purified subunit proteins were found to be relatively weak immunogens. Efficacy of subunit vaccines could be considerably improved by using adjuvants such as squalene, which represents an unsaturated aliphatic hydrocarbon (MF59), or by using liposomes [183,184]. A number of adjuvants for infectious diseases including influenza have been reviewed elsewhere [185]. Liposome-based virosome vaccines have been commercially available for a number of years, as described in the next chapter. Immunogenicity study with MF59 adjuvant and subunit vaccine demonstrated that a consistently higher immune response is observed in MF59-adjuvanted subunit vaccine as compared to nonadjuvanted subunit and split influenza vaccines [183]. MF59-adjuvanted vaccine was clinically well tolerated, also after re-immunization in subsequent influenza seasons. An MF59-adjuvanted inactivated influenza vaccine containing A/Panama/2007/99 (H3N2) induced broader serological protection against heterovariant influenza virus strain A/Fujian/411/02 (H3N2) than a subunit and a split influenza vaccine [186]. The results showed that, while less than 80% of elder people vaccinated with conventional vaccines had protective levels of antibodies against the A/Fujian/2002 heterovariant strain, those vaccinated with the MF59-adjuvanted vaccine had protective levels of antibodies in more than 98% of the cases. Importantly, it has been also shown that vaccination with a subunit influenza vaccine with the MF59 adjuvant neither induced antisqualene antibodies nor enhanced preexisting antisqualene antibody titers [187]. This is important, because the presence of antisqualene antibodies in some recipients of squalene-contaminated anthrax vaccine has been linked to the Gulf War syndrome [188].

Other adjuvant systems are also being evaluated in the context of peptides, subunit proteins, as well as inactivated influenza vaccines including adjuvants based on CD40 monoclonal antibody [189,190]. CD40 is a costimulatory receptor on B lymphocytes, and signaling through CD40 molecules greatly enhances lymphocyte activation in the presence of antigen receptor stimulation. Conjugates of CD40 monoclonal antibody were made with three potential influenza vaccines: a peptide-based vaccine containing T and B cell epitopes from virus HA, a killed whole-virus vaccine, and a commercial split-virus vaccine. CD40 mAb conjugates in each case were found to be more immunogenic [190]. Inoculation of mice i.n. with an anti-CD40 monoclonal antibody and NP366–374 peptide, corresponding to a CTL epitope on NP, encapsulated in liposome induced protective CTL responses against influenza A virus [191].

25.6.2 Influenza Virosome Vaccines

The combination of a liposome with subunits of the influenza virus is called avirosome. The viral antigens HA and NA can be anchored into the nanoparticle or liposome lipid layer, thus resembling natural influenza virus particle [192]. Commercial virosomal influenza vaccines are available in Europe such as Inflexal V licensed for all age groups. Inflexal V represents an inactivated vaccine that has virosomes in its formulation acting as carrier/adjuvant, and it is composed by highly purified surface antigens of influenza virus strains A and B, propagated in embryonated chicken eggs and inactivated with beta-propiolactone. The vaccine's antigen composition follows annual WHO recommendations and contains 15 μg of HA for each recommended strain. Inflexal V was originally introduced in 1997 and is registered now in more than 40 countries. The vaccine has a very good safety record. Neither formaldehyde nor thiomersal (thimerosal), an organomercury preservative, is contained within the vaccine. The manufacturing process results in minimal residual quantities of antibiotics, detergent, and chicken proteins compared with other influenza vaccines. Another virosomal influenza vaccine, Invivac, has been also evaluated clinically and has been on the market. Virosomal vaccines mostly target the elderly, as well as other populations with impaired immune responses to conventional influenza vaccines [182,193,194].

Virosomal nasal vaccine, NasalFlu, containing strong mucosal adjuvant, *Escherichia coli* heat-labile enterotoxin (LT), was also licensed in Switzerland but later withdrawn due to increased number of cases of Bell's palsy associated with vaccination [195]. Unfortunately, LT appears to be an important component of the vaccine, as phase I clinical study showed that the use of LT as a mucosal adjuvant is necessary to obtain a humoral immune response comparable to that with parenteral vaccination [196]. Currently, second-generation virosomes are being developed for various prophylactic and therapeutic indications. The inclusion of additional components to optimize virosome assembly, to stabilize the formulations, or to enhance the immunostimulatory properties have further improved and broadened the applicability of this platform [192].

25.6.3 Virus-Like Particles as Influenza Vaccines

During the course of infection in virus-infected host cells, structural proteins and genomic nucleic acids of influenza virus assemble into progeny virion particles, which are released from infected cells. Interestingly, influenza structural proteins maintain this intrinsic ability to self-assemble into influenza VLPs, following expression of the structural genes in cell culture systems [197–200]. The size and morphology of such self-assembled influenza VLPs resemble those of influenza virions. However, VLPs are noninfectious, because in the cell culture expression systems, viral proteins self-assemble in the absence of the viral genetic material. This ensures intrinsic safety of recombinant VLPs.

For many other viruses, VLPs have been also generated that closely resemble structures formed by their counterpart viruses, including hepatitis B viruses (HBVs) [201,202], human papilloma virus (HPV) [203,204], severe acute respiratory syndrome (SARS) coronavirus [205], Ebola and Marburg filoviruses [206], and other viruses [200,207]. Such noninfectious VLPs were found to be promising candidates for the production of vaccines against many diseases because highly repetitive structures of VLPs and high-density display of epitopes is very effective in eliciting strong immune responses. VLPs are often morphologically and antigenically indistinguishable from the respective viruses, and the epitopes within VLPs preserve native conformation found in the viruses. Furthermore, the size and particulate nature of VLPs may facilitate their uptake by dendritic cells, which is advantageous for the development of effective immunity [208].

During the past decade, significant progress has been achieved in expression of various types of VLPs for many viruses, as reviewed elsewhere [200]. In general, VLP-based vaccines for influenza can be divided into native VLPs and chimeric VLPs, depending on the approach used to generate VLPs and the way of inducing immunity against influenza virus [200].

25.6.3.1 Native VLPs as Influenza Vaccines

Native VLPs are formed by self-assembly of native, full-length structural proteins. In these cases, VLPs are generated that have morphological and antigenic characteristics similar, if not identical, to naturally occurring authentic viral and subviral particles. Because native VLPs are structurally and morphologically similar to the wild-type viruses, when administered to a susceptible host, such VLPs induce the immune responses that in many aspects resemble those induced by the respective cognate viruses.

In many cases, native VLPs have already proven successful as vaccines against the respective virus infections. For example, small envelope protein, HBsAg, of HBV forms in yeast or mammalian cells is secreted as subviral (22 nm) particles that

are essentially identical to a natural product of HBV infection and found in patient blood at levels greater than the virion itself. Both plasma-derived and recombinant 22 nm particles provided successful HBV vaccines. Similarly, expression of the L1 protein of HPV6 and HPV16 in cultured cells have lead to the assembly of VLPs [203,204] that are similar to the virus particles formed during papillomavirus replication, although the natural particles also contain the L2 protein. Gardasil, a recombinant quadrivalent VLP vaccine based on the L1 proteins of HPV (types 6, 11, 16, and 18), has been produced in yeast cell culture and licensed for the commercial use as an HPV vaccine. Recently, Cervarix vaccine produced in insect cell culture by using baculovirus expression system has also been licensed. The licensure of the HPV VLP vaccine will undoubtedly provide an impetus for further development of VLP-based vaccines for other viruses including influenza.

The expression of native VLPs for influenza virus poses some challenges. Unlike HBV or HPV viruses, influenza virus particle consists of numerous structural proteins. The protein–protein interactions and the roles of each protein, lipids, and host cell factors in the assembly and morphogenesis of VLPs are not fully understood. In order to achieve correct assembly of native influenza VLPs, proper interactions should take place between the inner influenza proteins, between the surface protein subunits within the lipid bilayer, and between the surface envelope and the inner influenza proteins.

Introduction of the efficient protein expression systems as well the reverse genetics methods have been critically important for efficient coexpression of influenza proteins including VLPs [209–213]. The examples of native influenza VLPs that have been reported in the literature are shown in Table 25.2. In the early experiments, in order to generate influenza VLPs, 10 influenza proteins were expressed in mammalian cell culture. For example, Mena et al. used COS-1 cells, which were initially infected with recombinant vaccinia virus expressing the T7 RNA polymerase and then transfected with plasmids to express all 10 influenza proteins [214]. Influenza VLPs

TABLE 25.2
Examples of Native Influenza VLP Vaccines

Strain	Genes	Size, nm	Expression	Testing, Model	References
H3N2	PB2, PB1, PA, NP, HA, NA, M1, M2, NS1, NS2	80–120	COS1	nt	[214,224]
H1N1	PB2, PB1, PA, NP, NA, NA, M1, M2, NS2	nt	293T	Mice	[199]
H3N2	HA, NA, M1, M2	~100	Sf9	nt	[197]
H9N2	HA, NA, M1	80–120	Sf9	Mice, rats, ferrets	[184,198]
H3N2	HA, M1	nt	Sf9	Mice	[218]
H5N1	HA, M1		Insect	Chickens	[221]
H1N1	HA, M1	80–120	Sf9	Mice	[219]
H3N2	HA, NA, M1	nt	Sf9	Mice, ferrets	[215]
H5N1	HA, NA, M1	nt	Sf9	Mice	[216]
H3N2	HA, NA	100	293T	nt	[222]
H1N1	NA, M1	100		Mice	[220]
H5N1	HA		Insect	Chickens	[221]

nt, not tested.

resembling wild-type influenza virus were found in the culture medium. Further, these VLPs were also capable of encapsidating a foreign gene and transferring it to fresh MDCK cells, in which expression of their foreign gene could be detected.

Watanabe et al. generated influenza VLPs using nine proteins in transfected 293T cells [199]. NS2-knockout VLPs were injected into mice, which later were challenged with antigenically homologous influenza virus. The protective effect of such influenza VLPs demonstrated in these experiments highlights the potential of such nonreplicating VLPs as a vaccine approach.

By using baculovirus expression system, native VLPs have been generated in insect Sf9 cells by using four structural proteins, HA, NA, M1, and M2 [197]. Furthermore, efficient formation of influenza VLPs have been achieved in Sf9 cells following expression of only HA, NA, and M1 proteins from a single baculovirus construct [184,198,215]. For example, in order to generate H9N2 influenza VLPs in Sf9 cells, the HA, NA, and M1 genes were derived from influenza A/Hong Kong/1073/99 (H9N2) virus and introduced into recombinant baculovirus (rBV), each gene within its own expression cassette, which included a polyhedrin promoter and transcription termination sequences (Figure 25.3). As expected, the HA was expressed as HA0 in Sf9 cells, and no significant processing into HA1 and HA2 was observed. Influenza VLPs were purified from culture media by sucrose gradient centrifugation, and the presence of HA, NA, and M1 in VLPs was confirmed by SDS-PAGE, Western blot, hemagglutination assay, and NA enzyme activity assay. Electron microscopic examination of negatively stained samples revealed the presence of H9N2 VLPs with a diameter of approximately 80–120 nm, which showed surface spikes, characteristic of influenza HA protein on virions (Figure 25.4a).

Such native VLPs of influenza virus composed of HA, NA, and M1 induced robust immune responses in preclinical studies [184,198,215]. Immunogenicity and protective

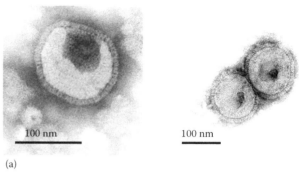

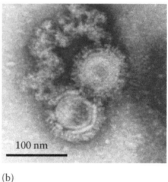

FIGURE 25.4 Negative staining electron microscopy of (a) H9N2 influenza native VLPs and (b) hybrid H7N1 VLPs. Influenza H9N2 VLPs were generated from influenza A/Hong Kong/1073/99 (H9N2) HA, NA, and M1 proteins, as shown in Figure 25.3. Influenza H7N1 hybrid VLPs were prepared by using HA gene from A/Shanghai/2/2013 (H7N9), while NA and M1 genes were derived from A/Puerto Rico/8/1934 (H1N1) virus. Bars represent 100 nm. For electron microscopy, VLPs were adsorbed on freshly discharged plastic/carbon-coated grids and stained with 2% sodium phosphotungstate, pH 6.5. Stained VLPs were observed by transmission electron microscope at magnifications ranging from 6,000× to 100,000× (Medigen, Inc.).

capacity of baculovirus-generated VLPs, subunit HA antigen, and liposome-adjuvanted VLP and HA antigens were also compared in BALB/c mice [184]. The results suggested that VLPs have advantage over recombinant subunit rH9 in terms of both immunogenicity and protection. Interestingly enough, addition of liposome adjuvant significantly improved characteristics of rH9 as a vaccine [184].

The protective capacity of H9N2 influenza VLPs was also confirmed in ferrets [184]. Ferrets are considered to be the most suitable animal model for preclinical evaluation of human influenza viruses. Ferrets received primary and booster vaccinations with 0.15, 1.5, or 15 μg of HA antigen within the H9N2 VLPs. Control animals received PBS only. Following vaccinations, animals that received 1.5 or 15 μg doses of VLPs were challenged with A/Hong Kong/1073/99 (H9N2) virus. Replication of the challenge H9N2 virus was determined by monitoring virus shedding in nasal washes after challenge. Infectious virus was detected in nasal washes of all animals on day 3 postchallenge. However, on day 5, only low titers of replicating virus were detected in the animals vaccinated with 1.5 or 15 μg of VLPs. In these groups, virus replication was undetectable on

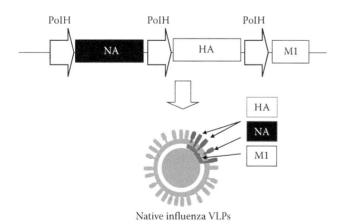

Native influenza VLPs

FIGURE 25.3 Baculovirus transfer vector for coexpression of influenza proteins and production of influenza native VLPs. Indicated are the polyhedrin promoter (PolH) and influenza genes. HA, hemagglutinin; NA, neuraminidase; M1, matrix protein. Positions of HA, NA, and M1 proteins on the surface of influenza VLPs are also indicated.

day 7, whereas control animals that received PBS showed over 3 $\log_{10}$ EID_{50}/mL of replicating influenza virus.

It has been shown also that influenza H3N2 VLPs composed of HA, NA, and M1 proteins induced broader immune responses than the whole virion inactivated influenza virus or recombinant HA vaccines [215].

We have also generated in a baculovirus expression system recombinant VLPs from the HA, NA, and M1 of H5N1 viruses [216]. In these cases, H5 proteins were engineered not to contain the polybasic cleavage site. Such cleavage-deficient HA proteins were capable of efficient assembling into VLPs in Sf9 insect cells. VLP vaccines were purified and administered to mice in either a one-dose or two-dose regimen, and the immune responses were compared to those induced by recombinant HA (rH5). Mice vaccinated with VLPs were protected against challenge regardless of whether the H5N1 clade was homologous or heterologous to the vaccine. However, rH5-vaccinated mice had significant weight loss and death following challenge with the heterologous clade virus. The association rate of antibody binding to HA correlated with protection and was enhanced using H5N1 VLPs, particularly when delivered i.n., compared to rH5 vaccines. The results showed that native H5N1 VLPs are effective influenza vaccine immunogens that elicit cross-clade protective immune responses to emerging H5N1 influenza isolates [216].

Three-protein VLPs composed of HA, NA, and M1 were also generated in COS-1 cells transfected with expression plasmids [217]. Influenza VLPs have also been successfully generated in Sf9 cells from the HA and M1 proteins only (Table 25.2), and such two-protein VLPs also induced detectable protective immune responses in mice [218,219]. Furthermore, VLPs were generated from NA and M1, and protective effect of such NA-M1 VLPs was demonstrated in mice [220]. Influenza VLPs have been also reported that contained the HA protein only [221].

In the experiments that involved Sf9 cells and rBVs, two strategies have been used for expression of influenza VLPs. In one strategy, influenza proteins were coexpressed from a single rBV [184,197,198,215]. The other strategy involved coinfection of the Sf9 cells with the two rBVs, one expressing the HA protein, whereas the other rBV expressed the M1 protein [218,219].

Most of the approaches for expressing influenza VLPs included M1 protein, because M1 is the most abundant protein in the virion and because it has been shown to be the driving force of influenza virus budding [217]. However, recently, two-protein influenza VLPs were also generated that consisted only of the HA and NA (Table 25.2) and did not include M1 [222]. HA protein, when expressed in noncytotoxic mammalian cell culture and treated with exogenous NA or coexpressed with viral NA, could be released from cells independently of M1. Incorporation of M1 into VLPs required HA expression, although when M1 was omitted from the VLPs, particles with morphologies similar to those of wild-type viruses were also observed.

Taken together, many studies have demonstrated that native influenza VLPs can be easily generated in eukaryotic cell culture expression systems by using combinations of HA, NA, M1, and other viral proteins. Furthermore, in preclinical and clinical studies, native influenza VLP vaccines have shown excellent characteristics as influenza vaccines. As mentioned earlier, influenza VLPs induced broad and robust immune responses as well as efficient protection against homologous and heterologous live influenza virus challenges.

The expression of native influenza VLPs also contributes to a better understanding of influenza virus assembly and morphogenesis. The influenza virion particle contains nine structural proteins. Studies with VLPs demonstrated that expression of M1 alone forms VLPs and that the expression of M1 can drive HA and NA into budding VLPs that are released into culture medium [197,217]. However, an alternative influenza assembly model suggests that envelope glycoproteins other than M1 control influenza virus budding by sorting to lipid raft microdomains and recruiting the internal viral core components. Recent demonstration of VLPs composed of HA and NA is consistent with the latter model [222].

VLPs composed of only two proteins, HA and M1, morphologically resemble wild-type influenza virions including characteristic surface spikes and are capable of inducing protection against lethal challenge with influenza in BALB/c mice [218,219]. The presence of additional influenza proteins, such as NA, M1, and M2, in the VLPs [197,198,215] may provide additional benefits including inducing broader immune responses, especially in outbred populations, because the NA, M1, and M2 proteins contain additional antigenic epitopes [73]. However, coexpression of several proteins in the cells during manufacturing process may reduce the overall levels of expression of VLPs due to promoter dilution effect or because of excessive metabolic burden [223]. Furthermore, overexpression of M2 may also dramatically decrease the yields of VLPs [224]. It was hypothesized that overexpression of M2, an ion channel protein, inhibited intracellular transport and drastically reduced accumulation of coexpressed HA and hence reduced the number of VLPs. It appears that the optimal composition of native VLPs still needs to be determined, which would ensure efficient and broadly protective characteristics and high levels of production of VLP-based influenza vaccines. Purification of VLPs may represent another potential challenge. Also, it remains to be determined if any molecules such as nucleic acids or host cell proteins are incorporated into the influenza VLPs and whether these play a role in the assembly and/or whether presence of these molecules may affect the use of VLPs as vaccines from the safety/regulatory points of view.

25.6.3.2 Engineering of Hybrid VLPs

Expression of native VLPs requires cloning and expression of multiple influenza proteins such as HA, NA, and M1, all derived from the same isolate of virus. This process is relatively time consuming and strain specific. For distinct viruses, new sets of genes need to be cloned and expressed. For example, in order to make a trivalent VLP vaccine, three genes (HA, NA, and M1) need to be expressed for each vaccine-relevant virus, which would require cloning and expression of nine genes to express VLPs for each virus. This can adversely affect timing and the cost of vaccine production. Furthermore, for distinct viruses, the levels of gene expression may vary, which can further complicate vaccine manufacturing. In order to simplify and standardize the VLP production process, hybrid VLPs were proposed [200]. Hybrid VLPs are defined as VLPs containing the full-length proteins from distinct viruses, for example, M1 is derived from one virus, while the HA and NA are derived from another virus [200]. Unlike chimeric VLPs, which are made of chimeric polypeptides containing epitopes or protein fragments from several viruses (as discussed below in Section 25.6.3.2), hybrid VLPs are prepared by using the unmodified wild-type proteins.

An example of hybrid influenza VLPs is shown on Figure 25.4b. In this case, HA was derived from A/Shanghai/2/2013 (H7N9) virus, whereas NA and M1 were derived from A/Puerto Rico/8/1934 (H1N1) virus. Expression of HA, NA, and M1 proteins resulted in hybrid H7N1, which were prepared and characterized at Medigen, Inc. (Frederick, MD). Electron microscopy examination confirmed presence of 100 nm particles morphologically resembling influenza virus (Figure 25.4b). In another study, hybrid VLPs were generated, which were formed by using retrovirus Gag protein core and the complete influenza HA, NA, and M2 proteins [225]. By using this approach, influenza-Gag VLPs were generated that contained retroviral core and the outer surface that mimicked properties of the influenza viral surface of two highly pathogenic influenza viruses of either H7N1 or H5N1 antigenic subtype. Such hybrid VLPs induced high-titer neutralizing antibodies in mice [225]. The phenomenon of generating influenza VLPs using core structure from unrelated viruses deserves further study and may provide important information regarding mechanism of influenza particle assembly and the specificity of interactions between the core and envelope proteins in such chimeric VLPs. Recently, it has been shown that VLPs of brome mosaic virus, a nonenveloped icosahedral plant virus, can be generated that contain spherical gold nanoparticle cores. Interestingly, variation of the gold core diameter provided control over the VLP structure, as the number of subunits required for a complete particle increased with the core diameter [226]. This technology could also be useful for studying, and possibly controlling, the assembly process of influenza VLPs.

A novel hybrid VLP platform has been reported that contained three HA subtypes colocalized within a VLP. Experimental triple-HA VLPs were designed to colocalize HA proteins derived from H5N1, H7N2, and H2N3 viruses, as well as NA and M1 from H5N1 avian influenza (Figure 25.5). Such triple-HA VLPs were immunogenic and protected ferrets from challenge from all three potentially pandemic influenza viruses (Figure 25.7a). Similarly, triple-HA VLPs containing HA subtypes derived from seasonal H1N1, H3N2, and type B influenza viruses (Figure 25.6) protected ferrets from three seasonal influenza viruses [227] (Figure 25.7b).

We also described a hybrid VLP that colocalized HA proteins derived from H5N1, H7N2, and H9N2 viruses, as well as NA and M1 derived from influenza PR8 strain [228].

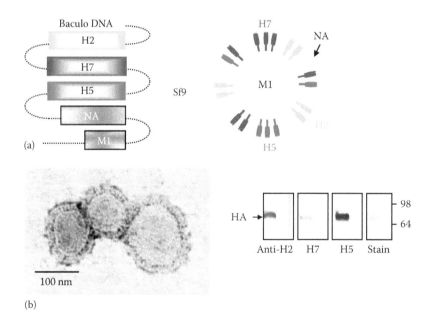

FIGURE 25.5 Preparation of triple-HA hybrid VLP vaccine expressing HA proteins derived from H2, H5, and H7 subtype viruses. (a) rBV and schematic representation of triple-HA pandemic VLPs expressed in Sf9 cells. Influenza HA gene sequences were derived from A/Swine/Missouri/4296424/2006 (H2N3), A/New York/107/2003 (H7N2), and A/Vietnam/1203/2004 (H5N1). NA and M1 gene sequences were derived from A/Indonesia/05/2005 (H5N1) virus. The H2, H7, H5, NA, and M1 genes were combined within recombinant rBV so that each gene was expressed from its own expression cassette. (b) Triple-HA hybrid VLPs, by transmission electron microscopy and Western blot. VLPs were harvested from the medium of rBV-infected Sf9 cells on day 3. VLPs were concentrated and purified by 20% step sucrose gradient ultracentrifugation. SDS-PAGE and Western blot is shown on the right. Locations of influenza HA proteins and molecular weight markers (kDa) are indicated. Western blot was done using ferret antisera against A/Mallard/NY/6750/1978 (H2N2), A/New York/107/2003 (H7N2), and A/Vietnam/1203/2004 (H5N1) viruses, followed by alkaline phosphatase-conjugated antispecies IgG (H + L). SDS gel was stained using GelCode Blue.

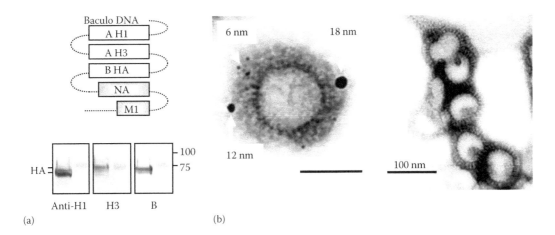

FIGURE 25.6 Triple-HA hybrid VLP vaccine expressing seasonal H1 and H3 influenza A and influenza B virus HA. (a) rBV for expression of triple-HA VLPs in Sf9 cells. HA gene sequences were derived from A/New Caledonia/20/1999 (H1N1), A/New York/55/2004 (H3N2), and B/Shanghai/361/2002. NA and M1 genes were from A/Indonesia/05/2005 (H5N1) virus. Western blot was done using ferret antisera against A/New Caledonia/20/1999 (H1N1), A/New York/55/2004 (H3N2), and B/Shanghai/361/2002 viruses. (b) Immunoelectron microscopy (left panel) was done using sucrose-gradient-purified VLPs. VLPs were probed with the mixture of primary antibodies against H1, H3, and type B influenza from rabbit, mouse, and guinea pigs, respectively. Secondary antibodies were donkey anti-rabbit labeled with 18 nm gold, anti-mouse labeled with 6 nm gold, and anti-guinea pig labeled with 12 nm gold particles. Transmission negative staining electron microscopy (right panel) was done using staining with 1% phosphotungstic acid. Bars, 100 nm.

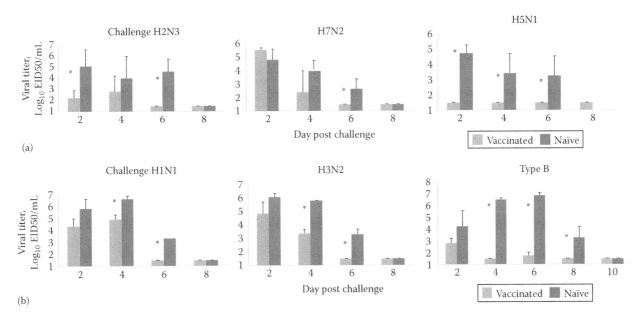

FIGURE 25.7 (a) Protective efficacy of triple-HA pandemic VLP vaccine after challenge with virulent strains of influenza A/Swine/Missouri/4296424/2006 (H2N3), A/New York/107/2004 (H7N2), and A/Vietnam/1203/2004 (H5N1) viruses. Animals were challenged i.n. with 10^6 EID50 of each influenza virus. Shown are the titers of influenza viruses at indicated days after challenges in the nasal washes of VLP-vaccinated (blue bars) and nonvaccinated (red bars) ferrets. Detection limit was 1.5 log10 EID50/mL of influenza virus. *p < 0.05. (b) Protective efficacy of triple-HA seasonal VLP vaccine in ferrets after challenge with influenza A/New Caledonia/20/1999 (H1N1), A/New York/55/2004 (H3N2), and B/Shanghai/361/2002 viruses. Animals were challenged i.n. with 10^6 EID50 of each influenza virus. Shown are the titers of influenza viruses at indicated days after challenges in the nasal washes of VLP-vaccinated (blue bars) and nonvac-cinated (red bars) ferrets. Detection limit was 1.5 log10 EID50/mL of influenza virus. *p < 0.05 between VLP-vaccinated and naive groups for each day postchallenge.

A baculovirus vector was configured to coexpress the H5, H7, and H9 genes from A/Vietnam/1203/2004 (H5N1), A/NewYork/107/2003(H7N2), and A/HongKong/33982/2009 (H9N2) viruses, respectively, as well as NA and matrix (M1) genes from A/Puerto Rico/8/1934 (H1N1) virus. Coexpression of these genes in Sf9 cells resulted in production of triple-HA VLPs containing HA molecules derived from the three influenza viruses. The triple-HA VLPs exhibited hemagglutination and NA activities and morphologically resembled influenza virions. Intranasal vaccination of ferrets with the VLPs resulted in induction of serum antibody responses and protection against experimental challenges with H5N1, H7N2, and H9N2 viruses [228]. Thus, triple-HA hybrid VLP technology may represent novel strategy for rapid development of trivalent seasonal and pandemic vaccines [200,227,228]. In summary, flexible influenza VLP technology offers promising alternatives to the conventional reassortment or reverse genetic platforms without concerns or limitations imposed by the manipulation of live viruses.

25.6.3.3 Chimeric VLPs Containing Influenza Epitopes

The structural components of native VLPs derived from many viruses have proven amenable to the insertion of foreign antigenic sequences, allowing protein engineering of *chimeric* VLPs that expose foreign antigen(s) on their surface. In other words, many VLPs can be used as *carriers* or *platforms* for the presentation of foreign epitopes and/or targeting molecules on chimeric VLPs [200–202,229–231]. This method can significantly improve immunogenicity of peptide epitopes. In some cases, foreign peptide epitopes or the entire proteins, or even nonprotein antigens, may be chemically conjugated to pre-formed VLPs [231]. Alternatively, foreign antigen may be exposed on the surface of VLPs via modification of the VLP gene sequences, such that fusion proteins composed of VLP protein and foreign antigen are assembled into VLPs during de novo synthesis [232]. In this case, it is important that foreign epitope is inserted

so that it does not interfere with the assembly process of the subunits of carrier VLPs [233]. Methods for mapping protein domains that are exposed on the surface or internalized within VLPs have been developed [230,234,235]. Also, 3D structures for many viruses have been determined by cryoelectron microscopy, x-ray crystallography, or a combination of both methods, which greatly facilitates the rational design of the proteins that assemble into chimeric VLPs [236–239].

Because chimeric VLPs consist of a core platform VLP as well as of the antigenic epitopes of interest, administration of such chimeric VLPs in vivo usually induces responses to the antigenic epitopes of interest as well as to the epitopes within carrier VLPs.

VLPs derived from several viruses have been configured to carry and display influenza antigenic epitopes, especially fragments derived from M2 protein. Extracellular part of M2 protein, approximately 23 aa residues, is highly conserved in known human influenza A strains. Therefore, M2-based vaccine may be able to protect from all human influenza A strains that could result in a *universal* influenza vaccine and a major improvement over currently used vaccines.

Universal vaccine targets have recently included the conserved stalk (stem) domain of the HA (Figure 25.8). Because of the conservation of the stalk domain, broadly protective vaccine candidates based on the epitopes in the stalk domain, for example, chimeric HA structures, are currently under development and show promising results in animals models [46,240].

As mentioned earlier, foreign epitopes can be fused to carrier VLP through either chemical conjugation or genetic fusion. An example of conjugated vaccine is represented by HPV VLPs, to which extracellular domain of influenza M2 protein was chemically conjugated. Conjugates comprised approximately 4000 copies of the antigenic peptide per VLP. Such chimeric M2-HPV VLPs were in average larger in size as compared to HPV VLP carrier alone. Such a conjugate

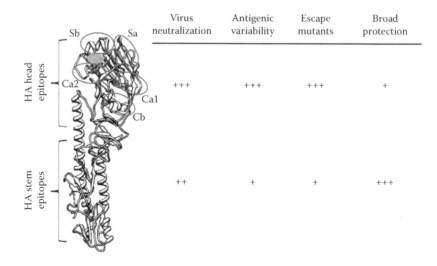

	Virus neutralization	Antigenic variability	Escape mutants	Broad protection
HA head epitopes	+++	+++	+++	+
HA stem epitopes	++	+	+	+++

FIGURE 25.8 Comparison of influenza HA globular head epitopes and HA stem region epitopes [240]. A structural view of the H1 HA molecule (monomer, A/Puerto Rico/8/34) showing the different globular head epitopes (red circles) and their localization close to the receptor-binding domain (brown).

vaccine has been formulated with adjuvant and administered to mice, in which it induced immune responses and protection against lethal challenge with influenza virus [231].

Genetic fusion of influenza M2 protein sequences with the core gene (HBcAg) of HBV provides an example of an alternative approach to generate chimeric M2-HBcAg VLPs. The M2 gene fragment corresponding to extracellular domain was fused to the HBcAg gene to create fusion gene coding for M2-HBc VLPs. Administration of M2-HBc VLPs purified from *E. coli* to mice provided up to 100% protection against a lethal virus challenge. The protection appeared to be mediated by antibodies, as it was transferable using serum [232]. However, it appears that such M2-specific antibodies neither bind efficiently to the free influenza virus nor neutralize virus infection but bind to M2 protein expressed on the surface of virus-infected cells. Therefore, the M2 antiserum does not prevent infection and only reduces disease at low challenge doses. At higher challenge doses, the M2 antiserum fails to protect mice [241]. Attempts to improve vaccine characteristics of chimeric M2-HBc VLPs were carried out by optimizing the antigen design by either displaying more copies of the M2 polypeptide or by inserting the M2 polypeptide into different sites of the HBcAg carrier, the amino-terminus or the immunodominant loop. Such engineered variants showed promise after i.n. coadministration with cholera toxin A1, a powerful vaccine adjuvant [111,242].

Further, characteristics of chemically conjugated chimeric M2-HBc VLPs were compared to those of genetically fused M2-HBc VLPs. The conjugated M2 appeared to be better accessible on the surface of HBc VLPs and induced stronger immune response than genetically fused M2 [243].

Tandem repeats of heterologous M2e sequences (M2e5x) derived from human, swine, and avian-origin influenza A viruses were expressed on influenza VLPs in a membrane-anchored form. Immunization of mice with M2e5x VLPs induced protective antibodies cross-reactive to antigenically different influenza A viruses and conferred cross-protection [244].

Furthermore, hybrid and chimeric approaches can be combined in the production of VLPs. For example, VLPs were made for SARS virus, which were composed of chimeric SARS spike (S) protein and influenza M1 protein. In this study, chimeric S protein was modified to contain the carboxy terminal region of influenza HA protein with the view to facilitate interaction with M1 and to improve production of SARS VLPs [200,245].

Thus, many types of chimeric VLPs expressing influenza antigens have been generated using protein engineering methods and showed promise as vaccines in preclinical models. However, the limited number and/or relatively small size of influenza antigens that can be incorporated into some VLPs may limit the practical utility of many chimeric influenza VLP approaches, while manufacturing and/or regulatory considerations may also prove to be a significant barrier to vaccine development. The first clinical study for a M2-based universal influenza vaccine has been announced [246].

25.6.3.4 Chimeric Influenza VLPs as a Carrier of Foreign Epitopes

The available data suggest that not only influenza epitopes can be displayed on the surface of carrier VLPs derived from unrelated viruses but also influenza VLPs themselves may also be engineered as a carrier for foreign antigenic epitopes. For example, it has been shown that influenza VLPs are capable of incorporating vesicular stomatitis virus (VSV) G protein [197]. In another study, *Bacillus anthracis* protective antigen (BPA) fragments, 90 or 140 aa in length, were inserted at the C-terminal flank of the HA signal peptide and expressed as the HA1 subunit. The chimeric proteins could be cleaved into the HA1 and HA2 subunits by trypsin and could be also incorporated into recombinant influenza viruses suggesting that viral envelope can tolerate foreign inserts without precluding assembly. The inserted BPA domains were maintained in the HA gene segments following several passages in MDCK cells or embryonated chicken eggs. Immunization of mice with either recombinant viruses or with DNA plasmids that express chimeric BPA/HA proteins induced antibody responses against both HA and BPA components of the protein [247]. Although VLPs were not generated in this study, these experiments suggests that similar modifications of HA protein with foreign epitopes may be compatible with the formation of chimeric influenza VLPs.

25.7 INFLUENZA AS A GENETIC VIRUS VECTOR

Using molecular engineering technologies, a concept of using influenza as a genetic vector has been applied to the development of vaccine candidates for several pathogens [248,249]. Both propagation-competent and propagation-incompetent influenza virus vectors have been described. Strategies for the construction of propagation-competent influenza vectors included the insertion of foreign antigenic epitopes into influenza virus glycoproteins [247], rescue of bicistronic genes into infectious viruses, and the expression of polyproteins. Influenza virus vectors have been obtained, which express both B and T cell epitopes from different pathogens. These constructs have been shown to induce in vaccinated animals systemic and local antibody responses and/or cytotoxic T cell responses against the expressed antigenic epitopes [248].

For example, vaccination of mice with recombinant influenza and vaccinia viruses expressing antigens from *Plasmodium yoelii* resulted in a significant protective immune response against malaria in this model. Mice immunized with recombinant influenza viruses expressing HIV epitopes generated long-lasting HIV-specific serum antibody response as well as secretory IgA in the nasal, vaginal, and intestinal mucosa [249].

Evaluation of propagation-incompetent influenza vectors has been also initiated. As mentioned in the previous chapters, in the COS-1 cells expressing all 10 influenza virus-encoded proteins, the transfected CAT RNA could be rescued into influenza VLPs that were budded into the supernatant fluids [214]. The released VLPs not only resembled influenza virions but also transferred the encapsidated CAT RNA to MDCK cell cultures. Such VLPs required trypsin treatment to deliver the RNA to fresh cells and could be neutralized by a monoclonal antibody specific for the influenza A virus HA. These data indicated that influenza VLPs are capable of encapsidating a synthetic virus-like RNA, which can be delivered to fresh cells for expression of foreign gene of interest. For other virus vectors, it has been shown that such *VLP vectors* encapsidate nucleic acids by utilizing the ability of viral structural proteins to recognize specific encapsidation signals within the nucleic acid sequences. As with other propagation-incompetent virus vectors, the advantage of such vectors is that when injected in a susceptible host, they can generate immune response almost exclusively against the foreign antigen expressed from the vector, whereas low, if any, immune response is generated against the structural proteins of the vector itself [158]. However, more experiments are needed to demonstrate the potential of influenza VLPs as a vaccine vector system.

25.8 CONCLUSION

Every year, influenza virus causes up to 500,000 vaccine-preventable deaths in the world [8]. Furthermore, a constant menacing threat of another influenza pandemic has the potential to cause even larger numbers of deaths and illnesses over a short period of time than any other known health threat. Because vaccination is a powerful and cost-effective countermeasure to the threat of either seasonal or pandemic influenza, vaccines against influenza viruses have been developed and licensed during the last decades in the United States, Europe, and around the world. The existing vaccines have reduced significantly the impact of influenza disease on public health and world economies.

However, several important issues still remain to be addressed in order to improve traditional influenza vaccines. As discussed in detail in the previous chapters, these issues include suboptimal efficacy of the existing vaccines in the young and elderly patients, reactogenicity in some people, and the difficulty to accommodate rapid changes in the vaccine composition in response to, and in order to match, the changing circulating influenza virus strains. Many of these issues stem from using relatively inefficient, chicken-egg-based technology for the production of the majority of licensed vaccines. Potentially, limitations in vaccine efficacy, safety, and manufacturing may cause, and have caused in the past, suboptimal vaccine efficacy and reduced protection of the population, as well as delays in supply, and even shortages of the vaccines. Such limitations may become insurmountable in the case of a pandemic, which can result in significant loss of human lives and massive economic damage.

Significant efforts have been undertaken by both vaccine manufacturers as well as the research community to address these important issues. Among successful strategies are the attempts to adopt cultured cells instead of chicken eggs for vaccine manufacturing, which could bring significant improvements to both the vaccine production and vaccine safety. Further, faster and newer methods of generating vaccine seed viruses are being implemented such as using the reverse genetics technology. In addition, steps are being taken for improvements in virus growth, virus inactivation, and vaccine purification methods. Introduction of quadrivalent influenza vaccines improves protection against multiple circulating influenza strains. Alternative vaccine inoculation schedules and advanced adjuvants are also being evaluated in order to improve immunogenicity, efficacy, and safety of influenza vaccines to make lower vaccine doses more effective. A promising way of improving vaccination against a respiratory virus such as influenza could be mucosal, such as i.n. vaccine administration. Inactivated influenza vaccines are currently administered i.m. to patients. In general, i.m. vaccinations result in the elicitation of effective virus-neutralizing serum antibody response and effective protection against potentially severe or even fatal consequences. Live-attenuated FluMist vaccine is administered i.n. and appears to induce a more balanced immune response. The advantage of i.n. administration could be the elicitation of effective mucosal IgA that may be able to bind and stop the incoming virus in the upper nasal pharyngeal area and thus prevent its passage to the lower respiratory tract.

In addition to improvement of current vaccines, new vaccine candidates are also being actively developed using molecular technologies including protein engineering. The first recombinant influenza vaccine FluBlok has been approved. Novel vaccine candidates range from synthetic peptides to the recently developed influenza VLPs, which mimic whole influenza viruses. Virtually, every new concept or technology is being applied or adapted for influenza research or developed specifically for influenza research and vaccine development. The impact of influenza virus and influenza disease on public health and economy as well as the significant market for influenza vaccines provide an impetus to these efforts.

This concerted research resulted in a considerable progress in our understanding of influenza virus and the disease as well as of influenza vaccines. This knowledge can be applied to further advance influenza research and to create vaccines that are safer and even more effective than the ones we are currently using. Importantly, the requirements for inducing effective protective immunity as well as critically important protective antigens have been identified during the past years of influenza vaccine research. The available data strongly suggest that HA, the major surface envelope glycoprotein, is the most important antigen in the currently used influenza vaccines. Presence of functionally active HA is the standard requirement for all currently used inactivated influenza vaccines, because eliciting of virus-neutralization response correlates with the presence of functionally active HA in the vaccine. It appears that other antigenic components of the vaccine are of lesser clinical importance for protection. For example, NA, another influenza surface antigen, appears not to significantly affect protection [180], although in other studies, some protective effect of NA has been implicated retrospectively [13,250].

The available comparative studies of different inactivated vaccines have demonstrated that different vaccines have varying immunogenic characteristics. For example, the whole-virus vaccines appear to have higher immunogenicity when compared to split vaccines [89], and virosome vaccines appear to have higher immunogenicity when compared to subunit vaccine [182]. It is plausible to suggest that the higher level of immunogenicity depends on the amount of correctly folded and antigenically active HA. Correct conformation of HA within a lipid membrane is also required for fusion activity [251]. Functional activity of HA, such as the ability to agglutinate erythrocytes, also depends on the correct folding of the HA protein. The higher is the chance of correctly folded HA in vaccine, the higher is the immunogenicity and protective capacity of HA. Our data demonstrated that baculovirus-generated influenza VLPs have higher immunogenicity profile in mice as compared to baculovirus-generated subunit recombinant HA [184]. Interestingly enough, immunogenicity of the subunit HA could be drastically improved by adding non-phospholipid liposomes [184]. Further, immunogenicity of split vaccines can also be significantly improved by adding lipid-containing liposomes or other lipid-based adjuvants, such as MF59, an oil-in-water emulsion containing unsaturated aliphatic hydrocarbon squalene [183]. Virosome vaccines representing liposome-embedded subunits of HA represent an example of highly immunogenic influenza vaccines. Thus, highly immunogenic influenza vaccines including whole-virus vaccines, VLPs, and virosomes have lipoprotein membrane or other lipid components as an integral part of the vaccine, suggesting that this plays an important role in the correct conformation and functional activity of HA. Presence of lipid components appears to result in higher overall immunogenicity and protective efficacy of the vaccines. The reasons for such importance of lipid components in the correct conformation of HA probably stem from the intimate involvement of lipid-containing cellular membranes in the life cycle of influenza virus including budding of progeny particles from the infected cells. In the influenza virus, HA and NA proteins are embedded in the outer envelope, which is derived from cell plasma membrane during budding. The plasma membrane is not homogeneous, and the lipids of the plasma membrane are distributed nonrandomly and can self-associate and organize into a liquid ordered phase. Sphingolipids and cholesterol contribute to cellular membrane organization by packing densely to form microdomains in the plasma membrane commonly called lipid rafts. These specialized membrane regions are involved in the budding of many enveloped viruses including influenza virus. Influenza virus HA and NA are known to associate with lipid rafts, and clustering of HA within the rafts is an intrinsic property of the HA protein. NA is also found sequestered within the same microdomains as HA, whereas the M2 ion channel protein does not concentrate within the raft-like microdomains.

Depletion of cholesterol from cells decreased the diameter of the HA clusters [252]. The involvement of plasma membrane in the distribution of HA in the membrane may also trigger the HA folding and/or stabilize correct conformation of functionally active HA. Alternatively, lipid membrane may contribute to the advantageous presentation of influenza antigens, which can result in better antigen processing [89]. Absence of lipid component in a vaccine, such as in the case of subunit vaccines, results in significant losses of immunogenic activity and require higher doses to induce protection. However, immunogenicity of the subunit vaccines can be restored by adding lipids, as in the case of liposomes/virosomes. Taken together, these observations strongly support the idea of important role of conformation and lipids for optimal preparation and immunogenicity of nanoparticle-based influenza vaccines. This phenomenon may require further studies and comparison of the activity of HA protein as well as of the protein and lipid content in the various vaccines. Future progress in our understanding of influenza virus will provide additional impetus for the development of novel influenza vaccines with improved safety and efficacy characteristics.

ACKNOWLEDGMENTS

We thank Prof. Paul Pumpens for the opportunity to work on this review. This work was supported in part by the USDA NIFA grant 2013-33610-21041. The views expressed in this chapter are those of the authors and do not necessarily reflect the views of the funding agencies.

REFERENCES

1. Potter, C.W., Chronicle of influenza pandemics. *Textbook of Influenza*, eds. R.G. Webster, K.G. Nicholson, and A.J. Hay. 1998, Oxford, U.K.: Blackwell Science.
2. Webby, R.J. and R.G. Webster, Emergence of influenza A viruses. *Philos Trans R Soc Lond B Biol Sci*, 2001. **356**(1416): 1817–1828.
3. Smith, W., C.H. Andrewes, and P.P. Laidlaw, A virus obtained from influenza patients. *Lancet*, 1933. **2**: 66–68.
4. Horsfall, F.L.J., E.H. Lennette, E.R. Rickard, C.H. Andrewes, W. Smith, and C.H. Stuart-Harris, The nomenclature of influenza. *Lancet*, 1940. **236**(ii): 413.
5. Cox, N.J. and K. Subbarao, Influenza. *Lancet*, 1999. **354**(9186): 1277–1282.
6. Brankston, G. et al., Transmission of influenza A in human beings. *Lancet Infect Dis*, 2007. **7**(4): 257–265.
7. CDC, Estimating Seasonal Influenza-Associated Deaths in the United States: CDC Study Confirms Variability of Flu. 2014. http://www.cdc.gov/flu/about/disease/us_flu-related_deaths.htm.
8. WHO, http://www.who.int/mediacentre/factsheets/2003/fs211/en/, accessed February 3, 2015. World Health Organization, Influenza Fact Sheet N211.
9. Taubenberger, J.K. et al., Integrating historical, clinical and molecular genetic data in order to explain the origin and virulence of the 1918 Spanish influenza virus. *Philos Trans R Soc Lond B Biol Sci*, 2001. **356**(1416): 1829–1839.
10. de Jong, M.D. and T.T. Hien, Avian influenza A (H5N1). *J Clin Virol*, 2006. **35**(1): 2–13.

11. Johnson, N.P. and J. Mueller, Updating the accounts: Global mortality of the 1918–1920 "Spanish" influenza pandemic. *Bull Hist Med*, 2002. **76**(1): 105–115.

12. Tumpey, T.M. et al., Characterization of the reconstructed 1918 Spanish influenza pandemic virus. *Science*, 2005. **310**(5745): 77–80.

13. Lipatov, A.S. et al., Influenza: Emergence and control. *J Virol*, 2004. **78**(17): 8951–8959.

14. Peiris, J.S., L.L. Poon, and Y. Guan, Emergence of a novel swine-origin influenza A virus (S-OIV) H1N1 virus in humans. *J Clin Virol*, 2009. **45**(3): 169–173.

15. Brockwell-Staats, C., R.G. Webster, and R.J. Webby, Diversity of influenza viruses in swine and the emergence of a novel human pandemic influenza A (H1N1). *Influenza Other Respi Viruses*, 2009. **3**(5): 207–213.

16. Centers for Disease Control and Prevention, Swine influenza A (H1N1) infection in two children—Southern California, March-April 2009. *MMWR: Morb Mortal Wkly Rep*, 2009. **58**(15): 400–402.

17. Cohen, J. and M. Enserink, Swine flu. After delays, WHO agrees: The 2009 pandemic has begun. *Science*, 2009. **324**(5934): 1496–1497.

18. Koopmans, M. et al., Transmission of H7N7 avian influenza A virus to human beings during a large outbreak in commercial poultry farms in the Netherlands. *Lancet*, 2004. **363**(9409): 587–593.

19. Gao, R. et al., Human infection with a novel avian-origin influenza A (H7N9) virus. *N Engl J Med*, 2013. **368**: 1887–1897.

20. CDC, Update: Influenza activity—United States and Worldwide, 2005–06 season, and composition of the 2006–07 influenza vaccine. *MMWR: Morb Mortal Wkly Rep*, 2006. **55**: 648–653.

21. WHO. Cumulative number of confirmed human cases for avian influenza A(H5N1) reported to WHO, 2003–2013. 2013 [cited 2014]; Available from: http://www.who.int/influenza/human_animal_interface/EN_GIP_20131210CumulativeNumberH5N1cases.pdf. Accessed February 3, 2015.

22. Peiris, M. et al., Human infection with influenza H9N2. *Lancet*, 1999. **354**(9182): 916–917.

23. To, K.K. et al., Emergence in China of human disease due to avian influenza A(H10N8)—Cause for concern? *J Infect*, 2014. **68**(3): 205–215.

24. Palese, P., Influenza: Old and new threats. *Nat Med*, 2004. **10**(12 Suppl): S82–S87.

25. Palese, P., Making better influenza virus vaccines? *Emerg Infect Dis*, 2006. **12**(1): 61–65.

26. Wright, P., Orthomyxoviruses. *Fields Virology*, eds. D.M. Knipe and P. Howley, pp. 1691–1740. 2002, Philadelphia, PA: Lippincott Williams and Wilkins.

27. Palese, P. and M.L. Shaw, Orthomyxoviridae: The viruses and their replication. *Fields Virology*, 5th edn., eds. D.M. Knipe and P.M. Howley, pp. 1647–1690. 2007, Philadelphia, PA: Lippincott Williams & Wilkins.

28. Arvin, A.M. and H.B. Greenberg, New viral vaccines. *Virology*, 2006. **344**(1): 240–249.

29. Glezen, W.P. et al., Epidemiologic observations of influenza B virus infections in Houston, Texas, 1976–1977. *Am J Epidemiol*, 1980. **111**(1): 13–22.

30. Hite, L.K. et al., Medically attended pediatric influenza during the resurgence of the Victoria lineage of influenza B virus. *Int J Infect Dis*, 2007. **11**(1): 40–47.

31. Moriuchi, H. et al., Community-acquired influenza C virus infection in children. *J Pediatr*, 1991. **118**(2): 235–238.

32. Manuguerra, J.C., C. Hannoun, and M. Aymard, Influenza C virus infection in France. *J Infect*, 1992. **24**(1): 91–99.

33. Matsuzaki, Y. et al., A nationwide epidemic of influenza C virus infection in Japan in 2004. *J Clin Microbiol*, 2007. **45**(3): 783–788.

34. Kimura, H. et al., Interspecies transmission of influenza C virus between humans and pigs. *Virus Res*, 1997. **48**(1): 71–79.

35. Joosting, A.C. et al., Production of common colds in human volunteers by influenza C virus. *Br Med J*, 1968. **4**(5624): 153–154.

36. Yamada, S. et al., Haemagglutinin mutations responsible for the binding of H5N1 influenza A viruses to human-type receptors. *Nature*, 2006. **444**(7117): 378–382.

37. Russell, R.J. et al., The structure of H5N1 avian influenza neuraminidase suggests new opportunities for drug design. *Nature*, 2006. **443**(7107): 45–49.

38. Horimoto, T. and Y. Kawaoka, Pandemic threat posed by avian influenza A viruses. *Clin Microbiol Rev*, 2001. **14**(1): 129–149.

39. Palese, P. et al., Characterization of temperature sensitive influenza virus mutants defective in neuraminidase. *Virology*, 1974. **61**(2): 397–410.

40. Matrosovich, M.N. et al., Neuraminidase is important for the initiation of influenza virus infection in human airway epithelium. *J Virol*, 2004. **78**(22): 12665–12667.

41. Hay, A.J. et al., The molecular basis of the specific anti-influenza action of amantadine. *Embo J*, 1985. **4**(11): 3021–3024.

42. Monto, A.S. and N.H. Arden, Implications of viral resistance to amantadine in control of influenza A. *Clin Infect Dis*, 1992. **15**(2): 362–367; discussion 368–369.

43. Palese, P. and R.W. Compans, Inhibition of influenza virus replication in tissue culture by 2-deoxy-2,3-dehydro-N-trifluoroacetylneuraminic acid (FANA): Mechanism of action. *J Gen Virol*, 1976. **33**(1): 159–163.

44. Kiso, M. et al., Resistant influenza A viruses in children treated with oseltamivir: Descriptive study. *Lancet*, 2004. **364**(9436): 759–765.

45. Le, Q.M. et al., Avian flu: Isolation of drug-resistant H5N1 virus. *Nature*, 2005. **437**(7062): 1108.

46. Krammer, F. and P. Palese, Influenza virus hemagglutinin stalk-based antibodies and vaccines. *Curr Opin Virol*, 2013. **3**(5): 521–530.

47. Doyle, T.M. et al., Universal anti-neuraminidase antibody inhibiting all influenza A subtypes. *Antiviral Res*, 2013. **100**(2): 567–574.

48. Zheng, M., J. Luo, and Z. Chen, Development of universal influenza vaccines based on influenza virus M and NP genes. *Infection*, 2013. **42**: 251–262.

49. Schultz-Cherry, S. et al., Influenza virus ns1 protein induces apoptosis in cultured cells. *J Virol*, 2001. **75**(17): 7875–7881.

50. Seo, S.H., E. Hoffmann, and R.G. Webster, Lethal H5N1 influenza viruses escape host anti-viral cytokine responses. *Nat Med*, 2002. **8**(9): 950–954.

51. Engel, D.A., The influenza virus NS1 protein as a therapeutic target. *Antiviral Res*, 2013. **99**(3): 409–416.

52. Jablonski, J.J. et al., Design, synthesis, and evaluation of novel small molecule inhibitors of the influenza virus protein NS1. *Bioorg Med Chem*, 2012. **20**(1): 487–497.

53. Hatta, M. and Y. Kawaoka, The NB protein of influenza B virus is not necessary for virus replication in vitro. *J Virol*, 2003. **77**(10): 6050–6054.

54. McElhaney, J.E., R.N. Coler, and S.L. Baldwin, Immunologic correlates of protection and potential role for adjuvants to improve influenza vaccines in older adults. *Expert Rev Vaccines*, 2013. **12**(7): 759–766.

55. Guillot, L. et al., Involvement of toll-like receptor 3 in the immune response of lung epithelial cells to double-stranded RNA and influenza A virus. *J Biol Chem*, 2005. **280**(7): 5571–5580.

56. Barchet, W. et al., Dendritic cells respond to influenza virus through TLR7- and PKR-independent pathways. *Eur J Immunol*, 2005. **35**(1): 236–242.

57. Hayden, F.G. et al., Local and systemic cytokine responses during experimental human influenza A virus infection. Relation to symptom formation and host defense. *J Clin Invest*, 1998. **101**(3): 643–649.

58. de Jong, M.D. et al., Fatal outcome of human influenza A (H5N1) is associated with high viral load and hypercytokinemia. *Nat Med*, 2006. **12**(10): 1203–1207.

59. Garcia-Sastre, A. et al., The role of interferon in influenza virus tissue tropism. *J Virol*, 1998. **72**(11): 8550–8558.

60. He, X.S. et al., T cell-dependent production of IFN-gamma by NK cells in response to influenza A virus. *J Clin Invest*, 2004. **114**(12): 1812–1819.

61. Nicholls, J.M., The battle between influenza and the innate immune response in the human respiratory tract. *Infect Chemother*, 2013. **45**(1): 11–21.

62. Abe, T. et al., Baculovirus induces an innate immune response and confers protection from lethal influenza virus infection in mice. *J Immunol*, 2003. **171**(3): 1133–1139.

63. Hervas-Stubbs, S. et al., Insect baculoviruses strongly potentiate adaptive immune responses by inducing type I IFN. *J Immunol*, 2007. **178**(4): 2361–2369.

64. Fiore, A.E., C.B. Bridges, J.M. Katz, and N.J. Cox, Inactivated influenza vaccines, *Vaccines*, eds. S.A. Plotkin, W. Orenstein, and P.A. Offit. 2012, Elsevier Health Sciences. pp. 257–293.

65. Dormitzer, P.R. et al., Influenza vaccine immunology. *Immunol Rev*, 2011. **239**(1): 167–77.

66. Rimmelzwaan, G.F. and J.M. Katz, Immune responses to infection with H5N1 influenza virus. *Virus Res*, 2013. **178**(1): 44–52.

67. Brown, D.M., E. Roman, and S.L. Swain, CD4 T cell responses to influenza infection. *Semin Immunol*, 2004. **16**(3): 171–177.

68. Powell, T.J. et al., CD8+ T cells responding to influenza infection reach and persist at higher numbers than CD4+ T cells independently of precursor frequency. *Clin Immunol*, 2004. **113**(1): 89–100.

69. Thomas, P.G. et al., Cell-mediated protection in influenza infection. *Emerg Infect Dis*, 2006. **12**(1): 48–54.

70. He, X.S. et al., Analysis of the frequencies and of the memory T cell phenotypes of human CD8+ T cells specific for influenza A viruses. *J Infect Dis*, 2003. **187**(7): 1075–1084.

71. Sasaki, S. et al., Comparison of the influenza virus-specific effector and memory B-cell responses to immunization of children and adults with live attenuated or inactivated influenza virus vaccines. *J Virol*, 2007. **81**(1): 215–228.

72. Zeman, A.M. et al., Humoral and cellular immune responses in children given annual immunization with trivalent inactivated influenza vaccine. *Pediatr Infect Dis J*, 2007. **26**(2): 107–115.

73. Bui, H.H. et al., Ab and T cell epitopes of influenza A virus, knowledge and opportunities. *Proc Natl Acad Sci U S A*, 2007. **104**(1): 246–251.

74. Scherle, P.A., G. Palladino, and W. Gerhard, Mice can recover from pulmonary influenza virus infection in the absence of class I-restricted cytotoxic T cells. *J Immunol*, 1992. **148**(1): 212–217.

75. O'Neill, E. and R.O. Donis, Generation and characterization of candidate vaccine viruses for prepandemic influenza vaccines. *Curr Top Microbiol Immunol*, 2009. **333**: 83–108.

76. Cox, M.M., Recombinant protein vaccines produced in insect cells. *Vaccine*, 2012. **30**(10): 1759–1766.

77. Cox, M.M., P.A. Patriarca, and J. Treanor, FluBlok, a recombinant hemagglutinin influenza vaccine. *Influenza Other Respir Viruses*, 2008. **2**(6): 211–219.

78. Harper, S.A. et al., Prevention and control of influenza. Recommendations of the Advisory Committee on Immunization Practices (ACIP). *MMWR Recomm Rep*, 2005. **54**(RR-8): 1–40.

79. McKeage, K., Inactivated quadrivalent split-virus seasonal influenza vaccine (Fluarix(R) quadrivalent): A review of its use in the prevention of disease caused by influenza A and B. *Drugs*, 2013. **73**(14): 1587–1594.

80. Toback, S.L. et al., Quadrivalent Ann Arbor strain live-attenuated influenza vaccine. *Expert Rev Vaccines*, 2012. **11**(11): 1293–1303.

81. Wood, J.M., Standardization of inactivated influenza vaccines. *Textbook of Influenza*, eds. R.G. Webster, K.G. Nicholson, and A.J. Hay. 1998, Oxford, U.K.: Blackwell Science.

82. Wood, J.M. and R.A. Levandowski, The influenza vaccine licensing process. *Vaccine*, 2003. **21**(16): 1786–1788.

83. van Essen, G.A. et al., Influenza vaccination in 2000: Recommendations and vaccine use in 50 developed and rapidly developing countries. *Vaccine*, 2003. **21**(16): 1780–1785.

84. CDC, Interim guidance for influenza outbreak management in long-term care facilities. 2014. http://www.cdc.gov/flu/professionals/infectioncontrol/ltc-facility-guidance.htm. Accessed February 3, 2015.

85. Huber, V.C., Influenza vaccines: From whole virus preparations to recombinant protein technology. *Expert Rev Vaccines*, 2013. **13**: 31–42.

86. Gerin, J.L. and N.G. Anderson, Purification of influenza virus in the K-II zonal centrifuge. *Nature*, 1969. **221**(5187): 1255–1256.

87. Barry, D.W. et al., Comparative trial of influenza vaccines. I. Immunogenicity of whole virus and split product vaccines in man. *Am J Epidemiol*, 1976. **104**(1): 34–46.

88. Gross, P.A. et al., A controlled double-blind comparison of reactogenicity, immunogenicity, and protective efficacy of whole-virus and split-product influenza vaccines in children. *J Infect Dis*, 1977. **136**(5): 623–632.

89. Hovden, A.O., R.J. Cox, and L.R. Haaheim, Whole influenza virus vaccine is more immunogenic than split influenza virus vaccine and induces primarily an IgG2a response in BALB/c mice. *Scand J Immunol*, 2005. **62**(1): 36–44.

90. van der Velden, M.V. et al., Safety and immunogenicity of a vero cell culture-derived whole-virus influenza A(H5N1) vaccine in a pediatric population. *J Infect Dis*, 2014. **209**(1): 12–23.

91. Keitel, W.A. and P.A. Piedra, Influenza A(H5N1) vaccines: Are we better prepared for the next pandemic? *J Infect Dis*, 2014. **209**(1): 1–3.

92. Müller, M. et al., Safety and immunogenicity of a cell-culture (Vero) derived whole virus H5N1 vaccine. *International Meeting on Emerging Diseases and Surveillance*. 2007. Vienna, Austria.

93. Davenport, F.M. et al., Comparisons of serologic and febrile responses in humans to vaccination with influenza a viruses or their hemagglutinins. *J Lab Clin Med*, 1964. **63**: 5–13.

94. Treanor, J.J. et al., Rapid licensure of a new, inactivated influenza vaccine in the United States. *Hum Vaccin*, 2005. **1**(6): 239–244.

95. CDC, *Cell-Based Flu Vaccines.* 2013. Available from: http://www.cdc.gov/flu/protect/vaccine/cell-based.htm. Accessed February 3, 2015.

96. Barker, C.I. and M.D. Snape, Pandemic influenza A H1N1 vaccines and narcolepsy: Vaccine safety surveillance in action. *Lancet Infect Dis*, 2013. **14**: 227–238.

97. Clegg, C.H., J.A. Rininger, and S.L. Baldwin, Clinical vaccine development for H5N1 influenza. *Expert Rev Vaccines*, 2013. **12**(7): 767–777.

98. GlaxoSmithKline, H5N1 vaccine approved by the U.S. FDA as pandemic influenza preparedness measure 2013. http://www.gsk.com/en-gb/media/press-releases/2013/h5n1-vaccine-approved-by-the-us-fda-as-pandemic-influenza-preparedness-measure/. Accessed February 3, 2015.

99. Baz, M. et al., H5N1 vaccines in humans. *Virus Res*, 2013. **178**(1): 78–98.

100. FDA, 2013. H5N1 Influenza Virus Vaccine, manufactured by Sanofi Pasteur, Inc. Questions and Answers. Available at: http://www.fda.gov/BiologicsBloodVaccines/Vaccines/QuestionsaboutVaccines/ucm080753.htm. Accessed February 3, 2015.

101. Cox, R.J., K.A. Brokstad, and P. Ogra, Influenza virus: Immunity and vaccination strategies. Comparison of the immune response to inactivated and live, attenuated influenza vaccines. *Scand J Immunol*, 2004. **59**(1): 1–15.

102. Belshe, R. et al., Safety, immunogenicity and efficacy of intranasal, live attenuated influenza vaccine. *Expert Rev Vaccines*, 2004. **3**(6): 643–654.

103. Maassab, H.F., C.A. Heilman, and M.L. Herlocher, Cold-adapted influenza viruses for use as live vaccines for man. *Adv Biotechnol Processes*, 1990. **14**: 203–242.

104. Murphy, B.R. and K. Coelingh, Principles underlying the development and use of live attenuated cold-adapted influenza A and B virus vaccines. *Viral Immunol*, 2002. **15**(2): 295–323.

105. Treanor, J.J. et al., Evaluation of trivalent, live, cold-adapted (CAIV-T) and inactivated (TIV) influenza vaccines in prevention of virus infection and illness following challenge of adults with wild-type influenza A (H1N1), A (H3N2), and B viruses. *Vaccine*, 1999. **18**(9–10): 899–906.

106. Capua, I. and S. Marangon, Control of avian influenza in poultry. *Emerg Infect Dis*, 2006. **12**(9): 1319–1324.

107. Pantin-Jackwood, M.J. and D.L. Suarez, Vaccination of domestic ducks against H5N1 HPAI: A review. *Virus Res*, 2013. **178**(1): 21–34.

108. Tian, G. et al., Protective efficacy in chickens, geese and ducks of an H5N1-inactivated vaccine developed by reverse genetics. *Virology*, 2005. **341**(1): 153–162.

109. Swayne, D.E. et al., Protection against diverse highly pathogenic H5 avian influenza viruses in chickens immunized with a recombinant fowlpox vaccine containing an H5 avian influenza hemagglutinin gene insert. *Vaccine*, 2000. **18**(11–12): 1088–1095.

110. CDC, Update: Influenza activity United States, 2004–05 season. *MMWR: Morb Mortal Wkly Rep*, 2005. **54**(8): 193–196.

111. Fiers, W. et al., A "universal" human influenza A vaccine. *Virus Res*, 2004. **103**(1–2): 173–176.

112. Goodwin, K., C. Viboud, and L. Simonsen, Antibody response to influenza vaccination in the elderly: A quantitative review. *Vaccine*, 2006. **24**(8): 1159–1169.

113. Wright, P.F., The use of inactivated influenza vaccine in children. *Semin Pediatr Infect Dis*, 2006. **17**(4): 200–205.

114. Beyer, W.E. et al., Antibody induction by influenza vaccines in the elderly: A review of the literature. *Vaccine*, 1989. **7**(5): 385–394.

115. Thompson, W.W. et al., Mortality associated with influenza and respiratory syncytial virus in the United States. *JAMA*, 2003. **289**(2): 179–186.

116. Katz, J.M. et al., Immunity to influenza: The challenges of protecting an aging population. *Immunol Res*, 2004. **29**(1–3): 113–124.

117. Munoz, F.M., Influenza virus infection in infancy and early childhood. *Paediatr Respir Rev*, 2003. **4**(2): 99–104.

118. Sullivan, S.J., R. Jacobson, and G.A. Poland, Advances in the vaccination of the elderly against influenza: Role of a high-dose vaccine. *Expert Rev Vaccines*, 2010. **9**(10): 1127–1133.

119. Kilbourne, E.D., Future influenza vaccines and the use of genetic recombinants. *Bull World Health Organ*, 1969. **41**(3): 643–645.

120. Lugovtsev, V.Y., G.M. Vodeiko, and R.A. Levandowski, Mutational pattern of influenza B viruses adapted to high growth replication in embryonated eggs. *Virus Res*, 2005. **109**(2): 149–157.

121. Widjaja, L. et al., Molecular changes associated with adaptation of human influenza A virus in embryonated chicken eggs. *Virology*, 2006. **350**(1): 137–145.

122. Medina, J. et al., Vero/CHOK1, a novel mixture of cell lines that is optimal for the rescue of influenza A vaccine seeds. *J Virol Methods*, 2013. **196C**: 25–31.

123. FDA, 2009. Influenza Vaccine Needs. As a result of the vaccine shortage, FDA had to identify additional sources of vaccine that could be made available under an FDA investigational new drug (IND) application. Available at: http://www.fda.gov/NewsEvents/Testimony/ucm113117.htm. Accessed February 3, 2015.

124. Auewarakul, P. et al., Antibody responses after dose-sparing intradermal influenza vaccination. *Vaccine*, 2007. **25**(4): 659–663.

125. Sambhara, S. et al., Heterosubtypic immunity against human influenza A viruses, including recently emerged avian H5 and H9 viruses, induced by FLU-ISCOM vaccine in mice requires both cytotoxic T-lymphocyte and macrophage function. *Cell Immunol*, 2001. **211**(2): 143–153.

126. Young, F. and F. Marra, A systematic review of intradermal influenza vaccines. *Vaccine*, 2011. **29**(48): 8788–8801.

127. Brown, L.E., The role of adjuvants in vaccines for seasonal and pandemic influenza. *Vaccine*, 2010. **28**(50): 8043–8045.

128. Asahi-Ozaki, Y. et al., Intranasal administration of adjuvant-combined recombinant influenza virus HA vaccine protects mice from the lethal H5N1 virus infection. *Microbes Infect*, 2006. **8**(12–13): 2706–2714.

129. Halperin, S.A. et al., Phase I, randomized, controlled trial to study the reactogenicity and immunogenicity of a nasal, inactivated trivalent influenza virus vaccine in healthy adults. *Hum Vaccin*, 2005. **1**(1): 37–42.

130. Atmar, R.L. et al., A dose-response evaluation of inactivated influenza vaccine given intranasally and intramuscularly to healthy young adults. *Vaccine*, 2007. **25**: 5367–5373.

131. Subbarao, K. and T. Joseph, Scientific barriers to developing vaccines against avian influenza viruses. *Nat Rev Immunol*, 2007. **7**(4): 267–278.

132. Horimoto, T. and Y. Kawaoka, Strategies for developing vaccines against H5N1 influenza A viruses. *Trends Mol Med*, 2006. **12**(11): 506–514.

133. Neumann, G., M. Ozawa, and Y. Kawaoka, Reverse genetics of influenza viruses. *Methods Mol Biol*, 2012. **865**: 193–206.

134. Neumann, G. et al., An improved reverse genetics system for influenza A virus generation and its implications for vaccine production. *Proc Natl Acad Sci U S A*, 2005. **102**(46): 16825–16829.

135. Forde, G.M., Rapid-response vaccines—Does DNA offer a solution? *Nat Biotechnol*, 2005. **23**(9): 1059–1062.

136. Sato, Y. et al., Immunostimulatory DNA sequences necessary for effective intradermal gene immunization. *Science*, 1996. **273**(5273): 352–354.

137. Klinman, D.M., G. Yamshchikov, and Y. Ishigatsubo, Contribution of CpG motifs to the immunogenicity of DNA vaccines. *J Immunol*, 1997. **158**(8): 3635–3639.

138. Robinson, H.L. et al., DNA immunization for influenza virus: Studies using hemagglutinin- and nucleoprotein-expressing DNAs. *J Infect Dis*, 1997. **176**(Suppl 1): S50–S55.

139. Robinson, H.L., L.A. Hunt, and R.G. Webster, Protection against a lethal influenza virus challenge by immunization with a haemagglutinin-expressing plasmid DNA. *Vaccine*, 1993. **11**(9): 957–960.

140. Webster, R.G. et al., Protection of ferrets against influenza challenge with a DNA vaccine to the haemagglutinin. *Vaccine*, 1994. **12**(16): 1495–1498.

141. Kodihalli, S. et al., DNA vaccine encoding hemagglutinin provides protective immunity against H5N1 influenza virus infection in mice. *J Virol*, 1999. **73**(3): 2094–2098.

142. Chen, Z. et al., Comparison of the ability of viral protein-expressing plasmid DNAs to protect against influenza. *Vaccine*, 1998. **16**(16): 1544–1549.

143. Ulmer, J.B. et al., Heterologous protection against influenza by injection of DNA encoding a viral protein. *Science*, 1993. **259**(5102): 1745–1749.

144. Tompkins, S.M. et al., Matrix protein 2 vaccination and protection against influenza viruses, including subtype H5N1. *Emerg Infect Dis*, 2007. **13**(3): 426–435.

145. Song, J.M. et al., DNA vaccination in the skin using microneedles improves protection against influenza. *Mol Ther*, 2012. **20**(7): 1472–1480.

146. Wang, S. et al., Hemagglutinin (HA) proteins from H1 and H3 serotypes of influenza A viruses require different antigen designs for the induction of optimal protective antibody responses as studied by codon-optimized HA DNA vaccines. *J Virol*, 2006. **80**(23): 11628–11637.

147. Laddy, D.J. and D.B. Weiner, From plasmids to protection: A review of DNA vaccines against infectious diseases. *Int Rev Immunol*, 2006. **25**(3–4): 99–123.

148. Donnelly, J., K. Berry, and J.B. Ulmer, Technical and regulatory hurdles for DNA vaccines. *Int J Parasitol*, 2003. **33**(5–6): 457–467.

149. Ulmer, J.B., U. Valley, and R. Rappuoli, Vaccine manufacturing: Challenges and solutions. *Nat Biotechnol*, 2006. **24**(11): 1377–1383.

150. Robertson, J.S. and K. Cichutek, European Union guidance on the quality, safety and efficacy of DNA vaccines and regulatory requirements. *Dev Biol (Basel)*, 2000. **104**: 53–56.

151. Smith, H.A. and D.M. Klinman, The regulation of DNA vaccines. *Curr Opin Biotechnol*, 2001. **12**(3): 299–303.

152. NIH. 2007. NIAID DNA Vaccine for H5N1 Avian Influenza Enters Human Trial. Available from: http://www.nih.gov/news/pr/jan2007/niaid-02.htm. Accessed February 3, 2015.

153. Ledgerwood, J.E. et al., Influenza virus h5 DNA vaccination is immunogenic by intramuscular and intradermal routes in humans. *Clin Vaccine Immunol*, 2012. **19**(11): 1792–1797.

154. Le Gall-Recule, G. et al., Importance of a prime-boost DNA/protein vaccination to protect chickens against low-pathogenic H7 avian influenza infection. *Avian Dis*, 2007. **51**(Suppl 1): 490–494.

155. Epstein, S.L. et al., Protection against multiple influenza A subtypes by vaccination with highly conserved nucleoprotein. *Vaccine*, 2005. **23**(46–47): 5404–5410.

156. Carroll, T.D. et al., Efficacy of influenza vaccination of elderly rhesus macaques is dramatically improved by addition of a cationic lipid/DNA adjuvant. *J Infect Dis*, 2014. **209**(1): 24–33.

157. Kopecky-Bromberg, S.A. and P. Palese, Recombinant vectors as influenza vaccines. *Curr Top Microbiol Immunol*, 2009. **333**: 243–267.

158. Pushko, P. et al., Replicon-helper systems from attenuated Venezuelan equine encephalitis virus: Expression of heterologous genes in vitro and immunization against heterologous pathogens in vivo. *Virology*, 1997. **239**(2): 389–401.

159. Hoelscher, M.A. et al., Development of adenoviral-vector-based pandemic influenza vaccine against antigenically distinct human H5N1 strains in mice. *Lancet*, 2006. **367**(9509): 475–481.

160. Draper, S.J., M.G. Cottingham, and S.C. Gilbert, Utilizing poxviral vectored vaccines for antibody induction-progress and prospects. *Vaccine*, 2013. **31**(39): 4223–4230.

161. Bublot, M. et al., Development and use of fowlpox vectored vaccines for avian influenza. *Ann N Y Acad Sci*, 2006. **1081**: 193–201.

162. Kreijtz, J.H. et al., Recombinant modified vaccinia virus ankara-based vaccine induces protective immunity in mice against infection with influenza virus H5N1. *J Infect Dis*, 2007. **195**(11): 1598–1606.

163. Schultz-Cherry, S. et al., Influenza virus (A/HK/156/97) hemagglutinin expressed by an alphavirus replicon system protects chickens against lethal infection with Hong Kong-origin H5N1 viruses. *Virology*, 2000. **278**(1): 55–59.

164. Weaver, E.A., Vaccines within vaccines: The use of adenovirus types 4 and 7 as influenza vaccine vectors. *Hum Vaccin Immunother*, 2013. **10**(3): 544–556.

165. Gao, W. et al., Protection of mice and poultry from lethal H5N1 avian influenza virus through adenovirus-based immunization. *J Virol*, 2006. **80**(4): 1959–1964.

166. Redfield, R.R. et al., Disseminated vaccinia in a military recruit with human immunodeficiency virus (HIV) disease. *N Engl J Med*, 1987. **316**(11): 673–676.

167. Velkov, T. et al., The antigenic architecture of the hemagglutinin of influenza H5N1 viruses. *Mol Immunol*, 2013. **56**(4): 705–719.

168. Meisel, H. et al., Fine mapping and functional characterization of two immuno-dominant regions from the preS2 sequence of hepatitis B virus. *Intervirology*, 1994. **37**(6): 330–339.

169. Bichko, V. et al., Epitopes recognized by antibodies to denatured core protein of hepatitis B virus. *Mol Immunol*, 1993. **30**(3): 221–231.

170. Sominskaya, I. et al., Tetrapeptide QDPR is a minimal immunodominant epitope within the preS2 domain of hepatitis B virus. *Immunol Lett*, 1992. **33**(2): 169–172.

171. Sominskaya, I. et al., Determination of the minimal length of preS1 epitope recognized by a monoclonal antibody which inhibits attachment of hepatitis B virus to hepatocytes. *Med Microbiol Immunol*, 1992. **181**(4): 215–226.

172. Ulrich, R. et al., Precise localization of the epitope of major BLV envelope protein. *Acta Virol*, 1991. **35**(3): 302.

173. Levy, R. et al., Fine and domain-level epitope mapping of botulinum neurotoxin type A neutralizing antibodies by yeast surface display. *J Mol Biol*, 2007. **365**(1): 196–210.

174. Ernst, W.A. et al., Protection against H1, H5, H6 and H9 influenza A infection with liposomal matrix 2 epitope vaccines. *Vaccine*, 2006. **24**(24): 5158–5168.

175. Kammoun, H. et al., Immunogenicity of live attenuated B. pertussis BPZE1 producing the universal influenza vaccine candidate M2e. *PLoS One*, 2013. **8**(3): e59198.

176. Vergara-Alert, J. et al., Conserved synthetic peptides from the hemagglutinin of influenza viruses induce broad humoral and T-cell responses in a pig model. *PLoS One*, 2012. **7**(7): e40524.

177. Krammer, F. et al., H3 stalk-based chimeric hemagglutinin influenza virus constructs protect mice from H7N9 challenge. *J Virol*, 2013. **88**: 2340–2343.

178. Wang, T.T. et al., Vaccination with a synthetic peptide from the influenza virus hemagglutinin provides protection against distinct viral subtypes. *Proc Natl Acad Sci U S A*, 2010. **107**(44): 18979–18984.

179. Nicholson, K.G. et al., Clinical studies of monovalent inactivated whole virus and subunit A/USSR/77 (H1N1) vaccine: Serological responses and clinical reactions. *J Biol Stand*, 1979. **7**(2): 123–136.

180. Treanor, J.J. et al., Safety and immunogenicity of a baculovirus-expressed hemagglutinin influenza vaccine: A randomized controlled trial. *JAMA*, 2007. **297**(14): 1577–1582.

181. Brands, R. et al., Influvac: A safe Madin Darby Canine Kidney (MDCK) cell culture-based influenza vaccine. *Dev Biol Stand*, 1999. **98**: 93–100; discussion 111.

182. Conne, P. et al., Immunogenicity of trivalent subunit versus virosome-formulated influenza vaccines in geriatric patients. *Vaccine*, 1997. **15**(15): 1675–1679.

183. Podda, A., The adjuvanted influenza vaccines with novel adjuvants: Experience with the MF59-adjuvanted vaccine. *Vaccine*, 2001. **19**(17–19): 2673–2680.

184. Pushko, P. et al., Evaluation of influenza virus-like particles and Novasome adjuvant as candidate vaccine for avian influenza. *Vaccine*, 2007. **25**: 4283–4290.

185. O'Hagan, D.T., M.L. MacKichan, and M. Singh, Recent developments in adjuvants for vaccines against infectious diseases. *Biomol Eng*, 2001. **18**(3): 69–85.

186. Del Giudice, G. et al., An MF59-adjuvanted inactivated influenza vaccine containing A/Panama/1999 (H3N2) induced broader serological protection against heterovariant influenza virus strain A/Fujian/2002 than a subunit and a split influenza vaccine. *Vaccine*, 2006. **24**(16): 3063–3065.

187. Del Giudice, G. et al., Vaccines with the MF59 adjuvant do not stimulate antibody responses against squalene. *Clin Vaccine Immunol*, 2006. **13**(9): 1010–1013.

188. Asa, P.B., R.B. Wilson, and R.F. Garry, Antibodies to squalene in recipients of anthrax vaccine. *Exp Mol Pathol*, 2002. **73**(1): 19–27.

189. Barr, T. et al., Antibodies against cell surface antigens as very potent immunological adjuvants. *Vaccine*, 2006. **24**(Suppl 2): S2-20-1.

190. Hatzifoti, C. and A.W. Heath, CD40-mediated enhancement of immune responses against three forms of influenza vaccine. *Immunology*, 2007. **122**: 98–106.

191. Ninomiya, A. et al., Intranasal administration of a synthetic peptide vaccine encapsulated in liposome together with an anti-CD40 antibody induces protective immunity against influenza A virus in mice. *Vaccine*, 2002. **20**(25–26): 3123–3129.

192. Moser, C. et al., Influenza virosomes as vaccine adjuvant and carrier system. *Expert Rev Vaccines*, 2013. **12**(7): 779–791.

193. Glück, R. et al., Immunogenicity of new virosome influenza vaccine in elderly people. *Lancet*, 1994. **344**(8916): 160–163.

194. de Bruijn, I.A. et al., Clinical experience with inactivated, virosomal influenza vaccine. *Vaccine*, 2005. **23**(Suppl 1): S39–S49.

195. Mutsch, M. et al., Use of the inactivated intranasal influenza vaccine and the risk of Bell's palsy in Switzerland. *N Engl J Med*, 2004. **350**(9): 896–903.

196. Glück, U., J.O. Gebbers, and R. Glück, Phase 1 evaluation of intranasal virosomal influenza vaccine with and without *Escherichia coli* heat-labile toxin in adult volunteers. *J Virol*, 1999. **73**(9): 7780–7786.

197. Latham, T. and J.M. Galarza, Formation of wild-type and chimeric influenza virus-like particles following simultaneous expression of only four structural proteins. *J Virol*, 2001. **75**(13): 6154–6165.

198. Pushko, P. et al., Influenza virus-like particles comprised of the HA, NA, and M1 proteins of H9N2 influenza virus induce protective immune responses in BALB/c mice. *Vaccine*, 2005. **23**(50): 5751–5759.

199. Watanabe, T. et al., Immunogenicity and protective efficacy of replication-incompetent influenza virus-like particles. *J Virol*, 2002. **76**(2): 767–773.

200. Pushko, P., P. Pumpens, and E. Grens, Development of virus-like particle technology from small highly symmetric to large complex virus-like particle structures. *Intervirology*, 2013. **56**(3): 141–165.

201. Pumpens, P. and E. Grens, HBV core particles as a carrier for B cell/T cell epitopes. *Intervirology*, 2001. **44**(2–3): 98–114.

202. Borisova, G.P. et al., Recombinant core particles of hepatitis B virus exposing foreign antigenic determinants on their surface. *FEBS Lett*, 1989. **259**(1): 121–124.

203. Sasagawa, T. et al., Synthesis and assembly of virus-like particles of human papillomaviruses type 6 and type 16 in fission yeast Schizosaccharomyces pombe. *Virology*, 1995. **206**(1): 126–135.

204. Kirnbauer, R., Papillomavirus-like particles for serology and vaccine development. *Intervirology*, 1996. **39**(1–2): 54–61.

205. Mortola, E. and P. Roy, Efficient assembly and release of SARS coronavirus-like particles by a heterologous expression system. *FEBS Lett*, 2004. **576**(1–2): 174–178.

206. Bosio, C.M. et al., Ebola and Marburg virus-like particles activate human myeloid dendritic cells. *Virology*, 2004. **326**(2): 280–287.

207. Noad, R. and P. Roy, Virus-like particles as immunogens. *Trends Microbiol*, 2003. **11**(9): 438–444.

208. Fifis, T. et al., Size-dependent immunogenicity: Therapeutic and protective properties of nano-vaccines against tumors. *J Immunol*, 2004. **173**(5): 3148–3154.

209. Luytjes, W. et al., Amplification, expression, and packaging of foreign gene by influenza virus. *Cell*, 1989. **59**(6): 1107–1113.

210. Garcia-Sastre, A. and P. Palese, Genetic manipulation of negative-strand RNA virus genomes. *Annu Rev Microbiol*, 1993. **47**: 765–790.

211. Neumann, G. and Y. Kawaoka, Reverse genetics of influenza virus. *Virology*, 2001. **287**(2): 243–250.

212. Schickli, J.H. et al., Plasmid-only rescue of influenza A virus vaccine candidates. *Philos Trans R Soc Lond B Biol Sci*, 2001. **356**(1416): 1965–1973.

213. St Angelo, C. et al., Two of the three influenza viral polymerase proteins expressed by using baculovirus vectors form a complex in insect cells. *J Virol*, 1987. **61**(2): 361–365.

214. Mena, I. et al., Rescue of a synthetic chloramphenicol acetyltransferase RNA into influenza virus-like particles obtained from recombinant plasmids. *J Virol*, 1996. **70**(8): 5016–5024.

215. Bright, R.A. et al., Influenza virus-like particles elicit broader immune responses than whole virion inactivated influenza virus or recombinant hemagglutinin. *Vaccine*, 2007. **25**: 3871–3878.

216. Bright, R.A., D.M. Carter, F.R. Toapanta, J.D. Steckbeck, K.S. Cole, N.M. Kumar, P. Pushko, G. Smith, T.M. Tumpey, and T.M. Ross, Cross-clade protective immune responses to pandemic H5N1 viruses elicited by an influenza virus-like particle. *Plos One*, 2007. **3**: e1501.

217. Gomez-Puertas, P. et al., Influenza virus matrix protein is the major driving force in virus budding. *J Virol*, 2000. **74**(24): 11538–11547.

218. Galarza, J.M., T. Latham, and A. Cupo, Virus-like particle (VLP) vaccine conferred complete protection against a lethal influenza virus challenge. *Viral Immunol*, 2005. **18**(1): 244–251.

219. Quan, F.S. et al., Virus-like particle vaccine induces protective immunity against homologous and heterologous strains of influenza virus. *J Virol*, 2007. **81**(7): 3514–3524.

220. Easterbrook, J.D. et al., Protection against a lethal H5N1 influenza challenge by intranasal immunization with virus-like particles containing 2009 pandemic H1N1 neuraminidase in mice. *Virology*, 2012. **432**(1): 39–44.

221. Choi, J.G. et al., Protective efficacy of baculovirus-derived influenza virus-like particles bearing H5 HA alone or in combination with M1 in chickens. *Vet Microbiol*, 2013. **162**(2–4): 623–630.

222. Chen, B.J. et al., Influenza virus hemagglutinin and neuraminidase, but not the matrix protein, are required for assembly and budding of plasmid-derived virus-like particles. *J Virol*, 2007. **81**(13): 7111–7123.

223. Roldao, A. et al., Modeling rotavirus-like particles production in a baculovirus expression vector system: Infection kinetics, baculovirus DNA replication, mRNA synthesis and protein production. *J Biotechnol*, 2007. **128**(4): 875–894.

224. Gomez-Puertas, P. et al., Efficient formation of influenza virus-like particles: Dependence on the expression levels of viral proteins. *J Gen Virol*, 1999. **80**(Pt 7): 1635–1645.

225. Szecsi, J. et al., Induction of neutralising antibodies by virus-like particles harbouring surface proteins from highly pathogenic H5N1 and H7N1 influenza viruses. *Virol J*, 2006. **3**: 70.

226. Sun, J. et al., Core-controlled polymorphism in virus-like particles. *Proc Natl Acad Sci U S A*, 2007. **104**(4): 1354–1359.

227. Pushko, P. et al., Influenza virus-like particle can accommodate multiple subtypes of hemagglutinin and protect from multiple influenza types and subtypes. *Vaccine*, 2011. **29**(35): 5911–5918.

228. Tretyakova, I. et al., Intranasal vaccination with H5, H7 and H9 hemagglutinins co-localized in a virus-like particle protects ferrets from multiple avian influenza viruses. *Virology*, 2013. **442**(1): 67–73.

229. Pumpens, P. et al., Evaluation of HBs, HBc, and frCP virus-like particles for expression of human papillomavirus 16 E7 oncoprotein epitopes. *Intervirology*, 2002. **45**(1): 24–32.

230. Pushko, P. et al., Analysis of RNA phage fr coat protein assembly by insertion, deletion and substitution mutagenesis. *Protein Eng*, 1993. **6**(8): 883–891.

231. Ionescu, R.M. et al., Pharmaceutical and immunological evaluation of human papillomavirus viruslike particle as an antigen carrier. *J Pharm Sci*, 2006. **95**(1): 70–79.

232. Neirynck, S. et al., A universal influenza A vaccine based on the extracellular domain of the M2 protein. *Nat Med*, 1999. **5**(10): 1157–1163.

233. Pumpens, P. and Grens, E., Artificial genes for chimeric virus-like particles. *Artificial DNA*, eds. Y.E. Khudyakov and H.A. Fields. 2003, New York: CRC Press. pp. 249–327.

234. Pushko, P. et al., Identification of hepatitis B virus core protein regions exposed or internalized at the surface of HBcAg particles by scanning with monoclonal antibodies. *Virology*, 1994. **202**(2): 912–920.

235. Luo, L. et al., Mapping of functional domains for HIV-2 gag assembly into virus-like particles. *Virology*, 1994. **205**(2): 496–502.

236. Rossmann, M.G. et al., Combining X-ray crystallography and electron microscopy. *Structure*, 2005. **13**(3): 355–362.

237. Zhang, Y. et al., Structure of immature West Nile virus. *J Virol*, 2007. **81**(11): 6141–6145.

238. Prasad, B.V. et al., X-ray crystallographic structure of the Norwalk virus capsid. *Science*, 1999. **286**(5438): 287–290.

239. Crowther, R.A. et al., Three-dimensional structure of hepatitis B virus core particles determined by electron cryomicroscopy. *Cell*, 1994. **77**(6): 943–950.

240. Ellebedy, A.H. and R. Ahmed, Re-engaging cross-reactive memory B cells: The influenza puzzle. *Front Immunol*, 2012. **3**: 53.

241. Jegerlehner, A. et al., Influenza A vaccine based on the extracellular domain of M2: Weak protection mediated via antibody-dependent NK cell activity. *J Immunol*, 2004. **172**(9): 5598–5605.

242. De Filette, M. et al., Improved design and intranasal delivery of an M2e-based human influenza A vaccine. *Vaccine*, 2006. **24**(44–46): 6597–6601.

243. Jegerlehner, A. et al., A molecular assembly system that renders antigens of choice highly repetitive for induction of protective B cell responses. *Vaccine*, 2002. **20**(25–26): 3104–3112.

244. Kim, M.C. et al., Multiple heterologous M2 extracellular domains presented on virus-like particles confer broader and stronger M2 immunity than live influenza A virus infection. *Antiviral Res*, 2013. **99**(3): 328–335.

245. Liu, Y.V. et al., Chimeric severe acute respiratory syndrome coronavirus (SARS-CoV) S glycoprotein and influenza matrix 1 efficiently form virus-like particles (VLPs) that protect mice against challenge with SARS-CoV. *Vaccine*, 2011. **29**(38): 6606–6613.

246. Medicalnews.com. 2007. Human Clinical Trials Have Begun On Universal FluVaccine. Available from: http://www.medical-newstoday.com/articles/77166.php. Accessed February 3, 2015.

247. Li, Z.N. et al., Chimeric influenza virus hemagglutinin proteins containing large domains of the *Bacillus anthracis* protective antigen: Protein characterization, incorporation into infectious influenza viruses, and antigenicity. *J Virol*, 2005. **79**(15): 10003–10012.

248. Garcia-Sastre, A. and P. Palese, Influenza virus vectors. *Biologicals*, 1995. **23**(2): 171–178.

249. Palese, P. et al., Development of novel influenza virus vaccines and vectors. *J Infect Dis*, 1997. **176**(Suppl 1): S45–S49.

250. Couch, R.B. et al., Antibody correlates and predictors of immunity to naturally occurring influenza in humans and the importance of antibody to the neuraminidase. *J Infect Dis*, 2013. **207**(6): 974–981.

251. Lai, A.L. et al., Fusion peptide of influenza hemagglutinin requires a fixed angle boomerang structure for activity. *J Biol Chem*, 2006. **281**(9): 5760–5770.

252. Leser, G.P. and R.A. Lamb, Influenza virus assembly and budding in raft-derived microdomains: A quantitative analysis of the surface distribution of HA, NA and M2 proteins. *Virology*, 2005. **342**(2): 215–227.

26 Retrovirus-Derived Virus-Like Particles

Bertrand Bellier, Charlotte Dalba, and David Klatzmann

CONTENTS

Despite the impact of vaccination programs, which have significantly reduced the incidence and mortality of infectious diseases worldwide, there is still a great need to develop safer vaccination strategies that induce long-lasting immunity. Recent advances in genomics and proteomics have identified a wide variety of potential new vaccine modalities, including recombinant proteins and synthetic peptides. While these clearly defined antigens offer important safety advantages, they are not always processed and presented effectively by the immune system. For efficacy, the particle form of a pathogen seems of major importance. Indeed, vaccine delivery systems that mimic pathogen size, shape, and surface molecule organization appear advantageous for the induction of protective immune responses, notably neutralizing antibodies, as exemplified by inactivated or attenuated virus vaccines, or vaccine formulations with particles in the viral or bacterial size range [1,2]. For safety, as much as possible of the pathogen genome should be deleted to avoid potential reversion or too severe side effects. In short, a desirable vaccine design features most properties of a pathogen, with the exception of causing disease. Genome-free, virus-like particles (VLPs) meet these criteria. Various VLPs have demonstrated safe and efficient induction of humoral and cellular immune responses in animal studies [3], and clinical trials [4–8], and some have already been marketed. VLPs may therefore be a potent platform to induce specific immune responses against antigens of choice. Many different VLP types have been adapted for this purpose, as illustrated in this book.

Retrovirus-like particles (retroVLPs) were initially developed as candidate vaccines against HIV [9,10]. In their simplest form, they consist solely of wild-type HIV Gag polyproteins that, requiring neither structural or enzymatic viral proteins nor a viral genome, suffice for assembly and budding of virions resembling their wild-type cognates [11–13]. In more complex forms, both HIV Gag and Env proteins are incorporated into the retroVLPs [14,15]. Above all, the latter enveloped type (eVLP) has proven to induce encouraging cellular and humoral immune responses, including neutralizing antibody responses, both in mouse [16,17] and in Rhesus macaque [18,19].

We pioneered the development of recombinant retroVLPs based on simple oncoretroviruses, the murine leukemia retrovirus (MLV), rather than complex lentiviruses [20,21]. MLV-based retroVLPs can be designed in many ways (Figure 26.1) [21]. Cellular expression of the MLV-Gag protein suffices to generate roughly spherical, pleomorphic, membrane-enclosed particles, with a diameter of 80–120 nm [22], and additional expression of Env glycoprotein, of viral or nonviral origin, and allows generation of replication-defective particles that are capable of cell entry but lack the viral genome [23,24]. Furthermore, MLV-based retroVLPs easily lend themselves to incorporation of foreign antigens, and they are small enough to be expressed from heterologous vectors. Whereas MLV-Gag readily forms VLPs when expressed in murine cells, HIV-Gag does not [25], making the mouse model inappropriate for studies of the immunogenicity of HIV-based retroVLPs expressed from genetic vectors. Beyond this technical difficulty, the use of HIV proteins in vaccines would render vaccinated people seropositive for HIV. Finally, compared to HIV, the MLV genome is much less complex, rendering its genetic engineering more straightforward.

26.1 PSEUDOTYPED OR ANTIGEN-DISPLAYING retroVLPs

The key to the success of VLPs as immunogens is thought to reside in their accurate mimicry of wild-type virus particles. The presentation of an antigen in a highly ordered, repetitive array, such as on many viruses and bacteria, normally provokes strong humoral responses, whereas the same antigen presented as a monomer usually is much less immunogenic [26]. The self-assembling core structures of

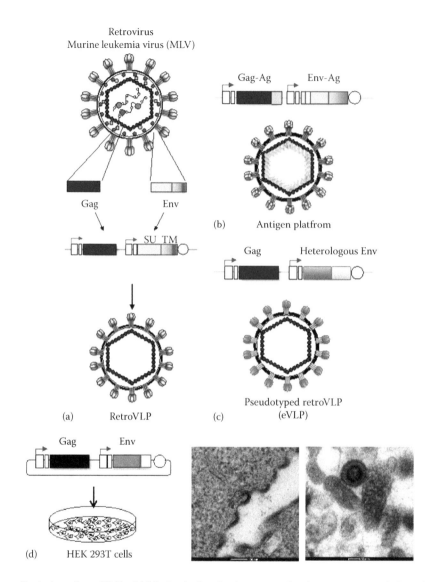

FIGURE 26.1   The versatile design of retroVLPs. (a) Murine leukemia viruses are simple structure, consisting of a core made of structural proteins, containing the viral genome and enzymes, surrounded by a lipid bilayer taken up from the cellular membrane during budding, and expressing envelops proteins at its surface. The sole expression of Gag proteins by mammalian cells is sufficient to produce a retroVLP consisting of Gag and the lipid bilayer, and also expressing Env proteins if they are coexpressed with Gag. Such retroVLP does not comprise the viral genome nor the enzymes necessary for viral replication, that is, protease, reverse transcriptase, and integrase. (b) Antigens can be incorporated at multiple locations in the retroVLPs. They can be fused to Gag (blue) and thus be expressed in approximately 2000 copies inside the retroVLP. They can be incorporated into Env (yellow), and thus be expressed at the surface of the particle. (c) The autologous retroviral envelope can be substituted for a heterologous envelope protein, generating pseudotyped particles (eVLPs). The heterologous envelope can be wild type or can be modified to improve its incorporation on the retroVLP. (d) Transfection of 293T cells results in eVLP production, with particles budding from the cell surface and then released. In the absence of the protease, the Gag precursor is not cleaved, and the retroVLPs retain the typical doughnut shape of immature viral particles rather than the condensed core found in mature viral particle.

many different viruses are amenable to pseudotyping; that is, they can be adapted by recombinant technology to incorporate entire or truncated versions of an envelope protein from a heterologous virus of interest. Pseudotyping serves two important functions: a means for ordered presentation of an antigenic determinant of interest and a means to enhance or target vector uptake by antigen-presenting cells (APCs). The cores of retroviruses have proven exceptionally accommodating for this type of *make-up* [23]. Indeed, retroVLPs can be pseudotyped with glycoproteins from most retroviruses [27] and from other families of enveloped viruses, including

lymphocytic choriomeningitis virus, spleen necrosis virus, vesicular stomatitis virus (VSV), hepatitis C virus, yellow fever virus, West Nile virus, influenza virus, HIV, and CMV [28–33], without harming their functionality. Obviously, this makes retroVLPs excellent vaccine candidates not only for retroviral diseases such as AIDS but also for a wide range of other viral diseases of clinical interest such as hepatitis C, yellow fever, West Nile fever, and CMV.

MLV Env glycoproteins can be modified to display small or large epitopes [34] or even entire proteins [35]. Using eVLPs fitted with such engineered envelopes in the mouse, we first showed that a dominant T-cell epitope displayed on the Env protein induced significantly stronger cellular immune responses than the same antigen molecule administered as recombinant peptide formulated in Freund's adjuvant [36].

Chimeric retroVLPs can be generated by conjugating or fusing target antigens to viral structural proteins that self-assemble into VLPs (Figure 26.1b). We previously demonstrated that MLV Gag could be easily modified to display vaccine antigens [22,37,38]. This strategy appears highly efficient to induce T-cell-specific immune responses, favoring the antigen cross-presentation to CD8+ T cells. Indeed, we observed in two different models that recombinant retroVLPs can be easily captured by DCs, which then efficiently prime CD8+ T cells [22,38].

26.2   ADJUVANT PROPERTIES OF retroVLPs

In vitro studies showed that retroVLPs efficiently induce dendritic cell activation and prime T cells, including human CD8+ T cells, using human monocyte-derived dendritic cells [22]. In vivo studies showed that retroVLPs induce activation of spleen DCs that upregulate costimulatory molecules CD83, CD80, and CD86 (not published). Moreover, transcriptome analysis on sorted CD11c+ cells from the spleen of C57Bl/6 mice 6 h after retroVLP intravenous injection showed significantly upregulated signatures related to immune response, proteasome activity, and viral processes, confirming that, in the absence of adjuvant, retroVLPs can elicit DC activation in vitro and in vivo (manuscript in preparation).

We evaluated the ability to improve the immunogenicity of retroVLPs by combining them with adjuvants. We evaluated the influence on the immune response of a noncoding viral-derived single-stranded RNA (ncRNA) that can be encapsidated into retroVLPs and should induce TLR7/8 signaling. We observed that ncRNA carried by eVLPs promotes Th1 responses and improves CD8+ T cell proliferation in vitro and in vivo in a MyD88-dependant manner [39]. Thus, ncRNA acts as a TLR7/8-acting adjuvant for triggering and modulation of immune responses against vaccine antigens carried by retroVLPs.

26.3   ALTERNATIVE retroVLP PRODUCTION STRATEGIES

While retroVLPs are attractive candidates to generate protective immune responses, their production may represent a limitation. In contrast to nonenveloped VLPs that are formed by expression and self-assembly of structural proteins, in vitro production of eVLPs requires expression of multiple proteins that do not self-assemble. DNA plasmids that express all components of eVLPs can be injected as vaccines in vivo, circumventing the need for in vitro eVLP production (Figure 26.2b). We previously demonstrated that eVLP-expressing plasmids induce significantly better antigen-specific responses and antiviral immune protection than plasmids bearing a single mutation preventing eVLP assembly [32,36]. This established

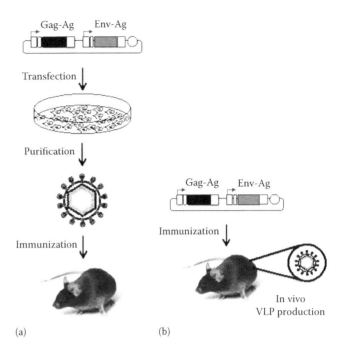

FIGURE 26.2   Modalities for eVLPs vaccination. (a) eVLPs can be produced in vitro by transfection of mammalian cells, purified, and then used for immunization. (b) The same type of plasmid used for in vitro eVLP production can be directly used as the vaccine product. Its injection will result in in vivo expression of the proteins and thus in in vivo eVLP production.

that eVLP-expressing DNA vaccines represent a notable improvement over *classical* DNA vaccines, and this strategy was used in several vaccine developments, including against HCV and HIV.

26.4   eVLPs HAVE DEMONSTRATED SUPERIOR EFFICACY

26.4.1   HCV

Retroviruses offer the opportunity to display complete glycoproteins in their native conformation on their surface. Many heterologous viral glycoproteins, including HCV glycoproteins, can be incorporated into retrovirus particles and mediate infectivity, a property indicating conserved conformation, allowing proper receptor binding and postbinding entry events [40,41].

We generated HCV-pseudotyped retroVLPs (HCV-eVLPs) and evaluated their immunogenicity. We observed that HCV-eVLPs induced poor antibody responses in mice, particularly against E1, even after several injections [41]. To increase integration of envelope glycoproteins at the surface of the particles and to improve their immunogenicity, structural modifications were performed. Replacement, in the C-terminal region, of the native HCV-E1 and HCV-E2 transmembrane and cytoplasmic domains with that of VSV (VSV-G) formed HCV-E1G and HCV-E2G, which were highly expressed and incorporated into VSV virions [42,43]. Similarly, we designed chimeric HCV-E1G glycoproteins that showed improved antigen incorporation into retrovirus-based VLPs and allowed to produce eVLPs pseudotyped with the E1G protein alone [41]. Combining eVLPs pseudotyped with either HCV-E1E2 or HCV-E1G, we observed that such recombinant eVLPs induced high-titer antibodies, including neutralizing antibodies, in both mouse and macaque, in prime-boost strategies using HCV-recombinant viral vectors for priming [41] or DNA vaccine [37]. The neutralizing antibodies generated after HCV-eVLP immunization were shown to cross-neutralize all of the HCV genotypes tested [41]. Importantly, HCV-E1-specific antibodies were detected when HCV-E1E2-eVLPs were injected concomitantly with HCV-E1G-eVLPs but not when injected alone, demonstrating that separation of E1 from E2 is required to elicit an anti-E1 antibody response. In these studies, eVLPs but not recombinant adenovirus and measles vectors were able to elicit significant levels of anti-E1 antibodies, demonstrating the superior potential of the retroVLP-based platform.

26.4.2 HIV

We designed HIV-specific eVLPs displaying HIV-GP140$_{TM}$ at their surface. Since HIV displays very low numbers of GP160 trimeric spikes at its surface [44], we developed a chimeric form of the envelope glycoprotein to increase its valency. The gp140$_{TM}$ sequence derives from the full-length gp160 from the JRFL strain (clade B) in which the transmembrane and cytoplasmic (TM, CT) domains have been replaced by those of the VSV glycoprotein (VSV-G). We quantified the HIV antigens on eVLPs and demonstrated that the amounts of antigens increased more than threefold in the HIV-GP140$_{TM}$-eVLPs compared to controls (manuscript in preparation), highlighting the interest to use the chimeric form of the HIV glycoprotein.

We measured the HIV-specific antibodies induced in rabbit in a prime-boost immunization protocol using either HIV-GP140$_{TM}$-eVLP-forming DNA vaccines followed by HIV-GP140$_{TM}$-eVLPs, or *classical* GP140 DNA vaccines followed by GP140 proteins, and observed significantly higher levels of HIV-specific antibodies with the former combination (unpublished data). Using the same set and combinations of immunogens, we also demonstrated the advantage of combining intranasal prime with intrarectal or intravaginal boost to induce HIV-specific mucosal immunity (manuscript in preparation). We next aim to perform mucosal challenge experiments in nonhuman primates to evaluate the protective efficacy of our vaccine strategy to prevent HIV infection in mucosa.

26.4.3 CMV

Human cytomegalovirus (HCMV) can be congenitally transmitted and then lead to neurologic impairments. Also, HCMV is the most common viral infection in organ transplant recipients. Neutralizing antibodies against HCMV, and notably those directed against glycoprotein B (gB), protect against infection [45,46]. Most vaccine candidates against HCMV failed to induce long-lasting neutralizing antibody responses that prevent infection of epithelial cells [47,48]. At EPIXIS,* we designed retroVLPs pseudotyped with gB (HCMV-eVLPs), either native gB or recombinant gB comprising a fusion between the transmembrane and cytoplasmic domains of the VSV-G protein and the extracellular domain of HCMV gB (gB-G). These eVLPs produced in mammalian cells induced potent neutralizing antibody responses, with the capacity to neutralize infection of both fibroblast and epithelial cells [33]. The magnitude of the neutralizing response was higher with eVLPs expressing gB-G, and above the threshold that is considered necessary for efficacy. This vaccine, developed by *VBI vaccines*, is scheduled to enter clinical development in 2015 (http://www.vbivaccines.com).

26.4.4 PERSPECTIVES

eVLPs have proven efficient at inducing neutralizing antibodies in many experimental settings. For HCV and CMV, they have succeeded to trigger responses that could not be achieved with most other vaccine platforms used. The versatile engineering of eVLPs capitalizes on decades of fundamental retrovirology and development of retroviruses as gene delivery vectors. They represent a flexible platform, able to accommodate a broad range of additional antigens and design improvements. eVLPs will soon be clinically evaluated in the field of HCMV. Positive results for this indication should unleash further efforts for developing retroVLPs for challenging settings, such as but not limited to HCV, HIV, RSV, flaviviruses, Ebola, and other hemorrhagic fever viruses.

REFERENCES

1. Jennings GT, Bachmann MF: Designing recombinant vaccines with viral properties: A rational approach to more effective vaccines. *Curr Mol Med* 2007, **7**(2):143–155.
2. Bachmann MF, Jennings GT: Vaccine delivery: A matter of size, geometry, kinetics and molecular patterns. *Nat Rev Immunol* 2010, **10**(11):787–796.
3. Boisgerault F, Moron G, Leclerc C: Virus-like particles: A new family of delivery systems. *Expert Rev Vaccines* 2002, **1**(1):101–109.
4. Harper DM, Franco EL, Wheeler C, Ferris DG, Jenkins D, Schuind A, Zahaf T, Innis B, Naud P, De Carvalho NS et al.: Efficacy of a bivalent L1 virus-like particle vaccine in prevention of infection with human papillomavirus types 16 and 18 in young women: A randomised controlled trial. *Lancet* 2004, **364**(9447):1757–1765.
5. Villa LL, Costa RL, Petta CA, Andrade RP, Ault KA, Giuliano AR, Wheeler CM, Koutsky LA, Malm C, Lehtinen M et al.: Prophylactic quadrivalent human papillomavirus (types 6, 11, 16, and 18) L1 virus-like particle vaccine in young women: A randomised double-blind placebo-controlled multicentre phase II efficacy trial. *Lancet Oncol* 2005, **6**(5):271–278.

---

* EPIXIS was a biotechnology company formed in 2003 to develop the eVLP vaccine platform (21) that was acquired in 2011 by VBI Vaccines.

6. Atmar RL, Bernstein DI, Harro CD, Al-Ibrahim MS, Chen WH, Ferreira J, Estes MK, Graham DY, Opekun AR, Richardson C et al.: Norovirus vaccine against experimental human Norwalk Virus illness. *N Engl J Med* 2011, **365**(23):2178–2187.

7. Chang LJ, Dowd KA, Mendoza FH, Saunders JG, Sitar S, Plummer SH, Yamshchikov G, Sarwar UN, Hu Z, Enama ME et al.: Safety and tolerability of chikungunya virus-like particle vaccine in healthy adults: A phase 1 dose-escalation trial. *Lancet* 2014, **384**(9959):2046–2052.

8. Low JG, Lee LS, Ooi EE, Ethirajulu K, Yeo P, Matter A, Connolly JE, Skibinski DA, Saudan P, Bachmann M et al.: Safety and immunogenicity of a virus-like particle pandemic influenza A (H1N1) 2009 vaccine: Results from a double-blinded, randomized Phase I clinical trial in healthy Asian volunteers. *Vaccine* 2014, **32**(39):5041–5048.

9. Young KR, Ross TM: Particle-based vaccines for HIV-1 infection. *Curr Drug Targets Infect Disord* 2003, **3**(2):151–169.

10. Doan LX, Li M, Chen C, Yao Q: Virus-like particles as HIV-1 vaccines. *Rev Med Virol* 2005, **15**(2):75–88.

11. Delchambre M, Gheysen D, Thines D, Thiriart C, Jacobs E, Verdin E, Horth M, Burny A, Bex F: The GAG precursor of simian immunodeficiency virus assembles into virus-like particles. *EMBO J* 1989, **8**(9):2653–2660.

12. Karacostas V, Nagashima K, Gonda MA, Moss B: Human immunodeficiency virus-like particles produced by a vaccinia virus expression vector. *Proc Natl Acad Sci U S A* 1989, **86**(22):8964–8967.

13. Jacobs E, Gheysen D, Thines D, Francotte M, de Wilde M: The HIV-1 Gag precursor Pr55gag synthesized in yeast is myristoylated and targeted to the plasma membrane. *Gene* 1989, **79**(1):71–81.

14. Haffar O, Garrigues J, Travis B, Moran P, Zarling J, Hu SL: Human immunodeficiency virus-like, nonreplicating, gag-env particles assemble in a recombinant vaccinia virus expression system. *J Virol* 1990, **64**(6):2653–2659.

15. Krausslich HG, Ochsenbauer C, Traenckner AM, Mergener K, Facke M, Gelderblom HR, Bosch V: Analysis of protein expression and virus-like particle formation in mammalian cell lines stably expressing HIV-1 gag and env gene products with or without active HIV proteinase. *Virology* 1993, **192**(2):605–617.

16. Yao Q, Bu Z, Vzorov A, Yang C, Compans RW: Virus-like particle and DNA-based candidate AIDS vaccines. *Vaccine* 2003, **21**(7–8):638–643.

17. Visciano ML, Diomede L, Tagliamonte M, Tornesello ML, Asti V, Bomsel M, Buonaguro FM, Lopalco L, Buonaguro L: Generation of HIV-1 Virus-Like Particles expressing different HIV-1 glycoproteins. *Vaccine* 2011, **29**(31):4903–4912.

18. Notka F, Stahl-Hennig C, Dittmer U, Wolf H, Wagner R: Accelerated clearance of SHIV in rhesus monkeys by virus-like particle vaccines is dependent on induction of neutralizing antibodies. *Vaccine* 1999, **18**(3–4):291–301.

19. Ross TM, Pereira LE, Luckay A, McNicholl JM, Garcia-Lerma JG, Heneine W, Eugene HS, Pierce-Paul BR, Zhang J, Hendry RM et al.: A polyvalent clade B virus-like particle HIV vaccine combined with partially protective oral preexposure prophylaxis prevents simian-human immunodeficiency virus infection in macaques and primes for virus-amplified immunity. *AIDS Res Hum Retroviruses* 2014, 30:1072–1081.

20. Dalba C, Bellier B, Kasahara N, Klatzmann D: Replication-competent vectors and empty virus-like particles: New retroviral vector designs for cancer gene therapy or vaccines. *Mol Ther* 2007, **15**(3):457–466.

21. Klatzmann D, Salzmann JL, Bellier B, Frisen C, Cosset FL: Synthetic viruses and uses thereof. Patent US20040071661, EP2338984.

22. Lescaille G, Pitoiset F, Macedo R, Baillou C, Huret C, Klatzmann D, Tartour E, Lemoine FM, Bellier B: Efficacy of DNA vaccines forming e7 recombinant retroviral virus-like particles for the treatment of human papillomavirus-induced cancers. *Hum Gene Ther* 2013, **24**(5):533–544.

23. Sandrin V, Muriaux D, Darlix JL, Cosset FL: Intracellular trafficking of Gag and Env proteins and their interactions modulate pseudotyping of retroviruses. *J Virol* 2004, **78**(13):7153–7164.

24. Sedlik C, Vigneron J, Torrieri-Dramard L, Pitoiset F, Denizeau J, Chesneau C, de la Rochere P, Lantz O, Thery C, Bellier B: Different immunogenicity but similar antitumor efficacy of two DNA vaccines coding for an antigen secreted in different membrane vesicle-associated forms. *J Extracell Vesicles* 2014, **3**:10.3402.

25. Mariani R, Rutter G, Harris ME, Hope TJ, Krausslich HG, Landau NR: A block to human immunodeficiency virus type 1 assembly in murine cells. *J Virol* 2000, **74**(8):3859–3870.

26. Bachmann MF, Rohrer UH, Kundig TM, Burki K, Hengartner H, Zinkernagel RM: The influence of antigen organization on B cell responsiveness. *Science* 1993, **262**(5138):1448–1451.

27. Sandrin V, Russell SJ, Cosset FL: Targeting retroviral and lentiviral vectors. *Curr Top Microbiol Immunol* 2003, **281**:137–178.

28. Miletic H, Bruns M, Tsiakas K, Vogt B, Rezai R, Baum C, Kuhlke K, Cosset FL, Ostertag W, Lother H et al.: Retroviral vectors pseudotyped with lymphocytic choriomeningitis virus. *J Virol* 1999, **73**(7):6114–6116.

29. Engelstadter M, Buchholz CJ, Bobkova M, Steidl S, Merget-Millitzer H, Willemsen RA, Stitz J, Cichutek K: Targeted gene transfer to lymphocytes using murine leukaemia virus vectors pseudotyped with spleen necrosis virus envelope proteins. *Gene Ther* 2001, **8**(15):1202–1206.

30. Bartosch B, Dubuisson J, Cosset FL: Infectious hepatitis C virus pseudo-particles containing functional E1-E2 envelope protein complexes. *J Exp Med* 2003, **197**(5):633–642.

31. Sandrin V, Cosset FL: Intracellular versus cell surface assembly of retroviral pseudotypes is determined by the cellular localization of the viral glycoprotein, its capacity to interact with Gag, and the expression of the Nef protein. *J Biol Chem* 2006, **281**(1):528–542.

32. Bellier B, Huret C, Miyalou M, Desjardins D, Frenkiel MP, Despres P, Tangy F, Dalba C, Klatzmann D: DNA vaccines expressing retrovirus-like particles are efficient immunogens to induce neutralizing antibodies. *Vaccine* 2009, **27**(42):5772–5780.

33. Kirchmeier M, Fluckiger AC, Soare C, Bozic J, Ontsouka B, Ahmed T, Diress A, Pereira L, Schodel F, Plotkin S et al.: Enveloped virus-like particle expression of human cytomegalovirus glycoprotein B antigen induces antibodies with potent and broad neutralizing activity. *Clin Vaccine Immunol* 2014, **21**(2):174–180.

34. Kayman SC, Park H, Saxon M, Pinter A: The hypervariable domain of the murine leukemia virus surface protein tolerates large insertions and deletions, enabling development of a retroviral particle display system. *J Virol* 1999, **73**(3):1802–1808.

35. Erlwein O, Buchholz CJ, Schnierle BS: The proline-rich region of the ecotropic Moloney murine leukaemia virus envelope protein tolerates the insertion of the green fluorescent protein and allows the generation of replication-competent virus. *J Gen Virol* 2003, **84**(Part 2):369–373.

36. Bellier B, Dalba C, Clerc B, Desjardins D, Drury R, Cosset FL, Collins M, Klatzmann D: DNA vaccines encoding retrovirus-based virus-like particles induce efficient immune responses without adjuvant. *Vaccine* 2006, **24**(14):2643–2655.

37. Huret C, Desjardins D, Miyalou M, Levacher B, Amadoudji Zin M, Bonduelle O, Combadiere B, Dalba C, Klatzmann D, Bellier B: Recombinant retrovirus-derived virus-like particle-based vaccines induce hepatitis C virus-specific cellular and neutralizing immune responses in mice. *Vaccine* 2013, **31**(11):1540–1547.

38. Sedlik C, Saron M, Sarraseca J, Casal I, Leclerc C: Recombinant parvovirus-like particles as an antigen carrier: A novel nonreplicative exogenous antigen to elicit protective antiviral cytotoxic T cells. *Proc Natl Acad Sci U S A* 1997, **94**(14):7503–7508.

39. Bellier B, Klatzmann D: Methods for improving immunogenicity of virus-like particles. Patent application US61/781,700.

40. Bartosch B, Bukh J, Meunier JC, Granier C, Engle RE, Blackwelder WC, Emerson SU, Cosset FL, Purcell RH: In vitro assay for neutralizing antibody to hepatitis C virus: Evidence for broadly conserved neutralization epitopes. *Proc Natl Acad Sci U S A* 2003, **100**(24):14199–14204.

41. Garrone P, Fluckiger AC, Mangeot PE, Gauthier E, Dupeyrot-Lacas P, Mancip J, Cangialosi A, Du Chene I, Legrand R, Mangeot I et al.: A prime-boost strategy using virus-like particles pseudotyped for HCV proteins triggers broadly neutralizing antibodies in macaques. *Sci Transl Med* 2011, **3**(94):94ra71.

42. Buonocore L, Blight KJ, Rice CM, Rose JK: Characterization of vesicular stomatitis virus recombinants that express and incorporate high levels of hepatitis C virus glycoproteins. *J Virol* 2002, **76**(14):6865–6872.

43. Flint M, McKeating JA: The C-terminal region of the hepatitis C virus E1 glycoprotein confers localization within the endoplasmic reticulum. *J Gen Virol* 1999, **80** (Part 8):1943–1947.

44. Schiller J, Chackerian B: Why HIV virions have low numbers of envelope spikes: Implications for vaccine development. *PLoS Pathog* 2014, **10**(8):e1004254.

45. Adler SP: Human CMV vaccine trials: What if CMV caused a rash? *J Clin Virol* 2008, **41**(3):231–236.

46. Plotkin SA: Is there a formula for an effective CMV vaccine? *J Clin Virol* 2002, **25**(Suppl 2):S13–S21.

47. Cui X, Meza BP, Adler SP, McVoy MA: Cytomegalovirus vaccines fail to induce epithelial entry neutralizing antibodies comparable to natural infection. *Vaccine* 2008, **26**(45): 5760–5766.

48. Wang D, Li F, Freed DC, Finnefrock AC, Tang A, Grimes SN, Casimiro DR, Fu TM: Quantitative analysis of neutralizing antibody response to human cytomegalovirus in natural infection. *Vaccine* 2011, **29**(48):9075–9080.

27 Cancer Therapy Applying Viral Nanoparticles

Kenneth Lundstrom

CONTENTS

27.1 INTRODUCTION

Inefficiency in drug delivery has been a serious problem in drug development for years. It significantly compromises the drug action at the target site/tissue, while it simultaneously increases the damage to normal tissue, causing serious adverse events in patients. Furthermore, diagnostic and monitoring capacity has been hampered by inefficient reagent delivery. One approach in addressing these problems has been to subject drug molecules and delivery vectors to chemical and biological modifications by applying nanotechnology [1]. This includes the employment of varied nanomaterials including synthetic materials and naturally occurring biomaterials [2]. In this context, various liposome- and polymer-based formulations have been engineered for drugs such as Doxil® [3] and triptorelin [4], respectively.

The extraordinary capacity of viruses to promote highly efficient host cell infection has made them attractive as delivery vehicles. Both VNPs and VLPs have been subjected to engineering efforts to carry drug molecules, quantum dots (QDs), other nanoparticles, or imaging reagents for efficient delivery of cargo [5]. Furthermore, introduction of targeting ligands on the external surface of the particles has enabled cell-specific delivery [6]. VNPs can be defined as particles carrying viral genomic information, while VLPs are genome free and possess no risk of infectious biohazard but capable of carrying other types of cargo. This chapter gives an overview of various VNPs and VLPs and their applications in cancer therapy.

27.2 VARIOUS FORMS OF VIRUS NANOPARTICLES AND VIRUS-LIKE PARTICLES

The VNPs and VLPs are generally derived from a broad range of viruses including bacteriophages and plant and animal viruses (Table 27.1). They possess an ability to self-assemble in vitro and provide efficient delivery of various foreign cargo to host cells [7]. Due to the different types of viruses used for generation of VNPs and VLPs, they represent different levels of complexity and structure. Generally, they are composed of one or several viral structural proteins, which can be produced applying bacterial, plant, yeast, insect, and mammalian expression systems. Typically, the particles lack full-length genomes, which render them noninfectious albeit they are able to generate one round of transduction of host cells and thereby deliver their cargo. VLPs can be classified depending on their structural composition as nonenveloped and enveloped. The nonenveloped VLPs exist in both icosahedral and helical forms depending on which type of virus is applied. Furthermore, they have been used as nonchimeric particles, which possess native unmodified viral structural proteins or as chimeric particles, which are subjected to modifications of the viral structural proteins by addition or substitution of heterologous components in the form of encapsidation signals, cell targeting, or immunological epitopes [8]. The engineering of VLPs based on enveloped viruses is more complex. Generally, the VLPs consist of a nucleocapsid surrounded by a host cell–derived

TABLE 27.1

Examples of VLPs Used in Diagnostics and Therapy

Viral Origin	Reagent	Application	References
Bacteriophages			
AP205 phage	GnRH, HIV, influenza A	Vaccines	[13]
fr phage	HaPV	Vaccine	[14]
Lambda phage	HIV	Vaccine	[10]
MS2 phage	HIV, malaria	Vaccines	[11,12]
	Antisense	Leukemia	[68]
P22 phage	Polymers, metals	Cargo delivery	[15]
PP7 phage	HIV, HPV	Vaccines	[77,78]
Qβ phage	HPV, influenza	Vaccines, cancer therapy	[17,18]
	Nicotine, melanoma	Smoking cessation, therapy	[19,56]
T7 phage	HBV peptide PreS1	HepG2 cell delivery	[57]
Plant viruses			
Alfalfa mosaic virus (AlMV)	Rabies G & N peptides	Vaccines	[24]
	RSV G peptide	Vaccines	[20]
Bamboo mosaic virus (BaMV)	FMDV, IBDV	Vaccines	[79,80]
Brome mosaic virus (BMV)	Nanogold	Cargo delivery	[21]
Cowpea chlorotic mottle virus (CCMV)	Polystyrene, nanometal	Cargo delivery	[22,23,81]
	RNA, Phthalocyanine	Nucleic acid, drug delivery	[22,82]
Cowpea mosaic virus (CPMV)	*P. aeruginosa*, MEV	Vaccines	[25,26]
	CPV		[80]
	Fullerenes	Cancer targeting	[72]
Cucumber mosaic virus (CMV)	DNA, fluorophore	Newcastle disease virus, HCV	[27,29]
CMV		Alzheimer's disease	[28]
Johnsongrass mosaic virus	*P. falciparum*	Vaccines	[84]
(JGMV)	JEV		[85]
Papaya mosaic virus (PapMV)	Influenza A	Vaccine	[86]
Plum pox virus (PPV)	CPV	Vaccine	[87]
Potato virus X (PVX)	HIV-1, Tuberculosis	Vaccines	[31,32]
	HPV		[33]
Potato virus Y (PVY)	HBV	Protein delivery	[34]
Tomato bushy stunt virus	Ricin toxin	Vaccines	[88]
Tobacco mosaic virus (TMV)	Poliovirus, malaria, HIV	Delivery, vaccines	[34,89,90]
	Rabies, MHV, FMDV		[36,91,92]
	CRPV, ROPV		[13,37]
Animal viruses			
Adenovirus	HER3 targeting	Breast cancer	[70]
	Integrin/factor IX	Tumor targeting	[71]
Human parvovirus B19	Parvovirus, dengue virus	Vaccines	[47,48]
	Anthrax		[49]
Infectious hypodermal and hematopoietic necrosis virus (IHHNV)	Antiviral agents	Antivirals	[93]
Goose parvovirus (GPV)	GPV	Vaccine	[94]
Porcine parvovirus (PPV)	Polio, porcine circovirus	Vaccines	[51,95]
	Malaria		[50]
Canine parvovirus (CPV)	Polio, rabies	Vaccines	[83,96,97]
Hepatitis E virus (HEV)	HEV, herpes simplex	Vaccines	[52,53]
Porcine encephalomyocarditis virus (EMCV)	EMCV	Vaccine	[98]
Foot-and-mouth disease virus	FMDV	VLPs from *E. coli*	[99]
Flock house virus (FHV)	FHV, anthrax toxin	Vaccine	[54]
Dragon grouper nervous necrosis virus (DGNNV)	DGNNV	VLPs from *E. coli*	[100]
Kunjin virus	GCSF	Intratumoral VLP delivery	[73]
Norwalk virus (NV)	NV	Vaccines	[101,102]

(Continued)

TABLE 27.1 (*Continued*)
Examples of VLPs Used in Diagnostics and Therapy

Viral Origin	Reagent	Application	References
Rabbit hemorrhagic disease virus (RHDV)	RHDV, FMDV	Vaccines	[103,104]
	Ovalbumin	Cancer therapy	[59]
Murine polyoma virus (MuPV)	HBV, PSA	Vaccines, cancer therapy	[58,105]
Hamster polyoma virus (HaPV)	HBV, hantavirus	Vaccines	[106,107]
	Mucin 1 epitope	Vaccine, cancer therapy	[108]
Simian virus 40 (SV40)	Fusion proteins	Cargo delivery	[38]
	Targeting peptides	Targeted delivery	[40,41]
Human papillomavirus (HPV)	HPV	Approved vaccine	[64]
Bovine papillomavirus (BPV)	HPV, amyloid-β	Vaccine, Alzheimer's disease	[55,109]
	MUC1 peptide	Cancer therapy	[110]
Bluetongue virus (BTV)	BTV, HBV preS2	Vaccines	[64,111]
	Influenza		[112]
Rotavirus (RT)	RT, RT-Norovirus	Vaccines	[113,114]
	GFP	Delivery, imaging	[42]
Hepatitis B virus (HBV) core	Influenza M2, malaria	Vaccines	[115,116]
HBV surface	HBV	Vaccines	[117,118]
Woodchuck hepatitis virus (WHV)	Influenza M2e	Vaccine	[119]
West Nile virus (WNV)	WNV	Vaccine	[120]
Dengue virus	Dengue virus	Vaccines, VLPs from yeast	[121,122]
Hepatitis C virus (HCV)	HCV	Vaccines	[123,124]
Japanese encephalitis virus (JEV)	Chimeric JEV-TEV	Vaccine	[125]
Semliki Forest virus	β-galactosidase	Liposome-SFV: cancer therapy	[74]
	Interleukin-12		[75]
Sindbis virus (SIN)	DNA, RNA, protein	Cargo delivery	[43–46]
	Fluorophore, gold particles		
Ross river virus (RRV)	RRV	Vaccine	[43]
Chikungunya virus (CHIKV)	CHIKV	Vaccine	[126]
Severe acute respiratory syndrome–related coronavirus (SARS-CoV)	SARS-CoV	Vaccines	[127]
	SARS-CoV-Influenza		[128]
Rift Valley fever virus (RVFV)	RVFV	Vaccine	[129]
Hanta virus (HTNV)	HTNV	Vaccine	[130]
Human immunodeficiency virus-1 (HIV-1)	HIV-1, pseudorabies	Vaccines	[131,132]
	HPV16 E7 protein		[133]
Human immunodeficiency virus-2 (HIV-2)	HIV-2	Vaccines	[134,135]
Simian immunodeficiency virus (SIV)	SIV, SHIV	Vaccines	[136,137]
Rous sarcoma virus (Rous SV)	RSV, RSV-hPRR	Vaccine, bioassays	[138,139]
Murine leukemia virus (MLV)	MLV	Alzheimer's disease, VSV	[140,141]
		Prion disease	[142]
Influenza A virus	Influenza	Vaccines	[143]
	Pseudotyped Gag		[144]
Lassa virus	Lassa	Vaccine	[145]
Newcastle disease virus	NDV-RSV	Vaccines	[146]
Respiratory syncytial virus (RSV)	NDV-RSV	Vaccines	[147]
Nipah virus (NiV)	NiV	Vaccine	[148]
Sendai virus (SeV)	SeV	VLPs, VLPs from yeast	[149,150]
Mumps virus (MuV)	MuV	VLPs, VLPs from yeast	[151,152]
Tioman virus (TioV)	TioV	VLPs from yeast	[153]
Measles virus (MV)	MV	VLPs from yeast	[154]
Ebolavirus (EBOV)	EBOV	Vaccine	[155]
	EBOV-MARV	Chimeric VLPs	[156]

lipid bilayer embedded with viral spike proteins [9]. The generation of these complex enveloped VLPs therefore relies on appropriate expression systems and cell lines as well as access to efficient purification methods. In the following text are presented examples of VLPs based on bacteriophages, plant viruses, and animal viruses (Table 27.1). Emphasis is put on the engineering of VLPs with different assembly structures and cargo and their applications for a number of indications including vaccinations against infectious agents.

27.3 BACTERIOPHAGES

A number of bacteriophage species have been subjected to VLP production. In this context, the lambda phage has been employed to generate HIV spikes on lambda scaffolds [10]. However, these chimeric VLPs did not provide any improved antibody response to HIV envelope. Furthermore, the MS2 phage was engineered to express HIV [11] and malaria [12] peptide epitopes for immunization studies. Additionally, peptides for angiotensin II, HIV Nef, gonadotropin-releasing hormone, and influenza A M2-protein were fused to the coat protein of the AP205 phage [13] and assembled into VLPs, which demonstrated strong immunogenicity in mice. Likewise, chimeric bacteriophage fr VLPs containing the immune-dominant C-terminal region of hamster polyoma-virus VP1 induced a strong VP1-specific antibody response in rabbits and mice [14]. Bacteriophages have also been applied as delivery vectors for various reagents. For instance, the P22 phage showed efficient packaging and delivery of a number of small molecules including metals and ligands as well as larger molecules such as polymers [15]. It was also demonstrated that pretreatment with P22 VLPs generated an enhanced specific response to the model antigen ovalbumin (OVA) chemically conjugated to the exterior of a small heat shock protein (sHsp) cage, which shows structural similarities to VLPs [16]. Qβ phage VLPs have found applications in immunizations against human papilloma virus (HPV) [17] and influenza virus [18] infections.

In another study, a nicotine derivative was chemically linked to the Qβ coat protein and VLPs generated for intramuscular immunization of 229 subjects in a 6-month phase II clinical trial on smoking cessation [19]. The vaccination was safe and induced antibody responses in all individuals. A statistically significant difference in abstinence was obtained after 2 months between subjects treated with nicotine-Qβ VLPs (47.2%) and the control group (35.1%). However, between 2 and 6 months, no significant difference was observed. Moreover, subgroup analysis demonstrated that individuals with low antibody levels did not increase abstinence in comparison to the placebo receivers.

27.4 PLANT VIRUSES

Self-assembled VLPs have been obtained from brome mosaic virus (BMV) coats surrounding functionalized gold core nanoparticles [21]. In this case, the gold particles can mimic the electrostatic behavior of nucleic acid in native virus assembly and provides an example of potential cargo delivery to host cells. Moreover, cowpea chlorotic mottle virus (CCMV) has been developed for delivery of various types of cargo including polystyrenes [22] and nanometals [23]. Self-assembly of CCMV VLPs and the anionic polymer poly(styrene sulfonate) (PSS) generated five molecular masses from 0.4 to 3.4 MDa and the size of the polymer suggestively affected the capsid size [22]. It was demonstrated that CCMV VLPs can be reversibly assembled and disassembled by adjusting the pH [23]. As the capsid disassembles into 90 capsid protein dimers at pH 7.5, addition of negatively charged polyelectrolytes or particles is mandatory, which restricts the encapsulation of other cargo molecules. The application of N-terminal histidine-tagged capsid protein proteins allows the formation of nanometer-sized protein particles or capsid-like structures stable at pH 7.5 containing multiple proteins expanding the application range of CCMV VLPs. Moreover, the packaging capacity of RNA in CCMV VLPs was investigated [22]. RNAs ranging from 140 to 12,000 nucleotides were completely packaged as long as the protein/RNA mass ratio was sufficiently high. In case of RNAs shorter than 3000 nucleotides, 24–26 nm nucleocapsids with two or more copies of RNA are generated. Longer RNAs (>4500 nucleotides) and single RNA molecules are assembled into capsids as large as 30 nm. Furthermore, when water-soluble phthalocyanine (Pc) was encapsulated into CCMV VLPs, water-soluble phthalocyanine (Pc) either Pc stacks or Pc dimers was obtained depending on assembly conditions. Potentially CCMV VLPs containing Pc will find applications as photosensitizer systems in photodynamic therapy.

A large number of plant virus–based VLPs have been applied for vaccine development (Table 27.1). For instance, rabies virus glycoprotein (G protein) and nucleoprotein (N protein) were fused to the core protein (CP) of Alfalfa mosaic virus (AlMV) and VLPs generated in *Nicotiana tabacum* [24]. Immunization of mice with VLPs resulted in protection against rabies challenges. Furthermore, as mice fed on virus-infected spinach (*Spinacia oleracea*) leaves were immune to lethal doses of rabies virus [24], volunteers previously immunized with rabies virus received virus-infected unprocessed raw spinach leaves responded specifically against the peptide antigen. Interestingly, rabies virus–nonimmune individuals showed significant antibody responses to either rabies or AlMV. In another application of AlMV, a 21-mer peptide of the respiratory syncytial virus (RSV) G protein was fused to the AlMV CP and tested in vitro in human dendritic cells (DCs) and in vivo in human nonprimates [20]. The human DCs generated solid CD4(+) and CD8(+) T-cell responses. Likewise, strong cellular and humoral immune responses were observed in the primates. Additional vaccine studies have employed cowpea mosaic virus (CPMV) by displacement of foreign peptides on the surface of VLPs [25]. In this context, immunization of mice with a peptide derived from the outer membrane protein F of *Pseudomonas aeruginosa* generated high titers of specific IgG antibodies and provided protection against different immunotypes of *P. aeruginosa*. In another study, a short linear epitope of the mink enteritis virus (MEV) VP2 capsid protein was expressed on the surface of CPMV VLPs in the black-eyed bean *Vigna unguiculata* [26]. A single subcutaneous administration of CPMV-MEV VLPs conferred protection against MEV challenges. Moreover, as MEV, canine parvovirus (CPV), and feline panleukopenia virus share the epitope used in this study, the same vaccine could be administered in mink, cat, and dog. Furthermore, subcutaneous or intranasal vaccination of NH mice with CPMV VLPs carrying a 17-mer peptide of CPV as monomers or dimers generated strong antibody responses, however, 10-fold lower than obtained with native plant virus.

The cucumber mosaic virus (CMV) coat protein was expressed from the potato virus X (PVX)-based vector in *Nicotiana benthamiana* plants generating VLPs [27]. Furthermore, engineering of epitopes from Newcastle disease virus (NDV) in the coat protein provided immunoreactive antibodies against CMV and NDV. In another study, the CMV was genetically modified to express amyloid β protein (Aβ)-derived fragments [28]. Out of six chimeric CMV-Aβ constructs, only one (Aβ-1-15-CMV), in position 248 of the coat protein, elicited humoral immune responses to Aβ1-42 antiserum, suggesting the potency as a vaccine candidate against Alzheimer's disease. Moreover, chimeric CMV VLPs (R9-CMV) expressing a 27 amino acid synthetic peptide from hepatitis C virus (HCV) induced humoral immune responses in rabbits fed with lettuce plants infected with the chimeric R9-CMV, indicating that this approach could provide the production of stable oral vaccine against HCV [30]. In another study on HCV, thw R9 peptide was fused directly to the PVX coat protein (PVX(R9)CP) or via the 2A ribosomal skip (PVX(R9-2A)CP) [30]. Only the latter construct yielded systemic infection in *N. benthamiana* and the presence of R9 peptide in infected plant extracts was demonstrated. BALB/c mice immunized with purified PVX(R9-2A)CP VLPs showed specific anti-R9 IgG titers of 1:50,000. Moreover, monoclonal antibodies for R9 were obtained from spleens of immunized mice. PVX(R9-2A)CP VLPs also induced a specific reaction to sera from chronically infected HCV patients in 35% of cases.

Similarly, the highly conserved ELDKWA epitope of HIV-1 glycoprotein 41 (gp41) expressed as an N-terminal fusion to the PVX coat protein elicited high levels of HIV-1 specific IgG and IgA antibodies after intraperitoneal and intranasal immunization of mice [31]. Additionally, specific human antibody responses against the gp41 epitope were obtained in severe combined immunodeficient (SCID) mice reconstituted with human peripheral blood lymphocytes, offering a novel perspective for the development of HIV vaccines. In another approach, the full-length tuberculosis (TB) ESAT-6 protein antigen was fused via an FMDV 2A peptide sequence to the PVX coat protein [32]. Expression studies revealed the production of free CP and ESAT-6 as well as the chimeric ESAT-6-2A-CP on the surface of VLPs, which could be used for vaccination. In a similar study, an epitope (amino acids 108–120) of the HPV-16 L2 minor capsid protein was fused to the PVX coat protein and expressed in

transgenic *N. benthamiana* [33]. Subcutaneous immunization of mice with the L2 108–120 PVX-CP VLPs resulted in antibodies against both PVX CP and the L2 108–120 epitope. It has also been demonstrated that N-terminal insertions into the potato virus Y (PVY) coat protein allow assembly of VLPs [34]. In this context, introduction of the foreign epitope preS1 and the full-length rubredoxin at the N-terminus generated filamentous VLPs with a packaging capacity of up to 71 amino acids. However, C-terminal fusions to the PVY coat protein produced mainly unstructured protein aggregates. Immunization studies in mice showed that chimeric PVY-preS1 CP VLPs elicited a strong anti-preS1 immune response. Moreover, selected malarial B-cell epitopes were inserted in the surface loop in the tobacco mosaic virus coat protein or fused to its C-terminus [35]. Chimeric VLPs were purified from tobacco plants showing a 20:1 ratio of wild-type TMV coat protein and chimeric VLPs.

In another study, two constructs with 10 (TMV-5B19) and 15 (TMV-5B19L) amino acids from the murine hepatitis virus (MHV) 5B19 epitope fused to the TMV coat protein were propagated in tobacco plants and immunogold labeling with the MAb5B19 monoclonal antibody showed expression and display of MHV epitopes on the surface of TMV VLPs [36]. Intranasal administration of TMV-5B19L elicited 5B19- and TMV CP-specific IgG and IgA antibodies in mice. Mice subcutaneously or intranasally vaccinated survived challenges with lethal doses of MHV. Similarly, the G5–24 epitope for Rabies virus glycoprotein was successfully displayed on the surface of TMV and virus produced in infected leaves of *Nicotiana tabacum* [24]. In vaccine evaluations for papilloma viruses, the L2 peptide (amino acids 94–122) from cottontail rabbit papillomavirus (CRPV) and rabbit oral papilloma virus (ROPV) was introduced into the TMV coat protein [37]. New Zealand rabbits were vaccinated with CRPVL2, ROPVL2, and combined CRPVL2 + ROPVL2 TMV VLPs and thereafter challenged with infectious CRPV and ROPV. Complete protection against CRPV was observed for animals immunized with CRPVL2 and CRPVL2 + ROPVL2. However, ROPVL2 vaccinated rabbits showed only a week response against CRPV challenges.

27.5 ANIMAL VIRUSES

A large number of mammalian viruses have been subjected to the engineering of VLPs (Table 27.1). Most applications have focused on vaccine development although additional approaches include cargo delivery and imaging. For instance, it was demonstrated that foreign proteins can be incorporated in the capsid structure of the simian virus 40 (SV40) [38]. EGFP was fused to the *C*-terminus of the VP2/3 capsid protein, and the resulting VLPs retained their infectivity. Furthermore, the yeast cytosine deaminase (yCD) was introduced into the VP2/3 capsid protein. CV-1 cells infected with the SV40-yCD VLPs became sensitive to 5-fluorocytosine-induced cell death in relation to the capacity of yCD modifying 5-fluorocytisine to 5-fluorouracil. In attempts to provide methods for imaging such events as in living cells, virus tracking, nanoparticle targeting, and drug delivery, QD-containing SV40 VLPs were engineered [39]. The SV40-QD VLPs showed homogeneity in size (24 nm) and solid stability. This approach allowed the monitoring of SV40-QD VLPs in living cells and demonstrated entrance by caveolar endocytosis, transport along microtubules, and accumulation in the endoplasmic reticulum as expected for the early steps of SV40 infection. Furthermore, display of foreign peptides within two surface loops in the SV40 major capsid protein VP1 has allowed cell-type-specific delivery of VLPs [40]. Introduction of RGD motifs in the VP1 resulted in integrin binding in vitro and in cell-specific targeting in an RGD-dependent manner. In contrast, insertion of Flag-tags in the same VP1 site prevented cell attachment. In another approach, SV40 VLPs were conjugated to human epidermal growth factor (hEGF) and human epithelial carcinoma A431 cells overexpressing hEGF were evaluated for infection by luciferase reporter expression [41]. Significant increase in EGF receptor-mediated endocytosis occurred.

Rotavirus VLPs have been subjected to studies on delivery of bioactive molecules to the gut [42]. Replication-deficient rotavirus-GFP VLPs were generated in a baculovirus expression system and showed strong fluorescence in MA104 cells. Furthermore, fluorescence imaging was performed in healthy and 2,4,6-trinitrobenzene sulfonic acid (TNBS)-treated mice after intragastric administration. Both GFP fluorescence and viral proteins were detected in intestinal samples. Additionally, alphaviruses have been subjected to VLP production for various cargo delivery [43]. Both Sindbis virus (SIN) and Ross River virus (RRV) nucleocapsids have been produced in a T7-based *Escherichia coli* expression system. The capsid protein oligomerization occurred only in presence of single-stranded but not double-stranded nucleic acid. Moreover, truncated SIN capsid was evaluated for assembly. Deletions of up to 18 N-terminal residues (CP 19–264) still provided fully competent VLP assembly. SIN VLPs have further been subjected to incorporation of short sequences of single-stranded DNA, RNA, small fluorescent-labeled oligonucleotides, and gold particles [44]. It was demonstrated that electrostatic interactions are driving VLP formation. The number of cargo molecules varied with different cargo, but the total negative charge remained constant. However, despite carrying a negative charge, L-Glu molecules could not be successfully incorporated into VLPs [45]. One favorable reason for using SIN particles for cargo delivery has been their systemic tumor targeting [46].

There are numerous examples of applying mammalian VLPs for vaccine development (Table 27.1). For instance, empty human B19 parvovirus capsid VLPs composed of either VP2 or VP1-VP2 combinations were expressed from baculovirus vectors and purified and evaluated in mice, guinea pigs, and rabbits [47]. The VLPs lacking VP1 or containing 4% of VP1 elicited poor virus neutralizing activity, whereas VLPs with more than 25% VP1 protein provoked strong neutralizing responses against B19 virus. Furthermore, parvovirus B19 VLPs carrying dengue-2 virus-specific epitopes were subjected to immunization of BALB/c mice, resulting in strong humoral responses [48]. The vaccinated

mice produced high and robust antidengue 2 titers and efficiently neutralized live dengue 2 virus in 50%-plaque-reduction neutralization (PRNT50) tests. In another application, parvovirus B19 VLPs carried a small-loop peptide of domain 4 of *Bacillus anthracis* protective antigen (PA) [49]. Anti-PA IgG titers up to 2.5×10^4 were obtained in immunized BALB/c mice. The vaccination further showed prevention of mortality of RAW264.7 mouse-macrophage cells from challenges with lethal doses of toxin. Vaccine development has also targeted porcine parvovirus (PPV) VLPs carrying the CD8(+) T cell epitope of the circumsporozoite (CS) protein from *Plasmodium yoelii* in the VP2 capsid protein to address malaria [50]. Administration of PPV-PYCS VLPs in combination with vaccinia virus expressing the full-length PYCS in BALB/c mice demonstrated strong specific CD8(+) T cell responses to CS, resulting in 95% reduction in parasites 2 days after challenges with sporozoite. However, vaccination with only PPV-PACS was ineffective. Poliovirus C3.T and C3.B epitopes introduced at the N-terminus of the PPV VP2 allowed VLP formation [51]. The chimeric VLPs with the C3.T epitope induced T cell responses. However, the PPV-C3.B VLPs did not elicit any peptide-specific antibody responses, suggesting that the B cell epitope was not exposed on the surface of VLPs. Recombinant hepatitis E virus (HEV) capsid protein generated VLPs in potato plants when plant expression cassettes were applied [52]. However, the assembly of VLPs was limited in potato tubers and oral immunization of mice failed to generate detectable antibody responses in serum. In another attempt, the 11 amino acid B cell epitope tag was fused to the C-terminal of the HEV coat protein and incorporated in HEV VLPs [53]. Oral administration of chimeric HEV VLPs elicited specific IgG and IgA responses in intestinal secretions already 2 weeks after immunization. Furthermore, the insect Flock House virus (FHV) VLPs displayed *Bacillus anthracis* AB-type toxin receptor binding protective antigen (PA) on the surface, demonstrating inhibition of lethal toxin action both in vitro and in vivo [54]. The potent toxin-neutralizing antibody response protected rats from lethal anthrax challenges after a single immunization. In attempts to develop vaccines against Alzheimer's disease, bovine papilloma virus (BPV) VLPs displaying repetitively nine amino acids of Aβ protein on the surface were engineered [55]. Particularly, the chosen Aβ peptide contains a functional B cell epitope but lacks known T cell epitopes. Mice and rabbits vaccinated with these BPV-Aβ VLPs induced high titer antibodies and inhibited assembly of Aβ (1–42) peptides into neurotoxic fibrils. Furthermore, deposits of Aβ were observed in the brain of immunized APP/presenilin 1 transgenic mice. Additionally, increased levels of Aβ peptide were detected in plasma.

27.6 CANCER THERAPY APPROACHES WITH VLPs

Current inefficiency and the presence of severe adverse events in cancer therapy are to a large part due to insufficient and/or inappropriate delivery of therapeutic agents. Again, numerous approaches have been taken to deliver different types of *therapeutics* with the aid of VLPs. In this context, the majority of applications have involved immunization approaches with VLPs to provide both therapeutic and prophylactic efficacies. For instance, when Qβ VLPs were covalently linked to a peptide from the melanoma self-antigen Melan-A, strong IgG1 and IgG3 antibody responses in vaccinated melanoma patients were observed [56]. However, the CD4(+) T-cell responses were primarily specific for Qβ and require additional optimization toward immunogenicity against the cargo peptide. In attempts to engineer targeted vectors, the preS1 region of the hepatitis B virus (HBV) envelope protein was introduced into the T7 coat structure [57]. The chimeric T7 VLPs were able to efficiently transfect HepG2 cells in a dose- and time-dependent manner and therefore potentially present an interesting delivery vehicle for the treatment of liver cancer.

To address cancer immunotherapy, murine polyomavirus (MuPV) VLPs were engineered to display the entire human prostate-specific antigen (PSA) [58]. The PSA-MuPV VLPs showed only marginal protection against tumor challenges in BALB/c mice, but loading onto murine DCs together with CpG adjuvant efficiently protected animals from tumor growth. In another study, ovalbumin (OVA) or OVA-derived CD4 and CD8 epitopes were chemically conjugated to rabbit hemorrhagic disease virus (RHDV) VLPs and induced high antigen-specific cytotoxicity in vivo [59]. Moreover, the growth of the aggressive B16.OVA melanoma was significantly delayed in mice vaccinated with chimeric VLPs. Also, VLPs derived from RHDV showed binding of the galactose-containing α-galactosylceramide adjuvant, forming composite particles [60]. Administration of these VLPs activated splenic iNKT cells and resulted in production of interferon-γ and interleukin-4, rendering prophylactic protection against subcutaneous tumor challenges. Furthermore, simian immunodeficiency virus (SIV) and simian-human immunodeficiency virus (SHIV) VLPs incorporated with the murine Trop2 glycoprotein were overexpressed in pancreatic cancer [61]. Immunization of C57BL/6 tumor-bearing mice demonstrated a significant decrease in tumor growth. Additionally, the vaccination led to Trop2-specific cytotoxic T lymphocytes, strong antibody response, and decrease in the population of regulatory T cells and myeloid-derived suppressor cells in tumor tissue. When VLP immunization was combined with gemcitabine treatment, a significant increase in the survival of tumor-bearing mice was observed. Likewise, hepatitis B virus core antigen (HBcAg) VLPs displaying the highly selective tumor-associated cell lineage marker claudin-18 isoform 2 (CLDN18.2) elicited antibody responses in mice and rabbits [62]. Furthermore, efficient killing of CLDN18.2 cells was observed in vitro and partial protection against challenges of mice with syngeneic tumor cells stably expressing CLDN18.2. In another study, bovine papillomavirus (BPV) VLPs displayed a polycationic MUC1 peptide in various regions of the BVP L1 protein and were produced from baculovirus vectors in insect cells [63]. Chimeric BPV-MUC1 VLPs induced robust activation of bone marrow-derived DCs, and immunization of transgenic human MUC1

mice delayed the growth of transplanted MUC1 tumors and showed complete tumor rejection in some animals. One of the most advanced applications of VLPs has certainly been the approval of Cervarix, a vaccine against cervical cancer caused by HPV [64]. In this context, the HPV L1 capsid protein was expressed in insect cells using baculovirus to generate VLPs for immunization studies [65].

In another study, a novel in vivo DNA packaging method was developed for the assembly of JC virus (JCV) VLPs consisting of the major structural protein VP1 [66]. VLPs carrying the GFP as a reporter gene and the herpes simplex virus thymidine kinase (tk) as a suicide gene were demonstrated to specifically target human colon carcinoma cells (COLO-320 HSR) in a nude mouse model. Furthermore, treatment with ganciclovir reduced the tumor volume significantly. The HBV capsid protein with a C-terminal truncation fused with p19 RNA binding protein was the basis for VLP assembly for the encapsulation of siRNA molecules [67]. Additionally, VLP tumor targeting was obtained through RGD peptides after intravenous tail vein injection in tumor-bearing mice. Prolonged in vivo circulation was observed for encapsulated siRNAs, which increased their in vivo potential. The effect of siRNA-based gene silencing was evaluated in red fluorescent protein (RFP)-expressing B16F10 cells. In vivo, the chimeric VLPs carrying RFP siRNAs targeted tumor tissue and efficiently suppressed RFP expression. Recently, adeno-associated virus (AAV) type 2 VLPs were produced in baculovirus-infected insect cells, purified by chromatography and coated with PEI [68]. Furthermore, siRNA sequences were packaged into the AAV2-VLPs and delivered at high transfection rates to MCF-7 breast cancer cells, demonstrating more than 60% cell death within 72 h. In another approach, the MS2 bacteriophage capsid proteins were applied for VLP-based delivery of antisense oligonucleotides to target p120 mRNA, a biomarker overexpressed in myelogenous leukemia cells [69]. Covalent attachment of transferrin on the VLP surface provided targeting of tumor cells expressing transferrin receptors. The difficulties of transfer from in vitro success to in vivo efficacy were illustrated by HER3 targeting of adenovirus to breast cancer xenografts in mice [70]. An epidermal growth factor-like domain of the human heregulin-α was introduced into the HI loop of the Adenovirus 5 (Ad5) fiber, which resulted in enhanced infection of tumor cells expressing the cognate receptors HER3/ErbB3 and HER4/Erb4. However, intratumoral injection of mice with breast cancer xenografts showed no improvement in tumor targeting compared to conventional adenovirus vectors. Similarly, Ad5 vectors were modified for targeting integrin expression in cancer cells and simultaneously detargeting the coxsackie-adenovirus receptor (CAR) to prevent interaction with factor IX (FIX)/C4b-binding protein (C4BP) [71]. This vector showed reduced viral toxicity. However, despite efficient tumor transduction, the modified Ad5 vector did not provide enhanced tumor regression capacity because of restricted intratumoral delivery. In another approach, C(60) fullerenes were conjugated to CPMV VLPs [72]. Dye-labeled VLP-PEG-C(60) complexes were efficiently taken up by a human cancer line that potentially provides the possibility for photoactivated tumor therapy. Furthermore, a noncytopathic RNA replicon system based on the flavivirus Kunjin encoding granulocyte colony-stimulating factor (GCFS) was developed for intratumoral VLP delivery [73]. The Kunjin-GCFS VLPs were able to cure more than 50% of established subcutaneous CT26 colon carcinomas and B16-OVA melanomas. Tumor regression correlated with the induction of anti-cancer CD8 T cells. Interestingly, treatment of subcutaneous CT26 tumors also provided regression of CT26 lung metastases. In another approach, replication-deficient Semliki Forest particles were encapsulated in liposomes to provide enhanced targeting of tumor cells [75]. Intraperitoneal delivery of encapsulated SFV-LacZ particles to immunodeficient mice with human prostate tumor xenografts demonstrated accumulation of β-galactosidase in tumor tissue. Furthermore, intravenous administration of liposome-SFV particles expressing interleukin-12 (IL-12) showed a good safety profile in melanoma and kidney carcinoma patients in a phase I clinical trial [75].

27.7 CONCLUSIONS AND FUTURE ASPECTS

In this review, numerous examples have been given on the production of VLPs for delivery of a variety of cargo such as nucleic acids, fluorescent reagents, drug molecules, and imaging reagents. Furthermore, incorporation of antigens in the surface structure of VLPs has provided means for efficient vaccine development. This approach has resulted in an FDA-approved vaccine against cervical cancer. Most interestingly, the repertoire of VLPs stretches from bacteriophages to plant and animal origin. It has allowed the opportunity to choose specifically VLPs for appropriate applications. It is easy to envision that numerous novel applications will emerge in the near future. Particularly, gene silencing approaches applying microRNA provide means for novel strategies in drug development and might generate excellent opportunities for the discovery of more efficient drugs.

REFERENCES

1. Wagner, V., Dullaart, A., Bock, A.K. et al. 2006. The emerging nanomedicine landscape. *Nat. Biotechnol.* 24: 1211–1217.
2. Yildiz, I., Shukla, S., and Steinmetz, N.F. 2011. Applications of viral nanoparticles in medicine. *Curr. Opin. Biotechnol.* 22: 901–908.
3. James, N.D., Coker, R.J., Tomlinson, D. et al. 1994. Liposomal doxorubicin (Doxil): An effective new treatment for Kaposi's sarcoma in AIDS. *Clin. Oncol.* 6: 294–296.
4. Lundstrom, K. 2009. Nanocarriers for delivery of peptides and proteins. In: Jorgensen, L. and Nielsen, H.M. (eds.), *Delivery Technologies for Biopharmaceuticals: Peptides, Proteins, Nucleic Acids and Vaccines.* Chichester, United Kingdom: John Wiley & Sons Ltd., pp. 193–205.
5. Pushko, P., Pumpens, P., and Grens, E. 2013. Development of virus-like particle technology from small highly symmetric to large complex virus-like particle structures. *Intervirology* 56: 141–165.
6. Pokorski, J.K. and Steinmetz, N.F. 2011. The art of engineering viral nanoparticles. *Mol. Pharm.* 8: 29–43.
7. Zlotnick, A. and Mukhopahyay, S. 2011. Virus assembly, allostery and antivirals. *Trends Microbiol.* 19: 14–23.
8. Pumpens, P. and Grens, E. 2002. Artificial genes for chimeric virus-like particles. In: Khudyakov, Y.E. and Fields, H.A. (eds.), *Artificial DNA: Methods and Applications.* Boca Raton, FL: CRC Press LLC, pp. 249–327.
9. Vaney, M.C. and Rey, F.A. 2011. Class II enveloped viruses. *Cell Microbiol.* 13: 1451–1459.
10. Mattiacio, J., Walter, S., Brewer, M. et al. 2011. Dense display of HIV-1 envelope spikes on the lambda phage scaffold does not result in the generation of improved antibody responses to HIV-1 Env. *Vaccine* 29: 2637–2647.
11. Peabody, D.S., Manifold-Wheeler, B., Medford, A. et al. 2008. Immunogenic display of diverse peptides on virus-like particles of RNA phage MS2. *J. Mol. Biol.* 380: 252–263.
12. Heal, K.G., Hill, H.R., Stockley, P.G. et al. 1999. Expression and immunogenicity of a liver stage malaria epitope presented as a foreign peptide on the surface of RNA-free MS2 bacteriophage capsids. *Vaccine* 18: 251–258.
13. Tissot, A.C., Renhofa, R., Schmitz, N. et al. 2010. Versatile virus-like particle carrier for epitope based vaccines. *PLoS One* 5: e9809.
14. Voronkova, T., Grosch, A., Kazaks, A. et al. 2002. Chimeric bacteriophage fr virus-like particles harboring the immunodominant C-terminal region of hamster polyomavirus VP1 induce a strong VP1-specific antibody response in rabbits and mice. *Viral. Immunol.* 15: 627–643.
15. Uchida, M., Morris, D.S., Kang, S. et al. 2012. Site-directed coordination chemistry with P22 virus-like particles. *Langmuir* 28: 1998–2006.
16. Richert, L.E., Servid, A.E., Harmsen, A.L. et al. 2012. A virus-like particle vaccine platform elicits heightened and hastened local lung mucosal antibody production after a single dose. *Vaccine* 30: 3653–3665.
17. Vasiljeva, I., Kozlovska, T., Cielens, I. et al. 1998. Mosaic Qbeta coats as a new presentation model. *FEBS Lett.* 431: 7–11.
18. Bessa, J., Schmitz, N., Hinton, H.J. et al. 2008. Efficient induction of mucosal and systemic immune responses by virus-like particles administered intranasally: Implications for vaccine design. *Eur. J. Immunol.* 38: 114–126.
19. Cornuz, J., Zwahlen, S., Jungi, W.F. et al. 2008. A vaccine against nicotine for smoking cessation: A randomized controlled trial. *PLoS One* 3: e2547.
20. Yusibov, V., Mett, V., Mett, V. et al. 2005. Peptide-based candidate vaccine against respiratory syncytial virus. *Vaccine* 23: 2261–2265.
21. Chen, C., Kwak, E.S., Stein, B. et al. 2005. Packaging of gold particles in viral capsids. *J. Nanosci. Nanotechnol.* 5: 2029–2033.
22. Cadena-Nava, R.D., Comas-Garcia, M., Garmann, R.F. et al. 2012. Self-assembly of viral capsid protein and RNA molecules of different sizes: Requirement for a specific high protein/RNA mass ratio. *J. Virol.* 86: 3318–3326.
23. Minten, I.J., Wilke, K.D., Hendriks, L.J. et al. 2011. Metal-ion-induced formation and stabilization of protein cages based on the cowpea chlorotic mottle virus. *Small* 7: 911–919.
24. Yusibov, V., Hooper, D.C., Spitsin, S.V. et al. 2002. Expression in plants and immunogenicity of plant virus-based experimental rabies vaccine. *Vaccine* 20: 3155–3164.
25. Brennan, F.R., Gilleland, L.B., Staczek, J. et al. 1999. A chimeric plant virus vaccine protects mice against bacterial infection. *Microbiology* 145: 2061–2067.
26. Dalsgaard, K., Uttenthal, A., Jones, T.D. et al. 1997. Plant-derived vaccine protects target animals against a viral disease. *Nat. Biotechnol.* 15: 248–252.

27. Natilla, A., Hammond, R.W. and Nemchinov, L.G. 2006. Epitope presentation system based on cucumber mosaic virus coat protein expressed from a potato virus X-based vector. *Arch. Virol.* 151: 1373–1386.

28. Vitti, A., Piazzolla, G., Condelli, V. et al. 2010. Cucumber mosaic virus as the expression system for a potential vaccine against Alzheimer's disease. *J. Virol. Methods* 169: 332–340.

29. Nuzzaci, M., Vitti, A., Condelli, V. et al. 2010. in vitro stability of Cucumber mosaic virus nanoparticles carrying a hepatitis C virus-derived epitope under simulated gastro-intestinal conditions and in vivo efficacy of edible vaccine. *J. Virol. Methods* 165: 211–215.

30. Uhde-Holzem, K., Schlösser, V., Viazov, S. et al. 2010. Immunogenic properties of chimeric potato virus X particles displaying the hepatitis C virus hypervariable region I peptide R9. *J. Virol. Methods* 166: 12–20.

31. Marusic, C., Rizza, P., Lattanzi, I. et al. 2001. Chimeric plant virus particles as immunogens for inducing murine and human immune responses against human immunodeficiency virus type I. *J. Virol.* 75: 8434–8439.

32. Zelada, A.M., Calamante, G., de la Paz Santangelo, M. et al. 2006. Expression of tuberculosis antigen ESAT-6 in Nicotiana tabacum using a potato virus X-based vector. *Tuberculosis (Edinb.)* 86: 263–267.

33. Cerovska, N., Hoffmeisterova, H., Moravec, T. et al. 2012. Transient expression of human papillomavirus type 16 L2 epitope fused to N- and C-terminus of coat protein of Potato virus X in plants. *J. Biosci.* 37: 125–133.

34. Kalnciema, I., Skarstina, D., Ose, V. et al. 2012. Potato virus Y-like particles as a new carrier for the presentation of foreign protein stretches. *Mol. Biotechnol.* 52: 129–139.

35. Turpen, T.H., Reinl, S.J., Charoenvit, Y. et al. 1995. Malarial epitopes expressed on the surface of recombinant tobacco mosaic virus. *Biotechnology (NY)* 13: 53–57.

36. Bendahmane, M., Koo, M., Karrer, E. et al. 1999. Display of epitopes on the surface of tobacco mosaic virus: Impact of charge and isoelectric point of the epitope on virus-host interactions. *J. Mol. Biol.* 290: 9–20.

37. Palmer, K.E., Benko, A., Doucette, S.A. et al. 2006. Protection of rabbits against cutaneous papillomavirus infection using recombinant tobacco mosaic virus containing L2 capsid epitopes. *Vaccine* 24: 5516–5525.

38. Inoue, T., Kawano, M.A., Takahashi, R.U. et al. 2008. Engineering of SV40-based nanocapsules for delivery of heterologous proteins as fusions with the minor capsid proteins VP2/3. *J. Biotechnol.* 134: 181–192.

39. Li, F., Zhang, Z.P., Peng, J. et al. 2009. Imaging viral behavior in mammalian cells with self-assembled capsid-quantum-dot hybrid particles. *Small* 5: 718–726.

40. Takahashi, R.U., Kanesashi, S.N., Inoue, T. et al. 2008. Presentation of functional foreign peptides on the surface of SV40 virus-like particles. *J. Biotechnol.* 135: 385–392.

41. Kitai, Y., Fukuda, H., Enomoto, T. et al. 2011. Cell selective targeting of a simian virus 40 virus-like particle conjugated to epidermal growth factor. *J. Biotechnol.* 155: 251–256.

42. Cortes-Perez, N.G., Sapin, C., Jaffrelo, I. et al. 2010. Rotavirus-like particles: A novel nanocarrier for the gut. *J. Biomed. Biotechnol.* 2010: 317545.

43. Tellinghuisen, T.L., Hamburger, A.E., Fischer, B.R. et al. 1999. In vitro assembly of alphavirus cores by using nucleocapsid cores in mature virus. *J. Virol.* 73: 5309–5319.

44. Mukhopadhyay, S., Chipman, P.R., Hong, E.M. et al. 2002. In vitro assembled alphavirus core-like particles maintain a structure similar to that of nucleocapsid cores in mature virus. *J. Virol.* 76: 11128–11132.

45. Cheng, F., Tsvetkova, I.B., Khuong, Y.-L. et al. 2012. The packaging of different cargo into enveloped viral nanoparticles. *Mol. Pharm.* 10: 51–58.

46. Tseng, J.C., Levin, B., Hurtado, A. et al. 2004. Systemic tumor targeting and killing by Sindbis viral vectors. *Nat. Biotechnol.* 22: 70–77.

47. Bansal, G.P., Hatfield, J.A., Dunn, F.E. et al. 1993. Candidate recombinant vaccine for human B19 parvovirus. *J. Infect. Dis.* 167: 1034–1044.

48. Amexis, G. and Young, N.S. 2006. Parvovirus B19 empty capsids as antigen carriers for presentation of antigenic determinants of dengue 2 virus. *J. Infect. Dis.* 194: 790–794.

49. Ogasawara, Y., Amexis, G., Yamaguchi, H. et al. 2006. Recombinant viral-like particles of parvovirus B19 as antigen carriers of anthrax protective antigen. *In Vivo* 20: 319–324.

50. Rodríguez, D., González-Aseguinolaza, G., Rodríguez, J.R. et al. 2012. Vaccine efficacy against malaria by the combination of porcine parvovirus-like particles and vaccinia virus vectors expressing CS of Plasmodium. *PLoS One* 7: e34445.

51. Sedlik, C., Sarraseca, J., Rueda, P. et al. 1995. Immunogenicity of poliovirus B and T cell epitopes presented by hybrid porcine parvovirus particles. *J. Gen. Virol.* 76: 2361–2368.

52. Maloney, B.J., Takeda, N., Suzaki, Y. et al. 2005. Challenge in creating a vaccine to prevent hepatitis E. *Vaccine* 23: 1870–1874.

53. Niikura, M., Takamura, S., Kim, G. et al. 2002. Chimeric recombinant hepatitis E virus-like particles as an oral vaccine vehicle presenting foreign epitopes. *Virology* 293: 273–280.

54. Manayani, D.J., Thomas, D., Dryden, K.A. et al. 2007. A viral nanoparticle with dual function as an anthrax toxin and vaccine. *PLoS Pathog.* 3: 1422–1431.

55. Zamora, E., Handisurya, A., Shafti-Keramat, S. et al. 2006. Papillomavirus-like particles are an effective platform for amyloid-beta immunization in rabbits and transgenic mice. *J. Immunol.* 177: 2662–2670.

56. Braun, M., Jandus, C., Maurer, P. et al. 2012. Virus-like particles induce robust human T-helper cell responses. *Eur. J. Immunol.* 42: 330–340.

57. Tang, K.H., Yusoff, K., and Tan, W.S. 2009. Display of hepatitis B virus PreS1 peptide on bacteriophage T7 and its potential in gene delivery into HepG2 cells. *J. Virol. Methods* 159: 194–199.

58. Eriksson, M., Andreasson, K., Weidmann, J. et al. 2011. Murine polyomavirus virus-like particles carrying full-length human PSA protect BALB/c mice from outgrowth of a PSA expressing tumor. *PLoS One* 6: e23828.

59. Peacey, M., Wilson, S., Perret, R. et al. 2008. Virus-like particles from rabbit hemorrhagic disease virus can induce an anti-tumor response. *Vaccine* 26: 5334–5337.

60. McKee, S.J., Young, V.L., Clow, F. et al. 2012. Virus-like particles and α-galactosylceramide form a self-adjuvanting composite particle that elicits anti-tumor responses. *J. Control Release* 159: 338–345.

61. Cubas, R., Zhang, S., Li, M. et al. 2011. Chimeric Trop2 virus-like particles: A potential immunotherapeutic approach against pancreatic cancer. *J. Immunother.* 34: 251–263.

62. Klamp, T., Schumacher, J., Huber, G. et al. 2011. Highly specific auto-antibodies against claudin-18 isoform 2 induced by a chimeric HBcAg virus-like particle vaccine kill tumor cells and inhibit the growth of lung metastases. *Cancer Res.* 71: 516–527.

63. Pejawar-Gaddy, S., Rajawat, Y., Hilioti, Z. et al 2010. Generation of a tumor vaccine candidate based on conjugation of a MUC1 peptide to polyionic papillomavirus virus-like particles. *Cancer Immunol. Immunother.* 59: 1685–1696.

64. Centers for Disease Control and Prevention (CDC). 2010. FDA licensure of bivalent human papillomavirus vaccine (HPV2, Cervarix) for use in females and updated HPV vaccination recommendations from the Advisory Committee on Immunization Practices (ACIP). *MMWR—Morb. Mortal Wkly. Rep.* 59: 626–629.

65. Senger, T., Schädlich, L., Gissmann, L. et al. 2009. Enhanced papillomavirus-like particle production in insect cells. *Virology* 388: 344–353.

66. Chen, L.S., Wang, M., Ou, W.C. et al. 2010. Efficient gene transfer using the human JC virus-like particle that inhibits human colon adenocarcinoma growth in a nude mouse model. *Gene Ther.* 17: 1033–1041.

67. Kyung-mi, C., Kwangmeyung, K., Ick, C.K. et al. 2013. Systemic delivery of siRNA by chimeric capsid protein: Tumor targeting and RNAi activity in vivo. *Mol. Pharm.* 10: 18–25.

68. Shao, W., Paul, A., Abbasi, S. et al. 2012. A novel polyethyleneimine-coated adeno-associated virus-like particle formulation for efficient siRNA delivery in breast cancer therapy: Preparation and in vitro analysis. *Int. J. Nanomed.* 7: 1575–1586.

69. Wu, M., Sherwin, T., Brown, T.L. et al. 2005. Delivery of antisense oligonucleotides to leukemia cells by RNA bacteriophage capsids. *Nanomedicine* 1: 67–76.

70. MacLeod, S.H., Elgadi, M.M., Bossi, G. et al. 2012. HER3 targeting of adenovirus by fiber modification increases infection of breast cancer cells in vitro, but not following intratumoral injection in mice. *Cancer Gene Ther.* 19: 888–898.

71. Coughlan, L., Vallath, S., Gros, A. et al. 2012. Combined fiber modifications both to target α(v)β(6 and detarget the coxsackievirus-adenovirus receptor improve virus toxicity but fail to improve antitumoral efficacy relative to adenovirus serotype 5. *Hum. Gene Ther.* 23: 960–979.

72. Steinmetz, N.F., Hong, V., Spoerke, E.D. et al. 2009. Buckyballs meet viral nanoparticles: Candidates for biomedics. *J. Am. Chem. Soc.* 131: 17093–17095.

73. Hoang-Le, D., Smeenk, L., Anraku, I. et al. 2009. A Kunjin replicon vector encoding granulocyte macrophage colony-stimulating factor for intra-tumoral gene therapy. *Gene Ther.* 16: 190–199.

74. Lundstrom, K. 2006. Alphavirus vectors for gene therapy applications. In: Hunt, K.K., Vorburger, S., and Swisher, S.G. (eds.), *Cancer Drug Discovery and Development: Gene Therapy for Cancer.* Totowa, NJ: Humana Press, Inc., pp. 109–119.

75. Ren, H., Boulikas, T., Lundstrom, K. et al. 2003. Immunogene therapy of recurrent glioblastoma multiforme with a liposomally encapsulated replication-incompetent Semliki Forest virus vector carrying the human interleukin-12 gene—A phase I/II clinical protocol. *J. Neurooncol.* 65: 191.

76. Spohn, G., Jennings, G.T., Martina, B.E. et al. 2010. A VLP-based vaccine targeting domain III of the West Nile virus E protein protects from lethal infection in mice. *Virol. J.* 7: 146.

77. Caldeira Jdo, C., Medford, A., Kines, R.C. et al. 2010. Immunogenic display of diverse peptides, including a broadly cross-type neutralizing human papillomavirus L2 epitope, on virus-like particles of the RNA bacteriophage PP7. *Vaccine* 28: 4384–4393.

78. Tumban, E., Peabody, J., Peabody, D.S. et al. 2011. A pan-HPV vaccine based on bacteriophage PP7 VLPs displaying broadly cross-neutralizing epitopes from the HPV minor capsid protein L2. *PLoS One* 6: e23310.

79. Yang, C.D., Liao, J.T., Lai, C.Y. et al. 2007. Induction of protective immunity in swine by recombinant bamboo mosaic virus expressing foot-and-mouth disease virus epitopes. *BMC Biotechnol.* 7: 62.

80. Chen, T.H., Chen, T.H., Hu, C.C. et al. 2012. Induction of protective immunity in chickens immunized with plant-made chimeric Bamboo mosaic virus particles expressing very virulent infectious bursal disease virus antigen. *Virus Res.* 166: 109–115.

81. Hu, Y., Zandi, R., Anivitarte, A. et al. 2008. Packaging of a polymer by a viral capsid: The interplay between polymer and length and capsid size. *Biophys. J.* 94: 1428–1436.

82. Brasch, M., de la Escosura, A., Ma, Y. et al. 2011. Encapsulation of phthalocyanine supramolecular stacks into virus-like particles. *J. Am. Chem. Soc.* 133: 6878–6881.

83. Nicholas, B.L., Brennan, F.R., Martinez-Torrecuadrada, J.L. et al. 2002. Characterization of the immune response to canine parvovirus induced by vaccination with chimeric plant viruses. *Vaccine* 20: 2727–2734.

84. Jagadish, M.N., Hamilton, R.C., Fernandez, C.S. et al. 1993. High level production of hybrid potyvirus-like particles carrying repetitive copies of foreign antigens in *Escherichia coli*. *Biotechnology* (*NY*) 11: 1166–1170.

85. Saini, M. and Vrati, S. 2003. A Japanese encephalitis virus peptide present on Johnson grass mosaic virus-like particles induces virus-neutralizing antibodies and protects mice against lethal challenge. *J. Virol.* 77: 3487–3494.

86. Savard, C., Laliberté-Gagné, M.E., Babin, C. et al. 2012. Improvement of the PapMV nanoparticle adjuvant property through an increase of its avidity for the antigen [influenza NP]. *Vaccine* 30: 2535–2542.

87. Fernandez-Fernandez, M.R., Martinez-Torrecuadrada, J.L., Casal, J.L. et al. 1998. Development of an antigen presentation system based on plum pox polyvirus. *FEBS Lett.* 427: 229–235.

88. Kumar, S., Ochoa, W., Singh, P. et al. 2009. Tomato bushy stunt virus (TBSV), a versatile platform for polyvalent display of antigenic epitopes and vaccine design. *Virology* 388: 185–190.

89. Haynes, J.R., Cunningham, J., von Seefried, A. et al. 1986. Development of a genetically engineered, candidate polio vaccine employing the self-assembling properties of the tobacco mosaic virus coat protein. *Biotechnology* (*NY*) 4: 637–641.

90. Sugiyama, Y., Hamamoto, H., Takemoto, S. et al. 1995. Systemic production of foreign peptides on the particle surface of tobacco mosaic virus. *FEBS Lett.* 359: 247–250.

91. Koo, M., Bendahmane, M., Lettieri, G.A. et al. 1999. Protective immunity against murine hepatitis virus (MHV) induced by intranasal or subcutaneous administration of hybrids of tobacco mosaic virus that carries an MHV epitope. *Proc. Natl. Acad. Sci. USA* 96: 7774–7779.

92. Wu, L., Jiang, L., Zhou, Z. et al. 2003. Expression of foot-and-mouth disease virus epitopes in tobacco by a tobacco mosaic virus-based vector. *Vaccine* 21: 4390–4398.

93. Hou, L., Wu, H., Xu, L. et al. 2009. Expression and self-assembly of virus-like particles of infectious hypodermal and hematopoietic necrosis virus in *Escherichia coli*. *Arch. Virol.* 154: 547–553.

94. Ju, H., Wei, N., Wang, Q. et al. 2011. Goose parvovirus structural proteins expressed by a recombinant baculovirus self-assemble into virus-like particles with strong immunogenicity in goose. *Biochim. Biophys. Res. Commun.* 409: 131–136.

95. Pan, Q., He, K., and Huang, K. 2008. Development of recombinant porcine parvovirus-like particles as an antigen carrier formed by the hybrid VP2 protein carrying immunoreactive epitope of porcine circovirus type 2. *Vaccine* 26: 2119–2126.

96. Rueda, P., Martinez-Torrecuadrada, J.L., Sarraseca, J. et al. 1999. Engineering parvovirus-like particles for the induction of B-cell CD4(+) and CTL responses. *Vaccine* 18: 325–332.

97. Feng, H., Liang, M., Wang, H.I. et al. 2011. Recombinant canine parvovirus-like particles express foreign epitopes in silk-worm pupae. *Vet. Microbiol.* 154: 49–57.

98. Jeoung, H.Y., Lee, W.H., Jeong, W. et al. 2011. Immunogenicity and safety of virus-like particle of the porcine encephalomyocarditis virus in pig. *Virol. J.* 8: 170.

99. Lee, C.D., Yan, Y.P., Liang, S.M. et al. 2009. Production of FMDV virus-like particles by a SUMO fusion protein approach in *Escherichia coli*. *J. Biomed. Sci.* 16: 69.

100. Lu, M.W., Liu, W., and Lin, C.S. 2003. Infection competition against grouper nervous necrosis virus by virus-like particles produced in *Escherichia coli*. *J. Gen. Virol.* 84: 1577–1582.

101. Santi, L., Batchelor, L., Huang, Z. et al. 2008. An efficient plant viral expression system generating orally immunogenic Norwalk virus-like particles. *Vaccine* 26: 1846–1854.

102. Atmar, R.L., Bernstein, D.I., Harro, C.D. et al. 2011. Norovirus vaccine against experimental human Norwalk Virus illness. *N. Engl. J. Med.* 365: 2178–2187.

103. Laurent, S., Vautherot, J.F., Madeleine, M.F. et al. 1994. Recombinant rabbit hemorrhagic disease virus capsid protein expressed in baculovirus self-assembles into virus-like particles and induces protection. *J. Virol.* 68: 6794–6798.

104. Crisci, E., Fraile, I., Moreno, N. et al. 2012. Chimeric calicivirus-like particles elicit specific immune responses in pigs. *Vaccine* 30: 2427–2439.

105. Skrastina, D., Bulavaite, A., Sominskaya, I. et al. 2008. High immunogenicity of a hydrophilic component of the hepatitis B virus preS1 sequence exposed on the surface of three virus-like particle carriers. *Vaccine* 26: 1972–1981.

106. Gedvilaite, A., Frömmel, C., Sasnauskas, K. et al. 2000. Formation of immunogenic virus-like particles by inserting epitopes into surface-exposed regions of hamster polyomavirus major capsid protein. *Virology* 273: 21–35.

107. Gedvilaite, A., Zvirbliene, A., Staniulus, J. et al. 2004. Segments of Puumala hantavirus nucleocapsid protein inserted into chimeric polyomavirus-derived virus-like particles induce a strong immune response in mice. *Viral. Immunol.* 17: 51–68.

108. Dorn, D.C., Lawatscheck, R., Zvirbliene, A. et al. 2008. Cellular and humoral immunogenicity of hamster polyomavirus-derived virus-like particles harboring a mucin 1 cytotoxic T-cell epitope. *Viral. Immunol.* 21: 12–27.

109. Slupetzky, K., Gambhira, R., Culp, T.D. et al. 2007. A papilloma-virus-like particle (VLP) vaccine displaying HPV 16 L2 epitopes induces cross-neutralizing antibodies to HPV11. *Vaccine* 25: 2001–2010.

110. Roy, P., Urakawa, T., Van Dijk, A.A. et al. 1990. Recombinant virus vaccine for bluetongue disease in sheep. *J. Virol.* 64: 1998–2003.

111. Belyaev, A.S. and Roy, P. 1992. Presentation of hepatitis B virus preS2 epitope on bluetongue virus core-like particles. *Virology* 190: 840–844.

112. Adler, S., Reay, P., Roy, P. et al. 1998. Induction of T cell response by bluetongue virus core-like particles expressing a T cell epitope of the M1 protein of influenza A virus. *Med. Microbiol. Immunol.* 187: 91–96.

113. Azevedo, M.S., Gonzales, A.M., Yuan, I. et al. 2010. An oral versus intranasal prime/boost regimen using attenuated human rotavirus of VP2 and VP6 virus-like particles with immunostimulating complexes influences protection and antibody secreting cell responses to rotavirus in a neonatal gnotobiotic pig model. *Clin. Vaccine Immunol.* 17: 420–428.

114. Blazevic, V., Lappalainen, S., Nurminen, K. et al. 2011. Norovirus VLPs and rotavirus VP6 protein as combined vaccine for childhood gastroenteritis. *Vaccine* 29: 8126–8133.

115. Neirynck, S., Deroo, T., Saelens, X. et al. 1999. A universal influenza A vaccine based on the extracellular domain of the M2 protein. *Nat. Med.* 5: 1157–1163.

116. Schödel, F., Wirtz, R., Peterson, D. et al. 1994. Immunity to malaria elicited by hybrid hepatitis B virus core particles carrying circumsporozoite protein epitopes. *J. Exp. Med.* 180: 1037–1046.

117. Engerix B: Summary for basis approval. FDA, 1988. http://www.fda.gov/down-loads/BiologicsBloodVaccines/Vaccines/ApprovedProduct/UCM110155.pdf. Accessed January 31, 2014.

118. Recombivax HB: Summary for basis approval. FDA, 1987. http://www.fda.gov/down-loads/BiologicsBloodVaccines/Vaccines/ApprovedProduct/UCM244544.pdf.

119. Ameiss, K., Ashraf, S., Kong, W. et al. 2010. Delivery of woodchuck hepatitis virus-like particle presented influenza M2e by recombinant attenuated Salmonella displaying a delayed lysis phenotype. *Vaccine* 28: 6704–6713.

120. Ohtaki, N., Takahashi, H., Kaneko, K. et al. 2010. Immunogenicity and efficacy of two types of West Nile virus-like particles different in size and maturation as a second-generation vaccine candidate. *Vaccine* 28: 6588–6596.

121. Zhang, S., Liang, M., Gu, W. et al. 2011. Vaccination with dengue virus-like particles induces humoral and cellular immune responses in mice. *Virol. J.* 8: 333.

122. Liu, W., Jiang, H., Zhou, J. et al. 2010. Recombinant dengue virus-like particles from *Pichia pastoris*: Efficient production and immunological properties. *Virus Genes* 40: 53–59.

123. Lechmann, M., Murata, K., Satoi, J. et al. 2001. Hepatitis C virus-like particles induce virus-specific humoral and cellular immune responses in mice. *Hepatology* 34: 417–423.

124. Elmowalid, G.A., Qiao, M., Jeong, S.H. et al. 2007. Immunization with hepatitis C virus-like particles results in control of hepatitis C virus infection in chimpanzees. *Proc. Natl. Acad. Sci. USA* 104: 8427–8432.

125. Yoshii, K., Goto, A., Kawakami, K. et al. 2008. Construction and application of chimeric virus-like particles of tick-borne encephalitis virus and mosquito-borne Japanese encephalitis virus. *J. Gen. Virol.* 89: 200–211.

126. Akahata, W., Yang, Z.Y., Andersen, H. et al. 2010. A virus-like particle vaccine for epidemic Chikungunya virus protects nonhuman primates against infection. *Nat. Med.* 16: 334–338.

127. Bai, B., Hu, Q., Hu, H. et al. 2008. Virus-like particles of SARS-like coronavirus formed by membrane proteins from different origins demonstrate stimulating activity in human dendritic cells. *PL papilloma-virus oS One* 3: e2685.

128. Liu, Y.V., Massare, M.J., Bernard, D.I. et al. 2011. Chimeric severe acute respiratory syndrome coronavirus (SARS-CoV) S glycoprotein and influenza matrix 1 efficiently form virus-like particles (VLPs) that protect mice against challenge with SARS-CoV. *Vaccine* 29: 6606–6613.

129. Habjan, M., Penski, N., Wagner, V. et al. 2009. Efficient production of Rift Valley fever virus-like particles: The antiviral protein MxA can inhibit primary transcription of bunyaviruses. *Virology* 385: 400–408.

130. Li, C., Liu, F., Liang, M. et al. 2010. Hantavirus-like particles generated in CHO cells induce specific immune responses in C57BL/6 mice. *Vaccine* 28: 4294–4300.

131. Tong, T., Crooks, E.T., Osawa, K. et al. 2012. HIV-1 virus-like particles bearing pure env trimers expose neutralizing epitopes but occlude nonneutralizing epitopes. *J. Virol.* 86: 3574–3587.

132. Garnier, L., Ravallec, M., Blanchard, P. et al. 1995. Incorporation of pseudorabies virus gD into human immunodeficiency virus type I Gag particles produced in baculovirus-infected cells. *J. Virol.* 69: 4060–4068.

133. Di Bonito, P., Grasso, F., Mochi, S. et al. 2009. Anti-tumor CD8+ T cell immunity elicited by HIV-1-based virus-like particles incorporating HPV-16 E7 protein. *Virology* 395: 45–55.

134. Luo, L., Li, Y., Cannon, P.M. et al. 1992. Chimeric gag-V3 virus-like particles of human immunodeficiency virus induce virus-neutralizing antibodies. *Proc. Natl. Acad. Sci. USA* 89: 10527–10531.

135. Kang, C.Y., Luo, L., Wainberg, M.A. et al. 1999. Development of HIV/AIDS vaccine using chimeric gag-env virus-like particles. *Biol. Chem.* 380: 353–364.

136. Yamshchikov, G.V., Ritter, G.D., Vey, M. et al. 1995. Assembly of SIV virus-like particles containing envelope proteins using a baculovirus expression system. *Virology* 214: 50–58.

137. Zhang, R., Zhang, S., Li, M. et al. 2010. Incorporation of CD40 ligand into SHIV virus-like particles (VLP) enhances SHIV-VLP-induced dendritic cell activation and boosts immune responses against HIV. *Vaccine* 28: 5114–5127.

138. Deo, V.K., Tsuji, Y., Yasuda, T. et al. 2011. Expression of an RSV-gag virus-like particle in insect cell lines and silkworm larvae. *J. Virol. Methods* 177: 147–152.

139. Tsuji, Y., Deo, V.K., Kato, T. et al. 2011. Production of Rous sarcoma virus-like particles displaying human transmembrane protein in silkworm larvae and its application to ligand-receptor binding assay. *J. Biotechnol.* 155: 185–192.

140. Leb, V.M., Jahn-Schmid, B., Kueng, H.J. et al. 2009. Modulation of allergen-specific T-lymphocyte function by virus-like particles decorated with HLA class II molecules. *J. Allergy Clin. Immunol.* 124: 121–128.

141. Bach, P., Kamphuis, E., Odermatt, B. et al. 2007. Vesicular stomatitis virus glycoprotein displaying retrovirus-like particles induce a type I IFN receptor-dependent switch to neutralizing IgG antibodies. *J. Immunol.* 178: 5839–5847.

142. Nikles, D., Bach, P., Boller, K. et al. 2005. Circumventing tolerance to the prion protein (PrP): Vaccination with PrP-displaying retrovirus-like particles induces humoral immune responses against the native form of cellular PrP. *J. Virol.* 79: 4033–4042.

143. Bright, R.A., Carter, D.M., Daniluk, S. et al. 2007: Influenza virus-like particles elicit broader immune responses than whole virion inactivated influenza virus or recombinant hemagglutinin. *Vaccine* 25: 3871–3878.

144. Haynes, J.R., Dokken, L., Wiley, J.A. et al. 2009. Influenza-pseudotyped Gag virus-like particle vaccines provide broad protection against highly pathogenic avian influenza challenge. *Vaccine* 27: 530–541.

145. Branco, L.M., Grove, J.N., Geske, F.J. et al. 2010. Lassa virus-like particles displaying all major immunological determinants as a vaccine candidate for Lassa hemorrhagic fever. *Virol. J.* 7: 279.

146. McGinnes, L.W., Gravel, K.A., Finberg, R.W. et al. 2011. Assembly and immunological properties of Newcastle disease virus-like particles containing the respiratory syncytial virus F and G proteins. *J. Virol.* 85: 366–377.

147. Murawski, M.R., McGinnes, L.W., Finberg, R.W. et al. 2010. Newcastle disease virus-like particles containing respiratory syncytial virus G protein induced protection in BALB/c mice, with no evidence of immunopathology. *J. Virol.* 84: 1110–1123.

148. Walpita, P., Barr, J., Sherman, M. et al. 2011. Vaccine potential of Nipah virus-like particles. *PLoS One* 6: e18437.

149. Sugahara, F., Uchiyama, T., Watanabe, H. et al. 2004. Paramyxovirus Sendai virus-like particle formation by expression of multiple viral proteins and acceleration of its release by C protein. *Virology* 325: 1–10.

150. Juozapaitis, M., Slibinskas, R., Staniulis, J. et al. 2005. Generation of Sendai virus nucleocapsid-like particles in yeast. *Virus Res.* 108: 221–224.

151. Li, M., Schmitt, P.T., Li, Z. et al. 2009. Mumps virus matrix, fusion, and nucleocapsid proteins cooperate for efficient production of virus-like particles. *J. Virol.* 83: 7261–7272.

152. Slibinskas, R., Zvirbliene, A., Gedvilaite, A. et al. 2003. Synthesis of mumps virus nucleocapsid protein in yeast *Pichia pastoris*. *J. Biotechnol.* 103: 43–49.

153. Petraityte, R., Tamosiunas, P.L., Juozapaitis, M. et al. 2009. Generation of Tioman virus nucleocapsid-like particles in yeast *Saccharomyces cerevisiae*. *Virus Res.* 145: 92–96.

154. Slibinskas, R., Samuel, D., Gedvilaite, A. et al. 2004. Synthesis of the measles virus nucleoprotein in yeast *Saccharomyces cerevisiae*. *J. Biotechnol.* 107: 115–124.

155. Warfield, K.L., Swenson, D.L., Olinger, G.G. et al. 2007. Ebola virus-like particle-based vaccine protect against lethal Ebola virus challenge. *I. Infect.* 196(Suppl.): S430–S437.

156. Swenson, D.L., Warfield, K.L., Negley, D.L. et al. 2005. Virus-like particles exhibit potential as a pan-filo-virus vaccine for both Ebola and Marburg viral infections. *Vaccine* 23: 3033–3042.

28 Alphaviral Vectors for Cancer Treatment

Anna Zajakina, Jelena Vasilevska, Tatjana Kozlovska, and Kenneth Lundstrom

CONTENTS

28.1 INTRODUCTION

28.1.1 OVERVIEW OF ALPHAVIRUS REPLICATION AND ASSEMBLY

Alphaviruses represent a large class of arthropod-borne viruses (Arboviruses) belonging to the Togaviridae family of viruses, which cause transient febrile illness or more severe diseases such as encephalitis. The most commonly used vectors were generated on the basis of three alphaviruses: Semliki Forest virus (SFV) (Liljestrom and Garoff 1991), Sindbis virus (SIN) (Xiong et al. 1989), and Venezuelan equine encephalitis virus (VEE) (Davis et al. 1989).

The biology of these alphaviruses is similar. Their genome comprises a 5′ end capped and 3′ end polyadenylated RNA molecule of approximately 12 kb in length. Since the RNA has a positive polarity, it is infectious, capable of initiation of replication and translation when introduced into the cytoplasm of host cells. Functionally, the genome is divided into two parts coding for the nonstructural and structural proteins, respectively (Figure 28.1). Two-thirds of the 5′ end of the RNA genome encodes a polyprotein that is processed into four viral nonstructural proteins responsible for the replication of the plus strand (42S) genome into full-length minus strands. These molecules then serve as templates for the production of new 42S genomic RNAs and subgenomic 26S RNAs. The latter is an approximately 4000 nucleotide long subgenomic RNA and is collinear with the last third of the genome. Its synthesis is internally initiated at the 26S promoter on the 42S minus RNA strand.

The subgenomic RNA codes for the structural proteins of the virus, which are also synthesized as a polyprotein precursor in the order C-E3-E2-6K-E1. Once the capsid (C) protein has been synthesized, it acts as an autoprotease, cleaving itself off the nascent chain (Hahn and Strauss 1990; Schlesinger and Schlesinger 2001). At the endoplasmic reticulum (ER) membrane, the nascent chain is cotranslationally translocated and cleaved further by a signal peptidase to the three structural membrane proteins, p62 (precursor of E3-E2), 6K, and E1. After synthesis, the C protein complexes with genomic RNA into nucleocapsid structures in the cell cytoplasm. Usually, only the genomic RNA is packaged due to the presence of the encapsidation signal within the nsP1 and nsp2 genes for SIN and SFV, respectively (Frolova et al. 1997).

The membrane proteins undergo extensive posttranslational modifications within the biosynthetic transport pathway of the cell. The precursor protein p62 is proteolytically cleaved during the transport to the cell surface to form the mature envelope glycoprotein E2 (Lobigs and Garoff 1990). The p62 forms a heterodimer with E1 in the ER (Barth et al. 1995). This dimer is transported to the plasma

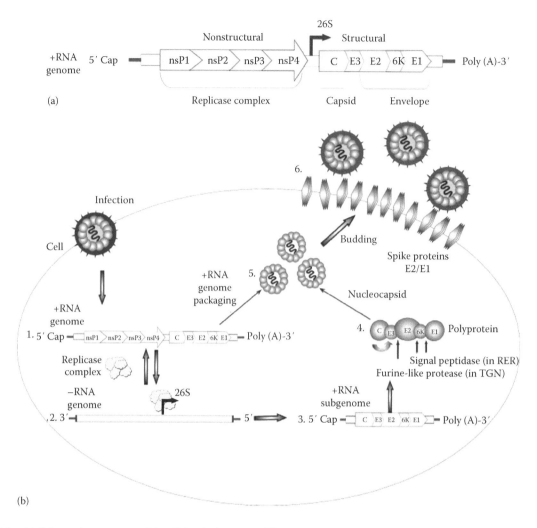

FIGURE 28.1 (a) Schematic structure of the alphaviral genome. The genome is a positive-strand RNA molecule that is capped and polyadenylated. It encodes two polyproteins indicated as nonstructural and structural parts of the genome. (b) Replication cycle of alphaviruses. After virus particle penetration and uncoating, the viral genomic RNA is released into the cytoplasm. The replication is initiated by translation of the nonstructural proteins (nsP1–4) and formation of the replicase complex (1). The positive strand genomic RNA serves as a template for a full-length complementary negative strand (2). The negative strand in turn serves as a template for the synthesis of subgenomic RNA expressed from 26S subgenomic minus RNA promoter (3). The structural proteins are translated from the subgenomic RNA and are processed posttranslationally into the individual proteins (4). The positive RNA genome is encapsidated by the capsid protein (5). Virus budding occurs on the cytoplasmic membrane via spike proteins and nucleocapsid interactions (6). RER, rough endoplasmic reticulum; TGN, trans-Golgi network; NC, nucleocapsid.

membrane, where virus budding occurs via spike nucleocapsid interactions (for review, see Jose et al. 2009). At a very late (post-Golgi) stage of transport, the p62 protein is cleaved to E3 and E2 by host furin-like proteases. This cleavage activates the host cell–binding function of the virion as well as the membrane fusion potential of E1. In the absence of p62 cleavage, virus particles are noninfectious. This feature was used for the construction of conditionally infectious particles (Salminen et al. 1992). In SFV, E3 remains part of the mature virion (Garoff et al. 1990), whereas it is shed from the spike in SIN (Welch and Sefton 1979).

Active alphaviral RNA replication triggers cell death in infected cells. The nonstructural region of the genome has been shown to be sufficient for the induction of apoptosis,

while the structural region can be replaced by a gene of interest (Urban et al. 2008). Although the precise mechanisms of cell death and virus persistence remain unclear, activation of double-stranded RNA-activated protein kinase R (PKR) has been proposed to contribute to blocking protein synthesis and induction of apoptosis in infected cells (Balachandran et al. 2000; Gorchakov et al. 2004; Venticinque and Meruelo 2010). Other factors, such as caspase cleavage (Nava et al. 1998), reduction in intracellular superoxide levels (Lin et al. 1999), bcl-2 downregulation (Scallan et al. 1997), and cyclin-dependent kinase 2 (CDK2) activation (Hu et al. 2009), may participate in alphavirus-mediated induction of apoptosis, suggesting that several pathways are involved in this potentially cell-specific process.

28.1.2 Structure of Alphavirus Vectors

The essential elements of the expression plasmids are shown in Figure 28.2. The expression system is based on the full-length cDNA clone of the corresponding alphavirus. The classical vectors were generated in such a way that the heterologous insert replaces the structural genes downstream of the 26S subgenomic promoter. Therefore, the vectors contain only the nonstructural coding region, which is required for the production of the nsP1–4 replicase complex, the 26S subgenomic promoter, and a multiple cloning site with several unique restriction sites for the foreign gene insertion. Because the RNA replication is dependent on short sequence elements located at the 5′- and 3′-ends of the genomic RNA (Kuhn et al. 1990), these regions are also included in the vector construct.

To generate infectious particles, the genes encoding structural proteins can be provided in trans. This is a central part of the alphavirus expression technology representing the packaging of recombinant RNAs into infectious particles using a helper construct encoding the viral structural genes. In this procedure, in vitro-made recombinant and helper RNAs are cotransfected into animal host cells (Figure 28.3). The recombinant RNA codes for the RNA replicase needed for the amplification of both incoming RNA species

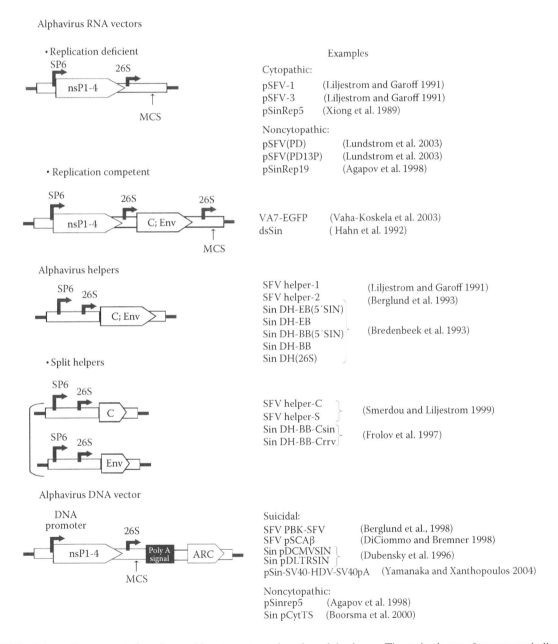

FIGURE 28.2 Schematic representation of recombinant constructs based on alphaviruses. The main classes of vectors are indicated: RNA vectors, including replication-deficient (with helper systems) and replication-competent vectors, and DNA/RNA layered vectors. Examples of each type of vector are indicated for Semliki Forest (SFV) and Sindbis (SIN) viruses. MCS, multiple cloning site; ARC, antibiotic resistance cassette.

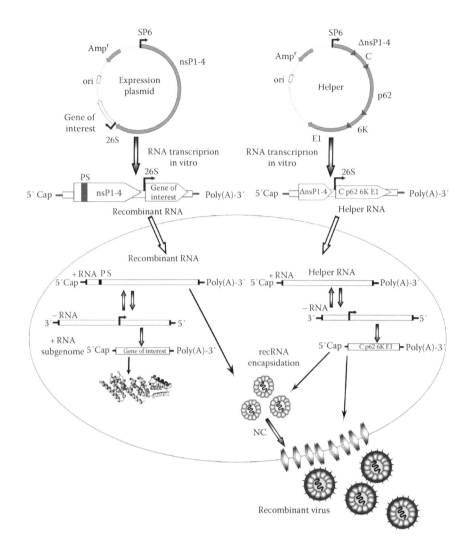

FIGURE 28.3 Recombinant alphavirus production. Two types of in vitro synthesized RNAs (recombinant RNA and helper RNA) are cotransfected into cells. Both RNAs are replicated by the alphavirus replicase complex (nsP1–4). The helper RNA provides the alphavirus structural proteins, which form recombinant virus particles with encapsidated (packaged) recombinant RNA. PS, packaging signal; NC, nucleocapsid.

and the gene of interest, whereas the helper RNA encodes the structural proteins for the assembly of new virus particles. The helper vector is constructed by deleting a large portion of the nonstructural genes, retaining the 5′ and 3′ end signals needed for RNA replication. Since almost the complete nsP region of the helper is deleted, RNA produced from this construct will not replicate in the cell, due to the lack of a functional replicase complex. When helper RNA is cotransfected with recombinant RNA, the helper construct provides the structural proteins in trans to assemble new virus particles, while the recombinant construct provides the nonstructural proteins for RNA replication of both recombinant and helper RNAs. The goal for this trans-complementation process is selective packaging of only recombinant RNAs into virus particles, because the helper vector lacks RNA packaging sequence signals recognized by the capsid protein. This packaging signal is located on the replicase coding sequence. The produced recombinant virus stock, therefore, contains only recombinant genomes, and when such virus particles are used

to infect animal host cells, no helper proteins are expressed, providing a one-step virus infection.

In order to reduce the chance of recombination and generation of replication-competent virus, a packaging system in which capsid and envelope genes are produced from separate vectors has been developed (Smerdou and Liljeström 1999). Since the capsid gene contains a translational enhancer (Sjoberg et al. 1994), this sequence was inserted in front of the spike sequence p62-6K-E1. On the other hand, to provide cotranslational removal of the enhancer sequence and normal biosynthesis of the spike complex, a sequence coding for the foot-and-mouth disease virus 2A autoprotease was inserted in frame between the capsid translational enhancer and the spike genes. Cotransfection of cells with both helper RNAs (SFV-helper-C and SFV-helper-S) and the SFV vector replicon carrying a foreign gene led to the production of recombinant particles with high titers (up to 8×10^8 particles per 10^6 cells). An empirical frequency of recombination and replication-competent virus appearance in this system would be very low (10^{-13}), emphasizing the high biosafety of the system based on two-helper RNAs. A similar strategy of the two-helper RNA system was developed on the basis of the Sindbis virus replicon (Frolov et al. 1997). In this system, the Sindbis spike genes were fused to the capsid gene of Ross River virus containing deletions in the RNA-binding domain, which maintained both the translation-enhancing and the self-cleaving activities. The same bipartite helper packaging system has also been described for VEE (Pushko et al. 1997), but in this case, the spike proteins were expressed without the capsid translation enhancer, which apparently is not needed in the VEE context.

In contrast to replication-deficient particles providing one round of infection, production of replication-competent virus can be achieved by cell transfection with a single construct (Hahn et al. 1992; Vaha-Koskela et al. 2003). These vectors may contain two 26S promoters, leading to the synthesis of two subgenomic mRNAs: one is responsible for the expression of the heterologous product, and the other for the synthesis of virus structural proteins (Figure 28.2, replication-competent vectors). Alternatively, the gene of interest can be inserted into nonstructural polyprotein genes and expressed as a fusion protein (Atasheva et al. 2007), or it can be expressed as a cleavable part of a structural or nonstructural polyprotein (Thomas et al. 2003; Tamberg et al. 2007). These vectors are self-replicating, produce infectious virus particles, and can spread from cell to cell in a manner similar to that of the parental virus. The obvious advantage of replication-competent vectors is the increased efficacy of in vivo gene delivery, which should allow for spreading in infected tissue and, therefore, enhancing the therapeutic efficacy. However, the disadvantage is the safety concern related to potential uncontrolled spread of infectious particles.

Another type of alphaviral vectors commonly used in vaccine development is based on DNA vectors. To allow direct application of plasmid DNA, the SP6 RNA polymerase promoter has been replaced by a DNA promoter (e.g., CMV IE, RSV LTR). In this case, transient transfection of plasmid DNA will result in expression of the gene of interest. DNA vectors could also contain a selection marker for stable transfection. Moreover, it is possible to cotransfect the DNA vector with a DNA-based helper vector, and to obtain recombinant particles or to use helper-producing cell lines. However, the virus titer has generally been significantly lower in this system than in the case of RNA-based particle production (Diciommo and Bremner 1998).

28.1.3 Gene Delivery by Alphavirus Vectors

Among recombinant viruses, alphaviral vectors are good candidates for cancer gene therapy due to their ability to mediate strong cytotoxic effects through the induction of p53-independent apoptosis, their ability to efficiently overcome immunological tolerance by the activation of innate antiviral pathways, and the subsequent triggering of cytotoxic T-lymphocyte responses against tumors (Lundstrom 2009; Quetglas et al. 2010; Osada et al. 2012). The advantages of alphaviral vectors also include a low specific immune

response against the vector itself and the absence of vector preimmunity in the major part of the population.

Alphaviruses have been widely accepted to display a broad tissue tropism and can efficiently infect and induce apoptosis in many types of cancer cells (Wahlfors et al. 2000; Rheme et al. 2005). In practice, the infectivity and cytotoxic properties of alphaviruses can vary significantly between different types of cancer cells. A low level of IFN-β production in some tumor cells has been shown to determine the susceptibility and oncolysis of tumors to SIN virus (Huang et al. 2012). Whether this correlation exists for other alphaviruses remains unknown. Although SFV, SIN, and VEE are closely related, substantial differences may exist in their tissue tropism, vector infectivity, and cytotoxicity (Wahlfors et al. 2000).

The mechanism of virus infection and possible regulation of entry, which is important for tumor targeting, is currently controversial. Different proteins have been proposed as candidate receptors for alphavirus infection (for review, see Kononchik et al. 2011; Leung et al. 2011). However, because alphaviruses infect genetically divergent cells, they probably utilize multiple proteins as receptors or alternative entry pathways in different cells. Thus far, only SIN has been considered to be capable of targeting tumors upon systemic injection into mouse models (Meruelo 2004; Tseng et al. 2004b, 2010; Unno et al. 2005). The mediastinal lymph nodes (MLNs) were shown as a site of early transient heterologous protein expression after intraperitoneal injection of SIN vectors, providing the generation of effector and memory CD8+ T cells against expressed tumor-associated antigen (Granot et al. 2014). In contrast, SFV vectors have been applied in most studies through intratumoral administration (Murphy et al. 2000; Maatta et al. 2007; Rodriguez-Madoz et al. 2007; Quetglas et al. 2012b). We have recently observed that SFV can target tumors upon systemic injection at a reduced viral dose, demonstrating that tumor-targeted delivery of the vector may be possible under certain conditions (Vasilevska et al. 2012).

Despite rapid induction of apoptosis in infected cells, treatment with natural oncolytic alphaviral vectors did not result in complete tumor regression (Chikkanna-Gowda et al. 2005; Smyth et al. 2005). The administration of immunomodulator genes, such as cytokines or growth factors, was more efficient and led to successful tumor inhibition or complete regression in animal models (Asselin-Paturel et al. 1999; Rodriguez-Madoz et al. 2005; Lyons et al. 2007). The use of replication-competent viruses is limited by safety restrictions, and most studies in recent years have focused on the use of suicidal replicons in the form of recombinant particles and DNA- or RNA-based vectors.

A promising new approach in vaccine development has been the use of plasmid DNA for immunization. In difference to conventional plasmid DNA, protein expression directed by alphaviral DNA vectors has a suicidal effect due to their strong cytopathic effect. Therefore, the use of these vectors eliminates the undesirable consequences of DNA integration into the host genome. Taking into consideration that the SFV replicon vector is based on in vitro transcribed RNA, the corresponding RNA (naked or encapsulated into transfection vesicles) can also be used as an alternative vector for cancer therapy. Self-replicating RNA can provide the same efficient cytoplasmic transgene expression and induction of apoptosis in host cells. It is safe for in vivo applications and does not induce antivector immunity (Cheng et al. 2001b; Vignuzzi et al. 2001; Saxena et al. 2009). Intramuscular injection of as little as 1 μg of naked SFV RNA provided complete tumor protection and extended the survival of treated mice when tumor cells were injected 2 days before immunization (Ying et al. 1999).

The issue of transgene distribution and persistence in vivo is of special importance when applications in gene and cancer therapy are considered. Therefore, in order to evaluate persistence and distribution of recombinant SFV particles, SFV replicon–based DNA plasmid and conventional DNA plasmid were compared after intramuscular injection mouse and chicken model (Morris-Downes et al. 2001). The presence of transgene was detected by RT-PCR. Recombinant SFV particles persisted for 7 days at the injection site, while SFV replicon–based plasmid and conventional DNA plasmid could be detected up to 93 and 246 days, respectively, at the injection site. In chickens, transgene could be detected up to 1 day at the injection site in case of recombinant SFV particles and up to 17 days for the SFV-based plasmid and 25 days for the conventional DNA plasmid. In mice lymph nodes, the recombinant SFV particles were detectable for 1 day, and both plasmids for 3 months. Similarly, the two plasmids were present up to 3 months in tissues distal from the site of injection, indicating dissemination.

Localization and persistence of replicon RNA is also dependent on the route of administration (Colmenero et al. 2001). Intravenous administration resulted in a systemic distribution, and the reporter gene was detectable in spleen and lymph nodes as well as in nonlymphoid tissues. Subcutaneous injection leads to a local distribution in the draining lymph nodes and skin surrounding the injection site, while intramuscular injection resulted in expression in local lymph nodes and at the injection site. This study confirmed the transient nature of SFV particles in vivo, since the reporter gene was almost undetectable by day 6 after injection by all examined administration routes. Intratumoral injection of SFV leads to localization of SFV-RNA in tumor cells and draining lymph node only (Colmenero et al. 2002). The short persistence renders the recombinant replicon particles as a safe vaccine tool, but not relevant for applications where prolonged gene expression in vivo is desired.

28.2 EFFICACY OF THERAPEUTIC AND PROPHYLACTIC ANTICANCER VACCINES BASED ON ALPHAVIRUS VECTORS

Mainly SFV, SIN, and VEE vectors have been tested for cancer treatment in animal models. Here we provide a detailed overview on different strategies of cancer treatment using immunogene delivery as naked RNA, plasmid DNA, and virus particles by replication-deficient vectors. Moreover, the possibility to employ oncolytic replication-competent alphaviruses is discussed.

28.2.1 MELANOMA

Melanoma is one of the most aggressive forms of skin cancer. Melanoma tumors arise from melanocytes and contain specific tumor-associated antigens (TAAs), which can be categorized as differentiation antigens such as Pmel17/gp100, p75/tyrosinase-related protein TRP-1, MART-1/Melan-A, and the retained intron in tyrosinase-related protein (TRP-2-INT2) as well as TAAs like MAGE or melanoma cell adhesion molecule (MUC18) (Pleshkan et al. 2011). These antigens act as ideal targets for melanoma immunotherapy because of their preferential expression in melanocytes and melanoma cells. Alphavirus vectors have demonstrated advances in targeted prophylactic and therapeutic immunotherapy of melanoma in several preclinical studies.

Tyrosinase is an essential enzyme involved in the initial stages of melanin biosynthesis in melanocytes and melanoma cells (Kumar et al. 2011). The ability to stimulate or augment an immune response against melanoma by alphaviral antitumor vaccines expressing either murine or human tyrosinase-related protein 1 (TRP-1) has been shown in two different studies. The efficacy of DNA-based SIN tumor vaccines pSIN-mTRP-1 and SIN-hTRP-1 was evaluated in a B16 mouse melanoma model (Leitner et al. 2003). It was one of the first demonstrations of prophylactic immunization with an alphavirus DNA vector where intramuscular injection was able to break immunological tolerance and provide immunity against melanoma, when inoculated 5 days prior to cancer cells challenge. Similarly, the high prophylactic potential of alphavirus-based vaccines was confirmed by demonstrating the ability of VEE virus-like particle (VLP) vectors encoding murine or human TRP-1 to induce strong immune responses and to provide a significant tumor growth delay in immunocompetent melanoma tumor–bearing mice (Goldberg et al. 2005).

In another study, the transmembrane melanosomal glycoprotein TRP-2 was applied as a therapeutic gene in controlling of melanoma growth (Avogadri et al. 2010). It has been shown that VEE-TRP-2 VLPs induced time-dependent tumor protection when vaccination was started as late as 5 days after tumor inoculation. Importantly, vaccination with VEE-TRP-2 was more effective than a combination of VEE-gp100 with VEE-tyrosinase vectors. Moreover, the efficacy of the combination of all three VEE vectors was not significantly better than VEE-TRP-2 alone.

The SIN DNA vector (SINCp) was used to express the murine melanoma cell adhesion molecule (MCAM/MUC18) for vaccination against murine melanoma (Leslie et al. 2007). MUC18 is expressed in late primary and metastatic melanoma, but hardly at all in healthy melanocytes. Immunization with this vector showed no antitumor effect against parental B16F10 cells, probably because of extremely low expression of murine MUC18 in those melanoma cells. To increase this antigen expression in tumor cells, new B16F10 cells transduced with MUC18 were obtained. Vaccination against MUC18 resulted in the induction of humoral and CD8+ T-cell immune responses against melanoma. In order to investigate the efficacy of recombinant alphavirus-based vaccines for the stimulation of human immune responses, SFV VLPs encoding MAGE-3 were administered in humanized BALB/c mice (Trimera murine model) (Ni et al. 2005). The results showed that the SFV vector elicited human MAGE–specific antibody and CTL responses in Trimera mice.

Several studies in melanoma therapy development have shown that expression of cytokine genes in tumor cells generally resulted in dramatic alteration of tumor cell growth and induction of tumor-specific immunity. For instance, a single intratumoral injection of recombinant SFV particles expressing IL-12 caused significant inhibition of melanoma growth in tumor-bearing mice (Asselin-Paturel et al. 1999). Intratumoral administration of SFV-IL12 led to dramatic tumor necrosis in all treated mice, resulting in 70%–90% tumor growth inhibition. However, complete tumor regression was not achieved in this study. To improve the efficacy of antimelanoma therapy, a combined strategy including administration of SFV-IL-12 VLPs and systemic costimulation with agonist anti-CD137 monoclonal antibodies has been explored (Quetglas et al. 2012a), showing the powerful synergistic effects. Briefly, immune system stimulation with agonist agents acting on CD137 expressed on primed T cells resulted in the enhancement of tumor-eradicating cytotoxic T-cell responses (Melero et al. 1997). In contrast to suboptimal therapeutic effect provided by intratumoral injection of SFV-IL-12 VLPs alone, combined administration of both SFV-IL-12 and CD137 mAb dramatically increased the therapeutic efficacy, inducing 50% and 75% of complete tumor remission, respectively.

The oncolytic potential of a replication-competent alphavirus vector was also investigated for the treatment of melanoma. In this context, the avirulent SFV strain A7 expressing EGFP was applied intravenously, intraperitoneally, and intratumorally as a therapeutic vaccine in human melanoma–bearing SCID xenografts (Vaha-Koskela et al. 2006). A single inoculation of the VA7 vector resulted in significant tumor regression, irrespective of the route of administration. The neurotropism of SFV did not restrict its ability to target tumors, as within 3 weeks, VA7 had caused regression of tumors to far below the starting volume. Despite the positive treatment dynamic, small isolated groups of dividing tumor cells were detected within strands of connective tissue, indicating the potential tumor remission in the future.

28.2.2 Breast Cancer

Worldwide, breast cancer is the most common cancer among women. The HER2/*neu* is a member of the tyrosine kinase receptor family overexpressed in 30%–40% of breast cancers, correlating with increased metastasis and poor prognosis, due to the increase of mitotic activity, mutation of the p53 gene, negative estrogen receptor status, and absence of bcl-2 (Banin Hirata et al. 2014). The HER2/*neu* has a high potential as tumor antigen in breast cancer therapy. A DNA-based SIN vector ELVIS-*neu* expressing the *neu* gene was used for intramuscular vaccination 14 days before injection of cancer cells overexpressing *neu* (Lachman et al. 2001). The results showed a strong protection of mice against tumor

development. Vaccination led to reduction of the incidence of lung metastasis from mammary fat pad tumors and also reduced the number of lung metastases resulting from intravenous injection of *neu*-overexpressing cells. Interestingly, intradermal vaccination also provided protection and required 80% less plasmid for a similar level of protection. The beneficial results of cancer vaccines based on the SIN *neu* vector to treat pre-existing tumors were also confirmed (Wang et al. 2005). It was shown that therapeutic efficacy of the pSINCP/*neu* vaccine depended on the order of vector and cancer cell injection, indicating that the prophylactic vaccine was effective only when administered before tumor challenge.

The therapeutic potential of immunotherapy with the pSINCP/*neu* DNA vaccine and the VEE/*neu* VLPs was enhanced by combination with the chemical anticancer agents doxorubicin and paclitaxel (Eralp et al. 2004). Administration of 5 mg/kg doxorubicin prior to pSINCP/*neu* DNA and VEE/*neu* VLPs vaccination led to a significant delay in tumor progression. Despite doxorubicin being established as a standard adjuvant therapy for breast cancer, mice receiving chemotherapy alone did not demonstrate reduced tumor growth. Interestingly, but in contrast to the previous results, combined treatment with paclitaxel (25 mg/kg) increased the effectiveness of only the VEE/*neu* VLP vaccine.

In another study, 36% of rat breast tumors were eliminated when a VEE/*neu* VLP vaccine was subcutaneously administered to animals with aggressive preexisting mammary tumors (Laust et al. 2007). When DC-based cancer immunotherapy was combined with VEE/*neu* VLP administration, induction of both cellular and humoral immunity against *neu* was observed in transgenic human breast tumor–bearing mice (Moran et al. 2007). Furthermore, the combination treatment led to significant inhibition of tumor growth. Not only tumor antigens have been used for immune system stimulation. Cytokines such as IL-12 have potential as cancer therapy agents because of their antitumor and antimetastatic activities. For instance, enhanced IL-12 expression has been established from SFV10-E VLPs as a potential treatment for breast cancer (Chikkanna-Gowda et al. 2005). The enhanced SFV10-E vector has shown up to 10 times higher expression levels of foreign genes as compared to the original SFV10 vector. Intratumoral administration of high titer SFV-E-IL-12 VLPs caused complete tumor regression in four out of six mice and noticeably reduced the amount of lung metastases.

The vascular endothelial growth factor receptor-2 (VEGFR-2) serves also as an attractive therapeutic target, because it is required for neovascularization within tumors and has been shown to be important for tumor growth, invasion, and metastasis. The requirements for therapeutic efficacy of VEGFR-2-expressing vectors are associated with the induction of an antibody response against VEGFR-2. It was demonstrated that both tumor growth and pulmonary metastatic spread were significantly inhibited in mice with preexisting tumors when subjected to five immunizations with SFV10-E VLP expressing VEGFR-2 (Lyons et al. 2007). Moreover, a significant tumor regression was observed after coimmunization of mice with SFV particles encoding VEGFR-2 and IL-4.

28.2.3 Lung Cancer

Lung cancer is the leading cause of cancer death in the industrialized world. The majority of patients are diagnosed at a locally advanced or metastatic stage, making systemic therapies the mainstay for treatment (Blanchon et al. 2006). In case of lung cancer, promising results were obtained only using direct intratumoral alphaviral vector administration. The first preclinical study using alphaviral vectors in lung cancer treatment was performed by direct intratumoral injections of SFV VLPs expressing EGFP into human non–small lung cancer xenografts (Murphy et al. 2000). The outcome was tumor growth inhibition, and in some cases, complete tumor regression was achieved. The effect was mediated by p53-independent apoptosis and necrosis, but required repeated intratumoral administration (three to six injections) and very high doses of the vector (1×10^{10} IU/mL).

In another lung tumor study, the synergistic effect of combined therapy with SFV-IL-12 VLP and anti-CD137 monoclonal antibodies was demonstrated (Quetglas et al. 2012a). Similar to B16 mouse melanoma, syngeneic TC-1 lung carcinoma was inhibited by a clinically feasible therapeutic combination involving intratumoral treatment with SFV-IL-12 and systemic costimulation with anti-CD137 monoclonal antibodies.

Oncolytic virotherapy with the attenuated replication-competent SFV vector VA7-EGFP showed a good safety profile and resulted in almost complete inhibition of tumor growth in human lung adenocarcinoma NMRI nu/nu mouse models upon intratumoral administration (Maatta et al. 2007). In contrast, systemic administration resulted in only delayed tumor growth (intravenous injection) or total absence of response (intraperitoneal injection).

28.2.4 Colon Cancer

Colon cancer is the second most common cancer in the European Union, and at least 50% of patients develop recurrences or metastases during their illness due to aggressive behavior of this cancer type (Kuipers et al. 2013). Several preclinical studies have demonstrated high potential of alphavirus vectors in the development of colorectal cancer therapy. For instance, an alphaviral DNA vector encoding LacZ (pSIN1.5-β-gal) was compared to a conventional CMV promoter–based β-gal plasmid in CT26.CL25 tumors (Leitner et al. 2000). It was shown that intramuscular immunization with plasmid DNA replicons of mice with preexisting tumors elicited immune responses at doses 100–1000-fold lower than when performed with conventional DNA plasmids. Mice bearing experimental tumors expressing the β-gal reporter antigen were effectively treated and resulted in significant prolongation of survival rates.

Several alphavirus vectors have been engineered to express different cytokines to enhance antitumoral immune responses against colon cancer. A single intratumoral injection of 10^7 and 10^8 VLP of SFV-IL-12 resulted in a complete tumor regression in 36% and 80%, respectively, in a

mouse colon adenocarcinoma model (Rodriguez-Madoz et al. 2005). Moreover, application of the modified SFV vectors with a natural capsid translation enhancer significantly increased IL-12 expression and tumor regression in treated mice. Six doses of high titer SFV10-E VLPs expressing IL-12 induced complete regression in all colon carcinoma tumor–bearing mice (CT26 model). During the treatment stage, tumor swelling occurred in relation to intratumoral necrosis and inflammation (Chikkanna-Gowda et al. 2005). In a similar study, colon carcinoma tumor–bearing mice were treated with six inoculations of high titer SFV10-E VLP expressing the murine IL-18 gene along with an Ig-kappa leader sequence (Chikkanna-Gowda et al. 2006). Although the growth of treated tumors was delayed, complete tumor regression was achieved only in 33% of treated mice, where the induction of avascular and suppurative necrosis was observed.

In order to investigate the potential of angiogenesis inhibition in primary colon carcinoma, mice were immunized with SFV10-E VLP expressing VEGFR-2 10 days prior to tumor cell injection (Lyons et al. 2007). Similar to 4T1 mouse breast carcinoma, the growth of CT26 colon carcinoma was inhibited in vaccinated mice. Microvessel density analysis showed that immunization with SFV-VEGFR-2 VLPs led to a significant inhibition of tumor angiogenesis. Moreover, coimmunization of mice with SFV VLPs encoding VEGFR-2 and IL-4 led to enhancement of mice survival and production of high titers of anti-VEGFR-2 antibodies in contrast to coimmunization with VEGFR-2 and IL-12, or VEGFR-2 alone.

The therapeutic potential of alphaviruses as oncolytic agents has also been studied. The virulent SFV4 strain and its derivative recombinant SFV-p62-6k vector, containing deletions of the capsid and E1 genes, were used to stimulate immunity in a CT26 model (Smyth et al. 2005). Direct intratumoral injection of replication-deficient VLPs or virulent SFV4 resulted in an immediate and intense inflammatory reaction and significant effect on survival. No differences were observed in inhibition of tumor growth between VLP- and SFV4-treated animals. However, the antitumor effect could be enhanced by preimmunization of animals with the VLP vector.

28.2.5 Ovarian Cancer

Ovarian cancer is the sixth most common malignancy in women. In advanced ovarian cancers, tumors spread throughout the peritoneal cavity and induce the production of ascites. Therefore, the high potential of systemic tumor targeting by SIN vectors is an important factor in treatment of this type of cancer. Efficacy of SIN vector application in ovarian cancer treatment has been evaluated using several strategies, such as immunotherapy, oncolytic virotherapy, and combined therapy with chemical agents. It has been shown that SIN VLP vectors have the ability to systemically and specifically target metastasized tumors within the peritoneal cavity, leading to significant suppression of tumor growth in ovarian tumor–bearing xenograft models (Tseng et al. 2004b). However, incorporation of antitumor cytokine genes such as IL-12 and IL-15 genes significantly enhanced the efficacy of the vector (Tseng et al. 2004a; Granot et al. 2011). Additionally, using C.B-17-SCID beige mice with selective impairment of natural killer (NK) cell functions, and C.B-17-SCID ovarian tumor–bearing mice, it was demonstrated that anticancer efficacy of SIN vectors is largely NK cell dependent and depletion of these cells caused a significant decrease in the therapeutic potential (Granot et al. 2011; Granot and Meruelo 2012). Because of low efficacy of penetration in tumor vascular structures, the SIN vectors were not able to reach and kill all tumor cells to ensure complete tumor regression. To solve this problem, a chemotherapeutic drug was used. Due to the ability of paclitaxel to inhibit tumor angiogenesis at low concentration and increase blood vessel permeability, the combined treatment with paclitaxel (taxol) at a concentration of 16 mg/kg and intraperitoneally inoculated SIN-LacZ VLPs dramatically enhanced therapeutic effects (Tseng et al. 2010).

SFV replicon vectors were also utilized for ovarian cancer immunotherapy with granulocyte-macrophage colony-stimulating factor (GM-CSF), which is an important hematopoietic growth factor and immune modulator (Klimp et al. 2001). It was shown that intraperitoneal injection of SFV-GM-CSF VLPs in murine ovarian tumor models provided rise in the number of macrophages and neutrophils. It resulted in a modest tumor growth inhibition, but with no survival benefit.

The potential of the oncolytic SIN AR339 strain was evaluated for ovarian cancer treatment (Unno et al. 2005). Although the SIN AR339 is a replication-competent vector, this strain has not been reported to cause any serious human disease. Intraperitoneal vector administration in the human ovarian xenograft mouse model provided significant suppression of ascite formation, an important therapeutic outcome in the treatment of ovarian cancer. In another study, mouse ovarian carcinoma xenografts were treated with an oncolytic SFV vector, using a readministration strategy with the same or another (vaccinia) viral vector (Zhang et al. 2010). In contrast to reinoculations of the same virus, the heterologous vector administration led to remarkably increased oncolysis and generation of antitumor immunity that significantly prolonged the survival.

28.2.6 Cervical Cancer

Human papilloma virus (HPV) E6 and E7 oncogenes are promising targets for cervical cancer vaccine development. SFV VLPs expressing the HPV16 E6 and E7 as separate proteins were applied as prophylactic vaccines in a mouse model for cervical cancer based on TC-1 cells expressing HPV16 E6E7 (Daemen et al. 2000). Preimmunization with three injections of 10^4 pSFV-E6E7 VLPs induced HPV-specific CTL response in 50% of the mice, whereas three inoculation with an increased virus dose of 10^6 resulted in CTL response in all treated mice. Furthermore, immunization with the highest dose of 5×10^6 of SFV-E6E7 VLPs protected 40% of

the mice from tumor challenges. To enhance oncogene production and improve the cellular immune responses against E6 and E7, a new vector encoding a fusion protein of E6 and E7 together with the SFV core translational enhancer (pSFV3enh-E6,7) was generated (Daemen et al. 2002). Immunizations with 5×10^6 SFV3-enhE6,7 VLPs protected four out of five mice from tumor development, and a second tumor challenge in tumor-free animals revealed complete long-term protection against tumor occurrence. In a further study, authors confirmed the high potential of the proposed SFV-enhE6,7 VLP vaccine by intravenous and intramuscular administrations (Daemen et al. 2004).

In addition to the promising prophylactic properties of the SFV3-enhE6,7 VLP vector, its therapeutic potential in tumor-bearing mice was also investigated (Daemen et al. 2003). Subcutaneous injections of the vector at the 2nd, 7th, and 14th day after TC-1 cell challenge resulted in rapid CTL response induction and efficient protection against fast-growing tumors (90%–100% of treated mice were protected even after a second tumor challenge). Importantly, the efficacy of using adenovirus Ad-E6,7 VLPs was dramatically lower (20%–40% of treated mice were protected) (Riezebos-Brilman et al. 2007). To improve the therapeutic efficacy, coadministration of SFV3-enhE6,7 VLPs and different doses of SFV-IL-12 VLPs was examined (Riezebos-Brilman et al. 2009). The results of coinoculation of both vectors depended significantly on the viral dose and injection schedule. Synergistic antitumor activity was observed only at a low dose of SFV-IL12. Furthermore, heterologous prime-boost immunization strategy was shown to provide advantages over single immunization (Walczak et al. 2011). Heterologous prime boost with SFV3-enhE6,7 VLPs and virosomes containing the E7 protein resulted in higher numbers of antigen-specific CTL in mice than applying homologous protocols. Nevertheless, the high number of CTL initially primed by the heterologous protocols did not correlate with enhanced antitumor responses in vivo.

SIN self-replicating RNA vectors were developed to induce E7-specific immunity in a TC-1 mice model (Cheng et al. 2001a). Intramuscular inoculation of RNA, encoding HPV E7 oncogene alone (SINrep5-E7), induced poor humoral and cellular immune responses and provided no protection against tumor challenge. However, another construct expressing E7 as a fusion with secretory Sig protein and lysosome-associated membrane protein-1 (LAMP-1) in the SINrep5-Sig/E7/LAMP-1 vector demonstrated E7-specific CD4+ helper T cells and CD8+ cytotoxic T cell activity and increased in vivo antitumor effect. Addition of the LAMP-1 endosomal/lysosomal sorting signal to the E7 protein significantly enhanced the oncogene processing and presentation in vivo in the case of uptake of apoptotic cells by APCs at sites of vector RNA inoculation.

The efficacy of prophylactic and therapeutic vaccines based on SIN VLPs expressing both E7 and calreticulin (CRT), an ER Ca2+-binding transporter participating in antigen processing and presentation with major histocompatibility complex (MHC) class I, was tested (Cheng et al. 2006).

The developed SINrep5-CRT/E7 VLP vector was able to generate antigen-specific immune responses, antiangiogenic effect, and a strong antitumor activity. Intramuscular vaccination with SINrep5-CRT/E7 VLPs 1 week prior to the challenge with TC-1 cells provided excellent protection of all treated animals. To determine the therapeutic potential of the vector in established tumors, both immunocompetent and nude mice were intramuscularly inoculated with SINrep5-CRT/E7 VLPs 2 days after tumor cell injection. Although this strategy did not provide complete tumor elimination as a prophylactic approach, it resulted in a significantly lower number of pulmonary tumor nodules in both mice groups.

Similar to other alphaviral vectors, VEE VLP-based vectors expressing E7 provided satisfactory results when used as a prophylactic vaccine. It was shown that two subcutaneous preimmunizations with VEE-E7 VLPs 2 weeks prior to cancer cell injection prevented tumor formation in mice (Velders et al. 2001). Moreover, mice challenged 3 months after immunization with cancer cells did not develop tumors, indicating induction of long-term memory responses by the vector. In contrast, the therapeutic approach was efficient only in 67% of treated tumor-bearing mice. In other studies, the efficacy of the VEE vector was increased by expression of both E6 and E7 oncogenes from the same vector (Eiben et al. 2002; Cassetti et al. 2004). To test the HLA-restricted capabilities of the vaccine, an HPV tumor model was established on the basis of HLA-A*0201 transgenic mice. In this case, preimmunization with VEE-E6E7 VLPs protected 100% of immune-competent and HLA-A*0201 transgenic mice from tumor development and induced specific T cell immune response against HLA-A*0201-restricted HPV16 epitopes. As a therapeutic vaccine, VEE VLPs were inoculated after a tumor challenge. Although the therapy did not completely eradicate tumors, approximately 90% of immune competent and transgenic mice demonstrated elimination of established tumors.

The anticancer potential of the oncolytic replication-competent SIN virus AR339 strain was also explored using different human cervical cancer xenografts in mice (C33A) (Unno et al. 2005). Therapeutic treatment by intratumoral or intravenous injection of the vector resulted in remarkable regression of tumor growth through induction of necrosis.

28.2.7 Prostate Cancer

Alphavirus-based gene therapy represents an attractive strategy for noninvasive treatment of prostate cancer, where current clinical interventions show limited efficacy. The first promising results in preclinical trials were obtained using apoptosis-resistant tumor models. Immunodeficient mouse models with established rat prostate tumors, overexpressing the *Bcl-2* oncogene, were treated by intratumoral injections of SFV VLPs encoding the proapoptotic gene *Bax*, which plays a key role in programmed cell death (Murphy et al. 2001). Expression of the *Bax* gene by the SFV1 vector enhanced its cytopathic potential and led to

a remarkable 47% reduction in tumor volume compared to the control. However, complete regression was not achieved in this study.

One alternative strategy for prostate cancer treatment is immune system stimulation against specific prostate cancer antigens, like the prostate-specific membrane antigen (PSMA), the six transmembrane epithelial antigen of the prostate (STEAP), and the prostate stem cell antigen (PSCA) (Naz and Shiley 2012). PSMA is a highly restricted prostate cell surface antigen. The VEE VLP vector, producing human PSMA, has demonstrated strong cellular and humoral immunities in mice upon subcutaneous inoculation (Durso et al. 2007). Although additional preclinical studies were not conducted, due to the absence of relevant PSMA tumor challenge models, the efficacy of VEE-PSMA VLP was studied in clinical trials (see Section 28.3).

The STEAP antigen is also an attractive target for immunotherapy, because it is predominantly expressed in prostate tissue, and is found to be upregulated in multiple cancer cell lines (Gomes et al. 2012). The potential of VEE VLPs expressing mouse STEAP was assessed in the context of prophylactic and therapeutic approaches (Garcia-Hernandez et al. 2007). Mice preimmunized with VEE-STEAP VLPs showed a specific induced immune response and significantly prolonged overall survival of TRAMPC-2 prostate tumor–bearing mice. The therapeutic effect of the VEE vector was tested by coadministration with the STEAP plasmid DNA vaccine, demonstrating short, but statistically significant delay in tumor growth. More beneficial results were obtained using a PSCA as a target, which *is* upregulated in a large proportion of localized and metastatic prostate cancers. Prophylactic vaccination of transgenic (TRAMP) mice with the PSCA-cDNA plasmid followed by VEE-PSCA VLP inoculation generated a specific immune response and antitumor protection in 90% of TRAMP mice.

28.2.8 Brain Cancer

In preclinical studies, intracranial injection of B16 mouse melanoma cells was applied to generate a mouse brain tumor model. This model was used to investigate the therapeutic potential of an SFV VLP vaccine–encoding mouse endostatin—a protein, possessing antiangiogenic properties (Yamanaka et al. 2001a). A significant reduction of intratumoral vascularization was observed in tumor sections after intratumoral injection of SFV-endostatin VLPs.

Another strategy of brain tumor therapy includes the immunization of mice with DCs isolated from bone marrow and transduced with SFV VLPs expressing cytokines or specific cDNAs from melanoma or glioma cells (Yamanaka et al. 2001b). It was shown that prevaccination with DCs transduced by the same type of cDNA as the tumor (SFV-mediated B16 complementary cDNA or SFV-mediated 203 glioma cDNA vectors for B16 and 203 glioma tumors, respectively) provided protection from tumor challenge. Moreover, therapeutic vaccination of brain tumor–bearing mice prolonged the overall survival.

Therapeutic immunization with DCs that have been pulsed with SFV IL-12 also significantly prolonged survival of B16 brain tumor–bearing mice (Yamanaka et al. 2002). A similar survival rate has been detected after stimulation of the immune system with DCs transduced with SFV-IL-12 VLPs in combination with systemically administered IL-18 (Yamanaka et al. 2003). Interestingly, that combination of DCs pulsed with SFV-IL-12 and systemic inoculation of IL-18 has increased the survival rate.

Human melanoma–associated antigen gp100 is a melanocyte differentiation antigen, which has been also detected in multiple glioma cancer cell lines (Liu et al. 2004). It was shown that vaccination with a plasmid DNA-based SIN vector expressing human gp100 and murine IL-18 induced specific antitumor CTL immune responses and provided antitumor protection (Yamanaka and Xanthopoulos 2005). Three prophylactic immunizations with both pSIN-hgp100 and pSIN-IL-18 DNA resulted in prevention of the formation of B16-hgp100-transfected tumors. Therapeutic vaccination of mice with established B16-hgp100 tumors showed significant survival prolongation (90 days) with both vectors, where median mice survival treated with either pSIN-hgp100 or pSIN-IL-18 DNA was 24–28 days.

The antitumor capacity of the oncolytic replication-competent SFV VA7 vector was investigated in immunocompetent rat glioma tumor models (Maatta et al. 2007). Neither intravenous nor intraperitoneal administration provided any positive therapeutic efficacy in glioma-bearing rats. In contrast, direct intratumoral injections of SFV VA7 led to a significant reduction of tumor growth. However, these beneficial results were followed by accelerated increase in tumor mass, leading to eventual death of the animals. Despite the promising results of oncolytic virotherapy in other cancer types described earlier, the SFV VA7 vector demonstrated insufficient efficacy in brain tumors in immunocompetent mice. Nevertheless, systemic inoculation of the SFV VA7 vector in nude mice caused complete subcutaneous brain tumor eradication while leaving healthy brain tissue unharmed (Heikkila et al. 2010). Furthermore, improved long-term survival was observed in 16 of a total of 17 animals.

28.2.9 HEPATOCELLULAR CARCINOMA

Hepatocellular carcinoma (HCC) is a liver cancer that has limited therapeutic options. In preclinical studies, woodchucks chronically infected with woodchuck hepatitis virus serve as a model for liver cancer therapy development. The therapeutic potential of SFV-E-IL12 VLPs was evaluated in woodchucks with hepatic tumors (Rodriguez-Madoz et al. 2009). The results indicated that a single intratumoral injection of vector provided partial, dose-dependent tumor regression in 58% of treated animals, leading to reduction in tumor volume of up to 70% 4 weeks after treatment. The promising therapeutic results were associated with a general activation of cellular immune responses against HCC. Nevertheless, tumor growth was restored thereafter. In a recent study, an L-PK/c-myc transgenic mice model was applied, providing

spontaneous appearance of hepatic tumors with latency, histopathology, and genetic characteristics similar to human HCCs (Rodriguez-Madoz et al. 2014). Intratumoral inoculation of SFV-IL-12 induced growth arrest in most tumors, providing 100% survival rate.

28.2.10 OSTEOSARCOMA

Osteosarcoma is the most common primary malignant bone tumor, which typically metastasizes into bones, lungs, and other soft tissues. The oncolytic SFV vector VA7 was tested as a virotherapy candidate against unresectable osteosarcoma. Subcutaneous human osteosarcoma nude mice xenografts were treated by three intratumoral injections of SFV VA7-EGFP (Ketola et al. 2008). Treatment with the oncolytic SFV was highly efficient, showing significant reduction of tumor size in comparison with the oncolytic adenoviral Ad5Δ24 vector. Additionally, a highly aggressive orthotopic osteosarcoma nude mouse model characterized by invasion to surrounding tissues, and emergence of hematogenous pulmonary metastases, was treated with VA7-EGFP. Intratumoral inoculations of oncolytic SFV significantly enhanced the survival rate in the orthotopic osteosarcoma model. However, none of the mice were eventually cured.

28.3 CLINICAL TRIALS

Following beneficial results of cancer treatment in preclinical trials, some therapeutic strategies have been evaluated in humans. The first phase I/II clinical study was performed using the SFV vector expressing the human IL-12 gene and encapsulated in cationic liposomes (LSFV–IL12) (Ren et al. 2003). To assess the biosafety and optimal dosage of the vector, LSFV–IL12 was intravenously administered in cancer patients with stage III or IV metastasizing melanoma or renal cell carcinoma every third day for 4 weeks in two different concentrations. The therapy demonstrated no toxicity or any significant changes in the function of internal organs. However, therapeutic potential was indicated by a 10-fold increase in IL-12 concentration in the peripheral blood of treated patients. In another phase I/II study, repeated inoculations of VEE VLPs expressing the carcinoembryonic antigen (CEA) induced clinically relevant CEA-specific T cell and antibody responses due to the ability of alphaviruses to infect DCs (Morse et al. 2010). The study which included patients with advanced or metastatic cases of lung, colon, breast, appendix, or pancreatic cancers were pretreated with multiple courses of chemotherapy and received up to four injections of VEE-CEA VLPs. The majority of patients showed a dramatically low rate of clinical responses after the therapy. Regression of liver metastasis in one patient with pancreatic cancer was detected. Moreover, two patients with no evidence of disease remained in remission and two patients were able to maintain stable disease. One of the most recent alphavirus-based clinical trials targeted prostate cancer

TABLE 28.1

Prophylactic Anticancer Vaccines Based on Recombinant Alphavirus Vectors

Vector	Expressed Transgene	Tumor Model	Vaccination and Vector Administration Types	Vaccination Efficacy	References
SIN DNA/RNA layered vectors					
SIN DNA	TRP-1	Mouse melanoma B16	Prophylactic (i.m.)	Immunity Tumor prevention (60%–70%)	Leitner et al. (2003)
SIN DNA	MUC18	Mouse melanoma B16	Prophylactic (s.c.)	Ineffective	MC Leslie et al. (2007)
		Mouse melanoma B16 transduced with MUC18	Prophylactic (s.c.)	Immunity	
				Tumor prevention (50%)	
SIN DNA	HER/*neu*	Mouse breast cancer A2L2	Prophylactic (i.m.), (i.d.)	Tumor prevention (80%) Reduction of metastasis Increase of survival	Lachman et al. (2001)
SIN DNA	HER/*neu*	Mouse breast cancer A2L2	Prophylactic (i.m.)	Tumor prevention (50%) Partial tumor reduction Reduction of metastasis	Wang et al. (2005)
SIN DNA	LacZ	Mouse colon cancer CT26.CL25, β-gal-expressing clone	Prophylactic (i.m.)	Complete tumor prevention (100%)	Leitner et al. (2000)
SIN DNA + SIN DNA	Gp100	Mouse brain cancer model B16	Prophylactic (i.m.)	Tumor prevention (40%)	Yamanaka et al. (2005)
	IL-18			Increase of survival	
SFV DNA	P1A	Mouse mastocytoma P815	Prophylactic (i.m.)	Tumor prevention (60%–70%)	Ni et al. (2004)
SIN RNA	Sig/E7/LAMP-1	Mouse cervical cancer model TC1	Prophylactic (i.m.)	Immunity Reduction of metastasis	Cheng et al. (2001a)
SIN viral particles vectors					
SIN VLP	CRT/E7	Mouse cervical cancer model TC1	Prophylactic (i.m.)	Complete tumor prevention (100%) Immunity	Cheng et al. (2006)
SIN VLP	P1A	Mouse mastocytoma P815	Prophylactic (i.p.)	Tumor prevention (80%)	Ni et al. (2004)
SFV and SFV-E viral particles vectors					
SFV VLP	HPV16 E6, E7	Mouse cervical cancer model TC1	Prophylactic (s.c.), (i.p.)	Tumor prevention (40%)	Daemen et al. (2000)
SFV-E VLP	HPV16 E6, E7	Mouse cervical cancer model TC1	Prophylactic (s.c.), (i.p.)	Complete tumor prevention (100%)	Daemen et al. (2002)
SFV10-E VLP	VEGFR-2	Mouse breast cancer 4T1	Prophylactic (s.c.)	Partial tumor reduction Reduction of metastasis	Lyons et al. (2007)
SFV-E VLP	VEGFR-2	Mouse colon cancer CT26	Prophylactic (s.c.)	Partial tumor reduction Increase of survival Angiogenesis inhibition	Lyons et al. (2007)
SFV-E VLP + IL-12	VEGFR-2	Mouse colon cancer CT26	Prophylactic (s.c.)	Inefficient	Lyons et al. (2007)
SFV-E VLP + IL-4	VEGFR-2	Mouse colon cancer CT26	Prophylactic (s.c.)	Immunity Increase of survival	Lyons et al. (2007)
SFV-E VLP	P1A	Mouse mastocytoma P815	Prophylactic (i.v.)	Immunity Tumor prevention (90%)	Colmenero et al. (2002)
SFV VLP + Ad VLP	P1A	Mouse mastocytoma P1.HTR3	Prophylactic (i.v.)	Increase of survival	Näslund et al. (2007)
VEE viral particles vectors					
VEE VLP	TRP-1	Mouse melanoma B16	Prophylactic (s.c.)	Immunity Partial tumor reduction	Goldberg et al. (2005)
VEE VLP	Gp100	Mouse melanoma B16	Prophylactic (s.c.)	Ineffective	Avogadri, et al. (2010)
VEE VLP	Tyr	Mouse melanoma B16	Prophylactic (s.c.)	Ineffective	Avogadri et al. (2010)
VEE VLP	TPR-2	Mouse melanoma B16	Prophylactic (s.c.)	Partial tumor reduction	Avogadri et al. (2010)
VEE VLP	HER/*neu*	Mouse breast cancer A2L2	Prophylactic (s.c.)	Complete tumor prevention (100%) Complete metastasis prevention (100%)	Wang et al. (2005)

(Continued)

TABLE 28.1 (*Continued*)

Prophylactic Anticancer Vaccines Based on Recombinant Alphavirus Vectors

Vector	Expressed Transgene	Tumor Model	Vaccination and Vector Administration Types	Vaccination Efficacy	References
VEE VLP	E7	Mouse cervical cancer C3	Prophylactic (s.c.)	Complete tumor prevention (100%)	Velders et al. (2001)
VEE VLP	HPV16 E6, E7	Cervical cancer model HLF16	Prophylactic (s.c.)	Complete tumor prevention (100%)	Eiben et al. (2002)
VEE VLP	HPV16 E6, E7	Mouse cervical cancer C3, TC1	Prophylactic (s.c.)	Complete tumor prevention (100%)	Cassetti et al. (2004)
VEE VLP	STEAP	Mouse prostate cancer TRAMPC-2	Prophylactic (s.c.)	Immunity Increase of survival	Garcia-Hernandez et al. (2007)
VEE VLP + cDNA PSCA	PSCA	Mouse prostate cancer TRAMPC-2	Prophylactic (s.c.)	Tumor prevention (76%)	Garcia-Hernandez et al. (2008)
DCs + VEE VLP	HER/*neu*	Human breast cancer NT2	Prophylactic (s.c.)	Immunity Partial tumor reduction	Moran et al. (2007)

(Slovin et al. 2013). The immunotherapeutic efficacy of VEE VLPs carrying PSMA was evaluated for patients with castration-resistant metastatic prostate cancer. The patients received a maximum of five subcutaneous injections in the deltoid region at two different vector concentrations. The vaccination did not cause adverse systemic or local toxicity and was generally well tolerated. However, the therapeutic effect of both immunization strategies was very low. No cellular immune response to PSMA was observed, and only a small number of patients demonstrated a humoral response to PSMA.

28.4 CONCLUDING REMARKS

The most interesting and promising applications of alphaviruses for cancer treatment are summarized in Tables 28.1 through 28.3. Alphavirus-based delivery platforms have numerous advantages, which render them attractive tools for immunotherapeutic vaccine and cancer therapy.

Safety: When the suicidal replication-deficient alphavirus particles are used, viral structural genes are not present and the infectious virus capable of infecting new target cells cannot generate virus progeny in immunized host cells. Particularly, the use of the second-generation helper vector (Berglund et al. 1993) and the split helper system (Smerdou and Liljeström 1999) prevents homologous recombination events and generation of replication-competent virus progeny. Moreover, RNA replicon–based vaccines are not prone to random integration into the host genome, thus avoiding the risk of cell transformation and development of tolerance or anti-DNA antibodies due to persistence, which present a limitation for the conventional DNA vaccines. Furthermore, the viral RNA is degraded within 5–7 days. Additionally, apoptosis induced by alphaviruses in transfected cells is another safety feature of vaccines based on suicidal alphavirus vectors.

No preexisting immunity: Alphaviruses generally possess no widespread immunity in the human and animal populations although some epidemics related to SFV, SIN, and VEE has been documented. This limitation observed for other viral expression systems does not prevent the use of alphaviruses for in vivo expression of heterologous genes.

Repeated administration: The viral structural genes are not intracellularly expressed. Therefore, alphavirus replicons can be repeatedly administered, since the host immune response to the vector itself does not cause rejection when subjected to booster immunizations, which has presented some serious limitations for other vector systems.

Stimulation of immune responses: Due to induction of apoptosis, gene expression is transient and lytic. The induced apoptosis assists in the uptake of transfected cells by DCs and, subsequently, facilitates activation and stimulation of these cells. Additionally, the viral double-stranded RNA molecules generated during alphavirus RNA replication provide an immunostimulatory effect on DCs and on innate immunity.

Application of alphavirus replicon systems induces broad and robust humoral and cellular immune responses to a wide array of tumor antigens and confers protection against tumor challenges as has been demonstrated in numerous studies employing several animal models. The demonstrated ability of alphavirus-based vaccines to break immunological tolerance to self-antigens is crucial for cancer therapy. Indeed, these vaccines have been found effective in cancer therapy models both in prophylactic and therapeutic settings. Plethora of successful preclinical studies already performed and future vector developments and improvements in vector delivery and targeting will contribute to widening the range of alphavirus vector applications and potentially paving the way for extremely versatile tools for future immunotherapy and cancer gene therapy.

TABLE 28.2

Therapeutic Anticancer Vaccines Based on Recombinant Alphaviral Vectors

Vector	Expressed Transgene	Tumor Model	Vaccination and Vector Administration Types	Vaccination Efficacy	References
SIN DNA/RNA layered vectors					
SIN DNA	HER/*neu*	Mouse breast cancer A2L2	Therapeutic (into the foot pad)	Ineffective	Eralp et al. (2004)
SIN DNA + DOX	HER/*neu*	Mouse breast cancer A2L2	Therapeutic (into the foot pad)	Partial tumor reduction	Eralp et al. (2004)
SIN DNA + PTX	HER/*neu*	Mouse breast cancer A2L2	Therapeutic (into the foot pad)	Ineffective	Eralp et al. (2004)
SIN DNA	HER/*neu*	Mouse breast cancer A2L2	Therapeutic (i.m.)	Ineffective	Wang et al. (2005)
SIN DNA + Ad vector	HER/*neu*	Mouse breast cancer A2L2	Therapeutic (i.m.)	Increase of survival	Wang et al. (2005)
SIN DNA	LacZ	Mouse colon cancer CT26.CL25	Therapeutic (i.m.)	Increase of survival	Leitner et al. (2000)
SIN DNA + SIN DNA	Gp100 IL-12	Mouse brain cancer model B16	Therapeutic (i.m.)	Increase of survival	Yamanaka et al. (2005)
SIN DNA	P1A	Mouse mastocytoma P815	Therapeutic (i.m.)	Tumor prevention (40%)	Ni et al. (2004)
SIN viral particles vectors					
SIN VLP	P1A	Mouse mastocytoma P815	Therapeutic (i.p.)	Tumor prevention (50%)	Ni et al. (2004)
SIN VLP	LacZ IL-12	Human ovarian cancer ES-2	Therapeutic (i.p.)	Partial tumor reduction	Tseng et al. (2004b)
SIN VLP	IL-12 IL-15	Human ovarian cancer ES-2	Therapeutic (i.p.)	Partial tumor reduction	Tseng et al. (2004a)
SIN VLP + paclitaxel	LacZ	Human ovarian cancer ES-2	Therapeutic (i.p)	Partial tumor reduction Increase of survival	Tseng et al. (2010)
SIN VLP	CRT/E7	Mouse cervical cancer model TC1	Therapeutic (i.m.)	Reduction of pulmonary nodules	Cheng et al. (2006)
SFV and SFV-E viral particles vectors					
SFV VLP	IL-12	Mouse melanoma B16	Therapeutic (i.t.)	Partial tumor reduction (70%–90%)	Asselin-Paturel et al. (1999)
SFV-VLP + CD137 mAb	IL-12	Mouse melanoma B16	Therapeutic (i.t.)	Complete tumor reduction (50%, 75%) Immunity	Quetglas et al. (2012)
SFV-IL-with CD137 mAb	IL-12	Mouse lung cancer TC1	Therapeutic (i.t.)	Partial tumor reduction	Quetglas et al. (2012)
SFV VLP	EGFP	Human lung cancer H358a	Therapeutic (i.t.)	Partial tumor reduction Complete tumor reduction (40%)	Murphy et al. (2000)
SFV VLP	GM-CSF	Mouse ovarian cancer MOT	Therapeutic (i.p.)	Partial tumor reduction Immunity	Klimp et al. (2001)
SFV VLP	BAX	Rat prostate cancer AT3	Therapeutic (i.t.)	Partial tumor reduction	Murphy et al. (2001)
SFV VLP	Endostatin	Mouse brain cancer model B16	Therapeutic (i.t.)	Partial tumor reduction Increase of survival	Yamanaka et al. (2001a)
SFV-E VLP	IL-12	Woodchucks liver cancer WCH17	Therapeutic (i.t.)	Partial tumor reduction	Rodriguez-Madoz et al. (2009)
SFV-E VLP	P1A	Mouse mastocytoma P815	Therapeutic (i.t.), (peritumoral)	Partial tumor reduction Complete tumor reduction (46%)	Colmenero et al. (2002)
SFV VLP	IL-12	Mouse mastocytoma P815	Therapeutic (i.t.), (peritumoral)	Partial tumor reduction Complete tumor reduction (53%)	Colmenero et al. (2002)
SFV VLP	IL-12	Mouse colon cancer MC38	Therapeutic (i.t.)	Complete tumor reduction (80%)	Rodriguez-Madoz et al. (2005)
SFV-E VLP	IL-12	Mouse colon cancer MC38	Therapeutic (i.t.)	Complete tumor reduction (92%)	Rodriguez-Madoz et al. (2005)

(Continued)

TABLE 28.2 (*Continued*)

Therapeutic Anticancer Vaccines Based on Recombinant Alphaviral Vectors

Vector	Expressed Transgene	Tumor Model	Vaccination and Vector Administration Types	Vaccination Efficacy	References
SFV-E VLP	IL-18	Mouse colon cancer CT26	Therapeutic (i.t.)	Partial tumor reduction Complete tumor reduction (33%)	Chikkanna-Gowda et al. (2006)
SFV10-E VLP	VEGFR-2	Mouse breast cancer 4T1	Therapeutic (i.t.)	Partial tumor reduction Reduction of metastasis	Lyons et al. (2007)
SFV-E VLP	HPV16 E6, E7	Mouse cervical cancer model TC1	Therapeutic (s.c.)	Complete tumor reduction (100%	Daemen et al. (2003)
SFV-E VLP + SFV VLP	HPV16 E6, E7 IL-21	Mouse cervical cancer model TC1	Therapeutic (s.c.)	Complete tumor reduction (28%)	Riezebos-Brilman et al. (2009)
VEE viral particles vectors					
VEE VLP	TPR-2	Mouse melanoma B16	Therapeutic (s.c.)	Immunity Reduction of metastasis	Avogadri et al. (2010)
VEE VLP	HER/*neu*	Rat breast cancer 13762 MAT B III	Therapeutic (s.c.), (i.m.)	Immunity Complete tumor reduction (50%)	Nelson et al. (2003)
VEE VLP	HER/*neu*	Rat breast cancer 13762 MAT B III	Therapeutic (s.c.)	Partial tumor reduction	Laust et al. (2007)
VEE VLP	E7	Mouse cervical cancer C3	Therapeutic (s.c.)	Partial tumor reduction	Velders et al. (2001)
VEE VLP	HPV16 E6, E7	Cervical cancer model HLF16	Therapeutic (s.c.)	Partial tumor reduction	Eiben et al. (2002)
VEE VLP	HPV16 E6, E7	Mouse cervical cancer C3 cell line HLF16 cell line	Therapeutic (s.c.) Therapeutic (s.c.)	Partial tumor reduction Complete tumor reduction (90%–100%)	Cassetti et al. (2004)
VEE VLP + STEAP cDNA	STEAP	Mouse prostate cancer TRAMPC-2	Therapeutic (s.c.)	Partial tumor reduction	Garcia-Hernandez et al. (2007)

TABLE 28.3

Oncolytic Anticancer Vaccines Based on Replication-Competent Alphavirus

Vector	Expressed Transgene	Tumor Model	Vaccination and Vector Administration Type	Vaccination Efficacy	References
Sindbis virus					
SIN AR339	EGPF	Human ovarian cancer OMC-3	Therapeutic (i.p.)	Suppression of ascites formation	Unno et al. (2005)
SIN AR339	Nonspecified	Human cervical cancer HeLaS3 and C33A	Therapeutic (i.t.)	Partial tumor reduction Increase of survival	Unno (2005)
SIN AR339	Nonspecified	Human cervical cancer C33A cell line	Therapeutic (i.v.)	Partial tumor reduction	Unno et al. (2005)
SFV virus					
SFV VA7	EGFP	Human melanoma A2058	Therapeutic (i.v.), (i.p.), (i.t.)	Partial tumor reduction	Vähä-Koskela et al. (2006)
SFV VA7	EGFP	Human lung cancer A549	Therapeutic (i.t.) Therapeutic (i.v.) Therapeutic (i.p.)	Almost complete tumor reduction Partial tumor reduction Ineffective	Maatta et al. (2008)
SFV VA7	EGFP	Rat brain cancer BT4C	Therapeutic (i.v.), (i.p.) Therapeutic (i.t.)	Ineffective Partial tumor reduction	Maatta et al. (2007)
SFV VA7	EGFP	Human brain cancer U87	Therapeutic (i.v.)	Complete tumor reduction (95%)	Heikkilä et al. (2010)
SFV VA7	EGFP	Human osteosarcoma Saos2LM7	Therapeutic (i.t.)	Partial tumor reduction	Ketola et al. (2008)
SFV + VV	Nonspecified	Mouse ovarian cancer MOSEC	Therapeutic (i.p.)	Increase of survival Immunity	Zhang et al. (2010)
SFV wild type	Nonspecified	Mouse fibrosarcoma WEHI-11	Prophylactic (i.p.)	Complete tumor reduction (80%)	Griffith et al. (1975)

ACKNOWLEGMENTS

The authors thank Henrik Garoff (Stockholm) for giving them the opportunity to work with alphaviruses and for the kind scientific support; and Paul Pumpens (Riga) for help with submission of the manuscript. Their work was supported by the Latvian National Research Programme BIOMEDICINE (2014–2017).

REFERENCES

Agapov EV, Frolov I, Lindenbach BD, Prágai BM, Schlesinger S, Rice CM. 1998. Noncytopathic Sindbis virus RNA vectors for heterologous gene expression. *Proc Natl Acad Sci U S A* 95(22):12989–12994.

Asselin-Paturel C, Lassau N, Guinebretiere JM, Zhang J, Gay F, Bex F, Hallez S, Leclere J, Peronneau P, Mami-Chouaib F, Chouaib S. 1999. Transfer of the murine interleukin-12 gene in vivo by a Semliki Forest virus vector induces B16 tumor regression through inhibition of tumor blood vessel formation monitored by Doppler ultrasonography. *Gene Ther* 6(4):606–615.

Atasheva S, Gorchakov R, English R, Frolov I, Frolova E. 2007. Development of Sindbis viruses encoding nsP2/GFP chimeric proteins and their application for studying nsP2 functioning. *J Virol* 81(10):5046–5057.

Avogadri F, Merghoub T, Maughan MF, Hirschhorn-Cymerman D, Morris J, Ritter E, Olmsted R, Houghton AN, Wolchok JD. 2010. Alphavirus replicon particles expressing TRP-2 provide potent therapeutic effect on melanoma through activation of humoral and cellular immunity. *PLoS ONE* 5(9):e12670.

Balachandran S, Roberts PC, Kipperman T, Bhalla KN, Compans RW, Archer DR, Barber GN. 2000. Alpha/beta interferons potentiate virus-induced apoptosis through activation of the FADD/Caspase-8 death signaling pathway. *J Virol* 74(3):1513–1523.

Banin Hirata BK, Oda JM, Losi GR, Ariza CB, de Oliveira CE, Watanabe MA. 2014. Molecular markers for breast cancer: Prediction on tumor behavior. *Dis Markers* 2014:513158.

Barth BU, Wahlberg JM, Garoff H. 1995. The oligomerization reaction of the Semliki Forest virus membrane protein subunits. *J Cell Biol* 128(3):283–291.

Berglund P, Sjöberg M, Garoff H, Atkins GJ, Sheahan BJ, Liljestrom P. 1993. Semliki Forest virus expression system: Production of conditionally infectious recombinant particles. *Bio/Technology* 11:916–920.

Berglund P, Smerdou C, Fleeton MN, Tubulekas I, Liljeström P. 1998. Enhancing immune responses using suicidal DNA vaccines. *Nat Biotechnol* 16(6):562–565.

Blanchon F, Grivaux M, Asselain B, Lebas FX, Orlando JP, Piquet J, Zureik M. 2006. 4-year mortality in patients with non-small-cell lung cancer: Development and validation of a prognostic index. *Lancet Oncol* 7(10):829–836.

Boorsma M, Nieba L, Koller D, Bachmann MF, Bailey JE, Renner WA. 2000. A temperature-regulated replicon-based DNA expression system. *Nat Biotechnol* 18(4):429–432.

Bredenbeek PJ, Frolov I, Rice CM, Schlesinger S. 1993. Sindbis virus expression vectors: Packaging of RNA replicons by using defective helper RNAs. *J Virol* 67:6439.

Cassetti MC, McElhiney SP, Shahabi V, Pullen JK, Le Poole I, Eiben GL, Smith LR, Kast WM. 2004. Antitumor efficacy of Venezuelan equine encephalitis virus replicon particles encoding mutated HPV16 E6 and E7 genes. *Vaccine* 22(3–4):520–527.

Cheng WF, Hung CF, Chai CY, Hsu KF, He L, Rice CM, Ling M, Wu TC. 2001a. Enhancement of Sindbis virus self-replicating RNA vaccine potency by linkage of *Mycobacterium tuberculosis* heat shock protein 70 gene to an antigen gene. *J Immunol* 166(10):6218–6226.

Cheng WF, Hung CF, Hsu KF, Chai CY, He L, Ling M, Slater LA, Roden RB, Wu TC. 2001b. Enhancement of Sindbis virus self-replicating RNA vaccine potency by targeting antigen to endosomal/lysosomal compartments. *Hum Gene Ther* 12(3):235–252.

Cheng WF, Lee CN, Su YN, Chai CY, Chang MC, Polo JM, Hung CF, Wu TC, Hsieh CY, Chen CA. 2006. Sindbis virus replicon particles encoding calreticulin linked to a tumor antigen generate long-term tumor-specific immunity. *Cancer Gene Ther* 13(9):873–885.

Chikkanna-Gowda CP, McNally S, Sheahan BJ, Fleeton MN, Atkins GJ. 2006. Inhibition of murine K-BALB and CT26 tumour growth using a Semliki Forest virus vector with enhanced expression of IL-18. *Oncol Rep* 16(4):713–719.

Chikkanna-Gowda CP, Sheahan BJ, Fleeton MN, Atkins GJ. 2005. Regression of mouse tumours and inhibition of metastases following administration of a Semliki Forest virus vector with enhanced expression of IL-12. *Gene Ther* 12(16):1253–1263.

Colmenero P, Berglund P, Kambayashi T, Biberfeld P, Liljestrom P, Jondal M. 2001. Recombinant Semliki Forest virus vaccine vectors: The route of injection determines the localization of vector RNA and subsequent T cell response. *Gene Ther* 8(17):1307–1314.

Colmenero P, Chen M, Castanos-Velez E, Liljestrom P, Jondal M. 2002. Immunotherapy with recombinant SFV-replicons expressing the P815A tumor antigen or IL-12 induces tumor regression. *Int J Cancer* 98(4):554–560.

Daemen T, Pries F, Bungener L, Kraak M, Regts J, Wilschut J. 2000. Genetic immunization against cervical carcinoma: Induction of cytotoxic T lymphocyte activity with a recombinant alphavirus vector expressing human papillomavirus type 16 E6 and E7. *Gene Ther* 7(21):1859–1866.

Daemen T, Regts J, Holtrop M, Wilschut J. 2002. Immunization strategy against cervical cancer involving an alphavirus vector expressing high levels of a stable fusion protein of human papillomavirus 16 E6 and E7. *Gene Ther* 9(2):85–94.

Daemen T, Riezebos-Brilman A, Bungener L, Regts J, Dontje B, Wilschut J. 2003. Eradication of established HPV16-transformed tumours after immunisation with recombinant Semliki Forest virus expressing a fusion protein of E6 and E7. *Vaccine* 21(11–12):1082–1088.

Daemen T, Riezebos-Brilman A, Regts J, Dontje B, van der ZA, Wilschut J. 2004. Superior therapeutic efficacy of alphavirus-mediated immunization against human papilloma virus type 16 antigens in a murine tumour model: Effects of the route of immunization. *Antivir Ther* 9(5):733–742.

Davis NL, Willis LV, Smith JF, Johnston RE. 1989. In vitro synthesis of infectious Venezuelan equine encephalitis virus RNA from a cDNA clone: Analysis of a viable deletion mutant. *Virology* 171(1):189–204.

Diciommo DP, Bremner R. 1998. Rapid, high level protein production using DNA-based Semliki Forest virus vectors. *J Biol Chem* 273(29):18060–18066.

Dubensky TW Jr, Driver DA, Polo JM, Belli BA, Latham EM, Ibanez CE, Chada S et al. 1996. Sindbis virus DNA-based expression vectors: Utility for in vitro and in vivo gene transfer. *J Virol* 70(1):508–519.

Durso RJ, Andjelic S, Gardner JP, Margitich DJ, Donovan GP, Arrigale RR, Wang X et al. 2007. A novel alphavirus vaccine encoding prostate-specific membrane antigen elicits potent cellular and humoral immune responses. *Clin Cancer Res* 13(13):3999–4008.

Eiben GL, Velders MP, Schreiber H, Cassetti MC, Pullen JK, Smith LR, Kast WM. 2002. Establishment of an HLA-A*0201 human papillomavirus type 16 tumor model to determine the efficacy of vaccination strategies in HLA-A*0201 transgenic mice. *Cancer Res* 62(20):5792–5799.

Eralp Y, Wang X, Wang JP, Maughan MF, Polo JM, Lachman LB. 2004. Doxorubicin and paclitaxel enhance the antitumor efficacy of vaccines directed against HER 2/*neu* in a murine mammary carcinoma model. *Breast Cancer Res* 6(4):R275–R283.

Frolov I, Frolova E, Schlesinger S. 1997. Sindbis virus replicons and Sindbis virus: Assembly of chimeras and of particles deficient in virus RNA. *J Virol* 71:2819.

Frolova E, Frolov I, Schlesinger S. 1997. Packaging signals in alphaviruses. *J Virol* 71(1):248–258.

Garcia-Hernandez ML, Gray A, Hubby B, Kast WM. 2007. In vivo effects of vaccination with six-transmembrane epithelial antigen of the prostate: A candidate antigen for treating prostate cancer. *Cancer Res* 67(3):1344–1351.

Garcia-Hernandez ML, Gray A, Hubby B, Klinger OJ, Kast WM. 2008. Prostate stem cell antigen vaccination induces a long-term protective immune response against prostate cancer in the absence of autoimmunity. *Cancer Res* 68(3):861–869.

Garoff H, Huylebroeck D, Robinson A, Tillman U, Liljestrom P. 1990. The signal sequence of the p62 protein of Semliki Forest virus is involved in initiation but not in completing chain translocation. *J Cell Biol* 111(3):867–876.

Goldberg SM, Bartido SM, Gardner JP, Guevara-Patino JA, Montgomery SC, Perales MA, Maughan MF et al. 2005. Comparison of two cancer vaccines targeting tyrosinase: Plasmid DNA and recombinant alphavirus replicon particles. *Clin Cancer Res* 11(22):8114–8121.

Gomes IM, Maia CJ, Santos CR. 2012. STEAP proteins: From structure to applications in cancer therapy. *Mol Cancer Res* 10(5):573–587.

Gorchakov R, Frolova E, Williams BR, Rice CM, Frolov I. 2004. PKR-dependent and -independent mechanisms are involved in translational shutoff during Sindbis virus infection. *J Virol* 78(16):8455–8467.

Granot T, Meruelo D. 2012. The role of natural killer cells in combinatorial anti-cancer therapy using Sindbis viral vectors and irinotecan. *Cancer Gene Ther* 19(8):588–591.

Granot T, Venticinque L, Tseng JC, Meruelo D. 2011. Activation of cytotoxic and regulatory functions of NK cells by Sindbis viral vectors. *PLoS ONE* 6(6):e20598.

Granot T, Yamanashi Y, Meruelo D. 2014. Sindbis viral vectors transiently deliver tumor-associated antigens to lymph nodes and elicit diversified antitumor CD8+ T-cell immunity. *Mol Ther* 22(1):112–122.

Hahn CS, Hahn YS, Braciale TJ, Rice CM. 1992. Infectious Sindbis virus transient expression vectors for studying antigen processing and presentation. *Proc Natl Acad Sci U S A* 89(7):2679–2683.

Hahn CS, Strauss JH. 1990. Site-directed mutagenesis of the proposed catalytic amino acids of the Sindbis virus capsid protein autoprotease. *J Virol* 64(6):3069–3073.

Heikkila JE, Vaha-Koskela MJ, Ruotsalainen JJ, Martikainen MW, Stanford MM, McCart JA, Bell JC, Hinkkanen AE. 2010. Intravenously administered alphavirus vector VA7 eradicates orthotopic human glioma xenografts in nude mice. *PLoS ONE* 5(1):e8603.

Hu J, Cai XF, Yan G. 2009. Alphavirus M1 induces apoptosis of malignant glioma cells via downregulation and nucleolar translocation of p21WAF1/CIP1 protein. *Cell Cycle* 8(20): 3328–3339.

Huang PY, Guo JH, Hwang LH. 2012. Oncolytic Sindbis virus targets tumors defective in the interferon response and induces significant bystander antitumor immunity in vivo. *Mol Ther* 20(2):298–305.

Jose J, Snyder JE, Kuhn RJ. 2009. A structural and functional perspective of alphavirus replication and assembly. *Future Microbiol* 4(7):837–856.

Ketola A, Hinkkanen A, Yongabi F, Furu P, Maatta AM, Liimatainen T, Pirinen R et al. 2008. Oncolytic Semliki forest virus vector as a novel candidate against unresectable osteosarcoma. *Cancer Res* 68(20):8342–8350.

Klimp AH, van der Vaart E, Lansink PO, Withoff S, de Vries EG, Scherphof GL, Wilschut J, Daemen T. 2001. Activation of peritoneal cells upon in vivo transfection with a recombinant alphavirus expressing GM-CSF. *Gene Ther* 8(4):300–307.

Kononchik JP Jr, Hernandez R, Brown DT. 2011. An alternative pathway for alphavirus entry. *Virol J* 8:304.

Kuhn RJ, Hong Z, Strauss JH. 1990. Mutagenesis of the 3′ nontranslated region of Sindbis virus RNA. *J Virol* 64(4):1465–1476.

Kuipers EJ, Rosch T, Bretthauer M. 2013. Colorectal cancer screening—Optimizing current strategies and new directions. *Nat Rev Clin Oncol* 10(3):130–142.

Kumar CM, Sathisha UV, Dharmesh S, Rao AG, Singh SA. 2011. Interaction of sesamol (3,4-methylenedioxyphenol) with tyrosinase and its effect on melanin synthesis. *Biochimie* 93(3):562–569.

Lachman LB, Rao XM, Kremer RH, Ozpolat B, Kiriakova G, Price JE. 2001. DNA vaccination against *neu* reduces breast cancer incidence and metastasis in mice. *Cancer Gene Ther* 8(4):259–268.

Laust AK, Sur BW, Wang K, Hubby B, Smith JF, Nelson EL. 2007. VRP immunotherapy targeting *neu*: Treatment efficacy and evidence for immunoediting in a stringent rat mammary tumor model. *Breast Cancer Res Treat* 106(3):371–382.

Leitner WW, Hwang LN, deVeer MJ, Zhou A, Silverman RH, Williams BR, Dubensky TW, Ying H, Restifo NP. 2003. Alphavirus-based DNA vaccine breaks immunological tolerance by activating innate antiviral pathways. *Nat Med* 9(1):33–39.

Leitner WW, Ying H, Driver DA, Dubensky TW, Restifo NP. 2000. Enhancement of tumor-specific immune response with plasmid DNA replicon vectors. *Cancer Res* 60(1):51–55.

Leslie MC, Zhao YJ, Lachman LB, Hwu P, Wu GJ, Bar-Eli M. 2007. Immunization against MUC18/MCAM, a novel antigen that drives melanoma invasion and metastasis. *Gene Ther* 14(4):316–323.

Leung JY, Ng MM, Chu JJ. 2011. Replication of alphaviruses: A review on the entry process of alphaviruses into cells. *Adv Virol* 2011:249640.

Liljestrom P, Garoff H. 1991. A new generation of animal cell expression vectors based on the Semliki Forest virus replicon. *Biotechnology* 9(12):1356–1361.

Lin KI, Pasinelli P, Brown RH, Hardwick JM, Ratan RR. 1999. Decreased intracellular superoxide levels activate Sindbis virus-induced apoptosis. *J Biol Chem* 274(19):13650–13655.

Liu G, Ying H, Zeng G, Wheeler CJ, Black KL, Yu JS. 2004. HER-2, gp100, and MAGE-1 are expressed in human glioblastoma and recognized by cytotoxic T cells. *Cancer Res* 64(14):4980–4986.

Lobigs M, Garoff H. 1990. Fusion function of the Semliki Forest virus spike is activated by proteolytic cleavage of the envelope glycoprotein precursor p62. *J Virol* 64(3):1233–1240.

Lundstrom K. 2009. Alphaviruses in gene therapy. *Viruses* 1(1):13–25.

Lundstrom K, Abenavoli A, Malgaroli A, Ehrengruber MU. 2003. Novel Semliki Forest virus vectors with reduced cytotoxicity and temperature sensitivity for long-term enhancement of transgene expression. *Mol Ther* 7(2):202–209.

Lyons JA, Sheahan BJ, Galbraith SE, Mehra R, Atkins GJ, Fleeton MN. 2007. Inhibition of angiogenesis by a Semliki Forest virus vector expressing VEGFR-2 reduces tumour growth and metastasis in mice. *Gene Ther* 14(6):503–513.

Maatta AM, Liimatainen T, Wahlfors T, Wirth T, Vaha-Koskela M, Jansson L, Valonen P et al. 2007. Evaluation of cancer virotherapy with attenuated replicative Semliki forest virus in different rodent tumor models. *Int J Cancer* 121(4):863–870.

Maatta AM, Makinen K, Ketola A, Liimatainen T, Yongabi FN, Vaha-Koskela M, Pirinen R et al. 2008. Replication competent Semliki Forest virus prolongs survival in experimental lung cancer. *Int J Cancer* 123(7):1704–1711.

Melero I, Shuford WW, Newby SA, Aruffo A, Ledbetter JA, Hellstrom KE, Mittler RS, Chen L. 1997. Monoclonal antibodies against the 4–1BB T-cell activation molecule eradicate established tumors. *Nat Med* 3(6):682–685.

Meruelo D. 2004. Systemic gene therapy by Sindbis vectors: A potentially safe and effective targeted therapy for identifying and killing tumor cells in vivo. *Discov Med* 4(20):54–57.

Moran TP, Burgents JE, Long B, Ferrer I, Jaffee EM, Tisch RM, Johnston RE, Serody JS. 2007. Alphaviral vector-transduced dendritic cells are successful therapeutic vaccines against *neu*-overexpressing tumors in wild-type mice. *Vaccine* 25(36):6604–6612.

Morris-Downes MM, Phenix KV, Smyth J, Sheahan BJ, Lileqvist S, Mooney DA, Liljestrom P, Todd D, Atkins GJ. 2001. Semliki Forest virus-based vaccines: Persistence, distribution and pathological analysis in two animal systems. *Vaccine* 19(15–16):1978–1988.

Morse MA, Hobeika AC, Osada T, Berglund P, Hubby B, Negri S, Niedzwiecki D et al. 2010. An alphavirus vector overcomes the presence of neutralizing antibodies and elevated numbers of Tregs to induce immune responses in humans with advanced cancer. *J Clin Invest* 120(9):3234–3241.

Murphy AM, Morris-Downes MM, Sheahan BJ, Atkins GJ. 2000. Inhibition of human lung carcinoma cell growth by apoptosis induction using Semliki Forest virus recombinant particles. *Gene Ther* 7(17):1477–1482.

Murphy AM, Sheahan BJ, Atkins GJ. 2001. Induction of apoptosis in BCL-2-expressing rat prostate cancer cells using the Semliki Forest virus vector. *Int J Cancer* 94(4):572–578.

Näslund TI, Uyttenhove C, Nordstrom EK, Colau D, Wrnier G, Jondal M, Van den Eynde BJ et al. 2007. Comparative prime-boost vaccinations using Semliki Forest virus, adenovirus, and ALVAC vectors demonstrate differences in the generation of a protective central memory CTL response against the P815 tumor. *J Immunol* 178(11):6761–6769.

Nava VE, Rosen A, Veliuona MA, Clem RJ, Levine B, Hardwick JM. 1998. Sindbis virus induces apoptosis through a caspase-dependent, CrmA-sensitive pathway. *J Virol* 72(1):452–459.

Naz RK, Shiley B. 2012. Prophylactic vaccines for prevention of prostate cancer. *Front Biosci (Schol Ed)* 4:932–940.

Nelson EL, Prieto D, Alexander TG, Pushko P, Lofts LA, Rauner JO, Kamrud KI et al. 2003. Venezuelan equine encephalitis replicon immunization overcomes intrinsic tolerance and elicits effective anti-tumor immunity to the 'self' tumor-associated antigen, neu in a rat mammary tumor model. *Breast Cancer Res Treat* 82(3):169–183.

Ni B, Gao W, Zhu B, Lin Z, Jia Z, Zhou W, Zhao J, Wang L, Wu Y. 2005. Induction of specific human primary immune responses to a Semliki Forest virus-based tumor vaccine in a Trimera mouse model. *Cancer Immunol Immunother* 54(5):489–498.

Ni B, LinZ, Zhou L, Wang L, Jia Z, Zhou W, Diciommo DP et al. 2004. Induction of P815 tumor immunity by DNA-based recombinant Semliki Forest virus or replicon DNA expressing the P1A gene. *Cancer Detect Prev* 28(6):418–425.

Osada T, Morse MA, Hobeika A, Lyerly HK. 2012. Novel recombinant alphaviral and adenoviral vectors for cancer immunotherapy. *Semin Oncol* 39(3):305–310.

Pleshkan VV, Zinov'eva MV, Sverdlov ED. 2011. Melanoma: Surface markers as the first point of targeted delivery of therapeutic genes in multilevel gene therapy. *Mol Biol (Mosk)* 45(3):416–433.

Pushko P, Parker M, Ludwig GV, Davis NL, Johnston RE, Smith JF. 1997. Replicon-helper systems from attenuated Venezuelan equine encephalitis virus: Expression of heterologous genes in vitro and immunization against heterologous pathogens in vivo. *Virology*, 239:389–401.

Quetglas JI, Dubrot J, Bezunartea J, Sanmamed MF, Hervas-Stubbs S, Smerdou C, Melero I. 2012a. Immunotherapeutic synergy between anti-CD137 mAb and intratumoral administration of a cytopathic Semliki Forest virus encoding IL-12. *Mol Ther* 20(9):1664–1675.

Quetglas JI, Fioravanti J, Ardaiz N, Medina-Echeverz J, Baraibar I, Prieto J, Smerdou C, Berraondo P. 2012b. A Semliki Forest virus vector engineered to express IFNalpha induces efficient elimination of established tumors. *Gene Ther* 19(3):271–278.

Quetglas JI, Ruiz-Guillen M, Aranda A, Casales E, Bezunartea J, Smerdou C. 2010. Alphavirus vectors for cancer therapy. *Virus Res* 153(2):179–196.

Ren H, Boulikas T, Lundstrom K, Soling A, Warnke PC, Rainov NG. 2003. Immunogene therapy of recurrent glioblastoma multiforme with a liposomally encapsulated replication-incompetent Semliki forest virus vector carrying the human interleukin-12 gene—A phase I/II clinical protocol. *J Neurooncol* 64(1–2):147–154.

Rheme C, Ehrengruber MU, Grandgirard D. 2005. Alphaviral cytotoxicity and its implication in vector development. *Exp Physiol* 90(1):45–52.

Riezebos-Brilman A, Regts J, Chen M, Wilschut J, Daemen T. 2009. Augmentation of alphavirus vector-induced human papilloma virus-specific immune and anti-tumour responses by co-expression of interleukin-12. *Vaccine* 27(5):701–707.

Riezebos-Brilman A, Walczak M, Regts J, Rots MG, Kamps G, Dontje B, Haisma HY, Wilschut J, Daemen T. 2007. A comparative study on the immunotherapeutic efficacy of recombinant Semliki Forest virus and adenovirus vector systems in a murine model for cervical cancer. *Gene Ther* 14(24):1695–1704.

Rodriguez-Madoz JR, Liu KH, Quetglas JI, Ruiz-Guillen M, Otano I, Crettaz J, Butler SD et al. 2009. Semliki forest virus expressing interleukin-12 induces antiviral and antitumoral responses in woodchucks with chronic viral hepatitis and hepatocellular carcinoma. *J Virol* 83(23):12266–12278.

Rodriguez-Madoz JR, Prieto J, Smerdou C. 2005. Semliki forest virus vectors engineered to express higher IL-12 levels induce efficient elimination of murine colon adenocarcinomas. *Mol Ther* 12(1):153–163.

Rodriguez-Madoz JR, Prieto J, Smerdou C. 2007. Biodistribution and tumor infectivity of semliki forest virus vectors in mice: Effects of re-administration. *Mol Ther* 15(12):2164–2171.

Rodriguez-Madoz JR, Zabala M, Alfaro M, Prieto J, Kramer MG, Smerdou C. 2014. Short-term intratumoral interleukin-12 expressed from an alphaviral vector is sufficient to induce an efficient antitumoral response against spontaneous hepatocellular carcinomas. *Hum Gene Ther* 25(2):132–143.

Salminen A, Wahlberg JM, Lobigs M, Liljeström P, Garoff H. 1992. Membrane fusion process of Semliki Forest virus. II: Cleavage-dependent reorganization of the spike protein complex controls virus entry. *J Cell Biol* 116:349–357.

Saxena S, Sonwane AA, Dahiya SS, Patel CL, Saini M, Rai A, Gupta PK. 2009. Induction of immune responses and protection in mice against rabies using a self-replicating RNA vaccine encoding rabies virus glycoprotein. *Vet Microbiol* 136(1–2):36–44.

Scallan MF, Allsopp TE, Fazakerley JK. 1997. bcl-2 acts early to restrict Semliki Forest virus replication and delays virus-induced programmed cell death. *J Virol* 71(2):1583–1590.

Schlesinger S, Schlesinger MJ. 2001. *Togaviridae*: The viruses and their replication, In *Fields' Virology*, 4th edn. DM Knipe and PM Howley, eds. Philadelphia, PA: Lippincott Williams and Wilkins, pp. 895–916.

Sjoberg EM, Suomalainen M, Garoff H. 1994. A significantly improved Semliki Forest virus expression system based on translation enhancer segments from the viral capsid gene. *Biotechnology* (*NY*) 12:1127.

Slovin SF, Kehoe M, Durso R, Fernandez C, Olson W, Gao JP, Israel R, Scher HI, Morris S. 2013. A phase I dose escalation trial of vaccine replicon particles (VRP) expressing prostate-specific membrane antigen (PSMA) in subjects with prostate cancer. *Vaccine* 31(6):943–949.

Smerdou C, Liljeström P. 1999. Two-helper RNA system for production of recombinant Semliki forest virus particles. *J Virol* 73:1092–1098.

Smyth JW, Fleeton MN, Sheahan BJ, Atkins GJ. 2005. Treatment of rapidly growing K-BALB and CT26 mouse tumours using Semliki Forest virus and its derived vector. *Gene Ther* 12(2):147–159.

Tamberg N, Lulla V, Fragkoudis R, Lulla A, Fazakerley JK, Merits A. 2007. Insertion of EGFP into the replicase gene of Semliki Forest virus results in a novel, genetically stable marker virus. *J Gen Virol* 88(Pt 4):1225–1230.

Thomas JM, Klimstra WB, Ryman KD, Heidner HW. 2003. Sindbis virus vectors designed to express a foreign protein as a cleavable component of the viral structural polyprotein. *J Virol* 77(10):5598–5606.

Tseng JC, Granot T, DiGiacomo V, Levin B, Meruelo D. 2010. Enhanced specific delivery and targeting of oncolytic Sindbis viral vectors by modulating vascular leakiness in tumor. *Cancer Gene Ther* 17(4):244–255.

Tseng JC, Hurtado A, Yee H, Levin B, Boivin C, Benet M, Blank SV, Pellicer A, Meruelo D. 2004a. Using sindbis viral vectors for specific detection and suppression of advanced ovarian cancer in animal models. *Cancer Res* 64(18):6684–6692.

Tseng JC, Levin B, Hurtado A, Yee H, Perez de Castro I, Jimenez M, Shamamian P, Jin R, Novick RP, Pellicer A, Meruelo D. 2004b. Systemic tumor targeting and killing by Sindbis viral vectors. *Nat Biotechnol* 22(1):70–77.

Unno Y, Shino Y, Kondo F, Igarashi N, Wang G, Shimura R, Yamaguchi T, Asano T, Saisho H, Sekiya S, Shirasawa H. 2005. Oncolytic viral therapy for cervical and ovarian cancer cells by Sindbis virus AR339 strain. *Clin Cancer Res* 11(12):4553–4560.

Urban C, Rheme C, Maerz S, Berg B, Pick R, Nitschke R, Borner C. 2008. Apoptosis induced by Semliki Forest virus is RNA replication dependent and mediated via Bak. *Cell Death Differ* 15(9):1396–1407.

Vaha-Koskela MJ, Kallio JP, Jansson LC, Heikkila JE, Zakhartchenko VA, Kallajoki MA, Kahari VM, Hinkkanen AE. 2006. Oncolytic capacity of attenuated replicative semliki forest virus in human melanoma xenografts in severe combined immunodeficient mice. *Cancer Res* 66(14):7185–7194.

Vaha-Koskela MJ, Tuittila MT, Nygardas PT, Nyman JK, Ehrengruber MU, Renggli M, Hinkkanen AE. 2003. A novel neurotropic expression vector based on the avirulent A7(74) strain of Semliki Forest virus. *J Neurovirol* 9(1):1–15.

Vasilevska J, Skrastina D, Spunde K, Garoff H, Kozlovska T, Zajakina A. 2012. Semliki Forest virus biodistribution in tumor-free and 4T1 mammary tumor-bearing mice: A comparison of transgene delivery by recombinant virus particles and naked RNA replicon. *Cancer Gene Ther* 19(8):579–587.

Velders MP, McElhiney S, Cassetti MC, Eiben GL, Higgins T, Kovacs GR, Elmishad AG, Kast WM, Smith LR. 2001. Eradication of established tumors by vaccination with Venezuelan equine encephalitis virus replicon particles delivering human papillomavirus 16 E7 RNA. *Cancer Res* 61(21):7861–7867.

Venticinque L, Meruelo D. 2010. Sindbis viral vector induced apoptosis requires translational inhibition and signaling through Mcl-1 and Bak. *Mol Cancer* 9:37.

Vignuzzi M, Gerbaud S, van der WS, Escriou N. 2001. Naked RNA immunization with replicons derived from poliovirus and Semliki Forest virus genomes for the generation of a cytotoxic T cell response against the influenza A virus nucleoprotein. *J Gen Virol* 82(Pt 7):1737–1747.

Wahlfors JJ, Zullo SA, Loimas S, Nelson DM, Morgan RA. 2000. Evaluation of recombinant alphaviruses as vectors in gene therapy. *Gene Ther* 7(6):472–480.

Walczak M, de Mare A, Riezebos-Brilman A, Regts J, Hoogeboom BN, Visser JT, Fiedler M et al. 2011. Heterologous prime-boost immunizations with a virosomal and an alphavirus replicon vaccine. *Mol Pharm* 8(1):65–77.

Wang X, Wang JP, Rao XM, Price JE, Zhou HS, Lachman LB. 2005. Prime-boost vaccination with plasmid and adenovirus gene vaccines control HER2/*neu*+ metastatic breast cancer in mice. *Breast Cancer Res* 7(5):R580–R588.

Welch WJ, Sefton BM. 1979. Two small virus-specific polypeptides are produced during infection with Sindbis virus. *J Virol* 29(3):1186–1195.

Xiong C, Levis R, Shen P, Schlesinger S, Rice CM, Huang HV. 1989. Sindbis virus: An efficient, broad host range vector for gene expression in animal cells. *Science* 243(4895):1188–1191.

Yamanaka R, Tsuchiya N, Yajima N, Honma J, Hasegawa H, Tanaka R, Ramsey J, Blaese RM, Xanthopoulos KG. 2003. Induction of an antitumor immunological response by an intratumoral injection of dendritic cells pulsed with genetically engineered Semliki Forest virus to produce interleukin-18 combined with the systemic administration of interleukin-12. *J Neurosurg* 99(4):746–753.

Yamanaka R, Xanthopoulos KG. 2004. Development of improved Sind bis virus-based DNA expression vector. *DNA Cell Biol* 23(2):75–80.

Yamanaka R, Xanthopoulos KG. 2005. Induction of antigen-specific immune responses against malignant brain tumors by intramuscular injection of sindbis DNA encoding gp100 and IL-18. *DNA Cell Biol* 24(5):317–324.

Yamanaka R, Zullo SA, Ramsey J, Onodera M, Tanaka R, Blaese M, Xanthopoulos KG. 2001a. Induction of therapeutic antitumor antiangiogenesis by intratumoral injection of genetically engineered endostatin-producing Semliki Forest virus. *Cancer Gene Ther* 8(10):796–802.

Yamanaka R, Zullo SA, Ramsey J, Yajima N, Tsuchiya N, Tanaka R, Blaese M, Xanthopoulos KG. 2002. Marked enhancement of antitumor immune responses in mouse brain tumor models by genetically modified dendritic cells producing Semliki Forest virus-mediated interleukin-12. *J Neurosurg* 97(3):611–618.

Yamanaka R, Zullo SA, Tanaka R, Blaese M, Xanthopoulos KG. 2001b. Enhancement of antitumor immune response in glioma models in mice by genetically modified dendritic cells pulsed with Semliki forest virus-mediated complementary DNA. *J Neurosurg* 94(3):474–481.

Ying H, Zaks TZ, Wang RF, Irvine KR, Kammula US, Marincola FM, Leitner WW, Restifo NP. 1999. Cancer therapy using a self-replicating RNA vaccine. *Nat Med* 5(7):823–827.

Zhang YQ, Tsai YC, Monie A, Wu TC, Hung CF. 2010. Enhancing the therapeutic effect against ovarian cancer through a combination of viral oncolysis and antigen-specific immunotherapy. *Mol Ther* 18(4):692–699.

29 Synthetic Virus-Like Particles in Vaccine Design

Arin Ghasparian and John A. Robinson

CONTENTS

29.1 INTRODUCTION

In this review, we discuss the *synthetic viruslike particle* (SVLP) technology, which was conceived as a next-generation delivery system for synthetically derived vaccine candidates. A characterizing feature of the SVLP technology is the use of rational structure-based methods for vaccine design, coupled with the tools of synthetic and medicinal chemistry for production, purification, and characterization. The SVLP technology exploits advances in our understanding of the mechanisms underlying innate and adaptive immunity, the structural biology of neutralizing immune responses, and peptide and protein engineering for the design of conformationally stable B-cell epitope mimetics.

SVLP-based vaccine candidates comprise a novel nanoparticle delivery system, which uses synthetic lipopeptide building blocks that spontaneously self-assemble in aqueous buffer into highly immunogenic nanoparticles in the 20–30 nm size range. Around 60–80 copies of the lipopeptide are incorporated into each nanoparticle. B-cell epitopes, in the form of conformationally stable proteins or synthetic antigen mimetics (SAMs), can be conjugated using suitable linkers to the lipopeptide building blocks (Figure 29.1). Alternatively, small molecule haptens (e.g., drugs) or synthetic oligosaccharides can also be linked to the SVLPs. After self-assembly, 60–80 copies of the B-cell epitope are then displayed across the outer surface of the nanoparticle and can effectively activate B cells through cross-linking of B-cell receptors (BCRs). However, SVLPs contain no genetic information and cannot replicate in cells.

The SAMs used to elicit B-cell responses are designed to adopt folded 3D structures that accurately mimic the target epitopes found, for example, on infectious viruses and bacteria. The design of SAMs is facilitated by the rapidly growing database of crystal and NMR structures of nAb fragments bound to their target antigens.[1,2] This information can be exploited for the rational structure-based design of natively folded SAMs.[3] Computer-aided design methods are now also being used to graft conformational epitopes onto alternative protein scaffolds that can be produced either synthetically or recombinantly.[4–15] Again, SVLPs represent an ideal vehicle for the delivery of such grafted epitopes to the immune system.

The lipopeptide building blocks of SVLPs also incorporate T-cell epitopes. After SVLP uptake by antigen-presenting cells (APCs) and processing, presentation of these epitopes with MHC-I or MHC-II proteins at the cell surface initiates stimulation of T-cell responses. An additional layer in SVLP design exploits the flexibility of synthetic chemistry to introduce into the lipopeptide building blocks, pathogen-associated molecular patterns (PAMPs) that can be recognized by pattern recognition receptors (PRRs).[16–20] By incorporating ligands for PRRs, SVLP nanoparticles are endowed with self-adjuvanting properties and can directly stimulate PRRs in cells of the innate and adaptive immune systems. For example, the lipid component of the lipopeptide building blocks can be related to the lipid anchors found in bacterially derived membrane lipoproteins, which are potent stimulators of mammalian Toll-like receptors (TLRs).[21–23] These aspects of SVLP design are discussed in more detail in the following.

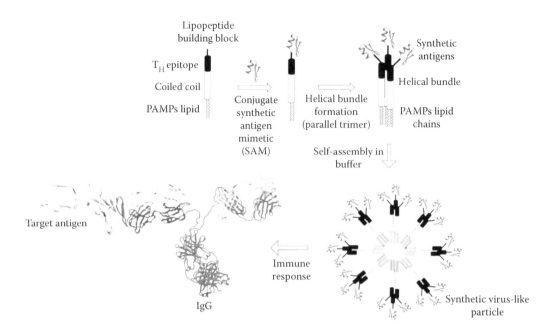

FIGURE 29.1 From lipopeptide design to a target-specific humoral immune response. SVLPs are formed from synthetic lipopeptide building blocks by a process of self-assembly in buffered aqueous solution. The lipopeptide contains a coiled-coil motif fused to a T-helper epitope and a lipid attached at one terminus. A structure-based, target-derived SAM is conjugated to the lipopeptide. An immune response (IgG) against the target antigen is elicited by SAMs displayed on the surface of the SVLP.

29.2 SELF-ASSEMBLY OF LIPOPEPTIDES INTO SVLPs

The lipopeptide building blocks required for SVLP formation contain a coiled-coil motif that plays a key role in the self-assembly process leading to nanoparticle formation. Two biophysical properties of the lipopeptide building blocks are exploited during self-assembly: firstly, the ability of the coiled-coil sequence to drive association of the lipopeptides into helical bundles of defined composition and secondly, the hydrophobic effect associated with the lipid tails, which drives self-association of the helical bundles, with burial of the lipid chains in the center of the micelle-like nanoparticle (Figure 29.1). The lipopeptide building blocks, therefore, typically contain (1) an appropriate coiled-coil sequence to direct formation of helical bundles, (2) one or more T-cell epitopes fused to the coiled coil, and (3) a lipid chain, such as one of the TLR2 ligands (Pam_2Cys or Pam_3Cys), or a phospholipid such as phosphatidylethanolamine, coupled to one terminus of the peptide. The immunological properties of the SVLPs are further tuned by incorporating PAMPs, which after nanoparticle uptake by APCs can interact with PRRs, and (4) a site and suitable conjugation chemistry for coupling to a B-cell epitope, such as a folded protein or a SAM (Figure 29.1).

29.2.1 Coiled-Coil Motifs for SVLP Engineering

The coiled coil is one of the most important and robust modules in contemporary peptide and protein engineering efforts aimed at the design of polypeptides with predictable folded architectures and functional properties.[24,25] Coiled coils adopt amphipathic α-helical structures that promote self-association, either within or between polypeptide chains, through formation of helical bundles.[26,27] Many coiled-coil helical bundles are found in nature, but the increasing understanding of the *rules* that govern coiled-coil association has enabled in some cases the creation of novel assembled nanostructures from primary sequence alone.[28] This opens new avenues to rational design and to *nonnatural* peptide sequences that self-assemble into nanostructures with predetermined architectures and are potentially of broad value in biomaterials science and synthetic biology.[29–33]

Coiled coils display great flexibility in mediating protein–protein interactions, in terms of the number, constitution, and orientation (parallel or antiparallel) of interacting helices (Figure 29.2).[34,35] Many coiled-coil domains contain easily recognizable, tandemly duplicated heptad motifs, in which patterns of hydrophobic (H) and polar (P) residues occur in a defined order $(HPPHPPP)_n$, often presented as $(abcdefg)_n$ to denote a tandemly repeated seven residue unit where residues a and d are hydrophobic and the rest are polar. This spacing of hydrophobic residues, combined with the helical periodicity of ca. 3.6 residues per helical turn, leads to amphipathic α-helical structures, where all the hydrophobic side chains lie on one face of the α-helix. The nonpolar faces are then buried upon helix–helix association. In naturally occurring proteins, helical bundles with 2, 3, 4, and 5 parallel helices have been defined crystallographically.[34,35] The slight mismatch between spacing and helical periodicity (3.5 vs. 3.6 residues/turn) leads to a preferred left-handed

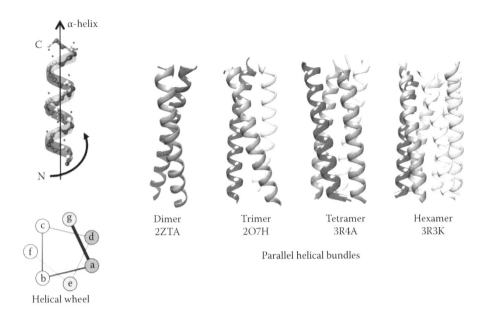

FIGURE 29.2 Naturally occurring and designed coiled coils assemble into helical bundles with defined oligomer number and orientation. Shown is a typical α-helical conformation, and a helical wheel showing how the residues of a heptad repeat (*abcdefg*) are displaced around an α-helix, with residues closer to the helix–helix interfaces shaded progressively darker. Crystal structures of a coiled-coil parallel dimer (2ZTA), trimer (2O7H), tetramer (3R4A), and hexamer (3R3K) are also shown.

supercoiling of the helices around each other, and a characteristic knobs-into-holes packing of hydrophobic side chains at the helical interface. Furthermore, the polar residues at positions *e* and *g* in each heptad repeat, which flank the hydrophobic core, offer possibilities for additional stabilizing interhelical contacts, for example, through salt bridging (Figure 29.2).

The types of residues at the *a*, *d*, *e*, and *g* positions in a heptad repeat play a dominant role in defining the nature of the helical interfaces in coiled-coil assemblies. Such information provides a starting point for the design of nonnatural and engineered peptide sequences that target predetermined oligomeric states, such as parallel dimer, trimer, or tetramer.[30,36,37] For example, designed sequences with the signatures *a* = *d* = Ile and *a* = Leu with *d* = Ile generally specify parallel trimeric and tetrameric coiled coils, respectively. However, in other cases, the preferred oligomeric state of a coiled coil can be very sensitive to the precise choice of residue at these key positions, showing that the energetic differences between different coiled-coil states can sometimes be very small.[30]

Many naturally occurring coiled coils form 2, 3, 4, or 5 helix bundles based on heptad repeat units.[34] However, a naturally occurring parallel right-handed coiled-coil tetramer contains an 11-residue repeat.[38] Only a few higher-order parallel helical bundles have so far been described. An interesting example is a *de novo* designed peptide that forms a regular hexameric helical bundle.[39] Starting from a known tetrameric coiled coil, residue changes were made to widen the interfacial angle between individual helices, and to extend the knobs-into-holes interactions to include not only the *a*–*d* side chains, but also the *e*, *g*, *b*, and *c* side

chains (Figure 29.2). The resulting peptide forms a unique parallel six-helix bundle with the anticipated regular classical coiled-coil packing arrangement. In another example, an engineered mutant of a GCN4-p1 leucine zipper peptide was shown to form a parallel seven-helix bundle.[40] The GCN4-p1 peptide itself is a parallel dimeric coiled coil, although various mutants are known to adopt parallel 2-, 3-, or 4-helix structures.[36] Finally, mention can be made of a 12-helix ring formed by part of the bacterial exporter TolC, comprising 6 antiparallel dimers.[41] As yet there are no known coiled coils with bundles of 8, 9, 10, or 11 helices.[25] In the case of synthetically derived coiled-coil sequences, there is also the possibility to modulate the properties of the helical bundles through the incorporation of nonnatural amino acids. Already coiled coils containing β-amino acids, as well as fluorinated and glycosylated α-amino acids, have been described.[42–45] These and other nonnatural modifications open an almost unlimited molecular space to explore in the optimization of coiled coils for SVLP design.

Of the naturally occurring coiled coils, those found in various viral fusion proteins also represent interesting starting points for use in SVLP design, since these structures often contain B- and T-cell epitopes that can elicit protective/neutralizing immune responses in humans.[13,46–48] For example, parallel trimeric coiled-coil sequences are found in the class I fusion proteins of influenza virus, SARS virus, Ebola virus, RSV, HIV-1, and parainfluenza viruses.[49] The trimeric coiled coil from the RSV F glycoprotein was exploited in an early example of SVLP design.[50] Progress is also being made in the development of hetero-oligomeric coiled coils. For example, peptide sequences have been designed that adopt stable parallel heterodimeric and heterotrimeric helical bundles.[51–53]

29.2.2 INCORPORATING T-CELL EPITOPES IN SVLP DESIGN

Acquired immunity involves both humoral immunity, resulting from the production of antibodies by B lymphocytes, as well as cell-mediated immunity by the activation of CD8[+] T lymphocytes. Also, CD4[+] T-helper lymphocytes are required for the B-cell-mediated production of class-switched, affinity-optimized, antibody responses to intact antigens. Whereas antibodies recognize intact soluble antigens, T cells only *see* their antigen (T-cell epitope) when it is presented on the surface of an APC in a complex with a major histocompatibility complex (MHC) molecule (human MHC molecules are also referred to as human leukocyte antigens [HLAs]).[54–56] The T-cell epitope is generally a peptide of 10–20 residues that has been derived from either *self-* or *nonself* proteins by proteolysis. The T-cell receptor (TCR) on the T cell recognizes and responds to foreign peptides bound to MHC class I or class II proteins on the surface of the APC.

Crystallographic data have revealed that the MHC class I and II molecules have similar overall structures, including two extended α-helices that sit above a floor created by strands of a β-pleated sheet.[55] This architecture forms a distinct groove between the two helices into which a complementary T-cell epitope can bind (Figure 29.3).[57] However, the peptide-binding groove of the class I molecule is more restricted in size than that of the class II molecule, which accounts for the observation that MHC-I-binding peptides are typically shorter (10 residues) than those that bind to MHC-II molecules (14–16 residues). Peptides bound to MHC-II can extend out of both ends of the binding groove. However, MHC class I–bound peptides can also be longer, and because of restrictions on the dimensions of the peptide-binding cleft, this can cause the peptide to bulge out of the binding pocket.

The TCR bears structural homology to the immunoglobulin family and typically consists of two disulfide-linked chains, α and β, each of which folds into two immunoglobulin-like domains.[56] Both the MHC-peptide complex and TCR present relatively flat interfacial surfaces, so when a typical TCR recognizes a MHC-peptide complex, the TCR interacts not only with the T-cell epitope, but also with the MHC protein (Figure 29.3). Indeed, MHC haplotype restriction demonstrates a direct requirement for MHC recognition by the TCR.

Although the structures of only a few TCR–peptide–MHC-II complexes have so far been determined, one well-studied example is the HA1.7 TCR and a so-called promiscuous CD4[+] T-cell epitope derived from influenza hemagglutinin (HA$_{306–318}$) bound to a HLA-DR1 molecule (Figure 29.3).[58–60] This particular T-cell epitope can bind to many different MHC-II haplotypes, and the HA1.7 TCR can promiscuously bind to different MHC-IIs, presenting a wide range of distinct peptide epitopes. In the HA1.7-HA$_{306–318}$-HLA-DR4 complex, the T-cell epitope binds in an extended, but not regular β-sheet conformation, with five side chains projected *up* to contact the TCR, and a similar number pointing toward the MHC-II molecule. Both electrostatic and hydrophobic interactions are seen between the peptide and its binding partners, but not in a regularly alternating pattern, as seen, for example, in a regular β-sheet or the heptad repeats in coiled-coil peptides. In the intact folded influenza hemagglutinin, this HA$_{306–318}$ sequence adopts a quite different conformation to that seen when bound to the MHC-II molecule.[61]

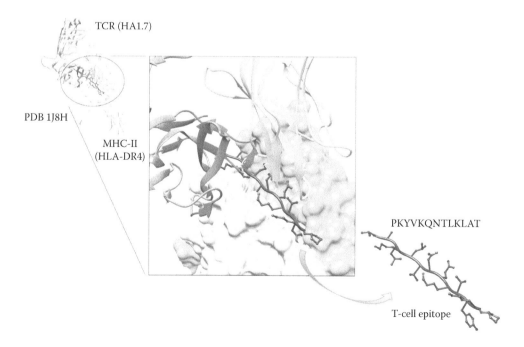

FIGURE 29.3 Structural anatomy of an MHC molecule (green/blue chains) and recognition by the TCR (yellow/orange chains) of a peptide–MHC complex (PDB 1J8H). The HA1.7 TCR and a *promiscuous* T-cell epitope (pink) derived from influenza hemagglutinin (HA$_{306–318}$) bound to a HLA-DR4 MHC-II molecule.

An important part of SVLP design is the incorporation of relevant T-cell epitopes into the key lipopeptide building blocks, by direct fusion with the coiled-coil motif. The high thermodynamic stability of coiled coils can be readily exploited to design fusions between coiled coils and T-helper epitopes, without affecting the formation of stable helical bundles. One example is illustrated in Figure 29.4.[62] Here, the coiled coil (IEKKIEA)$_4$ is known to form trimeric parallel helical bundles with high thermal stability.[63] This coiled coil was fused to a promiscuous CD4+ T-helper epitope (IEKKIAKMEKASSVFNVVNS) previously identified in the circumsporozoite (CS) protein of the malaria parasite *Plasmodium falciparum*. This epitope corresponds (with two Cys-to-Ala substitutions) to residues 379–398 of the CS protein, and is recognized by mouse and human T cells in association with a wide variety of different MHC class II molecules.[64,65] It is notable that fusion of the two sequences is possible while retaining the pattern of hydrophobic residues at *a* and *d* positions, well into the sequence of the T-cell epitope (Figure 29.4). The design of the lipopeptide building block was completed by addition of KKKC at the C-terminus, to allow conjugation of a B-cell epitope through the cysteine thiol. A lipid moiety was added at the N-terminus, comprising either a phospholipid related to phosphatidylethanolamine, or Pam$_2$Cys or Pam$_3$Cys moieties.

Similar design strategies can be followed to generate lipopeptide building blocks containing other T-cell epitopes. For example, in an SVLP-based infectious disease vaccine,

it may be advantageous to have both B- and T-cell epitopes from the same bacterial or viral pathogen target in the same nanoparticle.

29.2.3 INTEGRATING PAMPS INTO SVLPS

The stimulation of a strong immune response to vaccination requires an appropriate coactivation of the innate immune system.[16,66] Vaccines that incorporate signals that target innate immune receptors typically show significantly enhanced vaccine-induced immunity.[67] The PRRs are an important family of germ-line-encoded immunoreceptors found in cells of the innate and adaptive immune systems, including DCs and B and T lymphocytes.[16] PRRs detect invading pathogens by binding to *PAMPs*. It is, therefore, important to incorporate ligands (PAMPs) for PRRs into SVLP nanoparticles.

The different families of PRRs contain a small number of distinct protein folds, including leucine-rich repeats (LRRs) in TLRs and NOD-like receptors (NLRs), the DExH box helicase domain in RIG-like receptors (RLRs), and the C-type lectin (CTL) domain in the CTLs.[68] The PAMPs recognized by these receptors include bacterial lipopeptides, lipopolysaccharide, and oligosaccharides, as well as macromolecules such as viral or bacterial DNA, RNA, and pathogen-derived proteins such as flagellin. In addition, an increasing number of nonnatural, synthetic ligands are being developed that bind to PRRs and act as either

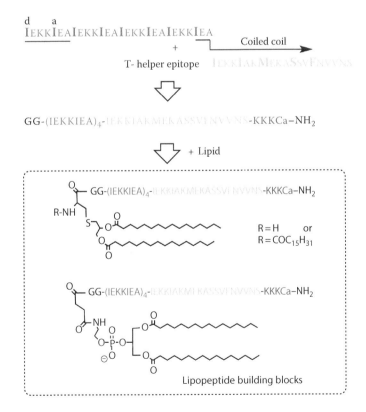

FIGURE 29.4 Design of typical lipopeptide building blocks. A coiled-coil sequence with the heptad repeat (IEKKIEA)$_4$ (blue) is fused to a promiscuous T-helper epitope (green). As lipid component, the TLR ligands Pam$_{2/3}$Cys and 1,3-dipalmitoyl-glycero-2-phosphoethanolamine are shown. SAMs are linked through the unique Cys residue at the C-terminus (red) of the lipopeptide.

agonists or antagonists, thereby influencing their downstream signaling pathways. These molecules may be useful as adjuvants to stimulate an adaptive immune response to a vaccine, and to influence the type of response, so that the most effective form of immunity for each specific pathogen is produced (e.g., Th1 versus Th2 responses, or by enhancing cross-presentation of T-cell antigens by DCs).[67,69] In the course of a natural viral or bacterial infection, multiple PAMPs will likely generate overlapping signals delivered by a combination of TLRs, NLRs, RLRs, and/or CLRs. However, overstimulation by some TLR agonists can cause unwanted systemic inflammatory responses.[70,71] Use of the SVLP nanoparticle delivery system with selected TLR agonists may be ideal for achieving the enhanced immunogenicity sought in a vaccine adjuvant, while avoiding safety concerns due to overly strong inflammatory responses.

The classes of synthetic PRR ligands of interest in SVLP design include lipid-based analogues of $Pam_{2/3}Cys$ and lipid A. Other lipids that interact with TLRs[72] or are involved in the activation of invariant natural killer T (iNKT) cells, synthetic derivatives of dsRNA and DNA that are recognized by TLRs and RLRs,[73,74] synthetic heterocycles that stimulate TLR7/8,[72] carbohydrate ligands that interact with CLRs,[75] and peptidoglycan fragments that bind to NLRs also play key roles in innate immune responses.[76] Glycolipids, such as α-GalCer, are important in the activation of iNKT cells, which in turn also provide both innate and adaptive forms of help to B cells.[77]

Crystal structures have revealed how the TLR1/2 heterodimer binds to lipopeptides with a triacylated N-terminus (Pam_3Cys), while the TLR2/6 heterodimer binds lipopeptides with a diacylated N-terminus (Pam_2Cys).[21,78,79] All TLRs share a common basic 3D architecture, with multiple LRRs arranged in a horseshoe- or crescent-shaped structure that together comprises the N-terminal ectodomain responsible for ligand binding.[79] A single transmembrane segment links the ectodomain to the C-terminal domain, which interacts with adapter proteins in the cytoplasm, an event that culminates in the activation of transcription factors and to altered patterns of gene expression.[80] Structural studies support the view that lipopeptides/lipoproteins containing Pam_2Cys or Pam_3Cys represent specific ligands of TLR2, recognized by heterodimers of TLR1/2 or TLR2/6 (Figure 29.5). All bacterial lipoproteins contain a glycerol moiety linked via a thioether to the side chain of an N-terminal Cys residue (S-[2,3-bis(acyloxy)-(2R)-propyl-(R)-cysteinyl), and acylated with two long chain fatty acids. Lipoproteins anchored in the membrane of Grampositive bacteria generally have a free N-terminus (Pam_2Cys), whereas in Gram-negative bacteria, lipoproteins contain three lipid chains (Pam_3Cys). The residues immediately downstream of the N-terminal Cys in bacterial lipoproteins are variable in sequence, but the first is typically Gly or has a small polar side chain. The following residues do not exert a strong influence on signaling and do not make significant contact to the TLR upon binding (Figure 29.5).[68,79] Structure-activity studies

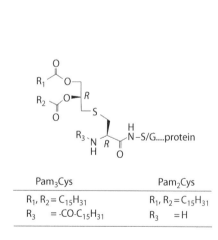

Pam_3Cys	Pam_2Cys
$R_1, R_2 = C_{15}H_{31}$	$R_1, R_2 = C_{15}H_{31}$
$R_3 = -CO-C_{15}H_{31}$	$R_3 = H$

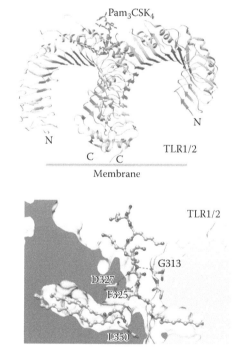

FIGURE 29.5 *Left*, Structures of lipoproteins Pam_2Cys and Pam_3Cys; *right*, crystal structure of the human heterodimeric TLR1/2-Pam_3CSK_4 complex (PDB 2Z7X); TLR2 (blue), TLR1 (green), lipid (red). View from the side with membrane below and a slice through the complex showing the buried binding sites for the lipid chains. The two ester-linked lipids project into a binding pocket in TLR2 (blue) and the amide-linked lipid into one in TLR1 (green). Hydrogen-bonding (HB) interactions (pink dotted) with the ligand head group are highlighted.

have revealed that the immune modulatory activity is strongly dependent on the fatty acid chain length and the correct absolute configuration of the natural $(2R)$-dihydroxypropyl-(R)-cysteine.[81-84] Also, replacement of the sulfur by CH_2 reduces biological activity significantly.[84]

The incorporation of $Pam_{2/3}Cys$ into one of the lipopeptide building blocks needed for SVLP self-assembly is straightforward. Most convenient is conjugation of the lipid at the N-terminus of the peptide, which can be performed *on-resin* after assembly of the peptide chain by solid-phase peptide synthesis. In the example shown in Figure 29.4, two Gly residues are included as linker between the $Pam_{2/3}Cys$ moiety and the start of the coiled-coil heptad repeats. During peptide synthesis, after each amino acid coupling, a capping step can be performed with acetic anhydride. This has the practical advantage that after completion of peptide assembly, and coupling of $Pam_{2/3}Cys$ to the free N-terminus, the HPLC retention time of the peptide is dramatically altered by lipidation, thus greatly facilitating HPLC purification of the desired lipopeptide. As an alternative to $Pam_{2/3}Cys$, lipopeptide building blocks containing a phospholipid moiety at the N-terminus (e.g., phosphatidylethanolamine) have also been described.[50,62] Attachment of the lipid moiety at the C-terminus of the peptide chain is also feasible, using suitable conjugation chemistry (unpublished work).

29.3 SVLPs FOR DELIVERY OF B-CELL EPITOPES

Most successful prophylactic vaccines to date rely for their protective effects on the production of neutralizing antibodies (nAbs) in the animal host. In recent years, structural biology has contributed enormously to our understanding of how antibodies recognize vaccine antigens in general, and neutralizing epitopes in particular. The crystal structure of a pathogen-derived antigen bound to its cognate nAb reveals the detailed 3D structure of the epitope, comprising often a unique surface region of a folded protein, against which the protective humoral response was elicited. This 3D structural information is of great value for the design of novel and improved antigens. For example, peptide and protein engineering approaches have recently been reported for epitope grafting onto suitable folded peptide or protein scaffolds.[6,7,9-15]

The nAbs often used for structural studies are typically derived from the late stages of an immune response, after multiple rounds of antibody somatic mutation and selection. Their production requires triggering of naive, perhaps rare BCRs, to prime for the nAb response. For this, an immune response focused on epitope mimetics with a native-like fold, and delivered in a multivalent format, with provision of promiscuous CD4$^+$ T-cell epitopes and appropriate PRR stimulation, may be particularly effective. Indeed, long-acting antibody responses, elicited after only one or two antigenic exposures, are predicted to occur when a multivalent antigen triggers strong B-cell activation with effective T-cell help.[85,86] One interesting example of a structure-based, epitope-focused vaccine design strategy was described recently, using structural information on an epitope on respiratory syncytial virus (RSV) targeted by the licensed, prophylactic nAb palivizumab (also known as Synagis) and an affinity-matured variant (motavizumab).[13]

SVLPs provide an ideal delivery vehicle for conformationally stable B-cell epitope mimetics. One recent example focused on the V3 loop in the envelope glycoprotein gp120 of HIV-1. The β-hairpin V3 loop in gp120 becomes exposed in the CD4-induced conformation of gp120, after the virus docks with the primary CD4 receptor on target cells.[87] The tip of the V3 loop is then able to interact with the chemokine coreceptor (CXCR4 or CCR5), which initiates virus entry into the cell. Several crystal structures of nAb fragments bound to linear peptides derived from the V3 loop have been determined.[88-92] V3-loop-derived linear peptides, however, are flexible in solution and do not adopt folded structures. Consequently, linear peptides cannot be expected to efficiently elicit antibodies specific to a folded V3 epitope. However, conformationally stable V3 loop mimetics have been designed, in the form of macrocyclic peptides containing a β-hairpin-stabilizing D-Pro-L-Pro template.[93,94] One of these cyclic peptides accurately mimics the V3 loop seen in a complex with the nAb F425-B4e8 (Figure 29.6).[95] This β-hairpin V3 loop mimetic was coupled to SVLPs through a unique Cys residue near the C-terminus of the lipopeptide building block.[93] A computer model of the resulting SVLPs is shown in Figure 29.6, which is supported by extensive biophysical data (see below). These V3-SVLPs elicited high IgG titers in rabbits, including antibodies that bind to gp120 and show HIV-1 neutralizing activity, although not against neutralization-insensitive tier-2 HIV-1 strains unless the viruses were first engineered by deleting the V1V2 loop region.[93] This is consistent with more recent immunological and structural studies showing that V3 loops on the viral surface are often inaccessible to antibody binding, due to active shielding by the V1V2 loops.[96-99] Indeed, the V2 loop has recently become a hot target for vaccine design, which may also be addressed using the SVLP technology.

New technologies are now emerging for the design of conformationally stable folded protein domains,[10,100,101] including peptides containing stapled helices,[102,103] and miniproteins with defined folds.[104] Such folded domains may be useful for epitope grafting and immunogen design. New technologies, such as native chemical ligation,[105] are also now allowing the efficient synthesis of large folded peptides and proteins by chemical synthesis.[106] We predict that as this field advances, the engineering of a large variety of conformationally stable B-cell epitope mimetics will become feasible. The SVLP technology provides an ideal platform for delivering folded epitope mimetics to the immune system.

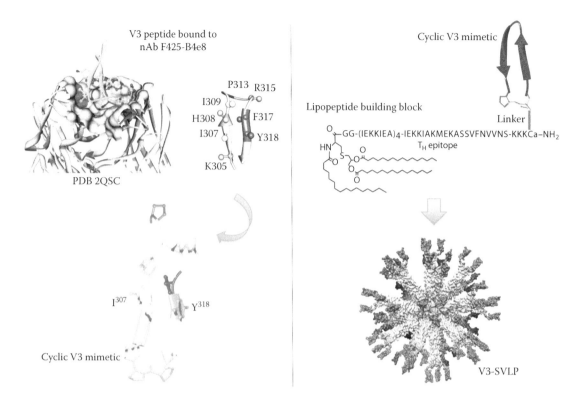

FIGURE 29.6 From structure-based SAM design (*top left*) to an SVLP-based HIV-1 vaccine candidate (*bottom right*).[93] *Left*, crystal structure (PDB 2QSC) showing the V3 peptide (green) bound to the HIV-1 nAb F425-B4e8; the Ab-bound V3 loop conformation is shown adjacent, with cross strand HB by dotted lines, and selected β-carbons as balls; *below*, NMR structure of a designed macrocyclic V3 mimetic linked to a D-Pro-L-Pro template (Ile, green; Phe, dark pink; Tyr, light pink; D/L-Pro, orange; Gly, yellow). *Right*, the V3 mimetic (red/orange) linked to a lipopeptide building block (Figure 29.4) and *below*, computer model of the resulting SVLP nanoparticle containing 24 helical bundles and 72 copies of the loop mimetic (coiled coil/T_H epitope in blue/green).[93] The lipid chains are buried in the core of the particle.

29.4 NANOSIZED SVLPs

When the lipopeptide building blocks are dissolved in phosphate buffer, the formation of homogeneous SVLP nanoparticles, with diameters in the 20–30 nm range, can be readily detected by dynamic light scattering (DLS).[50,62,93] Sedimentation equilibrium analytical ultracentrifugation (SE-AUC) analyses suggest that the nanoparticles have a mass close to 500–600 kDa, and that each particle contains 20–30 helical bundles, and so 60–90 copies of the epitope mimetic. The nanoparticles can be visualized by negative staining transmission electron microscopy. These experimental data support the computer model of a typical SVLP nanoparticle shown in Figure 29.6. The computer model contains 24 trimeric helical bundles, arranged manually in a spherical assembly with a diameter of ca. 30 nm, with the coiled coils radiating out with regular spacing from the central lipid core. In aqueous solution, however, thermal motion can be expected, so the particles should not possess a formal symmetry.

The computer model of a related SVLP was also assessed by small-angle x-ray scattering (SAXS) and small-angle neutron scattering (SANS) experiments.[62,107] SAXS analysis suggested that the average nanoparticle size in solution is slightly larger than predicted by the computer model, likely due to slight variations in particle size in the range 0.85×–2.0×. SANS studies were also performed with lipopeptide building blocks containing phosphatidylethanolamine (Figure 29.4) and fatty acid chains with protium (^1H) and with fully deuterated (^2H) lipid chains. Comparing the SANS scattering patterns observed with both SVLP samples showed clearly that the lipid chains are buried in the center of the particle.[62,107] The SANS data also indicated that the lipid chain packing density and aggregation number were significantly constrained by the coiled-coil motif of the peptide head groups, which are closely packed in this part of the micelle. In the presence of Ca^{2+} ions, the SANS scattering data were consistent with sequestration of the Ca^{2+} ions, with the phosphate groups within the interior of the lipopeptide micelles.[107] TEM micrographs of the Ca^{2+}-loaded SVLPs showed discrete spherically shaped electron-dense nanoparticles with a mean diameter of 22 ± 3 nm. Fluorescent spectroscopy studies indicated that small organic fluorophores such as pyrene and perylene could permeate through the peptide corona into the hydrophobic core of the calcium-free but not calcium-loaded SVLPs. Alternatively, the SVLPs could be first loaded with dye and then enclosed in the hydrophobic core by treatment with Ca^{2+} and calcium phosphate.[107]

29.5 SVLPs AND INTERACTIONS WITH B CELLS AND DENDRITIC CELLS

Dendritic cells (DCs) are among the first to encounter an invading pathogen or vaccine in peripheral tissues, an event that leads to rapid antigen uptake into endosomes or phagosomes, followed by maturation and migration of the DCs to draining lymph nodes.[108] Free microbial proteins, including VLPs, may be carried by the draining lymph fluid to the secondary lymphoid organs for uptake by resident DCs.[109] In either case, DCs in secondary lymphoid organs process endocytosed or phagocytosed antigens by proteolysis after fusion with vesicles containing newly synthesized MHC-II molecules. The antigen-derived peptide–MHC-II complexes are then trafficked to the cell surface. In the lymph nodes, a small fraction of the large pool of naive CD4+ T cells will express a TCR capable of strong or sustained binding to any given microbial peptide–MHC-II complex.[110] This interaction leads to activation of these rare T cells, the production of growth factors and cytokines, rapid cell division, which in turn produces several hundred thousand progeny effector T cells. During this process, the effector T cells differentiate into specific subsets, including T helper 1 (Th1), T helper 2 (Th2), T helper 17 (Th17), and T follicular helper cells.[111] Signals elicited by PRRs in various immune cells in response to PAMPs in the pathogen or vaccine have a major influence on this qualitative aspect of the immune response.[67]

DCs and macrophages take up extracellular proteins into endosomes by macropinocytosis or particles by phagocytosis. However, B cells are very inefficient at these processes and can efficiently internalize only antigens that bind to their BCR. After internalization, processing leads to antigen-derived peptide–MHC-II complexes that are trafficked to the surface, ready for interaction with TCRs. Activated CD4+ T cells provide *help* to activated B cells to undergo class switching and somatic hypermutation within specialized regions of the lymph nodes. Thus, engagement of BCR by cognate antigens presented by APCs in lymph nodes drives the first stages of B-cell activation.

Although B cells in lymph nodes can gain rapid access to soluble antigens arriving through afferent lymph vessels, larger antigens, such as viral aggregates tethered to the surface of APCs (such as macrophages, DCs or fDCs), are particularly efficient at inducing B-cell responses.[112] The interaction of B cells with antigens presented on APCs triggers the formation of an immunological synapse that facilitates the efficient extraction and processing of membrane-tethered antigens.[113] In a rapid actin-dependent membrane spreading response at the center of this synapse between the B cell and APC, multiple BCRs engage a cluster of antigens and are then surrounded by a ring of membrane-bound signaling molecules. This spreading response helps to increase the number of BCR–antigen complexes and coreceptors into a BCR microcluster. This is followed by a contraction phase in which antigen–BCR complexes are drawn into a central cluster, mediated by rearrangements of the cortical actin cytoskeleton of the B cell.[112] This process is intimately linked to the signaling capacity of the cell, and the amount of antigen that is accumulated at the synapse and available for efficient antigen extraction and processing. For this reason, a multivalent format of antigen delivery, such as with SVLPs, is important for inducing strong adaptive humoral immune responses.

BCR internalization with bound antigen is clathrin dependent and leads to rapid degradation of antigens and processing to form peptide–MHC-II complexes. Indeed good evidence exists that antigen degradation may already begin before internalization, at the immunological synapse, through the actions of secreted extracellular proteases and lipases.[114] After antigen encounter, internalization, and processing, B cells migrate in the lymph node toward the T-cell boundary, where contacts with cognate T-helper cells can be established. The interaction between B and T cells also leads to the formation of an immunological synapse, where bidirectional activation signals are exchanged between cells. After interacting with T cells, some B cells migrate to the interfollicular region of the lymph node, where they proliferate into short-term plasmablasts that contribute to the primary immune response by generating antibodies, typically of relatively low affinity. Other activated B cells migrate into the B-cell follicle, where they continue to proliferate and form germinal centers. Here, the B cells undergo affinity maturation and differentiate into either antibody-producing plasma cells or long-lived memory B cells.[112]

It is of special interest to understand how DCs interact with a complex antigen like an SVLP. Studies with recombinant proteins have shown that DCs degrade internalized protein antigens rather slowly, and thus retain antigen in lymphoid organs for extended periods, enhancing their ability to disseminate antigens throughout the immune system.[115] Fluorescently labeled SVLPs have been used to monitor their interaction with porcine monocyte-derived DCs.[116] These studies showed that SVLPs were rapidly bound (within seconds) by DCs, and cell uptake peaked around 30 min. Thereafter, most SVLPs localized into peripheral structures. Endocytosis was dominated by a rapid caveolin-independent route involving lipid raft polarization. Although the main uptake mechanism was macropinocytosis, other mechanisms seemed to operate in the background. Processing of SVLPs was much slower, involved fusion with EEA-1+ early endosomes in the periphery. Further studies showed that the degradation in DCs was relatively slow and antigen retention was high, rather like that observed with protein antigens. Slow antigen processing is likely beneficial and potentially allows peptides from SVLPs to enter sorting compartments for presentation to B lymphocytes.[117,118] Some colocalization with MHC-II-positive structures at the periphery could also be observed.[116] Further studies of SVLPs carrying TLR2 ligands showed that SVLPs activated DCs in a dose-dependent manner, leading to secretion of proinflammatory cytokines, such as IL6.[119]

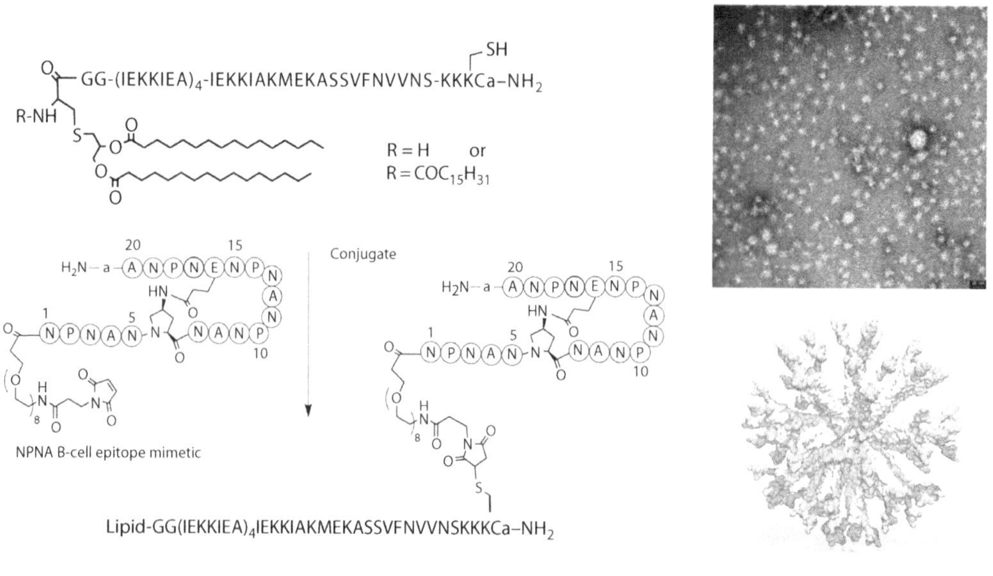

FIGURE 29.7 An SVLP-based synthetic malaria vaccine candidate. *Left*, the lipopeptide building block (Figure 29.4) is conjugated to a macrocyclic conformationally constrained B-cell epitope mimetic based upon the NPNA repeat region of the CS protein of the malaria parasite. *Right*, an EM image of the SVLP nanoparticles and a computer model of an idealized SVLP particle showing multivalent display of the NPNA mimetic (orange).

29.6 IMMUNE RESPONSES FROM SVLPs

The very high immunogenicity of SVLPs is also shown by studies on a potential malaria vaccine candidate.[62] In this work, the lipopeptide building block described earlier (Figure 29.4) was combined with a conformationally constrained B-cell epitope mimetic. The SAM comprised a macrocyclic peptide that was designed to mimic the immunodominant NPNA (Asn-Pro-Asn-Ala) epitope of the CS protein from *P. falciparum*.[120] This tetrapeptide motif is tandemly repeated almost 40 times in the CS protein. Antibodies targeting this region of the CS protein block infection of human liver hepatocytes by sporozoites. The NPNA motif adopts type-I β-turn conformations in crystals of a pentapeptide (Ac-ANPNA).[121] β-Turns were stabilized in the B-cell epitope mimetic by incorporating NPNA motifs into a macrocyclic structure (Figure 29.7). The resulting macrocyclic NPNA-mimetic was conjugated to the lipopeptide building block, which also contains a promiscuous T-helper epitope from the CS protein, using a maleimide linker and the unique Cys residue close to the C-terminus of the lipopeptide. A convergent synthetic strategy gave the final SVLP building block in good synthetic yield and purity.

This SVLP-based vaccine elicited very high titers of parasite-specific IgG in rabbits and mice. In rabbits, the response reached almost saturation after a single boost with an antigen dose of 10 μg. Significant improvements of the avidity index of the antibodies between the first and second antigen exposure supported the occurrence of affinity maturation during the immune response to the vaccine.[62] In a dose ranging study, strong titers were also observed at lower antigen doses (0.4 μg), although two

booster immunizations were then required to achieve complete seroconversion in all animals. These SVLPs were also highly immunogenic in mice. The response was of long duration with only a marginal decline in antibody titers over 40 weeks and IgG1/IgG2a ratios of 1.2–3.2. SAR studies showed that self-assembly of the building blocks into SVLP nanostructures, and the stability of the nanoparticles, has an important influence on immunogenicity. Deletion of the coiled-coil sequence from the lipopeptide building block abolished the IgG response, indicating that, in the absence of coformulated adjuvants or TLR ligands, a well-defined nanostructure is important for the strong antibody responses. Finally, the rabbit antibodies were also shown to bind *P. falciparum* sporozoites. Very high titers of sporozoite cross-reactive IgG could be detected by immunofluorescence assays in sera collected after the second and third immunizations.

The immunological properties of SVLPs have also been studied in pigs.[116] Pig and human DC subsets are closely related functionally, which make pigs an attractive model for immunological studies. Immunization of pigs with SVLPs carrying an epitope mimetic derived from foot and mouth disease virus, without additional adjuvant, elicited strong antibody and activated/memory T-cell responses, thereby confirming the high immunogenicity of SVLPs in these larger animals.

ACKNOWLEDGMENTS

The authors thank Dr. Kerstin Moehle for help in preparing the figures. The SVLP technology is being developed for commercial use by Virometix AG.

REFERENCES

1. P. R. Dormitzer, G. Grandi, and R. Rappuoli, *Nat. Rev. Microbiol.* 2012, **10**, 807.

2. P. R. Dormitzer, J. B. Ulmer, and R. Rappuoli, *Trends Biotechnol.* 2008, **26**, 659.

3. J. W. Back and J. P. M. Langedijk, In *Adv. Immunol.*; C. J. M. Melief, ed.; Elsevier Academic Press, Inc., San Diego, CA, 2012; Vol. 114, p. 33.

4. D. Corti and A. Lanzavecchia, *Annu. Rev. Immunol.* 2013, **31**, 705.

5. D. W. Kulp and W. R. Schief, *Curr. Opin. Virol.* 2013, **3**, 322.

6. B. E. Correia, J. T. Bates, R. J. Loomis, G. Baneyx, C. Carrico, J. G. Jardine, P. Rupert et al., *Nature* 2014, **507**, 201.

7. S. J. Fleishman, T. A. Whitehead, D. C. Ekiert, C. Dreyfus, J. E. Corn, E.-M. Strauch, I. A. Wilson, and D. Baker, *Science* 2011, **332**, 816.

8. J. S. McLellan, B. E. Correia, M. Chen, Y. P. Yang, B. S. Graham, W. R. Schief, and P. D. Kwong, *J. Mol. Biol.* 2011, **409**, 853.

9. M. L. Azoitei, Y. E. A. Ban, J. P. Julien, S. Bryson, A. Schroeter, O. Kalyuzhniy, J. R. Porter, Y. Adachi, D. Baker, E. F. Pai, and W. R. Schief, *J. Mol. Biol.* 2012, **415**, 175.

10. M. L. Azoitei, B. E. Correia, Y.-E. A. Ban, C. Carrico, O. Kalyuzhniy, L. Chen, A. Schroeter et al., *Science* 2011, **334**, 373.

11. B. E. Correia, Y. E. A. Ban, M. A. Holmes, H. Y. Xu, K. Ellingson, Z. Kraft, C. Carrico et al., *Structure* 2010, **18**, 1116.

12. J. Holm, M. Ferreras, H. Ipsen, P. A. Wurtzen, M. Gajhede, J. N. Larsen, K. Lund, and M. D. Spangfort, *J. Biol. Chem.* 2011, **286**, 17569.

13. J. S. McLellan, M. Chen, M. G. Joyce, M. Sastry, G. B. E. Stewart-Jones, Y. P. Yang, B. S. Zhang et al., *Science* 2013, **342**, 592.

14. J. S. McLellan, M. Pancera, C. Carrico, J. Gorman, J.-P. Julien, R. Khayat, R. Louder et al., *Nature* 2011, **480**, 336.

15. R. L. Stanfield, J.-P. Julien, R. Pejchal, J. S. Gach, M. B. Zwick, and I. A. Wilson, *J. Mol. Biol.* 2011, **414**, 460.

16. M. C. Michallet, G. Rota, K. Maslowski, and G. Guarda, *Curr. Opin. Microbiol.* 2013, **16**, 296.

17. S. Akira, *Phil. Trans. R. Soc. B* 2011, **366**, 2748.

18. H. Kumar, T. Kawai, and S. Akira, *Biochem. J.* 2009, **420**, 1.

19. H. Kumar, T. Kawai, and S. Akira, *Int. Rev. Immunol.* 2011, **30**, 16.

20. O. Takeuchi and S. Akira, *Cell* 2010, **140**, 805.

21. M. S. Jin and J.-O. Lee, *Immunity* 2008, **29**, 182.

22. I. Botos, D. M. Segal, and D. R. Davies, *Structure* 2011, **19**, 447.

23. T. Kawai and S. Akira, *Nat. Immunol.* 2010, **11**, 373.

24. A. L. Boyle and D. N. Woolfson, *Chem. Soc. Rev.* 2011, **40**, 4295.

25. D. N. Woolfson, G. J. Bartlett, M. Bruning, and A. R. Thomson, *Curr. Opin. Struct. Biol.* 2012, **22**, 432.

26. G. Grigoryan and A. E. Keating, *Curr. Opin. Struct. Biol.* 2008, **18**, 477.

27. D. N. Woolfson, In *Adv. Prot. Chem.*; A. D. P. David and M. S. John, eds.; Academic Press: 2005; Vol. 70, p. 79.

28. T. L. Vincent, P. J. Green, and D. N. Woolfson, *Bioinformatics* 2013, **29**, 69.

29. A. L. Boyle, E. H. C. Bromley, G. J. Bartlett, R. B. Sessions, T. H. Sharp, C. L. Williams, P. M. G. Curmi, N. R. Forde, H. Linke, and D. N. Woolfson, *J. Am. Chem. Soc.* 2012, **134**, 15457.

30. J. M. Fletcher, A. L. Boyle, M. Bruning, G. J. Bartlett, T. L. Vincent, N. R. Zaccai, C. T. Armstrong et al., *ACS Synth. Biol.* 2012, **1**, 240.

31. J. M. Fletcher, R. L. Harniman, F. R. H. Barnes, A. L. Boyle, A. Collins, J. Mantell, T. H. Sharp et al., *Science* 2013, **340**, 595.

32. T. H. Sharp, M. Bruning, J. Mantell, R. B. Sessions, A. R. Thomson, N. R. Zaccai, R. L. Brady, P. Verkade, and D. N. Woolfson, *Proc. Natl. Acad. Sci. USA* 2012, **109**, 13266.

33. H. Gradisar, S. Bozic, T. Doles, D. Vengust, I. Hafner-Bratkovic, A. Mertelj, B. Webb, A. Sali, S. Klavzar, and R. Jerala, *Nat. Chem. Biol.* 2013, **9**, 362.

34. A. N. Lupas and M. Gruber, In *Adv. Prot. Chem.*; A. D. P. David and M. S. John, eds.; Academic Press: 2005; Vol. 70, p. 37.

35. E. Moutevelis and D. N. Woolfson, *J. Mol. Biol.* 2009, **385**, 726.

36. P. B. Harbury, T. Zhang, P. S. Kim, and T. Alber, *Science* 1993, **262**, 1401.

37. P. B. Harbury, P. S. Kim, and T. Alber, *Nature* 1994, **371**, 80.

38. J. Stetefeld, M. Jenny, T. Schulthess, R. Landwehr, J. Engel, and R. A. Kammerer, *Nat. Struct. Mol. Biol.* 2000, **7**, 772.

39. N. R. Zaccai, B. Chi, A. R. Thomson, A. L. Boyle, G. J. Bartlett, M. Bruning, N. Linden, R. B. Sessions, P. J. Booth, R. L. Brady, and D. N. Woolfson, *Nat. Chem. Biol.* 2011, **7**, 935.

40. J. Liu, Q. Zheng, Y. Deng, C.-S. Cheng, N. R. Kallenbach, and M. Lu, *Proc. Natl. Acad. Sci. USA* 2006, **103**, 15457.

41. V. Koronakis, A. Sharff, E. Koronakis, B. Luisi, and C. Hughes, *Nature* 2000, **405**, 914.

42. J. L. Price, W. S. Horne, and S. H. Gellman, *J. Am. Chem. Soc.* 2010, **132**, 12378.

43. J. A. Falenski, U. I. M. Gerling, and B. Koksch, *Bioorg. Med. Chem.* 2010, **18**, 3703.

44. J. K. Montclare, S. Son, G. A. Clark, K. Kumar, and D. A. Tirrell, *ChemBioChem* 2009, **10**, 84.

45. B. C. Buer, R. de la Salud-Bea, H. M. Al Hashimi, and E. N. G. Marsh, *Biochemistry* 2009, **48**, 10810.

46. F. Krammer and P. Palese, *Curr. Opin. Virol.* 2013, **3**, 521.

47. J. S. McLellan, M. Chen, S. Leung, K. W. Graepel, X. L. Du, Y. P. Yang, T. Q. Zhou et al., *Science* 2013, **340**, 1113.

48. A. P. West, L. Scharf, J. F. Scheid, F. Klein, P. J. Bjorkman, and M. C. Nussenzweig, *Cell* 2014, **156**, 633.

49. R. A. Lamb and T. S. Jardetzky, *Curr. Opin. Struct. Biol.* 2007, **17**, 427.

50. F. Boato, R. M. Thomas, A. Ghasparian, A. Freund-Renard, K. Moehle, and J. A. Robinson, *Angew. Chem. Int. Ed.* 2007, **46**, 9015.

51. N. A. Schnarr and A. J. Kennan, *J. Am. Chem. Soc.* 2002, **125**, 667.

52. T. Mizuno, K. Suzuki, T. Imai, Y. Kitade, Y. Furutani, M. Kudou, M. Oda, H. Kandori, K. Tsumoto, and T. Tanaka, *Org. Biomol. Chem.* 2009, **7**, 3102.

53. F. Thomas, A. L. Boyle, A. J. Burton, and D. N. Woolfson, *J. Am. Chem. Soc.* 2013, **135**, 5161.

54. I. Ferrero, O. Michelin, and I. Luescher, *Encyclopedia of Life Sciences (eLS)*; John Wiley & Sons, Ltd.: 2007.

55. B. M. Baker, D. R. Scott, S. J. Blevins, and W. F. Hawse, *Immunol. Rev.* 2012, **250**, 10.

56. M. Bhati, D. K. Cole, J. McCluskey, A. K. Sewell, and J. Rossjohn, *Prot. Sci.* 2014, **23**, 260.

57. D. M. Templeton and K. Moehle, *Pure Appl. Chem.* 2014, **86**, 1435.

58. J. Hennecke, A. Carfi, and D. C. Wiley, *EMBO J.* 2000, **19**, 5611.

59. J. Hennecke and D. C. Wiley, *J. Exp. Med.* 2002, **195**, 571.

60. C. J. Holland, P. J. Rizkallah, S. Vollers, J. M. Calvo-Calle, F. Madura, A. Fuller, A. K. Sewell, L. J. Stern, A. Godkin, and D. K. Cole, *Sci. Rep.* 2012, **2**, 629.

61. D. Fleury, S. A. Wharton, J. J. Skehel, M. Knossow, and T. Bizebard, *Nat. Struct. Mol. Biol.* 1998, **5**, 119.
62. A. Ghasparian, T. Riedel, J. Koomullil, K. Moehle, C. Gorba, D. I. Svergun, A. W. Perriman, S. Mann, M. Tamborrini, G. Pluschke, and J. A. Robinson, *ChemBioChem* 2011, **12**, 100.
63. K. Suzuki, H. Hiroaki, D. Kohda, and T. Tanaka, *Prot. Eng.* 1998, **11**, 1051.
64. J. Kilgus, T. Jardetzky, J. C. Gorga, A. Trzeciak, D. Gillessen, and F. Sinigaglia, *J. Immunol.* 1991, **146**, 307.
65. F. Sinigaglia, M. Guttinger, J. Kilgus, D. M. Doran, H. Matile, H. Etlinger, A. Trzeciak, D. Gillessen, and J. R. L. Pink, *Nature* 1988, **336**, 778.
66. R. Suresh and D. M. Mosser, *Adv. Physiol. Educ.* 2013, **37**, 284.
67. R. L. Coffman, A. Sher, and R. A. Seder, *Immunity* 2010, **33**, 492.
68. J. A. Robinson and K. Moehle, *Pure Appl. Chem.* 2014, **86**, 1483.
69. O. P. Joffre, E. Segura, A. Savina, and S. Amigorena, *Nat. Rev. Immunol.* 2012, **12**, 557.
70. H. Hemmi, T. Kaisho, O. Takeuchi, S. Sato, H. Sanjo, K. Hoshino, T. Horiuchi, H. Tomizawa, K. Takeda, and S. Akira, *Nat. Immunol.* 2002, **3**, 196.
71. P. J. Pockros, D. Guyader, H. Patton, M. J. Tong, T. Wright, J. G. McHutchison, and T.-C. Meng, *J. Hepatol.* 2007, **47**, 174.
72. W. M. Hussein, T. Y. Liu, M. Skwarczynski, and I. Toth, *Exp. Opin. Therap. Pat.* 2014, **24**, 453.
73. E. J. Hennessy, A. E. Parker, and L. A. J. O'Neill, *Nat. Rev. Drug Discov.* 2010, **9**, 293.
74. J. Rehwinkel and C. R.-e. Sousa, *Curr. Opin. Microbiol.* 2013, **16**, 485.
75. B. Lepenies, J. Lee, and S. Sonkaria, *Adv. Drug Del. Rev.* 2013, **65**, 1271.
76. J. K. Krishnaswamy, T. Chu, and S. C. Eisenbarth, *Trends Immunol.* 2013, **34**, 224.
77. E. E. Vomhof-DeKrey, J. Yates, and E. A. Leadbetter, *Curr. Opin. Immunol.* 2014, **28**, 12.
78. N. J. Gay and M. Gangloff, *Annu. Rev. Biochem.* 2007, **76**, 141.
79. J. Y. Kang and J.-O. Lee, *Annu. Rev. Biochem.* 2011, **80**, 917.
80. D. H. Song and J.-O. Lee, *Immunol. Rev.* 2012, **250**, 216.
81. U. Buwitt-Beckmann, H. Heine, K.-H. Wiesmüller, G. Jung, R. Brock, S. Akira, and A. J. Ulmer, *Eur. J. Immunol.* 2005, **35**, 282.
82. U. Buwitt-Beckmann, H. Heine, K.-H. Wiesmüller, G. Jung, R. Brock, S. Akira, and A. J. Ulmer, *J. Biol. Chem.* 2006, **281**, 9049.
83. U. Buwitt-Beckmann, H. Heine, K.-H. Wiesmüller, G. Jung, R. Brock, and A. J. Ulmer, *FEBS J.* 2005, **272**, 6354.
84. R. Spohn, U. Buwitt-Beckmann, R. Brock, G. Jung, A. J. Ulmer, and K.-H. Wiesmüller, *Vaccine* 2004, **22**, 2494.
85. I. J. Amanna, N. E. Carlson, and M. K. Slifka, *N. Engl. J. Med.* 2007, **357**, 1903.
86. M. K. Slifka and I. Amanna, *Vaccine* 2014, **32**, 2948.
87. E. E. H. Tran, M. J. Borgnia, O. Kuybeda, D. M. Schauder, A. Bartesaghi, G. A. Frank, G. Sapiro, J. L. S. Milne, and S. Subramaniam, *PLoS Pathog.* 2012, **8**, 18.
88. V. Burke, C. Williams, M. Sukumaran, S.-S. Kim, H. Li, X.-H. Wang, M. K. Gorny, S. Zolla-Pazner, and X.-P. Kong, *Structure* 2009, **17**, 1538.
89. R. Pantophlet, R. O. Aguilar-Sino, T. Wrin, L. A. Cavacini, and D. R. Burton, *Virology* 2007, **364**, 441.
90. O. Rosen, M. Sharon, S. R. Quadt-Akabayov, and J. Anglister, *Proc. Natl. Acad. Sci. USA* 2006, **103**, 13950.
91. R. L. Stanfield, M. K. Gorny, C. Williams, S. Zolla-Pazner, and I. A. Wilson, *Structure* 2004, **12**, 193.
92. R. L. Stanfield, M. K. Gorny, S. Zolla-Pazner, and I. A. Wilson, *J. Virol.* 2006, **80**, 6093.
93. T. Riedel, A. Ghasparian, K. Moehle, P. Rusert, A. Trkola, and J. A. Robinson, *ChemBioChem* 2011, **12**, 2829.
94. A. Mann, N. Friedrich, A. Krarup, J. Weber, E. Stiegeler, B. Dreier, P. Pugach et al., *J. Virol.* 2013, **87**, 5868.
95. C. H. Bell, R. Pantophlet, A. Schiefner, L. A. Cavacini, R. L. Stanfield, D. R. Burton, and I. A. Wilson, *J. Mol. Biol.* 2008, **375**, 969.
96. P. Rusert, A. Krarup, C. Magnus, O. F. Brandenberg, J. Weber, A. K. Ehlert, R. R. Regoes, H. F. Gunthard, and A. Trkola, *J. Exp. Med.* 2011, **208**, 1419.
97. J.-P. Julien, A. Cupo, D. Sok, R. L. Stanfield, D. Lyumkis, M. C. Deller, P.-J. Klasse et al., *Science* 2013, **342**, 1477.
98. D. Lyumkis, J.-P. Julien, N. de Val, A. Cupo, C. S. Potter, P.-J. Klasse, D. R. Burton et al., *Science* 2013, **342**, 1484.
99. A. Bartesaghi, A. Merk, M. J. Borgnia, J. L. S. Milne, and S. Subramaniam, *Nat. Struct. Mol. Biol.* 2013, **20**, 1352.
100. C. Vita, J. Vizzavona, E. Drakopoulou, S. Zinn-Justin, B. Gilquin, and A. Ménez, *Biopolymers* 1998, **47**, 93.
101. D. R. Burton, *Proc. Natl. Acad. Sci. USA* 2010, **107**, 17859.
102. R. S. Harrison, N. E. Shepherd, H. N. Hoang, G. Ruiz-Gómez, T. A. Hill, R. W. Driver, V. S. Desai, P. R. Young, G. Abbenante, and D. P. Fairlie, *Proc. Natl. Acad. Sci. USA* 2010, **107**, 11686.
103. Y.-W. Kim, T. N. Grossman, and G. L. Verdine, *Nat. Protocols* 2011, **6**, 761.
104. D. J. Craik and A. C. Conibear, *J. Org. Chem.* 2011, **76**, 4805.
105. P. E. Dawson and S. B. H. Kent, *Annu. Rev. Biochem.* 2000, **69**, 923.
106. H. Hojo, *Curr. Opin. Struct. Biol.* 2014, **26**, 16.
107. A. W. Perriman, D. S. Williams, A. J. Jackson, I. Grillo, J. M. Koomullil, A. Ghasparian, J. A. Robinson, and S. Mann, *Small* 2010, **6**, 1191.
108. A. Teijeira, E. Russo, and C. Halin, *Sem. Immunopathol.* 2014, **36**, 261.
109. A. A. Itano and M. K. Jenkins, *Nat. Immunol.* 2003, **4**, 733.
110. M. L. Dustin, *Mol. Cell* 2014, **54**, 255.
111. J. Zhu, H. Yamane, and W. E. Paul, *Annu. Rev. Immunol.* 2010, **28**, 445.
112. M.-I. Yuseff, P. Pierobon, A. Reversat, and A.-M. Lennon-Dumenil, *Nat. Rev. Immunol.* 2013, **13**, 475.
113. F. D. Batista and N. E. Harwood, *Nat. Rev. Immunol.* 2009, **9**, 15.
114. M.-I. Yuseff, A. Reversat, D. Lankar, J. Diaz, I. Fanget, P. Pierobon, V. Randrian et al., *Immunity* 2011, **35**, 361.
115. L. Delamarre, M. Pack, H. Chang, I. Mellman, and E. S. Trombetta, *Science* 2005, **307**, 1630.
116. R. Sharma, A. Ghasparian, J. A. Robinson, and K. C. McCullough, *PLoS One* 2012, **7**, e43248.
117. D. Le Roux, A. Le Bon, A. Dumas, K. Taleb, M. Sachse, R. Sikora, M. Julithe, A. Benmerah, G. Bismuth, and F. Niedergang, *Blood* 2011, **119**, 95.
118. M. Wykes, A. Pombo, C. Jenkins, and G. G. MacPherson, *J. Immunol.* 1998, **161**, 1313.
119. R. Sharma, A. Ghasparian, T. Riedel, J. A. Robinson, and K. C. McCullough, 2014, manuscript in preparation.
120. S. L. Okitsu, U. Kienzl, K. Moehle, O. Silvie, E. Peduzzi, M. S. Mueller, R. W. Sauerwein et al., *Chem. Biol.* 2007, **14**, 577.
121. A. Ghasparian, K. Moehle, A. Linden, and J. A. Robinson, *Chem. Commun.* 2006, 174.

Index

T - #0553 - 071024 - C524 - 279/216/23 - PB - 9780367658779 - Gloss Lamination